WITHDRAWN BY THE
UNIVERSITY OF MICHIGAN

INTERNATIONAL SERIES IN
NATURAL PHILOSOPHY

General Editor: D. ter HAAR

VOLUME 103

Metal Vapours in Flames

Other Titles of Interest

BOUMANS
Line Coincidence Tables for Inductively Coupled Plasma Atomic Emission Spectrometry (2 volume set)

CHEN & KIRSH
Analysis of Thermally Stimulated Processes

EESLEY
Coherent Raman Spectroscopy

FUWA
Recent Advances in Analytical Spectroscopy

JALURIA
Natural Convection Heat and Mass Transfer

MEITES
An Introduction to Chemical Equilibrium and Kinetics

OHSE & BINDER
Thermodynamics and Transport Properties of Alkali Metals

SPALDING
Combustion and Heat Transfer

Journals*

Journal of Quantitative Spectroscopy and Radiative Transfer

Plasma Physics

Progress in Analytical Atomic Spectroscopy

Progress in Energy and Combustion Science

Progress in Reaction Kinetics

Spectrochimica Acta A and B

*Free specimen copy available on request.

A full list of titles in the International Series on Natural Philosophy follows the Index.

Metal Vapours in Flames

by

C Th J ALKEMADE
Tj HOLLANDER
W SNELLEMAN
P J Th ZEEGERS

Rijksuniversiteit Utrecht, The Netherlands

PERGAMON PRESS

OXFORD · NEW YORK · TORONTO · SYDNEY · PARIS · FRANKFURT

U.K.	Pergamon Press Ltd., Headington Hill Hall, Oxford OX3 0BW, England
U.S.A.	Pergamon Press Inc., Maxwell House, Fairview Park, Elmsford, New York 10523, U.S.A.
CANADA	Pergamon Press Canada Ltd., Suite 104, 150 Consumers Rd., Willowdale, Ontario M2J 1P9, Canada
AUSTRALIA	Pergamon Press (Aust.) Pty. Ltd., P.O. Box 544, Potts Point, N.S.W. 2011, Australia
FRANCE	Pergamon Press SARL, 24 rue des Ecoles, 75240 Paris, Cedex 05, France
FEDERAL REPUBLIC OF GERMANY	Pergamon Press GmbH, 6242 Kronberg-Taunus, Hammerweg 6, Federal Republic of Germany

Copyright © 1982 C. Th. J. Alkemade; T. Hollander; W. Snelleman; P. J. Th. Zeegers

All Rights Reserved. No part of this publication may be reproduced, stored in a retrieval system or transmitted in any form or by any means: electronic, electrostatic, magnetic tape, mechanical, photocopying, recording or otherwise, without permission in writing from the publishers.

First edition 1982

Library of Congress Cataloging in Publication Data

Main entry under title:
Metal vapours in flames.
(International series in natural philosophy ; v. 103)
Bibliography: p.
Includes index.
1. Flame. 2. Metal vapors. 3. Flame spectroscopy.
I. Alkemade, C. Theodorus J. (Cornelis Theodorus Joseph)
II. Series.
QD516.M55   1982   541.3'61   82-421
AACR2

British Library Cataloguing in Publication Data

Metal vapours in flames.—(International series in natural philosophy; v. 103)
1. Metal vapours
I. Alkemade, C. Th. J. II. Series
669'.94   QD171
ISBN 0-08-018061-2

In order to make this volume available as economically and as rapidly as possible the typescript has been reproduced in its original form. This method unfortunately has its typographical limitations but it is hoped that they in no way distract the reader.

Printed in Great Britain by A. Wheaton & Co. Ltd., Exeter

Preface

This book deals with the properties and behaviour of metal atoms and their compounds in the vapour phase in laboratory-scale combustion flames. Interest in this subject stems from the use of metal vapours in combustion research and diagnostics as well as from the application of flames to the fundamental study of metal species and to chemical analysis.

Many books have already been written about flames and combustion systems, atoms and molecules, and analytical flame spectroscopy. However, most of these books, if they deal with flames at all, focus on only one aspect of the flame, or metal vapours play only a subsidiary rôle in the discussion, or this rôle is considered merely from the analytical point of view. This book is rather unique in that it has a synoptic scope: it covers in one volume all the main aspects of metal vapours in flames, such as excitation, radiation, dissociation, ionization and diffusion. Owing to the strong interdependencies of these aspects such a wide view is mandatory for an unambiguous interpretation of observations in flames. Also, the results of flame studies are set in a wider perspective by comparing them with those obtained in other systems, as well as with theory. At the same time the authors have attempted to bridge possible communication gaps by bringing together in one book techniques and experimental results reported in journals, conference proceedings, etc., belonging to widely different areas of research and covering a large span of time.

This book is aimed mainly at the experimentalist, for whom it may serve as a research manual, state-of-the-art report and reference source. In addition, it may serve as an introduction to the field for the student or the non-specialist who wants to appraise the potentialities of the flame method.

The authors set out to make this book as self-contained as possible. The underlying theoretical concepts and relationships from Atomic and Molecular physics, Quantum mechanics, Radiation theory, Statistical mechanics and Kinetics, as well as the basic experimental notions and techniques are recapitulated. Thus the reader can acquire or brush up the background knowledge he or she needs to understand the more specialized chapters, without having to frequently consult various textbooks. In this recapitulation emphasis is on insight rather than completeness, and on the presentation of equations for practical use.

The level of treatment is generally such that graduate students in physics, chemistry and engineering should be able to follow the text without too much difficulty.

The authors have taken great pains to present an unambiguous nomenclature and a consistent classification of the main concepts and processes. They have adhered to international conventions where these exist, but in many other cases they had to choose a supplementary terminology. In any case, alternative terms encountered frequently in the literature have been added for orientation. Great care has also been taken in the selection of the symbols. We refer to Appendices A.3 and A.5 for a survey.

In choosing units for physical quantities we have interpreted liberally the Rules of the prevailing S.I. System. Decimal fractions of MKS units have been used throughout, including the Ångstrom (Å). Only in exceptional cases did we use incoherent units such as electronvolt (eV) and atmosphere (atm). We feel justified because this is still common practice in atomic physics and because most flames burn in the free atmosphere. Conversion factors for incoherent units are presented in Appendix A.1, while fundamental constants are expressed in different unit systems in Appendix A.2. Whenever equations involve electrical quantities, they are formulated both in the E.S. and in the rationalized Giorgi System.

The text has been interspersed with numerous references to current as well as to older literature. The purpose of these references is not only to suggest supplementary reading, support statements or draw attention to work done. It is also our aim to give due credit to original work or to prevent older but still useful publications from falling into oblivion. Our selection of references was not based on merit-rating, nor do we necessarily agree with the contents of the papers selected. The reader may notice in our selection a certain bias towards publications from our own laboratory. Although the main text was completed at the end of 1979, it has been possible to include a good many of the most relevant publications that appeared before the summer of 1980 and even a few more recent papers or theses. For easy identification the authors' surnames and the year of publication are included in the text, while the Bibliography contains the full details. Although the Bibliography contains only the references that have been actually cited in the text, the number of references approaches 2000; the Bibliography was therefore synthesized and printed by computer. As a special service to potential readers among spectro-

PREFACE

chemists, a separate bibliography of books on analytical flame spectroscopy has been added to Chapter I.

So that the book can be useful also as a reference work for incidental consultation, a detailed Subject Index has been compiled. This Index may be especially helpful when the reader is interested in a subject or item that does not belong to any particular chapter or section. We mention for example 'two-photon process', 'C_2H_2-N_2O flame' and 'shock-tube' as items that are mentioned at various places throughout the book.

To help the reader survey the contents we have made a rather detailed division of the subject matter into chapters, sections, subsections, etc. The main division has been made according to the following scheme.

In Chapter I the title of the book is explained and a general introduction is presented. The relevance and the advantages as well as disadvantages of flame studies are sketched.

In Chapter II some well-known basic concepts and relationships from theory are recapitulated, while the flame still remains in the background.

Chapter III deals with instrumental components and measuring methods that are basic to most experimental flame studies. The description of flame spectrometric components and arrangements is fairly detailed and is interspersed with various practical hints and measuring tricks. Several sections (III.4-7 and III.14-16) dealing with spectroscopic measurements in general and signal-to-noise theory are believed to be of wider interest to the experimentalist.

In the context of this book the flame acts as a 'solvent', 'milieu' or 'heat bath' for the metal species to be studied. The relevant properties and the different types of flames are therefore reviewed in Chapter IV.

The absolute or relative concentrations of metal species in the flame often play a key rôle in the interpretation of flame experiments. Chapter V is devoted to a systematic exposition of methods of determining these concentrations.

The remaining chapters (VI—X) deal with various aspects of the behaviour of metal vapours in flames and constitute the backbone of the book. In these chapters we not only give the results of flame experiments, but we also compare them with results obtained in metal vapour cells, atomic beams, shock-tubes, etc., and with theory if available. Collisional, radiative and chemical excitation as well as de-excitation processes are discussed at some length in Chapter VI. Table VI.1 presents a unique, updated list of quenching and fine-structure mixing cross sections as measured in flames and vapour cells for more than 400 transitions in metal atoms. This table may meet a long-felt need not only of flame spectroscopists but of all those interested in inelastic collisions of metal atoms with neutral gas particles at thermal energies. Chapter VII deals with the spectral profile of atomic metal lines, and in particular with the effects of collision line broadening and shift. Chapter VIII surveys the kinds of metal compounds and their formation reactions, and treats in some detail methods of determining dissociation energies. The dissociation energies listed in Tables VIII.1—4, which also contain non-flame results, may again be of wider interest. The state of ionization of metal atoms and compounds and, above all, the processes by which ions are formed or removed, are discussed in Chapter IX. Emphasis is on the determination of collisional ionization rate constants and on the theoretical interpretation of the values found in flames. Finally, Chapter X reviews the methods of measuring diffusion constants of metal vapours and lists the few results that have been obtained so far.

Nine appendices containing supplementary information or detailed mathematical derivations conclude the text.

For the benefit of the spectrochemist, the significance of each subject for the analytical application of flame spectroscopy has been summarized in the appropriate chapter.

Since the book covers an extended ground, selective reading has been made easier by the insertion of detailed cross-references to previous chapters or by repeating the essentials of previous discussions. This approach may make (parts of) this book also useful to those engaged in other fields, such as high-temperature research, reaction kinetics, plasma physics, astrophysics, chemical analysis and magneto-hydrodynamic power generation.

The fate of a book as ambitious in its scope as the present one is that it is likely to appear either too early or too late. If there is rapid progress in the field, the book may seem already outdated when released from the press. If a decline has set in and interest in the subject has dwindled, the book may appear less relevant. When the writing started many years ago, there was a boom in fundamental flame studies involving metal vapours. It was then hard to keep up with new developments, so several parts of the manuscript had to be rewritten more than once. In that period important papers continued to flow from a handful of very active Schools in East and West. Nobody will take offence if we mention here only the 'Sugden School', which had its cradle in Cambridge (U.K.). However, in more recent years the rate of progress has undeniably slowed down. A similar trend is to be observed in the development of analytical flame spectroscopy. By providing improved instrumentation or by discovering new effects, this branch of applied science has so often stimulated and even contributed to more fundamental studies. This trend, on the other hand, has made it easier for us to assemble and assess the final results of a long series of investigations undertaken by various groups of authors in different areas. Fortunately, one can now foresee a revival of flame studies thanks to the advent of tunable optical lasers and the renewed interest in combustion

PREFACE

systems in connection with rocket propulsion and power generation. We hope that this book may become an aid or even a stimulus to this revival.

In an over-optimistic mood the first author initially undertook to write this book all by himself. Before he had come half-way he realized that the rate of progress in the field was faster than his rate of writing. He was then fortunate enough to find three experienced co-workers in the Utrecht flame group who were willing to collaborate. Although each of them had his own specialization and preferences, the book as a whole should be considered as a product of common effort and responsibility. Of course, the joint experience of the authors in this wide field is only limited. They had to stick out their neck when treating complicated or even controversial matters with which they felt less familiar. We apologize in advance for any resulting shortcomings.

Acknowledgements. Of the many who have assisted us in completing this work we can mention only a few by name.

In the first place we are indebted to Dr D. TER HAAR (Oxford, U.K.) who, in his function as General Editor for Pergamon Press, invited us to write this book. His advice and interest have always been appreciated.

We are very grateful to SHEILA McNAB for her substantial and continuous help in improving and clarifying the English text. The authors must take full responsibility for any residual flaws in the definitive version.

Several secretaries in succession have been involved in typing a difficult manuscript, which demanded high accuracy and great perseverance. They managed to keep smiling. Our warmest thanks to all of them and particularly to DIET BOS for her work in the initial stage and to MARINA FELLER for her work in the arduous final stage.

At the request of Pergamon Press, INA GODWIN (North Moreton, U.K.) took on the gigantic task of preparing a camera-ready copy typescript. We are most appreciative of her painstaking efforts and her endless patience and we thank her for many useful suggestions.

We should mention too the assistance we received from GERRIT KUIPER (who sub-edited a large part of the manuscript), from HILDE ELBERSE and HENK van ZOEST (who drew many figures) and from CEES PRINS (who computer-programmed the bibliography). We were always able to call upon other people in the Fysisch Laboratorium too. We thank them all for their indispensable contributions.

Many members, past and present, of the Molecular Physics section of our laboratory have greatly obliged us by giving their critical comments, by supplying material or data, and by assisting with proofreading. Some are mentioned by name in the captions of the figures, spectrograms or tables concerned.

We also wish to thank several colleagues, in particular Dr D.R. BATES, Dr E. BAUER and Dr A. SCHADEE, who upon request sent us their expert comments on certain parts of the original draft.

Last but not least we acknowledge the cooperation of authors and publishers who gave us permission to reproduce or redraw figures from their publications. The sources have, of course, been duly cited in the captions.

Utrecht
June 1981

C.Th.J. Alkemade

Tj. Hollander

W. Snelleman

P.J.Th. Zeegers

Contents

CHAPTER I		INTRODUCTION	1

Flames (1). Metal vapours (2). Properties and interactions studied in flames (2). Comparison of metal vapour studies in flames and in other systems (4). Applications of metal vapour studies carried out in flames (6). Conclusion (10). Special bibliography of books dealing with analytical flame spectroscopy (11)

CHAPTER II		BASIC CONCEPTS AND GENERAL RELATIONSHIPS	13
1	INTRODUCTION		13
2	THE STRUCTURE AND SPECTRUM OF ATOMS AND MOLECULES (A SUMMARY)		15
	2a	The Structure of Atoms	15
		2a-1 Atomic Species and States	15
		2a-2 Atomic Energy Levels and Degeneracy	18
	2b	The Structure of Molecules	21
		2b-1 Molecular Species and States	21
		2b-2 Molecular Energy Levels and Degeneracy	26
	2c	The Atomic Line Spectrum	30
		General (30). Optical selection rules (31). Some consequences of the selection rules (32). (Hyper-) fine-structure (33)	
	2d	The Electronic Spectrum of Di-Atomic Molecules	34
		General (34). Optical selection rules (35). Description of band spectra (36)	
	2e	Continuous Spectra	39
3	THE EQUILIBRIUM STATE		41
	3a	Thermodynamic Equilibrium, Temperature, and Equipartition of Energy	41
	3b	Equilibrium Distribution Laws	43
		3b-1 The Maxwell Law of Velocity Distribution	43
		3b-2 The Boltzmann Law of Internal Energy Distribution	44
		The Boltzmann law (44). Atomic internal partition function (44). Molecular internal partition functions (46)	
		3b-3 The Mass-Action-Law	49

CHAPTER II (continued)

	3b-4	The Saha Law	53
	3b-5	The Planck Radiation Law	55
3c	Relaxation; Local and Partial Equilibria		57
	3c-1	General	57
	3c-2	Relaxation	58
	3c-3	Local and Partial Equilibria	60
		Local thermodynamic equilibrium (60). Partial equilibria (60)	

4 KINETIC CONCEPTS AND PROCESSES 63
  4a Rates and Rate Constants 64
    4a-1 Definitions 64
    4a-2 Detailed-Balance Relations between Rate Constants 68
    4a-3 Rate Constants and Relaxation 72
      First example (73). Second example (74). Other examples (75)
    4a-4 Experimental Methods of Determining Collisional De-Excitation Rate Constants for Metal Atoms 76
      General (76). Absolute fluorescence method (76). Relative fluorescence method (77). Population-depletion method (77). Saturation method (78). Lifetime method (78). Phase-shift method (78). Discussion (79)
  4b Cross Sections and Rate Constants 80
    4b-1 Cross Section for Gas-Kinetic Collisions and Related Concepts 80
    4b-2 Classical Expression for the Rate Constant; Activation Energy 82
    4b-3 General Definition of Effective Cross Section of Process 85
    4b-4 Microscopic-Reversibility Relation 88
  4c Activated Complex Theory 89
  4d Potential Energy Surfaces 91
  4e Pseudo-Crossing of Potential Energy Curves 93
  4f Processes through Potential-Curve Crossing 95
    General (95). Selection rules (96). Transmission coefficient (96). Activation energy (99). Effective cross section (99). Multiple crossings; ionic-curve crossing (99). Concluding remarks (102)
  4g Resonance Processes 102
  4h General Classification of Processes 103

5 THE INTENSITIES OF RADIATIVE TRANSITIONS 110
  5a Types of Radiative Transitions and their Rate Constants 110
    5a-1 Radiative Transitions and Einstein Radiation Coefficients for a Single Atomic Line 110
      Types of radiative transitions (110). Light scattering and fluorescence (113). Case of weak radiation field (116). Case of strong radiation field (117). Einstein radiation coefficients (117)
    5a-2 Oscillator Strength, Absorptivity, and other Parameters related to the Einstein Radiation Coefficients 123
      Radiative lifetime (123). Oscillator strength (124). Absorptivity (125). Net absorptivity (absorption and induced emission combined) (130)
    5a-3 Radiative Rate Constants for Atomic Multiplets and Molecular Bands 133
      Atomic multiplets (133). Molecular bands (134)
  5b Causes of Line Broadening and Shapes of Spectral Line Profiles 138
    5b-1 General 138

CONTENTS xiii

CHAPTER II (continued)

    5b-2 Causes of Line Broadening and Associated Spectral Distribution Functions 141
Natural broadening (141). Radiation Broadening (144). Collision broadening (145). Doppler broadening (152)

    5b-3 The Combination of Broadening Effects; the Voigt Line Profile 154
The combination of Lorentzian profiles (154). The combination of Lorentzian and Doppler profiles; the Voigt line profile (155)

  5c Spectral Line Intensity in Absorption, Emission, and Fluorescence 158
    5c-1 General 158
    5c-2 The Strength of Absorption 159
Absorption line profile; Beer's law (159). Absorption factor with arbitrary spectral source profile (164). Peak absorption (164). Integral absorption (166). Curve-of-growth (COG) (169). Saturation (173)

    5c-3 The Intensity of Thermal Emission and Self-Absorption; the Line-Reversal Temperature 175
Definition of thermal radiation (175). Self-absorption and radiation diffusion (176). Thermal emission intensity of atomic lines (without self-absorption) (178). Thermal intensity of molecular emissions (without self-absorption) (179). Thermal emission intensity of atomic lines (with self-absorption) (182). Self-reversal (187). Quantities determinable from thermal emission intensity (188). The line-reversal method (190). Conclusions (191)

    5c-4 The Intensity of Fluorescence 192
The efficiency of fluorescence (192). The intensity of fluorescence (general) (194). The intensity of fluorescence (without self-absorption) (195). The intensity of fluorescence (with self-absorption) (198). Case of continuum source (200). Case of narrow-line source (201). Intensity of fluorescence under saturation conditions (202). Conclusion (207)

CHAPTER III GENERAL INSTRUMENTAL ASPECTS AND EXPERIMENTAL METHODS 209

1 INTRODUCTION 209
2 BURNERS AND NEBULIZERS 211
  2a Introduction 211
  2b Premix Burners for Laminar Flames at 1 atm Pressure 212
    2b-1 General 212
    2b-2 Méker Burners 213
    2b-3 Porous Plate-Burners 220
    2b-4 Slot Burners 220
  2c Low-Pressure Burners 221
  2d Burner Mounting and Adjustment 223
  2e Nebulizers 224
    2e-1 Construction and Operation 224
    2e-2 Properties and Utility 226
    2e-3 Nebulizers for Low-Pressure Flames 229
  2f Preparation of Metal Salt Solutions 230
  2g Regulation and Calibration of Gas Flows 230
Regulation (230). Calibration (233)
  2h Overall Performance of Nebulizer-Burner Systems 234

CHAPTER III (continued)

3   SPECIAL TECHNIQUES OF SEEDING A FLAME 236
    3a   Introduction of Dry Metal Salt Aerosols (without Solvent) 236
         3a-1   Nebulization followed by Removal of Solvent 236
         3a-2   Direct Evaporation of Solid Metal Salts 237
    3b   Other Special Seeding Techniques 238
         Twin nebulizers (238). Hieftje droplet generator (238).
         Liquid bubble method (239). Introduction of halogens (239)

4   THE OPTICAL TRAIN 240
    4a   Introduction 240
    4b   Figures of Merit for Monochromators 241
    4c   Distortion of the Spectral Energy Distribution 248
    4d   Alignment and Area Selection 250
    4e   Polarization Effects 253
    4f   Special Optical Systems and Techniques 253
         Double-monochromator (253). Intensity-modulation (253).
         Double spatial-beam techniques (254). Double spectral-
         beam (or dual-wavelength) techniques (255). Derivative
         spectroscopy (256). Rapid-scan and recording methods (258).
         Multiplexing methods (259). Selective-modulation techniques
         and resonance monochromator (260)

5   PHOTODETECTORS AND NOISE 261
    5a   Introduction 261
    5b   General Characteristics of Photomultiplier Tubes 264
         Photo-emission (264). Secondary-electron emission (264).
         Dark current (266). Pulse counting (266)
    5c   Noise 268
         Shot noise; spectral noise power (268). Secondary-emission
         noise (271). Flicker noise (272). Johnson noise (272)
    5d   Practical Considerations 273
         Selection (273). Technique of measuring (274). Reduction
         of dark current (275). Mounting and operation (276)

6   THE ELECTRICAL MEASURING SYSTEM 277
    6a   Introduction 277
    6b   General Measuring Systems 278

7   LIGHT SOURCES FOR ABSORPTION AND FLUORESCENCE MEASUREMENTS 285
    7a   General 285
    7b   Spectral Continuum Sources 286
         7b-1   Tungsten Strip Lamp 286
         7b-2   Tungsten Brush Lamp 287
         7b-3   High-Pressure Xenon Arc Lamp 287
         7b-4   High-Pressure Mercury Arc Lamp 288
         7b-5   Other Continuum Sources 288
    7c   Spectral Line Sources 289
         7c-1   Hollow-Cathode Lamp (HCL) 290
         7c-2   Low-Pressure Metal Vapour Discharge Lamp 292
         7c-3   Electrodeless Discharge Lamp (EDL) 293
         7c-4   Lasers 296
         7c-5   Other Spectral Line Sources 296
    7d   Survey of Characteristics of Spectral Line and Continuum Sources 298

8   MEASUREMENT OF THE NEBULIZATION EFFICIENCY 302
    8a   General 302
    8b   Methods of Measurement 303

CHAPTER III (continued)

9	RELEASE OF METAL VAPOUR FROM AEROSOL	306
	9a Desolvation of Wet Aerosol	306
	9b Methods for Checking (In-)Complete Desolvation	307
	9c Volatilization of Dry Aerosol	309
	Formation of dry aerosols (310). Factors controlling the volatilization (311)	
	9d Methods for Checking (In-)Complete Volatilization	313
10	THE MEASUREMENT OF FLAME TEMPERATURE	315
	10a Equilibrium Considerations	315
	10b The Translational or 'True' Temperature	316
	10c The Excitation Temperature	317
	10c-1 The Two-Line Method (in Emission)	317
	10c-2 The Slope Method	318
	10c-3 The Emission-Absorption Method	318
	10c-4 Methods using Fluorescence	319
	10c-5 The Method of Line-Reversal	320
	The measurement of line-reversal temperature (320). Background light sources (323). Systematic errors (324). Random errors (325)	
	10d The Measurement of Temperature-Distributions	326
11	THE DETERMINATION OF THE RISE-VELOCITY OF THE BURNT FLAME GASES	329
12	THE MEASUREMENT OF THE FLAME THICKNESS AND OF THE DISTRIBUTION OF METAL VAPOUR IN THE FLAME	332
	Definition of flame thickness (333). Measurement of spatial distribution (333). Description of spatial distributions (334)	
13	METHODS FOR CHECKING THE LINEARITY OF A PHOTO-ELECTRIC MEASURING SYSTEM	336
14	THE MEASUREMENT OF EMISSION	339
	14a Introduction	339
	14b Geometrical Considerations in Emission Measurements	341
	14c The Optical Train	343
	14d Expressions for the Signal Strength	343
	14e Error Sources in Emission Measurements	344
	14f Optimization of the SNR	346
	14f-1 Introduction	346
	14f-2 D.C. Measurement in the Presence of Background Shot Noise	348
	14f-3 D.C. Measurement in the Presence of Background Flicker Noise	355
	14f-4 A.C. Measurement in the Presence of Background Noise	359
	14g The Measurement of Absolute Intensities and Intensity Ratios	362
	Absolute intensities (362). Intensity ratios (364). Radiation standards (365)	
15	THE MEASUREMENT OF ABSORPTION	368
	15a Introduction	368
	15b General Outline of Set-Ups used for Absorption Measurements	369
	15b-1 Single-Beam Set-Up	370
	15b-2 Double-Beam Set-Up	371
	15c Detailed Discussion of Components of the Experimental Set-Up	376
	15c-1 Light Sources	376
	15c-2 Optical Beam Formation	377
	15c-3 Modulation and Beam Splitting	378

CHAPTER III (continued)

	15c-4 Detection	379
	15c-5 Stability	380
	15c-6 Interference from Thermal Metal Emission and Flame Background	381
	15c-7 Systematic Error Sources	382
15d	Signal-to-Noise Ratio Considerations and their Application in the Optimization of Experimental Conditions in Absorption Measurements	383
	15d-1 Case I: Measurement of Integral Absorption of an Atomic Metal Line	384
	15d-2 Case II: Measurement of Absorption Factor with a Narrow-Line Source	387
16	THE MEASUREMENT OF FLUORESCENCE	390
16a	Introduction	390
16b	The Measurement of Fluorescence Efficiency	393
16c	Optimization of SNR	400

CHAPTER IV   TYPES AND PROPERTIES OF NONSEEDED FLAMES   413

1	INTRODUCTION	413
2	GENERAL TYPES OF FLAMES	417
3	THE STRUCTURE OF PREMIXED LAMINAR FLAMES	419
	Primary combustion or inner zone (419). Interzonal region or central region (422). Secondary combustion or outer zone (423)	
4	CALCULATION OF EQUILIBRIUM COMPOSITION AND ADIABATIC TEMPERATURE	424
4a	Introduction	424
4b	Methods of Calculating Equilibrium Composition and Adiabatic Temperature	425
	Equilibrium composition (425). Adiabatic temperature (427)	
4c	Results of Equilibrium-Composition and Adiabatic-Temperature Calculations	428
5	DEVIATIONS FROM FLAME EQUILIBRIUM COMPOSITION	434
5a	Introduction	434
5b	Methods of Evaluating Radical Concentrations	437
	5b-1 General	437
	5b-2 Concentration of H Radicals	440
	5b-3 Concentration of OH Radicals	443
	5b-4 Concentration of O Radicals	445
	5b-5 Conversion of Relative into Absolute Radical Concentrations	448
5c	Some Observations on Excess Radical Concentrations and their Relaxation Rates	451
6	MEASURED FLAME TEMPERATURES	455
6a	Causes of Differences between Calculated and Measured Flame Temperatures	455
6b	Inhomogeneities in Temperature	459
7	NATURAL FLAME IONIZATION	460
	General (460). Ion concentration (460). Identity, formation, and decay of ions (461)	
8	FLAME BACKGROUND SPECTRUM	465
	General (465). (Quasi-)continuum background spectrum (467)	
9	FLUCTUATIONS IN THE FLAME EMISSION	470
	General (470). Excess low-frequency noise and shot noise (471). Figure-of-merit (472). Experiments (475)	

CONTENTS xvii

CHAPTER V DETERMINATION OF METAL CONCENTRATION IN THE FLAME   479

1   INTRODUCTION AND DEFINITIONS   479
2   SPECTROMETRIC DETERMINATION OF ABSOLUTE METAL CONCENTRATION   481
    2a   The Absolute-Intensity Method   481
         Case of negligible self-absorption (481). Inhomogeneity
         of temperature and concentration (482). Case of strong
         self-absorption (484)
    2b   The Integral-Absorption Method   486
         Description (486). Comparison of the absolute-intensity
         and the integral-absorption method (489)
    2c   The Peak-Absorption Method   491
         Description (491). Comparison of the peak-absorption and
         the integral-absorption method (495)
    2d   Curve-of-Growth Methods (in Emission and Absorption)   495
         Description (495). Distortion of the experimental
         COG (499)
    2e   Duplication Methods   504
3   SPECTROMETRIC DETERMINATION OF CONCENTRATION RATIOS   508
    3a   Concentration Ratios of One Species   509
         3a-1   Intensity Measurements   509
         3a-2   Absorption Measurements   510
    3b   Concentration Ratios between Different Species   511
         3b-1   Intensity Measurements   512
         3b-2   Absorption Measurements   512
4   SPECIAL METHODS OF DETERMINING ABSOLUTE METAL CONCENTRATIONS   513
    4a   Use of the Saha Equation for Ionization Equilibrium   513
    4b   Other Methods   515
5   CALCULATION OF ABSOLUTE METAL CONCENTRATION FROM MEASURED SALT SUPPLY   515
    5a   Calculation of Metal Concentration from Measured Nebulization   515
         Efficiency
    5b   Calibration of Metal Supply with Weighed Beads of Salt   519
         and some Other Methods
6   THE COMPARISON METHOD AND ELEMENTS SUITABLE AS FULLY ATOMIZED   520
    STANDARDS
         General (520). Sodium as fully atomized standard element (523).
         Silver as fully atomized standard element (525). Copper as
         fully atomized standard element (526). Other elements (526).
         Conclusion (527)

CHAPTER VI   EXCITATION AND DE-EXCITATION OF METAL SPECIES   529

1   INTRODUCTION   529
         Survey and terminology (529). Importance of (de-)excitation
         studies (532). Advantages and disadvantages of flame studies
         (533)
2   PHYSICAL EXCITATION AND DE-EXCITATION PROCESSES   534
    2a   (De-)Excitation of Electronic States through Conversion of   536
         Translational Energy only
         2a-1   Quenching and Excitation by Collisions with Unexcited   536
                Atoms and Molecules
                Experiments relating to the lowest excited level (537).
                Experiments relating to higher excited levels (539).
                Theory (544). Conclusions (547)

CHAPTER VI (continued)

	2a-2	Mixing of Multiplet Components by Collisions with Unexcited Atoms	548
		Experiments (548). Theory (552). Self- and mutual mixing (555). Conclusions (556)	
	2a-3	Quenching and Excitation by Collisions with Free Electrons	556
2b		(De-)Excitation of Electronic States Involving Conversion of Internal Energy of Collision Partner	560
	2b-1	Introduction	560
	2b-2	Quenching and Excitation by Collisions with Molecules	561
		General (561). Experiments (565). Alkali atoms (565). Hg atoms (569). Other atoms (570). The participation of internal molecular energy (570). Resonance effects (577). Conclusions from the experiments (580). Theory (581)	
	2b-3	Mixing of Multiplet Components by Collisions with Molecules	593
		Experiments (593). Theory (596). Conclusions (598)	
	2b-4	Electronic-Excitation Transfer between Dissimilar Atoms upon Collision	599
		Experiments (600). Theory (602)	
		TABLE VI.1 Cross Sections of Collisional (De-)Excitation Processes	603
2c		Radiative (De-)Excitation of Electronic States	619
	2c-1	General Expressions for the Radiative (De-)Excitation Rates	619
	2c-2	General Dependence of Radiant Density on Excited-State Population (Flame as the only Source of Radiation)	622
	2c-3	General Dependence of Excited-State Population on Radiant Density and on Fluorescence Efficiency	624
	2c-4	Approximate Solution when the Flame is the only Source of Radiative Excitation	626
	2c-5	Approximate Solution for a Flame Irradiated by an External Light Source	628
		Case of conventional light sources (no saturation) (628). Case of laser sources (with saturation) (630)	
3		CHEMICAL EXCITATION AND DE-EXCITATION PROCESSES (CHEMILUMINESCENCE)	634
3a		Introduction	634
3b		Hard Chemiluminescence	635
	3b-1	Chemi-Excitation Reactions in which only Atomic Metal Species are Involved	635
	3b-2	Chemi-Excitation Reactions Involving Metal Ions	638
	3b-3	Chemi-Excitation Reactions Involving Metal Compounds	639
3c		Soft Chemiluminescence	641
	3c-1	Qualitative Considerations	641
	3c-2	Quantitative Considerations	643
		Rate constants of chemi-excitation reactions (643). Rate constants of chemiquenching reactions (647). Discussion of some experimental rate constants (649)	
	3c-3	Reactions Involving a Metal Compound	650
		Conclusion (652)	
4		NONTHERMAL RADIATION OF ATOMIC LINES	652
4a		On the Concept of 'Thermal Radiation'	652
4b		Relationship between Outgoing Radiation Intensity and Excited-State Population	654

CONTENTS   xix

CHAPTER   VI (continued)

    4c   Causes and Conditions of Nonthermal Radiation   657
        4c-1 Nonthermal Radiation due to a Relaxation in the   658
              Excited-State Population
        4c-2 Nonthermal Radiation caused by Nonequilibrated Physical   659
              (De-)Excitation Processes
        4c-3 Suprathermal Radiation caused by Nonequilibrated   663
              Chemiluminescent Reactions
                  Case of (induced) chemiluminescence through 3-body
                  association reaction (664). Factors influencing
                  suprathermal chemiluminescence (665). Case of (true)
                  chemiluminescence through 2-body exchange reaction
                  (667). Extra Doppler broadening (668)
    4d   Experimental Methods of Evaluating Nonthermal Radiation   668
    4e   Some Conclusions for the Analytical Application of Flame   673
        Emission Spectroscopy

CHAPTER VII   BROADENING AND SHIFT OF ATOMIC METAL LINES   677

1   INTRODUCTION   677

2   THEORETICAL   679
    2a Formulation of the Problem   679
    2b Collision Broadening   682
        2b-1 General Considerations   682
        2b-2 Survey of Collision Broadening Theories   683
        2b-3 Semi-Classical (SC) Theories of Collision Broadening   683
            2b-3.1 Impact Approximation   685
            2b-3.2 Quasi-Static (QS) Theories of Adiabatic Line   690
                    Broadening
            2b-3.3 General-Pressures Theories of Adiabatic Line   692
                    Broadening
        2b-4 Quantummechanical (QM) Theories of Collision Broadening   695
        2b-5 The Line Profile due to the Simultaneous Action of   700
            Natural, Doppler, Adiabatic and Diabatic (Foreign Gas)
            Collision Broadening
        2b-6 Influence of J- and M_J-Mixing Collisions on the Line   702
            Profile
        2b-7 (Integrated) Line Profiles with H.F.S.   704
        2b-8 The Occurrence of Satellite Bands   705
        2b-9 Extra Doppler Broadening of Nonthermal Emission Lines   708

3   EXPERIMENTAL   710
    3a Determination of Emission Line Profiles in Flames   710
        3a-1 Scanning Interferometer   710
        3a-2 Scanning Monochromator   712
    3b Determination of Absorption Line Profiles in Flames   712
        3b-1 Scanning Interferometer   712
        3b-2 Zeeman Scanning   714
        3b-3 Scanning Monochromator   715
        3b-4 Tunable Dye Laser   716
    3c Determination of Line Widths from Integrated Line Intensities   718
        General (718)
        3c-1 Emission Curve-of-Growth   719
        3c-2 Absorption Curve-of-Growth (Relative)   719
        3c-3 Duplication Curve   719
        3c-4 Emission Curve-of-Growth plus Absolute Integral Absorption   720
        3c-5 Absorption Curve-of-Growth (Absolute)   720
        3c-6 Emission Curve-of-Growth plus Duplication Curve   720

CHAPTER VII (continued)

	3c-7	Fluorescence 'Curve-of-Growth'	721
	3c-8	Integral Absorption plus Peak Absorption	722

4 RESULTS FROM FLAME EXPERIMENTS IN COMPARISON WITH NON-FLAME RESULTS AND WITH THEORY — 722

    4a Comparison of Line Profiles Measured in Flames with Theoretical Models — 723

        Broadening effect of J-mixing and quenching collisions in flames (723). Line-core and very near line-wing profiles (725). Far-wing profiles (729). Confrontation with theory (732). Conclusions about line wings (732). Satellites (733). Adequacy of the Voigt model (733). Validity of Kirchhoff's law (736)

    4b Comparison of some Nonflame Line-Broadening Experiments with Theoretical Models — 736

    4c Derivation of Force Constants of (Difference) Interaction Potentials from Line Profiles — 739

    4d a-Parameters and (Adiabatic) Line-Broadening Cross Sections Derived from Flame Experiments — 742

    4e Experimental a-Parameters Obtained in Analytically Useful Flames — 747

    4f Observations of the Effect of Hyper-Fine-Structure on Line Profile and Curve-of-Growth in Flames — 749

CHAPTER VIII FORMATION OF METAL COMPOUNDS — 751

1 INTRODUCTION — 751

2 METHODS OF DETERMINING DISSOCIATION ENERGIES (D_0) AND VALUES REPORTED — 754

    2a Methods of Determining D_0 — 754

        2a-1 I. Molecular Methods — 755

        2a-2 II. Thermochemical Methods — 760

    2b Reported Values of D_0 — 764

3 KIND AND ABUNDANCE OF METAL COMPOUNDS — 770

    3a Metal Oxides — 771

    3b Metal Hydroxides and Hydrides — 775

    3c Metal Halides — 779

4 FORMATION REACTIONS OF METAL COMPOUNDS — 783

    4a General Considerations of Metal Compound Formation — 784

        4a-1 General Kinetic Considerations — 784

        4a-2 Determination of Dominant Reaction Path from Variation of Association Factor with Height — 787

        4a-3 Determination of Dominant Reaction Path from Dependence of Association Factor on T and Flame Composition — 789

    4b Survey of Formation Reactions for Specific Metal Compounds — 791

        4b-1 Metal Oxides — 791

        4b-2 Metal Hydroxides — 793

        4b-3 Metal Hydrides — 794

        4b-4 Metal Halides — 795

5 FLAME SPECTROMETRIC DETERMINATION OF DISSOCIATION ENERGIES — 796

    5a Flame-Spectrometric Equilibrium Method — 797

    5b Some Typical Applications of the Flame-Spectrometric Equilibrium Method — 802

        5b-1 Determination of D_0(LiOH) — 802

        5b-2 Determination of D_0(BaCl) — 805

        5b-3 Determination of D_0(SrOH) — 806

        5b-4 Simultaneous Determination of D_0(CaOH) and $D_0\{Ca(OH)_2\}$ — 808

        5b-5 Simultaneous Determination of D_0(BaO) and D_0(BaOH) — 809

        5b-6 Conclusion — 812

CONTENTS xxi

CHAPTER IX   IONIZATION IN SEEDED FLAMES   813

1 INTRODUCTION   813
2 SURVEY OF IONIZATION PROCESSES   817
   2a Classification and Examples of Ionization Processes   817
   2b Factors Controlling the Probability of an Ionization Process   826
3 SOME USEFUL THEORETICAL RELATIONSHIPS   828
   3a Electric Conductivity   829
   3b Thermal Ionization   832
      3b-1 Case of a Single Metallic Element Forming no Compounds   832
      3b-2 Ionization Interference between Two Metallic Elements   833
      3b-3 Case of a Metallic Element Forming Neutral as well as Ionized Atoms and Compounds   834
   3c Ionization Relaxation   837
      Case A (838). Case B (843). Case C (844)
   3d Conditions for Thermal Ionization and Causes of Nonthermal Ionization   846
      Thermal ionization (846). Ionization relaxation (846). Physical and chemical disequilibrium (847). Conclusion (849)

4 EXPERIMENTAL TECHNIQUES   850
   4a Survey of Experimental Techniques   850
   4b Detailed Discussion   854
      Optical spectrometry (854). Ion mass spectrometry (855). Electrostatic probe technique (857). R.f. resonance technique (861). Cyclotron-resonance technique (862). Microwave-cavity resonance technique (863). Microwave attenuation technique (863)

5 EXPERIMENTAL RESULTS   864
   5a Measurements of Ionization and Ion-Recombination Rate Constants with Metallic Additives   864
      5a-1 Case A   866
      5a-2 Case B   869
      5a-3 Case C   876
   5b Survey of Further Ionization Experiments with Additives in Flames   880

6 DISCUSSION AND CONCLUSIONS   882
   6a Conclusions from and Intercomparison of Ionization and Recombination Rate Constants of Alkali Atoms Measured in Flames   882
      6a-1 Some General Conclusions; the Original Ladder Model of Hollander et al.   882
      6a-2 Intercomparison of Standardized Experimental $k_i(T)$ Values   886
      6a-3 Comparison of E_{ion} and E_A; T-Dependence of k_i and k_r 888
      6a-4 The Experimental Ionization Cross Sections   890
   6b Comparison of Experiment with Theory   892
      6b-1 A Theoretical Discussion of the Ladder Model and the Summation Rule in Multi-Component Flames   892
      6b-2 The Ionization Efficiency of Ar   898
      6b-3 Expected Sequence of $\sigma_{i,j}$ Values for Different Alkali Atoms   898
      6b-4 The Activation Energy and the Ionic-Curve Crossing Model of Ionization   899
      Qualitative considerations (899). Quantitative considerations (902)
   6c Consequences of Metal Ionization in Flames   903
      6c-1 Consequences for other Metal Vapour Studies   903
      6c-2 Consequences in Analytical Flame Spectroscopy   905

CHAPTER X	DIFFUSION	909
1	INTRODUCTION	909
2	THEORETICAL ESTIMATE OF DIFFUSION COEFFICIENTS	911
3	MODEL CALCULATIONS OF ATOMIC DISTRIBUTIONS	914
3a	Diffusion from a Point Source in a Laminar Flame	914
3b	Diffusion from a Line Source in a Laminar Flame	916
3c	The Validity of the Models in Actual Flame Experiments	917
4	MEASUREMENTS OF DIFFUSION COEFFICIENTS IN FLAMES	918
4a	Measurements in Simple Situations	918
4b	Measurements in More Complex Situations	919
5	THE IMPORTANCE OF DIFFUSION IN METAL VAPOUR EXPERIMENTS	921

APPENDICES 923

A.1	CONVERSION FACTORS OF INCOHERENT UNITS	923
A.2	SOME FUNDAMENTAL CONSTANTS	924
A.3	LIST OF SYMBOLS AND TERMS USED	925
A.4	IONIZATION ENERGIES AND SPECTRAL LINE CHARACTERISTICS OF SOME METAL ATOMS	939
A.5	GENERAL RADIANT QUANTITIES	943
A.6	THE CONCEPT OF OPTICAL CONDUCTANCE	945
A.6-1	Radiating Surface S of Arbitrary Shape in Combination with Stop D	945
A.6-2	Flat Surface S Radiating into Cone C	946
A.6-3	Sphere Radiating into Free Outside Space	946
A.7	DERIVATION OF EXPRESSIONS FOR SIGNAL-TO-NOISE RATIO (SNR)	947
A.7-1	General SNR Expression for Paired D.C. Meter Readings	947
A.7-2	SNR Expression for Paired Readings on a D.C. Meter with Exponential Step Response in the Case of 'White' Noise	949
A.7-3	SNR Expression for Paired Integrator Readings in the Case of 'White' Noise	950
A.7-4	SNR Expression for Paired Readings on a D.C. Meter with Exponential Step Response in the Case of Flicker Noise	950
A.7-5	SNR Expression for Paired Integrator Readings in the Case of Flicker Noise	951
A.8	SPECTROGRAMS OF THE VISIBLE BANDS OF Ca, Sr AND Ba IN FLAMES	953
A.9	DERIVATION OF RELAXATION EQUATION INCLUDING NEGATIVE-ION FORMATION	957

BIBLIOGRAPHY 959

SUBJECT INDEX 999

OTHER TITLES IN THE SERIES IN NATURAL PHILOSOPHY 1031

CHAPTER I

Introduction

This chapter gives a general introduction to the subject matter of this book. The use of flames in metal vapour studies will be explained in general and compared to related studies in other systems. The usefulness of data obtained in flame studies for other branches of science and technology will be reviewed; the significance of these data for the spectrochemical application of flame spectroscopy will be stressed. The basis of analytical flame spectroscopy is briefly described, and we give a bibliography of textbooks dealing with this method of analysis.

Flames. In common parlance the term 'flame' is used to describe a great variety of phenomena (see instances in Sect. IV.2 sub 1). We shall use this term in a strict sense as we are primarily interested in the common, hot chemical flame. A (purely) *chemical flame* can be defined as one that proceeds through a highly exoergic, autocatalytic chemical reaction in the gas phase, which propagates spontaneously and rapidly through a combustible mixture after ignition. The laboratory flames used in metal vapour studies are made stationary by means of a burner. Fuel and oxidant gases are continuously supplied (see Sect. III.2) and are usually premixed before being ignited above the outlet(s) of the burner head. The burner fixes the flame in space. Different zones can be distinguished in such a flame (see Sect. IV.2 and Fig. IV.1). The central zone, extending over a volume of the order of $1-10$ cm^3 above the primary combustion zone, is the most suitable one for studying the behaviour of metal vapours under (near-)equilibrium conditions. The physical state of the flame gas in this zone can be fairly well characterized by a more or less uniform temperature value, called the flame temperature (see Sect. IV.6). Flame temperatures range from approximately 1500 to 3000 K. Most studies are made in flames burning in the open air at 1 atm pressure, that is, without walls; if desired, the flame can be screened off from the ambient air by a flowing gas sheath.

The flow pattern of a flame is determined by pressure gradients, buoyancy, inertia and friction. Burners can be designed so that they produce, in good approximation, a one-dimensional flame flowing usually in a vertical direction. The height above the combustion zone then corresponds to the travel or residence time of the species contained in the rising flame gas. Typical rise velocities are 10^2 to 10^3 cm/s; relaxation studies with a time resolution better than 1 ms are possible.

Metal vapours. In the context of this book, the function of the flame is to provide a vapour of neutral or ionized metal atoms or metal compounds in a gas of moderately high temperature and with variable chemical composition. The metal vapour should be accessible to optical, electrical or mass-spectrometric measurements, and its (local) concentration should be constant during the observation and should be accurately controllable. The metal species are present in the flame in trace amounts, that is, their concentrations are low compared to those of the main flame constituents. The flame may thus be looked upon as a 'solvent' (or chemical bath) and as a 'thermostat' (or heat bath) for the metal species.

The term 'metal vapours', as used in this book, is taken in a broad sense. It does not only include the metallic elements such as sodium or copper, but also metalloids such as boron. Incidentally we shall also deal with the behaviour of non-metallic elements such as the halogens.

A convenient and common method of introducing metal species into the flame is by pneumatic nebulization (see Sect. III.2). A metal salt solution is aspirated and sprayed by a gas jet, obtained for example by forcing the air stream feeding the flame through a nozzle. The mist droplets thus formed usually pass through a spray chamber, where they (partly) evaporate and where the larger droplets are caught by the walls. The air stream containing the wet aerosol is led to the flame, where the desolvation is completed and the solute particles are volatilized (see Sect. III.9). Halogens can be directly introduced into a flame in the form of a vapour. In the vapour phase (part of) the metal salt molecules dissociate and (part of) the metal atoms may form new chemical compounds (e.g., hydroxides) by reacting with flame constituents (see Chapt. VIII). If the flame temperature is high enough, an appreciable fraction of the metal atoms may be ionized (see Chapt. IX). By changing the composition or the concentration of the metal salt solution, the kind and concentration of the metallic element(s) in the flame can be controlled in a simple and reproducible way.

Properties and interactions studied in flames. By introducing metal vapours into flames a great variety of properties of metal atoms or their compounds, and various kinds of interactions between these species and flame constituents can be quantitatively investigated. Examples of properties studied in flames are: the emission and absorption spectrum of metal species, the oscillator strength of their optical transitions, and the dissociation and ionization energy of metal compounds (see Chapts VIII and IX, respectively). Some of these studies are nowadays mainly of historical interest (such as the determination of oscillator strengths) and will not be explicitly dealt with.

Examples of interactions studied are: collisional broadening and shift of atomic spectral lines (see Chapt. VII), collisional de-excitation ('quenching') and doublet mixing of excited atoms and chemiluminescence (see Chapt. VI), collisional and chemical ionization processes (see Chapt. IX), formation reactions of metal compounds (see Chapt. VIII), and diffusion (see Chapt. X). Again, studies of mainly historical interest will not be mentioned; the interested reader is referred to the well-known book of Mitchell and Zemansky (1961), which describes some of the older flame experiments.[†]

Since free electrons and ions as well as metal species are generally present in flames in trace concentrations, the metal species mostly interact with a neutral flame molecule or radical. Only in ionization experiments are interactions with ionized species or electrons important. The situation in flames is thus typically different from that in electric arcs and other 'true' plasmas.

We shall treat only reactions that occur in the homogeneous gas phase. There is a growing interest in heterogeneous reactions involving metallic elements in flames. The latter reactions, which may be of certain fundamental or practical importance, are, however, much more complicated and less well-understood. For a quantitative interpretation of metal vapour studies it is, however, often necessary to know if the dry aerosol particles conveying the metal after desolvation are in fact completely volatilized in time. We shall therefore briefly deal with the volatilization processes and reactions which involve the condensed phase (see Sect. III.9).

Most metal vapour studies in flames are performed by optical methods including emission, absorption and fluorescence. The spectrum of the metal atoms or compounds is characteristic for the kind of species. Spectral intensities can usually be measured with a precision of one per cent or better by making use of photo-electric detectors. The advent of tunable dye lasers as an intense source of variable wavelength has considerably broadened the potentialities of the optical methods. The instrumentation and performance of optical intensity measurements in emission, absorption and fluorescence is amply described in Sects III.4 and III.14-16.

[†] See also the 'Bibliography on Flame Spectroscopy (1800 - 1966)' by R. Mavrodineanu, N.B.S. Miscellaneous Publ. 281, Washington, D.C., 1967.

Chapt. I INTRODUCTION

Other experimental means such as the ion mass spectrometer have been employed in special areas of flame studies only and will be discussed in the pertinent chapters.

Comparison of metal vapour studies in flames and in other systems. Most properties and interactions studied in flames with metal vapours have been investigated also in other systems, but under different conditions. We mention, for example, the experiments on the quenching of excited metal atoms and on the collisional broadening of atomic spectral lines in a closed, heated gas cell, with optical windows, containing the metal vapour. Collisional excitation and ionization of metal atoms are investigated also in shock-tubes or with the help of crossed-molecular beam apparatus, wherein a beam of metal atoms intersects a beam of the atomic or molecular collision partner. For a long time the emission spectra of metal atoms and metal compounds have been studied in hot plasmas such as the electric arc. Chemical reactions involving metal species are now also studied at elevated temperatures in, e.g., a high-temperature fast-flow reactor (see Fontijn et al. 1972, 1975, 1977) or a heat-pipe-oven reactor (see Hessel, Broida and Drullinger 1975, and Luria, Eckstrom and Benson 1976). Recombination rates of metal ions have been measured in after-glow systems. Electrothermal devices such as the heated graphite furnace have frequently been used in spectroscopic work with metal vapours.

Compared with these alternative systems, flames offer certain advantages, but have also disadvantages, as a milieu for studying metal vapours. When using the flame region downstream from the combustion zone, we have the following *advantages*:

(i)   Metal species can be simply and reproducibly introduced into the flame, with minimum flame disturbance, by nebulizing a metal salt solution; the density of metal species in the flame can be simply, quickly and precisely controlled by varying the solution concentration; combinations of different metallic as well as non-metallic species can be introduced by nebulizing mixed solutions or by choosing an appropriate solvent; the species thus introduced are usually contained in an extended flame region with fairly uniform temperature and gas composition.

(ii)   Flames can be made to have a fairly stable and controllable temperature, which can be accurately measured by optical means (see Sect. III.10); flame temperatures are intermediate between those of heated vapour cells and those of hot plasmas, e.g., electric arcs; flames have proved useful in extending the temperature range of measurements in heated vapour cells (see Sects VI.2a and 2b); the elevated temperature of the flame permits the study of metal species via their thermal emission.

INTRODUCTION

(iii) Open flames burning at 1 atm pressure have no solid walls to screen them from the surrounding air; these flames are therefore easily accessible to (optical and electrical) observations and the correct positioning of the flame section under investigation (e.g., with respect to the detector or background radiation source) is simple; there are none of the problems one finds with heated vapour cells caused by chemical reactions of metal species or gases like O_2 with the walls or windows.

(iv) In flames certain metal compounds such as CaOH and radicals such as H and OH, which are unstable at lower temperatures, occur as stable species in the gas phase; their properties and interactions can thus be conveniently studied.

(v) Flames can be selected that do not produce much background radiation or background ionization of their own; this minimizes interference in spectroscopic observations or in metal ionization studies.

(vi) Flames can be realized as a one-dimensional gas flow, which permits the study of relaxation effects in the millisecond range.

(vii) The composition and temperature of a well-designed flame can be derived with fair accuracy (see Sect. IV.4); the flame gas attains at some distance from the combustion zone a fair degree of physical and chemical equilibration; consequently one can apply certain equilibrium distribution laws when interpreting flame observations.

(viii) As an advantage from a practical point of view we note finally that burner-nebulizer systems with the associated gas handling are relatively cheap, simple to operate and easy to maintain; because usually the spectral background radiation of a flame is comparatively weak and the flame emission spectra of metal atoms are not crowded, a simple (and cheap) spectral selector often suffices for measuring atomic line intensities.

Some general *disadvantages* of flames are however:

(i) The flame is a multi-component system containing various combustion products and often an inert gas as well; thermal dissociation of, for example, H_2O molecules produces different species, viz. H_2, H, O_2, O and OH, even in a stoichiometric H_2-O_2 flame; the metal species studied may thus have a choice of interacting with different partners; this complicates the interpretation of collision or reaction rates measured.

(ii) The presence of chemically reactive species in the flame may lead to side-effects that complicate the interpretation of flame measurements.

Chapt. I   INTRODUCTION

(iii)   Elements that have a strong tendency to form (hydro-)oxides will often be present in only a very small fraction as free atoms.

(iv)   The limited temperature of the flame restricts the observations to neutral or singly ionized metal species.

(v)   The flame temperature cannot be varied over a wide range without at the same time changing the chemical composition.

(vi)   The total pressure of the flame gas cannot be varied unless it is enclosed in a tank; this invalidates some of the aforementioned advantages.

(vii)   Because of the low partial pressure of the metal vapour interactions between metal species themselves are hard to observe in flames.

(viii)   As in other bulk systems (vapour cell, furnace, shock-tube, arc), the kinetic data obtained about collisions and reactions are averaged values over the statistical distribution of relative velocity and over the internal states of the collision partner; this is in contrast to the data obtained in molecular beam experiments with velocity- and internal-state selection.

Applications of metal vapour studies carried out in flames. The information about properties of metal species and their interactions with other species, obtained from flame studies, may find and has in fact found application in several branches of science and technology. Here we summarize only some of these applications.

The study of metal atoms, ions or compounds may yield data on the state of the flame itself and on the processes occurring therein. The metal species can thus act as a 'probe' for acquiring information about flames and combustion systems in general (e.g., the temperature and radical concentrations; see Sects III.10 and IV.5).

Measurements of alkali ionization in flames may find application in magnetohydrodynamic power generators seeded with alkali vapour, as the physical conditions in the two systems are comparable (see, e.g., McCune 1963). These measurements may also find application in 'augmented flames', i.e. flames with alkali ions which are additionally heated by electric power dissipation. Furthermore, metal ionization studies may contribute to our understanding of ionization in rocket exhaust gases where the physical conditions are again comparable to those in flames. This ionization effect disturbs the transmission of radio signals from spacecraft upon re-entry into the atmosphere.

The study of the oxidation reactions of metal species such as Al in flames may contribute to the application of these metals as high-energy rocket fuels (see, e.g., Newman and Page 1971).

INTRODUCTION   Chapt. I

The study of chemi-excitation of metal atoms by radical recombination in flames may contribute to the development of a chemical laser (see, e.g., Zwillenberg 1975).

In order to better understand the anti-knock effect of Pb in combustion engines, a study has been made of the behaviour of this element in flames (see Sugden and Knewstubb 1956).

The operation of the H_2-diffusion flame detector in gas-chromatography is closely connected with ionization processes in flames. For the detection of P- and S compounds in the effluent of a gas-chromatograph the corresponding spectral emissions in a H_2-rich flame are utilized.

Knowledge of the spectral broadening of the Na-D lines in flames might help to understand the light emission in Na flares (see, e.g., Douda and Bair 1974).

There are various phenomena in the upper-atmosphere that involve excited metal atoms, which are directly related to physical and chemical excitation processes in flames. The outcome of Na fluorescence measurements in flames as well as in vapour cells may be of help to us to interpret the Na resonance-fluorescence signals induced by a probing laser beam in the atmosphere ('lidar' = light detection and ranging).

In the atmosphere of cool stars, where conditions may be comparable to those in flames, spectral bands of, e.g., Ti-, V-, Al- and Ba monoxides have been observed. Laboratory studies of the same bands in flames may thus be helpful in interpreting stellar spectra.

The most extensive application of these metal vapour studies is doubtlessly found in analytical flame spectroscopy.

Analytical flame spectroscopy is a spectral method of qualitative and quantitative chemical analysis of the elemental constituents (usually the metallic elements) of samples which are in or brought into the form of a solution. This solution is sprayed into a flame where the solvent and solute are vapourized and decomposed. The element to be analysed (called: analyte) may be present in the flame as free atoms or may form new bonds with components of the flame gas. The wavelengths of the atomic or molecular emission, of the atomic absorption or atomic fluorescence in the flame is characteristic for the kind of analyte. The intensity of these optical signals depends on the analyte concentration in the flame, which in turn depends on its solution concentration. Analytical flame spectroscopy is a relative method which is calibrated by measuring, usually by photo-electrical means, these intensities while spraying into the flame reference samples containing the analyte in known concentrations. In this way an empirical analytical curve is constructed, which enables one to derive analyte concentrations in unknown samples

Chapt. I   INTRODUCTION

from the corresponding relative intensities.

The three main branches of analytical flame spectroscopy are called: *flame emission spectroscopy* (FES), *flame atomic absorption spectroscopy* (FAAS), and *flame atomic fluorescence spectroscopy* (FAFS) (see Fig. I.1). In FAAS an auxiliary light source is used in order to measure the atomic absorption at the wavelength of a suitable resonance line of the analyte. Often this light source contains an atomic vapour of the element to be analysed (e.g., the hollow-cathode lamp); it then emits a spectral line at the same wavelength as the absorption line of the analyte in the flame. In FAFS an auxiliary light source is used to generate a fluorescence line of the analyte in the flame. A further, brief description of analytical flame spectroscopy will be given in Sects III.14a and III.16a.

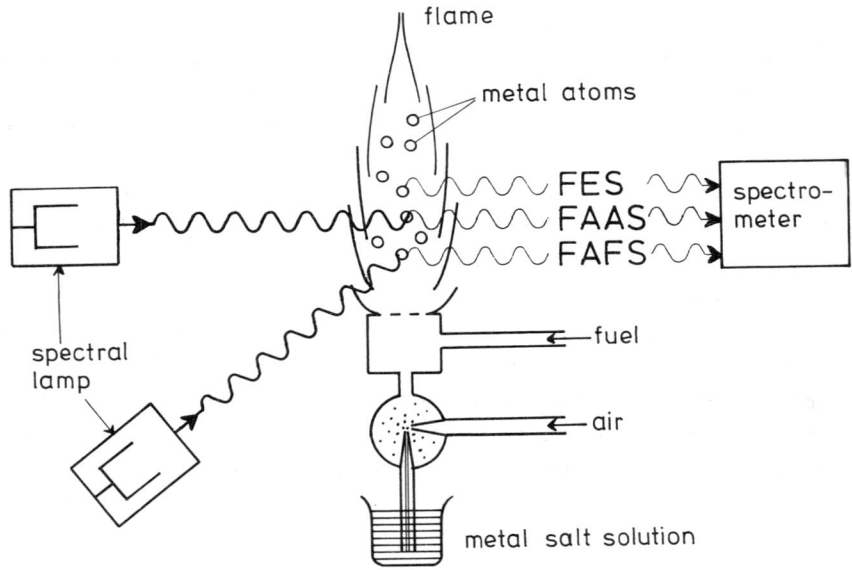

Fig. I.1 Block diagram of flame emission (FES), flame atomic absorption (FAAS) and flame atomic fluorescence (FAFS) spectrometer.

The analytical application of flame emission spectroscopy dates back to the last century; Kirchhoff and Bunsen proposed already in 1860 (Ann. Physik, **110**, p.161) to use FES for qualitative chemical analysis, whereas Champion, Pellet and Grenier in 1873 (C.R. Acad. Sci. Paris, **76**, p.707) applied FES as a quantitative method by measuring spectral line intensities by visual means. There was a real breakthrough in the analytical application of FES after spectrographic and later photo-electric detection techniques were introduced in the twenties and the thirties of this century, respectively.

Although atomic absorption of metal vapours had been studied much earlier in vapour cells, the sun's atmosphere and in flames, the analytical potentialities of FAAS were not recognized until Walsh (1955) and Alkemade and Milatz (1955), independently and simultaneously, introduced this new branch of spectrochemical analysis. It has undergone an explosive development since the mid-sixties and has now become a wide-spread, universal tool of analysis. In addition to flames other simple 'atomizers', such as the electric tube furnace, are being applied to convert the analyte in liquid or solid samples into an atomic vapour for atomic absorption analysis (AAS).

At an instigation of Alkemade (1963), Winefordner and Vickers (1964a) started to investigate systematically the analytical potentialities of FAFS and to demonstrate its applicability in real analyses. This third branch of analytical atomic spectroscopy has certain specific advantages, but in practice it is far from being a rival to AAS. We believe that ultimately these three branches of analytical (flame) spectroscopy will not replace but rather supplement each other.

The reader who requires a more complete historical description of analytical flame spectroscopy is referred to chapters in the books listed below, and to the reviews given by Alkemade (1973), Walsh (1975), and Winefordner (1975).

A full theoretical understanding is not a prerequisite for the practical application of flame spectroscopy as a method of analysis. The only requirement for practical application is that the relationship between optical signal and solution concentration is reproducible and unambiguous. However, in the search for optimum operating conditions relating to sensitivity, detection limit (i.e. lowest concentration detectable), linearity of the analytical curve, freedom from interferences by concomitants in the sample, etc., knowledge of the basic properties and interactions of metal species in the flame is most helpful. This knowledge is provided by physical and physico-chemical studies on metal vapours in flames. The furnishing of such knowledge has not been the only contribution of these fundamental studies to the development of analytical flame spectroscopy. The *physical methods* themselves employed in such studies have often been precursors of new methods of chemical analysis, as is demonstrated by the history of flame emission, atomic absorption and atomic fluorescence spectroscopy (see Alkemade 1973).[†] Also in other areas fundamental research in physics has often been the 'base-camp' for exploring new physical methods of analysis (e.g., mass spectrometry and Mössbauer spectroscopy).

[†] See footnote on p.10

Not only has analytical flame spectroscopy profited from the findings of fundamental flame studies, but the latter, conversely, have often also benefited from developments in the analytical instrumentation (e.g., hollow-cathode lamps and the realization of very hot flames). In addition, the experiences and even the discovery of new effects in analytical applications have often triggered off new fundamental studies. Examples are the mutual ionization interference between two alkali elements (see Sect. IX.6c-2), the suprathermal chemiluminescence of metal lines in fuel-rich flames with alcoholic solvents (see Sect. VI.3), the shape of the analytical curve, the appearance of unknown lines or bands in the flame spectra, and the peculiar effects encountered when a tunable dye-laser is used as a light source in FAFS. The frequent references in this book to the analytical literature testify to the indebtness of fundamental flame research to analytical flame spectroscopy.

For the convenience of the reader who is interested in analytical flame spectroscopy a special bibliography of books on this subject is added to this chapter.

Conclusion. Metal vapour studies in flames have been and are being pursued in various laboratories scattered over the world. Most of this research is concentrated in the U.S.A., the U.S.S.R. and Great Britain. In some laboratories these studies are linked with the development and understanding of analytical flame spectroscopy and of spectrochemistry in general.

> Research on metal vapours in flames in the Physical Laboratory of the University at Utrecht dates back to the early thirties when the measurement of oscillator strengths by emission methods was in the foreground (see, e.g., van der Held 1932). In the late forties links were formed with analytical flame emission spectroscopy, which were later extended to include flame atomic absorption and atomic fluorescence spectroscopy. Although these links still exist, the present investigations with flames in our laboratory form part of a more general research programme in atomic and molecular physics. This programme is aimed at the experimental study of the behaviour of metal atoms, ions and compounds in the gas-phase under thermal-energy conditions. The facilities used include — in addition to flames — a shock-tube, molecular-beam apparatus and various dye-laser systems. These facilities enable us to study an effect from different angles and under different conditions.

---

† The use of the ion mass spectrometer in ionization studies of metal vapours in flames suggests the possible application of a 'flame ionic mass spectrometer' in chemical analysis (see Alkemade 1973). An ion mass spectrometer combined with a plasma ion source at atmospheric pressure has recently been developed by Gray (1974, 1975, 1975a, 1975b) for the trace analysis of metals in solutions. The 'opto-galvanic effect' found in flames with metal vapour, illuminated by a strong laser beam tuned at a metal resonance line (see Green *et al.* 1976a, Alkemade 1977, and van Dijk 1978), is useful in chemical analysis (see Green *et al.* 1976).

INTRODUCTION   Chapt. I

SPECIAL BIBLIOGRAPHY OF BOOKS DEALING WITH ANALYTICAL FLAME SPECTROSCOPY
(E = emission-, A = absorption-, F = fluorescence-method)

Alkemade, C.Th.J. and Herrmann R.: 'Fundamentals of analytical flame
spectroscopy'. Bristol (U.K.): A. Hilger, 1979. (E,A,F)

Angino, E.E. and Billings, G.K.: 'Atomic absorption spectrometry in
geology', 2nd ed., Amsterdam: Elsevier, 1972. (A)

Burriel-Martí, F. and Ramírez-Muñoz, J.: 'Flame photometry, a manual
of methods and applications', 4th ed., Amsterdam: Elsevier, 1964. (E)

Christian, G.D. and Feldman, F.J.: 'Atomic absorption spectroscopy:
applications in agriculture, biology and medicine', New York: Wiley-
Interscience, 1970. (A)

Dean, J.A.: 'Flame photometry', New York: McGraw-Hill, 1960. (E)

Dean, J.A. and Rains, T.C. (Eds.): 'Flame emission and atomic absorption
spectrometry, I : Theory', New York: M. Dekker, 1969. (E,A)

Dean, J.A. and Rains, T.C. (Eds.): 'Flame emission and atomic absorption
spectrometry, II : Components and techniques', New York: M. Dekker, 1971. (E,A)

Dean, J.A. and Rains, T.C. (Eds.): 'Flame emission and atomic absorption
spectrometry, III: Elements and matrices', New York: M. Dekker, 1975. (E,A)

Dvorák, J., Rubeška, I. and Řezáč, Z.,: 'Flame photometry; laboratory
practice', London: Butterworths, 1971 (transl. from the Czech). (E)

Elwell, W.T. and Gidley, J.A.F.: 'Atomic absorption spectrophotometry',
2nd ed., Oxford: Pergamon, 1966. (A)

Herrmann, R. and Alkemade, C.Th.J.: 'Flammenphotometrie', Heidelberg etc.:
Springer Verlag, 2nd ed., 1960; 'Flame photometry', New York etc.:
Interscience, 2nd revised ed., 1963 (translated by P.T. Gilbert Jr.). (E)

Hoda, K. and Hasegawa, T.: 'Atomic absorption spectroscopic analysis',
Tokyo: Genshi Kyuko Bunseki, Kondanska, 1972. (A)

Kirkbright, G.F. and Sargent, M.: 'Atomic absorption and fluorescence
spectroscopy', London: Academic Press, 1974. (A,F)

L'vov, B.V.: 'Atomic absorption spectrochemical analysis', London:
A. Hilger, 1970 (translated from the Russian). (A)

Mavrodineanu, R. and Boiteux, H.: 'Flame Spectroscopy', New York:
Wiley, 1965. (E)

Mavrodineanu, R. (Ed.): 'Analytical flame spectroscopy; selected topics',
Philips Technical Library, London: McMillan, 1970. (E,A,F)

Parsons, M.L., Smith, B.W. and Bentley, G.E.: 'Handbook of flame
spectroscopy', New York: Plenum Press, 1975. (E,A,F)

Pietzka, G.: 'Flammenspektrometrie', Ullmann's Enzyklopädie der
technischen Chemie, Band 2/1, München-Berlin: Urban und Schwarzenberg,
1961. (E)

Pinta, M. (Ed.): 'Spectrométrie d'absorption atomique, I: Problèmes
généraux', and II: Application à l'analyse chimique', Paris: Masson,
O.R.S.T.O.M., 1971. English transl.: 'Atomic absorption spectrometry',
London: A. Hilger, 1975. (A)

Poluektov, N.S.: 'Techniques in flame photometric analysis', New York:
Consultants Bureau, 1961 (translated from the Russian). (E)

Chapt. I   INTRODUCTION

Price, W.J.: 'Spectrochemical analysis by atomic absorption', London:
Heyden, 1979.   (A)

Pruvot, P.: 'Spectrophotométrie de flammes', Paris: Gauthier-Villars,
1972.

Pungor, E.: 'Flame photometry theory', London: Van Nostrand, 1967
(translated from the Hungarian).   (E)

Ramírez-Muñoz, J.: 'Atomic absorption spectroscopy and analysis by
atomic absorption flame photometry', Amsterdam: Elsevier, 1968.   (A)

Reynolds, R.J. and Aldous, K.: 'Atomic absorption spectroscopy, a
practical guide', London: Griffin, 1970.   (A)

Rubeška, I. and Moldan, B.: 'Atomic absorption spectrophotometry',
London: Butterworths, 1969 (translated from the Czech).   (A)

Schuhknecht, W.: 'Die Flammenspektralanalyse', Stuttgart:
Enke Verlag, 1961.   (E)

Slavin, W.: 'Atomic absorption spectroscopy', New York: Wiley-
Interscience, 1968.   (A)

Sychra, V., Svoboda, V. and Rubeška, I.: 'Atomic fluorescence spectroscopy',
London: Van Nostrand Reinhold, 1975.   (F)

Welz, B.: 'Atom-Absorptions-Spektroskopie', Weinheim (Germany): Verlag
Chemie, 1972.   (A)

Winefordner, J.D. (Ed.): 'Spectrochemical methods of analysis', New
York: Wiley-Interscience, 1971.   (E,A,F)

Winefordner, J.D., Schulman, S.G. and O'Haver, T.C.: 'Luminescence
spectrometry in analytical chemistry', New York: Wiley-Interscience,
1972.   (F)

Added in proof

Omenetto, N., (Ed.): 'Analytical laser spectroscopy', New York: (A,F)
Wiley, 1979.

12

CHAPTER II

Basic Concepts and General Relationships

1. INTRODUCTION

This chapter presents a selection of fundamental concepts and general relationships that are basic to the understanding of the following chapters and the pertinent literature. Insofar as the material to be presented is common knowledge and can be found in the usual textbooks, only a recapitulation, without derivation or much explanation, will be given. The condensed exposition of this material is not suited for a first study; but as a survey it may aid the educated reader to brush up his general knowledge or to look up quickly the exact formulation and significance of any particular concept or law he may come across. References will be given to textbooks and review articles for further study.

The subject matter is divided into four sections, which are written in such a way that they can be read more or less independently of each other. Detailed cross-references are given wherever use is made of specific information contained in other sections. Section 2[†] surveys the general structure and optical spectrum of an isolated particle (atom or molecule) without going into the underlying quantum-mechanical theory. Although continuous spectra are normally associated with processes involving more than one particle, they are included here too. Section 3 deals with the statistical distribution laws for an ensemble of free particles or photons in thermodynamic equilibrium. The underlying statistical-mechanical theory is, again, not treated; only results important for use later are given. For the description of a system in equilibrium kinetic considerations are not needed. The state of an equilibrium system at given volume or pressure is characterized by a single macroscopic parameter, the temperature. The concept of temperature and its significance in partial- or near-equilibrium systems are discussed briefly; methods

---
[†] Section numbers quoted without a roman cipher refer to the same chapter.

and results of flame temperature measurements are treated in other chapters. However, the lack of complete thermodynamic equilibrium in flames, with or without metal vapours, makes it necessary to introduce kinetic concepts and to go into the general microscopic description of processes. This will be done in Section 4. Because flame studies derive their fundamental interest largely from the data they give on the mechanisms and rates of physical processes and chemical reactions, this section has turned out to be more extensive and detailed than the previous ones. The topics to be discussed belong to widespread areas of Physics and Physical Chemistry; this makes it difficult to provide a coherent presentation. Actual examples of processes and experimental results are not given; these will be found in other chapters. This section mainly provides a theoretical framework for discussing and classifying processes in general. It also summarizes the basis of various experimental methods of determining rate constants of de-excitation processes. We have elaborated in this section on the proper distinction between phenomenological and microscopic quantities while introducing a terminology which although perhaps cumbersome is unambiguous. This may facilitate the interpretation of the experimental results obtained in bulk systems (such as a flame) and their comparison with related data obtained, e.g., in molecular-beam experiments.

Intensity measurements in the optical spectrum are still the main channel through which we obtain information on the state of the flame gas and the metal vapour contained in it. Besides, they form the basis of quantitative analytical flame spectroscopy (see Chapt. I). Ample attention is therefore given in a special Section 5 to the kinds and rate constants of radiative processes in general (Subsection 5a) and to the derivation of quantitative expressions for the intensity of spectral lines in absorption, thermal emission, and fluorescence (Subsection 5c). This intensity may depend on the spectral profile of the line considered; we have therefore first treated spectral line broadening in Subsection 5b. Spectral line broadening in flames is also of interest in itself; Subsection 5b thus serves also as an introduction to Chapt. VII where this subject is treated in more detail. The application of lasers in the study of metal vapours in flames has prompted a theoretical discussion of the special effects of strong radiation fields on the atomic transitions.

Most flame spectrometric work has been done on atomic vapours; the major part of Section 5 deals therefore with optical transitions in free atoms. However, quite a number of the general equations presented hold also for molecular lines if they do not spectrally overlap with each other. The complicated detailed expressions for the thermal emission intensity of molecular lines and bands have been worked out explicitly.

2. THE STRUCTURE AND SPECTRUM OF ATOMS AND MOLECULES (A SUMMARY)

This section presents only a summary of generally known facts and relationships extracted from the quantum theory of atoms and molecules, as far as is useful within the context of this book or for reading the pertinent literature. The structure and the spectrum will be treated separately. This somewhat unorthodox division is made because, within the context of this book, the structure is relevant not only in relation to radiation but also in relation to collisional processes and reactions. For simplicity's sake we shall restrict ourselves, for the case of molecules, in the main to di-atomic molecules. We have kept to a minimum explanations and formulae that can easily be found in the appropriate textbooks. These textbooks are so numerous that there seems little point in recommending any particular ones. Nevertheless we would make an exception for the advanced standard works by Herzberg (1950, 1966, 1971), Rosen (1964), and Sobel'man (1979), the introductory chapters in the books by Thorne (1974) and Gaydon (1968, 1974), and the more extensive but still introductory theoretical chapters in the book by Mavrodineanu and Boiteux (1965). The last two books discuss atomic and molecular spectra in special relation to flame spectroscopy. In this section we follow closely the terminology used by Shore and Menzel (1968).

2a THE STRUCTURE OF ATOMS
2a-1 *ATOMIC SPECIES AND STATES*

An atom or atomic ion is characterized by its atomic number (Z) and mass number (A), and by its net charge expressed in elementary charge units. This information is expressed by writing for the symbol of, e.g., a singly ionized Na atom: $^{23}_{11}Na^+$, where the superscript number is A and the subscript is Z. Neutral, singly and doubly positively ionized atoms may also be denoted by adding I, II, III after the chemical symbol. Species of the same element having different A numbers are called *isotopes*.

A given species can exist in different internal *(quantum) states*. A state, or more precisely: an eigenstate of the operators of energy, angular momentum and projection of angular momentum, is characterized by assigning four quantum numbers to each of the electrons and by indicating how the electronic angular momenta are coupled with each other and with the nuclear spin.

As a set of *electron quantum numbers* we can take the integral *principal quantum number* n $(n = 1, 2, 3 \ldots \infty)$, the integral *orbital angular-momentum* (or *azimuthal*) *quantum number* l $(0 \leq l \leq n-1)$, the integral *orbital magnetic quantum number* m_l $(-l \leq m_l \leq +l)$, and the *spin magnetic quantum number* m_s. For m_s we have only two values: $m_s = \pm s$ where $s = \frac{1}{2}$ is the *spin quantum number*. The magnetic quantum numbers describe the quantization of the projection of the orbital and spin angular momentum, $\mathbf{l}$ and $\mathbf{s}$ respectively, along the z-axis; this z-axis may be chosen parallel to the direction of a magnetic field or to any other preferential direction present in the physical system being studied. We have the relations

$$\mathbf{l}^2 = l(l+1)\hbar^2 \quad \text{and} \quad \mathbf{s}^2 = s(s+1)\hbar^2 = \tfrac{3}{4}\hbar^2 \qquad (\text{II.1})$$

$$(\mathbf{l})_z = m_l \hbar \quad \text{and} \quad (\mathbf{s})_z = m_s \hbar \qquad (\text{II.2})$$

with $\hbar \equiv h/2\pi$ and $h =$ Planck's constant.

The value of n specifies the *shell* to which the electron belongs (K-, L- or M-shell for $n = 1, 2$ or 3, respectively; etc.). The value of l, at given n, specifies the *subshell* (s-, p-, d- or f-subshell for $l = 0, 1, 2$ or 3, respectively; etc.). The maximum number of electrons that can be stored in a given (sub-)shell is limited by *Pauli's exclusion principle*, according to which no two electrons in the same atom can have the same 4 quantum numbers. The distribution of the electrons over the subshells, that is, the *electron configuration*, is indicated by writing, e.g.: $(1s)^2 (2s)^2 (2p)^6 (3s)$ for the neutral Na atom in its ground state. The number in brackets indicates the n value and the following letter symbolizes the l value, while the superscript indicates the number of electrons (if larger than one) contained in the subshell considered.

For a given electron configuration we can further characterize the state of the atom as a whole by specifying the coupling of the angular momenta. In the *Russell-Saunders* or *L-S coupling scheme* the $\mathbf{l}_i$ vectors of the different electrons numbered by i add up to a *resultant orbital angular momentum*: $\mathbf{L} = \sum_i \mathbf{l}_i$; likewise the $\mathbf{s}_i$ vectors add up to a *resultant (electron) spin*: $\mathbf{S} = \sum_i \mathbf{s}_i$. The quantization of the magnitude of these resultant vectors is described by the (integral) *orbital quantum number* L and the (integral or half odd-integral) *spin quantum number* S for the electron cloud as a whole. The number S is a half odd-integral or an integral according to whether there is an odd or even number of electrons. The quantization of the projection of these resultant vectors along the z-axis is described by the *orbital* and *spin magnetic quantum numbers*, M_L and M_S respectively, for the electron cloud as a whole. M_L can assume all integral values between and including the limits: $-L$ and $+L$; M_S can assume either all integral or all half odd-integral values between and including the limits: $-S$ and $+S$ according to whether S is an integral or half odd-integral number. Atomic states with $L = 0, 1, 2$ or 3 are called S-, P-, D- or F-states, respectively, and so on. All electrons in a *closed* shell or subshell yield a resultant $L = S = 0$. The possible L- and S values of an atom are thus determined by the electrons that are found in open (sub-)shells. In optical spectroscopy these are usually the *valence electrons* as distinct from the *core electrons*. For atoms with only one valence electron, such as the alkalis, we have consequently: $L = l$, $M_L = m_l$, $S = s$ ($= \frac{1}{2}$) and $M_S = m_s$ ($= \pm \frac{1}{2}$).

The quantization rules for the vector addition of angular momenta are based on the following general addition rules for any pair of angular momenta: $\mathbf{a} + \mathbf{b} = \mathbf{c}$. Let a, b, c and m_a, m_b, m_c be the quantum numbers describing the magnitudes and projections of the corresponding vectors, respectively; we then have (cf. Eqs II.1 and 2)

$$\mathbf{c}^2 = c(c+1)\hbar^2 \text{ and } (\mathbf{c})_z = m_c \hbar \, ; \tag{II.3a}$$

$$c = |a-b|, |a-b|+1, \ldots, (a+b)-1, (a+b) \, ; \tag{II.3b}$$

$$m_c = -c, -c+1, \ldots, c-1, c \, . \tag{II.3c}$$

It follows that there are $(2b+1)$ different c values possible if $b \leqslant a$ and $(2a+1)$ values if $a \leqslant b$. For given c, there are $(2c+1)$ different m_c values. In general, a, b and c may assume positive integral or half odd-integral values or be zero.

The vectors **L** and **S** combine to form the *total electronic angular momentum* **J** defined by

$$\mathbf{J} \equiv \mathbf{L} + \mathbf{S} \tag{II.4}$$

and characterized by quantum number J and magnetic quantum number M_J.

According to the addition rules Eqs II.3a-c, consecutive values of J, which are either integral or half odd-integral, differ by unity and range from $|L-S|$ to $(L+S)$; consecutive values of M_J differ again by unity and range from $-J$ to $+J$. There are thus $(2S+1)$ different J values, for given S and L, if $L \geqslant S$; there are $(2J+1)$ different M_J values for each J. An atomic state having quantum numbers L, S and J is symbolized by adding the multiplicity $(2S+1)$ and J as superscript and subscript, respectively, to the capital letter symbolizing the L value. For example, the lowest excited Na state having $S = \frac{1}{2}$, $L = 1$ and $J = \frac{1}{2}$ is symbolized by: $^2P_{\frac{1}{2}}$.

When an atom is in a quantummechanical state with given J value and with equal expectation values or probabilities for any of the possible M_J values, the atom is unpolarized and unaligned. The atom is *polarized* when the expectation 'value' $\langle \mathbf{J} \rangle$ is unequal to $\mathbf{0}$. The atom then has a preference for being *oriented* in a certain direction. When $\langle \mathbf{J} \rangle$ equals $\mathbf{0}$ but $\langle J_x J_y \rangle - \langle J_x \rangle \langle J_y \rangle \neq 0$, **J** has a preference to lie parallel to a certain line and the atom is said to be *aligned*. For instance, when $J = M_J = 0$, no polarization or alignment is possible. When $J = \frac{1}{2}$ and the probability of finding $M_J = +\frac{1}{2}$ exceeds that for $M_J = -\frac{1}{2}$, or conversely, the atom is polarized; alignment is, however, not possible in this case. When $J = 1$ and the probabilities for $M_J = +1$ and $M_J = -1$ are equal but different from the probability for $M_J = 0$, the atom is aligned.

Finally, the *total atomic angular momentum* **F**, including the *nuclear spin* **I**, is obtained from

$$\mathbf{F} \equiv \mathbf{J} + \mathbf{I} \,. \tag{II.5}$$

The quantum number I associated with **I** can have integral or half odd-integral values and depends on the isotopic species considered. The quantization rules for F and M_F are analogous to those for J and M_J.

The $L-S$ coupling scheme works well for the lighter atoms. For the heavier atoms the (j,j) *coupling scheme* should be applied instead. Here $\mathbf{l}_i$ and $\mathbf{s}_i$ of each electron separately are coupled to form a resultant angular momentum $\mathbf{j}_i$ defined by

$$\mathbf{j}_i \equiv \mathbf{l}_i + \mathbf{s}_i \,. \tag{II.6}$$

The quantization rules for the corresponding *inner quantum number* j and associated magnetic quantum number m_j are again based on Eqs II.3a-c. The total electronic angular momentum of the atom is then obtained from

$$\mathbf{J} \equiv \sum_i \mathbf{j}_i \,. \tag{II.7}$$

The atomic state (without nuclear spin) is now characterized by writing up the complete set of j_i values and the resultant J value in the form: $(j_1, j_2 \ldots)_J$. There is formally a unique correspondence, for a given electron configuration, between this notation and the $L-S-J$ notation. The latter is often also used for the heavier elements in spectroscopic handbooks, although L and S are no longer good quantum numbers here. The (j,j) coupling scheme is again simplified by the fact that closed (sub-)shells always yield a resultant $J=0$.

Whichever coupling scheme applies, J, the internal energy E, and the parity are always well-defined quantities for an atomic eigenstate. The *parity* is odd (symbolized by superscript o) or even, according to whether the atomic wave function does or does not change its algebraic sign upon inversion of the electron coordinates. Configurations for which $\sum_i l_i$ is an even number lead to states with even parity; if this sum is odd, the states have odd parity (we disregard here the possibility of 'configuration mixing'). The electrons contained in closed (sub-)shells always contribute to $\sum_i l_i$ by an even number; the parity therefore depends on the valence electrons alone.

2a-2 *ATOMIC ENERGY LEVELS AND DEGENERACY*

The *internal energy* E of an atomic state is an important quantity in determining the position of the spectral lines (see Subsection 2c). For a given electron configuration, quantum states that differ among themselves only in total quantum number M_J have the same energy in the absence of a magnetic or an electric field; they are said to be degenerate (*spatial* or *'magnetic'* *degeneracy*). The

THE STRUCTURE AND SPECTRUM OF ATOMS AND MOLECULES Sect. II.2a

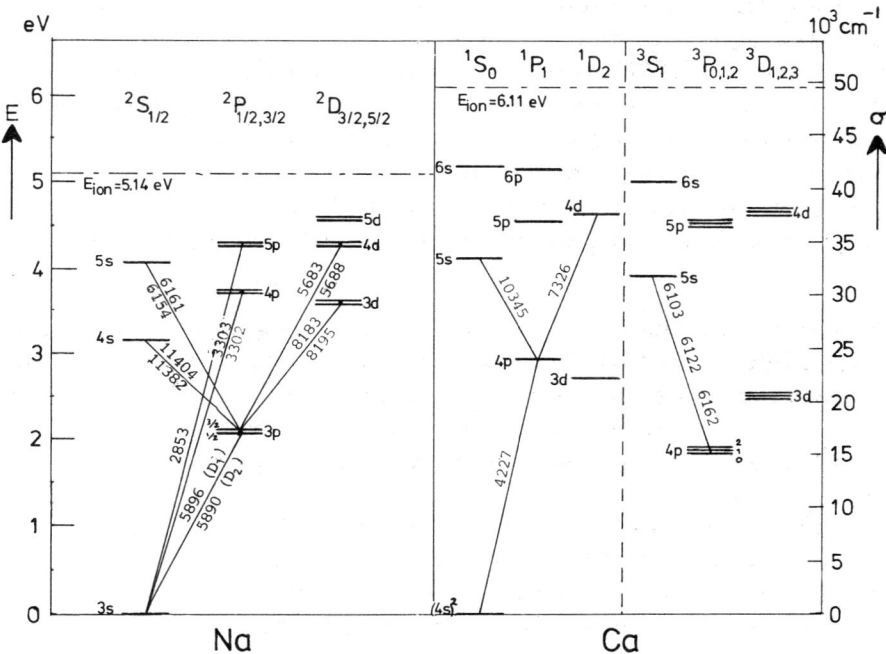

Fig. II.1 Energy level diagrams for neutral Na and Ca atoms. The ionization limit (E_{ion}) is indicated by —·—·—·. The zero-point of the energy scale is chosen at the ground level. The excitation energy E is expressed in eV and the corresponding wave number σ in cm^{-1}. Multiplet splittings are not to scale. The J numbers for the $3p\,^2P$ Na doublet and the $(4s)(4p)\,^3P$ Ca triplet are added. The configuration of the valence electron(s) in the ground state is indicated; in the case of Ca one of them remains in the 4s subshell. Lines connecting energy levels show some of the allowed optical transitions; the corresponding (multiplet) wavelengths are indicated in Å along these lines. (The data have been derived from D'Ans-Lax Taschenbuch für Chemiker und Physiker, Band III, Springer, Berlin, 1970.)

degeneracy (factor) g_J equals $2J+1$ as there are $(2J+1)$ different M_J values. These states together form an (*energy*) *level*.[†] In the *energy level diagram* the energy of each level is represented by the position of a horizontal bar along a vertical energy scale (see Fig. II.1). The energy of the lowest level, called *ground level*, or the (first) *ionization limit* may be taken as zeropoint of the energy scale; in this book we generally make the first choice. For an isolated

---

[†] The distinction between 'states' and 'levels' introduced here is not generally applied in textbooks (cf. Mavrodineanu and Boiteux 1965, for example). In the literature and in other places in this book these terms are often used in a more loose sense but should not cause confusion.

atom all levels below the ionization limit are discrete; those above it are continuously distributed and correspond to an ionized atom plus a free electron having a continuum of possible values of translational energy of relative motion.

We disregard here the existence of discrete, unstable auto-ionizing states involving the simultaneous excitation of two electrons, or of unstable negative-ion states resulting from the temporary binding of an extra electron ('resonance states'). These states lie above the ionization limit but play no role in flame spectroscopy.

Energy levels lying above the ground level are called *excited* or *excitation levels*. The energy difference between an excited level and the ground level is called the *excitation energy*; the latter is often expressed in electronvolt (eV) or converted to *wavenumber in vacuo*, $\sigma \equiv 1/\lambda$, and expressed in cm^{-1} (see Appendix A.1 for conversion into other units). The *excitation potential*, expressed in V, is numerically equal to the excitation energy expressed in eV. Since the principal quantum number n is unlimited, there is an infinity of excited levels; they crowd together when $n \to \infty$, that is, when the ionization limit is approached. In the level diagram the energy levels can be arranged in separate columns according to their $S-$ and $L-$ (and $J-$) values (see Fig. II.1). When the excited levels correspond to the promotion of one valence electron to a state of higher energy, the levels within each column can be distinguished by putting the quantum number n of this electron in front of the level symbol. One may also specify the complete configuration of the valence electron(s) in front of the level symbol.

The same distinction between ground and excited levels can also be made between *ground* and *excited states*.

The spatial degeneracy of an energy level with $J \neq 0$ is removed when a magnetic field is applied; the level is then split into $(2J+1)$ *Zeeman sublevels* (nondegenerate Zeeman states) of equal mutual separation which is proportional to the magnetic field strength. In an electric field a splitting into *Stark sublevels* with different $|M_J|$ values occurs. The mutual separation increases here in general proportionally to the *square* of the field strength in lowest-order approximation. Each Stark sublevel is, however, still doubly degenerate with respect to the sign of M_J, except for $M_J = 0$. A *linear* Stark effect is found for the H atom, where l-degeneracy exists, that is where the energy difference between levels with different l but same n is negligible.

When the $L-S$ coupling scheme applies, all levels with the same $L-$ and S value, for a given electron configuration, lie close together and are said to form a *multiplet (of levels)* or a *(spectroscopic) term*. If $L \geqslant S$ there are $(2S+1)$ such levels which have slightly different energy and are distinguished by their J values. We call $(2S+1)$ the *multiplicity* of the multiplet or term. For alkali atoms having $S = \frac{1}{2}$ the multiplicity is two and we speak of a *doublet* (see, e.g., in Fig. II.1 the ^2P doublets of Na, each consisting of a $^2P_{\frac{1}{2}}$ and a $^2P_{\frac{3}{2}}$ level). When $S=0$ or $S=1$, as in the alkaline-earth atoms, we have *singlets* or *triplets*, respectively (see the same figure). The *term value* or the position of the term along the energy scale in the *term diagram* is determined by the mean energy of the constituent levels. Because of its close connection to the atomic spectrum, the term value is often expressed in wave number σ (unit: cm^{-1}).

The existence of a multiplet of levels with slightly different energies is due to an interaction between the magnetic dipole moment associated with the orbital angular momentum and that associated with the electron spin; it is called *fine-structure* (f.s.). When $S=0$, that is, for singlet terms, or when $L=0$, that is, for S terms,[†] no fine-structure occurs (see Fig. II.1). The doublet splitting with the alkali atoms decreases with increasing quantum number n and with decreasing atomic number Z. The splitting of the lowest Na doublet (commonly called Na-D doublet) amounts to 2.1×10^{-3} eV or 17 cm^{-1} (see also Appendix A.4 presenting some atomic excitation energies).

A further splitting, called *hyper-fine-structure* (h.f.s.), arises as a result of the interaction between **J** and **I** (if $I \neq 0$). When $J \leq I$, there are $(2J+1)$ h.f.s. levels, each characterized by a different quantum number F; if $J \geq I$, there are $(2I+1)$ h.f.s. levels. Since this interaction is much weaker than the f.s. interaction, the energy splittings are also much less. For example, with Na having a nuclear spin $I = \frac{3}{2}$ the h.f.s. splitting of the $3^2S_{\frac{1}{2}}$ ground level amounts to 0.059 cm^{-1} whereas it is only 0.006 cm^{-1} for the lowest excited $3^2P_{\frac{1}{2}}$ level. Because the h.f.s. splitting is so small, it can be measured by radio-frequency methods; its magnitude is therefore often expressed in frequency units (see Appendix A.1 for conversion of units).

The energy of an atomic state depends weakly on the nuclear mass. As a consequence, there is an *isotope effect* in the values of the excitation energies.

An extensive tabulation of atomic energy levels is found in the publications by Moore (1949-1952-1958 and 1972) and by Bashkin and Stoner (1975); a large collection of atomic energy levels of interest in flame spectroscopy has been presented by Mavrodineanu and Boiteux (1965).

2b THE STRUCTURE OF MOLECULES
2b-1 *MOLECULAR SPECIES AND STATES*

Neutral or ionized molecules are specified by the number, kinds and geometrical configuration of the constituent nuclei and by their net charge. We distinguish between di-atomic and poly-atomic molecules, between homonuclear (e.g.

[†] The *symbol* S for $L=0$ states should not be confused with the *spin* quantum number S.

$^{14}N_2$) and heteronuclear (e.g. $^{14}N\,^{15}N$, NO) di-atomic molecules, between isotopic species, and between (co-)linear and nonlinear poly-atomics. Molecular species are represented by the chemical symbols (and mass numbers) of the constituent atoms and their net charge expressed in elementary charge units, for example: $H^{35}Cl$, $CaOH^+$ and $^{15}N_2$. Molecules, such as C_2, OH, CH, with unpaired electrons are often called *radicals*; they are chemically reactive and can therefore perform an important role in flames.

A given molecule can exist in different *(eigen-)states* with respect to the orbital motions and spins of the electrons and to the vibrational and rotational motions of the nuclei. We shall ignore here the nuclear spin as it is usually of little consequence in flame spectrometric work. Moreover, we shall restrict ourselves mainly to di-atomic molecules which have only one degree of freedom for vibration and two for rotation in a centre-of-mass system.

The over-all (eigen-)state of a di-atomic molecule is characterized in the first place by the *quantum numbers* S and Λ, which relate to the *resultant spin* and the *component of the resultant orbital angular momentum* along the internuclear axis of the valence electrons, respectively. The electrons in the cores of the constituent atoms do not play a role. In fact Λ corresponds to the resultant magnetic quantum number M_L in the atomic case, if we take the internuclear axis as z-axis. Defining $\mathbf{\Lambda}$ as the component of $\mathbf{L}$ along this axis, we have for its magnitude: $|\mathbf{\Lambda}| = |\Lambda|\hbar$. Since in a di-atomic molecule the field acting on the valence electrons is not central but has only axial symmetry, the resultant azimuthal quantum number L is no longer a good quantum number here; but M_L is still a good one. Because of the axial symmetry, only the absolute value $|\Lambda|$ matters.[†] States with $|\Lambda| = 0, 1, 2 \ldots$ are called $\Sigma, \Pi, \Delta, \ldots$ states. The multiplicity $(2S+1)$ is added as a superscript to this letter symbol, for example: $^2\Pi$ (compare with the notation ^2P in the atomic case). A molecule may have different electronic eigenstates with the same $|\Lambda|$ value; these are distinguished from each other by writing a Roman letter in front of the Greek letter symbol for $|\Lambda|$. The letter X refers to the ground state; capital letters A, B, ... refer to different excited states with the same $|\Lambda|$ and the same multiplicity as the ground state; small letters a, b, ... refer to states with a different multiplicity. We have, for example, the following electronic states of C_2: $X^3\Pi$ (ground state), $A^3\Pi$ and $a^1\Sigma$ (excited states).

Nondegenerate Σ states (which show no Λ-doubling; see below) can be further classified as *negative* (−) or *positive* (+), depending on whether the electronic eigenfunction changes or does not change its sign upon reflection in a

---

[†] Often one defines $\Lambda \equiv |M_L|$ instead of $\Lambda \equiv M_L$ as is done here.

plane through the internuclear axis (which is a plane of symmetry). This is marked by adding - or + as a superscript: Σ^- or Σ^+. When the two nuclei have identical charge (e.g., $^{12}C_2$ or $^{12}C^{13}C$), the electronic eigenfunction remains unaltered or changes only its sign upon reflection of the electron positions in the centre of the nuclear charges. This is marked by adding the subscript g (German: gerade = even) or u (German: ungerade = odd) to the letter symbol for the state considered. For the states of C_2 mentioned above we have: $X^3\Pi_u$, $A^3\Pi_g$ and $a^1\Sigma_g^+$.

The molecular state is further characterized by the *vibrational quantum number* v, which can take the values $0, 1, 2, 3 \ldots$. A di-atomic molecule can be conceived as a one-dimensional oscillator, that is, the internuclear distance r oscillates around an equilibrium value r_0. The two nuclei move with respect to each other along the line of their joining under the influence of an interaction force which can be described by a *potential energy function* $V(r)$ having a minimum at $r = r_0$ (see the *potential energy curves*, also called *potential curves*, depicted by way of illustration in Fig. II.2). This function is given by the dependence on

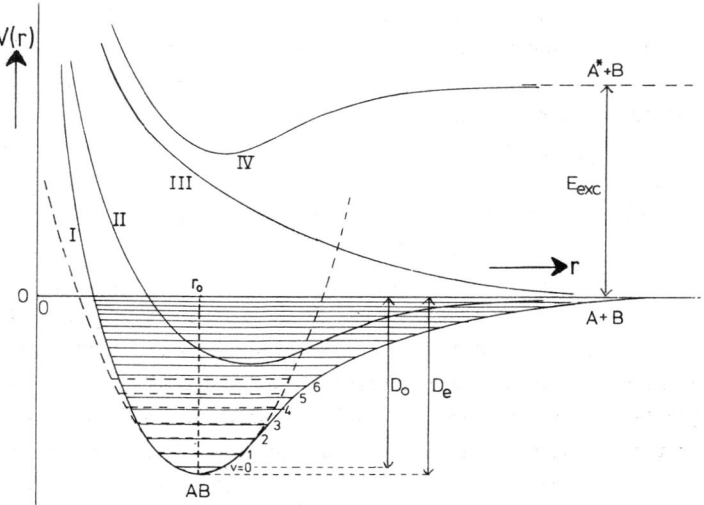

Fig. II.2 Potential energy curves for di-atomic molecule AB (schematic). Curve I corresponds to ground-state molecule, with vibrational levels $v = 0, 1, 2, \ldots$ being represented by horizontal, drawn lines and with equilibrium internuclear distance $= r_0$. Parabola matching best curve I is shown by dashed curve and some corresponding vibrational levels of the ideal oscillator are represented by dashed horizontal lines. D_0 and D_e are dissociation energies of $AB \rightarrow A + B$ as counted from the $v = 0$ level and from the minimum of curve I, respectively.
Curves II and IV belong to an electronically excited, stable molecule; E_{exc} is the excitation energy of excited atom A^* with which curve IV corresponds asymptotically for $r \rightarrow \infty$. Curve III belongs to an electronically excited, unstable molecule dissociating into ground-state atoms A and B.

r of the potential energy of the Coulomb repulsion between the nuclei, plus the total electronic (kinetic + potential) energy as calculated at *fixed* internuclear separation. Such a description is valid if the relative velocity of the nuclei is much smaller than the velocities of the electrons in their orbits. These orbits can then adjust themselves momentarily to the slowly varying internuclear distance (Born-Oppenheimer approximation; see also Subsection 2b-2). The zeropoint of the $V(r)$ scale corresponds to infinite separation of the atoms in their ground states and at rest with respect to each other. Different $V(r)$ functions are associated with different electronic states. Some of them, showing a minimum at finite r, may lead to a 'stable' molecule (apart from the possibility of radiative transitions to other states); others, showing a monotonous fall with increasing r, lead to an unstable molecule (see Fig. II.2). The $V(r)$ curve for the electronic ground state has the lowest minimum. When a molecule in a certain state can dissociate into free atoms, one or both of which are in an excited state, the corresponding $V(r)$ function approaches asymptotically the positive atomic excitation energy or sum of atomic excitation energies for $r \to \infty$ (see curve IV in Fig. II.2). A similar situation arises when a neutral molecule in a certain state can dissociate into a pair of oppositely charged atoms; here the difference between the ionization energy of the one and the electron affinity of the other atom determines the asymptotic value of $V(r)$.

The potential curves show a steep rise when r tends to zero because of the Coulombic repulsion of the electrons in the cores of the two atoms. On the other hand, the curves show a smooth course for large r, as the interaction between the atoms then becomes weak. The $V(r)$ curve belonging to a stable electronic state is therefore asymmetric with respect to its minimum and consequently deviates from a parabola. The vibration of the nuclei is thus anharmonic and the molecule is a vibrator rather than a pure oscillator. Only within a restricted interval around the minimum can the potential curve be approximated by a parabola (see dashed curve in Fig. II.2) and only then can the quantum theory for a harmonic oscillator be applied. Therefore, in the range of small v values only, one may expect equidistant vibrational energy levels; for larger v values the energy levels will crowd together towards the dissociation limit, called *convergence limit*[†] (see the horizontal bars marked $v = 0, 1, 2 \ldots$ in Fig. II.2). This has consequences for the vibrational energy as a function of v and for the frequencies of the vibrational bands in the spectrum (see below and Subsection 2d). The quantum number v still has a well-defined meaning but it has a finite upper limit; above this limit the molecule dissociates.

[†] The term convergence limit here does not imply that this limit is approached asymptotically for $v \to \infty$ (cf. the ionization limit in the atomic case); in fact v goes to a finite limit.

The orbital and spin angular momenta of the electrons combine with the angular momentum of nuclear rotation to form the *total molecular angular momentum* **J** with associated quantum numbers J and M_J (apart from the nuclear spin). J is a strict quantum number and plays an important role in the rotational structure of molecular spectra (see Subsection 2d). There are several ways in which the different angular momenta can be coupled to form the resultant **J**; we mention here only the two most important ones, called Hund's (coupling) case a and b.

In *Hund's case a* the vector **Λ** combines with the vector **Σ**, which is the component of **S** along the internuclear axis, to form a resultant vector **Ω**

$$\mathbf{\Omega} \equiv \mathbf{\Lambda} + \mathbf{\Sigma}. \tag{II.8}$$

Just as quantum number Λ was identified with M_L, we have here: $\Sigma \triangleq M_S$ and $|\mathbf{\Sigma}| = |\Sigma| \hbar$. Like **Λ** and **Σ**, **Ω** is parallel to the internuclear axis. As a consequence, the quantum number Ω which describes the magnitude of **Ω** by: $|\mathbf{\Omega}| = \Omega \hbar$, is obtained from: $\Omega = |\Lambda + \Sigma|$. For a $^3\Delta$ state (with $S = 1$ and $|\Lambda| = 2$), for example, there are three different Σ values: $-1, 0, +1$ and thus three different Ω values: 1, 2, 3. The vector **Ω** describes the coupling of the electronic spin to the electronic orbital motion in a way analogous to the vector **J** in the atomic case (cf. Eq. II.4). Different Ω values correspond to different energy values. The Ω value is marked as a subscript to the Greek capital letter symbolizing $|\Lambda|$; we thus distinguish in this example between $^3\Delta_1$, $^3\Delta_2$ and $^3\Delta_3$ states.[†] The total molecular angular momentum **J** is now obtained by adding **Ω** to the angular momentum **R** of the nuclear rotation

$$\mathbf{J} = \mathbf{\Omega} + \mathbf{R}. \tag{II.9}$$

The quantum number J can take only the values: $\Omega, \Omega+1, \Omega+2,...$; the fact that no J values inferior to Ω exist, may be intuitively expected as **R** is perpendicular to the internuclear axis and thus to **Ω**. It should be realized that J is always a strict quantum number while in Hund's case a Ω behaves, in good approximation, as a quantum number too. No such quantum number can be associated, however, with **R**, unless $\Omega = 0$.

In *Hund's case b* **Λ** first couples with **R** to form a resultant **N** with quantum number N[*]

$$\mathbf{N} = \mathbf{\Lambda} + \mathbf{R}, \tag{II.10}$$

where N can take only the values $|\Lambda|, |\Lambda|+1, |\Lambda|+2, ...$ (cf. above). **J** is then

---

[†] Also when $S > |\Lambda|$, there are still $(2S+1)$ *different* energy levels, which are distinguished from each other by marking them: $|\Lambda| + S$, $|\Lambda| + S - 1, ... , |\Lambda| - S$ (< 0) instead of using Ω.

[*] This quantum number was denoted by the letter K in the older literature.

obtained from

$$J = N + S, \tag{II.11}$$

with J being limited by: $N-S \leqslant J \leqslant N+S$ $(N \geqslant S)$. In this case J, N and Λ are good quantum numbers, but Σ and Ω no longer exist as such. For $^1\Sigma$ states with $S = \Lambda = 0$, however, a good quantum number can be associated with $\mathbf{R} \equiv \mathbf{J}$.

Case b holds when the electronic spin is relatively unimportant; it holds, with some exceptions, for all Σ states and for all singlet states. For singlet states we have: $J = N$ (called: *case b'*). Case b also holds approximately for molecules with only very light atoms, such as OH and MgH. Case a, being indistinguishable from case b when the spin S is equal to zero, holds when the fine-structure splitting due to the electronic spin is relatively large. Complications arise in the identification of molecular spectra when the upper and the lower level correspond to a different coupling case.

Each rotational level of a molecule in a given electronic state can be classified as *positive* (+) or *negative* (−), depending on whether the *total* eigenfunction remains unchanged upon reflection to the origin or just changes its sign, respectively. The *sign* of a molecular state corresponds to the parity of an atomic state; this sign must not be confused with the distinction between Σ^+ and Σ^- states. For an electronic Σ^+ state, levels with even N are always positive and with odd N negative; for a Σ^- state the reverse holds.

2b-2 MOLECULAR ENERGY LEVELS AND DEGENERACY

In the absence of applied electric or magnetic fields, molecular states that differ only in total magnetic quantum number M_J have exactly the same energy and show a *spatial degeneracy* $g_J = 2J+1$, just as in the atomic case.

A further two-fold degeneracy, called Λ-*doubling*, arises for states with $\Lambda \neq 0$ because the energy of orbital motion of the electrons remains the same when Λ changes sign (that is, when the sense of their rotation around the internuclear axis is reversed). However, this degeneracy is slightly removed owing to the coupling of the electronic orbital motion with the nuclear rotation. The resulting energy splitting† is usually quite small and will be ignored so that a degeneracy $g_\Lambda = 2$ for $|\Lambda| > 0$ and $g_\Lambda = 1$ for $\Lambda = 0$ (i.e. for Σ states) will be assumed in the following.

The *multiplicity* due to the resultant electron spin S leads practically to a further degeneracy of the molecular states: $g_S = 2S+1$ in Hund's case b. In case a, a larger fine-structure splitting may be found; it depends on the kind

---

† Note that the corresponding eigenstates are not functions labelled by $+|\Lambda|$ and $-|\Lambda|$ respectively, but are linear combinations thereof. These states have opposite sign.

of problem being studied whether one wants to treat the multiplet practically as a set of degenerate states with $g_S = 2S+1$, or not.

The *energy* of a *molecular level*, that is, a set of degenerate states or a single, nondegenerate state, is composed of contributions from the electronic, vibrational and rotational motions. In the *Born-Oppenheimer* (B.-O.) *approximation* the electronic and nuclear motions are treated as being independent of each other. In the most simple form, the molecular eigenfunction is considered as a product of three eigenfunctions describing the three (electronic + nuclear) motions separately. The total internal energy, E_m, of the molecule in a particular state is then the sum of an electronic, a vibrational, and a rotational term:

$$E_m = E_e + E_v + E_r. \tag{II.12}$$

There are, however, interactions of various types between these three kinds of internal motion in a molecule. We shall see presently how these interactions affect the three terms in Eq. II.12 and thus E_m.

The *electronic energy* E_e, including a possible splitting due to the electron spin in Hund's case a, is given by the minimum value of the potential energy function pertinent to the stable electronic state considered. Electronic excitation energies are usually in the range of 1 - 10 eV.

Taking into account the anharmonicity of the vibrational motion, we obtain for the quantized *vibrational energy* E_v in good approximation the expression

$$E_v \equiv hcG(v) = hc\omega_e(v+\tfrac{1}{2}) - hcx_e\omega_e(v+\tfrac{1}{2})^2 \quad (v = 0, 1, 2, 3, \ldots). \tag{II.13}$$

Herein ω_e is the fundamental vibrational wave number[†] for infinitesimal amplitude and x_e is the anharmonicity constant, which is small (≈ 0.01) and positive. The factor hc derives from the conversion of wave number into energy. $G(v)$ is referred to as the vibrational term (value), generally expressed in cm^{-1}. For low v numbers the vibrational terms are practically equidistant with separation ω_e.

The potential curve can often be closely approximated by a Morse potential (see textbooks) and we obtain the relation

$$x_e\omega_e = \omega_e^2/4D_e^*, \tag{II.14}$$

where D_e^* is the dissociation energy expressed as wave number and counted from the minimum of the potential curve (see Fig. II.2). We see from Eq. II.13 that the lowest vibrational energy, obtained for $v = 0$, is not zero but, in good approximation, equals $\tfrac{1}{2}hc\omega_e$, which is called the *zeropoint vibrational energy*. The lowest energy that a molecule can have in a given electronic state is therefore not E_e

---

[†] The choice of symbol is historical; it should not be confused with the circular frequency of vibration $\omega_v \equiv 2\pi\nu_v$, which differs from $\omega_e \equiv \nu_v/c$ (ν_v = vibrational frequency).

but $E_e + \frac{1}{2}hc\omega_e$. The actual, that is, physically determinable *dissociation energy*, D_0, is consequently related to D_e^* by

$$D_0/hc \equiv D_0^* = D_e^* - \tfrac{1}{2}\omega_e . \qquad (\text{II.15})$$

Since ω_e is usually in the $300-3000$ cm^{-1} range, this zeropoint energy, and, more generally, the vibrational-level separations, are of the order of 0.1 eV.

The values of ω_e and x_e, and thus the vibrational terms, depend in general on the electronic state, because the shape of the potential curve does. The effective shape of this curve is, moreover, influenced by the centrifugal effect of the molecular rotation.

This centrifugal effect can be formally taken into account by adding, for fixed J value, a positive 'centrifugal potential term' $\propto J(J+1)/\mu r^2$ (μ = reduced mass) to $V(r)$. The resulting curve is called the *effective potential curve* (see Gaydon 1968). The position of the minimum in the latter curve is shifted to larger r values when J increases. To the right of this minimum a local maximum, called *centrifugal barrier*, arises, which has to be surmounted when the molecule is to dissociate at given J value. At very high J values the effective potential curve may even show no minimum at all; the molecule is then 'rotationally unstable'.

Since the centrifugal potential term represents the kinetic energy of rotation, its effect on the effective potential curve is not great as long as this kinetic energy is of the order of kT (k = Boltzmann constant; T = absolute temperature) and $D_0 \gg kT$.

The expression for the *rotational energy* E_r depends on the way in which the electronic and nuclear angular momenta are coupled. In Hund's case a we have the approximate expression

$$E_r = hcB\{J(J+1) - \Omega^2\} , \qquad (\text{II.16})$$

where $B \equiv h/8\pi^2 Ic$ is called the *rotational constant*, I is the moment of inertia of the molecule in its equilibrium position with respect to an axis through the centre of mass, perpendicular to the internuclear axis, and c is the light velocity. We have: $I = \mu r_0^2$ where r_0 is the equilibrium separation of the nuclei, and μ is again the reduced mass. The value of B depends on the electronic state, which determines r_0 through the associated potential curve. The value of B diminishes slightly with increasing J as r_0 increases owing to centrifugal stretching. There is an interaction between rotation and vibration because the momentaneous value of the moment of inertia depends on r which changes periodically and comparatively rapidly with time owing to the vibration. This interaction can be taken into account by considering B in Eq. II.16 as a function of v.

It is noted from the preceding subsection that J cannot be less than Ω; the latter has a fixed value for a given electronic state in Hund's case a.

In some expressions for E_r a J-independent term $A\Omega^2$ is added to the right-hand side of Eq. II.16. The coefficient A is related to the moment of inertia of the electrons around the internuclear axis. Since Ω too is related to the motion of the electrons only, this extra term can be considered as being included in E_e instead of E_r.

In Hund's case b we have, instead of Eq. II.16,

$$E_r = hcB\{N(N+1) - \Lambda^2\}, \qquad (II.17)$$

with $N = |\Lambda|, |\Lambda|+1, \ldots$. For singlet states $\Omega = |\Lambda|$ and $J = N$, so Eqs II.16 and 17 become identical.

For $^1\Sigma$ states we have $\Omega = \Lambda = 0$. Then J relates to the nuclear rotation only and takes the values $0, 1, 2, \ldots$, while the expressions in Eqs II.16 and 17 coincide and read simply

$$E_r = hcBJ(J+1) \qquad (J = 0, 1, 2, \ldots). \qquad (II.18)$$

The rotational-energy separations, $E_r(J+1) - E_r(J)$, increase linearly with J and are much smaller than the vibrational-energy separations, as hcB is of the order of 10^{-4} eV.

Molecules containing hydrogen have small μ values and thus relatively high rotational constants and large rotational-energy separations.

The above described quantization of the rotational energy can be disturbed by *predissociation*. This effect occurs when a molecule in a stable electronic state, having acquired a certain amount of vibrational-rotational energy, can make a radiationless transition to an unstable state of the same energy. After this transition it dissociates spontaneously. When this transition takes place in a time which is shorter than the rotation period, the rotational energy levels become blurred; this need not be the case for the vibrational energy level, as the vibration period is usually much shorter.

At flame temperatures, usually a large number of rotational energy levels are populated, but only a few vibrational levels (cf. Sect. II.3). For large J (or N) the distinction between the expressions in Eqs II.16 and 17 vanishes as Ω or $|\Lambda|$ are usually not much larger than 1. Molecular states with $S > 3$ are rather the exception.

Energy level diagrams of *poly-atomic molecules* are much more complex because of their larger number of degrees of freedom. With nonlinear molecules it is possible to define three, mutually orthogonal principal axes of inertia with correspondingly three *principal moments of inertia* and *three rotational constants*. In this frame of reference the expression for E_r as a function of J takes its simplest form. Even in the absence of any coupling between the electronic and nuclear motions the expression for E_r of an asymmetric-top molecule (characterized by three different rotational constants) is still much more complex than Eq. II.18. There are, in addition, $3N-6$ *normal vibrations* or *modes* in which the vibrational motions of the N nuclei can be decomposed. Each mode is characterized by its own fundamental frequency; some of these modes may be degenerate. In first approximation, the contribution of each nondegenerate mode to E_v is similar to the leading term in Eq. II.13 for a di-atomic molecule; the over-all state of vibration is then characterized by a set of vibrational quantum numbers $v_1, v_2, v_3, \ldots$, each of which can take the values $0, 1, 2, \ldots$.

Linear poly-atomic molecules behave much the same as di-atomics with regard to their rotational and electronic degrees of freedom. They are again characterized by a single (principal) moment of inertia relating to an axis perpendicular to the internuclear axis. There are here $3N-5$ normal modes; for a tri-atomic linear molecule at least two of them are degenerate.

For detailed information about poly-atomic molecules we refer to, e.g., Mavrodineanu and Boiteux (1965) and Herzberg (1966)

Gaydon (1974) has tabulated molecular constants of di- and poly-atomics in various electronic states which are of interest for the flame background spectrum. Molecular constants of some metal compounds of interest in flame spectroscopy have been presented by Mavrodineanu and Boiteux (1965). Extensive tables of molecular data are found in Rosen (1970) and, for di-atomics, in Barrow (1979).

2c THE ATOMIC LINE SPECTRUM

General. The spectrum of a neutral atom is marked by Roman number I, that of the singly ionized atom by II, etc. The (central) *frequencies* of the spectral lines of an atomic species are determined by the differences in energy between the *upper* and *lower level*, $E_q - E_p$, between which an optical transition is 'allowed', according to

$$\nu = (E_q - E_p)/h , \qquad (II.19)$$

where h is Planck's constant. This relation follows from Bohr's postulate no. 2 (recoil effects due to the linear momentum of the photon are totally negligible in flame spectroscopy). Optical selection rules tell us whether an optical transition is allowed or not (see below). The *wavelength* of the line is obtained from: $\lambda = c/\nu$ and the *wave number* σ from: $\sigma \equiv 1/\lambda = \nu/c$, where c is the light velocity in vacuo. One may also define λ in air by multiplying the vacuum value by $1/n$, where n is the refractive index of air at the considered wavelength (see also end of Sect. III.4b). Wave numbers are always defined in vacuo so that they are directly related to energy differences. Although officially the nanometer (1 nm = 10^{-9} m) is recommended as the unit for λ, the ångström (1Å = 0.1 nm) is still in use as a suitable unit in the optical range. Wave numbers are usually expressed in cm^{-1}; the name 'kayser' for this unit is not officially approved. Frequencies are most often used in theoretical work. In experimental work (very) small spectral separations are often expressed in mÅ ($1\,mÅ = 10^{-4}$ nm $= 10^{-1}$ pm) or in frequency units. Atomic lines of interest in flame spectroscopy have photon energies in the $1-5$ eV range corresponding to a wavelength range of about $10^4 - 2500$ Å. We refer to Appendix A.1 for conversion factors of the various units in use.

The optical transitions observed in flame spectroscopy generally involve a jump of a single electron belonging to an open outer subshell (e.g., an open s subshell for the alkalis) or to a closed outer subshell when all other subshells are also closed (e.g., a closed s subshell for the alkaline-earths). If the configuration of all other electrons remains the same as in the ground level, it is sufficient to specify only the quantum numbers (n, l) of this one electron in the upper and lower level, in addition to the atomic quantum numbers. For example, the 4267 Å Ca emission line may be described as a $4s^1S_0 \leftarrow 4p^1P_1$ transition. For the alkali lines we have: $l = L$ and we can describe the 5890.0 and 5895.9 Å Na emission lines simply as $3^2S_{\frac{1}{2}} \leftarrow 3^2P_{\frac{3}{2}}$ and $3^2S_{\frac{1}{2}} \leftarrow 3^2P_{\frac{1}{2}}$ transitions, respectively, omitting l.

It is customary to represent an optical transition by writing the symbol of the lower level first and that of the higher level second, both being separated by a dash (—). An absorption transition is then denoted by an arrow pointing to the right (→), whereas an emission transition is denoted by an arrow in the reverse direction (←). In molecular spectroscopy the opposite notation is used. Because of this controversy and also because we shall have to deal with collisional transitions too, we shall not keep strictly to these conventions.

Optical selection rules. We are interested here in *electric-dipole radiation*, involving the emission or absorption of a *single* photon; this kind of radiation is the usual one that is observed in flame spectroscopy. In this case the number of allowed transitions is restricted by the following *general selection rules* that must be strictly obeyed:

1. the *parities* of the upper and lower level must be different;
2. the jump in l is restricted by: $\Delta l = \pm 1$ (or $\Delta l = +1$ when $l = 0$ in the initial level) if only one electron jumps;
3. the jump in J is restricted by: $\Delta J = 0$ or ± 1, but $J = 0 \to J = 0$ is not allowed; for the magnetic quantum number we have: $\Delta M_J = 0$ or ± 1, but $M_J = 0 \to M_J = 0$ is not allowed when $\Delta J = 0$;
4. the rules that apply for J and M_J also apply for F and M_F;
5. the jump in quantum number n, however, is unrestricted.

The selection rule for M_J has consequences for the Zeeman splitting of spectral lines in a magnetic field. However, when the magnetic field is very strong, other selection rules replace the rule for M_J.

The selection rule for M_J is also related to the *polarization* of the radiation (in the presence as well as absence of a magnetic field). When we choose, e.g., the z-axis parallel to the propagation direction of the observed radiation beam, transitions with $\Delta M_J = \pm 1$ produce radiations that are circularly polarized in opposite directions. When the z-axis is perpendicular to the beam direction, a transition $\Delta M_J = 0$ corresponds to linearly polarized radiation with polarization axis parallel to the z-axis. For a further theoretical treatment we refer to, e.g., Sobel'man (1972).

We have, in addition, the following *special selection rules* that hold only approximately, depending on whether the $L-S$ or (j,j) coupling scheme prevails. If $L-S$ coupling applies to both levels, we have the further restrictions:

6. $\Delta L = 0$ or ± 1;
7. $\Delta S = 0$.

When (j,j) coupling applies to both levels the last two selection rules are replaced by

8. $\Delta j = 0$ or ± 1, if only one electron is involved in the optical transition.

It is noted that this last selection rule is broader than the two previous special rules: all transitions allowed by Rules 6 and 7 are also allowed by Rule 8. The last rule, however, also allows transitions between levels with different multiplicities, that is: $\Delta S \neq 0$ (called *intercombination lines*). In mixed coupling cases these special selection rules are not strictly applicable.

In *two-photon* processes (see also Sect. II.5a-1), only transitions between levels of the same parity are allowed; we further have the selection rules: $|\Delta J| \leq 2$, and $|\Delta L| \leq 2$ in the case of $L-S$ coupling (see for a more detailed formulation of the selection rules Cagnac, Grynberg and Biraben 1973). In the case of alkali atoms, the selection rules allow only: $\Delta L = 0$ and $|\Delta L| = 2$.

Some consequences of the selection rules. Since Δn is unrestricted (Rule 5) and $\Delta l = \pm 1$ (Rule 2), various series of spectral lines are observed with the alkali atoms, which correspond to transitions between S- and P levels, P- and D levels etc. In this case the ground ^{2}S level can only combine with the ^{2}P levels. All spectral lines that arise from an allowed transition to or from the ground level are called *resonance lines*; the other lines are called *nonresonance lines*. The resonance line(s) originating from the *lowest* excited multiplet that can optically combine with the ground level, and vice versa, are called *first resonance line(s)*.[†]

[†] In some books, only these lines are called resonance lines.

The lines of the well-known yellow Na-D doublet are thus first resonance lines.

Rule 7 forbids transitions between the singlet and triplet terms of the alkaline-earth atoms (see Fig. II.1); there are therefore two separate systems in the spectra of these atoms: a singlet and a triplet system. Since the ground level is a singlet, no downward optical transitions are allowed for the lowest Ca triplet term ($4p^3P$); we call such levels *metastable*.

For the heavier atoms, such as Hg and Zn, the (j,j) coupling scheme applies and $\Delta S \neq 0$ is allowed. The strongest resonance line of Hg at 2536.5 Å combines the middle $6p^3P^0_1$ level of the lowest ^3P term with the singlet $6s^1S_0$ ground level (note that Rules 1, 2, 3 and 8 are obeyed in this transition). A transition from the two other levels of this triplet ($6p^3P^0_2$ and $6p^3P^0_0$) to the ground level is not allowed because of Rule 3; this pair of triplet levels is thus metastable.

(Hyper-)fine-structure. A *fine-structure* (f.s.) in the *spectrum* arises when optical transitions are allowed from or to different levels belonging to the same upper or lower multiplet. For example, for the alkali atoms transitions are allowed from each level of a $^2P_{\frac{1}{2},\frac{3}{2}}$ doublet to the single $^2S_{\frac{1}{2}}$ ground level. The resonance lines thus form *spectral doublets*; the spectral spacing of the doublet components is determined by the *energy* splitting of the ^2P doublet term. The f.s. splitting of the alkali first resonance doublets increases with increasing atomic number from 0.16 Å for Li (which is of a magnitude comparable to the line width in flames) to 422 Å for Cs. For a given alkali atom the splitting of the resonance doublets decreases with increasing principal quantum number n of the upper term.

For the alkalis there are (only) three optical transitions allowed between a $^2D_{\frac{3}{2},\frac{5}{2}}$ doublet term and a $^2P_{\frac{1}{2},\frac{3}{2}}$ term because of Rule 3 which tells us that $|\Delta J|$ cannot be larger than one. The existence of a third component (close to one of the others) leads to a *compound doublet*; the component with $\Delta J = 0$ is, however, weak.

Thallium is an example where the ground level ($6p^2P^0_{\frac{1}{2}}$) forms part of a doublet term; the other level ($6p^2P^0_{\frac{3}{2}}$) lies at 0.97 eV above the ground level. The latter is a metastable level as it cannot optically combine with the ground level because of Rules 1 and 2. Both levels combine with the single $7s^2S_{\frac{1}{2}}$ level lying at 3.28 eV; the resulting lines at 3776 and 5350 Å thus form a spectral doublet, although their spectral separation is large.

A *hyper-fine-structure* (h.f.s.) may be observed in the *spectrum* as a consequence of the h.f.s. splitting of the energy levels due to the *nuclear spin*. Neglecting the h.f.s. splitting of the Na ^2P levels, which is much smaller than that of the Na ^2S ground level (see Subsection 2a), we find each of the Na first resonance lines split into two h.f.s. components ($I = \frac{3}{2}$). This splitting amounts to about 0.02 Å and can barely be resolved in flames. In the Cu 3247 Å and In 4100 Å lines, however, a distinct hyper-fine-structure can be observed in flames.

A hyper-fine-structure in the spectrum can also result from the *isotope shift* when various isotopes of the same element are present. The isotope shift amounts to about 0.16 Å for the first resonance lines of ^7Li; this effect has been used for isotope analysis of $^7Li/^6Li$ mixtures.

For a detailed discussion of the effects of h.f.s. in analytical atomic spectroscopy, we refer, e.g., to Preobrazhenskii (1963), L'vov (1972, 1972a), and L'vov *et al.* (1976).

General tabulations of wavelengths of atomic lines are found in, e.g., Harrison (1948), Moore (1949-1952-1958 and 1972), Zaidel, Prokof'ev and Raiski (1961), Jaffe (1962), Mavrodineanu and Boiteux (1965), Striganow and Sventitskii (1968), Gilbert (1970/71), Parsons and McElfresh (1971), Meggers, Corliss and Scribner (1975), and Alkemade and Herrmann (1979).

2d THE ELECTRONIC SPECTRUM OF DI-ATOMIC MOLECULES

General. As with the atomic spectrum, the frequencies of the molecular spectral lines are determined by the positions of the energy levels, through Eq. II.19, and by the selection rules. Since there are many more energy levels in a molecule than in an atom, the molecular spectrum is much richer in lines, which are usually crowded into groups. When the spectral resolution of the spectrometer is insufficient or the molecular lines overlap because of spectral broadening (see Sect. II.5b), the individual lines within a group may not be resolved. For this reason these groups were called 'bands' to distinguish them from the sharp atomic 'lines'. We shall use the term band in a more general sense (see below), independently of whether it is spectrally resolved or not.

Because the energy-level spacing with di-atomic molecules containing one hydrogen atom, like CuH, is relatively large, the corresponding bands show an 'open' structure in flames. Tri-atomic molecules, like CaOH, usually exhibit unresolved band spectra in flames, even with good instrumental resolution (see Appendix A.8). An exception is, for example, BO_2. Most metal compounds of interest in flame molecular spectroscopy are di- or tri-atomics containing only one metal atom.

We shall be concerned mainly with molecular bands in the visible and u.v. regions of the spectrum, which are due to combined electronic, vibrational and rotational transitions. Transitions that involve only a change in vibrational and rotational energy give rise to bands in the (near-)infrared, whereas purely rotational spectra lie in the far infrared or microwave regions.

THE STRUCTURE AND SPECTRUM OF ATOMS AND MOLECULES   Sect. II.2d

In this subsection we shall restrict ourselves to the electronic band spectra of di-atomic molecules.

Optical selection rules. Electric-dipole transitions involving only one photon are restricted by the following *general selection rules:*

1. the molecular states between which the transition takes place must have opposite sign, that is: $+ \to -$ or $- \to +$ (this rule replaces Rule 1 in the atomic case);
2. $\Delta J = 0$ or ± 1, but $J = 0 \to J = 0$ is forbidden (identical to Rule 3 in the atomic case);
3. for molecules with nuclei of equal charge, even electronic states combine only with odd ones, and vice versa, that is: $g \to u$ or $u \to g$.

There are, in addition, *special selection rules* which depend on which of Hund's cases applies. If the same case applies to the upper and the lower level, we have additional restrictions; some of them are:

4. in cases a and b, $\Delta \Lambda = 0$ or ± 1, but a transition between a Σ^+ and a Σ^- state is forbidden;
5. in cases a and b, $\Delta S = 0$; however, in practice, weak transitions with $\Delta S = \pm 1$ do occur;
6. in case a, $\Delta \Sigma = 0$ and $\Delta \Omega = 0$ or ± 1;
7. in case b, $\Delta N = 0$ or ± 1, but $\Delta N = 0$ is forbidden for $\Sigma - \Sigma$ transitions.

There are no selection rules for the vibrational quantum number v in a combined electronic-vibrational transition. However, the (relative) strength of a transition between vibrational levels with quantum number v' and v'', belonging to two different electronic states, depends sensitively on the relative positions of the corresponding potential energy curves (see Fig. II.3). According to the *Franck-Condon principle*, the internuclear distance r will hardly change during the short time that an electronic transition takes. The electronic transition can therefore be represented by a *vertical* line between two points of the potential curves involved. When the molecule is found in a vibrational level, with quantum number v', belonging to the upper electronic state, downward transitions to those v'' levels are favoured, whose 'end points' along the r scale coincide with or lie in the vicinity of those of the initial v' level, and vice versa (see Fig. II.3). The end or turning points correspond to the minimum and maximum values of r that the molecule, considered as a classical vibrator, can reach. A harmonic classical vibrator spends most of its time in the vicinity of these turning points. In Quantummechanics the relative transition probabilities are described as a function

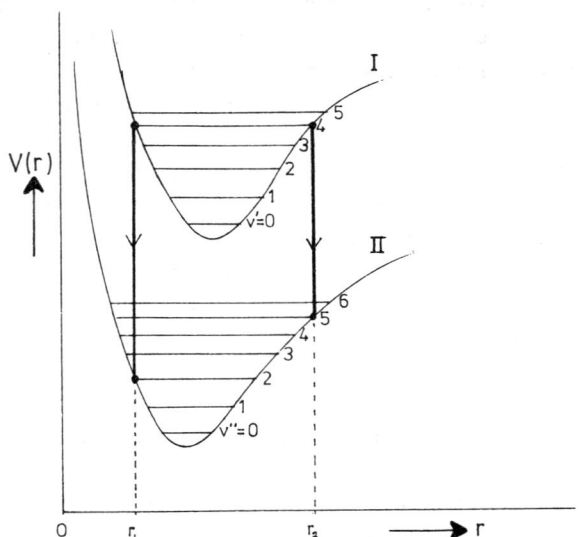

Fig. II.3 According to the *Franck-Condon principle* optical transitions from the vibrational level $v' = 4$ in the upper electronic state I take place preferably to those vibrational levels ($v'' = 2$ and 5) in the lower state II, that have turning points (r_1 and r_2) lying *vertically* under those of the initial level, and vice versa.

of (v', v'') by the *Franck-Condon factor*, which is the squared overlap integral of the vibrational wave functions involved (see also Sect. II.5a - 3).

Description of band spectra. A *band* comprises all molecular lines originating from a given vibrational and electronic transition. A band is thus associated with a pair of vibrational levels with quantum numbers v' and v'', belonging to a specific upper and lower electronic state, labelled by q and p, respectively. Each band has an *origin* that is determined by the difference in electronic plus vibrational energy of the upper and lower level (cf. Eqs II.12 and 13). All bands belonging to a given electronic transition constitute together a *band system*; a band system is characterized by the difference in electronic energy, $E_q - E_p$, and by the coefficients occurring in the expressions for $E_{v'}$ and $E_{v''}$ (cf. Eq. II.13). The bands belonging to the same electronic transition and having a given $\Delta v \equiv v' - v''$ form a *sequence*. The bands belonging to the same electronic transition and having a fixed v' (or v'') form a v''- (or v'-) *progression*.

Within a band we distinguish between the *P-, Q- and R branch* according to ΔJ ($\equiv J' - J''$) $= -1, 0$ and $+1$, respectively (cf. Rule 2); single and double primes denote here again the quantum numbers in the upper and lower electronic state, respectively. The positions of the lines within each branch are determined by the rotational constants B of the upper and lower vibrational level, according to Eqs II.16 or 17. Since these constants are in general different, the lines of a Q branch do not coincide. A Q branch does not always occur; for example, for a transition between two $^1\Sigma^+$ or two $^1\Sigma^-$ states (see Rule 4), ΔJ cannot be zero because of Rule 1 in combination with the fact that in this case the sign of the molecular level does not change when J is not altered. Note also from Subsection 2b-1 that J has a lower bound: $J \geqslant \Omega$ in coupling case a and that N has a lower bound $N \geqslant |\Lambda|$ in coupling case b. There may therefore be some 'missing' rotational lines in the range of low $J-$ or N values.

As a further consequence of the difference in rotational constants of the upper and lower level a number of rotational lines may be found to crowd together, thus forming a more or less sharp edge in the spectrum, called *band head*. The band head need not coincide with the band origin. When the rest of the band is found at the red (or violet) side of this head, the band is said to be *degraded* to the red (or violet).

A rotational line in a given (v', v'') band can be characterized by the capital letter (P, Q or R) which denotes the branch to which it belongs; this letter is followed by the quantum number N'' of the lower level, placed between parentheses. In case a, N is formally defined by: $N \equiv J - \Sigma$. When ΔN ($\equiv N' - N''$) $= -1$, 0 or $+1$, a superscript capital letter P, Q or R is added to the main letter symbol for ΔJ; for example, a transition: $\Delta N = +1$, $\Delta J = 0$ and $N'' = 13$ is denoted by: $^R Q(13)$. When $\Delta N = \Delta J$, the superscript for ΔN is omitted. Since there are $(2S+1)$ different J values for given N value, two cardinal numbers are added as subscripts to the ΔJ symbol in order to distinguish between them, one for the upper and another for the lower level. For example, in the case of a doublet the notation $^R Q_{21}(N'')$ indicates that $J' = N' - \tfrac{1}{2}$ and $J'' = N'' + \tfrac{1}{2}$.

In molecular spectroscopy, a transition is represented by writing first the symbols characterizing the upper level and second the symbols of the lower level, with a dash (—) between them (cf. Section 2c for atomic lines).

A *resonance band* is defined as a band whose lower electronic level is the ground level.

When $|\Lambda| \neq 0$ for both levels a splitting of each rotational line occurs because of Λ-doubling. This splitting is however very small and of no significance in flame spectroscopy. The same can be said of the small line splittings caused by the electronic spin when a transition between two Σ states occurs.

The rotational structure of a band may become diffuse or be suddenly broken off at or above a certain N value as a result of *predissociation* (see Subsection 2b-2). This effect has been utilized to determine dissociation energies (see Sect. VIII.2).

The spectral positions of the band heads and of the individual rotational lines within a band may be shifted as a result of *isotope effects* on the electronic, vibrational and rotational energy levels. The effect on the vibrational and rotational energy levels is greater than on the electronic energy level; it is recalled that the fundamental vibrational frequency and the rotational constant are inversely proportional to the square-root of the reduced mass and to the reduced mass, respectively. Isotope effects are especially strong when in a hydride molecule H is replaced by one of its isotopes.

Gaydon (1974) has tabulated and described molecular emission and absorption spectra in flames with and without additives and has given data on the terms and molecular constants involved. Mavrodineanu and Boiteux (1965) have presented spectrograms of molecular emission bands which they photographed in premixed stoichiometric C_2H_2-air and C_2H_2-O_2 flames with metal vapours. They have also tabulated the spectral features of the observed bands, the terms and the molecular constants involved. Spectrometric records of bands emitted by some metal compounds in flames are found in, e.g., Herrmann and Alkemade (1960, 1963) and in Alkemade and Herrmann (1979). General, extensive tabulations, etc., of molecular bands are found, e.g., in Herzberg (1950, 1966, 1971), Gatterer et al. (1957), and Pearce and Gaydon (1976).

Flame spectroscopy has contributed to our knowledge of molecular spectra, in particular with respect to the identification of their emitters. This is true especially for the bands of Ca, Sr and Ba emitted in flames; the emitter, term designation and excitation energy of some of these bands have been long disputed. A large variety of test experiments with alkaline-earth elements in flames, ovens, arcs, sparks and molecular beams have been done in the past. In particular, isotope substitution ($^{18}O_2$ for $^{16}O_2$; D for H) has been employed to identify the emitter. A survey of results obtained and of opinions expressed in the literature has been given by van der Hurk, Hollander and Alkemade (1973) for the visible Ca, Sr and Ba bands, and by Frank and Krauss (000) for the Ca bands (see also for more general band identifications Mavrodineanu and Boiteux 1965, and Gaydon 1974). The bands emitted in C_2H_2-air, H_2-air, moist CO-N_2-O_2 and CO-N_2O flames by Ca at 5540, 6020 and 6230 Å, and by Sr at 6060, 6470, 6690 and 6820 Å have been assigned by van der Hurk, Hollander and Alkemade (1973) to the monohydroxides. The same authors have assigned the diffuse Ba bands found in C_2H_2-air flames at 4870 and 5120 Å also to the monohydroxide, whereas other structures could be identified with known BaO bands. It might well be possible that under different experimental conditions bands from other emitters are found in the same spectral regions as the above mentioned bands (cf. Frank and Krauss 000). All alkaline-earth flame bands found so far are emitted by species that contain only one alkaline-earth atom.

In Appendix A.8 some records of the visible Ca, Sr and Ba bands in flames are reproduced.

2e CONTINUOUS SPECTRA

In addition to a discrete line or band spectrum, flames with or without additives may emit a *continuous spectrum* or *(spectral) continuum*. We speak of a continuous spectrum when the spectral intensity $I_\lambda(\lambda)$ (see definition in Appendix A.5) is a smooth, continuous function of wavelength λ over an extended wavelength range. We exclude here *apparent continua* that may be observed when a spectrum consisting of many close-lying lines is scanned by a spectrometer of insufficient resolution. We also exclude the case of a *quasi-continuum*[†] that arises when the individual lines in a banded spectrum lie so close to each other that they overlap because of (intrinsic) spectral line broadening.

A (real) spectral continuum can be emitted in the flame by incompletely volatilized particles because of *incandescence*. Elements such as Al, Cr and V may readily form involatile oxide particles when salt solutions are nebulized into the flame (see Sect. III.9). Their thermal radiation depends on the flame temperature (which also influences the rate of volatilization) and is found mainly in the (infra-)red part of the spectrum. The temperature of these particles can be raised above the flame temperature through the release of chemical energy in the recombination of excess H- and OH radicals on the particles' surfaces (see Kallend 1967); the resulting suprathermal radiation is called *candoluminescence* (see Gaydon 1974). These nonspecific continuous radiations enhance the spectral background and cause spectral interference in analytical flame spectroscopy when concomitants forming involatile compounds are present at high concentration in the solution (see textbooks cited in Chapt. I).

In flames, a continuum can also be emitted in the gaseous phase as a result of a *recombination process* between two particles, for example, $K^+ + e^- \rightarrow K + h\nu$ (*ion-recombination continuum*) and $Na + OH \rightarrow NaOH + h\nu$ (*chemical association continuum*). The energy released in the recombination process equals the sum of the discrete ionization or dissociation energy of the product particle and the initial translational energy of relative motion, which is continuously distributed. When one of the reactants is in an excited state, its discrete excitation energy is added to the total energy released in the recombination. The recombination energy, or part of it, can be radiated as a photon. On account of the continuous distribution of the translational energy, the frequency of this radiation — in a certain range — will also be continuously distributed. Besides, a discrete part of the

---

[†] The expression 'quasi-continuous' is also used in this book to denote that the spectral width of the radiation from an auxiliary light source (e.g. laser) is large compared to the width of the atomic absorption line in the flame. The equation describing the absorption of this radiation is then, virtually, the same as for a real continuum (see Sect. II.5c-2). On an absolute scale, the radiation may still be contained in a narrow spectral interval.

recombination energy may be converted into quantized internal energy of the product particle. This may cause discontinuities or 'edges' in the spectrum of the ion-recombination radiation, which correspond to different atomic excitation levels. Since the atomic levels converge towards the ionization limit, the edges that belong to high excitation levels become indistinct. The position of the edge with the highest frequency corresponds to the ionization energy counted from the ground level. Under certain conditions, a *short-wavelength cut-off* or *short-wave limit* may be found in the association continuum, which corresponds to the dissociation energy of the product molecule in the ground level.

Because of the principle of microscopic reversibility (see Sect. II.4b-4), the above processes can also occur in the reverse direction, that is, in absorption giving rise to a *photo-ionization* and *photo-dissociation continuum*. The latter process has been observed in flames with, e.g., alkali- and Sr halides (see Yoshimura et al. 1977, and Furuta et al. 1978). The long-wavelength limit of the photo-dissociation continuum is connected with the dissociation energy (see Sect. VIII.2a).

The occurrence of an association continuum can usually be described by means of potential curves (see also Sect. II.4). It may happen that two particles, in their ground states, approach each other on a mainly repulsive potential curve, forming a temporary, unstable collision complex. When no internal energy is removed from this complex, in the form of radiation or by energy transfer to a third collision partner, it will redissociate. The complex may be stabilized, however, when it can make a spontaneous transition to a lower lying, stable molecular state by emission of a photon with a continuous frequency-distribution. The probability of photon emission per collision depends on the ratio of the duration of the collision to the radiative lifetime (defined in Sect. II.5a-2).

It may also happen that the collision complex first passes, by curve-crossing, from the initial repulsive potential curve to another curve belonging to a stable, but electronically excited state, without photon emission. This step may be followed by an electronic transition to the ground state with emission of a photon. Because of the effect of predissociation (see Sect. II.2b-2), the emitted spectral radiation may have a continuous character or show diffuse bands in a certain range.

A different situation exists when the unstable collision complex first makes a radiationless transition to a stable, excited molecular state of discrete energy, which is induced by a collision with a third partner. When this step is followed by a radiative transition to, e.g., the ground state, the emitted spectrum will have a discrete, banded structure (see process no.39 in Table II.1 on page 108).

Whereas the significance of ion-recombination continua in flames has been questioned by James and Sugden (1958) and Alkemade (1959), the occurrence of chemical association continua of various kinds has been firmly established in the region above the combustion zone, although they are often weak in comparison to the atomic line and molecular band emissions. Flame radicals like H, OH and O, which may be present in excess over chemical equilibrium concentrations, usually participate in the radiative association process. Additives if present in large concentrations may produce observable continua, for example, by the following association reactions: $A + OH \rightarrow AOH$ (A = alkali atom; see James and Sugden 1958, Alkemade 1959, and McEwan and Phillips 1967), $H + Cl \rightarrow HCl$ and $Cl + Cl \rightarrow Cl_2$ in fuel-rich and oxygen-rich H_2 flames, respectively (see Phillips and Sugden 1961), or $SO + O \rightarrow SO_2$. Examples of reactions between flame constituents which may produce continua are: $CO + O \rightarrow CO_2$ and $NO + O \rightarrow NO_2$. We shall encounter several of these examples in other sections (e.g., Sects IV.5 and IV.8; see for a general review also Sugden 1962). In some cases there is still doubt whether a real continuum is produced or a quasi-continuous emission which is connected with third-partner effects.

Continua are associated with *processes* between two particles rather than with properties of individual species. They may yield information about the (repulsive) interaction potential which governs the process. On the other hand the long-wave limit of photo-dissociation continua can yield information about the dissociation energy of the product molecule. For a further discussion of spectral continua we refer to Gaydon (1968, 1974) and Thrush (1968).

3. THE EQUILIBRIUM STATE [†]

3a THERMODYNAMIC EQUILIBRIUM, TEMPERATURE, AND EQUIPARTITION OF ENERGY

When a system is in contact with a heat bath at uniform temperature, it will attain after a while a state of *thermodynamic equilibrium* if it is not disturbed from the outside. The heat bath may be a furnace which encloses the system (e.g., a gas mixture), or it may be an opaque gas of such large extensions that it screens any interior volume element from exchange of energy and particles with the surrounding world. When the 'system' is, e.g., a metal vapour present as a trace in a flame, the flame may be considered approximately as a heat bath which fixes the temperature of the system.

[†] For a more extensive discussion in relation to flames and hot gases see, for example, Unsöld (1955), Lewis and von Elbe (1961), Laidler (1963), Mavrodineanu and Boiteux (1965), Boumans (1966), Stevens (1967), Richter (1968), and Gaydon and Wolfhard (1970). The general theoretical background of this section is treated in, e.g., ter Haar (1954), Cambel, Duclos and Anderson (1963), and Fowler and Guggenheim (1965).

The particles contained in a gas mixture possess energy in various forms and may dissociate or ionize into fragments; they may also emit and absorb radiation at certain wavelengths. In thermodynamic equilibrium, the distribution of energy over the various translational and internal degrees of freedom and the distribution of dissociation and ionization products, as well as the spectral distribution of the radiant density, are governed by a single and universal parameter, T. This parameter is called <u>the temperature</u> of the system. At given value of T, these distributions are independent of the type and rate of the detailed mechanisms through which energy is exchanged between the various forms or through which dissociation and ionization are achieved. The equilibrium distribution depends on the temperature and the molecular properties of the individual species, not on the type of interactions between them.[†]

In equilibrium there is a certain partition of the average values of energy stored in different forms. For the translational energy the average value per degree of freedom (in the x-, y- and z-directions, respectively) and per particle amounts to $\tfrac{1}{2}kT$, where k is Boltzmann's constant and T is expressed in kelvin (K). When considered per mole of substance the average translational energy per degree of freedom amounts to $\tfrac{1}{2}RT$ where R is the gas constant (for numerical values of k and R see Appendix A.2). We thus have

$$\tfrac{1}{2} m\overline{v_x^2} = \tfrac{1}{2} m\overline{v_y^2} = \tfrac{1}{2} m\overline{v_z^2} = \tfrac{1}{2} kT \, , \qquad (\text{II}.20a)$$

$$\tfrac{1}{2} m\overline{v^2} = \tfrac{3}{2} kT \qquad (\text{II}.20b)$$

for the average translational energy of a particle with mass m, absolute velocity v, and velocity components v_x, v_y and v_z. This classical *equipartition law* holds for all forms of energy that have a continuous range of variation and that can be expressed as homogeneous quadratic forms of 'coordinates' or 'momenta'. Consequently we have for the average kinetic energy associated with the relative motion of any pair of particles with masses m_1 and m_2

$$\tfrac{1}{2} \mu \overline{g^2} = \tfrac{3}{2} kT \qquad (\text{II}.21)$$

where g is the absolute relative velocity and μ is the *reduced mass* defined by

$$\mu \equiv m_1 m_2 / (m_1 + m_2) \, . \qquad (\text{II}.22)$$

The average kinetic energy associated with the *component* of **g** along the line connecting the centres of m_1 and m_2 is given by $\tfrac{1}{2} kT$. The kinetic energy associated with the motion of the centre of mass, having velocity v_c, has an average value given by

$$\tfrac{1}{2} (m_1 + m_2) \overline{v_c^2} = \tfrac{3}{2} kT \, . \qquad (\text{II}.23)$$

[†] This holds at least if the interactions are sufficiently weak.

THE EQUILIBRIUM STATE   Sect. II.3b

The equipartition law does not hold for the quantized internal energy of atoms or molecules. However, it holds in good approximation for the molecular rotational energy if kT is much larger than the differences in energy between the rotational levels. This condition is satisfied at flame temperatures for the three rotational degrees of freedom of nonlinear poly-atomic molecules; for di-atomics or linear poly-atomics this condition is not satisfied for the degree of freedom associated with the rotation around the internuclear axis (characterized by an 'infinite' rotational constant).

Real gas systems are usually not in a state of strict thermodynamic equilibrium. We shall encounter in this book many examples of deviations from that state. For practical purposes, however, we can still speak of a (partial) equilibrium in a restricted and approximate sense (see Subsection 3c). The consequences for the definition of flame temperature will be discussed in Sect. III.10.

3b EQUILIBRIUM DISTRIBUTION LAWS

3b-1 *THE MAXWELL LAW OF VELOCITY DISTRIBUTION*

The fraction, $f(v_x)\, dv_x$, of particles with a *velocity component* along the x-axis in the infinitesimal interval $(v_x, v_x + dv_x)$ is, in equilibrium, given by the *Maxwell law of velocity distribution*

$$f(v_x)\, dv_x = (m/2\pi kT)^{\frac{1}{2}} \exp\left[-mv_x^2/2kT\right] dv_x \,. \tag{II.24}$$

The same expressions hold for the distribution of v_y and v_z, as the velocity distribution in equilibrium is isotropic. The integral of the distribution function between the limits $-\infty$ and $+\infty$ is normalized to unity.

For the *absolute velocity* v we have the related distribution function

$$f(v)\, dv = 4\pi v^2 \, (m/2\pi kT)^{\frac{3}{2}} \exp\left[-mv^2/2\, kT\right] dv \,. \tag{II.25}$$

From this expression one proves that the average absolute velocity $\bar{v}$ is given by

$$\bar{v} = (8\, kT/\pi m)^{\frac{1}{2}} \,. \tag{II.26}$$

For the *kinetic energy* $E_k = \frac{1}{2} mv^2$ we find from Eq. II.25 the distribution function

$$f(E_k)\, dE_k = (2/\pi^{\frac{1}{2}})\, E_k^{\frac{1}{2}}/(kT)^{\frac{3}{2}} \exp\left[-E_k/kT\right] dE_k \,. \tag{II.27}$$

The fraction of particles having a kinetic energy larger than a given value E_A follows from Eq. II.27, after substituting: $E_k/kT \equiv s$,

$$F(E_k > E_A) = (2/\pi^{\frac{1}{2}}) \int_{E_A/kT}^{\infty} s^{\frac{1}{2}} \exp\left[-s\right] ds \tag{II.28}$$

When $E_A/kT \gg 1$, this fraction varies virtually $\propto \exp[-E_A/kT]$; for $E_A/kT = 10$, it is of the order of 10^{-4}.

The equipartition law for the translational energy (see Subsection 3a) follows from Eqs II.24 and 25 or 27.

Similar distribution laws hold for the velocity, v_c, of the centre of mass of any pair of particles (with masses m_1 and m_2) and for their relative velocity g. One has only to replace in the foregoing equations m by (m_1+m_2) or by the reduced mass μ, respectively. In particular, one thus finds from Eq. II.26 for the average relative velocity

$$\bar{g} = (8kT/\pi\mu)^{\frac{1}{2}} . \tag{II.29}$$

In all the above equations the ratio m/k can be replaced by M/R where M is the molar mass and R the molar gas constant.

3b-2 THE BOLTZMANN LAW OF INTERNAL ENERGY DISTRIBUTION

The Boltzmann law. This law determines in equilibrium the fractional population of the various discrete levels of internal energy of a particle of a given species by

$$f(E_j) = g_j \exp[-E_j/kT]/Q . \tag{II.30}$$

Here E_j is the *energy* of the jth level with respect to that of the ground level ($E_0 = 0$) and g_j is the *statistical weight* of this level. The *internal partition function* or *state sum* Q is added to the denominator in order to normalize $\sum_{i=0}^{\infty} f(E_i)$ to unity; Q is therefore defined by

$$Q \equiv \sum_{i=0}^{\infty} g_i \exp[-E_i/kT] , \tag{II.31}$$

and is a function of T. When E_j is expressed in eV and T in K, we have numerically

$$\exp[-E_j/kT] = 10^{-5040\, E_j/T} . \tag{II.32}$$

The statistical weight of a degenerate energy level is determined by the (product of) degeneracy factor(s). For a non-degenerate level $g_j = 1$.

When the total number density (= number per unit volume) of particles of the considered species: $n_t \equiv \sum_{j=0}^{\infty} n_j$ is given, we have for the densities n_i and n_j of particles in the ith and j^{th} level ($i, j = 0, 1, \ldots$)

$$n_j = n_t f(E_j) , \tag{II.33a}$$

$$n_i/n_j = f(E_i)/f(E_j) . \tag{II.33b}$$

Atomic internal partition function. For neutral or ionized atoms whose energy levels are labelled by the quantum number, J, of the total electronic angular momentum we have: $g = 2J+1$ (see Sect. II.2a-2). When we disregard the

THE EQUILIBRIUM STATE   Sect. II.3b

fine-structure splitting, we have for the statistical weight of the multiplet as a whole, characterized by the total electronic spin and orbital quantum numbers S and L: $g = (2S+1)(2L+1)$. We have left out of account the nuclear spin because the h.f.s. splitting of the energy levels is usually negligible in flame spectroscopy; because of the nuclear spin (I) we would have to add a factor $(2I+1)$ in the numerator as well as in the denominator of the expression in Eq. II.30, which thus drops out.

Since the number of energy levels in an atom is infinite (the principal quantum number n is unlimited; see Sect. II.2a-1) and since the exponential factors in the expression for Q have a positive lower bound set by the ionization energy E_{ion}, Q would be infinite according to Eq. II.31. This divergency of the electronic partition function has long been a subject of theoretical discussion (see, e.g., Cambel, Duclos and Anderson 1963, Boumans 1966, Rouse 1967, Richter 1968, and Thorne 1974). This problem can be in practice solved by assuming an upper cut-off in the summation over the energy levels.

Several propositions have been made about the choice of the highest, finite n level that should be incorporated in the expression for Q. For example, levels that correspond to an electron orbit whose radius exceeds half the mean distance between two neighbouring atoms in the gas may be assumed to make no real contribution to Q. Another approximation is cutting-off all levels whose excitation energies approach the ionization limit within, say, kT. In a plasma containing also singly ionized atoms and free electrons, the atomic partition function may be made finite by cutting-off all levels above the *effective* ionization limit. The latter lies below the ionization limit of an isolated atom owing to the effect of Debije shielding (see Subsection 3b-4). Depending on which approximation is used, the location of the cut-off level may vary with temperature, total gas pressure or electron density. However, at the temperatures, pressures or electron densities expected in flames these different approximations will not give significant differences in the calculations.

At flame temperatures ($T \leqslant 3000$ K) reasonably accurate partition functions for atoms are obtained by including only levels with excitation energies up to about 2 eV (see de Galan, Smith and Winefordner 1968). The latter authors have expressed $Q(T)$ in a polynomial of 5th order in T, which is valid between 1500 and 7000 K; coefficients of this polynomial have been given for a large collection of neutral and singly ionized atoms. The same authors and Boumans (1968) give references to literature where atomic partition functions for higher temperatures have been calculated.

Atoms with a single ground level and whose lowest excitation level lies at 2 eV or higher have a partition function that is practically equal to g_0 at flame temperatures. In other words, in equilibrium the majority of, e.g., Na atoms in the flame is present in the $^2S_{\frac{1}{2}}$ ground level ($g_0 = 2$). For the lowest Na 3^2P doublet with $g = 2+4 = 6$ and $E_1 = 2.1$ eV we calculate: $f(E_1) = 8.7 \times 10^{-4}$ at $T = 3000$ K. The fractional population of this doublet is thus very small; it is a fortiori very small for all higher levels. In the case of atoms having a ground multiplet with an f.s. splitting $\leq kT$ and possibly some other, low-lying multiplets, we are not entitled to replace Q by g_0. An example is Sn having a ^3P ground multiplet with $^3P_0$ being the ground level and an f.s. splitting of 0.2 eV; the fraction of Sn atoms in the ground level ($g_0 = 1$) is thus markedly less than 1 at $T \geqslant 2000$ K.

The temperature-dependence of the fractional population of the excited levels is mainly governed by the exponential factor in Eq. II.30 (called: *Boltzmann factor*). The population thus grows strongly with increasing temperature. The *relative* increase for given ΔT (>0) is proportional to the excitation energy. The increase of $Q(T)$ with increasing flame temperature is usually weak or negligible.

Boumans (1968) has described to a first approximation the influence of $Q(T)$ on the T-dependence of $f(E_j)$ as an effective lowering of the excitation energy E_j. He has calculated effective correction terms for E_j that hold in various, broad ranges of temperatures, for a large number of neutral and singly ionized atoms. The largest correction term valid for $T = 2000 - 3000$ K was found for the tungsten atom and amounted to 0.3 eV. Most correction terms are 0.1 eV or less.

Molecular internal partition functions. In the Born-Oppenheimer approximation (see Sect. II.2b) we can factorize the total internal partition function Q_m according to the electronic and vibrational plus rotational degrees of freedom

$$Q_m = \sum_u Q_{r,u} Q_{v,u} g_u \exp[-E_u/kT] . \qquad (II.34)$$

Here u labels the electronic state while $Q_{r,u}$ and $Q_{v,u}$ are the rotational and vibrational partition functions, respectively, belonging to that state.

For heteronuclear di-atomic or linear poly-atomic molecules the *rotational partition function* is obtained from Eq. II.31 if we replace there E_i by the expression for the rotational energy given in Eq. II.18, and g_i by the rotational degeneracy factor: $g_J = 2J+1$ (see Sect. II.2b-2). Eq. II.18 is strictly valid for $^1\Sigma$ states only, but it is here used in the B.-O. approximation for other states too; the result obtained is usually good enough if T is not very low (see Tatum 1967). Since at flame temperatures the separation between adjacent rotational levels is small compared to kT, we can replace the summation in the expression for Q_r by

THE EQUILIBRIUM STATE   Sect. II.3b

an integration and obtain

$$Q_r = kT/hcB = 8\pi^2 kTI/h^2 . \tag{II.35}$$

Here I is the moment of inertia and B is the rotational constant (see Sect. II.2b-2), which, in general, depend on the electronic state and — to a weaker extent — also on the vibrational state. We find the fractional population of the J levels by inserting $2J+1$ for g_j, $E_r(J)$ for E_j and Q_r for Q into Eq. II.30. It is found that the highest population occurs for the level with $J \simeq (kT/2hcB)^{\frac{1}{2}}$, which is usually much larger than one.

For homonuclear di-atomics, however, we must take into account explicitly the nuclear spin in calculating the rotational level population (see the general literature cited at the beginning of this section). If $I = 0$, then of each pair of consecutive rotational levels one will not be populated at all.

For a nonlinear poly-atomic molecule we have instead of Eq. II.35

$$Q_r = \pi^{\frac{1}{2}}(kT/hc)^{\frac{3}{2}}/(ABC)^{\frac{1}{2}} \quad \text{with} \quad A \equiv h/8\pi^2 I_A c, \text{ etc.,} \tag{II.36}$$

where I_A, I_B, I_C are the principal moments of inertia.

Sometimes a symmetry number is included in the expression for Q_r. This is, however, not needed when Q_r is used to calculate through Eq. II.30 the fractional population of the energy levels of one species. A symmetry number is only appropriate in the expression for the dissociation constant (see Subsection 3b-3).

For a di-atomic molecule the *vibrational partition function* is obtained from Eq. II.31 if we replace there E_i by the expression for the vibrational energy given in Eq. II.13 while putting $g_i = 1$. If we retain in the latter equation only the term that is linear in v (i.e. if we neglect the anharmonicity) we get in good approximation for not too high temperatures

$$Q_v = \left(1 - \exp[-hc\omega_e/kT]\right)^{-1} = \left(1 - 10^{-0.625\,\omega_e/T}\right)^{-1} . \tag{II.37a}$$

In the latter expression we have to express ω_e in cm^{-1} and T in K. Since we have dropped the zeropoint energy in the expression for Q_v, we must do the same when we substitute the expression for E_j in calculating the fractional level population from Eq. II.30. When the quantum of vibrational energy, $hc\omega_e$, is small compared to kT, we can approximate Eq. II.37a by

$$Q_v \simeq kT/hc\omega_e . \tag{II.37b}$$

For poly-atomic molecules Q_v can be approximated as a product of similar factors as given by Eq. II.37a, each involving the ω_e value corresponding to one of the normal vibrations.

The *internal molecular partition function* of a *di-atomic heteronuclear* molecule is finally found by labelling the expressions for Q_r and Q_v from Eqs II.35 and 37a with u and by inserting them into Eq. II.34 where we also replace g_u by $(g_\Lambda)_u$. A considerable simplification is obtained when the summation over the electronic states u can be restricted to the ground multiplet, no other low-lying states being assumed. Since B as well as ω_e are the same for each state of the same multiplet, we then obtain

$$Q_m \simeq Q_{r,0} Q_{v,0} Q_{e,0} = (kT/hcB)\left(1 - \exp\left[-hc\omega_e/kT\right]\right)^{-1} \sum_{\Sigma_0} g_{\Lambda,0} \exp\left[-E_{\Sigma_0}/kT\right]. \tag{II.38}$$

Here $Q_{e,0}$ is the electronic partition function including only the ground multiplet levels, which are labelled by Σ_0; all other quantities relate to the ground multiplet. When all E_{Σ_0}'s are small compared to kT, $Q_{e,0}$ simply equals the statistical weight, g_0, of the ground multiplet, that is

$$Q_{e,0} \simeq g_0 \equiv g_{\Lambda,0}(2S_0 + 1). \tag{II.39}$$

When other low-lying multiplets are present, we may still use Eq. II.38, with the summation now including these multiplets, if B and ω_e are not too different for all multiplets involved. Then we have generally

$$Q_m \simeq Q_r Q_v Q_e, \tag{II.40}$$

with Q_r and Q_v given by Eqs 35 and 37a and with Q_e including the low-lying multiplets.

More precise, detailed expressions for the fractional populations of electronic, vibrational and rotational levels of homo- and heteronuclear di-atomics are found in, e.g., Tatum (1967).

For *poly-atomic* molecules we obtain under the same conditions as those underlying Eq. II.38 similar approximate expressions. For *linear* poly-atomic molecules we have only to replace $Q_{v,0}$ in Eq. II.38 by a product of similar factors, each for one vibrational mode. For *nonlinear* poly-atomic molecules we find by using Eq. II.36

$$Q_m = \left\{\pi^{\frac{1}{2}}(kT/hc)^{\frac{3}{2}}/(ABC)^{\frac{1}{2}}\right\}\prod_i\left(1 - \exp\left[-hc\omega_i/kT\right]\right)^{-1} Q_{e,0}, \tag{II.41}$$

where all quantities refer to the ground multiplet. In the presence of other, low-lying multiplets we may write

$$Q_m \simeq Q_r Q_v Q_e, \tag{II.42}$$

with Q_e including these multiplets if ABC and the ω_i's are about the same for all multiplets involved. When two vibrational modes have the same ω_i, say: $\omega_1 = \omega_2 \equiv \omega^*$, the corresponding factor with ω^* occurs twice in Eq. II.41. The vibrational level with energy $nhc\omega^*$ is then degenerate with a degeneracy factor equal to $n+1$. This

THE EQUILIBRIUM STATE

degeneracy factor should be substituted for g_j in the Boltzmann law, Eq. II.30, when calculating the population of the vibrational levels.

3b-3 THE MASS-ACTION-LAW

The mass-action-law relates the concentrations of the products of a chemical reaction to those of the reactants in the equilibrium state. Let us consider first the case of a balanced *dissociation* and reverse *association reaction* in the gas phase:

$$AB \rightleftharpoons A + B ,$$

where A and B may be atoms or molecules. Denoting the number densities of the chemical species by n_{AB} (= number of molecules AB per unit volume, summed over all energy levels), and so on, we have according to the *mass-action-law* at temperature T

$$n_A n_B / n_{AB} = K_d(T) . \qquad (II.43)$$

Here $K_d(T)$ is the *dissociation constant*, which depends only on T but not on the total pressure. $K_d(T)$ as defined here has the same dimension and is expressed in the same unit as the number density, e.g., in cm^{-3}.

Statistical Mechanics shows that $K_d(T)$ depends on the properties of the individual species involved, in particular on the dissociation energy of AB, but *not* on the specific path along which or the rate at which the actual dissociation reaction takes place (see also Sect. II.4a-2). $K_d(T)$ does not depend on whether the actual dissociation reaction proceeds in a single step or involves several, consecutive steps. It also remains the same when the species involved participate in other reactions as well. According to Fig. II.2, the *dissociation energy* D_0 is the positive energy required for the decomposition of an isolated molecule AB in its lowest (real) energy level ($v'' = 0$) into neutral fragments A and B; these fragments are supposed to be at their lowest energy level and to have zero kinetic energy of relative motion at infinite separation. When a potential barrier has to be surmounted in the decomposition, D_0 relates to the *net* energy required. D_0 is usually expressed in eV per molecule or in kJ (in older literature also in kcal) per mole.

The *statistical-mechanical expression* for $K_d(T)$ reads

$$K_d(T) = (2\pi k T/h^2)^{\frac{3}{2}} (m_A m_B / m_{AB})^{\frac{3}{2}} (Q_A Q_B / Q_{AB}) (s_{AB}/s_A s_B) \exp[-D_0/kT]. \qquad (II.44)$$

Here m_A is the mass of particle A, Q_A its internal partition function (see Subsection 3b-2), s_A its symmetry number, and so on. The *symmetry number* s reflects the symmetry of the geometrical structure of the particle when it is a molecule; molecules without any symmetry, e.g., heteronuclear di-atomics, have $s = 1$; homonuclear di-atomics and linear tri-atomics with two identical extreme nuclei have $s = 2$. For atoms, s has no meaning and we formally define $s \equiv 1$ for them. The factor $(2\pi kT/h^2)^{\frac{3}{2}}$ appearing in the expression of Eq. II.44 is a consequence of the role played by the translational partition functions in the dissociation equilibrium.

Since the internal partition functions are not, or are only weakly dependent on temperature and D_0 is often much larger than kT at flame temperatures, the T-dependence of $K_d(T)$ is mainly described by the exponential factor; the latter may be written as a power of 10 according to Eq. II.32 if D_0 is expressed as eV per molecule. When D_0 is taken as energy per mole, we must replace k by the gas constant R in the exponential factor in Eq. II.44. We conclude that K_d increases strongly with increasing T; the relative increase in K_d, at given positive ΔT, is proportional to D_0. A numerical expression for $K_d(T)$ for practical use will be found in Eq. VIII.36.

The *degree of dissociation*, β_d, of a metal compound AB (A = metal atom) is here defined by

$$\beta_d \equiv n_A/(n_A + n_{AB}) \; . \tag{II.45}$$

It does not depend on K_d only, but also on $(n_A + n_{AB})$ or n_B. In general β_d increases when $K_d(T)$ is increased, other things being kept constant. We refer for a further discussion to Chapt. VIII.

In the literature the mass-action-law for gas-mixtures is often formulated in terms of partial pressures. The *partial pressure*, p_X or $p(X)$, of a component X is the pressure (expressed, for instance, in atm) that this component would exert in the same volume if all other components of the gas-mixture were absent. The numerical relation between p and the number density n for an ideal gas will be given in Sect. V.1. We obtain from Eq. II.43

$$p_A p_B/p_{AB} \equiv K_d^*(T) = (kT/p_0)K_d(T) \; , \tag{II.46}$$

where p_0 stands for the numerical factor that converts the unit of 1 atm to the CGS unit of pressure if K_d is expressed in CGS units. Note that the dimensions and units of $K_d^*(T)$ and $K_d(T)$ are different. Equation II.46 holds if the gases behave as ideal gases; this condition is generally satisfied at flame temperatures. The use of Eq. II.46 has some advantages in calculations of the equilibrium composition of the flame at fixed total pressure (see Sect. IV.4).

Besides the statistical-mechanical expression for $K_d(T)$, given by Eq. II.44, we can also give a *thermodynamic expression* which relates $K_d(T)$ to the thermodynamical properties of the species involved. Let G be the *Gibbs free energy* or the *free enthalpy* defined by:

$$G \equiv U + pV - TS \; , \tag{II.47}$$

where U is the internal energy, p the pressure, V the volume, and S is the entropy of the gas considered. Defining the *enthalpy* H by

$$H \equiv U + pV \tag{II.48}$$

we also have

$$G = H - TS \tag{II.49}$$

The *change* in G, H and S that occurs when one mole of AB is dissociated into A+B in the standard state at temperature T is denoted by ΔG_T^0, ΔH_T^0 and ΔS_T^0. When G increases upon dissociation we take ΔG_T^0 etc. as being positive. As *standard state* one usually assumes that the pressure is 1 atm and that the gas behaves as an ideal gas. The differences between a real gas and an ideal gas are generally negligible at flame temperatures. Denoting the standard pressure by p_0, we express the concentrations of A, B and AB as fractions of p_0, defined by

$$\pi_A \equiv p_A/p_0, \text{ and so on.} \tag{II.50}$$

Note that, in contrast to the partial pressure p_A introduced earlier, π_A has dimension one. For a clear distinction we shall call π_A the *fractional pressure*, although π_A and p_A are numerically the same when the standard pressure p_0 is the unit of pressure. We can now formulate the thermodynamical expression for the mass-action-law at temperature T by

$$\pi_A \pi_B / \pi_{AB} = \exp\left[-\Delta G_T^0/RT\right] = \exp\left[\Delta S_T^0/R\right] \exp\left[-\Delta H_T^0/RT\right]. \tag{II.51}$$

The first equation follows from the fact that the Gibbs free energy of a system in equilibrium at given T and p is minimal. Since dissociation is an endo-ergic process, the *heat of reaction* (*in casu: dissociation*) at constant p_0 and T, given by ΔH_T^0, has a positive value. A large ΔH_T^0 represses the dissociation; on the other hand, the exponential factor with positive ΔS_T^0 favours the dissociation. The increase in entropy upon dissociation is essentially connected with the fact that two independent particles are formed out of one. The entropy factor explains why dissociation can be appreciable even when $\Delta H_T^0/RT \gg 1$, whereas the fractional population of an energy level according to the Boltzmann law (which does not involve an entropy factor) is always very small for $E_j/kT \gg 1$. The opposite effects of these two thermodynamic quantities reflect the general property of nature to tend to minimum energy as well as to maximum entropy.

Although π_A and p_A, and so on, are numerically the same, we prefer to formulate the latter equation in terms of fractional pressures instead of partial pressures. This is done to stress the point that the exponential factors in Eq. II.51 have dimension one and to recall that the 'concentrations' of A, B and AB should be expressed in conformity with the standard state to which the thermodynamic functions relate. It is not at all necessary that the standard state be defined as a certain pressure; it could just as well be defined as a certain volume.

Comparison of the thermodynamical expression in Eq. II.51 with the statistical-mechanical expression in Eq. 44 suggests that ΔH_T^0 is closely related to D_0, whereas ΔS_T^0 involves the internal plus translational partition functions of the species involved. In fact we have the equality: $D_0 = \Delta H_0^0$, where ΔH_0^0 refers to the system at zero kelvin in the ideal-gas state and both quantities are expressed in the same units.

If ΔH_θ^0 and ΔS_θ^0 are known at some standard temperature θ, usually 298 K, we can calculate $\Delta G_T^0/RT$, and thus the dissociation constant at T, from

$$\Delta G_T^0/RT = \Delta H_\theta^0/RT - \Delta S_\theta^0/R - \int_\theta^T T'^{-2} \left(\int_\theta^{T'} \Delta c_p' \, dT'' \right) dT'/R. \qquad (II.52)$$

Here $\Delta c_p'$ is the difference in molar heat capacity at constant p ($=p_0$) between the dissociation products on the one side and the dissociating compound on the other. This equation follows from the general relations derived from Thermodynamics

$$H_T = H_\theta + \int_\theta^T C_p \, dT', \qquad (II.53a)$$

$$S_T = S_\theta + \int_\theta^T (C_p/T') \, dT', \qquad (II.53b)$$

where C_p is the heat capacity at constant pressure. When in the transition from θ to T phase changes occur, the appropriate latent heats should be accounted for, as the thermochemical data at standard temperature and pressure usually refer to the substances that are in their normal phase at temperature θ. For pure (ideal) gases each degree of freedom contributes maximally $\frac{1}{2}R$ to the molar heat capacity at constant volume c_v' and thus to c_p' which equals $c_v' + R$. The full contribution $\frac{1}{2}R$ is obtained if the classical equipartition law (see Subsection 3a) holds. When at the temperature considered the quantization of the internal energy is important, the contribution will be less. Values of ΔS_θ^0 and ΔH_θ^0 and of c_p' as a function of temperature may be derived from thermodynamic tables (see, e.g., JANAF 1971).

A determination of the *relative* variation of the dissociation constant with T, i.e. $d(\ln K_d)/dT$, at known T yields also D_0 through application of Eq. II.44, provided that the relative variations of the partition functions with T are known; the latter variations are usually weak in comparison with that of the exponential factor containing D_0. Writing: $\Delta G_T^0/RT = -\ln K_d^*$, according to Eq. II.51, and differentiating the right-hand side of Eq. II.52 with respect to T, we find

$$d(\ln K_d)/dT = \Delta H_T^0/RT^2, \qquad (II.54a)$$

$$d(\ln K_d)/dT^{-1} = -\Delta H_T^0/R. \qquad (II.54b)$$

It is immaterial whether we write here K_d or K_d^*. Thus a determination of the

THE EQUILIBRIUM STATE   Sect. II.3b

relative variation of K_d with T yields directly ΔH_T^0 if T is known. From ΔH_T^0 thus determined one calculates its value at any other temperature, for example θ, by using according to Eq. II.53a

$$\Delta H_\theta^0 = \Delta H_T^0 + \int_T^\theta \Delta c_p' \, dT' \, , \tag{II.55}$$

if all species involved remain in gas phase. Under the latter condition one may also use Eq. 55, with $\theta = 0$, for calculating $\Delta H_0^0 \equiv D_0$.

An *absolute* determination of the dissociation constant at known T yields directly the value of ΔG_T^0 through Eq. II.51 and indirectly the value of D_0 through Eq. II.44, if all internal partition functions are known.

Once ΔG_T^0 has been determined, one finds ΔH_T^0 from: $\Delta H_T^0 = \Delta G_T^0 + T\Delta S_T^0$ (see Eq. II.49) if ΔS_T^0 is known. The latter can be calculated according to Eq. II.53b from

$$\Delta S_T^0 = \Delta S_\theta^0 + \int_\theta^T (\Delta c_p'/T') \, dT' \, , \tag{II.56}$$

if all species remain in the gas phase. This equation may also be used, with $\theta \equiv 0$, to calculate ΔS_T^0 from ΔS_0^0, which is zero according to the third law of Thermodynamics. The integral in this equation may be evaluated by using thermal data or data on the molecular structure, etc. Finally, one finds again $D_0 \equiv \Delta H_0^0$ from ΔH_T^0 by using Eq. II.55 with $\theta = 0$.

In the case of a balanced *binary exchange reaction* in the gas phase:

$$A + B \rightleftharpoons C + D$$

the mass-action-law tells that in equilibrium at temperature T

$$n_C n_D / n_A n_B = K(T) \, , \tag{II.57}$$

where K is called the *equilibrium constant* of the reaction and depends only on T. In this example, K has dimension one and is independent of whether concentrations or partial pressures are used in the left-hand side of the latter equation. The thermodynamical expression for $K(T)$ has the same form as that for the dissociation constant in Eq. II.51. The statistical-mechanical expression reads

$$K(T) = (m_C m_D / m_A m_B)^{\frac{3}{2}} (Q_C Q_D / Q_A Q_B)(s_A s_B / s_C s_D) \exp[-\Delta U_0^0/kT]. \tag{II.58}$$

Here $\Delta U_0^0 \, (= \Delta H_0^0 = D_0)$ is the energy of the reaction for one pair of particles A and B, defined in a similar way as D_0, that is, at zero temperature and in the ideal-gas phase.

3b-4   *THE SAHA LAW*

The Saha law relates the concentration of the neutral species to those of the ionization products in equilibrium. We consider the following *ionization* and reverse *recombination processes* in the gas phase:

53

Chapt. II   BASIC CONCEPTS AND GENERAL RELATIONSHIPS

$$A \rightleftharpoons A^+ + e^-.$$

The *Saha law* reads

$$[A^+][e^-] / [A] = K_i(T), \tag{II.59}$$

where $[A^+]$ denotes the number density of ions (summed over all energy levels), and so on, and $K_i(T)$ is called the *ionization constant*, which depends only on T. The ionization process is quite analogous to the dissociation process and the statistical-mechanical expression for $K_i(T)$ derives immediately from that for $K_d(T)$ given by Eq. II.44. Realizing that a free electron has a statistical weight equal to 2 because of its spin and that its mass m_e is very small compared to m_A, we get

$$K_i(T) = 2 (2\pi m_e kT/h^2)^{\frac{3}{2}} (Q_{A^+}/Q_A) \exp[-E_{ion}/kT]. \tag{II.60a}$$

Herein E_{ion} is the (positive) *ionization energy* (taken here per particle), which is defined formally in the same way as the dissociation energy D_0. The value of E_{ion} corresponds to the (first) ionization limit considered in Sect. II.2a for atoms. Expressing E_{ion} in eV, T in K and $K_i(T)$ in cm^{-3} we have the numerical expression

$$K_i(T) = 4.83 \times 10^{15} T^{\frac{3}{2}} (Q_{A^+}/Q_A) 10^{-5040\, E_{ion}/T}. \tag{II.60b}$$

When the concentrations of the species involved are expressed in partial pressures, we should replace $K_i(T)$ by $K_i^*(T) \equiv (kT/p_0) K_i(T)$, as in Eq. II.46, where p_0 is the numerical conversion factor for the unit of pressure chosen (which need not be coherent with the unit of number density). This replacement introduces an extra factor T in the statistical-mechanical expression for $K_i^*(T)$.

Similar considerations as given above for $K_d(T)$ apply again to $K_i(T)$. The smaller E_{ion} and the higher T are, the stronger the ionization will be, other things being kept constant, because ionization is an endo-ergic process. The degree of ionization β_i is defined in a way analogous to β_d in Eq. II.45

$$\beta_i \equiv [A^+] / ([A^+] + [A]). \tag{II.61}$$

The equilibrium degree of ionization depends not only on $K_i(T)$ but also on $([A^+] + [A])$ or $[e^-]$; this will be discussed in more detail in Sect. IX.3b.

The Saha law holds also for *negative ions*, formed by attachment of a free electron to a neutral particle. We have for the process:

$$A^- \rightleftharpoons A + e^-$$

the equilibrium relation

$$[A][e^-] / [A^-] = K_{det}(T). \tag{II.62}$$

The expression for the *equilibrium constant for detachment*, $K_{det}(T)$, is analogous to that for $K_i(T)$ in Eq. II.60a or 60b, with E_{ion} being replaced by the (positive) *electron affinity* E_{aff}; the latter is defined in a similar way as E_{ion} and represents the binding energy of the extra electron to the particle. Instead of the ratio (Q_{A^+}/Q_A) we now have to insert (Q_A/Q_{A^-}) in Eq. II.60a or 60b.

The value that should be filled in for E_{ion} in Eq. II.60a or 60b in the case of a *real* plasma may be somewhat lower than the value for an isolated atom. This *depression of the (effective) ionization energy*, ΔE_{ion}, is a consequence of the *Debije** *shielding* of the field of the positive ion; this shielding results from the local increase of the electron concentration caused by Coulomb attraction (see e.g., Cambel, Duclos and Anderson 1963, Richter 1968, and Thorne 1974). The electric field of the ion practically does not extend beyond a sphere with radius equal to the *Debije shielding distance* (or *Debije length* or *Debije radius*) ρ_D. The latter parameter is given by (in e.s. CGS units[†])

$$\rho_D = (kT/8\pi e^2 [e^-])^{\frac{1}{2}} \approx 5(T/[e^-])^{\frac{1}{2}}. \tag{II.63}$$

When an electron is removed from the atom, the former will no longer be bound to the parent ion when their distance exceeds ρ_D. Consequently, in a real plasma the minimum energy required to create a free electron-ion pair is actually less than E_{ion}, which was defined for infinite separation of an isolated electron-ion pair. The effective depression of the ionization energy, ΔE_{ion}, is of the order of e^2/ρ_D, and is numerically given by

$$|\Delta E_{ion}| \approx 3 \times 10^{-8} ([e^-]/T)^{\frac{1}{2}} \quad \text{(in eV)}, \tag{II.64}$$

where $[e^-]$ is expressed in cm^{-3}. In a typical flame with $T \approx 2500$ K and $[e^-]$ as large as 10^{14} cm^{-3} we have: $|\Delta E_{ion}| \approx 10^{-2}$ eV, which is quite negligible.

This depression of the actual ionization energy removes the divergency of the electronic partition function (see Subsection 3b-2).

3b-5 THE PLANCK RADIATION LAW

In thermodynamic equilibrium the *spectral volume density of radiant energy* $\rho_\lambda^b(\lambda,T)$, expressed, e.g., in erg cm^{-3} Å^{-1} (see also Appendix A.5), is given by the *Planck law*, which reads in the vacuum case

$$\rho_\lambda^b(\lambda,T) = 8\pi hc \lambda^{-5}/(\exp[hc/\lambda kT] - 1), \tag{II.65a}$$

where all symbols have their usual meaning and c is defined in vacuo. In order to express the spectral volume density per unit of frequency ν, we have to multiply the latter expression by $|d\lambda/d\nu|$ and to convert λ into $\nu = c/\lambda$; the result is

$$\rho_\nu^b(\nu,T) = (8\pi h\nu^3/c^3)/(\exp[h\nu/kT] - 1). \tag{II.65b}$$

---

* In the English literature the (Dutch) name Debije is customarily spelt as 'Debye'.

[†] In the S.I. system of units we should replace e^2 by $e^2/4\pi\varepsilon_0$ with ε_0 = the vacuum permittivity.

The latter quantity is expressed, e.g., in $erg\,cm^{-3}\,Hz^{-1}$. The radiant energy considered here is isotropically distributed and unpolarized; each independent polarization direction contributes an equal part. We note from these equations that the density of the radiation field depends only on λ (or ν) and T, but not on material properties.

The *spectral radiance*, $B_\lambda^b(\lambda,T)$, of a black body in thermodynamic equilibrium, defined as the radiant energy emitted per unit of time, per unit wavelength interval and per unit solid angle by a surface element of unit area, normal to the surface, follows from Eq. II.65a through the relation

$$B_\lambda^b(\lambda,T) = (c/4\pi)\,\rho_\lambda^b(\lambda,T) . \qquad (II.66)$$

A *black body* (or *full radiator*) has, by definition, an absorption factor equal to unity, that is, it completely absorbs any radiation incident on its surface, whatever the direction, wavelength or polarization of this radiation (see also Sects II.5c-2 and 5c-3). A full radiator may be realized by making a small opening in the wall of a cavity ('Hohlraum') that is kept at uniform temperature. Any radiation incident on this opening from the outside will have a negligible chance of re-escaping through the opening if the latter is small enough; it will thus be, virtually, completely trapped, i.e. absorbed, inside the cavity. From Eqs II.65a and 66 we obtain the *Planck law* in another formulation

$$B_\lambda^b(\lambda,T) = 2hc^2\lambda^{-5}/(\exp[hc/\lambda kT]-1) . \qquad (II.67a)$$

The latter quantity is expressed, e.g., in $erg\,s^{-1}\,sr^{-1}\,cm^{-2}\,\text{Å}^{-1}$. Alternatively we get from Eq. II.65b

$$B_\nu^b(\nu,T) = (2h\nu^3/c^2)/(\exp[h\nu/kT]-1) . \qquad (II.67b)$$

Defining the *radiation constants* c_1, c_2 and c_3 by[†] (for numerical values see Appendix A.2)

$$c_1 \equiv 2hc^2 , \quad c_2 \equiv hc/k \quad \text{and} \quad c_3 \equiv 8\pi hc , \qquad (II.68)$$

we can simplify Eqs II.65a and 67a by writing

$$\rho_\lambda^b(\lambda/T) = c_3\,\lambda^{-5}/(\exp[c_2/\lambda T]-1) , \qquad (II.69)$$

$$B_\lambda^b(\lambda/T) = c_1\,\lambda^{-5}/(\exp[c_2/\lambda T]-1) . \qquad (II.70)$$

Radiance is defined as the radiant flux *per unit solid angle* and per unit area perpendicular to the direction of radiation (see also Appendix A.5). As a

---

[†] The quantities c_1 and c_2 are usually called the *first* and *second* radiation constant, respectively. In the literature one also often defines $c_1 \equiv 2\pi hc^2$ instead of $\equiv 2hc^2$.

THE EQUILIBRIUM STATE   Sect. II.3c

consequence of this definition, B_λ^b is isotropic and independent of the size and shape of the black body's surface, whatever these may be. The spectral radiant flux (see Appendix A.5) emitted by a black body per unit area into a *hemisphere* is given by πB_λ^b (not: $2\pi B_\lambda^b$!).

For $\lambda T \leqslant 1.6$ cm K the exponential term in the denominator of the above expressions is larger than 100 and Planck's law may be approximated by *Wien's radiation law*, with an accuracy better than 1%,

$$B_\lambda^b (\lambda, T) \simeq c_1 \lambda^{-5} \exp[-c_2/\lambda T], \qquad (II.71a)$$

$$B_\nu^b (\nu, T) \simeq (2h\nu^3/c^2) \exp[-h\nu/kT]. \qquad (II.71b)$$

In flame spectroscopy the latter approximation is certainly a good one.

The wavelength, λ_m, at which $B_\lambda^b(\lambda, T)$ attains its maximum value as a function of λ, at given T, is determined by *Wien's displacement law*

$$\lambda_m T = 0.20 \, hc/k = 2.9 \times 10^7 \quad (\text{Å K}). \qquad (II.72)$$

At $T = 3000$ K we find: $\lambda_m \simeq 10^4$ Å.

When the radiation field is considered in a medium with refractive index n, the expressions for ρ_λ^b and B_λ^b in Eqs II.69 and 70 have to be multiplied by n^3 and n^2, respectively. In a flame the deviation of n from unity is, however, small (see Sect. II.5a-2).

The radiance of an arbitrary *thermal radiator* (not a black body) depends not only on λ and T, but also on material properties as expressed by *Kirchhoff's law*. We shall formulate this law in Sect. II.5c-3 (see Eq. II.293) after introducing the absorption and emission factors.

3c   RELAXATION ; LOCAL AND PARTIAL EQUILIBRIA
3c-1   *GENERAL*

A system attains thermodynamic equilibrium only if the following conditions are fulfilled: firstly the system must be 'closed', that is, it may only be in contact with a heat bath at uniform temperature (see Subsection 3a), but it should not exchange energy or particles with the surroundings; there should be no perturbations from the outside that 'drive' the system out of equilibrium (*adiabatic condition*). Secondly, the system, left to itself, should be given sufficient time to relax towards an equilibrium state, if it was not in that state initially (*relaxation condition*). We assume here that all degrees of freedom of the system (including ionization, dissociation and radiation) can interact, either directly or indirectly, with each other; the time rates of these interactions may, however, be small (but not zero) and different.

Thermodynamic equilibrium in the strict sense is either 'complete' or it simply does not exist. It implies *detailed balance*, that is, the rate of each particular process (collisional or radiative) should exactly balance that of the reverse process (see Sect. II.4a-2). When the distribution function for one degree of freedom is disturbed, the condition of detailed balance is no longer fulfilled and all other distribution functions are — in principle — affected, too. It is then — in principle — no longer possible to describe the complete state of the system by a few parameters only (e.g., temperature, pressure, total element concentrations, molecular constants); one has to specify then in detail the concentrations of each species and the distribution functions for each degree of freedom separately. Moreover, these specifications may vary with place or time; if they do not vary with time, one has a *stationary system* (not necessarily an equilibrium system). Whereas knowledge about the interaction *processes* is not required for a theoretical calculation of the state of a system in thermodynamic equilibrium, such detailed knowledge is essential for calculations with nonequilibrium systems.

The concepts of equilibrium and temperature may, however, be applied in practice in a less strict sense; but essentially this can only be done if we are prepared to accept some approximations (see below). The evaluation of these approximations involves the use of kinetic concepts, which are the subject of the following section. Here we shall present only a general description of the ways in which we can deal with the existence of deviations from thermodynamic equilibrium such as are expected in real systems like flames. The actual extent of these deviations in flames and their consequences for the temperature measurement, the state of ionization and dissociation, and the radiation intensity will be treated in other chapters.

3c-2   *RELAXATION*

When energy is supplied suddenly in some specific form to a system (e.g., combustion energy in the combustion zone of a flame), the system will tend to a new state of equilibrium that is characterized by an increased temperature. The initial distribution of the energy supplied over the various forms and species of the system does not conform to an equilibrium situation, and it will take some time for equipartition of energy[†] to be attained. Equilibration of the system involves a redistribution of the energy among all degrees of freedom (including dissociation and ionization). In flames this is achieved by collisions, chemical reactions, and the emission and re-absorption of radiation. The rates at which these processes proceed are not infinitely large. The system relaxes to the equilibrium state; the relaxation for the various degrees of freedom can be described by different relaxation times (see for a quantitative formulation Sect. II.4a-3). Expressions describing the relaxation in special cases will be found in Eqs II.105, IV.12-14 and VI.58, and in Sect. IX.3c.

[†] Equipartition is here used in a broader sense than in the purely classical equipartition law mentioned in Subsection 3a.

THE EQUILIBRIUM STATE

The temperature that the flame gas attains under the adiabatic condition, after relaxation is over, is called the *final* or *adiabatic flame temperature*. It has often a theoretical meaning only (see Sect. IV.6). As long as the adiabatic condition is fulfilled, the flame gas, considered itself as the system studied, is not in contact with a heat bath. The flame derives therefore its adiabatic temperature from the equipartition of the original heat of combustion over all available degrees of freedom (see Sect. IV.4).

The relaxation times for energy exchange may be roughly characterized by the average number of gas-kinetic collisions with other molecules that a given molecule must undergo for exchanging energy of one form into the same form or into another form. These numbers may be denoted by $\bar{N}_{TT}$, $\bar{N}_{RR}$, $\bar{N}_{VV}$ for the exchange of translational (T), rotational (R) and vibrational (V) energy within one form, and by $\bar{N}_{TR}$, etc., for the exchange between translational and rotational energy, etc. Denoting the *collision frequency* (= number of gas-kinetic collisions per second for one molecule) by ν, we find the corresponding relaxation times by calculating the ratios $\bar{N}_{TT}/\nu$, etc. For a flame gas at 1 atm pressure, ν is typically of the order of 10^9 s^{-1}. For particles of comparable mass Classical Mechanics tells us that $\bar{N}_{TT}$ is of the order of unity; the relaxation time for the establishment of a Maxwellian velocity distribution in a flame is thus of the order of $10^{-8}-10^{-9}$ s. Since $\bar{N}_{RR}$ is usually also of the order of unity, the rotational energy will be distributed over the rotational levels according to a Boltzmann-like function within about the same period. The equipartition between the translational and rotational degrees of freedom however takes somewhat more time; for N_2 we have $\bar{N}_{RT} \approx 10$, but for H_2 we have $\bar{N}_{RT} \approx 300$ because of its more widely spaced rotational levels (see Sect. II.2b).

Exchange of vibrational energy usually takes place in small steps. When the spacing of the vibrational levels is large compared to kT, the vibrational relaxation time may be expected to depend markedly on temperature. In general $\bar{N}_{VV}$ is much smaller than $\bar{N}_{VT}$; the relaxation time of VV exchange is not much longer than for RT exchange. In a pure N_2, CO or CO_2 gas at 1 atm pressure and $T \approx 2000$ K, relaxation times of the order of 100 μs have been found for VT exchange. In the presence of H_2O the latter relaxation time for N_2 is reduced by 1 or 2 orders of magnitude; this catalytic effect of H_2O is due to a ready VV exchange between N_2 and H_2O combined with a comparatively fast VT exchange for H_2O.

Because (at least) a trace of H_2O is always present in flames, its catalytic effect on the VT exchange for N_2 etc., which is the slowest process, causes equilibration to be attained within 10 μs not only for the translational and rotational energy but also for the vibrational energy. The flame gases rise with a velocity typically of the order of 10^3 cm s^{-1}. Consequently, the initial disturbance

Chapt. II   BASIC CONCEPTS AND GENERAL RELATIONSHIPS

in the distribution over the various forms of internal and translational energy, caused by the irreversible combustion reactions, will not persist over a height interval much larger than $10^{-5} \times 10^3 = 10^{-2}$ cm.

For a general theoretical analysis of the persistence of nonequilibrium conditions caused by chemical reactions we refer to Lewis, Pease and Taylor (1956). Energy-relaxation in gases and plasmas is furthermore discussed, theoretically and experimentally, by, e.g., Callear (1965), Stevens (1967), Taylor and Bitterman (1969), and in several chapters of Hochstim (1969a). In particular, the effect of selection rules and of the conservation of total linear and angular momentum on the relaxation times has been discussed by Callear (1965). Jenkins and Sugden (1969), and Gaydon and Wolfhard (1979) have briefly discussed the significance of these effects in flames and have given further references to the literature.

3c-3 *LOCAL AND PARTIAL EQUILIBRIA*

Local thermodynamic equilibrium. Actual flames are not truly adiabatic systems. The flame gases radiate heat to the environment and lose heat by conduction, while the entrainment of fresh air at the flame border may cause an additional supply of heat by secondary combustion, as well as cooling by convection. Concentration gradients of various flame constituents are usually found both in vertical and in horizontal directions. Consequently, a net transport of heat, radiation, and mass occurs throughout the flame. Under this condition a general, thermodynamic state of equilibrium, characterized by a single temperature, cannot exist in the flame. However, if the rate of these transport processes is comparatively slow with respect to the rate at which the energy is locally partitioned over the various degrees of freedom, the concept of a *local thermodynamic equilibrium* (LTE) characterized by a *local temperature* is meaningful. We refer for a more extensive, mathematical formulation of this concept to Richter (1968).

Partial equilibria. In one respect every flame with or without metal vapour obviously lacks thermodynamic equilibrium: the density of the radiation field. Although the flame may emit some strong bands of its own in the near ultraviolet and the infrared (see Sect. IV.8) and some strong resonance lines when supplied with metal vapour at high concentrations, its spectral radiance in most parts of the spectrum lies far below the Planck value at flame temperature[†]. This even applies to sooting flames, which display a comparatively strong continuous background emission. In principle, this deviation would upset flame equilibrium for all other

---

[†] If a cylindrical flame of 3 cm diameter and 20 cm height would radiate as a black body at a temperature of 2500 K, the total radiant flux would be about 30 kW; this is one order of magnitude more than the combustion power released in an average flame of this kind. (Unpublished calculations by Mr G. Kuiper, Physics Laboratory, Utrecht.)

degrees of freedom too. However, radiation plays only a minor part in the equilibration of the other forms of energy including dissociation and ionization. This equilibration is normally *collision-dominated*, even in low-pressure flames. The defect of radiant energy may only affect the equilibrium population of electronic excitation levels under certain conditions (see Sect. VI.4c). We also note that a considerable overpopulation of one or a few atomic levels may occur when a strong external beam is absorbed by a metal line (see Sect. VI.2c). However, in both cases the overall state of the flame gas is hardly affected. On the other hand, the intensity of the flame radiation can be sensitively influenced by possible deviations of the energy-distribution or chemical composition from the equilibrium state.

It may therefore well happen that all other degrees of freedom in the flame closely approach a state of thermodynamic equilibrium notwithstanding the lack of radiation equilibrium. In that case we shall speak of a state of (*local*) *thermal equilibrium*.[†] The temperature parameter that describes (locally) all these other distribution functions as well as the state of dissociation and ionization is then defined as the (local) flame temperature. In the case of thermal equilibrium, the flame emits *thermal radiation*, as distinct from the Planck radiation which corresponds to strict thermodynamic equilibrium (see for a more precise definition and a discussion Sect. VI.4a).

It may also happen that the physical degrees of freedom (translational motion; internal excitation) of all particles are distributed according to the Maxwell-Boltzmann law with a common T parameter, but that the degree of dissociation or ionization of some species does not conform to the mass-action-law or the Saha law. We shall then say that *physical equilibrium* exists but not full *chemical equilibrium*.

When a certain chemical reaction: $A + B \rightleftharpoons C + D$ proceeds fast in both directions so that it is practically balanced, one may find that the concentrations of A, B, C and D are related to each other in accordance with the mass-action-law for that specific reaction at flame temperature. When, for example, A and C also take part in another raction: $A + C \rightleftharpoons AC$ which does not proceed fast enough to be balanced too, the concentrations of A, C and AC will not conform to the mass-action-law for the latter reaction. We then say that a *partial chemical equilibrium* exists for the former reaction; the reaction is then *partially balanced*. It should be noted that this partial equilibrium and partial balance can exist only in approximation. For a quantitative description of such situations we need the use of kinetic concepts which will be introduced in Sect. II.4a-2. Here we note only

---

[†] Richter (1968) calls this a state of 'incomplete thermodynamic equilibrium'; other authors assume implicitly absence of radiation equilibrium when using the term local thermodynamic equilibrium (cf. *ibidem*).

that in full chemical equilibrium the chemical composition of a system can be calculated unambiguously by writing up any set of mass-action equations that connect the various chemical species to each other. The only requirement is that the number of equations is large enough to yield a unique solution. We need not know if the massaction equations correspond to reactions actually occurring in the system nor need we know the rates at which these reactions occur. This is, however, no longer true when some of the chemical equilibria are only partial in the above sense. One has then to know specifically which reactions are balanced and which are not, and what are the rate constants of the latter. We shall encounter actual examples of such deviations from chemical equilibrium and of their consequences in flames in Sects IV.5 and 6, VI.3, VIII.4 and IX.3d.

We may also have a *partial equilibrium* with respect to one of the physical forms of energy. It may happen that during or shortly after a disturbance of the system, e.g., caused by a combustion reaction, the amount of energy stored in the translational, rotational and vibrational degrees of freedom does not correspond to thermodynamic equilibrium. In other words, the exchange of energy *between* these different forms of energy may not be fast enough to ensure equipartition. The exchange of energy *within* one and the same form may, however, be fast enough to set up a Maxwellian- or Boltzmann-type distribution within each of these forms (see also Subsection 3c-2). The T parameters characterizing these separate distributions need then not be identical and one distinguishes between the *(effective) translational, rotational* and *vibrational temperature*. In a similar way we may also speak of an *(effective) electronic-excitation temperature*. After the disturbance is over, these different T parameters will tend to the same, final temperature value as the system relaxes to physical equilibrium. This relaxation is usually described by an exponential function of time (see Lewis, Pease and Taylor 1956).

The temperature parameters describing the distributions over each particular form of energy and for each atomic or molecular species can be separately determined in flames by spectroscopic methods (see Sects II.5c-3 and III.10). Their intercomparison yields information on how far equilibration of the physical degrees of freedom has proceeded (see also Sect. VI.4d).

Some authors also speak of an (effective) 'dissociation' or 'ionization temperature' which is to be substituted for T in the mass-action-law or Saha law in order to describe the *actual* state of dissociation or ionization. When an equilibrium state has not yet been reached owing to slow relaxation, the dissociation or ionization 'temperature' differs from the T parameter describing the physical equilibrium state. If for a particular ionization process the ionization temperature conforms to flame temperature, we shall say that *thermal ionization* exists for that process; if this applies for all ionization processes in the flame, *Saha equilibrium* exists (see Sect. IX.2a).

Whenever in a flame deviations from equilibrium exist for some specific form(s) of energy, one often defines the effective translational temperature as the flame temperature (see Sect. III.10). A concordant Maxwellian velocity distribution for all atomic and molecular species is expected to be most readily attained (see the above). For free electrons, having a discomparable mass, a deviating translational temperature, called *electron temperature*, might be found under certain flame conditions (see Sect. IV.7). We note that whenever deviations from thermodynamic equilibrium are found, the definition of the (flame) temperature is a matter of convention rather than of principle.

4. KINETIC CONCEPTS AND PROCESSES

In this section we introduce the main kinetic concepts and their relationships and present a general classification of the main processes that are important in the context of our book. The term *process* is used here in a broad sense, covering elastic and inelastic collisions as well as chemical reactions and physical and chemical ionization processes. Spontaneous emission of radiation is included only in so far as it competes with other de-excitation processes; detailed expressions for the rates of photon emission and absorption will be found in the following section. The emphasis lies generally on the rate or probability per second of inelastic collisions and reactions in the gas phase.

In Subsections 4a-1 to 4a-3 the concepts of rate and rate constant are defined and their significance for the equilibrium state (detailed balance) and for the relaxation to the equilibrium state are discussed. In Subsection 4a-4 the foregoing concepts and relationships are employed — for illustration as well as for use later, in Chapt. VI — in a survey of experimental methods of determining de-excitation rate constants in flames and other gaseous systems containing metal vapours. Methods of measuring rate constants of flame radical reactions and metal ionization processes will be mentioned separately in Sects IV.5 and IX.3c. The relationships between rate constant, (effective) cross section, mean free path and collision frequency are presented in Subsection 4b; in the latter subsection microscopic reversibility too is formulated in terms of effective cross sections. The subsequent Subsections 4c-4g provide an introduction into the theory of elementary processes, as far as is relevant for the subject of this book. Subsection 4h concludes this section with a general description and classification of processes. More detailed classifications and examples of specific types of processes will be given later in the appropriate chapters.

Although the scope of this section is more general, most concepts and

Chapt. II BASIC CONCEPTS AND GENERAL RELATIONSHIPS

relationships treated will be exemplified with (de-)excitation processes involving metal atoms. The section thus forms a theoretical background in particular to Chapt. VI.

The subject matter of this section is vast and diverse. It is concerned with macroscopic phenomena as well as microscopic mechanisms, and relates to Molecular Dynamics, Kinetic Chemistry, Quantum Mechanics, Statistical Mechanics and Thermodynamics as basic disciplines. There is a wealth of good textbooks available in this field. We mention for further study only the following, which have been chosen rather arbitrarily. Introductory textbooks are, e.g., Glasstone, Laidler and Eyring (1941), Laidler (1955, 1963), Cambel, Duclos and Anderson (1963) and Levine and Bernstein (1974). More advanced textbooks or chapters are, e.g., Lewis, Pease and Taylor (1956), Bates (1962), Moore (1962), Mott and Massey (1965), Fowler and Guggenheim (1965), Drawin (1968), Hartmann (1968), Hochstim (1969a), Massey, Burhop and Gilbody (1971), Nikitin (1974, 1975), and Bernstein (1979). For experimental data and methods of calculating rate constants of inelastic collisions and reactions see also, e.g., Johnston (1968), Heinrichs (1968), Nikitin (1968b) and Gilmore, Bauer and McGowan (1967, 1969). An extensive bibliography (up to and including 1966) has been published by Hochstim (1969). The books by Mavrodineanu and Boiteux (1965) and Gaydon and Wolfhard (1979) have sections on rate processes that relate to flames. Some publications on specific topics will be mentioned separately in the following subsections.

4a RATES AND RATE CONSTANTS

4a-1 *DEFINITIONS*

Consider, for example, the following three elementary processes in which a metal atom M^0 in the ground level is excited to a specific excitation level M^*, and vice versa:

$$M^0 + h\nu \rightleftharpoons M^* , \qquad (1)$$
$$M^0 + Z \rightleftharpoons M^* + Z , \qquad (2)$$
$$M^0 + X + Y \rightleftharpoons M^* + XY . \qquad (3)$$

Here $h\nu$ represents a photon with frequency ν, and X, Y and Z are atomic or molecular constituents of the flame gas. Because of the *principle of microscopic reversibility*, each elementary process can also proceed in the reverse direction. Thus for the forward photo-excitation process (1) there is a corresponding reverse photon emission process, represented by an arrow pointing from right to left; for the collisional excitation process (2) there is a corresponding collisional de-excitation process; for the excitation process (3) by chemical association there is a corresponding dissociative de-excitation process. We call, for example, M^0, X and Y the *reactants*, and M^* and XY the *products* of forward process (3); and conversely in the reverse process. In process (2) we call Z the *collision partner* of M^0 (or M^*); in forward process (3) we call X and Y *reaction partners*.

We now define the *rate*[†] (also called: *'velocity'*), v_n, of forward process (n) as the number of excited atoms M^* that are formed per second and per unit volume by the process considered. The rate of the reverse process is denoted by v_{-n}.

Forward process (1) involves the 'encounter' of an M^0 atom and a photon considered here as 'particle'; the reverse process (1) involves only a single M^* atom which spontaneously decays to the ground level by photon emission. The latter process is an example of a *unimolecular* process, and its rate, v_{-1}, is determined by the *unimolecular* (or: *monomolecular*) *rate constant* for spontaneous emission A (expressed in s^{-1}) according to

$$v_{-1} = A [M^*]. \tag{II.73}$$

Here $[M^*]$ denotes the number density (= number of particles per unit volume) of metal atoms in the excitation level considered. The rate constant in this example is called the *(Einstein) transition probability (per second) for spontaneous emission*; its significance for the emission intensity and its relationship to other radiative transition probabilities will be treated in Sect. II.5.

The *rate constant* (or: *rate coefficient*), k_n, of an elementary process (n), which consists only of a single step, can generally be defined by singling out the concentration factors[*]; so we have for the *bimolecular* (or *binary*) forward process (2) and the *termolecular* (or *ternary*) forward process (3) the relations

$$v_2 = k_2 [M^0][Z], \tag{II.74}$$

$$v_3 = k_3 [M^0][X][Y], \tag{II.75}$$

whereas for the bimolecular reverse processes (2) and (3) it holds that

$$v_{-2} = k_{-2} [M^*][Z], \tag{II.76}$$

$$v_{-3} = k_{-3} [M^*][XY]. \tag{II.77}$$

We note that k_2 and k_{-3} are expressed, e.g., in $cm^3 s^{-1}$ but k_3 is expressed in $cm^6 s^{-1}$. When concentrations are expressed in number of moles per cm^3, we express v_2 in $mol\, cm^{-3} s^{-1}$ and k_2 in $mol^{-1} cm^3 s^{-1}$. For a better distinction we may write the units for v_2 and k_2 in the former case as: molecule $cm^{-3} s^{-1}$ and molecule$^{-1} cm^3 s^{-1}$, respectively.

[†] For practical reasons the term 'rate' is used here in a sense that is different from that recommended by IUPAC (1970).

[*] The rate coefficients generally depend on the relative-velocity and internal-energy distribution of the reactants. In the absence of thermodynamic equilibrium these distributions might depend on the concentrations of the species involved. The rate coefficients are then no true constants of proportionality. In the following we shall consider k_n as to be independent of concentration (for a further discussion see Sect. II.4a-2).

Chapt. II   BASIC CONCEPTS AND GENERAL RELATIONSHIPS

In certain cases it may be convenient to include the concentration(s) of (some of) the partners in the rate constant, for example by defining the *apparent unimolecular rate constant*: $k^u_{-2} \equiv k_{-2}[Z]$ or: $k^u_{-3} \equiv k_{-3}[XY]$. In fact k^u_{-2} and k^u_{-3} denote the probability per second that a *given* M^* atom is de-activated through the reverse process (2) or (3), respectively. These constants have the same dimension $[t]^{-1}$ as A; they are, however, not really constant[†] as their values depend on the concentration of the partner(s) included.

The above concepts can easily be extended to other elementary collisional processes, chemical reactions and ionization processes.

The above forward processes (2) and (3) are first-order in $[M^0]$, $[Z]$, $[X]$ and $[Y]$; if $X \equiv Y$, process (3) is second-order in $[X]$. The over-all order of forward process (2) is two, and that of forward process (3) is three. The *order* of a process conforms here to the *molecularity*, discussed earlier, because the processes considered are elementary. If, however, an over-all process consists of several steps, its rate might vary with reactant concentration(s) to a power different from that that would follow from the over-all reaction equation (see Laidler 1963). To complex reactions it may even not always be possible to assign a certain order. The molecularity relates to the number of particles participating in an elementary reaction step and is always defined. See for the distinction between elementary and multi-step processes also Subsection 4h.

The rate v_2 of collisional excitation process (2) and the resulting rate constant k_2, as measured, e.g., in a cell containing metal atoms and N_2 molecules, are in fact weighted over the distributions of relative velocity, g, and of the internal energy, E_i, of N_2. For N_2 molecules in each energy level i we may define specifically a *microscopic rate constant* $k_2^{(i)}$, which is again weighted over the velocity distribution. Assuming that the normalized distribution function, $f(g)$, for the relative velocity does not depend on the internal energy, we can write

$$k_2^{(i)} = \int_0^\infty k_2^{(i)}(g) f(g) \, dg . \tag{II.78}$$

Here $k_2^{(i)}(g)$ is the *velocity-dependent microscopic rate constant*. Denoting the normalized distribution function for the molecular level populations by $f(E_i)$, we finally arrive at

$$k_2 = \sum_{i=0}^\infty k_2^{(i)} f(E_i) = \sum_{i=0}^\infty f(E_i) \int_0^\infty k_2^{(i)}(g) f(g) \, dg . \tag{II.79}$$

If we insert k_2 into Eq. II.74, we should then interpret $[Z]$ as a summation over all internal-energy levels. We accordingly distinguish between the *phenomenological* (or: *bulk*) *rate constant* k_2 and the *microscopic rate constant* $k_2^{(i)}(g)$ or $k_2^{(i)}$.

[†] The term rate *coefficient* would be more appropriate here.

When the distribution functions involved in the last equation conform to the Maxwell-Boltzmann laws at temperature T, we obtain the *thermal rate constant* $k_2(T)$, which is a function of T only. Note that the population distribution of the energy levels of M is not involved here; this is because we have defined k_2 as relating to the excitation of M from and to a *specific* energy level. The same holds for $k_{-2}(T)$.

When the experimental conditions are such that, for example, the excited atoms M^* have a *peaked* relative-velocity distribution with respect to the N_2 molecules, we can approximate $f(g)$ by a delta function. The measured de-excitation rate constant, $k_{-2}(g)$, is then a function of g. In Subsection 4a-4 we shall encounter a case of this kind. Crossed-molecular beam experiments with velocity-selection offer another possibility of measuring rate constants as a function of g.

When the rate of collisional excitation, etc., is measured in a multi-component gas, e.g. a flame, we obtain an *over-all* apparent unimolecular rate constant: $\overline{v_2} = \overline{k_2^u}\,[M^0]$. The bar denotes that the rate (constant) is a kind of weighted average over the different collision partners Z_j ($j = 1, 2, \ldots$) present. We can express this over-all rate constant in the *specific*, bimolecular rate constants $k_{2,j}$ as follows

$$\overline{k_2^u} = \sum_j k_{2,j}[Z_j] \equiv \overline{k_2}[Z]_t . \qquad (II.80)$$

In the second expression we have introduced: $[Z]_t \equiv \Sigma_j [Z_j]$; this expression may be used if we cannot or do not want to specify the $[Z_j]$'s. This equation is based on the additivity of the contributions from each species. Each of the specific (phenomenological) rate constants $k_{2,j}$ can, in turn, be expressed in the corresponding microscopic rate constants by Eq. II.79. If all $k_{2,j}$'s are thermal and all relative $[Z_j]$'s for which $k_{2,j} \neq 0$ conform to the mass-action-law, we may call $\overline{k_2^u}(T)$ thermal too. Note, however, that — in contrast to $k_{2,j}(T)$ — the thermal over-all rate constant depends not only on the temperature of the flame but also on its composition.

In order to find the thermal specific rate constants separately at given T, one measures $\overline{k_2^u}(T)$ in a number of isothermal flames with the same qualitative but different, known quantitative compositions: $[Z_1], [Z_2], \ldots [Z_N]$. One can thus obtain a set of N independent, linear equations in the N unknowns $k_{2,j}$. If one of the species Z_j is *a priori* known to give a negligible contribution to the excitation rate (because of its inherent inefficiency or small concentration), the number of equations can be reduced. It is often wise to provide for more independent, linear equations than there are unknowns in order to test the reliability of the computation.

Chapt. II BASIC CONCEPTS AND GENERAL RELATIONSHIPS

4a-2 *DETAILED-BALANCE RELATIONS BETWEEN RATE CONSTANTS*

The *principle of detailed balance* states that in thermodynamic equilibrium the rate at which each micro-process occurs exactly equals the rate of the reverse micro-process. In a *micro-process* the initial and final conditions of the interaction system, such as the relative velocity, the quantum numbers of the internal states, the impact parameter, are fully specified. This principle applies also after averaging over the relative velocity and those internal states that are not specified in the over-all process. This principle is a consequence of microscopic reversibility (see Subsection 4b-4) in combination with the equilibrium laws (see Sect. II.3b).

Application of this principle to the *excitation processes* (2) and (3), mentioned in the preceding subsection, leads to: $(v_2)_e = (v_{-2})_e$ and $(v_3)_e = (v_{-3})_e$, where the subscript e refers to the equilibrium state. Combination of these equalities with Eqs II.74-77 yields

$$(k_2)_e [M^0]_e [Z]_e = (k_{-2})_e [M^*]_e [Z]_e , \qquad (II.81)$$

$$(k_3)_e [M^0]_e [X]_e [Y]_e = (k_{-3})_e [M^*]_e [XY]_e . \qquad (II.82)$$

When different collision partners Z_j contribute to the (de-)excitation of a particle M, the relation in Eq. II.81 holds for each Z_j separately.

From these balance equations we obtain the following relations between the thermal rate constants of each process and its reverse

$$(k_2)_e / (k_{-2})_e = [M^*]_e / [M^0]_e , \qquad (II.83)$$

$$(k_3)_e / (k_{-3})_e = ([M^*]_e / [M^0]_e) [XY]_e / [X]_e [Y]_e . \qquad (II.84)$$

Here $(k_2)_e$ is equal to $k_2(T)$ as defined in the preceding subsection, and so on. Using now the Boltzmann law (Eq. II.30) for $[M^*]_e / [M^0]_e$ and the mass-action-law (Eq. II.43) for the dissociation equilibrium of: $XY \rightleftharpoons X + Y$, we arrive finally at the important *detailed-balance relations for conjugated rate constants*

$$(k_2)_e / (k_{-2})_e = (g^*/g_0) \exp[-E/kT] , \qquad (II.85)$$

$$(k_3)_e / (k_{-3})_e = (g^*/g_0) \exp[-E/kT] / K_d(T) . \qquad (II.86)$$

Here g^* and g_0 are the statistical weights of the excited and ground level, respectively, E is the excitation energy, and $K_d(T)$ the dissociation constant.

We note from Eq. II.81 that the concentration of the collision partner, $[Z]_e$, drops out of the detailed-balance equation for collisional (de-)excitation. This equation therefore remains valid even if $[Z]$ does not correspond to chemical equilibrium. However, the Maxwell-Boltzmann laws must be satisfied by Z

KINETIC CONCEPTS AND PROCESSES   Sect. II.4a

if Eqs II.81 and 85 are to hold. Deviations from the dissociation equilibrium of XY obviously render Eq. II.82 invalid; but then again if all reaction partners satisfy the Maxwell-Boltzmann laws, Eq. II.86 still remains valid.

Similar detailed-balance relations hold for the conjugated thermal rate constants of any process. The relation corresponding to *radiative process* (1) is given by Eq. II.151 in Sect. II.5a-1. For an arbitrary *chemical reaction*: $A + B + \ldots \rightleftharpoons C + D + \ldots$ with equilibrium constant $K_n(T)$ we have in equilibrium because of detailed balance

$$(k_n)_e [A]_e [B]_e \ldots = (k_{-n})_e [C]_e [D]_e \ldots , \qquad (II.87)$$

$$(k_n)_e / (k_{-n})_e = K_n(T) . \qquad (II.88)$$

In the case of a *dissociation reaction*: $AB + Z \rightleftharpoons A + B + Z$ we have in particular for the dissociation and association rate constants, k_d and k_a respectively,

$$(k_d)_e [AB]_e [Z]_e = (k_a)_e [A]_e [B]_e [Z]_e , \qquad (II.89)$$

$$(k_d)_e / (k_a)_e = K_d(T) . \qquad (II.90)$$

Note that the concentration of the 'third body' Z (see Subsection 4h) drops out of Eq. II.89; this equation and Eq. II.90 therefore remain valid also when [Z] does not conform to chemical equilibrium. A similar relation as Eq. II.90 holds also for the apparent unimolecular dissociation rate constant: $k_d^u \equiv k_d [Z]$ and the conjugated apparent bimolecular association rate constant: $k_a^b \equiv k_a [Z]$

$$(k_d^u)_e / (k_a^b)_e = K_d(T) . \qquad (II.91)$$

We again note that any deviation of [Z] from chemical equilibrium affects both k_d^u and k_a^b in the same way; consequently it does not affect their ratio, so that Eq. II.91 remains valid.

As another application of the principle of detailed balance we find for the collisional ionization process: $M + Z \rightleftharpoons M^+ + e^- + Z$ the equilibrium relations

$$(k_i')_e [M]_e [Z]_e = (k_r')_e [M^+]_e [e^-]_e [Z]_e , \qquad (II.92)$$

$$(k_i')_e / (k_r')_e = K_i(T) . \qquad (II.93)$$

Here k_i' and k_r' are the bimolecular and termolecular ionization and recombination rate constants, respectively, and $K_i(T)$ is the ionization constant (see Eq. II.59). In Eq. II.92 $[M]_e$ represents the equilibrium concentration of M summed over all its energy levels. Defining the apparent unimolecular ionization rate constant: $k_i \equiv k_i' [Z]$ and the apparent bimolecular recombination rate constant: $k_r \equiv k_r' [Z]$, we also find the equalities

69

$$(k_i)_e [M]_e = (k_r)_e [M^+]_e [e^-]_e \,, \qquad (II.94)$$

$$(k_i)_e / (k_r)_e = K_i(T) \,. \qquad (II.95)$$

The latter equalities remain valid when [Z] deviates from chemical equilibrium (compare with the dissociation case above).

The collisional-excitation, dissociation and ionization processes considered above are all endo-ergic. Consequently an exponential factor: $\exp[-E/kT]$, with E representing the excitation-, dissociation- or ionization-energy, occurs in the expressions for the ratio of conjugated thermal rate constants in Eqs II.85, 90 or 91, and 93 or 95 (see also Eqs II.44 and 60a for the dissociation and ionization constants). A high value for E/T makes the rate constant of the forward process very small relative to that of the reverse, exo-ergic process and thus reduces the equilibrium population of the excited level or the equilibrium degree of dissociation and ionization. We shall see in Subsection 4b-2 that the rate constant for the forward, endo-ergic process is the most affected by the value of E/T; when $E/T \to \infty$, this rate constant tends exponentially to zero.

We generally conclude that the thermal rate constant of any process can be calculated from that of the reverse process by the use of equilibrium laws only. In particular, the temperature-dependence of a thermal rate constant can be predicted through the above detailed-balance relations if the temperature-dependence of the rate constant for the reverse process is known.

It should be noted that the equilibrium laws themselves are *independent* of the magnitude of the rate constants. For example, the relative population $[M^*]_e / [M^0]_e$ is the same whether process (2) or process (3) prevails. The admittance of any additional process that contributes to the excitation of M will have no effect on the equilibrium value of $[M^*]$. This is, of course, a direct consequence of the principle of detailed balance. The enhanced rate at which M^* is formed because of this additional process is just balanced by the equally enhanced rate at which M^* is destroyed because of the additional reverse process. Consequently *no information whatsoever can be gained in regard to the type and rate of processes* that are operative in a system by studying its behaviour *in thermodynamic equilibrium* only.

In practice deviations from thermodynamic equilibrium may occur and one is faced with the problem of whether the above detailed-balance relations remain valid. Ultimately this problem must be solved by considering in detail the micro-processes that make up each bulk process. Generally, however, we can expect that the values of the rate constants will not be essentially different from those in equilibrium, as long as the internal and translational energy distributions of the particles involved are practically the same as in equilibrium (cf. Eq. II.79 and the ensuing

KINETIC CONCEPTS AND PROCESSES   Sect. II.4a

discussion on the thermal rate constants in Subsection 4a-1). This holds true at least if the rate constants are defined for an elementary process involving a single step only, and if all concentration factors are taken out. Any deviation of the *concentrations* of the reactants from their equilibrium values as such (caused, for example, by chemical disequilibrium or by photo-excitation by means of an external light beam) will then not affect the *rate constants*. Such deviations, however, do affect the *rates* and may, for instance, render Eqs II.81 and 82 invalid.

   The rate constant of a micro-process, where the reactants have a specified relative velocity and internal energy, will in general depend on the initial state. When the forward micro-process is not balanced by the same reverse micro-process, one might expect a depletion of reactants in those states for which the microscopic rate constants are the highest. The distribution of reactants over the different states will then deviate from equilibrium, and the phenomenological rate constant will be less than its thermal value (cf. Eq. II.79). However, when the rate of the process is much slower than the rate at which the reactants are redistributed over their various states by collisions with other molecules being in thermal equilibrium, the rate constant of the unbalanced process will still be close to its thermal value. It has been shown, for example, that the rate constant of an unbalanced dissociation reaction will practically equal its thermal value if kT is much less than the dissociation energy or, more generally speaking, is much less than the activation energy (see Rice 1961, Snider 1965, and Pyun 1968). In the special case when the reaction energy has to be supplied from the vibrational degrees of freedom of the molecular partners, the rate constant will be close to its thermal value if the vibrational relaxation rate greatly exceeds the reaction rate (see Kondratev and Nikitin 1967).

   If the distribution of the reactants over their internal states (e.g., the internal-energy distribution of a molecular collision partner Z or the population distribution over the components of an atomic multiplet M^0 or M^* in the case of process (2)) is affected by an unbalanced process, the phenomenological rate constants may depend on the degree of unbalance. The rate constants are then functions of the non-equilibrated reactant concentrations. In other words, k_2 and k_{-2} in Eqs II.74 and 76 can no longer be considered as constants of proportionality. When different collision partners exist together, the degree of unbalance and consequently the rate constant for each partner may then depend on the concentrations of all partners. Then the over-all rate constant is not a linear function of the $[Z_j]$'s and Eq. II.80 does not hold. When the degree of unbalance is also influenced by the occurrence of a competitive process (e.g., process (3)), the value of k_2 etc. may depend on k_3 etc.

   Fortunately, deviations from excitation, dissociation and ionization

71

equilibrium usually seem to have no marked effect upon the associated rate constants in the flame gas above the combustion zone. It is understood here that we have singled out all concentration factors in the expression for the rate constant.

We are now able to formulate more precisely the conditions for the existence of a *partial chemical equilibrium*, which was already mentioned in Sect. II.3c-3. Consider, for example, the chemical reaction: $A + B \rightleftharpoons C + D$, and suppose that not all of the species involved are in full chemical equilibrium with the rest of the system. Suppose, however, that the bimolecular rate constants, k_n and k_{-n}, of this reaction equal their thermal values (see for conditions the above). In general, the non-equilibrated species may also take part in other chemical reactions. Now if the rate at which the first mentioned reaction proceeds in *both* directions is high, whereas that of any of these other reactions is low, a state of balance between the rates v_n and v_{-n} of the former reaction will be rapidly approached (cf. next section). Then $v_n \simeq v_{-n}$, whereas the other reactions still remain unbalanced. Consequently we get

$$k_n(T)\,[A][B] \simeq k_{-n}(T)\,[C][D] \,, \qquad (II.96)$$

and because of Eq. II.88, with $(k_n)_e \triangleq k_n(T)$ etc.,

$$[C][D]/[A][B] \simeq k_n(T)/k_{-n}(T) = K_n(T) \,, \qquad (II.97)$$

where $K_n(T)$ is the equilibrium constant of the balanced reaction. We can conclude therefore that the concentrations of A, B, C and D are mutually related by the pertinent mass-action-law, although each of them individually may deviate from equilibrium. The rapidly established partial equilibrium for this particular reaction links the concentrations of the non-equilibrated species, without forcing them to attain their equilibrium values. When for example [A] and [C] are about equal to their equilibrium values, the balanced reaction forces only the *ratio* [B]/[D] to equal about its equilibrium value $[B]_e/[D]_e$. When B and D are both present in excess over their equilibrium concentrations, they may slowly relax to equilibrium, e.g., through the association reaction: $B + D + Z \rightarrow BD + Z$. The latter reaction is slow because it involves a 'third body' Z (see Subsection 4h). We shall encounter examples of such situations in flames in other chapters.

In a similar way a *partial ionization equilibrium* may be set up. It links the concentrations of the neutral and parent ionic species with that of the electrons, in accordance with the Saha law, notwithstanding the fact that the electrons may be present in excess (see Sect. IX.2a).

4a-3   *RATE CONSTANTS AND RELAXATION*

We shall derive, for illustration and use later, the quantitative

KINETIC CONCEPTS AND PROCESSES   Sect. II.4a

relationship between the rate constants and the relaxation time τ with which the system tends to a stationary state, in two special examples.

First example. Let the (de-)excitation of a metal atom having only a ground level (M^0) and one excited level (M^*) proceed exclusively through the radiative process (1) and the collision process (2) mentioned earlier. Suppose furthermore that the intensity of the radiation field is so low that photo-excitation virtually plays no role. This may be the case with a flame with low metal concentration and no external radiation field.

Suppose that by some means (e.g., a very short light pulse that is resonant with the optical transition) the initial concentration $[M^*]$ is raised to a high value $[M^*]_0$ at time $t=0$. The relaxation of $[M^*]$ towards the stationary state for $t>0$ will be described by the differential equation

$$d[M^*]/dt = (d[M^*]/dt)_1 + (d[M^*]/dt)_2 . \tag{II.98}$$

The terms on the right-hand side of this equation, labelled by 1 and 2, denote the contributions of processes (1) and (2) to the total time derivative. This time derivative is considered in a reference system at rest with respect to the metal vapour, which moves upward at the flame rise-velocity v_r. From the definitions of the rate constants A, k_2 and k_{-2} (see Subsection 4a-1) we get

$$(d[M^*]/dt)_1 = -A[M^*] , \tag{II.99}$$

$$(d[M^*]/dt)_2 = -k_{-2}[M^*][Z] + k_2[M^0][Z] , \tag{II.100}$$

$$d[M^*]/dt = -(A + k_{-2}[Z])[M^*] + k_2[M^0][Z] . \tag{II.101}$$

Now suppose that: $[M^*] \ll [M^0]$ and that the collisional rate constants are thermal (which are plausible assumptions under usual flame conditions). We can then treat $[M^0]$ as a constant and as equal to its equilibrium value, and apply the detailed-balance relation between k_2 and k_{-2}, given by Eq. II.83. We then get from the last equation

$$d[M^*]/dt = -(A + k_{-2}[Z])[M^*] + k_{-2}[M^*]_e[Z] . \tag{II.102}$$

The stationary value $[M^*]_{st}$ is found by putting: $d[M^*]/dt = 0$ in Eq. II.101

$$[M^*]_{st} = [M^*]_e k_{-2}[Z]/(A + k_{-2}[Z]) . \tag{II.103}$$

The deviation of $[M^*]$ from its stationary value is described by the differential equation

$$d([M^*] - [M^*]_{st})/dt = -(A + k_{-2}[Z])([M^*] - [M^*]_{st}) . \tag{II.104}$$

The solution is an exponential function of time

$$[M^*] - [M^*]_{st} = ([M^*]_0 - [M^*]_{st}) \exp[-(A + k_{-2}[Z])t] . \tag{II.105}$$

Here we have tacitly assumed that $k_{-2}[Z]$ and $[M^*]_e$ are independent of time (that is, independent of height in the flame). From the last equation we derive for the *relaxation time* τ

$$\tau = (A + k_{-2}[Z])^{-1}. \tag{II.106}$$

We can interpret τ also as the *actual lifetime* of the atom in the excited level, that is, the mean time the atom stays in this level before it becomes de-excited by radiation or collision (cf. the definition of radiative lifetime in Sect. II.5a-2). Usually τ is very short for electronic excitation levels under flame conditions so that a stationary state is attained virtually immediately (see also Sect. VI.4c-1).

In this example the stationary state does not conform to the equilibrium state as the *population factor*

$$P \equiv [M^*]_{st} / [M^*]_e \tag{II.107}$$

is smaller than unity according to Eq. II.103. There is thus a depletion of the excited-level population, which is all the stronger the larger the ratio $A/k_{-2}[Z]$ is. This deviation is due to the lack of detailed balance caused by the assumed absence of photo-excitation. If photo-excitation occurred in a Planck radiation field, the stationary excited-level population would exactly equal $[M^*]_e$ given by Boltzmann's law. Note from Eq. II.103 that in the above example the violation of detailed balance for the radiative process (1) also removes detailed balance for the collisional process (2), notwithstanding the fact that the collisional rate constants themselves are thermal and obey Eq. II.85. However, when $A \ll k_{-2}[Z]$, process (1) is unimportant compared to process (2), and detailed balance is restored for the latter, while the stationary level population approaches the Boltzmann value. We then have a state of thermal (not: thermodynamic) equilibrium, as defined in Sect. II.3c-3. The (de-)excitation is then collision-dominated (see for a further discussion Sects VI.2c-3 and 4c-2).

The above example clearly demonstrates the difference between *detailed* balance: $(d[M^*]/dt)_1 = (d[M^*]/dt)_2 = 0$ (a condition for *equilibrium*) and *total* balance: $(d[M^*]/dt)_{total} = 0$ (a condition for *stationarity*). The latter implies only that the total rate of de-excitation balances the total rate of excitation.

Whereas the equilibrium population $[M^*]_e$ does not depend on the rate constants involved, the stationary population outside equilibrium as well as the rate at which the equilibrium or stationary population is approached do depend on the rate constants. This dependence opens ways for measuring rate constants (see next subsection).

Second example. Suppose now that processes (2) and (3) are the only important processes for the excitation of M. Following the same procedure as in the first example, we can now prove that the relaxation time for the attainment of a stationary excited-level population equals $(k_{-2}[Z] + k_{-3}[XY])^{-1}$. The ratio $[M^*]/[M^0]$ in the *stationary state* is found to be

$$\left(\frac{[M^*]}{[M^0]}\right)_{st} = \frac{k_2[Z] + k_3[X][Y]}{k_{-2}[Z] + k_{-3}[XY]} \ . \tag{II.108}$$

It is easily verified by using the equilibrium relations Eqs II.85 and 86 and the mass-action-law Eq. II.43 that $([M^*]/[M^0])_{st}$ obeys the Boltzmann law, Eq. II.30, when the rate constants are thermal and there is dissociation equilibrium. The depletion effect due to radiative disequilibrium, found in the former example, does not appear here as we have assumed that process (1) is unimportant relative to processes (2) and (3). An excess concentration of dissociation products X and Y would, however, lead, in the stationary state, to a population of M^* in excess of its equilibrium value. The effect of dissociation disequilibrium on the population of M^* is the greater, the larger the rate of chemical process (3) is in comparison to that of collisional process (2) (see for a further discussion Sect. VI.4c-3).

Under usual flame conditions, the relaxation time for the attainment of a stationary excited-level population is very short compared to the residence time of the M atoms in the flame. On the other hand, the values of the quantities occurring in the expression for the stationary population may vary with the height of observation h_{obs} in the flame; this height is connected with the particle's travel time t through: $h_{obs} = v_r t$, where v_r is the constant vertical rise-velocity of the flame gas. In the example described by Eq. II.103 $[M^*]_e$, which depends exponentially on the temperature T, will vary with t when T varies with height. In the other example described by Eq. II.108 k_2 may vary when T varies, while $[X][Y]$ may vary because of a slow recombination of excess flame radicals (see Sect. IV.5). The actual value of $[M^*]/[M^0]$ at any height will then still be close to the *local* value of $([M^*]/[M^0])_{st}$ calculated by taking the quantities in the right-hand side of Eq. II.108 at the height considered. This holds true at least in the region above the combustion zone.

Other examples of relaxation equations involving rate constants will be found in Sects IV.5 and IX.3c relating to radical-recombination and metal-ionization processes. The relaxation rates for these processes may be considerably slower than for the (de-)excitation of electronic states. A stationary state may then not be attained, at least not at low height in the flame.

For a complete solution a diffusion term has to be included in the relaxation equation; an example of a complete ionization relaxation equation is given in Eq. IX.23.

4a-4 EXPERIMENTAL METHODS OF DETERMINING COLLISIONAL DE-EXCITATION RATE CONSTANTS FOR METAL ATOMS

General. We describe here the principles of some experimental methods that have found application in the determination of *collisional de-excitation rate constants*, also called: *quenching rate constants*, of electronic levels in atomic metal vapours. The data obtained by flame experiments will be discussed and compared with related experiments in vapour cells and shock-tubes in Chapt. VI, where also literature references are to be found concerning the experimental methods described below. For a review and critical discussion of some of the methods to be mentioned we refer also to, e.g., the book of Massey, Burhop and Gilbody (1971) and to the reports and dissertations of Lijnse (1972,1973), Siara (1972) and Earl (1973).

All methods employ essentially the existence or creation of some *deviation* from thermal equilibrium (cf. Subsection II.4a-2). We distinguish between *stationary methods* (nos 1-4) and *relaxation methods* (nos 5, 6) according to whether a stationary deviation or its relaxation is measured. For simplicity's sake we assume a two-level model for the atom M, which absorbs and emits only one resonance line at wavelength λ_0. We also assume that only the radiative and collisional processes (1) and (2) (see Subsection 4a-1) are responsible for (de-)excitation. Extensions to more general cases are easily made. In accordance with the usual experimental conditions we also assume that collisions between metal atoms are negligible because of their low density; so we shall only consider collisions between M and a partner of a different species Z (e.g., a N_2 molecule).

1. *Absolute fluorescence method.* When a flame with atomic metal vapour is irradiated by an external light source with a spectral component at λ_0, a fraction Y of the absorbed *primary* radiant flux will be re-emitted as *secondary* radiation, called (*resonance*) *fluorescence*, in all directions. The remaining fraction of the absorbed power is converted into heat when photo-excited atoms make a radiationless transition to the ground level as a result of collisional de-excitation. The fluorescence intensity is thus (partially) *quenched* by these collisions and we have therefore called the collisional de-excitation rate constant the *quenching rate constant*. This fraction Y is called the *efficiency of fluorescence*[†] (see for a more general definition Sect. II.5c-4) and equals, according to the *Stern-Volmer formula*, the ratio of the optical transition probability per second, A, to the total probability of de-excitation per second and per excited atom. Applying the definitions of the rate constants A and k_{-2} given in Subsection 4a-1 we can express the Stern-Volmer formula in the case when only the reverse processes (1) and (2) contribute to the de-excitation, by

---

[†] Also called yield factor of fluorescence in the literature.

KINETIC CONCEPTS AND PROCESSES   Sect. II.4a

$$Y = A/(A + k_{-2}[Z]) \equiv A/(A + k') . \qquad (II.109a)$$

Here k' (in s^{-1}) denotes the apparent *uni*molecular quenching rate constant $k_{-2}^u \equiv k_{-2}[Z]$. When all three reverse processes (1)-(3) contribute to the de-excitation, we have instead

$$Y = A/(A + k_{-2}[Z] + k_{-3}[XY]) \equiv A/(A + k') , \qquad (II.109b)$$

where now: $k' \equiv k_{-2}[Z] + k_{-3}[XY]$. When $k' \gg A$, virtually no fluorescence is observed ($Y \simeq 0$) because of the dominance of quenching collisions. When $k' \ll A$, Y attains its maximum = 1 and practically all absorbed primary radiation will be re-emitted as fluorescence.

An *absolute* measurement of Y (see Sect. III.16b) at known [Z] and A yields the bimolecular quenching rate constant k_{-2} through Eq. II.109a. For an accurate determination Y should not be too close to unity. This method has been applied mostly to flames, which usually contain several quenching species. One finds k_{-2} for each species separately by measuring Y in a number of isothermal flames with varying but known gas composition (see Subsection II.4a-1).

2.   *Relative fluorescence method*. This method is based upon a *relative* measurement of Y as a function of known [Z]. We have from the Stern-Volmer formula, Eq. II.109a,

$$Y^{-1} \propto A + k_{-2}[Z] . \qquad (II.110)$$

When Y^{-1} is plotted versus [Z] the absolute value of k_{-2} can be derived from the slope and intercept of the straight line obtained if A is known. This method has been applied mostly in metal vapour cells with a single quenching gas whose pressure can be easily controlled.

3.   *Population-depletion method*. As was discussed in the first example in Subsection 4a-3, the stationary density, $[M^*]_{st}$, of excited atoms may be below the Boltzmann value $[M^*]_e$ because of radiative dis-equilibrium (absence of adequate photo-excitation). From Eqs II.103 and 109a we find for the population factor P

$$P \equiv [M^*]_{st} / [M^*]_e = 1 - Y . \qquad (II.111)$$

As will be seen from Sect. III.10c, it is possible to measure directly the 'excitation temperature', T_{exc}, of the pertinent excitation level. This temperature value is defined implicitly by using formally the Boltzmann relation, Eq. II.30,

$$[M^*]_{st}/[M^0] \equiv (g^*/g_0) \exp[-E/kT_{exc}] . \qquad (II.112)$$

The corresponding equilibrium ratio $[M^*]_e / [M^0]_e$ is determined by the true gas temperature T through the Boltzmann law

$$[M^*]_e / [M^0]_e = (g^*/g_0) \exp[-E/kT] . \qquad (II.113)$$

Combining the latter pair of equations and putting $[M^0] = [M^0]_e$, we get

$$P \equiv [M^*]_{st} / [M^*]_e = \exp[(E/k)(T^{-1} - T_{exc}^{-1})]. \qquad (II.114)$$

If we have measured T_{exc} and we know T and E, we can derive Y from Eqs II.114 and 111. From the Y value thus determined we find k_{-2} as explained in Method 1 above.

The population-depletion method has been applied with metal vapours in a shock-heated gas whose true temperature can be accurately calculated from the shock-parameters (see Gaydon and Hurle 1963, and Tsuchiya 1964). This method works only when the gas after passage of the shock-front quickly reaches a state of thermal equilibrium; vibrational relaxation therefore should not occur. According to Eq. II.106 (with $A \approx 10^7 \, s^{-1}$) the metal vapour attains a stationary state of excitation in a very short time compared to the total duration of the shock pulse ($\approx 10^2 \, \mu s$).

4. *Saturation method.* In principle Y can also be determined from the saturation parameter; this parameter characterizes the spectral volume density of radiant energy that is required to saturate the atomic transition by laser irradiation tuned at the transition wavelength (see Sect. II.5c-4). Care should be taken that the laser-pulse duration is not too short compared to the relaxation time for atomic excitation. At radiation powers where saturation sets in, this time is about equal to that given by Eq. II.106; at higher radiation levels, this time is shortened under the influence of induced emission.

5. *Lifetime method.* One can determine the lifetime of an excited state after pulse excitation by measuring the decay of the fluorescence intensity as a function of time, e.g., by the delayed-coincidence technique. The quenching rate constant then follows from Eq. II.106 at known A and $[Z]$. This method can also be used to measure A in the absence of a quenching gas. The time resolution of the detection apparatus should be good, i.e. typically of the order of a few ns.

6. *Phase-shift method.* When the atoms are excited by a radiation beam with a.c. modulated intensity, a phase shift occurs between the a.c. components of the exciting and the fluorescent beams. This phase shift ϕ is determined by the modulation frequency f_{mod} and the relaxation- or lifetime τ of the excited state according to

$$\tan\phi = 2\pi f_{mod} \tau. \qquad (II.115)$$

One can obtain τ and thus the quenching rate constant by measuring ϕ at known f_{mod} and known A. In order to measure τ with good accuracy, the modulation period $(= f_{mod}^{-1})$ should be of the order of τ.

Discussion. There are a few other methods and other variants of the above methods which have not been mentioned here (see Massey, Burhop and Gilbody 1971). Besides, the excitation of the atomic vapour may be achieved not only by photo-excitation techniques, but also by sensitized-fluorescence (see Sect. II.5a-1) or electron bombardment.

When one of the above methods is applied to a bulk-gas containing atomic metal vapour, the relative-velocity distribution of the (excited as well as the unexcited) atoms will be Maxwellian. One then finds the thermal quenching rate constant, which is a function of *temperature*. With the *photo-dissociation technique* one employs a quartz cell containing vapour of a metal salt, e.g. NaI, and the quenching gas. This vapour is irradiated by an intense beam from a metal spark, emitting a spectral line of sufficiently short wavelength, or from a hydrogen arc, emitting a continuous spectrum from which a narrow u.v. band is selected by a monochromator. If the wavelength of this radiation is sufficiently far below the photo-dissociation limit of the metal salt (see Sect. II.2e), an *excited* Na atom and a free I atom in the ground state are formed (cf. process no. 27 in Table II.1). The excess photon energy is converted into kinetic energy of relative motion of the dissociation products. By varying the wavelength of the exciting radiation the initial velocity of the excited Na atoms relative to the quenching gas can be controlled. The actual relative-velocity distribution of the excited atoms at the moment of quenching may be broadened by secondary factors. In any case, the photo-dissociation technique combined, e.g., with Method 2 or Method 5 (with pulsed photo-dissociation) can be used to determine quenching rate constants as a function of mean *relative velocity* for a more or less sharply peaked velocity distribution. Results obtained and literature references will be given in Chapt. VI.

Systematic errors in the rate constants may arise from uncertainties in the optical transition probabilities, A. Besides, the fluorescence intensities of resonance lines have to be corrected for losses due to self-absorption (see Sect. II.4c-3). This correction may be obtained by measuring the fluorescence efficiency as a function of metal concentration and extrapolating to zero concentration (see Hooymayers and Alkemade 1966; see Sect. III.16b). Alternatively, one may apply a semi-empirical correction formula that is valid for not too high metal concentrations (see Jenkins 1966). In vapour-cell measurements radiation trapping or imprisonment of resonance lines may cause a systematic error by enlarging the effective optical lifetime of the excited state (see Sect. II.5c-3). This error may be kept small by choosing low metal concentrations. Furthermore, pressure-dependent collisional broadening and shift (see Chapter VII) of the absorption line may cause errors, when the exciting line is comparatively narrow. Through the latter effects the absorbed primary light intensity varies with pressure, which distorts the

Chapt. II. BASIC CONCEPTS AND GENERAL RELATIONSHIPS

dependence of relative fluorescence intensity on [Z] in Method 2. Neglect of these effects has introduced errors into the earlier quenching measurements with vapour cells. Polarization effects may also introduce an error, as the rate of depolarizing collisions depends on the gas pressure (see McGillis and Krause 1968, 1968a). Spurious polarization of the exciting light beam may be caused by the optical components (see Sect. III.4). When the quenching efficiency of the gas under investigation is small, molecular impurities in trace concentrations in the vapour cell may interfere by competition. Moreover, chemically active impurities may affect the concentration of free metal atoms in vapour cells.

> In vapour cells, fluorescence measurements are often done in a layer close to the observation window in order to minimize re-absorption or radiation trapping (see Krause 1966). Quenching collisions of excited atoms with the cell wall might then lead to errors. One expects, however, from the displacement that an excited atom with thermal velocity undergoes during its lifetime that wall collisions are normally unimportant.

The influence of the above error sources depends on the method used. For example, factors that affect the production rate of excited atoms or the atomic metal concentration are of no consequence with the relaxation methods (see Krause 1975). However, the reliability of the latter methods may be affected by radiation trapping, as are the stationary methods. In general, most error sources can be eliminated by working at low metal and quencher concentrations. The signal-to-noise ratio, however, may then become worse and a compromise has to be sought. Other error sources will be mentioned in Sect. VI.2a-2 in relation with doublet-mixing experiments. Sources of error connected with the measuring apparatus are discussed in Sect. III.16.

4b   CROSS SECTIONS AND RATE CONSTANTS
4b-1 *CROSS SECTION FOR GAS-KINETIC COLLISIONS AND RELATED CONCEPTS*

Gas-kinetic collisions between species M and Z are associated with the transfer of linear momentum and are closely related to the diffusion process (see Sect. X.2). The rate, v_0, of binary gas-kinetic collisions between M and Z, considered as hard spheres with radii r_1 and r_2, is determined by

$$v_0 = [M][Z]\,\pi(r_1+r_2)^2\,\bar{g}, \tag{II.116}$$

where $\bar{g}$ is the mean relative velocity. We now define: $\sigma_0 \equiv \pi(r_1+r_2)^2$ as the *cross section* for this process.[†] The order of magnitude of σ_0 for atoms or

---

[†] In the literature cross sections are sometimes defined as $(r_1+r_2)^2$ without a factor π.

molecules is $10^{-15}-10^{-14}$ cm^2; σ_0 is often expressed in Å^2 ($= 10^{-16}$ cm^2). Consequently we have for the rate v_0 and corresponding rate constant k_0

$$v_0 = [M][Z]k_0 = [M][Z]\sigma_0 \bar{g} . \tag{II.117}$$

In the case of Maxwell equilibrium $\bar{g}$ is given by Eq. II.29 and we get

$$v_0 = [M][Z]\sigma_0 (8kT/\pi\mu)^{\frac{1}{2}} . \tag{II.118}$$

In general the cross section σ_0, which is a measure for the average loss of linear momentum in the forward direction, depends on $\bar{g}$. It is therefore better to insert in the right-hand side of the former expression: $\overline{\sigma_0(g)g}$ instead of $\sigma_0\bar{g}$, or to define: $\sigma_0 \equiv \overline{\sigma_0(g)g}/\bar{g}$.

This cross section $\sigma_0(g)$ is closely related to the *cross section for scattering*, $\sigma_{sc}(g)$, which can be studied as a function of g in molecular-beam experiments with velocity selection. If the *differential scattering cross section*, which characterizes the dependence of the scattering probability as a function of scattering angle, θ, is independent of θ, the velocity-dependent cross sections $\sigma_0(g)$ and $\sigma_{sc}(g)$ are equal (see Drawin 1968).

The *collision frequency* ν (in s^{-1}) is defined as the average number of gas-kinetic collisions that a given particle of species M makes per second with an arbitrary Z particle. In fact ν equals the apparent unimolecular rate constant that is obtained by dividing v_0 by [M]. We thus find from Eq. II.118 in the case of Maxwell equilibrium

$$\nu \equiv v_0/[M] = [Z]\sigma_0 (8kT/\pi\mu)^{\frac{1}{2}} . \tag{II.119}$$

When M is a Na atom and Z a typical flame molecule such as N_2, ν amounts to about 3×10^9 s^{-1} at 1 atm pressure and at 2600 K. Because of the reduced-mass factor, $\mu^{-\frac{1}{2}}$, ν is much larger for electron-molecule collisions (see Sect. IX.3a). One also expects from these formulas that collisions between two Na atoms are roughly a factor $[N_2]/[Na]$ rarer than collisions between an Na atom and an N_2 molecule.

The *collisional lifetime* or: *intercollision time* τ_{ic} is the reciprocal of the collision frequency; it is the average time between two successive collisions of a given M particle with an arbitrary Z particle. The *duration* of a collision between a Na atom and a flame molecule is many orders of magnitude shorter than the collisional lifetime under usual flame conditions. Taking a typical value of 10^{-13} s for this duration and of 10^{-10} s for τ_{ic}, one estimates that the probability of a *ternary* collision between a given Na atom and two N_2 molecules is $10^{-10}/10^{-13} = 10^3 \times$ smaller than the probability of a *binary* collision with one molecule.

The *mean free path*, $\lambda_{M,Z}$, of a particle M with respect to gas-kinetic collisions with particles Z is the average distance covered by M before it

Chapt. II   BASIC CONCEPTS AND GENERAL RELATIONSHIPS

collides with Z. According to this definition we have the relation

$$\lambda_{M,Z} \equiv \bar{v}_M \tau_{ic} = ([Z]\sigma_0)^{-1}(\bar{v}_M/\bar{g}) \,, \qquad (II.120a)$$

or in the case of Maxwell equilibrium (see Eqs II.26 and 29)

$$\lambda_{M,Z} = ([Z]\sigma_0)^{-1}(\mu/m_M)^{\frac{1}{2}} \,. \qquad (II.120b)$$

Herein v_M and m_M are the absolute velocity and mass of M, respectively. It is noted that the expressions II.120a and 120b are only approximately valid as we have assumed herein that σ_0 is independent of g. Since in flames at 1 atm pressure [Z] is of the order of 10^{18} cm^{-3} for the major flame constituents and σ_0 is of the order of $10^{-15} - 10^{-14}$ cm^2, we calculate the mean free path to be equal to roughly $10^{-4} - 10^{-3}$ cm.

4b-2   *CLASSICAL EXPRESSION FOR THE RATE CONSTANT; ACTIVATION ENERGY*

A collision can lead to a *process* (e.g. chemical reaction or excitation of one of the collision partners) if sufficient energy is available in the collision system to overcome the required *activation energy* (or *threshold*) E_A of the process. If this energy has to be supplied from the kinetic energy associated with the component of the relative motion of the colliding partners along the line of their centres, the thermal rate v_n of process (n) is calculated from the Maxwell law to be

$$v_n(T) = P[M][Z]\sigma_0(8kT/\pi\mu)^{\frac{1}{2}}\exp[-E_A/kT] \,. \qquad (II.121)$$

The right-hand side of this equation differs from that of Eq. II.118 by an exponential factor and by the *steric factor* P. The former factor describes the fraction of collision pairs that have sufficient energy to overcome the energy barrier. The latter factor has been added in order to account for the fact that not every gas-kinetic collision with sufficient energy will actually lead to the process considered. The steric factor may be markedly less than unity when the process takes place only if the collision partners have a special orientation with respect to each other or with respect to their line of connection. More generally, a factor $P<1$ is included not only to take geometrical restrictions into account but also to adapt formally the classical expression to restrictions of quantum-mechanical origin. It is further noted that in some cases the process, e.g. a dissociation or ionization reaction, does not take place in a single step; it may actually consist of a succession of (smaller) steps involving particles in different excited levels (*ladder-mechanism*; see for a special example Sect. IX.6). The steric factor for the collisional dissociation of a di-atomic molecule includes, for example, the quantum of vibrational energy, because dissociation proceeds by 'climbing up the ladder' of vibrational levels (see, e.g., Stevens 1967).

According to the definition of the bimolecular rate constant (cf. Eq. II.74) we get from Eq. II.121 for the thermal rate constant of process (n)

$$k_n(T) \equiv v_n/[M][Z] = P\sigma_0 \, (8kT/\pi\mu)^{\frac{1}{2}} \exp[-E_A/kT] \,. \tag{II.122}$$

The factor preceding the exponential factor in the last expression is called the *pre-exponential factor* or *frequency-factor*; it has the same dimension as k_n. Only for unimolecular rate constants has the factor the dimension of frequency, $[t]^{-1}$.

When more degrees of freedom are available in the colliding system to overcome the activation energy (for example, the translational energy of relative motion perpendicular to the line of centres, or internal molecular energy), the right-hand side of Eq. II.122 should be extended. On condition that the internal molecular energy participating in the process is distributed quasi-continuously in accordance with the Boltzmann law and that also Maxwell's law applies, we have for the thermal rate constant

$$k_n(T) = P\sigma_0 \, (8kT/\pi\mu)^{\frac{1}{2}} \exp[-E_A/kT] \times$$
$$\times \left\{ 1 + (E_A/kT)^{s-1}/(s-1)! + (E_A/kT)^{s-2}/(s-2)! + \ldots \right\}. \tag{II.123}$$

Here $(2s-2)$ is the number of *additional* quadratic terms in the general expression for the available energy (in the centre-of-mass system). The expression between braces $\{\ldots\}$ holds for integer values of s; the series stops at the sth term. for example, in the case of atomic excitation by collisions with free electrons, $s = 2$ and the expression between braces reduces to: $(1 + E_A/kT)$.

When in a (limited) range of temperatures, the thermal rate constant $k_n(T)$ is observed to depend (approximately) exponentially on T^{-1}, one can express the experimental rate constant in the form of the *Arrhenius' equation*

$$k_n(T) = A \exp[-E_{Arrh}/kT] \,, \tag{II.124}$$

where A and the *Arrhenius' energy of activation* E_{Arrh} are constants in the temperature range considered. One has the relation

$$d(\ln k_n)/dT^{-1} = -E_{Arrh}/k \,. \tag{II.125a}$$

When we start from the classical expression for $k_n(T)$ in Eq. II.122, we find the relation

$$d(\ln k_n)/dT^{-1} = -(E_A/k + \tfrac{1}{2}T) \,. \tag{II.125b}$$

Comparing the latter pair of equations one sees that E_{Arrh} and E_A are not identical; in the case when Eq. II.122 applies we have the relation

$$E_{Arrh} = E_A + \tfrac{1}{2}kT \,. \tag{II.126}$$

In flames the term $\tfrac{1}{2}kT$ (≈ 0.2 eV) is relatively small when $E_A \gtrsim 1$ eV.

When $k_n(T)$ varies proportionally to a *negative* power of T, strict application of the Arrhenius' equation would lead to a negative value for E_{Arrh}. In such cases it is advisable to choose $E_{Arrh} = 0$ and to write the pre-exponential factor A as a negative power of T (see Schofield 1967a).

The activation energy E_A and the energy, ΔH_0^0, of a reaction (see Sect. II.3b-3) should be well distinguished. When $\Delta H_0^0 > 0$, we have $E_A \geqslant \Delta H_0^0 > 0$. But a positive E_A may also be found with an exo-ergic reaction for which $\Delta H_0^0 < 0$. In the latter case the 'reaction path' leads over an energy barrier. A similar case of an exo-ergic quenching process with positive E_A is depicted in Fig. II.5b on page 94.

Binary chemical reactions between atoms and di-atomic molecules in the gas phase at room temperature have generally frequency factors A of 10^{14} to 10^{13} cm^3 mol^{-1} s^{-1} (10^{-10} to 10^{-11} cm^3 molecule^{-1} s^{-1}). Frequency factors for slightly more complex reactants may be one or two orders of magnitude less (see Laidler 1963).

When E_A for the forward process is known, its value for the reverse process follows by applying the detailed-balance relation, Eq. II.88, between k_n and k_{-n}. The *difference* between the two E_A values is directly connected with the reaction energy. This relation is schematically shown in Fig. II.4.

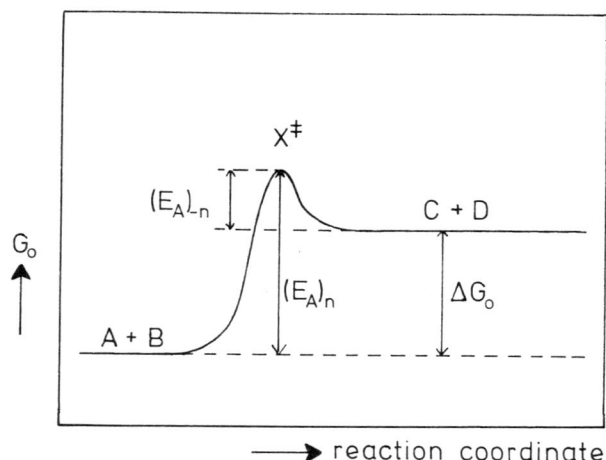

Fig. II.4 Schematic graph of the change of free enthalpy G_0 at 0 K during the course of a chemical reaction: $A + B \rightarrow X^{\neq} \rightarrow C + D$. The reaction proceeds through an 'activated complex' $X^{\neq}$ located somewhere at the top of the curve (which constitutes a 'potential barrier'). The activation energy of the forward and of the reverse reaction is denoted by $(E_A)_n$ and $(E_A)_{-n}$, respectively; their difference equals the free enthalpy of reaction ΔG_0.

4b-3 GENERAL DEFINITION OF EFFECTIVE CROSS SECTION OF PROCESS

In the preceding subsection the thermal rate constant of a process was expressed in the (velocity-independent) cross section for gas-kinetic collisions. The latter can be derived from diffusion experiments. If the activation energy is known and the steric factor estimated, an estimate of the thermal rate constant may be obtained. For a more accurate description, which may also be used in non-equilibrium conditions, we have to take into account explicitly the dependence of the probability of the process on relative velocity, initial state of the reactants, etc., and the corresponding distribution functions. In such a description the concept of effective cross section is useful.

Consider again a certain process (n) between two reactants, M and Z. The process can be decomposed into micro-processes for which the initial and final states of the collision system are specified in detail (see Subsection 4a-2). Let the velocity-dependent bimolecular rate constant of such a micro-process be denoted by: $k_n^{(i,f)}(g)$, where superscripts (i,f) stand for the symbols specifying the internal states of the reactants and of the products, respectively. We now <u>define</u> formally the (*microscopic*) *effective cross section*[†] of process (n)

$$\sigma_n^{(i,f)}(g) \equiv k_n^{(i,f)}(g)/g \,, \qquad (II.127)$$

by analogy with the relation: $\sigma_0 = k_0/\bar{g}$ in Eq. II.117. It is noted that the effective cross section has the dimension of area. The effective cross section, however, has no (direct) relationship with the dimensions of the particles involved, in contrast to the gas-kinetic cross section σ_0; the former cross section may be much larger or much smaller than the latter one. When we sum over all accessible internal states of the products, we obtain

$$\sigma_n^{(i)}(g) \equiv \sum_f \sigma_n^{(i,f)}(g) \,, \qquad (II.128)$$

which depends only on the initial state of the reactants. In the example considered in Eq. II.79, we have the relation: $k_2^{(i)}(g) = \sigma_2^{(i)}(g)g$.

It is noted that, e.g., in the collisional excitation process (2) considered in Subsection 4a-1, the ground-state metal atom is defined as 'reactant' (M^0) while the same metal atom in a certain energy level (M^*) is defined as 'product'. The summation over the final states of M^* here only includes the degenerate states of M^*. If the final state of the collision partner Z is not specified in the excitation process, the summation extends over all (accessible) internal states of the latter.

[†] An alternative symbol for the effective cross section used in this book is σ_{eff}.

Referring again to the special case considered in Eq. II.79, we obtain in general for the *phenomenological* bimolecular rate constant of process (n)

$$k_n = \sum_i f^{(i)} \int_0^\infty g\, \sigma_n^{(i)}(g)\, f(g)\, dg \,. \tag{II.129}$$

Here $f^{(i)}$ is the normalized distribution function for the internal states of the reactants taken together and $f(g)$ is the same for the relative velocity.

With collisional excitation process (2) the summation over the initial internal states is again restricted to the degenerate states of M^0, but extends over all available states of the collision partner Z. When we consider the rate of collisional ionization from the ground-state M^0 only: $M^0 + Z \rightarrow M^+ + e^- + Z$, the same restriction should be made in the summation; $\sigma_n^{(i)}$ then relates specifically to the collision pair $(M^0 + Z)$. However, when we consider the rate of ionization of M present in various internal states, the summation should include these states. In a chemical reaction between M and Z where the initial states are not specified the summation extends over all available internal states.

From the phenomenological rate constant k_n described by Eq. II.129 we obtain, by definition, the *phenomenological effective cross section*[†] $\langle \sigma_n \rangle$

$$\langle \sigma_n \rangle \equiv k_n / \bar{g} \,. \tag{II.130}$$

In contrast to the microscopic effective cross section, $\langle \sigma_n \rangle$ depends on the distribution of the reactants over their internal states and over their relative velocities. It depends therefore not only on the kind of process considered but also on the milieu wherein the process is studied. If these distribution functions obey the Maxwell-Boltzmann laws at temperature T we obtain the *thermal (phenomenological) effective cross section*, which is as a function of T

$$\langle \sigma_n \rangle(T) = k_n(T) / (8kT/\pi\mu)^{\frac{1}{2}} \,, \tag{II.131}$$

where $k_n(T)$ is the thermal rate constant as defined in Subsection 4a-1.

The thermal effective cross section is a phenomenological quantity which follows directly from experimental observations in a bulk system where a temperature can be defined. Measurements of $\langle \sigma_n \rangle(T)$ as a function of T in general do not provide detailed information about the microscopic effective cross sections, because of the complicated implicit relationship between these two kinds of cross sections.

The activation-energy factor and the steric factor, occurring in the classical expression Eq. II.122 or 123 for the rate constant, are included in $\langle \sigma_n \rangle(T)$.

[†] Called 'average total cross section' and denoted by the same symbol by Drawin (1968).

KINETIC CONCEPTS AND PROCESSES   Sect. II.4b

For a direct comparison of effective cross sections measured at different temperatures, the activation-energy factor is sometimes taken out of the expression for $\langle \sigma_n \rangle(T)$. The latter factor can be found experimentally as explained in the preceding subsection (cf. Eq. II.125b).

As an illustration and for use later in Sect. VI.2a-3 we show here how a direct relation between the thermal and the microscopic effective cross section can be found in a special case and under simplified conditions. We consider the excitation of a metal atom in the ground state by collision with a free electron $M^0 + e^- \rightarrow M^* + e^-$. We assume that the microscopic effective cross section has a constant value σ_{thr} for electron energies E_e above threshold ($= E_{exc}$) and is zero otherwise. Because of the large mass ratio between atom and electron, we may approximate the reduced mass μ by the electron mass m_e, and g by the electron velocity v_e. Assuming that the electron energy is distributed according to Maxwell's law, we have for the thermal cross section of excitation by electron impact through Eqs II.131, 129 and 25

$$\langle \sigma_e \rangle(T) = \sigma_{thr} 4\pi (m_e/2\pi kT)^{\frac{3}{2}} \int_{v_{thr}}^{\infty} \exp\left[-m_e v_e^2/2kT\right] v_e^3 dv_e / (8kT/\pi m_e)^{\frac{1}{2}} \quad (II.132a)$$

with $v_{thr} \equiv (2E_{exc}/m_e)^{\frac{1}{2}}$. The solution is

$$\langle \sigma_e \rangle(T) = \sigma_{thr} (1 + E_{exc}/kT) \exp\left[-E_{exc}/kT\right]. \quad (II.132b)$$

Note that we obtain the same result when we apply the classical expression, Eq. II.123 with $s = 2$, for $k_n(T)$ instead of Eq. II.129 and identify $E_A \triangleq E_{exc}$ and $P\sigma_0 \triangleq \sigma_{thr}$. When we extract the activation-energy factor from the expression for $\langle \sigma_e \rangle(T)$, the latter becomes identical to the constant microscopic cross section above threshold. It would not make much difference if the latter varied markedly with electron energy for energies a few times kT above threshold or higher, as the Maxwell distribution function drops sharply with increasing electron energy.

When several collision partners Z_j ($j = 1, 2, \ldots$) contribute independently to the process considered, we find from Eqs II.80 and 131 for the *over-all unimolecular rate constant*

$$\overline{k_n^u}(T) = \sum_j (8kT/\pi \mu_j)^{\frac{1}{2}} \langle \sigma_{n,j} \rangle(T) [Z_j]. \quad (II.133)$$

When measurements are done in a multi-component gas milieu and we cannot or do not want to specify the contributions from each component, we may use the *over-all thermal effective cross section* defined by

$$\langle \overline{\sigma_n} \rangle(T) \equiv \left\{ \overline{k_n^u}(T)/(8kT/\pi\bar{\mu})^{\frac{1}{2}} \right\} / [Z]_t, \quad (II.134)$$

where $\bar{\mu}$ is a weighted average of the μ's and $[Z]_t \equiv \Sigma_j [Z_j]$. As outlined in Subsection 4a-1, we can, however, find the specific effective cross sections $\langle \sigma_{n,j} \rangle(T)$

from Eq. II.133 by measuring $\overline{k_n^u}(T)$ in a number of isothermal flames with varying but known quantitative composition.

4b-4 *MICROSCOPIC-REVERSIBILITY RELATION* [†]

We consider a binary micro-process and the conjugated reverse binary micro-process which is obtained by time-reversal. Let $\sigma_n^{(i,f)}(g')$ and $\sigma_{-n}^{(f,i)}(g'')$ be the corresponding microscopic effective cross sections as defined in Subsection 4b-3, and let g' and g'' be the initial and final relative velocity, respectively, in the forward micro-process. According to *microscopic reversibility* we have the following symmetric relation between these two cross sections

$$(g')^2 \sigma_n^{(i,f)}(g') = (g'')^2 \sigma_{-n}^{(f,i)}(g'') . \tag{II.135}$$

This relation holds if the reactants and products are in nondegenerate states; if not, the combined statistical weight of the particles in the initial state should be added as a factor to the left-hand side and that of the particles in the final state should be added to the right-hand side.

The above microscopic-reversibility relation follows straightforwardly from Quantum Mechanics and is a consequence of the invariance for time-reversal of the Hamiltonian that describes the inelastic scattering process. In contrast to the principle of detailed balance, microscopic reversibility does not depend on statistical-mechanical considerations. Microscopic reversibility does not relate to the average rate of processes in a statistical ensemble, but relates to the quantum-mechanical probability of an individual micro-process. Microscopic reversibility is therefore not restricted to systems in equilibrium; this again is in contrast to the principle of detailed balance.

We can make use of Eq. II.135 in combination with the Maxwell-Boltzmann laws for the energy distribution to prove the *detailed-balance relation* between the thermal rate constants, $k_n(T)$ and $k_{-n}(T)$, of the forward process and reverse process. This derivation starts from the general relationship between k_n and $\sigma_n^{(i,f)}(g)$ which is obtained by combining Eqs II.128 and 129. Application of this general relationship to k_n as well as to k_{-n} and using microscopic reversibility, Eq. II.135, and inserting the Maxwell and Boltzmann laws for $f(g)$ and $f^{(i)}$ in Eq. II.129, we obtain the detailed-balance relation sought. In the case of collisional (de-)excitation process (2) we then find, in accordance with Eq. II.85

$$k_{exc}(T)/k_q(T) = (g^*/g_0) \exp[-E_{exc}/kT] . \tag{II.136}$$

Herein g^* and g_0 are the statistical weights of the atom in the excited and ground

---

[†] For further reading on this topic we refer to, e.g., Light, Ross and Shuler (1969), Morrissey (1975), and Mahan (1975).

state; E_{exc} is the excitation energy; $k_{exc}(T)$ and $k_q(T)$ may be taken either as the bimolecular excitation and quenching rate constants, respectively, or as the corresponding unimolecular rate constants. Applying Eq. II.130 we immediately find the same *relation* for the corresponding *thermal effective cross sections*

$$\langle \sigma_{exc}\rangle(T)/\langle \sigma_q\rangle(T) = (g^*/g_0)\exp[-E_{exc}/kT] . \qquad (II.137)$$

Note that this relation between effective cross sections is essentially a detailed-balance relation, whereas the relation in Eq. II.135 reflects only microscopic reversibility. Note also that detailed-balance relations such as those given in Eqs II.136 or 137 require only that the energy distributions that appear in the expression for the rate constant in Eq. II.129 obey the Maxwell-Boltzmann laws. In the above example the population of the energy levels of M need not obey the Boltzmann law. Equations II.136 and 137 therefore do not require complete equilibrium (in contrast to the principle of detailed balance itself; see Subsection 4a-2).

We have more generally for any bimolecular collisional process (n) in which M atoms are transferred from level i (with statistical weight g_i) to level j (with statistical weight g_j)

$$\langle \sigma_n\rangle(T)/\langle \sigma_{-n}\rangle(T) = (g_j/g_i)\exp[-(E_j-E_i)/kT] . \qquad (II.138)$$

4c   ACTIVATED COMPLEX THEORY[†]

The *activated-complex theory* is aimed at a more quantitative prediction of the absolute rate of chemical reactions than can be provided by the simple kinetic (or collision) theory described in Subsection 4b-2. The activated-complex theory explicitly accounts for the (quantized) structure of the reacting particles, which are no longer simply considered as hard spheres. The latter approximation works reasonably well only for simple molecules and atoms. In the theory, it is assumed that upon collision the reactants (A and B) form a temporary *activated complex* $X^{\ddagger}$ (*activation complex*, *transition complex* or *quasi-molecule*). This complex may be conceived as existing at the top of the energy barrier shown in Fig. II.4. The complex dissociates at a rate ν_d (s^{-1}) into the reaction products C and D, according to:

$$A + B \rightarrow X^{\ddagger} \rightarrow C + D$$

and *vice versa*. The concentration of the activated complex $[X^{\ddagger}]$ is assumed to be in equilibrium with the concentrations of the reactants according to the mass-action-law

$$[X^{\ddagger}]_e/[A]_e[B]_e = K^{\ddagger}(T) . \qquad (II.139)$$

[†] Occasionally also called: 'theory of absolute reaction rates'. For a fuller treatment we refer to, e.g., Glasstone, Laidler and Eyring (1941), Laidler (1955, 1963), Lewis, Pease and Taylor (1956), **Kondratev and Nikitin (1967)**, and Nikitin (1974).

Chapt. II BASIC CONCEPTS AND GENERAL RELATIONSHIPS

The equilibrium constant $K^{\ddagger}(T)$ is related to the changes $(\Delta S^{\ddagger})_T^0$ and $(\Delta H^{\ddagger})_T^0$ in entropy and enthalpy that arise when one mole of the activated complex is formed from the reactants in their standard states. (The standard state chosen conforms to the units in which the concentrations and k_n are expressed; see Sect. II.3b-3.) This relation is similar to that given by Eq. II.51.

The rate of the above process, labelled by n, is given by: $v_n = v_d [X^{\ddagger}]_e = v_d [A]_e [B]_e K^{\ddagger}(T)$, so that by analogy with Eq. II.74, $k_n = v_d K^{\ddagger}(T)$. It is assumed that the rate v_d at which a molecule $X^{\ddagger}$ is decomposed is equal to the frequency of the loose vibration which allows the complex to dissociate into $C + D$. This vibrational frequency also occurs in the statistical expression for $(\Delta S^{\ddagger})_T^0$ or $K^{\ddagger}(T)$ (cf. Eqs II.44 and 38). If we define $(\Delta S_{\ddagger})_T^0$ as the change in entropy when one mole $X^{\ddagger}$ is formed, <u>without</u> taking into account the contribution from this loose vibration, we find for the rate constant

$$k_n = \kappa (kT/h)\, \exp[(\Delta S_{\ddagger})_T^0/R]\, \exp[-(\Delta H^{\ddagger})_T^0/RT]\,. \tag{II.140}$$

In this expression, v_d has dropped out provided the condition that $hv_d \ll kT$ holds. The *transmission coefficient* κ is added in order to account for the possibility that, after completing one cycle of the loose vibration, the activated complex may not have actually disintegrated into the products of the reaction considered. The introduction of this coefficient is a consequence of the simplifying assumptions made in the theory. For nonadiabatic reactions, which involve a change in electronic state, κ may be much less than unity.

A comparison of Eqs II.122 and 140 immediately suggests a close relation between the enthalpy of activation $(\Delta H^{\ddagger})_T^0$ and the activation energy E_A. It is also seen that a negative change in entropy leads to a small steric factor P in the expression of the simple collision theory. A negative sign of $(\Delta S_{\ddagger})_T^0$ occurs in a bimolecular reaction where two molecules combine to form one activated complex. This reduces the number of degrees of freedom and thereby the entropy.

The applicability of the activated-complex theory (as well as of the simple collision theory) is restricted by the assumption of an equilibrium distribution of the reactants over their various degrees of freedom. Moreover, the internal-energy distribution of the activated complex itself should also conform to equilibrium at the temperature of the system. This implies that for this complex *inter*molecular energy exchange should be much faster than the conversion of the complex into products. At low pressures this condition may not be satisfied. Furthermore, an activated complex should be definable with well-determined molecular properties that allows a calculation of its thermodynamic functions. The theory works better for poly-atomic molecules with many degrees of freedom than for di-atomics. In the latter case dissociation reactions are better described by applying a 'strong-collision model' or a 'ladder-climbing model' (see Subsection 4b-2).

KINETIC CONCEPTS AND PROCESSES   Sect. II.4d

For a proper definition of the activated complex, and for a calculation of its thermodynamic functions and of the transmission coefficient, it is useful to consider the potential-energy surfaces of the collision complex (see the following subsections). This holds especially when reactions are considered that involve electronically excited states.

4d   POTENTIAL ENERGY SURFACES [†]

The potential (energy) surfaces are a multi-dimensional extension of the potential (energy) curve, $V(r)$, which was introduced in Sect. II.2b-1 (see also Fig. II.2) to describe the internal-energy levels of a di-atomic molecule. Potential energy surfaces are a useful means not only for the description of stable poly-atomic molecules but also for a pictorial representation or quantitative calculations of the course and rate of a process between two particles. This process may be an elastic collision or an inelastic one involving, e.g., electronic (de-)excitation of one of the particles, a chemical reaction, leading possibly also to electronic (de-)excitation, or an ionization process.

When two atoms or molecules, A and B, pass each other at short distance, they may interact strongly and form together a *temporary complex* (*collision complex* or *quasi-molecule*). (This complex need not have the special properties that are ascribed to the 'activated complex' considered in the preceding subsection.) When the lifetime of the complex is of the order of a normal vibration period ($\approx 10^{-13} - 10^{-14}$ s), we call it *short-lived*; when it is much longer, say 10^{-9} s, we speak of a *long-lived* complex.

The nuclei in the complex can be considered as moving in a field described by a *potential energy* which is a function of their mutual distances. The origin of this potential energy is similar to that in the di-atomic case considered in Sect. II.2b-1; this description is again valid under the condition of the Born-Oppenheimer approximation. The electronic eigenfunctions and eigenenergies then contain the internuclear distances as parameters. The complex molecule, just like a normal, stable molecule, can exist in different states of excitation. The potential energy of interaction is usually a different function of the internuclear distance(s) for each electronic state.

In the simplest case of two colliding atoms, the curve representing the potential energy $V(r)$ as a function of internuclear distance r was called the (one-dimensional) *potential* (*energy*) *curve* (see Fig. II.2). When an atom and a

---

[†] For a further study of the theoretical problems involved in this and the following subsections, we refer to, e.g., Glasstone, Laidler and Eyring (1941), Laidler (1955), Bates (1962), Mott and Massey (1965), Landau and Lifshitz (1965), Slater (1968), Bauer (1969), Margenau and Kestner (1971), Polanyi (1972), Torrens (1972), Nikitin (1974), and Janev (1976).

Chapt. II   BASIC CONCEPTS AND GENERAL RELATIONSHIPS

di-atomic molecule form a linear complex molecule (for example Na-H_2), V is a function of the distance r_{Na-H} between the Na nucleus and the nearest proton as well as of the distance r_{H-H} between the two protons. The function $V(r_{Na-H}, r_{H-H})$ is then graphically represented by a two-dimensional *potential (energy) surface*. This surface depends in general on the orientation of the H_2 molecule with respect to the line connecting the Na atom and the centre of H_2. When more variables are required to specify the mutual positions of the nuclei, the potential energy is in general represented by a *potential (energy) hyper-surface*.

There exist certain *correlation rules* that predict the various possible spectroscopic designations of the complex molecule that is formed from the reactants when they are in specified electronic states. For example, an excited Na atom in a ^2P state may combine with an H_2 molecule in the $^1\Sigma_g^+$ ground state to form an NaH_2 complex in either a $^2\Sigma^+$ or a $^2\Pi$ state. The combination of two species in their electronic ground states need not necessarily lead to a molecule in the ground state. A Ca atom having a ^1S ground state may combine with an oxygen atom having a ^3P ground state to form a CaO molecule in an excited electronic state.

The detailed course of a chemical reaction can now be traced by a *trajectory* (or *reaction path*) on one or more of the (hyper-)surfaces. Consider, for example, the exchange reaction: $D + H_2 \rightarrow DH + H$, where all species are in their ground states, and assume that the complex is a linear tri-atomic molecule. The trajectory starts somewhere at 'infinite' distance between D and H_2 where the H atoms are close together; it ends at 'infinite' distance between DH and the remaining H atom. Since both H_2 and DH in their ground states are stable molecules, characterized by the existence of a minimum in their potential curves, the potential surface of the D−H−H complex shows two valleys that are connected by a saddle point. At this point the distance between D and the nearest proton, and that between the two protons are of the same order of magnitude. The activation energy of the reaction is minimum when the trajectory followed by the system passes through this saddle point. The height of this point relative to the depth of the valley at infinite distance between D and H_2 is in fact the threshold or activation energy, E_A, of the reaction (cf. Subsection 4b-2). Because of the *principle of microscopic reversibility* the inverse reaction follows the same trajectory but in the reverse direction.

In this example the *activated complex*, whose thermodynamic functions enter into the expression, Eq. II.140, for the rate constant, may be readily located in the saddle point. For more complex reactions involving surfaces of different electronic species, the activated complex is usually more difficult to locate. In some processes no activation energy is involved at all and the position of the activated complex cannot be connected with that of a local maximum or saddle point on the potential surface.

4e   PSEUDO-CROSSING OF POTENTIAL ENERGY CURVES

It may happen that during the course of a collision or reaction the system jumps from one potential surface to another. When a reaction or collision results in the formation or destruction of an excited electronic state or in the production of a pair of ions (*nonadiabatic* or *diabatic process*), the *input* and *output channels* of the process correspond to different potential energy surfaces of the complex; a transition between these surfaces must then have taken place either directly, or indirectly via other surfaces. We consider here, for example, the quenching collision: $A^* + B \rightarrow A + B$ between an electronically excited A atom and a B atom. In this process, electronic excitation energy is converted into the translational energy of the relative motion of A and B. In Fig. II.5a as well as II.5b, two potential energy curves are sketched; at infinite separation one of these curves corresponds to a free excited A^* and a free B atom, and the other to a free A and a free B atom, both in their ground states. The difference in height between the asymptotes of both curves for $r \rightarrow \infty$ is equal to the electronic excitation energy of the free A^* atom. Each of the combinations $A^* + B$ and $A + B$ leads to a collision complex of a certain electronic species; this species is designated by the resultant orbital angular momentum $|\Lambda|$ and spin S of the electrons in both atoms (see Sect. II.2b-1) and by the symmetry properties. In the lowest-order approximation, the combination $A^* + B$ is associated with a potential energy curve that is composed of branches I and IV connected through the intermediate dotted line (see Figs II.5a and 5b). Similarly, the combination of $A + B$ yields, in the lowest-order approximation, the potential energy curve that is composed of branches II and III. In the example chosen in Fig. II.5a the former curve has a minimum at a finite value of r, so A^* and B are attracted to each other in some range of r at the right-hand side of this minimum. A local maximum, called centrifugal barrier (see Sect. II.2b-2), may arise in branch I of this curve when the collision is not head-on. The collision is not head-on when the *impact parameter* b differs from zero; this parameter is the distance between A and the (extrapolated) line of initial motion of B relative to A. When b and the initial relative velocity are not zero, the system has an angular momentum of relative nuclear motion with quantum number $J_n > 0$. The value of J_n determines the height of the barrier; the larger J_n is, the higher the barrier.

The shape chosen for the curve II-III is typical for the case when the interaction between A and B has a repulsive character. The same applies for the shape of the curve I-IV in Fig. II.5b.

We have assumed that in the lowest-order approximation the two potential curves (I-IV and II-III) cross each other somewhere at $r = r_c$. However, at this *crossing point*, an interaction between the two electronic states of the collision

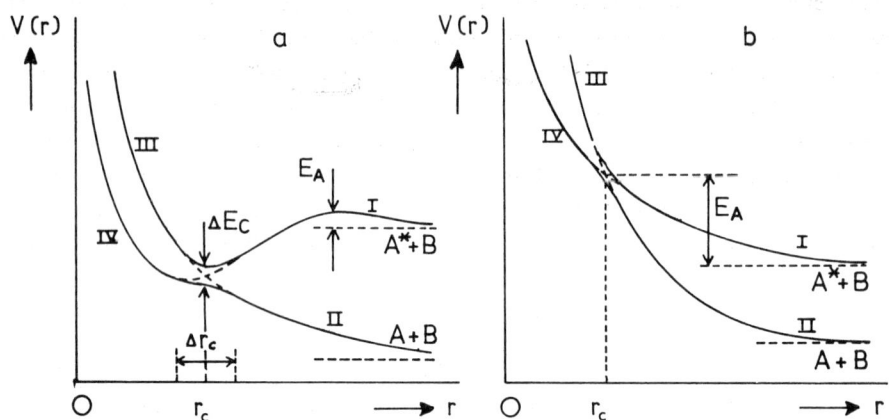

Fig. II.5 The potential energy V of a collision complex consisting of atoms A and B is plotted as a function of internuclear distance r. The upper curve corresponds to a free, excited atom A^* and a free ground-state atom B when $r \to \infty$. The lower curve corresponds to free atoms A and B in their ground states when $r \to \infty$. The vertical distance between the dashed asymptotes of both curves, for $r \to \infty$, equals the excitation energy of A^*. In lowest-order approximation, the combination A + B yields the curve that is composed of branches II and III, connected by the intermediate dotted line, whereas the combination $A^* + B$ yields the curve containing the other two branches. These diabatic curves cross each other at $r = r_c$. In higher-order approximation, resonance splitting near the crossing point must be taken into account and the curves will 'repel' each other, so that crossing is avoided (see solid curves). The minimum difference in V between the latter, adiabatic curves is found at $r = r_c$ and equals ΔE_c. During collision a radiationless transition between the upper and lower curve is possible in a range Δr_c around the point of pseudo-crossing.

Fig. 5a is typical for the case where the diabatic curve for $A^* + B$ shows a minimum at some finite value of r. In contrast, Fig. 5b shows the case where both diabatic curves are repulsive (no minimum) in the whole range considered; here a high threshold or activation energy E_A for the quenching collision may exist. A threshold for quenching in the case of Fig. 5a may arise from the existence of a centrifugal barrier when the collision is not head-on. (From Alkemade and Zeegers 1971.)

complex may occur, leading to a mutual separation of the potential curves (called *resonance splitting*). As a result, we obtain in higher-order approximation the two curves drawn as solid lines in the figures. The upper solid curve comprises branches I and III; the lower one contains the two other branches. Because of this resonance splitting, the potential curves are said to have a *pseudo-* or *quasi-crossing* (also called *avoided crossing*). The interaction at the crossing point is strong when the state of the complex formed from $A^* + B$ is of the same electronic species (e.g., $^1\Sigma$) as that resulting from $A + B$. In such a case the resonance separation, that is, the minimum difference ΔE_c in energy between the upper and lower solid curves, is relatively large. For states of different electronic species, the resonance separation ΔE_c is small.

According to quantum theory, the resonance splitting is determined by

$$\Delta E_c = 2|V_{12}(r_c)| , \qquad (II.141)$$

where $V_{12}(r_c)$ is the matrix element of the perturbation energy with respect to the zero-order eigenstates corresponding to the curve I-IV and II-III, taken at $r = r_c$. In the zero-order Born-Oppenheimer approximation, the coupling between the electronic and nuclear motions was disregarded. It is this coupling, however, that introduces in higher-order approximation a perturbation term in the Hamiltonian of the complex molecule (see Coulson and Zalewski 1962, Landau and Lifshitz 1965, and Henry and Kasha 1968). Due to this coupling, a mixing of the two zero-order eigenstates occurs, which is especially important near the crossing point.

The solid curves depicted in Figs II.5a and 5b are called *adiabatic* potential-energy curves. The collision complex follows these curves when it moves adiabatically, that is, when the relative nuclear motion is so slow that the electron orbitals of the quasi-molecule are allowed to readjust themselves momentaneously at each internuclear separation. The other curves (I-IV and II-III) are called the *diabatic* potential-energy curves. The collision complex follows the latter curves when the nuclear motion is comparatively so fast that readjustment of the electron orbitals becomes too sluggish. The interaction force experienced by the collision complex during its motion thus depends on whether its motion is adiabatic or diabatic.

Two-dimensional potential surfaces may cross each other along a crossing curve, and so on for multi-dimensional surfaces. Here again we distinguish between adiabatic and diabatic surfaces. In the case of a reaction between, e.g., K and Br_2, the coupling of K with the internal motion of Br_2 is included in the adiabatic surfaces, but not in the diabatic ones.

4f PROCESSES THROUGH POTENTIAL-CURVE CROSSING

General. Potential energy surfaces are useful for gaining an insight into

Chapt. II BASIC CONCEPTS AND GENERAL RELATIONSHIPS

the detailed course, the probability, the possible activation energy, and so on, of processes that are based on short-range interaction forces. This applies to inelastic collisions as well as chemical reactions and ionization processes. The examples given in the following basic discussion relate mainly to processes which are relevant in the collisional (de-)excitation of metal atoms by heavy particles; an occasional excursion is made into ionizing collisions. For a process to occur through curve- or surface-crossing, there must be sufficient energy available to reach the crossing region while the crossing probability must not be too low. We first consider a two-state model where the two potential curves happen to have one pseudo-crossing. Processes involving multiple crossings between more than two potential curves will be briefly discussed thereafter.

Selection rules. In the examples shown in Fig. II.5 the de-excitation reaction: $A^* + B \rightarrow A + B$ involves a radiationless, diabatic transition of the collision complex from the upper adiabatic potential curve to the lower one. When a transition is allowed, it usually takes place near the point of pseudo-crossing, that is, at $r = r_c$. A large resonance splitting ΔE_c, however, results in a low transition probability. The *no-crossing rule* for di-atomic molecules forbids transitions between states of the same electronic species; in this case the resonance splitting is large (see Subsection 4e). The probability of a radiationless transition may be further restricted by selection rules. The *Wigner spin-conservation rule* requires that the resultant spin S of the reactants be the same as that of the products. When the spins of the separate reactants cannot be combined to the same quantum number as the spins of the products, the process (inelastic collision or reaction) is 'forbidden'. Exceptions to this rule are possible, in particular when one of the atoms shows j,j coupling. A violation of this rule may also occur when spiralling collision trajectories occur, so that the duration of the collision is long enough to allow a spin-forbidden transition (see Gilmore, Bauer and McGowan 1967).

Another selection rule says that the change in orbital angular momentum Λ of the collision complex should satisfy: $\Delta\Lambda = \pm 1$ or 0 (cf. the optical selection rule 4 in Sect. II.2d). It should be noted that there are no selection rules for the transitions that the individual collision partners can undergo.

Transmission coefficient. Even when the collision complex reaches the crossing point and the transition is not forbidden by selection rules, the process considered may not always take place. The probability that it actually takes place is described by the *transmission coefficient* κ (cf. Subsection 4c). The transmission coefficient depends on the radial component of the relative velocity: $g_r \equiv |dr/dt|$ of the collision partners near the crossing point; we shall denote this value by g_c.

The probability of a jump between two adiabatic potential curves is the largest at or near the crossing point where the potential energy gap is smallest and equal to ΔE_c. According to the Heisenberg uncertainty relation (cf. Eq. II.211), the uncertainty in the energy of the collision complex when it spends a time τ in the crossing region Δr_c (see Fig. II.5) is of the order of $\hbar/\tau$. If the amount of uncertainty in energy is much smaller than ΔE_c, the complex remains with a large probability on the original adiabatic curve, that is, it behaves adiabatically. If, on the other hand, the uncertainty is larger than ΔE_c, the complex may undergo a diabatic transition to the other potential curve. Since the time spent in the crossing region is determined by: $\tau \simeq \Delta r_c/g_c$, we can characterize the probability of a transition by the *adiabaticity parameter*, also called *Massey parameter*, ξ which is here defined by (see Massey, Burhop and Gilbody 1971, and Levine and Bernstein 1974)[†]

$$\xi \equiv \Delta E_c \Delta r_c / \hbar g_c . \qquad (II.142)$$

This parameter plays generally an important role whenever a system undergoes a jump in (quantized) energy during a limited time interval. When $\xi \gg 1$ the collision complex behaves adiabatically.

The value of ΔE_c occurring in the Massey parameter is given by Eq. II.141, while Δr_c can be expressed in ΔE_c and in the slopes F_1 and F_2 of the tangents of the two diabatic curves crossing each other at $r = r_c$ (see the dotted lines in Fig. II.5) according to: $\Delta r_c \approx \Delta E_c / |F_1 - F_2|$. From the expression for the Massey parameter thus obtained one can derive the *Landau-Zener formula* for the transmission coefficient

$$\kappa = 2W(1-W) \qquad (II.143)$$

with

$$W = \exp[-4\pi^2 |V_{12}(r_c)|^2 / (\hbar g_c |F_1 - F_2|)] . \qquad (II.144)$$

The quantity W is the probability that the complex, which is initially in the state $A^* + B$, makes a diabatic transition from branch I to branch IV during a single passage of the crossing point $r = r_c$ or vice versa. After the first passage of the crossing point, the system reaches a turning point at some minimal distance r_{min} ($< r_c$) and subsequently passes the crossing point a second time. De-excitation results when the system moves from branch I to IV at first passage, and from branch IV to II at second passage of the crossing point. Since the probability for the first transition is W, the probability for the second transition equals $(1-W)$. When these transitions can be assumed to occur independently, the combined probability equals $W(1-W)$. De-excitation is, however, also achieved when the system

---

[†] Some authors use h instead of $\hbar \equiv h/2\pi$ in the denominator of this expression.

first moves from branch I to III, and then from III to II. The probability for this sequency of transitions is $(1-W)W$. When the collision complex can actually reach the transition region at $r \simeq r_c$, the over-all probability that quenching of A^* results is thus the sum of these identical product probabilities, in accordance with Eq. II.143.

It is easily seen that the over-all probability is maximum for $W = \tfrac{1}{2}$, and it drops to zero for either $W \to 0$ or $W \to 1$. The value of W depends, through the matrix element V_{12}, on the perturbation energy and on the electronic species of the states of $(A^* + B)$ and $(A + B)$ of the quasi-molecule (see the above). The effect on W of a small value of $|F_1 - F_2|$, that is, an acute angle between the tangents of the zero-order potential curves at $r = r_c$, is similar to that of a large resonance splitting. One finds that g_c must be of the order of $\Delta E_c \Delta r_c / \hbar$ to make $W = \tfrac{1}{2}$, that is, to attain a maximum transmission coefficient. This corresponds to a value $\xi = 1$ for the Massey parameter.

The Landau-Zener formula has been derived under conditions some of which strongly restrict its applicability. One condition is that only two electronic states need be considered (two-level model). Another condition is that the kinetic energy, E_k, associated with the velocity of approach, g_c, should be large compared to ΔE_c. Coulson and Zalewski (1962) and Henry and Kasha (1968) have examined the implications and restrictions of the Landau-Zener theory in more detail. The former authors have shown, for some realistic assumptions, that the validity of the above formula is restricted by the inequality: $3(E_k/\mu)^{\tfrac{1}{4}} \ll \Delta r_c \ll 3\,\text{Å}$. Here E_k is expressed in electronvolts, the reduced mass μ of the collision partners in atomic mass units, and the length Δr_c characterizing the range of $|V_{12}(r)|$ in Å units. This inequality condition will hardly be met in practice. The same authors have derived a more general expression for W which replaces Eq. II.144

$$W = \exp\left[-\left|\int_{-\infty}^{\infty} V_{12}(r)\,dr\right|^2 / \hbar^2 g_c^2\right]. \qquad (\text{II.145})$$

This equation is valid for $E_k > 1\,\text{eV}$. It will be difficult, however, to evaluate the integral of the matrix element in actual cases (see Henry and Kasha 1968). We refer for a further discussion of the applicability and limitations of the Landau-Zener formula to Bates, Johnston and Stewart (1964), Heinrichs (1968), and Bauer (1969), and to the books mentioned at the beginning of this section. However, even when the conditions underlying the derivation of this formula are not strictly fulfilled, it may often be used at least for a semi-quantitative or qualitative description.

In the derivation of the above formulae for W, semi-classical approximations are made for the motions of the nuclei. According to Coulson and Zalewski, in practice these approximations will be valid in most cases.

KINETIC CONCEPTS AND PROCESSES   Sect. II.4f

Activation energy. In the example shown in Fig. II.5a the initial kinetic energy of relative motion of A^* and B must be larger than the centrifugal barrier in order to reach the crossing region. As explained in Subsection 4e, the height of this barrier depends on the quantum number, J_n, of the angular momentum of relative motion. By averaging over the statistical distribution of J_n one can calculate the (effective) activation energy for the quenching process.

When a repulsive interaction exists between A^* and B as well as between A and B, the crossing point of the potential curves may lie well above the asymptotes of both curves for $r \to \infty$ (see Fig. II.5b). The activation energy is then determined by the height of the crossing point relative to the initial asymptote of the system. Quantum-mechanical tunnelling, however, may allow a transition to take place below the crossing point at a value of r slightly larger than r_c (see Nikitin 1966).

The activation energy of the reverse, excitation process: $A + B \to A^* + B$ is in general obtained by adding the excitation energy of a free A^* atom to the activation energy of the forward quenching process (cf. Fig. II.4).

Effective cross section. When the transmission factor κ is about unity, we may estimate the effective cross section, σ_{eff}, of the curve-crossing process to be of the order of πr_c^2 apart from a possible activation-energy factor (see Subsection 4b-2).

A more quantitative approach for calculating σ_{eff} is to use the expression for κ given by the Landau-Zener formula, Eqs II.143 and 144. According to this formula, κ depends on the radial component, g_c, of the relative velocity at the crossing point. The value of g_c, in turn, as well as the possibility for the collision complex to reach the crossing region over the top of the centrifugal barrier (see Fig. II.5a) depend on the initial relative velocity g_0 and the impact parameter b (see for definition Subsection 4e). When, for <u>given</u> g_0, the value of b is so large ($\geqslant b_{max}$) that the barrier cannot be overcome, the probability of a quenching process will be zero (in classical approximation). The effective quenching cross section, $\sigma_{eff}(g_0)$, as a function of g_0 is therefore obtained from

$$\sigma_{eff}(g_0) = \int_0^{b_{max}} \kappa (2\pi b) \, db \; , \qquad (II.146)$$

wherein b_{max} as well as κ are functions of g_0 while κ depends also on b. Consequently, $\sigma_{eff}(g_0)$ will be at most equal to πb_{max}^2 (corresponding with $\kappa \equiv 1$). The thermal effective cross section, $\sigma_{eff}(T)$, is obtained as a function of temperature T by averaging $\sigma_{eff}(g_0)$ over a Maxwellian velocity distribution.

Multiple crossings; ionic-curve crossing. For more complicated collision systems involving different electronically excited states it may be necessary to

Chapt. II BASIC CONCEPTS AND GENERAL RELATIONSHIPS

consider crossings between more than two potential hyper-surfaces. For poly-atomic (quasi-)molecules the no-crossing rule does not hold, so the number of allowed crossings is increased. In such systems various reaction paths along different potential surfaces are feasible, and the transmission coefficient must be evaluated for each possible crossing if the probabilities of crossings are independent.

In general, when a certain process can take place through different 'channels', the overall probability, at given initial velocity, is not simply the sum of the probabilities for each channel. One has, instead, to compute by Quantum Mechanics the complex 'scattering amplitude' for each channel separately. The overall probability is then obtained by squaring the absolute value of the *sum* of these separate *amplitudes*. Such a procedure may give rise to interference effects between different channels, which has no analogue in a classical description.

$A + B$ and $A^* + B$ may have no pseudo-crossing, at least not in an accessible energy range, when, e.g., A and A^* are an alkali atom in its ground and first excited state, respectively, and B is a halogen molecule or an H_2 molecule in the electronic ground state. However, in these examples both surfaces intersect the 'attractive' *ionic* (or *polar*) *surface* that represents a *charge-transfer state*, $A^+ + B^-$, and whose shape is, roughly speaking, determined by the Coulomb potential of a positive A^+ ion combined with a negative B^- ion (see Fig. II.6). When B is a di-atomic molecule, the occurrence of a pseudo-crossing of the ionic surface with each of the covalent $A + B$ and $A^* + B$ surfaces may depend on the configuration of the tri-atomic quasi-molecule (see also Sect. VI.2b-2). A *quenching collision* then proceeds through the formation of an ionic complex as an intermediate state and involves two successive crossings. In the situation sketched in Fig. II.6, one possible course of the quenching process is that the system moves from branch I (corresponding to the initial, excited state) through branch II (corresponding to the ionic state) to branch III, whereupon it makes a final transition to branch IV of the ground state. It may, however, also happen that the system retraces its path by going back from branch III, through II, to the initial state at branch I. Then quenching is not achieved. The probability of a return to the initial state will be reduced if the collision complex becomes 'stabilized' during its motion in the potential well formed by the branches II and III. Stabilization occurs, for example, when part of the energy associated with this motion is converted into vibrational energy of the molecular collision partner. The complex will then be 'trapped' in this potential well, where it may perform several vibrations until it crosses over to the lower potential curve containing branch IV.

Collisional ionization of atom A may occur when the translational energy of relative motion of A and B is sufficient to surmount the difference between the ionization energy, E_{ion}, of A and the electron affinity, E_{aff}, of B (see also

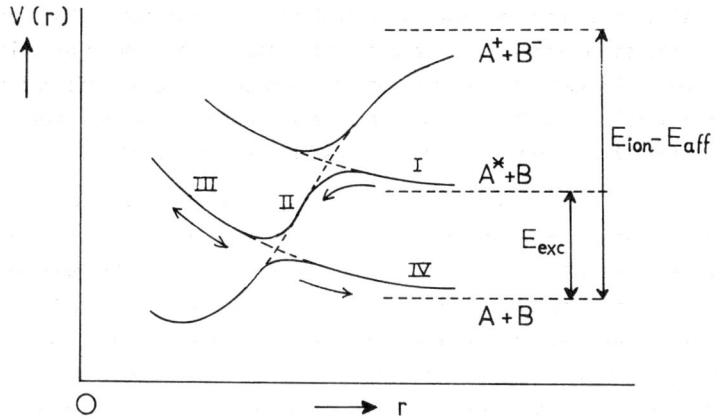

Fig. II.6 Potential energy curves $V(r)$ are shown schematically for the collision complex A + B (see also the caption for Fig. II.5). The upper curve corresponds, for $r \to \infty$, to a free positive A^+ ion and a free negative B^- ion in their electronic ground states. The covalent curves for A + B and $A^* + B$ are assumed to have no pseudo-crossing; they both have, however, pseudo-crossings with the steep, attractive branch of the ionic curve $A^+ + B^-$. Radiationless transitions are possible near the points of pseudo-crossing. The arrows along the branches I, II, III, and IV indicate, in this order, a possible course of a quenching collision. When atoms A and B collide with sufficient translational energy, they may end up as a pair of free ions A^+ and B^-.

Sect. IX.2). This difference determines the position of the asymptote of the ionic curve as well as the location of the crossing points. In some cases a stable free B^- ion may not exist (e.g., when B is an N_2 molecule; see Sect. VI.2b-2).

A charge-transfer state is formed when, for example, a Br_2 molecule picks up a valence electron from a K atom when they approach each other. This may happen at relatively large distance as these species have a large E_{aff} and a low E_{ion}, respectively. Once an ion pair has been formed, the Coulomb force brings the particles together and a chemical reaction can take place. This mechanism is known as the '*harpoon model*', because the electron is ejected as a harpoon to haul in the Br_2 molecule.

Quenching of alkali atoms in their lowest excited state by collisions with di-atomic molecules through ionic-curve crossing plays an important role in flames (see Sect. VI.2b-2). According to the principle of microscopic reversibility the reverse process is important for the collisional excitation of alkali atoms in flames. The participation of the internal molecular energy complicates the quantitative description of these processes (see *ibidem*).

Chapt. II BASIC CONCEPTS AND GENERAL RELATIONSHIPS

Concluding remarks. In de-excitation collisions between atoms, electronic excitation energy is converted into translational energy of relative motion. The conservation law of total linear momentum determines how this translational energy is distributed among the atoms after the collision. When molecules are involved in an inelastic collision, the distribution over the translational and rotational degrees of freedom after the collision should conform to the conservation laws of the total linear as well as angular momentum.

Lack of accuracy in the calculation of potential energy curves often hampers the theoretical derivation of rate constants. In some simple cases, successful attempts have been made, however, as with alkali-noble gas systems. We shall return to this problem in Chapt. VI, which deals with (de-)excitation processes.

The theoretical description of processes between atoms and molecules is often simplified by application of *semi-classical methods*. The motion of these heavy particles under the influence of the potential energy of their interaction is calculated classically; but the transitions between different internal states is calculated quantummechanically. The angular momentum, J_n, of relative motion of, e.g., a He and a Na atom, with an impact parameter of, say, 3 Å, is typically of the order of 100 at $T \approx 2500\,\text{K}$. This high J_n value justifies the semi-classical approach.

Measurements of velocity-dependent differential cross sections for elastic scattering of, e.g., a ground-state or an excited alkali atom by noble-gas atoms, with crossed molecular beams, are a sensitive check for calculated interaction potentials (see, e.g., Saxon, Olson and Liu 1977, and van Deventer 1980).

4g RESONANCE PROCESSES

The processes considered in the preceding subsection were based on transitions between different electronic states, and are made possible by pseudo-crossings of potential energy surfaces in the temporary complex molecule. Surface crossing is usually connected with the short-range interaction between the species involved. Thus the probability of a (de-)excitation process is related to the reactivity of the collision partner. The rate constant of the process is then not expected to depend critically on the difference in total internal energy, $\Delta E(\infty)$, of both partners at infinite separation, in their initial and final states, respectively. The actual transition takes place near the crossing point, where the potential energy of the system may deviate markedly from its value in the initial and (or) final state.

However, effective transfer of internal energy between two collision partners of different species may also proceed without potential surface crossing if there is *close resonance* or *quasi-resonance*, that is, $\Delta E(\infty) \approx 0$. Suppose that a long-range interaction exists, so that transitions are possible even when the partners are a relatively large distance R from each other. The effective cross section of the transfer process will approach its maximum value, for given relative velocity g, when the Massey parameter in the transition region is about equal to unity or is less. The Massey parameter ξ is here approximated by: $\xi \simeq \Delta E(R) R / \hbar g$, where $\Delta E(R)$ is the jump in internal energy that occurs when the transition takes place at $r \approx R$ (cf. Eq. II.142). If R is large enough to make $\Delta E(R)$ about equal to $\Delta E(\infty)$, the effective cross section will be larger the smaller the *resonance defect* $\Delta E(\infty)$; this statement is known as *Franck's rule*. For a given value of ΔE, the effective cross section attains a maximum value, as a function of g, when ξ becomes equal to about unity.

Resonance processes and processes occurring by curve crossing are, in a certain sense, complementary. It may well happen that the same collisional excitation process behaves as a quasi-resonance process at high relative velocities, whereas at low velocities it takes place through curve-crossing.

Resonance processes are usually described by semi-classical theory, but in simple cases a fully quantummechanical theory has also been presented (see, e.g., Nakamura 1968); the latter theory confirms, at least qualitatively, the significance of the Massey parameter as a criterion for quasi-resonant energy transfer. For a recent exposition of the various theories in use we refer to the standard book by Nikitin (1974). Other expositions of earlier date have been given by Callear (1967), Gallagher (1968), Bates and Estermann (1968), and Massey, Burhop and Gilbody (1971).

In the literature one frequently finds that a distinction is made between (near-)adiabatic and non-adiabatic (diabatic) resonance processes. This distinction is related to whether the Massey parameter is much larger than or less than unity. The use of this distinction has been criticized, however, by Laidler (1955).

4h GENERAL CLASSIFICATION OF PROCESSES

In Table II.1 a general classification is presented of processes in the gas phase that may be relevant inside and outside the primary combustion zone of flames with and without metal vapours. The table includes also a few processes that have been studied in other gaseous milieus (such as metal vapour cells) and to which reference is made in this book. According to the principle of microscopic reversibility, to each process a reverse process can be associated. When both are relevant in flames, they are classified separately; otherwise only the relevant one is classified. Examples of processes in flames are found in other sections of this book.

This classification is not exhaustive but is believed to include the main types of processes that are involved in the (de-)excitation of metal species and combustion products in flames. A more extensive and detailed classification of radiation and ionization processes in flames will be given in Sects II.5a and IX.2a, respectively. The purpose of the general classification presented here is to provide a survey and general terminology that may help in the reading of this book and, more generally, of the pertinent literature.

A categorization of (in-)elastic collisions and reactions, involving also excited species, ions, electrons as well as photons, can be found in the bibliography by Hochstim (1969) and in the book of Massey, Burhop and Gilbody (1971). Our classification and terminology is tailored to the more restricted needs of our book. Since a generally accepted terminology is not yet available, differences in the names of processes are to be noted. Alternative names are occasionally added in the table.

The general term *process* covers (in-)elastic collisions, chemical reactions, and ionization and radiation processes. Two main categories of processes can be distinguished: *physical* and *chemical*.[†] In the latter category new chemical bonds are formed and/or old bonds are broken; in the former the chemical bonds remain unaltered. In both categories momentum, internal and translational energy, and/or electric charge may be exchanged between the reactants or between a reactant and a radiation field. The term reactant is here taken in the broad sense as defined in Subsection 4a-1.

A further, practical distinction made is that between a *single-step* (or *elementary*) process and a *multi-step* process. We shall speak of a single-step process when the initial and final states of the interacting system succeed each other without a distinct intermediate state or when the lifetime of the intermediate state (temporary collision complex) is of the order of 10^{-13} s or less; this interval roughly corresponds to the vibration period of a normal di-atomic molecule or to the duration of a gas-kinetic collision between two neutral atoms (see Subsection 4b-1). When one or more intermediate states are discernable with a lifetime of the order of 10^{-10} s or longer, we shall consider the formation and destruction of these states as as many single steps. The (over-all) process is then said to be a multi-step (e.g., 2-step) process. Note that the inter-collision time in a flame at 1 atm pressure is of the order of $10^{-9} - 10^{-10}$ s (see Subsection 4b-1). Note also that the optical lifetime for spontaneous emission (see Sect. II.5a-2) of an electronically excited atom is of the order of 10^{-8} s or longer.

The distinction between one- and two-step radiation processes involving the absorption and re-emission of one photon or the absorption of two photons, is more

---

[†] Alkemade and Zeegers (1971) have classified ionization as a chemical process because a particle (electron) is exchanged. We now call an ionization process chemical or physical according to whether new chemical bonds are formed or not (see Table II.1).

subtle and will be discussed later in some detail in Sect. II.5a-1.

Finally we can distinguish between *uni-*, *bi-* and *termolecular* elementary processes, as explained in Subsection 4a-1. Bi- and termolecular processes are also called: 2- and 3-body processes (collisions; reactions). When a chemical reaction takes place between two particles in the presence of another particle whose chemical identity is not changed, we call the latter particle a '*third body*' or '*third partner*' (see reactions no. 17, 20, 22, 23 in Table II.1). This partner serves often to carry off (part of) the chemical energy that is released in an exo-ergic association reaction. The reaction product may thereby be stabilized, since it would otherwise disintegrate again into fragments immediately ($\simeq 10^{-13}$ s) after its formation. Stabilization occurs if the internal energy of the association product is reduced to below its dissociation energy. After the exo-ergic reaction is completed, the third body and/or the reaction product(s) may be found in an excited but 'stable' electronic state (see reactions no. 22 and 23). The excited particle may thereupon emit radiation at discrete wavelength(s) (see two-step processes no. 38 and 39), or it may subsequently be quenched by an inelastic collision of the 2nd kind (see process no. 4).

The association product of two particles can also be stabilized by the simultaneous emission of a photon with a continuous spectrum (see photochemical reaction no. 26, and Sect. II.2e for further discussion).

Stabilization of an association product can also be brought about by ionization of the product (process no. 28) or by ionization of a third body (process no. 30). The latter process can also occur in two steps (process no. 44).

An analogous situation exists in the stabilization of an ion-electron recombination by a third body (process no. 12) or by the emission of a photon with continuous spectrum (process no. 14).

Whenever *three* particles are produced in a *bi*molecular chemical reaction (no. 21) or collisional ionization process (no. 11), the entropy of the system is increased. This increase enhances the equilibrium constant $K(T)$ of the reaction or the ionization constant $K_i(T)$ (cf. Eq. II.51 and the ensuing discussion). It also enhances the thermal rate constant of the forward process in comparison to that of the reverse process since their ratio equals the equilibrium constant (see Eq. II.88).

As we have seen in Subsection 4b-1, the probability of a bimolecular collision between a given particle AB and a flame molecule is roughly 10^3 times larger than that of a termolecular collision between AB and two flame molecules in flames at 1 atm. Termolecular reactions of type no. 17 are thus expected *a priori* to be $10^3 \times$ less frequent than bimolecular reactions of type 16 or 15 when the reaction partners of AB are major flame constituents. For this reason, reactions that are bimolecular in both directions (such as no. 15 and 16) are, generally speaking, more quickly balanced in flames than reactions such as no. 20 and 21. We have, however, here left out of account the influence on the reaction velocity of other factors such as the activation energy and the steric factor. A quantitative discussion, including these factors, of the relative probabilities of the production of AB through the bimolecular reverse exchange reaction no. 15 and the termolecular association reaction no. 20 will be given in Sect. VIII.4a for a particular case (see also Sugden 1956).

One has to be careful in drawing general conclusions from such *a priori* estimates. It appears, for instance, that the 2-body radiative association process no. 26 often produces less radiation than the 2-step, 3-body association process no. 39. Thrush (1968) has explained the greater efficiency of the 3-body chemiluminescence process by the absence of rigid restrictions as to near-resonance of energy levels and conservation of angular momentum; such restrictions do hold for the 2-body chemiluminescence process.

Fluorescence and chemiluminescence in flames are treated quantitatively and in more detail in Sects VI.2 and VI.3, respectively. The emission intensities are determined by the rates of the processes that lead to the production and destruction of the excited states involved. Various processes may concur in parallel or in succession, which complicates the calculations, especially in the primary combustion zone. The effect of self-absorption (process no. 37) on the fluorescence intensity will be considered in Sect. II.5c-4.

Under certain conditions, the rate of excitation of, for example, a metal atom M through two-step association process no. 41 with $M \hat{=} D$, could be larger than when the metal atom is directly excited as a third body in the same association reaction, process no. 23, with $M \hat{=} C$ (see Alkemade 1963). Suppose that the rate constant or the efficiency of producing a vibrationally excited N_2^* molecule, acting as a third body in the first step of process no. 41, is the same as that of producing directly an excited metal atom in the corresponding association process, reaction no. 23. In a flame where N_2 is the major constituent, the rate of production of N_2^* per unit volume will then exceed the production rate of M^* through reaction no. 23 by a factor $x \equiv [N_2]/[M]$. Suppose that the probability of a collision between N_2^* and another N_2 molecule is larger by about the same factor x than that between N_2^* and a metal atom M^0 (i.e. we assume equal effective cross sections and mean relative velocities for both types of collisions). Let η_1 be the probability that the excess vibrational energy of N_2^* is degraded upon collision with another N_2 molecule. The reciprocal value (η_1^{-1}) is the *persistence factor* of the (excess) vibrational energy. Let η_2 be the probability that this vibrational energy is converted into electronic excitation energy when N_2^* collides with M^0. It is then simply found that the rate of producing M^* through two-step process no. 41 exceeds the production rate through direct process no. 23 by the factor $x(\eta_2/x\eta_1) = \eta_2/\eta_1$. A molecule excited in one of the higher vibrational levels by process no. 41 gives up its excess energy by small quanta only, at each collision with another molecule. Hence it may retain sufficient vibrational energy to excite the metal atom, after it has undergone a number of collisions. Consequently the persistence factor η_1^{-1} may be markedly larger than unity (see also Sect. II.3c-2). If η_2 is of the order of unity, two-step process no. 41 may then be more effective than the corresponding direct process, reaction no. 23.

NOTES TO TABLE II.1

* A general classification of processes in the gas-phase that are discussed in this book is given. A more detailed classification of purely radiative processes and ionization processes is given in Sects II.5a-1 and IX.2a, respectively.

† A, B, ... stand for an atom, molecule, radical (or electron) in an unspecified internal state; a particle in the ground state is denoted by superscript 0; a particle in a specific state of excitation is denoted by * or **; a singly ionized particle is denoted by +; e^- represents a free electron; $h\nu_d$, $h\nu_1$ and $h\nu_2$ represent a photon of discrete energy; $h\nu_c$ represents a photon from a spectral continuum.

†† When the reverse process is of the same class, it is listed as 'id'.

(1) The particles exchange only linear momentum **p**.

(2) M_J is the magnetic quantum number of atom A, which remains here in the same energy level (see also Sects II.5b-2, VI.1, and VII.2b-6 and 4f).

(3) The product atom A may also be found in an excited level below that of A^* that does not belong to the same multiplet as A^*.

(4) A^{**} and A^* are in different levels of the same multiplet.

(5) See Sect. II.5a-1.

(6) The product particle A may also be found in an excited energy level below that of A^*.

(7) Only of importance in flames when an intense laser beam is employed.

(8) See Sect. IX.2a.

(9) A may be in the ground or in an excited electronic state.

(10) See Sect. II.2e.

(11) When the reaction is exo-ergic, the reaction product(s) may be found in an excited electronic state.

(12) This reaction may also produce BC^* in an excited electronic state.

(13) See Sect. VI.1.

(14) Atoms A in the first and second step are different particles of the same species, present in the same radiating medium (see Sect. II.5c-3).

Chapt. II   BASIC CONCEPTS AND GENERAL RELATIONSHIPS

Table II.1

General Classification of Processes*

No.	Process[†]	Name(s)	Foot-note	No. of reverse process[††]	Kind of process
		Single-step processes			
		Physical			
1.	$A(p_1) + B(p_2) \to A(p_3) + B(p_4)$	elastic collision; elastic scattering	(1)	id.	elastic collision
2.	$A(M_J) + B \to A(M'_J) + B$	M_J-mixing collision	(2)	id.	
3.	$A^0 + B \to A^* + B$	collisional excitation; collision of the 1st kind		4.	
4.	$A^* + B \to A^0 + B$	collisional de-excitation; quenching collision; collision of the 2nd kind; superelastic collision	(3)	3.	collisional (de-)excitation
5.	$A^{**} + B \to A^* + B$	(fine-structure or intra-multiplet) mixing collision	(4)	id.	
6.	$A^* + B^0 \to A^0 + B^*$	excitation transfer collision		id.	
7.	$A^* \to A^0 + h\nu_d$	(de-excitation by) spontaneous emission	(5),(6)	9.	
8.	$A^* + h\nu_d \to A^0 + 2h\nu_d$	(de-excitation by) induced emission	(5),(7)		radiative (de-)excitation
9.	$A^0 + h\nu_d \to A^*$	photo-excitation; (photon) absorption	(5)	7.	
10.	$A^0 + 2h\nu_d \to A^*$	two-photon absorption	(5),(7)		
11.	$A^{(*)} + B \to A^+ + e^- + B$	collisional ionization	(8),(9)	12.	collisional ionization
12.	$A^+ + e^- + B \to A^{(*)} + B$	3-body ion-electron recombination	(8),(9)	11.	
13.	$A + h\nu_c \to A^+ + e^-$	photo-ionization	(10)	14.	radiative ionization
14.	$A^+ + e^- \to A^{(*)} + h\nu_c$	radiative ion-electron recombination; ion-recombination continuum emission	(10)	13.	
		Chemical			
15.	$AB + C \to A + BC$	2-body exchange reaction	(11)	id.	
16.	$AB + CD \to AC + BD$	2-body exchange reaction	(11)	id.	
17.	$AB + CD + E \to AC + BD + E$	3-body exchange reaction	(11)	id.	
18.	$A^* + BC \to AB + C$	chemiquenching by 2-body exchange		19.	
19.	$AB + C \to A^* + BC$	(true) chemi-excitation by 2-body exchange	(12)	18.	chemical reaction and chemi-(de-)excitation
20.	$A + B + C \to AB + C$	3-body association (of A and B)		21.	
21.	$AB + C \to A + B + C$	collisional dissociation (of AB)		20.	
22.	$A + B + C \to AB^* + C$	(true) 3-body associative excitation			
23.	$A + B + C \to AB + C^*$	(induced) 3-body associative excitation		24.	
24.	$AB + C^* \to A + B + C$	2-body dissociative quenching		23.	
25.	$AB + h\nu_c \to A + B$	photodissociation; photolysis	(10)	26.	
26.	$A + B \to AB + h\nu_c$	radiative association; association-continuum emission	(10)	25.	photochemical reaction
27.	$AB + h\nu_c \to A^* + B$	excitation by photolysis	(13)		

Table II.1 (cont.)

28.	$A + B \rightarrow AB^+ + e^-$	2-body associative ionization	(8)	29.	
29.	$AB^+ + e^- \rightarrow A + B$	2-body dissociative neutralization	(8)	28.	chemi-ionization and -neutralization
30.	$A + B + C \rightarrow A^+ + e^- + BC$	3-body associative ionization	(8)	31.	
31.	$A^+ + e^- + BC \rightarrow A + B + C$	3-body dissociative neutralization	(8)	30.	

		Multi-step processes		No. of 1-step process	
32.	$A + BC \rightarrow AB + C$ $AB + DE \rightarrow ABD + E$ $\overline{A + BC + DE \rightarrow ABD + C + E}$	over-all 3-body exchange reaction		15. 15.	chemical reaction
33.	$A + B + C \rightarrow AB^* + C$ $AB^* + D \rightarrow AB^0 + D$ $\overline{A + B \rightarrow AB^0}$	over-all 2-body association reaction		22. 4.	
34.	$A^0 + h\nu_1 \rightarrow A^*$ $A^* + h\nu_2 \rightarrow A^{**}$	stepwise photo-excitation	(13)	9. 9.	
35.	$A^0 + h\nu_d \rightarrow A^*$ $A^* \rightarrow A^0 + h\nu_d$	photoluminescence; resonance fluorescence	(13)	9. 7.	photo-excitation and -luminescence
36.	$A^0 + h\nu_1 \rightarrow A^{**}$ $A^{**} \rightarrow A^* + h\nu_2$	photoluminescence; non-resonance fluorescence	(13)	9. 7.	
37.	$A^* \rightarrow A^0 + h\nu_d$ $h\nu_d + A^0 \rightarrow A^*$	self-absorption; re-absorption	(14)	7. 9.	
38.	$A + B + C \rightarrow AB + C^*$ $C^* \rightarrow C^0 + h\nu_d$	(induced) chemiluminescence by 3-body association		23. 7.	
39.	$A + B + C \rightarrow AB^* + C$ $AB^* \rightarrow AB^0 + h\nu_d$	(true) chemiluminescence by 3-body association		23. 7.	
40.	$A + B + C \rightarrow AB^* + C$ $AB^* + D^0 \rightarrow AB + D^*$	2-step chemi-excitation (of D) by 3-body association		23. 6.	chemi-excitation and -luminescence
41.	$A + B + C \rightarrow AB + C^*$ $C^* + D^0 \rightarrow C + D^*$	2-step chemi-excitation (of D) by 3-body association		23. 6.	
42.	$AB + CD \rightarrow AC^* + BD$ $AC^* + E^0 \rightarrow AC + E^*$	2-step chemi-excitation (of E) by 2-body exchange		16. 6.	
43.	$A^0 + B \rightarrow A^* + B$ $A^* + C \rightarrow A^+ + e^- + C$	cumulative collisional ionization	(8)	3. 11.	
44.	$A + B + C \rightarrow AB^* + C$ $AB^* + D \rightarrow AB + D^+ + e^-$	over-all 3-body associative ionization	(8)	22. 11.	ionization
45.	$A^0 + h\nu_d \rightarrow A^*$ $A^* + B \rightarrow A^+ + e^- + B$	process producing an opto-galvanic effect	(7),(8)	9. 11.	

Chapt. II BASIC CONCEPTS AND GENERAL RELATIONSHIPS

5. THE INTENSITIES OF RADIATIVE TRANSITIONS [†]

5a TYPES OF RADIATIVE TRANSITIONS AND THEIR RATE CONSTANTS

5a-1 *RADIATIVE TRANSITIONS AND EINSTEIN RADIATION COEFFICIENTS FOR A SINGLE ATOMIC LINE*
Types of radiative transitions. A free atom with two energy levels labelled p and q, with discrete energy values, E_p and E_q ($E_q > E_p$), may undergo optical transitions in a very narrow frequency band around ν_0 ($=c/\lambda_0$) $= (E_q - E_p)/h$. The energy levels may be degenerate with degeneracies or statistical weights g_p and g_q, respectively.

The atom may be raised from the lower state p to the upper state q by *absorption*[*] of a photon from a radiation field with frequency $\nu \approx \nu_0$. The atom may conversely undergo a downward transition from state q to state p by *spontaneous emission* of a photon or by *induced emission* (also called *stimulated emission*) of a photon under the action of a radiation field with frequency $\nu \approx \nu_0$. The radiation emitted in the latter process is coherent with the inducing radiation field, i.e., it has the same direction of propagation, frequency ν and phase.

Multi-photon processes in which two or more photons act simultaneously to achieve one optical transition are mentioned here for completeness' sake only (for a general introduction see, e.g., Bloembergen and Levenson 1976, and for detailed calculations of two-photon excitation processes see Nienhuis and Schuller 1978). Two-photon excitation of Cd-, Zn- and Na atoms have been observed in flames irradiated by a single, tunable pulsed dye-laser (Fraser and Winefordner 1972, van Dijk and Alkemade 1980, van Dijk *et al.* 1978). The occurrence of a two-photon process was established by observing the fluorescence originating from the energy level with excitation energy equal to twice the photon energy and from other levels populated therefrom in cascade. The two-photon excitation rate grows proportionally to the square of the field intensity, as distinct from the normal one-photon process. Also, the selection rules are different for two- and one-photon transitions (see Sect. II.2c). With conventional radiation sources the former process can be disregarded. A deviation from the dependence on the squared intensity occurs when the two-photon transition becomes saturated, as has been observed by van Dijk *et al.* 1978, e.g., for the Na($3^2S \rightarrow 5^2S$) transition in an H_2-O_2-Ar flame (see Fig. II.7).

[†] For an introduction to this subject see also Penner (1959), Mitchell and Zemansky (1961), Garbuny (1965), Shore and Menzel (1968), and Thorne (1974).

[*] The term absorption is ambiguous (see the discussion in Sect. II.5a-2 in relation to the term absorptivity). Here the term is used to denote a certain quantum process, irrespective of whether this process is followed by a radiative or radiationless transition.

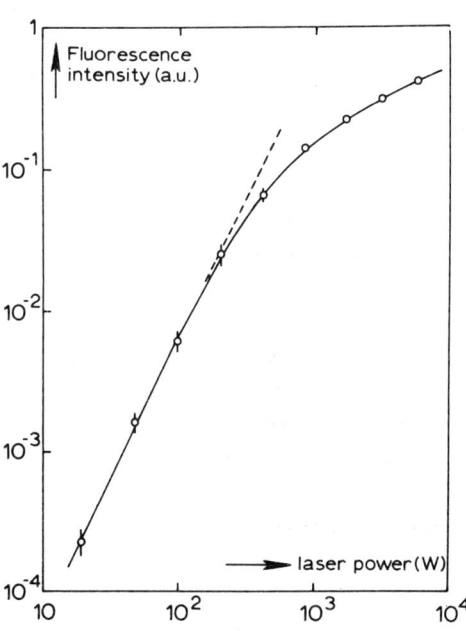

Fig. II.7 Double-photon excitation rate of Na(5s) level in H_2-O_2-Ar flame, monitored by fluorescence signal from 4p→3s transition (cf. Fig. II.1), is plotted as a function of broad-band pulsed-laser power at 6022 Å. Dashed line shows the initial asymptote which corresponds to a square dependence of the signal on the laser power. (From van Dijk et al. 1978.)

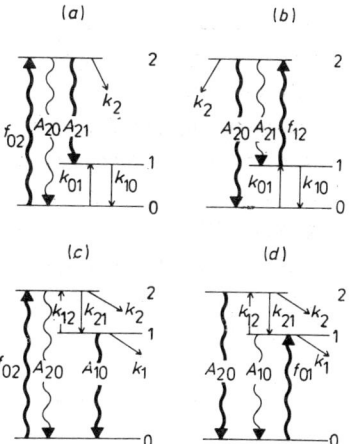

Fig. II.8 Schematic representation of various types of nonresonance atomic fluorescence with the aid of a level diagram. The number 0 denotes ground level; the numbers 1 and 2 denote excited levels. Wavy lines with arrows indicate radiative transitions; straight lines with arrows denote collisional transitions. Absorption transitions and observed fluorescence transitions are represented by heavy wavy arrows pointing upwards and downwards, respectively. The oscillator strength f, spontaneous transition probability A or unimolecular collisional rate constant k are specified for each transition. Case (a) and (b) refer to *direct line fluorescence*; case (c) and (d) refer to *stepwise line fluorescence* and are distinguished as *Stokes* and *anti-Stokes fluorescence*, respectively. (From Alkemade 1970.)

When the gas pressure is sufficiently high, radiationless transitions between states p and q due to collisions may also become important. Collisions which excite the atom are called *collisions of the first kind*; those which remove the excitation energy from the excited atom (de-excitation) are called *collisions of the second kind*. When an atom absorbs a photon and subsequently undergoes a collision of the second kind, the final result is that radiant energy is converted into kinetic (translational, vibrational or rotational) energy. When, however, photon absorption is followed by spontaneous photon emission, one speaks of *photoluminescence* or *fluorescence*. When the same spectral line is absorbed and re-emitted, *resonance fluorescence* is said to exist, which is related to elastic *scattering* (see below).

When the frequencies of the absorbed (*primary*) radiation and of the re-emitted (*secondary*) radiation correspond to different atomic transitions, one speaks of *nonresonance fluorescence*. Several cases can be distinguished here (see Fig. II.8; see also Omenetto and Winefordner 1972). *Direct line fluorescence* exists when the transitions involved in the absorption and subsequent re-emission process have a common upper level.[†] When the upper levels are different for the two transitions, *stepwise line fluorescence*[*] occurs. In this process atoms excited to one upper level by photon absorption are transferred, usually by collisions, to the other level from which the fluorescent line is emitted. In the case when the wavelength of the latter line is shorter than the wavelength of the absorbed radiation, we are dealing with *anti-Stokes fluorescence*. The deficit in photon energy is made up by thermal energy and the term *thermally assisted* (*stepwise line*) *fluorescence* can be used here. In the more usual case we have *Stokes fluorescence*.

Sensitized fluorescence occurs when (donor) atoms are excited by means of an external radiation beam and collide with (acceptor) atoms of another species while transferring their excitation energy to the acceptor atom which thereupon undergoes a radiative de-excitation. This effect is not important in flames (see Winefordner and Mansfield 1967, and Alkemade 1968), but it may become significant in vapour bulb experiments (cf. Sect. VI.2b-4). *Sensitized luminescence* in general occurs when the donor atoms are excited by radiation or any other means (e.g., collisions of the first kind).

[†] Note that 'resonance fluorescence' as defined by Inaba (1976) includes both 'direct line fluorescence' and 'resonance fluorescence' as defined in our text.

[*] Also called 'cross fluorescence' or 'sensitized fluorescence' in some papers; the latter term will be used here with a different meaning (see tne following). The terminology followed here conforms to Omenetto and Winefordner (1972) and IUPAC (1976a). We distinguish it from *stepwise radiative excitation* which consists of two separate radiative excitation steps in succession (see also Sect. VI.1).

THE INTENSITIES OF RADIATIVE TRANSITIONS   Sect. II.5a

Light scattering and fluorescence. The distinction between these terms, whether or not they are preceded by the term resonance, is not always clear in the literature and different authors have often used them with different meanings (for a critical discussion see Inaba 1976 and the references cited there). We here leave out of account light scattering by particulate matter such as *Mie scattering*, which occurs when the dimensions of the particle are close to or larger than the wavelength. Atomic or molecular light scattering may be *elastic* (*Rayleigh scattering*, with no exchange of energy between atom and radiation field) or *inelastic* (*Raman scattering*, in which a discrete amount of energy is exchanged, corresponding to an atomic excitation or de-excitation). We shall restrict ourselves here to a comparison of elastic scattering and resonance fluorescence in gases; these two re-emission processes have in common the fact that after the process the atom is again found in the same energy level (but not necessarily in the same quantum state) as before. The theory of re-emission has not yet been worked out fully for arbitrary intensity and spectral characteristics of the exciting radiation and for arbitrary collisional interactions of the radiating atom with the surrounding gas molecules.[†] A consistent and exact distinction between light scattering and resonance fluorescence, which can be applied unambiguously in all feasible experimental situations, can therefore hardly be given. We shall content ourselves with presenting only a provisional, over-all description which may suffice within the context of our book.[*]

One point of distinction that can be made is that *elastic light scattering* is, in general, not linked with a particular intermediate energy level and is found, though to a weak extent, at any frequency of the incident radiation. The intensity of the scattered radiation is proportional to the square of an absolute amplitude which is a summation of partial complex amplitudes each of which describes the interaction of the radiation field with one of the atomic excited states. The magnitude of a partial amplitude decreases with increasing difference between photon energy and excitation energy. When the frequency of the incident radiation comes close to or coincides with one of the atomic transition frequencies (case of *resonant light scattering* [‡]), the particular excited state involved plays a dominant

---

[†] A general treatment of the quantummechanical theory of resonant light scattering (for arbitrarily strong, coherent and incoherent excitation and without collisions), a critical discussion of previous theories and an extensive bibliography have been given by Kimble and Mandel (1976, 1977); a quantummechanical theory including collisions for weak excitation has been given by Nienhuis and Schuller (1977a, 1977b) for nondegenerate levels and Nienhuis (1978) for degenerate ones.

[*] Inaba (1976) has given a somewhat different description in connection with laser monitoring of the atmosphere.

[‡] The term resonant light scattering does not necessarily imply that the scattered photons have (about) the same frequency as the incident photons; we can thus speak of resonant Raman scattering as well. This is in contrast with resonance fluorescence where the same atomic line is absorbed and re-emitted (see definition earlier in this subsection).

113

Chapt. II   BASIC CONCEPTS AND GENERAL RELATIONSHIPS

role in determining the intensity, polarization and angular distribution of the scattered radiation. The intensity of resonant scattering is substantially higher than in the case of nonresonant scattering.

Furthermore, scattering is to be conceived as a *single absorption-re-emission act* according to classical as well as quantummechanical theory (see, e.g., Heitler 1954). One cannot unambiguously define the intermediate state in which the atom (separate from the radiation field) is found after the absorption and before the re-emission act. Scattering is a two-photon process leading to the destruction and creation of a photon in a single step. In the case of nonresonant scattering this destruction and creation take place at the same instant, and collisions-of-the-second-kind have no effect. In the case of resonant scattering these collisions may shorten the delay time between photon absorption and photon emission. Apart from Doppler broadening, the spectral distribution of the re-emitted photons is solely determined by the spectral profile and the intensity of the exciting radiation if the initial state is not affected by collisions (see also below).

Resonance fluorescence, and fluorescence in general, are usually thought to involve a 'real' excitation of the atom as such (considered separately from the radiation field) to a particular energy level. This implies that the frequency of the incident radiation matches or is in the vicinity of a particular atomic transition frequency. The rate of excitation varies with detuning according to the collision- and Doppler-broadened absorption line profile.

When the wings of two neighbouring atomic lines overlap partially (as in the case of the Na-D doublet), and the frequency of the exciting radiation lies somewhere between the line centres, two distinct fluorescence transitions can take place in different atoms. The resulting fluorescence is then simply the sum of the separate fluorescence effects; we disregard here complications due to saturation in strong fields (see Sect. II.5c-4). This is in contrast to the case of scattering where complex amplitudes (not: intensities) have to be added together and where quantummechanical interference effects may appear.

The fluorescence process can be considered as a sequence of *two separate radiation acts*: absorption of a photon followed, after a while, by spontaneous emission of another photon. In between the atom is found in a particular energy level from which it may 'choose' to be transferred to another level by emission of radiation or by a collision. The probability of the overall fluorescence process is the **product** of the probability of photon absorption and the fractional probability of photon emission (see also Sect. II.5c-4).

These features of the fluorescence process are essentially a consequence of collisional interactions † that result in a (partial) 'loss of memory' on the part

---

† The term collision-induced fluorescence has therefore also been in use.

THE INTENSITIES OF RADIATIVE TRANSITIONS   Sect. II.5a

of the excited atom with regard to the phase and frequency, direction and polarization of the exciting radiation. These collisions cause the fluorescence radiation to be incoherent with respect to the exciting radiation. The spectral profile of the fluorescence line is broadened by adiabatic and diabatic (or quenching) collisions in rather the same way as the thermal emission line (see Sect. II.5b-2).[†] When the collisions are frequent enough to thermalize the velocity of the excited atom, the fluorescence line has the same centre and spectral profile as the thermal emission line, irrespective of the profile and detuning of the exciting beam. However, in an exceptionally strong radiation field line splitting or radiation broadening may occur (see below).

Resonant scattering and resonance fluorescence as described here are aspects of the re-emission process that dominates in the absence and presence of frequent collisions, respectively. A gradual transition between these extreme aspects will be found, depending on the pressure and kind of gas applied. They may be discriminated through their different polarizations (see Vriens 1977). For atomic metal lines in flames at 1 atm pressure resonance fluorescence, which is collision-induced and incoherent, is usually dominant (see also Sects II.5c-4 and VI.1).

Jongerius *et al.* (1979) have found in an H_2-O_2-Ar flame at 1 atm that the ratio of Rayleigh scattering to resonance fluorescence was about 0.7% and 1.1% when the laser was detuned by about 10 Å to the red from the centre of the Na-D_1 line and to the blue from the centre of the D_2 line, respectively. The re-emission was observed at right angles to the direction of the laser beam and to its linear-polarization direction. In an $H_2-O_2-N_2$ flame these ratios appeared to be about a factor of 5 larger as a result of the stronger fluorescence quenching in this flame (cf. Sect. VI.2) and of the independence of Rayleigh scattering of collisions. When the laser polarization direction was chosen parallel to the observation direction, the relative Rayleigh scattering was strongly reduced, as expected (see van Dijk 1977). Chan and Daily (1978, 1979a) have made similar measurements in a methane-air flame, and Hosch and Piepmeier (1978) in an acetylene-air flame.

Raman and Rayleigh scattering as well as resonance fluorescence have been investigated as a possible means for monitoring the local temperature and composition of flames and combustion systems with the aid of a monochromatic laser beam (see Lapp and Penney 1974, Daily 1976, and Eckbreth 1978).

For further clarification we give here a brief description of some

---

[†] This holds at least in the impact approximation (see Omont, Smith and Cooper 1972); a more general theory has been derived by Nienhuis and Schuller (1977a).

Chapt. II   BASIC CONCEPTS AND GENERAL RELATIONSHIPS

characteristics of the scattered radiation. These characteristics depend, in general, on whether the incident radiation is monochromatic (spectral width ≪ reciprocal radiative lifetime; see Sect. II.5b-2) or (quasi-)continuous, whether it is weak or strong, on the perturbation by collisions and on the atomic velocity distribution.

(i)   *Case of weak radiation field*. When the radiation field is so weak that it does not noticeably affect the lifetime of the excited state (as is expected for conventional radiation sources), the scattering intensity is linear in the field intensity (as is the fluorescence intensity; see Sect. II.5c-4). Since Rayleigh scattering is coherent, there is a close correlation in frequency and phase between the absorbed and emitted radiation considered in a reference frame moving with the radiating atom; the frequency distribution of the emitted radiation is determined by that of the incident radiation. A monochromatic radiation beam produces scattered radiation with the same narrow frequency distribution if the atoms are stationary.[†] The relative variation in intensity of the scattered radiation with frequency of the incident radiation is determined by the natural line-broadening profile (see Sect. II.5b-2). In a (quasi-)continuous field (with spectral width ≫ reciprocal radiative lifetime) the width and spectral distribution of the scattered radiation conform to the natural line-broadening profile.[*]

Resonant scattering (and under certain conditions also resonance fluorescence; see Sect. VI.1) may show an anisotropic angular distribution and a certain degree of polarization that depend on the direction and polarization of the incident beam, and on the quantum number, J, of the atomic states involved (see Fano and Macek 1973). We assume here that no magnetic field is present and that the ground-state atoms are not polarized by the weak radiation beam. When the scattering is observed at right angles to the unpolarized-beam direction, a certain degree of linear polarization may be found along an axis perpendicular to the beam direction. Atomic lines having an upper level with $J = \frac{1}{2}$ are, however, not polarized.

In flames the atoms have a Maxwellian velocity-distribution and the above frequency relation between the incident and the <u>observed</u> scattered radiation will be perturbed due to the Doppler effect (see also Sect. II.5b-2). The frequency of the scattered radiation is smeared out over the full Doppler width when the radiation is observed at right angles to the incident beam.

[†] We consider here the case when the initial state is the ground state. Strictly speaking, the scattered radiation is spectrally broadened by an amount that corresponds to the inverse lifetime of the ground state, which is however small because of the low absorption rate in weak fields (see also Sect. II.5b-2).

[*] An enlightening discussion on the effect of coherence on the emission line profile is found, e.g., in the Appendix of Holstein's (1951) paper.

THE INTENSITIES OF RADIATIVE TRANSITIONS   Sect. II.5a

(ii) *Case of strong radiation field*. A strong laser beam cannot only affect the populations of the upper and lower level to such an extent that saturation of the atomic transition sets in (see Sect. 5a-2). It can, under certain conditions also influence the spectral characteristics of the scattered radiation, the polarization and the velocity-distribution of the ground-state atoms (see Sect. 5a-2).

The spectrum of the radiation scattered by atoms in a strong, monochromatic laser beam consists of a sharp, coherent Rayleigh peak and an incoherent, broadened peak at the same frequency, which is flanked by two other incoherent peaks (*Rabi effect* or *a.c. Stark effect*; see Mollow 1969, 1972, 1973, Jacquinot 1976, Grove, Wu and Ezekiel 1977, Zoller 1978, and Hertel and Stoll 1977). The spectral separation of the latter, symmetrically located, side peaks increases proportionally to the square-root of the beam intensity if the frequencies of the radiation beam and the atomic line coincide. At very high intensities the incoherent components become dominant over the coherent component.

In a strong, broad-band, i.e. quasi-continuous laser beam, the scattered radiation shows a Lorentzian spectral profile whose width increases linearly with the spectral beam intensity at the atomic line centre λ_0 (*radiation broadening*; see Sect. II.5b-2). The peak of the scattered-line profile coincides with λ_0, if the spectral beam intensity is symmetrically distributed with respect to λ_0. If it is not, a shift of this peak will be found that is proportional to the spectral beam intensity (*light shift*; see Maassen 1976, Cohen-Tannoudji 1977). Both the Rabi effect and radiation broadening are caused by a direct influence of the radiation field on the atomic energy levels.

The quantitative theoretical description of these effects is complicated by collisions (see the analysis by Nienhuis 1980). Moreover, collisional and Doppler line broadening make these effects hard (but not impossible) to observe even with laser sources producing a high spectral irradiance in the flame.

Einstein radiation coefficients. The intensities of the different radiative transitions are determined by the *Einstein coefficients of radiation*. Let n_p and n_q be the (number) densities (= number of particles per unit volume) of atoms in state p and q. For simplicity's sake we assume either that the atoms are randomly oriented (unpolarized) or that the radiation field in which they are found is isotropic and unpolarized. These assumptions are justified in most spectroscopic flame studies or applications (see end of this subsection for further discussion). When the spectral[†] radiant energy density of a *continuous* radiation field at $\nu \simeq \nu_0$ is given by ρ_{ν_0} (radiant energy per unit of volume and per unit of frequency interval; see Appendix A.5), we have

---

[†] See footnote on the next page

$B_{pq}\,\rho_{\nu_0}\,n_p$ = number of absorption transitions per unit volume and per second, (II.147)

$B_{qp}\,\rho_{\nu_0}\,n_q$ = number of induced-emission transitions per unit volume and per second, (II.148)

$A_{qp}\,n_q$ = number of spontaneous-emission transitions per unit volume and per second. (II.149)

The coefficients B_{pq}, B_{qp} and A_{qp} are called the *Einstein (probability) coefficients* or *Einstein radiation coefficients* or *transition probabilities for absorption*, *induced emission*, and *spontaneous emission*, respectively. They are, strictly speaking, not probabilities, but probability densities. Note that A_{qp}, for example, is expressed in s^{-1} and B_{pq} in $cm^3\,erg^{-1}s^{-1}\,Hz$. The Einstein coefficients are characteristic for the optical transition considered. They are constants, as long as the wave functions of the atomic states involved are not altered by electric or magnetic fields or by neighbouring particles, and as long as the refractive index does not deviate markedly from unity (see also the end of this subsection and of Sect. 5a-2).

It follows, for example, from Eq. II.147 that the probability of photon absorption by an atom in the lower state during time interval Δt is given by $B_{pq}\,\rho_{\nu_0}\,\Delta t$. It appears, however, from the quantummechanical theory of radiative transitions that the proportionality between this probability and Δt is restricted by two conditions:

$$B_{pq}\,\rho_{\nu_0}\,\Delta t \ll 1 \,, \qquad (i)$$

$$\Delta t > \Delta\nu^{-1} \,, \qquad (ii)$$

where $\Delta\nu$ is the width of the spectral range in which $\rho_\nu \approx \rho_{\nu_0}$. Condition (i) justifies the use of the underlying first-order perturbation theory. Condition (ii) may be interpreted as a consequence of the uncertainty relation between Δt and $\Delta\nu$; when $\Delta t \leqslant \Delta\nu^{-1}$, the *spectral* energy density is not well-defined. Both conditions together can be satisfied only if $B_{pq}\,\rho_{\nu_0} \ll \Delta\nu$. We refer to Cohen-Tannoudji (1962) and Nienhuis (1977) for a more detailed discussion.

[†] Optical quantities which are defined *per unit of frequency interval* are preceded by the adjective '*spectral*' and distinguished by a subscript ν. In general, they may be a function of ν. This may be made explicit by writing, for example, $\rho_\nu(\nu)$. When defined per unit of wavelength interval, the subscript λ is used. The dimension of these quantities changes when ν is replaced by λ. We always assume that the spectral density, etc., of a *continuous* radiation field is practically constant over the spectral width of the atomic line. This also applies to a *quasi-continuous* field whose spectral density varies smoothly only in a limited frequency interval that is, however, broad in comparison with the atomic line width. (See also footnote on page 39.)

THE INTENSITIES OF RADIATIVE TRANSITIONS   Sect. II.5a

Because of the assumed isotropy, the radiant flux $d\Phi_{sp}$ (energy per second; see Appendix A.5) emitted spontaneously by a volume element dV into a solid angle Ω is given by

$$d\Phi_{sp} = A_{qp}\, n_q\, (\Omega/4\pi)(h\nu_0)\, dV\,. \qquad (II.150)$$

The volume element should be chosen so small that self-absorption (see Sect. II.5c-3) may be neglected.

When instead of ρ_ν the spectral radiance B_ν (see Appendix A.5; $B_\nu = \rho_\nu c/4\pi$ with c = velocity of light) or ρ_λ is used in Eqs II.147 and 148, the definitions and dimensions of the Einstein B coefficients will be changed accordingly.

Simple *relations between the Einstein coefficients* can be derived by thermodynamic equilibrium considerations. We assume that a gas of atoms is enclosed in a cavity or 'Hohlraum', for example a furnace, and that this system, including the continuous radiation field, is in thermodynamic equilibrium at temperature T. Because of the principle of detailed balance (see Sect. II.4a-2), the rate of upward optical transitions equals that of the inverse, downward optical transitions, i.e.,

$$B_{pq}\, \rho^b_{\nu_0}\, (n_p)_e = A_{qp}\, (n_q)_e + B_{qp}\, \rho^b_{\nu_0}\, (n_q)_e \qquad (II.151)$$

or

$$(n_q)_e / (n_p)_e = B_{pq}\, \rho^b_{\nu_0} / (A_{qp} + B_{qp}\, \rho^b_{\nu_0})\,. \qquad (II.152)$$

Here $\rho^b_{\nu_0}$ is the spectral density of a black-body radiation field at temperature T, given by Planck's law (see Eq. II.65b). The ratio of equilibrium number densities, $(n_q)_e/(n_p)_e$, is given by Boltzmann's law (see Eq. II.30) and is independent of the way(s) in which the atoms become excited or de-excited. Substituting these equilibrium laws in Eq. II.152 and putting: $E_q - E_p = h\nu_0$, one finds the following relations between the Einstein coefficients

$$A_{qp}/B_{qp} = 8\pi h\nu_0^3/c^3\,, \qquad (II.153)$$

$$B_{qp}/B_{pq} = g_p/g_q\,. \qquad (II.154)$$

Since these coefficients are not dependent on the existence of equilibrium, these relations are more generally valid. From Eqs II.153 and II.154 we derive by substituting $\nu_0/c = 1/\lambda_0$ the numerical relations (in CGS units)

$$B_{qp} = 6.01 \times 10^{24}\, \lambda_0^3\, A_{qp}\,, \qquad (II.155)$$

$$B_{pq} = 6.01 \times 10^{24}\, \lambda_0^3\, (g_q/g_p)\, A_{qp}\,. \qquad (II.156)$$

Note that B_{pq}/A_{qp} increases proportionally to λ_0^3.

It follows from Eqs II.148, 149 and 153, and from Planck's law (Eq. II.65b) that in thermodynamic equilibrium

119

Chapt. II   BASIC CONCEPTS AND GENERAL RELATIONSHIPS

$$\frac{\text{rate of spontaneous emission}}{\text{rate of induced emission}} = \frac{A_{qp}(n_q)_e}{B_{qp}\rho_{\nu_0}^b(n_q)_e} = \exp[h\nu_0/kT] - 1 \ . \quad (\text{II.157})$$

For optical frequencies $(h\nu_0 \geqslant 1\,\text{eV})$ and under thermal flame conditions $(kT \approx 0.2\,\text{eV})$, induced emission may always be neglected in comparison with spontaneous emission.[†]
It follows from Eqs II.147, 148 and 154 that generally

$$\frac{\text{rate of absorption}}{\text{rate of induced emission}} = \frac{B_{pq}\rho_{\nu_0}n_p}{B_{qp}\rho_{\nu_0}n_q} = \frac{g_q}{g_p} \cdot \frac{n_p}{n_q} \ . \quad (\text{II.158})$$

For Boltzmann equilibrium, this ratio equals $\exp[h\nu_0/kT]$, and again we can conclude that induced emission may be neglected in comparison with absorption, at least under thermal flame conditions. Henceforth in the normal case we shall only consider spontaneous-emission and absorption transitions; in the case of a strong laser field, however, induced emission will be taken into account, too (see Sects 5a-2 and 5c).

The neglect of induced emission is equivalent to substituting Wien's approximate radiation law (see Eq. II.71) for Planck's law. In fact, if we had applied Wien's law in Eq. II.152, the coefficient B_{qp} would have turned out to be zero.

From the principle of detailed balance it also follows that in equilibrium the photons emitted and those absorbed from a continuous radiation field must have the same *spectral distribution*[*], $S_\nu(\nu)$. Here $S_\nu(\nu)d\nu$ is defined as the fraction of photons with frequencies in the interval $(\nu, \nu+d\nu)$. This function $S_\nu(\nu)$ has dimension $[t]$ and is normalized by

$$\int_{\text{line}} S_\nu(\nu)\, d\nu \equiv 1 \ , \quad (\text{II.159})$$

where the integral extends over the whole spectral line. The principle of detailed balance is also valid within any frequency interval $(\nu, \nu+d\nu)$ and thus in equilibrium the rates of emission and absorption processes within this interval must also be equal. This equality of the spectral distributions is the very basis of Kirchhoff's law which will be used in Sect. II.5c-3. It should be noted that this conclusion, reached from detailed balance, is not necessarily true in situations that are out of equilibrium. The spectral distribution function, in contrast to

---

[†] For the first resonance doublet of Cs, with $h\nu_0 = 1.4\,\text{eV}$, and at a flame temperature of 3000 K we have: $\exp[h\nu_0/kT] = 270$, so that the relative error made in this neglect is less than $\tfrac{1}{2}\,\%$.

[*] See also footnote on page 118.

THE INTENSITIES OF RADIATIVE TRANSITIONS   Sect. II.5a

the Einstein coefficients, depends, in general, sensitively on milieu factors and on the prevailing (de-)excitation mechanisms. Deviations from equilibrium might affect this function differently for emission and absorption (see Sects II.5b-2 and VI.4c-3).

The equality of the spectral distribution for emission and absorption does not necessarily imply that the precise frequency of a spontaneous emission act must also equal that of the preceding absorption act. Collisional perturbations and Doppler effects may destroy the coherence between the absorbed and re-emitted radiation as observed by a stationary observer (see the above). The conclusion drawn above from the principle of detailed balance in thermodynamic equilibrium has only a statistical meaning. It links two probability distributions for a large ensemble of emission and absorption acts, but does not connect the individual emission and preceding absorption acts.

It follows from the above that in a continuous radiation field in thermodynamic equilibrium, the absorption rate expressed per second and per unit volume in frequency interval $(\nu, \nu+d\nu)$ is given by: $B_{pq} S_\nu n_p \rho_{\nu_0}^b d\nu$. The product $B_{pq} S_\nu$ may be called the *spectral Einstein coefficient for absorption*. The same rate expression can also be applied for an *arbitrary* continuous radiation field if we replace $\rho_{\nu_0}^b$ by the spectral energy density ρ_{ν_0} (if S_ν does not depend on the radiation field). Thus we have more generally for a *continuous* field:

$$\text{number of absorptions in } d\nu \text{ per second and per unit volume} = B_{pq} S_\nu n_p \rho_{\nu_0} \quad (II.160)$$

In a *monochromatic* radiation field peaked at some frequency ν near ν_0 and with energy density $\rho(\nu)$[†] we have:

$$\text{number of absorptions per second and per unit volume} = B_{pq} S_\nu n_p \rho(\nu). \quad (II.161)$$

The latter expression follows by replacing $\rho_\nu d\nu$ in Eq. II.160 by $\rho(\nu)$. Under usual flame conditions, $S_\nu(\nu)$ is independent of the radiation field (see Sect. II.5b-2).

The *quantummechanical expression for the Einstein radiation coefficients* depends on the multipole character of the optical transition involved. The spectral lines observed are usually connected with electric-dipole transitions. In this case, the Einstein coefficient of spontaneous emission is given by (see, e.g., Herzberg 1950, Heitler 1954, Richter 1968, Unsöld 1968, and Shore and Menzel 1968)

$$A_{qp} = (64\pi^4/3h\lambda_0^3)(S_{qp}/g_q) = 3.14 \times 10^{29} S_{qp}/(g_q \lambda_0^3), \quad (II.162)$$

[†] Note the difference between energy density ρ and spectral energy density ρ_ν (see also Appendix A.5).

Chapt. II   BASIS CONCEPTS AND GENERAL RELATIONSHIPS

where the numerical factor refers to e.s. CGS units [†]. S_{qp} ($= S_{pq}$) is the sum of the squared *dipole matrix elements* (or: *transition moments*) for all possible transitions between the sublevels of the degenerate upper and degenerate lower level. The degeneracy or statistical weight g is given by the quantum number J through: $g = 2J+1$ (see Sect. II.2a). The theoretical quantity S_{qp} is called the *line strength* and has the dimension of the square of an electric dipole moment; it is frequently expressed in atomic units, $e^2 a_0^2$, where e is the elementary charge and a_0 is the radius of the first Bohr orbit in the hydrogen atom ($e^2 a_0^2 = 6.46 \times 10^{-36}$ e.s. CGS units $= 7.2 \times 10^{-59}$ C² m²). The value of S_{qp} can be calculated if the wave functions of the states involved are known with sufficient accuracy. Under usual flame conditions and in the absence of strong fields, we may use here the wave functions of the isolated atom. For strong dipole transitions, as in the case of the first resonance lines of the alkali atoms, A_{qp} is of the order of 10^8 s^{-1}. For certain combinations of quantum numbers, expressed by selection rules (see Sect. II.2c), S_{qp} is zero and electric-dipole transitions are not 'allowed'. In that case, magnetic-dipole transitions characterized by the magnetic-dipole matrix element, may still be possible. They yield much lower transition probabilities, typically of the order of 10^3 s^{-1}. An excited atomic state that cannot be connected to the ground state or any other lower state by an electric-dipole transition is called a *metastable state*. It has a long radiative lifetime (see Sect. II.5a-2).

Making use of Eqs II.153 and 154 for the relation between the Einstein coefficients one finds from the above expression for A_{qp} the theoretical expression for the Einstein coefficient for absorption (in e.s. CGS system of units)

$$B_{pq} = (8\pi^3/3h^2)(S_{pq}/g_p) = 1.88 \times 10^{54} S_{qp}/g_p . \qquad (II.163)$$

It is noted that Eq. II.162, with λ_0 representing the wavelength in vacuo, holds strictly speaking under the condition that the refractive index n equals unity.

In principle, the quantummechanical expressions for A_{qp} and B_{pq} are modified when the radiating atoms are embedded in a continuous medium with *refractive index* $n \neq 1$ or when the density of the atoms becomes so high that anomalous dispersion is noticeable at $\nu \simeq \nu_0$. The latter effect will be discussed in Section 5a-2 in relation to the absorptivity. Nienhuis and Alkemade (1976) have derived expressions for the Einstein coefficients when the radiating atoms are embedded at low density (no anomalous dispersion) in a non-dissipative, weakly dispersive continuous medium. The medium affects through its refractive index not only the wavelength factor in Eq. II.162, but also, through its permittivity, the quantum-

---

[†] In the rationalized Giorgi system of units a factor $4\pi\varepsilon_0$ should be added to the denominator of expressions II.162 and 163.

THE INTENSITIES OF RADIATIVE TRANSITIONS
Sect. II.5a

mechanical operators for the electric field and the magnetic induction. Assuming that the permeability is virtually the same as in vacuo, we find that A_{qp} is proportional to n, whereas B_{pq} as well as B_{qp} vary as n^{-2} for negligible dispersion. The absorptivity $k(\nu)$ integrated over the atomic line profile (see later in Sect. 5a-2) turns out to vary as n^{-1}. In flame gases at 1 atm pressure $(n-1)$ is well below 10^{-3} in the optical frequency range and the above effects can thus usually be totally neglected.

It should be realized that the above expressions hold in the case of an *unpolarized, isotropic* radiation field. For a radiation beam traversing a gas of atoms in *one direction* and *plane-polarized* along the z-axis, we should replace the squared dipole matrix element contained in S_{qp} by three times the squared matrix element of the z-component. However, if the absorbing atoms have no preferential orientation and assuming absence of electric or magnetic fields, this replacement makes no difference. This holds because under these conditions the squared matrix elements of all three components of the dipole moment are the same, while their sum equals the squared matrix element of the total dipole moment. Thus we conclude that under usual flame conditions where the orientation of the atoms is randomized by collisions the absorption-rate equations given hold true regardless of the directional distribution and polarization of the radiation field. It is only the (spectral) energy density that matters. Some reservation should perhaps be made in the case of strong laser beams.

5a-2   *OSCILLATOR STRENGTH, ABSORPTIVITY, AND OTHER PARAMETERS RELATED TO THE EINSTEIN RADIATION COEFFICIENTS*

Radiative lifetime. The reciprocal of A_{qp} has the dimension of time and is called the *radiative* (or *optical*) *lifetime*, $(\tau_r)_{qp}$, of the excited state q with respect to spontaneous radiative transitions to the state p. We thus have

$$(\tau_r)_{qp} = A_{qp}^{-1} \ . \tag{II.164}$$

For strong lines with $A_{qp} \approx 10^8 \, \text{s}^{-1}$, this time is thus of the order of 10^{-8} s. In the absence of other (de-)excitation processes induced by radiation or collisions and if re-absorption of photons (cf. Sect. II.5c-4) is neglected, the decay of the population, $N_q(t)$, of state q after pulse excitation at $t=0$ is given by (cf. Eq. II.105)

$$N_q(t) = N_q(0) \exp[-At] = N_q(0) \exp[-t/\tau_r] \ . \tag{II.165}$$

Under these conditions, τ_r does indeed appear to be the actual lifetime of the excited state, i.e., the average time an atom stays in the excited state. Measurements of this decay have been used to determine A (see, e.g., Karstensen and Schramm 1967; cf. Sect. II.4a-4).

When spontaneous transitions to several lower states are possible from the excited state q, the reciprocal radiative lifetime equals the sum of the separate reciprocal lifetimes, that is

$$(\tau_\mathbf{r})_q^{-1} = \sum_p (\tau_\mathbf{r})_{qp}^{-1} = \sum_p A_{qp} \ . \tag{II.166}$$

Oscillator strength. The *oscillator strength*, f_{pq}, *for absorption*, or the *f-value*, represents the adjustment of the classical theory of radiation absorption to the quantummechanical theory. Here we simply define the f-value by (see, e.g., Richter 1968, Mitchell and Zemansky 1961, and Unsöld 1968)

$$f_{pq} \equiv (hm_e\nu_0/\pi e^2) B_{pq} = 2.50 \times 10^{-25} B_{pq}/\lambda_0 \ , \tag{II.167}$$

where m_e and e are the mass and absolute value of the charge of an electron, respectively, while the numerical factor refers to electrostatic CGS units [†]. Expressing B_{pq} into A_{qp} through Eqs II.153 and 154, we also find (see, e.g., Richter 1968)

$$f_{pq} = (m_e c^3/8\pi^2 e^2)(g_q/g_p) A_{qp}/\nu_0^2 = 1.50\,(g_q/g_p)\,\lambda_0^2 A_{qp} \ , \tag{II.168}$$

$$(g_p f_{pq})/(g_q A_{qp}) = 1.50\,\lambda_0^2 \ , \tag{II.169}$$

where the numerical factor refers to CGS units. The products $g_p f_{pq}$ and $g_q A_{qp}$ are called the *weighted oscillator strength* and *weighted transition probability*, respectively. Combining Eq. II.164 with the latter equation one also has (in CGS units)

$$f_{pq}(\tau_\mathbf{r})_{qp} = 1.50\,(g_q/g_p)\,\lambda_0^2 \ . \tag{II.170}$$

The relation between f_{pq} and the line strength follows from Eqs II.168 and 162

$$f_{pq} = (8\pi^2 m_e/3he^2)\,\nu_0 S_{qp}/g_p \ , \tag{II.171}$$

which has the same form in the e.s. CGS as in the rationalized Giorgi system of units because S_{qp} includes a factor e^2.

The use of the f-value has a great practical advantage over that of the corresponding Einstein coefficient B, because f is a quantity with dimension one and is typically of the order of unity for strong spectral lines. This follows directly from Eq. II.168 by substituting: $A \approx 10^8\,\text{s}^{-1}$ and $\lambda_0 \approx 0.5 \times 10^{-4}$ cm, since

---

[†] The form of Eq. II.167, and of all equations derived therefrom, holds in the (rather historical) e.s. CGS system of units where $\varepsilon_0 = 1$ and no rationalization is applied. In order to formulate them in the rationalized Giorgi system one has to replace e^2 by $e^2/4\pi\varepsilon_0$.

THE INTENSITIES OF RADIATIVE TRANSITIONS   Sect. II.5a

g_q/g_p is usually of the order of unity.

f-values of atomic spectral lines are obtained more often by experimental methods than by theory. A survey of experimental methods (including measurements of emission, absorption, anomalous dispersion and optical lifetime) can be found in Mitchell and Zemansky (1961), Unsöld (1968), and Boumans (1966). f-values or gf-values have been tabulated by, e.g., Corliss and Bozman (1962), Allen (1963), Wiese, Smith and Glennon (1966), Corliss (1967), Wiese, Smith and Miles (1969), and L'vov (1970) (see also Miles and Wiese 1970 for a bibliography). A collection of gf-values of some atomic metal lines that may be relevant in flame spectroscopy is presented in Appendix A.4. The (weighted) Einstein coefficients of spontaneous emission are easily obtained from the (weighted) f-values through Eq. II.169. The accuracy with which absolute f- or A values are known should not be over-estimated. Errors up to a factor of 2 might be expected in the N.B.S. tables published by Corliss and Bozman (1962) and revised by Corliss (1967) for transitions with $\lambda < 2450$ Å (see Boumans 1966). The f-values for the first resonance lines of the alkali atoms have been thoroughly investigated. The expected accuracies are 3% for Na and Li, 8% for K (see Wiese, Smith and Glennon 1966, and Wiese, Smith and Miles 1969) and probably somewhat worse for Rb and Cs. The accuracy of the absolute f-value for the Cu 3247 Å resonance line is about 2% (see Hannaford and McDonald 1978). Relative f-values are usually much better known, since they are less affected by systematic errors in the calibration of atomic density, radiation intensity, temperature of the source, etc., involved in their measurement. In Sects II.5c-2 and 5c-3 we shall briefly mention methods of determining f- or A values from spectrometric measurements in absorption or emission that are applicable to metal vapours in flames (see also Sect. V.1). Determination of relative or absolute f- or A values by flame spectrometric methods are described in, e.g., van der Held (1932), Heïerman (1937), James and Sugden (1955), Hinnov (1957), Hinnov and Kohn (1957), Mavrodineanu and Boiteux (1965), Ostroumenko and Rossikhin (1965), L'vov (1970), Brown and Parsons (1978), Reif et al. (1978), and in 'Optical Transition Probabilities' (1962, 1963). Some of these determinations have only historical interest.

Absorptivity. A related quantity of direct experimental significance is the absorptivity $k(\nu)$ (often also called absorption coefficient; see, however, Appendix A.5) which describes the absorption of a monochromatic radiation beam in an infinitesimal layer as a function of frequency. Consider a radiation beam travelling along the x-axis through a gas of atoms and having a sharply peaked spectral distribution at some frequency ν (near ν_0). The beam flux $\Phi(x,\nu)$ (expressed, e.g., in erg per second) is a function of x, as the beam is attenuated by absorption in the gas. We shall disregard here the contribution from spontaneous and induced emission to the beam flux (cf. later in this section). The relative attenuation of the flux per unit of length in the beam direction is now defined as the *absorptivity*, $k(\nu)$,

$$k(\nu) \equiv -\frac{d\Phi(x,\nu)}{dx} / \Phi(x,\nu) \ . \tag{II.172}$$

Note that $k(\nu)$ has dimension $[l]^{-1}$ and is expressed, e.g., in cm^{-1}.

Strictly speaking, the expression: *absorption of radiation in a medium* means the conversion of radiant energy into heat. In a gas, this conversion involves the excitation of an atom by photon absorption and the subsequent loss of excitation energy by a collision-of-the-second-kind. As the collision partner undergoes further inelastic collisions at random, this energy is ultimately converted into heat.

The expression: *absorption of a photon by an atom* means only the radiative transition of an atom to a higher state of excitation. It is irrelevant whether the excited atom is subsequently transferred to another state by a radiative or by a collisional process. Conversion of photon energy into heat is thus not necessarily implied herewith.

Extinction of a radiation beam traversing a medium can be caused by (real) absorption or by scattering (in particular, fluorescence) of radiation. The scattered or fluorescent radiation is usually distributed over all space directions and is thus 'lost' from the radiation beam, at least if the beam subtends an infinitesimal solid angle. When nonresonance fluorescence occurs, a 'loss' in beam intensity at the original frequency always occurs, irrespective of the beam divergence.

The *absorptivity* and related absorption quantities are defined here with regard to the extinction of a parallel beam propagating within an infinitesimal solid angle, as a result of (real) absorption and fluorescence. Rayleigh scattering by the flame gas molecules can safely be neglected. The contribution from induced emission, which is always in the beam direction, will be dealt with separately on page 130 ff.

The relation between the (local) values of $k(\nu)$ and of B_{pq} or f_{pq} in the absorbing gas may be derived by considering an infinitesimally thin volume element, $dV = O\,dx$, at x, with O = surface area perpendicular to the x-axis and dx = thickness (see Fig. II.9). Let $\Phi(x,\nu)$ be the monochromatic radiation flux incident on O, then the radiant energy absorbed in dV in time interval dt is determined by

$$\text{absorbed energy in } (dV, dt) = k(\nu)\,\Phi(x,\nu)\,dx\,dt\ . \tag{II.173}$$

On the other hand, if we assume dV so small to make the radiant flux and density of atoms, n_p, uniform inside dV, it follows from Eq. II.161 also that

THE INTENSITIES OF RADIATIVE TRANSITIONS   Sect. II.5a

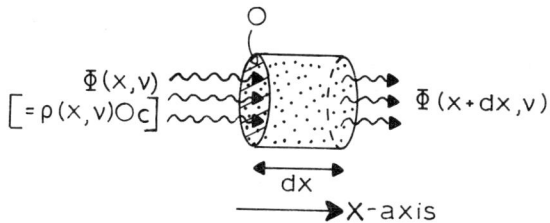

Fig. II.9 Monochromatic radiation beam propagating along the x-axis through an absorbing medium. Absorbed power in volume element $dV = O dx$, containing $n_p dV$ atoms in the absorbing state, is calculated to find the relationship between absorptivity $k(\nu)$ and Einstein coefficient B_{pq} (see text).

absorbed energy in $(dV, dt) = B_{pq} S_\nu n_p \rho(x,\nu) h\nu_0 O dx\, dt$.   (II.174)

The volume density, $\rho(x,\nu)$, of the monochromatic radiant energy at place x is related to $\Phi(x,\nu)$ by

$$\rho(x,\nu) = \Phi(x,\nu)/Oc .$$ (II.175)

Combining Eqs II.173, 174 and 175 we finally get

$$k(\nu) = B_{pq} S_\nu h \nu_0 n_p/c .$$ (II.176)

Note that $k(\nu)$ does not explicitly depend on Φ and that it varies with frequency through the function $S_\nu(\nu)$. If neither S_ν nor n_p depend on the radiation density, $k(\nu)$ is independent of Φ. Multiple photon absorption (see Section 5a-1), radiation broadening (see Sect. II.5b-2), and induced emission (see page 130 ff. in this subsection) have been ignored here. It is also seen that $k(\nu)$ is strictly proportional to the density of atoms in the lower state of the optical transition. When this density varies with position, $k(\nu)$ will vary with position too. In general, $k(\nu)$ is a *local* property of the absorbing medium, and independent of the beam direction and beam polarization for unpolarized atoms.

It also appears from Eq. II.176 that the spectral variation of the absorptivity, i.e. the spectral profile of the absorption line for an optically thin layer, is thus identical to the spectral distribution of the photons emitted or absorbed by the individual atoms. A layer of absorbing gas is *optically thin* (at frequency ν), if the product of $k(\nu)$ and the thickness, l, of the layer in the beam direction is small compared to unity. The relative attenuation of the radiation beam passing this layer is then small too and practically equal to $k(\nu)l$. In Sect. II.5c-2 we shall see that the latter conclusions do not hold for an optically thick absorbing layer. The quantity $k(\nu)l$ is called the *optical thickness* at the considered frequency and has dimension one. When the absorptivity depends on the

x-coordinate along the line of observation, the optical thickness is given by:

$$\int_0^l k(x,\nu)\,dx\ .$$

The relation between $k(\nu)$ and f_{pq} or A_{qp} is found by combining Eqs II.176, 167, 153 and 154. One finds (in CGS units)

$$k(\nu) = (\pi e^2/m_e c)\, n_p S_\nu f_{pq} = 2.65 \times 10^{-2} n_p S_\nu f_{pq}\ , \tag{II.177}$$

$$k(\nu) = (c^2/8\pi\nu_0^2)(g_q/g_p)\, n_p S_\nu A_{qp}. \tag{II.178}$$

Equation II.177 also follows directly from classical dispersion theory, if we substitute there $n_p f_{pq}$ for the density of classically absorbing atoms (see Garbuny 1965 and Unsöld 1968).

After integration over the line profile, one finds from Eqs II.159 and 177 (in CGS units)

$$\int_{\text{line}} k(\nu)\,d\nu = (\pi e^2/m_e c)\, n_p f_{pq} = 2.65 \times 10^{-2}\, n_p f_{pq}\ . \tag{II.179}$$

The dimension of the latter integral is $[l]^{-1}[t]^{-1}$. The important conclusion from Eq. II.179 is that the relationship between the *integrated absorptivity* and the f-value is <u>independent</u> of the line shape. The integrated absorptivity determines the radiant power absorbed per unit path length from a *continuous* radiation beam with a constant spectral flux, Φ_{ν_0}, at $\nu \approx \nu_0$. In contrast to the integrated value, the peak value, $k_{\max}$, does depend on the line shape. Defining the *effective spectral width*, $\Delta\nu_{\text{eff}}$, of the normalized $S_\nu(\nu)$ profile by

$$\Delta\nu_{\text{eff}} \equiv \int_{\text{line}} S_\nu(\nu)\,d\nu/S_\nu(\nu_0) = 1/S_\nu(\nu_0)\ , \tag{II.180}$$

where $S_\nu(\nu_0)$ is the maximum value of $S_\nu(\nu)$, we immediately have for $k_{\max}\,[\equiv k(\nu_0)]$

$$k_{\max} = \int_{\text{line}} k(\nu)\,d\nu/\Delta\nu_{\text{eff}} = 2.65\times 10^{-2}\, n_p f_{pq}/\Delta\nu_{\text{eff}}\ , \tag{II.181}$$

$$k(\nu)/k_{\max} = S_\nu(\nu)\,\Delta\nu_{\text{eff}}\ . \tag{II.182}$$

It appears that the broader the line, the smaller $k_{\max}$ for given $n_p f_{pq}$ value.

When wavelengths are used instead of frequencies, we get, replacing S_ν by S_λ (see also Eqs II.207 and 209), for the integrated absorptivity (in CGS units)

$$\int_{\text{line}} k(\lambda)\,d\lambda = (\pi e^2/m_e c^2)\,\lambda_0^2 n_p f_{pq} = 8.83 \times 10^{-13} \lambda_0^2\, n_p f_{pq}\ . \tag{II.183}$$

The latter integral has dimension one, in contrast to the corresponding integral over frequency. Similarly we find for $k_{\max}\ [\equiv k(\lambda_0)]$ the numerical expression (in CGS units)

THE INTENSITIES OF RADIATIVE TRANSITIONS   Sect. II.5a

$$k_{max} = 8.83 \times 10^{-13} \lambda_0^2 n_p f_{pq}/\Delta\lambda_{eff} \; . \qquad (II.184)$$

Referring, for example, to the Na resonance line at $\lambda_0 = 5.9 \times 10^{-5}$ cm with $f = 0.65$ and assuming only Doppler broadening at flame temperatures ($\Delta\lambda_{eff} \approx 5 \times 10^{-10}$ cm; see Sect. II.5b-2) and a moderately high ground-state density $n_0 = 10^{11}$ cm^{-3}, one estimates from Eq. II.183: $k_{max} \approx 0.4$ cm^{-1}.

The proportionality between $k(\nu)$ and n_p leads naturally to the definition of the *atomic absorptivity* $\kappa(\nu)$ through

$$\kappa(\nu) \equiv k(\nu)/n_p \; . \qquad (II.185)$$

This proportionality constant having the dimension of $[l]^2$ may also be interpreted as the *optical cross section for absorption*, when we visualize the photon flux at frequency ν as a stream of particles moving with the speed of light and captured by atomic targets with cross-sectional area equal to $\kappa(\nu)$ (cf. the general definition of effective cross section in Sect. II.4b-3). From Eqs II.177 and 185 we have for $\kappa(\nu)$ and $\kappa_{max} [\equiv \kappa(\nu_0)]$ (in CGS units; see also footnote on page 124)

$$\kappa(\nu) = (\pi e^2/m_e c) S_\nu(\nu) f_{pq} = 2.65 \times 10^{-2} S_\nu(\nu) f_{pq} \; , \qquad (II.186)$$

$$\kappa_{max} = (\pi e^2/m_e c) f_{pq}/\Delta\nu_{eff} = 2.65 \times 10^{-2} f_{pq}/\Delta\nu_{eff}$$

$$= 8.83 \times 10^{-13} \lambda_0^2 f_{pq}/\Delta\lambda_{eff} \; . \qquad (II.187)$$

For a resonance line with only natural broadening $(\Delta\nu_N)_{eff} = A_{qp}/4$ (see Eqs II.214 and 216), one finds through Eq. II.168 that κ_{max} is of the order $(10^{-2} - 10^{-1}) \times \lambda_0^2$ if $f \approx 1$. In the above example of a Doppler-broadened Na resonance line, κ_{max} is about $10^{-3} \times \lambda_0^2 = 4 \times 10^{-12}$ cm^2 and thus much larger than the gas-kinetic cross section.

The above equations for the absorptivity $k(\nu)$ hold true if the refractive index n equals unity. At high atomic density *anomalous dispersion* (see Born and Wolf 1959 and Thorne 1974) may become perceptible around the line centre. Classical dispersion theory tells us that $k(\nu)$ in Eq. II.177 should be replaced by $n(\nu)k(\nu)$ (see Garbuny 1965 and Unsöld 1968). The absolute deviation $|n-1|$ due to anomalous dispersion attains a maximum on either side of line centre just where the absorptivity has dropped to half its peak value. This maximum deviation is given by (in the e.s. CGS units system)

$$|n-1|_{max} \simeq n_p f_{pq} e^2/4\pi m_e \nu_0 \delta\nu_L \qquad (II.188)$$

for an absorption line with Lorentzian spectral profile and half-intensity width $\delta\nu_L$ (cf. Sect. II.5b-2). Assuming that the wavelength of the line centre is less than 10^{-4} cm, that the oscillator strength is at most unity, and $\delta\nu_L/\nu_0 \approx 5 \times 10^{-6}$ (which is a reasonable estimate for atomic lines in flames at 1 atm), one

Chapt. II BASIC CONCEPTS AND GENERAL RELATIONSHIPS

calculates that this maximum deviation is less than 10^{-2} for $n_p < 2 \times 10^{14}$ cm^{-3} (corresponding to a partial pressure less than 10^{-4} atm) (see also van Trigt, Hollander and Alkemade 1965). The latter condition for n_p is usually fulfilled in flames (see Sect. III.2h). Although classical theory may not describe the anomalous dispersion exactly, we may safely neglect its effect on the atomic absorption of metal vapours in flames.

Net absorptivity (absorption and induced emission combined). When under the action of a strong laser field the population of the upper level q becomes comparable with that of the lower level p, *induced emission* has to be taken into account. As a result, *saturation* of the optical transition sets in.[†] Since the induced radiation is emitted in the same direction and at the same frequency as the inducing radiation beam, the loss in beam intensity due to absorption is (partly) compensated by the gain due to induced emission. For the *net absorptivity* $k'(\nu)$ which allows for both absorption and induced emission, we have from Eqs II.176 and 148

$$k'(\nu) = (S_\nu h\nu_0/c)(B_{pq} n_p - B_{qp} n_q) \ . \tag{II.189}$$

Using Eq. II.154 for the relationship between the Einstein coefficients B_{pq} and B_{qp} we immediately get from the latter equation

$$k'(\nu) = (B_{pq} S_\nu h\nu_0/c)\{n_p - (g_p/g_q) n_q\} \ , \tag{II.190}$$

$$k'(\nu) = (n_p B_{pq} S_\nu h\nu_0/c)\{1 - (g_p/g_q)(n_q/n_p)\} \ . \tag{II.191}$$

Quantitative expressions for the excited-state population as a function of field intensity will be derived in Sect. VI.2c-5 by kinetic considerations. It will be seen that $(g_p/g_q)(n_q/n_p)$ tends asymptotically towards unity when the field intensity at $\nu \simeq \nu_0$ goes to infinity (see Eq. VI.34b). The net absorptivity then drops to zero and the atomic vapour becomes *bleached*, that is, transparent for radiation at $\nu \simeq \nu_0$. The optical transition is then said to be *fully saturated*. Saturation sets in when the de-excitation rate due to induced emission becomes comparable with the collisional plus spontaneous-emission de-excitation rate. For given A_{qp} the collisional de-excitation rate is related to the fluorescence efficiency Y; the onset of saturation thus depends on Y (see Eq. VI.38a). Rate constants for de-excitation by collision plus spontaneous emission are typically

---

[†] Piepmeier (1972, 1972a) has been the first one to discuss the possibility of inducing saturation in an atomic vapour in flames by a monochromatic laser beam. Omenetto, Benetti, Hart, Winefordner and Alkemade (1973) have extended the theoretical analysis to the case of a broad-band laser beam, which effectively acts as a quasi-continuum source.

of the order of 10^8-10^9 s^{-1} for atomic lines in flames at 1 atm pressure. For a 2-level atom with total atomic density $n_a \equiv n_p + n_q$ we have in the limit of full saturation: $n_p \to n_a/(1+g_q/g_p)$ and $n_q \to n_a(g_q/g_p)/(1+g_q/g_p)$. For $g_p = g_q$, one half of the atoms is then in the lower state and the other half in the upper state.

A negative $k'(\nu)$ value, i.e. *net amplification* of an incident radiation beam, is obtained when n_q exceeds $n_p(g_q/g_p)$. This can be realized by selective population of the upper level, for example an excited Ne level, in an electric discharge through a suitable He-Ne mixture. This *population inversion* is the basis of laser action ('laser' = light amplification by stimulated emission of radiation).

The possibility of obtaining inversion in excited radicals produced by combustion reactions in low-pressure flames was investigated by Bleekrode and Nieuwpoort (1965a).

The use of the Einstein coefficients B_{pq} and B_{qp} makes the above equations valid only on condition that the atoms are not noticeably polarized by the laser beam (see Sect. II.5a-1). In other words, collisional mixing of the M_J sublevels should be sufficiently strong to preclude selective (de-)population of any sublevel by the laser beam. This condition may be fulfilled in flames at 1 atm pressure (cf. Sect. VI.1). If it is not fulfilled, the equation for $k'(\nu)$ becomes more complicated and its value will depend on the polarization and on the direction of observation with respect to the laser beam direction.

Another condition as to the validity of Eqs II.189, 190 and 191 is that the saturated absorption line is homogeneously broadened. *Homogeneous broadening* exists when for all atoms in the lower state the probability for absorbing a photon within a specified time interval Δt_0 is the same function of frequency ν (see Hercher 1967). In other words, all these atoms should contribute equally to the absorption of radiation at any frequency. When this is not the case, we speak of an *inhomogeneously* broadened line. Which case applies still depends on Δt_0. Suppose that the absorption line is mainly broadened by the Doppler effect (see Sect. II.5b-2) and that Δt_0 is so short that the velocity of an atom remains unchanged during Δt_0. Then only those atoms that happen to have a velocity component v_x in the beam direction satisfying the Doppler relation: $v_x = c(\nu - \nu_0)/\nu_0$, can contribute to the absorption and induced emission at frequency ν. The fraction of these atoms is determined by the velocity-distribution and by the width of the spectral line that would be observed if all atoms were stationary. For this Δt_0 value the absorption line is to be considered as inhomogeneously broadened. If, however, Δt_0 is much larger than the relaxation time for the velocity-distribution, i.e. the average time needed for an atom to change appreciably its velocity, the probability for absorbing a photon with frequency ν during this Δt_0 would be the same

Chapt. II   BASIC CONCEPTS AND GENERAL RELATIONSHIPS

for each atom in the vapour. This probability is simply given by the Doppler profile of the spectral line. We have then the case of homogeneous broadening.

A similar distinction can be made when collision broadening (see Sect. II.5b-2) dominates in the wings of the line profile. The existence of homogeneous or inhomogeneous broadening then depends on whether Δt_0 is long or short compared to the duration of the disturbance of the energy levels by neighbouring flame molecules, respectively (see also Sharp and Goldwasser 1976). Natural line broadening (see Sect. II.5b-2) is always homogeneous.

The distinction between homogeneous and inhomogeneous broadening has consequences for the spectral profile of the net absorptivity when saturation is achieved by means of a strong, *monochromatic* laser beam with frequency ν_L. Consider again the case when Doppler broadening is dominant. Let κ_0 be the optical cross section for absorption of a photon by an atom whose velocity component v_x^0 satisfies the above Doppler relation at $\nu = \nu_L$. Let E be the irradiance of the laser beam, expressed in number of quanta per second and per unit area (see Appendix A.5). Then $(\kappa_0 E)^{-1}$ is the average time such an atom needs to become excited by photon absorption or de-excited by induced emission. For simplicity, we here assume $g_p = g_q$. When the latter time is short compared with the velocity-relaxation time, inhomogeneous broadening occurs. For the class of atoms whose velocity components v_x satisfy the Doppler relation, the population difference, $dn_p - dn_q$, is more saturated than for the other atoms which do not have the suitable velocity to interact directly with the laser beam. A dip or *'Bennett hole'* is said to be *'burnt'* in the curve representing the distribution of $(dn_p - dn_q)$ over v_x (see Fig. II.10b). For a Doppler broadened line the $k'(\nu)$ profile, observed in the laser beam direction, conforms to this distribution curve and thus shows *spectral hole-burning* at $\nu = \nu_L$ (see Fig. II.10c; see also Hercher 1967 and Shimoda 1976a).

The latter dip, however, does not show up when $k'(\nu)$ is measured in a direction perpendicular to the laser beam, as the velocity components in two mutually perpendicular directions are statistically uncorrelated.

Inhomogeneous broadening also exists when the duration of the monochromatic laser pulse is short compared to the velocity-relaxation time.

In the inhomogeneous-broadening case saturation is thus not spread uniformly over the whole line profile and Eqs II.189, 190 and 191 lose their meaning. However, in flames at 1 atm pressure one rather expects homogeneous broadening to occur because (i) the velocity-relaxation time is sufficiently short (typically 10^{-9} s; see Daily 1979), and (ii) the line width due to collision broadening is comparable to the Doppler width.

With the use of *broad-band* lasers all velocity-classes experience the same saturation effect. We then have homogeneous broadening, and Eqs II.189, 190 and

THE INTENSITIES OF RADIATIVE TRANSITIONS Sect. II.5a

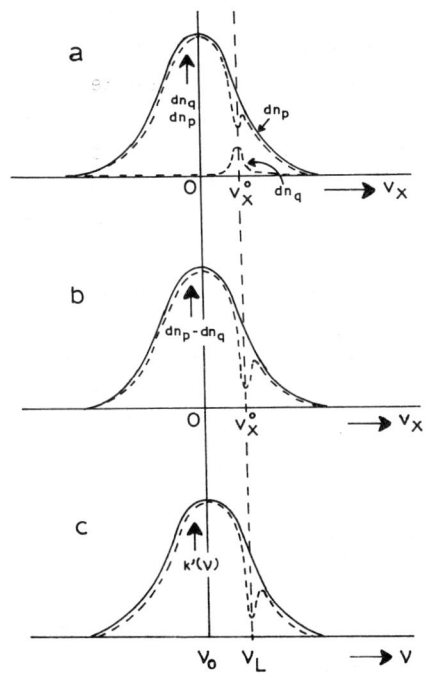

Fig. II.10 Effect of hole-burning with monochromatic laser beam (frequency ν_L) passing through an absorbing medium in the case of inhomogeneous line broadening (schematic). Figure a represents the differential populations, dn_p and dn_q, of the lower and upper level of the atomic absorption line as a function of velocity component, v_x, in the beam direction. Here, and in the other figures, the solid and dotted curves refer to absence and presence of saturating laser radiation, respectively. Figure b shows the difference $(dn_p - dn_q)$ as a function of v_x, which is directly related to the net absorptivity $k'(\nu)$ — measured in the beam direction — as a function of frequency (see Figure c) through the Doppler relation. For simplicity, g_p/g_q is put equal to 1.

191 are applicable.

For an arbitrary spectral profile of the laser beam, expressions for saturation have been presented by Greenstein and Bates (1975); Sharp and Goldwasser (1976) have considered the case of a Gaussian spectral profile.

5a-3   *RADIATIVE RATE CONSTANTS FOR ATOMIC MULTIPLETS AND MOLECULAR BANDS*

Atomic multiplets. The spectral components of an atomic multiplet are often so close to each other that they are not resolved by the spectrometer or that their spectral profiles even overlap intrinsically. Under certain conditions and in certain respects, it is then appropriate to define f- or A values for the multiplet as a whole. We shall assume that the multiplet levels of the upper as well as of the lower term are so close to each other that they are occupied in proportion to their statistical weights $g = 2J + 1$ (see Sects II.2a-2 and 3b-2). That is to say, we neglect the effect of the small energy difference between these levels on the thermal population distribution. These assumptions are reasonably accurate for most alkali doublets in flames. In the following, we also neglect the small differences in wavelength between the multiplet components.

By assigning a statistical weight $\sum_p (2J_p + 1)$ to the lower term of the

Chapt. II   BASIC CONCEPTS AND GENERAL RELATIONSHIPS

multiplet transition (p refers here to the different levels of this term) and correspondingly a weight $\sum_q (2J_q+1)$ to the higher term, we can define f- and A values for the multiplet as a whole by

$$f^{\text{mult}} \equiv \sum_{p,q} (2J_p+1) f_{pq} \bigg/ \sum_p (2J_p+1) \, , \qquad (II.192)$$

$$A^{\text{mult}} \equiv \sum_{p,q} (2J_q+1) A_{qp} \bigg/ \sum_q (2J_q+1) \, . \qquad (II.193)$$

The *integrated* absorptivities, which are directly proportional to the oscillator strengths according to Eq. II.179, can be summed in the same way as in Eq. II.192. Thus the power absorbed per unit length from a continuous radiation beam by the multiplet as a whole is the same as for a single line with $f = f^{\text{mult}}$ and with lower-state density $= \sum_p n_p$. It is not essential whether the multiplet components actually overlap or not. This is in general not true for the *peak* absorptivities; the latter may be summed only if the multiplet components fully overlap (i.e., if their intrinsic line widths are large compared to the fine-structure splitting).[†]

Similarly, the spontaneous emission of the multiplet as a whole is the same as that of a single line with $A = A^{\text{mult}}$ and with upper-state density $= \sum_q n_q$.

For the components of the first resonance doublets of the alkali atoms, the (S_{qp}/g_q) values occurring in Eq. II.162 are, virtually, the same and so are their A values, apart from slight differences because of the λ_0^3 factor for the heavier alkali atoms. It then follows from Eq. II.168 that the corresponding f-values are in the ratio of the statistical weights of the doublet levels, i.e., $4 : 2 = 2$. For an atom with one valence electron, the correspondence with classical radiation theory leads to the *sum rule*: $\sum_q f_{0q} = 1$ (for transitions from the ground state), where the summation extends over all excitation levels q. Since the f-values for transitions to excitation levels above the lowest ones are relatively small (see also Appendix A.4), one expects the f-values of the components of the first resonance doublet to be $\frac{2}{3}$ and $\frac{1}{3}$, respectively. This expectation is reasonably fulfilled.

Molecular bands. The transition probabilities for a spectral line of a molecular band can be expressed in the line strength according to the same Eqs II.162 and 163 as were presented for an atomic line. The line strength is defined in the same way as in the atomic case, if we replace the atomic eigenfunctions

---

[†] A distinction must be made between *apparent* overlap of neighbouring lines or (h.)f.s. components in the *observed* spectrum due to insufficient spectral resolution of the monochromator etc., and *intrinsic* overlap in the true spectrum of the source. The latter overlap would also be observed with a monochromator with infinite spectral resolution.

THE INTENSITIES OF RADIATIVE TRANSITIONS   Sect. II.5a

occurring in the matrix elements by the total molecular eigenfunctions and also
take into account the contribution of the nuclear charges to the total dipole
moment.

We shall consider here a *di-atomic molecule* which undergoes a combined
electronic, vibrational and rotational radiative transition (see also Sect. II.2d).[†]
The electronic upper and lower states are characterized by q, Σ' and p, Σ'', whereas the vibrational and rotational quantum numbers of the upper level are denoted by
v' and J', respectively, and by v'' and J'' in the lower level (see Sect. II.2b).[*]
The nuclear spin is disregarded; the optical transitions are summed over the h.f.s.
components. From Eqs II.162 and 163 we immediately have for the transition probabilities of a *rotational line* characterized by the transition $p,\Sigma'' \leftrightarrow q,\Sigma'$, $v'' \leftrightarrow v'$
and $J'' \leftrightarrow J'$, when we equalize the degeneracy of the upper level to the product of
the electronic degeneracy $g_{\Lambda'}$ and the rotational degeneracy $(2J'+1)$, and that of
the lower level to $g_{\Lambda''}(2J''+1)$ (see Sect. II.2b-2), in the e.s. CGS system of
units [‡]

$$A_{qp,\,\Sigma'\Sigma'',\,v'v'',\,J'J''} = (64\pi^4/3\,h\lambda_0^3)\,S_{qp,\Sigma'\Sigma'',v'v'',\,J'J''}\,/\,g_{\Lambda'}(2J'+1), \quad (II.194)$$

$$B_{pq,\,\Sigma''\Sigma',\,v''v',\,J''J'} = (8\pi^3/3h^2)\,S_{pq,\,\Sigma''\Sigma',\,v''v',\,J''J'}\,/\,g_{\Lambda''}(2J''+1). \quad (II.195)$$

We find from Eq. II.171 for the oscillator strength of the rotational line

$$f_{pq,\,\Sigma''\Sigma',\,v''v',\,J''J'} = (8\pi^2 m_e/3\,he^2)\nu_0\,S_{pq,\,\Sigma''\Sigma',\,v''v',\,J''J'}\,/\,g_{\Lambda''}(2J''+1). \quad (II.196)$$

The two S factors occurring herein are equal and are called the *line strength* (see
for a general definition Section 5a-1). We note that for heteronuclear di-atomics
the line strength of a line consisting of two unresolved Λ-doubling components is
twice the strength of either component, whereas the corresponding transition probabilities are equal.

[†] The theory of molecular transition probabilities is rather intricate, even for
di-atomic molecules. Here we present only a condensed survey. For practical
application and a more detailed understanding of the following formulas the
reader is referred to, e.g., Penner (1959), Nicholls and Stewart (1962), Bingel
(1967), Tatum (1967), Kovács (1969), Hougen (1970), Whiting and Nicholls (1974),
Kuznetsova et al. (1974), and Schadee (1978). For poly-atomic molecules and radicals the books of Herzberg (1954, 1971) are especially recommended.

[*] Note that, in contrast to Sects II.2b and 2d, we have here taken out Σ' and Σ''
from the labels q and p for the electronic states. In Hund's coupling case b
(see Sect. II.2b) Σ is not a quantum number but is only used as a number labelling the $(2S+1)$ levels of the multiplet.

[‡] In the rationalized Giorgi system of units a factor $4\pi\varepsilon_0$ should be added to
the denominator of the expressions II.194 and 195 whereas the expression for f
remains formally unchanged (cf. Eq. II.171).

Chapt. II   BASIC CONCEPTS AND GENERAL RELATIONSHIPS

In the Born-Oppenheimer approximation, the total molecular eigenfunction is written as a product of electronic and vibrational plus rotational eigenfunctions (see Sect. II.2b). To this approximation we can resolve the line strength as a product of the *squared electronic dipole matrix element* S^{el}_{qp}, the *Franck-Condon factor* $F_{v'v''}$ (see Sect. II.2d), and the *Hönl-London factor* H which includes the rotational quantum numbers.[†] The last two factors have dimension one. It appears, however, upon closer examination that an electronic matrix element that depends only on the electronic states involved cannot be defined because of the variation of this matrix element with the internuclear distance. Through this variation the electronic and vibrational transition probabilities are coupled. The introduction of a separate electronic matrix element is therefore of no practical use (see Schadee 1967 and Thorne 1974). Following Whiting and Nicholls (1974) and Schadee (1978) we resolve the line strength as a product of only <u>two</u> factors including the Hönl-London factor

$$S_{qp,\,\Sigma'\Sigma'',\,v'v'',\,J'J''} = \left|R^{v'v''}_{qp}\right|^2 H^{\Lambda'\Sigma'J'}_{\Lambda''\Sigma''J''}, \qquad (II.197)$$

where $\left|R^{v'v''}_{qp}\right|$ is the root-mean-square of the transition moments of the electronic subtransitions due to both Λ-doubling and spin multiplicity. One can now define the *band strength*

$$S^{v'v''}_{qp} \equiv h_\Lambda (2S+1)\left|R^{v'v''}_{qp}\right|^2, \qquad (II.198)$$

being the product of $\left|R^{v'v''}_{qp}\right|^2$ and the number of electronic subtransitions: $h_\Lambda(2S+1)$ with $h_\Lambda \equiv 1$ for transitions between Σ states and $\equiv 2$ for all other transitions. We have assumed that the upper and lower electronic states have the same multiplicity $(2S+1)$, according to selection rule no. 5 in Section II.2d. The band strength determines the *band transition probability*, which characterizes[*] the probability of spontaneous transition from a <u>given</u> (q,v',J') level to <u>all</u> accessible levels with given (p,v''), through

$$A^{v'v''}_{qp} \equiv \frac{\sum_{\Sigma'\Sigma''}\sum_{J'J''} A_{qp,\Sigma'\Sigma'',v'v'',J'J''}}{2S+1} = (64\pi^4/3h\lambda_0^3)\,S^{v'v''}_{qp}/g_{\Lambda'}(2S+1).$$

(II.199)

The second part of this equation follows from Eqs II.194, 197 and 198, by making use of the *sum rule* or *normalization* of the Hönl-London factors (see e.g. Whiting

---

[†] Other symbols used for the Franck-Condon and Hönl-London factors are: $q_{v'v''}$ and $S_{J'J''}$ (or $S_{\Sigma J}$), respectively (for detailed expressions for the Hönl-London factor see also Schadee 1971).

[*] The precise, operational significance of this formal quantity will appear from the expression for the band radiance given by Eq. II.287 in Sect. II.5c-3.

THE INTENSITIES OF RADIATIVE TRANSITIONS   Sect. II.5a

and Nicholls 1974)
$$\sum_{\Sigma'\Sigma''}\sum_{J''} H^{\Lambda'\Sigma'J'}_{\Lambda''\Sigma''J''} = h_\Lambda \, (2S+1)(2J'+1) \,, \qquad (II.200)$$

with h_Λ as defined above. For convenience the label Λ denotes here and in the following the *absolute* value of Λ as defined in Sect. II.2b. The *total transition probability*, $A^{v'}_{qp}$, which is a characteristic for the spontaneous transition from a given (q,v',J') level to <u>all</u> accessible (v'',J'') levels in the lower electronic state p, is given by

$$A^{v'}_{qp} \equiv \sum_{v''} A^{v'v''}_{qp} \,. \qquad (II.201)$$

We can also define a *band oscillator strength* by a similar summation of the f-values of the absorption transitions from a <u>given</u> (p,v'',J'') level to <u>all</u> accessible levels with given (q,v') and find

$$f^{v''v'}_{pq} \equiv (8\pi^2 m_e /3he^2) \, \nu_0 \, S^{v'v''}_{qp}/g_{\Lambda''} \, (2S+1) \,. \qquad (II.202)$$

One may formally take out the Franck-Condon factor $F_{v'v''}$ in the expression for the band strength and hence write

$$A^{v'v''}_{qp} \equiv F_{v'v''} \, A^{el}_{qp} \,, \qquad (II.203)$$

$$f^{v''v'}_{pq} \equiv F_{v'v''} \, f^{el}_{pq} \,. \qquad (II.204)$$

The quantities labelled by 'el' depend, however, markedly on the vibrational transition (see Schadee 1967). They are thus of little practical use for the description of the strength of a band system as a whole (see also Eqs II.290 and 291 in Sect. II.5c-3). Only for a given sequence of bands (see Sect. II.2d) are these quantities reasonably constant and is the variation of the transition probability, etc., with v' (or v'') mainly described by the Franck-Condon factor. For the latter factor separately we have the *sum rule*

$$\sum_{v''} F_{v'v''} = 1 \,. \qquad (II.205)$$

The *line* oscillator strength, given by Eq. II.196, is directly related to the *absorptivity* integrated over the *line*, for unit density of molecules in the absorbing level, just as in the atomic case (cf. Eq. II.179). The absorptivity integrated over an absorption *band* $(p,v'' \to q,v')$ for unity density of molecules in the (p,v'') level is, in a similar way, related to the *band* oscillator strength given by Eq. II.202. This holds at least if we can neglect the variation of ν_0 over the absorption band. We note from Eq. II.202 that the band oscillator strength is proportional to ν_0 and <u>independent</u> of J''. We refer for an

Chapt. II   BASIC CONCEPTS AND GENERAL RELATIONSHIPS

enlightening and critical discussion on the operational significance and manipulation of these concepts to Tatum (1967) and Thorne (1974). Expressions for the integrated absorptivity of a rotational line or a vibrational band of a di-atomic molecule are found in Tatum (1967) and Kuznetsova et al. (1974). We note from Eqs II.194 and 196 that the emission intensity (including the photon energy $h\nu_0$) of a given rotational line varies $\propto \nu_0^4$, whereas the integrated absorptivity of the line varies $\propto \nu_0$, all other factors being kept the same.

Except for the (far) u.v. bands, $f_{pq}^{v''v'}$ is usually small for optical transitions, typically of the order of 10^{-3} or less (see for an explanation Herzberg 1950, p.385), whereas f is of the order of unity for the strongest atomic resonance lines. This distinction in f-values makes the *peak* absorptivities of the rotational lines very small in comparison with those of an atomic line, at equal populations of the absorbing level. At equal *total* molecular and atomic densities, the peak absorptivity of nonoverlapping rotational lines is further reduced because the molecules are distributed over many more levels than the atoms. Consequently (self-)absorption (see Sect. II.5c) is — as a rule — much more difficult to observe in the flame spectra of metal compounds than with atomic resonance lines. Transition probabilities or oscillator strengths for molecular bands of metal compounds are much less well-known than for atomic metal lines. For the BaO ($A^1\Sigma - X^1\Sigma$) band system, for example, $A_{qp}^{v'}$ values varying between $1.5 \times$ and 3.0×10^6 s^{-1} have been found by Wentink and Spindler (1972) (see also Johnson 1972). For a critical selection of reliable experimental values we refer to Kuznetsova et al. (1974), who have also presented a review of theoretical expressions for the line strength, etc., and of experimental methods for determining them.

5b   CAUSES OF LINE BROADENING AND SHAPES OF SPECTRAL LINE PROFILES
5b-1 *GENERAL*

When an atomic line spectrum is recorded through a spectrometer, each separate 'monochromatic' line shows a spectral profile with finite width. Unless the spectrometer has a very high spectral resolution, the width and profile observed are mainly determined by the properties of the spectral apparatus, as will be discussed in Sect. III.4c. But even in the case of infinitely high resolution, the spectral line will still appear to be broadened, its profile reflecting the spectral distribution of the photons emitted (or absorbed) by the source. This 'true' spectral line profile will be the subject of this section. We assume that the source is *optically thin*, i.e., that $k(\nu)l$ is small compared to unity at any frequency ν near line centre ν_0. Here $k(\nu)$ is the absorptivity and l the thickness of the source in the direction of observation (see Section II.5a-2). The distortion and additional broadening of the line profile in an optically thick medium will be discussed in Sect. II.5c. Any apparent line broadening resulting from an unresolved (hyper-)fine-structure of the line (see Sect. II.2c) will be ignored. Throughout this section we shall consider the case of an isolated, single atomic spectral line. The results obtained are also applicable to an isolated line of a molecular band. In molecular spectra, however, overlap of the individual line profiles often occurs (see Penner 1959, and Arnold, Whiting and Lyle 1969).

A general introduction into spectral line broadening and in particular collision broadening is found in numerous books and review articles, e.g. Mitchell and Zemansky (1961), Ch'en and Takeo (1957), Penner (1959), Unsöld (1968), Traving (1960, 1968), Kuhn (1969), L'vov (1970), and Hindmarsh and Farr (1972), and an extensive treatment has been given by Breene (1961).

In Sect. II.5a-1 we have introduced the *spectral distribution function*, $S_\nu(\nu)$ or $S_\lambda(\lambda)$, of emitted or absorbed photons, normalized by

$$\int_{\text{line}} S_\nu(\nu) \, d\nu = 1 \,, \tag{II.206}$$

$$\int_{\text{line}} S_\lambda(\lambda) \, d\lambda = 1 \,. \tag{II.207}$$

Because
$$d\nu = \left|\frac{d\nu}{d\lambda}\right|_{\lambda_0} d\lambda = (c/\lambda_0^2) \, d\lambda \,, \tag{II.208}$$

$$S_\nu = S_\lambda(\lambda_0^2/c) \,. \tag{II.209}$$

In this section we shall mostly use frequencies, because the expressions for the line broadening become more simple then and their theoretical interpretation more transparent. Conversion from frequencies into wavelengths can easily be made through Eqs II.208 and 209.

The *effective width*, $\Delta\nu_{\text{eff}}$ or $\Delta\lambda_{\text{eff}}$, is defined by

$$\Delta\nu_{\text{eff}} \equiv 1/S_\nu(\nu_0) \quad \text{or} \quad \Delta\lambda_{\text{eff}} \equiv 1/S_\lambda(\lambda_0) \,. \tag{II.210}$$

Here $S_\nu(\nu_0)$, etc., is the peak value. The effective width is thus the width of an equivalent rectangular line profile with the same peak value and the same area ($\equiv 1$).

In addition, it is customary to define the *half-intensity width*, $\delta\nu$ or $\delta\lambda$, (also called: *full width at half maximum*, FWHM) as the full difference in frequency or wavelength between the points where $S_\nu(\nu)$ or $S_\lambda(\lambda)$ has dropped to half its maximum value. The relation between $\Delta\nu_{\text{eff}}$ and $\delta\nu$ depends on the particular line shape.

In some branches of spectroscopy, the half-intensity width or half-width is defined as half the distance between the points of half intensity. Where confusion might arise, the term *full* half-intensity width should be used for the former definition. Anyway, the ambiguous term 'half-width' should be avoided.

Chapt. II　　　　　　　　　　　　　　BASIC CONCEPTS AND GENERAL RELATIONSHIPS

The spectral profile of the radiation emitted or absorbed (from a continuum source) by an optically thin (homogeneous) medium as a whole, is the same as the spectral distribution, $S_\nu(\nu)$, of the photons emitted or absorbed by the individual atoms inside the medium. For an absorption line this is made clear by the linear relation between $k(\nu)$ and $S_\nu(\nu)$ in Eq. II.176 or 177. For an emission line, this follows from the fact that all photons generated inside an optically thin source leave the source with negligible chance of re-absorption (see also Sect. II.5c-3). All conclusions about S_ν reached in the following sections thus hold equally well for an emission as for an absorption line in a homogeneous, optically thin medium.

A discussion of the different causes of line broadening and the associated distribution functions is essential for the calculation of the radiation intensity in an optically thick medium (see Sect. II.5c). The optical thickness itself depends on the line broadening for given atomic density and thickness of the source. Also an insight into the various line-broadening effects enables us to estimate the narrowness of, and possible overlap between atomic (and molecular) lines occurring in a flame. A knowledge of the line profile (and line shift) is furthermore appropriate for calculating the signal obtained in atomic absorption spectroscopy when a background source producing a narrow line of the same atomic transition is used. The study of spectral line profiles and line shifts may also yield information on the interactions between the radiating atom and the surrounding flame molecules. This topic will be more extensively dealt with in Chapt. VII. Under certain conditions, the spectral line profile may, finally, yield information about the translational temperature in the source.

Attention will be given mainly to those broadening effects that are relevant in flames.

Broadening of the S_ν profile should be well distinguished from *saturation broadening* which is observed when a tunable, strong and monochromatic laser beam is scanned over the absorption line (see, e.g., Greenstein and Bates 1975). When we measure the dependence of the laser absorption or of the laser-induced fluorescence as a function of detuning of the laser frequency ν_L with respect to ν_0, we find a curve whose shape depends on the laser intensity because of saturation (cf. Sect. II.5a-2). Equation II.190 describes the dependence of the net absorptivity $k'(\nu)$ on the population difference, $\{n_p - (g_p/g_q)n_q\}$, of the lower and upper levels. (We assume here the case of homogeneous line broadening and the absence of atomic polarization; see Sect. II.5a-2.) Under (partial) saturation conditions, this population difference depends, in turn, on the laser intensity as well as on the detuning $\nu_L - \nu_0$ (see Sect. VI.2c-5 for a quantitative treatment). For a given laser intensity, the degree of saturation will be maximal if $\nu_L = \nu_0$. Consequently, the reduction of $k'(\nu)$ caused by saturation will be maximal at line centre, whereas it will be practically vanish at the line wings. As a result the profile of the curve representing the laser absorption as a function of detuning will appear to be flattened around the line centre, and thus broadened, when compared with the profile measured in the absence of saturation. This so-called saturation broadening is more marked the higher the laser intensity, i.e., the stronger the saturation near line centre (see Omenetto et al. 1980 for a detailed study in flames).

Saturation broadening is a well-known effect in microwave spectroscopy where saturation is more readily attained than in optical spectroscopy.

Figure II.11 shows the effect of saturation broadening on the laser-excited fluorescence intensity when the laser frequency is scanned over one of the Na-D lines in a flame.

It should be noted that saturation broadening does <u>not</u> affect the spectral profile of $k'(\nu)$ itself, which is determined by $S_\nu(\nu)$ contained as a factor in Eqs II.189, 190 and 191. If $k'(\nu)$ were measured with the aid of an auxiliary, monochromatic but weak radiation beam, the same $k'(\nu)$ profile would be found as in the absence of saturation (at least under the above general assumptions). The only effect of saturation is here a frequency-independent, overall reduction of $k'(\nu)$.

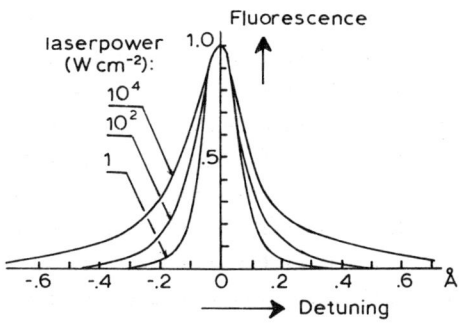

Fig. II.11 Normalized excitation function of Na fluorescence at 3 different power values of pulsed dye laser with spectral width of 0.14 Å. The curve at lowest laser power relates to non-saturation conditions; the increased width of the other curves is due to saturation broadening. (From Alkemade 1977 and van Dijk et al. 1978.)

In the following discussion of line broadening we shall thus disregard saturation broadening.

5b-2 *CAUSES OF LINE BROADENING AND ASSOCIATED SPECTRAL DISTRIBUTION FUNCTIONS*

Natural broadening. Even the line emitted or absorbed by an atom at rest without external disturbances is not infinitely sharp. This is a direct consequence of the finite radiative lifetime, τ_r, of the excited state in combination with the *Heisenberg uncertainty relation* which relates τ_r to the minimal uncertainty in energy, ΔE, of the state according to

$$\Delta E \tau_r \simeq \hbar \, (\equiv h/2\pi) \, . \qquad (II.211)$$

The uncertainty in energy of the upper and lower state is, in turn, directly related to the uncertainty in frequency ν through the relation: $\nu = (E_q - E_p)/h$. When

Chapt. II BASIC CONCEPTS AND GENERAL RELATIONSHIPS

$(\tau_r)_q$ and $(\tau_r)_p$ refer to the upper and lower state, respectively, we obtain for the uncertainty in frequency or half-intensity width

$$\delta\nu_N = \{(\tau_r)_q^{-1} + (\tau_r)_p^{-1}\}/2\pi . \tag{II.212}$$

The radiative lifetimes are determined by the Einstein transition probabilities of all possible spontaneous emissions from states q and p, respectively, through Eq. II.166. For resonance lines, the radiative lifetime of the lower state, i.e. ground state, can be put equal to infinity, so that we obtain

$$\delta\nu_N = 1/2\pi(\tau_r)_q^{-1} \quad \text{(for a resonance line)}. \tag{II.213}$$

For a transition from the lowest excited state to the ground state (first resonance line) we simply have

$$\delta\nu_N = A_{q0}/2\pi \quad \text{(for the first resonance line)}, \tag{II.214}$$

where A_{q0} is the transition probability of the line. Expressing A_{q0} in f_{0q} through Eq. II.168: $f_{0q}A_{q0}^{-1} = 1.5\,(g_q/g_0)\lambda_0^2$, and putting $f \approx 1$ for strong lines and $g_q/g_0 \approx 1$, we find from Eq. II.214: $\delta\lambda_N = \delta\nu_N |d\lambda/d\nu|_{\nu_0} \approx 3 \times 10^{-12}$ cm or 0.3 mÅ, independent of λ_0. For lower f-values the natural line width will be even smaller. Under flame conditions natural broadening is negligible compared to the other broadening effects.

The normalized spectral profile associated with natural broadening is given by

$$S_{\nu,N} = \frac{2/(\pi\delta\nu_N)}{1 + \{2(\nu_0 - \nu)/\delta\nu_N\}^2} . \tag{II.215}$$

This so-called *dispersion* or *Lorentzian profile* is symmetric with respect to the line centre and falls off $\propto (\nu_0 - \nu)^{-2}$ in the line wings, i.e., at distances $|\nu_0 - \nu| \gg \delta\nu_N$. The peak value is $2/(\pi\delta\nu_N)$, from which the effective natural line width, $(\Delta\nu_{eff})_N$, follows through Eq. II.210

$$(\Delta\nu_{eff})_N = (\pi/2)\delta\nu_N . \tag{II.216}$$

The derivation of Eqs II.212 and 215 from the Heisenberg relation, Eq. II.211, is not straightforward. The latter applies strictly for the standard deviation of a Gaussian distribution. In fact, Eq. II.215 shows that the broadening of the line and thus of the energy level due to spontaneous emission is described by a Lorentzian distribution function. The standard deviation, $\sqrt{(\nu - \nu_0)^2}$, for a Lorentzian distribution function even appears to be infinite [since $(\nu - \nu_0)^2$ appears in the denominator of the distribution function !].

The line profile due to natural broadening can be derived more consistently in the framework of classical radiation theory as follows.

Classically, the spontaneous emission of a photon corresponds to the conversion of atomic oscillator energy into radiation energy. This conversion has a damping effect on the oscillation (*radiation damping*). As a consequence, the oscillation

$$x(t) = A(t) \exp[j\omega_0 t] , \qquad (II.217)$$

(with A = complex amplitude including a phase factor $\exp[j\eta]$, $\omega_0 = 2\pi\nu_0$, t = time, and j = imaginary unit) decays in time. This decay is described by an exponential t-dependence of A

$$A(t) = A(0) \exp[-t/2\tau_r] , \qquad (II.218)$$

if $A(0)$ is the initial amplitude of the oscillator excited at $t = 0$. Since the energy, $E(t)$, of the oscillator is proportional to the squared absolute amplitude, we also have

$$E(t) \propto \exp[-t/\tau_r] . \qquad (II.219)$$

This function is similar to the decay function of the excited state population in the quantum description (see Eq. II.165). The reciprocal of the Einstein transition probability, A^{-1}, is thus equal to the exponential decay time of the classical oscillator energy, and to half the decay time of the oscillator amplitude (see Unsöld 1968).

For a statistically stationary ensemble of independent, identical oscillators excited in a random fashion, the normalized *autocorrelation function*, $\phi(s)$, of the amplitude can be defined by (see Traving 1968)

$$\phi(s) \equiv \frac{\overline{A^*(t)A(t+s)}}{\overline{A^*(t)A(t)}} , \qquad (II.220)$$

where the asterisk denotes the complex conjugate and the bar denotes the average over the ensemble, which is independent of t. Note that $\phi(0) = 1$ due to normalization. It can be shown that the shape of the function $\phi(s)$ is the same as the decay function for the amplitude of a single oscillator. Thus we have from Eq. II.218

$$\phi(s) = \exp[-s/2\tau_r] . \qquad (II.221)$$

In addition, we may define the *correlation time* τ through

$$\tau \equiv \int_0^\infty \phi(s)\,ds/\phi(0) = \int_0^\infty \phi(s)\,ds . \qquad (II.222)$$

This quantity is a measure for the time during which appreciable correlation exists; it is connected to the *coherence time* or *coherence length* of the radiation field. In our case τ appears to equal $2\tau_r$ if we calculate the integral

Chapt. II BASIC CONCEPTS AND GENERAL RELATIONSHIPS

in Eq. II.222.

The spectral distribution of the radiation power emitted by a random ensemble of identical oscillators can generally be derived from the autocorrelation function, $\phi(s)$, through a Fourier transformation

$$S_\nu(\nu) = \text{Re } 2\int_0^\infty \phi(s) \exp[-j2\pi(\nu-\nu_0)s] \, ds, \qquad (II.223)$$

where Re denotes the real part. When $\phi(s)$ is a real function, as in the above case, we also have

$$S_\nu(\nu) = 2\int_0^\infty \phi(s) \cos\{2\pi(\nu-\nu_0)s\} \, ds . \qquad (II.224)$$

Substituting for $\phi(s)$ the expression given by Eq. II.221, we find for $S_\nu(\nu)$ the dispersion function given by Eq. II.215, with $\delta\nu_N = 1/2\pi\tau_r$.

A strict proof of the natural line profile can also be given on the basis of the Dirac theory of radiation through the Weisskopf-Wigner formula (see Heitler 1954, and Penner 1959).

Radiation broadening. The radiative lifetime of a state may also be shortened when optical transitions from this state (absorption or emission) are induced by a strong radiation field. This causes an additional broadening of the energy levels and thus of the spectral line: *radiation* or *power broadening*, increasing in proportion to the spectral radiant energy density (see Sect. II.5a-1). With conventional light sources and in normal experimental conditions in flame spectroscopy, this effect is not expected to be noticeable, but with strong, pulsed lasers tuned at the atomic line centre the effect may be observable (see below).

A strong radiation field causes radiation broadening only in the *incoherent case*, that is, when the coherence time of the field oscillation is short compared with the radiative lifetime of the states considered (see Ya'akobi 1972, and Maassen 1976). In other words, the frequency bandwidth of the radiation field must be large compared with the rate constant (in s^{-1}) of the induced atomic transitions.

In the opposite *coherent case* no broadening, but a line splitting results. This is the a.c. Stark or Rabi effect mentioned in Sect. II.5a-1.

An estimate of radiation broadening in comparison with natural broadening may be obtained as follows. Assume that a (quasi-)continuous radiation field is present which induces absorption and emission transitions between the ground state 0 and the upper state q. We now compare the corresponding rate constants (i.e. number of transitions per second and **per atom** in the initial state) with that of the spontaneous emission, A_{q0}. These rate constants or inverse radiative lifetimes are proportional to the corresponding line-broadening effects (cf. Eq. II.214). The rate constants of absorption and induced emission are about equal, as g_q/g_0 is of the order of unity (see Eq. II.158). The ratio of the rate constants of induced and spontaneous emission follows from Eq. II.157 if we replace there $\rho^b_{\nu_0}$ by ρ_{ν_0} of the radiation field present. We can now formally express ρ_{ν_0} in terms of an effective radiation temperature, T_{eff}, by equalizing: $\rho^b_{\nu_0}(T_{eff}) \equiv \rho_{\nu_0}$. A Planck radiation field at T_{eff} would yield the same absorption and induced-emission rates as the actual field (which need not be isotropic, however). Substituting T_{eff} for T and E_{exc} for $h\nu_0$ in Eq. II.157 we get an expression for the ratio of the induced and spontaneous rate constants, that is, of the corresponding line-broadening effects. Taking an excitation energy of 2.1 eV (as for the first resonance line of Na) and putting T_{eff} equal to 5000 K (comparable with the radiance temperature of a high-pressure xenon lamp; see Eq. II.301 and Prugger 1969), we find the latter ratio to be of the order 10^{-2}. It should be realized that in order to obtain a spectral energy density equal to $\rho^b_{\nu_0}(5000\ K)$, the radiation from the xenon lamp must be focussed under the maximum possible solid angle of 4π sr into the flame. This is practically impossible to realize. So even with this intense conventional light source induced radiative transitions have a negligible effect on the spectral line width.

It is not essential in the above derivation that the lower state of the optical transition considered is the ground state. The conclusion drawn is more generally valid.

In order to observe radiation broadening in flames, the induced emission and absorption rates (in s^{-1}) per atom in the excited and ground state, respectively, must be comparable to the frequency width (in Hz) of the spectral line. The latter may exceed the natural line width by two orders of magnitude (see also below). One estimates that for the Na-D lines in an H_2-O_2-Ar flame at about 1800 K the spectral energy density should exceed roughly 2×10^{-10} erg cm^{-3} Hz^{-1} in order to make radiation broadening important (see Alkemade 1977 and Nienhuis 1980). Van Calcar (1980) has indeed observed in a similar flame for the first time radiation broadening of the Na-D fluorescence lines excited by a pulsed dye-laser beam with 30 kW/cm^2 peak irradiance and 350 mÅ bandwidth.

Collision broadening. Under *collision broadening* we group together all broadening effects due to interactions of the radiating or absorbing atom with neighbouring atoms or molecules. The term 'collision' is to be understood here in a broad sense. It is not restricted to the sudden impact of two particles, but describes any interaction, pulse-like or continuous, of the radiating atom with one or possibly more particles in its vicinity.[†] We may distinguish these broadening

---

[†] The term 'encounter' would here be more appropriate than 'collision'.

Chapt. II BASIC CONCEPTS AND GENERAL RELATIONSHIPS

collisions according to the *type of interaction*: *diabatic* or *adiabatic*, or to the *kind of collision partner*. Following the classification as used by Margenau and Lewis (1959) and by Townes and Schawlow (1955) (in microwave spectroscopy), we characterize diabatic broadening collisions as those by which the atom is transferred to another quantum state with lower or higher energy. Adiabatic broadening collisions, on the other hand, leave the atom in the same quantum state or transfer it to another state belonging to the same degenerate energy level (M_J mixing; see also Sect. VII.2b-6). In classical language, diabatic broadening collisions terminate the oscillation or change definitely the amplitude of oscillation of the atom. Adiabatic collisions only shift the phase of oscillation and/or change the orientation of the oscillator or its frequency within a narrow range. These changes have a statistical character.

Both types of collision cause a spectral line broadening which increases with increasing density of perturbers. For this reason, collision broadening is often referred to as *pressure broadening*. It depends, however, not only on the pressure of the perturbing gas molecules but also on the temperature and, of course, on the kind of perturbers. A study of collision broadening may yield information about the interaction potential between the perturber and the radiating atom in the states between which the optical transition takes place. This will be discussed more extensively in Sect. VII.4c.

When the perturbing particles are of a different kind than the radiating atom, the term *foreign gas broadening* is used. When the perturbers are of the same species, *resonance-*, *Holtsmark-* or *self-broadening* is said to occur. *Stark broadening* occurs when the perturbers are ions, electrons, or molecules with a strong permanent electric dipole moment. These particles generate a statistically distributed electric field at the position of the radiating atom, which causes the atomic levels and consequently the spectral line to be broadened as a consequence of the Stark effect. Stark broadening is a special type of *field broadening* and is of considerable influence in plasmas with a high degree of ionization. Resonance and Stark broadening are relatively unimportant for atomic metal lines in flames because of the low density of metal atoms (the partial pressure is usually below 10^{-4} atm; see Sect. III.2h) and of ionized species (see Behmenburg 1964, van Trigt, Hollander and Alkemade 1965, and Alkemade 1968; see also Sects IV.7 and IX.1). The justification of the neglect of resonance and Stark broadening in flames will be discussed in more detail in Sect. VII.2a.

> The nomenclature used in the literature, in particular the definition of Lorentz broadening, is sometimes ambiguous. Some authors apply this term to inelastic broadening collisions as in the original Lorentz theory (see, e.g., Mavrodineanu and Boiteux 1965); others use it for elastic broadening collisions

THE INTENSITIES OF RADIATIVE TRANSITIONS   Sect. II.5b

(see, e.g., Mitchell and Zemansky 1961). We therefore prefer the less contaminated term collision broadening covering both types of collisions. Also the term adiabatic collision is preferred to elastic collision. The small random deviation, $\Delta\nu$, in frequency of the emitted or absorbed photons, with respect to the frequency ν_0 of the undisturbed atom, implies that a small amount of energy, $h\Delta\nu$, is converted during the collision process. This process is thus not strictly elastic. The term interaction broadening (see de Galan and Wagenaar 1971) cannot be used as a substitute for collision broadening either, as there are also interactions with a radiation field which may lead to radiation broadening (see above).

A semiclassical *theory* of spectral line broadening including *diabatic collision broadening* has been presented by Anderson (1949) (in the impact approximation) and by Nienhuis (1973) (see also Sect. VII.2b-4). Following Hooymayers and Alkemade (1966a) we shall derive here only an estimate of its effect by simply treating it as we treated natural broadening, for both broadening effects are connected with the finite lifetime of the states involved in the transition.

Diabatic collisions which transfer the atom to another energy level have the effect of shortening the lifetime of the initial state. Due to the statistical character of these collisions, the corresponding decay of excited-state population in time, e.g., after pulse excitation at $t=0$, is described by an exponential function similar to that occurring in Eq. II.165 for the radiative decay. Instead of the radiative lifetime τ_r, we now have the *collisional lifetime* $\tau_{c'}$. Its reciprocal value $\tau_{c'}^{-1}$ is the mean number of diabatic collisions per second and per excited atom, which can be expressed in a bimolecular rate constant $k_{c'}$ according to: $\tau_{c'}^{-1} = k_{c'} n_X$, where X denotes the perturber (see Sect. II.4a-1). Thus, replacing $\delta\nu_N$ in Eqs II.213 and 215 by the corresponding quantity for diabatic collision broadening, we find for a resonance line†

$$\delta\nu_{c'} = 1/(2\pi\tau_{c'}), \qquad (II.225)$$

$$S_{\nu,c'} = \frac{2/(\pi\delta\nu_{c'})}{1 + \{2(\nu_0 - \nu)/\delta\nu_{c'}\}^2}. \qquad (II.226)$$

For a first resonance line with known Einstein transition probability A, the value of $\tau_{c'}$ can be obtained from a measurement of the quantum efficiency of fluorescence Y (see Sect. II.5c-4; Hooymayers and Alkemade 1966a). Diabatic collision broadening and quenching of fluorescence radiation are both connected with

---

† The rate constant for the **reversed** diabatic collision process (per atom in the **ground** state) is much smaller and can thus be neglected (see Sect. II.4a-2).

inelastic collisions of the excited atom. For this reason, this type of broadening is also called *quenching broadening*. The relation between $\tau_{c'}$ and Y follows from Eq. II.306 (in which the unimolecular rate constant k' should be identified with $\tau_{c'}^{-1}$) and reads

$$\tau_{c'}^{-1} = A(1-Y)/Y \ . \tag{II.227}$$

Since in flames at 1 atm pressure, Y may range from a few per cent up to about 50 per cent for the alkali resonance lines, $\delta\nu_{c'}$ ranges here from about 1 to $100 \times$ the natural half-intensity width which is related to A through Eq. II.214. In flames diluted with Ar, for example, Y may be as large as 0.5 and we expect $\delta\lambda_{c'} \approx 0.3$ mÅ in flames diluted with N_2 showing much smaller Y values, $\delta\lambda_{c'}$ may be of the order of 10 mÅ. When the gas pressure is lowered, $\delta\nu_{c'}$ is reduced in the same proportion, because of the linear dependence of $\tau_{c'}^{-1}$ on n_x (see above). Diabatic collision broadening is, however, usually overshadowed by adiabatic collision broadening, as we shall see presently (see also Sect. VII.2b-5).

In the above heuristic derivation the spectral line was considered in emission. Theory shows that quenching broadening has the same effect in emission as in absorption (see Nienhuis 1973).

The *theory* of *adiabatic collision broadening* has been worked out extensively for atomic lines, classically as well as quantummechanically, while numerous experiments have been performed to verify the theory and to obtain detailed information about the adiabatic interactions. Experiments are especially needed to check the validity of the various approximations made in the theoretical calculations. Here only a short introduction will be given, whereas references to the literature and a more elaborate treatment will be presented in Sect. VII.2b-3 in connection with the interpretation of flame and other experiments.

The main effects of adiabatic interactions are:
(i)   a broadening of the spectral line,
(ii) a shift of the line centre,
(iii) an asymmetry between the far line wings at either side of the centre.

For a full quantitative understanding of these effects, the detailed interaction of the atom in the excited as well as the ground state with neighbouring particles must be examined. We shall leave the discussion of effects (ii) and (iii) to Sect. VII.2 and give here only an elementary and simplified description of the broadening effect (i) in the framework of *classical* theory.

We assume, as an approximation, that the density of the perturbers, n_x, is low enough to make the duration of the interaction small compared to the mean time between the collisions. The interaction of the radiating atom with more than one perturber at the same time may then be neglected. The duration of interaction is also supposed to be small in comparison to the duration of the radiation process.

THE INTENSITIES OF RADIATIVE TRANSITIONS   Sect. II.5b

These assumptions are the basis of the *impact approximation*. A further simplification may be obtained by assuming the relative velocity of the collision partners to be constant (g_0). Each collision results, after its completion, in a shift, $\Delta\eta(b)$, of the *phase* of oscillation,[†] which is a function of impact parameter b. The absolute value of the (complex) amplitude of oscillation is assumed to remain unchanged. By an integration over the statistical distribution of b, the autocorrelation function of the complex amplitude can be expressed in the function $\Delta\eta(b)$. This autocorrelation function describes the correlation in complex amplitude {including the phase factor $\exp[j\eta(t)]$; j = imaginary unit; $\eta(t)$ = time-dependent phase of oscillation}, taken at two different instants t' and t'', as a function of their difference $s \equiv t'' - t'$ (for more detail, see below). When the oscillation would have an undisturbed, strictly constant phase and frequency ν_0, the correlation would be complete for any value of s. The spectral distribution would consist of an infinitely sharp line located at ν_0. When phase disturbances occur, noticeable correlation only exists over a limited time interval of the order of the correlation time $\tau_{c''}$. The frequency-distribution of the line will then have a certain width of the order $\tau_{c''}^{-1}$. Thus the line becomes the broader the shorter $\tau_{c''}$ is, i.e., the more frequently phase disturbing collisions occur. Fourier transformation of the autocorrelation function yields the spectral line shape, which is again given by a dispersion or Lorentzian function

$$S_{\nu,c''} = \frac{2/(\pi\delta\nu_{c''})}{1 + \{2(\nu_0 - \nu)/\delta\nu_{c''}\}^2} . \qquad (II.228)$$

Here the half-intensity width for adiabatic collision broadening is given by

$$\delta\nu_{c''} = 1/(\pi\tau_{c''}) . \qquad (II.229)$$

In addition to this broadening, also a shift of the line centre away from ν_0 may occur, which has been neglected for simplicity's sake in this discussion. An example of a purely collision broadened line profile is presented in Fig. II.12.

It is customary to interpret the correlation time as the average time between two successive 'strong' collisions. A 'strong' collision is defined as one which destroys the phase memory of the oscillator. That is, the phases before and after such collisions are completely uncorrelated. The reciprocal of $\tau_{c''}$ is thus the average number of strong collisions per second, which can be expressed by: $\tau_{c''}^{-1} = g_0 \sigma_{c''} n_x$ (cf. Sect. II.4b-1). The quantity $\sigma_{c''}$ is called the *optical cross section for adiabatic collision broadening*[*] and is thus related through Eq. II.229

[†] We disregard here, for simplicity's sake, the effect of collisions that result in a re-orientation of the atomic oscillator.

[*] Not to be confused with the optical cross section for absorption introduced in Sect. II.5a-2.

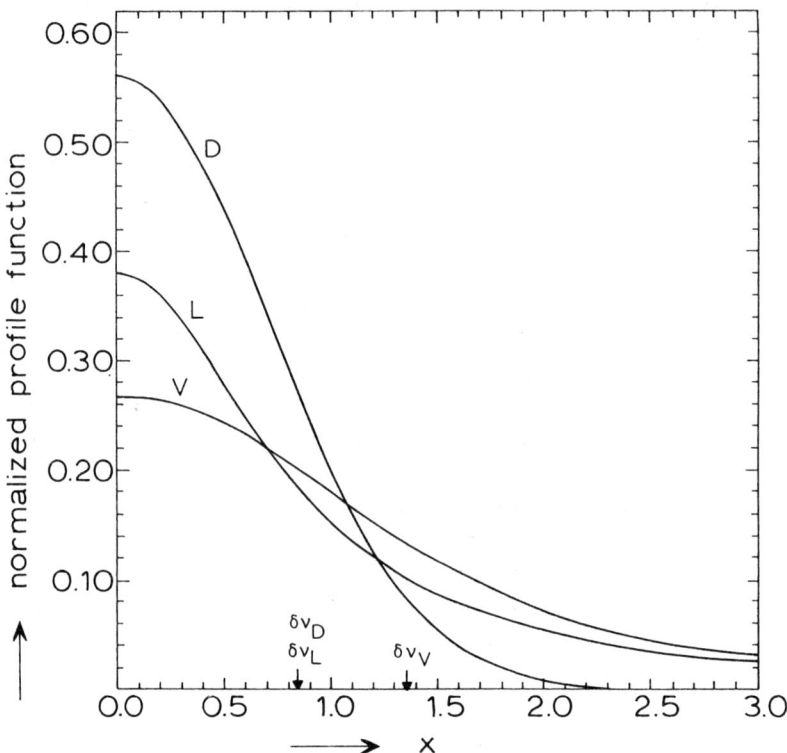

Fig. II.12 Normalized theoretical Doppler (D), Lorentzian (L), and Voigt (V) line profiles are shown as functions of $x \equiv 2\sqrt{\ln 2}\,(\nu-\nu_0)/\delta\nu$ where $\delta\nu = \delta\nu_D = \delta\nu_L$ (or $a = 0.84$). The ordinate values are given by Eqs II.239, 228, and 245, respectively, multiplied by $\delta\nu/(2\sqrt{\ln 2})$, where $\delta\nu_C{''} = \delta\nu_L$. The profiles are symmetric with respect to the line centre ν_0 (corresponding to $x=0$); the areas of the half profiles shown are equal to $\frac{1}{2}$ in accordance with the normalization condition Eq. II.206. Arrows indicate the positions of the points at half peak-intensity. (According to calculations performed by Dr B. Jansen at the Fysisch Laboratorium, State University, Utrecht.)

THE INTENSITIES OF RADIATIVE TRANSITIONS   Sect. II.5b

to $\delta\nu_{c''}$ according to [†]

$$\delta\nu_{c''} = g_0 \sigma_{c''} n_x / \pi \ . \qquad (II.230)$$

For a gas with a Maxwellian velocity distribution we find $\sigma_{c''}$ from $\delta\nu_{c''}$ by replacing g_0 in Eq. II.230 by the average relative velocity (see Eq. II.29). This yields

$$\delta\nu_{c''} = (2/\pi)(2kT/\pi\mu)^{\frac{1}{2}} \sigma_{c''} n_x \ , \qquad (II.231)$$

where μ is the reduced mass of the collision partners and $\sigma_{c''}$ is assumed to be independent of g_0.

For collisions with foreign, neutral gas molecules or atoms, $\sigma_{c''}$ may be larger by an order of magnitude than the gas-kinetic cross section. It should be noted that $\sigma_{c''}$ is, in general, a function of velocity and thus of temperature, although it does not vary very strongly with temperature (see Sect. VII.2b-3.1). For Na atoms perturbed by N_2 in flames of 1 atm pressure and at $T \approx 2500$ K, we get from $\sigma_{c''} \approx 50 \text{Å}^2$ a value $\delta\lambda_{c''} \approx 3.5 \times 10^{-2}$ Å. This width thus exceeds by a large amount the natural line width and, albeit usually to a lesser extent, also the diabatic collision width. For resonance broadening the optical cross sections may be very high, say of the order of 10^4 Å^2 or higher.

The relation between half-intensity width and effective line width is the same as for natural broadening (see Eq. II.216), because of the similarity in line profile.

The adiabatic collision broadening can be classically described by considering the oscillation (see Eq. II.217)

$$x(t) = A(t) \exp[j\omega_0 t] \ , \qquad (II.232)$$

where $A(t)$ is the complex amplitude including the time-dependent phase factor $\exp[j\eta(t)]$. Since the phase disturbances through adiabatic collisions are random, the phase $\eta(t)$ is a fluctuating function of time, whereas the absolute value $|A(t)|$ is constant (radiation damping and inelastic collisions are now neglected). Using the definition of the autocorrelation function, $\phi(s)$, of the complex amplitude $A(t)$ (see Eq. II.220) we have

$$\phi(s) = \overline{\exp[j\eta(t+s) - j\eta(t)]} \ . \qquad (II.233)$$

Here the bar denotes an average over time t, which is identical with an ensemble average for a statistically stationary system. It can be proved that $\phi(s)$ is an exponential function of s (see Traving 1968)

[†] The formal relation between line width and cross section in Eq. II.230 conforms to the (semi-)classical expressions presented by Traving (1968) and Behmenburg (1968). It differs by a factor 2 from Baranger's (1958, 1962) quantummechanical expression. The explanation of this apparent discrepancy has been given by Baranger (1958).

Chapt. II BASIC CONCEPTS AND GENERAL RELATIONSHIPS

$$\phi(s) = \exp\left[-s/\tau_{c''}\right], \quad (II.234)$$

where $\tau_{c''}$ is the correlation time. We assume that $\tau_{c''}$ is a real quantity; when $\tau_{c''}$ is complex, there is also a shift in the central frequency, which is here neglected. The autocorrelation function in Eq. II.234 is of the same type as that for natural broadening (see Eq. II.221). Applying Fourier transformation of $\phi(s)$ (see Eq. II.224) we thus find again a dispersion function for the spectral line profile with half-intensity width $\delta\nu_{c''} = 1/(\pi\tau_{c''})$, in accordance with Eqs II.228 and 229.

The difference of a factor 2 in the expressions for the half-intensity width of natural broadening and adiabatic collision broadening, respectively, should be noted (compare Eqs II.213 and 229). This difference arises from the fact that the radiative lifetime τ_r is, in fact, the correlation time for the energy of the oscillator, whereas $\tau_{c''}$ is the correlation time for the amplitude (compare also Eqs II.218 and 219). Since the energy is proportional to the square of absolute amplitude, a difference of a factor 2 in the exponents in Eqs II.221 and 234 results.

Doppler broadening. This broadening effect is caused by the statistical distribution of the velocities of the emitting or absorbing atoms along the line of observation. Up till now the atoms were considered to be at rest with respect to the observer, so that the oscillation frequency of the radiating atom and the observed frequency were the same. Due to the *Doppler effect*, the observed frequency, ν, of a light wave emitted by an atom oscillating at ν_0 and moving with velocity v_x towards the observer, is given by

$$(\nu-\nu_0)/\nu_0 = v_x/c, \quad (II.235)$$

where c is the light velocity. If we assume that the velocity v_x of the atom while radiating does not change (see for a more strict formulation of this condition Eq. II.242), we find that the frequency distribution of the observed spectral line is directly related to the v_x distribution through Eq. II.235. In thermal equilibrium a Gaussian distribution of velocities v_x does exist according to Maxwell's law (see Eq. II.24). The profile of a Doppler broadened line is thus described by a Gaussian function with peak at ν_0

$$S_{\nu,D} = \{1/\sqrt{\pi\overline{(\Delta\nu)^2}}\}\exp\left[-(\nu-\nu_0)^2/\overline{(\Delta\nu)^2}\right]. \quad (II.236)$$

Here $\overline{(\Delta\nu)^2}$ is the mean square deviation of ν from ν_0, which is determined by $\overline{(\Delta v_x)^2}$ through Eq. II.235

$$\overline{(\Delta\nu)^2} = \overline{(\Delta v_x)^2}(\nu_0/c)^2 = 2kT\nu_0^2/mc^2. \quad (II.237)$$

In the latter equation use is made of the equipartition law for the translational motion of an atom with mass m at temperature T (see Eq. II.20). The half-intensity width of this spectral distribution is given by

$$\delta\nu_D = 2\sqrt{(\ln 2)\overline{(\Delta\nu)^2}} = 2\sqrt{2(\ln 2)kT/m}\,(\nu_0/c)\,. \tag{II.238}$$

Expressing $\sqrt{\overline{\Delta\nu^2}}$ in $\delta\nu_D$ through the latter relation we find from Eq. II.236

$$S_{\nu,D} = (2\sqrt{\ln 2}/\delta\nu_D\sqrt{\pi})\exp[-4(\ln 2)(\nu-\nu_0)^2/\delta\nu_D^2]\,. \tag{II.239}$$

We have numerically

$$\delta\nu_D/\nu_0 = \delta\lambda_D/\lambda_0 = 7.16\times 10^{-7}\sqrt{T/M_r} \tag{II.240}$$

where T is expressed in kelvin and M_r is the relative mass of the emitting species. For the first resonance line of Na and at temperature $T = 2500\,\text{K}$, we find: $\delta\lambda_D = 4.5\times 10^{-2}$ Å. An example of a pure Doppler line profile is shown in Fig. II.12.

Using the definition in Eq. II.210, we find for the effective line width (= reciprocal peak value of $S_{\nu,D}$) through Eqs II.236 and 238

$$(\Delta\nu_{\text{eff}})_D = (S_{\nu,D})_{\max}^{-1} = (\sqrt{\pi}/2\sqrt{\ln 2})\,\delta\nu_D = 1.06\,\delta\nu_D\,. \tag{II.241}$$

Thus the integrated Doppler line profile equals the product of peak value and half-intensity width within an error of about 6%.

The validity of the above expressions for the spectral line shape and line width is restricted by the condition that the radiating atom does not noticeably change its velocity within a time interval of about $(2\pi\delta\nu_D)^{-1}$ (see Traving 1968). Using the Doppler relation Eq. II.235 one can prove that this condition is equivalent to the restriction (see Traving 1968, and Nienhuis 1973)

$$2\pi\lambda_{\text{free}} \gg \lambda_0\,, \tag{II.242}$$

where λ_{free} is the mean free path between two gas-kinetic collisions. In flames at 1 atm pressure, $2\pi\lambda_{\text{free}}$ is of the order of 20 μm. When the latter condition is not fulfilled, however, a reduction of Doppler width (*collisional line-narrowing* or *Dicke narrowing*) and a deviation from the Gaussian profile result (see Dicke 1953, and Shimoda 1976a; see also Sect. VII.2a).

When excited atoms are produced by a chemiluminescent reaction, the excess of chemical energy liberated may be partly converted into extra translational energy of the excited atom. On the other hand, excited atoms with extra high translational energies may have a greater probability to be quenched by the reverse chemical reaction. In thermodynamic equilibrium, the combined effects of these opposite chemical reactions do not upset the Maxwellian velocity distribution for the excited atoms. If, however, deviations occur from chemical equilibrium for the reactants and (or) products of this reaction, the velocity distribution of the excited atoms as well as the Doppler width may be affected. When the reactants of the chemiluminescent reaction are present in excess concentrations, the Doppler width of the emission

Chapt. II BASIC CONCEPTS AND GENERAL RELATIONSHIPS

line may exceed the equilibrium value (see Alkemade 1963; see also Sect. VII.2b-9).

5b-3   *THE COMBINATION OF BROADENING EFFECTS; THE VOIGT LINE PROFILE*

The broadening effects discussed in the foregoing section usually occur in combination. The resulting total line profile and line width depend on the contributions from the separate broadening effects. Roughly speaking, the relative importance of these contributions can be judged from the corresponding half-intensity widths. From the examples given, we expect that under flame conditions (at 1 atm pressure) adiabatic collision broadening and Doppler broadening are dominant, while diabatic collision broadening might compete only when Y is of the order 10^{-2} or lower (see Eq. II.227; see also Jansen 1976). Here we want to discuss how and under which restricting conditions the separate broadening effects must be combined to yield the total line profile. In this discussion, we shall neglect any shift or asymmetry of the line profile. We shall thus use the idealized, symmetrical Lorentzian and Gaussian line shapes for collision and Doppler broadening, respectively, given by Eqs II.226, 228 and 236 or 239. Under flame conditions, these approximations are mostly valid in calculations of integrated line intensities, as will be shown in Sect. VII.2b-7.

The combination of Lorentzian profiles. Assuming that natural and collision broadening are mutually independent, we obtain for the combined line profile again a Lorentzian function with half-intensity width, $\delta\nu_L$, given by

$$\delta\nu_L = \delta\nu_N + \delta\nu_C \ . \tag{II.243}$$

Here $\delta\nu_C$ refers to (adiabatic and/or diabatic) collision broadening. We have grouped both types of collision broadening together; they do not act independently of each other, so that their half-intensity widths cannot simply be added (see Nienhuis 1973). Interferences between adiabatic and diabatic collisions are, however, not expected when in a gas mixture one type of perturber causes (mainly) adiabatic, and another (mainly) diabatic line broadening.

The result expressed by Eq. II.243 can be simply understood on account of a general rule for independent broadening processes. This rule states that the autocorrelation function for the combined line profile is the product of the autocorrelation functions for the separate profiles (see Traving 1968). Since in the case considered the correlation functions are exponential (see Eqs II.221 and 234), the product correlation function is again exponential. The reciprocal of the total correlation time is the sum of the separate reciprocal correlation times. This leads immediately to Eq. II.243 because of the relation between half-intensity width and correlation time.

The combined effect of natural and collision broadening on the line shape

THE INTENSITIES OF RADIATIVE TRANSITIONS   Sect. II.5b

is thus described by the general Lorentzian function

$$S_{\nu,L} = \frac{2/(\pi \delta \nu_L)}{1 + \{2(\nu_0 - \nu)/\delta \nu_L\}^2} \, , \tag{II.244}$$

where $\delta \nu_L$ is given by Eq. II.243. Under flame conditions, we usually have that $\delta \nu_C \gg \delta \nu_N$ so that, virtually, $\delta \nu_L = \delta \nu_C$.

The combination of Lorentzian and Doppler profiles; the Voigt line profile.
A simplification is usually introduced by assuming that collision and Doppler broadening are mutually independent processes. But this approximation is not obvious: an excited atom moving with high velocity (in the direction of observation) might have a different chance to undergo a broadening collision as a slowly moving atom. A general theory has been given by Galatry (1961) for the combined effects of Doppler and collision broadening by taking into account simultaneously the perturbation of phase and change in velocity (see also the more refined and recent theories presented by Rautian and Sobel'man 1967, Edmonds 1968, Gersten and Foley 1968, Smith et al. 1971, 1971a, Nienhuis 1973, Ward, Cooper and Smith 1974, Herbert 1974, and Berman 1975). For simplicity's sake, we shall here assume mutual independence of both broadening effects. The neglect of mutual dependence is permitted, anyway, as far as the central portion of the line profile is concerned (see Mizushima 1967; see also Sect. VII.2b-6). The combined line profile is then found by folding mathematically the normalized Doppler and Lorentzian profiles given by Eqs II.236 and 244 respectively.

The procedure of *folding* or *convolution* may be visualized as follows (see Fig. II.13; see also Eq. III.17 in Sect. III.4c). The frequency scale is divided into a large number of infinitesimal intervals $d\nu$ and the Lorentzian profile is correspondingly decomposed into an equally large number of narrow rectangular segments with width $d\nu$ and height $S_{\nu,L}$ (some of these segments are shown in Fig. II.13). As a result of Doppler broadening, each of these segments obtains a Gaussian profile with a peak at the same frequency and with the same area = $S_{\nu,L} d\nu$. The contributions of all these, partly overlapping elemental Doppler curves are added for a given frequency ν. The result of these additions, represented as a function of ν, describes the combined or convoluted line profile. One easily recognizes that the integrated combined line profile meets again the normalization condition, i.e., is equal to unity, since the original Lorentzian profile was normalized. The same result would have been obtained, of course, when one starts by decomposing the original Doppler profile, etc. In general, the normalized combined line profile will be broader than either of the two original profiles and its peak will be lower than the peaks of either original profile.

Mathematically the result of this folding is described by the following

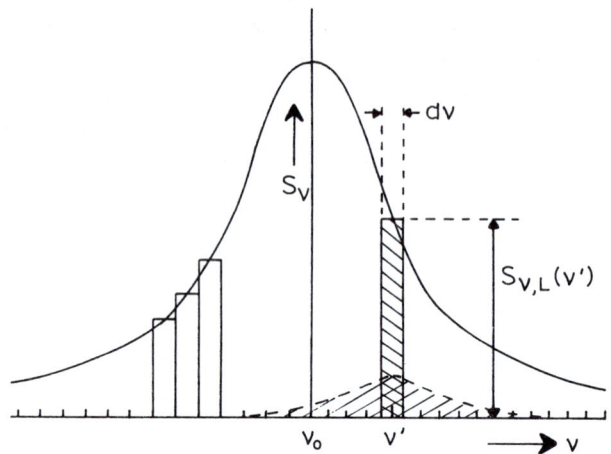

Fig. II.13 Illustration of the procedure of folding or convolution of a normalized Lorentzian profile $S_{\nu,L}(\nu)$ with a normalized Doppler profile $S_{\nu,D}(\nu)$ (see text). One infinitesimal segment of the former profile at frequency ν' and with (shaded) area $S_{\nu,L}(\nu')d\nu$ is shown, that is to be replaced by the broken, Gaussian shaped curve representing $S_{\nu,D}(\nu_0 - \nu' + \nu) \times S_{\nu,L}(\nu')d\nu$ as a function of ν; the (shaded) area under the latter curve is the same and it is centred at ν'. Summation of all such elemental Doppler curves, obtained for each segment $d\nu$, yields the convolution of the two line profiles as a function of ν. The actual result of such a convolution is shown in Fig. II.12 for a special example.

integral expression for the normalized combined line profile

$$S_{\nu,V} = (2\sqrt{\ln 2}/\pi\sqrt{\pi})(a/\delta\nu_D) \int_{-\infty}^{\infty} \frac{\exp[-y^2]dy}{a^2 + (v-y)^2}, \qquad (II.245)$$

which is called the *Voigt distribution function* (see Mitchell and Zemansky 1961, Traving 1968, and Unsöld 1968). The parameters a and v with dimension one are defined by

$$a \equiv (\delta\nu_L/\delta\nu_D)\sqrt{\ln 2} = 0.84\ \delta\nu_L/\delta\nu_D, \qquad (II.246)$$

$$v \equiv (\nu - \nu_0)\ 2\sqrt{\ln 2}/\delta\nu_D, \qquad (II.247)$$

while y is an integration variable with dimension one. Note that the factor preceding the integral in Eq. II.245 equals the peak value of the normalized *Doppler* profile (given by Eq. II.241) multiplied by a/π. The symmetry of the Voigt function with respect to ν_0 is easily recognized from Eq. II.245. The Voigt profile is neither pure Gaussian nor pure Lorentzian; only in the limit $a \to 0$ or $a \to \infty$ does it approach either of these functions asymptotically.

THE INTENSITIES OF RADIATIVE TRANSITIONS   Sect. II.5b

The exponential form of the numerator in the integrand stems from the original Doppler function, while the appearance of quadratic terms in the denominator is a reminescent of the original Lorentzian function. The Voigt distribution function is fully determined by $\delta\nu_D$ and the a-*parameter* which will be shown in Sect. II.5c-2 to play an important role for the curves-of-growth. When we convert the frequency scale into a v scale according to Eq. II.247 and express the distribution function accordingly per unit of v interval, the Voigt profile solely depends on the a-parameter. This parameter varies as some moderate power of temperature, since the Doppler and collision widths do so (see Sect. VII.2b).

It should be noted that in the older literature an a-parameter has occasionally been used that is <u>twice</u> the a-parameter given by Eq. II.246. The latter a-parameter is identical to the a-parameter defined by Unsöld (1968).

Neither the exact width nor the exact peak value of the Voigt profile can be simply expressed as a function of a and $\delta\nu_D$. Numerical values for the ratio $(S_{\nu,\mathbf{v}}/S_{\nu,D})$ at $\nu = \nu_0$ for different a-parameters can be found in the literature (see, e.g., Posener 1959, Mitchell and Zemansky 1961, Armstrong 1967, and L'vov 1970). The following approximate expressions can be given for this ratio within an error limit of 10%

$$(S_{\nu,\mathbf{v}}/S_{\nu,D})_{\nu=\nu_0} \simeq (1+1.2\,a)^{-1} \quad \text{for} \quad 0<a<2, \qquad (II.248)$$

$$(S_{\nu,\mathbf{v}}/S_{\nu,D})_{\nu=\nu_0} \simeq 0.56/a \quad \text{for} \quad a>2. \qquad (II.249)$$

The half-intensity width, $\delta\nu_V$, of the Voigt profile cannot be found by simply adding $\delta\nu_L$ and $\delta\nu_D$. For all a-values one has approximately within a 1% error (see Whiting 1968; for further empirical approximations see Olivero and Longbothum 1977)

$$\delta\nu_V \simeq \tfrac{1}{2}\delta\nu_L + \sqrt{\tfrac{1}{4}\delta\nu_L^2 + \delta\nu_D^2}. \qquad (II.250)$$

Using the approximated $\delta\nu_V$ from Eq. II.250, one finds the peak value of the Voigt distribution within a 1% error limit from (see Whiting 1968)

$$(S_{\nu,\mathbf{v}})_{\nu=\nu_0} \equiv \frac{1}{\Delta\nu_{\text{eff}}} \simeq \frac{1}{\delta\nu_V[1.065 + 0.447(\delta\nu_L/\delta\nu_V) + 0.058(\delta\nu_L/\delta\nu_V)^2]}.$$
$$(II.251)$$

For the calculation of the Voigt function computational programs have been worked out (see Young 1965a, Hummer 1965, Armstrong 1967, and Herbert 1974), while numerical values have been tabulated in the literature (see, e.g., Penner 1959, Posener 1959, Mitchell and Zemansky 1961, and Unsöld 1968; see for a complete list of references Olivero and Longbothum 1977). In special ranges of a and (or) v, auxiliary mathematical functions and expansions can be used to facilitate the

Chapt. II   BASIC CONCEPTS AND GENERAL RELATIONSHIPS

approximative calculation of the Voigt function (see, e.g., Penner 1959, Mitchell and Zemansky 1961, Young 1965a, Unsöld 1968, Traving 1968, and Whiting 1968).

Generally speaking, the behaviour of the Voigt function in the line wings is dominated by the Lorentzian function, as the latter drops appreciably more slowly with increasing distance from the line centre than does the exponential Gaussian function. This behaviour of the line wings is even found in cases where the Lorentzian width is appreciably smaller than the Doppler width. On the other hand, the Voigt function approaches more closely a Gaussian function near the line centre for $a<1$ (see Unsöld 1968). An example of a Voigt profile is presented in Fig. II.12 for $a=0.84$, i.e. for $\delta\nu_L = \delta\nu_D$ (see Eq. II.246), together with the separate pure Lorentzian and Doppler profiles. Values of the a-parameter found in flames at 1 atm pressure are of this order of magnitude (see Sect. VII.4d).

Whenever extensive and precise calculations of the Voigt profile are made, the premises in the derivation of the Voigt formula, Eq. II.245, should be well borne in mind: mutual independence of collision and Doppler broadening, neglect of deviations from the Lorentzian profile, and the fulfilment of Eq. II.242 as a condition for the assumed Doppler profile. In flames, the condition of a Maxwellian velocity distribution for the existence of a Gaussian Doppler profile is usually fulfilled.

5c   SPECTRAL LINE INTENSITY IN ABSORPTION, EMISSION, AND FLUORESCENCE
5c-1 *GENERAL*

In the foregoing sections, the rate (Section 5a) and spectral distribution (Section 5b) of optical transitions were considered for single atoms or for an infinitesimal volume element of the radiating gas. Now we shall deal with the intensity of spectral lines observed in absorption, emission or fluorescence in a hot gas (e.g., a flame) of arbitrary optical thickness. The practical problems of measuring these intensities and the influence of the spectral apparatus on the observed spectral line shape will be discussed in Sects III.4 and 14-16. Here we are interested only in the absorbing and radiating properties of the hot gas itself. Certain simplifications as to the geometrical and optical conditions and the spatial distribution of temperature and radiating species will be introduced. Then the basic dependence of the line intensity on factors such as atomic density, flame thickness, oscillator strength, temperature and line broadening will become more obvious. At the same time, possibilities of deriving these parameters from intensity measurements will be explained briefly and reference made to following chapters containing more detailed discussion. The theoretical dependence of the line intensity on atomic density explains the shape of the analytical curves in

THE INTENSITIES OF RADIATIVE TRANSITIONS   Sect. II.5c

atomic spectroscopy. The emission intensity will be treated under thermal equilibrium conditions only. Nonthermal emission in flames will be considered in Sect. VI.4 in relation to the excitation mechanisms.

In the following, mainly atomic line intensities will be considered. Molecular band intensities in emission will be discussed briefly in Sect. 5c-3. Molecular absorption and fluorescence play only a subordinate role for metal vapours in flames and will not be considered here.

Absorption will be considered first (Sect. 5c-2), so that self-absorption of emission lines can then be treated (Sect. 5c-3) in a straight-forward way. Fluorescence is a combined effect of absorption and emission, and its discussion in Sect. 5c-4 is a logical conclusion to this section.

To ensure an unambiguous interpretation of the quantitative expressions of line intensities, the nomenclature and main symbols used for radiant quantities are summarized in Appendix A.5. In practice it is, however, often unnecessary to consider what radiant quantity is really measured, or to refer in a discussion to a particular radiant quantity when its specification is either evident or irrelevant. Often relative radiation measurements are expressed in arbitrary units, for example, scale divisions. Following the recommendations of IUPAC (1972, 1976a) we shall use in these cases the indefinite expression *intensity*, I, for the relative strength of radiation in a loose sense. Note, however, from Appendix A.5 that the term 'radiant intensity' with same symbol I has a well-defined meaning and dimension. It will usually be clear from the context whether I is used in a loose or in a strict sense.

In Appendix A.4 and in the following sections, spectral positions are characterized by wavelengths in vacuo rather than frequencies. This is because the intensity expressions to be presented will be quoted frequently in the chapters dealing with experimental situations where the use of wavelengths is more customary. Spectral quantities described in terms of wavelength can easily be translated in terms of frequency through Eqs II.208 and 209.

5c-2   *THE STRENGTH OF ABSORPTION*

Absorption line profile; Beer's law. When a radiation beam from a background source is focused into a flame containing metal vapour, loss of beam intensity may occur due to resonance absorption (see Sect. II.5a-2). Reflection at the flame boundaries and nonresonant scattering by free atoms or molecules can here be fully neglected. The effect of induced emission on the net absorption will be treated separately in this subsection. The contribution of resonance fluorescence to the beam intensity will be ignored (see Sect. II.5a-2) as will be the contribution from the thermal radiation of the absorbing gas, which can be eliminated, e.g., by a modulation technique (see Sect. III.15).

Chapt. II   BASIC CONCEPTS AND GENERAL RELATIONSHIPS

The strength of absorption is expressed by the *absorption factor* α (also called *absorptance*) being the ratio of absorbed intensity I_A to incident intensity I_0

$$\alpha \equiv I_A/I_0 \ . \tag{II.252}$$

The term intensity is here used in a loose sense and may stand for the radiant flux or any other quantity that specifies the strength of the radiation beam. Denoting the transmitted intensity by I_t, we define the *transmission factor* τ (also called *transmittance*) by

$$\tau \equiv I_t/I_0 \ . \tag{II.253}$$

Because $I_t = I_0 - I_A$, we have

$$\alpha = 1 - \tau \ . \tag{II.254}$$

The (*decadic*) *absorbance* A or A^{abs} is defined by[†]

$$A^{abs} \equiv -\log_{10} \tau = -\log_{10}(1-\alpha) \ . \tag{II.255}$$

Note that the absorbance ranges from 0 to infinity, whereas the absorption factor ranges from 0 to 1. In analytical atomic absorption spectroscopy the quantity A is preferred to α because one obtains with the former quantity straight analytical curves (see also below).

By assuming a strictly monochromatic radiation beam with variable wavelength, we can define $\alpha(\lambda)$, etc., as a function of wavelength λ (or frequency ν). When a spectral line is recorded in absorption against a background source with continuous spectrum, a dip is seen with a peak at central wavelength λ_0. If we assume infinite spectral resolution of the spectrometer, the shape of this dip is described by the function $\alpha(\lambda)$, which is called the (*true*) *absorption line profile*. The clue to the quantitative description of this profile is given by Beer's law which connects the absorption factor $\alpha(\lambda)$ to the absorptivity $k(\lambda)$ defined in Sect. II.5a-2.

In the derivation of Beer's law, we consider the ideal case of a parallel, strictly monochromatic radiation beam traversing a slab of gas with thickness l in the beam direction (here chosen as x-axis). The (irradiated) section of the gas is assumed to have a uniform temperature T, whereas the density, $n_p(x)$, of atoms in the absorbing state may be a function of x. The beam is assumed to be sufficiently narrow to consider n_p at any x as practically constant over the beam cross section. Excluding absorption by other atoms or molecules as well as scattering by nonvolatilized metal compounds (see Sect. III.9), we find the transmitted intensity

---

[†] The superscript 'abs' may be dropped when there is no risk of confusion with the Einstein probability A.

THE INTENSITIES OF RADIATIVE TRANSITIONS   Sect. II.5c

$I_t(\lambda)$ by integrating $dI(x,\lambda)/dx$ from $x=0$ to $x=l$ with the help of Eq. II.172. The result obtained is called Beer's law (or Beer-Lambert's law)

$$I_t(\lambda) = I_0(\lambda) \exp[-\int_0^l k(x,\lambda)dx]. \qquad (II.256)$$

Here use is made of the boundary condition: $I(0,\lambda) = I_0(\lambda)$. According to Eq. II.185 the absorptivity $k(x,\lambda)$ depends on x through: $k(x,\lambda) = \kappa(\lambda) n_p(x)$, where $\kappa(\lambda)$ is the atomic absorptivity which is independent of atomic density n_p.

In the above integration, $k(x,\lambda)$ was assumed to be independent of $I(x,\lambda)$. A variation of absorptivity with intensity could result from a possible dependence of n_p or of the spectral profile of $\kappa(\lambda)$ on intensity. However, it was shown in Sect. II.5b-2 that absorption (and emission) transitions induced by a radiation field are unlikely to contribute noticeably to the optical line broadening under flame conditions. Since absorption lines observed in flames usually originate from the ground state or a low-lying excitation state $(E_{exc} < 1 \text{ eV})$ with a relatively high thermal population, n_p too will hardly be affected by the absorption transitions and thus by the beam intensity. An exception must be made for the case when a strong laser beam causes saturation (see the end of this subsection).

In the following, we shall suppose for simplicity's sake that n_p does not depend on x nor on I. Writing $k(\lambda) = \kappa(\lambda) n_p$ and using the definitions of $\alpha(\lambda)$ and $\tau(\lambda)$, we then obtain

$$\alpha(\lambda) = 1 - \tau(\lambda) = 1 - \exp[-\kappa(\lambda) n_p l], \qquad (II.257)$$

and with the help of Eq. II.255

$$A^{abs}(\lambda) = (\log_{10} e) \kappa(\lambda) n_p l = 0.434 \kappa(\lambda) n_p l. \qquad (II.258)$$

Using the numerical expression for $\kappa(\lambda)$ which is obtained from Eq. II.186 with the aid of Eq. II.209, we finally get (in CGS units)

$$\alpha(\lambda) = 1 - \exp[-8.83 \times 10^{-13} \lambda_0^2 S_\lambda(\lambda) f_{pq} n_p l], \qquad (II.259)$$

$$A^{abs}(\lambda) = 3.78 \times 10^{-13} \lambda_0^2 S_\lambda(\lambda) f_{pq} n_p l. \qquad (II.260)$$

The spectral profile of the absorption line is related to the wavelength dependence of $S_\lambda(\lambda)$. For a given λ, the strength of absorption depends further only on the <u>product</u> of the oscillator strength f_{pq}, the lower-state density n_p, and the thickness l. A doubling of the thickness is thus equivalent to a doubling of the atomic density or the f-value. Since l may vary with the beam direction, $\alpha(\lambda)$ depends on direction, too. For resonance lines the lower-state density n_0 is, virtually, equal to the total density of free atoms n_a, if there are no low-lying excited states (see Sect. II.3b-2). When there are low-lying excited states, we have: $n_0 = (g_0/Q)n_a$, where the partition function Q is only weakly dependent on temperature T (see Sect. II.3b-2). A significant exponential dependence of n_p on T exists, according to the Boltzmann law, when the lower state is at least a few times kT above the ground state. The strong decrease in the Boltzmann factor with increasing excitation energy explains why in flames atomic absorption is mostly found for resonance lines only. However, in hot flames ($T \approx 3000$ K), absorption lines originating from levels as high as about 1 eV above the ground level may become noticeable (see Alkemade and Herrmann 1979).

We shall now consider more closely the spectral profiles of $\alpha(\lambda)$ and $A^{abs}(\lambda)$. From Eq. II.260 the spectral profile of the absorbance appears to be identical to the profile function, $S_\lambda(\lambda)$, for the absorptivity, for any value of n_p. Under flame conditions, $S_\lambda(\lambda)$ is usually described by a Voigt function which depends on the Doppler width $\delta\lambda_D$ and a-parameter (see Eq. II.245). When n_p is enhanced by an arbitrary factor, the absorbance is enhanced by the same factor at any λ.

The situation is different for $\alpha(\lambda)$. In the range of n_p values at which the optical thickness $k(\lambda)l$ is small compared to unity for any $\lambda \approx \lambda_0$, we may approximate Eqs. II.257 and 259 by series expansion

$$\alpha(\lambda) \to k(\lambda)l = \kappa(\lambda) n_p l = 8.83 \times 10^{-13} \lambda_0^2 S_\lambda(\lambda) f_{pq} n_p l \quad \text{for} \quad n_p \to 0. \quad (II.261)$$

In the limit of vanishing optical thickness, the spectral profile of $\alpha(\lambda)$ is thus again identical to $S_\lambda(\lambda)$ (see Fig. II.14). In this limit, $\alpha(\lambda)$ is also proportional to n_p at any λ and differs from the absorbance by a constant factor, 0.434. However, it follows from Eq. II.257 that $\alpha(\lambda)$ grows less than proportionally with increasing n_p when the optical thickness becomes of order unity. In the limit of high optical thickness, $\alpha(\lambda)$ approaches the plateau value unity (see curve labelled ∞). The closest approach to this plateau is found at the line centre, where the optical thickness as a function of λ is maximum. This results in a flattening of the absorption line profile near the line centre and consequently in an increase in half-intensity width (see Fig. II.14). For high optical thickness near the line centre, the absorption line develops in breadth rather than in height upon further increase of atomic density. This additional broadening effect due to a large optical thickness is sometimes called *concentration-broadening* and should be well distinguished from the broadening effects discussed in Sect. II.5b for the $k(\lambda)$ profile. The width of the latter profile is independent of atomic density and equals the absorption line width in the low-density limit only.

The exact profile of $\alpha(\lambda)$ depends in rather a complicated way on $\delta\lambda_D$ and on the a-parameter which together determine the Voigt function through an integral expression (see Eq. II.245). A simplification in the numerical calculations is

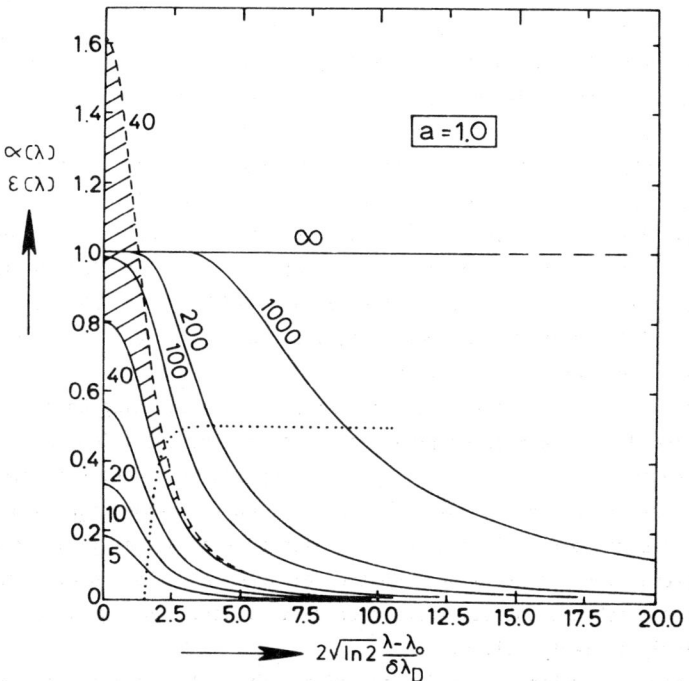

Fig. II.14 Theoretical spectral profiles (full curves) of the absorption factor $\alpha(\lambda)$ [which is identical to the emission factor $\varepsilon(\lambda)$] calculated for a resonance line with a-parameter = 1.0 for different atomic densities, n_0, in the ground state. Only the right half of the symmetrical profiles is shown; wavelength differences have been divided by the Doppler width $\delta\lambda_D$ and multiplied by $\sqrt{\ln 2}$. The figures indicated on each profile represent $(n_0 fl/b)$ in cm^{-2} s (for an explanation of symbols see the caption to Fig. II.15). The dashed curve shows the profile one would expect for $n_0 fl/b = 40$ cm^{-2}s were there no 'concentration- (or self-absorption) broadening'; this curve has the same shape as the absorption profile $k(\lambda)$. The hatched area between this dashed curve and the corresponding full curve represents half of the loss in total emission intensity due to self-absorption. The dotted curve is the locus of the points on the full curves where $\alpha(\lambda)$ [or $\varepsilon(\lambda)$] has dropped to half its peak value. Note that the half-intensity width is concentration-dependent and that the central portion of the profiles flattens out at high densities. For $n_0 \to \infty$, the half-intensity width tends asymptotically to infinity; in this limit the thermal flame emission behaves like a black body.
(From Jansen 1976; note that a factor 2 should be added to the expression for x in Fig. 2 of Jansen 1976.)

Chapt. II   BASIC CONCEPTS AND GENERAL RELATIONSHIPS

obtained by scaling all wavelengths by the Doppler half-intensity width (see also Sect. II.5b-3 and Fig. II.14). The profile expressed as a function of $(\lambda - \lambda_0)/\delta\lambda_D$ is independent of $\delta\lambda_D$ for given a-parameter.

Absorption factor with arbitrary spectral source profile. Atomic absorption experiments are usually not aimed at the measurement of α as a function of λ. The narrowness of absorption lines in flames (of the order of 0.01 to 0.1 Å) would require a high spectral resolution or special spectral scanning techniques (see Sect. VII.3b). In atomic absorption spectroscopy, one usually measures α (or A^{abs}) for a background source with a fixed spectral intensity distribution around λ_0. Let the spectral intensity of the incident source radiation (possibly after passing a spectrometer with a relatively broad bandwidth $\Delta\lambda_M$) be described by $(I_\lambda)_0$ as a function of λ. The total incident intensity I_0 which is actually measured in the absence of absorption is obtained by integrating $(I_\lambda)_0$ over wavelength. For a given atomic absorption line, the absorption factor measured is then given by

$$\alpha \equiv \frac{I_0 - I_t}{I_0} = 1 - \frac{\int (I_\lambda)_0 \exp[-\kappa(\lambda) n_p l] d\lambda}{\int (I_\lambda)_0 d\lambda}, \qquad (II.262)$$

where the integration extends over the whole wavelength interval in which $(I_\lambda)_0$ exists. The latter expression follows directly from Beer's law (Eq. II.257). The variation of α with lower-state density n_p now involves, in general, the spectral characteristics of the background source. This variation is typically different in the case of a 'narrow-line source' (measurement of peak absorption) and a 'continuum source' (measurement of integral absorption). We shall consider these two borderline cases more closely now. L'vov (1970) has given expressions for some typical intermediate cases (source line profile equal to that of absorption line; source line with arbitrary Gaussian profile; source line split into two monochromatic components). Van Gelder (1970) has presented extensive calculations of atomic flame absorbance for different emission line profiles that may be expected with gas-discharge sources.

Peak absorption. A *narrow-line source* is defined as one which radiates only within a wavelength interval that is narrow in comparison with the width of the $k(\lambda)$ profile. It does not suffice that the source line width is small compared to the width of the $\alpha(\lambda)$ profile [the latter may appreciably exceed the width of the $k(\lambda)$ profile at high optical thickness; see Fig. II.14]. Some spectral line sources emitting the same atomic line as the absorption line may meet this condition approximately (see Sect. III.7c). If we disregard a possible absorption line shift resulting from collisions with flame molecules, the centres of the source-emission and flame-absorption lines coincide at λ_0. Then $(I_\lambda)_0$ occurring

THE INTENSITIES OF RADIATIVE TRANSITIONS   Sect. II.5c

in Eq. II.262 differs from zero only in a narrow wavelength interval around λ_0 in which $\kappa(\lambda)$ can be considered as constant and equal to its maximum value $\kappa(\lambda_0)[\equiv \kappa_m]$. Making this approximation in the evaluation of the integral for I_t in Eq. II.262, we find for the *peak absorption factor* $\alpha(\lambda_0)$

$$\alpha(\lambda_0) = 1 - \frac{\exp[-\kappa_m n_p l] \int (I_\lambda)_0 \, d\lambda}{\int (I_\lambda)_0 \, d\lambda} = 1 - \exp[-\kappa_m n_p l]. \quad (II.263)$$

For the *peak absorbance*, one immediately finds by putting $\lambda = \lambda_0$ in Eq. II.258

$$A^{abs}(\lambda_0) = 0.434 \, \kappa_m n_p l, \quad (II.264)$$

which shows a strictly linear dependence on n_p. If n_p varies proportionally to the concentration of the metal salt in the nebulized solution, a linear analytical curve is thus obtained for $A^{abs}(\lambda_0)$ in analytical atomic absorption spectroscopy. When $\alpha(\lambda_0)$ is plotted versus concentration, a convex analytical curve approaching a horizontal plateau is obtained in the concentration range where $\kappa_m n_p l \geqslant 1$. This shows the advantage of using absorbance values instead of absorption factors in analytical absorption spectroscopy with narrow-line sources.

In practice, deviations from strict linearity may occur for the absorbance curves (see, e.g. Christian and Feldman 1970, L'vov 1970, Pinta 1971, de Galan and Wagenaar 1971, and Alkemade and Herrmann 1979). The validity of Eq. II.264 was based on the following suppositions which may not be fully realized in practice: all light rays travel over the same distance, l, through the absorbing flame, the source line has a negligible spectral width, and the spectral line has no unresolved (hyper-)fine-structure. The effect of finite source line width on the analytical curve has been considered by Winefordner and Vickers (1964) and van Gelder (1970), while that of hyper-fine-structure has been considered by the same authors and by Vidale (1960). The effect of the line shift will be discussed in Chapt. VII.

For *low optical thickness*, we may again approximate $\alpha(\lambda_0)$ by a series expansion of the exponential term in Eq. II.263. Using Eq. II.187 we find (in CGS units)

$$\alpha(\lambda_0) \rightarrow \kappa_m n_p l = 8.83 \times 10^{-13} \lambda_0^2 f_{pq} n_p l / \Delta\lambda_{eff} \quad \text{for} \quad n_p \rightarrow 0. \quad (II.265)$$

The latter equation shows clearly the effect of line broadening on the peak absorption. For a Doppler profile, $\Delta\lambda_{eff}$ follows from Eqs II.240 and 241, if the flame temperature is (approximately) known. For a Voigt profile, $\Delta\lambda_{eff}$ can be calculated from Eq. II.251 in combination with Eq. II.208. It is recalled that in flames self-broadening is usually negligible (see Sect. II.5b-1), so that $\Delta\lambda_{eff}$ in Eq. II.265 may be considered as being independent of the atomic metal density.

Chapt. II   BASIC CONCEPTS AND GENERAL RELATIONSHIPS

Equation II.265 can be used in analytical atomic absorption spectroscopy to estimate the *characteristic concentration*, i.e., the solution concentration that produces an absorption factor of 1% or an absorbance of 0.0044 (see IUPAC 1976a). The relation between density of atoms in the absorbing state (usually the ground state) and solution concentration must then be known (see Sect. V.5).

It is possible to derive *f-values* or *atomic densities* from peak-absorption measurements, if the effective line width is known (see Russel, Shelton and Walsh 1957, and L'vov 1965, 1970a and 1972). In the determination of relative f-values from peak absorption measurements for lines of the same atom, only relative n_p values and relative effective line widths are required. When absorption takes place from low-lying excited states, relative n_p values can be safely obtained from Boltzmann's law, if the flame temperature is known. The experimental determination of atomic densities (or f-values) from absorption measurements in flames will be discussed in Chapt. V.

Integral absorption. When a *continuum source* (e.g., a tungsten strip lamp) is used as background source in atomic absorption spectroscopy, we may consider its spectral intensity to be constant over the whole width of the absorption line (i.e. $I_\lambda = I_{\lambda_0}$). In this second border-line case, the evaluation of the integral in Eq. II.262 again becomes simple. We shall now consider the absorbed intensity, $I_A \equiv I_0 - I_t$, rather than the absorption factor, I_A/I_0, since the former does not contain explicitly the bandwidth of the spectral apparatus transmitting the continuous radiation.[†] It is assumed that this bandwidth is sufficiently large to encompass virtually the whole absorption line profile. We then have from Eq. II.262, using $I_A = \alpha I_0$ and $I_0 = (I_{\lambda_0})_0 \int d\lambda$,

$$I_A = (I_{\lambda_0})_0 A_t^{(\lambda)} = (I_{\nu_0})_0 A_t^{(\nu)} , \qquad (II.266)$$

where the *integral absorption* A_t is defined by[*]

$$A_t^{(\lambda)} \equiv \int \alpha(\lambda) d\lambda = \int \{1 - \exp[-\kappa(\lambda) n_p l]\} d\lambda , \qquad (II.267)$$

$$A_t^{(\nu)} \equiv \int \alpha(\nu) d\nu = (c/\lambda_0^2) A_t^{(\lambda)} . \qquad (II.268)$$

The integral extends over the whole wavelength interval around λ_0 where $\alpha(\lambda)$ differs significantly from zero. In fact, $A_t^{(\lambda)}$ is the area under the $\alpha(\lambda)$ curve

---

[†] Compare Eq. II.266 with Eq. II.271. Of course, when the slit width, s, of the monochromator is varied, I_{λ_0}, considered at the exit of the monochromator, will be changed in proportion to s and thus implicitly in proportion to the bandwidth if the latter is determined by s (see Sect. III.4b).

[*] The superscripts (λ) or (ν) may be dropped when they are evident from the context.

THE INTENSITIES OF RADIATIVE TRANSITIONS   Sect. II.5c

presented in Fig. II.14. Note that $A_t^{(\lambda)}$ has dimension [l] and should be expressed in the same unit as the wavelength. In contrast, $A_t^{(\nu)}$ has dimension $[t]^{-1}$ and is expressed in Hz. The integral absorption $A_t^{(\nu)}$ equals the *total absorption*, A_G, as defined by Mitchell and Zemansky (1961), divided by 2π. The integral absorption is also known as the *equivalent width* of the absorption line, especially in the astrophysical literature. In the following, we shall drop the superscript λ in $A_t^{(\lambda)}$.

Using the numerical expression for $\alpha(\lambda)$ in Eq. II.259, the integral absorption can be computed as a function of the product $(f_{pq}\,n_p l\,\lambda_0^2)$, if the Voigt profile is known. Before presenting results of these calculations, we shall first consider two asymptotic expressions for A_t that hold in the limits of low and high optical thickness at the line centre, respectively.

At *low optical thickness* or low atomic densities, we obtain from the asymptotic expression for $\alpha(\lambda)$ in Eq. II.261 through Eqs II.267 and 183

$$A_t \to \int \kappa(\lambda)\,n_p l\,d\lambda = (\pi e^2/m_e c^2)\,\lambda_0^2 f_{pq}\,n_p l, \quad \text{for} \quad n_p \to 0 \qquad (II.269)$$

or numerically (in CGS units)

$$A_t \to 8.83 \times 10^{-13}\,\lambda_0^2 f_{pq}\,n_p l \quad \text{for} \quad n_p \to 0. \qquad (II.270)$$

At low optical thickness, A_t is thus **independent** of the shape and width of the spectral line and is proportional to n_p. It is thus possible to determine either the f-value or the atomic density from a measurement of A_t without knowledge of the line width (see also Sect. V.2b). On the other hand, a comparison of integral absorption and peak absorption (see Eq. II.265), in the limit of low densities, yields a value for the effective line width.

The experimental determination of A_t through Eq. II.266 requires an absolute calibration of the spectral bandwidth of the monochromator used. If the effective bandwidth, $\Delta\lambda_M$, of the monochromator (see Sect. III.4b) is small enough to treat $(I_\lambda)_0$ as independent of wavelength, we can write: $I_0 = (I_{\lambda_0})_0\,\Delta\lambda_M$. One then obtains for $\alpha_c \equiv I_A/I_0$ through Eq. II.266

$$\alpha_c = A_t/\Delta\lambda_M. \qquad (II.271)$$

A measurement of α_c and $\Delta\lambda_M$ thus yields A_t. The practical problem of measuring α_c values will be dealt with in Sect. III.15. Note that in the limit $n_p \to 0$, α_c is smaller by a factor $\Delta\lambda_M/\Delta\lambda_{eff}$ than the peak absorption factor $\alpha(\lambda_0)$ (compare Eqs II.270, 271 and 265). Eq. II.271 ceases to be valid when $\Delta\lambda_M$ becomes comparable to the width of the $\alpha(\lambda)$ profile.

The incipient deviation from the asymptotic linear relationship with n_p (Eq. II.269) can be estimated as follows (see Alkemade and Zeegers 1971). Retaining the linear and quadratic terms in the series expansion of the exponential term in Eq. II.267, one gets for $\kappa(\lambda) n_p l \ll 1$

$$A_t \simeq \int \kappa(\lambda) \, n_p l \, d\lambda - \tfrac{1}{2} \int \kappa(\lambda)^2 n_p^2 l^2 \, d\lambda . \tag{II.272}$$

Defining the weighted average value of $\kappa(\lambda)$ by

$$\overline{\kappa(\lambda)} \equiv \int \kappa(\lambda)^2 d\lambda / \int \kappa(\lambda) d\lambda , \tag{II.273}$$

we have to second-order approximation in n_p

$$A_t \simeq n_p l \int \kappa(\lambda) \, d\lambda \{1 - \tfrac{1}{2} \overline{\kappa(\lambda)} n_p l\} . \tag{II.274}$$

The relative deviation from the linear term is thus of first order in n_p. The extent of this deviation may be estimated by putting $\overline{\kappa(\lambda)} \approx \kappa_m$ and applying Eq. II.187 (for the yellow Na line in usual flames at 1 atm pressure, $\kappa_m / \overline{\kappa(\lambda)} \approx 1.4$; see Alkemade 1970). It then becomes apparent that the smaller $\Delta\lambda_{\text{eff}}$ is, the lower is the n_p value at which this deviation becomes noticeable. Thus we see the influence of the line width on the behaviour of A_t (see also below).

Also in the opposite case of *high optical thickness*, i.e., high atomic density, a simple asymptotic expression can be derived for A_t (see Hinnov 1957, Mitchell and Zemansky 1961, and Unsöld 1968) which reads, in CGS units,

$$A_t \rightarrow \sqrt{\left(\frac{2\pi e^2}{m_e c^2}\right) \lambda_0^2 \delta\lambda_L \, f_{pq} n_p l} = 1.33 \times 10^{-6} \lambda_0 \sqrt{\delta\lambda_L \, f_{pq} n_p l} \quad \text{(for large } n_p\text{)} . \tag{II.275}$$

In the range of high atomic densities, A_t thus appears to vary $\propto \sqrt{n_p}$, while its value for given $f_{pq} n_p l$ depends on the half-intensity width of the Lorentzian component of the Voigt profile. The Doppler width does not show up in the high-density limit.

The latter conclusion may be understood by considering the $\alpha(\lambda)$ profiles presented in Fig. II.14 for high optical thickness. Under this condition, the central part of the profile has practically reached the plateau and does not change anymore when n_p is increased further. The increase of A_t, i.e. the area under the profile, with increasing density is then brought about mainly by an expansion in breadth of the profile. This expansion depends on the behaviour of the Voigt distribution function in the (far) line wings. But this behaviour is dominated by the Lorentzian component (see Sect. II.5b-3). It is essentially the characteristic wavelength dependence in the *wings* of the Lorentzian profile (i.e., $S_{\lambda,L} \propto (\lambda - \lambda_0)^{-2}$ for $|\lambda - \lambda_0| \gg \delta\lambda_L$; see Eq. II.244) that brings about the asymptotic square-root dependence of A_t on n_p.

In a double-logarithmic plot of A_t as a function of n_p, the parabolic, high-density branch (called the *square-root-branch*) as well as the *linear branch* at low densities appear as two straight asymptotes with slopes arctan $\tfrac{1}{2}$ and arctan 1, respectively. The value A_t^* of the integral absorption at the *intersection* of these lines is related to $\delta\lambda_L$ simply by (see Hinnov 1957, Hinnov and Kohn 1957, and Hofmann and Kohn 1961)

$$A_t^* = 2\delta\lambda_L \ . \tag{II.276}$$

This relation directly follows from equating the right-hand sides of Eqs II.269 and 275. For the density n_p^* in the intersection point one obtains (in CGS units)

$$n_p^* = \left(\frac{2\,m_e c^2}{\pi e^2}\right) \delta\lambda_L/\lambda_0^2 f_{pq} l = 2.26 \times 10^{12}\, \delta\lambda_L/\lambda_0^2\, f_{pq} l \ . \tag{II.277}$$

Thus $\delta\lambda_L$ can be obtained through Eq. II.276 by an absolute measurement of the integral absorption (see also Eq. II.271) for sufficiently low and high atomic densities, respectively. Only the ratio of atomic densities need be known. Once $\delta\lambda_L$ is known, $n_p^* f_{pq}$ follows through application of Eq. II.277 (see also Winefordner, Vickers and Remington 1965).

In the intermediate range of n_p values, an exact analytical expression cannot be given for A_t and its value must be computed numerically from Eq. II.267 with the help of tabulated values of the Voigt distribution function. Numerous graphical presentations as well as tabulations are found in the literature (see, e.g., van der Held's original curves in Mitchell and Zemansky 1961; Penner 1959 and Unsöld 1968 where there is also an extensive discussion; Hollander 1964, Young 1965a, van Trigt, Hollander and Alkemade 1965 who give computer calculations with an accuracy of about $1:10^4$; Penner and Kavanagh 1953, Hinnov 1957, Posener 1959, Hummer 1965, Armstrong 1967, Goldman 1968, Whiting 1968, and van Trigt 1968). A simple, approximate expression has been given by Rodgers and Williams (1974), and by Hill (1979).

Curve-of-growth (COG). The graphical presentation of the integral absorption as a function of atomic density is usually given in the form of universal *curves-of-growth* by plotting $A_t^{(\nu)} \sqrt{\ln 2}/\delta\nu_D$ vs. $f_{pq} n_p l \sqrt{\ln 2}/\pi\delta\nu_D$ on a double-logarithmic scale (see Fig. II.15). The ordinate values have dimension unity, whereas the abscissa values have dimension $[l]^{-2}[t]$ and are here expressed in $cm^{-2}s$. The position and shape of the curves thus obtained are determined solely by the a-parameter, and their linear asymptotes at the low end of the abscissa scale coincide. These curves-of-growth depend solely on the a-parameter since all frequencies are scaled by the Doppler width (see Sect. II.5b-3). Conversion to wavelengths is obtained by replacing on the ordinate axis: $A_t^{(\nu)} \sqrt{\ln 2}/\delta\nu_D$ by $A_t^{(\lambda)} \sqrt{\ln 2}/\delta\lambda_D$, and on the abscissa axis: $f_{pq} n_p l \sqrt{\ln 2}/\pi\delta\nu_D$ by $f_{pq} n_p l\, \lambda_0^2 \sqrt{\ln 2}/\pi c\delta\lambda_D$. In this replacement, use is made of the relations in Eqs II.268 and 208.

The shape of all curves-of-growth with $a > 0$ in the asymptotic ranges of low or high densities is the same. In the intermediate region, the shape shows a characteristic dependence on the a-parameter for $a \leqslant 1$. For $a \geqslant 1$ the slope of the curve decreases monotonically with increasing abscissa value to the final value

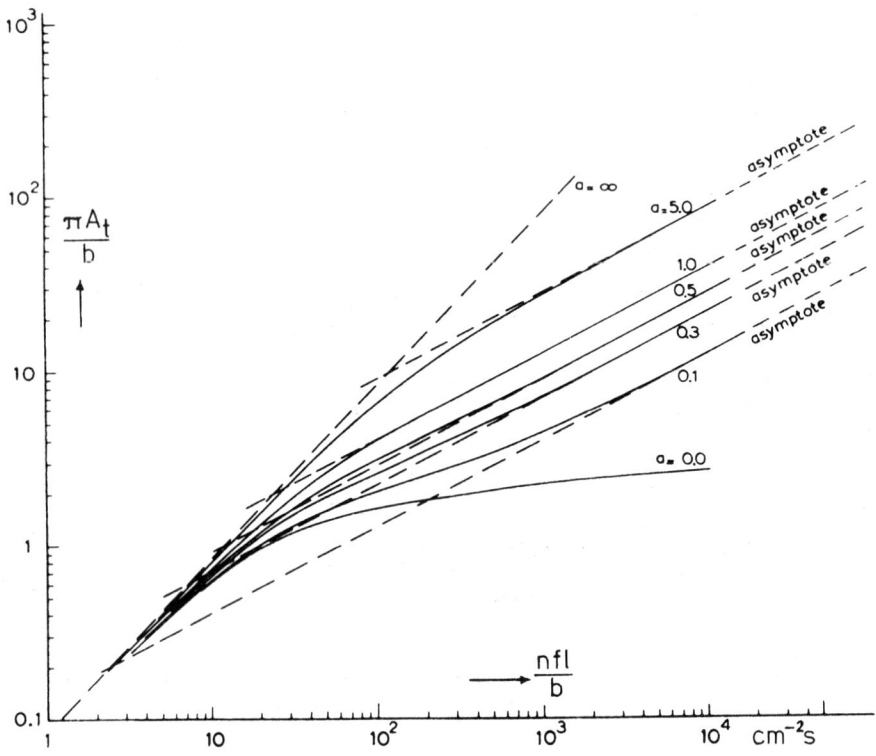

Fig. II.15 Double-logarithmic plot of theoretical curves-of-growth based on a Voigt profile with various a-parameters as indicated in the figure. Legends: A_t = integral absorption (also called: 'equivalent width', and equal to $A_G/2\pi$ with A_G = 'total absorption' as defined originally by Mitchell and Zemansky 1961; see for definition Eq. II.268); $b = \pi\delta\nu_D/\sqrt{\ln 2}$ with $\delta\nu_D$ = half-intensity width of Doppler profile (given by Eq. II.238); n = uniform number density of atoms in the absorbing state; f = oscillator strength of the absorption considered; l = flame depth measured along the line of observation; a is parameter defined by Eq. II.246. Dotted lines are asymptotes.

THE INTENSITIES OF RADIATIVE TRANSITIONS   Sect. II.5c

arctan $\frac{1}{2}$. For $a < 1$, however, an inflection point appears, i.e. the slope attains a minimum value below arctan $\frac{1}{2}$ somewhere in the transition region between the two asymptotes. The value of this minimum decreases markedly with decreasing a-parameter. A peculiar flat-shaped curve is obtained for $a = 0$, i.e., for the case of pure Doppler broadening. The steep slope in the wings of the Gaussian Doppler profile means that the $\alpha(\lambda)$ profile expands but slowly in breadth with increasing n_p. Since A_t represents the area under the latter profile, it also grows but slowly with n_p in the high-density limit.

It is possible to derive a-*parameters* experimentally from the shape of the curves for $a \leqslant 1$ by plotting the integral absorption in relative measure versus the ratio of atomic densities (see Sect. VII.3). Once the a-parameter is known, one can find $f_{pq} n_p l / \delta \nu_D$ in absolute value by shifting the relative experimental curve into coincidence with the corresponding theoretical curve-of-growth. By measuring l and calculating $\delta \nu_D$ from Eq. II.240, one then finds the f-*value* if n_p is known, or the *atomic density* n_p at any point of the curve if the f-value is known (see Sect. V.2d).

An experimental curve-of-growth can be determined by spraying a series of metal salt solutions with varying concentrations into the flame. The density, n_p, of atoms in the absorbing state must attain sufficiently high values to make the flame optically thick. Typical threshold values for n_0 are of the order of 10^{11} cm^{-3} for the yellow Na resonance line for an assumed value of $l = 1$ cm. This condition can usually be met only with resonance lines. In flame atomic absorption spectroscopy, analytical curves have been found with a continuum source which do indeed show a parabolic shape at high atom concentrations (see for a survey, Alkemade 1968).

The many *presuppositions* underlying the simple *curve-of-growth theory* should be borne in mind in experimental applications. First of all, the absorption line should be simple and well isolated from other absorbing lines. When the line shows an instrumentally unresolved (hyper-)fine-structure with known intensity ratios, the theory can be extended simply if these components do not intrinsically overlap[†] (see Alkemade 1954, van Trigt, Hollander and Alkemade 1965; see also Sect. VII.2b-7). For the unresolved first resonance doublets of Na, K, Cs and Rb which show no intrinsic spectral overlap in flames, we obtain the asymptotic expressions by replacing f by $(f_1 + f_2)$ in Eq. II.269 and by $(\sqrt{f_1} + \sqrt{f_2})^2$ in Eq. II.275, respectively (see van Trigt, Hollander and Alkemade 1965). Here f_1 and f_2 are the f-values of the doublet components. The factor 2 in Eq. II.276 for the intersection point of

---

[†] See also footnote on page 134.

Chapt. II   BASIC CONCEPTS AND GENERAL RELATIONSHIPS

the asymptotes should correspondingly be replaced by the factor 3.89, if $f_1 = \frac{1}{3}$ and $f_2 = \frac{2}{3}$. These substitution rules hold if the a-parameter is the same for both components, which is a reasonable assumption.

When intrinsic spectral overlap does occur, the theoretical calculation may become more difficult. Expressions and asymptotic expansions for the integral absorption, which depends on the degree of overlap, have been given by Matossi, Mayer and Rauscher (1949), Preobrazhenskii (1963), Rukosueva (1964), Sakai and Stauffer (1964), van Trigt, Hollander and Alkemade (1965), Plass (1965), van Trigt (1968), and Demayo, Hunter and Kruus (1968). The *degree of overlap* γ is defined as the ratio of the distance of neighbouring h.f.s. components to the half-intensity width (this width is usually assumed to be the same for all components). Partial overlap of neighbouring rotational lines may also be important in calculations of the integral absorption for molecular spectra (see Penner 1959).

The assumption of a Voigt profile for the absorptivity in the curve-of-growth theory is another idealisation. The Voigt profile was calculated by assuming independence of Doppler and collision broadening effects (see Sect. II.5b-3). Besides, collisions may bring about a shift and asymmetry of the line profile (see Chapt. VII). In contrast to the case of a narrow-line source, a shift of line centre does not affect the integral absorption measured with a continuum source. The asymmetry occurring in the line wings might affect the course of the curve-of-growth at high atomic densities (see Jansen 1976, and Jansen and Hollander 1977). Experimental curves-of-growth (especially those measured with high precision in emission; see Sect. II.5c-3) generally show the predicted parabolic behaviour in the range of high densities in flames. Asymmetry apparently does not produce serious deviations here; this is in agreement with numerical calculations of the curve-of-growth based on an actually observed, asymmetric line profile (see Sect. VII.4a).

The parabolic asymptote usually found for metal lines in flames up to high atomic densities also proves that the condition of a constant a-parameter is fulfilled here. Self-broadening would make the a-parameter dependent on atomic density, but it has apparently no noticeable effect under flame conditions (see also Sect. II.5b-1). The effect of self-broadening on the integral absorption at high densities has been treated by Rice and Ragone (1965).

Further restricting conditions as to the applicability of the curve-of-growth theory are the uniformity of the a-parameter (and therefore of temperature and gas composition) and of the atomic density in the flame section under observation. When the atomic density varies with position along the line of observation, $k(\lambda)l$ should simply be replaced by $\int_0^l k(x,\lambda)dx$, as it was done in Eq. II.256. It is essential, however, that this integral has the same value for all light rays of

THE INTENSITIES OF RADIATIVE TRANSITIONS   Sect. II.5c

the radiation beam traversing the flame. When absorption measurements are made with converging beams subtending a large solid angle, this condition may not be fulfilled and one obtains for the curve-of-growth a sort of weighted average. Finally, in measurements of integral absorption, the bandwidth of the monochromator should be large enough to encompass the wings of the absorption line just as well as the line centre. One should bear in mind that the absorption line broadens considerably at large optical thickness (see Fig. II.14).

Saturation. When saturation of the optical transition occurs in a strong laser beam, we have to replace $k(\nu)$ by the *net* absorptivity $k'(\nu)$ given by Eqs II.189, 190 and 191. (We assume here and in the following that the general conditions underlying these equations are fulfilled.) As a result the absorption factor will be reduced; it also becomes dependent on the incident beam intensity as well as on the efficiency of fluorescence Y, which is contained in the saturation parameter (see Eqs VI.38a and b). This reduction is found not only in the direction of the saturating beam, but also in other directions. The flame with metal vapour then behaves like a saturable absorber or nonlinear filter (see Hercher 1967) for frequencies around the line centre. The flame ultimately becomes transparent when the beam intensity far exceeds the saturation parameter. This holds true how-ever large the atomic density and flame depth are.

The dependence of the transmission factor on the laser irradiance for various Na concentrations in the sprayed solution has been measured by Kuhl, Neumann and Kriese (1973) in a propane-air flame.

In the general case, quantitative expressions for the net absorption factor, absorbance or integral absorption cannot be derived easily under (near-)saturation conditions. This holds because the degree of saturation, and thus $k'(\nu)$, depend on the intensity and spectral distribution of the *local* radiation field. The latter parameters depend on the attenuation of the laser beam inside the flame, which, in turn, is related to these absorption quantities. Additional complications arise when the incident laser beam is convergent and/or not uniform over its cross section, or when the beam diameter is smaller than the flame width.

To simplify matters we shall assume that the spectral density of the beam, the efficiency of fluorescence, and the atomic density are uniform along the line of observation. The degree of saturation and $k'(\nu)$, which is proportional to $\{n_p - (g_p/g_q)n_q\}$, are then uniform too (cf. Eqs II.190 and VI.37, 38a and 39a). These assumptions may be approximately met when the optical depth of the metal vapour at $\nu \simeq \nu_0$ in the laser-beam direction is small and a homogeneous flame is used that is uniformly irradiated by the laser. A special choice of flame geometry and observation direction may also justify these assumptions (see Sect. II.5c-4).

The *net* absorption factor, etc., denoted by adding a prime to the appropriate symbol, is then found simply by substituting the expression: $\{n_p - (g_p/g_q)n_q\}$ for n_p in Eqs II.257, 258 and 267. The reason becomes clear if one compares Eq. II.190 for $k'(\nu)$ with Eq. II.176 for $k(\nu)$, remembering that $k(\nu) = \kappa(\nu) n_p$. The dependence of $\{n_p - (g_p/g_q)n_q\}$ on the local intensity of the laser beam is described by Eq. VI.37, to be derived in Sect. VI.2c-5.

For example, let us apply this substitution to demonstrate the occurrence of saturable absorption. When the optical depth in the direction considered is small (low atomic density and/or strong saturation) we can approximate $\alpha'(\nu)$ by $k'(\nu)l$. According to Eqs VI.37 and 39a, to be derived in Sect. VI.2c-5, $k'(\nu)$ decreases proportionally to the inverse laser intensity when saturation is strong. Consequently, $\alpha'(\nu) \to 0$ when the laser intensity $\to \infty$. In this limit, the absolute absorbed laser power (= absorption factor times incident laser power) is independent of the laser power. This also leads to a saturation in the yield of fluorescence as a function of laser intensity (see Sect. II.5c-4).

Saturable absorption in the flame produced a peculiar effect when a flame fed by a concentrated Na solution was placed inside the cavity of a pulsed dye-laser pumped by a flash-lamp with laser-pulse duration of about 0.5 μs (see van Dijk, Alkemade and Zeegers 1978). When the laser cavity was initially tuned by a Fabry-Pérot interferometer at a wavelength near the centre of one of the Na-D lines, laser action at this wavelength was prohibited because of radiative loss inside the cavity. The laser may then be forced to oscillate first in an adjacent mode at another resonance frequency of the interferometer. After a short delay, however, laser oscillation in the original (dominant) mode re-appeared, whereas that in the adjacent mode was extinguished. When at given (high) Na concentration the laser cavity was tuned too closely to the atomic line centre, no oscillation appeared at all in the dominant mode.

The effects were explained by considering the kinetic equations that couple the two modes. One has to take into account that the radiation loss due to Na absorption diminishes with growing oscillation amplitude in the dominant mode because of the saturation effect. Laser oscillation in the dominant mode may then ultimately be restored.

At *low* Na concentration, the presence of Na vapour inside the cavity may not inhibit laser action at the wavelength of the Na line. Slightly above threshold the laser output power will, however, depend critically on the absorption and thus on the concentration of Na vapour because of the nonlinear operation of the laser. This fact may be utilized to detect small concentrations of metal atoms in a vapour cell or flame and thus to improve the detection limit by several orders of magnitude in analytical atomic absorption spectroscopy (see, e.g., Thrash, Weyssenhoff and Shirk 1971, and Kuhl, Marowski and Torge 1972).

THE INTENSITIES OF RADIATIVE TRANSITIONS   Sect. II.5c

Applying the above substitution for n_p in Eq. II.267 we find for the *net integral absorption* A'_t (defined as an integral over wavelength)

$$A'_t = \int \left\{ 1 - \exp\left[-\kappa(\lambda)\, n_p l \{1 - (g_p/g_q)(n_q/n_p)\}\right] \right\} d\lambda \,. \tag{II.278}$$

In thermal equilibrium we have: $(g_p/g_q)(n_q/n_p) = \exp[-(E_q - E_p)/kT]$ according to Boltzmann's law. Since the exponential factor is very small for optical lines in flames [see also footnote † on page 120], we can always replace A'_t by A_t under thermal equilibrium conditions.

5c-3   *THE INTENSITY OF THERMAL EMISSION AND SELF-ABSORPTION;*
       THE LINE-REVERSAL TEMPERATURE

Definition of thermal radiation. The term thermal radiation is sometimes used in the literature in rather a confused and ambiguous sense. Therefore, we present here a definition of this term that seems to be best adapted to flame spectroscopy (see Alkemade 1963, 1970a, and Alkemade and Zeegers 1971). The radiation of a spectral line emitted by atoms M in a certain portion of the flame is called *thermal* if the total (i.e., spectrally integrated) line radiance as well as the spectral shape of the line conform to Maxwell-Boltzmann equilibrium. We here assume that the temperature in the flame part considered is uniform. The existence of thermal radiation implies that the emission factor $\varepsilon(\lambda)$ and the absorption factor $\alpha(\lambda)$ are equal for any value of λ within the spectral line profile (see Eq. II.293). This equality exists if the relative population of the upper and lower levels of the optical transition with respect to the ground level conforms to the Boltzmann law. Moreover, the velocity distribution of the atoms in both levels should conform to the Maxwell law, so that the Doppler half-intensity width of the line is given by Eq. II.238.

The intensity of thermal radiation emitted in a given spectral line can be calculated if the temperature and the atomic density as a function of position in the flame are known. Thermal radiation should be clearly distinguished from black-body radiation which, according to Planck's law, depends only on temperature and wavelength. Black-body radiation is independent of the material properties of the source, unlike thermal radiation as defined above. Black-body radiation would only be found if all degrees of freedom, including the radiation field, were in full thermodynamic equilibrium. The fact that thermal radiation can deviate from Planck's law implies that the Boltzmann and Maxwell distribution laws are not obeyed exactly either, if there is an interaction between the atoms and the radiation field. The concept of thermal radiation can thus be applied only in an approximate sense, that is, on condition that this interaction is relatively unimportant for the state of the atoms. We refer to Sect. VI.4a for a further discussion of this concept and of the conditions relating to its applicability. In the following we shall assume that these conditions are fulfilled.

Self-absorption and radiation diffusion. Not all photons generated inside a gaseous light source contribute to the outgoing radiation intensity. Photons emitted by atoms undergoing a downward transition from state q to state p may be re-absorbed on their way out by other atoms of the same element that happen to be in state p and that are thereupon transferred to state q. This effect is found – in principle – with atomic as well as molecular lines and is called *self-absorption*. Self-absorption results in a loss of outgoing radiation intensity as well as in a broadening of the emission line profile (see below).

The probability for a photon to escape the light source without re-absorption is determined by the optical thickness of the source measured from the place of photon generation in the direction of observation. The optical thickness again depends on the precise frequency of the photon. In flames with metal vapour, a high optical thickness and thus marked self-absorption is mainly found near the centre of atomic resonance lines at high atomic densities. In principle, it is sufficient to know the absorptivity and photon-emission rate as functions of position and frequency in order to calculate the outgoing radiation power, inclusive of the effect of self-absorption. Such calculations, however, become very complicated for nonuniform light sources that are optically thick near the line centre and in which the population of the upper state depends on the local radiant energy density (see Sect. VI.2c). At low pressures, complications may also arise when the spectral distribution of emitted photons depends on the spectral distribution of the local radiation field due to coherence between absorbed and re-emitted photons (see Sect. II.5a-1). Through such interdependences the rates and spectral distributions of photon generation at different places in the source are linked; these distributions may also deviate from the spectral profile of the absorptivity $k(\nu)$.

When the efficiency of fluorescence Y (see Sect. II.5c-4) is close to unity as, for example, in low-pressure sources, a given photon may be re-absorbed and re-emitted many times in succession inside the source, each time changing its direction and frequency within the spectral line width. Since the absorptivity $k(\nu)$ depends on the frequency ν, the probability that the photon will be re-absorbed per unit path length will vary each time, too. The reciprocal quantity, $k(\nu)^{-1}$, can be interpreted as a mean free path length and the process of multiple re-absorption and re-emission as a *radiation diffusion*. When the major part of the photons, before reaching the boundaries of the source, is converted into heat through a collision of the second kind (see Sects II.5a-1 and 5a-2), we speak of *radiation trapping* or *imprisonment* (see, e.g., Pringsheim 1949).

A consequence of radiation trapping is that the intensity ratio of fluorescent resonance and nonresonance lines decreases as the atomic density increases (see, e.g., Held and Stephens 1975).

Radiation diffusion will lead to an increase in the *effective radiative lifetime* which is roughly given by: $(\tau_r)_{\text{eff}} = (pA)^{-1}$ (see Copley and Krause 1969). Here A is the transition probability for spontaneous emission of the resonance line and p the average probability for an emitted photon to escape the vapour cloud without being re-absorbed. This probability, which is averaged over the spectral line profile and over space, decreases with increasing ground-state density and with increasing shortest dimension of the vapour cloud (see also the discussion following Eq. VI.29 in Sect. VI.2c-4). The above expression for $(\tau_r)_{\text{eff}}$ can be visualized by saying that a photon jumps p^{-1} times between neighbouring metal atoms before escaping, and resides on each of these atoms during, on the average, A^{-1} seconds. This diffusion effect must be taken into account, or, if possible, avoided, when A is to be determined by lifetime measurements (see Sect. II.5a-2). When $(\tau_r)_{\text{eff}}$ exceeds the quenching lifetime of the excited state (that is, the reciprocal quenching rate constant; cf. Eq. II.106), radiation trapping occurs.

The calculation of the total radiation 'output' and its spectral distribution, for given 'input conditions' (collisional excitation rate in a light source; incident radiation flux in a fluorescence cell) is done through the application of the *theory of radiative transfer* (or *transport*). Such calculations are often cumbersome and have to be done by computer if Y is close to unity. In border-line cases use can be made of approximate expressions. There is special interest in such calculations in astrophysics and in the field of gas-discharge sources.[†]

For flames at 1 atm pressure with $Y \leqslant 0.5$, closed analytical expressions have been presented by Hooymayers (1966) and Hooymayers and Alkemade (1966a) in order to calculate the deviation from thermal emission due to radiative disequilibrium in successive orders of approximation (see Sects VI.2c and 4c-2). When the flame gas is mainly composed of molecular species, so that Y is considerably smaller than unity (see Sect. VI.2b), every re-absorbed photon is practically lost, that is, it is converted into heat. The (de-)excitation of the atoms is then collision-dominated (see Sect. II.3c-3) and the thermal emission intensity, including the effect of self-absorption, can be calculated in a straightforward manner (see below). Nonthermal radiation in flames will be further treated in Sect. VI.4

[†] See, for instance, the numerous publications in the Journal of Quantitative Spectroscopy and Radiative Transfer and the Special Report no. 174: 'Research in space science' on the 'Formation of spectral lines', Proc. 2nd Havard-Smithsonian Conference on Stellar Atmospheres, May 17, 1965, Cambridge, U.S.A.; see also Penner (1959), Richter (1968), Goulard, Scala and Thomas (1968), van Trigt (1969, 1970 and 1971), and Falk, Becker-Ross and Schiller (1971).

Chapt. II   BASIC CONCEPTS AND GENERAL RELATIONSHIPS

in connection with a discussion of the (de-)excitation mechanisms in Sects VI.2 and 3.

Thermal emission intensity of atomic lines (without self-absorption). A quantitative expression for the intensity of thermal radiation *in the absence of self-absorption* (i.e., for optical thickness at line centre much smaller than unity) follows directly from Eq. II.150. We find from this equation the radiance, B, of an *atomic line* as observed at a given position on the surface of the source and in a given direction by integrating over the line of observation (see also Fig. III.20 in Sect. III.14b)

$$B = (A_{qp} hc/4\pi\lambda_0) \int_0^l n_q(x)\, dx\ . \tag{II.279}$$

Here l represents the thickness of the source along the line of observation (taken as the x-axis), while $n_q(x)$ describes the dependence of excited state density on x in the interval $(0, l)$. A_{qp} is the Einstein coefficient of spontaneous emission, and hc/λ_0 is the photon energy.

Expressing $n_q(x)$ in terms of the atomic density $n_a(x)$ through Boltzmann's law (Eq. II.30), we also have

$$B = (A_{qp} hc/4\pi\lambda_0)(g_q/Q) \exp[-E_q/kT] \int_0^l n_a(x)\, dx\ , \tag{II.280}$$

where g_q is the statistical weight of the excited level, Q the atomic (internal) partition function, and E_q the excitation energy. When the atoms are distributed uniformly over the observation volume, Eq. II.280 reduces to (in CGS units)

$$B = (hc/4\pi\lambda_0)(g_q/Q) \exp[-E_q/kT]\, A_{qp} n_a l =$$
$$= 1.58 \times 10^{-17} (A_{qp} g_q/\lambda_0 Q) \exp[-E_q/kT]\, n_a l\ . \tag{II.281}$$

When self-absorption is negligible, B appears to be proportional to the product $(n_a l)$. For given $(n_a l)$, B depends strongly on temperature through the exponential Boltzmann factor, since for optical lines in flames $E_q/kT \geq 4$. For atoms with low-lying excitation levels, Q may depend slightly on T, too (see Sect. II.3b-2). Disregarding the latter temperature dependence and keeping $(n_a l)$ constant, one finds from Eq. II.281 through differentiation with respect to T

$$\frac{T}{B}\left(\frac{dB}{dT}\right) \simeq \left(\frac{E_q}{kT}\right)\ . \tag{II.282}$$

Thus in a flame of 2500 K, a relative increase of 1% (i.e., 25 deg) in T causes a relative increase in B equal to $(E_q/kT) \approx 10\%$ for the yellow Na line.

For resonance lines we have: $E_q = hc/\lambda_0$. Keeping all other quantities in Eq. II.281 constant, we find for the intensity ratio of two resonance lines at

THE INTENSITIES OF RADIATIVE TRANSITIONS   Sect. II.5c

λ_0 = 6000 and 3000 Å, respectively, a value of about 10^4 at T = 2500 K. This high
ratio explains why in flame atomic emission spectroscopy ultraviolet lines appear
to be comparatively weak.

Close-lying multiplet components have relative intensities proportional to
their $(g_q A_{qp})$ values. These again are proportional to the $(g_p f_{pq})$ values by reason of Eq. II.168, or to the f_{pq}-values if these components happen to have a
common lower level (as for the alkali doublets).

We note that the spectral line shape does not appear in Eq. II.280 or 281
for the radiance in the absence of self-absorption. The spectral line shape is
described by the function $S_\lambda(\lambda)$ defined in Sect. II.5b-1, and the spectral radiance
is therefore given by

$$B_\lambda(\lambda) = B S_\lambda(\lambda) . \qquad (II.283)$$

In flames, $S_\lambda(\lambda)$ can be reasonably well described by a Voigt function (see Eq.
II.245). The same function $S_\lambda(\lambda)$ describes the wavelength dependence of $k(\lambda)$ or
the absorption line profile in the case considered of low optical thickness.

Thermal intensity of molecular emissions (without self-absorption). For
di-atomic molecules the expression for the thermal radiance of a *rotational line*,
in the absence of self-absorption, follows directly from Eq. II.279 when we replace there the atomic transition probability A_{qp} by $A_{qp,\Sigma'\Sigma'',v'v'',J'J''}$ given by
Eq. II.194 and n_q by the expression given by Eq. II.33a and the Boltzmann law,
Eq. II.30, with $g_j \triangleq g_{\Lambda'}(2J'+1)$ [†]. We then obtain *

$$B_{line} = (16\pi^3 c/3\lambda_0^4) S_{qp,\Sigma'\Sigma'',v'v'',J'J''} n_m l \exp[-(E_q + E_{v'} + E_{r'})/kT]/Q_m ,$$
(II.284)

where n_m is the total (uniform) density of molecules, $(E_q + E_{v'} + E_{r'})$ the sum of
the electronic, vibrational and rotational energies in the upper level,[‡] and Q_m
the total internal partition function of the molecule (with the exclusion of the
nuclear partition function, which drops out). The full expression for Q_m is given
by Eq. II.34 in combination with Eqs II.35 and 37a. When the molecule has a ground
multiplet with negligible f.s. splitting and no other low-lying electronic states,

---

[†] A more detailed, critical discussion of methods for calculating the relative population of di-atomic molecular energy levels and absolute intensities of molecular lines and bands has been given by Tatum (1967) and Kuznetsova *et al.* (1974).

* This expression holds for a rotational line of a heteronuclear molecule whose rotational levels show degeneracy as to M_J and Λ-doubling.

[‡] For simplicity's sake we assume that the fine-structure splitting due to the electronic spin is small compared to kT; this is a reasonable assumption at flame temperatures.

Chapt. II   BASIC CONCEPTS AND GENERAL RELATIONSHIPS

Q_m may be approximated by (see Eqs II.38 and 39)

$$Q_m \simeq g_0 Q_{v,0} Q_{r,0} \,, \qquad (II.285)$$

where g_0 is the degeneracy of the ground multiplet, and $Q_{v,0}$ and $Q_{r,0}$ are the vibrational and rotational partition functions belonging to this multiplet. The dependence of the line radiance on T, at given $(n_m l)$, is governed mainly by the exponential Boltzmann factor in Eq. II.284 and, to a lesser extent, by the dependence of $(Q_{v,0} Q_{r,0})$ on T. At flame temperatures, the latter product varies $\propto T$ or $\propto T^2$ according to whether the vibrational energy quantum is large or small compared to kT, respectively (see Eqs II.35, 37a and 37b). The relative intensities of the rotational lines within one band are proportional to:

$$H^{\Lambda' \Sigma' J'}_{\Lambda'' \Sigma'' J''} \exp[-E_{r'}/kT]$$

(see Eq. II.197).

By plotting on a logarithmic scale the relative intensities of the rotational lines belonging to the same band, divided by ν_0^4 times the Hönl-London factor, versus $E_{r'}$ one finds T from the slope of the straight line obtained (see Herzberg 1950, Penner 1959, and Gaydon 1974). When no general thermal equilibrium exists, the T value thus found is called the *rotational temperature* T_r.

The total radiance of a *band* $v' \to v''$, belonging to an electronic transition $q \to p$, follows from Eq. II.284 after summation over all allowed $\Sigma', J' \to \Sigma'', J''$ transitions. Using Eqs II.197 and 198 and the sum rule expressed by Eq. II.200 we find

$$B_{band} = (16\pi^3 c/3\lambda_0^4) \, S^{v'v''}_{qp} \, n_m l \, \exp[-(E_q + E_{v'})/kT] \, (Q_{r,q}/Q_m) \,, \qquad (II.286)$$

or, using Eq. II.199,

$$B_{band} = (hc/4\pi\lambda_0) \, A^{v'v''}_{qp} \, n_m l \, \exp[-(E_q + E_{v'})/kT] \, g_{\Lambda'} (2S+1)(Q_{r,q}/Q_m) \,. \qquad (II.287)$$

In these expressions we have neglected the variation of λ_0 over the band considered. Here $Q_{r,q}$ is the rotational partition function belonging to state q. The dependence of the band radiance on T, at given $(n_m l)$, is governed mainly by the exponential Boltzmann factor and, to a lesser extent, by the temperature dependence of the ratio of partition functions. Approximating Q_m according to Eq. II.285, one finds that at flame temperature this ratio varies $\propto T^0$ or $\propto T^{-1}$ according to whether the vibrational energy quantum is large or small compared to kT, respectively. The temperature dependences of $Q_{r,q}$ and $Q_{r,0}$ (contained in Q_m) cancel each other at flame temperatures (see Eq. II.35). The relative band intensities belonging to a given electronic transition are proportional to: $S^{v'v''}_{qp} \exp[-E_{v'}/kT]$,

THE INTENSITIES OF RADIATIVE TRANSITIONS   Sect. II.5c

where the band strength $S_{qp}^{v'v''}$ includes the Franck-Condon factor (cf. Eq. II.203). This holds, if we neglect the variation of the rotational constant B' with v' due to vibration-rotation interaction as well as the variation of λ_0^4 among the bands. The Franck-Condon factors depend in a critical way on the relative positions and shapes of the potential energy curves in state q and p, respectively (see Sect. II.2d). In contrast to the Hönl-London factors which are universal functions of the quantum numbers involved, the F.-C. factors are characteristic for the molecular species considered.

From a logarithmic plot of relative band intensity divided by v_0^4 times band strength versus $E_{v'}$, one finds the *vibrational temperature* T_v in a similar way as described above for the rotational temperature. This method is more difficult to apply than the determination of the rotational temperature (see Gaydon 1974; a detailed discussion of this method has been presented by Smit 1950).

We find the total radiance of a *band system* belonging to a given electronic transition $q \to p$ from Eq. II.286 by summation over all possible vibrational transitions

$$B_{\text{syst}} = (16\pi^3 c/3\bar{\lambda}_0^4) \left\{ \sum_{v'v''} S_{qp}^{v'v''} \exp[-E_{v'}/kT] \right\} n_m l \exp[-E_q/kT] (Q_{r,q}/Q_m) \tag{II.288}$$

or, using Eq. II.287,

$$B_{\text{syst}} = (hc/4\pi\bar{\lambda}_0) \left\{ \sum_{v'v''} A_{qp}^{v'v''} \exp[-E_{v'}/kT] \right\} n_m l \exp[-E_q/kT] g_{\Lambda'}(2S+1)(Q_{r,q}/Q_m). \tag{II.289}$$

When the band system spreads over an appreciable wavelength region, the definition of $\bar{\lambda}_0$ may include some ambiguity. It is a sort of weighted average which may shift with temperature when the spectral intensity distribution varies markedly with T.

The formal similarity of the expressions for a molecular band system (Eq. II.289) and for an atomic line (Eq. II.281) becomes more evident when we define an *effective electronic transition probability* $\bar{A}_{qp}^{\text{el}}$ by

$$\bar{A}_{qp}^{\text{el}} \equiv \sum_{v'v''} A_{qp}^{v'v''} \exp[-E_{v'}/kT]/Q_{v,q}, \tag{II.290}$$

where $Q_{v,q}$ is the vibrational partition function in state q. If A_{qp}^{el}, defined by Eq. II.203, were independent of v' and v'' (which, however, is not the case; see Schadee 1967), we would find, by using the sum rule Eq. II.205, that $\bar{A}_{qp}^{\text{el}} \equiv A_{qp}^{\text{el}}$. Approximating Q_m by the expression in Eq. II.285, we obtain from Eqs II.289 and 290

$$B_{\text{syst}} = (hc/4\pi\bar{\lambda}_0) \bar{A}_{qp}^{\text{el}} n_m l \exp[-E_q/kT](g_q Q_{v,q} Q_{r,q}/g_0 Q_{v,0} Q_{r,0}) \tag{II.291}$$

Chapt. II   BASIC CONCEPTS AND GENERAL RELATIONSHIPS

with $g_q = g_{\Lambda'} (2S+1)$. When we also approximate the atomic Q occurring in Eq. II.281 by g_0, the latter equation for the atomic line radiance and Eq. II.291 for the band-system radiance become quite similar. Since the ratios of the statistical weights and of the partition functions occurring in these two equations are of order unity, we see that the atomic line radiance and band-system radiance are in the ratio of A_{qp} to $\bar{A}_{qp}^{el}$ for equal electronic excitation energy, number density and (average) wavelength. It should be stressed that $\bar{A}_{qp}^{el}$ is not a constant for a given electronic transition, but varies with temperature. Eq. II.291 is therefore of little use in actual calculations; it may serve for obtaining a qualitative insight into the main factors determining the total intensity of a band system in comparison with the case of an atomic line. At high concentrations the intensity of atomic resonance lines may be depressed by self-absorption (see below), which is negligible for molecular bands (see the discussion at the end of Sect. II.5a-3).

Thermal emission intensity of atomic lines (with self-absorption). In the presence of *self-absorption*, the (spectral) line radiance can be calculated by integrating the emission intensity of a volume element over the line of observation (as was done in Eq. II.279) while now taking into account the loss in intensity due to re-absorption of photons inside the flame. This will be done in Sect. VI.4b for thermal as well as nonthermal radiation. The *thermal* (spectral) line radiance can be found more directly by applying Kirchhoff's law for a thermal radiator. We define the *emission factor*, $\varepsilon(\lambda)$, of a thermal radiator at temperature T, as the ratio of its spectral radiance $B_\lambda(\lambda)$ to that of a full radiator (or black body), $B_\lambda^b(\lambda,T)$, taken at the same wavelength and same temperature. Thus we have

$$\varepsilon(\lambda) \equiv B_\lambda(\lambda)/B_\lambda^b(\lambda,T) . \tag{II.292}$$

Kirchhoff's law states that at any λ (see, e.g., Penner 1959 and Unsöld 1968)

$$\varepsilon(\lambda) \equiv \alpha(\lambda) , \tag{II.293}$$

where the absorption factor $\alpha(\lambda)$ has been defined in Sect. II.5c-2 and should be considered along the same line of observation. We thus find quite generally, i.e. for arbitrary optical thickness and at any λ, from Eqs. II.292 and 293

$$B_\lambda(\lambda) = \alpha(\lambda)\, B_\lambda^b(\lambda,T) = \alpha(\lambda)\, B_{\lambda_0}^b(T) \tag{II.294}$$

and

$$B = \int_{\text{line}} B_\lambda \, d\lambda = A_t\, B_{\lambda_0}^b(T) , \tag{II.295}$$

where the integral absorption A_t has been define by Eq. II.267. In practice the variation of $B_\lambda^b(\lambda,T)$ over the width of the narrow spectral line can be disregarded. We are therefore justified in replacing $B_\lambda^b(\lambda,T)$ by the value $B_{\lambda_0}^b(T)$ at the

THE INTENSITIES OF RADIATIVE TRANSITIONS   Sect. II.5c

line centre.

In these equations for B_λ and B the (small) contribution of the induced emission is neglected. One obtains an exact expression for B by applying the exact Planck law for $B_{\lambda_0}^b(T)$ and using A_t' as given by Eq. II.278, while substituting there the Boltzmann ratio for $(g_p/g_q)(n_q/n_p)$ (see Omenetto, Winefordner and Alkemade 1975). As was argued before (see also footnote † on page 120 of Sect. II.5a-1), the neglect of induced emission is warranted under thermal flame conditions. But if we use A_t instead of A_t' we should also consistently replace Planck's law by the approximate Wien law given by Eq. II.71a (see Omenetto, Winefordner and Alkemade 1975).

The strict applicablity of Eq. II.295 for the line radiance is limited not solely by the conditions put forward already in Sect. II.5c-2 with respect to the integral absorption A_t. For the factor $B_{\lambda_0}^b(T)$ in the expression for the radiance to make sense, the source temperature should also be strictly uniform in the region where the radiating and absorbing atoms are found. Another condition underlying the applicability of Eqs II.294 and 295 is that the relative population of the excited state conforms to Boltzmann equilibrium. This condition is not fulfilled (exactly), if the radiation field deviates from that of a black body at the temperature of the source (see Sect. VI.4c-2 sub iv). There may also be other causes of deviation from Boltzmann equilibrium; these will be discussed in Sect. VI.4c.

Equations II.294 and 295 show the intimate relation that exists between the spectral characteristics and the strengths of a spectral line seen in emission and absorption, respectively. The appearance of the exponential $B_{\lambda_0}^b(T)$ factor in the expression for the emission line accounts for the additional excitation energy that must be raised in the thermal emission process.

According to Eqs II.293 and 294 the variation of B_λ or $\varepsilon(\lambda)$ with wavelength around the line centre is identical to that of $\alpha(\lambda)$ which is described by Eq. II.259 (see Fig. II.16). All discussions in Section 5c-2 pertaining to the absorption line profile can be transferred directly to the profile of a self-absorbed emission line. It is therefore expected that with growing optical thickness (or atomic density) the *thermal* spectral radiance will approach asymptotically an upper limit, $B_{\lambda_0}^b(T)$, given by Wien's law, *which it can never exceed*. When the optical thickness of an isothermal flame becomes large at the line centre, the flame radiates effectively as a black body of the same temperature in the central part of the line profile. As a result of self-absorption, the emission line profile becomes more and more distorted and broadened the higher the atomic density, that is to say the larger the optical thickness at line centre (see Fig. II.17). This broadening effect is sometimes called *self-absorption broadening* or *concentration broadening*. Note that this effect is **not** present in the $k(\lambda)$ profile, but

183

Chapt. II   BASIC CONCEPTS AND GENERAL RELATIONSHIPS

only in the $\alpha(\lambda)$ or in the $\varepsilon(\lambda)$ profile.

The radiance B depends on atomic density in the same way as the integral absorption A_t (see Section 5c-2). In the limit of low densities (small optical thickness at line centre), B therefore increases proportionally to density and self-absorption plays no role. In the limit of high densities, B grows as the square root of density. This behaviour is typical for self-absorbed emission lines in the presence of collision broadening. The course of the full intensity-density curve follows from the curve-of-growth theory and is determined by the a-parameter and the Doppler half-intensity width (see Fig. II.15).

The onset of self-absorption is determined by the value of $\frac{1}{2}\overline{\kappa(\lambda)}\,n_p l$, as can be seen from the asymptotic expression for A_t in Eq. II.274. The latter expression with $\overline{\kappa(\lambda)} \simeq \kappa_m$ can be used to estimate the error of approximation that is made in the application of Eqs II.279-281, derived while neglecting self-absorption. The deciding factors contained in $\kappa_m n_p l$ are: $f_{pq}\,n_p$ (see Eq. II.187), since the order of magnitude of l and of $\Delta\lambda_{eff}/\lambda_0^2$ usually does not vary very much in flame atomic spectroscopy. When f is of the order of unity and most atoms are in the ground state ($n_0 \approx n_a$), self-absorption of resonance lines is expected to

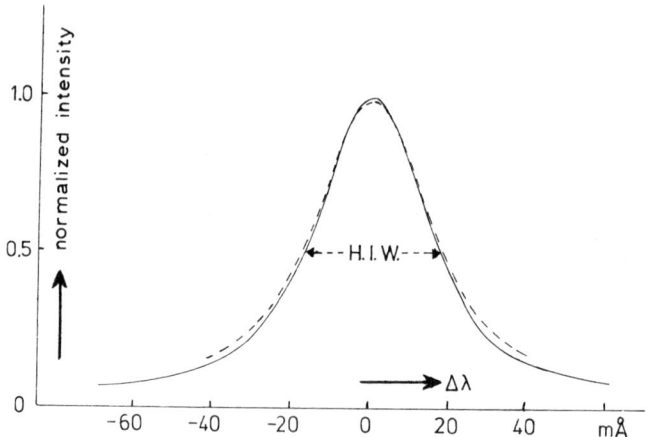

Fig. II.16 Normalized emission (dashed curve) and absorption (solid curve) line profiles of the resonance line of Sr (4607.33 Å) measured with a Fabry-Pérot interferometer and a Zeeman-scanning technique, respectively, in an C_2H_2-air flame at 1 atm and 2275 K with a low atomic Sr density. The emission line profile was averaged over six repeated scans. The arrow indicates the half-intensity width (H.I.W.). The zero point on the $\Delta\lambda$ scale corresponds to the peak of the line, which showed a collisional shift of about 9 mÅ.
(Derived from Jansen 1976 with permission of the author.)

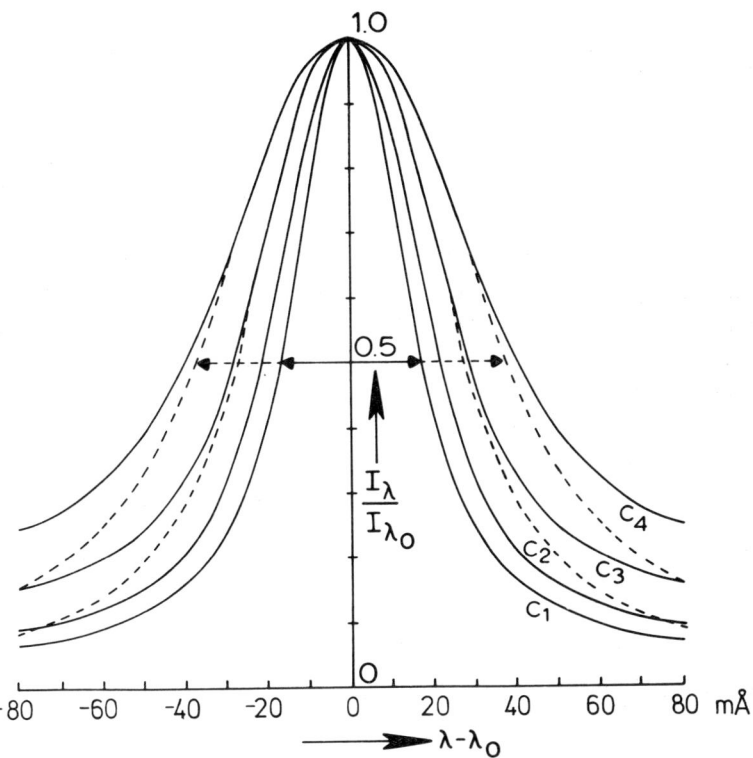

Fig. II.17 Normalized spectral profiles of the emission resonance line of Sr (4607.33 Å) measured with a Fabry-Pérot interferometer in a shielded, isothermal C_2H_2-air flame at 1 atm and 2275 K for different Sr solution concentrations (c_1 = 320; c_2 = 1280; c_3 = 2560 and c_4 = 5120 mg/l). The instrumental broadening of the interferometer was negligible. The dashed curves show the wings of the profiles for c_3 and c_4 after correction for spectral overlap of successive orders of the interferometer. Note the increase of half-intensity width (indicated by arrows at I_λ/I_{λ_0} = 0.5) with increasing concentration as a result of self-absorption. The profile for the lowest solution concentration (c_1) is practically free from self-absorption broadening. The experimental a-parameter (= 1.25) is only slightly larger than the value a = 1.0 assumed in the theoretical $\alpha(\lambda)$ profiles in Fig. II.14, so that both sets of profiles can be meaningfully compared.
(From Jansen 1976.)

occur in flames for $n_a l > 10^{11}\,\text{cm}^{-2}$, as has often been demonstrated by the convex curvature of the corresponding analytical curves (see Alkemade and Herrmann 1979). We note from Eq. II.295 that it is not the absolute radiance as such that is a measure for the occurrence of self-absorption. Ultraviolet resonance lines, which are weak in flames because of a very small Planck factor, may still show appreciable self-absorption because of a large $f_{oq}n_0$ value. The reasons that molecular resonance lines of metal compounds are usually hard to observe in absorption in flames (see Sect. 5c-2) and also show little or no self-absorption in emission are one and the same.

The *temperature dependence* of B is mainly governed by the exponential factor in $B_{\lambda_0}^b(T)$ and, for nonresonance lines, also by the exponential Boltzmann factor contained in n_p. Nonresonance lines with $E_p \gtrsim 1\,\text{eV}$ (and thus $n_p \ll n_a$), however, are practically free from self-absorption in flames and Eq. II.280, with $E_q = E_p + hc/\lambda_0$, can be applied to describe their temperature dependence. For self-absorbed resonance lines with $E_q = hc/\lambda_0$, the exponential factor contained in $B_{\lambda_0}^b(T)$ is just the same as that in Eq. II.281 which holds in the absence of self-absorption. An additional but weak dependence on T may exist for self-absorbed resonance lines through the variation of A_t with T (see Alkemade 1954). The latter variation is brought about by the temperature dependence of collision and Doppler broadening (see also Sects VII.2 and 4d). Disregarding the slight variation of A_t with T, we can conclude that the intensity of the same resonance line with and without self-absorption varies with temperature through the same exponential factor. Eq. II.153 thus describes the relative variation of intensity with T also for self-absorbed resonance lines.

The effect of self-absorption for lines with an intrinsically unresolved (hyper-)fine-structure or for overlapping (molecular) lines can be treated in the same way as the corresponding integral absorption (see Sect. 5c-2; see also Penner 1959, Hunt and Sibulkin 1967, and Arnold, Whiting and Lyle 1969).

Equations II.294 and 295 are valid for arbitrary optical thickness and thus also hold in the limiting case of vanishing optical thickness. It can easily be proved that Eq. II.295 includes Eq. II.281 as a border-line case in the limit of negligible self-absorption, i.e., for $n_p \to 0$. Using Eqs II.269 and 71a we find from Eq. II.295

$$B \to (2\pi e^2 h/m_e) f_{pq} n_p l \exp[-hc/\lambda_0 kT]/\lambda_0^3 \quad \text{for } n_p \to 0. \quad \text{(II.296)}$$

Applying Boltzmann's law for n_p and writing: $E_q = E_p + hc/\lambda_0$, we find from this

$$B \to (2\pi e^2 h/m_e) f_{pq} g_p n_a l \exp[-E_q/kT]/\lambda_0^3 Q \quad \text{for } n_a \to 0. \quad \text{(II.297)}$$

Finally, using the general relation between $(f_{pq} g_p)$ and $(A_{qp} g_q)$ in Eq. II.168, we find that Eq. II.297 is identical to Eq. II.281, as expected.

THE INTENSITIES OF RADIATIVE TRANSITIONS   Sect. II.5c

At high atomic densities where the square-root-branch of the curve-of-growth develops (see Sect. 5c-2) we find an expression for the radiance of a resonance line by inserting the asymptotic expression for A_t (Eq. II.275) and Wien's law for $B^b_{\lambda_0}(T)$ (Eq. II.71a) in Eq. II.295. Using $hc/\lambda_0 = E_q$ and expressing f_{oq} in A_{qo} by means of Eq. II.168, we obtain for $n_0 \to \infty$

$$B \to (hc^{\frac{3}{2}}/\pi^{\frac{1}{2}} \lambda_0^3) \exp\left[-E_q/kT\right]\{\delta\lambda_L(g_q/g_0) A_{qo} n_0 l\}^{\frac{1}{2}} . \qquad (II.298)$$

From this we derive, by expressing n_0 in terms of the atomic density n_a through Boltzmann's law, for $n_a \to \infty$

$$B \to (hc^{\frac{3}{2}}/\pi^{\frac{1}{2}} \lambda_0^3) \exp\left[-E_q/kT\right] \{\delta\lambda_L(g_q/Q) A_{qo} n_a l\}^{\frac{1}{2}} . \qquad (II.299)$$

Self-reversal. The absorptivity, $k(\lambda)[= \kappa(\lambda) n_p]$, may be interpreted as the reciprocal penetration depth for photons of wavelength λ. The main contribution to B_λ comes therefore from excited atoms that lie within a distance $k(\lambda)^{-1}$ from the front surface of the source, measured along the line of observation (see also Eq. VI.48 in Sect. VI.4b). Since $k(\lambda)$ has a peak value at the line centre, the central portion of the emission line profile originates from emitters which are — on the average — closer to the front surface than the emitters which contribute to the line wings. When $k(\lambda_0)^{-1}$ becomes very small compared to the flame depth l, the central line emission originates from a relatively thin layer near the front surface. When the flame temperature drops towards the surface (see Sect. IV.6), the spectral radiance in the central portion of the line corresponds effectively to a lower temperature than the spectral radiance in the line wings. A dip may even appear at the centre of the emission line profile, which becomes more pronounced the larger the atomic density and the steeper the temperature gradient. This special self-absorption effect which occurs in sources with a relatively cooler outer layer is called *self-reversal*. In the spectral recording of the emission line profile the central dip appears as a *'reversed line'*. Self-reversal is often found for metal resonance lines at high atomic densities in unshielded flames. It may also occur in gas-discharge spectral lamps in which most of the excitation is confined to the inner part of the atomic cloud (see Alkemade and Herrmann 1979). When self-reversal occurs, a deviation from the square-root-branch in the intensity-density curve is found at high densities. The intensity, integrated over the line profile, then grows more slowly than the square-root of the density.

The self-reversal dip is due basically to the drop in temperature near the border of the flame. The population, n_q, of the excited state still obeys Boltzmann's law at the local temperature. This dip should therefore not be confused with the dip that appears at the centre of a strongly self-absorbed resonance line in a flame with uniform temperature where n_q is below the

Chapt. II BASIC CONCEPTS AND GENERAL RELATIONSHIPS

Boltzmann value at the border (see Fig. VI.8). This deviation from Boltzmann equilibrium occurs as a result of radiative disequilibrium in flames with insufficient collisional excitation (see Sect. VI.4c-2). The intensity-density curve is also distorted by this infrathermal radiation effect, but in a typically different way from the above case of self-reversal (see *ibidem*).

Quantities determinable from thermal emission intensity. In the expressions II.281 (or II.297), 295 and 299 for the atomic line radiance with negligible, arbitrary, and strong self-absorption, respectively, several quantities occur, such as the Einstein transition probability A or the related f-value, the uniform atomic density n_a or the density, n_p, of atoms in the lower state of the optical transition, the temperature T, the excitation energy E_q, and the a-parameter. The latter parameter is contained in the A_t factor and only plays a role, if self-absorption is noticeable. In principle, it is possible to determine each of these quantities, or combinations thereof, by measurements of the thermal line intensity. In all theoretical expressions, the quantities f (or A), n_a and source thickness l occur only in combination, that is, as a product: $(fn_a l)$ or $(An_a l)$. The value of l may be easily assessed experimentally. Thus all spectroscopic methods of determining (relative or absolute) atomic densities from intensity measurements can also be applied to determine f- or A *values*. If the one quantity is known, the other follows. In Chapt. V we shall discuss in more detail those methods of determining n_a which have found application in flame spectroscopy. Here only a summary is presented.

When *atomic densities* are determined from line intensities in the presence of self-absorption, the a-parameter and Doppler width must be known. The latter can be calculated easily by means of Eq. II.240 if the flame temperature is (approximately) known. Methods of determining the *a-parameter* from line-intensity measurements will be dealt with in Chapt. VII. It should be recalled from Eq. II.295 that the total line intensity is proportional to the integral absorption A_t. We can therefore obtain information about the a-parameter by comparing the shape of the intensity versus density curve (or experimental emission curve-of-growth) with the theoretical curves-of-growth in the range where self-absorption is significant (see Sect. 5c-2).

Instead of measuring curves-of-growth in emission for the determination of the a-parameter and/or atomic densities, we may find it more advantageous to measure the *duplication curve* or *Gouy's curve* (see also Sect. V.2e). The duplication curve relates the *duplication factor* D to the atomic density. D is defined as the relative increase in line intensity I, or integral absorption, caused by a doubling of the product $(fn_a l)$

THE INTENSITIES OF RADIATIVE TRANSITIONS Sect. II.5c

$$D \equiv \frac{I(2fn_a l) - I(fn_a l)}{I(fn_a l)} \ . \qquad (II.300)$$

D may be measured by duplicating the flame, i.e., by placing a second, identical flame behind the original flame, or by means of a mirror placed behind the flame (see Gouy 1879 and Alkemade 1954; see also Sect. V.2e). The theoretical duplication curve relating D to $(fn_p l \sqrt{\ln 2}/\pi \delta \nu_D)$ is obtained directly from the theoretical curve-of-growth and is usually plotted double-logarithmically (see Fig. II.18).

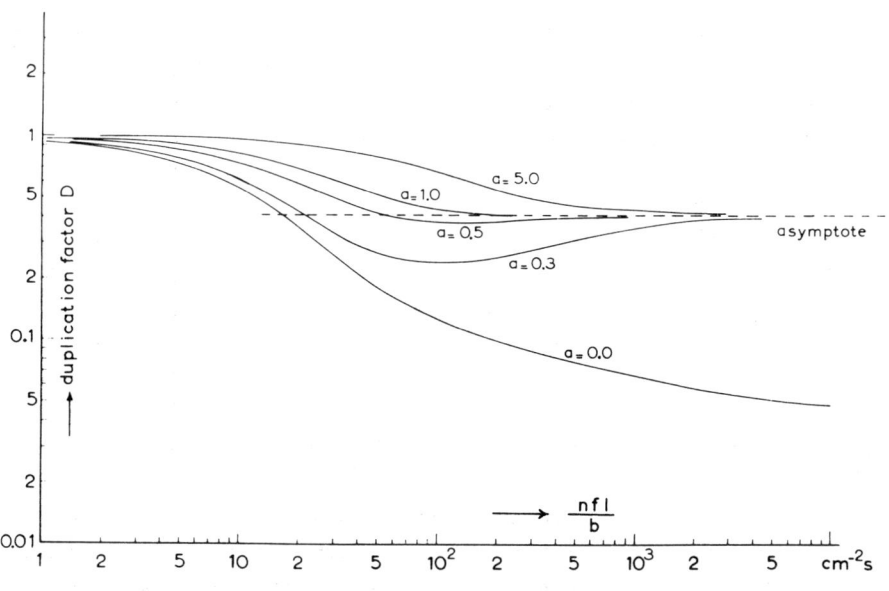

Fig. II.18 Duplication curves. See for meaning of symbols the caption of Fig. II.15. (From Hollander 1964.)

Its position and shape depend only on the a-parameter. Roughly speaking, the behaviour of D reflects the change in slope of the curve-of-growth shown in Fig. II.15 as a function of atomic density. The linear and square-root branches of the curve-of-growth for $n_p \to 0$ and $n_p \to \infty$, respectively, correspond to two horizontal asymptotes of the D-curve with asymptotic values $D = 1$ and $D = \sqrt{2}-1 = 0.415$, respectively (see Fig. II.18). A local minimum is found in the D-curves for $a \leqslant 1$. The depth of this minimum is related to the a-value and can be used to measure this parameter (see Chapt. VII). Once a and f are known, D measurements can be used to determine atomic densities (see Sect. V.2e).

The *excitation energy* E_q can be determined through Eq. II.281 by an absolute measurement of B in the absence of self-absorption, if the flame temperature and the other quantities in this equation are known. Methods of measuring absolute

Chapt. II BASIC CONCEPTS AND GENERAL RELATIONSHIPS

intensities will be treated in Sect. III.14g. E_q can also be determined from relative intensity measurements through Eq. II.282 by varying the known flame temperature. A possible variation of $(n_a l)$ and Q with temperature must be corrected for.

When E_q and the other parameters in Eq. II.281 are known, the *flame temperature* T_f can be determined by an absolute measurement of B in the absence of self-absorption. When the integral absorption A_t can be measured separately (see Sect. III.15), the temperature can also be derived through Eq. II.295 from an absolute measurement of B in the presence of self-absorption. In the latter method, neither the f-value nor the atomic density need be known.

The line-reversal method. An elegant, commonly used spectroscopic method of measuring flame temperatures is the method of line-reversal. It does not involve absolute intensity measurements nor does it require knowledge of the f-value, atomic density, etc. However, there must be an auxiliary continuum source available with a variable, calibrated radiance temperature at λ_0. The *radiance temperature*, T_r, of a continuum source having a spectral radiance $B^c_{\lambda_0}$ at λ_0 is implicitly defined by equating

$$B^c_{\lambda_0} \equiv B^b_{\lambda_0}(T_r) \ , \tag{II.301}$$

where the black-body radiance $B^b_{\lambda_0}(T_r)$ as a function of T_r is given by Planck's or Wien's law. For a thermal radiator (like an incandescent source) where $B^c_{\lambda_0} = \varepsilon(\lambda) B^b_{\lambda_0}(T)$, the *radiance* temperature is fully determined by its *true* temperature T and its emission factor at given λ (cf. Eq. II.292). When $\varepsilon(\lambda) \equiv 1$, $T_r \equiv T$ at any λ; if $\varepsilon(\lambda) < 1$, we have $T_r < T$.

The line-reversal method is based on the following principle. The integrated radiance B_E of an atomic (non)resonance line at λ_0 emitted by the flame is given by (see Eq. II.295)

$$B_E = B^b_{\lambda_0}(T_f) A_t \ . \tag{II.302}$$

Here $B^b_{\lambda_0}(T_f)$ is the spectral radiance of a black body at flame temperature T_f, while A_t is the integral absorption. One assumes the existence of thermal emission and a homogeneous flame temperature.

When, on the other hand, a continuum source with spectral radiance B^c_λ at $\lambda \simeq \lambda_0$ is placed immediately behind the flame or imaged upon it, the radiance of this source observed through the flame will be weakened at $\lambda \simeq \lambda_0$ because of absorption. According to Eq. II.266 the absorbed radiance B_A, integrated over the whole width of the line, is given by

$$B_A = B^c_{\lambda_0} A_t \ . \tag{II.303}$$

Now if the continuum source happens to have a radiance temperature T_r

equal to T_f at wavelength λ_0, we immediately see from Eqs II.301-303 that the thermally emitted radiance of the flame just compensates for the absorbed radiance. This is true regardless of the concentration of the atomic vapour. The continuous spectrum observed with the background source will then not be altered when the flame with atomic vapour is placed between it and the spectroscope. However, when $B_{\lambda_0}^c < B_{\lambda_0}^b(T_f)$, the balance ceases and there will be a net emission of radiation at λ_0. The atomic line then appears as an emission peak superimposed on the continuous spectrum. When $B_{\lambda_0}^c > B_{\lambda_0}^b(T_f)$, the reverse is true and an absorption dip appears in the continuous background spectrum. When $B_{\lambda_0}^c$ is gradually increased from some value below $B_{\lambda_0}^b(T_f)$, a transition from an absorption dip to a superimposed emission peak will be seen at λ_0. The point of transition or '*line reversal*' occurs when $B_{\lambda_0}^c \equiv B_{\lambda_0}^b(T_r)$ just goes through the value $B_{\lambda_0}^b(T_f)$, i.e., when $T_r = T_f$.

The application and limitations of the line-reversal method for measuring flame temperatures will be dealt with in Sect. III.10c-5.

Conclusions. To conclude this section, we list some criteria for recognizing the actual occurrence of self-absorption in light sources (flame, discharge lamp).

1. When the atomic density can be varied by known ratios (e.g., by spraying solutions of known concentrations into the flame), the presence of self-absorption can be recognized from the shape of the intensity versus density curve plotted on relative scales (see Fig. II.15). The relative loss of intensity due to self-absorption can be evaluated by comparing the measured intensity with the extrapolated low-density asymptote which represents the intensity in the absence of self-absorption.

2. When the atomic density is not proportional to the solution concentration (e.g., owing to a concentration-dependent degree of ionization or nebulization efficiency; see Sects IX.3b and III.8), we can apply, instead of criterion 1, the method of flame duplication with a back mirror. The latter method can also be recommended when the metal is not sprayed as a salt solution into the flame.

3. The intensity ratios of the components of a multiplet are affected by self-absorption when their $g_p f_{pq}$-values are different. The alkali doublet components have an intensity ratio equal to 2 in the absence of self-absorption, and equal to $\sqrt{2}$ for strong self-absorption. This criterion may be useful in the case of spectral-line sources where the foregoing criteria cannot be applied. However, since uniform excitation conditions do not often exist in gas-discharge sources, one can hardly draw quantitative conclusions from the multiplet ratios in the presence of self-absorption.

4. An additional peak absorption measurement with a spectral-line source as background may also give information about the occurrence of self-absorption. If (self-)absorption is weak, this measurement can be used to estimate, with the help of Eq. II.274, the relative loss in intensity due to self-absorption. When precise, quantitative conclusions are wanted, the a-parameter and the collisional line shift must be known.

5. Spectral scanning of the emission line profile and comparison with the calculated Voigt or Doppler profile expected at low optical thickness may also be useful (see Fig. II.14; see also Jansen 1976). This method, however, requires a spectral apparatus with high resolution and one must have some knowledge about the relevant broadening parameters.

6. Broadening of an atomic emission line and thus self-absorption may also be checked — in a rather indirect way — by measuring the absorption factor when the line radiation passes through an auxiliary flame containing the same atomic vapour at a moderate density. The absorption factor measured is then compared with the factor measured in the same absorbing flame for another spectral-line source which is known to have a spectral width smaller than the absorption line width. If both absorption factors turn out to be about equal, the spectral width of the investigated line source must also be smaller than, or at most about equal to the absorption line width. When the investigated line source, however, is strongly broadened by self-absorption, a lower absorption factor is measured. Instead of measuring absorption factors, one may also compare the intensities of fluorescence induced by the two line sources in a suitable auxiliary flame. The advantage of measuring fluorescence intensities is that these are not found from a (small) difference between two large signals, as is often the case with absorption measurements. This fluorescence method may be useful as a qualitative check for self-absorption with spectral lamps; this method is especially sensitive to the occurrence of self-reversal.

7. The occurrence and extent of self-absorption in a homogeneous source may also be evaluated by theoretical calculations with the help of the equations or theoretical curves-of-growth presented in the preceding section. The f-value, atom density, source thickness, Doppler width, and a-parameter should be known.

5c-4 *THE INTENSITY OF FLUORESCENCE*

The efficiency of fluorescence. Fluorescence stems from the re-emission of photons absorbed from a primary radiation beam (see Sect. II.5a-1). The fluorescence intensity depends on the absorbed primary radiation intensity, the *quantum efficiency of fluorescence* Y (or Y_q), i.e., the probability that an absorbed

THE INTENSITIES OF RADIATIVE TRANSITIONS   Sect. II.5c

photon will be re-emitted, and on the extent of self-absorption of the fluorescence radiation.

The *Stern-Volmer formula* provides a kinetic expression for the quantum efficiency of fluorescence

$$Y = \frac{A_{qp}}{\text{total probability of de-excitation of state } q \text{ per second}} . \quad (II.304)$$

It is assumed here that atoms are raised from some lower state to state q by absorption of primary photons and that fluorescence is observed specifically at the wavelength of the optical transition $q \to p$ with transition probability A_{qp}. De-excitation of an atom in state q may occur either by radiative transitions to state p and any other lower state, or by radiationless transitions to any state below or above state q induced by inelastic collisions or chemical reactions. These collisions and reactions *quench* the fluorescence and reduce the Y value; one speaks here of *physical* and *chemical quenching*, respectively. Let the unimolecular rate constant k' denote the total probability per second of a radiationless transition of an atom initially in state q (cf. Sect. II.4a-1). The total probability per second of radiative de-excitation is equal to the reciprocal radiative lifetime, $(\tau_r)_q^{-1}$, of state q (see Eq. II.166). Therefore Eq. II.304 can also be written as

$$Y = A_{qp} / \{k' + (\tau_r)_q^{-1}\} . \quad (II.305)$$

When only one optical transition from state q is allowed, Eq. II.305 reduces to

$$Y = A/(k' + A) = 1/(k'/A + 1) , \quad (II.306)$$

where A is the probability of spontaneous emission per second.

The *power efficiency of fluorescence* Y_p (defined as re-emitted radiant power divided by absorbed power) is obtained by multiplying Y by the ratio of fluorescence photon energy to absorbed photon energy. For resonance fluorescence Y_p equals Y.

In flames at 1 atm pressure, Y may range, for example, from a few percent to roughly 50% for the first resonance lines of the alkali atoms. A determination of Y through fluorescence measurements (see Sect. III.16) yields the value of k', if A is known. In this way, effective cross sections can be derived for quenching of electronic states by collisions (see Sects VI.2a and 2b). Experimental methods of measuring fluorescence intensities and discriminating them from thermal emission in flames will be dealt with in Sect. III.16. Besides playing a role in de-excitation studies, atomic fluorescence has also found limited application in spectrochemical analysis (see Chapt. I and Sect. III.16a).

The discussion in this section will be restricted to resonance fluorescence of atomic resonance lines. In flames significant absorption of primary radiation is usually limited to atomic resonance lines. Molecular fluorescence from elements nebulized into a flame is more difficult to observe, because molecular absorption is much weaker than atomic absorption (see Sect. II.5c-2). With a xenon lamp as primary light source fluorescence from alkaline-earth compounds and PO has been observed in cool hydrogen flames having weak background emission (see Jenkins 1969a, Human and Zeegers 1975, and Haraguchi et $al.$ 1976). However, with intense laser sources strong fluorescence band spectra from CaOH, SrOH, MnO, CrO and YO have been observed in various flames including the C_2H_2-air flame (see Weeks, Haraguchi and Winefordner 1978, Blackburn, Mermet and Winefordner 1978, and Wijchers et $al.$ 1980). Fluorescence from flame radicals has also been observed with conventional as well as laser sources (see references cited in Sect. IV.8). The theoretical treatment of molecular fluorescence is complicated by the existence of many radiative and radiationless transitions from the excited state to other states (see end of page 207). Similar complications also arise, although to a much lesser extent, for atoms with low-lying multiplet states or excited to high-lying states; this matter will be discussed in Sects VI.2a-1 and 2c-5.

The intensity of fluorescence (general). When a resonance line is observed in fluorescence, the fluorescence intensity may be weakened by self-absorption just as in the case of thermal emission. The extent of self-absorption depends on the optical thickness of the flame in the direction of observation. There is a certain probability that a photon generated by fluorescence inside the flame will be reabsorbed on its way out by other metal atoms in the ground state. When Y is close to unity, most re-absorbed photons will again be re-emitted as fluorescence, and so on. A given photon may then 'jump' many times from one atom to another before it leaves the flame or is converted into heat as a result of a quenching collision. The general theory of radiative transfer must then be applied to describe this radiation diffusion or imprisonment (see Sect. II.5c-3). In flames at 1 atm Y is generally below 0.5; we shall therefore neglect here tertiary and subsequent photon emission steps. The approximate expressions to be presented are thus valid to first order of Y only.

We make the following assumptions in treating steady-state fluorescence (see Daily 1979 for transient fluorescence effects in flames).

1. We assume that the absorptivity as a function of wavelength is described by a Voigt profile (see Sect. II.5b-3) and that the curve-of-growth theory is applicable. Line shift is ignored.

2. Coherence between the absorbed and re-emitted photons is assumed to be negligible (see Sect. II.5a-1). The re-emitted photons are assumed to be distributed isotropically and to show no preferential polarization. Their

THE INTENSITIES OF RADIATIVE TRANSITIONS   Sect. II.5c

spectral distribution is given by the same Voigt function as that which describes the wavelength dependence of the absorptivity.

In an H_2-air flame of 1 atm, a degree of polarization of the order of 10% has been found for the Cd line at 2288 Å by Chenevier and Lombardi (1972), notwithstanding the large rate of depolarizing collisions expected at 1 atm. A possible explanation for this outcome might be the simultaneous occurrence of quenching collisions which tend to counteract the effect of the (adiabatic) depolarizing collisions (see Barrat et al. 1966). The polarization of the Cd line appeared to decrease with increasing magnetic field parallel to the observation direction; this is due to the Hanle effect (see, e.g., Thorne 1974). Similar experiments have also been done in a low-pressure flame by Chenevier and Lombardi (1974). In most studies, however, no strong polarization of the fluorescence radiation has been found in flames at 1 atm pressure (see Sect. VI.1).

3. Tertiary emission of re-absorbed photons is neglected (see above).

4. The density of atoms, n_0, in the ground state (= lower state of the optical transition) is not affected by the primary radiation field. The case of partial saturation by a laser beam will be treated separately at the end of this subsection.

5. The temperature, atomic density, a-parameter and fluorescence efficiency are assumed to be uniform throughout the flame region observed.

6. The flame region observed is assumed to be brick-shaped with volume $V = l\,LH$ (see Fig. II.19). The whole left-hand side of this region is irradiated by a homogeneous, parallel beam perpendicular to the side. The beam direction is chosen as the x-axis; the position of the irradiated side corresponds to $x = 0$. Fluorescence is observed from the whole irradiated flame volume. When self-absorption is significant, we assume that the fluorescence is observed under a small solid angle in a direction perpendicular to the front face (y-axis; see Fig. II.19).

7. It is assumed that thermal emission of the atomic vapour is eliminated in the fluorescence intensity measurements (see Sect. III.16). The contribution of induced emission in the case of partial saturation will be considered separately at the end of this subsection.

The intensity of fluorescence (without self-absorption). We first consider the border-line case of *negligible self-absorption*, i.e., of low optical thickness in the y-direction ($n_0 l \to 0$).

The radiant flux, Φ_F, of fluorescence observed under an arbitrary solid angle Ω_F and in an arbitrary direction is then simply given by

Chapt. II BASIC CONCEPTS AND GENERAL RELATIONSHIPS

$$\Phi_F = \Phi_A Y \Omega_F / 4\pi . \qquad (II.307)$$

Here Φ_A is the total absorbed primary radiant flux. In this equation use is made of the isotropic distribution of the fluorescence radiation and of the equality of power efficiency Y_p to quantum efficiency Y (see the above). Eq. II.307 provides the basis for the experimental determination of Y (see Sect. III.16b).

When the primary radiation source is a *continuum source*, we have from Eq. II.266

$$\Phi_A = (\Phi_{\lambda_0})_0 A_t = E_{\lambda_0} H l A_t , \qquad (II.308)$$

where $(\Phi_{\lambda_0})_0$ and E_{λ_0} are the incident spectral radiant flux and spectral irradiance at λ_0, respectively. The integral absorption A_t is to be considered in the direction of the incident beam (see Fig. II.19). When $n_0 L \to 0$, we can use the asymptotic expression for A_t given by Eq. II.269 and find from Eqs II.307 and 308

$$\Phi_F \to 7.05 \times 10^{-14} \lambda_0^2 f n_0 E_{\lambda_0} Y V \Omega_F \quad \text{for} \quad n_0 L \to 0 , \qquad (II.309)$$

where all quantities are expressed in CGS units.

When a *narrow-line source* emitting a spectral line precisely at λ_0 is used, we have

$$\Phi_A = \Phi_0 \alpha(\lambda_0) = E H l \alpha(\lambda_0) . \qquad (II.310)$$

Here $\alpha(\lambda_0)$ is the peak absorption factor of the flame, considered in the direction of the incident beam, and E is the irradiance (integrated over the relatively narrow emission line profile). Using the asymptotic expression for $\alpha(\lambda_0)$ given by Eq. II.265, we find from Eqs II.307 and 310

$$\Phi_F \to 7.05 \times 10^{-14} \lambda_0^2 f n_0 (E/\Delta\lambda_{eff}) Y V \Omega_F \quad \text{for} \quad n_0 L \to 0 , \qquad (II.311)$$

where all quantities are expressed in CGS units. The effective width, $\Delta\lambda_{eff}$, of the Voigt profile depends on the a-parameter and the Doppler half-intensity width (see Sect. II.5b-3). By comparing Eqs II.309 and 311 one sees that one obtains the same fluorescence flux from a narrow-line source as from a continuum source if $E = E_{\lambda_0} \Delta\lambda_{eff}$. Values of E or E_{λ_0} obtainable with practical radiation sources may be estimated by multiplying the (spectral) radiance values quoted in Sect. III.7 by the solid angle under which the source is imaged onto the flame[†] (see also Sect. VI.2c-5).

Equations II.309 and 311 are also applicable in the case of an arbitrary flame shape and nonuniform atomic density. One has then simply to replace the product $(n_0 V)$ by the total number of ground-state atoms in the (uniformly) irradiated

---

[†] This holds because the (spectral) radiance of the source image is the same as that of the source itself, if we disregard optical losses (see Appendix A.6).

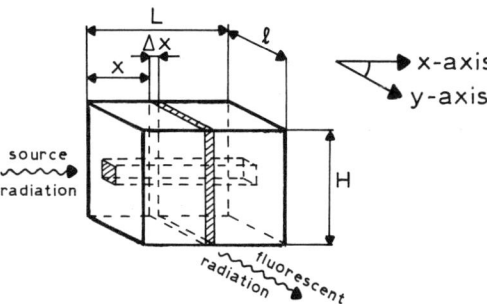

Fig. II.19 Dimensions of brick-shaped flame volume considered in the derivation of the fluorescence intensity. The whole of the left-hand side is supposed to be illuminated by the source, whereas the whole of the front is observed in fluorescence. (From Alkemade 1970.)

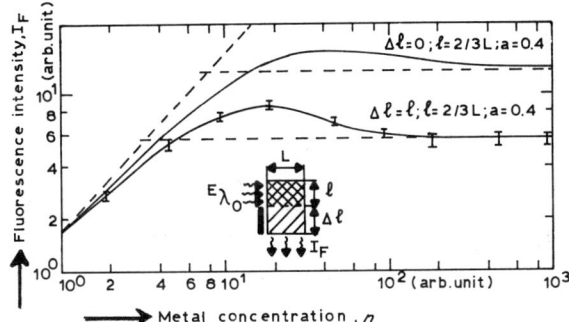

Fig. II.20 Fluorescence intensity I_F as a function of metal concentration n for the flame volume considered in Fig. II.19 with a continuum light-source. The upper curve was calculated for a fully illuminated flame volume with $a = 0.4$ (see Eq. II.319). The lower curve was calculated for a partially illuminated flame volume (see insert) with the same a-parameter. Experimental values with their spreads are represented by vertical bars and refer to the 2852 Å Mg line in a premixed acetylene-air flame with a xenon lamp as exciting light-source. Their relative positions are to be compared with the lower theoretical curve. (According to calculations and measurements by Zeegers and Winefordner 1971, reported also in Alkemade 1970.)

Chapt. II   BASIC CONCEPTS AND GENERAL RELATIONSHIPS

flame volume. When the (spectral) irradiance is not uniform over the beam cross section, Eq. II.309 or 311 can only be applied to an infinitesimally thin bar parallel to the x-axis (see bar with hatched end-face in Fig. II.19). The total fluorescence flux is then obtained by integrating over all similar bars contained in the radiation beam.

It follows from Eqs II.309 and 311 that the fluorescence flux, in the absence of self-absorption, grows proportionally to n_0, as is to be expected. If Y is considerably less than unity (i.e., $A \ll k'$ in Eq. II.306), it appears from Eq. II.306 that $Y \propto (A/k') \propto (f/k')$. The fluorescence flux is then proportional to (f^2/k'), whereas for $1-Y \ll 1$ it is proportional to f. One should note that the absorbed radiation flux and the thermal emission flux are always proportional to f in the limit of $n_0 \to 0$. In atomic fluorescence spectroscopy, the f-value may therefore play an even more important role than in atomic absorption or emission spectroscopy.

The intensity of fluorescence (with self-absorption). In the presence of *self-absorption*, the calculation of the fluorescence intensity becomes more complicated. Under certain simplifying conditions, expressions have been derived in the literature which include also the case when only part of the irradiated flame volume is observed in fluorescence or when the primary irradiance does not extend over the whole flame width l (see Fig. II.19; see Hooymayers 1968, Zeegers, Smith and Winefordner 1968, Alkemade 1970, and Zeegers and Winefordner 1971[†]). When the observed fluorescence radiation has to pass through a flame layer that contains metal vapour but is not irradiated by the primary beam, a kind of self-reversal effect occurs. The absence of photo-excitation in this layer has the same effect as the large reduction in thermal excitation in the relatively cool outer layer of a flame (see Sect. II.5c-3). Consequently an additional reduction of fluorescence intensity occurs at high atomic densities and a typical self-reversal dip may appear in the fluorescence line profile. In the following we shall exclude this self-reversal effect by assuming that the flame is uniformly irradiated over its full width l by the primary beam. The theory can, however, be simply extended to the case of partial illumination as is shown in the insert of Fig. II.20 (see Alkemade 1970). The same figure also shows the good agreement between the theoretically predicted fluorescence intensity-density curve and the experimental curve found by Zeegers and Winefordner (1971) in the latter case.

A further simplification is obtained here by considering specifically the

---

[†] The expressions derived in the earlier papers of McCarthy, Parsons and Winefordner (1967) and Jenkins (1967) are not complete.

THE INTENSITIES OF RADIATIVE TRANSITIONS   Sect. II.5c

radiant intensity, I_F, of fluorescence radiation emitted by the whole front face in
a direction perpendicular to this face (see Fig. II.19). Assuming also that all
the above conditions 1 to 7 are applicable, we shall derive here, following Alkemade
(1970), an expression for I_F which allows us to discuss the dependence of I_F on
n_0 directly on the basis of the curve-of-growth theory.

Let $n_1(x)$ denote the local density of those atoms in the excited state
that are brought into this state by absorption of primary photons. The atoms excited by collisions are here disregarded, as they do not contribute to the fluorescence emission. Since the exciting radiation is gradually weakened on its way
through the flame along the x-axis, the density n_1 will, in general, be a function
of x. Inside a thin slab of flame gas with thickness Δx (see Fig. II.19) n_1 may,
however, be considered to be uniform because the exciting radiation is assumed to be
uniform over the cross section of the beam; the beam attenuation is also uniform
here, as n_0 does not depend on position. We can then apply Eq. VI.57b, as anticipated from Sect. VI.4b, which generally describes the radiance of a resonance line
with self-absorption under *nonthermal* excitation conditions. The radiance $B_F(x)$
of the fluorescence line observed at distance x from the irradiated face in a
direction perpendicular to the front face is then found to be

$$B_F(x) = C_1 \{n_1(x)/n_0\} A_t(n_0 l) . \qquad (II.312)$$

$A_t(n_0 l)$ denotes the value of A_t for a flame with uniform ground-state concentration
n_0 and depth l which is measured along the line of observation. The constant C_1
follows from Eq. VI.57b: $C_1 = (2hc^2/\lambda_0^5)(g_0/g_1)$. For n_1 we have the local balance
equation

$$An_1(x) = Yv_1(x) , \qquad (II.313)$$

where $v_1(x)$ is the number of excitations per second and per cubic centimetre,
caused by primary photon absorption, as a function of x. This balance equation
follows directly from the definition of the quantum efficiency Y. Combination of
Eqs II.312 and 313 yields

$$B_F(x) = (C_1 Y/A) \, v_1(x) A_t(n_0 l)/n_0 . \qquad (II.314)$$

The radiant intensity ΔI_F (in erg s^{-1}sr^{-1}) of the fluorescence emitted by the
designated flame slab with surface area $H\Delta x$, follows from Eq. II.314

$$\Delta I_F(x) = B_F(x) H \Delta x = (C_1 YH/A) \, v_1(x) \{A_t(n_0 l)/n_0\} \Delta x . \qquad (II.315)$$

The total radiant intensity I_F of the fluorescence radiation in the chosen direction of observation is obtained by integration over x

$$I_F = C_2 \{A_t(n_0 l)/n_0\} \int_0^L v_1(x) \, dx \qquad (II.316)$$

with: $C_2 \equiv C_1 YH/A$. The integral in the latter equation equals the number of

Chapt. II BASIC CONCEPTS AND GENERAL RELATIONSHIPS

primary photons absorbed per second in an arbitrary bar with length L parallel to the x-axis and with a cross section of 1 cm^2 (see baulk with hatched end-face in Fig. II.19). The combined dependences of this integral and of $A_t(n_0 l)/n_0$ on n_0 determine the variation of I_F with n_0, at least under the idealized geometrical conditions imposed. In two extreme cases with respect to the spectral profile of the source radiation, Eq. II.316 leads to a simple expression, as we shall see presently.

Case of continuum source. Consider first the case of a continuum source which produces a spectral irradiance E_{λ_0} at the surface of the flame ($x=0$) at $\lambda \simeq \lambda_0$. The power absorbed in the whole bar shown in Fig. II.19 is then given by (cf. Eq. II.266)

$$\int E_{\lambda_0} \alpha(\lambda)\,d\lambda = E_{\lambda_0} A_t(n_0 L) ,\qquad (II.317)$$

as the bar has a length L. Thus we have

$$\int_0^L v_1(x)\,dx = E_{\lambda_0} A_t(n_0 L)/h\nu_0 \qquad (II.318)$$

and with the aid of Eq. II.316

$$I_F = C_3 A_t(n_0 l)\, A_t(n_0 L)/n_0 \qquad (II.319)$$

with $C_3 \equiv C_1 Y H E_{\lambda_0}/Ah\nu_0 = (2cg_0/\lambda_0^4 g_1)(Y/A)HE_{\lambda_0}$. Note from Eq. II.304 that Y/A is the actual lifetime of the excited state (which is determined mainly by the rate constant of collisional de-excitation if $Y \ll 1$).

The dependence of I_F on n_0 (which determines the shape of the analytical curve in analytical atomic fluorescence spectroscopy) is thus directly related to the curve-of-growth through Eq. II.319. We note that the factor $A_t(n_0 L)$ in Eq. II.319 accounts for the absorbed source radiation as a function of n_0 and that the factor $A_t(n_0 l)/n_0$ accounts for the loss of fluorescence radiation due to self-absorption. The possibility of describing these absorption effects by means of two separate factors is connected with the special geometry of the flame and the special optical conditions chosen.

For a given a-parameter the shape of the analytical curve, plotted double-logarithmically, depends only on the ratio $\zeta \equiv L/l$. In other words, for a given ζ value we can always bring the analytical curve into coincidence with the curve describing $\log_{10}[A_t(\eta)A_t(\zeta\eta)/\eta]$ as a function of $\log_{10}\eta$ through appropriate shifts parallel to the coordinate axes.

The asymptotic behaviour of the analytical curve follows directly from Eq. II.319 if we recall the general asymptotic behaviour of the curve-of-growth shown in Fig. II.15. When $n_0 \to 0$, $A_t(n_0 l)$ and $A_t(n_0 L)$ vary as $(n_0 l)$ and $(n_0 L)$, respectively. Thus we expect a linear asymptote

$$I_F \propto n_0^2 lL/n_0 = n_0 lL \quad \text{for} \quad n_0 \to 0. \qquad (II.320)$$

THE INTENSITIES OF RADIATIVE TRANSITIONS Sect. II.5c

For large n_0 values, $A_t(n_0 l)$ and $A_t(n_0 L)$ vary as $(n_0 l)^{\frac{1}{2}}$ and $(n_0 L)^{\frac{1}{2}}$, respectively, (see Eq. II.275) and we have

$$I_F \propto (n_0 l)^{\frac{1}{2}} (n_0 L)^{\frac{1}{2}} / n_0 = (lL)^{\frac{1}{2}} \quad \text{for} \quad n_0 \to \infty, \quad (II.321)$$

which is independent of n_0. With a continuum source, the analytical curve therefore approaches a horizontal asymptote in the high-density range. Fig. II.20 shows a calculated fluorescence intensity-density curve with a continuum source together with its linear and horizontal asymptotes for an assumed value $a = 0.4$.

The appearance of a local maximum in the curve of Fig. II.20 should be noted. This is connected with the inflection that occurs in the curve-of-growth for $a \leqslant 1$ (see Sect. 5c-2 and Fig. II.15). Because of this inflection there is an intermediate range of densities where the derivative γ of $\log_{10} A_t$ as a function of $\log_{10} n_0$ is <u>smaller</u> than the final asymptotic value $\gamma = \frac{1}{2}$. Upon differentiating $\log_{10} I_F$ as a function of $\log_{10} n_0$ one gets from Eq. II.319

$$(d \log_{10} I_F / d \log_{10} n_0) = \gamma + \gamma' - 1, \quad (II.322)$$

where $\gamma \equiv d \log_{10} A_t(n_0 l) / d \log_{10} n_0$ and $\gamma' \equiv d \log_{10} A_t(n_0 L) / d \log_{10} n_0$. When for some (finite) value of n_0, $(\gamma + \gamma')$ just equals unity, an extremum (here: a maximum) will occur in the analytical curve. If, for example, $l = L$, this maximum is found at that value of n_0 at which the tangent of the curve-of-growth, plotted double-logarithmically, has a slope $\tan \beta = \frac{1}{2}$. Since for $l = L$, $\gamma = \gamma' = \tan \beta$, we have here $\gamma + \gamma' = 1$. It is possible to determine the a-parameter (if $a \leqslant 1$) from the ratio of maximum intensity to the plateau value. This parameter can also be found from the abscissa value of the intersection point of the two asymptotes in Fig. II.20 (see Hooymayers 1968).

In the limit $n_0 \to \infty$, both γ and γ' tend towards $\frac{1}{2}$ and a second extremum is attained. This extremum is identical to the constant plateau which the curve approaches asymptotically. This (asymptotic) extremum occurs for all a-values > 0.

Case of narrow-line source. We now consider the other extreme case of a narrow-line source with same central wavelength as the absorption line and with a spectral line width that is small compared to that of the absorption line. The irradiance of the source line at the surface of the flame is denoted by E. The fraction of source radiation absorbed per centimetre path length may thus be assumed to equal the peak absorptivity k_m. According to Beer's law the number of primary photons absorbed per second in the bar considered in Fig. II.19 is

$$\int_0^L v_1(x) dx = (E/hv_0)(1 - \exp[-k_m L]). \quad (II.323)$$

For I_F we then get through Eq. 316

$$I_F = C_4 A_t(n_0 l)(1 - \exp[-k_m L]) / n_0 \quad (II.324)$$

with $C_4 \equiv C_2 E/h\nu_0$.

Considering the asymptotic behaviour of A_t and expanding the exponential function in a series, we find for small n_0 values approximately

$$I_F \propto n_0 l k_m L/n_0 = k_m lL \quad \text{for} \quad n_0 \to 0, \tag{II.325}$$

where k_m is proportional to n_0. Thus there is again an initial linear asymptote $(I_F \propto n_0)$. In contrast to the continuum case, the slope of this asymptote now depends on k_m and thus on the spectral width of the absorption line (see Eq. II.187) with $\kappa_{max} \equiv k_m/n_0$).

For high densities the exponential term in Eq. II.324 drops to zero, that is, practically all the primary photons become absorbed while $A_t(n_0 l)$ behaves as $(n_0 l)^{\frac{1}{2}}$ (for $a > 0$). We then get from Eq. II.324: $I_F \propto 1/n_0^{\frac{1}{2}}$. The position of the high-density branch (with negative slope) of the analytical curve is insensitive to changes in L, but it still depends on l.

Since the curve rises at low densities and decays to zero for high densities, a maximum will occur for all values of the a-parameter. This maximum is positioned at some intermediate n_0 value where the fluorescence intensity becomes markedly affected by self-absorption while $k_m L$ is no longer small compared to unity. In a double-logarithmic plot the analytical curve has thus an initial asymptote with positive slope $\tan\beta = 1$, and a final asymptote with negative slope $\tan\beta = -\frac{1}{2}$. The shape of the curve and the point of intersection of these asymptotes depend on the a-parameter (see the curves calculated by Hooymayers 1968).

Intensity of fluorescence under saturation conditions. Up till now *induced emission* and *saturation* of the optical transition were disregarded. These effects may, however, become important when a strong tunable dye-laser is used to excite the atomic fluorescence.

The advantage of laser sources in fundamental as well as analytical atomic fluorescence spectroscopy stems firstly from the considerable gain in fluorescence signal and the resulting improvement in detection limit (see definition in Sect. III.14f-1). Secondly, when full saturation is attained, the fluorescence signal becomes insensitive to variations in the laser power. The saturated-fluorescence intensity then also becomes independent of the fluorescence efficiency, that is, of the flame gas composition, whereas it still remains proportional to the atomic concentration. Under (near-)saturation conditions self-absorption of the fluorescence radiation is reduced and the initial linear range of the analytical curve is extended to higher concentrations. However, when resonance fluorescence is applied, scattering of the laser radiation by unevaporated droplets or particles in the flame may interfere with analytical applications. Because the fluorescence signal saturates at increasing laser power but the scattering signal does not, this interference worsens as the laser power is increased. This interference does not occur when nonresonance fluorescence is observed. We refer to the textbooks cited in Chapt. I and to Allkins (1975) for further information on the use of lasers in analytical atomic spectroscopy with flames.[†] With strong pulsed laser sources one can also induce two-photon absorption in atomic vapours in flames, as has been demonstrated for Cd, Zn and Na by observing the fluorescence from levels with $E_{exc} = 2h\nu$ (see Sect. 5a-1).

The occurrence of (partial) saturation affects the net absorption of the primary radiation beam as well as the self-absorption of the fluorescence radiation. We have to replace $k(\nu)$ by the net absorptivity $k'(\nu)$ given by Eqs II.189, 190 and 191. Partial saturation also brings about a nonlinear dependence of excited-state population on beam intensity. In general no simple expression can be given for the fluorescence intensity as a function of the primary (spectral) laser irradiance and atomic density. Complications arise because the local net absorptivity depends on the local beam intensity which, in turn, depends on the absorption of the beam in the flame.

However, in the limiting cases of low atomic density or strong saturation, i.e. for a nearly transparent flame, we can again apply Eq. II.309 for Φ_F if we also assume the beam intensity and flame composition to be uniform. We now only have to substitute $n_0\{1-(g_0/g_1)(n_1/n_0)\}$ for n_0 according to Eq. II.191. Expressing n_0 in terms of the total atomic density $n_a \equiv n_0 + n_1$, we then find for a *broad-band* laser from Eq. II.309 (in CGS units)

$$\Phi_F = 7.05 \times 10^{-14} \lambda_0^2 f\, YV\, \Omega_F E_{\lambda_0} n_a \frac{1-(g_0/g_1)(n_1/n_0)}{1+(n_1/n_0)}, \qquad (II.326)$$

which holds if $n_a\{1-(g_0/g_1)(n_1/n_0)\}$ is sufficiently low. According to Eq. VI.34b, anticipated from Sect. VI.2c-5, we have: $(g_0/g_1)(n_1/n_0) = z/(1+z)$ with $z \equiv (E_{\lambda_0}/E_{\lambda_0}^S) g_0/(g_0+g_1)$ and *saturation parameter* $E_{\lambda_0}^S \equiv 1.52 \times 10^{-4}\, g_0/\lambda_0^5 Y(g_0+g_1)$ (in CGS units; see also Eqs VI.39b and 38b). Consequently we get from Eq. II.326 (in CGS units)[*]

$$\Phi_F = 7.05 \times 10^{-14} \lambda_0^2 f\, YV\Omega_F E_{\lambda_0} n_a/\{1+(E_{\lambda_0}/E_{\lambda_0}^S)\}. \qquad (II.327)$$

In the limit of $E_{\lambda_0} \to \infty$ we find that Φ_F tends to a constant value given by (in

---

[†] Denton and Malmstadt (1971) and Fraser and Winefordner (1971, 1972) were the first to investigate laser-excited atomic fluorescence in flames for analytical purposes. For a recent survey see also Omenetto and Winefordner (1979a).

[*] Compare with Omenetto, Benetti, Hart, Winefordner and Alkemade (1973) who derived essentially the same result in a somewhat different way.

Chapt. II BASIC CONCEPTS AND GENERAL RELATIONSHIPS

CGS units)

$$(\Phi_F)_{max} = 7.0 \times 10^{-14} \lambda_0^2 f \, YV \, \Omega_F n_a E_{\lambda_0}^s =$$
$$= 1.07 \times 10^{-17} \lambda_0^{-3} f \, V\Omega_F \, n_a g_0 / (g_0 + g_1) \, . \quad (II.328)$$

Making use of the relation between f and A in Eq. II.168 we can also write

$$(\Phi_F)_{max} = 1.60 \times 10^{-17} \lambda_0^{-1} A \, V\Omega_F \, n_a g_1 / (g_0 + g_1) \, . \quad (II.329)$$

The curve representing the function $\Phi_F(E_{\lambda_0})$ is called the *saturation curve*, which approaches asymptotically a plateau for $E_{\lambda_0} \to \infty$ (see Fig. II.21). It is seen from Eq. II.327 that a plot of Φ_F^{-1} versus $E_{\lambda_0}^{-1}$ should yield a straight line. For $z = g_0/(g_0 + g_1)$, i.e., for $E_{\lambda_0} = E_{\lambda_0}^s$, Φ_F attains half its plateau value $(\Phi_F)_{max}$. From a plot of relative Φ_F versus absolute E_{λ_0} we can therefore determine $E_{\lambda_0}^s$ from which Y follows (see also Omenetto, Benetti, Hart, Winefordner and Alkemade 1973). For the same z value we also find that the net absorptivity $k'(\nu)$ is reduced to half the absorptivity in the absence of the laser beam. From an absolute measurement of $(\Phi_F)_{max}$ and of the geometrical factors V and Ω_F we can find the product (An_a) through Eq. II.329; neither the source parameters nor the flame temperature nor the fluorescence efficiency need be known here. Once n_a is known, we can also find n_1, as $n_1/(n_1)_{max} = \Phi_F/(\Phi_F)_{max}$ with $(n_1)_{max} = n_a g_1/(g_1 + g_0)$.

The idealizing assumptions underlying the above theoretical conclusions should be kept well in mind. The irradiation of the flame volume observed may not be uniform, especially when a pencil-like laser beam traverses a broad flame. With pulsed dye-lasers, E_{λ_0} varies continuously with time, and the time-integrated fluorescence pulse will in general not simply obey the above relation. Attenuation of the laser beam through absorption in the flame and self-absorption of the fluorescence radiation at high atomic densities were not taken into account. However, as long as the atomic vapour is nearly saturated in the whole observed flame volume, (self-)absorption becomes unimportant. A 2-level model was assumed in the above derivation (saturated fluorescence for a 3-level model was treated by Omenetto and Winefordner 1979, 1979a). A special case is when the upper or lower term is a doublet (see Bolshov *et al.* 1977, Olivares and Hieftje 1978, Boutilier, Blackburn *et al.* 1978, and van Calcar *et al.* 1979; see page 632). When the collisional doublet mixing is strong and the doublet separation small, one may count the doublet effectively as one level in calculating n_0 or n_1 in full saturation while replacing g_0 or g_1 by the doublet weight factor. Attempts have been made by Kuhl, Neumann and Kriese (1973), Sharp and Goldwasser (1976), and Smith, Winefordner and Omenetto (1977) to measure the saturation curve (that is, the fluorescence intensity as a function of laser power) with Na vapour in flames. However, with pulsed lasers a plateau did not show up properly in the range

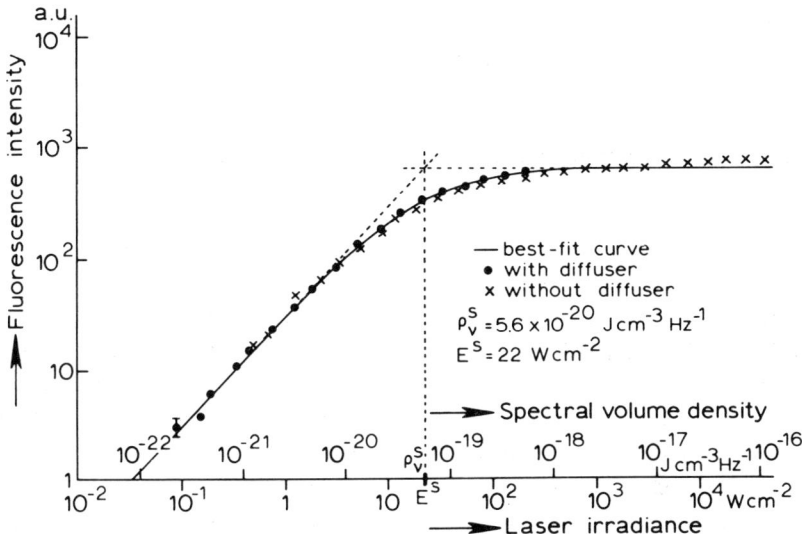

Fig. II.21 Saturation curve is shown representing the Na-D fluorescence peak-intensity in a H_2-O_2-Ar flame as a function of absolute intensity of a pulsed, broad-band dye-laser tuned at one of the D-components. Experimental points obtained with and without a diffuser in the laser beam are shown; a diffuser was used to improve the spatial uniformity of the laser intensity (at the cost of the maximum intensity obtainable). The drawn curve represents the best fit of the experimental points obtained with diffuser to the theoretical equation. The experimental value of the saturation parameter E^S or ρ_ν^S (indicated in the figure) agrees within 25% with the value calculated from the known efficiency of fluorescence Y. (From van Calcar et al. 1979.)

where the saturation parameter was far surpassed; the fluorescence intensity appeared to rise steadily with laser power. This can be explained, on the one hand, by the nonuniform intensity distribution across the laser beam when the beam cross section is small compared to the flame thickness (see Rodrigo and Measures 1973, Alkemade 1977, and Daily 1978). On the other hand, the fact that the dye-laser pulse does not abruptly terminate but decays smoothly with time can explain why a plateau is absent when the *integrated* fluorescence pulse is measured (see Alkemade 1977). When the laser-beam cross section was made comparable to the flame thickness and time-resolved fluorescence measurements were performed or a c.w. dye laser was used, a saturation curve was obtained for Na that accurately followed the theoretical curve (see van Calcar et al. 1979).

The general assumptions underlying the applicability of Eq. II.191 in the

Chapt. II   BASIC CONCEPTS AND GENERAL RELATIONSHIPS

above analysis, viz. homogeneous line broadening and absence of atomic polarization, are reasonably well fulfilled in flames at 1 atm pressure (see Sect. 5a-2). Under these conditions we can also find analogous expressions for the effect of (partial) saturation on Φ_F in the case of a *narrow-band* laser. We now have to start by making the same substitution for n_0 in Eq. II.311 as we made before in Eq. II.309. The appropriate expression to be filled in for z is, however, now different (see Sect. VI.2c-5).

When, even at partial saturation, the flame is optically thick, a simple expression for the self-absorbed fluorescence intensity is again obtained under the special geometrical conditions shown in Fig. II.19. In other words, we have to assume uniform illumination of the flame and uniform atomic density. The laser power absorbed in the whole flame, can, however, now not be simply expressed in an integral-absorption factor as before. We shall consider therefore only the fluorescence flux that is radiated from a thin flame slab, parallel to and at distance x from the illuminated side. (See hatched area with thickness Δx in Fig. II.19 where Δx is assumed to be small enough to treat the density of the laser beam as uniform within this slab.)

The fluorescence radiance $B_F(x)$ of this slab, as observed in a direction perpendicular to the laser beam, is again found from Eq. II.312 if we substitute $n_0\{1-(g_0/g_1)(n_1/n_0)\}$ for n_0 (see also Omenetto, Winefordner and Alkemade 1975).[†] We thus have, under partial-saturation conditions,

$$B_F(x) = (2hc^2/\lambda_0^5) A'_t n_1 (g_0/g_1)/\{n_0 - (g_0/g_1) n_1\} \, , \quad (II.330)$$

where A'_t refers to the *net* integral absorption in the observation direction (see Eq. II.278). The quantities n_1, n_0 and A'_t are functions of x, as the irradiance of the flame slab by the laser beam and thus the degree of saturation are generally dependent on x. From Eq. VI.36 in Sect. VI.2c-5 we have:

$$n_1(g_0/g_1)/\{n_0 - (g_0/g_1) n_1\} = \tfrac{1}{2} E_{\lambda_0}(x)/E^s_{\lambda_0} \, ,$$

in the case of a broad-band laser producing a spectral irradiance $E_{\lambda_0}(x)$ at position x inside the flame. Using Eq. VI.38b for the saturation parameter $E^s_{\lambda_0}$ we find from Eq. II.330 with the help of the latter expressions (in CGS units)

$$B_F(x) = 0.079 \, Y E_{\lambda_0}(x) A'_t \, . \quad (II.331)$$

Here A'_t is a function of $E_{\lambda_0}(x)$ and of the product $(n_a l)$, which can be found from Eq. II.278 in combination with Eqs II.190 and VI.37. The radiant intensity, $\Delta I_F(x)$ of the considered flame slab as a whole is obtained by multiplying $B_F(x)$ by its

---

[†] The justification for this substitution is more clearly seen by inspecting the derivation of this equation in Sect. VI.4b.

THE INTENSITIES OF RADIATIVE TRANSITIONS   Sect. II.5c

surface area $H\Delta x$ (see Fig. II.19).

If we keep n_a and thus $E_{\lambda_0}(x)$ constant, A'_t varies proportionally to $\sqrt{l}$ when self-absorption is strong. For fixed $E_{\lambda_0}(x)$, i.e. close to the irradiated side ($x \approx 0$), A'_t then also varies proportionally to $\sqrt{n_a}$. This does not hold, however, for arbitrary x, as $E_{\lambda_0}(x)$ and thus the degree of saturation then also decrease when n_a is increased. All that can be said is that saturation, in general, shifts the onset of self-absorption of the fluorescence radiation to higher $(n_a l)$ values. Saturation has thus a straightening effect on the fluorescence intensity-density curve or on the analytical curve in atomic fluorescence spectroscopy.

Conclusion. The conditions under which the equations presented are applicable in the case of self-absorbed fluorescence radiation are much more stringent than in the case of self-absorbed thermal radiation. The shape of the analytical curves found in atomic fluorescence spectroscopy with flames of arbitrary size and under nonuniform irradiation by a source with arbitrary spectral width may deviate markedly from the simple theory given here (for a more detailed discussion see Alkemade 1970). Nevertheless, the typical features predicted by this theory are at least in qualitative agreement with the shape of experimental analytical curves with a continuum- as well as spectral-line source (see, e.g., Winefordner and Staab 1964, Veillon *et al.* 1966, and Winefordner and Mansfield 1967; for a survey see also Alkemade and Herrmann 1979). This also applies to the theoretically predicted extension of the initial linear branch of the analytical curve to higher concentrations as a result of saturation by a laser source (see Omenetto, Hart, Benetti and Winefordner 1973, and Omenetto and Winefordner 1979).

In the experimental determination of Y values it is advisable to work under conditions of negligible or weak self-absorption (see Sect. III.16). In the derivation of the a-parameter from the shape of the fluorescence intensity-density curve with a continuum source (see Townsend *et al.* 1971), great care should be given to the geometrical and optical conditions in the measurement. It has proved possible, however, to realize experimental conditions that permit simple theoretical equations to be applied (see Zeegers and Winefordner 1971; see also Fig. II.20, lower curve). This may be concluded from the consistent Y- and a values derived by these authors from the fluorescence intensity-density curves, when compared to the values derived independently by other means.

Approximative total-radiance expressions for the steady-state fluorescence of *molecules* excited by a broad- and narrow-band source were presented by Boutilier, Blackburn *et al.* (1978) and Boutilier, Winefordner and Omenetto (1978). More detailed and exact expressions were derived by Chan and Daily (1979) and Wijchers (1981).

CHAPTER III

General Instrumental Aspects and Experimental Methods

1. INTRODUCTION

Flame spectroscopy is an experimental method that can be used for obtaining quantitative data about properties of metal vapours and about processes involving metal species. In addition, flame spectroscopy can serve as a method for quantitative spectrochemical analysis. In neither case can reliable results and optimum working conditions be obtained without adequate understanding and judicious use of the instrumental components and of the spectroscopic techniques. In studies of metal vapours we must also know, by measurement, the relevant flame properties if we are to arrive at a quantitative interpretation of the data.

In this chapter we give a description of the operation, design and performance of the instrumental components (Sects 2-7 [†]). Methods of checking their performance and of measuring the most important flame properties are thereupon treated in Sects 8-13. The chapter concludes with rather an extensive discussion of how relative or absolute optical signals in emission, absorption and fluorescence can be accurately measured in metal-seeded flames (Sects 14-16). Especially the first and last parts of this chapter may be a guide to the selection of the best type of instrumentation and measuring technique in a particular case. Experimental techniques that are important only in a special context (such as spectral line broadening or ionization) will be discussed separately in the appropriate chapters.

Sections 2 and 3 deal with components that are specific for a flame spectrometer, whereas Sects 4-7 are more generally relevant to any optical spectrometer.

[†] Section numbers quoted without a roman cipher refer to the same chapter.

Chapt. III   GENERAL INSTRUMENTAL ASPECTS AND EXPERIMENTAL METHODS

There are, of course, numerous specialized textbooks that deal extensively with the subjects treated in Sects 4-7, and references to them will be given at appropriate points in the text. We intend to review here only that body of knowledge that may be of direct interest in flame spectrometric work. We shall keep to a minimum those generally known facts and formulae that can easily be found in textbooks on Optics and Electronics. Lesser known aspects of the optical and electronic instrumentation that are relevant in the context of our book are treated in more detail or are elaborated in Appendices.

The formal aspects of noise are not dealt with in a separate section but are discussed along with photodetectors in Sect. 5. Although noise is a more general phenomenon, the major contribution to the inaccuracy of spectrometric measurements with photomultiplier tubes stems from fluctuations in the photo- or dark current. The general noise formulae presented will be used in the derivation of signal-to-noise ratios in optical measurements in Sects 14-16. The noise expressions will also be used in the description of the flame emission noise in Sect. IV.9.

Noise is not always simply a nuisance and a cause of trouble. Electric noise measurements have been utilized on one occasion to measure directly the electron temperature in flames (see Sect. IX.4a).

As an illustration of how many aspects and parameters of a flame spectrometric set-up may be actually involved in a single experiment, we mention here the determination of the relaxation of Na ionization towards Saha equilibrium in a given flame of uniform temperature. As will be discussed more fully in Sects IX.3c and 5, this relaxation can be determined by measuring the fall of atomic Na concentration with increasing height of observation above the combustion zone; this fall is caused by the gradual increase in Na ion formation due to relaxation. In this experiment the relative intensity of the Na resonance line is measured in emission or absorption at several well-defined loci in the flame (Sects 14 or 15 and 7). The Na line has to be spectrally isolated (Sect. 4) and detected (Sects 5 and 6); the linearity of the measuring system has to be checked in order to find true relative Na concentrations (Sect. 13). The height of measuring is varied by means of a calibrated vertical shift of the burner mounting (Sect. 2). During the measurement Na salt at a suitable concentration has to be introduced at a constant rate into the flame (Sect. 2). One has to make sure that the spray droplets are completely desolvated and that the salt particles are completely volatilized at the lowest height of observation (Sect. 9); otherwise a spurious height-dependence will be found. The flame should behave as a one-dimensional flow system so that there is a unique correspondence between height of measuring and travel time of the Na vapour; this makes certain demands on the burner design (Sect. 2). The rise-velocity of the

BURNERS AND NEBULIZERS   Sect. III.2a

burnt flame gas must be known in order to convert the height scale into a time scale (Sect. 11). The thickness of the metal-seeded flame part should be constant or its relative variation with height allowed for (Sect. 12). Last but not least, the flame temperature should be measured and its independence of height of measuring checked (Sect. 10).

Instrumental aspects will be discussed primarily in their general relation to metal vapour studies in flames. Most of these aspects are also relevant in analytical flame spectrometry and they are usually dealt with in the analytical textbooks cited in Chapt. I. There is, however, a clear distinction between these two sorts of flame spectrometry as regards the aims and constructional requirements (see also the separate sections). This circumstance entails a shift of emphasis in the choice of instrumental topics to be discussed in Sects 2-7. On the other hand, technical developments in analytical spectrometry, for example with respect to spectral line sources, have often greatly aided fundamental flame studies (see Alkemade 1973); the numerous references in the following sections to the analytical literature demonstrate this most clearly.

2. BURNERS AND NEBULIZERS

2a INTRODUCTION

In this section we shall discuss various types of burners and nebulizers that are in use in studies of metal vapours in flames. First of all we shall define what we understand by *burner* and *nebulizer*.

The function of a *burner* is to fix in space a continuously burning flame and more specifically its shape (height, diameter, etc.). A *nebulizer* serves as a tool for the reproducible introduction of metal species into the flame, usually as an aerosol of a metal-salt solution.

We can distinguish *premix burners* combined with chamber-type nebulizers, and *direct-injection burners*. With the premix burner, fuel and oxidant gases are thoroughly mixed inside the burner housing before they leave the burner ports and enter the primary combustion zone of the flame. This type of burner, which usually produces a more or less laminar flame, is used in combination with a *separate* unit for nebulizing the sample. The metal-salt solution is usually nebulized pneumatically by the oxidant (+ inert) gas which emerges as a jet stream from a nozzle into a spray chamber. In the spray chamber the gas jet is slowed down and homogeneously mixed with the mist droplets. In contrast, the direct-injection burner combines the function of nebulizing the sample and maintaining the flame. Here fuel and oxidant emerge from separate ports and are mixed above the burner through their turbulent motion. The flame produced is usually turbulent. It is mainly the oxidant that is

Chapt. III   GENERAL INSTRUMENTAL ASPECTS AND EXPERIMENTAL METHODS

used for nebulizing the sample at the tip of the burner. With this type of burner, the introduction of sample solution often affects strongly the flame (temperature, shape, homogeneity), in contrast to the premix burner with chamber-type nebulizer.

The premixed, laminar flames are (sometimes) *shielded* with a colourless (i.e. metal-free) *mantle flame* of the same composition and temperature as the inner (metal-seeded) flame, in order to promote the homogeneity (temperature, composition) of the inner flame. For some applications an additional *sheath* of flowing *inert gas* or a *silica tube* is used for preventing infusion of unwanted air, dust, etc. from the surroundings or for eliminating the secondary combustion zone (see Sect. IV.3).

In fundamental flame studies the premixed, laminar flames have a definite advantage over the turbulent flames produced by a direct-injection burner developed by Gilbert (1955). The latter flames have some important advantages in analytical flame spectroscopy (see Sect. IV.2, and for a detailed description of the latter type of burner see the textbooks on analytical flame spectroscopy cited in Chapt. I). In fundamental flame studies we require a number of special flame properties: the metal-seeded (inner) flame should be homogeneous and well defined as to composition, temperature, thickness, metal content, and rise-velocity of flame gases; at the same time the partial pressures of the flame gas components should be computable for given total gas pressure p_t (which may occasionally be below 1 atm); the height above the combustion zone should be related in a simple way to the travelling time through the rise-velocity of the flame gases to permit a study of processes as a function of time (see Eq. IV.1); the combustion zone and burnt gas region should be well separated. These demands can be reasonably met in premixed, laminar, shielded flames. Moreover, in these flames the introduction of an aerosol of a metal-salt solution by a separate nebulizer with spray chamber hardly affects the flame properties.

There will be no further discussion here of the direct-injection burner.

Although high-pressure flames ($p_t \approx 40$ atm) may be several hundreds of degrees hotter than the corresponding atmospheric-pressure flames (see Diederichsen and Wolfhard 1956), they have seldom been used in studies of metal vapours. Therefore they will not be considered here. Low-pressure flames have been used occasionally in such studies (see, e.g., Calcote and King 1955, Fristrom and Raezer 1956, Knewstubb and Sugden 1958, 1960, Bulewicz and Padley 1969, and Bleekrode 1970). The special burner-nebulizer systems required will be discussed in Sect. 2c.

2b   PREMIX BURNERS FOR LAMINAR FLAMES AT 1 ATM PRESSURE
2b-1   *GENERAL*

We shall discuss here the design of premix burners used for the production of laminar, stable and homogeneous flames. The flame cross section may be circular or rectangular. Additional (metal-free) mantle flames and/or screening sheaths of

BURNERS AND NEBULIZERS   Sect. III.2b

flowing inert gases are sometimes provided. The burner heads are sometimes (water-) cooled. Usual fuel gases are C_2H_2, H_2, and CO, whereas air, N_2O, and O_2 diluted by inert gases like N_2, Ar and CO_2 are used as oxidant gases. Because of the hazard of flash-back pure O_2 is seldom used here as an oxidant gas.[†]

Apart from the well-known simple Bunsen burner (see, e.g., Gaydon and Wolfhard 1970), which nowadays is only rarely used in flame studies of metal vapours, we may mention as the chief representative types of premix burners: Méker burners, porous-plate burners, and slot burners.

Méker burners may have a burner head (also called a burner tip or top) in which a regular array of small holes (also called ports or canals) is drilled (*grid burner*), or the burner head may consist of a bundle of capillaries acting as exit pipes (*capillary burner*). Depending on the distance between the exit pipes and the gas-flow conditions the primary combustion zone consists here of a large number of small separate combustion cones (typically a few mm high; see also Sect. IV.3); or the combustion cones may merge to produce a relatively flat combustion zone.

Porous-plate burners have a porous metal or ceramic plate acting as a burner head. The combustion zone is essentially flat. Measurements in the burnt gas region can be extended to an area immediately above the combustion zone; this is not possible with the Méker burner where measurements below the top of the individual cones are usually not very meaningful.

Slot burners are developed especially to provide a long optical path in atomic absorption measurements. They have a slot-shaped burner mount that produces a 50-100 mm long flame in a horizontal direction. The flames produced on slot burners are less stable because of the stretched shape of the combustion zone and the thinness of the flame. Furthermore the properties of these flames are quite similar to those produced on Méker burners. Sometimes *multi-slot burners* with several parallel slots are used to broaden the flame or to provide for a mantle flame.

2b-2   *MÉKER BURNERS*

We first discuss the conventional Méker grid-burner as a basic form from which other types may be derived. The design of the burner depends on a number of considerations; these considerations apply not only to the Méker burner but also to premix burners generally.

(i)   In order to produce the required flame, oxidant and fuel are here premixed in the burner housing (cf. above), while the flow of gas mixture that leaves the exit ports of the burner is made laminar. The base of the combustion cones (see

---

[†] It should be noted that the feasibility of premix burners for the production of $C_2H_2-O_2$ and H_2-O_2 flames has been reported by Fassel and Golightly (1967), Fiorino, Kniseley and Fassel (1968), and Frank and Krauss (1974a).

213

Chapt. III   GENERAL INSTRUMENTAL ASPECTS AND EXPERIMENTAL METHODS

Sect. IV.3) should be near to the burner head in order to improve stabilization. The height of the cones should be kept low, since high cones usually give rise to unstable flames and the vertical distance to the combustion zone would then be ill-defined. The presence of a mantle flame may be required to prevent self-reversal in temperature measurements (see Sect. III.10) or to stabilize the central flame (see below). When the mixing of the gases is not satisfactory inside the burner housing an additional external premixing chamber with anti-explosion device may be used (see below).

When measurements are made that depend sensitively on the gas composition (for example, measurements of the fluorescence efficiency, which is strongly affected by molecular gases), an additional sheath of a flowing noble gas is used for screening the flame from the surroundings.

(ii) The minimum supply of gas mixture required is often determined indirectly by the supply of (mainly diluent and/or oxidant) gases needed to operate the separate nebulizer.

(iii) Flash-back of the combustion zone into the burner housing and explosions inside the burner have to be prevented. The burner should be of firm construction, and have a safety outlet and gas leakages should be avoided by using heat-resistant gaskets. Flash-back can be prevented if one chooses the diameter of the exit ports below a critical value called the *quenching diameter* (depending on the qualitative composition of the combustible gas mixture) at which flash-back cannot take place regardless of the fuel/oxidant ratio and the velocity of the gas mixture. Values of quenching diameters are found in, e.g., Lewis and Von Elbe (1961), Gaydon and Wolfhard (1970), Hieftje (1971), Aldous, Bailey and Rankin (1972), and Suddendorf and Denton (1974). With rapidly burning gas mixtures the quenching diameter may be too small for practical use because of the difficulty of machining a multiport burner and of clogging (see below sub v). Whenever the port diameter exceeds the quenching diameter, the conditions for safely operating the gas flows must be determined and attended to (see, e.g., Saturday and Hieftje 1977). It is evident that the burner head with exit ports should be exchangeable when another kind of gas mixture with a different quenching diameter is to be burnt (see below). Furthermore, the number and arrangement of the exit ports and their diameter should be such that:

(iv) the total exit area equals F_u/v_u, where F_u denotes the volume of unburnt gas mixture supplied per unit of time at 1 atm, and v_u the required flow speed of the gases in the exit ports;

(v)   no stoppages should occur due to deposition of salt supplied to the flame.

The *diameter* of the exit ports is on the one hand determined by the safety

BURNERS AND NEBULIZERS   Sect. III.2b

requirements with respect to flash-back, which determine an upper limit, and on the other hand by the considerations connected with requirements (i), (ii) and (v). It should further be noted that the burning velocity v_b (see Sect. IV.1) and the flow speed, v_u, of the unburnt gas mixture are related to the apical angle, 2α, of the combustion cones (see Fig. IV.1 and Eq. IV.2). Since, in practice, the burning velocity is a given quantity and v_u may be chosen in a certain range, the diameter of the exit ports more or less determines the height of the combustion cones. As we want to keep this height as low as possible (see above sub i), we have the diameter of the exit ports as an adjustable parameter. Furthermore, turbulence in the exit ports should be avoided in order to promote the stability of the flame; this sets a lower limit on the port diameter; the same holds for demand (v): the ports should not be too narrow. In practice the optimum diameter is found by trial-and-error by achieving a compromise between the various demands. For a detailed discussion of burner-design criteria the reader is referred to Rann (1968, 1969), Gaydon and Wolfhard (1970), Hieftje (1971), and Aldous, Bailey and Rankin (1972).

The length, l, of the exit ports is linked to the diameter, $2r$, through an empirical relation $l/2r > 10$ in order to obtain a laminar gas flow (according to Mavrodineanu and Boiteux 1965; see also Alkemade and Herrmann 1979).

In practice the distance between the exit ports of a grid-type burner is chosen larger than the diameter of the port. This is done because too small a distance causes the undesirable merging of (some of) the separate cones into one big unstable cone (not to be confused with the merging of the cones discussed in connection with the capillary burner; see also below).

Figure III.1 shows the general design of a Méker grid burner being developed in the Physics Department of the State University at Utrecht (see Heiërman 1937, Vendrik 1949, Alkemade 1954, Hollander 1964, and Hooymayers 1966). This burner is made of nickel-plated brass and provides a mantle flame and/or sheath of flowing inert gas.[†]

This burner can be made suitable for burning several kinds of fuel gases (C_2H_2, CO, C_3H_8, H_2) with air, N_2O and mixtures of O_2 and inert gases (N_2, noble gases) by simply interchanging the burner heads with exit ports of appropriate diameters. The burner housing consists of two concentric cylindrical chambers, completely separated from each other and providing for the inner flame and mantle flame, respectively (see Fig. III.1). Between the base plate and the burner housing a teflon gasket of about 1 mm thickness is inserted.

[†] Slevin, Muscat and Vickers (1972) have investigated this type of burner for use in analytical atomic fluorescence spectroscopy. They conclude that this sheath burner is markedly better than (commercial) standard Méker-type burners, although the former was not specially designed for analytical use.

Chapt. III   GENERAL INSTRUMENTAL ASPECTS AND EXPERIMENTAL METHODS

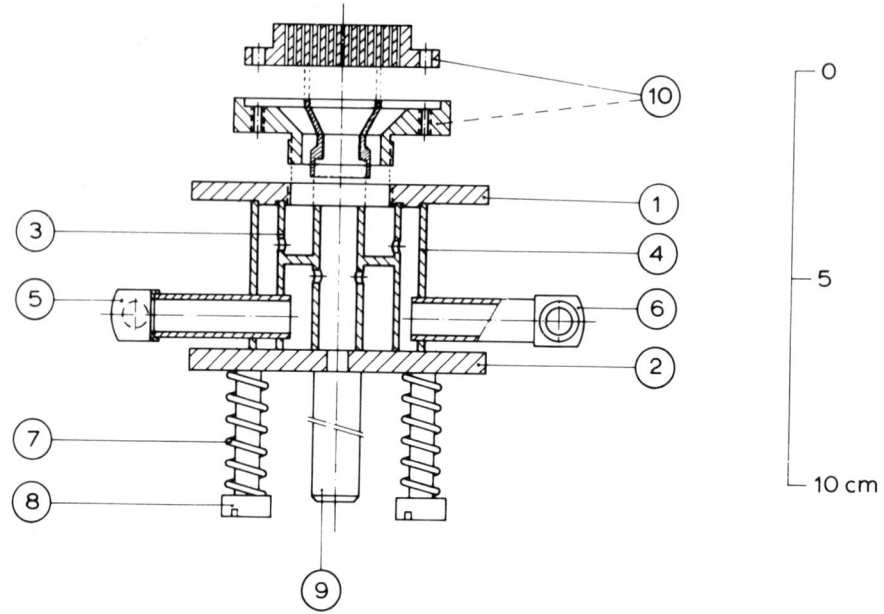

Fig. III.1   Design of cylindrical burner (Méker type).
1 =upper plate; 2 = base plate (serving as safety-outlet);
3 =wall separating outer and inner chamber; 4 =outer wall;
5 and 6 are inlet tubes for supply gases of central flame
and mantle flame respectively; 7 = spring pressing base
plate (2) against burner walls (3) and (4) (gasket between
base plate and burner walls is not shown); 8 =screw adjusting spring (7) and fixed in plate (1); 9 =holder; 10= burner head with exit ports.   (From Hollander 1964.)

The brass material of this burner is particularly suitable because of its unbreakableness and easily reproducible machining. It is easy to drill, (silver-)solder and handle (other burner materials are discussed in detail by Mavrodineanu and Boiteux 1965). Corrosion of the brass material due to metal solutions sprayed appeared to be but slight. Even at high flame temperatures where burner-head temperatures of $300°C$ were reached no trouble occurred from impurities such as copper atoms from the burner head in the flame.

For the C_2H_2-, C_3H_8- and H_2 flames a chimney proved to be superfluous when a mantle flame round the central flame was used. With a sheath of flowing inert gas an additional metal ring between flame and inert gas sheath sometimes seemed to be necessary to prevent the outer cones from being blown off (partial flame lifting, see below).

Some typical values for the diameter and length of the exit ports of Méker-

BURNERS AND NEBULIZERS   Sect. III.2b

type burners are the following: CO-air 1.2 mm diameter and 20 mm length; C_2H_2-air 0.8 mm diameter and 10 mm length; and for C_3H_8-air 2.0 mm diameter and 20 mm length. The figures for H_2-air flames are cited in Fig. III.2.

It should be noted that with the burner shown in Fig. III.1 a large variety of flames with different shapes (i.e., with circular, rectangular or square cross section) and dimensions can be produced by simply interchanging the burner top and using suitable adaptors between burner housing and burner top.

The burner described produces shielded flames and can, when necessary, supply a sheath of flowing inert gas. The length and diameter of the exit ports and the distance between them are equal for central flame and mantle flame, which have usually equal compositions. The thickness of the mantle flame is prescribed by the experimental conditions required. As a rule of thumb the thickness chosen is about half the diameter of the central flame. In order to avoid turbulence the gas flow per port is made equal for central and mantle flame.

Sometimes the outer row of cones of the mantle flame tends to blow off. This can generally be prevented by improving the premixing of the gases flowing to the mantle flame. For instance one can use extra premixing chambers outside the burner housing or place a metal ring round the mantle flame as high as the cones. The latter measure is recommended for gas mixtures such as CO-air which have a relatively low burning velocity (see van der Hurk 1974).

The flame can be surrounded by a sheath of inert gas that emerges from the exit ports of a hollow metal ring placed round the burner top (see Fig. III.2). These exit ports may be about 1-2 mm in diameter and about 2 mm apart. Usually there are 3-5 rows of exit ports.

To prevent the sheath of flowing gas from lifting off the flame the openings of the exit ports of the sheath are positioned somewhat higher than those of the flame exit ports. The difference in height is again of the order of the height of the combustion cones. The metal ring then also serves as a chimney.

Figure III.2 shows an example of the burner head for a sheathed H_2-O_2-inert gas flame with mantle flame. This burner head fits the Méker burner housing shown in Fig. III.1.

Cooling of the burner head may be necessary, especially when the cones are small and close to the head and cause a large heat flow to the burner. Cooling prevents deformation of the burner head and burning off of the head material. Cooling also stabilizes the cones. Water cooling is commonly used (see Rann 1968). However, cooling by, for instance, the sheath gas flowing through the hollow metal ring round the burner head may also be satisfactory in many cases (see Fig. III.2).

One can vary the flame temperature for a given combustible gas mixture

Chapt. III GENERAL INSTRUMENTAL ASPECTS AND EXPERIMENTAL METHODS

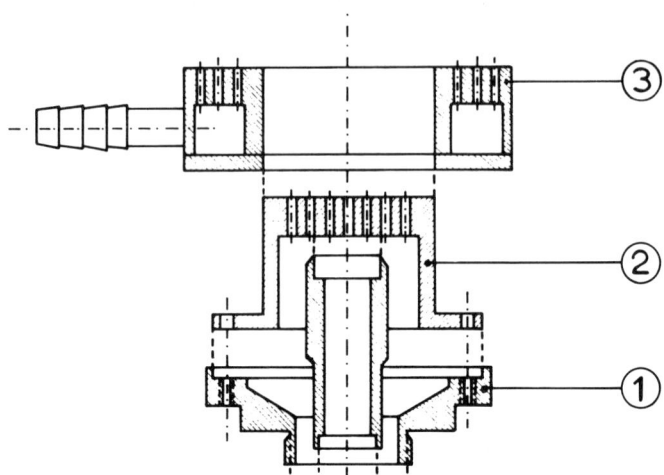

Fig. III.2 Exploded design of cylindrical Méker burner head used for sheathed H_2-O_2-inert gas flame with mantle flame; 1 = lower part that screws on to burner housing shown in Fig. III.1; 2 = burner grid for central and mantle flame; 3 = ring-shaped grid for flowing gas sheath.
(From van der Hurk 1974.)

at a given burner head over several hundreds of degrees centigrade by regulating the water cooling. For an air-acetylene flame a temperature variation of about 200 degrees was obtained (see Hollander and Broida 1969). In practice the temperature of a nickel-plated brass burner head can easily reach 300°C (see Snelleman 1965). By using a burner head that is insulated from the burner housing (layer-cake type burner head; see Hollander 1964, and Snelleman 1965) one may further increase the flame temperature by 80 degrees. Electrical preheating of the burner housing may also increase the temperature by a similar amount (see Hollander 1964).

The material used for the burner head discussed above is nickel-plated brass, but other materials may be used for special applications. (For a detailed discussion of this topic the reader is referred to, e.g., Mavrodineanu and Boiteux 1965.) Stainless steel is used when corrosion of the burner house or head is expected to occur, for example, when metal halides in relatively high concentrations are introduced by directly evaporating the solid material (see Sect. III.3a-2; see also Hollander 1964, Kalff 1971, and van der Hurk, Hollander and Alkemade 1973). Use of SyndaniO or wonderstone (both ceramic materials) is preferred when one is interested either in increasing the flame temperature or in the electrical insulation of the burner head (as in the case of electrical ionization measurements; see Sect. IX.4b).

BURNERS AND NEBULIZERS
Sect. III.2b

With the *capillary burner* the burner top consists of a bundle of stainless steel (or silica) capillaries. For the original design of this burner type (also called the Sugden burner by Schofield and Broida 1968) the reader is referred to Padley and Sugden (1958, 1959). Modifications of the basic form have been reported by Hofmann and Kohn (1961), Bulewicz, Phillips and Sugden (1961), Behmenburg (1964), Gurvich and Ryabova (1964a), Jenkins (1966), and McEwan and Phillips (1967).

The capillary burner produces a laminar flow for flames with a wide range of burning velocities up to several meters per second. Small, separate primary cones emerge from each capillary; these are usually about 0.5 mm high as in the case of $H_2-O_2-N_2$ or H_2-O_2-Ar flames. The tops of the cones tend to merge to produce a relatively flat combustion zone.

The inner flame is generally surrounded by a mantle flame of the same composition. The use of an additional gas sheath has been reported by Jenkins (1966) (see also Kelly and Padley 1972); all the possibilities and advantages of shielding and sheathing the inner flame discussed above apply, of course, to the capillary burner as well.

All the other considerations as to the diameter and length of exit ports discussed above in relation to the grid-type burner head also apply here. The only difference is that with the capillary burner the distance between the ports is very small so that a really flat combustion zone can be obtained. One real advantage of the grid-burner head is the simple exchangeability when other gas mixtures are to be burnt, whereas the capillary burner is less versatile in this respect. A special difficulty with the grid burner is the drilling of the holes, especially when the diameter is, for instance, only 0.4 mm (for H_2-air flames) and the bore length is about 10 mm, and when the tip material is tough as is the case with stainless steel.

In practice the bundle of stainless steel capillaries may be force-fitted into a perforated brass block or bar (see Hofmann and Kohn 1961, and Behmenburg 1964) or simply welded or soldered together (see Padley and Sugden 1959, Gurvich and Ryabova 1964a, and Jenkins 1966). Instead of stainless steel, nickel tubes (see Gurvich and Ryabova 1964a) and silica tubes have been used (see Bulewicz, Phillips and Sugden 1961). Silica is a suitable material when halogens are used as in that case metals corrode easily. The size and shape of the flame can be controlled just as simply as with the grid-type burner head.

The burner head may be cooled by water (see Padley and Sugden 1958, and Jenkins 1966). Behmenburg (1964), however, has used instead thin copper strips that were inserted between the capillaries and that conduct heat from the capillaries to the surroundings.

Chapt. III   GENERAL INSTRUMENTAL ASPECTS AND EXPERIMENTAL METHODS

2b-3   *POROUS PLATE-BURNERS*

The burner top consists here of a metal or ceramic porous plate which is generally water-cooled.

This burner has been developed by Kaskan and co-workers (see Kaskan 1958, 1965 for a detailed description) from the Botha-Spalding burner (see Botha and Spalding 1954; see also Schofield and Broida 1968). This burner can also be used at reduced pressure (see Sect. III.2c). Fristrom (1958) has used a modification which forms spherical flames.

Flames have been produced on this type of burner with burning velocities in the range of 4-40 cm s^{-1}. The production of H_2-air and CO-air flames (or H_2 plus CO-air flames) has been reported with this porous-plate burner by, e.g., Kaskan (1958, 1965), Fristrom (1958), Fenimore and Jones (1958), Fristrom and Westenburg (1965), and Carabetta and Kaskan (1967).

Owing to the noticeable heat transfer (dependent on gas flow rate) the flame temperature is lowered upon increasing the gas flow rate. A special complication may arise from the uncertainty about which fraction of the (metal-containing) aerosol is filtered out by the porous metal plug (see Kaskan 1965).

The flames produced may be surrounded by a metal-free mantle flame as usual (see Carabetta and Kaskan 1967). Additional sheathing has not been reported so far, but can easily be carried out.

2b-4   *SLOT BURNERS*

In the literature several types of slot burners have been reported (see, e.g., Fiorino, Kniseley and Fassel 1968, Kirkbright, Semb and West 1968, Willis 1970, and Rossi, Benetti and Omenetto 1971).

The widths of the slots are prescribed by the gas mixture to be burnt. Some typical data are the following: the slot length for coal gas, C_2H_2, H_2 and C_2H_2 with air as oxidant gas is usually 100 mm; when N_2O is used the length is reduced to 50 mm. The slot widths for the above fuel gases in the same sequence are 1.5, 1.5, 1.0 and 0.5 mm, respectively, for air as oxidant. This width is also 0.5 mm for C_2H_2-N_2O flames (see also Willis 1970).

As can be seen from a comparison of these data with those given for the Méker burner head, usually the width of the slots is 20-25% lower than the corresponding diameter of the exit ports with the Méker burner head (except for H_2-air flames).

Different modifications of the simple one-slot burner (see Amos and Willis 1966, and Fiorino, Kniseley and Fassel 1968) are known. The first modification is to shield the one-slot central flame with a mantle flame and thus produce a three-slot burner (*Boling burner*; see Herrmann 1971).

A variant, called a circular slot burner (see Rossi, Benetti and Omenetto

BURNERS AND NEBULIZERS   Sect. III.2c

1971) produces a metal-seeded inner flame with a circular cross section which is
surrounded by a cylindrical sheath of flowing argon. A circular slot burner with a
chimney of silica has also been reported (see Kirkbright, Semb and West 1968).
Generally speaking, the flames produced on slot burners are less stable,
homogeneous and 'quiet' than those produced on Méker or porous-plate burners. But
slot burners are primarily used in analytical atomic absorption spectroscopy, where
low detection limits are obtained by increasing the absorption path length.

2c   LOW-PRESSURE BURNERS

In this section we focus on low-pressure burners of the Méker type and on
porous-plate burners producing metal-seeded, shielded flames. We do not discuss
the various types of burners for the production of low-pressure, nonseeded flames
used for flame-radical recombination studies, flame-ionization measurements, etc.
(see, e.g., Calcote and King 1955, Fristrom and Raezer 1956, Fenimore and Jones
1958, Bulewicz and Padley 1969, Gaydon and Wolfhard 1979 and Bleekrode 1970).

Low-pressure flames are mostly used as a hot gas milieu where the number of
(in-)elastic collisions that a metal atom undergoes per second with flame gas molecules is reduced by a factor up to 1000. There is particular interest in such
flames in fluorescence quenching experiments (see Bleekrode and van Benthem 1969)
and line-broadening experiments (see Chapt. VII), as well as in those experiments
where the behaviour (e.g., chemi-ionization) of metal vapours in the primary combustion zone is studied. The latter experiments are facilitated by the increased
width of the combustion zone (see below).

The first device of this kind has been developed by Kaskan and co-workers
(see Kaskan 1958, 1965, and Carabetta and Kaskan 1967). We shall discuss this
(Kaskan-)burner later on in this section.

The low-pressure burner set-up shown in Fig. III.3 has been used for absorption measurements in C_2H_2-air, H_2-air, and C_2H_2- and H_2 flames burnt with mixtures
of (maximum) 50% O_2 + 50% N_2 (see Hollander and Broida 1969). It can also be used
for emission and fluorescence measurements; four quartz-glass windows at the same
height of observation, under right angles (see Fig. III.3), are therefore mounted
in the bell jar.

The burner housing is essentially the same as that shown in Fig. III.1.
The burner here consists of a system of long, concentric pipes at the top of which
a Méker type burner head is (silver) soldered. The burner head is water-cooled.
The simultaneous soldering of the concentric pipes brings about a number of technical problems, but soft soldering of the parts appeared to be dangerous (when cooling stops, the burner head falls apart). The burner produces laminar, shielded
flames with a sheath of flowing inert gas. For the introduction of metal vapour

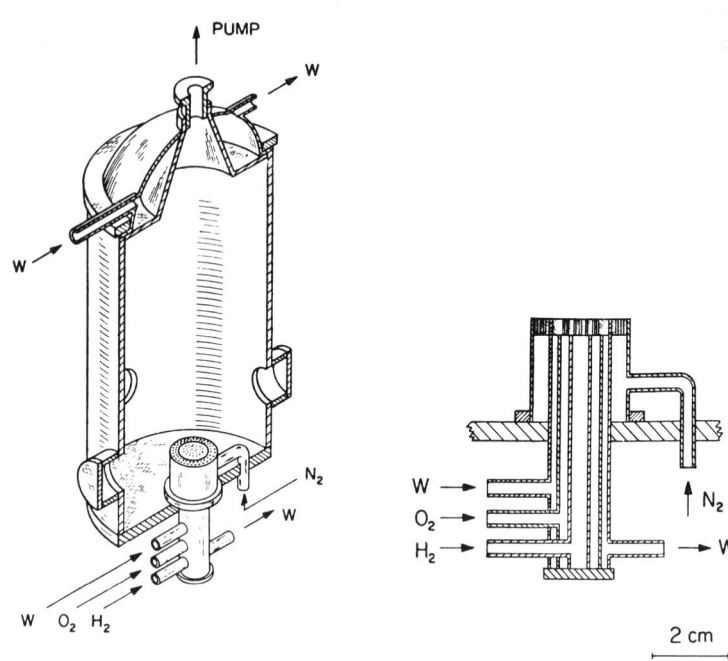

Fig. III.3 Low-pressure burner with water cooling (W). Flames are ignited with a sparking device. Note the sheath of flowing N_2 gas around the shielded flame, which keeps out dust and other contaminations. The height of observation can be varied by changing the position of the burner top with respect to the windows.
(From Hollander and Broida 1969.)

into the central flame see Sect. III.2e.

When the flame gas is at reduced pressure the danger of flash-back is much less. Therefore the diameter of the exit ports of the burner head is less critical. The diameter, $2r$, chosen for the exit ports in Fig. III.3 was 0.8 mm, i.e. the same value as for C_2H_2-air flames at 1 atm (see Sect. 2b-1), but greater diameters could be used as well. The thickness of the combustion zone increases by about the same factor by which the pressures decreases (see, e.g., Gaydon and Wolfhard 1970). At sufficiently low pressure the tops of the combustion cones tend to merge forming a disc-like combustion zone the thickness of which may become of the order of 1 cm (depending on the kind of gas mixture and pressure).

The flame burns inside a bell jar, which may be of pyrex or metal. If it

BURNERS AND NEBULIZERS   Sect. III.2d

is of metal, the mounting of the quartz glass windows requires some care. Since it proved necessary to cool the top of the bell jar (or to reduce the length of the flame; cf. below), a double-walled metal top with water cooling over its whole surface (see Fig. III.3) appeared to be the most convenient.

The pressure inside the bell jar is controlled with a rotary vacuum pump (pumping speed in the 100 to 500 ℓ/min range) and metered with a pressure gauge. Ignition of the flame can be carried out by using a pilot flame (see Kaskan 1958) which in turn is ignited with a sparking device. We found it simplest to reduce the pressure first to about 0.5 atm and then ignite the flame with the spark (see Hollander and Broida 1969). It appeared advisable to switch on the sheath gas supply after the flame was ignited (easier ignition of the flame).

With this set-up the flames produced appeared to remain laminar, cylindrical and well distinguished from the mantle flame even at pressures as low as 1 Torr. The combustion zone is then clearly disc-like. Typical temperatures obtained at the height of maximum temperature in C_2H_2- and H_2-air flames at pressure of about 50 Torr were 200-300 degrees lower than those of the corresponding atmospheric flames (see also Gaydon and Wolfhard 1979).

The original low-pressure burner described in detail by Kaskan (1958) is very similar to the one shown in Fig. III.3. A glass bell jar is used and a metal grid serves to limit the flame height, making cooling of the jar unnecessary. The burner head is a water-cooled porous plate (metal or ceramic). The flames produced are shielded with a mantle flame. Metals are introduced with a pneumatic nebulizer.

Kaskan and co-workers used $H_2-N_2-O_2$, H_2-CO-air and $CO-O_2-N_2$ flames at pressures down to 50-100 Torr. Typical flame temperatures for $H_2-O_2-N_2$ flames at pressures from 50 to 100 Torr are reported to range from 1250-1750 K. Typical rise-velocities are in the order of 10^2 cm s^{-1} (see Carabetta and Kaskan 1967).

2d   BURNER MOUNTING AND ADJUSTMENT

One can create a simple burner mounting by connecting a holding bar to the bottom of the burner house (see Fig. III.1) and by fixing this bar in a holder (hollow rod) mounted on a rail or an optical bench. The height of burners producing flames at 1 atm can be adjusted roughly by simply moving the burner-holding bar inside the holder and by fixing the height of the burner (with respect to the optical light path) by means of the screw in the holder. There is a pointer on the burner house which enables one to read the adjusted height. The height adjustment of low-pressure burners was discussed in Sect. III.2c. Precise height adjustment is feasible when micrometers are used.

Fine adjustment of the height of observation in the flame only makes sense

Chapt. III   GENERAL INSTRUMENTAL ASPECTS AND EXPERIMENTAL METHODS

when the image of the (vertical) flame is turned through 90° (see Sect. III.4d). The slit width, the usable solid angle (see Sect. III.14b) and the height of the combustion cones then limit the effective height resolution obtainable, when 'height' is related to the vertical distance to the combustion zone.

A vertical and/or horizontal, calibrated displacement of the burner is important when we are interested in the dependence on flame height or off-axis distance of parameters, such as temperature, metal content or ionic content. This applies, for instance, when time effects are involved, as in ionization-relaxation measurements (see Sect. IX.3c) or radical-recombination measurements (see Sect. IV.5). It should be noted that the height of observation is directly related to the travel time of the observed particles in the flame (see Eq. IV.1).

When the holding bar on the optical bench is replaced by a horizontal sliding support furnished with micrometer adjustment, off-axis measurements with good resolution are feasible (see, e.g., King and Scheurich 1966 who measured ion profiles in flames with a mass spectrometer; Gurvich and Ryabova 1964a, Mavrodineanu and Boiteux 1965, and Snelleman 1965). Vertical and horizontal positioning of the burner in the case of low-pressure flames is more cumbersome to carry out.

2e   NEBULIZERS

In this section only chamber-type nebulizers will be discussed. These nebulizers can be divided into three main categories:

(i)   pneumatic nebulizers,
(ii) ultrasonic nebulizers,
(iii) electrostatic nebulizers.

These three types of nebulizers are distinguished according to the source of energy used for nebulization: compressed gas, mechanical vibrations of electric field. One can distinguish between suction, gravity-fed, controlled-flow, modulated and reflux-type nebulizers according to the way in which the liquid is taken up. We refer to the general textbooks on analytical flame spectroscopy listed in Chapt. I for a more detailed description and for further references to the literature.

We shall discuss consecutively the pneumatic, the ultrasonic and the electrostatic nebulizers, compare their properties and finally draw conclusion about their utility.

2e-1   *CONSTRUCTION AND OPERATION*

(i)   With a *pneumatic* nebulizer usually driven by a stream of compressed air (or inert gas) emerging from a nozzle, the solution is aspirated from the sample container through the suction pressure of the gas jet and nebulized into a mist or aerosol of fine droplets. By desolvation, i.e., evaporation of the solvent from the droplets, this mist is converted into a dry aerosol consisting of a suspension

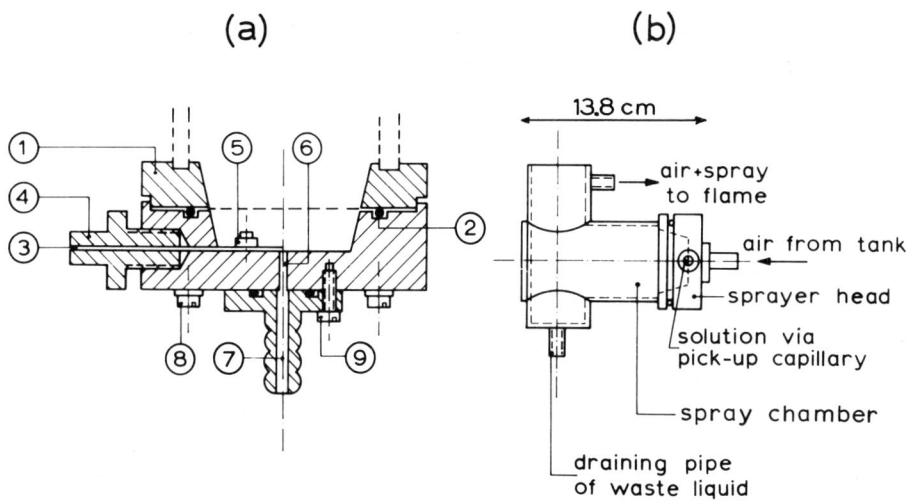

Fig. III.4 Design of angular-type pneumatic nebulizer (in perspex). (a) shows the sprayer head in cross section. (b) shows the nebulizer as seen from the side. 1 = end of spray chamber; 2 = groove filled with package material; 3 = adjustable stainless-steel nozzle for pick-up of liquid sample; 4 = holder of (3); 5 = screw fixing (3); 6 = inlet nozzle for air jet; 7 = air inlet (from tank); 8 = screw for fixing sprayer head to the spray chamber; 9 = screw for fastening (7) to the sprayer head. (From Hollander 1964.)

of solid particles of the solute (possibly in the form of hydrate compounds). In the hot flame milieu, volatilization of these particles follows: they are transformed into the gaseous state as free atoms and/or molecules (see Sect. III.9).

In the chamber-type nebulizer, the nebulizing gas jet expands from the nozzle (i.e., the part of the nebulizer where the aspirated liquid is disrupted by the impinging gas-jet into a spray of fairly coarse droplets) into a *spray chamber* (also called expansion chamber) where the gas jet is slowed down and homogeneously mixed with the mist droplets. Part of these droplets may evaporate, coalesce, or be caught by the chamber walls and drained off to waste.

The drainage of the spray chamber is usually controlled automatically by siphon action so that the liquid to be drained off is prevented from reaching too high a level inside the chamber. Care should be taken that there is always liquid in the waste siphon to prevent gas leakage.

One can distinguish angular-type nebulizers from concentric-type nebulizers (also called split-ring nebulizers) by the relative position of the nozzles for the nebulizing gas and the aspirated liquid. In Fig. III.4 an angular-type pneumatic nebulizer is shown.

Chapt. III   GENERAL INSTRUMENTAL ASPECTS AND EXPERIMENTAL METHODS

The specific data of the nebulizer shown in Fig. III.4 are the following: the nebulizer is made of perspex; air and (liquid) inlet nozzle have diameter of about 0.4 mm, the inner diameter of the (glass) liquid pick-up capillary may vary typically from 0.1 to 0.4 mm. The air pressure over the air capillary is typically between 1.5 and 2.0 atm. The aspiration rate (see definition later on in this section) varies from 0.5 to 4.0 cm^3 min^{-1}, depending on the diameter of the pick-up capillary. The height of the liquid level in the sample solution container with respect to the liquid inlet nozzle may influence this aspiration rate. Automatic drain-off is achieved by connecting a U-shaped tube to the draining pipe, the outlet of which is positioned a few centimetres higher than the outlet of the draining pipe.

(ii)   In the *ultrasonic* nebulizer the sample solution is placed on an ultrasonically vibrating plate, where it is broken up, transformed into a mist of fine droplets, and carried off to the flame by one of the gases feeding the flame. Denton and Malmstadt (1972) have made some modifications of the burner to increase the nebulization efficiency (see definition in Sect. III.8) of their burner-nebulizer system.

(iii)   *Electrostatic* nebulizers are realized by applying an electric field between the liquid nozzle and an electrode in the spray chamber. The liquid is broken up into small droplets, which are again carried off by one of the gases feeding the central flame. Tensions of up to about 1 kV are required to supply a sufficiently strong electric field.

Generally the nebulizers are operated continuously. However, intermittent and modulated sample introduction has been reported in the literature; this facilitates a clearer distinction between metal signal and flame background signal and thus increases the useful signal-to-background ratio (see, e.g., Lang 1966a, 1966b, and Bojovic and Antic-Jovanovic 1972). For further information we refer the reader again to the general textbooks cited in Chapt. I.

2e-2   *PROPERTIES AND UTILITY*

When comparing the properties of the three main categories of nebulizers we focus on the following aspects:[†]
(a)   simplicity of construction;
(b)   ease of operation (adjustment, source of energy used for nebulization, cleaning, liquid intake, drainage);

[†] For a critical comparison of these categories of nebulizers see also the detailed discussion in Dunken *et al.* (1964).

BURNERS AND NEBULIZERS   Sect. III.2e

(c)   prevention of contamination between different samples;
(d)   *nebulization efficiency*, ε_n, being the ratio of the amount of metal salt
        entering the flame as an aerosol to the amount of metal salt aspirated as
        a solution (for a more extensive discussion see Sect. III.8a); this effi-
        ciency is closely related to the size-distribution of the droplets formed
        by the sprayer;
(e)   reproducibility of operation and independence of ε_n of concentration of
        solute in the aspirated solution (*linearity of nebulization*);
(f)   corrosion effects due to the sample aspirated;
(g)   sample consumption, i.e., *aspiration rate* (also called *rate of liquid up-
        take*), F_1, usually expressed in $cm^3 min^{-1}$ or $cm^3 s^{-1}$.

The electrostatic nebulizer has, on the one hand, the advantage of relatively small sample consumption but, on the other hand, the serious drawback that
the rate of nebulization depends on the dipole moment of the molecules in the
liquid to be sprayed. Distilled water behaves badly and quite differently from
aqueous salt solutions. The latter even differ depending on their concentration
and composition. Moreover, many flames behave unstably as soon as large amounts of
charge carriers are introduced. It is clear from the above considerations that
where one is interested in a linear, versatile nebulizer which can be operated
simply, the electrostatic nebulizer is the least suitable (see also Herrmann 1971).

As a serious disadvantage of the ultrasonic nebulizer we may mention the
contamination from one sample to the next, when the metal salt concentrations
differ strongly in the two samples. This contamination can be overcome (but not
completely eliminated) by introducing distilled water between each sample (see
Herrmann 1971). Furthermore, the operation of an ultrasonic nebulizer requires an
electric power source that can produce sine waves or square waves in the frequency
range between roughly 10 kHz and 10 MHz; it is therefore more complicated and
costly to operate than the relatively simple pneumatic nebulizer, where one of the
compressed gases feeding the flame is used. It should be noted that for the production of spray droplets of about the same size as those leaving the spray chamber
of a pneumatic nebulizer, a generator frequency of more than 500 kHz must be used.
The droplet size is here directly related to the frequency applied (see Dunken *et
al*. 1964).

The angular-type pneumatic nebulizers are easier to construct than the concentric ones. With the special design shown in Fig. III.4 the mutual position of
the capillaries can be adjusted by simply disconnecting the sprayer from the spray
chamber. This adjustment can also be made during operation (i.e., with burning
flame) with the glass nebulizer developed by Alkemade (1954), where the adjustment
is carried out by rotating the slightly eccentric (glass) nozzles with respect to

Chapt. III   GENERAL INSTRUMENTAL ASPECTS AND EXPERIMENTAL METHODS

each other until an optimal position is achieved. With the concentric-type nebulizers this adjusting is usually more difficult (see Willis 1967). A disadvantage of the angular type is that the aspiration rate and mean droplet size depend much more critically on the relative position of the nozzles than in the case of the concentric-type nebulizer. An increase in ε_n can be obtained by increasing the air flow through the nebulizer, if the aspiration rate is held constant by changing the liquid pick-up capillary.

For comparison we shall give some representative data on nebulization efficiency for different types of nebulizers as reported in the literature (see also Alkemade 1970, Herrmann 1971, Alkemade and Herrmann 1979). The data refer to experimental conditions without preheating of gases, of spray chamber, etc.

With the angular- and concentric-type pneumatic nebulizer, ε_n may vary between roughly 1 and 15% for aqueous solutions (see Hollander 1964, Willis 1967, and Alkemade 1970). The aspitation rate F_1 is of the order of 1 cm^3 min^{-1}. When organic solvents are used, ε_n may exceed the value for water by a factor of 3 (see Pungor 1967). For ultrasonic nebulizers one may expect ε_n values for aqueous solutions of 59% at $F_1 = 0.1$ cm^3 min^{-1} and 32% at $F_1 = 0.3$ cm^3 min^{-1} (see Stupar 1969).

For electrostatic nebulizers ε_n values do not seem to have been reported. Since the aspiration rate for various (nonaqueous) solvents is very low here (roughly between 7×10^{-4} and 2×10^{-2} cm^3 min^{-1}; see Straubel 1954) the amount of metal salt actually reaching the flame is anyway much lower than for the other types of nebulizers. This is another reason why the electrostatic nebulizer is not used in analytical flame spectrometry.

The measurement of the nebulization efficiency ε_n will be dealt with in Sect. III.8b.

Preheating of the spray chamber and of the gases may considerably increase the nebulization efficiency. We shall discuss this topic in Sect. III.3a-1.

An important feature of the nebulizer is, of course, its reproducibility and linearity. The reproducibility of operation depends critically on the 'shock-proof' adjustment of the spray nozzles. With pneumatic nebulizers the pressure drop over the gas nozzle should be kept constant, and stoppages or clogging (for example, as a result of dust threads or deposits of salt) should be avoided. The temperature of the nebulizer should be kept constant (ε_n depends rather critically on temperature; see Sect. III.3a-1). Furthermore, the liquid pick-up capillary, liquid nozzle and spray chamber should be frequently cleaned by spraying intermittently with distilled water. With these precautions the spray production is reproducible within $\approx 1\%$ over a period of, say, 1 hour, if the gas operating the sprayer is supplied at a constant rate and temperature.

BURNERS AND NEBULIZERS   Sect. III.2e

The concentration range where the nebulizer operates linearly, i.e. where ε_n is independent of solution concentration, may differ from one (type of) nebulizer to another. It is not quite clear which concentration-dependent physical properties of the solution sprayed (viscosity, density and/or surface tension) are responsible for the onset of nonlinearity at high solution concentrations. The increased mass of the dry aerosol particles after desolvation at high solution concentrations may also play a role. For instance, in pneumatic nebulizers nonlinearity may start to show up with solution concentration of the order of 0.1-1 mol/ℓ.

Methods for checking the linearity of the nebulizer performance are discussed in Sect. III.8a.

Salt solutions must be handled very carefully to prevent contamination. Frequent (intermittent) nebulizing of distilled water and wiping of the liquid drops from the inlet of the liquid pick-up capillary is helpful in this respect. Deposits of big droplets or salt particles in the tubing and burner should be avoided for 'clean work'. Intermittent nebulization of organic solvents, such as alcohol, acetone, etc., has been shown to remove these deposits effectively. Especially when high solution concentrations have been used, frequent and thorough cleaning of nebulizer, tubing and burner are necessary.

Corrosion effects due to the chemical properties of the sample aspirated may show up in the case of stainless steel capillaries (see Fig. III.4) when copper solutions are aspirated (see Hollander 1964). With glass nebulizers corrosion effects are expected only for some fluorides; in the case of perspex nebulizers one should be careful with organic solvents such as acetone, alcohol, etc., that may eventually attack the perspex material.

As can be concluded from the above, the pneumatic nebulizers are more suitable for use in analytical applications and fundamental studies, and can be operated relatively simply with good reproducibility and linearity. However, they do not yield the highest nebulization efficiency.

2e-3   *NEBULIZERS FOR LOW-PRESSURE FLAMES*

Finally we want to mention some aspects about the use of nebulizers in low-pressure flame work. Only a few low-pressure flame devices with nebulizers have been reported in the literature. Carabetta and Kaskan (1967) have introduced alkali metals into low-pressure flames with a pneumatic nebulizer; Hollander and Broida (1969) have used a pneumatic nebulizer for the introduction of Na- and Zn solutions into their low-pressure flames, whereas King (1963) has reported a nebulizer operating at 20 Torr at very low flow rates. Finally Bouckaert, D'Olieslager and de Jaegere (1972) have described a direct-injection nebulizer that can be built into a low-pressure flame set-up.

The main difficulty with nebulizers operated in combination with low-pressure flames is that air from the environment can easily leak into the flame via the liquid pick-up capillary, resulting in changes in flame gas composition (see Hollander and Broida 1969). This problem can be circumvented by keeping the spray chamber at atmospheric pressure. This can be achieved by inserting an additional capillary resistance or needle valve between nebulizer outlet and burner. A serious disadvantage is that in this way considerable spray losses in the capillary or valve occur. When the spray is preheated, this loss is much less serious (see Bouckaert, D'Olieslager and de Jaegere 1972).

A different approach is to make use of the low pressure inside the chamber to suck the liquid from outside via the pick-up capillary. When this is done, the nebulizer should be continuously supplied with distilled water when no test solutions are sprayed, in order to avoid air leakage.

2f PREPARATION OF METAL SALT SOLUTIONS

The test solutions used are preferably obtained by diluting concentrated stock solutions that can be prepared with an accuracy of better than 0.1%. Deionized water or aqua bidest of good quality is recommended as a diluent. The quality of the distilled water is easily checked by spraying a blank sample into the flame to see if any metal emission occurs.

Glass bottles can be used for storing most of the (stock) solutions, but for alkaline-earth metal solutions polyethylene bottles are required (see Hollander 1964, and Hofmann and Kohn 1961). Storing of diluted test solutions for a longer time is, however, not to be recommended.

To prepare a stock solution, we use a very pure salt having an exactly known amount of crystal water. A stock solution of high concentration is obtained by weighing the salt and carefully dissolving it in distilled water. (It should be noted that for these stock solutions there is the hazard of concentration change resulting from evaporation of the solvent; thus the bottles must be provided with good fitting screw-caps.) As a rule of thumb, the stock solution concentration should be at most 50% of the concentration of the saturated salt solution at room temperature.

2g REGULATION AND CALIBRATION OF GAS FLOWS

Regulation. The gas flow is usually regulated by means of a reducing valve thus ensuring constant working pressure, and a needle valve, which permits precise adjustment of the flow (see, e.g., Hollander 1964, Fristrom and Westenberg 1965, Mavrodineanu and Boiteux 1965, Snelleman 1965, Gaydon and Wolfhard 1979, and the textbooks cited in Chapt. I). For reproducible adjustment of the gas flows one uses flowmeters rather than pressure gauges (for a detailed discussion see, e.g.,

BURNERS AND NEBULIZERS   Sect. III.2g

Mavrodineanu and Boiteux 1965).

The gases are stored in high-pressure gas tanks (cylinders), usualy under 150-200 atm; an exception is acetylene, which is stored at 15-20 atm; some gases, such as C_3H_8, are stored in liquid form. Usually the safety outlet of the burner (see Fig. III.1) effectively stops explosions caused, for instance, by flash-back; in the case of acetylene, however, some further precautions are recommended: the use of anti flash-back valves inserted behind the high-pressure regulator (see also Mavrodineanu and Boiteux 1965) and the use of nickel or stainless steel pipes instead of brass. (Brass is supposed to promote self-ignition of the acetylene gas.) Furthermore it is recommended that the C_2H_2 tanks should not be emptied completely because at low tank pressures some acetone vapour is found in the gas (see Sect. IV.6a).

Some practical hints about how to handle the combustion gases safely may be useful. When air or N_2O, etc. is used as oxidant, first the oxidant gas is supplied and then the fuel gas in the case of hydrocarbons. For CO and H_2 the reverse sequence is recommended. The gases are always switched off in reverse order. When a mixture of O_2 and diluent gas is used as oxidant, it is safe to premix these gases outside the burner. In other respects the switching on and off of the gas supplies is just the same as described above. When O_2 and diluent gas is used one should be very careful to check the diluent-gas supply in order to avoid the possibility of having pure oxygen as oxidant gas. If the oxidant supply is insufficient when hydrocarbons are burnt, soot forms and dust collects.

The normal *flowmeters* are either of the *capillary type* or they are *rotameters*. The former type is a closed U-shaped manometer which measures the difference in pressure over a capillary through which the flow to be measured is led. The liquid in the flowmeter may be water, mercury, xylol or dibutylphtalate (see Gaydon and Wolfhard 1970). The choice of liquid and capillary is based on the measuring range desired; different ranges may be obtained by switching or shunting different capillaries. Volatility and purity are additional factors that determine the choice of water, xylol or dibutylphtalate. Water is likely to evaporate and is easily contaminated, so a little soap or detergent is often added to prevent contamination. A drop of oil on top of the water columns may help to reduce the evaporation of the liquid. The mercury inside the flowmeter also tends to get dirty easily. The remedy is to renew the mercury after the flowmeter has been flushed with HNO_3, $KMnO_4$, etc.

A rotameter consists of a vertically mounted tapered glass or perspex tube containing a float made of metal or some other material. The volumetric flow rate is measured from the height of the float in the tube. A change in measuring range is achieved by varying the length (and width) of the tube or by changing the

Chapt. III   GENERAL INSTRUMENTAL ASPECTS AND EXPERIMENTAL METHODS

specific weight or size of the float.

The indication on the flowmeter is in first approximation independent of the (variable) stream resistance that may occur in the burner, nebulizer or gas tubing outside the flowmeter (see Alkemade 1954, and Hollander 1964). It should be noted, however, that when strong variations do occur (i.e., when the nebulizer capillaries are re-adjusted, or when the burner head is clogged), the reading on the flowmeter alters.

Usually the flowmeters metering the oxidant, (the diluent) and the fuel gas do not much influence each other despite their joint exits.

Corrections of the flowmeter reading have to be made as a result of variations in barometric pressure or because of variations in the outlet pressure due to burner-head clogging and nebulizer re-adjustment, etc. (see for detailed discussion Hollander 1964, and Snelleman 1965). It is advisable to check the pressure at the outlet of the flowmeter with an extra open-tube manometer; possible variations in the pressure, as discussed above, can then be easily detected. The variation in flowmeter reading caused by a pressure variation of, say, 10 Torr, may amount to about 1%. By inserting an additional needle valve between flowmeter and the open-tube manometer one can deliberately change the outlet pressure and read the changes in flow rate due to the (known) pressure change. In this way corrections for outlet-pressure variations can be derived quite simply.

Corrections of the flowmeter reading for temperature variations under normal room conditions (15-25 °C) are about equal to the attainable precision of the flowmeter reading ($\approx$ 1%) (see Snelleman 1965). Corrections for temperature variations depend on the viscosity and density of the gases to be metered. These effects may be calculated from literature data on the temperature dependence of viscosity and density: a variation in temperature of, say, 10 degrees may cause a flow rate variation of about 3%. If correction for temperature variation is necessary, one may follow the method of Ower and Pankhurst (1966) for rotameters. With modern rotameter designs, and for nearly all gases, the rotameter reading is virtually independent of the viscosity of the gas and it is only the temperature dependence of the density, ρ, of the gas that must be accounted for. So the volumetric flow rate F_1 at temperature T_1 is found from that at T_2 through the relation: $F_1/F_2 = \rho_2^{\frac{1}{2}}/\rho_1^{\frac{1}{2}}$ (see also Kirkbright and Sargent 1968).

Generally speaking one may in practice obtain a precision and reproducibility of gas flow adjustment typically of 0.5-1% with the capillary flowmeters, and 1-2% with the rotameters (see also Snelleman 1965, and Kalff 1971). These values depend on a number of factors, such as length of liquid column, kind of liquid, quietness of the float, etc.

It should be noted that the needle valve should be inserted between

BURNERS AND NEBULIZERS   Sect. III.2g

reducing valve and flowmeter, and not downstream from the flowmeter, since in the
latter case the calibration of the meter depends on the pressure change at the
meter outlet brought about be adjusting the needle valve (cf. above). However,
this effect does not influence the reproducibility of the meter reading. With
commercial flowmeters with built-in needle valves, one has to pay attention to this
effect (see also Mansfield and Winefordner 1968, and Kirkbright and Sargent 1968).

Calibration. Flowmeters may be calibrated either directly, or by comparison with a standard flowmeter for which a calibration curve is available for each of the gases used. Absolute calibration of the gas flow may be carried out directly as follows:

(i)   By the soap-bubble or the soap-film method (for a description the reader is referred to Gaydon and Wolfhard 1979).

(ii) By the pressure-rise method in a large vessel (see Gaydon and Wolfhard 1979).

(iii) By the 'water-repelling' method (see Alkemade 1954, and Hollander 1964); this simple method has the advantage that the calibration is carried out at <u>constant</u> outlet pressure.

When a standard flowmeter is available for which the corrections for temperature and working pressure are known, the calibration may be carried out by simply placing this standard flowmeter upstream from the meter to be calibrated: *method of series connection*. We have to correct the reading of the standard meter for the pressure drop over the meter that is to be calibrated. (With air flowmeters placed upstream from the nebulizer, we have to correct for the pressure drop over flowmeter + air nozzle.) A commercial integrating gas meter (i.e. the volumetric measuring device used for the metering of city gas) can be used as standard flowmeter. This gas meter has the advantage that the calibration does not depend on the kind of gas. Furthermore, the pressure drop over this meter is negligible. The meter may show systematic errors with gas supplies lower than, say, 0.5 ℓ min^{-1} (see Hollander 1964).

In another arrangement an extra needle valve + standard flowmeter are fitted in parallel to the needle valve + flowmeter to be calibrated: *substitution method* (see Fig. III.5). At the joint outlet an extra capillary resistance or needle valve (to create the required pressure drop) + an open-tube manometer are inserted in the tube leading to the burner. The needle valves, parallel to each other, are opened alternatively and the meter deflections are compared at the same deflection of the open-tube manometer. Additional corrections have to be made for the pressure drop over the extra capillary + open-tube manometer. This method is

233

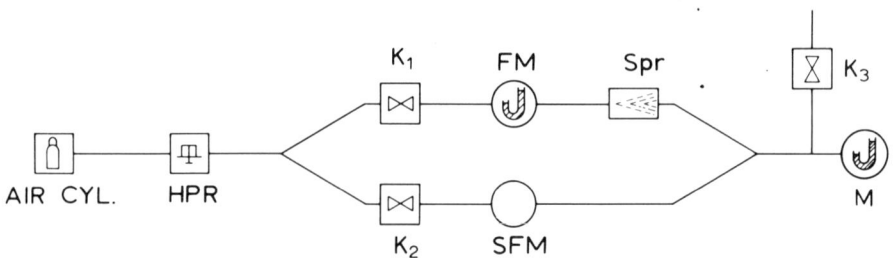

Fig. III.5 The standard flowmeter SFM is connected in parallel to the flowmeter FM to be calibrated, and sprayer Spr. K_1 and K_2 are needle valves. M is an open-tube manometer (filled with water), K_3 is a leakage valve which by-passes M. HPR is a pressure regulator. (From Hollander 1964.)

generally used when the standard flowmeter is a closed U-manometer filled with mercury and thus the pressure drop over this meter is comparatively high, so that the series-connection method is less suitable.

2h OVERALL PERFORMANCE OF NEBULIZER-BURNER SYSTEMS

In analytical flame atomic spectroscopy one is primarily interested in the efficiency with which the analysis element in the solution is converted into an atomic vapour. The *atomization efficiency*, ε_a, is defined as the ratio of the amount of element passing per second through the flame cross section at the observation height as free neutral atoms to the amount of element aspirated per second (see IUPAC 1976a). The atomization efficiency is a composite figure-of-merit given by: $\varepsilon_a = \varepsilon_n \beta_s \beta_v \beta_a$, where ε_n is the nebulization efficiency (see Sect. III.8a), β_s is the fraction desolvated (see Sect. III.9a), β_v is the fraction volatilized (see Sect. III.9c) and β_a is the fraction atomized (see Sect. V.1). For a given solvent the nebulization efficiency is characteristic for the nebulizer-burner system, β_s depends on the type and properties of nebulizer and flame, whereas β_v and β_a depend also on the properties of the element(s) present in the solution.

In fundamental flame studies it is sometimes convenient to define the *conversion factor* $K \equiv n_t/c$ where n_t is the total density of element in the gaseous phase in the flame and c is the molar solution concentration. K is a characteristic constant for the nebulizer-burner-flame system as a whole, which is — in first approximation — independent of the element if we assume complete volatilization ($\beta_v = 1$). K can be determined by measuring the absolute atomic concentration in the flame for a fully atomized element with $\beta_a \equiv 1$ (see Sects V.2 and 6). The value of K depends on the solution aspiration rate F_1 (see Sect. III.2e), the nebulization efficiency ε_n and the fraction desolvated β_s (which is usually unity for chamber-type nebulizers; see Sect. III.9b). Since F_1 and ε_n

BURNERS AND NEBULIZERS									Sect. III.2h

depend on the kind of solvent used, K is usually defined for an aqueous solution. Expressions for K can be easily derived from Eqs V.17 or V.18.

Table III.1

Some Conversion Factors K Reported in the Literature

$K \times 10^{-13}$ cm^{-3} per (mol ℓ^{-1})	Flame	F_u cm^3 s^{-1}	Nebulizer	Reference
2.3	H_2-O_2-N_2	140	chamber-type	Bulewicz and Padley (1971)
4	H_2-O_2-N_2	140	chamber-type	Jensen and Padley (1966)
5	CO-air	130	chamber-type	Hollander (1964)
5	C_2H_2-air	300	chamber-type	Zeegers and Winefordner (1971)
20	H_2-O_2-N_2	150	chamber-type	Gurvich and Ryabova (1964a)
60	C_2H_2-air	330	chamber-type	de Galan and Winefordner (1967)
100	C_2H_2-air	190	chamber-type	Rann (1968)
120	C_2H_2-air	190	chamber-type	Willis (1971)
200	H_2-O_2	225	direct-injection	Simon (1962)

For general orientation we have collected in Table III.1 some typical conversion factors reported. Most of them refer to premixed laminar flames at 1 atm pressure with chamber-type nebulizers having a rate of solution aspiration, F_1, of the order of a few times 0.01 cm^3 s^{-1}. The conversion factor is not critically dependent on F_1 since the product $\varepsilon_n F_1$ usually increases only slightly with increasing rate of solution aspiration (see Alkemade and Herrmann 1979). The flow rate F_u of the unburnt gas mixture has been included in the table.

As expected, it appears that direct-injection burners yield total element densities in the flame that are one order of magnitude larger than those with chamber-type nebulizers. In both cases, however, the partial pressure of the metal vapour in a flame is still very small ($\leqslant 10^{-4}$-10^{-3} atm), even when a metal solution containing 1 mol ℓ^{-1} is nebulized.

Of course, the widely varying values listed in the table cannot simply be transferred to other flames and nebulizers, and are given for illustrative purposes only. Even with a given nebulizer-burner combination the conversion factor may easily vary with time. Unnoticed changes in temperature, partial blocking of the liquid aspiration, a slow drift in flowmeter calibration, and so on, may account for this. It is for similar reasons that in analytical flame spectrometry reference solutions are repeatedly nebulized in succession with the sample solutions. Drift effects are then corrected for by taking the ratio of the spectrometer readings for sample and reference solution.

Chapt. III GENERAL INSTRUMENTAL ASPECTS AND EXPERIMENTAL METHODS

3. SPECIAL TECHNIQUES OF SEEDING A FLAME

In this section we shall discuss special techniques of seeding a flame with metal vapour and of introducing special components (such as chlorine) into the flame.

Formally we can distinguish three different seeding techniques, depending on the state of the material to be introduced, which may be solid salts or metals (without solvent), liquids, or vapours.

3a INTRODUCTION OF DRY METAL SALT AEROSOLS (WITHOUT SOLVENT)

In some investigations the addition of an aqueous solvent is undesirable or complicates matters (as, for example, in the study of the formation or emission of alkaline-earth oxides in 'dry' CO-air flames) (see Chapt. VIII). In the case of 'dry' CO flames metal vapours must be introduced without solvents into the flame. This may be achieved either by nebulization of a metal solution followed by quantitative removal of the solvent, or by direct introduction of dry metal (salt) by evaporating the solid material.

3a-1 *NEBULIZATION FOLLOWED BY REMOVAL OF SOLVENT*

Most of the cases we treat here involve aqueous solutions. When other solvents are used the introduction method is not essentially different. The solution is nebulized in the usual way, but the spray chamber is heated either by heating tape (see Hell *et al.* 1968, Veillon and Margoshes 1968, Venghiatis 1968, and Kalff 1971) or by infrared radiation (see Shifrin, Hell and Ramirez-Muñoz 1967, Slavin 1967, and Uny *et al.* 1971). The gases carrying off the aerosol droplets may also be preheated (see Rawson 1966, and Kalff 1971). The spray droplets are then quickly evaporated in the heated spray chamber and the gas and dry aerosol and solvent vapour is then led through a multi-stage condensing system. The first stage of this system is a water-cooled surface (called 'cooler'), where the largest bulk of the solvent is trapped before being introduced into the burner. The next cooling step can consist in putting one or more 'cold fingers' behind the (water) cooler (see Veillon and Margoshes 1968). In addition, the burner itself can be heated (see Hollander 1964, Rawson 1966, and Kalff 1971).

It should be noted that freezing out the solvent without using the above mentioned first stage of the condensing system and without heating the spray chamber does not work, since snow particles are then formed in the cold fingers and carried off, so that solvent is still transferred to the flame (see Hollander 1964).

Preheating methods are sometimes used in analytical spectrometry in order to enhance the nebulization efficiency of spray-chamber nebulizers (see Uny *et*

SPECIAL TECHNIQUES OF SEEDING A FLAME   Sect. III.3a

al. 1971). In the paper by Uny *et al*. a complete description of the set-up is
presented. The authors, however, do not give a quantitative answer about the
extent to which water can be removed from the gases. Veillon and Margoshes
(1968) claim that the amount of water introduced is quite negligible (as verified via OH emission) whereas the nebulization efficiency ε_n amounts to 35%,
which is about 10 times higher than without heating and water removal. It should
be noted also that with the method described here a constant and reproducible
supply of metal vapour can be achieved. In this respect the method is to be
preferred to methods involving direct evaporation of solid material (see Sect.
3a-2).

The experimental methods of determining the residual supply of H_2O to the
flame involve the optical measurement of LiOH-, CuOH-, (CuH-) or OH concentrations in the flame gases. The LiOH concentration is derived from emission or
absorption measurements on the Li resonance line (see Sect. VIII.5b), whereas
CuH-, CuOH- or OH concentrations can be derived from molecular-band intensity
measurements. Absorption measurements are mandatory when suprathermal chemiluminescent band emissions show up (see Sect. VI.3c-3).

3a-2   *DIRECT EVAPORATION OF SOLID METAL SALTS*

For an extensive survey of this topic (and especially the older techniques)
the reader is referred to Mavrodineanu and Boiteux (1965), Winefordner (1971), and
Dean and Rains (1969, 1971).

The first method to be discussed here is the *introduction of dry powder* by
means of an *ultrasonic stirring device* (see Rössler 1968, and Kalff 1971). The
powders were obtained from dry-frozen aqueous solutions (see Strachan Woods 1968)
from which salt crystals of about 1 μm diameter could be produced. These powders
were whirled up by means of the stirring device and carried off to the flame by
the oxidant. A serious disadvantage appeared to be that the flow of the aerosols
could not be kept sufficiently steady for a few minutes (see Kalff 1971). Clogging of various parts of the circuit between stirring device and burner appeared
seriously to hamper the measurements.

A second method is the use of a *solid-aerosol generator* (see Winge, Fassel
and Kniseley 1971): solid metal samples are transformed into aerosols of fine
metal particles. The sample serves as the cathode in a low-current d.c. arc discharge. A flowing gas stream transports the aerosol particles produced by cathodic
sputtering to the flame. The aerosol can be supplied intermittently by modulating
the electric power supply to the gas discharge. The authors report a reproducibility of 1.7%.

A third method is based on the *sublimation of solid metal* or *metal salts*
by heating (see Hollander 1964, and Fells and Harker 1967). Fells and Harker (1967)

describe a furnace ($T \leqslant 1300$ K) in which metals or metal salts are melted. An inert-gas stream is led over the bath of molten salts and picks up the (salt)vapour, which crystallizes into a dry aerosol of extremely fine particles. This aerosol can be transported over several metres without losses before being injected into the flame. The method is quantitative and reproducible. A disadvantage of this method is the relatively slow response of the metal content to variations of the current heating the furnace.

The evaporating system reported by Hollander (1964) is very similar to that described by Fells and Harker. The former author used an evaporation chamber in which several pots with different molten salts were placed. To fight corrosion due to metal halides, the chamber was made of stainless steel. The pots (made of Al_2O_3) were heated by placing them in small cylindrical furnaces. The temperature of each furnace could be controlled separately up to about 1500 K. With this system several metal salts could be introduced simultaneously and independently into the flame.

Other seeding techniques belonging to this class involve disruption of a spark between two electrodes made from the metal under investigation (see Mavrodineanu and Boiteux 1965) or the introduction of metal salt by placing a bead on a wire in the flame. The use of the latter technique for absolute calibration of the metal supply to the flame by measuring the loss of weight of the bead will be discussed in Sect. V.5b. A disadvantage of these (seldom used) techniques is that the rate of salt supply is not constant and is hard to control.

3b   OTHER SPECIAL SEEDING TECHNIQUES

Twin nebulizers. With this technique two separate (ordinary) pneumatic nebulizers are placed in parallel in the gas conduit; in other words, the gas stream is divided into two parts, each passing through one of the two nebulizers which have a joint outlet (see Alkemade and Jeuken 1957). In this way two different solutions can be introduced simultaneously but separately into the flame. When, for example, the effect of one component (a metal or an anion) in varying concentrations is studied on another metal, we need not prepare 'combined' solutions. With this technique (see Sect. III.9d) possible effects of reactions in the condensed phase between the two components when present in the same solution can be excluded.

Twin nebulizers are also used in experiments where ionization of the metal atoms under investigation is undesirable. The second nebulizer supplies the flame with a sufficient amount of an easily ionizable element, e.g. Cs or Rb, which suppresses the ionization of the metal introduced by the other nebulizer.

Hieftje droplet generator. This droplet generator developed by Hieftje and Malmstadt (1968) produces single droplets of known, constant size which can be controlled over a wide range (10-100 μm). This generator is used for the study of the

SPECIAL TECHNIQUES OF SEEDING A FLAME   Sect. III.3b

rate of processes or the sequence of events (desolvation, volatilization) that occur when a single droplet of a given solution travels through the flame (see Sect. III.9). It should be noted that the range of droplet sizes is here appreciably larger than in cases when a chamber-type nebulizer is used. In the latter case droplets may range in size from 1-10 μm (see also Clampitt and Hieftje 1972).

Liquid bubble method. Hieftje and Bystroff (1975) reported a special bubbling device for the introduction of Na vapour into the flame. This method of nebulization was chosen because of its simplicity, stability and low rate of solution consumption. The latter characteristic was important because of the extended time required to obtain a noise power spectrum. In this device, bubbles issuing from a dispersion tube immersed in a 200 μg/cm³ sodium solution rise to the surface and burst to produce small droplets which are efficiently delivered to the flame by the flowing gas.

Introduction of halogens. Halogens have been introduced into the flame, e.g., to study ionization processes (halogens act as efficient electron acceptors because of their favourable electron affinity) and the inhibition of the $CO + O_2$ combustion reaction (see Hall and Pierson 1969), and particularly in the study of the occurrence and dissociation energy of metal halides (see Sects VIII.3c and 5b).

In practice the corrosive action of (some of) these gases and their toxicity hamper the introduction of halogens in gaseous form. Therefore the use of halogen compounds in liquid form is widely advocated in the literature.

The commonest method for the introduction of halogens makes use of thermostatic bubblers where the saturated vapour is carried off by the inert-gas stream to the burner (see Bulewicz, James and Sugden 1956, Ryabova and Gurvich 1964, 1965a, and Gurvich and Ryabova 1964). Halogen concentrations smaller than 1% in the flame will not disturb the flame (see Bulewicz, James and Sugden 1956; see also Ryabova and Gurvich 1964, 1965a). This technique leads to the same experimental results as when the halogens are introduced in vapour form (see Bulewicz, James and Sugden 1956). The following liquid compounds have been used for the introduction of halogens: CCl_4 (see Ryabova and Gurvich 1965a); $CHCl_3$ (see Bulewicz et al. 1956); $(C_2F_5)_3$ (see Ryabova and Gurvich 1964); and hydrochloric acid (see Alkemade 1954). Gurvich and Ryabova (1964) have carried out an absolute calibration of the amount of CCl_4 introduced into the flame by weighing the loss in liquid.

When halogen gases have to be used, they are generally premixed with the inert gas or air before introduction into the burner, because of their possible reactions with, e.g., H_2. It should further be noted that when chlorine gas, etc., is supplied to the flame, its corrosive action on metal surfaces can be reduced by drying it before admitting it to the gas conduit.

Chapt. III   GENERAL INSTRUMENTAL ASPECTS AND EXPERIMENTAL METHODS

The introduction of halogens by nebulizing pure liquid-halogen compounds (through a second nebulizer, see above) is inferior to the methods described above. In practice, it is hardly possible to keep the halogen content in the flame low enough to avoid flame disturbance (see Hollander 1964).

4. THE OPTICAL TRAIN

4a   INTRODUCTION

There are a number of textbooks that deal thoroughly with imaging and dispersive systems used in spectral analysis (see, for instance, Stone 1963, Longhurst 1967, Fowles 1968, and Lipson and Lipson 1969). In this section we shall discuss some general aspects only, such as the relation between optical conductance and resolution, the distortion of spectral line shapes by the instrumental profile and the usefulness of auxiliary equipment such as double-beam instrumentation. We are not concerned here with fundamental aspects of information theory relating to spectrometric systems. The reader seeking information on this subject is referred to the important papers by Kaiser (1970); see also Fitzgerald and Winefordner (1975).

The purpose of the optical train in spectrometric apparatus is fourfold, viz.

1. To obtain an adequate wavelength selection for the measurements at hand so that no systematic errors appear in the final result. Possible errors include spectral interference of some neighbouring line or continuous background in the measurement of atomic line intensity, or the distortion of spectral profiles in a recorded spectrum.

2. To collect as much light as possible from the light source being investigated so that the random error in the measurement is minimized or so that weak signals can be measured in the presence of extraneous noise.

3. To select a well-defined volume element of the light source and (less frequently) a well-defined solid angle of acceptance.

4. To increase the signal-to-noise ratio by auxiliary (mechanical, optical or electronic) equipment. Such equipment is useful when the precision of the measurements is not limited by shot noise but by drift or excess low-frequency fluctuations (see Sect. III.14e). It is also useful in multiplex measurements.

The first two aims, characterized by the resolution R and the optical conductance L, will be treated together; the third aim is straightforward and requires little discussion. The fourth aim is wider in scope and will involve a discussion of a variety of special tricks; some attention will also be given to

THE OPTICAL TRAIN   Sect. III.4b

special detection methods and data processing.

4b   FIGURES OF MERIT FOR MONOCHROMATORS

Optical dispersive systems can be characterized by a figure of merit P_1 for line spectra and a related figure P_c for continuous spectra. On this basis a meaningful comparison can be drawn between prism monochromators, grating monochromators and interferometers including interference filters (see Jacquinot 1954).. If P_1 is to be a characteristic constant for a well-designed instrument, some requirements must be met: e.g., the available apertures should be completely filled, the width and height of entrance and exit slits (in slit instruments) should be matched, and slits and diaphragms should be large enough to prevent diffraction effects from playing a dominant role. In the description of the figures of merit the following terms will be used:

(i) $\tau(\lambda)$, the *transmission function* of the spectral apparatus, is a slowly varying function of wavelength λ which describes the transmission factor (see Appendix A.5) including reflection and absorption losses in the apparatus.

(ii) $A(\lambda_m - \lambda)$, the *normalized instrumental profile*, describes the relative distribution over wavelength of the radiation from a wavelength-independent continuous spectrum that is transmitted through the monochromator, when λ_m is the wavelength-setting. Normalization is obtained by defining $A(0) \equiv 1$. The profile is determined by dispersion, slit-widths, diffraction effects and optical imperfections. For example, in the case of rather large, mutually matched exit and entrance slit-widths, A is the well-known triangular function (see also Sect. III.4c).†

(iii) $\Delta\lambda_{eff} \equiv \int A(\lambda_m - \lambda) d\lambda$, the *effective spectral width*,* is a measure for the spectral resolution of the instrument.

The instrumental profile can be determined by recording the signal $S(\lambda_m)$ from a narrow emission line source with fixed central wavelength λ_0 as a function of wavelength setting λ_m. One then measures the *convolution* (see also Sect. II.5b-3) of the line profile $I^s_\lambda(\lambda)$ of the source and the instrumental profile A

$$S(\lambda_m) \propto \int_0^\infty I^s_\lambda(\lambda') A(\lambda_m - \lambda') d\lambda' . \qquad (III.1)$$

---

† Also in use, more specifically for spectrographic measurements, is the term '(entrance) slit function' (see de Galan and Winefordner 1968). This function describes the intensity distribution along the wavelength scale of a monochromatic image on the photographic plate. Convoluted with the (mostly rectangular) 'exit-slit function' it yields the 'total slit function' which is, after normalization, identical to our instrumental profile $A(\lambda_m - \lambda)$.

* This definition is analogous to that of the effective *line* width given by Eq. II.180; $\Delta\lambda_{eff}$ is also called *spectral bandwidth* by de Galan and Winefordner (1968).

Chapt. III   GENERAL INSTRUMENTAL ASPECTS AND EXPERIMENTAL METHODS

When the emission line profile can be described by a delta-function, Eq. III.1 reduces to $S(\lambda_m) \propto A(\lambda_m-\lambda_0)$. Calibration of the wavelength setting can be performed as described in Sect. III.4d.

When $\Delta\lambda_{eff}$ is measured to estimate only the spectral resolution (see Eq. III.7 below), the required accuracy is but moderate. However, in the calibration of absolute line intensities by means of a continuous standard source (see Sect. III.14g), a higher accuracy will be required.

If the half-intensity width of the spectral distribution under investigation is small compared to the effective spectral width of the instrument, we shall speak in this context of a '*line*' spectrum. In this case the figure of merit P_1 is the product of the resolution and the optical conductance. For a slit instrument this can be shown as follows (see also Fig. III.6):

The radiant flux emitted by a 'line' source at λ_0 and passing through the instrument is given by

$$\Phi_1 = BL\tau , \qquad (III.2)$$

where B is the (uniform) radiance (for explanation of terms see Appendix A.5) of

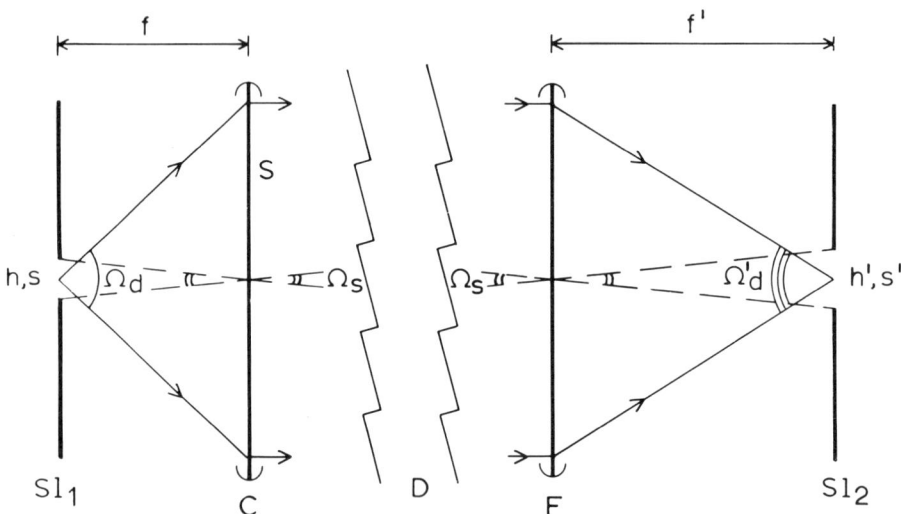

Fig. III.6 Quantities relating to the optical conductance of a monochromator. h and s are the height and width of the entrance slit Sl_1, S is the useful area of the collimator C (with focal length f), Ω_d and Ω_s are the solid angles subtended by the collimator C and the entrance slit, as viewed from Sl_1 and C, respectively. D is the dispersing element (grating, prism). F is the focusing lens (with focal length f') and Sl_2 is the exit slit.

THE OPTICAL TRAIN   Sect. III.4b

the 'line' source in the plane of the entrance slit, L the optical conductance and
τ the value of the transmission function at $\lambda = \lambda_0$. The *optical conductance*, L,
of the beam is (approximately) equal to the product of cross-sectional area and
solid angle of the radiation beam that passes through the instrument, if the beam
divergence is not too large (see Appendix A.6 for a more exact description and
further explanation).[†] This product is a constant along the beam.

The optical conductance of an optical component or a combination of components, such as a monochromator, is defined as the value of L obtained when the passing beam fills maximally all apertures present in the system. For the part of the monochromator that consists of the entrance slit Sl_1 (with width s and height h) and the collimator C we thus find L by considering a beam that completely fills Sl_1 and the useful area S of C. (The useful area may be limited by the dimensions of the dispersive element). The optical conductance of this beam follows from its cross section, $h \times s$, at the entrance slit and corresponding solid angle Ω_d (see Fig. III.6)

$$L = hs\Omega_d \, , \qquad (III.3)$$

if Ω_d is not too large. Since the optical conductance is invariant along the beam, we obtain the same value of L when we consider the cross section and solid angle of the beam at the place of the collimator

$$L = S\Omega_s \, . \qquad (III.4)$$

Here Ω_s is the solid angle subtended by Sl_1 when viewed from C (see Fig. III.6). For Ω_s we can therefore write

$$\Omega_s = \alpha \beta \, , \qquad (III.5)$$

where $\alpha = s/f$ and $\beta = h/f$ are the corresponding *angular* width and height of the slit, respectively (f is the focal length of C). One easily checks by geometrical considerations the equality of the expressions given by Eqs III.3 and III.4.

Similar expressions can be given for the part of the monochromator that consists of the focusing lens F (with focal length f' and same useful area S) and the exit slit Sl_2 (with dimensions s' and h'). If at the wavelength of the monochromatic radiation beam Sl_1 is exactly imaged upon Sl_2 (diffraction effects being neglected), we have: $h's' = hs(f'/f)^2$. All monochromatic radiation passing Sl_1 and the useful area of C also passes the exit slit, disregarding the transmission losses, etc., which are implicitly taken into account by the transmission factor τ in Eq. III.2. Because of the above relation between $h's'$ and hs, and

---

[†] The French term 'étendue' is more suitable to denote the L value for a radiation *beam*. The term 'conductance' is more appropriate to denote L as a property of an optical *component* or *system*.

Chapt. III   GENERAL INSTRUMENTAL ASPECTS AND EXPERIMENTAL METHODS

because $\Omega'_d = \Omega_d (f/f')^2$, we find (compare Eq. III.3) that the optical conductance of the two parts of the monochromator is then the same and simply equal to the optical conductance of the monochromator *as a whole*.

When, however, the image of the entrance slit in the focal plane of F does not match the exit slit, the optical conductance of the collimator part will be *different* from that of the focusing lens plus exit slit. The optical conductance of the whole monochromator is then equal to the smallest value of L of either part. When, for example, the exit slit is broader than the image of the entrance slit, the latter forms the *limiting* aperture and determines the optical conductance of the whole monochromator.

Similar considerations apply to the optical conductance of an optical train consisting, e.g., of light source, imaging lens, monochromator, photodetector, each of which can be characterized by a separate L value. The optical conductance of the whole train cannot exceed the smallest separate L value.

In the following we shall assume that the entrance and exit slits are mutually matched (neglecting again diffraction effects), and that the angular apertures are small enough to warrant the use of the above approximate equations for L. The optical conductance of the monochromator is then determined by Eq. III.4, which equally applies to the collimator plus entrance slit or to the focusing lens plus exit slit. From Eqs III.4 and III.5 we have $L = S\alpha\beta$. Expressing α by the relations

$$\alpha = \Delta\lambda_{\text{eff}} D \, , \qquad \qquad (\text{III.6})$$

$$R = \lambda_0 / \Delta\lambda_{\text{eff}} \, , \qquad \qquad (\text{III.7})$$

where R is the *resolution* and D the *angular dispersion*, one obtains

$$L = S\beta\lambda_0 D/R \, . \qquad \qquad (\text{III.8})$$

It follows from Eq. III.8 that the product of R and L,

$$P_1 \equiv RL = S\beta\lambda_0 D \, , \qquad \qquad (\text{III.9})$$

is a constant, with dimension of area, for a given apparatus and wavelength, λ_0. One may consider P_1 as a *figure of merit* of the instrument for spectral '*lines*'.

Bousquet (1969) following Jacquinot, showed that similar expressions for P_1 can be given for slit-monochromators as for (flat-plate) Michelson and Fabry-Pérot interferometers including interference-filters. He compared the values of P_1 for various instruments and found the ratio of the figure of merit of a grating monochromator to that of a prism monochromator equal to $\lambda \frac{dn}{d\lambda}$ (n = refractive index of the prism material) amounting to a factor 3 to 30. The corresponding ratio for an interferometer and a grating monochromator was found to be equal to $2\pi/\beta$, i.e. about a factor of 100, assuming the effective areas of the dispersing

THE OPTICAL TRAIN   Sect. III.4b

elements (grating, prism, mirrors) of the instruments to be equal. The factor $2\pi/\beta$ indicates that the axial symmetry which can be realized in an interferometer is the reason for the high optical conductance of such an instrument for a given value of R.

For sources with a *continuous* spectrum, i.e. a spectrum of which the half-intensity width exceeds significantly the effective spectral width of the apparatus, the radiant flux passing through the spectral instrument is given by (cf. Eq. III.1)

$$\Phi_c = \int_0^\infty B_\lambda(\lambda)\, S\alpha\beta\tau(\lambda)\, A(\lambda_m-\lambda)\, d\lambda = B_{\lambda m} S\alpha\beta\tau \Delta\lambda_{eff}, \quad (III.10)$$

when the spectral radiance $B_\lambda(\lambda)$ (for its definition see Appendix A.5) and $\tau(\lambda)$ can be considered to be constant over the interval $\Delta\lambda_{eff}$ and equal to their values $B_{\lambda m}$ and $\tau(\lambda_m) \equiv \tau$ respectively at λ_m. Inserting again the relations III.6 and III.7 we find

$$\Phi_c = \frac{B_{\lambda m} S\beta \lambda_m^2 \tau D}{R^2} = B_{\lambda m} L'\tau \quad (III.11)$$

in which $L' \equiv L\Delta\lambda_{eff}$. The product of L' and R^2 is now a constant of the apparatus at given λ_m and we define it as a *figure of merit* for *continuous* spectra

$$P_c \equiv L'R^2 = S\beta\lambda_m^2 D. \quad (III.12)$$

From Eq. III.9 with $\lambda_0 \equiv \lambda_m$ it follows that

$$P_c = P_1 \lambda_m. \quad (III.13)$$

It should be remembered that an atomic spectral line with a half-intensity width of, e.g., 0.1 Å in this context is to be considered as a 'line' when a prism monochromator of effective spectral width of 1 Å is used, but as a 'continuum' when a Fabry-Pérot interferometer of effective spectral width of 0.01 Å is used.

It follows from Eqs III.11 and III.12 that

$$\Phi_c = B_{\lambda m} P_c \tau/R^2, \quad (III.14)$$

and from Eqs III.2 and III.9 that

$$\Phi_1 = B P_1 \tau/R. \quad (III.15)$$

The constancy of P_c and P_1 demonstrates, for example, the well-known fact that widening of the slits (i.e. decreasing R) by a factor b increases Φ_c by a factor b^2, whereas it increases Φ_1 by only a factor b. An increase in resolution R by enlargement of the plate separation in an interferometer affects the light flux differently in the two cases. In the former case L and thus Φ_1 (see Eq. III.2) decrease linearly with R^{-1}, as LR is constant. In the latter case L' and thus Φ_c (see Eq. III.11) vary $\propto R^{-2}$, as $L'R^2$ is constant.

Some newer types of spectral instruments escape the consequence of decreasing

L with increasing resolution. The confocal Fabry-Pérot interferometer (see Connes 1956) and the field-widened Michelson-interferometer (see Bouchareine and Connes 1963) have a constant figure of merit for 'line' spectra equal to L/R. These instruments are suitable only in studies of very high resolution.

In Table III.2 the order of magnitude of P_1 (for a 'line' source) and of R_{max} are given for some spectral instruments.

Table III.2

Figure of Merit P_1, Maximum Resolution R_{max}, and Optical Conductance L at R_{max}

Spectral instrument	$P_1 = RL$ (cm² sr)	R_{max}	L (cm² sr) at R_{max}
glass absorption filter	-	20	-
interference filter	100	100-1000	0.1-1
prism monochromator	0.1	10^4	10^{-5}
grating monochromator	1	$(1-5) \times 10^5$	$(2-10) \times 10^{-6}$
flat F.-P. interferometer	100	$5 \times 10^5 - 5 \times 10^6$	$2 \times 10^{-4} - 2 \times 10^{-5}$
confocal F.-P. interferometer	-	$10^6 - 10^8$	-

The figures in the table do not allow for other differences between the instruments, such as difference in effective area S and transmission function $\tau(\lambda)$. For example, a grating of 100 cm² is quite feasible, whereas interferometer mirrors of this size and of the desired flatness cannot normally be obtained. On the other hand, gratings, when not properly blazed, have a lower transmission than the other instruments.

Interference filters can be considered as narrow-spaced Fabry-Pérot systems. The angular dependence of the transmission of such a filter has the same origin as the appearance of the rings and the central spot in a Fabry-Pérot pattern. Therefore the angle of acceptance for a monochromatic light beam falling on an interference filter is limited. The transmission decreases with increasing angle of incidence, whereas the wavelength of peak transmission shifts to lower values (see also James and Sternberg 1969). A glass absorption filter only exhibits the effects of increased reflection at the surfaces and a variation of effective thickness when a beam enters obliquely. Values for P_1 and L are therefore omitted in Table III.2.

The expressions for the figures of merit P_1 and P_c are helpful in solving various practical questions as to the optimal choice of a dispersive instrument, and the trade-off between R and L under different measuring conditions. Some examples are:

THE OPTICAL TRAIN   Sect. III.4b

1.   If for a given application a moderate resolution suffices, the use of an instrument with large P value still has the advantage of providing for large signal strengths (see Eqs. III.14 and III.15).

2.   Since   product of R and L is proportional to S, an increase in grating area by a factor b can be utilized to increase either R or L by the same factor, or R and L each by a factor $\sqrt{b}$ only.

3.   The focal lengths of a monochromator are only of importance with respect to tthe avoidance of optical and mechanical aberrations. The independence of P of the focal length expresses the fact that the (spectral) radiance of the source image cannot be increased by (de)magnification (see also Appendix A.6).

   If the radiance could be increased in (de)magnification (while keeping the refractive index in the image plane the same as in the object plane), one would be able to transport energy by thermal radiation from a body of a certain temperature to another body of higher temperature. Such a temperature lever, however, contradicts the second law of thermodynamics (see, e.g., Pippard 1966).

4.   The ratio of the signals from a given 'line' source and a given 'continuum' source is proportional to R, i.e. dependent on the slit widths. When the radiation from a line source is to be calibrated with the aid of a known continuum source (see Sect. III.14g), R should be known.

   Apart from R and L there are of course other aspects that characterize an instrument; we mention, for example, the overlap of orders in instruments based on diffraction. When in the Nth order (characterized by an optical path difference of $N\lambda$ between successive constructively interfering beams) radiation of wavelength λ_N is detected, then radiation of wavelength $\lambda_{N+1} = \frac{N}{N+1}\lambda_N$ is detected at the same focus in the spectrum of the $(N+1)^{st}$ order. The difference between λ_N and λ_{N+1} equals $\lambda_N/(N+1)$ and is called the *free spectral range* (FSR). This means, for example, that a visible spectrum in the first order does not overlap the same spectrum in the second order. However, the higher the order, the smaller the FSR becomes. In the 15th order, which may be realized with grazing incidence on a grating, the FSR in a visible spectrum is only about 400 Å. In a Fabry-Pérot interferometer with a plate separation of 1 cm, $N \approx 40,000$ and the FSR is only 1/8 Å. In a prism monochromator the problem does not exist, since here the spectrum is formed by refraction in the zeroth order only.

   The influence of the number of rulings per cm in a grating can be discussed by considering the theoretical *resolving power* R_{th} (see, e.g., Fowles 1968)

$$R_{th} = nN \,, \tag{III.16}$$

where n is the total number of rulings and N is the spectral order considered.

Chapt. III   GENERAL INSTRUMENTAL ASPECTS AND EXPERIMENTAL METHODS

When the number of rulings per cm is doubled (so that the total number becomes $2n$), the spectral order found for a given wavelength at the same place in the spectrum is halved, since the path difference between successive constructively interfering beams is halved. Therefore neither R_{th} nor L is changed by doubling the number of rulings per cm, assuming equally effective blazing, but the FSR is doubled. Overlap of orders is thus the less, the finer the ruling of the grating.

When a spectral interval broader than the FSR has to be investigated, one may place two or more instruments with decreasing resolution, increasing FSR and constant optical conductance in series. Examples are the Pepsios interferometer (see Mack et al. 1963), consisting of three Fabry-Pérot interferometers with a grating monochromator as 'predisperser'. In the latter case care should be taken that the monochromator can accommodate all the light that is conducted through the interferometer.

When, for example, one uses an interferometer with $R = 10^6$ and $RL = 10^2$ cm² sr (and thus $L = 10^{-4}$ cm² sr), in series with a monochromator with $RL = 10^{-1}$ cm² sr, the maximum usable resolution in the monochromator is only 10^3. This means that the slits have to be opened so wide that $R = 10^3$ in order to accommodate the light flux that is conducted through the interferometer.

If *wavelength calibration* of a spectrometer is required, a well-known atomic spectrum with many lines, often the spectrum of Fe, is used as a (secondary) standard of wavelengths. In addition, an interpolation procedure for intermediate values can be applied. Since the refractive index of air deviates from unity, for example, by about 3×10^{-4} in the visible part of the spectrum, the actual wavelengths measured in air are slightly smaller than those in vacuo to which the wavelength standards usually refer. A correction is therefore needed only in the (rare) case when one is interested in the actual wavelength in the medium.

4c   DISTORTION OF THE SPECTRAL ENERGY DISTRIBUTION

The shape of a recorded spectrum, $I_r(\lambda_m)$ is a convolution of the spectral intensity distribution in the source, $I_\lambda^s(\lambda)$ (see definition in Appendix A.5), and the normalized instrumental profile A (cf. Eq. III.1)

$$I_r(\lambda_m) = \int_0^\infty I_\lambda^s(\lambda') A(\lambda_m - \lambda') \, d\lambda' . \qquad (III.17)$$

When the effective bandwidth, $\Delta\lambda_{eff}$, of the instrument exceeds considerably the width of the spectral line in the source, the recorded spectrum is seriously distorted: it reflects the instrumental profile rather than the source line profile. However, this distortion is of no consequence, if one is not interested in the shape of the spectral line but, e.g., only in its integrated intensity or

THE OPTICAL TRAIN   Sect. III.4c

central wavelength.

The normalized instrumental profile of a monochromator is (approximately) triangular if the widths of the entrance slit and exit slit are mutually matched. This implies that the entrance slit is precisely imaged on the exit slit (apart from slit-diffraction effects) if the monochromator setting, λ_m, is tuned at the wavelength, λ_0, of the monochromatic radiation illuminating the entrance slit. When the slits are not matched the instrumental profile has the shape of a trapezium. The use of unmatched slit widths has no advantage in relation to signal strength with a 'line' source, whereas it yields a loss in resolution, since a broader wavelength interval is accepted by the exit slit. This loss in resolution may deteriorate the signal-to-noise ratio when the line emission is superimposed on a background continuum. An argument in favour of the use of unmatched slits might be that this reduces the effect of a possible drift in λ_m upon the measurements. Another advantage of a broader exit slit may be that one is surer of measuring the *integrated* source-line intensity (see Reif, Fassel and Kniseley 1976).

Conversely, if $\Delta\lambda_{eff}$ is much less than the spectral width of the structures in the source spectrum, the instrumental profile has approximately the shape of a Dirac delta-function and one obtains

$$I_r(\lambda_m) = \int_0^\infty I_\lambda^s(\lambda') \Delta\lambda_{eff} \delta(\lambda_m - \lambda') d\lambda' = I_\lambda^s(\lambda_m) \Delta\lambda_{eff}. \qquad (III.18)$$

The recorded spectrum $I_r(\lambda)$ is then practically identical in shape to the true spectrum $I_\lambda^s(\lambda)$.

For arbitrary $\Delta\lambda_{eff}$ the integrated source-line intensity, $\int I_\lambda^s(\lambda') d\lambda'$, is proportional to $I_r(\lambda_0)$; the factor of proportionality depends on the instrumental profile as well as on the source-line profile $I_\lambda^s(\lambda)$ (see de Galan and Winefordner 1968). The integrated source-line intensity may also be obtained by scanning $I_r(\lambda)$ and measuring the total area under the curve. This area is independent of the source-line profile (see Penner 1959).

Solutions of the convolution integral in Eq. III.17 have been tabulated by de Galan and Winefordner (1968) for various instrumental profile functions and spectral line shapes. Also expressions have been given for the integrated 'total slit function' and its peak value, from which $\Delta\lambda_{eff}$ can be directly derived. Penner (1959) has given theoretical expressions and graphical representations of distorted spectral line profiles for different instrumental conditions.

When high spectral resolution is needed, for example in the study of hyperfine-structure or rotational spectra and in the measurement of line profiles, the spectral instrument is generally used at its maximum resolution. Then R is limited, theoretically, by diffraction effects and, practically, also by optical

249

and mechanical imperfections. The instrumental profile is not triangular anymore but can be deduced from diffraction theory or it has to be determined experimentally. An empirical method of checking the occurrence of distortion is to make a series of runs with progressively smaller slit widths. If the shape of $I_r(\lambda)$ is nearly the same for two markedly different slit settings, distortion can be considered to be negligible when maximum resolution is not yet attained.

When the resolving power is insufficient for the problem at hand, the obvious solution is to use instruments with higher resolving power (see Table III.2). Another way out is to try to extract the real spectrum $I_\lambda^s(\lambda)$ by *deconvolution* from the recorded spectrum $I_r(\lambda)$ with the use of the known instrumental function $A(\lambda_m-\lambda)$. Deconvolution can be performed by a number of mathematical methods (see Lochte-Holtgreve and Richter 1968) and with graphical, analog or digital means. Although the principles, often based on Fourier transformation, are well understood, the gain in resolution achieved in practice is not much more than a factor of three (see Robaux and Roizen-Dossier 1970). The effective loss in L, which necessarily accompanies the deconvolution, varies exponentially with the gain in R (see, for example, Despain and Bell, 1970).

Often one wants to measure the half-intensity width of a spectral line instead of its complete profile. Instead of applying deconvolution one can avail oneself then more simply of certain (approximate) addition rules for the half-intensity widths involved. If the profiles of the source line and of the instrument are both Gaussian, quadratic addition of their half-intensity widths yields the corresponding width of the recorded line profile (compare with the well-known addition law for statistically uncorrelated measurement errors). If the two profiles are Lorentzian, a simple linear addition law holds. For intermediate cases, one can give only approximate rules that lie inbetween those for the cases mentioned or one can make use of computer studies (see Kusch, Röndigs and Wendt 1977).

It is to be noted that an enlargement of the monochromator slit-height not only linearly increases the optical conduction, L, but that it may also worsen the optical aberrations that occur for obliquely incident beams, resulting in a loss in resolution R. The use of a slit height of 2 cm in a $\frac{3}{4}$ meter grating monochromator may reduce R by a factor of 2 as compared to that with a slit height up to about 4 mm. When appropriately curved slits are used, slit heights up to 2 cm can be employed without loss of resolution. However, the alignment is more critical and it depends on the size of the source image whether indeed full use can be made of such a large slit height.

4d   ALIGNMENT AND AREA SELECTION

The complete filling of the available angular aperture Ω_d (or the effective

THE OPTICAL TRAIN   Sect. III.4d

area S, see Fig. III.6) and the optical alignment of the whole train is sometimes a problem in the case of complicated set-ups. A recommended procedure for alignment is to start with the first component (the light source) and the last component (the detector) and have as few components as possible in between, next to maximize the output, and then to insert the other components one by one in the optical path. A rapid check of the light path can afterwards be made by putting a point-like light source in place of the detector and by tracing the light path through the optical train back to the light source. In this way one can also determine accurately which part of the light source contributes to the signal.

A check of the alignment can also be made by replacing the light source by a laser and tracing the narrow laser beam through the optical train.

In this context two possible ways of imaging the light source should be mentioned. In the first one (see Fig. III.7) a small part P of the light source (flame) is imaged onto the restricting stop (e.g., the slit S1) so that every point of P contributes to the signal with a weight proportional to the solid angle Ω_1. In the second one (see Fig. III.8) the source is imaged with a lens L, close to S1, onto the collimating lens C (or equivalent component) of the dispersing element. In this way a larger part P of the light source contributes to the signal, but with a smaller weight proportional to Ω_2 ($< \Omega_1$). In a correctly dimensioned system the image of the flame part to be used just fills the area S of the collimating lens C. As long as the slit S1 and the collimator C are completely filled with light and the flame radiance is uniform, there is neither a difference in signal strength nor in resolution between the two cases. The only variables to choose are the area of the selected flame part P and the solid angle Ω of the useful light beam from each point of that part; their product remains constant, however.

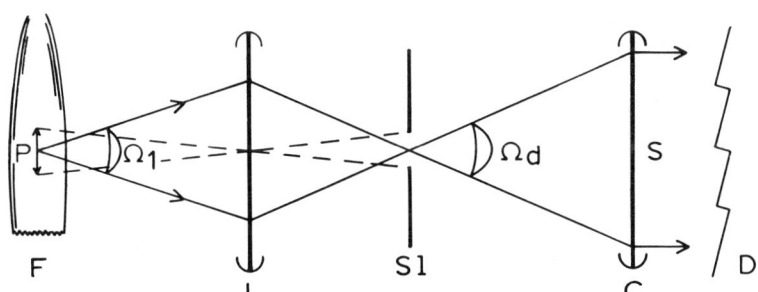

Fig. III.7 The flame F is imaged by lens L upon the plane of the entrance slit S1 of a monochromator with collimator lens C and dispersing element D. The part P of the flame that contributes to the signal measured is restricted by the slit dimensions; the solid angle Ω_1 under which the flame radiation is received is restricted by the solid angle Ω_d subtended by the (useful) area S of the collimator.

Chapt. III   GENERAL INSTRUMENTAL ASPECTS AND EXPERIMENTAL METHODS

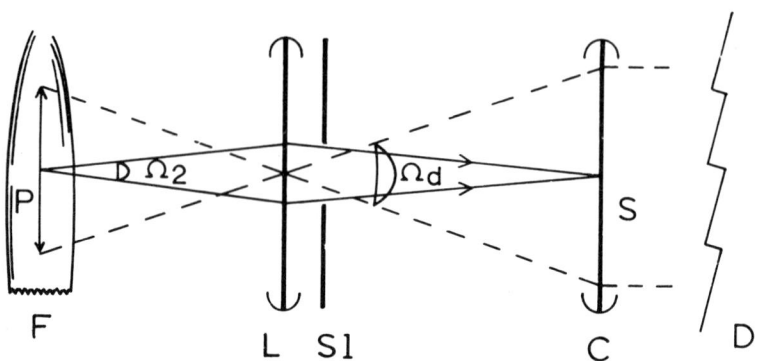

Fig. III.8 The flame is imaged by lens L, placed close to S1, upon the collimator lens. The observed flame part P is now restricted by Ω_d, and the solid angle Ω_2 under which the flame radiation is received is restricted by the slit dimensions.

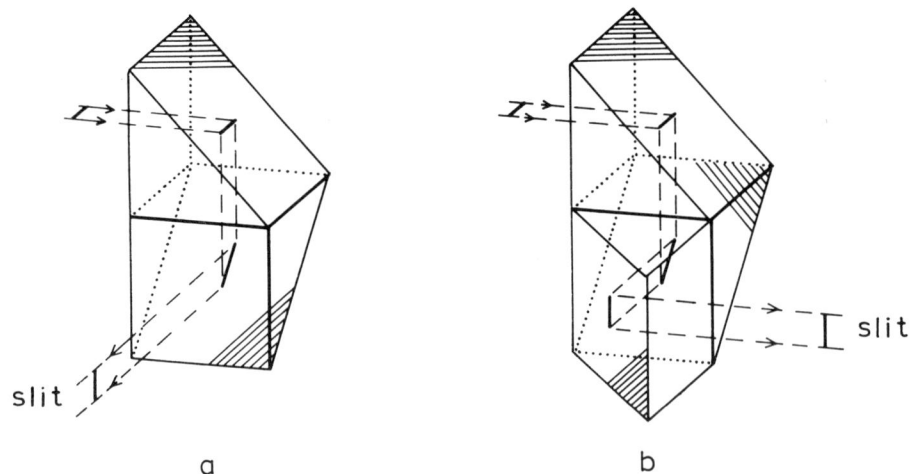

Fig. III.9 Prism-combinations that rotate a light beam through 90° around the optical axis. Note that in case a the optical axis is deflected over a right angle, whereas in case b a parallel shift occurs. Partly hatched areas are reflecting prism surfaces.

THE OPTICAL TRAIN   Sect. III.4f

There are circumstances in which one desires a high spatial resolution as to the height of measurement in the flame. In order to use in the set-up of Fig. III.7 the full height of the (generally vertical) entrance slit one can rotate the light beam around the optical axis through 90°. Fig. III.9 shows two prism-combinations that yield the desired rotation.

4e   POLARIZATION EFFECTS

Often an optical instrument or set-up partially polarizes the transmitted light beam because of oblique incidence of some beam rays on one or more reflecting or refracting surfaces. In intensity measurements one has thus to reckon with a polarization dependent transmission factor of the instrument. In polarization measurements a correction should be made for the polarizing effect of the instrument. Such a correction can be done with the aid of a strictly unpolarized light source and a rotatable linear polarizer. The partial polarization may amount to 30% in prism monochromators as well as in grating monochromators (see James and Sternberg 1969).

4f   SPECIAL OPTICAL SYSTEMS AND TECHNIQUES

Often extra components are built into the optical train and special techniques are applied to avoid or minimize disturbances that deteriorate the quality of the optical signal. Here we discuss briefly some of them that may be of use in flame spectroscopy.

1.   A *double-monochromator*, consisting in effect of two identical monochromators in series with the same wavelength setting, can be used to suppress stray light. '*Stray light*' is here defined as light of wavelengths outside the selected wavelength range that unintentionally passes through the exit slit of the monochromator. Stray light may be caused by dust and irregularities on the optical surfaces and by multiple reflections inside the instrument (see Larson *et al.* 1976). A reduction in stray light level by some orders of magnitude can generally be obtained. The resolution is increased by a factor of two, but the transmission factor is squared and thus lowered.

2.   *Intensity-modulation*, i.e. periodic variation of the light intensity as a function of time, can be achieved in a variety of ways. A common way is to interrupt the light beam mechanically by means of a chopper in the form of a rotating disc with equally spaced openings, blades or mirrors (if used in reflection). Vibrating (tuning-fork) choppers are also used; they have the advantage of ruggedness and compactness, and cause little disturbance through air draught (flame!) or mechanical vibrations. A disadvantage can be the limited cross-sectional area of the chopped beam. In general, mechanical choppers should be placed preferably where

the beam cross section is minimal to keep the dimensions of the moving parts as small as possible. Besides, the chopper should be placed close to the source of the beam to be modulated in order to prevent as far as possible modulation of spurious light signals (arising, e.g., from the surroundings).

The intensity of electric-discharge lamps can also be modulated (or pulsed) by modulating (or pulsing) the feeding current (see Sect. III.7).

The general advantages of the technique of intensity-modulation combined with a.c. detection tuned at the modulation frequency, f_{mod}, will be described in Sect. III.6b. The dependence of the signal-to-noise ratio on f_{mod} will be treated in Sect. III.14f. In the case of dominant flicker noise in an (unmodulated) background signal or dark current, f_{mod} should be chosen as high as possible. With normal mechanical choppers modulation frequencies up to about 1000 Hz can be realized in practice. A special chopper consisting of an air-turbine handpiece and a disk with 36 holes of 0.044 cm diameter has been designed for modulating a focused laser beam up to 300 kHz (see Selzer and Yen 1976).

Special electro-optical devices exist that permit the chopping of a light beam at much higher frequencies. A Pockels cell in combination with a polarizer and an analyzer can be used as a high-speed shutter below 1 ns. The interested reader is referred to the literature, for example, Fowles (1968).

3. *Double spatial-beam techniques*. One way of comparing the intensities of two light fluxes using a *single-beam instrument* is to measure them one after the other. As an example we mention the measurement of the transmission factor of a flame with metal vapour at the wavelength of an absorption line; such measurement involves the comparison of the intensities of a spectral line source as seen through the flame with and without metal vapour, respectively (see Sect. III.15b). When during the measuring time the source intensity or detector sensitivity varies because of drift or instability, the measurement may become inaccurate, especially when the transmission factor is near to unity (i.e. at weak absorption).

This disadvantage can be overcome by using a *double-beam technique* in which the two signals are detected simultaneously or in rapidly alternating succession and combined by a differential amplifier or ratio-meter.[†] When the two light beams

---

[†] When an instrument or experimental set-up has different channels that are read-out separately, without being combined, we shall speak of a *multi-channel instrument*. In contrast, a '*double-beam instrument*' combines the output signals of two beams automatically into one read-out signal corresponding, for example, to the difference, ratio or logarithmic ratio of the two beam intensities. Multi-channel spectrometers are in current use for multi-element analysis in analytical absorption, emission or fluorescence spectroscopy (see, e.g., Winefordner, Fitzgerald and Omenetto 1975, and Winefordner *et al.* 1976). In flame experiments an extra channel is sometimes used for monitoring the constancy and reproducibility of the flame conditions and of the supply of metal vapour (see Sects III.15b-1 and III.15c-5).

are distinguished by their wavelengths, the *double spectral-beam technique* can be applied (see below). When they occur at identical wavelengths or in identical wavelength intervals and are separated spatially, we speak of a *double spatial-beam technique*. In the above example of transmission measurements, one beam, called the *probing beam*, is directed through the flame, whereas the *reference beam* by-passes the flame (see Sect. III.15b-2).

In the d.c. mode each beam is detected by a separate photodetector and the detector output currents are subtracted or ratioed. In the a.c. mode, a common photodetector may be used if the spatially separated beams are encoded by a.c. modulation at opposite phases or at different frequencies. The modulation frequency should be high enough to reduce substantially the effect on the read-out of those low-frequency instabilities in the set-up that are shared by the two light beams (see also Sect. III.14f-4). Such instabilities may arise from the common light source or photodetector.

For a further discussion of the advantages and different modes of realization of double-beam techniques in flame spectroscopy we refer to Sects III.14 and III.15.

4. *Double spectral-beam (or dual-wavelength) techniques.* The intensities of two light beams originating from the same source but having different (central) wavelengths, e.g., a Na and a Li emission line, can be compared by separating the beams by means of two wavelength selectors (two filters or a monochromator with two exit slits). Spectral separation can also be used in order to correct the measured intensity of a spectral line for the contribution of a relatively strong underlying background continuum; one selector is then set at the wavelength of the spectral line, the other at a nearby wavelength (see Visser, Hamm and Zeeman 1976a). When an instrument incorporating such wavelength selectors permits a direct reading of the ratio or difference of the corresponding beam intensities, we shall speak of a *double spectral-beam* or *dual-wavelength technique*. In the d.c. mode this may be achieved by providing each selector with a separate photodetector, the outputs of which are fed to a differential amplifier or ratio-meter (compare above). In a.c. measurements one common photodetector can be used which is alternately exposed to the one and the other beam. This may be achieved by means of a rotating pair of filters or by alternately opening and closing the two exit slits. In effect this is a special case of wavelength-modulation where the transmitted (central) wavelength is a periodic, block-shaped function of time (see also below). The same a.c. measuring techniques can be applied and similar advantages are obtained as described above for a double spatial-beam set-up. Examples of applications in flame emission and absorption measurements will be given in Sect. III.14 and 15.

The *reference-element technique*, sometimes employed in analytical flame

Chapt. III   GENERAL INSTRUMENTAL ASPECTS AND EXPERIMENTAL METHODS

spectroscopy (see general literature listed in Chapt. I), is a special example of the dual-wavelength technique. It reduces, for example, the effect of drift in the performance of the nebulizer on the analyte signal.

5.   *Derivative spectroscopy*. Normally a spectrum is measured by recording the intensity, $I(\lambda)$, transmitted through a monochromator as a function of wavelength setting, λ. When an emission or absorption spectrum with detailed structure is to be scanned in the presence of a relatively strong, (quasi-)continuous background spectrum, it may be advantageous to scan directly the *derivative*, $dI(\lambda)/d\lambda$, as a function of λ. This is achieved by instrumental means in the technique known as *derivative spectroscopy* (see Evans and Thompson 1969, and Snelleman *et al.* 1970). It often leads to a better signal-to-noise ratio insofar as the latter is determined by instabilities in the background spectrum (see also Cahill 1980).

Derivative spectroscopy is essentially based on measuring the variation, $\Delta I(\lambda)$, in transmitted intensity when λ is varied over a small, ideally infinitesimal, wavelength interval $\Delta\lambda$. This may be achieved by measuring the difference of intensities transmitted through two, close-lying exit slits (compare with the dual-wavelength technique described above). More usually it is performed by modulating λ as a sinusoidal function of time t with amplitude $\tfrac{1}{2}\Delta\lambda$. This *wavelength-modulation* can be realized by mounting some vibrating element in the light path which varies periodically the wavelength transmitted. Such an element may be a laterally vibrating monochromator slit or a rotationally vibrating quartz plate that shifts the spectrum periodically over the stationary exit slit (see Fig. III.10a and b).

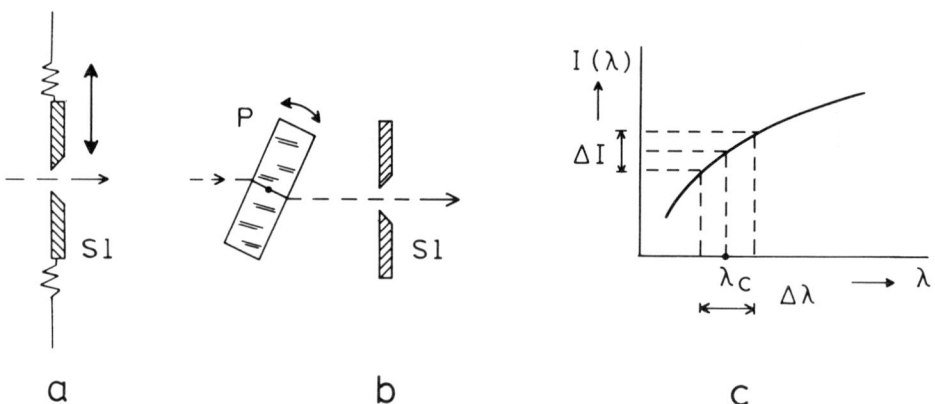

Fig. III.10 Figures a and b show how the wavelength, λ, of the radiation passing the exit slit Sl can be modulated by (a) lateral vibration of Sl or (b) rotational vibration of a thick quartz plate P in front of a stationary exit slit Sl. Figure c illustrates the effect of wavelength-modulation with amplitude $\tfrac{1}{2}\Delta\lambda$ around λ_c on the intensity $I(\lambda)$ of the transmitted radiation.

THE OPTICAL TRAIN   Sect. III.4f

The wavelength-modulation can be described by

$$\lambda = \lambda_c + \tfrac{1}{2}\Delta\lambda \sin 2\pi f_{mod} t \qquad (III.19)$$

where λ_c is the central wavelength and f_{mod} the frequency of modulation. Expansion of the transmitted intensity into a Taylor series as a function of wavelength around $\lambda = \lambda_c$ yields (see Fig. III.10c)

$$I(\lambda) = I(\lambda_c) + (dI/d\lambda)_{\lambda_c}(\lambda-\lambda_c) + \tfrac{1}{2}(d^2I/d\lambda^2)_{\lambda_c}(\lambda-\lambda_c)^2 + \ldots \qquad (III.20)$$

By comparing Eqs III.19 and III.20 one obtains I as a function of time t

$$I(t) = I(\lambda_c) + \tfrac{1}{2}(dI/d\lambda)_{\lambda_c} \Delta\lambda \sin 2\pi f_{mod} t +$$

$$+ \tfrac{1}{16}(d^2I/d\lambda^2)_{\lambda_c} \Delta\lambda^2 (1-\cos 4\pi f_{mod} t) + \ldots \ldots \qquad (III.21)$$

When synchronous detection (see Sect. III.6b) tuned at f_{mod} is applied, the meter measures only the first derivative, $dI/d\lambda$, at $\lambda = \lambda_c$. When the detection is tuned at the <u>double</u> frequency, $2f_{mod}$, the meter measures the <u>second</u> derivative, and so on. This holds true in fair approximation if $\Delta\lambda$ is not too large; when $\Delta\lambda$ is less than half the bandwidth $\Delta\lambda_{eff}$, the error made is less than about 5% (see Evans and Thompson 1969). When a slow steady λ_c-scan is superimposed on the λ-modulation, one thus measures directly the first- or second-derivative spectrum. In this way any component in the background spectrum that is a constant or a slowly varying function of wavelength is suppressed. This may be specially advantageous when the background intensity is not stable during the scanning period. Fig. III.11 illustrates the improvement in signal-to-noise ratio obtained by recording the second-derivative spectrum for a weak absorption doublet with a continuous background source. In this way a detection limit (for definition see Sect. III.14f-1) of the order of 0.01% can be reached for absorption measurements in flames, as can be deduced from Fig. III.11.

When one does not want to determine the profile of a spectral line but only its intensity in the presence of a relatively strong background spectrum, it may be profitable to measure the second derivative, with λ_c set at the line centre λ_0. The value of this derivative, which can be measured by tuning the synchronous detector at $2f_{mod}$, is proportional to the line intensity (in emission or absorption); it is unaffected by the magnitude of the background intensity and of its first derivative with respect to wavelength at λ_0. It is not essential in this respect that the effective bandwidth, $\Delta\lambda_{eff}$, of the monochromator or the modulation amplitude, $\tfrac{1}{2}\Delta\lambda$, are small as compared to the true line width. However, for a line spectrum the largest signal is obtained when $\Delta\lambda \approx 2\Delta\lambda_{eff}$. This technique has been used profitably in photoelectric line-reversal measurements of flame temperatures (see Sect. III.10c-5) and in the measurement of weak atomic absorption signals with a continuum background source (see Sect. III.15c-1).

Chapt. III GENERAL INSTRUMENTAL ASPECTS AND EXPERIMENTAL METHODS

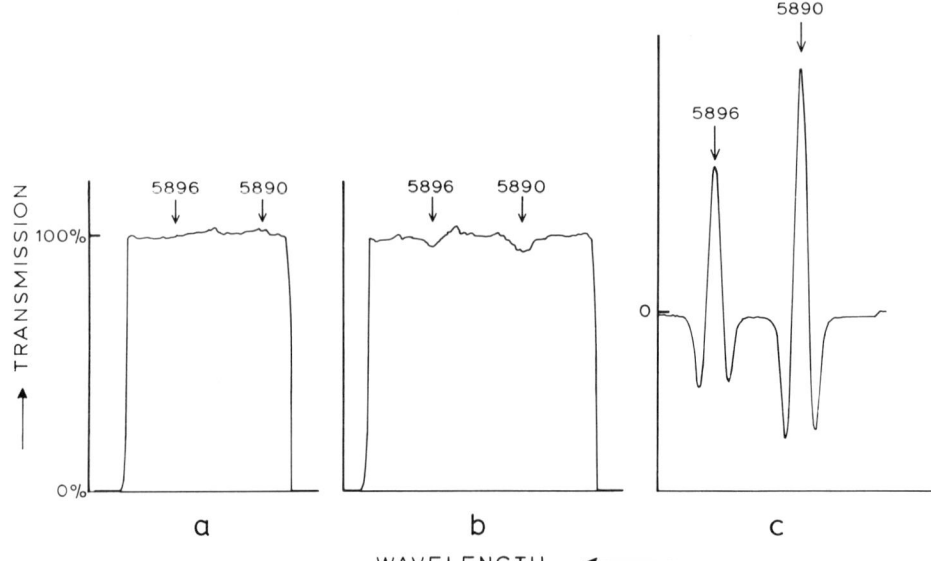

Fig. III.11 Comparison of signal-to-noise ratios obtained when a weak Na-D doublet in an air-hydrogen flame is scanned in absorption with a high-pressure xenon arc lamp as background source. Figure (a) refers to flame without sodium; figures (b) and (c) refer to flame with identical Na concentrations. In the case of figure (a) and (b) a mechanical chopper with modulation frequency 300 Hz was placed between lamp and flame; the lock-in meter was tuned at this modulation frequency. In the case of figure (c) the chopper was removed, but wavelength-modulation was applied (with the aid of a vibrating quartz plate) with modulation frequency 150 Hz and amplitude 0.5 Å, while the lock-in meter was tuned at twice this frequency (300 Hz). The intensity scale in figure (c) is amplified 20 X as compared to that in figures (a) and (b).
    Figure (c) clearly shows the gain in signal-to-noise ratio obtained by recording the second-derivative spectrum. The meter response time was in all three cases about 1 s, while $\Delta\lambda_{eff}$ amounted to 1 Å. (According to measurements made by Dr W. Snelleman at the Physics Laboratory of the University at Utrecht.)

The combined measurements of the zeroth, second, fourth, etc., derivative spectra can be utilized for a deconvolution procedure in order to retrieve the true atomic line profile. One has to know the instrumental profile and the kind of function by which the line profile can be described.

6. *Rapid-scan and recording methods.* For transient phenomena, use can be made of rapid-scanning monochromators that scan some hundreds of Å in a time of the order of a millisecond (Babrov et al. 1968, and Pimentel 1968). The common approach involves a rapid rotation of both the dispersing element and the mirror that

THE OPTICAL TRAIN   Sect. III.4f

focuses the radiation on the detector. A different approach (see Golightly, Kniseley and Fassel 1970) uses an image dissector tube (see Sect. III.5d under no. 3) in combination with a rapidly varying, axial magnetic field.

Many spectral elements can be rapidly and simultaneously recorded in systems incorporating multiple detectors. Such systems are multislit-spectrometers (see Mavrodineanu and Hughes 1968) and electronic image devices (see Sect. III.5a). Also short-time spectrography with mechanically or electronically operated shutters can be applied (see Lochte-Holtgreve and Richter 1968). Time-resolved photographic recording of transient spectra can also be obtained by sweeping the (small) image of the source along the length of the entrance slit of a spectrograph with the aid of a rotating mirror. On the photographic plate the time axis is then perpendicular to the wavelength axis (see Walters and Malmstadt 1965; see also Piepmeier and Malmstadt 1969).

Rapid scan and recording methods may also be used for stationary phenomena in order to avoid drift effects. Analog or digital averaging of a number of repetitive scans or recordings may help to improve the signal-to-noise ratio (see also Sect. III.14f, and Ernst 1965).

7. *Multiplexing methods*. It has been recognised for some time that any monochromator is rather wasteful in radiant energy, since in a complex spectrum only one spectral element at a time is measured (see, e.g., Winefordner $et\ al.$ 1976). A spectrograph provides greater informing power in this respect (see Sect. III.5a). There are two other procedures in use to measure a large number of spectral elements simultaneously: Fourier spectroscopy and Hadamard spectroscopy.

In *Fourier spectroscopy* (see Bell 1972) use is made of a Michelson interferometer, one end-mirror of which is moved at a constant velocity along the direction of the beam. The fringes in the interference pattern of the combined (monochromatic) beams pass the stationary exit diaphragm of the interferometer with a frequency, f, proportional to the velocity, v, of the mirror and to the optical frequency, ν, of the radiation: $f = 2v\nu/c$. Thus the amplitude of the radiation from every spectral element in a compound spectrum is modulated with a different frequency. The intensity of the outgoing beam is the square of the superposition of these modulated amplitudes. A record of this intensity as a function of time is called an *interferogram*, which can be Fourier-transformed into the original spectrum.[†]

[†] Work has been done in the Physics Laboratory at Utrecht in connection with the construction of a Fourier spectrometer for visible light with a maximum path difference of 120 cm (see Emonds 1981). A Fourier spectrometer for spectrochemical measurements from the mid-infrared to the ultraviolet has been described by Horlick and Yuen (1978).

Chapt. III   GENERAL INSTRUMENTAL ASPECTS AND EXPERIMENTAL METHODS

In *Hadamard spectroscopy* each spectral element from a conventional grating spectrometer is modulated by a rotating cylinder with transparent and opaque zones in a checker board-like pattern placed in the focal plane. The radiation from these thus encoded elements is simultaneously collected on a single photodetector and recorded as a function of time; it can be decoded afterwards into the original spectrum. The spectral resolution is determined by the spatial resolution of the modulation pattern.

A gain in signal-to-noise ratio of $(N/8)^{\frac{1}{2}}$ (with N = number of resolved elements) can be expected (see Treffers 1977) compared to a conventional scanning spectrometer. This holds if dark-current noise is the limiting noise source, as is often the case in infrared spectroscopy. When noise in the photosignal itself limits the precision of measurement, the gain in signal-to-noise ratio depends upon the form of the spectrum and may even turn into a loss; but drift is removed. There is then little advantage in these methods over the methods discussed under 6. However, Fourier spectroscopy, although technically more complicated, still has the advantage over conventional methods of a large optical conductance and offers the possibility of high resolving power.

A general discussion of the usefulness of these methods in atomic spectroscopy has been given by Plankey *et al.* (1974), Winefordner, Fitzgerald and Omenetto (1975), and Winefordner *et al.* (1976).

8. *Selective-modulation techniques and resonance monochromator.* Spectral selection of high resolution at discrete wavelengths can be obtained with nondispersive *selective-modulation techniques*. An atomic resonance line in the spectrum of a radiation beam can be selected from other (adjacent) spectral lines or a spectral continuum by modulating the transmission of the radiation beam through an atomic vapour cloud of the same element which selectively absorbs only the resonance line. This may be achieved by using a *pulsating vapour cloud* generated by an electrically modulated sputtering process in a hollow-cathode discharge (see Bowman, Sullivan and Walsh 1966, and Sullivan and Walsh 1968). Selective modulation may also be performed by modulating the wavelength of the absorption line in a periodically varying magnetic field as a result of the *Zeeman effect* (see Koizumi and Yasuda 1976). In the latter case the cell containing the vapour cloud is placed between the poles of an electromagnet. With the *negative-filter technique* of Alkemade and Milatz (1955) essentially the same effect is obtained by means of a double spatial-beam set-up. The radiation containing the resonance line to be selected, alternatively passes and by-passes the selectively absorbing vapour cloud. The atomic vapour may be generated, for example, by spraying a solution of the element concerned into an auxiliary flame. The experimental set-up resembles the double-beam set-up for absorption measurements in which the two beams are

modulated by a common light chopper at opposite phases (see Fig. III.25 in Sect. III.15b-2). By using synchronous detection (see Sect. III.6b) only the resonance line, but not the other spectral components in the radiation beam, is detected. The main advantage of these techniques is that the effective spectral resolution is determined by the very small width of the atomic absorption line. Atomic vapour cells at reduced pressure are thus especially advantageous in this respect. Besides, the wavelength setting is fixed by the properties of the absorbing atoms and thus cannot drift away.

Alkemade and Milatz (1955) were able to demonstrate by the negative-filter technique the occurrence of self-reversal (see Sect. II.5c-3) at the very centre of the Na-D lines in an unshielded C_2H_2-air flame at high Na concentrations. Because the auxiliary flame was fed by an Na solution with low concentration, it absorbs mainly the central portion of the much broader, self-reversed emission line emitted by the former flame.

Similar advantages are obtained with a *resonance monochromator* (also called resonance detector). The atomic vapour generated by cathodic sputtering re-emits part of the selectively absorbed resonance line as fluorescence, the intensity of which is measured (see Sullivan and Walsh 1965). By modulating the incoming radiation beam and by using a.c. detection the d.c. emission of the resonance monochromator itself is eliminated.

Of the variants described above the negative-filter technique has up till now found no application in atomic emission spectroscopy. An analogous technique has been used, however, in molecular spectroscopy as a molecule-specific detector. The term 'negative filter' indicates that spectral selection is not obtained through a narrow transmission band, like ordinary filters, but through a narrow absorption band. For a historical survey see Alkemade (1980).

In analytical atomic absorption spectroscopy Zeeman modulation has been applied for correcting the analyses for an underlying continuous background absorption (see Koizumi and Yasuda 1976).

5. PHOTODETECTORS AND NOISE

5a   INTRODUCTION

Most flame spectroscopic investigations in the near-infrared to far-ultraviolet part of the spectrum are done with the aid of a photo-electric detector, but occasionally a photographic plate is used (for a general survey, see, e.g., Herrmann and Alkemade 1960, 1963, Mavrodineanu and Boiteux 1965, and Veillon 1971). The advantage of the spectrographic method is its high 'informing power' (see

Kaiser 1970) but its drawbacks are nonlinearity and complicated processing. In flame spectrometry, direct-reading instruments equipped with photo-electric detectors are the most popular. One distinguishes between photo-emissive vacuum tubes (or phototubes) and solid-state photocells.

Solid-state photocells can be sub-divided into photodiodes (back-biased p-n junctions), photovoltaic cells (producing a phototension without being biased), and photoconductors (showing increased conductivity upon illumination). Examples of this category are the Si photodiode, the selenium barrier-layer and InSb photovoltaic cells, and the CdS photoconductor. The selenium cell and more rarely a photoconductive cell are used in simple analytical flame spectrometers equipped with spectral filters; these detectors are not expensive, they are reasonably responsive, simple to operate and are of sturdy construction (see, e.g., Winefordner, Schulman and O'Haver 1972). Photodiodes and phototransistors are attractive for simultaneous measurement at adjacent wavelengths with multi-channel spectrometers as they can be easily positioned in an array because of their small size (see Boumans and Brouwer 1972). The spectral response of various types of solid-state photodiodes and photoconductors may extend from the ultraviolet into the 1-10 μm region of the spectrum. Silicon photodiodes have a high quantum efficiency (up to 80%), broad spectral range (0.2 to 1.1 μm), wide dynamic response and short response time. When internal avalanche multiplication is applied they can compete with photomultiplier tubes in the 1 μm range (see Ingle and Crouch 1972b, and Keyes and Kingston 1972).

A combination of the main advantages of the photographic plate (high informing power) and of a photo-electric detector (direct electronic read-out) is offered by the utilization of an *electronic image sensor* placed in the focal plane of a dispersive instrument. The development of these sensors has been highly stimulated by their use as TV cameras. In spectroscopy they offer the possibility of simultaneous sensing or integrating part of the spectrum, and of scanning it rapidly and repeatedly (see, e.g., Olson 1972, Horlick and Codding 1973, Lowrance and Zucchino 1974, Knapp et al. 1974, Chester et al. 1976, and Horlick 1976). The possibility of automatic subtraction of background spectra is also noteworthy. In particular, silicon-diode array camera tubes (silicon vidicons) and secondary-electron conduction image tubes (both operating with a scanning electron beam in vacuo) and integrated solid-state devices like the self-scanning silicon-photodiode array have been tried for analytical applications in emission, absorption or fluorescence spectroscopy with flames (for a brief survey see Haraguchi et al. 1976). Electronic image sensors are in general inferior to photomultiplier tubes as regards responsivity and detection limits in analytical atomic spectroscopy (see Chester et al. 1976).

PHOTODETECTORS AND NOISE   Sect. III.5a

A detailed intercomparison of these detectors as to signal-to-noise ratios in
analytical atomic and molecular luminescence spectrometry has been given by
Cooney, Boutilier and Winefordner (1977).

   Photo-emissive vacuum tubes (or *phototubes*) are characterized by a good
linear response over a wide dynamic range, short response time (typically 10^{-8} s
or even less for photomultiplier tubes) and/or extended spectral response in the
ultraviolet through visible regions.[†] *Single-stage photo-emissive tubes* have,
however, rather poor responsivity (1-100 mA per W incident radiant flux at the
wavelength of their maximum response). But they are stable, inexpensive, easy to
operate, and have a very short response time (about 1 ns). Since this detector
is an almost ideal 'current source', a high load resistance (typically 10^{10} Ω) can
be used for efficient electronic amplification which may make up for the poor responsivity. However, a high load resistance may introduce insulation problems and
limit the speed of response because of the associated long RC-time. The distinc-
tive feature of *multiplier photo-emissive tubes* (commonly called *photomultiplier
tubes* or *photomultipliers*) is their very high and almost noise-free internal
amplification which is achieved by means of secondary-electron emission. This
offsets the need for a high load resistance and for elaborate electronic amplification, and even permits the counting of single photon pulses. It is interesting
to note that with the best photomultiplier tubes available the number of incident
photons producing, on the average, one pulse is 100 × less than the number of
photons required to darken a single silver grain in a good photographic emulsion.
From the point of view of signal-to-noise ratio, photomultipliers are in general
preferable to vacuum or solid-state photodiodes, especially at low radiation
levels where photon counting techniques are appropriate. The reverse may be true,
however, at high radiation levels (see Ingle and Crouch 1971, 1972b).

   In the following, attention will be paid mainly to the photomultiplier
tube, as it is nowadays very widely used in flame spectrometric investigations.
For a more detailed discussion of the physical basis and practical aspects of
photo-electric detectors, the reader should consult the literature, for example,
Benson (1958), Zworykin (1958), Lion (1959), Spicer and Wooten (1963), Young (1969,
1974), E.M.I. Photomultiplier Tubes (1970, 1972), and R.C.A. Photomultiplier Manual
PT-61.[*]

[†] *Solar-blind* photomultiplier tubes having a spectral response restricted to wavelengths below 3000 Å only are sometimes used without a spectral apparatus, e.g.,
in atomic fluorescence measurements, because they do not respond to the flame
background radiation in the visible and near-ultraviolet region.

[*] The reader who is interested in the practical aspects of photomultiplier tubes
is referred particularly to Young (1974).

Chapt. III   GENERAL INSTRUMENTAL ASPECTS AND EXPERIMENTAL METHODS

5b   GENERAL CHARACTERISTICS OF PHOTOMULTIPLIER TUBES

Photo-emission. Photo-emission is based upon the excitation of a single electron by absorption of a single photon near the surface of the photo-emitter; the electron is thus able to escape over the potential barrier at the surface into the vacuum. The photo-electrons thus released are guided to another electrode (first dynode with a photomultiplier tube, or anode with a vacuum photodiode) biased positively with respect to the photo-emitter (called *photocathode*). The photocathode may consist of a semi-transparent layer deposited on the inside of a window at the end of the tube (end-on type). In an alternative configuration, it is situated in the interior of the tube at some distance from a side window (side-window type). The responsivity of the photo-emitter to radiation of wavelength λ depends on the *quantum efficiency* $q(\lambda)$. The latter is defined as the ratio of the number of photo-electrons to the number of incident photons of wavelength λ. The function $q(\lambda)$ together with the wavelength-dependent transmission factor of the window material determine the *spectral response* (see Fig. III.12). The quantum efficiency drops to zero above the *long-wavelength threshold* which is determined by the photo-electric work function of the cathode; this work function depends on the composition of the cathode. An outstanding feature of the S-1 cathode is its spectral response that extends into the near-infrared (see Fig. III.12) because of its comparatively low work function.

Secondary-electron emission. The tension between the first dynode and photocathode is such that each accelerated photo-electron impinging on this dynode releases several *secondary electrons*. The average number of secondary electrons produced per incident primary electron is called the *secondary-emission coefficient* δ. These secondary electrons are accelerated by an electric field towards the next dynode where they, in turn, produce secondaries, and so on. When the number of dynode stages is n, each having the same secondary-emission coefficient, the *total gain factor*, G, of the photomultiplier tube is given by

$$G = C(g\delta)^n . \qquad (III.22)$$

Here the factor C accounts for the collection efficiency of the first dynode with regard to photo-electrons and for the non-zero probability that an impinging photo-electron produces no secondary electrons (C may be as high as 90%). The factor g describes the transfer efficiency of electrons between successive dynodes (g is usually close to 100%). Taking a typical case with $n = 10$ and $g\delta = 4$, one calculates G to be of the order of 10^6. By increasing the number of dynodes, gain factors up to about 10^8 can be realized. The amplified electron current that is finally collected by the anode may thus exceed the cathode current by many orders of magnitude, resulting in a responsivity up to 10^5 A per W radiant flux at the

264

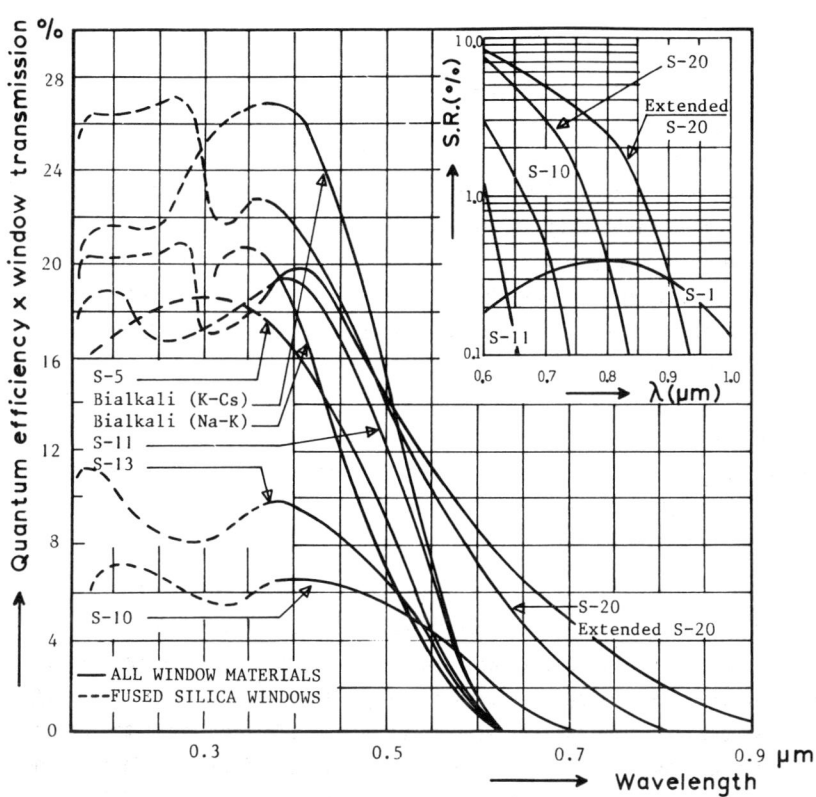

Fig. III.12 The spectral quantum response (S.R. ≡ quantum efficiency times window transmission) is plotted as a function of wavelength λ for various kinds of photocathodes, with indication of window material. (Redrawn with some modifications from E.M.I. Photomultiplier Tubes 1970, 1972.)

wavelength of maximum response. It is this huge gain in responsivity which gives the photomultiplier tube a unique advantage over the single-state photo-emissive tube.

Most photomultipliers require a tension of the order of 100 V per dynode stage for efficient secondary emission. The total tension required is thus of the order of 1000 V. The high-tension supply should be well stabilized, as a relative variation of 1% in tension produces a relative variation of roughly 0.7% in δ. According to Eq. III.22 this causes a relative variation of $n \times 0.7\% \approx 7\%$ in the total gain factor.

At high anode currents, departures from linearity between anode and cathode currents may occur due to heating of the last dynodes (which results in a drop of δ) and to space-charge effects. The maximum rating of the (mean) anode current specified by the manufacturer should be carefully observed. One must note too the maximum rating of the cathode current, as the collection efficiency C may deteriorate at high cathode currents. Another cause of departure from linearity may be the change in the division of the high-tension over the stages which occurs when high currents are drawn off at the last stage(s); in one case a departure of positive sign was found (see Sauerbrei 1972). A possible remedy is to use lower resistances in the divider circuit or to use tension stabilizers. A systematic investigation into the linear operation of photomultiplier tubes has been made by Land (1971).

Dark current. When a photomultiplier tube is operated in the dark, a *dark current* is usually observed. At low tension per stage, this current may be mainly due to ohmic leakage. At medium tension values, thermionic emission of electrons from the cathode (and the first dynode), which is amplified by secondary emission, may be the dominant component, at least at room temperature and for photocathodes with a low work function. Photocathodes which are sensitive in the near-infrared because of a low work function have intrinsically a relatively large thermal dark current. This component can be substantially reduced by cooling the cathode. The choice of a photomulitplier with a small cathode area may also help to reduce this component, as it is proportional to this area (compare also Sect. III.5d sub 3). At low temperatures, a residual dark current may be found which is due to luminescence and nuclear radiation effects caused, e.g., by the presence of radioactive ^{40}K in the window material. At high tension values, positive-ion currents, due to ionization of the residual gas, and field emission may cause an additional increase in dark current.

Pulse counting. If the time-resolution of the electronic apparatus following the photomultiplier is sufficiently good, single photo-electrons may be observed

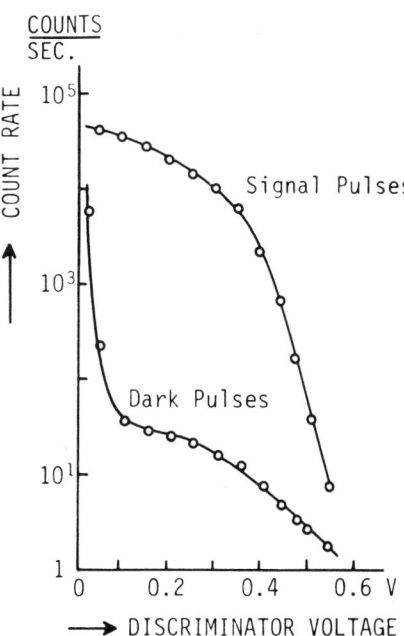

Fig. III.13 Integral pulse-height distributions of photon- and dark pulses for an IP 28 photomultiplier tube. The ordinate scale indicates the number of pulses per second whose height exceeds the discriminator voltage. (From Franklin, Horlick and Malmstadt 1969.)

as separate current *pulses* at low light levels. This allows the application of counting techniques for measuring weak light intensities (for a review see Eberhardt 1964, Morton 1968, Young 1969, 1971, 1974, and Malmstadt, Franklin and Horlick 1972). The pulse height is not constant but shows a statistical distribution that depends on the operating conditions. For some tubes there appears to be a disproportionately large number of small pulses in the dark current when compared to the photocurrent (see Fig. III.13). This might be due to thermionic emission of electrons from the first dynode. These electrons are multiplied by a smaller gain factor than those emitted by the photocathode. The ratio of signal to dark current can then be improved by eliminating these small pulses with the help of a suitably rated pulse-height discriminator. There may also be other causes of electron emission from the dynodes which produce an excess of small pulses both in the dark current and in the photocurrent (see Eberhardt 1964, and Young 1969). The time-resolution of the photomultuplier tube itself is limited by the duration (≈ 10 ns) or the rise time of the pulses as they appear at the anode. Advantages of pulse counting are the easy realization of long-term signal integration, the reduction in effective dark current through pulse-height discrimination, the direct digital presentation of the signal, and the fact that variations in the gain factor do not affect — in the first instance — the number of pulses, only their heights. These

Chapt. III GENERAL INSTRUMENTAL ASPECTS AND EXPERIMENTAL METHODS

properties make photon pulse counting especially advantageous at low light-levels and result in a slightly better signal-to-noise ratio than in the case of current measurements if shot noise (see Sect. 5c) is predominant (see Ingle and Crouch 1972a). A disadvantage is the nonlinear response at higher light levels where pile-up of pulses becomes statistically significant (see Smit and Alkemade 1963, and Ingle and Crouch 1972). In order to keep the deviation from linearity below 1%, the high-frequency cut-off of the measurement system must be at least 25 times larger than the average pulse rate (see Malmstadt, Franklin and Horlick 1972). Pulse counting is less affected by variations in the gain factor if the gain is sufficiently high.

Counting techniques are insensitive to variations in the gain factor only if the discriminator level is set so low that practically all pulses are counted. When in measuring weak signals the discriminator is set at a higher level in order to reduce the number of dark pulses (see above), variations in gain factor may have similar effects in counting and in current-measuring techniques (see Young 1974a).

5c   NOISE [†]

Shot noise; spectral noise power. Even under strictly constant operating conditions (in the dark or under constant illumination) the number N of electrons emitted by the photocathode in a given time interval τ is not constant but obeys Poisson statistics. When this number is measured repeatedly for a large series of equal time intervals, its *standard deviation* is found to be

$$\sigma_N \equiv \sqrt{\overline{(N-\overline{N})^2}} = \sqrt{\overline{N}} ,\qquad (III.23)$$

where the bar denotes the average value. The relative standard deviation is thus given by

$$\sigma_N/\overline{N} = 1/\sqrt{\overline{N}} = 1/\sqrt{\overline{n}\tau} ,\qquad (III.24)$$

if $\overline{n}$ is the mean number of electrons emitted per second. This relative standard deviation is thus inversely proportional to $\sqrt{\tau}$ and $\sqrt{\overline{n}}$. The transport of electrical charge, Q, in time interval τ is related to N through

$$Q = -Ne ,\qquad (III.25)$$

where e is the elementary charge. Combining Eqs III.23 and III.25 we get

---

[†] For an introduction to noise theory and analysis see Malmstadt, Enke and Crouch (1974) and Cova and Longoni (1979), and for an extensive treatment Bendat (1958), Blackman and Tukey (1959), MacDonald (1962), Bendat and Piersol (1966), and van der Ziel (1954, 1970, 1976). A brief survey of noise expressions for various types of photodetectors with pulse counting or current measurement is found in Young (1969), Robben (1971), amd Keyes and Kingston (1972). See also Sect. IV.9 for a brief description of the measurement of noise spectra in flames.

PHOTODETECTORS AND NOISE   Sect. III.5c

$$\sigma_Q = \sqrt{-e\bar{Q}} \; . \tag{III.26}$$

For the absolute value of the current as averaged over a time interval τ, we have: $i_\tau = |\bar{Q}|/\tau$, and for its standard deviation we obtain through Eq. III.26

$$\sigma_{i_\tau} = \sqrt{e\bar{i}/\tau} \; . \tag{III.27}$$

When instead of averaging the fluctuating current over the time interval τ, we transmit it through an electrical filter whose frequency-response is characterized by an (*effective*) *noise bandwidth* Δf and whose peak transmission is unity, we obtain from Eq. III.27 by replacing τ with $1/(2\Delta f)$ (see references cited in footnote on page 268)

$$\sigma_i = \sqrt{2e\bar{i}\Delta f} \; . \tag{III.28}$$

The quantity $2e\bar{i}$, which represents the mean square deviation of the current after passing through such a filter with $\Delta f = 1\,\text{Hz}$, is called the *spectral noise power*. The current fluctuations considered here are due to the random emission of single electrons and are called *shot-effect fluctuations* or simply *shot noise*. This type of noise has a '*white*' *noise spectrum*, i.e. its spectral noise power is independent of frequency f, or

$$W_{sh}(f) \equiv \sigma_i^2/\Delta f = 2e\bar{i} \; . \tag{III.29}$$

This means that σ_i in Eq. III.28 is independent of the peak frequency, f, of the transmission filter (it only depends on the bandwidth of the filter). For d.c. measurements this peak frequency equals, effectively, zero.[†] Shot noise varies typically as the square-root of the mean current, which is a direct consequence of the Poisson statistics.

Mathematically Δf is found by integrating the <u>square</u> of the filter transmission factor (expressed as a fraction of the peak transmission) over frequency (see Cath 1970). For an RC-filter or for a d.c. meter with an exponential step response with time constant τ_c we have (see Robben 1971, and Malmstadt, Enke and Crouch 1974)

$$\Delta f = \frac{1}{4RC} = \frac{1}{4\tau_c} \; . \tag{III.30}$$

For an arbitrary fluctuating signal $x(t)$ we can more generally define the

---

[†] This is strictly true only if the duration of the d.c. measurement is infinite. For a finite duration one has to be careful in defining an effective peak frequency when the noise spectrum rises indefinitely as f tends to zero. In the latter case the mathematical treatment presented in Appendix A.7 is appropriate. Besides, Eq. III.29 does not hold for 'infinite' frequencies; a roll-off of the spectral noise power sets ultimately in at very high frequencies where transit-time and quantummechanical effects become important (see MacDonald 1962).

Chapt. III   GENERAL INSTRUMENTAL ASPECTS AND EXPERIMENTAL METHODS

spectral noise power, $W_x(f)$, as a function of frequency, by using the concept of noise bandwidth Δf, through

$$W_x(f) \equiv \lim_{\Delta f \to 0} \sigma_x^2/\Delta f \; . \tag{III.31}$$

It is understood herewith that the peak frequency of the a.c. filter is kept constant and equal to f while Δf goes to zero. The dimension of $W_x(f)$ follows from Eq. III.31 and is $[x]^2[t]$, where $[x]$ denotes the dimension of x, etc. When x is an electric current, the spectral noise power is expressed, for example, in A²s or A²Hz⁻¹.

From Eq. III.31 we immediately get for the *variance* σ_x^2 measured with a filter peaked at f_0 and having a noise bandwidth Δf

$$\sigma_x^2 \simeq W_x(f_0) \Delta f \; . \tag{III.32}$$

For a noise spectrum with an arbitrary shape, Eq. III.32 holds only approximately if Δf is not too large [i.e. if $W_x(f)$ does not vary much within Δf]. For a white noise spectrum, Eq. III.32 is generally valid for any Δf.

When $y(t)$ is obtained from $x(t)$ by multiplication by a constant factor G we have

$$W_y(f) = G^2 W_x(f) \; . \tag{III.33}$$

Further relationships concerning the spectral noise power are to be found in Appendix A.7.

According to Eq. III.28 one can determine indirectly the photocurrent $\bar{i}_p$ by measuring its shot-noise power, $2e\bar{i}_p B$, in a broad frequency band B by means of an a.c. power meter (see Pao, Zitter and Griffiths 1966, and Pao and Griffiths 1967). In the literature the term a.c. method has been used for this method (see Inaba, Shimizu and Tsuji 1975) but the method must not be confused with the a.c. modulation techniques discussed in Sects III.4f and 6b. In the absence of other noise sources the relative error made in the indirect measurement of $\bar{i}_p$ through its shot-noise power is of the order of $1/\sqrt{B\tau}$ where $\tau (\geqslant B^{-1})$ is the measuring time, if B is smaller than the mean rate, $\bar{n}_p$, of photo-electron pulses (see Chapt. 13 in van der Ziel 1954). By increasing B, the relative error can be reduced until for $B \geqslant \bar{n}_p$ it becomes of the order of $1/\sqrt{\bar{n}_p \tau}$. The relative error is then of the same order as that made in the conventional d.c. and pulse-counting methods at similar measuring or counting times (cf. Eq. III.24). A detailed theoretical and experimental comparison of signal-to-noise ratios obtained with these three methods of measuring has been given by Inaba, Shimizu and Tsuji (1975) for weak light signals detected by a photomultiplier tube in the presence of dark-current noise.

PHOTODETECTORS AND NOISE   Sect. III.5c

Secondary-emission noise. The noise, σ_{i_a}, in the anode current of a photomultiplier tube is found by multiplying the shot noise in the cathode current by the total gain factor G (see Eq. III.22) while taking into account the collection efficiency C of the first dynode and the additional noise contribution introduced by the dynode multiplication process. Assuming that the number of secondary electrons released per incident electron and collected by the next dynode varies according to Poisson statistics (which may not necessarily be true; see Young 1969), we find [†]

$$\sigma_{i_a} = \sqrt{2e(g\delta)^n \bar{i}_a \{g\delta/(g\delta-1)\} \Delta f} \ . \qquad (III.34)$$

The factor between braces in this expression accounts for the *secondary-emission noise*. Since this factor is close to unity (say $\simeq 1.3$), the internal amplification is practically noise-free. For the mean anode current we can write: $\bar{i}_a = G(\bar{i}_p + \bar{i}_d)$ where $\bar{i}_p$ and $\bar{i}_d$ are the mean photo- and dark current, respectively, emitted by the cathode.

A lower limit is set for the detectability of weak light signals ($\bar{i}_p \ll \bar{i}_d$) by the inevitable shot noise in the dark current. Although the d.c. component of the dark current can be eliminated from the reading by applying a modulation technique, its shot-noise component cannot be eliminated in this way, as its spectral noise power is independent of frequency (compare also Sect. 5d sub 2). Improvement of the detection limit can be achieved, however, by reducing the dark current, by inserting a pulse-height discriminator (see Sect. 5b), or by decreasing Δf (i.e. by extending the observation time) when the photosignal is stationary (see Sect. III.14f).

For comparatively large photosignals ($\bar{i}_p \gg \bar{i}_d$), the inherent shot noise of the photocurrent itself limits the relative precision, $\sigma_{i_a}/\bar{i}_a$, of the measurement. This relative precision varies in inverse proportion to the square-root of the photosignal.

When thermionic emission from the dynodes also contributes to the dark current, Eq. III.34 would yield too high a value for the dark-current noise if we substituted there for $\bar{i}_a$ the dark current as observed at the anode (see O'Haver and Winefordner 1968, and Young 1969). However, the resulting error in the calculated σ_{i_a} value is not likely to exceed 10%. Equation III.34 also ceases to be valid when the other components of the dark current mentioned in Sect. 5b become noticeable.

[†] When only a small number of dynode stages was used with an E.M.I. 9558 S 20 photomultiplier tube in the measurement at high light levels, σ_{i_a} was found to exceed the theoretical value predicted by this equation by about a factor 2.6 (see Alkemade *et al.* 1979).

Chapt. III   GENERAL INSTRUMENTAL ASPECTS AND EXPERIMENTAL METHODS

Flicker noise. In the range of low frequencies photodetectors, especially solid-state photocells, may exhibit *flicker noise*.[†] In contrast to the shot-noise component, the spectral noise power (see Eq. III.31) of the flicker noise depends on frequency and varies approximately as $1/f$, while it is proportional to $\bar{i}^2$. Thus, for flicker noise, we have

$$W_{\mathrm{fl}}(f) = A\,\bar{i}^2/f \ . \tag{III.35}$$

The latter thus becomes increasingly important in comparison to shot noise, or to white noise in general, when f decreases and/or $\bar{i}$ increases. Any noise that behaves in a similar way will be called here *excess low-frequency (e.l.f.) noise*. The standard deviation of the combined current noises is found by adding quadratically the flicker-noise component $(\sigma_i)_{\mathrm{fl}}$ and the shot-noise component $(\sigma_i)_{\mathrm{sh}}$, which are mutually uncorrelated,

$$(\sigma_i)_{\mathrm{comb}} = \sqrt{(\sigma_i)^2_{\mathrm{fl}} + (\sigma_i)^2_{\mathrm{sh}}} \ . \tag{III.36}$$

The current- and frequency-dependence of the combined dark-current noises can thus be written as (see Eq. III.32 in combination with Eqs III.29 and III.35)

$$(\sigma_i)_{\mathrm{comb}} = \sqrt{(A\bar{i}_{\mathrm{d}}^2/f + B\bar{i}_{\mathrm{d}})\Delta f} \tag{III.37}$$

where A and B are constants. The first term on the right-hand side can be made small in comparison to the second term by applying a modulation technique at a sufficiently high modulation frequency f_{mod} (see also Sect. III.14f). The (minimum) modulation frequency required depends not only on the constants A and B but also on $\bar{i}_{\mathrm{d}}$.

The origin of flicker noise is usually not clear; it may be related to random instabilities of the surface conditions or temperature of the photocathode or of the high-tension supply. A similar kind of flicker noise or, more generally, excess low-frequency noise, may also exist in other components of the flame spectrometer (e.g., in the thermal flame emission, see Sect. IV.9, in the light source, and in components of the electronic amplifier, see Sect. III.6).

Johnson noise. When the output current of a photomultiplier tube is to be amplified electronically, a load resistor is inserted into the anode circuit. This

---

[†] Flicker noise has been found in the dark current of photomultiplier tubes by Boeschoten, Milatz and Smit (1954), Smit, Alkemade and Muntjewerff (1963), and Robben (1971). Young (1969, 1974) has summarized and critically discussed experiments relating to flicker noise in the dark current and photocurrent of this type of photodetector. The observability of flicker noise against an underground of shot noise depends, however, at given frequency on the mean cathode current (see Eq. III.37). The magnitude of this current should thus always be stated explicitly whenever flicker noise is reported to be present or absent.

PHOTODETECTORS AND NOISE   Sect. III.5d

introduces an additional noise component, as any ohmic resistor R produces a thermal noise current (or noise tension) as a result of the random thermal motion of the charge carriers inside the resistor. The effect of this so-called *Johnson noise* can be computed mathematically by adding an imaginary noise-current generator (of infinitely high internal impedance) parallel to the resistor. The standard deviation of the current fluctuations associated with this generator is given by the *Nyquist formula*

$$(\sigma_i)_{th} = \sqrt{4kT\Delta f/R} \, , \qquad (III.38)$$

where k is the Boltzmann constant and T the absolute temperature of the resistor. This equation holds universally for any linear dissipative system in, and often also away from thermodynamic equilibrium, irrespective of the prevailing microscopic conduction mechanism.

Since Johnson noise and shot noise are statistically uncorrelated, one has to add them quadratically to find the total current fluctuations in the anode circuit

$$(\sigma_i)_{tot} = \sqrt{(\sigma_i)^2_{sh} + (\sigma_i)^2_{th}} \simeq \sqrt{2e\Delta f/R} \sqrt{\bar{i}_a RG + 2kT/e} \, , \qquad (III.39)$$

where we have approximated $(g\delta)^n \simeq G$ and neglected secondary-emission noise. The Johnson noise of the load resistor at room temperature T_r is negligible if $\bar{i}_a RG \gg 2kT_r/e \simeq 0.05\,V$. This inequality determines the minimum $\bar{i}_a$- and R value required for the accuracy of measurement to be limited by shot noise only. When a photo-emissive tube without internal amplification but with the same cathode current $\bar{i}_c = \bar{i}_a/G$ is used instead of a multiplier tube, the load must be enhanced by a factor G^2 (typically of the order of 10^{12} or higher) in order to make shot noise dominant. Such an enhancement, however, is not feasible in practice.

A review of noise and signal-to-noise ratios for photomultiplier tubes has been presented by Kovaleva *et al.* (1966) and Ingle and Crouch (1972, 1972a) who also considered single-stage photo-emissive tubes and photon counting techniques. Rolfe and Moore (1970) and Robben (1971) have compared actual signal-to-noise ratios obtained by counting and d.c. techniques, respectively, for a number of photomultiplier tubes.

5d   PRACTICAL CONSIDERATIONS
      Here we shall give some additional, practical advice concerning the selection and operation of photomultiplier tubes.

1.   *Selection.* A wide range of photomultiplier tubes is available; these vary as to spectral response, quantum efficiency, maximum gain factor, cathode area, dark current, structure, size, maximum rating of cathode and anode currents, maximum

and minimum operating temperatures, time-response, price, etc. Tubes have been developed to meet special demands in various applications. When making a selection one often has to reach a compromise between the properties required, e.g., low dark current and yet good response to the red, ease of handling, and cost. Tubes most frequently employed or recommended for use in (analytical) flame spectrometric work have been listed by Menis and Rains (1970), Winefordner and Smith (1970), Veillon (1971), Ramirez-Muños (1971), and Fuwa (1971). In particular, the RCA-IP28 tube with S-5 photocathode has been used in flame spectrometry for a long time. Since various samples of the same type often show — within certain limits — different characteristics (see, e.g., Boeschoten, Milatz and Smit 1954, and Youngbluth 1970 as regards variations in gain factor, dark current, and quantum efficiency of the IP28 tube and in fatigue effects of the CsSb end-on tube), it may be advisable to select the best sample from a batch of similar tubes. When a close coupling between a diffuse light source and the photocathode is required or when the distance between source and photocathode is to be determined accurately, end-on photomultiplier tubes have an advantage over side-window tubes. In some applications (see, e.g., Sects III.14g and 16b) the variation in local responsivity over the surface of the photocathode may be troublesome. Youngbluth (1970) found a variation of about 20% with a CsSb end-on tube; the variation can be minimized by an optimal choice of focusing electrode tension. Reif, Kniseley and Fassel (1970) advise a careful selection of the most uniform area of the photocathode for accurate measurements. Other means of eliminating errors in the measurement of intensity ratios will be mentioned in Sect. III.16b.

2. *Technique of measuring.* A special feature of photomultiplier tubes is that they allow direct application of a digital measuring technique. This technique has special advantages in the detection of weak light signals (see Sect. 5b). Care should be taken that pile-up effects of the random photon and dark-current pulses do not spoil the linearity of the measurement (see Sect. 5b). The electronic equipment used should thus be sufficiently fast. Correction procedures and techniques for dead-time compensation can be applied to extend the dynamic range of operation (see Ingle and Crouch 1972). Since the photomultiplier tube has a very short response time, it lends itself easily to the detection of fast transient signals as well as modulated signals with high modulation frequencies. An advantage of modulation is that the d.c. component of the dark current, which may slowly drift, and its possible flicker-noise component can be eliminated in the measurements (see Sect. III.14f). This advantage also applies to stray light that is not modulated. Modulation combined with synchronous detection can be applied with analog as well as digital measuring systems (see Sect. III.6b).

When the dominant noise spectrum is white ('quantum-noise limited case'),

PHOTODETECTORS AND NOISE   Sect. III.5d

modulation of the light flux to be measured is not advantageous with respect to the signal-to-noise ratio (see also Sect. III.14f-4). This ratio is even slightly deteriorated as symmetrical chopping of the light beam discards half of the incoming photons. When the dark- or background current drifts slowly and steadily, occasional dark- or background readings, instead of a.c. modulation, may suffice to correct for their contribution to the readings. When low-frequency noise in excess of photon noise is present in the signal itself, modulation will not help either, as this excess noise is modulated and de-modulated together with the signal to be measured. Such excess noise may be caused, e.g., by random variations in the gain factor. The noise introduced by these variations is called *multiplicative noise* — as distinct from *additive noise*, which arises, e.g., from the Johnson noise.

3. *Reduction of dark current.* Several ways of reducing the dark current were already mentioned in Sect. 5b. Cooled photomultiplier housings are available for this purpose. Cooling to about $-40\,°C$ often suffices or is optimal; cooling down to liquid-nitrogen temperature may be advantageous only for S-1 photocathodes. However, the spectral response near the threshold wavelength may be deteriorated by cooling to liquid-nitrogen temperature (see Boeschoten, Milatz and Smit 1954, and Kovaleva *et al.* 1966). The signal-to-noise ratio in photon pulse counting does not always improve upon cooling (see Malmstadt, Franklin and Horlick 1972). Anyway, excessive cooling should be avoided because of possible damage to the tube. Condensation of water vapour on the window and on the insulating parts between the high-tension leads during cooling should, of course, be prevented. For a given photomultiplier tube the dark current at room temperature can also be reduced by confining the effective area of the photocathode through magnetic defocusing of the electrons emitted by the cathode. The effect of placing an annular magnet near the end-on photocathode is that only electrons emitted by the central part of the cathode arrive at the first dynode. In order to take full advantage of this reduction, the radiant flux to be detected should be focused onto a small spot at the centre. The dark-current to photocurrent ratio can also be reduced by an electrostatic suppressor grid mounted close to the photocathode, as the thermionic electrons are emitted with lower mean kinetic energy than the photo-electrons (see Eberhardt 1964). However, the threshold wavelength may be affected by this measure, too.

Special photomultiplier tubes have been devised in which the electron emission from a restricted section of the photocathode is 'imaged' on a slit between cathode and first dynode by means of a magnetic 'lens' (see Eberhardt 1964, Harber and Sonnek 1966, and Golightly, Kniseley and Fassel 1970). This arrangement may help to reduce the dark current. Another advantage is that the electron emission

Chapt. III   GENERAL INSTRUMENTAL ASPECTS AND EXPERIMENTAL METHODS

can be scanned over the cathode surface by varying the magnetic field (*image dissector tube*). When a detail of an optical spectrum is imaged on the photocathode, this detail can thus be scanned magnetically.

4.   *Mounting and operation*. The gain factor depends on the distribution of the high-tension over the chain of dynodes, in particular on the tension between cathode and first dynode. The manufacturer's recommendations should be consulted and checks made regularly to see that they are being adhered to. The responsivity of the photomultiplier can easily be controlled by varying the total high-tension applied to the dynode system. The absolute intensity calibration of the spectrometer including the photodetector will be discussed in Sect. III.14g. Care should be taken that any material in contact with the tube envelope is at cathode potential. A mu-metal shield (at cathode potential) is recommended in order to minimize interference by stray magnetic fields. One should see to it that the photomultiplier tube is screened from stray light and protected against dust, mechanical vibrations (caused, e.g., by the light chopper), and — especially when a flame is close by — against heat and chemical influences.

In some applications it is advisable not to focus the radiation beam onto a tiny spot on the cathode surface, but to irradiate the whole (effective) surface diffusely, as the sensitivity of the photocathode may vary markedly with place. When the cathode is slightly overfilled by the diffuse radiation beam, a small displacement of the cathode with respect to the incident beam will then have a minimum effect on the output current. It is not always realized that the responsivity of the photocathode may also depend on the direction of incidence of the radiation beam. The optical coupling of the photocathode to the exit slit of a monochromator may be facilitated with the aid of a light-guide (fibre-optics). Application of a liquid film (e.g., xylol) between the end-face of the light-guide and the photomultiplier window, in combination with a large angle of incidence, may improve the responsivity by a factor two (see Gunter, Grant and Shaw 1970). However, the dependence of the responsivity on the polarization of the light beam may be enhanced at oblique incidence on the photocathode. At normal incidence polarization effects may range from 1 to 10%; at large angles of incidence, effects up to 50% have been found (see Young 1974).

Fatigue and saturation effects of different kinds can be avoided by keeping the (mean) anode and cathode current far below the maximum ratings stated by the manufacturer (see Keene 1963). For stable operation it is advisable to expose the photomultiplier tube for some time to radiation before the actual measurements are taken. Departures from linearity may be checked by one of the methods to be discussed in Sect. III.13.

6. THE ELECTRICAL MEASURING SYSTEM

6a   INTRODUCTION

The electrical signal delivered by the photodetector is processed by the electrical measuring system and displayed on a read-out device. The system usually incorporates an electronic amplifier whose input and output impedances should be matched to the impedances of the photodetector and the read-out device, respectively. The signal may be processed either in analog form (as a current or tension whose magnitude represents the signal strength) or in digital form (as a train of pulses the time-rate of which corresponds to the signal strength and the number of which corresponds to the time-integral of the signal). An analog signal may be read out in analog form by a meter, pen recorder or oscilloscope. If one uses an analog-to-digital converter, the signal can also be read out in digital form. This may improve the reading precision (no pen drag or interpolation errors with scale reading). A digital signal may be directly read out in digital form by a counter and timing circuit, but it can also be presented in analog form by a rate meter. Some advantages of digital methods were mentioned in Sect. III.5b in connection with photon pulse counting. Digital methods are often used in more sophisticated or automated measuring systems, especially when computer operations are involved.

In this section we shall not go into details of the circuitry or components of electronic measuring systems. We shall confine ourselves to the more functional aspects of some general classes of measuring systems commonly used in flame spectroscopy. For a more detailed discussion, especially in relation to spectroscopic applications, the reader is advised to consult Malmstadt, Enke and Toren (1962), Hieftje (1972a), and Malmstadt, Enke and Crouch (1974). An introduction to this subject may be found in Cath (1970), Winefordner and Smith (1970), Veillon (1971), and Winefordner, Schulman and O'Haver (1972).

In general, a measuring system should meet the following demands, depending on the particular measuring problem under consideration:
- sufficient amplification,
- linear response over a sufficiently wide dynamic range,
- absence of drift (in the zeropoint as well as in the sensitivity),
- sufficiently fast response,
- rejection, as far as possible or is necessary, of the noise present in signal or background, or generated in the amplifier itself.

In addition, other properties of a more practical kind may be important:
- sturdy construction and easy operation,
- insensitivity to environmental interferences,
- versatility with respect to different measuring tasks,
- low cost and long life.

Chapt. III   GENERAL INSTRUMENTAL ASPECTS AND EXPERIMENTAL METHODS

In flame spectroscopic investigations, we are mostly dealing with the measurement of a quasi-stationary or slowly varying radiant flux. When this flux is uninterrupted, it generates a d.c. current in the photodetector, and a d.c. measuring system is required. Under certain circumstances, however, it may be advantageous to apply intensity-modulation of the radiation beam to be measured at a suitable fixed frequency (see also Sect. III.5d). This can be achieved by mechanically chopping the beam by means of a rotating disk with regularly placed, identical openings, or by modulating the electrical current feeding the background light source or by applying an a.c. feeding current. An a.c. measuring system must then be used. D.c. and a.c. measuring systems can in general be realized for analog as well as digital signals. We shall discuss these systems below.

When fast, transient signals are to be measured as a function of time in the presence of noise, as in fluorescence-decay or pulsed-laser experiments, more complicated measuring systems are needed. When these signals are repetitive, a single-channel 'box-car' integrator with time-scanning or a multi-channel sampling and integrating device can be used to improve the signal-to-noise ratio (see Veillon 1971, Winefordner, Schulman and O'Haver 1972, and Hieftje 1972b). These special techniques and the auto-correlation techniques for retrieving a stationary signal from background noise (see Hieftje 1972b) will not be considered. For a general introduction see, e.g., Malmstadt, Enke and Crouch (1974), and Cova and Longoni (1979).

6b   GENERAL MEASURING SYSTEMS

The simplest type of *d.c. measuring system* for analog signals consists of a d.c. amplifier followed by a low-pass RC-filter whose output is connected to a d.c. meter. The d.c. output current of the amplifier is normally proportional to the d.c. input current. The dynamic behaviour of the whole system is controlled by the time constant (= RC-time) of the filter, if it is appreciably longer than the response time of the meter. The use of an RC-filter is equivalent to averaging the signal plus noise over a time $2RC$ and results in a noise bandwidth $\Delta f = 1/(4RC)$ (see also Sect. III.5c). The signal-to-noise ratio can be improved by enlarging the RC-time, but this is accompanied by an increase in the response time of the system (see Sect. III.14f).

The *response time*, τ_r, of a measuring system is the time needed to reach a deflection that is a specified fraction of the final, steady-state deflection. For a system with an exponential step response (see Sect. III.14f-2 and Eq. III.67) with time constant τ_c, 99.0% of the final deflection is attained after $4.6\,\tau_c$ seconds, whereas after $2\pi\tau_c$ seconds a deflection larger than 99.8% is attained.

An operational amplifier with a capacitor as feedback element can be used as an *integrator* of the d.c. current for a given period of time.

THE ELECTRICAL MEASURING SYSTEM   Sect. III.6b

Electrometer amplifiers are noted for their relatively high input impedance and are especially suited for amplifying the d.c. photocurrent of a single-stage vacuum phototube (see Sect. III.5a). A very high input resistance, however, in combination with the stray capacity of the input circuit etc., may lead to an undesirably long input time constant.

Advantages of d.c. amplifier systems are their easy operation and comparatively low cost; electrometer amplifiers offer especially good sensitivity for photodetectors that operate as a current-source. A disadvantage is that they may be liable to drift effects and may also amplify undesirable d.c. signals (e.g., due to dark current or stray light) as well as the flicker noise produced by the photodetector or generated in the first amplifying stage. This flicker- or $1/f$ noise often appears — for reasons not well understood — in excess over the inevitable Johnson- and shot noise at low frequencies (see Sect. III.5c). Slow fluctuations or drift may be partly eliminated by making the response time appreciably shorter than the total observation time available, while repeating the measurements with closed and open phototube, respectively (see Sect. III.14f-3). This procedure is often more effective than if one minimizes the noise bandwidth by choosing the response time equal to the total observation time.

Drift effects of the amplifier itself may be reduced by applying chopper-stabilization. The d.c. current to be amplified is chopped by a switch which periodically opens and closes the input channel, and is thus converted to an a.c. current at the chopping frequency. This a.c. current is then amplified by a drift-free a.c. amplifier and finally converted again to a d.c. current by a second chopper operating synchronously with the first one (see also the discussion on lock-in amplifiers below). It should be realized that this procedure does not eliminate the drift or low-frequency noise occurring in the circuit (e.g., due to the dark current) before chopping takes place.

Counting the photomultiplier pulses during a fixed time interval is essentially equivalent to using a d.c. integrating system. In both cases, the signal to be measured is not discriminated against other, spurious signals caused by the dark current, stray light, etc. However, a digital system employing a wide-band a.c. amplifier, pulse-height discriminator, pulse-shaper, etc., is less sensitive to amplifier drift and noise, and can be operated easily for long counting (i.e. integration) times (see also Sect. III.5b).

A.c. measuring systems for analog signals in combination with light modulation can be realized in two different modes. The simplest form of a.c. measuring system consists of a *narrow-band* or *frequency-sensitive a.c. amplifier* which is followed by a linear or quadratic rectifier and a low-pass filter. Amplification

Chapt. III   GENERAL INSTRUMENTAL ASPECTS AND EXPERIMENTAL METHODS

is restricted to a narrow band around the modulation frequency by means of a selective a.c. filter. The selectivity can be improved by inserting a selective filter in the negative-feedback loop (*tuned amplifier*). The a.c. component of the modulated photosignal with modulation frequency is amplified and rectified. The tuned amplifier rejects any d.c. bias (caused, for example, by the dark current or unmodulated stray light) as well as all a.c. signals or noise components lying outside the transmission band of the filter. Flicker noise of the dark current and of the amplifier can thus be suppressed by choosing a sufficiently high modulation frequency (say $\geqslant 100\,\text{Hz}$; see Sect. III.14f-4). Besides, a.c. amplifiers can easily be made drift-free. The response time of the whole system is controlled by the time constant of the low-pass filter or by the response time of the meter. The bandwidth of the tuned filter, although small compared to the modulation frequency, is for practical reasons usually markedly larger than 1 Hz.

The noise transmitted through the tuned a.c. filter affects the reading in two different ways. On the one hand, this noise is rectified together with the a.c. signal and thus causes a zero-offset of the meter. Since the spectral noise power (see Sect. III.5c) may be considered to be independent of frequency within the small bandwidth of the tuned filter, this zero-offset is proportional to $\sqrt{\Delta f}$. On the other hand, the statistical nature of the noise also produces a fluctuation in the meter deflection whose standard deviation (or root-mean-square value) is proportional to $1/\sqrt{\tau_r}$, where τ_r is the response time of the system. When the noise spectrum is 'white' in the whole frequency range down to 0 Hz, nearly the same signal-to-noise ratio is obtained for a d.c. and an a.c. measuring system for the same response time (see Sect. III.14f). A slight deterioration occurs with the a.c. system, since the chopper throws away half of the incoming signal. However, with the d.c. system flicker noise may deteriorate this ratio markedly. On the other hand, the zero-offset produced by the rectified noise with the a.c. system may be a nuisance. Its magnitude is roughly a factor $\sqrt{\Delta f \tau_r}$ larger than the standard deviation of the meter fluctuations (if $\Delta f \gg \tau_r^{-1}$). This zero-offset can be reduced by narrowing the bandwidth Δf. However, the narrower the bandwidth, the more critical are slight deviations of the modulation frequency from the peak frequency of the filter.

It is not often realized that the zero-offset by the noise also has a disturbing effect on the dependence of the meter deflection on the signal strength. Let us assume, as before, a linear rectifier whose output is proportional to the mean absolute value of the a.c. input current. Then the d.c. output current i_t due to the rectified signal-plus-noise is related to the d.c. output current i_s due to the signal alone in the absence of noise, and to the noise zero-offset i_n in the absence of signal, approximately according to (see Alkemade 1954, and Alkemade and

THE ELECTRICAL MEASURING SYSTEM   Sect. III.6b

Lavèn 1957)

$$i_t \simeq \sqrt{(i_s)^2 + (i_n)^2} \ . \tag{III.40}$$

Thus it appears that i_t (i.e. the actual d.c. meter deflection) is no longer proportional to the signal strength (i_s) when the latter becomes comparable to, or smaller than the zero-offset (i_n). Simple subtraction of i_n from i_t (e.g. by shifting the zeropoint on the reading scale) does not help to restore a proportional relationship between meter deflection and signal strength. On the contrary, for a relatively large signal (say: $i_s > 10\,i_n$) a proportional relationship is in fact more closely approximated when the zeropoint is not corrected at all for this zero-offset. This holds since in the above expression i_n^2 is then practically negligible ($< 1\%$) compared to i_s^2.

An a.c. measuring system incorporating a *lock-in amplifier* [†] avoids the above practical difficulties. With this second type of a.c. system, the amplified a.c. component of the modulated photosignal (input signal) is converted to a strictly proportional d.c. current by means of *synchronous* or *phase- and frequency-sensitive rectification*. Essentially, synchronous rectification is based on some sort of 'multiplication' (beating) of the input signal by a periodic, noise-free reference signal or gate function of constant amplitude, that has the same frequency and phase as the modulated signal (see also Cath 1970, Hieftje 1972a, and Fisher 1977). As a result, a d.c. current is produced whose magnitude is proportional to the amplitude of the modulated signal. In general, the expression for the d.c. output current also contains a $\cos\phi$ factor, where ϕ is the phase difference between the modulated and the reference signal. This explains the phase-sensitivity of the synchronous rectifier. In addition to the d.c. current, higher harmonics of the modulation frequency are produced. But these harmonics can be easily removed by a low-pass RC-filter following the synchronous rectifier. When the input signal has a frequency unequal to that of the reference signal, no d.c. component will appear at the output of the rectifier. Only when the reference signal is not purely sinusoidal (e.g., when a square-wave gate function is used), may higher harmonics in the input signal also produce a d.c. output.

The mathematical basis of the synchronous rectifier can easily be understood by writing for the input signal: $A_s \cos\omega_s t$ (t = time; $\omega_s = 2\pi f_s$) and for the reference signal: $A_r \cos(\omega_r t + \phi)$. Multiplication yields for the output signal $i(t)$

$$i(t) \propto A_s A_r \cos\omega_s t \cdot \cos(\omega_r t + \phi) = \tfrac{1}{2} A_s A_r \cos\{(\omega_s - \omega_r)\,t - \phi\} + $$
$$+ \tfrac{1}{2} A_s A_r \cos\{(\omega_s + \omega_r)\,t + \phi\} \ . \tag{III.41}$$

[†] Also called synchronous, heterodyne or coherent detector, phase-lock amplifier, or phase- and frequency-sensitive rectifier.

Chapt. III   GENERAL INSTRUMENTAL ASPECTS AND EXPERIMENTAL METHODS

Only if $\omega_s = \omega_r$, do we obtain a d.c. component $\propto \frac{1}{2} A_s A_r \cos \phi$ (in addition to the second harmonic, which is discarded). When $\omega_s \neq \omega_r$, an a.c. beat signal at the difference frequency is produced (in addition to an a.c. current at the sum frequency).

When the reference signal has a square-wave form, it can be expanded in a Fourier series consisting of the fundamental wave (with frequency f_r) and the odd harmonics (with frequencies $3f_r$, $5f_r$, ...). The rectifier output current is then obtained by applying Eq. III.41 to each of these Fourier terms and adding the results. A d.c. output current is then produced for a.c. input currents with frequency f_r, $3f_r$, $5f_r$,... (see also Cath 1970 [†]).

Multiplication of two signals can be realized physically by means of a 'mixer' as in heterodyne radio receivers. A synchronous switch which simply transmits the input signal during one half period of the reference signal, and reverses its sign during the other half, has essentially the same effect. In the latter case the input signal is effectively multiplied by a square-wave gate function which assumes periodically the values +1 and −1, respectively.

The reference signal should be well synchronized with the modulator in respect of frequency and phase. When light modulation is achieved by means of a rotating chopper, this synchronization is most easily realized with the aid of an auxiliary, constant light beam falling on a separate photocell (e.g., a phototransistor) after being chopped by the same chopper. The a.c. output current of this photocell is then used to generate the reference signal. If the reference signal is used to drive a synchronous switch, (slight) variations in its amplitude are of no consequence. The phase of the reference signal can be adjusted with respect to that of the signal to be measured by a phase shifter or by varying the position of the auxiliary light beam along the circumference of the chopper disk. Any shift in frequency or phase of the modulator now affects the signal to be measured and the reference signal in the same way and is thus cancelled out. In other words, the modulator and reference signal are exactly locked to each other.

An early but unusual form of synchronous detection is the *a.c. galvanometer* which contains an electromagnet fed by an a.c. current of the same frequency as the modulation frequency (see Milatz *et al.* 1948). The deflection of the moving coil depends on the product of the current through this coil and the magnetic field strength. A constant deflection is obtained only for currents with the same frequency as that of the a.c. magnetic field. This device shows essentially the same characteristics as the synchronous rectifier described above.

[†] There is a misprint in Eq. (27) of this reference.

THE ELECTRICAL MEASURING SYSTEM
Sect. III.6b

The frequency bandwidth of this detector is about equal to the reciprocal response time of the moving coil (which is of the order of 1 Hz). This instrument has been used occasionally in flame emission as well as in double-beam flame atomic absorption measurements (see Alkemade 1954, and Alkemade and Milatz 1955).

Noise components having frequencies around the reference frequency produce low-frequency fluctuations near 0 Hz as a result of the beating effect of the synchronous rectifier. However, they do not produce a constant zero-offset as in the case of a linear rectifier (see Eq. III.41). The noise bandwidth Δf is again determined by the RC-time of the low-pass filter or by the response time of the meter if the latter happens to be the longer. The standard deviation of the meter fluctuations is proportional to $\sqrt{\Delta f}$ (see Sect. III.14f-4). The signal-to-noise ratio is about the same for a lock-in amplifier as for a tuned a.c. amplifier if the modulation frequency and response time are the same (disregarding the noise zero-offset in the latter case).[†]

Usually an a.c. filter centered at the modulation frequency is inserted before the synchronous rectifier. It prevents the amplifier or rectifier from being overloaded by noise. It need not be as sharply tuned as for the case of the tuned a.c. amplifier system, since the synchronous rectifier does not produce a noise zero-offset.

To summarize, the main advantages of a lock-in amplifier over a narrow-band or tuned a.c. amplifier are:
- the modulation frequency and the reference frequency driving the synchronous rectifier can be easily and accurately locked together,
- the a.c. filter preceding the rectifier may have a moderately large bandwidth and is thus not critical,
- no zero-offset by noise occurs,
- the linear response of the system is not affected by noise.

There is an additional advantage of a synchronous measuring system when a null-balance method is applied with a double-beam instrument (see also Alkemade and Milatz 1955, and Alkemade and Lavèn 1957). Suppose that the two light beams are modulated in opposite phase by the same chopper and that a single phototube is used for the detection of both beams. The use of a single phototube instead of two separate phototubes for each beam improves the stability of the null balance. In order to obtain null balance with the tuned a.c. amplifier, the amplitudes of the

---

[†] Signal-to-noise ratios obtained by a lock-in amplifier and by auto- and cross-correlation techniques in atomic fluorescence measurements have been compared and discussed on a qualitative basis by Hieftje, Bystroff and Lim (1973). The correlation techniques were found to be less susceptible to 'impulse noise' (arising, e.g., from incidental sparks or flashes) than the lock-in amplifier.

two modulated signals must be made equal while their phases must remain exactly opposite to each other. With a phase-sensitive lock-in amplifier these two conditions need not be fulfilled separately. Here only the **components** of the modulated signals **in phase** with the reference signal must be balanced. In other words, with a tuned amplifier two variables must be adjusted precisely to obtain null balance, whereas with a lock-in amplifier only one variable need be adjusted (see Sect. III.15c-3).

The greater instrumental complexity of the lock-in amplifier (requiring an additional device for generating the reference signal) may be a slight disadvantage. Also, care must be taken to maintain a constant, optimal phase-relationship between the modulator and the reference signal. Relative phase variations have minimal effect when we have: $\cos \bar{\phi} = \pm 1$ for the mean relative phase $\bar{\phi}$. Under this condition, the magnitude of the d.c. output signal is maximum. An inexpensive lock-in amplifier for frequencies ranging from 1 Hz to 5 MHz has been described by Caplan and Stern (1971).

The technique of synchronous rectification combined with modulation can be applied not only to analog signals, as described, but also to *digital*, i.e. pulse-like signals (see Smit 1961, Arecchi, Gatti and Sona 1966, Robben 1971, and Tebra and Visser 1972). A reference signal synchronized with the light modulator controls a gate circuit in such a way that the photon- plus background pulses, and the background pulses alone are counted alternately by two separate counters. If the two alternate count periods are exactly equal, the difference between the accumulated counts during the total measuring time τ_m is a measure for the photosignal as averaged over τ_m. Essentially the same result (as to the signal-to-noise ratio and the rejection of d.c. dark current, etc.) is now obtained as with a lock-in amplifier followed by an RC-filter with $RC = \frac{1}{2} \tau_m$. At sufficiently high modulation frequency or low counting rate, the noise is dominated by the shot effect (see Sect. III.5c) of the pulses. Amplifier noise spikes may contribute to the background noise, but their contribution can be markedly reduced by a suitable pulse-height discriminator setting. Instead of using two separate counters and subtracting their accumulated counts, one can use a single bidirectional or up-down counter, synchronized with the modulation frequency. During one half period of the reference signal this counter counts 'up' the incoming pulses, whereas it counts them 'down' during the other half period.

The optimal ratio $\Re$ of up- and down-periods giving the highest signal-to-noise ratio for fixed total measuring time is different in the weak-signal limit (where $\Re \to 1$) and in the strong-signal limit (where $\Re \to \infty$). The general case has been treated by Young (1974).

7. LIGHT SOURCES FOR ABSORPTION AND FLUORESCENCE MEASUREMENTS

7a GENERAL

In absorption and fluorescence measurements, a background light source is needed to produce the radiation which is to be absorbed by the metal atoms (and/or molecules). Light sources are in general distinguished as spectral continuum sources and spectral line sources, although there are also intermediate types, i.e., sources emitting a broad continuum with a superimposed line spectrum.

Examples of purely *spectral continuum sources* are the tungsten strip lamp and the anode of a low-current d.c. carbon arc. These lamps radiate light as a result of the high temperature of the strip or anode heated by electric current. They belong to the class of thermal radiators.

Spectral line sources can be realized either by an electric discharge with the aid of internal electrodes through a gas, a metal vapour or a mixture of both at low pressure, or by radio-frequency (r.f.) or microwave (MW) excitation of a metal vapour in a low-pressure inert gas (electrodeless discharge lamps, EDL). Examples of the former group are the hollow-cathode lamp (HCL) and its modifications, and the metal-vapour discharge lamps. The HCL is a *glow*-discharge lamp, whereas the metal-vapour discharge lamps belong to the class of *arc*-discharge lamps.

Examples of the *intermediate-type sources* are the arc discharge lamps; the most commonly used types are the high-pressure Xe-, Hg-, or (Hg+Xe) lamps. For convenience we shall discuss this type of source along with the continuum sources. Other background light sources, which are discussed briefly in this section, are flames, arcs, sparks, lasers, and atomic-beam light sources.

In the following subsections the characteristics of the various types of light sources will be discussed in relation to the demands of atomic absorption and fluorescence measurements (see Sects III.15 and III.16). The requirements for continuum sources used in line-reversal measurements will be dealt with in Sect. III.10. Since the demands may be different for continuum and line sources, both types will first be discussed in separate sections. Thereafter the characteristics of both types will be compared (see Sect. 7d). Quantitative data relating to (spectral) radiance, radiance temperature, size of the radiant area, available solid angle, line half-intensity width, short-term and long-term instability, etc, are summarized in Fig. III.15 and Table III.4 at the end of this section.

The precision of absorption and fluorescence measurements may depend on the stability and emission noise of the light source (see Sects III.15 and III.16). We shall distinguish between 'long-term instability' (including drift) and 'short-term instability' only from a merely practical view-point to describe the performance of a light source. By the former term we understand variations of the light emission

Chapt. III GENERAL INSTRUMENTAL ASPECTS AND EXPERIMENTAL METHODS

over periods of, say, one minute or longer. By the latter, we mean variations with periods typically of the order of 1 - 3 seconds.

In the low-frequency part of the emission noise spectrum (see Sect. III.5c) one often finds a noise component in excess of the inevitable shot noise component. When the spectral power of the excess noise component varies approximately as $1/f$ with frequency f, we call this excess component flicker- or $1/f$ noise (see Sect. III.5c). Experimental data on the noise characteristics of emission sources are scarce (see Snelleman 1968, Belyaev *et al.* 1968[†], Mansfield *et al.* 1968, and Hunziker 1971; see also the literature on flame-noise experiments, cited in Sect. IV.9). Data on the stability and emission noise will be presented in the following subsections.

Only a few technical data concerning the light sources will be dealt with here. For further details (construction, power supplies, etc.) the reader is referred to the following literature: Cann (1969), Dean and Rains (1969, 1971), Mavrodineanu (1970), Rubeška and Moldan (1969), Yakovlev and Shishatskaya (1969), Batrakov (1970), L'vov (1970), Winefordner, Schulman and O'Haver (1972), Winefordner (1971a), Elenbaas (1972), and Boumans (1972).

7b SPECTRAL CONTINUUM SOURCES

7b-1 *TUNGSTEN STRIP LAMP*

This incandescent source is widely used in line-reversal measurements (see Sect. III.10) in the visible and infrared region. Its emission is not sufficiently intense in the ultraviolet region (see Fig. III.15). Use of a quartz glass envelope or window is therefore of little practical value. The representative area of the strip amounts to about 20 mm². The true temperature of the strip may be as high as 3000 K, whereas the maximum radiance temperature (see definition in Sect. II.5c-3) of this source is about 2600 K and is somewhat dependent on λ. The spectral radiance of this thermal radiator follows directly from Planck's Law and the λ-dependent emission factor $\varepsilon(\lambda)$ (cf. Eqs II.293 and 294). For example, $\varepsilon(\lambda)$ for a flat tungsten strip varies from 0.43 to 0.45 in the λ-range from 6250 to 4700 Å. Each uncoated glass surface introduces a radiation loss of 4% (window plus focusing lens give a total loss of about 15%).

Excellent stability and reproducibility (better than 0.01%) can be obtained with an adequate d.c. power supply with optical feedback stabilization, or a battery. This stability is far greater than can be obtained with gas-discharge or arc lamps. Typical ratings are 20 - 30 V and 15 - 20 A maximum. Light-modulation

---

[†] Belyaev *et al.* (1968) have measured the frequency spectra (120 - 70000 Hz) of the emission fluctuations of a number of light sources (HCL, EDL, flames, arcs and plasma jets). Low-frequency fluctuations appeared in all of these cases.

LIGHT SOURCES FOR ABSORPTION AND FLUORESCENCE MEASUREMENTS   Sect. III.7b

is only feasible by mechanical chopping.

7b-2   *TUNGSTEN BRUSH LAMP*
This is a modification of the former lamp, i.e. the strip is replaced by a bundle of fine tungsten wires viewed end-on (see Quinn and Barber 1967). The brush structure enhances the emission factor from 0.44 (flat tungsten) to 0.95. The maximum obtainable radiance temperature is here 3000 K at $\lambda = 6500$ Å (see Snelleman 1969). This lamp has found little use.

7b-3   *HIGH-PRESSURE XENON ARC LAMP*
These lamps with pressures over 10 atm have found application in absorption and fluorescence measurements, especially in the ultraviolet (and to a lesser extent in line-reversal measurements, see Sect. III.10). They consist of two tungsten electrodes in a small quartz envelope. Most types have an arc gap as small as a few millimetres and are referred to as short or compact arcs. Xenon arc lamps are available in a wide range of input powers. Typical arc tensions are between 15 and 30 kV, while arc currents may range from 5 to more than 100 A, depending primarily on the arc length, which may vary from 1 to 9 mm. The starting tension of the lamp is between 20 and 30 kV; warming-up time is of the order of minutes.

In general the arc width also increases with arc power, so that the arc current density does not increase appreciably with power. The resulting increase in total radiation output of high-power lamps is thus primarily due to the increased arc cross-sectional area. The intensity is mostly concentrated near the cathode, known as the 'hot spot'. A typical size for this hot-spot for the 150 W xenon lamp is 2×2 mm^2. In order to obtain maximum radiant flux in a narrow beam, a special xenon lamp has been constructed with a built-in parabolic reflector (150 W Eimac source; see Bratzel, Dagnall and Winefordner 1970) giving a practically parallel beam of $4 - 5$ cm^2 cross section. The radiant flux from this source is about ten times higher than that from the 'normal' 150 W xenon lamp. This gain is mainly due to an increase in useful solid angle (see Table III.4).

Since the total radiative power efficiency of a xenon arc lamp is typically 30 - 50%, the lamp walls have to be cooled. With low-powered lamps natural convection cooling is usually sufficient, but at higher power forced cooling is necessary. Because of possible ozone formation, cooling with N_2 is preferable to air cooling.

For greatest stability and longest life xenon arc lamps are almost always operated on d.c. The instability of the arc lamp, including 'arc wander', is mainly due to the instabilities in the power supply. The (short-term) stability of the Xe lamps can be improved greatly by optical-feedback stabilization of the power supply (see Redfield 1961, Schurer and Stoelhorst 1968, Rutgers 1971, and Human, Zeegers and van Elst 1974). This holds for d.c. as well as for square-wave modulated power

Chapt. III   GENERAL INSTRUMENTAL ASPECTS AND EXPERIMENTAL METHODS

supplies. In the case of a.c. power supply arc wander causes extra instability. In the case of d.c. operation, polarity is important, as is also the case for other high-pressure d.c. arcs: the lamps should be operated with the anode facing upwards. The useful life ranges from 200-1000 hours. For special purposes, xenon flash lamps are available (see Winefordner, Schulman and O'Haver 1972). In flame studies the lamp output is either modulated with a mechanical chopper or by means of a square-wave power supply (see, e.g., Human, Zeegers and van Elst 1974; for pulsed operation of Xe lamps see also Johnson, Plankey and Winefordner 1974).

The Xe lamps show a smooth continuum spectrum between 2000 and 3000 Å (see Winefordner, Schulman and O'Haver 1972). Xenon lines appear superimposed on it in the near infrared. The visible continuum emission of a 500 W Xe arc lamp corresponds closely to that of a grey thermal radiator with temperature of 7000 K and emission factor of 0.06. Radiance temperatures of these lamps range from about 5000 to 6000 K at maximum ratings.

Some typical data on spectral radiance, stability, etc., will be presented in Sect. 7d.

For a detailed discussion on the feasibility of reducing the effects of instabilities from a continuum source in atomic absorption measurements, the reader is referred to Snelleman (1968). Further information about Xe arc noise has been given by Hunziker (1971).[†]

7b-4   *HIGH-PRESSURE MERCURY ARC LAMP*

Most of the properties of the high-pressure Hg lamps are similar to those of the Xe lamps. Their radiance temperature is also about 6000 K. The spectrum, however, consists of a number of very intense Hg lines superimposed on a continuum background (see Fig. III.14). Although more intense in the u.v. region than the Xe arc lamp of comparable input power, the mercury arc lamp may be less useful just because of its less smooth spectral distribution.

In Fig. III.14 the spectral distributions of a typical high-pressure Hg lamp and of an Xe lamp are compared.

7b-5   *OTHER CONTINUUM SOURCES*

The quartz halide lamp[*] with a radiance temperature of about 3000 K, which can be used down to 3000 Å in atomic absorption measurements (see Butler and Brink

---

[†] Hunziker shows an emission noise spectrum of a 150 W Xe lamp in the frequency range from 10 to 10^5 Hz. He found that at frequencies above 10^3 Hz shot noise is dominant, whereas in the lower frequency range excess fluctuations are produced by an inadequately filtered power supply.

[*] The quartz halide lamp or halogen lamp has tungsten coils in a quartz envelope filled with gas to which iodine has been added.

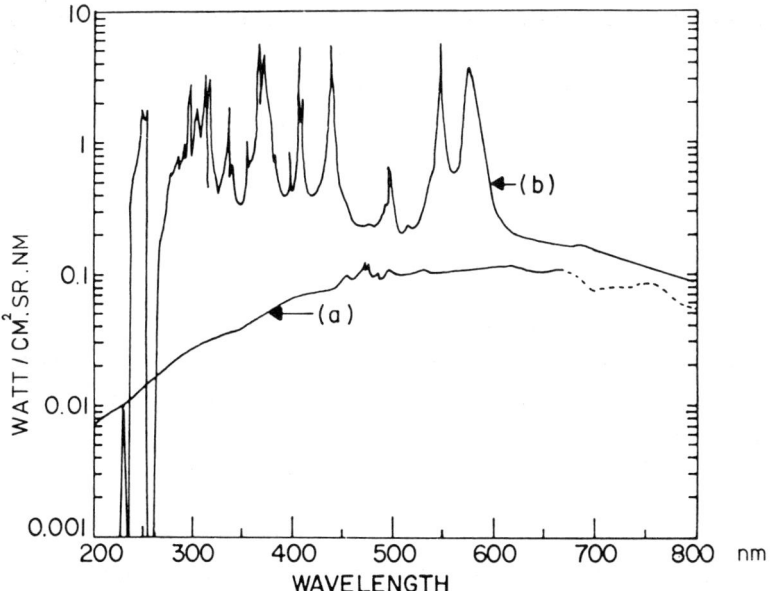

Fig. III.14 Spectral distribution of two typical sources: curve (a), PEK 75 xenon arc lamp (manufactured by PEK, Sunnyvale, Calif); curve (b), Osram HBO 100 mercury arc lamp. (Derived from Winefordner, Schulman and O'Haver 1972.)

1971, and Elenbaas 1972) and the anode of a low-current d.c. carbon arc with a radiance temperature 3800 K (see Snelleman 1969) are only mentioned here for completeness' sake.

7c   SPECTRAL LINE SOURCES

In atomic (peak) absorption measurements we often demand background source lines that are as narrow as possible (see Sects II.5c-2 and III.15). The need for a narrow emission line, in order to obtain maximum absorption factors, implies absence of noticeable self-absorption and self-reversal (see Sect. II.5c-3). In addition, a high radiance, a large useful solid angle, and low excess noise are often required.

When in absorption and fluorescence measurements a.c. detection or pulse detection is used, the light from the line source is mechanically chopped or electronically modulated by modulating the input power. In fluorescence measurements, where a high primary radiant flux in the flame is often essential, electronic modulation is often preferred, since a mechanical chopper may restrict the available solid angle (see Sect. III.16).

Chapt. III   GENERAL INSTRUMENTAL ASPECTS AND EXPERIMENTAL METHODS

7c-1   *HOLLOW-CATHODE LAMP (HCL)*

This lamp is a low-pressure glow-discharge lamp which, in its simplest form, consists of a cylindrical (glass or quartz glass) envelope filled with a low-pressure carrier gas,[†] mostly a noble gas, in which a hollow cathode and an anode are mounted. Typical properties of the HCL are its low gas temperature (350-400 K) and low gas pressure, so that narrow spectral lines can be produced. For construction details, manufacture, dimensions, etc., the reader is referred to numerous books and papers (see, e.g., Jones and Walsh 1960, Rubeška and Moldan 1969, L'vov 1970, Willis 1970, and Butler and Brink 1971).

In this lamp accelerated electrons generate positive ions upon collisions with atoms of the carrier gas. These ions gain energy in the electric field and collide with the cathode. Atoms of the cathode material are released by these collisions (cathode sputtering) and are excited in the discharge by inelastic collisions with electrons. They then radiate their atomic and ionic spectral lines.

The lamps are called sealed or demountable depending on whether the cathode is sealed within the lamp enclosure or can be exchanged by dismantling the lamp. (Demountable HCL's are mostly water-cooled.) Cathodes may be composed of one element or of several elements (see Butler and Brink 1971).

Although the demountable HCL's have the disadvantage of requiring a complex vacuum system (see Butler and Brink 1971), they do have the following advantages over the sealed ones: easily exchangeable cathodes for different elements and a longer life (see Rossi and Omenetto 1967). The latter advantage is due to the fact that the flow of fresh carrier gas promotes 'clean-up'. Since the shape and the dimensions of the cathode, as well as the other dimensions of the demountable HCL can easily be varied, the HCL can be optimized during operation. This may explain why with these lamps higher intensity at the same current and less self-reversal of the lines are obtained when compared with a sealed HCL (see Rossi and Omenetto 1967, and Butler and Brink 1969).

Typical ratings for the normal HCL are: starting d.c. tension 400 V and d.c. working current 10 - 40 mA at 150 - 350 V. In order to obtain a lamp stability of 0.5% or better the d.c. feeding current stability should be better than 0.2% (see L'vov 1970; see Table III.4). A lamp stability of better than 0.01% was obtained by optical stabilization of the d.c. power supply.[*]

Several modifications of the basic type HCL are known; we mention the d.c. boosted HCL (see Cartwright, Sebens and Slavin 1966, Davies 1967, and Butler and

---

[†] When the lamp is sealed, the term *fill-gas* is also in use.

[*] According to measurements by G. Bosman at the Physical Laboratory of the University at Utrecht on emission noise spectra from a Ba HCL (personal communication).

Brink 1971) where a second discharge applied between two auxiliary electrodes is used to excite the atomic vapour which is sputtered from the cathode by the primary discharge. For this boosting we can also use a microwave discharge; in this case a microwave antenna fed by a 2450 MHz, 120 W diathermy unit is generally applied (see also Subsection 7c-3). Booster currents are typically in the range 100 - 400 mA (see Cartwright, Sebens and Slavin 1966, and Davies 1967) while the MW boosting requires 30 - 90 W (see Human 1970, and Human, Zeegers and van Elst 1974).

A special narrow-line source for Cu- and Ba lines has been developed by van Gelder (see van Gelder 1970, and Butler and Brink 1971). In this source, a positive-column discharge is passed through the centre of a ring electrode. This electrode has a negative potential with respect to the positive column discharge. By restricting the positive column before and after the ring electrode by means of a constriction in the glass tube, a high current density is created immediately in front and behind the supplementary ring electrode. Excitation in the restricted plasma by (secondary) electrons takes place on the tube axis, where the atomic density is low (resulting in reduced self-absorption and self-reversal), but the electron density is high; in this way intense, but narrow spectral lines are obtained. Lowe (1971) describes a similar source for Ag, Cu, Au, Fe, Co, Ni and Cr.

The HCL and its modifications can easily be modulated by electronic sine-wave, square-pulse or square-wave modulation of the primary lamp current, or of the booster supply if there is one (see Goodfellow 1967, Scott and Butler 1967, Human 1970, Kielkopf 1971, Lowe 1971, Prugger, Grosskopf and Torge 1971, and Human, Zeegers and van Elst 1974). For further information on power supplies (stabilization; d.c.- or modulated operation, etc.) the reader is referred to, e.g., Scott and Butler (1967), Burgers (1969), L'vov (1970), Lowe (1971), and Human, Zeegers and van Elst (1974).

The useful life of the normal HCL's amounts to 1000 - 2000 hours of operation when used at moderate currents, or is equivalent to 3 ampere · hours (see L'vov 1970).

Data on stability of HCL's are presented in Sect. 7d (see Table III.4) together with radiance temperatures, (spectral) radiances (see Fig. III.15) and some other relevant characteristics.

One important feature of these sources remains to be discussed: the spectral shape of the source emission line, in particular its half-intensity width and, in connection with that, the possible occurrence of self-absorption and self-reversal. Several measurements of line widths and spectral line shapes have been reported in the recent literature (see Davis 1967, Human 1970, Bruce and Hannaford 1971, Jansen, Hollander and Franken 1974, and Wagenaar, Novotny and de Galan 1974; for a review see also Alkemade and Herrman 1979). (In these papers the first resonance lines of Ca, Sr, Ba, Cu and Ni were investigated.) Especially the work of Bruce and Hannaford (1971) has revealed that the main cause of extra line broadening in normal HCL's is self-absorption (see Sect. II.5c-3), and that collision broadening (see Sect. II.5b-2) does not play an important role. The following conclusions may be derived from these, mainly interferometric measurements:

(i) In a normal nonboosted HCL, the line width corresponds at low current (< 2 mA) to the Doppler width at $350-400$ K, and is independent of current in this range (see Bruce and Hannaford 1971).

(ii) At higher currents (roughly up to 15 mA), the lines are broadened by roughly a factor of $2-2.5$, and this broadening is mainly caused by self-absorption; the influence of collision broadening due to carrier gas atoms can be completely neglected.

(iii) At still higher currents (up to the maximum rating) the lines become self-reversed (see Sect. II.5c-3): the effective half-intensity width usually does not exceed $3-4$ times the Doppler width at $350-400$ K.

(iv) Square-wave or pulse modulation of the lamps sometimes narrows the line width by at most a factor of $2-3$, probably because of a reduction in self-absorption (see L'vov 1970).

(v) The main effect of boosting (d.c. or MW) on the line profile is the elimination of self-reversal; the line does not become narrower, but the central dip disappears. Furthermore, the line becomes considerably more intense because of extra excitation through the boosting (see Davies 1967, Human 1970, and Human, Zeegers and van Elst 1974). Both effects together easily yield a gain factor of $10-20$ in the lamp intensity. The van Gelder source, on the other hand, emits Cu resonance lines with a width that remains practically constant and presumably equal to the Doppler width at, say, $350-400$ K[†] with currents up to 50 mA, while the maximum permissible current is 75 mA. The gain in light intensity, as compared to conventional Cu HCL's, is here a factor of $5-15$. No directly measured line width values are yet available for this source (see also Lowe 1971).

7c-2   *LOW-PRESSURE METAL VAPOUR DISCHARGE LAMP*

These lamps (Osram-, Philips-, PenRay-lamps) are operated by an arc discharge at low vapour pressures. The lamp is filled with an inert gas, while the metal vapour is produced from the volatile metallic element by the thermal effect of the discharge. These lamps are commercially available for Hg, Cd, Zn and the alkali metals (for operation, construction, etc., see L'vov 1970, and Butler and Brink 1971). Similar lamps are also available for the noble gases.

[†] According to J. Lodde (personal communication)

LIGHT SOURCES FOR ABSORPTION AND FLUORESCENCE MEASUREMENTS   Sect. III.7c

A special modification of the low-pressure Hg lamp is the PenRay lamp (see Fluorescence News 1969, p.6), which is especially suited for wavelength calibration in the u.v., the determination of monochromator spectral resolution, etc. Because of its compact size, the source can easily be used in Zeeman-scanning experiments (see Sect. VII.3).

The lamps are mostly fed by an a.c. power supply; the working tension is 30 - 40 V and the starting tension 400 - 500 V (the power-supply specification is quoted in Rubeŝka and Moldan 1969, and Butler and Brink 1971). The currents are of the order of 1 A. These lamps can also be operated at d.c. power$^+$; source modulation can be achieved by square-wave or pulse modulation (see Lijnse and Elsenaar 1972). Typical values for the source life are 1000 hours of operation.

The lines are strongly self-reversed already at low current, which affects the line shape. The only way to prevent self-reversal (self-absorption cannot be avoided in these sources!) is to limit the current to 0.5 - 0.7 A, which is markedly less than the 'normal' operating current of 1.1 A or more. Remarkably enough quite often the light intensity is then also higher than at higher currents. Air-cooling of the lamp may prevent self-reversal and reduce the line half-intensity width considerably, i.e. by a factor of 2 or more. The only available line-width values were reported by Snider (1967) for a K lamp. The effective width of the heavily self-reversed line profiles were found to range from 78 to 151 mÅ at lamp currents from 0.5 to 1.3 A, i.e. at least one order of magnitude larger than the Doppler half-intensity width. For a more detailed discussion see Snider (1967) and Rubeŝka and Moldan (1969).

7c-3   *ELECTRODELESS DISCHARGE LAMP (EDL)*

These lamps generally have a cylindrical quartz glass envelope (diameter $\sim$ 12 mm; length 25 - 30 mm; these dimensions seem to be optimal at least for MW excited lamps; see Cooke, Dagnall and West 1971, and Jansen, Hollander and Franken 1974). The lamps contain inert gas (mostly Ar, Ne, He) at low pressure and some volatile pure metal or metal halide. Typical gas pressures are in the range 0.1 - 10 Torr, while the amount of metal (salt) ranges from 0.1 - 20 mg (see Dagnall, Thompson and West 1967, 1967a, Zacha, Bratzel, Winefordner and Mansfield 1968, Jansen, Hollander and Franken 1974, and Michel, Coleman and Winefordner 1978).

These sources are easily processed, although the reproducibility of the processing is problematic. The main problems such as those connected with light output, line width, stability and shelf life, arise from (yet unknown) chemical corrosion effects with the walls, incomplete outgassing due to the sealing off,

---

$^+$ According to Dr P.L. Lijnse (personal communication)

Chapt. III   GENERAL INSTRUMENTAL ASPECTS AND EXPERIMENTAL METHODS

role of carrier gas (for detailed information see Cooke, Dagnall and West 1971, 1971a, and Jansen, Hollander and Franken 1974). The manufacture of the lamps has therefore been the subject of many paper (see, e.g., Dagnall, Thompson and West 1967, 1967a, Dagnall and West 1968, Zacha, Bratzel, Winefordner and Mansfield 1968, Aldous *et al*. 1970, Woodward 1970, Cooke, Dagnall and West 1971, 1971a, Gleason and Pertel 1971, and Michel, Coleman and Winefordner 1978) and will not be discussed here further.

Excitation of the EDL's takes place by a radio-frequency field (see, e.g., Krause 1975) or more commonly by microwave power (usually a 2450 MHz, 120 W diathermy unit; see, e.g., Winefordner and Smith 1970, and Butler and Brink 1971). The discharge is initiated with a Tesla coil. In the case of MW excitation the lamps are excited with the aid of an antenna (see Zacha, Bratzel, Winefordner and Mansfield 1968, Mansfield *et al*. 1968, Winefordner and Smith 1970, and Butler and Brink 1971) or by placing them in $\frac{1}{2}$- or $\frac{3}{4}$-wave cavities (see Hollander and Broida 1967, 1969, Dagnall and West 1968, Mansfield *et al*. 1968, McCarroll 1970, Cooke, Dagnall and West 1971a, and Jansen, Hollander and Franken 1974; see also Fehsenfeld, Evenson and Broida 1965). The cavities have to be tuned and therefore the use of a reflected-power meter is recommended (see Hollander and Broida 1969; Jansen, Hollander and Franken 1974).

The discharge produced in the inert gas by the r.f. or MW field generates electrons which by collisions excite the metal (or noble gas) atoms in a 'cold plasma'. Very little is known about the actual processes that go on in this discharge; especially the question of the '*effective Doppler temperature*'[†] of the EDL's has still not been settled (see below).

Many attempts have been made to stabilize the light output of the EDL's. Long-term stability better than 0.5% can be obtained by preventing reflected power from being fed back towards the magnetron (see Jansen, Hollander and Franken 1974). The short-term stability can be improved by a factor of ten (see Brandenberger 1971) by replacing the mercury rectifiers in the power supply by silicon diodes. Mansfield *et al*. (1968) have measured the maximum source intensity, source fluctuations, and drift for a number of EDL's containing some 15 different metals. The short-term stabilities found ranged from 0.01% to 3% and the long-term ones from 0.01% to 5% depending on the metal used (see also Belyaev *et al*. 1968).

Square-wave modulation or a.c. modulation of the EDL output has been communicated by, e.g., Wildly and Thompson (1970), Hobbs, Kirkbright and West (1971), and Phillips (1971) (see Table III.4).

[†] By 'effective Doppler temperature' we understand the temperature value derived from the line half-intensity width through Eq. II.240, if and only if the line shape is found to be (at least approximately) Gaussian.

LIGHT SOURCES FOR ABSORPTION AND FLUORESCENCE MEASUREMENTS   Sect. III.7c

Very little is known about the shelf-life of EDL's. Pure metallic lamps seem to be much better in this respect than lamps filled with metal salts. Gleason and Pertel (1971) have produced a Hg EDL with Ar fill gas with a stability of better than 0.1%, that could be operated continuously over a period of 50 hours. Especially the properties of the alkali and alkaline-earth lamps are rather unpredictable and may show large variations from sample to sample (see Silvester and McCarthy 1970, and Jansen, Hollander and Franken 1974). It should be noted that several methods are being tried out to improve the stability and light output of the lamp by heating the lamp walls with heating tape (see Hollander and Broida 1969), by insulating the lamps either by surrounding them with a vacuum jacket (see Winefordner and Smith 1970), or simply by wrapping them in glass wool (see Mansfield et al. 1968, and Winefordner and Smith 1970). Browner and Winefordner (1973) have reported a new thermostat technique which makes use of heated Ar or N_2 gas flowing through the 'vacuum jacket'.

EDL's are especially suitable as primary light sources in Zeeman-scanning experiments because of their compactness (see Sect. VII.3).

We know but little about the line shapes and widths of the EDL's: only a few interferometric measurements have been reported in the literature (see Kirkbright and Sargent 1970, and Jansen, Hollander and Franken 1974). There are indications that the Doppler temperature of the EDL's is not higher than about 400 K, but in practice self-absorption makes the lines 2 - 8 times broader than the Doppler width that corresponds to this temperature value (see Kirkbright and Sargent 1970, and Jansen, Hollander and Franken 1974).

As an example, we mention the results obtained for a series of Sr lamps containing different Sr salts and different fill gases. Line profile measurements with an interferometer (apparatus half-intensity width ≈ 3 mÅ) revealed that the profile of the 4607.33 Å resonance line varied from a Doppler profile at 400 K ($\delta\lambda < 7$ mÅ) to heavily self-reversed profiles with a half-intensity width between 50 and 60 mÅ, depending on the metal content, microwave power and cooling conditions (see Jansen, Hollander and Franken 1974; cf. also Kirkbright and Sargent 1970 for similar observations on the Ca 4227 Å line). Self-reversal shows up in the EDL's when self-absorption has broadened the line to much larger widths (about 50 mÅ) than in the case of HCL's (about 15 mÅ). Relevant parameters appeared here to be the amount of metal salt in the lamp, the melting point of the salt, and the cooling of the lamp (see also Browner et al. 1972). A reduction of the line half-intensity width by about 35% could be obtained by firm air cooling without noticeable loss in intensity. Self-reversal of the lines could be prevented by keeping the metal content in the tubes below about 1 mg (see Jansen, Hollander and Franken 1974) and by firm cooling (see also Dagnall and West 1968).

Chapt. III   GENERAL INSTRUMENTAL ASPECTS AND EXPERIMENTAL METHODS

With the above series of Sr lamps the light output was about equal to that of a HCL at 10 mA; the line width was about 8—10 mÅ (compared to 10 mÅ for a HCL). Increase of the MW power did not noticeably affect the line width as long as firm air-cooling was applied and the amount of metal was small enough (see above). The light output of the EDL's could be improved by two orders of magnitude while the half-intensity width increased only to about 50 mÅ at which point self-reversal just sets in (see Jansen, Hollander and Franken 1974). This gain in light output was in an incidental case achieved by increasing the metal content in the EDL from below 1 mg to about 10 mg.

In Sect. 7d we shall summarize some characteristics of the EDL's. It should be kept in mind, however, that it is much more difficult to characterize the EDL than the HCL or any other line source.

7c-4   *LASERS*

The use of lasers, especially tunable dye lasers, in atomic fluorescence spectrometry has been reported by, e.g., Martin and Thomas (1967), Denton and Malmstadt (1971), Fraser and Winefordner (1971, 1972), Loth, Astier and Meyer (1972), and Kuhl, Marowsky and Torge (1972). For an introduction and survey of analytical laser spectroscopy see Omenetto (1979). The dye laser cell is pumped by a continuous (c.w.) or pulsed Ar$^+$ laser, a pulsed N_2 laser or a Xenon flashlamp (see Loth, Astier and Meyer 1972, Kuhl, Marowsky, Kunstmann and Schmidt 1972, and Kuhl, Marowsky and Torge 1972).

The tuning range of dye lasers is roughly between 3500 and 7000 Å, depending, inter alia, on the kind of dye. Spectral half-intensity widths as low as 1 mÅ can be attained by using etalons, but the dye lasers commonly used in atomic fluorescence measurements possess half-intensity widths of the order of 1-10 Å.

A comparison of total radiation flux values from dye lasers with those from other sources is given in Sect. 7d. It should be noted that only approximate values of the solid angle are known; Loth *et al.* have reported an angular beam divergence of 1 mrad. The pulsed dye lasers applied in fluorescence measurements may have a pulse duration as low as 2 ns, while the peak radiant flux of the system can easily exceed 10 kW (see Fraser and Winefordner 1971). For illustration we compare some typical pulsed tunable dye lasers in Table III.3.

7c-5   *OTHER SPECTRAL LINE SOURCES*

A *flame* may serve as a simple primary light source for absorption measurements (see Alkemade 1954, 1968, James and Sugden 1955, Hollander 1964, and Rann 1968). For fluorescence measurements its intensity is too low: a flame seeded with high metal concentrations may attain at most a spectral radiance in the centre of the atomic resonance lines corresponding to that of a black body at the flame

Table III.3

Some Characteristics of Pulsed Dye Lasers [+]

	N_2-pumped laser [1]	Xe-flashlamp pumped laser [2]	Xe-flashlamp pumped laser [3]
Peak output power (kW)	10	0.5	5.9 [‡]
Mean output power (mW)	0.5 – 1 mW (at 10 Hz repetition frequency)	0.5 – 1 mW (at 10 Hz repetition frequency)	1 mW (at 1 – 3 Hz repetition frequency)
Pulse duration (s)	$(2-8) \times 10^{-9}$	2×10^{-7}	$(3-10) \times 10^{-7}$
Rate (Hz)	1 – 25	1 – 20	Normally 1 – 3; possibly up to 100 [*]
Spectral width (Å)	1 – 10	1 – 2	$(3-5) \times 10^{-3}$
Tuning range (Å) (with different dyes)	3600 – 6500	4500 – 4700; 5700 – 6000	4500 – 4700; 5700 – 6000
Beam divergence (mrad)	⪆ 5	≈ 1	≈ 1
Beam cross section (mm²)	⪆ 40	≈ 40	≈ 40

[+] The characteristics listed should not be considered as fixed figures; they depend on the properties of the spectral filters inside the laser cavity, etc. For example, Xe-flashlamp pumped lasers have been operated at a higher peak power (≈ 100 kW), a lower repetition rate (≈ 1 Hz) and with an intermediate spectral width of ≈ 150 mÅ (see van Dijk 1977, and van Calcar et al. 1979).

[*] When repetition frequencies up to 100 Hz are used, the peak output power will drastically decrease as compared to frequencies of 1 – 3 Hz.

[‡] This (Zeiss) laser is available now with 100 – 120 kW peak power at 1 – 2 Å bandwidth and 1 Hz repetition rate. All power data quoted in Table III.3 can thus be multiplied by a factor of 20 for this particular source.

[1] Fraser and Winefordner (1971).

[2] Loth, Astier and Meyer (1972).

[3] Schmidt (1970), Kuhl, Marowsky, Kunstmann and Schmidt (1972), and Kuhl, Marowsky and Torge (1972).

temperature (see Sect. II.5c-3).

It should be noted that the primary combustion zone of a nonseeded flame has been used as a primary source for, e.g., OH absorption measurements.[†]

An advantage of the flame is that we can easily control the half-intensity width of its self-absorbed resonance lines by simply varying the concentration of the metal solution sprayed (see James and Sugden 1955, and Alkemade 1968; see also Sect. II.5c-3). In this way this half-intensity width can be made to exceed that of the absorption line by a factor of the order of ten; the flame then acts, virtually, as a quasi-continuum source. A disadvantage is, of course, that a lower limit is set for the half-intensity width of the emission line by the relatively large Doppler width at the flame temperature.

Carbon arcs and *sparks* have also been used as primary light sources. For information on these sources the reader is referred to the literature (see, e.g., Butler and Brink 1971, and Boumans 1966, 1972).

Finally we mention the use of resonance fluorescence from an *atomic beam* as a background line source for absorption measurements. This source has a very poor light output, depending on the primary light source used. When the fluorescence radiation from the source is observed perpendicular to the beam direction, extremely small half-intensity widths can be obtained, which are 1 or 2 orders of magnitude smaller than the Doppler width. The application of the atomic beam to absorption measurements is mostly restricted to those cases where photon counting is available for the detection of the weak signals.

7d SURVEY OF CHARACTERISTICS OF SPECTRAL LINE
AND CONTINUUM SOURCES

In Fig. III.15 the spectral radiances of various spectral continuum and line sources are shown. For the line sources a half-intensity width of 30 mÅ is assumed in order to reduce their measured radiances to *spectral* radiances (see Appendix A.5) for comparison with those of the continuum sources.

The spectral radiance of the high-pressure Hg-arc lamp as a function of λ can be derived from Fig. III.14. The radiant flux from the Eimac 150 W Xe-arc lamp (see above) is about ten times higher.

The absolute spectral radiances of the various modifications of the normal HCL can easily be estimated from the relative intensity data cited in Sect. 7c-1 in combination with the absolute values shown in Fig. III.15 for the normal HCL.

The *mean* output power of the N_2-laser pumped and the Xe flashlamp pumped pulsed dye lasers corresponds to about 1 mW (see Table III.3). The spectral bandwidth used is mostly about 2 Å and the solid angle of the laser beam about 10^{-6} sr.

[†] According to Dr H.P. Broida (personal communication).

LIGHT SOURCES FOR ABSORPTION AND FLUORESCENCE MEASUREMENTS Sect. III.7d

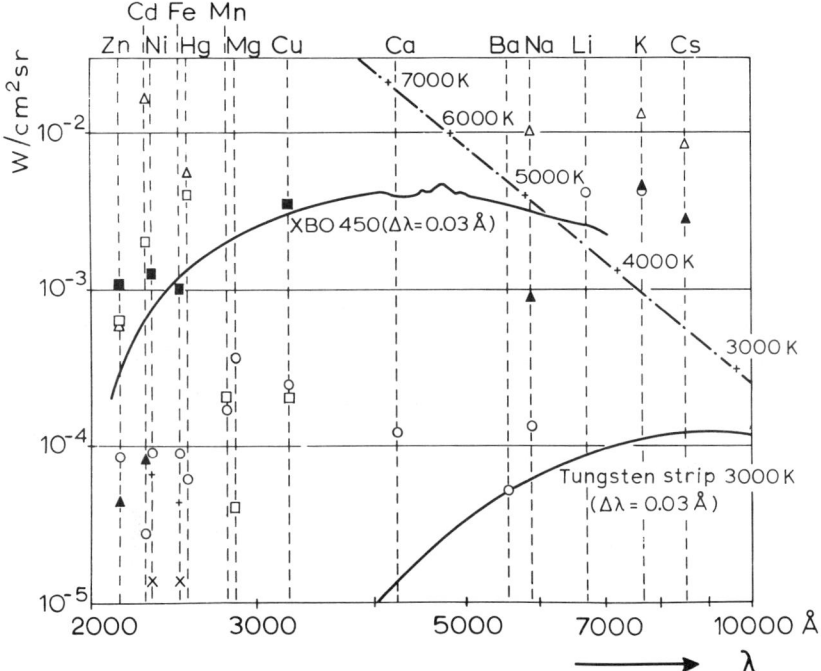

Fig. III.15 Comparison of radiances of some HCL's (0 - 0),'high-intensity' lamps (= modified HCL) without (X - X) and with (+ - +) electric boosting, low-pressure metal vapour discharge lamps with specified current (Δ - Δ) or reduced current (▲ - ▲) of 0.8 A with Cd, Cs, K and Na, and 1.3 A with Zn. Concentrated glow-discharge (■ - ■) and EDL (□ - □). For comparison the radiance of two typical continuum sources are presented (with Δλ = 0.03 Å): a black-body at the wavelength of maximum emission (+ - · - · +), a Xe-arc XBO 450, and a tungsten strip lamp at 3000 K. (Derived from Prugger 1969.)

The radiant flux of a 450 W Xenon lamp emitted at 5000 Å under the same solid angle and within the same spectral bandwidth is calculated to be about 2×10^{-4} mW (see Fig. III.15). Since, however, the useful solid angle of the Xe lamp may be as large as 1 sr, we can increase the latter value by roughly 10^6, so that its total radiant flux may still exceed the mean laser flux by a factor of about 200. However, during the pulse (ranging from 2 to 1000 ns) the output power of the laser exceeds that of the Xe lamp by some orders of magnitude (see the peak output-power data in Table III.3).

Tunable dye lasers, pumped with a c.w. Ar-ion laser, Xe flashlamp or N_2 laser, are also available with a much smaller half-intensity width of about 3 mÅ as compared to the value of 2 Å quoted above (see, e.g., Walther and Hall 1970, Hänsch,

Chapt. III GENERAL INSTRUMENTAL ASPECTS AND EXPERIMENTAL METHODS

Varsanyi and Schawlow 1971, Kuhl, Marowsky, Kunstmann and Schmidt 1972, Schieder, Walther and Wöste 1972, and Wallenstein and Hänsch 1975).

For the c.w. dye laser the output power may range from about 1 mW to 0.5 W, depending on the spectral bandwidth. With pulsed dye lasers one may attain 100- 120 kW peak power (for a pulse duration of 1 µs) at a bandwidth of the order of 0.1 - 1 Å. When reducing the bandwidth by a factor of, say, 100 - 1000, one reduces this peak power by roughly the same factor (see Wallenstein and Hänsch 1974, and Hänsch 1972).

The radiant flux of c.w. lasers again compares favourably with that of the most intense conventional continuum and line sources.

The very high peak power of N_2-pumped lasers, being a consequence of the very small pulse duration (typically of the order of 1 ns), may enhance the 'useful radiant flux' of these sources by several orders of magnitude when adequate electronic detection, such as a boxcar integrator or sample-and-hold circuit, is available (see Fraser and Winefordner 1971, 1972). A considerable gain in signal-to-noise ratio can then be obtained.[†]

In Table III.4 we summarize some useful characteristics of the sources discussed.

NOTES TO TABLE III.4

[†] The data represent the best obtainable with d.c. or a.c. operation. The Xe lamp and Eimac lamp data refer, however, to their 'normal' use, i.e. without optical-feedback stabilization. Most figures quoted are order-of-magnitude estimates.

[‡] P - square-wave and S - sine-wave modulation, as reported in the literature.

[*] The data refer to square-wave modulation.

[§] The shape of this area is not specified.

[⊕] The data given (except useful radiating area) are also valid for arc lamps operated at higher powers.

[⊗] According to Dr G.M. Hieftje (personal communication 1972) an increase in signal-to-noise ratio of a factor of 4 is obtained in absorption measurements with electronic square-wave modulation as compared to sine-wave modulation.

[†] Weide and Parson (1972) have reported a greatly improved signal-to-noise ratio in atomic fluorescence flame spectrometry of Zn with a pulsed hollow-cathode lamp in combination with a boxcar integrator read-out. The gain in signal-to-noise ratio as compared to the conventional HCL system was found to be a factor of 70 - 100.

Table III.4

Some Characteristics of Light Sources used in (Atomic) Spectroscopy

Type	Long-term[+] instability (%)	Short-term[+] instability (%)	Electronic modulation			Useful solid angle (sr)	Useful radiating area[5] (mm^2)
			Type[‡]	Maximum frequency (kHz)[*]	Maximum depth (%)		
Tungsten strip lamp	0.01	0.01	-	-	-	≈ 0.5	20
(Commercial) Xe-arc lamp⊕ (150 W)	2 – 5	1	P	5	75	≈ 0.5	4
(Commercial) Eimac Xe-arc lamp⊕ (150 W)	2 – 3	1	P	5	75	6	4
(Commercial) Hg-arc lamp⊕ (150 W)	2 – 3	1 – 2	⊗P,S	-	100	≈ 0.5	10
Normal HCL	1	0.1	P	2	90 – 100	≈ 0.5	25
D.c. boosted HCL	1	0.1	P	2	100	≈ 0.5	200
Microwave boosted HCL	1	0.1	P,S	-	-	≈ 0.5	200
Van Gelder HCL	0.5	0.5	-	-	-	≈ 0.5	25
Metal Vapour lamp	1	0.1	P,S	>10	100	≈ 0.5	100
EDL	1	0.1 – 0.5	P,S	20	50 – 100	≈ 0.5	250
Pulsed dye laser	≥ 1	≥ 1	pulsed (see Table III.3)			10^{-6}	10
C.w. laser (Ar$^+$-pumped)	0.1	≤ 0.1	-	-	-	10^{-6}	0.25

Chapt. III GENERAL INSTRUMENTAL ASPECTS AND EXPERIMENTAL METHODS

8. MEASUREMENT OF THE NEBULIZATION EFFICIENCY

8a   GENERAL

In fundamental flame studies a knowledge of the total number density, n_t, of a metallic additive in the flame is sometimes required. Here n_t refers to the presence of the element in the gaseous state in any form, that is, as free neutral atoms, ions, or molecular compounds. This total density can be derived from the amount of element supplied to the flame per second. Formulas relating n_t to the supply rate of element will be presented in Chapt. V (see Eqs V.17-19). In this section we shall be concerned with the experimental determination of the supply rate in the case when a metal salt solution is introduced as an aerosol through a pneumatic nebulizer (see Sect. III.2e). We shall discuss in a separate section which fraction of the element introduced is actually transformed into the gaseous state at the observation height in the flame (see Sect. III.9).

The rate at which an element is supplied to the flame would be simple to calculate by multiplying the *rate of solution aspiration* F_1 (in cm^3 s^{-1}) by the concentration, c, of the element in the solution if all the solution aspirated by the nebulizer actually reached the flame. With direct-injection burners (see Sect. III.2a) all the aspirated solution does indeed leave the burner port, but there may be losses of aerosol droplets that are thrown beyond the flame because of the turbulent motion of the gases leaving the burner (see Sect. V.5a). In chamber-type nebulizers (see Sect. III.2e), a large part of the mist droplets is usually caught by the walls of the spray chamber and then drains away. Droplets may also be deposited on the inner walls of both the gas conduit to the burner and the burner house. In this section we shall focus attention on the latter type of nebulizer, which is most commonly used in flame studies involving metal additives.

To describe these aerosol losses quantitatively, the *efficiency of nebulization* ε_n is defined as the ratio of the amount of metal salt entering the base of the flame per second to the amount of metal salt aspirated per second as a solution into the nebulizer. The supply rate of metallic element to the flame is then given by the product: $\varepsilon_n F_1 c$. It is mainly the factor ε_n which introduces an uncertainty into the supply rate. Since one cannot calculate this factor theoretically, one must determine it experimentally (see Sect. 8b).

It should be noted that in the literature different definitions have often been used, and confusion exists as to the terminology and choice of symbol. The definition adopted here conforms to the IUPAC recommendations (IUPAC 1976a) in that it does not refer to the amount of __solvent__ but to the amount of __solute__ reaching the flame. In our definition, ε_n also includes the aerosol losses in the gas conduit and in the burner house.

MEASUREMENT OF THE NEBULIZATION EFFICIENCY   Sect. III.8b

The value of ε_n depends on the design and operating conditions of the nebulizer-burner system as well as on the kind of solvent. These factors determine the original size-distribution of the mist droplets formed at the tip of the sprayer, the change in droplet diameter through subsequent evaporation and/or recombination (coalescence) of droplets, the fraction of droplets caught by the walls, etc. These effects may depend on the viscosity, surface tension and saturation pressure of the solvent vapour. In aqueous solutions, the kind or concentration of solute appears to have virtually no influence on ε_n if the concentration is well below 0.1-1 mol/ℓ (see Hinnov 1957, and Hofmann and Kohn 1961). This implies that the nebulization efficiency is independent of solution concentration for low concentration values; the nebulizer is then said to operate *linearly*, i.e., the supply rate of metal salt to the flame is proportional to the solution concentration. A possible nonlinear operation in the high-concentration range may distort the curve that relates, e.g., the metal emission intensity to the solution concentration. The linearity of the nebulizer can be checked by adding to the solutions a second element at a constant, low concentration and by observing its emission or absorption signal in the flame. This method works only if the two elements do not interfere with each other in the flame.

The factors ε_n and F_1 are in general mutually dependent. When a wider or shorter inlet capillary is chosen to increase the rate of aspiration, the efficiency of nebulization is often found to deteriorate. The supply rate of element to the flame should thus not be considered as being simply proportional to the liquid aspiration rate.

For a more comprehensive discussion of the nebulization process, its efficiency and methods of improving the efficiency, the reader is referred to the literature (e.g., Dean 1960, Mavrodineanu and Boiteux 1965, Christian and Feldman 1970, Alkemade 1970a, Herrmann 1971, and Alkemade and Herrmann 1979), where references to the original papers are also found.

8b   METHODS OF MEASUREMENT

Basically, there are three kinds of methods for determining ε_n by experiment:

1.   *Reflux methods* in which the amount of metal salt in the solution drained from the spray chamber is determined.

2.   *Trapping methods* in which the aerosol leaving the spray chamber or the burner ports (with unlit flame) is trapped and its metal content is determined.

3.   Derivation of the metal supply from an independent *absolute calibration* of the atomic metal concentration in the flame.

Chapt. III   GENERAL INSTRUMENTAL ASPECTS AND EXPERIMENTAL METHODS

The latter method is based on the inversion of Eqs V.17-19 which relate the total metal concentration in the flame to the supply of metal salt from the nebulizer. The application of the method requires that the test element be fully atomized (the metal should be present only in atomic form in the flame; see Sect. V.6) and that the aerosol be completely desolvated and volatilized (see Sect. III.9). A discussion of this method, including the absolute determination of metal concentrations in flames, will be given in Chapt. V.

In order to find ε_n by the reflux method, the difference between the metal salt content in the drained solution and the amount of metal salt aspirated is determined. With the trapping method one calculates the ratio of metal salt content in the trapped aerosol to the amount aspirated. The major difficulty in the former method is that the final result is usually obtained as a relatively small difference between two experimental quantities of about equal magnitude (ε_n is often 10% or less). The major problem with the second method may be an insufficient trapping efficiency (see below). Care should be taken with the insertion of the trapping device in the gas stream: one does not want to alter the properties of the aerosol to be trapped. In the reflux method and also when the aerosol is trapped directly behind the spray chamber, corrections should be made for the additional losses of aerosol in the gas conduit and the burner house. This can be done by washing the interior parts of the conduits and the burner and by analyzing the metal concentration in the washings. This analysis may be conveniently carried out by flame spectroscopy or by weighing the residue of metal salt after evaporation of the washings when a concentrated test solution is used (see Willis 1970).

The reflux method is usually applied by measuring the volume of a test solution drained continuously from the spray chamber during a known time interval (see Alkemade 1954, Hollander 1964, de Galan and Winefordner 1967, and Willis 1967, 1970). Because of partial evaporation of the drained solution inside the spray chamber, its metal concentration may be larger than in the original solution aspirated and a correction must be made (see Alkemade 1954, and Rann 1968). This may be done by analyzing the drained solution as described above for the washings. Since the draining does not usually proceed smoothly, a long draining period is recommended for improving the statistical error. A test element should be chosen which is not contaminated by the glassware and which provides good sensitivity in flame spectroscopic analysis. Mg-, Cd- and Mn solutions have been recommended in the literature.

As a special variant of the reflux method, Rann (1968) has applied recirculation of a given volume of test solution. The drained solution is aspirated again, and this process is repeated until the volume is reduced to a small fraction of the original. The increase in concentration in the recirculation solution due to partial evaporation (which may be monitored during the experiment

MEASUREMENT OF THE NEBULIZATION EFFICIENCY   Sect. III.8b

by flame spectroscopy) is compensated for by adding small amounts of solvent. From the difference between the quantity of metal salt in the original test volume and that in the remaining volume as well as in the washings of the gas conduits and burner house, the quantity of metal supplied to the flame is calculated. Comparison of this quantity with the quantity aspirated in the same time interval yields ε_n. An advantage of the recirculation method is that ε_n is now found from a relatively large difference between two experimental quantities.

Trapping methods can be realized simply with filters of glass wool or $CaCl_2$ traps (see Gurvich and Ryabova 1964a). To ensure maximum trapping efficiency, however, more elaborate equipment has been recommended. Electrostatic precipitation of the aerosol droplets has been applied successfully by Cotton and Jenkins (1969). When the aerosol is carried by a combustible gas mixture or when inflammable solvents are used, this method is not without danger. Aerosol particles can also be very efficiently trapped by steam nucleation and subsequent hot water gas-scrubbing (see Johnson and Smith 1972). This is achieved by mixing the gas stream containing the aerosol with steam (which condenses on the aerosol particles) and by letting the gas bubble through a hot water column. The metal salt concentration in the eluate is then analyzed by flame spectroscopy.

In a special case the metal content in the aerosol trapped by a filter has been determined by measuring the radioactivity of the filter after a salt solution of ^{60}Co or ^{59}Fe was nebulized (see Uny and Spitz 1970). This method requires additional equipment and special precautions.

Nebulization efficiencies reported in the literature are usually within the range of a few per cent to 15% for chamber-type nebulizers operated at room temperature and for aqueous solutions (see the literature cited in Alkemade 1970a; see also Sect. III.2). Table III.1 lists *conversion factors* (total number density of metal in the flame corresponding to 1M solution concentration) reported in the literature for some typical nebulizer-flame systems. Accidental errors as low as a few per cent in the determination of ε_n by reflux or trapping methods have been claimed (see de Galan and Winefordner 1967, Uny and Spitz 1970, and Johnson and Smith 1972). However, the systematic error here might be appreciably greater, depending on the method and equipment used and the skill of the experimenter. The systematic error may be checked by comparison with the result obtained by method no. 3. With the latter method, however, other systematic errors may arise. Systematic errors are unlikely, however, if all three methods yield the same results with different test elements.

Chapt. III   GENERAL INSTRUMENTAL ASPECTS AND EXPERIMENTAL METHODS

9. RELEASE OF METAL VAPOUR FROM AEROSOL

To obtain a metal vapour in the flame, wet aerosol droplets, produced by the nebulizer, must first be desolvated by evaporation of the solvent. Next the dry aerosol consisting of a suspension of solid or molten particles of the solute must be volatilized.

Desolvation and volatilization will be discussed separately in this section and methods for checking (in-)complete desolvation or volatilization in practice will be dealt with.

9a   DESOLVATION OF WET AEROSOL

When the desolvation of aerosol droplets is not complete in the flame, the fraction of desolvated aerosol should be taken into account when calculating the total metal concentration in the gaseous state for a given solution concentration (see Sect. V.5a). The *fraction desolvated*, β_s, is defined as the ratio of the amount of metal element passing the total horizontal flame cross section per second at the observation height in the desolvated state (i.e. either as dry aerosol or as vapour) to the total amount of metal element passing (see IUPAC 1976a). When β_s varies with observation height we speak more appropriately of the *local* fraction desolvated. This fraction may depend on the size distribution of the aerosol droplets entering the flame, the rate of desolvation which is a function of the size and composition of the droplets, the height of observation, and the rise-velocity of the flame gas. The two latter factors determine the time spent by a droplet in the flame before it enters the observation volume. In chamber-type nebulizers, the relevant size distribution may differ appreciably from the initial distribution at the tip of the sprayer inside the nebulizer because of subsequent evaporation (and possibly coalescence) of the droplets in the spray chamber, etc., and because the grosser droplets are caught by the chamber walls. For this type of nebulizer, the mean droplet size is usually so drastically reduced by these effects (coalescence plays only a subordinate role) that desolvation of the aerosol in the flame can be considered as complete, in contrast to the case of direct-injection burners. For a more comprehensive treatment of the desolvation process and the fraction desolvated, the reader is referred to the literature (e.g., Mavrodineanu and Boiteux 1965, Zeegers, Smith and Winefordner 1968, Hieftje and Malmstadt 1968, 1969, Alkemade 1969, 1970a, Clampitt and Hieftje 1972, and the general textbooks on analytical flame spectroscopy listed in Chapt. I).

The extent of desolvation in flames is governed by the rate at which heat is transferred from the ambient flame gas (with temperature T_f) to the boiling droplet (with boiling temperature T_b). Under certain simplifying conditions, we have for the loss of droplet mass m per unit time the theoretical relation (see

RELEASE OF METAL VAPOUR FROM AEROSOL   Sect. III.9b

van der Held 1932, and Williams 1965)

$$-dm/dt = (4\pi r\lambda/c_p)\ln\left\{1+c_p(T_f-T_b)/L\right\},\qquad(\text{III.42})$$

where r = droplet radius, c_p = specific heat capacity of solvent vapour at constant pressure, λ = thermal conductivity of the flame gas, and L = specific heat of vaporization. Since m is proportional to r^3, it follows by integrating this equation that the total time required for complete desolvation is proportional to the initial droplet surface area.

Calculations show that water droplets with initial diameter equal to 1 µm require approximately a time of 3 µs for complete evaporation in a flame of median temperature. If the rise-velocity of the flame gas equals 10^3 cm s^{-1}, all droplets with diameter of at most 1 µm will be completely evaporated before they have travelled over a distance of 0.03 mm. However, water droplets with initial diameter exceeding 30 µm, will not release their metal content when the height of observation is less than 30 mm. This example shows the order of magnitude of the initial droplet size required for efficient desolvation.

9b METHODS FOR CHECKING (IN-)COMPLETE DESOLVATION

(i) Using Eq. III.42 we can calculate, in principle, the fraction of aerosol that is desolvated at a given height in the flame, if the drop-size distribution at the flame base and rise-velocity are known. The size distribution may be measured be collecting the droplets leaving the burner (with flame off) on a glass slide coated, e.g., with magnesium oxide or on a film containing silver nitrate (see Ueno and Sato 1971; for a general discussion of experimental methods see Lapple 1960). The diameters of the droplets collected can be derived from their impact impressions on the slide, which can be measured with a microscope (see Dean and Carnes 1962). From the size distributions of water droplets measured with a direct-injection burner by Dean and Carnes (1962) it can be calculated that an appreciable fraction of the aerosol passes the first few centimetres above the burner tip without being completely evaporated. On the other hand, the diameters of the droplets emerging from a Méker burner combined with a chamber-type nebulizer have been reported to be below 1 µm (see Alkemade 1954), so that complete desolvation of the aerosol may be expected close to the flame base (see also Sect. 9a). Probably an appreciable fraction of these droplets may already be evaporated during their passage through the (thin) primary combustion zone. This may have important consequences when the released metal vapour takes part in nonequilibrated reactions occurring in this zone.

(ii) A more direct, experimental method to check the (in-)completeness of desolvation is based on the observation of scattering of an intense light beam by the

Chapt. III   GENERAL INSTRUMENTAL ASPECTS AND EXPERIMENTAL METHODS

unevaporated droplets in the flame. It should, however, be noted that only droplets with diameters exceeding the wavelength of the light contribute significantly to this scattering. This method has been applied to demonstrate the presence, and even to estimate the fraction of unevaporated droplets in unpremixed turbulent flames produced by a direct-injection burner (see Gibson, Grossman and Cooke 1963, Püschel, Simon and Herrmann 1964, and Parsons and Winefordner 1966). Also the droplets that are hurled outside the turbulent flame at the rim of the burner can be detected in this way. Light scattering by unevaporated droplets in the flame is generally absent when a chamber-type nebulizer is used (see Slevin, Muscat and Vickers 1972). It also appears to be markedly reduced with direct-injection burners when a volatile organic solvent is used or when the dispersion characteristics of the nebulizer are improved. These findings have practical consequences in analytical flame fluorescence spectroscopy where scattering of the primary radiation may cause erroneous signals.

(iii)   Indirect experimental evidence about the (in-)completeness of desolvation can be gained by measuring the relative variation of atomic metal concentration with height in the flame by one of the methods outlined in Chapt. V. In a homogeneous flame with uniform thickness and rise-velocity, the gradual progression of desolvation would make the atomic metal concentration increase with height if $\beta_s < 1$. It is here assumed that the metal is fully atomized (see Sect. V.6) and that no delay in the volatilization occurs (see below). When the radial distribution of the metal concentration varies also with height, one should consider the integral of the atomic metal concentration over the flame cross section as a function of height. Measurements of this kind with premixed flames combined with a chamber-type nebulizer have shown that practically all sodium nebulized into the flame is released in the gaseous state at the flame base (see Rann and Hambly 1965, and Alkemade 1970a). Atomic Na concentration distributions measured in an unpremixed $H_2 - O_2$ flame with a direct-injection burner have revealed a marked increase of β_s with height for an aqueous solution (see Simon 1960).

(iv)   Conclusions as to the state of desolvation in turbulent flames with a direct-injection burner have also been drawn from absolute measurements of the concentration distribution of a fully atomized element with suppressed ionization (see Simon 1960, and Püschel, Simon and Herrmann 1964). The shortage of free atoms found in the flame when compared to the known absolute quantity of metal salt nebulized was ascribed to incomplete desolvation of the droplets and to ejection of droplets outside this turbulent flame (see also Sect. V.5a). The concentration of the salt solution nebulized should be low enough to ensure that no delay in volatilization does occur.

RELEASE OF METAL VAPOUR FROM AEROSOL   Sect. III.9c

9c   VOLATILIZATION OF DRY AEROSOL

Dry aerosol particles will volatilize in the hot flame gases. Since volatilization takes time, it may be incomplete at the height of observation. The (*local*) *fraction volatilized*, β_v, is defined as the ratio of the amount of metal element passing through the flame cross section at the observation height in the gaseous state to the total amount of metal element passing in the desolvated state (see IUPAC 1976a). Incomplete volatilization may be the result of the low volatility of the dry aerosol particles, their large initial dimensions and insufficient time spent in the flame. With chamber-type nebulizers the gross mist droplets are separated, only the finer droplets containing relatively small quantities of salt enter the flame. Then volatilization is favoured by the smaller dimensions of the dry aerosol particles. In direct-injection burners, which feed all of the aspirated solution into the flame, the larger average dimensions of the dry aerosol particles and often the shorter time spent in the flame may prevent complete volatilization.

Incomplete volatilization may have some consequences in fundamental as well as in analytical flame spectrometry. For instance, the total metal content in the vapour phase calculated from the measured efficiency of nebulization (see Sect. V.5) will be too high. Nonequilibrated chemical processes in the inner zone are prevented from affecting directly the initial state of the metal vapour in the flame. Experimental curves-of-growth in the range of high salt concentrations will be distorted. The analytical curves for nonresonance lines and molecular bands, which are free from self-absorption, will still show a convex curvature in the high-concentration range. Furthermore, deterioration of the analytical detection limit may result. In flame emission spectrometry continuum emission of incandescent solid particles may cause spectral interference. In flame atomic absorption and especially in flame atomic fluorescence spectrometry, scattering of the primary radiation may lead to erroneous results. When the volatility of the dry aerosol particles depends on the chemical composition of the aspirated solution, solute-volatilization interferences can occur. A well-known example of this interference is the strong depression of the alkaline-earth emission or absorption signal by the addition of phosphor or aluminium compounds to the solution (see Alkemade and Jeuken 1957, and Alkemade and Voorhuis 1958). Solid particles may also interfere with the ionization of metal atoms, as has been suggested for Cr particles in H_2 flames by Jensen and Padley (1967).

This subsection will be confined to a brief qualitative discussion of the formation of dry aerosol particles in general and of the factors that control their volatilization. A more complete description and explanation of volatilization effects is found in L'vov (1966, 1970), Alkemade (1966, 1969, 1970a), Rains (1969), Rubeška (1971), and Alkemade and Herrmann (1979).

Chapt. III GENERAL INSTRUMENTAL ASPECTS AND EXPERIMENTAL METHODS

Formation of dry aerosols. In the simple case of an aqueous NaCl solution, one NaCl particle will be formed from each desolvated mist droplet. Since the melting and boiling points of NaCl (1075 and 1685 K, respectively) are lower than the temperature of most flames, these NaCl particles will be converted readily into the gaseous state, either as NaCl molecules or directly as atoms.

In the general case, the composition and state of the dry aerosol particles may depend on the form and concentration in which the element was present in the solution, as well as on the kind of solvent and concomitants in the solution, and the temperature and composition of the flame gas. In complex solutions competitive chemical reactions and fractional distillation in the aqueous phase might also be important.

For instance, an aqueous solution of $MgCl_2$ may lead to particles consisting of $MgCl_2 \cdot H_2O$ that can be decomposed after dehydration into solid $MgCl_2$ (with boiling point 1685 K) as well as into involatile solid MgO (see Halls and Townshend 1966). Magnesium nitrate decomposes to MgO at 594 K in the solid state. However, if the nitrate is heated rapidly, it may melt and volatilize before it forms oxide.

The extent of oxide formation in the condensed phase still depends on the environmental conditions in the flame. Under reducing flame conditions the formation of oxides is inhibited and the transformation of the metal solute into atomic vapour is enhanced when the oxide is involatile. In the hot, fuel-rich $C_2H_2 - N_2O$ flame, molybdenum oxide may be partly reduced by hydrocarbon radicals and hydrogen to molybdenum carbide and/or pure metal in the condensed phase (see Rubeška 1975). Since for this element the metal and the carbide are less volatile than the oxides, this reduction may hamper its volatilization even in hot flames.

The sequence of processes occurring when a particle is heated to flame temperature is still largely unknown. There is little information available about the nature, size and state of the dry aerosol particles under flame conditions (see also the discussion by Bulewicz following the paper of Soundy and Williams 1965).

We shall not endeavour to discuss in detail the (pyro-)chemical and physical processes that may play a role in the formation of dry aerosols. Such processes may include, e.g., hydrolysis, complex formation, ion pairing, precipitation and surface adsorption (see Thomas and Pickering 1971). The discussion of these processes is considerably complicated by their fast evolution in the flame. Here equilibrium considerations will hardly apply.

More experimental evidence is wanted. This may be gained, e.g., from electron microdiffraction analysis of particles captured from the flame. Shock-wave studies may be especially helpful since they enable observations to be made at a much shorter time scale than in flames. They can thus yield additional information on the rate of volatilization of solid particles at high temperatures. It was found that a noticeable fraction of NaCl was volatilized within a few microseconds after passage of the shock front (see Clouston, Gaydon and Glass 1958). Complete volatilization of a colloidal dust of pure Na seems to be attained within 30 µs (real gas time) after passage of the shock front. This follows from the observation that the Na vapour concentration remains constant in the shocked gas after that period (see Tsuchiya 1964). The latter observations might help to solve the question whether an aerosol containing a solid metal compound could volatilize during the very short time of its passage through the primary combustion zone. For premixed flames burning at 1 atm pressure this time is of the order of 10 µs. The release of metal vapour inside the combustion zone may be enhanced when the spray droplets on their way from the spraychamber to the burner outlet have got enough time to be desolvated before entering this zone. In low-pressure flames the combustion zone may be thicker by several orders of magnitude and metal vapour is thus expected to be released more completely in this zone.

A time delay of the order of 1 ms between complete desolvation and the appearance of Na radiation has been experimentally found for a droplet of an aqueous 10 ppm NaCl solution with an initial diameter of 50 µm, in a C_2H_2-air flame (see Hieftje and Malmstadt 1968, 1969). A special droplet generator was used to inject and observe isolated droplets of uniform size in the flame (see Sect. III.3b).

Factors controlling the volatilization. The temperatures of sublimation, melting and boiling may be considered as a first indication for the volatility of the compound formed. However, they need not necessarily be lower than the flame temperature for complete volatilization. Even when the flame temperature is considerably lower than the melting point, complete volatilization of dry aerosol particles is still possible, if the final vapour pressure of these particles remains below the saturation pressure.

The saturation pressure of the atomic vapour at flame temperature could set an upper limit to the free metal content in the flame as in the case of rhenium (see Gilbert 1966). But over-saturation might be possible.

In the simple case of a molten particle with radius r and mass m, the rate of volatilization, if diffusion-controlled, can be represented by the *Langmuir equation* (see L'vov 1966, 1970, and Alkemade 1970a)

$$-dm/dt = 4\pi r M D p_s/(RT). \tag{III.43}$$

Here t = time, M = molar mass (m and M in g) of the evaporating substance, D is the coefficient of diffusion of the vapour in the ambient flame gas, R is the gas constant, T is the absolute flame temperature, and p_s is the saturation pressure of the vapour at the surface of the particle. This equation holds in fair approximation if, on the one hand, p_s is markedly below 1 atm (i.e. temperature at particle surface below boiling point), whereas, on the other hand, the vapour pressure at large distance from the drop is negligible compared to p_s. The value of p_s is determined by the temperature at the surface of the drop. The latter will be only slightly lower than the flame temperature if p_s is at most of the order of 1-10 Torr (see L'vov 1966, 1970). A further condition for the validity of this equation is that r should be large compared to the free-path length of the molecules. In a flame at 1 atm pressure the latter may be of the order of 1 μm. If r is small compared to the mean free-path, dm/dt varies proportionally to r^2 (see Zung 1967, Okuyama and Zung 1967, and Bastiaans and Hieftje 1974).

It follows from this equation by integration over time that the total time required for complete volatilization of a particle is proportional to its initial surface area. The fraction of aerosol that is volatilized at a given height in the flame thus depends on the initial size distribution of the aerosol particles and on the travelling time from the flame base to the height considered. A larger height of observation thus corresponds to a higher fraction volatilized. This fraction increases too, when the average initial size of the particles is lowered. This may be brought about, for a given nebulizer, by a lowering of the solution concentration, or, for a given solution concentration, by a finer dispersion of mist droplets. Besides, a higher flame temperature as such also results in a more complete volatilization, owing to its influence on p_s. Also, convection may contribute to the transport of vapour from the particle surface and so enhance the rate of volatilization.

L'vov (1966, 1970) estimates on account of Eq. III.43 that in a $C_2H_2-O_2$ flame with $T \approx 3100$ K and for an Al solution concentration of 0.1%, complete volatilization can be expected to occur within 5×10^{-4} s for a molten drop of Al_2O_3 (melting point 2320 K; boiling point 3250 K) formed from a mist droplet of initial diameter of 20 μm. During this period the flame gas may have travelled over a height interval of the order of 0.5 cm.

When the dry aerosol particles attain their boiling point in the flame, the rate of volatilization is controlled by heat transport and can presumably be represented by an equation similar to Eq. III.42 for a boiling water droplet. In this case volatilization may be readily completed if the particle diameter is not excessively large.

When the aerosol particles are in the solid state, the calculation of their rate of volatilization may possibly be complicated by the nonspherical shape or porosity of the particles. In compound particles fractional volatilization may also complicate matters.

Under all practical conditions the total heat required per second to volatilize the dry aerosol is usually insignificant when compared to the heat of combustion released per second (see Alkemade 1969).

9d   METHODS FOR CHECKING (IN-)COMPLETE VOLATILIZATION

The poor state of our theoretical knowledge about the formation and volatillization of dry aerosol particles necessitates us to seek for *experimental methods* by which the occurrence of (in-)complete volatilization can be tested directly or indirectly in practical cases (see also Alkemade 1970a).

(i)   If the concentration of atomic alkali vapour is independent of the kind of anion of the alkali salt solution nebulized, complete volatilization is likely (see James and Sugden 1955, and discussion by Bulewicz in Soundy and Williams 1965).

(ii)   If the amount of metal vapour passing per second the total horizontal cross section of the flame appears to be independent of the height of this cross section above the burner, again complete volatilization seems probable. We assume that other factors such as diffusion losses are excluded or corrected for.

Independence of total metal vapour transport with respect to height has been found by Simon (1960) for Na in turbulent H_2-O_2 flames with a direct-injection burner in a height interval from 2 to 5 cm above the burner tip. A similar result has been found by Gibson, Grossman and Cooke (1963) for Na in an acetone solution nebulized into a similar flame. In thick premixed laminar flames with a spray chamber the total free Na concentration is usually found to be independent of height down to the bottom of the flame, for moderate Na solution concentrations at least (for example, see Hollander 1964).

On the other hand the depression of Ca emission by added phosphoric acid or aluminium salt has been found to decrease with increasing height of observation. Ca and P (or Al) are assumed to form an involatile compound particle, under participation of oxygen, that is gradually volatilized with increasing time spent in the flame. The Ca vapour concentration thus increases with height. This depression effect does not occur when Ca salt and phosphoric acid or aluminium salt are fed separately by means of two nebulizers in parallel into the same flame (see Alkemade and Jeuken 1957, Alkemade and Voorhuis 1958, Schuhknecht and Schinkel 1958, Fukushima 1959, and Poluektov 1961). This proves conclusively that the depression of Ca by P or Al is to be explained by processes in the condensed phase and not in the gaseous phase.

(iii)   If volatilization is incomplete, the metal vapour concentration increases by less than a factor of two when the metal solution concentration is doubled. At higher solution concentrations the average radius of dry aerosol particles will increase and consequently the total time required for complete volatilization will increase too.

This procedure has been used for checking that in a turbulent H_2-O_2 flame with direct-injection burner volatilization is complete for $KCl-$, $MgCl_2-$ and $CaCl_2$ solution concentrations up to $1\,g/l$ K and Mg, and to $0.1\,g/l$ Ca respectively (see Fischer and Kropp 1960, and Kropp 1960). As expected, the hotter $C_2H_2-O_2$ flame appeared to be more favourable in this respect than the cooler H_2-air flame.

A similar test for complete volatilization is obtained by checking whether the curve-of-growth obeys the square-root-law at high solution concentrations for self-absorbed atomic resonance lines (see Sect. II.5c-3). Furthermore when the band emission of a molecule, which shows practically no self-absorption, is plotted versus metal solution concentration, incomplete volatilization in the high concentration range causes a downward curvature. This curvature has successfully been applied, e.g., by Hollander (1964) and Hooymayers (1966) to correct metal vapour concentrations for incomplete volatilization.

(iv) When by application of the comparison method (see Sect. V.6) one finds that the atomic metal concentration is (much) lower than can possibly be explained by ionization or molecule formation, incomplete volatilization might be the cause.

(v) The occurrence of nonvolatilized particles in the flame may also be manifest from their incandescence. Sometimes separate luminous streaks can be discerned rising straight up from the burner orifices (see Kallend 1967). In fluorescence work Rayleigh scattering from these particles may be found when the flame is irradiated by a strong light source. The spectral continuum emission occurring when U- or Al solutions are nebulized is ascribed to incandescent oxide particles, which are known to have low volatility. In hydrogen flames the incandescence of metal oxide particles appeared to be enhanced by surface reactions of excess H- and OH radicals (see Kelly and Padley 1967, and Kallend 1967).

Rayleigh scattering of incident radiation from solid particles following a λ^{-4} dependence has been found when Ca, Ba, Ga, and Ni were nebulized into an unpremixed H_2-O_2 flame (see Veillon *et al.* 1966). When a polarizer was inserted into the incident radiation beam, the scattered radiation proved to be polarized, as expected. Scattering of a more complex type has also been found in flames (see Omenetto, Hart and Winefordner 1972, and Larkins and Willis 1974).

(vi) Spherical, hollow CrO particles, irregularly shaped, or fragments thereof,

THE MEASUREMENT OF FLAME TEMPERATURE   Sect. III.10a

have been collected by Kelly and Padley (1967) on carbon-coated copper grids passed through hydrogen flames into which chromic acid has been sprayed. These fragments might result from explosive decrepitation in the last stage of desolvation where a core of solvent is surrounded by the solute in the molten state. The size of the nondisintegrated spherical particles was roughly as expected from the number of Cr atoms contained in an average spray droplet.

In conclusion we emphasize that observations on volatilization made by one author under his particular experimental conditions can hardly be transferred to other conditions of measuring. Many independent, often unknown factors control the extent of volatilization of desolvated aerosol. In one case Cr is reported to be present partly as solid particles, presumably as oxides (see Padley 1959, Kelly and Padley 1967, and Jensen and Padley 1967). In another case, however, comparison of the atomic concentrations found by de Galan and Winefordner (1967) for Cr and Ca (after correction for the different degrees of dissociation of their oxides) shows that no appreciable volatilization loss for Cr occurs, assuming that Ca is completely volatilized.†

As a general rule fuel-rich, hot flames, such as the $C_2H_2-N_2O$ flame, are recommended whenever difficulties due to incomplete volatilization of refractory metal oxide particles are feared. Also a low solution concentration, a fine dispersion of spray droplets, and a large observation height in the flame may be recommended.

10. THE MEASUREMENT OF FLAME TEMPERATURE

10a   EQUILIBRIUM CONSIDERATIONS

As explained in Sect. II.3c-3, the concept of temperature requires a state of thermodynamic equilibrium. Only in that state does the parameter T have the same value for all distribution functions and all partial equilibria as described by the Maxwell law, the Boltzmann law, the Saha law, the mass-action-law, and by Planck's radiation law. In an open flame the main deviation from thermodynamic equilibrium above the combustion zone occurs in the density of radiation: only in the centre of strong spectral lines, and even then only in the inner parts of the flame, does the density of radiation obey Planck's law at the temperature T which describes (approximately) the other equilibria in the flame. Therefore, this radiation formula cannot really be used to determine the flame temperature.

If only Planck's law is not obeyed, we have a state of *thermal equilibrium*

---

† Personal communication by Dr P.J. Kalff.

Chapt. III   GENERAL INSTRUMENTAL ASPECTS AND EXPERIMENTAL METHODS

(see Sect. II.3c-3) with the common parameter T occurring in all the other equilibrium laws being defined as the flame temperature. To allow for spatial temperature variations we further specify the state of the gas by speaking of *local thermal equilibrium* (LTE; see also footnote † on page 61).

Local thermal equilibrium in flames can never be attained completely. In the first place, the radiation leak to the surroundings causes the population of the excited states of the atoms and molecules to be somewhat lower than the equilibrium value; the amount of the depletion depends on the ratio of the collisional to the radiative de-excitation rates (see Sect. VI.4c-2 sub point iv). In the second place, the processes of ionization and dissociation (and the reverse processes) are rather slow so that relaxation phenomena may be observed in flames (see Sects IX.3c and 3d for ionization relaxation and Sect. VIII.4a for dissociation relaxation).

The best temperature indicator is therefore the velocity distribution of the particles, as the relaxation time for the equilibration of the velocity distribution is very short (see Sect. II.3c-2). Above the combustion zone and under certain conditions the ratio of the populations of atomic and molecular levels, measured by means of the intensity of emission-, absorption- and/or fluorescence spectral lines may be a good indicator too. It should be realized that population ratios can be derived from thermal intensity measurements just because the radiation from the flame does deviate from that of a black body.

10b   THE TRANSLATIONAL OR 'TRUE' TEMPERATURE

When there is doubt about the equilibration of all degrees of freedom (apart from the density of radiation) the value of T describing the Maxwellian velocity distribution is often regarded or simply defined as the true flame temperature. This *translational* temperature can be determined optically by a measurement of the Doppler half-intensity width of a spectral line. The temperature then follows from the equation (see Eq. II.240)

$$\delta \nu_D / \nu_0 = 7.16 \times 10^{-7} \sqrt{T/M} \ . \qquad (III.44)$$

There are, however, two disadvantages in using the Doppler temperature. Firstly, the square-root dependence on temperature is weak. Secondly, other mechanisms which broaden spectral lines, especially collisional broadening, have effects of the same order of magnitude as Doppler broadening has in atmospheric-pressure flames (see Sect. II.5b). Therefore, although it is possible to correct for these additional broadening effects (see Sects VII.2b-5 and VII.4), the translational temperature is hardly ever used in flame measurements. Lück and Müller (1977), Lück and Thielen (1978) have measured Doppler temperatures in flames by spectral scanning of a single OH line using a tunable laser. It is, however, widely used as a concept for the true flame temperature from which other 'temperatures' may deviate; in this way deviations from thermal equilibrium (see Sect. II.3c-3) are simply defined.

It should be noted that suprathermal chemiluminescent lines might — under certain conditions — show an extra Doppler broadening (see Sect. VII.2b-9) which makes them unsuitable for translational temperature measurements.

Pitz *et al.* (1976) measured Doppler temperatures of a line profile obtained by Rayleigh scattering of laser light; although rather inaccurate, the method avoids the complication of collisional line broadening and yields local temperatures (see Sect. 10d).

10c THE EXCITATION TEMPERATURE

Expressions for the intensity of spectral lines in emission, absorption and fluorescence have been given in Sect. II.5c. These expressions contain T as a parameter and can thus in principle be used for temperature measurements, if the other parameters occurring therein are known or are eliminated. The density n of the particles involved is often unknown, the a-parameter sometimes uncertain and our knowledge of transition probabilities and partition functions is limited. The practical problems involved in an absolute intensity calibration of the apparatus (see Sect. III.14g) and the difficulty in determining the particle density (see Chapt. V) can generally be avoided at one and the same time if we measure the ratio of the intensities of two or more spectral lines or compare the thermal emission intensity and absorption strength of one and the same transition as in the line-reversal method (see below).

Since in these measurements the population (ratio) of excited levels is measured, we speak here of the *excitation temperature* pertaining to the spectral line(s) used. If in emission measurements the radiation is thermal (for definition see Sects II.3c-3 and VI.4a), the excitation temperature equals the flame temperature. The various possible methods are described below (see Bradshaw *et al.* 1979).

10c-1 *THE TWO-LINE METHOD (IN EMISSION)*

The ratio of the integral intensities of two thermal emission lines of the same element in the absence of self-absorption is given by (see Eq. II.280)

$$\frac{B_1}{B_2} = \frac{A_1 \lambda_2 g_1}{A_2 \lambda_1 g_2} \exp\left[(E_2 - E_1)/kT\right]. \tag{III.45}$$

We can therefore deduce the temperature by measuring this ratio, if we know (the ratio of) the quantities A, g, λ and E for both lines. Re-writing and differentiating Eq. III.45 we have

$$\frac{dT}{T} = \frac{kT}{(E_2 - E_1)} \frac{A_2 g_2 B_1 \lambda_1}{A_1 g_1 B_2 \lambda_2} d\left(\frac{A_1 g_1 B_2 \lambda_2}{A_2 g_2 B_1 \lambda_1}\right), \tag{III.46}$$

which shows that the accuracy of the result is served by a large energy difference between the upper levels of the lines used. The accuracy is mainly limited by the

Chapt. III   GENERAL INSTRUMENTAL ASPECTS AND EXPERIMENTAL METHODS

uncertainty in the values of the relative transition probabilities. It is difficult to make a general statement about these uncertainties (see Sect. II.5a-2 and Appendix A.4).

Equation III.45 does not allow for self-absorption. Therefore the concentration of the atomic species should be kept low when resonance lines are involved. Furthermore the relative spectral response of the apparatus should be calibrated (see Sect. III.4b) and its linear intensity-response should be tested (see Sect. III.13).

10c-2 *THE SLOPE METHOD*

One can enhance the accuracy of the two-line method by using a number of lines of the same element. By plotting the logarithm of the relative intensity divided by the gA/λ factor versus the energy of the upper level one finds the temperature T from the slope of the graph.

Reif, Fassel and Kniseley (1976) have recently investigated this method in detail. They selected groups of transitions of Fe atoms and measured their relative intensities in premixed $C_2H_2-O_2$ and $C_2H_2-N_2O$ flames at atmospheric pressure well above the combustion zone. From these data they calculated temperatures using Eq. III.45 on the basis of relative transition probabilities taken from six different sources. They also measured reversal temperatures (see below) on the Fe line at 3720 Å. The results showed that in both flames the calculated temperatures were scattered over a range of about 150 K[†] that included the reversal temperature. The two probable sources of error are the uncertainty of the transition probabilities and the nonuniform temperature distribution. If the former is taken to be the cause of the spread in the results, the error of 5% in temperature would indicate errors of the order of 10 to 20% in the relative transition probabilities.

One can also apply a similar method to molecular emission bands and thus determine the (effective) *rotational* or *vibrational temperature* as indicated in Sect. II.5c-3. The large number of rotational lines available makes up for the small differences between the energy levels used. The advantages of using rotational and vibrational temperatures are that they do not require the introduction of any additives into the flame and that for rotational lines the relative transition probablities are simple to calculate (see Sect. II.5c-3).

10c-3 *THE EMISSION-ABSORPTION METHOD*

The intensity of an emission line or band or of a continuous part of the

---

† See Note (12) in Appendix A.3.

THE MEASUREMENT OF FLAME TEMPERATURE   Sect. III.10c

flame spectrum can be compared to the absorption of the flame *in the same spectral
interval*. The relation between absorption and emission follows from Kirchhoff's
law (see Eq. II.295)

$$B = A_t B^b_{\lambda_0}(T_f) \quad (III.47)$$

where B is the radiance in the line or group of lines, A_t is the integral absorption in the same spectral interval, and $B^b_{\lambda_0}(T_f)$ is the black-body spectral radiance at flame temperature T_f. The integral absorption can be measured as a percentage value as described in Sect. III.15. The value of B is measured next and compared to $B^b_{\lambda_0}(T)$ of a black body of known but arbitrary temperature T. All measurements have to be made in the same spectral interval, i.e. with the same setting of the spectral apparatus. With the aid of Planck's law one then finds the temperature T_f for which Eq. III.47 holds. In effect the quantity measured is the population ratio of the upper and lower level of the same transition(s). Since the emission signal, which may be affected by self-absorption, and the integral-absorption signal depend on the density of the species involved in the same way, the density can be chosen arbitrarily. Measurements have been made, for example, on the infrared resonance bands of H_2O and CO_2 in flame gases (see Tourin 1966).

10c-4 *METHODS USING FLUORESCENCE*

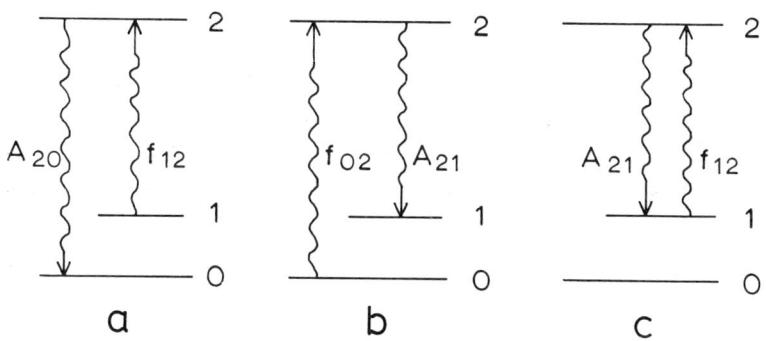

Fig. III.16 Schematic representation of the types of fluorescence used in temperature measurements. (a): Anti-Stokes fluorescence; (b): Stokes fluorescence; (c): resonance fluorescence of a nonresonance line.

Intensity ratios of atomic fluorescence lines (for their measurement see Sect. III.16) can also be used to measure excitation temperatures (see Omenetto, Benetti and Rossi 1972). We consider the relevant types of fluorescence in Fig. III.16. The ratio of the anti-Stokes to the Stokes fluorescence intensities F_{20} and F_{21} (expressed in photon flux) is given by, according to Alkemade (1970),

319

Chapt. III GENERAL INSTRUMENTAL ASPECTS AND EXPERIMENTAL METHODS

$$\frac{F_{20}}{F_{21}} = \frac{E_{\lambda 12}}{E_{\lambda 02}} \left(\frac{\lambda_{12}}{\lambda_{02}}\right)^5 10^{-5040E_1/T_f} \,, \qquad (III.48)$$

where $E_{\lambda 12}$ and $E_{\lambda 02}$ are the spectral irradiances (from a continuous source) at the wavelengths λ_{12} and λ_{02} of the spectral lines, E_1 the energy (in eV) of level 1 and T_f the flame temperature. Measurements of $E_{\lambda 12}/E_{\lambda 02}$ and F_{20}/F_{21} then yield, with the knowledge of the wavelengths of the spectral lines and the energy of level 1, the flame temperature. In essence, in this method the population of level 1 is compared to that of level 0 via excitation of level 2. The method is analogous to the emission-absorption method in which no knowledge of transition probabilities is required either. For detailed discussion we refer to Bradshaw et al. (1979, 1980).

Another possibility is to compare the resonance fluorescence, F_{res}, between levels 1 and 2 (with level 1 being populated thermally) and the Stokes fluorescence, F_{St}, from level 2 to level 1 (with level 2 being photo-excited from level 0). This ratio can be shown to be

$$\frac{F_{res}}{F_{St}} = \frac{E_{\lambda 12}}{E_{\lambda 02}} \left(\frac{\lambda_{12}}{\lambda_{02}}\right)^3 \frac{g_1 f_{12}}{g_0 f_{02}} 10^{-5040E_1/T_f} \,, \qquad (III.49)$$

where g_0 and g_1 are the statistical weights of levels 0 and 1, and f_{12} and f_{02} are the oscillator strengths. This method essentially compares the absorptions from level 0 and 1; therefore the relative oscillator strengths are needed in the calculation.

A variant of the first method is (formally) to assign a temperature T_c to the continuum background source so that the ratio $E_{\lambda 12}/E_{\lambda 02}$ corresponds to that for a black body at temperature T_c. It can then be shown that, when $F_{20}=F_{21}$, the temperatures T_c and T_f are equal.

The accuracy of the methods decreases when the height of level 1 above the ground level 0 decreases. This effect limits the choice of elements to be used. Measurements were made (see also Haraguchi and Winefordner 1977) with thallium and indium in air-H_2, air-C_2H_2 and N_2O-C_2H_2 flames. The temperatures obtained were in agreement with reversal-measurements (on a Tl line) within 1-2%. It seems therefore that these temperature measurements can yield reliable results. Moreover, they have the specific advantage that *spatial temperature distributions* can be measured directly (see Zizak, Cignoli and Benecchi 1979 and Omenetto 1979).

10c-5 *THE METHOD OF LINE-REVERSAL*

The measurement of line-reversal temperature. The principle of this method has been described in Sect. II.5c-3. In this method (in contrast to the methods mentioned in Subsections 10c-3 and 10c-4 which also employ the upper and lower levels of one transition) the absorption of a continuous background lamp by the

THE MEASUREMENT OF FLAME TEMPERATURE
Sect. III.10c

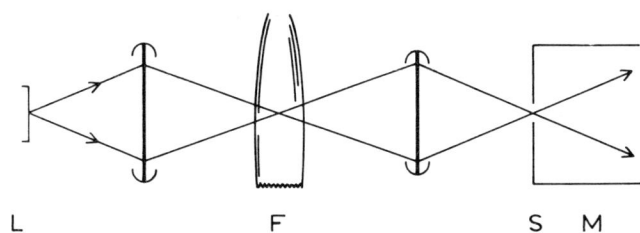

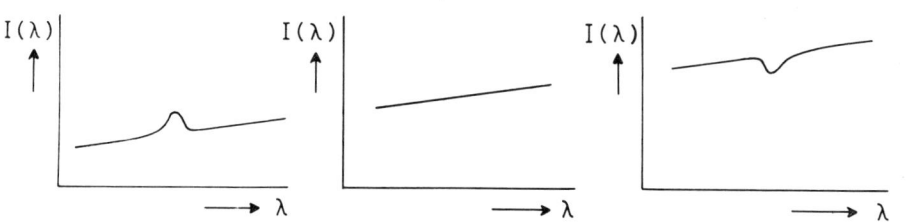

Fig. III.17 General set-up for a reversal-temperature measurement (upper part). A background lamp L with a continuous spectrum of controllable spectral intensity at wavelength λ_0 of the atomic line is imaged on a flame F containing metal atoms. Lamp and flame are imaged on a slit S of the monochromator M. The resulting spectra (lower part) from left to right are for $T_r < T_f$, $T_r = T_f$ and $T_r > T_f$, respectively, where T_r = radiance temperature of lamp and T_f is flame temperature.

lower level and the emission from the upper level are *added simultaneously* and *optically*. The 'reversal point' is attained when these signals add up to zero (see Fig. III.17). The accuracy of the method stems from the fact that it is a *null-method* that shifts the calibration problem to the background lamp, and that strong resonance lines can be used because self-absorption does not introduce an error.

The method has been employed extensively for flame temperatures (see Gaydon and Wolfhard 1979). In the early days the method involved viewing visually through a spectroscope the spectrum of a tungsten strip lamp imaged on the slit of the spectroscope. The flame, seeded with (most often) Na vapour, was placed between the lamp and the spectroscope. The current through the lamp is increased from a low starting value until the Na spectral line just disappears from the 'emission side' (current too low); next the same procedure is followed from the 'absorption side' (current too high). The mean of these two current values yields the current at the reversal point.

In photo-electric measurements one generally scans the combined spectrum of lamp and flame around the spectral line used at a number of different lamp currents.

A plot of residual line intensity versus lamp current then yields the reversal point by interpolation. Advantages of photo-electric measurements over visual ones are that one can use spectral lines for which the eye is not (optimally) sensitive, and that a photo-electric set-up can be made to have better temporally and spatially integrating properties than the eye, so that a better accuracy can be obtained (see below).

A measurement of the reversal point as described above takes several minutes. However, an increase in time-resolution can be obtained when the spectral apparatus is fitted for fast repetitive wavelength-scanning at frequency f_0 (of the order of 100 Hz) by a vibrating mirror or glass or quartz plate (see Sect. III.4f). When a small part $\Delta\lambda$ of the spectrum around the spectral line used is repetitively scanned (see Fig. III.18), the photo-electric signal will contain — apart from a d.c. signal due to the intensity of the lamp spectrum at λ_0 and an a.c. signal at f_0 due to the slope of the lamp spectrum — an a.c. signal of frequency $2f_0$ due to the superimposed spectral line. Phase-sensitive detection of the signal at frequency $2f_0$ directly indicates the occurrence of a net emission line, a net absorption line (opposite phase as compared to the emission line), or the reversal point. This method of derivative spectroscopy is further described in Sect. III.4f.

The concentration of metal vapour that is chosen for use in the flame is a compromise between two possible errors: too low a concentration offsets the accuracy, since the line is invisible or undetectable over too wide a range of lamp currents (see below); too high a concentration might introduce a systematic error since then the temperature of the edge of the flame is measured, and this is generally lower. The latter error can, however, be greatly reduced by surrounding

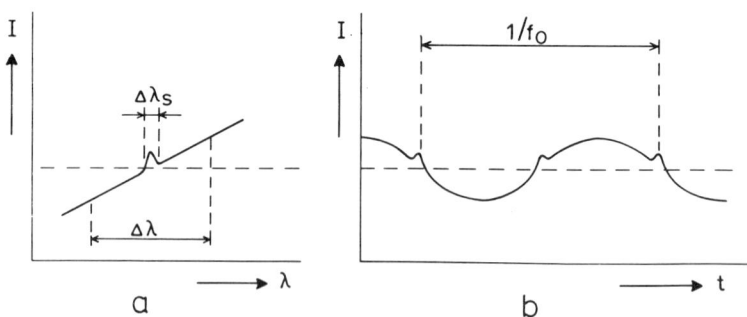

Fig. III.18 (a): The spectrum near the reversal point, which is scanned repetitively over an interval $\Delta\lambda$ with a slit width $\Delta\lambda_s$ and frequency f_0. (b): The time-dependent signal obtained from the exit slit.

THE MEASUREMENT OF FLAME TEMPERATURE   Sect. III.10c

the flame proper by a flame of equal composition into which no metal vapour is
introduced (see Sect. III.2a).

Background light sources. As explained in Sect. II.5c-3, the radiance
temperature of the background light source should be variable and at least equal to
the temperature to be measured. In practice the temperatures of some flames such
as H_2-O_2 and $C_2H_2-N_2O$ flames exceed the radiance temperature that is attainable
with the light source commonly used, namely the tungsten strip lamp (see Sect.
III.7b-1), which has a maximum $T_r \simeq 2600$ K at 6000 Å. For short periods higher
radiance temperatures can be attained, but the strip then noticeably evaporates,
invalidating its calibration. Use has been made of tungsten arc lamps which do
attain a higher temperature, but these cannot be accurately calibrated (see Gaydon
and Hurle 1963).

Another way of measuring higher reversal temperatures is to decrease the
flame emission while keeping the absorption constant. This can be achieved (see
Snelleman 1967) by *modulating* the flame emission *inefficiently*, e.g. by chopping
the emission with a 10% 'off' + 90% 'on' chopper placed between flame and detector.
The absorption is modulated at the same frequency with a normal 50% 'off' + 50% 'on'
chopper placed between background light source and flame; the 50% 'on' period
falls within the 90% 'on' period of the other chopper. The reversal point detected
at the modulation frequency, f_{mod}, then occurs at a flame temperature which is
higher (in practice up to 500 K) than the radiance temperature of the calibrated
background source.

Yet another way of measuring higher flame temperatures is to use a strong
background source such as a xenon lamp with a central radiance temperature of about
5000 K (see Sect. III.7). The disadvantage of that kind of lamp is the inhomogeneity of the temperature across the lamp. However, this fact need not invalidate a reversal measurement as long as the radiation, *integrated* over the area of
the slit of the monochromator, is used in the measurement *as well as* in the
calibration.

The procedure is to image the xenon lamp on the flame and then on the slit
of the monochromator in the usual way, and next reduce the radiation from the
xenon lamp (with neutral filters between the lamp and the flame) until the reversal point is measured. Finally, the reduced radiation is calibrated against a
tungsten lamp by means of a linear detector. The condition for a correct
measurement is that the radiation from the tungsten lamp used in the calibration
of the xenon lamp and the emission from the flame should subtend the same solid
angle and cover the same slit area. Further conditions (to eliminate the
influence of possible inhomogeneities in the surface of the detector) are that

Chapt. III   GENERAL INSTRUMENTAL ASPECTS AND EXPERIMENTAL METHODS

the light beams from the xenon lamp, the tungsten lamp and the flame should illuminate the limiting angular aperture uniformly and that this aperture but nnot the exit slit of the monochromator should be imaged on the detector.

Systematic errors. A number of effects may cause systematic errors in the measured reversal temperature. They are described below (see also Snelleman 1967).

The main source of error occurs when the solid angle or the area of the lamp radiation detected is smaller than that of the flame. The flame emission is then unduly favoured and the measured reversal temperature is too high.

The radiance temperature pertains to the *image* of the background lamp *in the flame*. Reflection losses at windows and lenses prior to incidence in the flame should therefore be corrected for.

The radiance temperature of the tungsten lamp is dependent on wavelength. The data for the emissivity of tungsten (see, e.g., de Vos 1954) can be used to allow for this dependence.

Light which is emitted by the flame in the direction of the tungsten-strip lamp may be reflected from a lens, a window or the strip itself, into the solid angle used in the measurements. Adjustment to the reversal point then requires a higher level of background radiation with this reflection than without it. Consequently the reversal point is found at too high a temperature. By a slight rotation of the strip lamp one can notice and avoid this effect, without invalidating the calibration of the lamp.

The flame emits light, in the spectral line used, over a much larger area and in a much larger solid angle than is used in the reversal measurement; the flame light used is limited by the optical conductance (slit area × aperture; see Sect. III.4b) of the monochromator. Stray light, especially flame light that has strayed at the lens between flame and monochromator, therefore tends to increase the detected flame radiation and to yield too high reversal temperatures. Clean lenses and the use of stops can adequately suppress the effects of stray light.

Real differences between the reversal temperature and the true flame temperature can be caused by the depletion of the upper level of the transition (see Sect. VI.4c-2 sub point iv). Such differences are generally small, but can be quite large ($\approx$ 100 K) in flames consisting of poor quenchers, e.g., the stoichiometric H_2-O_2-Ar flame. In such cases the measurement of the Y-factor or measurements with a series of different metal concentrations can lead to the correct value of T. The depletion decreases with increasing self-absorption in the spectral line, but remains finite. Therefore, extrapolation of the measured reversal temperatures to zero concentration and the application of a temperature correction given by the Y-factor are recommended.

THE MEASUREMENT OF FLAME TEMPERATURE   Sect. III.10c

Random errors. The accuracy with which the reversal point can be
established is governed by the detectability of a weak net emission or absorption
line superimposed upon a strong background continuum. The question of random errors
in emission measurements is dealt with in Sect. III.14f. We use here a result of
that section, assuming that the prevailing noise is shot noise in the radiation of
the continuous background source. Using Eq. III.78 (in Sect. III.14f-2) for the
signal-to-noise ratio (SNR) in the detection of a relatively weak spectral line
superimposed on a spectral continuum we have

$$\text{SNR} = \frac{B^s(\lambda_0)}{\sqrt{2\pi e B_\lambda^{bg}(\lambda_0)}} \sqrt{\frac{L\tau(\lambda_0)\,s(\lambda_0)\,\tau_r}{\Delta\lambda_{eff}}} \ . \tag{III.50}$$

Here $B^s(\lambda_0)$ is the net radiance of the spectral line used in the line-reversal
measurement, $B_\lambda^{bg}(\lambda_0)$ is the spectral radiance of the continuous background source
at λ_0, e is the elementary charge, L the optical conductance of the optical train
(lenses + monochromator; see Sect. III.4b), $\tau(\lambda_0)$ the transmission factor of the
optical train, $s(\lambda_0)$ the responsivity of the photocathode, $\Delta\lambda_{eff}$ the effective
bandwidth of the monochromator (see Sect. III.4b), and τ_r the response time of the
meter. The above equation applies in the case of a paired reading (background reading followed by signal + background reading τ_r seconds after application of the
signal; see Sect. III.14f-2). The *net* radiance, $B^s(\lambda_0)$, is the difference between
the radiance of the thermal emission line and the radiance absorbed from the source
radiation in the same spectral line. Using Eq. III.47 for the thermal line emission and Eq. II.303 for the absorbed radiance we get for arbitrary atom concentration

$$B^s(\lambda_0) = A_t B_{\lambda_0}^b(T_f) - A_t B_\lambda^{bg}(\lambda_0) \ . \tag{III.51}$$

Expressing the spectral radiance of the background source in its radiance temperature T_r according to Eq. II.301, we can rewrite Eq. III.51

$$B^s(\lambda_0) = A_t \left\{ B_{\lambda_0}^b(T_f) - B_{\lambda_0}^b(T_r) \right\} \ . \tag{III.52}$$

Near the reversal point we have: $\Delta T \equiv T_f - T_r \ll T_f$; we can thus also write

$$B^s(\lambda_0) = A_t \left\{ \frac{d B_{\lambda_0}^b(T)}{dT} \right\}_{T_f} \Delta T \ . \tag{III.53}$$

Assuming that the detection limit, $\widetilde{\Delta T}$, in the reversal-temperature measurement corresponds to a SNR = 3, we find from Eqs III.50 and III.53

$$\widetilde{\Delta T} = 3 \left\{ \frac{d B_{\lambda_0}^b(T)}{dT} \right\}_{T_f}^{-1} \left(\frac{1}{A_t}\right) \sqrt{\frac{2\pi e B_{\lambda_0}^b(T_f)\Delta\lambda_{eff}}{L\tau(\lambda_0)\,s(\lambda_0)\,\tau_r}} \ , \tag{III.54}$$

325

where we have approximated: $B_\lambda^{bg}(\lambda_0) \simeq B_{\lambda_0}^b(T_f)$, as $T_r \simeq T_f$. It is noted that essentially the same result is obtained in d.c. and in a.c. measurements (cf. Table III.6).

Equation III.54 shows that the absolute precision of the reversal-temperature measurement increases proportionally to A_t; the use of a resonance line and high atom concentrations, resulting in strong self-absorption, thus favours the precision. The precision also increases proportionally to $\sqrt{\tau_r}$; a slow meter response or a long measuring time thus favours the precision; the same applies when the measurement is repeated a number of times (see Sect. III.4f). Under usual measuring conditions we further have that both L and $\Delta\lambda_{eff}$ vary proportionally with the monochromator slit width (see Sect. III.4b); *the slit width then drops out of the expression for* $\widetilde{\Delta T}$!

When we take as an example: $T_f = 2500$ K, $\lambda_0 = 6000$ Å and $\tau_r = 1$ s and assume reasonable values of $\Delta\lambda_{eff}/L$ and $\tau(\lambda_0)$ for a medium prism monochromator and of $s(\lambda_0)$ for usual photocathodes in the visible range, we find for a markedly self-absorbed resonance line with $A_t = 0.1$ Å a detection limit: $\widetilde{\Delta T} = 0.4$ K. This value is in reasonable agreement with the experimental detection limit of about 1 K found by Snelleman and Smit (1968) in a C_2H_2-air flame of about 2500 K into which a Na solution of about 500 mg/l was sprayed. Visual line-reversal measurements are in general less precise because of the short integration time (≈ 0.2 s) and the smaller quantum efficiency of the eye at $\lambda_0 \approx 6000$ Å.

In summary, the (mostly systematic) errors mentioned can be eliminated to a large extent leaving a residual total error of the order of 5 K. This value is roughly equal to the accuracy generally attributed to the calibration of tungsten strip lamps in this temperature range.

10d   THE MEASUREMENT OF TEMPERATURE-DISTRIBUTIONS

In emission and absorption measurements the radiation is integrated along the line of sight. When the temperature along this line is not constant, one deduces from the measurements some sort of average temperature which may lead to erroneous conclusions (see Sect. IV.6 and Reif, Fassel and Kniseley 1973, 1974). In such cases one needs to measure *local temperatures*. Some methods are discussed here.

1.   Nonoptical methods such as those involving the use of *thermocouples* avoid line-of-sight integration. Such methods are extensively used in industrial flames and furnaces. The disadvantages of using these methods in laboratory flames are that they disturb the (generally) small flames too much and that the reliability decreases at higher temperatures (> 2000 K).

THE MEASUREMENT OF FLAME TEMPERATURE   Sect. III.10d

2. When a narrow (laser) beam is used for excitation of *fluorescence* and we observe the fluorescence intensity from a small section of the laser-illuminated flame part, we can masure the local flame temperature by one of the methods outlined in the previous sections and especially in Subsection 10c-4.

Haraguchi *et al.* (1977) used atomic lines of indium and thallium for temperature profile measurements. The values found in very inhomogeneous flames were consistent within 20 K in the centre and within 50 K at the edge of the flame. Lapp (1974) used laser-Raman-scattering to derive local temperatures from vibrational-rotational band profiles of nitrogen and hydrogen in a H_2-air flame. He found agreement within about 30 K with thermocouple measurements. The use of band peak ratios yielded a slightly better result than a complete curve-fitting procedure.

3. Temperature measurements have been made by *locally seeding* the flame with the thermometric species. This has been done, e.g., by Kohn (1914). She introduced a pearl of salt into the flame, which yielded metal vapour downstream and (by diffusion) a short distance upstream from the pearl. Local temperatures were then measured by line-reversal, showing however the local temperature disturbance by the pearl. Fernandez and Bastiaans (1979) have measured local flame temperatures by means of a single-droplet generator (see Sect. III.3b).

4. If the flame is cylindrically symmetric and no self-absorption occurs, the relation between the local radiation intensity $I(r)$ as a function of radial distance r and the distribution of the radiance $B(x)$ measured by scanning the flame in the horizontal x direction perpendicular to the optical axis (which is parallel to the y direction) is given by the *Abel integral equation* (cf. Fig. V.1)

$$B(x) \propto 2 \int_x^{r_0} I(r) \, r(r^2-x^2)^{-\frac{1}{2}} \, dr \, , \qquad (III.55)$$

where r_0 corresponds to the flame border. *Abel inversion* yields

$$I(r) \propto -\frac{1}{\pi} \int_r^{r_0} \frac{dB}{dx} (x^2-r^2)^{-\frac{1}{2}} \, dx \, . \qquad (III.56)$$

Generally $B(x)$ is not an analytical function and, moreover, only certain forms of $B(x)$ lead to an integral that can be evaluated. Therefore, several numerical methods have been devised (see Frie 1963, and Lochte-Holtgreven 1968). Mostly they consist of a 'peeling' procedure, in which the contributions of consecutive annular shells, starting from the outside, are subtracted from the measured profile $B(x)$. In a simple form, the contribution within each shell is considered to be constant; a more sophisticated procedure assumes a form: $I(r) = a r_i^2 + b_i$ within the ith

shell and the continuity of $I(r)$ and dI/dr as functions of r.

The main problems in the Abel inversion are the selection of the shell thickness and the criteria for distinguishing signal from noise (see Kock and Richter 1969, and ter Heerdt 1979), since the errors made in the subtraction of the contributions from all previous shells propagate into the following steps of the procedure.

Once the local radiation intensity has been derived, one can apply one of the emission methods from Subsection 10c to determine the local temperature. When the concentration of the thermometric species is constant as a function of r, the local intensity-distribution yields straightforwardly the relative temperature-distribution. It can be made absolute when the temperature in one place is known from other data. The temperature-distribution is directly found when the local intensity ratios of two or more transitions of one species are measured; these methods were described before and referred to as the two-line method and the slope method. The density-distribution need then not be known. The density distribution of the species can finally be recovered from the temperature and the emission when both are determined as a function of r (see Sect. V.2a).

When self-absorption occurs, the Abel inversion can only be applied in special cases (see Elder, Jerrick and Birkeland 1965). One first has to measure the transmission profile by scanning the flame in the x direction. When the background light source has intensity I_0, the measured transmission profile is (see Eq. II.256)

$$I(x) = I_0 \exp\left[- \int_{-y_0}^{y_0} k(x,y)\,dy \right]. \qquad (III.57)$$

Equation III.57 holds only if $k(x,y)$ is independent of wavelength within the spectral bandwidth used. In determinations of the atom distribution (see Sect. V.2c) one can use a hollow-cathode lamp (with its narrow spectral line profile) to measure the absorption in the centre of a spectral line only, where k is, virtually, constant. However, the problem at hand is the determination of local emission as well as local absorption in the same narrow spectral band to deduce local temperatures. While the correct local absorption can be obtained with the use of a hollow-cathode lamp, the local emission in the same narrow spectral band cannot be obtained so easily. It would require, e.g., the use of a Fabry-Pérot interferometer to select a small spectral bandwidth around the line centre. With the aid of the measured local transmission one can apply an Abel inversion to the emission profile with self-absorption and deduce the local temperature using one of the methods discussed before. Consequently local temperature measurements cannot be made easily when self-absorption occurs. The only case in which a temperature

THE DETERMINATION OF THE RISE-VELOCITY OF THE BURNT FLAME GASES   Sect. III.11

measurement (after Abel inversion) can be performed is when the spectrum is a real continuum (ter Heerdt 1979); this situation does not occur in flames.

11. THE DETERMINATION OF THE RISE-VELOCITY OF THE BURNT FLAME GASES

The rise-velocity of the flame gases downstream from the primary combustion zone is an essential parameter when observations made as a function of height in the flame have to be interpreted as time functions, e.g., in dissociation and ionization relaxation measurements.

Experimentally, rise-velocity determinations in flames are most commonly performed by particle track methods, although sometimes other methods have been tried (for a review see Fristrom and Westenberg 1965). Particles are brought into the flame by means of a sprayer (see Rann 1967), a settling chamber (see Fristrom and Westenberg 1965), a rotating pipe cleaner (see Lewis and von Elbe 1961) or simply by tapping the burner or the gas tubes in which some powder has been put (see Alkemade 1954, and Hollander 1964). The particles may be detected either by their incandescence, especially in flames with $T > 2200$ K, e.g., in the case of aluminium powder (see Hollander 1964) or solid carbon (see Rann 1967), or by their scattering of light from an external light source, e.g., in the case of bentonite clay or MgO (see Kumar and Pandya 1970). The radiated or scattered light is detected either photo-electrically (see Rann 1967) or photographically (see Lewis and von Elbe 1961). The rise-velocity can also be determined when light from a laser, having been deflected by a moving particle or phase object, is isolated by a schlieren aperture and then allowed to interfere with an unperturbed beam. The moving particle induces a Doppler shift in the wavelength, which can be measured by counting the number of interference fringes that pass a small opening (see Schwar and Weinberg 1969). Kleine (1973) has reported a variant of this method in which two laser beams are scattered by the same particle. Interference in the scattered light is detected by a frequency analyzer. These methods can be applied, without introducing particles, in milieus with steep refractive-index gradients. Particles have to be introduced in order to measure the velocity of a homogeneous gas flow.

For photo-electric measurements two horizontal slits are placed vertically above each other a certain distance apart. Behind the pair of slits a single photomultiplier is placed and its output is presented on an oscilloscope (see Alkemade 1954), or, alternatively, a photodetector is placed behind each slit, of which the output is presented on a dual-beam oscilloscope (see Rann 1967). The time taken for the particle to ascend the distance between the two slits can be derived from the horizontal distance between two pips in the oscilloscope trace. This method yields instantaneous results, but it has the disadvantage that it cannot

Chapt. III GENERAL INSTRUMENTAL ASPECTS AND EXPERIMENTAL METHODS

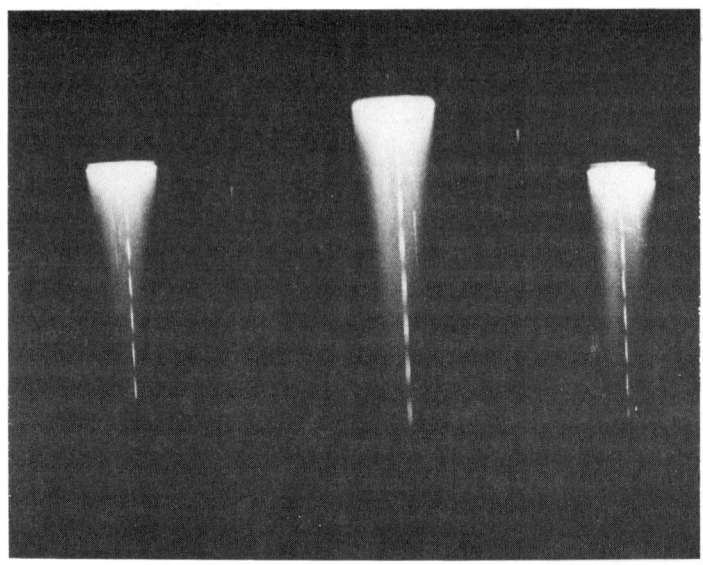

Fig. III.19 Picture made by means of a stationary camera of tracks caused by incandescent aluminium particles carried off by the rising gases of a CO-air flame. Between flame and camera a rotating chopper was inserted with regularly spaced openings. One opening of the chopper corresponds to a time interval of exposure of 1/625 s. The picture shows the three images of the flame: two (sharp) images via plane mirrors placed behind the flame, and between them the (unsharp) image of the flame directly. (From Hollander 1964.)

distinguish between particles conveyed in different layers of the flame, nor can it reveal whether the same particle is observed through either slit.

Photographically, the rise-velocity can be measured in various ways. The particles can be illuminated by a stroboscope (see Lewis and von Elbe 1961). The rise-velocity is found from the number of photographed track segments per cm and the frequency of the stroboscope. Also, the length of a single track can be determined after photographing, with a known (short) exposure time, incandescent particles or particles illuminated by an uninterrupted light source. In the latter methods difficulties brought about by the shutter (irregular or irreproducible movement) can be avoided by photographing through a disk with regularly spaced openings, rotating with a constant, known speed. Each particle track is presented as a broken line on the picture (see Fig. III.19). With this method the rise-velocity can be determined in a similar way as with the stroboscopic method.

330

THE DETERMINATION OF THE RISE-VELOCITY OF THE BURNT FLAME GASES   Sect. III.11

A central compound shutter has the disadvantage that it does not work instantly. Therefore the tracks will show vague ends. A focal-plane shutter, when applied with the shutter moving perpendicular to the direction of the particle, does not have this drawback. However, the shutter motion during the exposure may not be constant and reproducible (see Hollander 1964).

The contrast of the pictures may be improved by using a yellow filter to cut off the blue part of the flame spectrum and a didymium filter to eliminate light from evaporated atoms of Na which may have polluted the particles. In the pictures this pollution shows up as vague clouds. These photographic methods have one advantage in common: the whole gas-flow pattern can be made visible, and the velocity distribution over the flame volume can be simply determined. Three-dimensional positioning of each track can be achieved when the flame is photographed via two plane mirrors placed behind the flame under a suitably chosen angle (see Hollander 1964). Then the picture shows three images: two sharp images of the flame via the mirrors and, between them, the less sharp image of the flame (see Fig. III.19).

The mean accidental error in the rise-velocities measured by particle track methods may be as low as 0.5% (see Fristrom and Westenberg 1965). However, an appreciably larger systematic error may arise because of the gravitational force, which will induce a difference in velocity, Δv, of the particle and the flame gas.[†] It follows from Stokes' law that for spherical particles in a laminar gas flow: $\Delta v = 0.22 \, \rho g r^2/\eta$, where ρ is the particle density, g is the acceleration of gravity, r is the radius of particle, and η is the (average) gas viscosity. In practice, the particles will not be spherical, and consequently the velocity defect will be different. Alkemade (1954) has determined experimentally the velocity defect to be at most $25 \, \text{cm s}^{-1}$ for Al particles with a maximum radius of 6×10^{-3} cm. Usually smaller particles are used. Since the rise-velocity of most laminar flames is $500 \, \text{cm s}^{-1}$ or higher, gravity may cause a systematic error of 5% maximum.

It takes time for a particle to accelerate to its final velocity. The relaxation time involved is given by: $\tau = \Delta v/g$ (see Kuang-pang Li 1976, and Borgers 1978) and amounts to 1 to 10 ms, depending on the size and density of the particle. The height above the combustion zone where a particle reaches its final velocity within 5% lies between 1 and 3 cm for slow flames ($500 \, \text{cm s}^{-1}$) and between 10 to 30 cm for fast flames ($1500 \, \text{cm s}^{-1}$). Therefore, the height of observation should be chosen carefully. This relaxation effect has been studied theoretically and experimentally by Borgers (1978) and Borgers, Jongerius and Hollander (1980).

---

† The effects of gravitational and inertial forces in the application of the particle-track method to an electric arc have been described theoretically as well as experimentally by van Stekelenburg (1943).

Chapt. III   GENERAL INSTRUMENTAL ASPECTS AND EXPERIMENTAL METHODS

L'vov et al. (1976) have discussed the particle track method applied to flames produced on a slot burner. Such flames have a region (just above the combustion zone) where the gas velocity has also an appreciable horizontal component. In that region an additional systematic error may be expected.

The systematic errors connected with the mass of the solid particles can be avoided when one measures the rise-velocity of metal-vapour plumes that are produced by repetitively introducing tiny, individual droplets of a metal solution into the flame (see Boss and Hieftje 1978). The droplet generator described by Hieftje and Malmstadt (1968) can be used for this purpose, while the rise-velocity of these plumes can be determined by the two-slit method. In this way spatially resolved rise-velocities can be obtained. A disadvantage is that the injected droplets require a minimum time to begin releasing atoms in the flame..

Finally, the rise-velocity, v_r, of homogeneous, laminar flames can also be calculated from the relation: $v_r = \zeta(T_f/298) F_u/S$, where ζ is the molar expansion factor, F_u is the volume of unburnt gas mixture supplied per unit time at room temperature and 1 atm, S is the cross-sectional area of the flame, and T_f is the flame temperature. The factor ζ can be calculated from the compositions of the unburnt and burnt gas mixtures (for calculation see Sect. IV.4) and from the initial and final pressures. For cylindrical flames shielded with a flowing gas, calculated and measured rise-velocities have been found to agree within 5% (see Reid 1961, and Zeegers 1966). For a flame burning without such a protecting sheath the measured rise-velocity appeared to be about 70% of the calculated value (see Alkemade 1954); this discrepancy has been explained by assuming that a sheath of the unprotected flame is appreciably cooled and thus lost, due to direct contact with the ambient air.

12. THE MEASUREMENT OF THE FLAME THICKNESS AND OF THE DISTRIBUTION OF METAL VAPOUR IN THE FLAME

In optical measurements with metal vapours the thickness of the metal vapour along the line of observation often plays a role. This thickness, for example, determines the extent of self-absorption in emission measurements, or its variation in height has to be accounted for when the height-dependence of the (mean) metal concentration is to be derived from atomic absorption measurements. In the derivation of the absolute (mean) metal concentration from thermal-emission or absorption measurements this thickness has to be known (see Chapt. V). In flames that are not shielded by a metal-free flame mantle (see Sect. III.2a) and where the metal vapour is more or less uniformly distributed over the entire flame cross section, the flame thickness l is usually taken as a measure for the thickness of

THE FLAME THICKNESS AND DISTRIBUTION OF METAL VAPOUR   Sect. III.12

the observed vapour cloud.

The cross-sectional area of the flame also plays a role in the calculation of the vertical rise-velocity from the supply rate of combustible gas mixture (see Sect. III.11).

Definition of flame thickness. The flame thickness, or more precisely the cross-sectional dimensions of the flame, cannot be defined unambiguously. One may define the thickness by the visually perceived limits. However, the choice of the visual part of the spectrum as the measurement criterion is rather arbitrary. This criterion fails conspicuously in the case of almost invisible flames, such as the H_2-air flame, as well as in the case of luminous i.e. soot-carrying flames. In the latter the radiation from the soot particles makes other kinds of radiation, e.g., molecular radiation, hardly detectable. Besides, the stream lines of the particles may deviate from the flow lines of the flame gas.

A better criterion for determining flame thickness may be based on the horizontal temperature profile of the flame. One might then define the edge of the flame as the point at which the temperature has dropped appreciably, say half-way between maximum flame temperature and ambient temperature. Most flame processes that are relevant in metal vapour studies have virtually stopped at that temperature. Another way of defining flame thickness that has been used by Willis (1970) is to measure the envelope of the absorption region of OH radicals at 3064 Å.

The flame thickness is dependent on height because of the diffusive or turbulent mixing of the flame gases with the surrounding air. An increase or decrease in the thickness may occur, depending on whether the initial gas mixture was fuel-rich or lean on leaving the burner top.

Measurement of spatial distribution. In some experiments the nonuniform spatial distribution of the metal vapour[†] has to be known in two or three dimensions (in the horizontal plane and along the flame axis). Examples are: the determination of the diffusion coefficient (see Chapt. X), the establishment of complete or incomplete desolvation and volatilization (see Sect. III.9), and the calculation of the absolute metal concentration on the flame axis from the salt supply (see Eq. V.19). In analytical flame spectroscopy information about the spatial distribution of the species detected has been used for achieving the best operating conditions; we refer here to the pertinent textbooks cited in Chapt. I and to Hambly and Rann (1969).

An obvious method is to measure the emission intensity or absorbance of the

---

[†] Often called *concentration profile* in analytical flame spectroscopy.

Chapt. III   GENERAL INSTRUMENTAL ASPECTS AND EXPERIMENTAL METHODS

species of interest as a function of vertical and horizontal displacement of the flame in a plane perpendicular to the optical axis (for a discussion of the methods of measurement and results obtained see also Hambly and Rann 1969). In order to exclude the effect of a varying temperature on the excitation, most measurements of atomic distributions are done by absorption methods (see Sect. III.15 and Chapt. V). For most molecular species, however, only the emission method can be applied (see Sect. III.14 and Chapt. V). The spatial resolution is improved when one restricts the cross section and angular aperture of the light beam detected (see also Fig. III.20 in Sect. III.14). This avoids erroneous smoothing of the concentration profile. However, absorption and emission measurements yield concentrations that are integrated along the optical beam. In circularly symmetric flames one can determine the *local* concentration as a function of distance r from the flame axis by measuring the integrated concentration as a function of off-axis distance x of the optical beam (see Fig. V.1). Inversion of Abel's integral equation, Eq. V.2, then yields the distribution function in the horizontal plane (see Sects III.10d and V.2a). With prolongated flames burning on a slot burner one tries to avoid end effects while measuring the absorption (or emission) along a line parallel to the long side of the slot. The concentration may then be assumed uniform along this line. By varying the distance of this line to the median plane of the flame one directly finds the distribution in a horizontal plane at a given height.

Franklin, Baber and Koirtyohann (1976) used a linear photodiode array behind and parallel to the exit slit of a stigmatic monochromator while the flame was imaged on the entrance slit. This enabled them to obtain a rapid, accurate recording of the vertical or — with application of a dove prism — horizontal distribution of light of a given wavelength. The array-reading sequence was displayed on an oscilloscope screen or recorded in digital form.

With the advent of (dye) lasers local sensing of the metal vapour distribution by fluorescence has become feasible (see Omenetto 1975, and Lück and Müller 1977). According to this technique the saturated fluorescence caused by the narrow laser beam is viewed at a certain locus in a direction perpendicular to the laser beam, care being taken to avoid re-absorption of the fluorescent radiation. The saturated fluorescence intensity is insensitive to variations in laser intensity as well as fluorescence quenching, while it is proportional to the metal density (see Sect. II.5c-4).

Description of spatial distributions. In usual *flames on Méker-type burners* the horizontal distribution of metal vapour is fairly homogeneous. Each burner hole forms a point-like source. The central part of the flame quickly shows a uniform horizontal distribution because of the short distance between the holes, the

THE FLAME THICKNESS AND DISTRIBUTION OF METAL VAPOUR   Sect. III.12

expanding flow pattern just above the holes, and the diffusion of the metal vapour.

A different distribution is found in a flame on a Méker burner into which metal vapour is introduced through only one, central hole. This flame was used to study diffusion. This topic will be discussed in Chapter X.

In these flames, losses sideways to the surrounding air seem to be quite small, as test experiments on the height-dependence of the concentration of easily vapourized elements have shown (see also Sect. III.9). There is some difference, however, between shielded and unshielded flames. Measurements of the Na concentration in a C_2H_2-air flame (see Snelleman 1965) indicate that the diffusion of metal vapour from the central flame into the shielding flame is more marked than diffusion from the edges of an unshielded flame into the surrounding air.

The case of *slot-burner flames* is very different. Measurements of the absorbance patterns of many different atoms in a C_2H_2-air flame (see Rann and Hambly 1965) all show a distinct area of maximum absorbance, the shape of which differs from element to element and for rich and lean flames. In these flames the effects of convective transport of aerosol particles, of desolvation, volatilization and compound formation, and of diffusion interact to produce the pattern observed. The effect of volatilization has been investigated in these flames (see Willis 1970) by integration of the absorbance signal over the width of the flame at various heights. After an initial increase as a function of height, a plateau is reached for some metals (Na, Li, Rb), whereas for others the integrated absorbance decreases due to compound formation. The measurements of the horizontal distribution showed that in no case did the metals extend to the edge of the flame as defined by the presence of OH radicals. Further results showed that atoms of elements forming stable oxides exist mainly in the central region (of highest temperature), whereas those elements whose oxides are readily decomposed persist in the atomic form much nearer the edge.

Further investigations have been made by L'vov et al. (1976). The gases from a slot burner expand sideways by a factor of 10 to 20 after having left the burner. The aerosol particles do not attain the same horizontal velocity component as the flame gases and consequently have a distribution that is narrower than the flame width by a factor of 2 to 5. In the vertical direction the aerosol acquires a velocity component that is somewhat less than the gas velocity. After volatilization the resulting vapour exhibits a noticeable outward diffusion, which is still dependent on mass, i.e. on the kind of atom and possible compound formation. It is clear that a distribution dependent on all effects mentioned is hard to describe quantitatively. In some simple situations (in which diffusion processes dominated) quantitative results have been obtained (see Chapt. X).

Chapt. III   GENERAL INSTRUMENTAL ASPECTS AND EXPERIMENTAL METHODS

In analytical flame spectroscopy Koirtyohann and Pickett (1968) have reported *lateral-diffusion interference* with slot-burner flames; this interference is caused by a change in the transverse horizontal distribution (perpendicular to the slot) of the analyte by the concomitant. West, Fassel and Kniseley (1973) have explained this change by the delay in complete volatilization of the analyte in the presence of concomitant. As a consequence, the free analyte atoms or molecules have less time available to diffuse outwardly and their concentration in the centre of the flame is increased, once volatilization is completed. L'vov *et al.* (1976), who studied theoretically and experimentally the flow pattern of the aerosol particles and burnt gas above a slot burner, have, however, arrived at a different explanation for the lateral-diffusion interference of Pb observed by them with excess Al nitrate. Their explanation comes close to that originally suggested by Koirtyohann and Pickett (1968). For a discussion of this effect and similar interference effects we refer to Alkemade and Herrmann (1979).

13. METHODS FOR CHECKING THE LINEARITY OF A
    PHOTO-ELECTRIC MEASURING SYSTEM

Measurements of relative and absolute intensities are commonly performed with a linear photo-electric measuring system providing a reading that is proportional to the radiant flux received by the photo-electric detector. For example, curve-of-growth measurements in emission (see Sect. II.5c-3) require not only a linearly operating nebulizer (see Sect. III.8a) but also a linear photo-electric measuring system.

Departures from strict linearity may originate in the photo-electric detector itself (nonlinear photocathode emission; nonlinear internal amplification; pile-up effects in photon counting techniques; see Sect. III.5b), in the amplifier and, especially with a.c. detection, in the rectifier (see Sect. III.6b), or in the indicating instrument. By an appropriate graduation of the reading scale, nonlinearities in the deflection of the meter (and in the rectifier characteristic) can be corrected for. However, the nonlinearity arising for certain rectifiers in the low-signal range due to rectified noise (see Sect. III.6b) is not so easily redressed. Linear electronic amplification can be easily realized by appropriate feedback. Besides, the demand for linearity of the amplifier and meter can be dropped if a null-balance method is applied with a carefully calibrated balancing circuit. However, unless balancing is effectuated automatically, this procedure is more time-consuming. Nonlinear operation of the photomultiplier tube can be avoided by restricting the irradiance at the photocathode and/or the internal amplification factor (see Sect. III.5b). When an optical null-balance method is applied with a

THE LINEARITY OF A PHOTO-ELECTRIC MEASURING SYSTEM   Sect. III.13

ccalibrated attenuation wedge in the reference light beam (see Sect. III.15b-2), nonlinearity of the photodetector does not play a role. Because of its complications, the latter method is but rarely applied in flame spectrometry.

In order to check the linearity of the photo-electric measuring system as a whole or to construct a correction graph for the deviation from linearity, the following methods can be applied.

1.   In the *two-hole method*, a diaphragm with two holes is placed in the light beam. The reading obtained with both holes open is compared with the sum of the readings obtained when one or the other hole is covered, respectively. For a linear system, the two results should be identical. The size of the holes need not be equal or be known. However, the precision attained in this comparison is improved when the signals obtained with one or the other hole open are about equal. By repeating this procedure after changing the total light flux through the two holes, the linearity can be checked over an extended range of signal strengths. The position of the diaphragm with holes should be such that the light passing through either hole fills the same area of the photocathode. This may be realized, for example, by placing this diaphragm close to a lens that images the flame on the photocathode.

When the light beam is modulated by a mechanical chopper in combination with a.c. detection, the phases of the a.c. signals generated with one or the other hole open should be considered too. In case the rectifier is phase-insensitive, both phases (and waveforms) should be made exactly equal. The dimensions and positioning of the holes with respect to the chopper blades are then important. With a phase-sensitive rectifier these demands can, strictly speaking, be dropped, but for practical reasons both phases should be fairly close to the optimal phase (see also Sect. III.6b).

This simple and inexpensive method is used primarily to check whether linearity exists or not. It can be extended, however, to construct a complete calibration graph to correct for deviations from linearity. The accuracy attainable depends on the precision of the scale readings involved (and possibly of the phase adjustments). No extra calibrated components are required. In practice, an accuracy of about 1% may be readily attained.

2.   A well-known method of calibrating a photo-electric measuring system is based on the *inverse-square dependence on distance* of the irradiance by a sufficiently small-sized source. In its simplest form, the source directly irradiates a constant area of the photocathode (without insertion of any other optical component). The distance between source and photocathode should be kept sufficiently large in comparison with the dimensions of the source and the irradiated area. When we cannot measure the position of the photocathode accurately because the

Chapt. III   GENERAL INSTRUMENTAL ASPECTS AND EXPERIMENTAL METHODS

latter is not easily accessible, we can insert a diaphragm between source and phototube and a lens which makes an image of the diaphragm opening on the photocathode. The irradiance of the photocathode is then inversely proportional to the square of the distance between the point source and the diaphragm. This holds true if this distance is again sufficiently large and if all light rays passing through the diaphragm are caught by the lens. It is essential in this method that the irradiated area of the photocathode does not vary with distance from the point source. An accuracy of about 1% may be readily attained.

3.   The linearity can also be checked with *calibrated light attenuators* which are inserted between a constant light source and the photodetector. This check can be realized by means of a series of 'grey' filters whose transmission factors are known in the spectral region of interest and which are introduced apart or in combinations into the optical path. However, it is difficult to attain an accuracy of better than 1% in this way because of the following additional requirements (see Bennett 1966): insertion of the filter(s) should not markedly defocus the light beam because of a change in optical path length; the filter should be insensitive to any polarization introduced by the light source or by the optical system; the transmission factor should not vary significantly with wavelength within the bandwidth of the spectral selection device used; the transmission factor should be insensitive to the convergence angle of the light beam passing the filter; and errors by (multiple) reflections should be negligible.

Attenuation of a light beam by an accurately known factor can also be achieved by a system of two linear polarizers of high quality. A simple theoretical relation exists between the attenuation factor and the angle between the polarization axes of the polarizers. Difficulties, however, may arise when the emission from the light source is (slightly) polarized and the optical system or photodetector is also (slightly) sensitive to the direction of polarization. These difficulties can be avoided by using three linear polarizers in series; the polarization axes of the outer ones are kept constant and parallel, and only the middle one is rotated (see Bennett 1966). If polarizers of the highest quality are used, an accuracy of better than 0.1% can thus be attained, regardless of the polarization characteristics of the source and spectrometer used.

4.   A *calibrated tungsten strip lamp* (whose spectral radiance as a function of wavelength and lamp current is known; see Sects III.10c-5 and 14g) can also be used for checking the linearity. Since the spectral distribution of the emission varies with source temperature, a monochromator with sufficiently small bandwidth should be inserted in order to eliminate the wavelength-dependence of the detection system. Relative radiances may be calculated as a function of lamp current with an accuracy of about 1-2% for a given wavelength. When the spectral emission

THE MEASUREMENT OF EMISSION   Sect. III.14a

varies significantly with wavelength within the transmission bandwidth of the monochromator, the spectral characteristics of the latter should also be taken into account. Comparison of the calculated radiance values with the meter deflections then yields a check for the linearity of the whole system.

5. The linearity of the electronic measuring system alone may be checked by introducing *calibrated electrical signals* into the (pre-)amplifier input. This may be done conveniently by inserting a calibrated (low-impedance) voltage-source, V, in series with the load resistor, R, of the photodetector. This is equivalent to inserting a (high-impedance) current-source $i = V/R$ parallel to this resistor. In this way one can also accurately calibrate or check the sensitivity steps of the amplifier.

6. The linearity of a photodetector can be sensitively checked by *double a.c. modulation* of the incident light flux at two different frequencies, f_1 and f_2, that are not in the ratio of simple integral numbers (see Sauerbrei 1972). A deviation from linearity as small as $1:10^3$ can thus easily be detected by observing beat signals with frequencies $(mf_1 \pm nf_2)$ where m and n are integers.

14. THE MEASUREMENT OF EMISSION

14a INTRODUCTION

Emission measurements are performed for a number of purposes. The wavelengths of the spectral emission lines yield information about the structure of the atom or molecule. If the medium containing the metal vapour is in thermal equilibrium, the relative intensities of the spectral lines and bands can be used to determine either the relative vapour density, or the relative transition probabilities, or the temperature, when two of these quantities are known. Absolute measurements yield accordingly information on, e.g., absolute transition probabilities. Deviations from thermal equilibrium, i.e. deviations of the excited-level population from the Boltzmann value (see Sect. VI.4d) can be found from relative emission measurements. The spectral shape of emission lines may yield information on, e.g., collision processes and temperature. In analytical flame spectroscopy, the measurement of line or band intensities is the basis for quantitative analysis. The connection with absorption and fluorescence measurements to be discussed in the following sections is very strong, the choice between them being determined sometimes by the problem at hand and sometimes by the available apparatus and know-how.

The application of flame emission spectroscopy (FES) to *chemical analysis* of metal salt solutions dates far back into the past century. The appearance of the famous paper by G. Kirchhoff and R. Bunsen entitled: 'Chemische Analyse

Chapt. III   GENERAL INSTRUMENTAL ASPECTS AND EXPERIMENTAL METHODS

durch Spektralbeobachtungen' in Pogg. Ann. Physik. Chem. 110, 161-189 in 1860 marked the general acceptance of FES as a tool for qualitative analysis of metallic elements. Quantitative chemical analysis by flame emission spectrography was perfected by H. Lundegårdh who, in 1929 and 1934, published Parts I and II of his classical book: 'Die quantitative Spektralanalyse der Elemente'. Since World War II flame emission spectrometry, incorporating a photo-electric detector and a direct-reading device, has largely replaced flame spectrography.

As distinct from atomic emission spectroscopy by arcs and sparks (see Boumans 1966), FES provides a simple, cheap and often more precise tool for chemical analysis, especially for the alkali elements. Compared to atomic absorption spectroscopy (AAS; see Sect. III.15a) and atomic fluorescence spectroscopy (AFS; see Sect. III.16a), FES has certain advantages in the simultaneous determination of several elements in the same sample. Moreover, elements whose atoms do not readily emit or absorb optical radiation in flames can often still be detected in FES via the band emission of their compounds. So chlorine can be determined indirectly via the emission of InCl which is formed in the flame upon addition of indium. One disadvantage of FES compared to AAS and AFS is that the applicability of FES is limited by the condition of thermal excitation, i.e. the excitation energy must not be too high and/or the flame temperature too low. The introduction of the direct-injection burner by P.T. Gilbert, Jr., in 1951 (see Sect. III.2) for safely producing hot, unpremixed acetylene- or hydrogen-oxygen flames, and later the introduction of the hot, premixed acetylene-nitrous oxide flame by Willis and Amos (1966) have made it possible nowadays to determine more than 70 elements by FES. Also the exploitation of suprathermal chemiluminescence effects (see also Sect. VI.3) has helped to overcome the limitations imposed by the thermal-excitation condition in special cases.

The reader who wants further information is referred to the textbooks on analytical flame spectroscopy cited in Chapt. I.

The preceding sections contain information on the properties of the separate parts of a flame spectrometer. This information will be used here in discussing the optimization of intensity measurements in the spectrum emitted by a metal vapour in a flame. The goal of such optimization is essentially the enhancement of the signal-to-noise ratio (SNR) and the avoidance of systematic errors.

In emission measurements the first goal, viz. enhancement of the signal-to-noise ratio, requires the main attention. Except in absolute measurements, systematic errors do not usually pose a serious problem. To make clear the role of the various quantities and parameters involved in the SNR, we shall first consider the the role of the flame geometry (Sect. 14b) and the optical train (Sect. 14c),

THE MEASUREMENT OF EMISSION   Sect. III.14b

and then derive general expressions for the signal strength (Sect. 14d) and the error sources in emission measurements (Sect. 14e). Next the signal-to-noise ratio will be considered and recommendations given for its improvement (Sect. 14f). Finally the measurement of absolute intensities and intensity ratios will be discussed in Sect. 14g.

14b   GEOMETRICAL CONSIDERATIONS IN EMISSION MEASUREMENTS

The emitting species in a flame is located in a three-dimensional flame volume. An optical measurement integrates the emission along the optical path through the flame. Inhomogeneities in temperature or concentration, and self-absorption make the evaluation of the measurements more complicated than in the simple case of a uniform, optically thin medium.

Furthermore, for large solid angles of acceptance the direction of the outer 'rays' of the cone of light differ significantly from the direction of the optical axis. As a consequence, the area of the flame perpendicular to the optical axis that is contributing to the signal may be considerably larger than the area that is actually focused on the following stop (see Fig. III.20). There may therefore be a contradiction in trying to select a <u>small</u> flame volume, using a <u>large</u> angle of acceptance.

When in a cylindrically symmetric flame the intensity per unit volume of the emitted light depends — through temperature or concentration — only on the radial distance from the flame axis, this dependence can be found through an Abel-inversion of the transverse intensity profile (see Sect. V.2a).

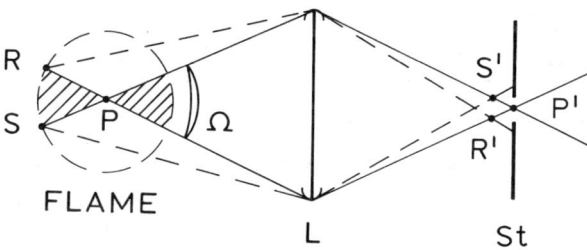

Fig. III.20 Illustration of the consequence of observing an extensive flame section under a large solid angle Ω. The centre, P, of the circular flame cross section is imaged by lens L in the centre, P', of stop St (for instance, monochromator entrance slit). All light emitted at P and collected by L passes St. Only a fraction of the light emitted at out-of-focus points R and S (imaged in R' and S') and collected by L passes St. Note that the area of flame section between R and S that contributes (partly) to the light flux passing St is larger than the area of flame section around P that contributes (fully) to this flux.

341

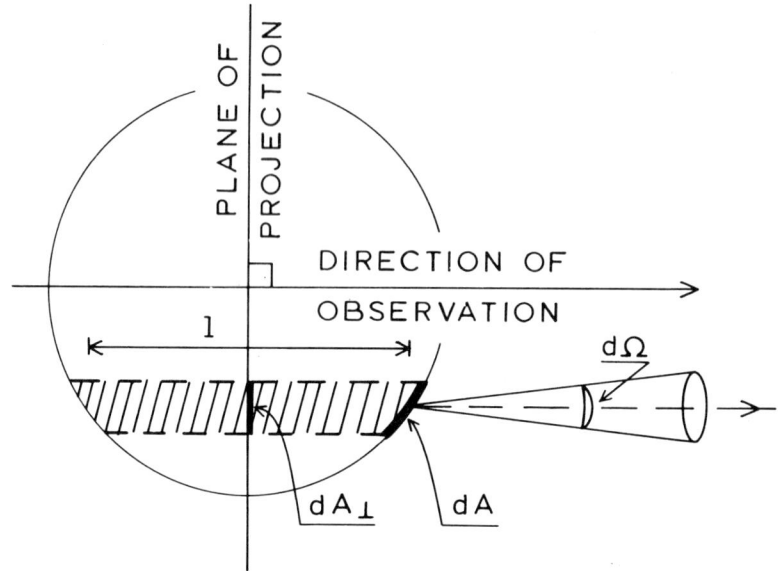

Fig. III.21 The light flux $d\Phi$ emitted by a flame through a surface element with projected area $dA_\perp$ and within a solid angle $d\Omega$ is given by: $d\Phi = Bd\Omega dA_\perp$, where the radiance B depends on the flame depth l. It appears that B varies with the direction of observation as well as with the position of the projected surface element in the plane of projection. The hatched flame section contributes to $d\Phi$.

In characterizing the absolute emission strength of a flame we shall often make use of the radiance B or spectral radiance B_λ. This quantity describes the (spectral) radiant flux emitted by the flame per unit of solid angle and per unit area of projected surface element perpendicular to the direction of observation (see also Appendix A,5). Fig. III.21 illustrates this concept for the simple case of a uniform, cylindrical flame, which is observed under an infinitesimal solid angle $d\Omega$ in a direction perpendicular to the flame axis. The light flux $d\Phi$ emitted within $d\Omega$ through a surface element with projected area $dA_\perp$ is given by: $d\Phi = Bd\Omega dA_\perp$. In the case of a uniform flame B is simply proportional to the depth l of the flame in the direction of observation (if we disregard self-absorption); evidently B depends on the direction of observation as well as on the position of the surface element. The usefulness of the (spectral) radiance stems from the fact that it is, apart from losses due to absorption, reflection and malfocus, invariant to optical (de-)magnification. In using this concept we have replaced in effect the three-dimensional flame volume by a radiating plane

THE MEASUREMENT OF EMISSION   Sect. III.14d

perpendicular to the direction of observation.

14c   THE OPTICAL TRAIN

The optical conductance (see Eq. III.4) of a given combination of flame, optics and detector is determined by the smallest value of the product ΩS that occurs in the set-up. Generally, for economic reasons, the optics, and more specifically the dispersing element, limit the optical conductance. An example of a different situation occurs in the measurement of a height-dependent phenomenon in a flame. The height of the usable flame area is then restricted to, say, 0.1 cm, while the beam divergence in the vertical plane is also restricted to, say, 0.1 rad (see Sect. 14b). Assuming a flame width of 2 cm and a beam divergence in the horizontal plane of 0.3 rad, the optical conductance is 0.06 cm² sr. This value is smaller than the optical conductance of an interference filter by an order of magnitude (see Sect. III.4b): the dispersing element cannot be filled with light from the source in the available solid angle and therefore is no longer the limiting factor in the conductance.

Another example is the common preference for photomultiplier tubes rather than solid-state detectors in emission measurements: the small area of the latter would become the limiting factor in the optical conductance. The disadvantage of a small photosensitive area, however, does not exist in the newer surface-junction detectors.

It will often occur that the optical train transmits light from a spectral line and a spectral continuum simultaneously. Referring to Sect. III.4, it is worth recalling here that the expression for the figure of merit of a dispersing element is different for these two types of spectral emissions, although the optical conductance is the same.

14d   EXPRESSIONS FOR THE SIGNAL STRENGTH[†]

Suppose that the flame emits radiation in a spectral line at central wavelength λ_0 with radiance $B(\lambda_0)$, superimposed on a continuum background with a spectral radiance $B_\lambda(\lambda)$. It is assumed that the effective bandwidth of the spectral selection device is appreciably larger than the spectral line width. The quantities $B(\lambda_0)$ and $B_\lambda(\lambda)$ contain parameters such as metal concentration, oscillator strength, flame depth, temperature, composition, etc., which are dealt with in other chapters.

The detected flux from the spectral line emission is given by Eq. III.2

[†] See also Boutilier, Pollard *et al.* (1978).

Chapt. III   GENERAL INSTRUMENTAL ASPECTS AND EXPERIMENTAL METHODS

$$\Phi_1 = B(\lambda_0) L\tau(\lambda_0) \ , \qquad (III.58)$$

where L is the optical conductance of the optical train, which is usually limited by the dispersing element (see Sect. III.4b), and $\tau(\lambda_0)$ is the transmission factor at λ_0, accounting for the reflective and absorptive losses of all optical components in the optical train. The corresponding expression for the flux from the continuous part of the spectrum follows from Eqs III.10, III.4 and III.5

$$\Phi_c = \int B_\lambda(\lambda) \, L\tau(\lambda) \, A(\lambda_0 - \lambda) \, d\lambda \ , \qquad (III.59)$$

where $A(\lambda_0 - \lambda)$ is the normalized instrumental profile of the monochromator with wavelength setting $\lambda_m = \lambda_0$ (see definition in Sect. III.4b). The corresponding photo-electric currents for a linear photodetector are

$$i_1 = B(\lambda_0) L\tau(\lambda_0) \, s(\lambda_0) \ , \qquad (III.60)$$

$$i_c = \int B_\lambda(\lambda) \, L\tau(\lambda) \, A(\lambda_0 - \lambda) \, s(\lambda) \, d\lambda \ , \qquad (III.61)$$

where $s(\lambda)$ is the responsivity of the photocathode (photocurrent per unit radiant flux). With monochromators and interferometers $B_\lambda(\lambda)$, $\tau(\lambda)$ and $s(\lambda)$ may often be considered to be constant over the spectral range in which $A(\lambda)$ deviates noticeably from zero. Using the effective bandwidth $\Delta\lambda_{eff}$ of the monochromator or interferometer as defined in Sect. III.4b, Eq. III.61 reduces to (cf. Eq. III.11)

$$i_c = B_\lambda(\lambda_0) \, L\tau(\lambda_0) \, s(\lambda_0) \, \Delta\lambda_{eff} \ . \qquad (III.62)$$

14e   ERROR SOURCES IN EMISSION MEASUREMENTS

Suppose we want to measure a constant light flux with a linear detector. The signal actually detected may deviate from the ideal one, i.e. a non-fluctuating signal that is proportional to the flux to be measured, in three respects. In the first place, a spurious signal, caused, e.g., by flame-background radiation, may be part of the measured signal. In the second place, the quantum nature of emission and detection of radiation causes fluctuations for which the term *shot noise* is colloquial (see Sect. III.5c). This noise adds an (only statistically predictable) error signal to the desired signal. In the third place, there are processes often of a technical nature which cause an additional scatter in the signal measured. Since they occur predominantly at low frequencies, they are called *excess low-frequency (e.l.f.) noise*. A special and important case of such noise sources is the one in which the spectral noise power (see Sect. III.5c) is inversely proportional to frequency. For this special case the term *flicker noise* is in use. These three types of noise may not only be found in the light signal to be measured, but also in the flame background signal, the dark current, and in some components of the amplifier. In the following we shall disregard the amplifier noise, which is

THE MEASUREMENT OF EMISSION   Sect. III.14e

usually unimportant when photomultiplier tubes are used. For simplicity's sake we shall also disregard the secondary-emission noise in the latter (see Sect. III.5c).

1. *Systematic errors* in the measurement of atomic line intensities may arise with d.c. methods from spurious signals produced by, e.g., flame background, stray light and dark current. Subtraction of these spurious signals, which can be found from a solvent blank measurement as a shift of the zero point, eliminates this error. The fluctuations or a steady drift of the spurious signal are, however, not eliminated in this way.

A more elegant and efficient way to eliminate the d.c. component and steady drift of spurious signals is the use of a.c. modulation techniques, which reject the signals that are not modulated. For example, when a periodic light chopper is placed in front of the flame and only the a.c. component of the light signal from the flame is detected ('intensity-modulation'; see Sect. III.4f), stray light and dark current are rejected, but the flame background continuum is not. When wavelength-modulation (see Sect. III.4f) is applied, the spectral line is detected, but the flame background continuum which is independent of wavelength in a restricted wavelength interval, as well as stray light and dark current are rejected.

Another source of systematic errors may be the deviation of the photodetector response, etc., from linearity; methods to check and correct for this deviation are discussed in Sect. III.13.

2. *Random errors* may be caused by the occurrence of *shot noise* in the signal to be measured as well as in spurious signals. In Sect. III.5c it was shown that the shot noise spectrum is 'white' and that the corresponding spectral noise power is described by (see Eq. III.29)

$$W_{sh}(f) = 2e\bar{i} \quad (\text{in } A^2/Hz) , \qquad (III.63)$$

where e = the elementary charge and $\bar{i}$ = mean absolute value of the current emitted by the photocathode. The total spectral noise power due to the shot effect in the photocurrents produced by several light fluxes and the dark current is found by simple addition

$$W_{sh}(f) = 2e \sum_j \bar{i}_j , \qquad (III.64)$$

where $\bar{i}_j$ refers to a component of the photocurrent or the dark current.

3. *Random errors* may also be caused by *e.l.f. noise* in the flame emission and the detector. The spectral noise power of this type of noise is usually proportional to the square of the mean photocurrent or dark current. We generally have

$$W_{elf}(f) = \sum_j C_j(f) \bar{i}_j^2 , \qquad (III.65)$$

Chapt. III   GENERAL INSTRUMENTAL ASPECTS AND EXPERIMENTAL METHODS

where $c_j(f)$ describes the frequency-dependence of the e.l.f. noise in the current component $\bar{i}_j$. A special case of e.l.f. noise is the *flicker noise* which is characterized by (see Eq. III.35)

$$c_j(f) = A_j/f \ . \qquad (III.66)$$

The constant A_j with dimension unity can be regarded as an inverse figure of merit for the low-frequency stability of the light source in question or of the photoelectric and thermionic emission process in the photocathode.

A different dependence of e.l.f. noise on frequency may be found, e.g., in solid-state detectors (due to generation-recombination noise; see van der Ziel 1970) and in the thermal emission of quasi-laminar flames; in the noise spectrum of the latter some rather sharp peaks around 10 - 30 Hz may occur (see Sect. IV.9).

4. Further *random errors* may be introduced by the so-called *quantizing noise* due to the limited precision of the meter reading or signal recording or to the finite resolution of the analog-to-digital conversion (see Malmstadt, Enke and Crouch 1974). We shall here disregard this source of error, as we can easily make it insignificant in comparison with the other noise sources by choosing a sufficiently high electronic amplification or by using scale expansion combined with zero-suppression (see also Sect. III.15b-2).

14f   OPTIMIZATION OF THE SNR
14f-1   *INTRODUCTION*

The *signal-to-noise ratio*, SNR, is connected with the precision with which a light flux can be measured. It also determines the lowest level of light flux that can be detected with given statistical confidence in the presence of background noise. The *limit of detection* is here defined as the smallest light flux that produces a reading which exceeds the standard deviation of the background reading by a factor of 3.

This definition corresponds to the limit of detection as defined in analytical flame spectroscopy with regard to the smallest solution concentration that can be determined in emission, absorption or fluorescence spectroscopy (see Kaiser and Menzies 1968 and IUPAC 1976, 1976a; see also the books on analytical flame spectroscopy listed in Chapt. I). This concept plays an important part in the assessment of an analytical method or instrument for trace analysis. In studies of metal vapours in flames the detection limit is of less importance, as there is usually a wide range of metal salt concentrations that can be chosen freely.

THE MEASUREMENT OF EMISSION
Sect. III.14f

Here we shall discuss how the experimental set-up can be optimized with respect to the SNR in emission measurements with a light source of given (spectral) radiance. One obvious way to do this is to reduce the e.l.f. noise sources as far as possible. This may be achieved, for instance, by choosing a more stable region of observation in the flame, a more stable mounting of the optical parts or by selecting a less noisy photodetector or electronic component. But once the best observation region and the best components have been selected, we have still to consider the dependence of the SNR on such parameters as the optical conductance, the spectral resolution, the frequency of detection (in case of a.c. modulation), the noise bandwidth, the sampling time or the response time of the measuring apparatus. In the following we shall mainly pay attention to these functional dependencies and not so much to the exact magnitude of numerical factors of order unity. A survey of the various characteristic times used in this and the following Sects 15 and 16 is given in Table III.5.

Table III.5
Characteristic Times used in Sections III.14 - 16

Symbol	Term	Meaning	Explained in Section
t	time	general	
τ	sampling time	time interval between introduction of sample into flame and meter reading	III.14f-2 (Fig. III.22)
τ_c	time constant or RC-time	time needed for meter to approach final deflection within (100/e) %	III.14f-2 (Eq. III.67)
τ_r	response time	$\tau_r \equiv 2\pi\tau_c$	III.14f-2
τ_m	measuring time	total time taken to measure a signal	III.14f-2
τ_i	integration time	time interval over which flame signal is integrated	
f_{mod}	modulation frequency	frequency at which flame signal is modulated	
τ_{mod}	modulation period	$\tau_{mod} = f_{mod}^{-1}$	
Δf	noise bandwidth	$\Delta f = \dfrac{1}{4\tau_c}$	III.5c

As a starting point we shall make use of the general equations III.60 or 62 for the signal current, and III.64-66 for the shot and flicker noise. All currents to be considered here refer to *primary* photo- or dark current emitted by the photocathode. Secondary-emission noise and amplifier noise will be disregarded.

Chapt. III   GENERAL INSTRUMENTAL ASPECTS AND EXPERIMENTAL METHODS

The SNR of interest, however, relates not directly to these currents, but to the final *readings* on the meter, recorder or counter. The whole measuring procedure thus enters essentially into the calculation of the SNR and must be judiciously specified lest these calculations become meaningless.[†] The specification of the measuring procedure includes such aspects as the method of correcting for spurious background or dark-current signals, total measuring time, sampling or integration time, step response of the read-out system and the number of repeated readings. Because of the great variety of measuring procedures and noise sources, we shall restrict ourselves to a few simple examples, laying more stress on basic insight than on completeness or exactness. The expressions to be derived are not only relevant to emission measurements, but — with some minor changes — are relevant also to absorption and fluorescence measurements (see Sects III.15 and 16).

Expressions for the SNR in emission measurements have been reported in the literature (for example, see, McCarthy 1971, Winefordner *et al.* 1976, Boutilier, Pollard *et al.* 1978, and Alkemade, Snelleman *et al.* 1978, 1980). Some expressions dealing with flicker noise should be considered critically.

14f-2   *D.C. MEASUREMENT IN THE PRESENCE OF BACKGROUND SHOT NOISE*

We assume here as a special case that the emission of an atomic metal line (to be called 'signal') is to be measured in the presence of a comparatively large *flame background emission with predominant shot noise*. The fluctuations in the line emission itself are assumed to be negligible. The measurements are supposed to be made with a single-beam set-up with one wavelength channel only. The sample (metal vapour) is introduced into the flame at time t_0 and the reading of the meter deflection is done τ seconds later; we call τ the *sampling time*. During this time interval the signal current i_s is assumed to be constant (see Fig. III.22a). The signal current is superimposed on a fluctuating d.c. current, $i_b(t)$, generated by the flame background emission to which the photodetector is continuously exposed. This background current has a mean value $\bar{i}_b$ and a fluctuating component $di_b(t) \equiv i_b(t) - \bar{i}_b$.

Case 1. We first consider the special case when a simple *current meter* is employed to measure the photocurrent after d.c. amplification. The signal current causes a meter deflection which is described by $x_s(t)$ as a function of time t (see Fig. III.22b). The meter deflection caused by the flame background alone is described

---

[†] For example, when flicker noise is the dominant noise source in the low-frequency range, the SNR cannot be simply equated to the ratio of the signal deflection to the standard deviation of the meter fluctuations. The standard deviation is, formally, infinite because integration of the spectral noise power (given by Eq. III.66) from $f = 0$ yields infinity. See Sect. III.14f-3 for a proper handling of this problem.

348

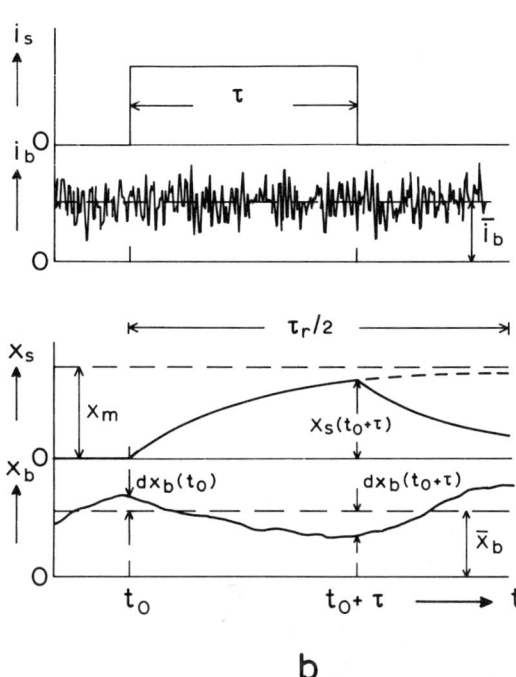

Fig. III.22 Figure a shows the signal photocurrent, i_s, and fluctuating background photocurrent, i_b, as functions of time t in the case of d.c. measurements with a current meter. The resulting meter deflections x_s and x_b, respectively, are shown in figure b as functions of t. The sample producing the signal is introduced at an arbitrary moment t_0; the sampling time is denoted by τ. The response time of the meter deflection is indicated by τ_r.

Chapt. III GENERAL INSTRUMENTAL ASPECTS AND EXPERIMENTAL METHODS

by the function $x_b(t)$ with mean value $\bar{x}_b$ and fluctuating component $dx_b(t)$ (see Fig. III.22b). Since the photodetector, amplifier and meter are assumed to operate as linear systems, the resulting total photocurrent (or total meter deflection) is simply found by adding the signal and background currents (or deflections).

The *step response*[†] of the meter is supposed to be described by an exponential function

$$x(t) = x_m \{1 - \exp[-(t-t_0)/\tau_c]\} \quad (\text{for } t \geq t_0) \ . \tag{III.67}$$

Here x_m is the maximum meter deflection which $x(t)$ approaches asymptotically when t becomes very large, while τ_c is the *time constant* of the meter. The latter may be controlled by a low-pass RC-filter connected to the terminals of the meter, in which case we have $\tau_c = RC$. We shall here define $\tau_r = 2\pi\tau_c$ as the *response time* of the meter; after τ_r seconds the meter approaches its final deflection within 0.2%.

In order to avoid systematic errors we have to subtract the mean background deflection, $\bar{x}_b$, from the meter readings (see also Sect. 14e). Since only a single wavelength channel is supposed to be available, this must be done by a separate background reading before (or after) the signal reading. In principle one could register the background deflection over an extended time interval and determine its precise mean value once and for all by averaging out its fluctuations. However, this would work only if the background emission were a truly stationary process. But in practice the flame background is likely to show a slow systematic drift due, for example, to a steadily drifting flame temperature or flame composition. This implies that $\bar{x}_b$ is not strictly a constant when measurements are made over a long period. In order to suppress the effect of this steady drift on the signal readings one can apply *paired readings* (see Kaiser and Menzies 1968): the instantaneous background deflection is read just before the metal vapour is introduced into the flame and its value is subtracted from the signal-plus-background reading made τ seconds later. The difference reading, Δx, is now taken as a corrected measure for the signal strength, and we have

$$\Delta x \equiv x_{s+b}(t_0 + \tau) - x_b(t_0) \ . \tag{III.68}$$

This method of paired readings is particularly useful when the signal is small compared with the mean background in which case scale expansion combined with zero suppression is appropriate. Assuming that the drift in $\bar{x}_b$ over the sampling time

---

[†] The step response describes the variation of the meter deflection with time t ($\geq t_0$) when a constant current is suddenly supplied to the meter at $t = t_0$ (Heaviside's step function).

THE MEASUREMENT OF EMISSION   Sect. III.14f

τ is still negligible, we get from Eq. III.68

$$\Delta x = x_s(t_0 + \tau) + \{dx_b(t_0 + \tau) - dx_b(t_0)\} \; , \tag{III.69}$$

where $dx_b(t)$ is the statistical fluctuation in the background deflection (see also Fig. III.22b). The SNR is then the ratio of the signal reading, $x_s(t_0 + \tau)$, to the standard deviation, $\sigma_{\Delta x}$, in the **difference** of the background fluctuations occurring τ seconds after each other. Thus we have in the case of paired d.c. meter readings

$$(\text{SNR})_1^{\text{pair}} = \frac{x_s(t_0 + \tau)}{\sigma_{\Delta x}} \tag{III.70}$$

with

$$\sigma_{\Delta x} \equiv \sqrt{\overline{\{dx_b(t_0 + \tau) - dx_b(t_0)\}^2}} \; . \tag{III.71}$$

The bar in expression III.71 denotes an average over time t_0 with τ being held constant.[†]

It should be realized that in Eq. III.70 both the numerator and the denominator depend in general on τ. The measuring procedure can be optimized by choosing τ such that their ratio becomes maximal. In Appendix A.7 (Eq. B.12) a general expression for the SNR in the case of shot noise is presented. It appears from this expression that for a meter with given response time, τ_r, the SNR is virtually optimized when the sampling time τ is at least equal to τ_r. In this section we shall derive an approximate expression for the SNR that holds under this special condition ($\tau \geqslant \tau_r$) only.

If $\tau \geqslant \tau_r$ the signal deflection x_s practically equals its asymptotic value x_m (see Eq. III.67) which can generally be written as

$$x_m = G i_s \; , \tag{III.72}$$

where G represents the d.c. sensitivity of the electronic measuring apparatus. Since the shot noise in the *photocurrent* shows practically no time-correlation (i.e. its fluctuations at any two distinct moments are statistically uncorrelated), the time-correlation of the fluctuations $dx_b(t)$ in the background *deflection* is fully determined by the sluggishness (or 'memory') of the meter as described by Eq. III.67. When $\tau \geqslant \tau_r$ ($\equiv 2\pi\tau_c$), however, the correlation in the meter fluctuations has practically died out, so we may write instead of Eq. III.71

$$\sigma_{\Delta x} = \sqrt{\overline{dx_b(t_0 + \tau)^2} + \overline{dx_b(t_0)^2}} = \sqrt{2}\,\sigma_b \; , \tag{III.73}$$

[†] For a stationary noise process the average over time for a single system is the same as the average over an ensemble of macroscopically identical systems at any fixed time (see MacDonald 1962).

Chapt. III   GENERAL INSTRUMENTAL ASPECTS AND EXPERIMENTAL METHODS

where σ_b is the standard deviation in the background fluctuation $dx_b(t)$. Making use of the relation between standard deviation and noise bandwidth Δf for shot noise (see Eq. III.29) we find

$$\sigma_b = G\sqrt{W_{sh}(0)\Delta f} = G\sqrt{2e\overline{i}_b \Delta f} \quad , \tag{III.74}$$

where the sensitivity G takes into account the conversion of photocurrent to meter deflection. In Eq. III.74 we have substituted for the spectral shot-noise power at zero frequency, $W_{sh}(0)$, the expression given by Eq. III.63. For a meter with time constant τ_c it holds that (see Eq. III.30)

$$\Delta f = 1/(4\tau_c) = \pi/(2\tau_r) \quad , \tag{III.75}$$

so that Eq. III.74 can be re-written as

$$\sigma_b = G\sqrt{\pi e \overline{i}_b / \tau_r} \quad . \tag{III.76}$$

Substituting this expression into Eq. III.73 and using Eqs III.70 and III.72 we find in the case of shot noise and for paired meter readings (with $\tau \geqslant \tau_r$)

$$(SNR)_1^{pair} = i_s \sqrt{\tau_r/(2\pi e \overline{i}_b)} \quad . \tag{III.77}$$

In the latter equation i_b may also refer to any other spurious photocurrent or dark current with dominant shot noise.

When the signal to be measured refers to a spectral $line$ emission, we can express i_s by Eq. III.60, whereas Eq. III.62 should be used for $\overline{i}_b$ in the case of a (quasi-)continuum background emission. As a result one gets from Eq. III.77

$$(SNR)_1^{pair} = \underbrace{\frac{B^s(\lambda_0)}{\sqrt{2\pi e B_\lambda^{bg}(\lambda_0)}}}_{source} \times \underbrace{\sqrt{\frac{L\tau(\lambda_0)}{\Delta\lambda_{eff}}}}_{optics} \times \underbrace{\sqrt{s(\lambda_0)}}_{\substack{photo-\\detector}} \times \underbrace{\sqrt{\tau_r}}_{\substack{read-\\out}} \quad . \tag{III.78}$$

The first factor in the right-hand side of Eq. III.78 is determined by the properties of the flame and the supply of metal vapour; these properties determine the radiance, $B^s(\lambda_0)$, of the line to be measured and the spectral radiance, $B_\lambda^{bg}(\lambda_0)$, of the background at $\lambda = \lambda_0$. The second factor describes the dependence of the SNR on the optical conductance, L, the transmission factor, $\tau(\lambda_0)$, and the effective spectral bandwidth, $\Delta\lambda_{eff}$, of the optical train at wavelength λ_0. When both L and $\Delta\lambda_{eff}$ are limited by the monochromator slit width and vary proportionally to it (see Eqs III.4−6), we see from Eq. III.78 that the SNR becomes <u>independent of the slit width</u>. A large slit width may, however, be preferable in order to make the other noise sources in the photodetector and measuring apparatus sufficiently small in comparison with the (inevitable) shot noise in the flame background. On the other hand, too large a slit width may lead to the flicker noise in the flame background becoming dominant over its shot noise, which would invalidate Eq. III.78 and

THE MEASUREMENT OF EMISSION   Sect. III.14f

deteriorate the SNR (compare Eq. III.96). In practice a compromise has to be found between these contrasting demands as to slit width.

Finally, the third and fourth factor in the expression III.78 show that the SNR increases proportionally to the square root of the photocathode responsivity and of the response time.

Since the sampling time, τ, does not explicitly occur in Eq. III.78 if $\tau \geqslant \tau_r$, it is most economical to choose τ as short as permitted, i.e. $\tau \approx \tau_r$. The total *measuring time*, τ_m, taken for a single pair of readings is then equal to $2\tau_r$, as it takes again about τ_r seconds before the signal deflection 'dies out' and the meter is again ready for the next measurement.

When the metal emission to be measured is a spectral *continuum* [with spectral radiance $B_\lambda^s(\lambda_0)$], we have to use Eq. III.62 for expressing both i_s and $\bar{i}_b$ in the parameters of the source and the optical train. The resulting expression for the SNR is given in Table III.6 on page 367.

By repeating the same measurement N times we can improve the SNR by a factor $\sqrt{N}$, as the fluctuations in the consecutive difference readings, Δx, are statistically independent. Since each single measurement takes $2\tau_r$ seconds, the total measuring time τ_m then equals $2N\tau_r$. The same improvement in SNR would also have been obtained in a single measurement if we had increased the response time τ_r by the same factor N. Thus we may replace more generally in Eq. III.78 the factor $\sqrt{\tau_r}$ by $\sqrt{\frac{1}{2}\tau_m}$, where τ_m either denotes the measuring time for a single measurement (with $\tau_m = 2\tau_r$) or the total measuring time spent in N repeated measurements (with $N = \tau_m/2\tau_r$). However, when τ_m is so long that the drift in the d.c. background emission during τ_m is no longer negligible, it is recommended to make $\tau_r = \tau_m/2N$ and to repeat the measurements N times. The effect of the background drift is then reduced.

The independence of the SNR of the number, N, of repeated measurements for a fixed total measuring time in the case of 'white' noise has also been generally proved by Léger et al. (1976).

It thus appears that with dominant shot noise the SNR generally increases proportionally to the square root of the (total) measuring time. This holds on condition that we use up the available measuring time most economically, i.e. we should choose the sampling time τ not longer than the response time τ_r.

In effect, 'chopping' the total measuring time, τ_m, into N consecutive time segments of equal length and taking the difference in meter deflections with signal 'off' and 'on' during each time segment is equivalent to 'modulating' the signal current, but not the background current, at a 'frequency' N/τ_m. A series of periodically repeated, paired d.c. readings eliminates a slowly drifting background as effectively as an a.c. measurement with signal modulation at

353

Chapt. III   GENERAL INSTRUMENTAL ASPECTS AND EXPERIMENTAL METHODS

a frequency $f_{mod} = N/\tau_m$ (see also Sect. III.14f-4).

Case 2. In the case when an *integrator* is used to measure the photocurrent after d.c. amplification, we identify the sampling time τ in Fig. III.22a as the *integration time* τ_i.

The integrator output, y_s, for a constant signal current i_s, is given by

$$y_s = G i_s \tau_i \ , \qquad (III.79)$$

where G is now the electronic amplification factor. The integrator output, y_b, due to the fluctuating background current $i_b(t)$, is given by

$$y_b = \int_{t_0}^{t_0+\tau_i} G i_b(t) \, dt = G \bar{i}_b \tau_i + \int_{t_0}^{t_0+\tau_i} G di_b(t) \, dt \ , \qquad (III.80)$$

where $di_b(t)$ is the shot-effect fluctuation in $i_b(t)$ as before. In order to eliminate the contribution of the mean background current we again apply the method of paired readings, i.e. we take the difference, Δy, between a pair of consecutive integrator readings, one for the background alone and one for the signal plus background τ_i seconds later. Thus we have for Δy

$$\Delta y = y_s + \int_{t_0}^{t_0+\tau_i} G di_b(t) \, dt - \int_{t_0-\tau_i}^{t_0} G di_b(t) \, dt \ . \qquad (III.81)$$

The signal-to-noise ratio is again defined as

$$(SNR)_2^{pair} = \frac{y_s}{\sigma_{\Delta y}} \ , \qquad (III.82)$$

with

$$\sigma_{\Delta y}^2 = 2 \left\{ \int_{t_0}^{t_0+\tau_i} G di_b(t) \, dt \right\}^2 \ . \qquad (III.83)$$

The factor 2 in Eq. III.83 appears because the fluctuations $di_b(t)$ during the two consecutive τ_i intervals are uncorrelated while the standard deviations of their integrals are the same. We can find the latter standard deviation by writing

$$\int_{t_0}^{t_0+\tau_i} di_b(t) \, dt \equiv dN_b e \qquad (III.84)$$

and

$$\bar{i}_b \tau_i \equiv \bar{N}_b e \ , \qquad (III.85)$$

where e = the elementary charge and N_b = the number of photo-electrons collected during the integration time τ_i due to the background emission. Because shot-effect fluctuations obey Poisson statistics we have for the standard deviation of

THE MEASUREMENT OF EMISSION   Sect. III.14f

dN_b according to Eq. III.23

$$\sigma_{N_b} = \sqrt{\bar{N}_b} \; . \tag{III.86}$$

From the latter equation and Eqs III.83, 84 and 85 we immediately find

$$\sigma_{\Delta y} = G\sqrt{2e\bar{i}_b \tau_i} \tag{III.87}$$

and hence with the help of Eqs III.82 and 79

$$(SNR)_2^{pair} = \frac{Gi_s \tau_i}{G\sqrt{2e\bar{i}_b \tau_i}} = i_s \sqrt{\tau_i/(2e\bar{i}_b)} \tag{III.88}$$

Comparing the SNR expressions given by Eqs III.88 and 77 for case 2 and 1, respectively, we see that an improvement of a factor $\sqrt{\pi} = 1.8$ in SNR is obtained with a pure integrator at a given measuring time $\tau_m \; (= 2\tau_i)$.

The dependence of $(SNR)_2^{pair}$ on the source parameters and instrumental characteristics is again given by Eq. III.78, if we replace there the factor $\sqrt{\tau_r}$ by $\sqrt{\pi\tau_i}$ (see also Table III.6). The same gain in SNR is obtained by repeating the integration measurements as found in case 1.

14f-3   *D.C. MEASUREMENT IN THE PRESENCE OF BACKGROUND FLICKER NOISE*

Here we shall consider the case of d.c. measurements in the presence of a comparatively large *flame background emission with predominant flicker noise*. The same measuring procedure as described in Sect. III.14f-2 for case 1 will be assumed here.

The main difference between the case of shot noise and that of flicker noise is that the presence of excess low-frequency components in the latter introduces a statistical correlation between the *photocurrent* fluctuations over a long time range. This makes itself also felt, of course, in the fluctuations of the meter deflection. These correlation effects have to be taken into account explicitly in the calculation of the standard deviation, $\sigma_{\Delta x}$, for paired readings (see Eq. III.71). In contrast to the case of shot noise, we can here no longer neglect correlation effects if the sampling time τ exceeds the meter response time τ_r; in fact we cannot do so for any τ however large. This greatly complicates the mathematical derivation. We shall present here only the general final expression for the variance $\sigma_{\Delta x}^2$ for __arbitrary__ values of τ and τ_r; for a full derivation of this expression the reader is referred to Appendix A.7-4 (see also Alkemade 1975, and Alkemade, Snelleman *et al.* 1978).

Using Eqs III.65 and 66 for the spectral power of the flicker noise in the background photocurrent i_b (with $A \equiv A_j$) and again assuming an exponential meter response (see Eq. III.67), the variance $\sigma_{\Delta x}^2$ is found to be

$$\sigma_{\Delta x}^2 = 2G^2 A \bar{i}_b^2 \int_0^\infty \frac{(1-\cos 2\pi f \tau)}{f(1+f^2 \tau_r^2)} df \ . \quad (III.89)$$

The factor $1/f$ in the integral expression stems from the $1/f$ dependence of the flicker noise; the factor $(1+f^2\tau_r^2)^{-1}$, with $\tau_r = 2\pi\tau_c$, describes the smoothing effect on the meter fluctuations brought about by the sluggishness of the meter response; the factor G contains, as before, the electronic amplification factor and the d.c. sensitivity of the read-out meter. Combining the latter equation with Eqs III.67 and 72 we find from Eq. III.70

$$(SNR)_1^{pair} = i_s (1-\exp[-2\pi\tau/\tau_r]) \Big/ \Big\{ 2A\bar{i}_b^2 \times \int_0^\infty \frac{(1-\cos 2\pi f \tau)}{f(1+f^2\tau_r^2)} df \Big\}^{\frac{1}{2}}. \quad (III.90)$$

This equation lends itself only to numerical computation. We can, however, prove directly from Eq. III.90 that the SNR is only dependent on the <u>ratio</u> τ/τ_r and not on τ and τ_r apart. This proof can be given by introducing two new variables with dimension unity, viz.

$$\beta \equiv 2\pi\tau/\tau_r \quad \text{and} \quad z \equiv 2\pi f \tau \ . \quad (III.91)$$

Equation III.90 then reduces to

$$(SNR)_1^{pair} = (i_s/\bar{i}_b \sqrt{2A})(1-\exp[-\beta]) \Big/ \Big\{ \int_0^\infty \frac{(1-\cos z)}{z(1+z^2/\beta^2)} dz \Big\}^{\frac{1}{2}} \equiv (i_s/\bar{i}_b \sqrt{2A}) \times f(\beta) .$$
$$(III.92)$$

The factor $f(\beta)$ in Eq. III.92 depends indeed only on β, that is on the ratio $2\pi\tau/\tau_r$, but not on τ and τ_r apart. Fig. III.23 shows the function $f(\beta)$ as obtained by computer calculations; it attains a maximum of about 0.88 at $\beta \simeq 0.8$, i.e. $\tau/\tau_r \simeq 1/8$, whereas it falls monotonically to zero if β goes to zero or infinity.

For given i_s, $\bar{i}_b$ and A the best SNR is thus obtained when we choose a sampling time τ equal to about $\tau_r/8$, which is roughly equal to the time constant τ_c ($=\tau_r/2\pi$) of the meter. Thus we have in the case of pure flicker noise for a single pair of readings at <u>optimal</u> $\tau/\tau_r \simeq 1/8$

$$(SNR)_{1,opt}^{pair} \simeq 0.62 \times i_s/(\bar{i}_b \sqrt{A}) \ . \quad (III.93)$$

This outcome is <u>independent</u> of how large τ or τ_r ($=8\tau$) are. Evidently the smoothing effect of an increased response time (compare Eq. III.77 for the case of shot noise) is fully compensated here by the rise in noise effect when longer time intervals (and thus flicker noise components with lower frequencies) are involved. A similar conclusion follows from the approximated calculations made by Léger et al.

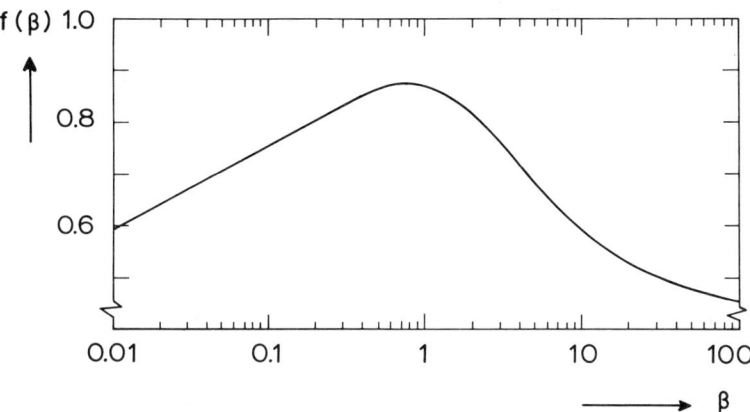

Fig. III.23 Single-logarithmic plot of the function $f(\beta)$ defined in Eq. III.92 with $\beta \equiv 2\pi\tau/\tau_r$. One can prove that $f(\beta)$ tends asymptotically to zero as $1/\sqrt{\ln \beta}$ for $\beta \to \infty$, and as $1/\sqrt{\ln(\beta^{-1})}$ for $\beta \to 0$.

(1976) in the case of pure flicker noise.

Roughly speaking, we can regard a d.c. meter with time constant τ_c as an 'integrator' with an integration period $\tau_i \approx \tau_c$. When we employ a *pure integrator* with paired readings, i.e. when we take the difference of two consecutive integrations — one over the background alone and one over the signal-plus-background — the SNR is expected to be of the same order of magnitude as that given by Eq. III.93. An exact calculation shows that the SNR obtained with an integrator is indeed almost the same (see Appendix A.7-5). Note that the SNR for flicker noise is independent of the integration time τ_i.

When the total measuring time, τ_m, is fixed, it is thus advantageous to make the sampling (or integration) time τ as short as possible and to repeat the measurements N times while turning the signal (but *not* the background) 'on' and 'off' alternately. The SNR is then enhanced by a factor $\sqrt{N}$. Since we have: $N \propto \tau_m/\tau$, the optimal SNR for repeated measurements within a total measuring time τ_m becomes

$$(SNR)^{rep}_{opt} \propto \sqrt{\tau_m/\tau} \times (SNR)^{pair}_{opt} \quad (\text{if } \tau_m \gg \tau) \ . \tag{III.94}$$

The same conclusion has been reached by Léger et al. (1976) for pure flicker noise by an approximated calculation.

Strictly speaking, an improvement of the SNR by a factor $\sqrt{N}$ occurs only if the statistical errors introduced by the flicker noise in the repeated readings are statistically uncorrelated. A similar problem exists as to the improvement of the SNR by time-averaging procedures based on successive scans of a reproducible, transient signal. It has been proved in the latter case that the SNR increases exactly according to the square-root of the number of scans if the background noise spectrum is of the type $1/f^\lambda$ with $0 \leqslant \lambda \leqslant 1$ including 'white' and flicker noise (see Ernst 1965).

The improvement in SNR obtained by repeated d.c. measurements is essentially similar to that obtained in an a.c. measurement with modulated signal but unmodulated background. In the case of repeated d.c. measurements we can interpret the reciprocal sampling time, τ^{-1}, as a 'modulation frequency', f_{mod}. We can then write the gain factor $\sqrt{\tau_m/\tau}$ in Eq. III.94 as $\sqrt{\tau_m f_{mod}}$, and from Eqs III.93 and 94 we find for the SNR

$$(SNR)_{opt}^{rep} \propto i_s \sqrt{\tau_m} \bigg/ \sqrt{\frac{A \overline{i_b^2}}{f_{mod}}} \qquad (III.95)$$

The expression under the square-root sign in the denominator is easily recognized as the spectral flicker noise power at frequency f_{mod} (compare Eqs III.65 and 66). The gain in SNR obtained by repeated d.c. measurements is obviously connected with the fact that the main contribution to the measurement error comes from the flicker-noise components with frequencies around f_{mod}. A similar situation exists in the case of a.c. measurements (see Sect. III.14f-4). A large f_{mod} (or a small τ in repeated d.c. measurements) is favourable as the spectral flicker-noise power decreases with increasing frequency.

By substituting for i_s and $\overline{i_b}$ in Eq. III.93 the expressions given by Eqs III.60 and 62 respectively, we again find the SNR as a function of source characteristics and instrumental parameters, in the case when a spectral *line* emission is to be measured. We then find

$$(SNR)_{opt}^{pair} = 0.62 \left\{ \frac{B^s(\lambda_0)}{B_\lambda^{bg}(\lambda_0)\sqrt{A}} \right\} \times \left(\frac{1}{\Delta\lambda_{eff}} \right) . \qquad (III.96)$$

Comparing the latter equation with Eq. III.78 holding for pure shot noise, we see that in the case of flicker noise all instrumental parameters drop out except the effective bandwidth of the spectral apparatus; the latter should be made as small as possible.

When a spectral *continuum* is to be measured, Eq. III.62 should be used for expressing i_s; the resulting expression for the SNR is given in Table III.6 on page 367.

In practice background flicker and background shot noise often concur. Since they are statistically uncorrelated, the combined standard deviation can be

THE MEASUREMENT OF EMISSION   Sect. III.14f

found by square addition of their respective standard deviations. The relative importance of flicker noise over shot noise increases proportional to $\sqrt{\bar{\imath}_b}$ (see also Eq. III.37).

Equations III.77 (88) and 93 (94) also apply when flicker noise in the *dark current*, i_d, of the phototube is predominant; we have only to replace $\bar{\imath}_b$ by $\bar{\imath}_d$.

14f-4   *A.C. MEASUREMENT IN THE PRESENCE OF BACKGROUND NOISE*

We now suppose that the flame emission signal is modulated (e.g., by a rotating chopper) at frequency f_{mod} and that only the a.c. component of the photocurrent is amplified and detected (see Sect. III.6b). The SNR will again be considered in the special cases when shot or flicker noise in the background current is dominant. Signal noise as well as amplifier noise are again assumed to be negligible. The a.c. amplifier + read-out system is assumed to be *linear*.[†] We shall not concern ourselves about numerical factors of order unity in the SNR expressions. The exact value of such factors, generally denoted by F, may depend on the type of a.c. detection used, the modulation waveform (and the reference waveform in the case of lock-in detection; see Sect. III.6b), the details of the reading procedure, etc.

The SNR depends essentially on whether or not the background with noise is also modulated. When the noise originates from the flame background emission which is modulated by a rotating chopper together with the emission signal, essentially the same SNR expressions apply as in the case of d.c. measurements. This holds because the same l.f. noise components (near 0 Hz) that interfere with the measurement of the d.c. signal are now transferred, together with the signal, to a frequency band around f_{mod}. Thus in the case of pure shot noise Eq. III.77 applies also in the a.c. case, apart from a numerical factor F, while τ_r now refers to the response of the read-out meter of the a.c. system. With flicker noise we essentially obtain the same SNR expression as in the d.c. case (see Eq. III.93), if we again choose an optimal ratio of sampling to response time. The modulation frequency is irrelevant in these cases. Modulation is thus not advantageous as far as the SNR is concerned; often it may even cause a slight deterioration, as the chopper throws away part of the emission signal that is available in the flame. There may be other reasons of a practical nature, however, for preferring a.c. measurements to d.c. measurements (see Sect. III.6b).

A different situation exists if only the signal to be measured is modulated but not the fluctuating background. This occurs for example when the noise

---

[†] That is, the meter deflection caused by the sum of two input currents is supposed to be the sum of the meter deflections caused by each input current separately.

Chapt. III   GENERAL INSTRUMENTAL ASPECTS AND EXPERIMENTAL METHODS

originates from the dark current of the photodetector or when the supply of metal-salt solution to the flame is modulated but not the flame background (see Sect. III.2e). Such a situation may also be found with double-beam set-ups in which the modulated emission signal is effectively differentiated from an underlying unmodulated background spectrum.

One can differentiate an emission line from an underlying continuum spectrum by using a *dual-wavelength set-up* (see Sect. III.4f-4) in which each wavelength channel is opened alternately, or by using similar methods of *derivative spectroscopy* (see Sect. III.4f-5). The same effect may be obtained — without any spectrally dispersive element — by the *negative-filter technique* (see Sect. III.4f-8).

It should be noted that with double-beam (a.c. or d.c.) techniques, only the d.c. component and certain e.l.f. noise components of the background emission can be suppressed, but not the shot-noise components. This holds because pure shot-effect fluctuations in any two light beams, even if they originate from the same source, are statistically uncorrelated.

If $G(f_{mod})$ denotes the sensitivity of the a.c. amplifier plus read-out meter for an a.c. input current with frequency f_{mod} we find for the SNR in the case of a single meter reading with sampling time τ equal to the response time τ_r

$$SNR = F \frac{G(f_{mod}) \, i_s}{\sqrt{G^2(f_{mod}) \, W_{i_b}(f_{mod}) \Delta f}} = F \frac{i_s}{\sqrt{W_{i_b}(f_{mod}) \Delta f}} \,. \qquad (III.97)$$

The denominator in the middle expression describes the standard deviation in the meter deflection, with $W_{i_b}(f)$ = the spectral noise power at frequency f of the (unmodulated) background current i_b and Δf = the noise bandwidth of the whole measuring system ($\Delta f \ll f_{mod}$; see also Eqs III.32 and 33); i_s denotes the RMS value of the a.c. component of the (modulated) input signal current with frequency f_{mod}. F is a numerical factor of order unity.

When a lock-in amplifier followed by a d.c. meter with exponential time-response is used, we have $\Delta f = 1/(4\tau_c) = \pi/(2\tau_r)$ as in the d.c. case (compare Eq. III.75). If the background current exhibits pure *shot noise*, we then find from Eqs III.63 and 97 for a single reading

$$SNR = F \frac{i_s}{\sqrt{2e\bar{i}_b(\pi/2\tau_r)}} = F i_s \sqrt{\tau_r/(\pi e \bar{i}_b)} \qquad (III.98)$$

When we want to eliminate a spurious modulated signal that contributes to the meter deflection, we can again apply paired readings (as explained in Sect. III.14f-2), with the consequence that an extra factor ½ appears under the square-root sign in

THE MEASUREMENT OF EMISSION   Sect. III.14f

Eq. III.98 (since $\tau = \tau_r$, there is no statistical correlation between the noises at the paired readings). We then get

$$(SNR)^{pair} = Fi_s \sqrt{\tau_r/(2\pi e \bar{i}_b)} . \qquad (III.99)$$

Apart from a noninteresting factor F, the latter outcome appears to be formally identical to that of Eq. III.77 which holds for paired readings in the d.c. case with $\tau = \tau_r$. It should be recalled that Eq. III.77 as well as Eq. III.99 generally hold for <u>any</u> d.c. background current i_b in the photodetector that produces shot noise. If i_b is generated by an unmodulated flame background emission, we have to apply Eq. III.98 or 99 for the SNR in the a.c. case. If the background emission is modulated together with the flame emission signal we should apply the same equation as in the d.c. case, i.e. Eq. III.77. But the outcome is essentially the same. Thus we conclude in general that shot noise — just because it has a 'white' noise spectrum (see Sect. III.5c) — can, in principle, not be suppressed by applying modulation techniques.

If *flicker noise* in the (unmodulated) background current i_b is dominant, we find by inserting $W_{i_b}(f_{mod}) = A\bar{i}_b^2/f_{mod}$ in Eq. III.97, while using Eq. III.75 for Δf

$$SNR = F'' i_s \sqrt{\tau_r f_{mod}}/(\bar{i}_b \sqrt{A}) , \qquad (III.100)$$

where we have included a factor $(\pi/2)^{-\frac{1}{2}}$ in F''. The latter equation again holds for a single reading with $\tau = \tau_r$. If we apply paired readings, then the SNR deteriorates by a factor $\sqrt{2}$ [†] resulting in

$$(SNR)^{pair} = F'' i_s \sqrt{\tau_r f_{mod}}/(\bar{i}_b \sqrt{2A}) . \qquad (III.101)$$

It is interesting to compare Eq. III.101 with Eq. III.93 which holds for a single pair of d.c. readings with optimal (τ/τ_r) ratio in the case of dominant flicker noise. The gain factor obtained in the SNR by modulating the signal, but not the background, thus appears to be of the order of $\sqrt{\tau_r f_{mod}}$. If we choose $\tau_r = 10$ s and $f_{mod} = 10^3$ Hz, a gain factor of about 100 is obtained. A similar gain factor would have been obtained if we had repeated the paired d.c. measurements $\sqrt{\tau_m f_{mod}}$ times during a total measuring time τ_m equal to the response time τ_r in the a.c. case (see the discussion pertinent to Eqs III.94 and 95 in Subsection 14f-3).

[†] The noise in the paired meter readings is statistically uncorrelated when $\tau \simeq \tau_r$. Any long-term correlation that exists in the original flicker noise is lost in the demodulation process which transforms the frequency band around f_{mod} to a band near 0 Hz (see also the discussion following Eq. III.94 in Subsection 14f-3).

When the flicker noise originates from the flame background emission, we see that the SNR can be greatly improved by modulating only the emission to be measured but not the flame background. The gain factor increases proportionally to $\sqrt{f_{mod}}$, since the flicker noise power that contributes to the meter fluctuations decreases $\propto 1/f_{mod}$ when f_{mod} is raised. Signal modulation leads generally to an improved SNR whenever the spectral power of the e.l.f. noise in the unmodulated background (flame background emission or dark current) near 0 Hz is markedly higher than around the modulation frequency.

Some further SNR expressions for the a.c. case are collected in Table III.6 on page 367. The gain factor $\sqrt{\pi}$ for integration measurements that is listed in the table for the a.c. case is the same as the gain factor for the d.c. case for shot noise. As in the d.c. shot-noise case, we can formally derive the SNR expression for a.c. integration measurements from the corresponding expression for a.c. current measurements by letting the meter response time τ_r tend to infinity while keeping the sampling (= integration) time τ fixed (see Subsection 3 in Appendix A.7). After demodulation the original flicker-noise components around f_{mod} are transformed to a noise current with a practically white spectrum in a frequency band close to 0 Hz. The reading meter following the demodulator 'sees' therefore only *white* noise in the relevant low-frequency band: $0 < f < \tau^{-1}$ where $\tau^{-1} \ll f_{mod}$. The effect on the SNR when $\tau_r \to \infty$ for fixed τ therefore is here the same as in the d.c. case with white shot noise (see Subsection 14f-2).

14g THE MEASUREMENT OF ABSOLUTE INTENSITIES AND INTENSITY RATIOS

Absolute intensities. If the transmission factors, geometrical dimensions, etc., of all components of the optical train as well as the responsivity of the photo-detector and the sensitivity of the measuring instrument were known, the meter readings could be converted to absolute radiant quantities by calculation only. However, such calculations are tedious and the results are in general not expected to be very accurate. The usual procedure for measuring *absolute intensities* is therefore to calibrate the whole set-up by means of a secondary radiation standard (e.g., a tungsten strip lamp; see below) which in turn is calibrated by comparison with a primary standard (e.g., a cavity at the melting point of gold, which radiates as a black body through a small opening; see Heusinkveld 1966). Absolute intensity measurement of the flame emission is then in fact performed by comparing the readings from the flame and from the radiation standard. In this comparison a linear response of the measuring system and identical measuring conditions are assumed. Identical measuring conditions, as far as the optics are concerned, are realized most simply by replacing the flame by the radiation standard or by imaging the latter on a plane through the flame axis as in the line-reversal set-up (see

THE MEASUREMENT OF EMISSION   Sect. III.14g

Sect. III.10c-5). In the latter case optical losses in the imaging system should be taken into account. If the limiting stop and angular aperture of the optical train are uniformly filled with light in both cases, the optical conductance, L, of the radiation beam passing the optical train is identical. One has then the advantage that the actual value of L drops out when the flame emission has to be calibrated in terms of radiance by means of a standard with known (spectral) radiance.

In the following we shall restrict ourselves to the calibration of the (spectral) flame radiance, from which, e.g., the concentration of emitting species can be derived (see Sect. V.2a). Besides, radiation standards incorporating a well-defined radiating surface (e.g., an incandescent tungsten strip) are usually calibrated in terms of spectral radiance. The flame radiance can be simply converted to other radiant quantities (such as radiant intensity, total radiant flux or radiant flux per unit of flame volume; see definitions in Appendix A.5) if the dimensions of the radiating flame volume considered are known.

Let us assume that a continuum source with known spectral radiance $B_\lambda^s(\lambda)$ is used as a standard and that the above measuring conditions as to L, etc., are fulfilled. We also assume that the spectral radiance does not vary significantly within the effective bandwidth, $\Delta\lambda_{eff}$, of the spectral apparatus (see for definition Sect. III.4b). We first consider the case when the flame emission to be calibrated is also a spectral continuum with spectral radiance $B_\lambda(\lambda)$. We then have from Eq. III.62 for the corresponding photocurrents, i_c^s and i_c, respectively,

$$\frac{i_c}{i_c^s} = \frac{B_\lambda(\lambda_0)}{B_\lambda^s(\lambda_0)} \ . \tag{III.102}$$

Through Eq. III.102 we find the absolute value of $B_\lambda(\lambda_0)$ at the wavelength setting λ_0 of the spectral apparatus, if we determine i_c/i_c^s from the corresponding ratio of meter readings. However, when the variation of $B_\lambda^s(\lambda)$ with wavelength within $\Delta\lambda_{eff}$ cannot be disregarded, one must resort to Eq. III.61 and take into account explicitly the instrumental profile function $A(\lambda_0 - \lambda)$. When the spectral radiance of the flame varies noticeably within $\Delta\lambda_{eff}$, only some sort of weighted average of $B_\lambda(\lambda)$ within this interval is found, unless the wavelength dependence of the latter is known. We refer to Bezemer (1976) for a further discussion of such complicating situations.

When the absolute radiance $B(\lambda_0)$ of a spectral line with half-intensity width much less than $\Delta\lambda_{eff}$ is to be determined, application of Eq. III.60 for the line emission and Eq. III.62 for the continuum standard emission yields

$$\frac{i_l}{i_c^s} = \frac{B(\lambda_0)}{B_\lambda^s(\lambda_0) \Delta\lambda_{eff}} \ . \tag{III.103}$$

363

Eq. III.103 enables us to determine the absolute line radiance if $\Delta\lambda_{eff}$ is known (see Sect. III.4b for the measurements of $\Delta\lambda_{eff}$; see also de Galan and Winefordner 1968, and Reif, Fassel and Kniseley 1976 for the special case when a monochromator with exit slit wider than the image of the entrance slit is used).

Systematic errors in the absolute measurement of intensities may arise from:

1. the contribution of unnoticed stray light to the signal readings, especially in the case of continuous radiation (see Sect. III.4f);

2. the occurrence of polarization effects in the standard lamp and in the spectral apparatus (see Sect. III.4e);

3. the nonuniform filling of the limiting stop and angular aperture of the optical train by the flame or the standard lamp; differences in the filling of the photocathode area when the sensitivity of the latter varies from place to place (see Sects III.5d and 16b);

4. deterioration of the standard lamp by ageing and irreproducibility of its operating conditions (see below);

5. possible deviations from the conditions underlying Eq. III.102 or 103, especially when the monochromator bandwidth is so large that the spectral radiance cannot be taken as constant or so small that the bandwidth becomes comparable to the flame line width (see de Galan and Winefordner 1968 for a discussion of the errors made in the latter case);

6. any systematic errors in the calibration of the standard lamp (see below) or in $\Delta\lambda_{eff}$;

7. deviation from linearity of the spectrometer response, especially if the optical signals from the flame and standard lamp are widely different (see also Sect. III.13);

8. any unnoticed variation in the spectrometer sensitivity in the course of the calibration procedure.

In general, an accuracy of better than, say, 10% may be considered as feasible in the absolute measurement of flame intensities in the visible and near-u.v. regions of the spectrum. This holds at least if certain precautions and check measurements are made and if the measurements are not affected by noise or spurious signals (see also Sect. 14e).

Intensity ratios. Intensity ratios for one and the same spectral line are directly given by the ratios of the corresponding spectrometer readings, if we assume a linear detector response. When the intensity ratio for two different spectral lines at λ_1 and λ_2 respectively is to be determined, we have to take into

THE MEASUREMENT OF EMISSION   Sect. III.14g

account the wavelength dependence of the properties of the spectrometer. According to Eq. III.60 we find the ratio of the radiances, $B(\lambda_1)/B(\lambda_2)$, from the spectrometer readings if we know the relative variation of transmission factor $\tau(\lambda)$ times detector responsivity $s(\lambda)$ with wavelength. We assume here that the optical conductance L of the radiation beam transmitted by the spectrometer does not depend on λ. This holds, for instance, in the case when L is determined by the slit width and angular aperture of the monochromator used (see Eq. III.3). When the slit width is varied by a known factor, L varies by the same factor if slit-diffraction effects can be neglected.

The relative variation of the product $\tau(\lambda) s(\lambda)$ with wavelength can be determined by means of a standard lamp emitting a continuous spectrum with known spectral radiance $B_\lambda^s(\lambda)$. We then see from Eq. III.62 that the relative variation of $\tau(\lambda) s(\lambda)$ can be found if we scan the spectrum of the standard lamp while keeping L constant and correcting for the variation of effective bandwidth $\Delta\lambda_{eff}$. For a fixed slit width, the latter variation is easily calculated from the known dispersion of the monochromator (if slit-diffraction effects can be disregarded; see Eq. III.6).

It is recalled from Sect. III.4b that the above considerations hold only if the width of the spectral lines is small compared to $\Delta\lambda_{eff}$. When this condition is not fulfilled, one can determine the ratio of the integrated intensities of two different lines by scanning them over wavelength and measuring the ratio of the areas under the curves. The latter ratio is independent of the spectral profiles of the lines (see Sect. III.4b).

Radiation standards. The standard most often used in flame spectrometry is the tungsten strip lamp. A flat strip of tungsten of, say, $2 \times 20\,mm^2$ is heated electrically by a current of 10 - 20 A in a glass or quartz casing containing a vacuum or, preferably, argon gas to suppress evaporation. The strip may attain temperatures of up to 2800 K, corresponding to a radiance temperature (see Sect. II.5c-3 and Eq. II.301) of about 2500 K, which is somewhat dependent on wavelength. The lamp is calibrated with the flat glass (or quartz) window against a primary radiation standard, generally at a wavelength of 6500 Å. The accuracy of the calibration is roughly 2% in flux and is somewhat dependent on wavelength and temperature. For the calculation of the real temperature of the strip from the calibrated radiance values and the calculation of radiances at wavelengths other than those used in the calibration, one uses tables of the emission factor (see Appendix A.5) of tungsten as a function of wavelength and temperature (see, e.g., de Vos 1954). The validity of the calibration of such lamps performed by standard laboratories is affected in the long run by the use of high temperatures which cause changes inside the strip and evaporation of the tungsten from the strip which is then deposited

Chapt. III   GENERAL INSTRUMENTAL ASPECTS AND EXPERIMENTAL METHODS

on the viewing window. The useful wavelength range is limited in the u.v. at about 3000 Å.

Greater spectral radiance throughout the spectrum, and especially in the u.v., can be obtained with the use of the anode of a carbon arc (radiance temperature about 3800 K; see Schurer 1968) and a high-pressure xenon lamp (radiance temperature about 5000 K; see ter Heerdt 1979), but the stability, the uniformity (see Snelleman 1968) and the reproducibility of these sources are inferior to those of the strip lamp.

Further details about the design and operation of these calibration sources are found in Sect. III.7 (see also Stair, Schneider and Jackson 1963, and Young 1974a).

NOTES TO TABLE III.6

† Expressions for signal-to-noise ratio (SNR) are given when the noise in a stationary background current i_b (photocathode current due to flame-background radiation or to thermionic emission) is dominant. Paired readings are assumed, i.e. the signal is found from the difference between a background reading and a consecutive signal-plus-background reading. i_s represents the d.c. photocathode current due to the emission signal in the d.c. case, and the RMS value of the corresponding a.c. component in the a.c. case. The expressions hold for a linear amplifier + read-out system with an exponential step response. See Table III.5 on page 347 for the meaning of sampling time τ, response time τ_r, integration time τ_i and f_{mod}, and see the text in Sect. III.14 for the meaning of the other parameters.

‡ The SNR for paired integration measurements is found by multiplying the SNR tabulated at the left-hand side by the gain factor, and by replacing τ_r by τ_i.

* Only the emission signal is assumed to be modulated, not the background-with-noise; if the latter is also modulated, the same SNR expressions apply as in the d.c. case apart from a numerical factor of order unity.

¶ F, F', F'', ... denote numerical factors of order unity that depend on the waveform of modulation, the type of a.c. detection, etc.

THE MEASUREMENT OF EMISSION   Sect. III.14g

Table III.6
Theoretical Expressions for SNR in Emission Measurements[+]

D.C. / a.c.	Specification (signal; background)	Background shot-noise		Background flicker-noise	
		SNR at $\tau = \tau_r$	gain factor for integration ($\tau_i \hat{=} \tau_r$)[‡]	SNR at optimal τ ($= \frac{1}{8}\tau_r$)	gain factor for integration (any τ_i)[‡]
d.c.	general case	$i_s \sqrt{\tau_r/(2\pi e \bar{i}_b)}$ (Eq. III.77)	$\times \sqrt{\pi}$	$0.62\, i_s/(\bar{i}_b \sqrt{A})$ (Eq. III.93)	$\times 1.0$ (Appendix A.7-5)
	metal line on background continuum	$B^s_{\lambda_0} \sqrt{\dfrac{L\tau(\lambda_0)s(\lambda_0)\Delta\lambda_{\mathrm{eff}}\,\tau_r}{2\pi e B^{bg}_{\lambda_0}\Delta\lambda_{\mathrm{eff}}}}$ (Eq. III.78)	$\times \sqrt{\pi}$	$0.62\, B^s/(B^{bg}_{\lambda_0}\Delta\lambda_{\mathrm{eff}}\sqrt{A})$	$\times 1.0$
	metal continuum on background continuum	$B^s_{\lambda_0} \sqrt{\dfrac{L\tau(\lambda_0)s(\lambda_0)\Delta\lambda_{\mathrm{eff}}\,\tau_r}{2\pi e B^{bg}_{\lambda_0}}}$	$\times \sqrt{\pi}$	$0.62\, B^s_{\lambda_0}/(B^{bg}_{\lambda_0}\sqrt{A})$ (Eq. III.96)	$\times 1.0$
a.c.[*]	general case	SNR at $\tau = \tau_r$	gain factor for integration ($\tau_i \hat{=} \tau_r$)[‡]	SNR at $\tau = \tau_r$	gain factor for integration ($\tau_i \hat{=} \tau_r$)[‡]
		$F i_s \sqrt{\tau_r/(2\pi e \bar{i}_b)}$ (Eq. III.99)	$\times \sqrt{\pi}$	$F''(i_s/\bar{i}_b)\sqrt{\tau_r f_{\mathrm{mod}}/(2A)}$ (Eq. III.101)	$\times \sqrt{\pi}$
	metal line on background continuum	$F' B^s_{\lambda_0} \sqrt{\dfrac{L\tau(\lambda_0)s(\lambda_0)\Delta\lambda_{\mathrm{eff}}\,\tau_r}{2\pi e B^{bg}_{\lambda_0}\Delta\lambda_{\mathrm{eff}}}}$	$\times \sqrt{\pi}$	$F'''\left(\dfrac{B^s_{\lambda_0}}{B^{bg}_{\lambda_0}\Delta\lambda_{\mathrm{eff}}}\right)\sqrt{\dfrac{\tau_r f_{\mathrm{mod}}}{2A}}$	$\times \sqrt{\pi}$
	metal continuum on background continuum	$F' B^s_{\lambda_0} \sqrt{\dfrac{L\tau(\lambda_0)s(\lambda_0)\Delta\lambda_{\mathrm{eff}}\,\tau_r}{2\pi e B^{bg}_{\lambda_0}}}$	$\times \sqrt{\pi}$	$F'''\left(\dfrac{B^s_{\lambda_0}}{B^{bg}_{\lambda_0}}\right)\sqrt{\dfrac{\tau_r f_{\mathrm{mod}}}{2A}}$	$\times \sqrt{\pi}$

Chapt. III   GENERAL INSTRUMENTAL ASPECTS AND EXPERIMENTAL METHODS

15. THE MEASUREMENT OF ABSORPTION

15a   INTRODUCTION

Absorption measurements may be carried out for several reasons: to determine (relative or absolute) number densities of metal atoms (ions, molecules) in the flame (see Sects V.2b, 2c and 2d); to derive line profiles (see Sect. VII.3b); to check on the existence of under- or overpopulation of excited atomic (or molecular) levels (see Sect. VI.4); to record molecular (or atomic) absorption spectra; to determine line-broadening parameters from measurements of the integral absorption and/or peak absorption factor (see Sect. VII.3b), etc. Quite often absorption measurements are preferable to emission measurements which could in principle yield the same information. For example, when there is doubt as to whether the excited level is thermally populated, the information on the number densities ought to be extracted from absorption measurements of ground-level population (see also McEwan 1965, and Zeegers 1966); when the excitation energy of the atomic level studied is relatively high or when the flame temperature is relatively low, its population may be too low to yield precise information from emission measurements; when height-dependent or off-axis measurements are involved, absorption measurements are affected much less by temperature variations. It may also happen that a number density has to be determined in absolute measure and that (relative) absorption measurements are more easily performed than the corresponding (absolute) emission measurements. Of course, for absorption measurements one needs an auxiliary light source (with known spectral characteristics). As examples of absorption measurements we mention the measurement of the absorption factor, α, or the absolute integral absorption, A_t (see Sect. II.5c-2 and Sect. III.15d-1) and the determination of absorption line profiles or peak absorption factors (see Sect. III.15d-2).

In the following we shall concern ourselves with the practical aspects of absorption measurements. Additional information on measuring techniques, analytical applications, etc., in atomic absorption spectrometry is to be found in numerous textbooks (e.g. Rubeška and Moldan 1969, Dean and Rains 1969, 1971, Christian and Feldman 1970, L'vov 1970, Willis 1970, Winefordner 1971a, Kirkbright and Sargent 1974, and Alkemade and Herrmann 1979); and articles (e.g., Simon 1962, Walsh 1966, de Galan and Winefordner 1967, McGee and Winefordner 1967, Rann 1968, Snelleman 1968, Fiorino, Kniseley and Fassel 1968, Zeegers, Townsend and Winefordner 1969, Hollander *et al.* 1970, and Smyly *et al.* 1971).

First of all we shall outline some practical set-ups for absorption measurements (Sect. 15b), then we shall discuss the various components and characteristics of these set-ups (Sect. 15c) and try to demonstrate from two examples how in the measuring practice optimization of experimental conditions can be realized (signal-to-noise considerations; Sect. 15d). In the latter section the absolute measurement

THE MEASUREMENT OF ABSORPTION   Sect. III.15b

of the integral absorption A_t (Sect. 15d-1) and of the wavelength-dependent absorption factor $\alpha(\lambda)$ (Sect. 15d-2) of a metal line is discussed, and consideration is given to the special problems involved in atomic absorption line-profile measurements.

Very weak absorption lines can be detected by placing the flame with atomic vapour inside the cavity of a dye laser emitting a continuous spectrum with a bandwidth of, say, 100 Å. The laser action at the wavelength of the absorption line depends sensitively on the radiation loss due to absorption and may even be quenched. An 'amplified' absorption line then appears in the output spectrum of the laser. Dye-laser amplified atomic absorption spectroscopy of Na, Sr, Ba$^+$ and rare-earth lines in flames has been demonstrated by Thrash, von Weyssenhoff and Shirk (1971), Konjevic and Konjevic (1973) and Maeda *et al.* (1976). Special methods such as these will not be considered in this section, which is restricted to usual, linear absorption spectroscopy. These methods have proved to be also useful in detecting (weak) absorption bands of, e.g., BaO and CuH (see Thrash, von Weyssenhoff and Shirk 1971) and monoxides of rare-earth elements (see Maeda *et al.* 1976).

15b   GENERAL OUTLINE OF SET-UPS USED FOR ABSORPTION MEASUREMENTS

Measurement of absorption factors, α, is basically a relative measurement, i.e. the beam intensity transmitted through the absorbing metal vapour in the flame is compared with that passing in the absence of metal vapour. If we denote the photosignals generated in the detecting system by the two beams by i_t and i_0, respectively, we generally have: $\alpha = (i_0 - i_t)/i_0 = 1 - i_t/i_0$.

In a *single-beam* set-up, i_t and i_0 are read consecutively either by feeding the flame with metal vapour or not (see Subsection 15b-1). The contribution of the flame background emission to the photocurrent detected has to be corrected for, or eliminated. The *double-beam* set-up is more complicated but provides possibilities for reducing the effect of instabilities or low-frequency noise in the common irradiating light source, and also in the photodetector if a common photodetector is used. This is of particular importance when measuring weak absorptions which are derived from a relatively small difference between two large signals.

There are essentially two types of double-beam set-ups (see also Sect. III.4f). In the one set-up there are two beams at the same wavelength or with the same spectral composition which are separated in *space*, the probing beam intersecting the flame, the reference beam by-passing it. We shall call this a *double spatial-beam set-up*. In the *dual-wavelength* or *double spectral-beam set-up*, two spatially coinciding beams of different wavelengths are selected, both of which

Chapt. III GENERAL INSTRUMENTAL ASPECTS AND EXPERIMENTAL METHODS

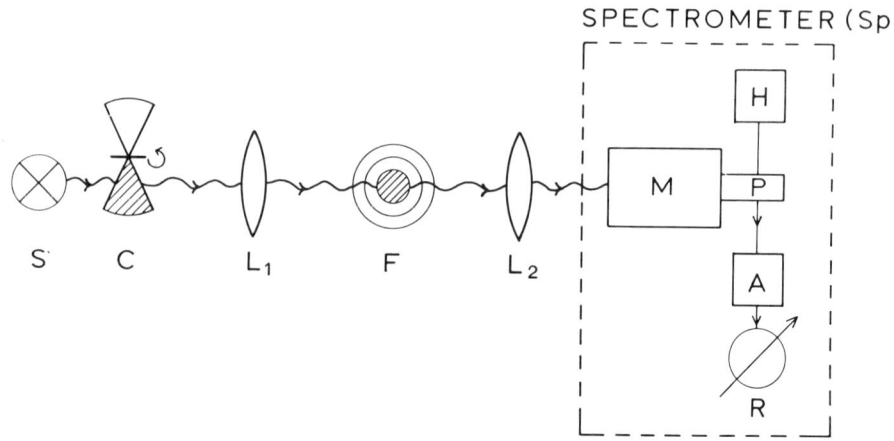

Fig. III.24 Block diagram of single-beam set-up for absorption measurements. S = irradiating light source; C = mechanical chopper; L_1, L_2 = lense; F = flame; M = monochromator; P = photomultiplier tube; H = high-tension supply; A = lock-in amplifier, and R = read-out system.

intersect the same flame. The flame with metal vapour only absorbs the probing beam at the one wavelength but not the reference beam at the other wavelength. The latter beam serves to monitor the variation with time of i_0; it is thus essential that the photocurrent generated by the latter beam, i_{ref}, is proportional to i_0. There are different ways of achieving this monitoring automatically by combining electrically the photosignals of the two beams (see Subsection 15b-2).

15b-1   *SINGLE-BEAM SET-UP*

In Fig. III.24 a typical example of a single-beam set-up is shown as a block diagram.

In order to eliminate the d.c. component of the flame emission and reduce the drift of the zeropoint, a mechanical chopper is placed between the light source and flame to modulate the light beam before it passes the flame. The a.c. component of the photocurrent with modulation frequency is amplified and detected by a lock-in amplifier (see Sect. III.6b). By controlling, e.g. the amplifier gain, the scale reading may be set to 100% transmission (= zero % absorption) when i_0 is measured. In this way a direct reading of the transmission or absorption factor is obtained. When drift in the source intensity, etc., is suspected, the 100% setting should be re-checked at regular intervals or the output signal should be recorded.

When an absorption line profile is being scanned or an extended absorption spectrum is being recorded over a long period of time, long-term instabilities of

THE MEASUREMENT OF ABSORPTION   Sect. III.15b

the light source may be corrected for by monitoring its intensity with a separate
detection + read-out channel. Such a system is in fact an intermediate form between
a single-beam and a double-beam set-up; in the latter, monitoring is carried out
automatically by appropriate combination of the reference and probing signals (see
Subsection 15b-2). An extra detection channel may also be useful in such experiments with single-beam or double-beam set-ups when one has to monitor the absorbing-
flame conditions, e.g., the constant delivery of absorbing species to the flame.
The extra channel may be used, e.g., to monitor the peak absorption of a spectral
line whose profile is being scanned, or the intensity of a particular emission or
absorption line (or band) during recording of an absorption spectrum.

In the range of low absorption factors *scale expansion* can be applied
successfully and results in a gain in sensitivity and reading precision. The zeropoint of the meter deflection is removed from the scale by electronic means (*zero-
suppression*) while the amplifier gain is increased. A given absorption in the
flame produces a larger meter deflection, which in fact now measures directly the
difference between i_0 and i_t.

Other single-beam set-ups may be devised from that shown in Fig. III.24 by
changing the components (see the other sections of this chapter for the choice of
optical and electrical components and modulation and detection techniques).

15b-2   *DOUBLE-BEAM SET-UP*

Figure III.25 shows block diagrams of two particular examples of a double
spatial-beam set-up, incorporating an optical attenuator (= optical wedge) in the
reference beam. In block diagram (1) the two beams are detected by one common
photomultiplier tube. In order to distinguish electronically the two components of
the photocurrent generated by the reference and probing beam, respectively, the two
beams are either modulated at a certain frequency but at opposite phase (as shown
in the figure; see Jansen, Hollander and Alkemade 1977b) or at different frequencies (see Elser and Winefordner 1972).

Block diagram (2) shows a set-up with two photodetectors, one for each beam.
The outputs of the beams are compared by a difference or ratio meter. The latter
diagram may be used with or without modulation of the light beams. The beam intensities can either be modulated mechanically by a rotating (or vibrating) chopper
placed between light source and beam splitter, or by electronic a.c., square-pulse
or square-wave modulation of the current feeding the common light source. When
a.c. modulation is applied, the a.c. components of the photocurrents have the same
frequency and phase. When absorption of a pulsed laser beam at a low repetition
rate is to be measured, set-up (2) without chopper is appropriate.

Numerous variations of the double-beam set-ups shown are possible if one
uses other components or rearranges the optical components (mirrors, lenses)

(1)

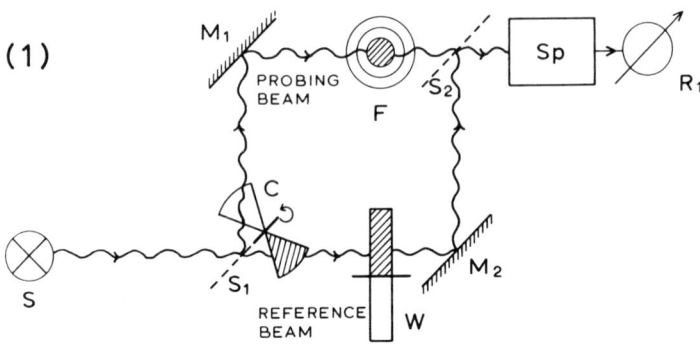

(2)

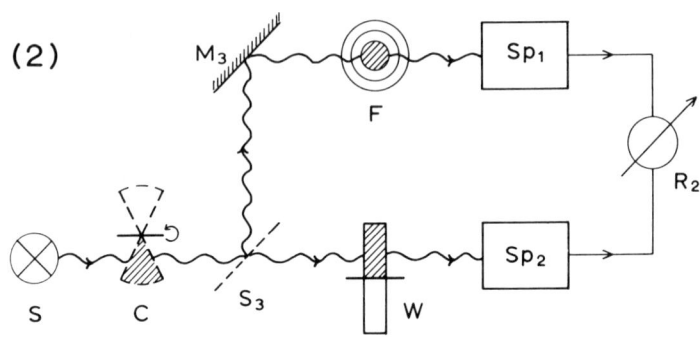

Fig. III.25 shows block diagrams of two typical double-beam set-ups for absorption measurements (see also caption of Fig. III.24); both set-ups possess optical compensation. Block diagram (1) shows the following components: S_1, S_2, S_3 are semi-transparent mirrors; M_1, M_2, M_3 are plane mirrors; W is a (circular) optical-attenuation wedge; Sp = spectral selection system + photodetector + amplifier; the remaining symbols have the same meaning as in Fig. III.24. In block diagram (2) one can discern a different chopper position and two separate selection + detection systems Sp_1 and Sp_2. With the common-detection system (1) the read-out system R_1 measures signal <u>differences</u>, whereas with the system (2) of two separate detectors R_2 measures either <u>differences</u> or <u>ratios</u>.

THE MEASUREMENT OF ABSORPTION   Sect. III.15b

producing the two light beams.

In analytical atomic absorption spectroscopy various double-beam spectrophotometers are now commercially available, although most spectrometers are still of the single-beam type.

An extra detection and read-out channel may again be used to monitor the constancy of flame conditions during prolonged absorption measurements. In their spectral line-scanning experiments Hollander, Jansen, Plaat and Alkemade (1970) monitored the flame conditions by measuring the atomic absorption of alkaline-earth vapour at a fixed (peak) wavelength of the resonance lines, using a separate background line source.

As to the *dual-wavelength* or *double spectral-beam* set-ups (see Sect. III.4f) the two different wavelengths applied may either be different lines from the same light source or different Zeeman components of the same line from the source (see Stephens and Ryan 1975; Uchida and Hattori 1975; Koizumi and Yasuda 1975, 1976, and Visser, Hamm and Zeeman 1976). One may also select two different parts of the spectrum from a spectral continuum source.

As we stated in Sect. III.4f, it is possible to modulate the wavelength instead of modulating the intensity of the background light source. Thus one obtains the advantages of a dual-wavelength method in a single-beam set-up. Elser and Wineforder (1972) combined wavelength modulation with intensity modulation of a continuum background source in flame atomic absorption spectrometry. They claimed an SNR improvement as compared to wavelength modulation alone.

In general, double-beam set-ups may be operated in three different modes for measuring absorption factors α:

(i) null-balance mode, using:

$$i_0 = i'_{ref} ; \quad i_t = i''_{ref} ; \quad \alpha = 1 - \frac{i''_{ref}}{i'_{ref}} ;$$

(ii) difference mode, using:

$$i_0 = i_{ref} ; \quad \alpha = (i_{ref} - i_t)/i_{ref} ;$$

(iii) ratio mode, using:

$$i_0 = i_{ref} ; \quad \alpha = 1 - i_t/i_{ref} .$$

(i) The *null-balance mode* is based on nulling the difference between the photosignals generated by the probing and reference beam, respectively. First the reference signal i_{ref} is adjusted to equalize i_0 which is generated by the probing beam in the absence of absorption by the flame. This adjustment may be accomplished by *optical balancing* of the two beam intensities, e.g., with the aid of an adjustable optical attenuator ('wedge'). It may also be accomplished by an

Chapt. III   GENERAL INSTRUMENTAL ASPECTS AND EXPERIMENTAL METHODS

electrical balancing circuit which compares the photosignals generated by the two beams. The proper optical balancing can be judged in the case of block diagram (1) from the 'nulling' of the lock-in output. In the case of the block diagram (2) the optical or electrical balancing can be adjusted by making use of a difference-detector that compares the two separate photosignals. Scale expansion can again be profitably used here to improve the precision of the null-adjustment (see Sect. 15b-1).

Next, the absorbing species is introduced into the flame and the reference-beam signal is again balanced against the (now attenuated) probing beam signal i_t, to the effect that $i''_{ref} = i_t$. The ratio $i_t/i_0 \equiv 1 - \alpha$ is then equal to the ratio i''_{ref}/i'_{ref}, which can be read from the calibrated scale of the optical or electrical balancing circuit.

In the case of block diagram (1) incorporating one common photodetector, electrical balancing is still possible if one separates the photo-electric current pulses released alternately by either beam by means of a synchronously gated circuit. During one half of the modulation period one gate is open and transmits, e.g., the reference signal to a separate channel. During the other half period a second gate is opened and the first one closed, and thus the probing signal is transmitted to a second channel. The integrated outputs of these channels may then be handled separately and compared with a differential or ratio meter read-out.

A practical example of the latter measuring device is the 'dual-beam-sampler' used for Zeeman scanning experiments by Jansen, Hollander and Alkemade 1977b). In order to improve the signal-to-noise ratio it proved necessary to correct 'beforehand' for the strong flame background radiation. Apart from sampling the two light beams as discussed above, one has also a third channel (referring to closed position of the chopper wheel) in which this flame signal is stored. Essentially one samples during a fraction of a quarter modulation period consecutively reference beam, flame signal, probing beam, and again flame signal. The electric pulse height is read into the channels, and after time-integration and electric subtraction of i_{bg} from i_{ref} and i_t, respectively, the two resulting, corrected signals are fed into a ratio meter that directly produces the signal ratio.

(ii)   In the *difference mode* one first balances the reference signal against the signal of the non-attenuated probing beam so that $i_{ref} = i_0$, as in the null-balance mode. Balancing is again achieved when the lock-in meter or the differential meter in the case of block diagram (1) and (2), respectively, reads zero. The (absolute) absorption signal $(i_0 - i_t)$ is then equal to the difference $(i_{ref} - i_t)$

374

which can be directly read from the deflection of the lock-in meter (see Alkemade and Milatz 1955) or the differential meter in case (1) and (2), respectively. The relative absorption, i.e. the absorption factor α, is derived by dividing the difference $(i_{\text{ref}} - i_t)$ by i_{ref}. The reference signal i_{ref} can be found by blocking the probing beam; for small α values one has to use different meter sensitivities at the separate readings of difference and reference signals, respectively.

(iii) In the *ratio mode* one employs block diagram (2) in which a divider circuit now compares the output currents of the two phototubes. By optical or electrical balancing one adjusts the meter to read unity ratio when no absorbing species is present in the flame, i.e. $i_{\text{ref}}/i_0 = 1$. When absorption occurs, the absorption factor is found from: $\alpha = 1 - (i_t/i_{\text{ref}})$, the ratio i_t/i_{ref} being directly read from the meter. When a logarithmic divider circuit is used, one obtains directly the absorbance as output (see Malmstadt, Enke and Crouch 1974).

It should be noted that this ratio mode cannot be realized directly in diagram (1). This mode can then only be effected by dual-beam sampling techniques.

The following *advantages and disadvantages* of the three different double-beam modes and their modifications are noted.

The *null-balance* can be adjusted very accurately (by using a high amplifier gain) and this mode yields a very smooth and simple measuring procedure. It is suitable for automation. In a non-automated mode the measuring process takes a relatively long time: the null-balance adjustment is a slow procedure. When one common detector is used, one has the advantage that instabilities and flicker noise in source and photodetector response drop out, at least partially (see also Sect. 15c-5). With two photodetectors, instabilities and flicker noise in the responsivity of the photodetectors are not eliminated at all. The application of electric compensation methods is rather obvious in this mode.

The *difference mode* with one common photodetector can be applied very accurately if a lock-in amplifier in combination with a chart paper recorder is used (see Hollander, Jansen, Plaat and Alkemade 1970). The signal i_{ref} is recorded at a relatively low sensitivity of the recorder, whereas the difference signal $(i_{\text{ref}} - i_t)$ is recorded with higher sensitivity of the recorder. This modification of the method of scale expansion (see Sect. 15c-5) permits absorption signals to be measured rapidly and accurately. In practice one can measure an absorption factor $\alpha = 0.001$ in flames with a relative accidental error of 10%.

In the difference mode instabilities and flicker noise from the source have a strongly reduced effect on the zero-point $(\alpha = 0)$. The same holds for the instability of the photodetector when one common detector is used. This mode is faster than mode (i).

Chapt. III   GENERAL INSTRUMENTAL ASPECTS AND EXPERIMENTAL METHODS

The *ratio mode* is simple and rather rapid. Instabilities and flicker noise from the source are completely eliminated in this mode, whereas instabilities from the two independent photodetectors are not reduced.

15c   DETAILED DISCUSSION OF COMPONENTS OF THE EXPERIMENTAL SET-UP

In this section we discuss some essential components of the experimental set-ups for absorption measurements.

15c-1   *LIGHT SOURCES*

There is detailed discussion of light sources in Sect. III.7. Here we shall consider the selection of light sources, and at the same time we shall give some technical hints.

The question whether a spectral continuum or a line source should be used, and, to a lesser degree, whether d.c., a.c. or pulsed operation is advisable, depends on the specific problem under consideration. We summarize here a number of advantages and disadvantages of either type of source (see also textbooks mentioned in Sect. 15a, and Nitis, Svoboda and Winefordner 1972, and Winefordner and Vickers 1972).

A continuum source in combination with a monochromator, etc., is versatile with respect to choice of wavelength, whereas a line source is wavelength-specific and often requires only a simple filter for selection of the one wavelength wanted. The wavelength setting in the former case may show instrumental drift; in the latter case it is absent. With the continuum source we have competition between the bandwidth of the wavelength-selector (a narrower bandwidth entails a larger absorption factor in atomic absorption measurements) and sufficiently high radiance. Of course a tunable dye laser with sufficiently small bandwidth combines versatility (in a limited λ range) with high radiance, and overcomes this difficulty. Its bandwidth can be chosen larger or smaller than the absorption line width, while its radiant power is orders of magnitude larger than that of the combination of conventional source + monochromator (see Sect. III.7). With line sources for peak absorption measurements the width of the emission line should be sufficiently small compared to the absorption line width in order to enhance the absorption factor as far as possible (i.e., absence of self-reversal and self-absorption, see Sect. III.7). In some applications (part of) a band spectrum of an auxiliary flame may be used as background source in absorption measurements of the same bands in the flame to be analyzed (see, e.g., Zeegers 1966).

The choice of the source may further depend on requirements like (spectral) radiance, wavelength or wavelength-range available, stability, etc. But also more constructional aspects may play a role. As an example we mention the preference of an electrodeless-discharge lamp to a hollow-cathode lamp in Zeeman-scanning

THE MEASUREMENT OF ABSORPTION   Sect. III.15c

measurements (see Sect. III.7).

In general the available solid angle under which the source emits its radiation is mostly not a factor of importance in absorption measurements.

15c-2 *OPTICAL BEAM FORMATION*

Special demands may be imposed on the shape and course of the light beam traversing the flame, depending on the kind of absorption measurements to be done and the goal of these measurements. We have two extreme approaches:

(i)   *Converging* beam: the background source is focused under a large solid angle on the centre of the absorbing flame, and this centre is again imaged on the entrance plane of the wavelength-selector.

(ii)   The beam travelling through the flame should be as parallel as possible.

In the choice between (i) and (ii) or intermediate cases several considerations may play a role; we distinguish between the following practical cases:

(1)   Demand of maximum radiance, for example, when the strong flame background emission disturbs the absorption measurements. One then has to focus the source under a maximum solid angle, Ω, in the flame ('converging beam').

(2)   When height-dependent or off-axis absorption measurements are required, one uses narrow parallel beams ('pencil beams'). The maximum acceptable Ω depends inter alia on the diameter of the flame (see Sect. III.14b).

(3)   When a precise value of absorptivity k is to be derived from the experimental α value in a uniform flame, the beam should be parallel but not necessarily narrow. The demand here is that the absorption path length inside the flame be equal for each ray within the light beam.

> It should be noted that the use of polarizers, interference filters and quarter-wave plates (phase-retardation plates) implies that the beams traversing these components are either parallel or slowly converging (up to $7°$ divergence). The diameter of quarter-wave plates available limits the beam diameter to, say, 1 cm.

(4)   When dye lasers are used as irradiating sources, the degree of saturation (see Sect. II.5c-4) can be varied by varying the beam cross section or by weakening the beam intensity. Variation of beam cross section can be obtained without appreciable loss of beam parallelity.

(5)   When weak absorption in a low-pressure flame requires the use of a multiple-pass system, slowly converging or parallel beams are preferred (see Bleekrode 1968a, and Gaydon and Wolfhard 1970). This holds especially when one wants to calibrate absorption measurements by measuring the absorption factor as a function

of the number of multiple passes through the flame (see Zeegers 1966).

The simplest way of obtaining 'pencil beams' is to place two diaphragms at some distance from each other in the light beam, but this leads to a considerable loss of intensity. With laser sources one makes successful use of beam narrowers without loss of total radiant power.

The demands of high beam flux and of high spatial resolution often conflict. In the case of pencil beams the optical conductance of the beam in the flame, L_{fl}, is usually smaller than the conductance of the optical train, L_{opt}. Conversely, when one wants to achieve a high beam flux by using converging beams, L_{opt} is usually the limiting factor. We refer to Sect. III.14b for a discussion of the optimal measuring conditions. The role of L_{fl} or L_{opt} here is analogous to that in emission measurements.

15c-3 *MODULATION AND BEAM SPLITTING*

Electronic source-modulation allows one to choose different types of wave forms: square-wave, sine-wave, square-pulse, etc. For a systematic discussion of the optimal waveform the reader is referred to Hieftje (1972). With mechanical beam modulation one has often square-wave modulation.

In block diagram (1) of Fig. III.25 mechanical modulation of the two beams at opposite phase is achieved by a rotating chopper which is so positioned that when one beam is blocked by one of the chopper blades the other is transmitted through an opening between two blades, and vice versa. The phase difference between the a.c. signals of the two beams may critically depend on the position of the chopper with respect to the two beams. However, when a lock-in detector is used, the relative phase, β, of the two a.c. signals need not be exactly equal to 180° in order to achieve a zero reading for equal amplitudes. This holds because only the components of the two a.c. signals that are in phase with the a.c. reference signal of the lock-in contribute to the reading of the lock-in meter (see Fig. III.26; see also Sect. III.6b). By turning the chopper motor around its axis, the phases of the resulting two a.c. signals with respect to the lock-in reference signal can always be made such that the meter deflection becomes zero. Zero reading is achieved if the angles δ_1 and δ_2 in Fig. III.26 are made equal.

When a nonsynchronous, ordinary a.c. meter is used instead of a lock-in meter, however, zero reading can never be obtained as long as the phase difference β of the two a.c. signals is not exactly equal to 180°. This holds because the a.c. meter measures the amplitude of the resultant a.c. signal (represented by vector OE in Fig. III.26), which can only be made zero if the two a.c. signals have equal amplitudes and opposite phases. Therefore in double-beam set-ups where two beams, modulated at opposite phase, are to be balanced with one common photodetector, synchronous measuring devices are strongly recommended.

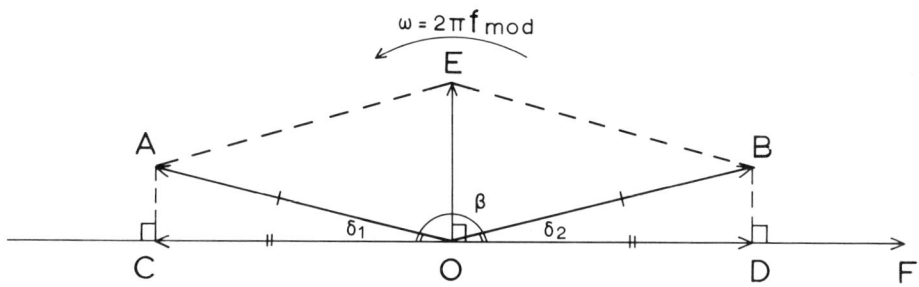

Fig. III.26 OA and OB are vectors representing the a.c. signals from the two modulated beams. Vector OF represents the a.c. reference signal of the lock-in meter. These three vectors rotate with angular velocity $\omega = 2\pi f_{mod}$ (f_{mod} = modulation frequency) around O. The meter reading is proportional to the difference of the in-phase components of OB and OA, represented by the projections OD and OC, respectively, of these vectors on the OF vector. When OA = OB, zero reading is obtained if the angles δ_1 and δ_2 are equal. The resultant vector OE is then perpendicular to the direction of vector OF, i.e. the resultant a.c. signal is then 90° out of phase with respect to the lock-in reference signal.

The formation and modulation of two light beams at (exactly) opposite phases can be achieved much more readily if one replaces the beam splitter plus blocking chopper blades in block diagrams Fig. III.25 by highly reflecting chopper blades (mirrors), a so-called butterfly mirror. The first beam is transmitted through the openings between the blades, whereas the second beam is split from the original incoming beam by reflection against the mirror blades. The rotation of the blades causes the two beams to be modulated at opposite phase irrespective of the relative position of the rotating disk on which the openings and mirrors are arranged in a symmetrical array.

15c-4 *DETECTION*

The detection and evaluation of the signal depends on whether d.c., a.c. modulated, or pulsed light sources are used, whether single- or double-beam set-ups are applied, and whether the two beams are modulated at the same or opposite phase (see Subsections 15b-1 and 15b-2).

With repetitively pulsed sources the signal can be detected with a single-channel box-car integrator with time-scanning, or a multi-channel sampling and integrating device. In these detection systems the signal can be time-resolved. We refer to Sect. III.6b for a further discussion and for references.

With single-beam set-ups the application of zero-suppression and scale expansion may be advisable when the absorption factor to be measured is relatively low (see Subsection 15b-1).

With double-beam set-ups one must realize that the d.c. mode necessarily

Chapt. III   GENERAL INSTRUMENTAL ASPECTS AND EXPERIMENTAL METHODS

implies the use of two separate detection systems; beam modulation by chopper with opposite phase requires only one detection system. For a further discussion on the feasibility of one common or two separate detection systems the reader is referred to Subsection 15b-2.

15c-5 *STABILITY*

Here we consider the mechanical and electronic stability of the various components and parts in a set-up for absorption measurements. Noise and drift effects will be considered separately in Subsections 15d-1 and 15d-2.

The instabilities considered here may be due to internal and external causes; for example, the xenon source may show arc wander (see Sect. III.7), the chopper motor may vibrate, causing vibrations of lenses, mirrors, etc. When distinguishing between the error sources due to instabilities in the various components, it is useful to discuss single-beam and double-beam set-ups separately.

In the *single-beam* set up of Fig. III.24 we have to deal with instabilities in the source, chopper, flame, wavelength-selector, photodetector response, amplifier and read-out system. As to the light source, two types of instabilities may show up: spatial and purely temporal ones. The spatial instability, i.e. a change in position of the radiating surface with respect to the optical axis ('arc wander') may lead to variations in the signal to be detected. Furthermore phase fluctuations in the a.c. component of the photocurrent may occur when mechanical modulation is applied. The remedy for arc wander is to collect the light from the whole radiating surface on the photodetector, or through the monochromator slit. In this case the source should not be imaged on slit or detector! When the phase-adjustment of the lock-in is optimal (i.e. $\cos\phi = 1$), the effect of phase fluctuations is minimal; this effect is, of course, absent when the detector is not phase-sensitive. Slow temporal instability of the source intensity can be taken into account by a separate monitoring channel (see Sect. 15b) or by improving the stability of the source power supply.

The vibrations of the chopper motor may cause extra modulation of the passing light beam and lead to frequency beats in the signal to be recorded. These vibrations may also cause the other components of the set-up to vibrate. The latter problem is circumvented either by solidly mounting the chopper or by mechanical damping of the vibrations. Additional disturbances may be caused by air turbulence produced by the chopper blades.

Since absorption measurements are essentially relative measurements (see the definition of absorption factor), spatial inhomogeneity of the photocathode surface is not a critical factor if the distribution of illumination over the photocathode remains the same.

THE MEASUREMENT OF ABSORPTION   Sect. III.15c

Instabilities may stem from irregularities in the metal vapour supply or metal atomization, turbulence, wavering of the flame due to draught, etc. Monitoring of the flame conditions (see above) may be helpful.

With the *double-beam* set-ups, a number of instabilities and drift effects are automatically cancelled out, when both beams are equally affected and treated. This holds especially when the two beams are modulated in opposite phase and a single photodetector with lock-in amplifier is used. It requires extreme care, however, to produce two practically identical light paths that will be affected in exactly the same way by all external disturbances. Vibrations of the chopper motor may easily affect the two beams differently and should therefore be eliminated. Careful attention should be given to the symmetric treatment of the two light beams, especially in the case of mechanical modulation.

In the double-beam set-ups shown in Fig. III.25 one beam passes through the flame, while the other passes through an optical wedge (= attenuator) for balancing. It is clear that because of this asymmetry in the two light beams, instabilities in the position of either the flame or the wedge will not be compensated for.

There is a difference in the reactions of set-ups (1) and (2) in Fig. III.25 to instantaneous instabilities in the source. With set-up (2) the two photodetectors register these instabilities simultaneously, as both beams are either modulated at the same phase or are not modulated at all. The zero reading will then not be affected, provided the two optical paths are identical. With set-up (1), however, source instabilities that occur within one half of the modulation period are obviously not balanced. This has consequences for the influence of flicker noise in the source emission on the zero reading (see Sect. 15d).

The effects of instabilities and flicker noise in the responsivity and in the dark current of the photodetector on the zero reading are reduced when double-beam set-up (1) with one photodetector is used. However, with double-beam set-up (2) uncorrelated instabilities and flicker noise in the responsivity of the two detectors are not reduced by applying a.c. modulation. A.c. modulation of the source only helps to reduce the effect of flicker noise in the dark currents.

15c-6 *INTERFERENCE FROM THERMAL METAL EMISSION AND FLAME BACKGROUND*

When modulated light sources are used, the d.c. components of the radiation from the flame background and from the thermal emission of the metal vapour are eliminated automatically because of the a.c. detection used. Only those a.c. components of the fluctuations that are within the bandpass of the a.c. measuring system remain. A.c. measuring techniques are therefore not affected by drift in these emissions; their flicker noise components can be reduced substantially if one chooses a sufficiently high modulation frequency (see Sect. 15d). With d.c.

Chapt. III   GENERAL INSTRUMENTAL ASPECTS AND EXPERIMENTAL METHODS

measuring techniques one can correct for the drift in these emissions either by re-checking them at regular time intervals or by recording them via a separate channel. Note that the flame background may be influenced by the supplying of a metal salt solution.

Whenever the flame background causes serious interference, it is advisable to use a smaller spectral bandwidth, a stronger light source or to apply an a.c. measuring technique. When the thermal emission of the atomic line itself interferes, only the last two possibilities are helpful. Dual-wavelength techniques, with either d.c. or a.c. detection, may also be used to reduce the d.c. and e.l.f. components of the background emission.

In the far-u.v. the flame background may also appear in absorption (see Sect. IV.8). This interference is, of course, not eliminated by source modulation, but it can be reduced by a dual-wavelength technique. Suppression of background interference has been investigated extensively in connection with the analytical application of atomic absorption spectroscopy (see the textbooks cited in Chapt. I).

15c-7 *SYSTEMATIC ERROR SOURCES*

In discussing systematic errors in absorption measurements we should make a distinction between the errors made in the absorption measurements proper and in the interpretation or evaluation of these measurements. Systematic errors of the latter kind may occur, for example, when one measures an absorption factor, α, of an atomic line with a narrow-line source in order to derive in that way the peak absorption factor. If the collisional line shift (see Sect. VII.2b-3) and the absorption line profile are not well-known, the peak absorption factor calculated from the measured α value will be erroneous.

Another example of systematic error sources is the derivation of the integral absorption, A_t, by means of Eq. II.271 from the α value measured with a continuum source and a monochromator. When the monochromator wavelength-setting is not properly tuned at the absorption line and its bandwidth is not accurately known or large enough compared to the absorption line width, a systematic error in A_t will result.

As a third example we mention the derivation, through the application of Beer's law, Eq. II.257, of the atomic absorptivity, $\kappa(\lambda)$, from the measured $\alpha(\lambda)$ value. When the absorption path length, l, is not the same for all light rays passing through the flame, or when the atomic density is not uniform, a systematic error will again arise.

Further examples will be given in Chapts V and VII where the derivation of metal densities and spectral line profiles from absorption measurements is

THE MEASUREMENT OF ABSORPTION   Sect. III.15d

discussed. Some of the above systematic error sources may be eliminated when one is interested in ratios rather than in absolute values of the quantities derived.

Sources of systematic errors in the measurement of absorption factors can be summarized as follows.

1. Contribution of thermal line emission or flame background to the readings (see preceding subsection).

2. Contribution of stray light to the readings, especially in the case of continuous radiation sources (see Sect. III.4f-1).

3. Scattering of the incident radiation by unevaporated particles in the flame (see Sect. III.9).

4. Nonlinearity of the photodetector or measuring apparatus (see Sect. III.13); unless optical null-balance is applied, special care should be taken when an intense pulsed laser beam is used.

5. Systematic errors in the wavelength-dependent calibration of the optical attenuator used in the optical null-balance mode (see Subsection 15b-2); systematic errors may also arise in this mode when the reference beam passing through the optical wedge is strongly divergent or polarized, or when its relative position on the wedge is not well-defined because the beam cross section is too large.

6. Insufficient spectral isolation of the emission line from concomitant emission lines or continuum radiation in the spectral line source.

7. The occurrence of saturation of the absorption transition if one uses an intense laser beam and one is interested in the undisturbed absorption factor.

8. Systematic variation of the central wavelength or spectral width of a pulsed laser beam during the pulse.

15d   SIGNAL-TO-NOISE RATIO CONSIDERATIONS AND THEIR
      APPLICATION IN THE OPTIMIZATION OF EXPERIMENTAL
      CONDITIONS IN ABSORPTION MEASUREMENTS

Optimization of the experimental conditions with respect to the signal-to-noise ratio (SNR) is aimed either at making the absorption measurements more precise or at detecting weaker absorption signals. Optimization is not only of interest in analytical atomic absorption spectroscopy (where the SNR problem has been studied extensively[†]), but also in studies of metal vapours in flames. Here

---
[†] See, e.g., Snelleman (1968), Rubeška and Moldan (1969), McCarthy (1971), L'vov (1970), Parsons and McElfresh (1972), Nitis, Svoboda and Winefordner (1972), and Kirkbright and Sargent (1974).

Chapt. III   GENERAL INSTRUMENTAL ASPECTS AND EXPERIMENTAL METHODS

we shall discuss the SNR and its optimization in two special cases of absorption measurements. Case I concerns the determination of the integral absorption, A_t, off an atomic metal line with the aid of a continuum background source (see Subsection 15d-1). Case II concerns the determination of the wavelength-dependent absorption factor, $\alpha(\lambda)$, with the aid of a narrow-line background source (see Subsection 15d-2).

15d-1   *CASE I: MEASUREMENT OF INTEGRAL ABSORPTION OF AN ATOMIC METAL LINE*

In the simplest case we have a single-beam set-up as shown in Fig. III.24. A continuum light source is used in combination with a monochromator. The expression for the integral absorption of the metal line when d.c. measurements are used is (see Eq. II.271; see also Hinnov and Kohn 1957, and Hofmann and Kohn 1961)

$$A_t^{(\lambda)} = \Delta\lambda_{eff} \frac{(i_0 - i_e + i_1)}{i_0} . \qquad (III.104)$$

Here $A_t^{(\lambda)}$ is the integral absorption (in Å; see Eq. II.267), and i_1, i_0 and i_e are the photocurrents corresponding to the atomic line emission, nonattenuated source emission, and atomic line plus attenuated source emission, respectively; the flame background emission can be included in i_1 and i_e, but drops out of the equation; $\Delta\lambda_{eff}$ is the effective monochromator bandwidth in Å (see Sect. III.4b); we assume that $\Delta\lambda_{eff} \gg \Delta\lambda_{abs}$, where $\Delta\lambda_{abs}$ = absorption line width.

With a.c. modulation measurements the equation reduces to

$$A_t^{(\lambda)} = \frac{\Delta\lambda_{eff}(i_0 - i_t)}{i_0} , \qquad (III.105)$$

where i_0 and i_t are the photocurrents corresponding to the nonattenuated and attenuated source emission, respectively.

In our SNR calculation we assume that the background light source is focused in the flame without optical losses, and that this image and the flame are seen by the spectrometer under the same solid angle. We also assume that only that part of the flame that is irradiated by the source is seen by the spectrometer. Under these conditions the relative contribution of the flame background and atomic line emission to the total noise is minimized. Furthermore we assume that noise arising from instabilities in the optical components, photodetector responsivity, amplifier gain, etc., is absent (for their influence see Subsection 15c-5).

When the background light source is intense enough and the absorption is comparatively weak (i.e. near the detection limit), the noise in the source radiation will be dominant. The differences $(i_0 - i_e + i_1)$ and $(i_0 - i_t)$, occurring in the numerator of expressions III.104 and 105, respectively, are then small compared to i_0. The relative error in the measurement of $A_t^{(\lambda)}$ is then mainly determined by the relative error in the measurement of these small differences and not by the

THE MEASUREMENT OF ABSORPTION   Sect. III.15d

error made in the measurement of i_0, which occurs in the denominator of these expressions. The SNR for $A_t^{(\lambda)}$ is thus, virtually, the same as the ratio of the absorbed light flux to the noise in the relatively large background-continuum flux transmitted by the spectrometer. Expressions for the SNR can thus be obtained immediately from those listed in Table III.6 for the analogous case of a weak emission line superimposed on a relatively strong flame background. We only have to replace there the radiance, B^s, of the atomic emission line by the __absorbed__ radiance, B_A^{ls}, of the light source, and $B_{\lambda_0}^{bg}$ by the nonattenuated spectral radiance, $B_{\lambda_0}^{ls}$, of the same source. For B_A^{ls} we can write (see Eq. II.303)

$$B_A^{ls} = B_{\lambda_0}^{ls} A_t^{(\lambda)} \; . \tag{III.106}$$

The resulting SNR expressions are listed in Table III.7.

Table III.7

Theoretical Expressions for SNR in Integral-Absorption Measurements*

D.c./a.c.	Shot noise SNR at $\tau = \tau_r$	Flicker noise SNR at optimal $\tau \; (\simeq \frac{1}{8} \tau_r)$
d.c.	$A_t^{(\lambda)} \sqrt{\dfrac{B_{\lambda_0}^{ls} L \tau(\lambda_0) s(\lambda_0) \tau_r}{2\pi e \Delta \lambda_{eff}}}$	$\dfrac{0.62 \, A_t^{(\lambda)}}{\Delta \lambda_{eff} \sqrt{A}}$
a.c.	$F' A_t^{(\lambda)} \sqrt{\dfrac{B_{\lambda_0}^{ls} L \tau(\lambda_0) s(\lambda_0) \tau_r}{2\pi e \Delta \lambda_{eff}}}$	$F''' \left(\dfrac{A_t^{(\lambda)}}{\Delta \lambda_{eff}}\right) \sqrt{\dfrac{\tau_r f_{mod}}{2A}}$

*
Expressions are given for the signal-to-noise ratio (SNR) in the measurement of the integral absorption, $A_t^{(\lambda)}$, of an atomic line when the noise in the background continuum source (with spectral radiance $B_{\lambda_0}^{ls}$) is dominant and the absorption is relatively weak. The expressions hold for paired readings with a linear amplifier + read-out system with response time τ_r and for a sampling time τ as specified in the table. For a further explanation of symbols and measuring conditions see the _Legend_ for Table III.6 on page 367 and the text. When integration measurements are made, the gain in SNR is the same as stated in Table III.6.

The expressions listed for the 'd.c. case' apply when a single beam set-up with or without intensity modulation is used (see Fig. III.24). Note that the SNR is not essentially changed by the application of intensity-modulation, as absorption signal and source fluctuations are then __both__ modulated (and de-modulated) at the same frequency. The expressions for the 'a.c. case' apply when a double-beam

Chapt. III   GENERAL INSTRUMENTAL ASPECTS AND EXPERIMENTAL METHODS

set-up is used in which the reference- and probing beam are modulated at opposite phases, as shown in diagram (1) in Fig. III.25. When a double-beam set-up is used in which these two beams are not modulated or are modulated at the same phase [see diagram (2) in Fig. III.25], source fluctuations which affect both beam intensities to the same extent are — in principle — cancelled altogether in the zero-reading. Only the shot-noise components remain, as they are statistically uncorrelated for the two beams; the SNR expressions listed for the 'd.c. shot-noise case' then apply. Of course, any flicker noise arising in the detection apparatus is not reduced unless the source radiation is modulated.

When double spectral-beam or dual-wavelength methods are applied in the absorption measurement, similar SNR considerations hold as given above for double spatial-beam methods. The only condition is that the excess-low-frequency fluctuations in the continuum source affect equally strongly the intensities of the probing beam (at $\lambda \simeq \lambda_0$) and of the reference beam at an adjacent wavelength.

The following conclusions as to optimization of the SNR can be drawn from Table III.7:

(i)   When $\Delta\lambda_{eff}$ is proportional to the slit width and thus to the optical conductance L, the SNR is **independent** of slit width in the case of dominant shot noise (see also Sect. III.14f-2).

(ii)   In the case of flicker noise a reduction of $\Delta\lambda_{eff}$ improves the SNR (as long as $\Delta\lambda_{eff} \gg \Delta\lambda_{abs}$). However, when $\Delta\lambda_{eff}$ is reduced at the expense of i_0, shot noise will ultimately become dominant over flicker noise (see Sect. III.14f-2).

(iii) The spectral radiance of the light source plays a role in the case of shot noise, but not in the case of flicker noise or any other e.l.f. noise whose RMS value is proportional to the source intensity (cf. Eq. III.65).

(iv)   In the case of shot noise, d.c. and a.c. measuring techniques give essentially the same SNR. In the case of flicker noise the gain in SNR obtained by applying a.c. modulation is about a factor $\sqrt{\tau_r f_{mod}}$.

For a discussion of the influence of other factors occurring in the SNR expressions we refer to Sect. III.14f.

When the background light source is not intense enough, when the absorption is strong or when $\Delta\lambda_{eff}$ is not much larger than $\Delta\lambda_{abs}$, other noise sources may become important, too. For example, the flame-background noise may be dominant in hot, incandescent flames; its contribution to the SNR can be read easily from the expressions in Table III.6 containing $B^{bg}_{\lambda_0}$, if we again replace therein B^s by B^{ls}_A. Shot noise in the thermal emission of the atomic line can always be neglected when the radiance temperature of the continuum background source exceeds the flame temperature and $\Delta\lambda_{eff} \gg \Delta\lambda_{abs}$; the source radiation received by the detector is

THE MEASUREMENT OF ABSORPTION   Sect. III.15d

then much stronger than the atomic line radiation. When the light source is so
weak that i_0 is comparable with or less than the dark current of the photodetector,
flicker noise in the latter current may become important; its effect can be reduced
by the application of a.c. modulation. When the absorption is relatively strong,
the fluctuations in A_t due to a fluctuating atomic density or flame depth may become noticeable. These fluctuations may be (partially) correlated with the fluctuations in the thermal line intensity because of the relation between the latter
intensity and A_t, expressed by Eq. II.295. The RMS values of the respective noise
contributions should then not be added quadratically!

In general, when several noise sources have, at least partially, a common
cause one should be careful when calculating their combined RMS value. For
example, temperature fluctuations affect not only the thermal excitation of the
flame background or metal line, but also the atomic density and thus the atomic
absorption through a varying degree of dissociation.

In the measurement of flame temperatures by the line-reversal method (see
Sects II.5c-3 and III.10c-5), the effects of a fluctuating metal supply on the
noise in the atomic emission and the atomic absorption signals are fully correlated and cancel each other out at the reversal point.

Snelleman (1968) has reported on his atomic absorption measurements with a
500 W xenon lamp applying derivative spectroscopy to a single spatial beam (compare Sect. III.4f sub 5). He has calculated the minimum detectable integral
absorption $(A_t)_{min}$ for his set-up, assuming that only shot noise was present.[†]
His measurements on the copper 3247 Å line appeared to be in agreement with his
calculations. The minimum integral absorption found at a measuring time of 1 s
corresponded in his case to an absorption factor $\alpha = 3 \times 10^{-4}$ for the source
radiation as transmitted by the spectrometer having a bandwidth $\Delta\lambda_{eff} = 0.3$ Å.
This absorption was obtained for a number density of 4×10^9 atoms per cm^3 in a
flame with a depth of 1.0 cm.

15d-2   *CASE II: MEASUREMENT OF ABSORPTION FACTOR WITH A NARROW-LINE SOURCE*

We assume that a background light source is used which emits a single emis-
sion line at wavelength λ, having a width that is small compared with $\Delta\lambda_{abs}$. The
use of a monochromator is not essential here as the spectral selectivity is determined by the width of this source emission line. Spectral apparatus may be required
however to reduce the continuous flame background or to select the emission line
wanted from the source spectrum.

[†] $(A_t)_{min}$ was calculated from a formula similar to that given in Table III.7 for
the shot-noise case, while the SNR was put equal to unity.

Chapt. III GENERAL INSTRUMENTAL ASPECTS AND EXPERIMENTAL METHODS

In the simplest case we have a single-beam set-up as shown in Fig. III.24. A more complicated double-beam set-up as shown in diagram (1) of Fig. III.25, with a manually operated or automatic optical balance and a lock-in amplifier, may be appropriate when one has to measure an absorption line profile at low atomic density using a Zeeman-scanning technique (see Hollander, Jansen, Plaat and Alkemade 1970; see also Sect. VII.3a-4).

In d.c. measurements we find $\alpha(\lambda)$ from

$$\alpha(\lambda) = (i_0 - i_e + i_1)/i_0 \ . \tag{III.107}$$

In a.c. modulation measurements $\alpha(\lambda)$ follows from

$$\alpha(\lambda) = (i_0 - i_t)/i_0 \ . \tag{III.108}$$

The meaning of the symbols is the same as in Eqs III.104 and 105.

In our SNR calculations we make the same assumptions about the spatial and angular apertures of the optical system as we did in the preceding subsection. If the source intensity is great enough and if $\alpha(\lambda) \ll 1$, the SNR is determined mainly by the relative error in the numerator of the above two expressions. In other words, the SNR in the determination of $\alpha(\lambda)$ is, virtually, the same as the ratio of the absorbed light flux to the noise in the relatively large, unattenuated flux from the background source. We find this SNR from the analogous general expressions in Table III.6 (see page 367) when we replace in it i_s by the current difference corresponding to the <u>absorbed</u> radiation flux, and $\bar{i}_b$ by the mean value of the unattenuated flux from the source. Applying Eq. III.60 and writing B^{1s} for the radiance of the line source (integrated over its spectral profile) we obtain the SNR expressions listed in Table III.8.

The remarks made in Subsection 15c-1 also hold for the applicability of the SNR expressions with different d.c. and a.c. measuring techniques. Similar conclusions can also be drawn about the optimization of the SNR if we exclude the role played by $\Delta\lambda_{eff}$. When $\alpha(\lambda)$ is of order unity, the noise in the atomic line emission or background emission of the flame may play a role and the given expressions will then cease to be valid. For $\alpha(\lambda) \simeq 1$ the transmitted source radiation and its noise are almost completely suppressed.

In studies of line broadening mechanisms in flames (see Chapt. VII) one is interested in the spectral profile of the *absorptivity* $k(\lambda)$ $[\equiv \kappa(\lambda) n_p$; see Eq. II.185] rather than in the $\alpha(\lambda)$ profile. The two profiles are equal only in the optically thin case, i.e. for $\alpha(\lambda) \ll 1$ (see Eq. II.261). However, according to the expressions in Table III.8 the SNR will be poor at low $\alpha(\lambda)$ values. One can find the $k(\lambda)$ [or $\kappa(\lambda)$] profile from absorption measurements made at arbitrary optical thickness if one converts $\alpha(\lambda)$ into absorbance $A^{abs}(\lambda)$ through Eq. II.255.

THE MEASUREMENT OF ABSORPTION   Sect. III.15d

Table III.8

Theoretical Expressions for SNR in Absorption Factor Measurements*

D.c./a.c.	Shot noise SNR at $\tau = \tau_r$	Flicker noise SNR at optimal τ ($\simeq \frac{1}{8}\tau_r$)
d.c.	$\alpha(\lambda) \sqrt{\dfrac{B^{1s} L\tau(\lambda) s(\lambda) \tau_r}{2\pi e}}$	$\dfrac{0.62\, \alpha(\lambda)}{\sqrt{A}}$
a.c.	$F'\alpha(\lambda) \sqrt{\dfrac{B^{1s} L\tau(\lambda) s(\lambda) \tau_r}{2\pi e}}$	$F'''\alpha(\lambda) \sqrt{\dfrac{\tau_r f_{mod}}{2A}}$

* Expressions are given for the signal-to-noise ratio (SNR) in the measurement of the absorption factor $\alpha(\lambda)$ with a narrow-line source at wavelength λ (with radiance B^{1s}) when the noise in the latter is dominant and $\alpha(\lambda) \ll 1$. For a further explanation see the Legends for Tables III.7, p. 385, and III.6, p. 367, and the text.

The latter quantity is strictly proportional to $\kappa(\lambda)$ according to Eq. II.258 if Beer's law holds. However, in order to obtain precise $A^{abs}(\lambda)$ values, $\alpha(\lambda)$ should be neither too small (see above) nor too close to unity (cf. Eq. II.255). When an absorption line profile is to be scanned, the density, n_p, of atoms in the absorbing state should therefore not be too high for precise measurements near line centre, and not too low for precise measurements at the line wings. One can circumvent these conflicting demands by measuring $\alpha(\lambda)$ at each chosen λ value for a series of widely varying solution concentrations, c, and by plotting $A^{abs}(\lambda)$ versus c at each λ. When n_p varies proportionally to c, the slope of the straight 'Beer plots' thus obtained varies with λ in strictly the same way as $\kappa(\lambda)$.

Such Beer plots have actually been applied in an accurate measurement of $\kappa(\lambda)$ profiles of the alkaline-earth resonance lines in flames by Hollander, Jansen, Plaat and Alkemade (1970). They used a Zeeman-scanning technique in a double-beam set-up as described above. Absorption factors of the order of 10^{-3} could be measured at an SNR ≈ 10 with this set-up incorporating an electrodeless-discharge lamp and a mechanical chopper at 50 Hz. When the source emission line is not much narrower than the absorption line profile, a deconvolution of the experimental absorption profile with respect to the source line profile should be made in order to find the true absorption profile (see Sect. VII.3a-4).

Chapt. III GENERAL INSTRUMENTAL ASPECTS AND EXPERIMENTAL METHODS

16. THE MEASUREMENT OF FLUORESCENCE

16a INTRODUCTION

Fluorescence measurements may be performed to study the transfer of energy from optically excited atoms (and molecules) to other particles in the flame upon collision (see Sect. VI.2) or for the purpose of flame diagnostics (see Sects III.10c-4 and V.2a). In analytical fluorescence spectroscopy the intensity of an atomic (or molecular) fluorescence line is measured to determine the concentration of the element in the solution sprayed into the flame (see below).

The potentialities of fluorescence studies in flames are strongly enhanced by the advent of intense, tunable dye lasers. The high spectral peak power of pulsed lasers even enables us to observe fluorescence from the higher atomic levels which are excited by a two-photon process (see, e.g., van Dijk et $al.$ 1978). An essential advantage of lasers over conventional excitation sources (high pressure xenon lamp or hollow-cathode lamp) is that saturation of optical transitions in the atomic vapour can be reached (see Sect. II.5c-4). The fluorescence intensity is then no longer proportional to the source intensity and at high saturation levels may even become independent of it.

Because most fundamental work as well as analytical work has been and is still being done with conventional light sources and because the theoretical description is simpler in the absence of saturation effects, we shall confine ourselves in this section to these kind of light sources. As a further, simplifying restriction we shall consider only the case of resonance fluorescence (see definition in Sect. II.5a-1) of an atomic resonance line.[†] We shall not deal here with the goals and results of flame fluorescence spectroscopy but only with its experimental and instrumental aspects.

The determination of the efficiency of fluorescence Y (see definition in Eq. II.307) is the experimental starting-point for most of the fundamental fluorescence studies. The avoidance of systematic errors is here more essential and often more problematical than the reduction of accidental errors. In Sect. 16b we shall describe what precautions have to be taken and how the experimental set-up can be optimized with respect to possible systematic errors in the measurement of Y. It should be stressed that Y values are only meaningful if they are measured in a flame region with a homogeneous and well-known gas composition (and temperature).

In analytical applications, however, one is primarily interested in a low limit of detection (expressed as minimal detectable solution concentration; see also Sect. III.14f-1) or high precision. The flame need not be homogeneous nor its properties be known, as this, like most other spectroscopic methods of

---

[†] References to molecular fluorescence spectroscopy in flames with conventional as well as laser sources are given in Sects II.5c-4 and IV.8.

THE MEASUREMENT OF FLUORESCENCE   Sect. III.16a

analysis are calibrated by means of reference solutions. Only constancy and reproducibility of the experimental conditions are important.

In contrast to fundamental flame studies, where one usually works with simple, well-known solutions, the presence of unknown or variable concomitants in the analytical sample may cause troublesome interferences. These interferences could give rise to systematic errors if they are not attended to.

In the analytical literature much attention has therefore been paid to the reduction of accidental errors, in particular to the enhancement of the signal-to-noise ratio (SNR) (see, e.g., Shull and Winefordner 1971, Omenetto, Fraser and Winefordner 1973, Winefordner et al., 1976, and Boutilier et al. 1977). The SNR also imports in fundamental work, but the experimental boundary conditions are often different here from those in spectrochemical analysis. Therefore the sometimes conflicting demands of a large SNR and a small systematic error will usually lead to a different compromise in the two cases.[†] In Sect. 16c we shall derive SNR expressions for fluorescence-intensity measurements under rather idealized experimental conditions. Although one hardly meets these conditions in practice these expressions may serve as a rough guide for optimizing the experimental set-up with respect to the SNR.

Since fluorescence is essentially an emission process preceded by photon absorption, the method of its measurement has many features in common with the methods of emission and absorption measurements. This also implies that the experimental arrangement for fluorescence measurements is — at least basically — more composite than in the two other cases. We need not repeat here what was already treated in the two preceding sections, but shall discuss only points that are specially relevant to fluorescence measurements.

A striking point of difference between fluorescence measurements and thermal-emission or absorption measurements in flame diagnostics is that in the two latter cases one probes the flame over the whole line of viewing, whereas local probing is possible in fluorescence. In emission or absorption measurements one obtains a (weighted) average of, e.g., the flame temperature or atomic density, when these parameters vary along the line of viewing. But when one measures the fluorescence from a small section of the flame zone irradiated by a pencil-like (laser) beam, local sensing of these parameters becomes feasible[*] (see, e.g., Omenetto 1975, Daily 1976, Lück and Müller 1977, Daily and Chan 1978, Haraguchi

---

[†] Nevertheless the sheathed Méker-grid burner described in Sect. III.2b-2 and shown in Fig. III.1 appeared to work equally satisfactorily for fundamental atomic fluorescence measurements and for analytical atomic fluorescence spectroscopy (see Slevin, Muscat and Vickers 1972).

[*] This is strictly true only if radiation diffusion (see Sect. II.5c-3) is negligible; this condition is certainly met in normal flames at 1 atm pressure.

Chapt. III GENERAL INSTRUMENTAL ASPECTS AND EXPERIMENTAL METHODS

and Winefordner 1977, Smith, Winefordner and Omenetto 1977, and Omenetto and Winefordner 1979, 1979a). This possibility may be of special interest for investigating inhomogeneous flames that are used in analytical flame spectroscopy.

The *analytical applications* of atomic fluorescence spectroscopy (AFS), atomic absorption spectroscopy (AAS) and atomic emission spectroscopy (AES) have certain advantages in common. The use of an auxiliary spectral-line source emitting a resonance line of the element to be determined makes both AFS and AAS element-specific. If this light source is intense enough or if its radiation is intensity-modulated at a sufficiently high frequency, no high demands are made on the resolution of the spectral apparatus in order to overcome spectral interference from the flame background or from concomitants in the flame. Interference from wavelength-independent scattering can be suppressed by wavelength-modulation and synchronous detection (see Sect. III.4f sub 5) or, more simply, by wavelength-scanning of the source emission. The latter technique was applied with a tunable c.w. dye laser for the Ba and Na lines in flames (see Green, Travis and Keller 1976, and Keller and Travis 1979).

Unlike the emission method, the absorption and fluorescence method can be applied to resonance lines with a relatively high excitation energy, even in cool flames, because thermal excitation plays no role in the case of AFS. In contrast to AAS, the fluorescence signal for low metal concentrations is not found as a small difference between two relatively large signals. Good limits of detection may therefore be expected with AFS, especially if the efficiency of fluorescence is high and the auxiliary light source is intense enough. The workable concentration range can therefore be extended to very low concentration values and the analytical curve is linear over many decades of the concentration axis. In AES the limit of detection is often set by fluctuations in the reading of the flame background emission which is subtracted from the analyte readings.

The analytical application of AFS, proposed by Alkemade (1963), has been greatly stimulated and improved by the extensive investigations carried out by Winefordner and co-workers (see, e.g., Winefordner, Schulman and O'Haver 1972). AFS has greatly profited from the technical achievements in AAS, especially with respect to intense light sources and nonflame atomizers. Nevertheless, the great success of the atomic absorption method, which was fully developed earlier, explains at least partly why AFS has hitherto found only limited application in actual analyses. The reader who wishes further information is referred to review articles (e.g., Winefordner 1971, Winefordner and Elser 1971, West and Cresser 1973, and Omenetto and Winefordner 1979a) and the textbooks on AFS listed in Chapt. I.

16b THE MEASUREMENT OF FLUORESCENCE EFFICIENCY

The efficiency of resonance fluorescence Y, defined in Sect. II.5c-4, can be determined by measuring the ratio of the fluorescence flux Φ_F, observed under solid angle Ω_F, to the absorbed primary-radiation flux Φ_A. If the fluorescence is emitted isotropically and if it is not weakened by self-absorption, Y follows from Eq. II.307

$$Y = (\Phi_F/\Phi_A)(4\pi/\Omega_F) = (u_F/u_A)\, q(4\pi/\Omega_F). \qquad (III.109)$$

Here u_F and u_A are the spectrometer readings corresponding to Φ_F and Φ_A, while the factor q takes account of the different spectrometer sensitivities, attenuation factors, etc., used in the two measurements. If self-absorption is not negligible, the (apparent) Y value found from Eq. III.109 is concentration-dependent. We can find the true Y value by plotting the apparent Y value as a function of solution concentration and extrapolating to zero concentration (see Hooymayers and Alkemade 1966). It appears, both experimentally and theoretically, that for low self-absorption the apparent Y value decreases linearly with increasing concentration (see Hooymayers and Alkemade 1966). This extrapolation procedure also corrects for the possible effect of radiation trapping (i.e, iterated re-absorption and re-emission of photons; see Sect. II.5c-3) on the measured Y value.

One can also correct for self-absorption by a semi-empirical method which relates the relative decrease of Y caused by self-absorption to the absorption factor measured for the primary beam (see Jenkins 1966). This relationship, however, strongly depends on the spectral profiles of the source radiation and the absorption line, the flame geometry, and other experimental factors. The coefficient appearing in this relationship is therefore determined empirically.

The set-up for *absolute* measurements of Y should meet the following demands:

(i) The solid angle Ω_F under which the fluorescence flux Φ_F is detected should be well-defined and equal for all elements of the fluorescing flame zone.

(ii) The whole fluorescing flame zone should be viewed by the spectrometer with uniform sensitivity.

(iii) In the measurement of the absorbed flux Φ_A all rays of the primary beam that pass the absorbing flame should be accepted by the spectrometer.

(iv) Differences in the spatial and angular distributions of the primary-radiation beam and the fluorescence beam should not affect the measurement of their intensity ratio.

Chapt. III   GENERAL INSTRUMENTAL ASPECTS AND EXPERIMENTAL METHODS

(v)   The contribution from thermal emission of the atomic line and flame background to the fluorescence and absorption signals should be negligible or corrected for.

(vi) Scattering of the primary beam by aerosol droplets or nonvolatilized dry aerosol particles as well as stray light in the optical system should be suppressed.

(vii) Noise from the thermal flame emission should be reduced as far as possible.

In the measurement of *relative* Y values some of these demands may be relaxed. Relative Y values are involved in the determination of doublet-mixing cross sections (see Sects VI.2a-2 and 2b-3). Relative Y measurements may also be done in order to eliminate Φ_F in the absolute determination of Y for an unknown metal line by comparing its fluorescence intensity with that of another metal line for which Y is known.

Figure III.27 shows a block diagram of a set-up for measuring Y values. The fluorescence is usually observed at right angles to the direction of the primary beam emitted by the source (see the 'fluorescence mode' in Fig. III.27). The absorbed primary flux is measured by the same spectrometer by rotating the optical

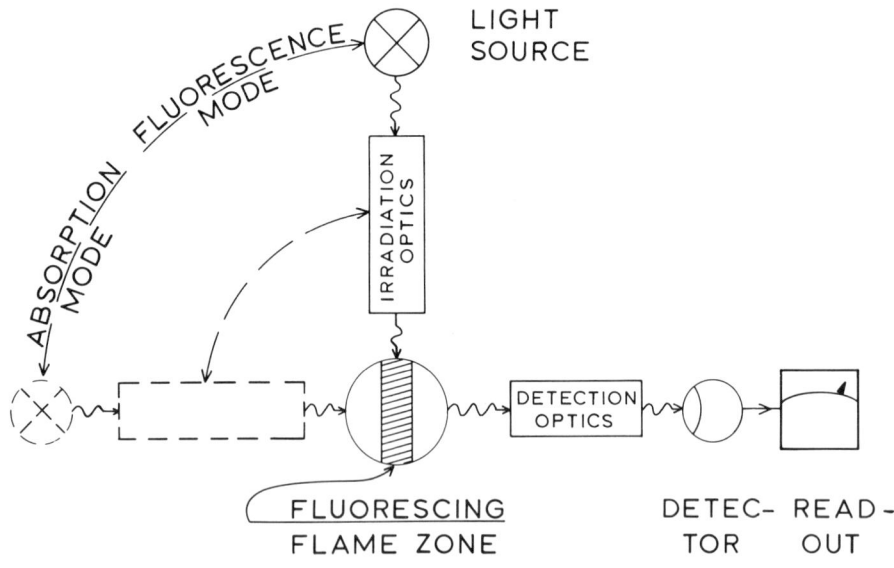

Fig. III.27 Block diagram of set-up for measuring fluorescence efficiency in flames. Bench with source and irradiation optics can be rotated over 90° around flame axis in order to change from 'fluorescence mode' to 'absorption mode'.

THE MEASUREMENT OF FLUORESCENCE   Sect. III.16b

bench with source and irradiation optics over 90° around the flame axis (see the 'absorption mode' in Fig. III.27). Absorption measurements have been dealt with in Sect. III.15. Rotation of this optical bench also allows one to check the isotropic distribution of the fluorescence in the horizontal plane. Details of the set-up will be discussed now with respect to the above demands.

The irradiation optics usually images the source on the flame. The breadth of this image should not exceed the breadth of the metal-seeded flame; the angular aperture of the primary beam in the flame should not exceed that of the detection optics (see below). The intensity of the primary beam is usually modulated by mechanical or electronic means in combination with a.c. detection (see Sect. III.4f-2) in order to eliminate the d.c. thermal flame emission (see demand v) and to reduce its noise (see demand vii). The modulation frequency should be sufficiently high to make the excess or flicker noise of the flame emission (see Sect. IV.9) negligible compared to its inevitable shot noise under the prevailing experimental conditions (see also Sect. 16c). The usual modulation frequencies are low enough to assure that stationary-state conditions are fully attained in the atomic vapour during each half cycle.

With respect to demands v and vi it is recommended that the primary flux within the spectral width of the absorption line be made as large as possible. When a continuum source is used, a monochromator in the irradiation optics helps to restrict scattering effects (see demand vi). Scattering in general can be reduced by using a nebulizer with spray-chamber producing a fine aerosol, a low solution concentration, a volatile solute and sometimes by choosing the observation volume not too close to the base of the flame (see also Sect. III.9d). Scattering by aerosol droplets can be checked by spraying a solvent blank, while scattering by dry aerosol particles can be detected by illuminating the metal-seeded flame at a different, adjacent wavelength. Stray light can be suppressed by using baffles and light traps in the irradiation and detection optics.

Scattering by dry aerosol particles may cause an unwanted matrix effect in the analytical application of atomic fluorescence with samples containing a high concentration of an involatile concomitant. Scattering by spray droplets is found especially with direct-injection burners. Several techniques can be applied to eliminate or correct for such interferences (e.g., the dual-wavelength technique; see Sect. III.4f-4). For a further discussion we refer to the pertinent textbooks on analytical flame spectroscopy listed in Chapt. I.

A spectral filter or monochromator may be inserted into the irradiation optics to select the line or multiplet component chosen for excitation. The limited optical conductance of the monochromator (see Sect. III.4b) may, however,

Chapt. III GENERAL INSTRUMENTAL ASPECTS AND EXPERIMENTAL METHODS

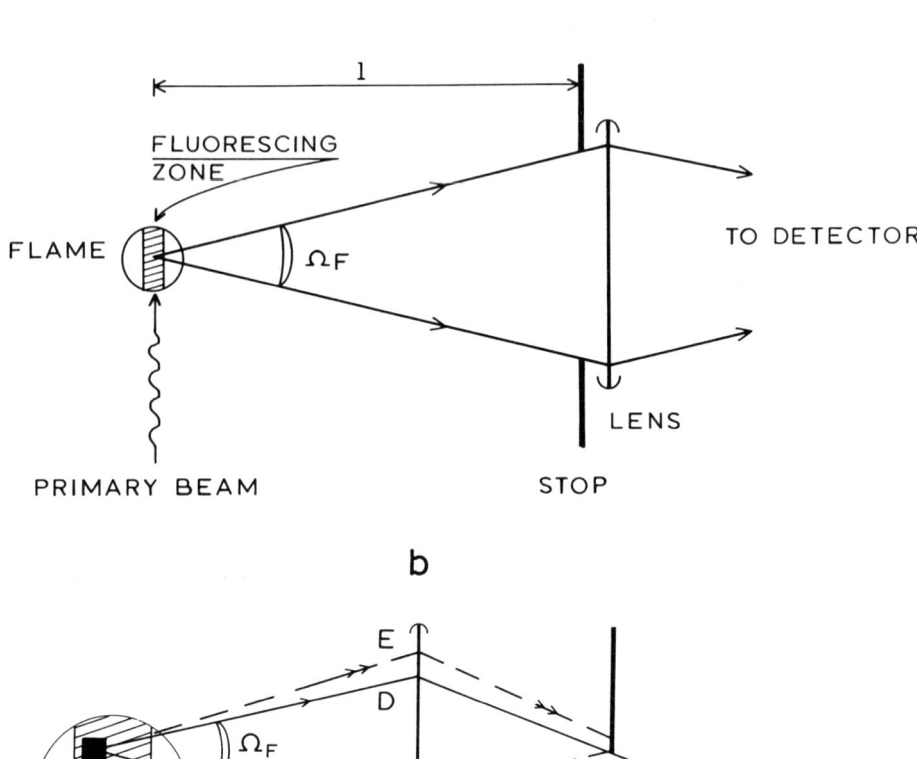

Fig. III.28 In figure (a) the stop placed in front of the lens defines a uniform solid angle Ω_F for all volume elements of the fluorescing flame zone if the dimensions of this zone are small compared to distance l between flame and stop. In figure (b) the stop placed at focal length f behind the lens ensures that Ω_F is independent of position of volume element P. Note that light ray PE outside cone with solid angle Ω_F does not pass through this stop. Because PD//OB and PC//OA, the magnitude of Ω_F is determined by the aperture of the stop and by f.

THE MEASUREMENT OF FLUORESCENCE   Sect. III.16b

have an adverse effect on the beam intensity in the flame and thus on the fluorescence signal.

The measurement of Φ_A may be contaminated by the fluorescence flux that is radiated in the same direction as the primary beam (see Pearce, de Galan and Winefordner 1968). If Y is close to unity, this contribution of fluorescence is about equal to $(\Omega_F/4\pi)\Phi_A$ (see Eq. III.109). If the solid angle Ω_A subtended by the primary beam through the flame is markedly less than Ω_F, the use of a smaller stop in the detection optics can reduce the contribution of fluorescence in the absorption measurements. This contribution can always be corrected for if one knows the fluorescence emitted at right angles to the primary beam, assuming that the fluorescence emission is isotropic.

The detection optics should be carefully designed to provide for a well-defined Ω_F (in order to avoid systematic errors; see i) as well as large Ω_F (in order to improve the signal-to-noise ratio; see Sect. 16c). This can be done by inserting a limiting stop with a well-known aperture at a well-defined place in the optical train. Care should be taken that all light rays passing through this stop do actually reach the photocathode. This stop may be placed close to the first lens that collects the fluorescence beam (see Fig. III.28a). If the spatial dimensions of the fluorescing flame zone are small compared to the flame—lens distance, the fluorescence from all volume elements of this zone will be accepted under practically the same solid angle Ω_F. If the dimensions of the fluorescing flame zone are not relatively small, a well-defined and uniform Ω_F value can be realized by placing a stop at focal length, f, behind the lens (see Fig. III.28b; see Jenkins 1966). For any volume element at position P the ultimate rays, PD and PC, that will just pass the stop are parallel to OB and OA, respectively. Thus Ω_F is independent of position P and determined only by the aperture of the stop and f. Of course the lens diameter should be large enough to collect all radiation emitted by the whole flame zone within the solid angle Ω_F.

Steps should be taken to ensure that the effective solid angle of acceptance is not diminished by the strong angular dependence of the spectral transmission of any interference filter(s) used (see Sect. III.4b). Interference filters should therefore be placed in that part of the optical train where the beam divergence is (made) minimal (see Fig. III.29).

Interference from the thermal flame emission can be minimized by placing in the image plane of the flame an extra stop that screens off the flame parts above and below the fluorescing zone (see demands v and vii). Care should be taken that this stop does not partially screen off the fluorescence beam, or the primary beam when the set-up is placed in the absorption mode.

Whether it is the fluorescing flame zone or the Ω_F-stop which is imaged on

Chapt. III   GENERAL INSTRUMENTAL ASPECTS AND EXPERIMENTAL METHODS

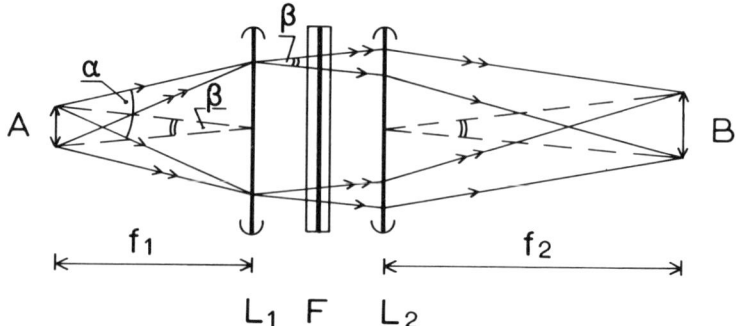

Fig. III.29 If one wants to collect radiation from object A under a large angle α, it is advisable to place interference filter F between two lenses L_1 and L_2, with focal lengths f_1 and f_2, which image A upon B. The angle of divergence, β, of the beam passing through F can then be kept small.

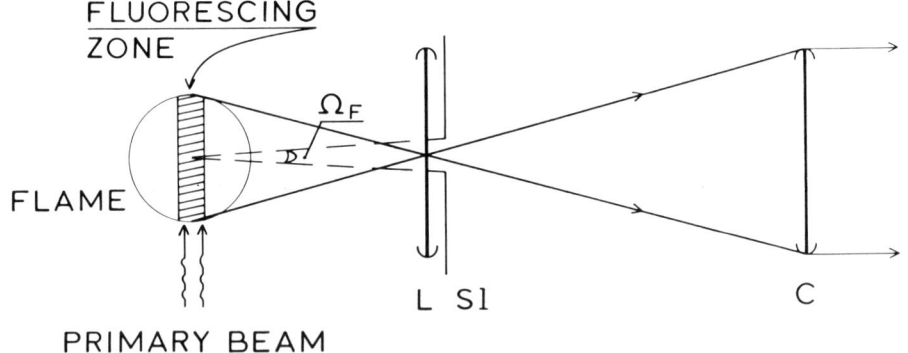

Fig. III.30 The whole fluorescing flame zone is imaged on the collimator lens, C, of the monochromator by means of a lens L placed in front of the entrance slit Sl.

THE MEASUREMENT OF FLUORESCENCE   Sect. III.16b

the photocathode, one should make sure that all fluorescence rays emitted by any element of the irradiated zone in any direction within Ω_F can reach the photosensitive surface of the cathode (see demand ii). With a monochromator one should therefore not image the fluorescing flame zone on the plane of the entrance slit, as the image is likely to be broader than the slit. It is better to image this zone on, and within the boundary of the collimator inside the monochromator by means of a lens in front of the entrance slit (see Fig. III.30). The solid angle Ω_F is then determined by the slit dimensions and the flame—slit distance. It is herewith assumed that the exit slit matches the image of the entrance slit and that slit-diffraction effects can be neglected. The narrowness of the slit severely restricts the magnitude of Ω_F and thus the fluorescence signal. When absolute Y values are to be measured, problems also arise when in the absorption mode the solid angle Ω_A subtended by the primary beam exceeds Ω_F determined by the slit dimensions. However, when only relative Y values are to be determined, demand (ii) does not apply (see Lijnse, Zeegers and Alkemade 1973).

Some investigators imaged the fluorescing flame zone or light source across the entrance slit while taking into account the loss of fluorescence or primary-radiation flux falling beside the slit by applying a geometrical correction factor (see Pearce, de Galan and Winefordner 1968). This factor is determined by the ratio of the slit area to the area of the image of the fluorescing flame zone or source. This correction procedure is justified only if the radiance of the image formed is uniform. However, the radiance of the fluorescing flame zone depends on the spatial distribution of the atoms and of the primary-beam intensity inside the flame and on the extent of self-absorption (thus on the atomic density and flame geometry). Unless measuring conditions are ideal a uniform fluorescence radiance cannot be expected.

When one applies the optical arrangement presented in Fig. III.30 to a circular flame with diameter of, say, 1 cm and a collimator lens with diameter of 6 cm and focal length $f = 30$ cm, the optimal distance l between flame and entrance slit is equal to $(1/6) \times 30 = 5$ cm. Such a short distance might be a nuisance in practice. This difficulty can be avoided, e.g., by forming a 1:1 image of the flame at a distance of 5 cm from the slit by means of an extra lens between flame and slit. The solid angle subtended by this lens as viewed from the slit should then be at least equal to that subtended by the collimator.

In *analytical* flame fluorescence spectrometry the fluorescing flame zone is usually imaged on the plane of the entrance slit when a monochromator is used (see also Jenkins 1967, and Cotton and Jenkins 1970). The fact that the monochromator then does not view the whole fluorescing flame zone is of no consequence, as this analytical method is calibrated by means of reference solutions.

Chapt. III   GENERAL INSTRUMENTAL ASPECTS AND EXPERIMENTAL METHODS

When the fluorescing flame zone is elongated, a gain in fluorescence signal may be obtained by viewing this zone at an oblique angle (see Cotton and Jenkins 1970) or by rotating the flame image over 90° around the optical axis (see Jenkins 1967; see also Sect. III.4d).

A systematic error in the measurement of Φ_F/Φ_A can arise when the sensitivity of the photocathode varies locally over its surface (see also Sect. III.5d). However, if both the fluorescence and the primary-radiation beams illuminate uniformly the same photocathode area, no error results. This may be the case if the stop that limits Ω_F is imaged on the photocathode and if, in addition, in the absorption mode the primary beam just fills and illuminates this stop uniformly. In practice the latter requirement may adversely restrict the magnitude of Ω_F. Another solution may be found by placing in front of the photocathode a diffusor (e.g., a ground quartz plate) which scatters the incoming light over the whole photocathode (see Jenkins 1966). In general, when the angular and spatial distributions of the primary and/or fluorescence beams are disparate, a nonuniform sensitivity of the photocathode poses a serious problem (see iv).

Nonlinearity of the spectrometer response due to the comparatively strong primary radiation received in the absorption mode can be avoided by attenuating this beam by means of a calibrated optical filter (see also Sect. III.13).

In actual measurements it will often be difficult to eliminate or at least to estimate all possible systematic errors. Besides, one often has to compromise, when choosing experimental parameters and components (such as Ω_F, spectral selector, light source), between the conflicting demands of minimal systematic and minimal accidental errors. It is therefore generally advisable to repeat the Y measurement in a given flame under widely varying experimental conditions and check the mutual agreement of the Y values found within their accidental-error limits.

It is advisable to check the condition of isotropic fluorescence (see Eq. III.109), especially if the primary radiation beam is polarized and the flame is operated at reduced pressure. An apparent anisotropy may be found when self-absorption is important.

Since Y may depend critically on the flame gas composition, use of a sheathed flame (see Sect. III.2a) is recommended, especially when a noble gas is employed as diluent.

16c   OPTIMIZATION OF SNR

The precision and limit of detection (see Sect. III.14f-1) in fluorescence measurements are connected with the signal-to-noise ratio (SNR). We shall here

THE MEASUREMENT OF FLUORESCENCE   Sect. III.16c

discuss how the SNR can be optimized, especially in the case where a relatively weak fluorescence signal is to be detected in the presence of background noise. Such a situation may occur in the measurement of the fluorescence efficiency Y, when the atom concentration must be kept low in order to reduce self-absorption and/or when Y is small. In analytical fluorescence spectroscopy such a situation may occur when the concentration of a trace element in a solution is to be determined.

Optimization is more complicated here than in the case of flame emission measurements (see Sect. III.14f), as more noise sources may be involved in the fluorescence measurements. Besides there is here a larger number of experimental parameters and optical components that can be varied. Not only are the photodetector and the associated optics relevant here, but so are the primary light source and the irradiation optics.

We shall assume starting-points and measuring conditions similar to those introduced in Sect. III.14f-1 in the discussion of the SNR in emission measurements. Again it is not our aim to present exact, complicated equations that would enable one to calculate the SNR in any concrete case. We are interested rather in the functional dependency of the SNR on certain experimental parameters in some special, idealized cases. This may give us a semi-quantitative insight in the relevancy of these parameters in SNR considerations. Numerical constants of order unity will be omitted in the expressions to be presented.

Expressions for the SNR in fluorescence measurements have been reported in the literature (for example, by Winefordner, Parsons, Mansfield and McCarthy 1967a, Winefordner 1971, McCarthy 1971, Winefordner, Fitzgerald and Omenetto 1975, Winefordner et al. 1976, and Boutilier, Pollard et al. 1978). Some of the expressions dealing with flicker-noise should be considered critically. For a survey we refer to Omenetto and Winefordner (1979).

The following idealistic and simplifying assumptions will be made.

(i)   One side of a flame with rectangular cross section is illuminated by a uniform radiation beam at normal incidence; the fluorescence is observed at right angles to this primary beam. The dimensions of the irradiated and observed flame zones and of the total flame cross section are as given in Fig. III.31a and b.

(ii)   The solid angles Ω_A, subtended by the primary beam, and Ω_F, subtended by the observed fluorescence beam, are small enough to treat the irradiated and observed flame zones as approximately brick-shaped.

(iii)   The atomic density, flame temperature, flame-background emission, fluorescence efficiency, and the intensity of scattered radiation, if any, are

401

Chapt. III GENERAL INSTRUMENTAL ASPECTS AND EXPERIMENTAL METHODS

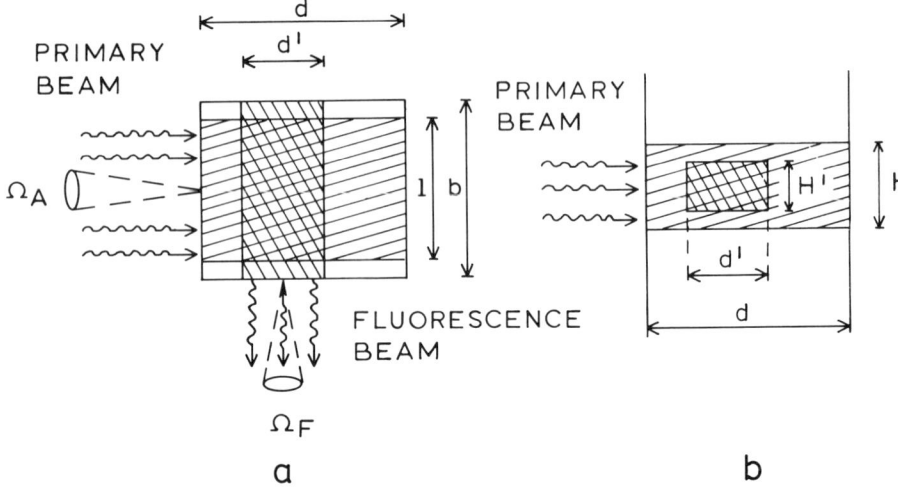

Fig. III.31   Strongly schematized top view (a) and front view from spectrometer (b) of irradiated and observed, brick-shaped flame zones. The flame is seeded by a metal vapour over its whole cross section $(b \times d)$. The fluorescence is observed from the doubly hatched flame zone $(l \times d' \times H')$. The solid angles Ω_A, subtended by the primary beam, and Ω_F, subtended by the observed fluorescence beam, are assumed to be small enough to treat these zones as brick-shaped.

assumed to be uniformly distributed over the flame zones in question.

(iv)   The intensity of the primary beam is uniformly distributed over its cross section ($l \times H$ in Fig. III.31).

(v)   The primary light source emits either a very narrow spectral line coincident with the centre λ_0 of the atomic line in the flame, or a (quasi-) continuous spectrum; the (spectral) radiance B (B_{λ_0}) of the source is assumed to be uniform.

(vi)   The density, n_0, of metal atoms in the ground-state is so low that both the absorbed primary flux Φ_A and the observed fluorescence flux Φ_F are proportional to n_0, while self-absorption is negligible; the relative attenuation of the primary-beam intensity in the flame is virtually negligible too.

We suppose that the intensity of the primary beam is modulated as usual and that a.c. detection is applied (see also Sect. 16b). For the rest the measuring procedure and meter characteristics are the same as described in Sect. III.14f.

THE MEASUREMENT OF FLUORESCENCE   Sect. III.16c

Thus we again suppose that paired readings are taken, so that the fluorescence signal is found from the difference between a background-plus-fluorescence reading and a reading of the background alone, separated by the sampling time τ. We choose the ratio of τ to the meter response time τ_r to be optimal (as in Sect. III.14f).

Separate SNR expressions will be derived for the cases that the predominant background noise arises from:
(1)   thermal emission of the atomic line considered, or
(2)   flame background emission with continuous spectrum, or
(3)   scattered radiation from the modulated primary beam inside the flame.
In practice these noise sources are often found to be the most important ones when one is working near the limit of detection.

As typical noise components we shall consider separately pure shot noise and pure flicker noise (see also Sect. III.14e). When both components are present together, we can calculate the resulting SNR expression by square addition of their respective standard deviations.

For noise sources (1) and (2) we have to apply the general SNR expressions given by Eqs III.99 and III.101, in the case of shot and of flicker noise, respectively. These expressions hold for a modulated signal current i_s and an unmodulated 'background' current i_b (see also Table III.6). Noise source (3) is modulated together with the fluorescence signal and the corresponding SNR is thus essentially the same as when no modulation of the primary beam was used. Therefore we can here apply the general SNR expressions given by Eqs III.77 and III.93, which hold for the d.c. case. The SNR expressions listed in Table III.9a and 9b (see pages 409-411) for the different cases considered are thus found by substituting for i_s and $\bar{i}_b$ the appropriate expressions for the photocurrent released by the fluorescence flux Φ_F, the thermal-emission flux, Φ_E, of the atomic line, the flame-background flux Φ_B, and the scattered-radiation flux Φ_S received by the photocathode, respectively. The expressions for Φ_F and Φ_S depend on whether a line or a continuum light source is used for excitation; we shall distinguish between these cases by writing Φ_F^l, Φ_S^l and Φ_F^c, Φ_S^c respectively. The radiant flux is converted into photocurrent by multiplying the former by the responsivity, $s(\lambda_0)$, of the photocathode (see also Sect. III.14d). For simplicity's sake it is assumed that the spectral radiance of the continuum source and of the flame background as well as the responsivity are practically λ-independent within the bandpass of the spectral selector used.

We find Φ_F^l from Eq. II.311 by multiplying its right-hand side by the transmission factor $\tau_d(\lambda_0)$ of the detection optics, by substituting for V the volume, $H'd'l$, of the observed fluorescence zone (see Fig. III.31), and by writing: $E = B\Omega_A \tau_i(\lambda_0)$. Here B is the radiance of the line source and $\tau_i(\lambda_0)$ the

403

Chapt. III GENERAL INSTRUMENTAL ASPECTS AND EXPERIMENTAL METHODS

transmission factor of the irradiation optics. Introducing the optical conductances $L_A = \Omega_A H l$ of the primary beam inside the flame and $L_F = \Omega_F H'd'$ of the observed fluorescence beam we finally get (in CGS units)

$$\Phi_F^1 = 7 \times 10^{-14} \, (\lambda_0^2 f_{pq}/\Delta\lambda_{eff}) \, \tau_i(\lambda_0) \, \tau_d(\lambda_0) \, n_0 \, YBL_A L_F/H \, , \qquad (III.110)$$

where f_{pq} = oscillator strength and $\Delta\lambda_{eff}$ = effective width of the atomic line in the flame.

In a similar way we get the expression for Φ_F^c from Eq. II.309 in the case of a continuum source (in CGS units)

$$\Phi_F^c = 7 \times 10^{-14} \lambda_0^2 f_{pq} \, \tau_i(\lambda_0) \, \tau_d(\lambda_0) n_0 \, YB_{\lambda_0} L_A L_F/H \, . \qquad (III.111)$$

The expressions for Φ_F hold on condition that the spectral bandwidth, $\Delta\lambda_d$, of the detection optics largely exceeds $\Delta\lambda_{eff}$ of the atomic line.

For the thermal-emission flux of the atomic line we find from Eq. II.281 (in CGS units)

$$\Phi_E = 1.6 \times 10^{-17} (A_{qp}/\lambda_0)(g_q/Q) \exp\left[-E_q/kT\right] \tau_d(\lambda_0) n_0 b L_F \, , \qquad (III.112)$$

if we equalize $n_a = n_0$ (see Sect. II.5c-3 and Fig. III.31 for meaning of symbols). It is here assumed that the thermal emission is observed from a brick-shaped flame zone with dimensions $H' \times d' \times b$, b being the depth of the whole flame in the direction of observation.

The flame-background flux Φ_B follows straightforwardly from its spectral radiance $B_{\lambda_0}^{bg}$ in the direction of observation

$$\Phi^{bg} = B_{\lambda_0}^{bg} \Delta\lambda_d \, \tau_d(\lambda_0) L_F = (B_{\lambda_0}^{bg}/b) \, \tau_d(\lambda_0) \, b \Delta\lambda_d L_F \, . \qquad (III.113)$$

The ratio $B_{\lambda_0}^{bg}/b$ occurring in this equation represents the radiant intensity (see Appendix A.5) of the flame background at λ_0 <u>per unit of flame volume</u>; this ratio is thus independent of the flame geometry and determined only by the temperature and gas composition of the flame.

The scattered-radiation flux has the same dependence on geometrical factors as the fluorescence flux considered above. For a line source we thus get

$$\Phi_S^1 = \xi(\lambda_0) \, n_{sc} \, \tau_i(\lambda_0) \, \tau_d(\lambda_0) BL_A L_F/H \, , \qquad (III.114)$$

where $\xi(\lambda_0)$ is a measure for the average scattering power at λ_0 of a single scatterer and n_{sc} is the number density of scatterers in the flame. In fact, the product $\xi(\lambda_0)n_{sc}$ represents the fraction of incident flux scattered per sterad and per cm length along the beam axis. Again for simplicity's sake we have assumed that the scattering is isotropic and that multiple scattering is negligible.

We obtain the scattered-radiation flux, Φ_S^c, for a continuum source simply

404

THE MEASUREMENT OF FLUORESCENCE   Sect. III.16c

by substituting $B_{\lambda_0}\Delta\lambda_i$ for B in Eq. III.114 (if $\Delta\lambda_i \leq \Delta\lambda_d$; otherwise $\Delta\lambda_d$ should be used here instead of $\Delta\lambda_i$).

The SNR expressions listed in Table III.9b and derived under idealized experimental conditions may guide us to optimize the set-up with respect to the SNR. We shall discuss briefly the influence of the separate experimental parameters and measuring conditions on the SNR in the twelve cases considered.

The influence of τ_r and f_{mod} is essentially the same as in the case of emission measurements (cf. Sect. III.14f).

The influence of the parameters contained in the coefficients $c_1, \ldots, c_5$ is seen from the definitions in Table III.9a and the expressions in Table III.9b. It appears that in cases 7-12 the SNR is independent of $\tau_d(\lambda_0)$ and $s(\lambda_0)$, and in cases 9 and 10 is also independent of $\tau_i(\lambda_0)$. In general the SNR improves with increasing c_1 and c_2 (i.e. with increasing fluorescence signal), and with decreasing c_3, c_4 and c_5. The former two coefficients depend on the atomic-line properties which are either fixed or variable in a restricted range only. Note that in cases 7 and 10 f_{pq} and A_{qp} drop out, as these optical transition coefficients are proportional to each other (see Eq. II.168). Lowering of the flame temperature considerably decreases c_3 and c_4, and may thus be advantageous in cases 1, 2, 4, 5, 7, 8, 10 and 11. In hydrogen flames the flame-background coefficient c_4 is comparatively small (see also Sect. IV.8) and this type of flame is in general suitable for fluorescence measurements. The magnitude of c_5 is connected with the occurrence of nonvolatilized solute particles or nondesolvated spray droplets. We refer to Sect. III.9 for a discussion of possible measures to minimize scattering effects; in this respect a direct-injection burner is generally not suited to fluorescence work.

The influence of ground-state density n_0 and fluorescence efficiency Y is trivial. Note, however, that n_0 drops out of the flicker-noise expressions 7 and 10, and occurs under the square-root sign in cases 1 and 4. When in cases 7 and 10 n_0 is made arbitrarily small, other noise sources may become dominant over the thermal-emission noise. It may then also happen that the shot-noise in the thermal-line emission (appearing in expressions 1 and 4) becomes dominant over the corresponding flicker noise. This is connected with the fact that the <u>relative</u> shot noise increases $\propto n_0^{-\frac{1}{2}}$ with decreasing n_0, whereas the <u>relative</u> flicker noise is independent of n_0 (see also Sect. IV.9). When n_0 is raised too high, self-absorption may set in and the SNR expressions become invalid.

In *analytical* flame fluorescence spectroscopy the detection limit (expressed in solution concentration c) can be improved by increasing the conversion factor n_0/c (see Sect. III.2h) by introducing the sample more efficiently into the flame. However, this does not hold in cases 7 and 10. A direct-injection burner can be used for increasing the conversion factor, but scattering noise may then become dominant.

The analytical detection limit can also be improved in all the cases listed if one works with argon-diluted flames, which generally give higher Y values (see Veillon et al. 1966, and Pearce, de Galan and Winefordner 1968). These high Y values are connected with the very small quenching cross section of Ar atoms (see Sect. VI.2a-1). An extra argon flame sheath is sometimes applied in order to prevent the surrounding air from penetrating into the flame (see also Sect. III.2a).

In all cases except for no. 9 and 12 the SNR is improved by enhancing the (spectral) radiance of the primary light source. In cases 9 and 12 the source intensity drops out, as both the fluorescence and scattering signal increase in proportion to the source intensity while the flicker noise in the scattering signal is a constant fraction of it. This fraction, determined by $A^{\frac{1}{2}}$, may depend not only on the excess low-frequency noise in the scattering process itself but also on the excess noise in the source radiation. Therefore, in cases 9 and 12 one should select a light source not on the basis of its absolute radiance but rather on the basis of its stability. A more stable light source would also make the fluorescence signal itself more stable. Fluctuations in the light source (e.g., a c.w. dye laser) can also be suppressed by monitoring the source intensity with an extra photodetector and ratio-ing the fluorescence (or scattering) signals versus the monitor output (see Green, Travis and Keller 1976).

The geometrical factors that appear explicitly in the SNR expressions are the height, H, of the irradiated flame zone and the depth, b, of the whole flame in the direction of observation (see also Fig. III.31). Their optimization cannot be discussed without also taking into consideration the optical parameters L_A and L_F, which may possibly be connected with these geometrical factors as well as with the spectral bandwidths $\Delta\lambda_i$ and $\Delta\lambda_d$. We recall that $L_A = \Omega_A l H$ and $L_F = \Omega_F d'H'$ with $H' \leqslant H$ and $l \leqslant b$. Since b is only found in the denominator of expressions 1, 2, 4, 5, 7, 8, 10 and 11, optimization requires that b be made equal to its lower limit l and that l be made as small as possible. But when b becomes too small, the flame will become unstable or inhomogeneous with respect to temperature and composition because of indrawn air. When Y values are to be measured accurately in a homogeneous portion of the flame, it is advisable to make b somewhat larger than l.

When a metal-free flame sheath (see Sect. III.2a) is applied to improve the homogeneity of the observed flame zone, the value of b in expressions 2, 5, 8 and 11 exceeds the value to be inserted for b in expressions 1, 4, 7 and 10. This holds because both the flame sheath and the central zone contribute to the flame

THE MEASUREMENT OF FLUORESCENCE   Sect. III.16c

background emission.

Whenever the minimal b value is fixed for practical reasons, the value of l ($\leq b$) is irrelevant as long as L_A remains the same. The SNR, however, can then be improved by decreasing H, which occurs in the denominator of all expressions except those in cases no. 9 and 12. H (and l) can be reduced by choosing other components in the irradiation optics. As long as these components do not restrict the spatial and angular extensions of the primary beam passing through them, the optical conductance, L_A, of the beam will not be changed (see Sect. III.4b and Appendix A.6). The only consequence of reducing H and l is then that Ω_A becomes larger. It should be realised, of course, that for too small H- and l values and thus for a too large Ω_A value the condition of an approximately brick-shaped irradiation zone will be violated so that the SNR-expressions become invalid. A reduction of H may ultimately entail a reduction of H' ($\leq H$) which is contained in L_F. However, by adapting the detection optics (i.e., by enlarging Ω_F) one can then still keep L_F constant. If this adaption is not made and if H' equals H, a further reduction of H will have no effect on the SNR in cases 3 and 6. Anyway, whenever a light source with a rectangular radiating surface (e.g., a tungsten-strip lamp) is imaged on the flame, this source should be positioned so that its shorter side stands vertically, thus making H minimal for given irradiation optics and L_A value.

In all cases where L_A and/or L_F appear in the SNR expressions, optimization means making these factors as large as possible. It is noted that L_F has only an influence on the SNR in cases 1-6 where shot noise prevails. The optical conductance, L_A, of the primary beam, however, may be limited by the finite optical conductance of the light source itself. In the case of a tungsten-strip lamp for example, the optical conductance is limited by the size of the strip and the solid angle subtended by the window as seen from the strip. A similar limitation may exist for the observed fluorescence beam whose optical conductance cannot exceed that of the photodetector including its housing. The available optical components (lenses, spectral selector, etc.) may impose a further restriction on the optical conductance of the irradiation and detection optics and thus on L_A and L_F. This restriction may be severe when a slit-instrument such as a monochromator is used (see Table III.2). Nondispersive spectral selectors are generally to be preferred for isolating the fluorescence line of interest, when the SNR does not depend on the spectral bandwidth. For fluorescence lines below 300 nm, where the background emission of most flames is low, a solar-blind photomultiplier (see Sect. III.5a) without any spectral selector has been applied successfully, e.g., by Vickers and Vaught (1969), Elser and Winefordner (1971), and Larkins and Willis (1974).

With monochromators the connection between optical conductance and spectral

Chapt. III   GENERAL INSTRUMENTAL ASPECTS AND EXPERIMENTAL METHODS

bandwidth should be taken into consideration when both quantities appear together in the SNR expressions (see cases 2, 5 and 6). According to Eq. III.9 the product, P_1, of optical conductance and spectral resolution is a constant (i.e. independent of slit width) for a given monochromator and wavelength. Consequently, if L_A or L_F are determined by the optical conductance of the monochromator, the ratio $L_A/\Delta\lambda_i$ (occurring in expression 6) or $L_F/\Delta\lambda_d$ (occurring in expressions 2 and 5) can only be improved by using another spectral selector with a greater P_1 value (see Table III.2). Of course, the values of L_A and $\Delta\lambda_d$, which are not interconnected, can be optimized independently (see cases 8 and 11). One should bear in mind that the SNR expressions given hold only on condition that $\Delta\lambda_d$ is sufficiently large compared to the effective width of the atomic line considered.

Special optical means have been used to increase Ω_A or Ω_F, and thus L_A or L_F. The simplest way is to place a concave mirror behind the flame on the axis of the irradiation or the detection optics (see, e.g., Benetti, Omenetto and Rossi 1971). Shull and Winefordner (1971) have chosen the most radical approach. They placed the light source (an electrodeless discharge lamp) in one focus of an elliptical cylinder and the flame in the other focus. Thereby they achieved roughly a tenfold increase in SNR. The increase of Ω_A and Ω_F is, of course, theoretically limited by the condition: $\Omega_A + \Omega_F \leq 4\pi$. When a maximum fluorescence signal is aimed at only, one should, ideally, make $\Omega_A = \Omega_F = 2\pi$ (see Alkemade 1963). However, when an optimal SNR is desired, it is in most cases more profitable to enlarge Ω_A rather than Ω_F (see Table III.9b). Although for large Ω_A- and Ω_F values the SNR expressions given are no longer strictly applicable, this general conclusion still remains valid.

Whenever the optical conductance of the detection optics is limited by the properties of a given monochromator (see Eq. III.3), optimization of the optical arrangement can only be achieved by making the (spectral) radiance of the light beam filling the entrance slit as high as possible (cf. Eqs III.2 and III.11). When the light source is focused on the plane of the slit or when the arrangement depicted in Fig. III.30 is used, the radiance at the slit cannot exceed the radiance of the source itself (see, e.g., Keitz 1971 and Appendix A.6). By placing a mirror behind the flame one can achieve at most a doubling of this radiance, if there are no losses due to reflection or re-absorption in the source, etc. Special optical systems, such as a Cassegrain mirror placed just in front of the entrance slit of a monochromator (see Mitchell 1970), or a combination of an ellipsoidal and a spherical mirror (see Benetti, Omenetto and Rossi 1971), cannot do better than a well chosen mirror behind the flame.

Table III.9a

Explanation of Coefficients Occurring in Table III.9b [†]

Symbol	Explanation
c_1	$= 7 \times 10^{-14} (\lambda_0^2 f_{pq} / \Delta \lambda_{\text{eff}}) \tau_i(\lambda_0) \tau_d(\lambda_0) s(\lambda_0)$ (in CGS units), where $\tau_i(\lambda_0)$ and $\tau_d(\lambda_0)$ is the transmission at λ_0 of the irradiation and the detection optics, respectively; $s(\lambda_0)$ is the responsivity of the photocathode at λ_0 and the other constants are the same as in Eq. II.311.
c_2	$= 7 \times 10^{-14} (\lambda_0^2 f_{pq}) \tau_i(\lambda_0) \tau_d(\lambda_0) s(\lambda_0)$ (in CGS units).
c_3	$= 1.6 \times 10^{-17} (A_{qp} g_q / \lambda_0 Q) \exp[-E_q/kT] \tau_d(\lambda_0) s(\lambda_0)$ (in CGS units), where the atomic constants are the same as in Eq. II.281 and T is the flame temperature.
c_4	$= \{B_{\lambda_0}^{\text{bg}} / b\} \tau_d(\lambda_0) s(\lambda_0)$, where $B_{\lambda_0}^{\text{bg}}$ is the spectral radiance of the flame-background continuum at λ_0 and b is the flame depth in the observation direction (see Fig. III.31).
c_5	$= \xi(\lambda_0) n_{\text{sc}} \tau_i(\lambda_0) \tau_d(\lambda_0) s(\lambda_0)$, where n_{sc} is the density of scatterers in the flame and $\xi(\lambda_0)$ characterizes the (average) scattering power of a single scatterer at λ_0.

[†] Numerical factors of order unity have been omitted (compare Table III.6 on page 367. The symbols c_1, c_2 and c_3 should not be confused with the radiation constants.

Chapt. III GENERAL INSTRUMENTAL ASPECTS AND EXPERIMENTAL METHODS

Table III.9b
Theoretical Expressions for SNR in Fluorescence Measurements[†]

Noise source \ Noise component	Shot-noise SNR × $(2\pi e/\tau_r)^{\frac{1}{2}}$				Flicker-noise SNR × $(2A)^{\frac{1}{2}}$		
	spectral line source		spectral continuum source		spectral line source		spectral continuum source
Thermal line emission *	$\frac{c_1}{\frac{c_2^2}{c_3}} \cdot \frac{n_0^{\frac{1}{2}} YBL\mathbf{A}}{H} \left(\frac{L_\mathbf{F}}{b}\right)^{\frac{1}{2}}$	1	$\frac{c_2}{c_3} \cdot \frac{n_0^{\frac{1}{2}} YB\lambda_0 L\mathbf{A}}{H}\left(\frac{L_\mathbf{F}}{b}\right)^{\frac{1}{2}}$	4	$\frac{c_1}{c_3} \cdot \frac{YBL\mathbf{A}}{Hb}(\tau_r f_{mod})^{\frac{1}{2}}$	7	$\frac{c_2}{c_3} \cdot \frac{YB\lambda_0 L\mathbf{A}}{Hb}(\tau_r f_{mod})^{\frac{1}{2}}$	10
Flame background emission *	$\frac{c_1}{\frac{c_2^2}{c_4}} \cdot \frac{n_0 YBL\mathbf{A}}{H}\left(\frac{L_\mathbf{F}}{b\Delta\lambda_d}\right)^{\frac{1}{2}}$	2	$\frac{c_2}{c_4} \cdot \frac{n_0 YB\lambda_0 L\mathbf{A}}{H}\left(\frac{L_\mathbf{F}}{b\Delta\lambda_d}\right)^{\frac{1}{2}}$	5	$\frac{c_1}{c_4} \cdot \frac{n_0 YBL\mathbf{A}}{Hb\Delta\lambda_d}(\tau_r f_{mod})^{\frac{1}{2}}$	8	$\frac{c_2}{c_4} \cdot \frac{n_0 YB\lambda_0 L\mathbf{A}}{Hb\Delta\lambda_d}(\tau_r f_{mod})^{\frac{1}{2}}$	11
Scattered radiation ¶	$\frac{c_1}{\frac{c_2^2}{c_5}} \cdot n_0 Y\left(\frac{BL L_\mathbf{F}}{H}\right)^{\frac{1}{2}}$	3	$\frac{c_2}{c_5} \cdot n_0 Y\left(\frac{B\lambda_0 L\mathbf{A} L_\mathbf{F}}{H\Delta\lambda_i}\right)^{\frac{1}{2}}$ ‡	6	$\frac{c_1}{c_5} \cdot n_0 Y$	9	$\frac{c_2}{c_5} \cdot \frac{n_0 Y}{\Delta\lambda_i}$ ‡	12

410

THE MEASUREMENT OF FLUORESCENCE   Sect. III.16c

NOTES TO TABLE III.9b

† Expressions for the signal-to-noise ratio (SNR) are given for dominant shot- or flicker noise in the thermal line emission, the flame background emission or the scattered radiation in the flame. The primary-beam intensity is supposed to be modulated at frequency f_{mod}. Paired readings are assumed (compare Sect. III.14f-2 and caption for Table III.6 on page 367). The characteristics of the amplifier + read-out system are the same as those assumed in Table III.6; τ_r is the response time and τ is the sampling time (see Table III.5). In all cases τ is assumed to equal τ_r, except for cases no. 9 and 12 where optimal $\tau = \frac{1}{8}\tau_r$. The flicker-noise component is characterized by a constant, A, defined in Eq. III.66.

Legend: n_0 = density of metal atoms in the ground-state; B and B_{λ_0} are the radiance and spectral radiance at line centre λ_0 of the irradiating spectral line source and continuum source, respectively; Y = efficiency of resonance fluorescence; L_F and L_A are the optical conductance of the fluorescence beam accepted by the detection optics and of the primary beam incident on the flame, respectively; H = height of irradiated flame zone (see Fig. III.31); b = depth of flame in the direction of observation; $\Delta\lambda_i$ and $\Delta\lambda_d$ are the effective spectral width of the irradiation optics and of the detection optics, respectively; e = elementary charge.

The coefficients $c_1, \ldots, c_5$ are explained in Table III.9a.

‡ It is here assumed that $\Delta\lambda_i \leqslant \Delta\lambda_d$.

* It is assumed that the thermal line emission and the flame background emission are accepted by the detection optics with the same optical conductance, L_F, as the fluorescence radiation.

¶ Scattering of the primary-radiation beam in the flame is assumed to be isotropic.

CHAPTER IV

Types and Properties of Nonseeded Flames

1. INTRODUCTION

In this book chemical flames and combustion are not in themselves subjects for study. They are considered here only as a means for studying metal vapours at moderately high temperature. The behaviour and state of metal vapour depend on the flame properties such as temperature and composition, which can be varied within wide limits. These properties have to be known when one wants to interpret the observations on metal species or to select the best experimental conditions for studying them. They are also important for the understanding of interference effects in analytical flame spectrometry or for selecting the optimum analytical flame. Since the metal vapour is usually present as a trace in the flame gas, the flame properties are to a large extent independent of its presence. In this chapter we shall therefore describe the types (Sect. 2)[†], structure (Sect. 3), and properties (Sects 4 to 9) of commonly used laboratory flames without metal seeding.

Most attention will be paid to *stationary, premixed, atmospheric-pressure flames*, that is, flames burning in the free atmosphere at a constant total pressure, p_t, equal to 1 atm. These flames are the ones most often used in studies of metal vapours as they can be made with reasonably uniform and well-defined properties, are stable and easy to operate (see also Sect. III.2a). In contrast to propagating flames (explosions), stationary flames are fed by a continuous flow of combustible gas mixture and are fixed in space by a burner. The latter flames are stationary with respect to the observer; but each element of the burnt gas moves upwards with a vertical velocity component called the *rise-velocity* v_r (see for methods of

---
[†] Section numbers quoted without a roman cipher refer to the same chapter.

Chapt. IV   TYPES AND PROPERTIES OF NONSEEDED FLAMES

measuring Sect. III.11). The state of the gas (and of the metal species contained in it) changes with *travel time* t_{tv}, that is the time that has elapsed since the gas entered the flame. When v_r is constant, the relation between t_{tv} and the height of observation, h_{obs}, above the base of the flame is given by

$$h_{obs} = v_r t_{tv} \ . \tag{IV.1}$$

This relation enables us to study relaxation processes having relaxation times of the order of 1 ms or longer, as v_r is of the order of 10^3 cm s^{-1} or less.

Combustion of a *fuel* like C_2H_2 by an *oxidant* like O_2 in the presence or absence of an *inert gas* like N_2 (see Sect. 2) is a highly exo-ergic chemical reaction which proceeds very fast in the *flame front* at the base of the premixed flame (see Sect. 3). If the (uniform) combustible gas mixture were stagnant and not bounded by walls, the flame front would propagate through it at a constant velocity perpendicular to the surface of the front. This velocity is called the *burning velocity* and is, for a particular fuel-oxidant-inert gas combination, a function of their mixing ratios, initial temperature, total pressure and humidity. The burning velocity is also influenced by possible turbulence of the gas mixture. With CO flames this velocity depends sensitively on humidity and on the presence of H_2 as impurity; these flames might not even burn at all if hydrogen (compounds) were completely absent (see Sect. 5a). The maximum value of the burning velocity may vary within wide limits for different combinations of gases, as is seen from Table IV.1. When there are walls nearby, for example, inside a tube, the burning velocity will be reduced. Propagation of the flame front is even prohibited, that is, the combustion is *quenched* when the diameter of the tube is less than the quenching diameter (see Sect. III.2b-2).

A self-sustained combustion reaction can proceed only if the mixing ratio of fuel and oxidant falls between a certain lower and higher limit, called *limits of inflammability*. These limits are usually expressed in percentage volume of gaseous fuel in the total volume of the gaseous mixture. These limits depend on the kind of gas mixture, the experimental conditions and the type of ignition.

Another parameter of practical significance is the *spontaneous-ignition temperature*. Its value does not vary much with the dimensions of the vessel containing the combustible gas mixture or with the fuel/oxidant ratio (within the limits of inflammability). For H_2, CO or C_2H_2 in air or O_2 spontaneous-ignition temperatures lie between 300 and 600°C.

The most important flame properties in metal vapour studies are the *temperature* and *chemical composition*. Both properties are interrelated, as will be shown in Sect. 4 where calculated equilibrium compositions and theoretical (adiabatic) temperatures are presented for various combustible gas mixtures. In Sect. 6 experimental flame temperatures are discussed and compared with theoretical values.

INTRODUCTION Sect. IV.1

Table IV.1

Theoretical Flame Temperatures [1] and Maximum Burning Velocities

Gas mixture	Stoichiometric		Optimum[2]		Maximum burning velocity[7]	
	Mixing ratio[3]	Temperature (K)[4]	Mixing ratio[3]	Temperature (K)[4]	v_b (cm/s)	Ref.
H_2-air	2.36	2385	2.22	2401	320-350 [8]	1-3
C_3H_8-air	23.65	2267	22.5	2279	36-48.5	1-5
C_2H_2-air	11.82	2537	9.22	2606	158-170	1-3
CO-air	2.36	2382	1.96	2417	18-20	3;6
H_2-O_2	0.5	3080	0.49	3081	1120-1190 [8]	1;2
C_3H_8-O_2	5.0	3094	4.4	3103	370-390	7
C_2H_2-O_2	2.5	3343	1.75	3430	1130;1140	1;2
CO-O_2	0.5	2971	0.45	2973	100	2
C_2N_2-O_2			1.0	4810 [5] 4850 [6]	270	1
H_2-N_2O	1.0	2960	0.95	2964	380-390	1
C_3H_8-N_2O	10.0	2932	8.3	2963		
C_2H_2-N_2O	5.0	3148	2.9	3254	160;285	1;7
CO-N_2O	1.0	2873	0.83	2881	~40	8

NOTES
[1] Calculated adiabatic flame temperatures at 1 atmosphere pressure for gas mixtures that were initially at room temperature.
[2] For the optimum gas mixture the theoretical temperature reaches its maximum.
[3] The mixing ratio is the mole ratio of the oxidant gas (air, etc.) to the fuel gas.
[4] Temperatures as calculated by the authors of this book.
[5] From reference 1.
[6] From reference 2.
[7] See reference 9 for a critical survey of literature data.
[8] See reference 10 for recent calculations and a comparison with experimental values.

REFERENCES

1 Willis (1968)
2 Gaydon and Wolfhard (1979)
3 Günther and Janish (1971)
4 Edmondson and Heap (1970)
5 Yumlu (1968)
6 Yumlu (1967)
7 Aldous, Bailey and Rankin (1972)
8 Kalff and Alkemade (1972)
9 Andrews and Bradley (1972)
10 Warnatz (1978)

Chapt. IV   TYPES AND PROPERTIES OF NONSEEDED FLAMES

Methods of measuring flame temperature have been discussed in Sect. III.10. It is seen from Table IV.1 that the highest attainable temperatures in flames at 1 atm pressure are nearly 5000 K. The temperature increases somewhat with increasing total pressure (see Sect. III.2a). Temperatures as low as about 1400 K can be realized with H_2 flames to which an excess of inert gas is added.

The concept of flame temperature is not fully unambiguous as a flame is not a closed thermodynamical system and there is only limited time available for the burnt gas to attain its equilibrium state. Inside the flame and between flame and surroundings a transport of heat, mass and radiation takes place continually (see for further discussion Sects II.3c and III.10a).

Flames are usually multi-component systems; even a flame produced by a pure, stoichiometric H_2-O_2 mixture contains some H_2, O_2, H, OH and O, in addition to H_2O, as a result of thermal dissociation. This can hamper the interpretation of observations on collision processes (excitation, ionization, diffusion, etc.; see Chapts VI, IX and X). The contribution of each (major) flame constituent to a certain collision process can, however, be assessed by varying the quantitative flame composition in a known way and measuring the resulting effect on the collisional rate (see Sect. II.4a-1). In Sect. 4 the methods of calculating the equilibrium composition will be treated and results given. The actual flame composition, however, may deviate from the calculated composition owing to trivial effects (e.g., indrawn air from the surroundings, which is not accounted for) or to fundamental relaxation effects in the establishment of flame-radical equilibria. The trivial effects can in most cases be overcome by appropriate instrumental means (see Sect. III.2). Deviations from radical-equilibria are especially noted in the cooler flames; they will be discussed in Sect. 5.

Although the flame as a whole is electrically neutral, *free electrons and ions* may be produced in it, be it at relatively low partial pressures. These charged species may, however, interfere with the ionization of the metal vapour and should be taken into account when the metal ionization is investigated (see Chapt. IX). Sect. 7 gives an account of the state of ionization in nonseeded flames.

In spectroscopic studies of metal vapours or in analytical applications of flame spectroscopy the spectral emission (or absorption) of the flame itself, called the *flame background*, may be a limiting factor. We shall therefore also deal with the flame background in Sect. 8.

Flames are, of course, not perfectly stable and we shall deal in Sect. 9 with their *fluctuations*. These fluctuations may limit the precision of intensity measurements or the detectability of weak emission signals when metal vapours are introduced. Spectral noise analysis (see Sect. III.5c) of the flame-background fluctuations enables us to calculate the signal-to-noise ratio in emission,

GENERAL TYPES OF FLAMES	Sect. IV.2

absorption or fluorescence measurements by using the theoretical expressions presented in Sects III.14-16.

The description of flames and combustion processes in this chapter is of limited scope and therefore rather superficial. The reader who wants a more extensive and basic treatment may consult a wide variety of well-documented, authoritative textbooks or review articles, e.g., Lewis and von Elbe (1961), Mavrodineanu (1961), Weinberg (1963), Fenimore (1964), Fristrom and Westenberg (1965), Mavrodineanu and Boiteux (1965), van Tiggelen (1968), Lawton and Weinberg (1969), Gaydon (1974), and Gaydon and Wolfhard (1979). The Proceedings of the series of International Symposia on Combustion also contain much useful information (see the pertinent references in the Bibliography under 'Proceedings').

2. GENERAL TYPES OF FLAMES

There is a great variety of laboratory-scale flames. They can be distinguished and classified from different points of view:

1. The *kind of fuel, oxidant* and *inert gas* (if any is present). The gases most frequently used in atomic vapour studies have been listed in Sect. III.2. In some cases combinations of different fuel, oxidant or inert gases are used.

Unusual flames applied in fundamental or analytical flame spectroscopy are, e.g., H_2-halogen flames (used for studying metal halides), the $C_2N_2-O_2$ flame (producing the highest temperature, see Table IV.1; see Vallee and Baker 1955), and the methylacetylene-propadiene-N_2O flame (combining a high temperature of about 3000 K with a low rise-velocity; see Mansell 1970). In analytical flame spectroscopy a liquid organic fuel or a combustible powder has been tried out incidentally (see references in Alkemade and Herrmann 1979). When an organic solvent, such as acetone, is nebulized into an (ordinary) flame, its vapour may contribute to the combustion (see Herrmann and Alkemade 1960, 1963).

Some high-temperature sources bearing the name 'flame' like the Langmuir- or atomic-hydrogen flame, spark-in-flame, augmented flame, or plasma flame are not genuine (purely) chemical flames and will not be discussed here (see Herrmann and Alkemade 1960, 1963, and Mavrodineanu and Boiteux 1965).

2. Besides the qualitative composition of the combustible gas mixture, the quantitative *mixing ratios* are important too for characterizing a flame. We distinguish sooting incandescent flames (only with organic fuels), fuel-rich, stoichiometric, and fuel-lean flames, depending on the fuel/oxygen ratio ('mixture strength'). The definition of a stoichiometric flame is given in Sect. 4c.

Chapt. IV   TYPES AND PROPERTIES OF NONSEEDED FLAMES

Reducing or fuel-rich flames are generally characterized by their ability to dissociate refractory metal oxides, and are frequently used in analytical spectroscopy. The reducing conditions in the $C_2H_2-O_2$ flame are most satisfactory when the ratio of total elemental C concentration to total elemental O concentration is about unity; higher ratios make this flame sooty and do not much improve its reducing properties (see L'vov 1970; see also Sect. 4c).

3. As to the *manner of mixing* the fuel gas with the oxidant we distinguish *premixed* and *unpremixed* flames. In the former type, mixing takes place inside the burner housing before the gases enter the flame (see Sect. III.2a). When the flow of combustible gas mixture leaving the burner port(s) is made laminar, a more or less laminar flame can be obtained.[†] In contrast, in an unpremixed flame the fuel and oxidant gases flow out of the burner through separate ports and are mixed in the flame itself. This mixing may take place smoothly through molecular diffusion or vehemently through turbulence. Turbulence is promoted by discharging the gases at high speed from the narrow exit ports in the direct-injection burner (see Sect. III.2a). Most flames with moderately high burning velocity (below, say, $500\,\mathrm{cm\,s^{-1}}$; see Table IV.1) can easily be premixed safely (see also Sect. III.2b-1). This distinguishes the flames burning with N_2O from the corresponding, not much hotter but more explosive flames burning with pure O_2.

Turbulent unpremixed flames produced above the tip of a direct-injection burner have found wide, successful application in analytical flame emission spectroscopy. With this burner highly explosive gas mixtures (H_2 or C_2H_2 with pure O_2) can be easily burnt safely while analyte solutions can be reproducibly introduced. For a more detailed description of this type of flame and for a comparison with premixed flames, we refer to the textbooks on analytical flame emission spectroscopy cited in Chapt. I. These flames are less suited for quantitative investigations of metal vapours (see Sect. III.2a). Because of their high temperature these flames do, however, produce rich atomic-line spectra and are therefore well-suited for spectral recordings.

Unpremixed H_2 diffusion flames burning with oxygen from the ambient air are characterized by their low temperature and favourable reducing property. Temperatures typically range between about 1000 and 2000 K. The hydrogen gas is often premixed with an inert gas, N_2 or Ar. These flames are occasionally

---

[†] The laminarity of a 10 cm long premixed cylindrical C_2H_2-air flame burning on a Méker burner as shown in Fig. III.1 was demonstrated by Snelleman (1965) by photographing the perfectly linear tracks of soot particles on the flame axis (see Fig. X.3). At the edge of the flame, where it comes into contact with the stagnant ambient air, eddies are formed which develop at greater heights into strong turbulence that finally extinguishes the flame (see also Alkemade 1954).

THE STRUCTURE OF PREMIXED LAMINAR FLAMES   Sect. IV.3

used in analytical flame spectroscopy, for example, for the determination of non-metals, such as P or S, by means of suprathermal chemiluminescence (see textbooks cited in Chapt. I).

4. As to the (*final*) *total gas pressure* we distinguish low-pressure, atmospheric-pressure, and high-pressure flames burning at $p_t \ll 1$, $=1$ and $\gg 1$ atm, respectively (see Sect. III.2a).

5. The *geometrical form*, in particular the shape of the cross section and general appearance of the flame, depend largely on the positioning of the exit port(s) or slit(s) in the burner head; they also depend on the absolute flow velocity at which the unburnt gas mixture is discharged from the burner. The possible presence of a chimney as in the Smithells separator (see Sect. 3) plays a role, too. One distinguishes rectangular, conical, cylindrical, bulbous, planar (as produced by a slot burner), flat, lifted and split flames (see ibidem).

3. THE STRUCTURE OF PREMIXED LAMINAR FLAMES

Premixed laminar flames show a distinct structure which can usually be perceived with the eye. Three adjacent regions can be distinguished, as is shown in Fig. IV.1. for the simplest case of a premixed, laminar and nonluminous *Bunsen flame* burning above a Bunsen burner with one, large, circular port. We present here a brief, somewhat phenomenological description of these zones but refer the reader to later sections or chapters for a more detailed description.

1. *Primary combustion zone or inner zone.*[†] This zone, found at the base of the flame and constituting the flame front, settles just above the rim of the burner outlet port. In this zone combustion of the fuel with the oxidant available in the premixed gas mixture takes place within a very short time (typically 10 μs at 1 atm). Consequently, this zone is very thin (typically 0.01 - 0.1 mm), at least when the gas flow is laminar. The thickness varies, roughly speaking, inversely with the pressure. Low-pressure flames are thus better suited for exploring this

---

[†] Also called inner, internal or blue cone when it has approximately the shape of a cone (as for the Bunsen flame), primary reaction zone, or tongue. The term reaction zone might be misleading as chemical reactions, considered on a microscopic scale, do occur throughout the whole flame; the difference is that in the combustion zone the reactions are (far) from balanced (they proceed 'irreversibly'), whereas they are (nearly) balanced in the interzonal region ('dynamic equilibrium'; see sub point 2). The actual occurrence of chemical reactions in the latter region can be deduced from their effects on the (de-)excitation and ionization rates of metal species (see Sects VI.3 and IX.2 and 3). The term reaction-free zone is therefore not appropriate for the latter region.
   Note that Fenimore (1964) and others denote the primary combustion zone as 'flame' and the zone above it as 'post-flame gas'.

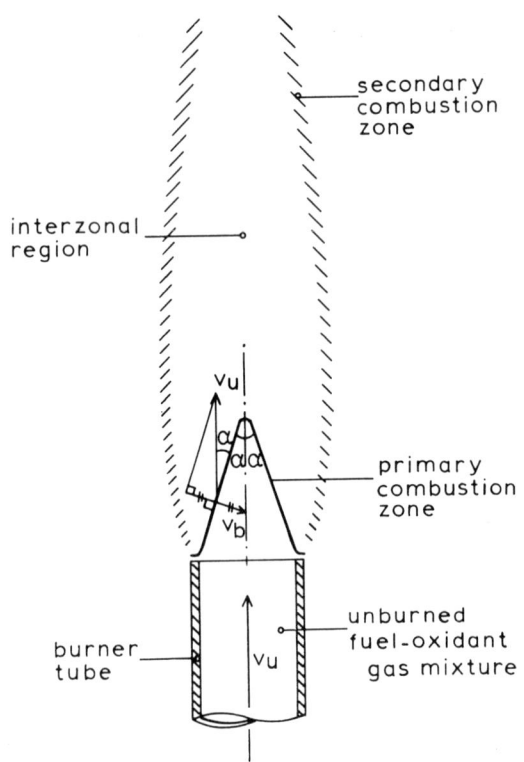

Fig. IV.1 Schematic section of a premixed laminar Bunsen flame in a plane through the axis of the cylindrical burner tube; the section shows the three regions of the flame discussed in the text. The magnitude of the burning velocity v_b equals the component of the unburned-gas velocity v_u, which is perpendicular to the flame front or primary combustion zone. (Redrawn from Alkemade and Herrmann 1979.)

zone (by mass spectrometry, electron spin resonance, electrical-probe measurements or optical spectroscopy) than atmospheric-pressure flames. The colour of this zone is blue-green in C_2H_2-air flames and blue-white in $C_2H_2 - N_2O$ flames. The origin of the emission of this zone is connected with the presence of excited radicals in excess concentrations and with chemical association reactions. The initiation and course of the combustion reactions producing these radicals will be discussed in Sect. 5a. There are also excess concentrations of free electrons and ions in the inner zone of hydrocarbon flames, as will be discussed in Sect. 7.

Upon closer inspection we can discern in the inner zone of hydrocarbon flames two layers: the bottom layer whose emission is due mainly to C_2, and an upper layer where CH and CHO are mainly responsible for the emission. The emission of OH radicals is observed throughout the entire inner zone. The radiation from the inner zone of hydrogen flames is mainly in the ultraviolet and stems from suprathermally excited OH radicals; this zone is hardly visible. A more detailed description of these emissions will be given in Sect. 8. Before entering the combustion zone the gas mixture passes through the *preheating zone* where it is heated

THE STRUCTURE OF PREMIXED LAMINAR FLAMES   Sect. IV.3

through conduction and where exothermic reactions are not yet significant.

When metal vapours are present in the inner zone, strong, suprathermal atomic emissions due to chemiluminescence may be noted (see Sect. VI.3). This phenomenon has been exploited in flame chemiluminescence spectroscopy for analytical applications (see Sect. VI.4e).

The geometrical shape of the inner zone is determined by the burning velocity v_b (see Sect. 1) and the velocity v_u at which the combustible gas mixture emerges from the burner port. We assume for simplicity that the latter velocity is uniform over the cross section of the gas stream and that it is directed vertically. The flame front tends to propagate in a normal direction and with speed v_b towards the unburnt gas mixture. For the flame front to be fixed in space the component of v_u normal to the front has to be just balanced against v_b at each point. This balancing of velocities (see arrows marked with two strokes) is shown for an arbitrary point on the front in Fig. IV.1. Assuming that v_b is constant along the whole front we thus find that the primary combustion zone with a round burner port is a *cone* with an apical angle 2α determined by

$$\sin \alpha = v_b/v_u . \qquad (IV.2)$$

The greater v_u is in comparison to v_b, the sharper the cone will be; with a sharper cone the flame will be stiffer — that is, it will flicker less. For large values of v_u/v_b the flame will be lifted or even blown off; if $v_u < v_b$ flashback can occur. Usually one adjusts v_u/v_b at a value larger than 5. The stability of the flame front and hence of the entire flame is governed not only by the value of $\sin \alpha$; other, secondary effects arising at the upper rim of the burner port may play a role, too. Note that at the apex and at the bottom of the inner cone Eq. IV.2 loses its meaning; other effects become important here, and as a consequence the inner zone becomes bell-shaped rather than cone-shaped.

The state of the gas in the inner zone is far from equilibrium. The chemical energy released in this zone is initially taken up by the direct products of the combustion reactions. These are formed in excess concentrations and with an excess amount of internal and translational energy. The excess energy is redistributed over the other components of the gas mixture by subsequent collisions. As long as the energetic and irreversible combustion reactions are not yet terminated, no thermodynamic equilibrium can exist. Thus the state of the gas in the inner zone cannot be described by a general temperature value. In particular, the state of electronic excitation, ionization and dissociation in this zone may strongly deviate from the equilibrium state to be expected at the final flame temperature. However, since the time of passage through this zone is considerably longer than the relaxation time for establishing a partial Maxwell equilibrium (see Sect. II.3c), a local 'translational temperature' may be meaningfully defined here for the bulk

Chapt. IV   TYPES AND PROPERTIES OF NONSEEDED FLAMES

gas components. Its value is a rapidly varying function of place inside this zone. It can be measured with a thermocouple in low-pressure flames where the inner zone is sufficiently thick.

In the inner zone, even the rotational and possibly the vibrational temperatures (see Sect. II.3c) of the inert gas components may approach rather closely the translational temperature at 1 atm. In fast flames, such as the $C_2H_2-O_2$ flame, having a thin combustion zone and a high burning velocity at 1 atm, the vibrational temperature probably lags behind the rapidly increasing translational temperature. We recall from Sect. II.3c that the conversion of vibrational energy upon collision is much less efficient than the conversion of rotational energy, which is, in turn, slower than the exchange of translational energy. The rotational temperature of OH (for its experimental determination see Sect. II.5c-3) was found to be equal to 9000 K in an $C_2H_2-O_2$ flame at $p_t = 1.5$ Torr, and about equal to the expected flame temperature of 3300 K at $p_t = 1$ atm (see Lewis, Pease and Taylor 1956, p.156). These different outcomes are to be explained by the much lower collisional rate in the low-pressure flame, which prevents the attainment of equilibrium for the rotational energy.

2.   *Interzonal region or central region.* After leaving the primary combustion zone the burnt gases have attained for the most part a state of equilibrium; the temperature is close to its equilibrium value and the concentrations of the major species correspond approximately to chemical equilibrium. The expansion and heating of the flame gas takes place mainly in the thin inner zone where a steep gradient in temperature and refractive index (observable by Schlieren-photography) is present. Because of this rapid expansion the volume elements of the gas passing the flame front are accelerated and their flow lines are broken and diverge. The course of the gas flow and the vertical rise-velocity (usually $10^2 - 10^3$ cm s^{-1}) can be established by means of the particle-track method (see Sect. III.11).

The more or less homogeneous region which the flame gases enter after leaving the inner zone is called the *interzonal region*, *interconal zone* or *central zone*. This region extends over a height of the order of 1 to 10 cm, depending on the burner design, gas flow, etc.; it is bounded by the inner and outer zones which are conically shaped in a Bunsen flame (see also sub point 3 and footnote on page 419).

The attainment of (near-)equilibrium in this region is explained by the high rates of the elastic and inelastic collisions and of the major bimolecular reactions. Even at a rise-velocity as high as 10^3 cm s^{-1}, a time of 100 µs is needed for travelling over a vertical distance of 1 mm. This time is long compared to the relaxation times for the translational and internal degrees of freedom of the flame molecules at 1 atm. As a result the translational, rotational and vibrational

THE STRUCTURE OF PREMIXED LAMINAR FLAMES						Sect. IV.3

degrees of freedom are in equilibrium like the concentrations of the major chemical species. An exception should be made for the concentrations of the flame radicals and flame ions which are minor species. They may be present in excess downstream from the combustion zone owing to the slowness of their recombination reactions; this will be separately discussed in Sects 5 and 7. Because of the overall equilibrium state, a local temperature can be meaningfully defined in the interzonal region (see also Sect. III.10a).

The flame background emission in the interzonal region is much weaker than in the inner zone (where strong suprathermal chemiluminescence is found). The interzonal region of hydrogen flames even appears dark to the eye. This fact and the relatively large extension and reasonable homogeneity of the interzonal region make it the most suitable zone for spectroscopic observations on metal vapours and for analytical flame emission spectroscopy.

3. *Secondary combustion zone or outer zone.* At the flame border where the burnt gases come into contact with the ambient air, a further, slow oxidation of CO and H_2 to CO_2 and H_2O occurs. This oxidation is accompanied by the release of additional heat (see Sect. 6) and by a weak chemiluminescent light emission (see Sect. 8). The rather diffuse zone where this takes place is called the *secondary combustion zone* or *outer zone*.[†] In hydrocarbon and carbonmonoxide flames this zone is recognizable by its pale blue-violet colour. It becomes more distinct the richer the fuel/oxidant mixture, i.e. the less complete the primary combustion is. With oxygen-rich mixtures secondary combustion practically disappears.

Oxygen and nitrogen penetrate into the flame from the surroundings by molecular diffusion or turbulent mixing. The outer zone has therefore some characteristics in common with an unpremixed flame. The additional heat released may have a stabilizing effect on the whole flame and may make up for the heat losses due to conduction and entrainment of cold air at the flame border. Its radiation, however, enhances the flame background emission, which has to be heeded in spectroscopic observations of metal vapours or in analytical applications. This adverse effect can be eliminated by shielding the flame from the ambient air by means of an inert-gas sheath or a silica tube surrounding the flame (see Sect. III.2a). In the latter case secondary combustion, if there is any, takes place at the top of the tube, as in the 'Smithells flame separator' (see Gaydon and Wolfhard 1979); one then speaks of a *split flame*.

The structure of premixed, laminar *flames burning on a Méker, porous-plate*

[†] Other names encountered in the literature are: secondary reaction zone, outer cone, panache or plume (see also footnote on page 419).

Chapt. IV   TYPES AND PROPERTIES OF NONSEEDED FLAMES

or *slot burner* (see Sect. III.2) is basically the same as described above for a
Bunsen flame. Only the geometrical forms of the three zones may be different. The
flame burning on a Méker burner with its close array of many narrow exit ports can
be conceived as a collection of small Bunsen flames whose burnt gases merge in a
broad, cylindrical interzonal region surrounded by a common outer zone. Above each
of the exit ports a tiny inner cone is usually formed which closely resembles the
cone depicted in Fig. IV.1. The division of the one, wide port as used in the
Bunsen burner into many, narrow ports such as are found in the Méker burner has the
advantage that the height of the inner cone is greatly reduced. This reduction,
again, diminishes the spread in real travel time that corresponds to a given height
of observation above the base of the flame (cf. Eq. IV.1). This improves the
resolution in travel time when measurements are made at varying height of observation as in relaxation studies.

In unpremixed, turbulent flames with a *direct-injection burner* the different
zones are blurred and not easily recognizable. Also, the physical and chemical
state of the gas inside each zone is not well-defined. The inner zone is irregular, ragged and erratic; this causes the flame to make a loud hissing noise.
The strong mixing with ambient air, that starts already at the bottom of the
flame, lowers the attainable flame temperature and makes the flame quite inhomogeneous with respect to temperature, composition and velocity, and also to metal
content when a metal salt solution is sprayed. We refer for a more detailed
description to the books on analytical flame spectroscopy cited in Chapt. I.

4. CALCULATION OF EQUILIBRIUM COMPOSITION AND ADIABATIC TEMPERATURE [†]

4a   INTRODUCTION

Premixed flames burning at 1 atm pressure approach a state of overall thermal equilibrium above the primary combustion zone. It follows, therefore, that the
concentrations of the major flame species, for example, H_2O, CO, CO_2, H_2 (in fuel-rich) or O_2 (in fuel-lean flames), can be calculated reasonably well from the
chemical-equilibrium laws at known flame temperature. Due to relaxation in the
establishment of flame-radical equilibria just downstream from the combustion zone,
the actual concentrations of the flame radicals such as H, OH and O may often
exceed their equilibrium value (see Sect. IV.5). Since these radicals are mostly

---

[†] For a more detailed treatment see Gaydon and Wolfhard (1970). The method of
calculating the equilibrium composition presented in this section is based
largely on their method.

CALCULATION OF EQUILIBRIUM COMPOSITION AND ADIABATIC TEMPERATURE   Sect. IV.4b

minor constituents in the flame gas, they do not strongly affect the overall properties of the flame.

Thermal equilibrium is said to exist if the concentrations of all species (atoms, molecules, ions) as well as the distribution over the internal and translational energies conform to the equilibrium laws (see Sect. II.3c-3). If after combustion thermal equilibrium is attained without an exchange of energy and mass with the surroundings (i.e., no convection or radiation of heat, no entrainment of air), the flame attains what is called the *adiabatic temperature*. This adiabatic temperature is solely determined by the qualitative and quantitative composition and initial temperature of the combustible mixture and by the total gas pressure, which is here assumed to be constant before and after combustion. The adiabatic temperature can be calculated accurately by using equilibrium laws. The actual flame temperature may deviate from the adiabatic value because of heat losses, secondary combustion due to entrainment of air, or deviations from chemical equilibrium (see Sect. IV.6a).

In Sect. 4b we describe methods of calculating the equilibrium composition and adiabatic flame temperature. Theoretical results for some typical flames are presented in Sect. 4c. These results are of interest not only because they often describe the overall state of the flame gases in first approximation, but also because much can be learned from the residual discrepancies between these calculated values and the values actually measured.

4b   METHODS OF CALCULATING EQUILIBRIUM COMPOSITION
       AND ADIABATIC TEMPERATURE

Equilibrium composition. First we consider methods by which the equilibrium composition can be calculated at the known or assumed flame temperature and total, constant pressure from the initial temperature and composition of the combustible gas mixture. Concentrations will be expressed here in partial pressures (for definition see Sect. II.3b-3).

If s elements form r gaseous species, the equilibrium composition of the mixture can be calculated from r equations: namely, $(s-1)$ mass-balance equations (expressing the ratios of number of moles of each element supplied), 1 equation for the constant total pressure p_t, and $(r-s)$ mutually independent chemical-equilibrium equations with known equilibrium constants. Since some of the chemical-equilibrium equations are nonlinear, it is not easy to solve this set of equations in a straightforward manner. However, Gaydon and Wolfhard (1970) have described a trial-and-error iteration procedure that converges quickly.

For *hydrocarbon flames* one first assumes trial values for the partial pressure of H_2O, $p(H_2O)$, and for the ratio $p(CO_2)/p(CO)$. Then one computes the gas

Chapt. IV TYPES AND PROPERTIES OF NONSEEDED FLAMES

gas composition without using the ratio, $[H]_t / [O]_t$, of total hydrogen and oxygen concentrations known from the initial gas supplies and the known total pressure p_t. The known values of these latter quantities are used to check the trial assumptions by comparing them with the corresponding values following from the computed gas composition. When no agreement is found within a properly chosen error interval, a next-order approximation of the gas composition is computed by starting again from better estimates for $p(H_2O)$ and $p(CO_2)/p(CO)$. The calculated total pressure depends mainly on the assumed value of $p(H_2O)$, and that of $[H]_t / [O]_t$ depends mainly on $p(CO_2)/p(CO)$. Therefore, a next-order estimate of $p(H_2O)$ can be based on the discrepancy between calculated and true total pressure only. For the same reason the next trial value for $p(CO_2)/p(CO)$ can be estimated from the discrepancy between the calculated and true value for $[H]_t / [O]_t$. It appears that the true values of p_t and $[H]_t / [O]_t$ may be approximated within 0.01% after less than ten iteration steps, provided the initially assumed values are not far outside the right order. Reasonable first-order estimates may be found from the simple chemical reaction equation in which only CO_2, CO, H_2O, H_2 or O_2 and the inert gases appear as products. For atmospheric-pressure flames with a temperature lower than 2000 K this simple estimate yields for the partial pressures of the major constituents values that are correct to within about 5% of their true values. From these estimates and the chemical equilibrium equations the equilibrium partial-pressures of the minor constituents (H, OH, O, NO) can also be calculated within 5%. For low-temperature flames, this rough calculation thus yields the equilibrium composition with reasonable accuracy.

In the case of a *hydrogen* or a *carbon monoxide flame*, it suffices to start with one trial value, $p(H_2O)$ or $p(O_2)$, respectively.[+] Now only the total pressure is used to check the initial estimate.

A disadvantage of the above equilibrium-constant method is that a specific computation scheme has to be set up for every set of reaction products. For instance, the computation scheme for an $C_2H_2-O_2$ flame is different from the scheme for an C_2H_2-air flame, because in the latter N_2 and NO also occur. Also a different computation scheme is required for very fuel-rich mixtures which produce solid carbon in equilibrium.

The above method may be modified to obtain a set of linear equations in parameters that are related to the partial pressures to be calculated (see Brinkley and Lewis 1952, and Weinberg 1957).

A more general method is to calculate the equilibrium composition of an

---

[+] The choice of two trial values as suggested by Gaydon and Wolfhard (1979) is not necessary for these types of flames.

CALCULATION OF EQUILIBRIUM COMPOSITION AND ADIABATIC TEMPERATURE   Sect. IV.4b

arbitrary mixture by minimizing the free enthalpy G_T (see Sect. II.3b-3) under the boundary conditions set by the mass-balance equations. In principle such a calculation may be carried out quite simply by using Lagrange multipliers. Since the free enthalpy of an individual component is a linear function of the logarithm of its mole fraction, minimization leads to a set of exponential equations which cannot be solved easily. White, Johnson and Dantzig (1958) have developed an iteration procedure which starts with the assumption of a trial composition. After some algebra they obtain a set of $(s+1)$ linear equations in s Lagrange multipliers and in one additional parameter derived from the total mole number (s = number of elements involved). The Lagrange multipliers solved from these equations and the initially assumed composition yield an improved value for the concentration of the individual components. These improved values may be used as a starting point for the next computation cycle. The iteration is repeated until successive solutions achieve the desired accuracy. This general computation method is more versatile than the equilibrium-constant method, but it is less suitable to use for computing the equilibrium composition by hand when one wants only an approximation within 5%.

Computer programs have been written for the free-enthalpy method (see Chester, Dagnall and Taylor 1970 and Wittenberg et al. 1979) and for the equilibrium-constant method (see Neumann and Knoche 1963, and Harker and Allen 1969).

In both methods the gases are assumed to behave ideally and use is made of the same thermodynamic data, i.e. the free enthalpy of the individual components, either directly as in the free-enthalpy method or indirectly as in the computation of the equilibrium constants.[†] Consequently, whatever the method of calculation used, the uncertainty in the free enthalpies determines the error in the equilibrium composition. Using thermodynamic data from various compilations (JANAF 1960, and Vedeneyev et al. 1966) Zeegers (1966) found the calculated equilibrium compositions of an C_2H_2-air flame to agree within 5%.

Adiabatic temperature. In order to calculate the adiabatic flame temperature one first calculates the equilibrium composition at an estimated temperature, T_0. Formally, the combustion process is considered to take place in two stages. In the first stage, the unburnt gas mixture is thought to be converted at initial temperature T_b (usually room temperature) and under constant total pressure into a mixture that has the same composition as that calculated in equilibrium at T_0. The heat of reaction, ΔH_{T_b}, liberated in this conversion (see Sect. II.3b-3) is

---

[†] It is not true to say (see Chester, Dagnall and Taylor 1970) that the equilibrium-constant method is less generally applicable than the free-enthalpy method. In the equilibrium-constant method the same thermodynamic data are required, and knowledge of the specific flame reactions is irrelevant for the calculation of an equilibrium composition

Chapt. IV   TYPES AND PROPERTIES OF NONSEEDED FLAMES

found from the difference in enthalpy of reactants and products at T_b. In the second stage, the products are heated from T_b up to T_0 at the same total pressure. The heat, $H(T_0 - T_b)$, consumed in this process is calculated from the difference in enthalpy of the gas mixture between T_b and T_0. From the discrepancy between ΔH_{T_b} and $H(T_0 - T_b)$, a better approximation for the adiabatic flame temperature, T_1, is estimated. Then the whole procedure starts again: calculation of the equilibrium composition at T_1, calculation of ΔH_{T_b} and $H(T_1 - T_b)$. This iteration is stopped when T_n and T_{n+1} equal each other within a properly chosen interval (see Snelleman 1968).

The accuracy of the value of the adiabatic flame temperature is determined by the uncertainty in the flame composition, in the heats of formation at 298 K, and in the specific heat capacities of the gas components. By estimating the errors in the thermodynamic quantities from a comparison of various literature sources, Snelleman (1968) has found the total error in the adiabatic temperature of a stoichiometric acetylene-air flame to be equal to about 5 K. There is no reason to expect a very different result for other flames in which the same products appear. The situation may be different in the case of flames containing solid particles. Adiabatic temperatures have been calculated by L'vov (1970) for flames in which solid carbon occurs, but no error limits have been stated.

4c   RESULTS OF EQUILIBRIUM-COMPOSITION AND
     ADIABATIC-TEMPERATURE CALCULATIONS

The equilibrium composition of C_2H_2-, $CO-$ and H_2 flames with either air or N_2O as oxidant gases and at 1 atm pressure are presented as a function of the fuel/oxidant ratio in Figs IV.2a-f. The fuel/oxidant ratio of the stoichiometric mixture is normalized to unity. The stoichiometric ratio is here defined as the fuel/oxidant ratio that would yield only CO_2 and/or H_2O as combustion products if no dissociation of these compounds into H, OH, O, CO, H_2 or O_2 were to occur. The compositions were calculated at the adiabatic temperature obtained for an initial temperature of 298 K. Since metal vapours are usually introduced into the flame by spraying aqueous solutions, an appropriate, small amount of H_2O was included in the calculation. For all flames the ratio of the H_2O- and oxidant-gas supplies was kept constant. Along with Snelleman (1965), we assumed that all H_2O entered the flame as vapour. Consequently, the influence of the desolvation of the aerosol on the adiabatic temperature was not taken into account (see also Alkemade 1969).

The normalized fuel/oxidant ratios range from fuel-lean (0.5) to fairly fuel-rich (2.3). The upper limit is lower than the ratio at which solid carbon is expected to form in C_2H_2 flames. Equilibrium compositions of very fuel-rich

CALCULATION OF EQUILIBRIUM COMPOSITION AND ADIABATIC TEMPERATURE Sect. IV.4c

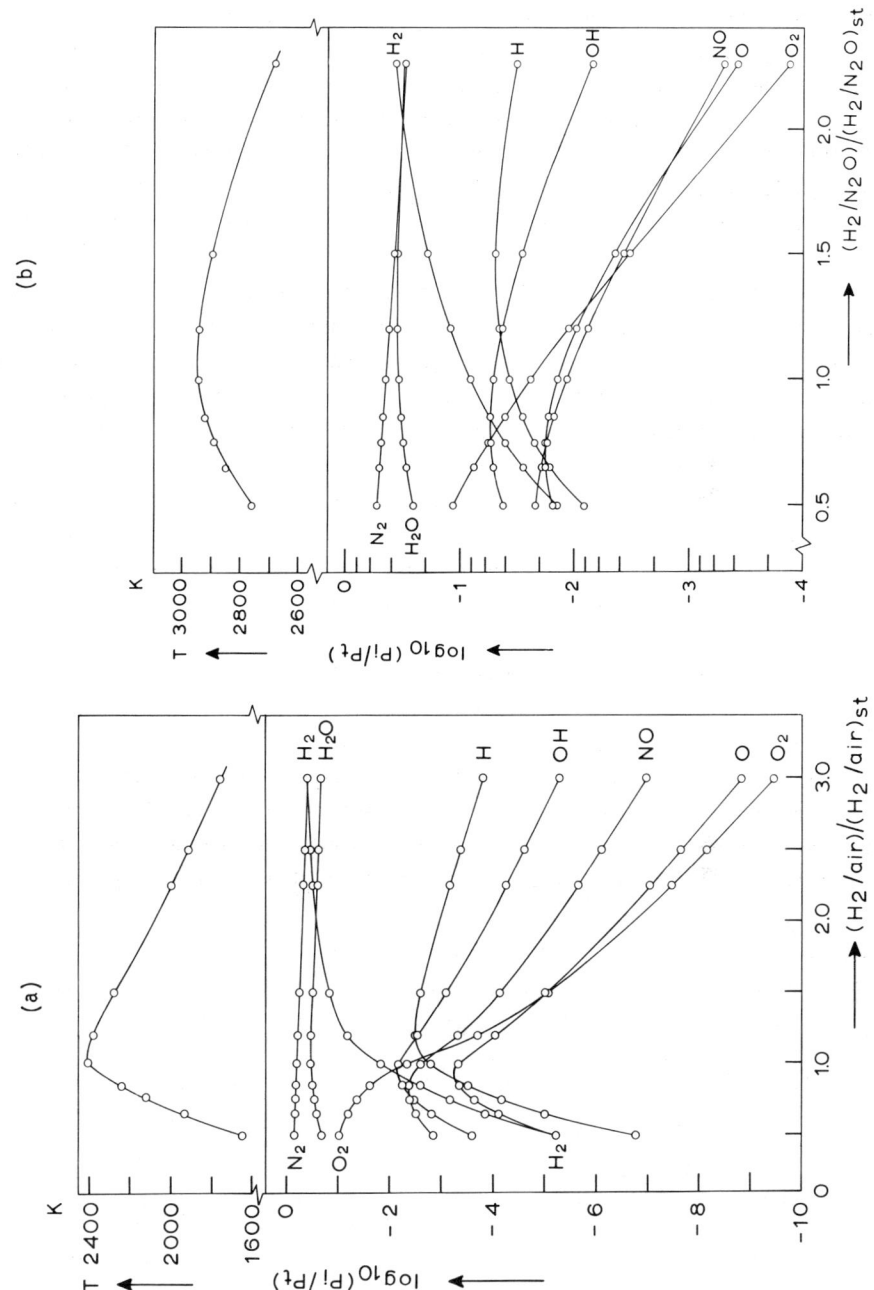

Chapt. IV   TYPES AND PROPERTIES OF NONSEEDED FLAMES

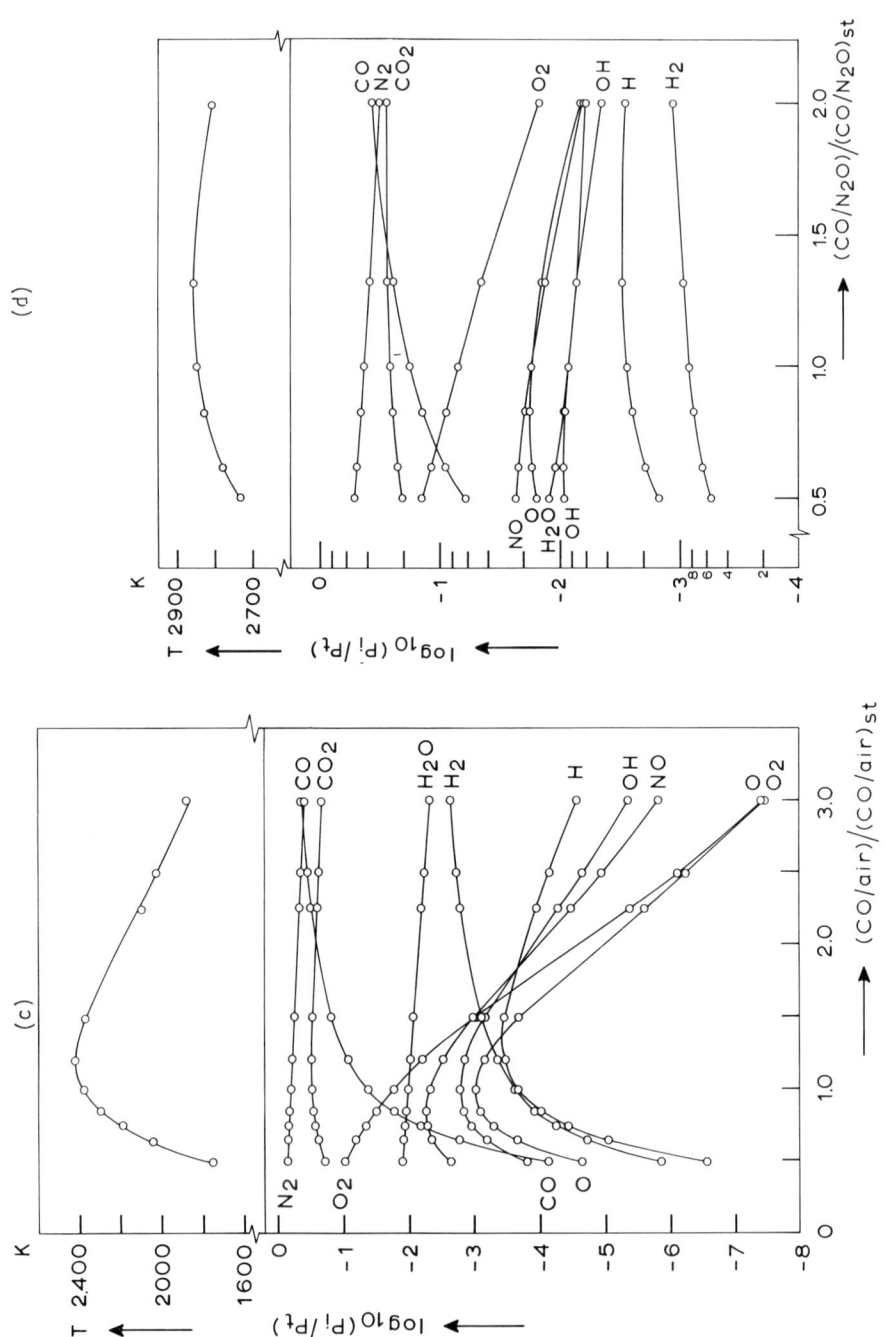

430

CALCULATION OF EQUILIBRIUM COMPOSITION AND ADIABATIC TEMPERATURE   Sect. IV.4c

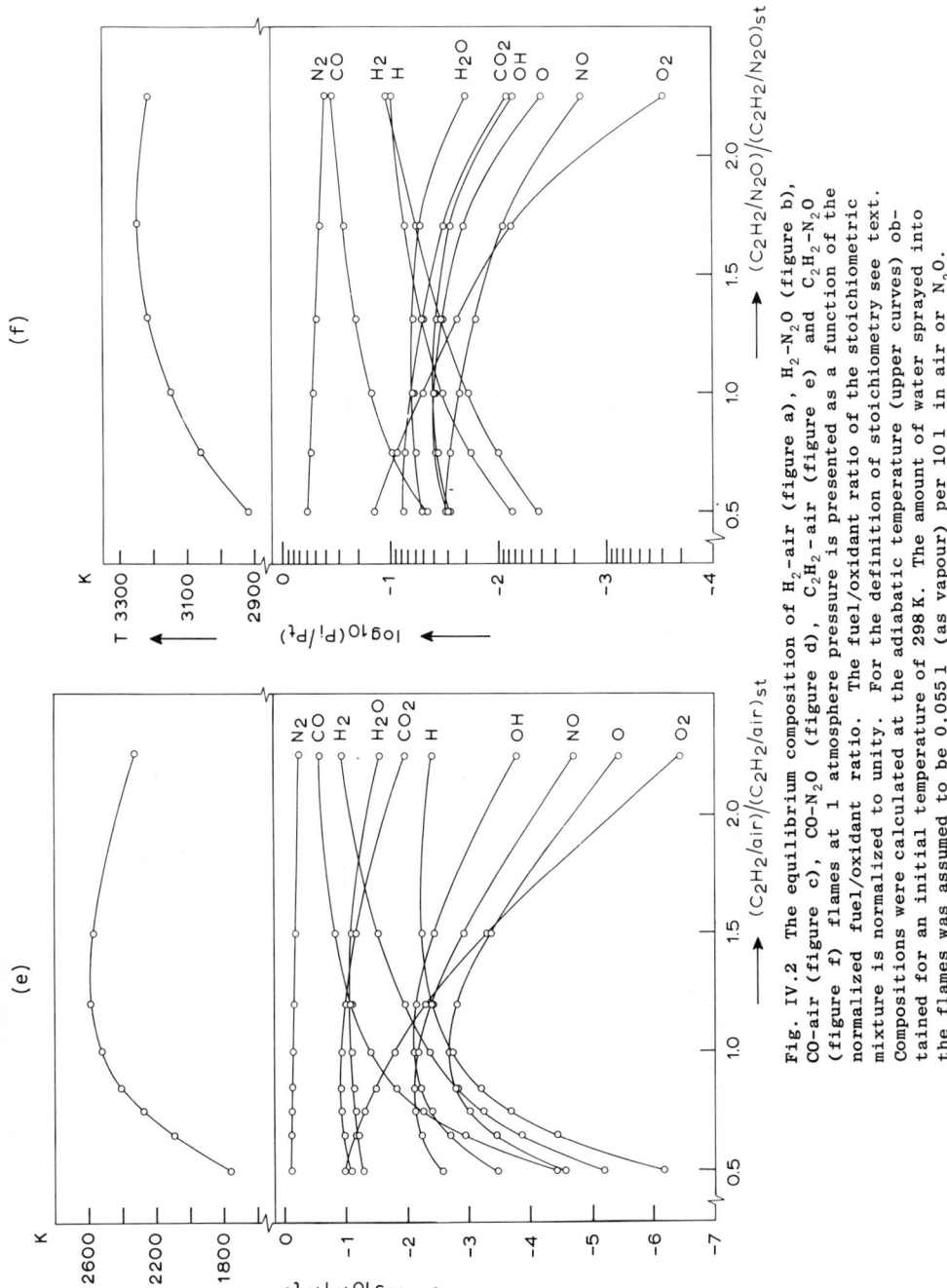

Fig. IV.2 The equilibrium composition of H_2-air (figure a), H_2-N_2O (figure b), CO-air (figure c), CO-N_2O (figure d), C_2H_2-air (figure e) and C_2H_2-N_2O (figure f) flames at 1 atmosphere pressure is presented as a function of the normalized fuel/oxidant ratio. The fuel/oxidant ratio of the stoichiometric mixture is normalized to unity. For the definition of stoichiometry see text. Compositions were calculated at the adiabatic temperature (upper curves) obtained for an initial temperature of 298 K. The amount of water sprayed into the flames was assumed to be 0.055 l (as vapour) per 10 l in air or N_2O.

Chapt. IV   TYPES AND PROPERTIES OF NONSEEDED FLAMES

flames ($[C_2H_2]/[O_2] > 1$) in which solid carbon is present have been given by L'vov (1970).[†] Equilibrium compositions of various kinds of flames have been given by Edwards, Smith and Brinkley (1953), Smith, Manton and Brinkley (1954), Lewis and von Elbe (1961), Jenkins and Sugden (1969), and Gaydon and Wolfhard (1979).

Figures IV.2a-f also show the influence of the mixing ratio on the temperature of some flames. They demonstrate clearly that a stoichiometric mixture does not yield the maximum flame temperature (see also Table IV.1 on page 415). In all cases the maximum flame temperature is reached for a more fuel-rich mixture, because of the fact that the main combustion products such as CO_2 and H_2O are partly dissociated at flame temperatures; this dissociation withdraws heat from the flame gases.

The adiabatic temperature increases with increasing initial temperature. But the specific heat of the flame and the degree of dissociation of its components also increase with increasing temperature. As a result the rise in the adiabatic temperature is smaller than the increase in initial temperature and the hotter the flame, the more important this deviation becomes. For instance, when the unburnt mixture is pre-heated by 100 K,[‡] the adiabatic temperature in a stoichiometric $H_2-O_2-N_2$ flame increases only by 50 K from about 2400 to 2450 K, and in a stoichiometric $C_2H_2-N_2O$ flame only by 30 K from 3140 to 3170 K.

The atomic oxygen concentration in very fuel-rich flames is of special interest, since it determines the reducing properties of these flames and consequently the equilibrium degree of dissociation of the metal oxides and, indirectly, also of the metal hydroxides (see Chapt. VIII). Fig. IV.3 presents the partial pressure of oxygen atoms in an $C_2H_2-N_2O$ flame as a function of the fuel/oxidant ratio. The $\log_{10}(p_O/p_t)$ curve is constructed from data given by Chester, Dagnall and Taylor (1970) and by L'vov et al. (1976). The adiabatic temperature of mixtures with a normalized fuel/oxidant ratio smaller than 2.3 was calculated; the adiabatic temperature of mixtures with a normalized fuel/oxidant ratio larger than 1.67 was taken from L'vov et al. (1976). When the normalized fuel/oxidant ratio is below 2.3, $p(O)$ depends in the first place on the fuel/oxidant ratio and to a lesser extent on the temperature. When this ratio is above 2.6, $p(O)$ is almost entirely determined by the saturated vapour pressure of carbon and therefore by the temperature, which, of course, depends again on the fuel/oxidant ratio.

From the few data given by L'vov (1970) we assume that for $C_2H_2-O_2$ flames a similar graph will be found (see Fig. IV.3), the only difference being a small shift along the $p(O)$ axis to higher values because of the higher temperatures

---

[†] Concentrations in Table 4.5 of this work are not expressed in **partial pressures** (as stated there) but in molar fractions (L'vov 1972).

[‡] See Note (12) in Appendix A.3.

CALCULATION OF EQUILIBRIUM COMPOSITION AND ADIABATIC TEMPERATURE Sect. IV.4c

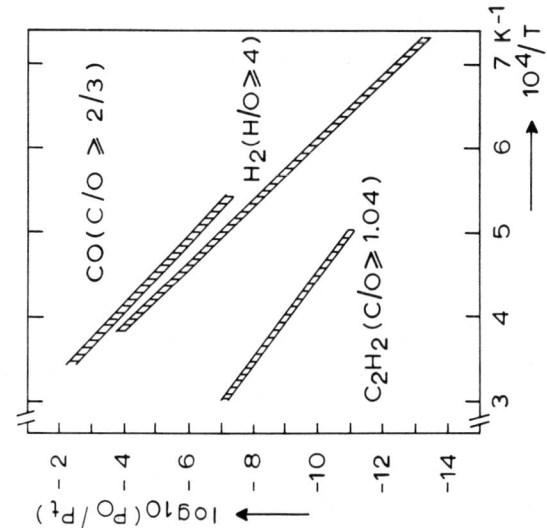

Fig. IV.4 For very fuel-rich flames an estimate (within a factor of about 2) of the atomic oxygen fractional pressure p_O/p_t can be read from this figure if the temperature T is known. The normalized fuel/oxidant ratio (see caption of Fig. IV.3) has to be larger than 2 for hydrogen and carbon monoxide flames, and larger than 2.6 for acetylene flames. The kind of oxidant gas and the kind of inert gas have, virtually, no influence on the position of the hatched regions.

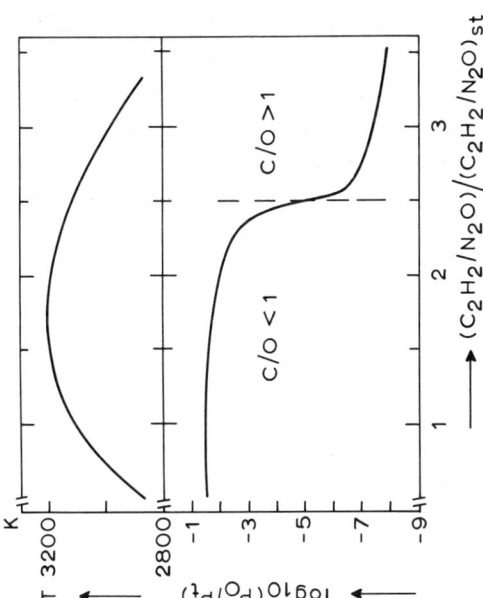

Fig. IV.3 The fractional pressure of atomic oxygen p_O/p_t (lower part) and the calculated temperature (upper part) in $C_2H_2-N_2O$ flames is presented as a function of the normalized fuel/oxidant ratio, i.e. $(C_2H_2/N_2O)/(C_2H_2/N_2O)_{stoich}$. The dotted line marks the normalized fuel/oxidant ratio (i.e. C/O = 1) above which atomic carbon vapour and solid carbon may be present in the flame. The data are taken from Chester, Dagnall and Taylor (1970) and L'vov, Katskov, Kruglikova and Polzik (1976); temperatures for flames with normalized fuel/oxidant ratio smaller than 2.3 were calculated by the authors. The figures refer to flames into which 0.055 l H_2O (as vapour) per 10 l N_2O was introduced.

433

Chapt. IV   TYPES AND PROPERTIES OF NONSEEDED FLAMES

obtained in these flames. In very fuel-rich H_2- and CO flames, $p(0)$ is again in first approximation determined by the temperature, although the fuel/oxidant ratio exerts some influence here, too. It is possible to estimate $p(0)$ in these flames within about a factor of two from the temperature only. Estimates of equilibrium values for $p(0)$ in very fuel-rich C_2H_2-, H_2- and CO flames can be read from Fig. IV.4, if the actual temperature of the flame is known. The lower limit of the range of fuel/oxidant ratios where these graphs are valid is indicated in the figure. The adiabatic flame temperatures corresponding to these lower limits (at assumed initial temperature of 300 K) terminate the graphs at the left-hand side of the $1/T$ axis. The type of oxidant gas, be it O_2, N_2O or air, and the type of inert gas have, virtually, no influence on the position of the curves in Fig. IV.4. The rather large difference between $p(0)$, at a given temperature, in very fuel-rich C_2H_2 flames on the one hand and in very fuel-rich H_2- and CO flames on the other hand is caused by the large difference in dissociation energy, D_0, of the respective main combustion products, namely: CO in C_2H_2 flames with $D_0 = 11.1\,\text{eV}$, CO_2 in CO flames with $D_0 = 5.4\,\text{eV}$, and H_2O in H_2 flames with $D_0 = 5.0\,\text{eV}$.

5. DEVIATIONS FROM FLAME EQUILIBRIUM COMPOSITION

5a   INTRODUCTION

In Sect. IV.4 methods of calculating the equilibrium composition of a gas mixture were described. However, in practice, chemical equilibrium is not always fully established everywhere downstream from the combustion zone. Therefore the actual concentrations may by different from the calculated equilibrium values. Here we shall discuss the deviation from the equilibrium concentration of certain species and describe some methods of determining the actual concentrations of the radicals H, OH and O, which often play an important role in the behaviour of metal vapours in flames.

Combustion proceeds in three stages. In the *first stage* the combustion reaction is initiated by radicals just below the combustion zone. Radicals are generated either by some relatively slow process, e.g., thermal decomposition of H_2 and/or O_2 (see Lewis and von Elbe 1961), or they may diffuse from the burnt gas (see Gaydon and Wolfhard 1979). In the *second stage*, which takes place in the primary combustion zone, the radicals multiply rapidly by chain(-branching) reactions. These are in H_2 flames (see Lewis and von Elbe 1961)

$H + O_2 \rightarrow OH + O$ (i) $OH + H_2 \rightarrow H_2O + H$ (iii)

$O + H_2 \rightarrow OH + H$ (ii) $OH + OH \rightarrow H_2O + O$,   (iv)

DEVIATIONS FROM FLAME EQUILIBRIUM COMPOSITION   Sect. IV.5a

and additionally in hydrocarbon and moist CO flames (see Brokaw 1967, and Dean and
Kistiakowski 1970)

$$CO + OH \rightarrow CO_2 + H \quad . \quad (v)$$

The rate constants of reactions (i) - (iv) are about the same at $T \approx 2000$ K (see
tables in Baulch et al. 1972, Jensen and Jones 1978, Dixon-Lewis and Williams (1977).

There is still some doubt whether it is possible to obtain a flame from
completely dry CO and hydrogen-free oxidant gas. Sulzmann, Meyers and Bartle
(1965, 1965a) suggest a reaction mechanism in which hydrogen plays no part.
Their suggestion is based on the study of the oxidation of CO by molecular oxygen in shock-tubes, in which they claimed to have at most approximately 1 ppm of
hydrogeneous impurities. However, Brokaw (1967) has shown that oxidation of CO
can only proceed by the reaction mechanism favoured by Sulzmann, Myers and Bartle
if the relevant rate constants have unusually high values. Therefore, Brokaw
favours an 'impurity' mechanism (reactions i - v) for which hydrogeneous impurity
concentrations as low as 7 ppm are sufficient to catalyze the combustion (see
Fishburne, Bilwakesh and Edse 1967).

Also in this second stage a rapid decomposition of hydrocarbons such as
C_2H_2, CH_4, C_3H_8 occurs by reaction with radicals (see Lewis and von Elbe 1961, and
Fenimore 1964). Typical decomposition and fragmentation products occurring in
excess concentrations are C_2, CH, CHO, CN, NH, in the ground state or in an excited
state (see Hand and Kistiakowski 1962, Arrington et al. 1965, 1965a, Becker and
Bayes 1968, and Gaydon and Wolfhard 1970). The second stage is completed when the
forward and backward rates of the reactions (i - v) have become equal. Then the
concentrations of the major combustion products, CO_2, CO, H_2O and H_2 (in fuel-rich
flames) or O_2 (in fuel-lean flames) will have reached their chemical equilibrium
values within a few percent. However, the concentrations of the radicals and, in
very fuel-rich flames, also the concentrations of the fragmentation products may
well exceed their chemical equilibrium values, which prevents the major combustion
products from attaining full equilibrium. In hydrocarbon flames burning with
enough oxygen to convert all carbon into CO, no C_2, CH, CN and NH are found beyond the combustion zone. In the interzonal region of very fuel-rich flames in
which also solid carbon is formed, emission bands of these fragmentation products
have been detected (see Zeegers 1966). Since concentration measurements of these
fragments in the region outside the combustion zone have not been reported so far,
conclusions about deviations of these species from equilibrium cannot yet be drawn.

The presence of fragmentation products in the combustion zone can be concluded from the appearance of their emission bands (see Gaydon and Wolfhard
1979). From absorption measurements in the combustion zone of low-pressure

Chapt. IV
TYPES AND PROPERTIES OF NONSEEDED FLAMES

$C_2H_2 - O_2$ flames ($O_2/C_2H_2 = 0.8-1.6$; final temperature = 2500 K, and pressure = 7.3 Torr), Bleekrode and Nieuwpoort (1965) and Bleekrode (1966) have detected the presence of C_2 and CH in the ground state at partial pressures of 10^{-5} to 10^{-6}, and 10^{-7} atm, respectively. These values exceed the equilibrium values calculated at the final flame temperature of 2500 K by factors of the order 10^{15} and 10^7, respectively (see Alkemade and Zeegers 1971). Baronavski and McDonald (1977) have measured the C_2 concentration in an $C_2H_2 - O_2$ flame at atmospheric pressure by laser-induced fluorescence. In their flame C_2 appeared to be present at a partial pressure of about 1.5×10^{-3} atm in accordance with the value found by Bleekrode and Nieuwpoort (1965).

In the *third stage* the chain reactions are terminated by recombination reactions that result in a decrease in the total number of radicals. In H_2 flames these recombination reactions are (for rate constants see Jensen and Jones 1978)

$$H + H + X \rightarrow H_2 + X \quad \text{(vi)}$$
$$H + OH + X \rightarrow H_2O + X \quad \text{(vii)}$$
$$O + O + X \rightarrow O_2 + X \quad \text{(viii)}$$

Additionally we have in hydrocarbon and carbon monoxide flames (see *ibidem*)

$$CO + O + X \rightarrow CO_2 + X \quad \text{(ix)}$$

All recombination reactions are strongly exo-ergic and require a third body, X, to accept (part of) their reaction energy for stabilization of the recombination product. Being termolecular these recombination reactions are relatively slow.

In this stage of the combustion, the fast binary reactions (i - v) are balanced at any place in the flame, while the slower recombination reactions (vi-ix) continue to reduce the concentration of H, OH and O to their appropriate equilibrium value. These balanced binary reactions bring about a partial equilibrium (see Sect. II.3c-3) between these radicals, through which their excess concentrations are interrelated (see Bulewicz, James and Sugden 1956). These relations, involving also bulk flame constituents, hold anywhere beyond the combustion zone, regardless of whether overall chemical equilibrium is established or not. From the set of five reactions (i) to (v) one finds four independent partial-equilibrium relations (equilibrium constants have been tabulated by Jensen and Jones 1978)

$$[OH][O] / [H][O_2] = K_i \quad \text{(IV.3)}$$

$$[OH][H] / [O][H_2] = K_{ii} \quad \text{(IV.4)}$$

$$[H_2O][H] / [H_2][OH] = K_{iii} \quad \text{(IV.5)}$$

$$[CO_2][H] / [CO][OH] = K_v \quad \text{(IV.6)}$$

with $K_{iv} = [H_2O][O] / [OH]^2 = K_{iii}/K_{ii}$. Here [OH] denotes the number density of OH radicals, etc. The equilibrium constants $K_i - K_v$ are determined by the local

DEVIATIONS FROM FLAME EQUILIBRIUM COMPOSITION   Sect. IV.5b

flame temperature. From these partial-equilibrium equations one can calculate the concentrations of any pair of radicals if one knows the experimental value of the concentration of the third radical and the (near-)equilibrium concentration of H_2O and H_2 (or CO_2 and CO) in fuel-rich flames at given temperature. In fuel-lean flames, the concentrations can be calculated by using the (near-)equilibrium concentration of O_2 instead of H_2 (see Bulewicz, James and Sugden 1956). Starting from measured $p(H)$ values, one can often use such calculations to check the validity and consistency of experimental methods to determine $p(O)$ and $p(OH)$. In stoichiometric flames, where neither H_2 nor O_2 is a bulk constituent, the gas composition, including $p(O_2)$ and $p(H_2)$, can still be found by calculation if the excess concentrations of two radicals are known (see Zeegers and Alkemade 1965, and Zeegers 1966).

Knowledge of the actual radical concentrations is required to understand the suprathermal chemiluminescence (see Sect. VI.3) and chemi-ionization (see Sect. IX.2) of metal species, as well as the spectral continua emitted by recombination reactions (see below and Sect. IV.8). The decrease of excess radical concentrations towards equilibrium explains the initial rise in temperature downstream from the combustion zone (see Sect. IV.6).

5b   METHODS OF EVALUATING RADICAL CONCENTRATIONS
5b-1 *GENERAL*

Table IV.2 summarizes the most common optical methods of determining radical concentrations in flames, the processes on which they are based, as well as some limitations in the practical application of such methods.

There are also nonoptical methods of determining radical concentrations in flames. We mention the local microprobe sampling of the flame gas and subsequent quantitative analysis of H and OH by *mass spectrometry*. This method has generally been applied in low-pressure flames (see Foner and Hudson 1953, and Fristrom 1963, 1963a) and occasionally also in flames at 1 atm (see Boers 1963, and Milne and Greene 1965, 1966). Some special mass-spectrometric methods in which $p(H)$ or $p(OH)$ are derived from the appearance of stable molecules and ions have been described by Fenimore and Jones (1958) and by Hayhurst and Kittelson (1972). *Paramagnetic-resonance methods* are also applied for the detection of flame radicals that possess unpaired electrons such as H (see Dixon-Lewis, Wilson and Westenberg 1966, Westenberg and Fristrom 1965, and Wilson and Westenberg 1967). Mass-spectrometric and paramagnetic-resonance methods require an elaborate apparatus. The conversion of the experimental data into absolute flame concentrations might be complicated by dubious instrumental factors. The spatial resolution of paramagnetic-resonance methods is poor, because the resonance cavity cannot be made much smaller than 1 cm (for a review of the above methods see Pungor 1967, and Schofield and Broida 1968).

Chapt. IV TYPES AND PROPERTIES OF NONSEEDED FLAMES

Table IV.2
Summary of Optical Methods for the Evaluation of Radical Concentrations in Flames

Method	Basic (overall) process	Radical determined	Relative or absolute	Quantity observed	Measurement and calibration required	Restrictions in the application
LiOH	$LiOH + H \rightleftarrows Li + H_2O$	H	Absolute	$[Li]/[Li]_t$	Na-comparison method	Not in cool O_2-rich flames; $T > 1600$ K, unless absorption measurements are used
LiOH	$LiOH + H \rightleftarrows Li + H_2O$	H	Relative	$\propto [Li]/[Li]_t$ (relative)	Relative emission or absorption measurements	
LiOH	$LiOH + H \rightleftarrows Li + H_2O$	H	Absolute	$\propto [Li]/[Li]_t$ (relative)	Adaptation of relative [H] to theoretical absolute [H]$_e$ at near-equilibrium	
NaCl	$Na + HCl \rightleftarrows NaCl + H$	H	Absolute	$[Na]_0/[Na]$	Relative Na line intensities and absolute calibration of Cl supply	H_2-rich and stoichiometric flames only
NaCl	$Na + HCl \rightleftarrows NaCl + H$	H	Relative	$[Na]_0/[Na]$	Relative Na line intensities at constant Cl supply	
CuH	$\begin{cases} Cu + H + X \rightleftarrows CuH + X \\ CuH + X \rightleftarrows CuH^* + X \end{cases}$	H	Relative	CuH band at 4280 Å	Relative thermal intensity measurement	Not in O_2-rich flames
NaOH	$Na + OH \rightarrow NaOH + h\nu$	OH	Relative	Continuum at 4000 Å	Relative intensity measurement	Correction for NaO$_2$ in O_2-rich flames
OH-absorption	$OH + h\nu \rightarrow OH^*$	OH	Absolute	OH band at 3064 Å	Absorption measurement of single rotational line with known f-value	Only for sufficiently strong OH concentrations

DEVIATIONS FROM FLAME EQUILIBRIUM COMPOSITION Sect. IV.5b

Table IV.2 (cont.)

OH-absorption	$OH + h\nu \rightarrow OH^*$	OH	Relative	OH band at 3064 Å	Absorption measurement of whole band, calibrated in relative [OH] by variable-pass technique	Only for sufficiently strong OH concentrations
OH-absorption	$OH + h\nu \rightarrow OH^*$	OH	Absolute	OH band at 3064 Å	Adaptation of relative [OH] to theoretical absolute $[OH]_e$ at near-equilibrium	
OH-fluorescence	$\{OH + h\nu \rightarrow OH^*$ $\{OH^* \rightarrow OH + h\nu$	OH	Relative	u.v. OH band	Intensity measurement of non-saturated fluorescence band excited by pulsed dye laser	Concentrations of major fluorescence-quenching species in flame must be known
IO	$I + O + X \rightleftharpoons IO + X$ [†]	O	Relative	IO band at 5307 Å	Relative intensity measurements of thermal radiation or supra-thermal chemiluminescence	In H_2-rich flames correction for underlying continuum required
NO	$NO + O \rightarrow NO_2 + h\nu$	O	Relative	Continuum (at 5000 Å)	Relative intensity measurement with added NO	
CO	$CO + O(+X) \rightarrow CO_2(+X) + h\nu$	O	Relative (Absolute)	(Quasi-)continuum (visible and near u.v.)	Relative or absolute intensity measurements with constant or known [CO]	Correction for thermal CO_2 radiation at near-equilibrium
Initial temperature rise	$CO + O + X \rightarrow CO_2 + X + $ heat	O	Relative Absolute	Temperature	Temperature measured as function of height	Only in CO flames where $CO + O$ is the dominant recombination reaction for O

[†] The observed IO emission may also be explained by the overall chemiluminescence reaction: $I + OH + (O)H \rightarrow IO^* + H_2(O)$ (see Sect. 5b-4).

Chapt. IV   TYPES AND PROPERTIES OF NONSEEDED FLAMES

5b-2   *CONCENTRATION OF H RADICALS*

Since its introduction by Bulewicz, James and Sugden (1956) the *LiOH-method* has found wide application for the determination of relative and absolute H concentrations. The method is based on the balanced exchange reaction: $LiOH + H \rightleftharpoons Li + H_2O$, described by the equilibrium constant K_{Li}

$$K_{Li} = [Li][H_2O] / [LiOH][H] . \qquad (IV.7)$$

The dissociation of LiOH proceeds through the nearly thermo-neutral reaction: $LiOH + H \to Li + H_2O$ and not through the endo-ergic reaction: $LiOH + X \to Li + OH + X$, as will be discussed in Sect. VIII.4a.

K_{Li} can be calculated at the local flame temperature from the known dissociation energy $D_0(LiOH)$ (see Eqs VIII.38 and 39). By substituting for $[H_2O]$ the local equilibrium concentration, which is virtually unaffected by deviations of the radical concentrations, one finds [H] in *absolute* measure from a determination of $\phi_{Li} \equiv [LiOH]/[Li]$. This ratio can be found from the measured ratio [Li]/[Na] for equimolar Li- and Na solutions (see Sect. V.3b). If ionization is suppressed, Na is present mainly as free atoms and only for a minor, but not always negligible, part as NaOH (see Sect. VIII.3b; see Cotton and Jenkins 1969, and Kelly and Padley 1971). Taking into account NaOH formation, one finds ϕ_{Li} according to Kalff (1971) and van der Hurk, Hollander and Alkemade (1973) from

$$\phi_{Li} = ([Na]/[Li] - 1)/(1 - [Na] K_{Li} / [Li] K_{Na}) , \qquad (IV.8)$$

when equimolar solutions of Li and Na are sprayed into the flame and ionization is suppressed. Here K_{Na} stands for the equilibrium constant of the balanced exchange reaction: $NaOH + H \rightleftharpoons Na + H_2O$.

We note that [H] will not be altered significantly upon introduction of a trace of Li or Na into the flame. The main advantage of Li as a test element is that the LiOH bond is fairly strong, whereas that of NaOH is but weak. Consequently, the deviation of [Na]/[Li] from unity and thus ϕ_{Li} can be measured precisely. For the same reason, the accuracy of the determination of ϕ_{Li} is only weakly affected by uncertainties in the ratio K_{Li}/K_{Na}, that is, by the uncertainty in the difference: $D_0(LiOH) - D_0(NaOH)$. Another advantage is that ionization of Li is generally insignificant.

For practical use, K_{Li} and K_{Na} are tabulated for some temperature values in Table IV.3, as calculated through Eqs VIII.38 and 39. The uncertainties in $D_0(LiOH)$ and $D_0(NaCl)$ are mentioned in Sect. VIII.5b.

In fuel-lean flames, formation of LiO_2 is expected to be negligible (see McEwan and Phillips 1967), but NaO_2 might form in appreciable amounts (see Kaskan 1965, and McEwan and Phillips 1966; see Sect. VIII.3a). Consequently, relative

DEVIATIONS FROM FLAME EQUILIBRIUM COMPOSITION   Sect. IV.5b

Table IV.3

Calculated[†] Equilibrium Constants $K_{Li}=[Li][H_2O]/[LiOH][H]$,
$K_{Na} = [Na][H_2O] / [NaOH][H]$ and $K_{NaCl}=[NaCl][H]/[Na][HCl]$
at Various Temperatures for Use in [H] Measurements

$T(K)$	1500	1750	2000	2250	2500	2750	3000
K_{Li}	23	10	5.3	3.4	2.3	1.7	1.3
K_{Na}	4.2×10^4	5.9×10^3	1.4×10^3	4.4×10^2	1.8×10^2	8.9×10^1	4.6×10^1
K_{NaCl}	0.67	0.91	1.2	1.4	1.6	1.9	2.1

[†] Assumed dissociation energies are: D_0 (LiOH → Li + OH) = 4.52 eV for a linear LiOH molecule (cf. Table VIII.5), D_0(NaCl) = 4.24 eV and D_0(HCl) = 4.43 eV.

[H] measurements by the LiOH-method can be performed just as well in fuel-rich as in fuel-lean flames; for absolute [H] measurements Na cannot be used as a reference element in fuel-lean flames. Some possible causes of systematic errors in emission measurements have to be mentioned. Firstly, in cool flames ($T < 2000$ K) with a large relative excess of radicals, suprathermal chemiluminescence may arise (see Sect. VI.3). Secondly, in flames with a high fluorescence efficiency of the Li or Na resonance lines (for instance, flames with noble gases as diluent), infrathermal population of the excited states may occur when the atomic concentrations are low (see Sect. VI.4). In such cases the line-intensity ratio of Li and Na is no longer simply related to [Li]/[Na]. Complications arising from nonthermal population of the excited states can be avoided by determining [Li] / [Na] from atomic absorption measurements (see McEwan and Phillips 1965, Zeegers 1966, Halstead and Jenkins 1967, and Cotton and Jenkins 1969; see also Sect. V.3b).

Relative [H] measurements are easily performed by the LiOH-method if the total Li concentration $[Li]_t$ ($\equiv [Li] + [LiOH]$), the H$_2$O concentration, and the temperature T remain, virtually, constant along the flame axis. This case applies when [H] is to be measured as a function of height in a thick, cylindrical flame over a limited height interval. Since for large ϕ values: $[LiOH] \simeq [Li]_t$, we see from the expression for K_{Li} that $[Li] \propto [H]$. Then a measurement of relative Li emission or absorption suffices to yield relative [H] values. When T varies with height, a correction should be made for the corresponding variation of $K_{Li}(T)$ (see Table IV.3), of the equilibrium concentration of H$_2$O, and of the thermal excitation of the Li line in case of emission measurements.

The *NaCl-method* of determining [H] after Bulewicz, Phillips and Sugden (1961) is based on the relative decrease in atomic Na concentration [Na], when a

Chapt. IV TYPES AND PROPERTIES OF NONSEEDED FLAMES

known amount of chlorine (in any form) is supplied to the flame, according to the partial equilibrium

$$Na + HCl \rightleftharpoons NaCl + H \qquad (x)$$

with equilibrium constant

$$K_{NaCl} = [NaCl][H]/[Na][HCl] \ . \qquad (IV.9)$$

If one makes the simplification that Na forms no other compounds than NaCl (cf. above) and that practically all chlorine is present in the flame as HCl, the ratio $[Na]_0/[Na]$ of atomic Na concentrations in absence and in presence of chlorine, respectively, is given by

$$[Na]_0/[Na] = 1 + K_{NaCl}[Cl]_t/[H] \ , \qquad (IV.10)$$

where $[Cl]_t$ stands for the total chlorine concentration. It follows from Eq. IV.10 that $[Na]_0/[Na]$ is a linear function of $[Cl]_t$. When $[Na]_0/[Na]$ is plotted as a function of the known concentration $[Cl]_t$, the *absolute* value of [H] can be calculated from the slope of the straight line obtained, if K_{NaCl} is known. For practical purposes, values of K_{NaCl} calculated at various temperatures are listed in Table IV.3 on p.441; the assumed values for $D_0(NaCl)$ and $D_0(HCl)$ are given in Table IV.3 and the uncertainty in $D_0(NaCl)$ is mentioned in Sect. VIII.2b. The other molecular constants used in this calculation were taken from the JANAF-tables (JANAF 1966) or from Herzberg (1950). If Na is also present as NaOH (compare above), then Eq. IV.10 will no longer hold. A detailed discussion of this matter is given in Sect. VIII.3c.

Methods of determining $[Na]_0/[Na]$ ratios will be discussed in Sect. V.6. Suprathermal chemiluminescence does not affect the experimental results obtained, if [H] is independent of $[Cl]_t$.

In flames with a low hydrogen content (e.g., fuel-lean flames or nearly dry CO flames) not all chlorine, added in quantities up to 0.5 volume percent, will be bound to HCl. Moreover, the presence of chlorine up to these amounts may interfere markedly with the flame kinetics involving H, thus affecting the H concentration to be determined. The conditions underlying Eq. IV.10 are no longer fulfilled, and the ratio $[Na]_0/[Na]$ ceases to be a linear function of $[Cl]_t$. Then [H] should be determined from the initial slope of this plot found for $[Cl]_t \rightarrow 0$, but the result is less accurate.

In H_2-rich flames absolute [H] values determined by the NaCl- and LiOH-method have been found in mutual agreement within 10%, if $D_0(NaCl)$ and $D_0(LiOH)$ are taken as listed in Tables VIII.4 and 2 on pages 769 and 768, respectively (see McEwan and Phillips 1967).

In practice the absolute calibration of $[Cl]_t$ requires some care. When

DEVIATIONS FROM FLAME EQUILIBRIUM COMPOSITION   Sect. IV.5b

[H] is studied as a function of height in *relative* measure only, $[Cl]_t$ need not be known if it can be assumed to be constant over the height interval studied. If the temperature too is independent of height, $([Na]_0/[Na]-1)$ is simply proportional to $[H]^{-1}$.

The *CuH-method* is based on the partial equilibrium

$$Cu + H + X \rightleftharpoons CuH + X \qquad (xi)$$

(see Sect. VIII.4a; see also Bulewicz and Sugden 1956a, and Bulewicz and Sugden 1958). Since [CuH] is usually small relative to [Cu], the thermal emission intensity of the CuH(0,0) band head at 4280 Å is proportional to [H] for a given Cu salt concentration in the sprayed solution. *Relative* [H] values as a function of height h can thus be obtained by following the band intensity with varying h. When T varies with h, the temperature-dependence of the partial equilibrium constant (see Fig. VIII.6 on p.790) and of the thermal excitation of the resonance band has to be accounted for. However, it has been found that these temperature-dependences practically compensate each other. Thus the ratio of CuH emission intensity to [H] is insensitive to variations in T (see Reid and Sugden 1962). The method may be applied in H_2-rich flames. In fuel-lean H_2 flames, the weak CuH band is swamped by a comparatively strong emission which might be due to CuO or CuOH (see McEwan and Phillips 1967).

Finally, [H] can also be determined by measuring the ratio of Ca^+ (or Sr^+) to $CaOH^+$ (or $SrOH^+$) concentrations (see Note 9 to Table IX.1 on p.825).

5b-3   *CONCENTRATION OF OH RADICALS*

The *OH-absorption method* utilizes the fact that OH shows a measurable absorption spectrum in most flames. No additives to the flame are required here. The OH band emission cannot safely be used, since it may be affected by nonthermal chemiluminescence effects in flames containing H- and OH radicals in excess (see Zeegers and Alkemade 1965a, Zeegers 1966, and Davies 1968).

The *absolute* OH concentration can be determined from the absorption of radiation at a single rotational line of an OH band. These determinations have been performed with a background spectral-line source, i.e. a discharge through water vapour, emitting the same OH band (see Kaskan 1958, and Carabetta and Kaskan 1967), and with a background continuum light source, i.e. a high-pressure xenon arc (see Zeegers, Townsend and Winefordner 1969). Both methods require a wavelength-selection device with a high resolution and knowledge of the absolute oscillator strength of the line observed (for values of oscillator strengths, see Dieke and Crosswhite 1962, and Learner 1962). In measurements with a line source, the spectral line profile of the background emission line as well as of the absorption line

Chapt. IV   TYPES AND PROPERTIES OF NONSEEDED FLAMES

has to be known. Uncertainties could arise here when the spectral profile of the
OH rotational lines (determined by the Doppler width and a-parameter; see Sect.
II.5b-3) is not precisely known as a function of flame temperature and composition.

Relative [OH] can be determined with low spectral resolution by absorption
measurements of the whole (0,0) band, while using a discharge through water vapour
or an auxiliary flame that emits the same OH band as a background source. This
method has been applied by Zeegers and Alkemade (1965) and Zeegers (1966) to measure relative [OH] as a function of h in an C_2H_2-air flame with a homogeneous
radial distribution of OH. The absorption factors found were converted into relative [OH] by measuring the absorption of the radiation from the background flame
when this radiation passes once, twice, and three times through the analysis flame
at a given height. In order to construct a relative calibration curve the three
absorption factors thus obtained were plotted versus the number of times the radiation from the background flame passed through the analysis flame.

The *NaOH-method* for measuring *relative* [OH] is based on the continuum
emission between 3000 and about 6000 Å arising from the recombination reaction:
$Na + OH \rightarrow NaOH + h\nu$ (see James and Sugden 1958). For given [Na], the emission
intensity is proportional to [OH]. In H_2 flames high Na concentrations are
required to give a measurable continuum emission. In acetylene and carbon monoxide
flames the situation is expected to be even worse because of their relatively high
background emission. This method is thus based on the rate of a process, and not
on the steady-state concentration of the product formed. The rate of the exo-ergic
reaction does not depend strongly on T; a variation of only a factor of two in a
T interval of 500 K for a given OH concentration has been observed (see James and
Sugden 1958). In cool O_2-rich flames allowance should be made for the slow recombination of free Na and O_2 to NaO_2 (see Sect. VIII.4b). This recombination
may also give rise to a continuum emission with $\lambda > 4400$ Å according to:
$Na + O_2 \rightarrow NaO_2 + h\nu$ (see McEwan and Phillips 1967). Therefore, one should be careful when applying this method in cool O_2-rich flames.

Recently Morley (1980) has attempted to monitor relative OH concentrations
in atmospheric-pressure hydrogen, hydrocarbon and carbon monoxide flames by
means of fluorescence-intensity measurements. A frequency-doubled N_2-pumped dye
laser tuned to a u.v. absorption band was used to excite the observed fluorescence band. The relative dependence of the fluorescence-quenching efficiency on
the flame composition was determined and accounted for. Saturation was avoided
in order to simplify the interpretation of the measurements. This method, which
has the advantage of a good spatial resolution, was also applied by Muller *et al.*
(1979), Stepowski and Cottereau (1979), and Cottereau and Stepowski (1980).

DEVIATIONS FROM FLAME EQUILIBRIUM COMPOSITION   Sect. IV.5b

5b-4   *CONCENTRATION OF O RADICALS*

The *NO-method* for determing *relative* [O] by adding NO gas to the flame according to James and Sugden (1955b) and Bulewicz and Sugden (1958a) employs the emission of the greenish continuum brought about by the reaction: $NO + O \rightarrow NO_2 + h\nu$ (see Gaydon 1944, 1946). Its intensity provides a measure for [O]. However, it is known that NO catalyses the recombination of H radicals (see Bulewicz and Sugden 1964; see also Sect. 5c). Since [H] and [O] are interrelated by the partial equilibria (i) and (ii) (see Eqs IV.3 and IV.4), this catalysis also affects [O]. Schofield and Broida (1968) have therefore concluded that the NO-method yields unreliable results in flames. We recommend that relative [O] be determined from the initial slope of the plot of continuum emission versus relative [NO], since for [NO] → 0 the catalytic effect on the radical recombination will vanish.

On the other hand, in pollution research the above method is used to determine [NO]. Here O atoms, generated by a microwave discharge through O_2, and NO enter a reaction vessel through separate inlets. Rapid mixing occurs and the above chemiluminescent reaction is observed (see Fontijn, Sabadell and Ronco 1970; see for a review Baulch, Drysdale and Home 1970).

The *CO-method*. In nonluminous hydrocarbon and carbon monoxide flames with a temperature lower than about 2000 K the intensity of the visible and u.v. quasi-continuum background emission between 3500 and 6000 Å is proportional to the product [CO][O] (see Kaskan 1959, and Gutman and Schott 1967). If [CO] is known, the relative [O] can be determined. Usually the flame background emission within a small wavelength range is measured. Good results have been obtained in a 10 Å interval at 4345 Å (see Kaskan 1959) and in a 40 Å interval at 4360, 4040 and 4010 Å (see Gutman and Schott 1967). Clyne and Thrush (1962) have found mutual agreement between the results of the NO and CO methods of determining [O].

If [O] is known by other methods, measurement of the background intensity yields [CO] (see Clyne and Thrush 1962).

In flames with a temperature higher than about 2000 K the thermal emission of CO_2 contributes significantly to the flame background emission (see Zeegers and Alkemade 1965, and Zeegers 1966) and the simple dependence on [CO][O] is lost. Then the *excess* emission, i.e. the difference between total emission and thermal emission, is proportional to $([CO][O] - [CO]_e[O]_e)$ (see Sect. IV.8). For flames in which the actual CO concentration, [CO], equals the equilibrium value $[CO]_e$, the excess emission is proportional to $([O] - [O]_e)$, i.e. to the excess atomic oxygen concentration.

The *IO-method* has been proposed to determine *relative* [O] (see Phillips and Sugden 1961). It entails observing the emission of the (0,4) band of IO at

Chapt. IV TYPES AND PROPERTIES OF NONSEEDED FLAMES

5307 Å produced when iodine, either as element or as methyl-iodide, is added in quantities up to 1 volume percent of the unburnt gas mixture. Apparently, addition of iodine in these quantities does not affect the radical concentrations; at least no such effects have been reported. In fuel-rich hydrogen flames a continuum arising from the overall process $H + I \rightarrow HI + h\nu$ overlap the IO band, and the intensity of the former must be subtracted. This correction can be performed approximately by measuring the continuum emission intensity adjacent to the IO band. Phillips and Sugden (1961) suggest that IO is thermally excited and that its concentration is determined by the supposedly balanced reaction

$$I + O + X \rightleftharpoons IO + X . \quad \text{(xii)}$$

The concentration of IO is relatively small, its dissociation energy being about 2.5 eV. Consequently, [I] may be assumed to be virtually unaffected by IO formation. The IO band emission is therefore proportional to [O] at a given temperature. However, the IO emission as observed by Phillips and Sugden (1961) may be explained equally well by a second mechanism proposed by the same authors, namely the balanced reaction

$$I + OH \rightleftharpoons IO + H \quad \text{(xiii)}$$

followed by suprathermal chemiluminescent excitation through

$$IO + H + (O)H \rightarrow IO^* + H_2(O) . \quad \text{(xiv)}$$

In both cases (reaction xii in combination with thermal emission, or reactions xiii plus xiv), the IO emission is proportional to $[H]^2$, which is proportional to [O] because of the interrelations of the radical concentrations (see Eqs IV.4 and 5). According to Phillips and Sugden (1961) the rate constant of reaction (xiv) in the second case would have to be at least 100 times larger than any previously measured rate constant of a ternary chemiluminescent reaction. Moreover, according to these authors, only rather stable molecules ($D_0 > 4$ eV) tend to show chemiluminescent band emission. Therefore, chemiluminescence of IO (with D_0 about 2.5 eV) seems unlikely. Phillips and Sugden (1961) conclude that the general trend of their observations points to the prevalence of reaction (xii) followed by thermal emission of IO. However, if reaction (xii) is effectively balanced, as assumed above, then the termolecular forward reaction (xii) should also have an unusually high rate constant, as pointed out by the same authors.

Using Phillips and Sugden's data: $T = 2200$ K, $[X] = 3 \times 10^{18}$ cm^{-3} , $[O] = 3 \times 10^{13}$ cm^{-3} , and assuming that reaction (xii) is effectively balanced within 2×10^{-5} s (corresponding to a height interval of about 0.1 mm), we calculate the lower limit of the termolecular rate constant of forward reaction (xii) to be 4×10^{-27} cm^6 s^{-1}. Termolecular rate constants are usually found in the range of

DEVIATIONS FROM FLAME EQUILIBRIUM COMPOSITION   Sect. IV.5b

10^{-30} to 10^{-32} cm^6 s^{-1} and rarely of the order of 10^{-29} cm^6 s^{-1} (see Gilmore, Bauer and McGowan 1967, Schofield 1967, and Jensen and Jones 1978).

Besides, a discrepancy of a factor of 1.8 is found when for given iodine supply the ratio of IO intensity to [O] in an H_2-rich flame is compared to the corresponding ratio in an O_2-rich flame at the same temperature (see McEwan and Phillips 1967). This discrepancy may be explained by assuming that reactions (xiii) and (xiv) prevail. In the latter case the intensity of the suprathermal IO band emission is expected to be dependent on flame composition, because the quenching of IO* depends on composition (see Sect. VI.3). If, however, IO is thermally excited, then the IO emission for given O- and I concentrations depends on temperature only and not on the composition.

Contrary to Schofield and Broida (1968), we conclude that it is still not certain whether the IO-method yields reliable results for [O]. The same holds for the (0,6) band of BrO at 4270 Å, which has also been used to measure relative [O] (see McEwan and Phillips 1967).

In *CO flames* where O is the dominant radical, *excess O concentrations* can be determined from the *initial rise of flame temperature* with height just above the combustion zone. This rise is caused by the heat that is gradually liberated in the slow recombination reactions: $CO + O + X \rightarrow CO_2 + X$ and: $O + O + X \rightarrow O_2 + X$. The latter reaction is usually of minor importance. To the extent that [O] approaches its equilibrium value $[O]_e$ with increasing height, T will approach the final flame temperature expected in full chemical equilibrium. When heat losses in the flame are corrected for, the difference between the actual value of T at a given height and the equilibrium value is proportional to the value of $[O] - [O]_e$ at that height. The proportionality constant includes the heat capacity of the flame gas and can be calculated. The value of [O] can be determined in *absolute* measure as a function of height, if $[O]_e$ is known by calculating the equilibrium composition (see Sect. IV.4). This method has been applied by Hollander (1964) and Zeegers (1966) to CO flames where [O] could not be calculated from measured values of [H]. Doubt about the reliability of [H] / [H_2O] determinations in nearly dry $CO - N_2O$ flames was cast by Kalff and Alkemade (1973) and van der Hurk, Hollander and Alkemade (1973). They did not find agreement between the concentration ratios measured and the equilibrium ratios calculated from Eqs IV.3 to IV.6 for flame regions where chemical equilibrium was expected to be fully established.

To our knowledge there is still no satisfactory general method of measuring [O] in flames. In hydrocarbon and hydrogen flames, the actual [O] can best be calculated from measured [OH] and theoretical equilibrium values $[OH]_e$ and $[O]_e$ by using the relation

Chapt. IV   TYPES AND PROPERTIES OF NONSEEDED FLAMES

$$[O] = ([OH]/[OH]_e)^2 [O]_e \ . \qquad (IV.11)$$

This relation follows directly from the balanced reaction (iv) and the reasonable approximation: $[H_2O] \approx [H_2O]_e$ if H_2O and H_2 are present in relative excess compared to H and OH (see Zeegers, Townsend and Winefordner 1969).

Finally, we stress that all optical methods discussed make sense only if the radicals are homogeneously distributed over the flame section viewed. Care should be taken with unshielded flames where secondary combustion at the flame border could cause radial inhomogeneity in the radical concentrations (see Zeegers 1966). With some methods, such as the CuH-method but not with the LiOH-method, we should account for the variation of flame thickness with height when deriving radical-concentration ratios from measured emission or absorption ratios.

5b-5   *CONVERSION OF RELATIVE INTO ABSOLUTE RADICAL CONCENTRATIONS*
It may occur that in a flame with constant T over the height interval studied, the plot of relative [H] (or [OH]) versus height approaches closely a horizontal asymptote higher up in the flame. Assuming that for this asymptote $[H]=[H]_e$ (or $[OH]=[OH]_e$), we can obtain a *conversion of relative into absolute concentration values* by calculating $[H]_e$ or $[OH]_e$ according to Sect. IV.4.

A similar absolute calibration of the relative [H] curve is also possible when T varies with height in a known way. The procedure, however, is much more complicated and involves theoretical expressions for the decay of excess radicals with h. We consider a particular case which has actually been observed by Zeegers and Alkemade (1965) and Zeegers (1966) in an C_2H_2-air flame where T drops uniformly with h after an inital sharp rise (see Fig. IV.5). The corresponding variation in absolute equilibrium concentration $[H]_e$ is also depicted in this figure and shows likewise a uniform fall downstream from the point of maximum temperature. The experimental *relative* [H] values obtained by the LiOH-method are also depicted in the figure. It is seen that at greater heights the experimental curve falls off with h at the same relative rate as the theoretical $[H]_e$ curve. For an absolute calibration of the experimental points we must not shift the experimental log [H] curve so far in a vertical direction that its right-hand tail coincides with that of the absolute log $[H]_e$ curve. The reason is that the actual [H] will lag constantly behind the continuously decreasing $[H]_e$ because of the relaxation in the establishment of radical equilibria. On account of this relaxation, the right-hand tails of both logarithmic plots will remain separated by a finite and practically constant vertical distance. This means that at great heights in the flame where $[H]_e$ drops uniformly the actual [H] will exceed $[H]_e$ by a constant factor. The absolute position of the experimental [H] curve can be found by calculating this factor by means of a theoretical formula.

448

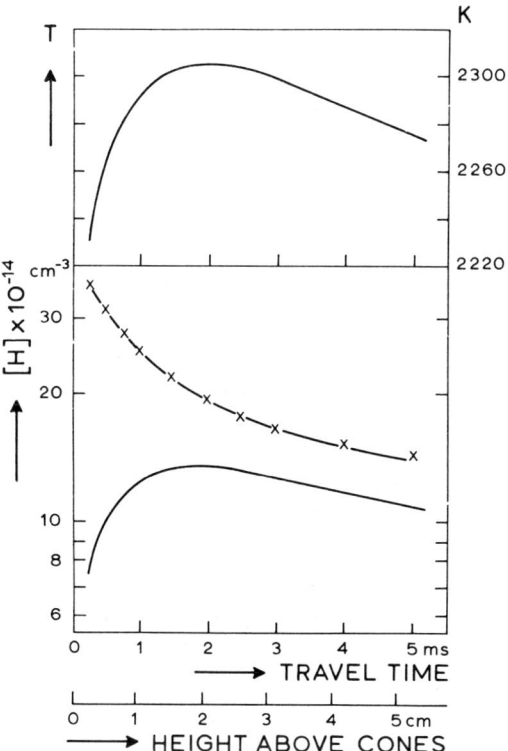

Fig. IV.5 The actual H concentration (× × ×) and the calculated equilibrium concentration (full curve) in an C_2H_2-air flame are plotted against travel time. The upper part gives the corresponding variation of the flame temperature T. (From Zeegers 1966.)

The mathematics involved in deriving this formula are straightforward, but rather laborious. This holds especially in C_2H_2-flames where four independent radical-recombination reactions may be operative at the same time, and the radical concentrations are linked with one another and with those of the bulk flame constituents by four partial-equilibrium relations (see Eqs IV.3 to 6). Zeegers and Alkemade (1965) and Zeegers (1966) have derived a formula that enables one to calculate the factor by which [H] deviates from $[H]_e$ at any point in the right-hand tail of the curve in Fig. IV.5. This formula involves the relative rate of decay of [H] and the derivative of T with respect to h, as well as the ratio of OH concentrations at two distinct heights. All these values can be found independently by experiment. The coefficients occurring in the formula can be calculated from the known chemical equilibrium constants, from calculated values of $[H_2O]_e$ and $[CO_2]_e$, and from roughly estimated values of [H] and $[H_2]$. The rate constants of the radical recombination reactions themselves need not be known, however.

It has proved possible to calculate $[H]/[H]_e$ with 5% accuracy in the

Chapt. IV   TYPES AND PROPERTIES OF NONSEEDED FLAMES

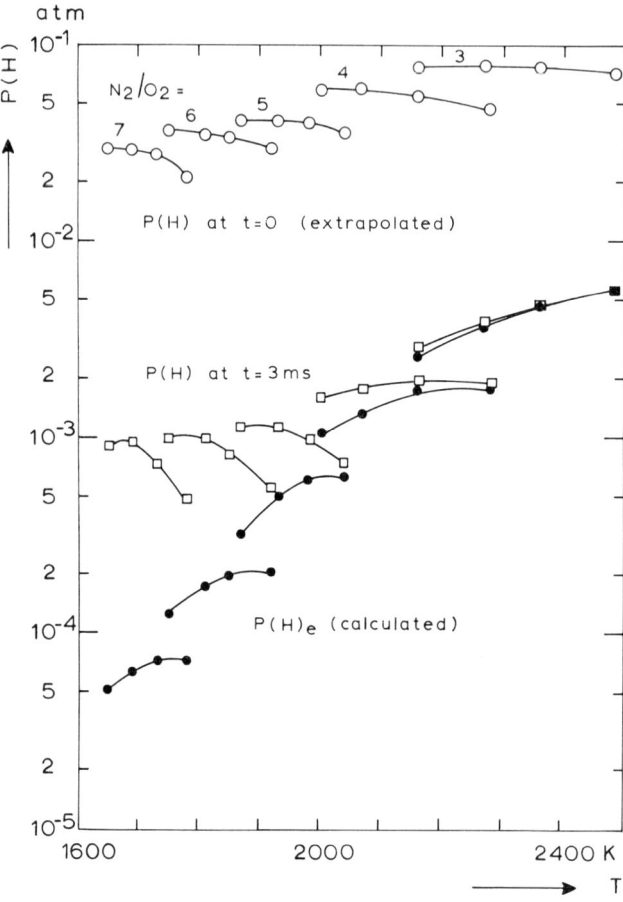

Fig. IV.6 The actual H concentration at the top of the combustion zone (i.e. at travel time $t = 0$) and at $t = 3$ ms as well as its equilibrium value at the maximum flame temperature T, marked by ○, □ and ●, respectively, are plotted against T for various 'families' of flames. The value of the ratio N_2/O_2, which characterizes a family of four flames, is indicated for the uppermost curves only, but it is the same for all curves lying on top of each other. The values of the ratio H_2/O_2 within one family are 4.0, 3.5, 3.0 and 2.5; the lowest value corresponds to the highest temperature within each family. (This figure is constructed from data measured by Padley and Sugden 1959.)

DEVIATIONS FROM FLAME EQUILIBRIUM COMPOSITION   Sect. IV.5c

right-hand tail of the curve shown in Fig. IV.5. Once this ratio is known and $[H]_e$ has been calculated in absolute measure at one height, absolute [H] values at all other heights follow then immediately from the experimental [H] curve.

This method of absolute standardization of the [H] curve through adaptation of the relative experimental curve to the absolute theoretical equilibrium curve works better, the closer radical equilibrium is approached. It can, then, give more accurate absolute concentrations than does the LiOH-method. This holds particularly when one is primarily interested in the slight deviation of actual [H] from $[H]_e$ under near-equilibrium conditions.

For OH a similar conversion procedure has been used by the same authors. Their results have been checked by an absolute [OH] determination from OH absorption measurements (see Zeegers, Townsend and Winefordner 1969; see also Sect. 5b-3); mutual agreement within 10% was found.

5c   SOME OBSERVATIONS ON EXCESS RADICAL CONCENTRATIONS
      AND THEIR RELAXATION RATES

H- , OH- and O concentrations at the top of the combustion zone have been estimated by Padley and Sugden (1959), who extrapolated radical concentrations measured as a function of vertical distance from this zone. Radical concentrations beyond the combustion zone have been determined by several authors who followed one of the methods outlined above (see, e.g., Bulewicz and Sugden 1964, and Zeegers and Alkemade 1965). Fig. IV.6 gives the results for [H] at the top of the combustion zone (see Padley and Sugden 1959) and at a point downstream from this zone corresponding to a travel time of 3 ms, as well as the equilibrium value, $[H]_e$, calculated at the final flame temperature by Bulewicz and Sugden (1964) for various 'families' of flames. H concentrations are plotted versus flame temperature. A flame 'family' is a set of flames having different H_2/O_2 unburnt-gas ratios but a fixed N_2/O_2 ratio (see also Sect. V.6). The value of N_2/O_2 is given in the figure. The points on each curve correspond to H_2/O_2 ratios of 2.5, 3.0, 3.5, and 4.0, respectively, in the direction of decreasing temperature.

The results for [OH] in the combustion zone are given in Fig. IV.7. The actual [H] and [OH] in one family behave differently, because in fuel-rich flames [OH] is strongly dependent on the fuel/oxidant ratio, whereas [H] is not. It is interesting to note that in these fuel-rich flames the actual [H] and [OH] in the combustion zone spread by no more than a factor of 5 and 10, respectively, whereas $[H]_e$ and $[OH]_e$ in the coolest flame (1650 K) are about a factor of 10^3 lower than in the hottest flame (2480 K). Consequently, the relative, but not the absolute excess of [H] and [OH] above their equilibrium values is much higher in cool flames than in hot flames.

Chapt. IV TYPES AND PROPERTIES OF NONSEEDED FLAMES

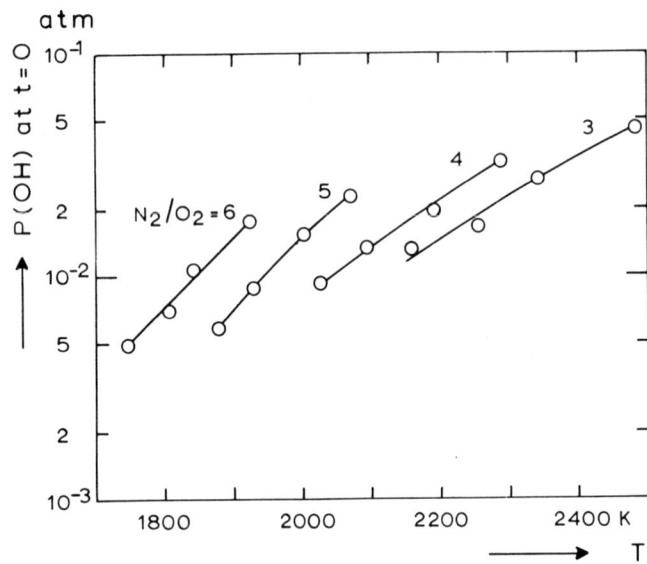

Fig. IV.7 The OH concentration at the top of the combustion zone found by extrapolation towards $t = 0$ is plotted against maximum flame temperature T for four families of four flames (see also Fig. IV.6). (Redrawn from Padley and Sugden 1959a.)

In fuel-rich hydrogen flames in which $[H_2]$, $[H_2O]$ and T may be assumed to be practically constant everywhere beyond the combustion zone the decrease of $[H]$ with travel time t is determined by the termolecular reactions (vi) and (vii) and the fast balanced reaction (iii). The decay rate is given by (see Bulewicz and Sugden 1958)

$$-d[H]/dt = 2(k_H + \alpha k_{OH})[X]([H]^2 - [H]_e^2)/(1+\alpha), \qquad (IV.12)$$

in which α stands for $K_{iii}^{-1}[H_2O]/[H_2]$ which equals $[OH]/[H]$ (see Eq. IV.5); X stands for the third body carrying off the recombination energy; k_H and k_{OH} stand for the rate constants of reaction (vi) and (vii), respectively. In the derivation of this equation it is assumed that the rate constants have the same value as in equilibrium and that only one kind of third body is effective (cf. below). Furthermore, the reverse reaction rate $k_{-H}[H_2][X]$ is set equal to $k_H[H]_e^2[X]$, which is correct when $[H_2] \simeq [H_2]_e$ as assumed above. Integrating Eq. IV.12 while putting $[H] = [H]_0$ at $t = 0$, we get

$$(1+\alpha)\ln\left\{\frac{[H]-[H]_e}{[H]+[H]_e} \cdot \frac{[H]_0+[H]_e}{[H]_0-[H]_e}\right\} = 2(k_H + \alpha k_{OH})[H]_e[X]t. \qquad (IV.13)$$

452

DEVIATIONS FROM FLAME EQUILIBRIUM COMPOSITION   Sect. IV.5c

For $[H] \gg [H]_e$ Eq. IV.13 reduces to

$$(1+\alpha)\left\{[H]^{-1} - [H]_0^{-1}\right\} \simeq 2(k_H + \alpha k_{OH})[X]t . \qquad (IV.14)$$

Along the same lines an analogous relation for $[OH]^{-1}$ can be obtained. Under the conditions mentioned, curves of $[H]^{-1}$ and $[OH]^{-1}$ versus travel time should be straight lines. Figs IV.8 and IV.9 show such curves for some $H_2-O_2-N_2$ flames, as measured by Padley and Sugden (1958). By extrapolating towards $t=0$ one finds $[H]$ in the combustion zone. The slope of the curve is a measure for the effective rate constant, i.e. the rate constant as averaged over the different contributions of several kinds of third bodies acting in the flame. A more exact expression is obtained by replacing $k_H[X]$ in Eq. IV.14 by: $\Sigma_j k_H^j[X_j]$. Here k_H^j stands for the specific rate constant of reaction (vi) with species X_j as third body. For k_{OH} a similar equation can be written.

It follows that the slopes in Figs IV.8 and 9 are in general not only dependent on α, but also on the concentrations of N_2, H_2O and H_2, being the major species in these flames. Values of $k_{(O)H}^j$ have been tabulated by Baulch et al. (1972), Jensen and Jones (1978), Dixon-Lewis and Williams (1977); for most third bodies k_H lies between 10^{-31} and 10^{-33} cm^6 s^{-1} and k_{OH} between 10^{-30} and 10^{-32} cm^{6}s^{-1}. In O_2-rich flames, where reaction (viii) also plays a role, and in hydrocarbon flames, where all reactions (vi) to (ix) contribute, the final equation for $d[H]/dt$ is much more complicated. In the radical-recombination region of most flames the temperature rises because of the heat released during the recombination (see, e.g., Padley and Sugden 1958, and Zeegers 1966). Consequently, there exists a temperature gradient downstream from the combustion zone. The partial equilibria of reactions (i) to (v) are therefore re-adjusted continuously at the varying local flame temperature. This implies that K_i and K_v (and thus also α in Eq. IV.12) are no longer constants. Under these circumstances, the simple equation IV.12 no longer holds.

Zeegers and Alkemade (1965) and Zeegers (1966) have derived a more general formula for $d[H]/dt$ that allows for all recombination reactions occurring in an C_2H_2-air flame, for the variation in the concentration of the major species with height, and for the effect of the variation in temperature on the equilibrium constants. However, their expression for $d[H]/dt$ (and $d[OH]/dt$) cannot be analytically integrated. The evaluation of the experimental results is then much more complicated.

Third bodies do not always act as energy acceptor only. They may play a role as an intermediary in the reaction kinetics as well. In that case *catalyzed radical recombination* is said to occur. The recombination of H and OH may be

453

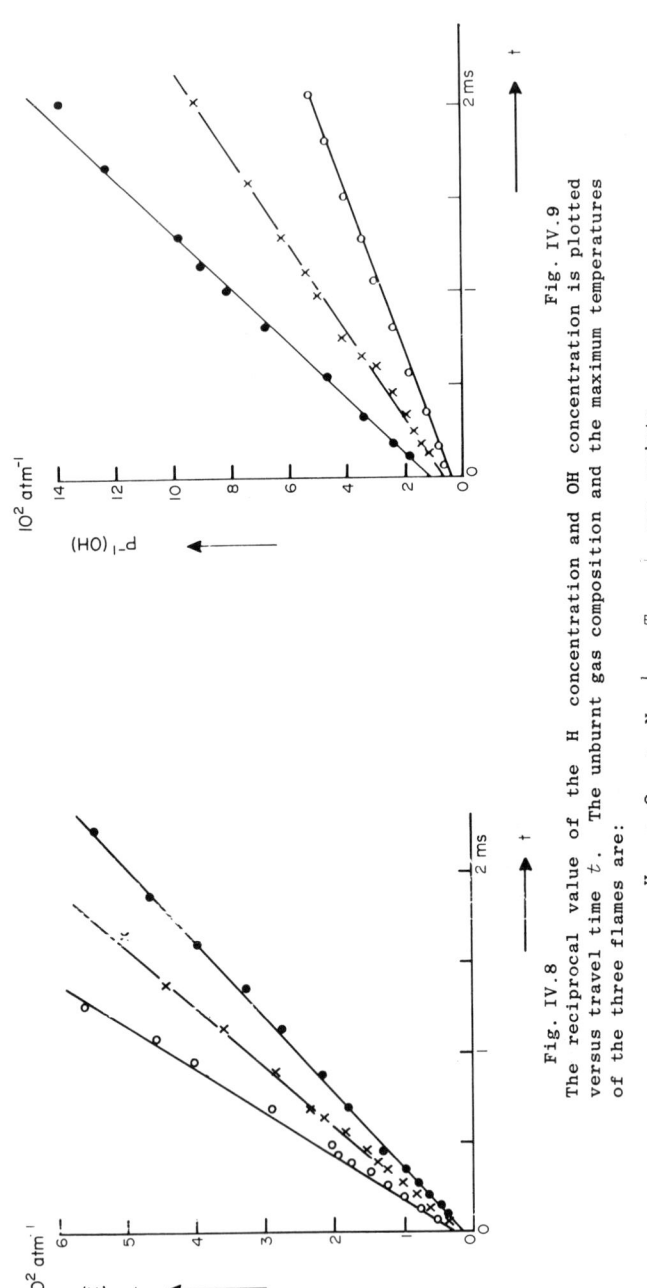

Fig. IV.8
Fig. IV.9
The reciprocal value of the H concentration and OH concentration is plotted versus travel time t. The unburnt gas composition and the maximum temperatures of the three flames are:

H_2 : O_2 : N_2	T	exp. points
2.5 : 1 : 5	2070 K	○
3.0 : 1 : 5	2000 K	×
3.5 : 1 : 5	1930 K	●

(Redrawn from Padley and Sugden 1959a.)

MEASURED FLAME TEMPERATURES   Sect. IV.6a

catalyzed, for instance, by NO (see Clyne and Thrush 1962, Bulewicz and Sugden 1964, Halstead and Jenkins 1968, and Smith 1972), by SO_2 (see Kallend 1967, 1972, Halstead and Jenkins 1969, and Durie, Johnson and Smith 1971), by Mg, Cr, Mn, Sn, U (Bulewicz and Padley 1971a), by Ca, Sr and Ba (Cotton and Jenkins 1971) and by K (Jensen, Jones and Mace 1979). Catalysis occurs by homogeneous reactions (with K, NO, SO_2) or by recombination of radicals on the surface of nonvolatilized particles (e.g., oxides of Sn and Cr). When determining radical concentrations one should be very careful with additives in high concentrations because they may affect the recombination rate and consequently the excess radical concentrations (see Bulewicz and Padley 1973). On the other hand, when studying the effect of excess radical concentrations on metal vapours, one should determine the radical concentration in a flame with metal additive in the same concentration as is used in the actual experiments.

6. MEASURED FLAME TEMPERATURES

In this section we discuss actual temperature measurements in flames. They will be compared to flame temperatures that have been calculated from thermodynamic data, as discussed in Sect. IV.4. The experimental methods have been discussed in Sect. III.10. From these foregoing sections it is recalled that the accuracy of calculated flame temperatures is about 5 K, possible exceptions being flames burning with very exotic fuels and flames containing solid carbon. The accuracy of measured temperatures is about 5 K for reversal measurements in homogeneous flames, but may be of the order of 100 K or even worse for, e.g., vibrational temperatures and local temperatures in very inhomogeneous flames. However, calculated or measured temperature differences of less than 5 K may still be meaningful when the variation of flame temperature with some specific quantity is studied.

6a   CAUSES OF DIFFERENCES BETWEEN CALCULATED
     AND MEASURED FLAME TEMPERATURES

In the calculation of temperatures the use of computers has virtually solved the problems caused by the necessary iterative procedures. There may still be differences between the calculations of various authors because different sets of thermodynamic data have been used. However, the JANAF-tables (JANAF 1960) constitute a widely used, consistent set of data.

Other differences in results may be due to different initial temperatures of the gases (these are not always stated) and the neglect of minor components of the initial mixture, e.g., water vapour, or of the reaction products, e.g., NO.

Measured flame temperatures, however, spread more widely than 5 K. Even if proper care is taken that the excitation temperature has been correctly determined,

Chapt. IV   TYPES AND PROPERTIES OF NONSEEDED FLAMES

there are still a number of factors that cause differences in the results of measurements on flames with the same fuel/oxidant/diluent gas ratio and the same temperature of the initial mixture:

1.   The geometry of the burner has a great influence on the conductive heat flow to the burner. The size and the number of holes or slots, the area of the burner top between holes or slots, the heat conductivity of the material, the velocity of the gas flow and the shape of the burner all influence the amount of heat that is lost through the burner. Part of this heat is restored to the flame by the preheating of the unburnt gases in the burner top. However, a net loss of a few percent of the total available heat, still dependent on the total gas flow, may occur, resulting in a drop in temperature of the order of 50 K. The use of insulating material can diminish the losses to less than one percent or 10 K (see Snelleman 1968). Cooling of the burner top, e.g., by a water flow system, increases the heat losses considerably.

2.   The flame temperature generally rises in the lower parts of the flame. This rise is caused by the slow chemical equilibration of the combustion products, as described in Sect. IV.5, which increases the thermal energy of the flame gases. This rise in temperature has been found to be of the order of 100 K. The final temperature is reached after a few milliseconds rise time; in fuel-rich flames this time is somewhat longer. Of course, the attainment of a constant temperature is only an indication that chemical equilibrium for the major components has been reached. For minor components one has to rely on the methods described in Sect. IV.5, since the equilibration of such components has little effect on temperature.

3.   The hot flame gases continuously lose energy by radiation in the infrared bands of mainly H_2O and CO_2 (at 2.7 μm and 4.4 μm). An energy loss equivalent to 10 K/ms has been measured in an C_2H_2-air flame by Snelleman (1968).

The well-known maximum in the temperature of some flames that occurs about 3 ms after the gases have left the burner, even in well shielded flames, is due to the combined effects of slow chemical equilibration and infrared radiation loss. The effect of turbulent and diffusive mixing with the surrounding air causes a further temperature drop; eventually, the top of the flame is no longer in a well-defined state.

4.   As discussed in Sect. VI.4c, sub point iv, the relative population of the upper level(s) of the spectral line(s) used may be decreased as a result of radiative disequilibrium. In flames containing good quenchers such as N_2 and CO_2 the depletion is small and leads to errors of only a few degrees. However, in flames containing only poor quenchers such as Ar and H_2O, e.g., the stoichiometric $H_2 - O_2$ - Ar flame, the depletion of the upper level of a transition may be

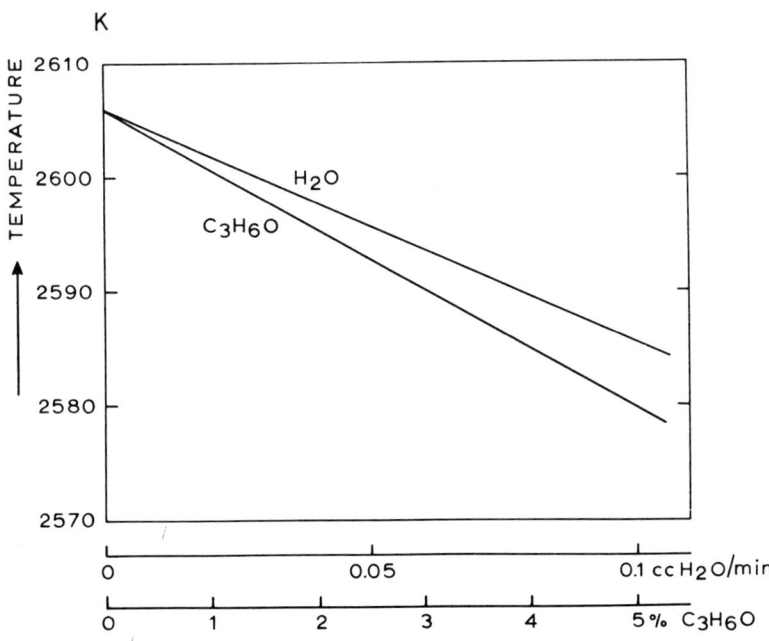

Fig. IV.10 The calculated influence of the amount of water introduced (as liquid), and the acetone content of the acetylene gas on the temperature of an air-acetylene flame (50 litres/min air, optimum mixing ratio). (From Snelleman 1969.)

considerable and differences between true temperature and reversal temperature of the order of 100 K may occur (see Hooymayers and Lijnse 1969). For optically thick lines the depopulation error is less.

5. The introduction of water decreases the flame temperature. Figure IV.10 gives an example of calculations of this decrease upon introduction of water vapour; it is recalled that with chamber-type nebulizers most of the water evaporates in the spray chamber, tubings and burner (see Sect. III.9a).

6. Acetylene from an almost exhausted cylinder yields lower flame temperatures than acetylene from a full cylinder since the percentage of acetone is higher in the former (see Fig. IV.10). Calculations and measurements on the influence of acetone agree only qualitatively, probably indicating that the acetone vapour is not saturated on leaving the cylinder.

When these effects are quantitatively allowed for in a particular flame, the calculated and the measured temperature agree within about 5 K (see Snelleman 1968). This result shows that the flame does in fact attain a fairly good state of equilibrium with regard to its major constituents. It also shows the validity of

Chapt. IV TYPES AND PROPERTIES OF NONSEEDED FLAMES

calculated temperatures. Cases in which measured temperatures exceed the calculated ones point to a deviation from equilibrium when the measuring procedure and the homogeneity of the temperature are sufficiently checked. Suprathermal chemiluminescent radiation may occur not only in the primary combustion zone but also in the interzonal region of the flame (see Sect. VI.3).

A graphical summary of the effects described is given in Fig. IV.11.

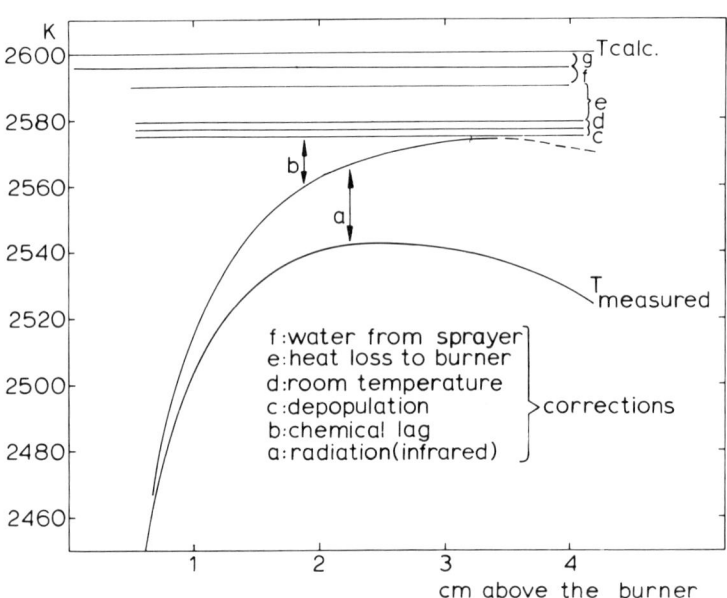

Fig. IV.11 A composite graph of the various steps in the comparison of measured (reversal) and calculated temperatures. Corrections: a: radiation loss between burner top and height of measurement. b: incomplete equilibration of the gases. c: difference between gas temperature and measured excitation temperature. d: effect of room temperature on flame temperature. e: net heat loss to the burner (conduction and radiation, minus heat restored to the flame). f: cooling effect of water sprayed into the flame. The remaining discrepancy (g) may be attributed to a difference between the practical and the thermodynamic temperature scale. (From Snelleman 1968a.) *Note*: the practical scale has since been revised (see Comité Int. des Poids et Mesures 1968); at 2500 K an increase of 4.5 K resulted.

458

6b INHOMOGENEITIES IN TEMPERATURE

When the temperature of the flame gases is not constant along the (normally horizontal) line of sight, an erroneous temperature is measured unless certain precautions are taken. Firstly, the edge of the flame is generally colder than the flame body because of mixing effects with the surrounding air. In emission methods, e.g., the two-line method (see Sect. III.10) the spectral lines used should be free from self-absorption. Then, the rather thin cold outer layer does not markedly affect the measurement of line intensity ratios. However, in line-reversal measurements and emission-absorption measurements optically thick spectral lines (and bands) are often used. In that situation one virtually measures the temperature of the outmost layer of the flame, facing the detector, where the thermometric species is still present.

Secondly, in nonshielded fuel-rich flames there is a secondary combustion zone at the boundary of the flame and the surrounding air, which can be hotter than the central flame (see Snelleman 1965, and Haraguchi et $al.$ 1976). This effect may also lead to erroneous results. This and the aforementioned effect can be avoided to a large extent by surrounding the flame proper by a mantle flame in which none of the spectral lines used in the temperature determination are emitted or absorbed. Generally this means that no metal solution is sprayed into the mantle flame. One should, however, be aware that in the upper parts of the flame, metal vapour can reach the outer parts of the flame by diffusion (see Chapt. X), while cold air diffuses inward.

Thirdly, the situation may occur that marked inhomogeneity in temperature is found also in the restricted region where the thermometric species is present. This complex situation has been treated in detail by Reif, Fassel and Kniseley (1973, 1974, 1975, 1976). These authors have pointed out the pitfalls in defining an 'average' temperature in such situations and show that, even with negligible self-absorption, the measured two-line temperature is neither the average temperature nor the weighted (as to density of thermometric species) average temperature. They also show that the measured temperature is dependent on the height of the energy levels of the optical transitions used and that the result is therefore dependent on the spectral line used; this holds even in the case when the population of the energy levels obeys the Boltzmann law locally. Computer calculations by Reif, Fassel and Kniseley (1974) yielded differences of up to 800 K when various plausible distributions of temperature and atoms, and different heights of energy levels were assumed; and differences up to 200 K when only different heights of energy levels were assumed while the temperature and atom distribution were kept fixed. The authors recommend the use of the Abel inversion procedure in inhomogeneous flames that are cylindrically symmetric (see Sect. III.10d).

For direct methods of spatially resolved measurements see Sect. III.10d.

Chapt. IV   TYPES AND PROPERTIES OF NONSEEDED FLAMES

7. NATURAL FLAME IONIZATION

General. It has been known since the beginning of the last century that flames may conduct electricity and therefore exhibit ionization. However, our understanding of ionization phenomena in flames is still not complete. On the one hand, modern experimental methods such as those based on microwave absorption, Langmuir probes, and ion mass spectrometry have considerably enriched our knowledge about the kinds, abundance and spatial distributions of charge carriers in flames of various compositions and pressures. We shall treat the experimental methods in Sect. IX.4 as part of our discussion on the ionization of metal vapours. On the other hand, theoretical analysis of the kinetics of production and removal of (positive and negative) ions has yielded a multitude of possible mechanisms. The fact that so many authors hold different or even conflicting views demonstrates how quickly this branch of flame and combustion research has grown. We shall not go into details and mathematical formulations; we wish only to summarize in a qualitative way our knowledge about the *natural flame ions* that are produced in absence of metal vapours. We refer the interested reader for detailed literature references and for further information to, e.g., the books by Miller (1968), Lawton and Weinberg (1969), and Gaydon and Wolfhard (1979), and to the reviews by Calcote (1965), Calcote and Jensen (1966), Feugier (1970), Sugden (1965, 1971), Fontijn (1972, 1974), Page (1973), and Hayhurst (1974). For a review of older work see Chapt. II in Shuler and Fenn (1963). The published Proceedings of the biennial International Symposia on Combustion and of the A.G.A.R.D. meeting on 'Fundamental studies of ions and plasmas' (see Wilsted 1965) are also recommended for consultation.

Ion concentration. In nonseeded hydrocarbon flames ionization is strongest at the top of the primary combustion zone. Here the ion concentration may be as high as 10^{12} cm^{-3} at a flame pressure of 1 atm, and 10^9 cm^{-3} at 2 Torr. These high ion concentrations as well as their pronounced decay with increasing distance from the combustion zone point to suprathermal ionization. The combustion products including the radicals OH, C_2 and CH have high ionization energies (> 12 eV); NO is an exception, having an ionization energy of 9.25 eV. Equilibrium calculations based on the Saha law (see Eq. II.59) yield much lower ion concentrations ranging from roughly 10^7 to 10^{11} cm^{-3} for normal atmospheric-pressure flames with about 1 mole percent NO and temperatures ranging from 2000 to 3000 K (see Gaydon and Wolfhard 1970).

Ion concentrations found in hydrogen and carbon monoxide flames are several orders of magnitude lower than in hydrocarbon flames. The ionization found is often due to metallic or organic impurities. The ion concentration in premixed

NATURAL FLAME IONIZATION   Sect. IV.7

H_2-air flames at 1 atm is at most of the order of 10^8 cm^{-3} (see Hayhurst and Sugden 1966, and Bulewicz and Padley 1969). In pure H_2 flames some residual ionization close to thermal equilibrium may be found, involving NO^+ and H_3O^+ ions. The latter ion is formed by the recombination reaction: $H + H + OH \rightarrow H_3O^+ + e^-$ (see Hayhurst and Telford 1972, 1975). In a pure CO-air flame the electron concentration was found to be below 5×10^7 cm^{-3} (see Borgers 1978).

Since the flame is electrically neutral (see Sect. IX.1), the total concentration of (singly charged) positive ions equals the total concentrations of (singly charged) negative ions plus free electrons. The free electrons usually outnumber the negative ions (see Hayhurst and Sugden 1966). The latter may become important when a large quantity of halogen is supplied (see Sect. IX.2); they may also be important near the preheating zone just below the primary combustion zone and in the cold border of the flame (see Sugden 1971).

In sooting flames appreciable amounts of free electrons can be produced by thermal ionization of carbon particles (see Lawton and Weinberg 1969). This cannot account, however, for the excessive ionization mentioned above.

Identity, formation, and decay of ions. Investigations with the ion mass spectrometer have revealed the simultaneous occurrence of many kinds of positive and negative ions in the primary combustion zone of hydrocarbon flames, with mass numbers up to at least 100 (see Hayhurst and Kittelson 1978). A few examples of species that have been identified are: C^+, CH_3^+, CHO^+, $C_2H_3O^+$, $C_3H_3^+$, C_3HO^+, H_3O^+, NO^+, and C^-, C_2^-, OH^-, O^-, O_2^-. In H_2- or CO flames to which small amounts of C_2H_2 and NO were added similar species were found. The measured dependence of ion concentrations on place in and above the thick combustion zone of low-pressure flames has been taken as a starting point to deduce the various possible production and decay mechanisms. Most ions are believed to be formed secondarily by charge exchange, polymerization or chemical reactions involving primary ions. The primary ions arise directly as products of an energetic chemical reaction; this formation process is called *chemi-ionization*. In the literature there is general agreement now that the most important primary ion is CHO^+.[†] This species is thought to be produced by the following reaction between neutral reactants

$$\text{CH (or CH}^*) + O \rightarrow CHO^+ + e^- . \qquad (i)$$

In fuel-rich hydrocarbon flames $C_3H_3^+$ may also be an important primary ion (see Feugier 1970 for a general discussion on the importance and formation reaction of this ion); it is formed by

---

[†] See Blades (1976) for a general critical discussion, and Semenov and Sokolik (1970) for a discussion of some other primary chemi-ionization routes and products.

Chapt. IV
TYPES AND PROPERTIES OF NONSEEDED FLAMES

$$CH^* + C_2H_2 \rightarrow C_3H_3^+ + e^- \ . \qquad (ii)$$

In both cases the rate of ion formation depends on the presence of CH radicals in the ground state, in an electronically excited state [$CH^*(A^2\Delta)$ or $(B^2\Sigma)$], or in a vibrationally excited state. This radical is known to be present in excess concentration in the combustion zone of hydrocarbon flames where it may be formed in the ground or excited state by reactions such as

$$C_2 + OH \rightarrow CO + CH \qquad (iii)$$

$$C_2H + O \rightarrow CO + CH \qquad (iv)$$

$$CH_3 + O \rightarrow CH + H_2O \ . \qquad (v)$$

Considerations based on the heat of formation of CHO^+ show that reaction (i) with CH in the ground state is energetically allowed in flames; this reaction is probably more important than the corresponding reaction with CH^* as the latter is present in smaller concentrations than CH (for further evidence and quantitative data see Fontijn 1974).

The most important secondary ion is H_3O^+, which is generally believed to be formed by

$$CHO^+ + H_2O \rightarrow H_3O^+ + CO \ . \qquad (vi)$$

It is normally the dominant ionic species in the last stage of the primary combustion process (that is, in the upper part of the combustion zone) and in the burnt gas. The long persistence of this ion away from the combustion zone is connected with the slowness of the dissociative neutralization reaction through which it is removed

$$H_3O^+ + e^- \rightarrow H + H + OH \text{ (or other neutral products)}. \qquad (vii)$$

Evidence for the appearance of $H + H + OH$ as reaction products was put forward by Hayhurst and Telford (1975). These authors also showed that reaction (vii) may be nearly balanced in pure H_2 flames and from this they derived a value of 7.2 eV for the proton affinity of H_2O, in agreement with values obtained in other ways.

Instead of an electron an OH^- ion may also participate in a similar dissociative neutralization reaction (see Calcote 1965, and Feugier 1970). The relative importance of the latter reaction depends, of course, on the relative abundance of OH^-.

NO^+ is another ion that can persist in excess concentration in the burnt gas of $C_2H_2 - O_2 - N_2$ flames; in some flames it may even be the principal positive flame ion (see Hayhurst and Kittelson 1978). It can be formed by charge transfer from primary ions to NO in the combustion zone and by dissociative charge transfer

NATURAL FLAME IONIZATION   Sect. IV.7

from H_3O^+ according to

$$NO + H_3O^+ \rightarrow NO^+ + H_2O + H \ . \qquad \text{(viii)}$$

The decay of the excess concentration of this ion downstream from the combustion zone occurs through the dissociative recombination reaction (see Hayhurst and Kittelson 1978)

$$NO^+ + e^- \rightarrow N + O \ . \qquad \text{(ix)}$$

In flames where reaction (vii) with participation of an electron is dominant, the rate of ion decay can be described by a second-order law with a recombination rate constant k_r that is independent of pressure and gas composition (see Calcote, Kurzius and Miller 1965, Kelly and Padley 1970, and Hayhurst and Telford 1974). Values of k_r between $0.5 \times$ and 5.0×10^{-7} cm^3 molecule^{-1} s^{-1} have been measured under different flame conditions at temperatures $T \simeq 2000-2500$ K and by different methods (see for a survey Kelly and Padley 1970, and Hayhurst and Telford 1974). Little agreement exists on the T-dependence of k_r. Measurements with a rotating single probe by Kelly and Padley (1970) in $(H_2 + C_2H_2) - O_2 - N_2$ flames yielded $k_r = (4.0 \pm 1.0) \times 10^{-7}$ cm^3 molecule^{-1} s^{-1} at $T = 2000$ K and at 1 atm pressure, and a temperature dependence $\propto T^{-1.6}$. Hayhurst and Telford (1974) and Hayhurst (1974) using an ion mass spectrometer reported $k_r = (4.1 \pm 1.0) \times 10^{-7}$ cm^3 molecule^{-1} s^{-1} at $T = 2250$ K with hardly any or at most a slightly positive temperature-dependence.

Porter (1970) observed a moderate third-partner effect on k_r by CO, N_2 and Ar, and a strong effect by H_2 in his low-pressure hydrocarbon flames. It is not certain, however, whether all his measurements relate to reaction (vii).

The rate constant of dissociative recombination of NO^+ by reaction (ix) is of the same order of magnitude as that of reaction (vii) for H_3O^+ (see Hayhurst and Kittelson 1978).

Negative ions are also formed in the combustion zone of hydrocarbon flames. C_2^- is believed to be the primary species and is formed by dissociative attachment of an electron to $C_2H_2O^*$ (see Calcote, Kurzius and Miller 1965). Species like OH^- and O^- may also be formed by simple or by dissociative attachment of an electron to OH, H_2O, O or O_2. Unlike the C_2^- ions, the OH^- ions, having a relatively high electron affinity of 1.8 eV, may persist away from the combustion zone. The interpretation of the experimental results obtained by different methods and in different flame conditions is not always unambiguous. In any case natural negative ions are believed to play only a minor role in the interzonal region of flames at 1 atm pressure. For a further discussion we refer to Hayhurst and Kittelson (1978).

Excessive ionization in the combustion zone may also be produced through

Chapt. IV TYPES AND PROPERTIES OF NONSEEDED FLAMES

collisional ionization of neutral molecules by impact of '*hot*' *electrons* having a suprathermal kinetic-energy distribution. Electrons, initially formed by some kind of physical collision, may acquire on the average extra energy when they make collisions of the second kind with vibrationally highly excited radicals or molecules under suprathermal conditions (see Cozens and von Engel 1964, von Engel and Cozens 1965, von Engel 1967, and Bell and Bradley 1970). Free electrons with excess kinetic energy could also be directly formed in a nonequilibrated, exo-ergic chemi-ionization process such as reaction (i) with CH^*. The importance of such energetic electrons as primary agent in the flame ionization has been stressed by von Engel and collaborators, and has given rise to considerable discussion (see also Lawton and Weinberg 1969). Electrons having energies slightly above the ionization energy are generally known to be very efficient at ionizing an atom or molecule upon collision. The central problem is whether the electron temperature T_e, describing the kinetic-energy distribution of the free electrons, can exceed the flame temperature T_f and if it does, to what extent. No definite and generally valid answer seems to have been provided so far in the literature.

The theoretical feasibility of T_e exceeding T_f as a result of collisions of electrons with suprathermally excited CO_2 molecules, etc., has been proved by Bell and Bradley (1970). The uncertainty about the real occurrence of elevated electron temperatures in and outside the combustion zone is connected with the possibility of systematic errors in the interpretation of the measurements. Inadequate spatial resolution may also cause uncertainties. The use of double, symmetrical electrostatic probes with not too large a diameter is recommended by Bradley and Ibrahim (1975). These authors have critically examined the reported electron temperature values that were measured with double probes in low-pressure hydrocarbon flames. They conclude that the true maximum elevation of T_e above T_f is in the range of only a few hundred K for N_2-diluted hydrocarbon flames with pressures between 20 and 100 Torr and temperatures up to 2000 K. Elevations over 1000 K must be considered as improbably large for this class of flames. The time needed for the relaxation of T_e towards T_f, if the cause of elevation of T_e were suddenly removed, was estimated by Bradley and Sheppard (1970) to be very short $(\approx 10^{-8}\,s)$ at pressures above 100 Torr

A few examples of literature reports on elevated electron temperatures will be given. Taran and Tverdokhlebov (1966) and Taran, Nesterko and Tsikora (1973) using a double-probe technique measured a maximum elevation of several hundred K in the combustion zone of a low-pressure C_2H_2-air flame; downstream from this zone T_e was found to equal T_f. Porter (1970) using a similar technique found elevations of at most about 200 K in the combustion zone of C_2H_2 flames at 18 Torr. Upstream from this zone, however, the electron temperature did not

FLAME BACKGROUND SPECTRUM
Sect. IV.8

decrease nearly as rapidly as the gas temperature, probably owing to the diffusion of vibrationally 'hot' molecules to the cooler, unburnt gas mixture. Attard (1966) determined T_e by measuring the electrical Johnson noise (cf. Sect. IX.4a) in an $C_2H_2-O_2$ torch flame at 1 atm. Assuming that the electrical conductivity stems mainly from the mobility of the free electrons, he identified the noise temperature with the electron temperature. Electron temperatures more than 1000 K in excess of the flame temperature of about 2500 K were found in the combustion zone; this outcome was in agreement with double-probe measurements by Cozens (1965) and von Engel and Cozens (1965) on the same flame. The latter measurements also showed that $T_e \simeq T_f$ in the interzonal region.

In the discussion about the origins of the excessive natural ionization in flames (chemi-ionization and/or 'hot' electrons), it is relevant to note that excessive ionization is also found in flames that do not show an elevated electron temperature. Moreover, an enhancement of electron temperature brought about by application of a strong d.c. field did not result in a rise in ion production. Lawton and Weinberg (1969) conclude from this that more than one important ionizing mechanism may be active in different flames.

Efforts have been made to compute the state of ionization in and outside the combustion zone by a mathematical model. So Ay, Ong and Sichel (1975) introduced a simplified mathematical model for calculating the spatial profile of the ion concentration in a one-dimensional flame, including ambipolar diffusion (see Sect. X.4b). Positive ions were supposed to be produced by reactions (i) and (vi), whereas negative ions and 'hot' electrons were disregarded; ion recombination was assumed to occur through reaction (vii). Using some additional empirical relationships, these authors obtained reasonable agreement between calculated and experimental ion profiles in low-pressure as well as atmospheric-pressure flames.

8. FLAME BACKGROUND SPECTRUM

General. The emission of flames without metal additives may have different origins, depending on the kind of flame, the fuel/oxidant ratio, and the place in the flame where the radiation is produced. In most flames there is a clear distinction between the radiation from the primary combustion (or inner) zone, the secondary combustion (or outer) zone, and the interzonal (or central) region. Hydrocarbon flames show most clearly the various possible types of background emission. Some strong bands of, e.g., OH, O_2 and C_2 may also appear in absorption (see Laidler 1955, Fiorino, Kniseley and Fassel 1968, and Mossholder, Fassel and Kniseley 1973). Fluorescence of OH radicals has been observed by Hooymayers and Alkemade (1967) in hydrogen flames with a Bi (3067.72 Å) hollow-cathode lamp as primary light source;

fluorescence of CH radicals has been observed near 4315 Å in an $C_2H_2-O_2$ flame by Barnes et al. (1973) and by Bonczyk and Shirley (1979); fluorescence of C_2 radicals has been observed in C_2H_2 flames by Becker, Haaks and Tatarczyk (1974) and by Blackburn, Mermet and Winefordner (1978), both groups using tunable dye-lasers, and fluorescence of PO, NH, CN and OH radicals in an C_2H_2-air flame by Fowler and Winefordner (1977) using a xenon arc lamp. Laser-induced fluorescence from C_2 radicals in flames has been observed by Baronavski and McDonald (1977a) too [†].

The blue-green colour of the *inner zone* is due mainly to the well-known emission bands of C_2 ($A^3\Pi_g \rightarrow X^3\Pi_u$, 4350 − 5800 Å) and of CH ($A^2\Delta \rightarrow X^2\Pi$, 3900 − 4350 Å). In the near-ultraviolet part of the spectrum the hydrocarbon flame bands (CHO or Vaidya bands, 3200 − 3900 Å) and OH bands ($A^2\Sigma \rightarrow X^2\Pi$, 2600 − 3200 Å) are found. In the far ultraviolet, bands of CO ($A^1\Pi \rightarrow X^1\Sigma^+$, 1200 − 2000 Å) have been observed (see Alder, Thompson and West 1970).

The nature of the flame emission in the *central region* depends on the fuel/oxidant ratio. When a very fuel-rich mixture is burned, a bright luminous flame is obtained. The spectrum of this flame is chiefly continuous and stems from glowing carbon particles. This nonspecific emission will not be discussed here.

When the fuel/oxidant ratio is lowered somewhat, the spectrum of the central region shows the same bands as that of the inner zone, although in the latter case the intensity of the bands is much higher.

When sufficient oxygen is available for a (nearly) complete combustion, the central region emits a spectrum that appears as quasi-continuum radiation when a spectral analyzer with moderate resolving power is used. Some distinct bands (OH and the Schumann-Runge bands: O_2 ($B^3\Sigma_u^- \rightarrow X^3\Sigma_g^-$, 2300 − 4000 Å) may be superposed on this quasi-continuum. Strong infrared bands of CO_2 at 4.4 μm and H_2O at 2.7 μm may appear in emission.

For a detailed survey of the discrete spectrum of the flame background emission and its origin, the reader is referred to the literature, for example, Mavrodineanu and Boiteux (1965), Gaydon and Wolfhard (1979), Alkemade and Zeegers (1971), and Gaydon (1948, 1974). Spectra of background emissions have been published by Naudeix (1971) for CH_4, city-gas and commercial-propane flames [*] burning with air; by Vickers et al. (1972) for the H_2-O_2-Ar and H_2-N_2O flames; by Sheinson and Williams (1973) for acetaldehyde and n-butane flames. Here we restrict ourselves to the continuum and quasi-continuum radiation, as most metal-vapour studies are done in the central flame region at such fuel/oxidant ratios that the discrete spectra from radicals like C_2, CH or CHO are here absent.

[†] For a recent study of C_2, CH, OH, CN in C_2H_2 flames see Fujiwara et al. (1979).
[*] City-gas: CH_4 30%, H_2 40%. Commercial-propane: C_3H_8 65%, C_3H_6 30%.

FLAME BACKGROUND SPECTRUM   Sect. IV.8

Apart from the discrete OH bands between 2600 and 3200 Å, the flame background
spectrum then consists mainly of a (quasi-)continuum.

In analytical flame spectroscopy, the flame background may often deteriorate
the limit of detection as the background intensity is, of course, not strictly
constant but may show fluctuations superposed on a slow drift. Moreover, its
intensity may be influenced by the presence of concomitants in the sample solution. For further information we refer to the textbooks cited in Chapt. I.

(Quasi-)continuum background spectrum. The (quasi-)continuum background
spectrum in flames may consist either of vibrational bands with unresolved, narrowly
spaced, rotational lines or of really continuum (recombination) radiation. When
radicals recombine, the stabilization of the association complex takes place mainly
by collisions with a third body, but in some cases stabilization by radiation also
is possible. The influence of the various ways of stabilization on the nature of
the emitted radiation was discussed in Sect. II.2e.

There are reported in the literature a few reactions that give rise to a
real continuum. Padley (1960) has observed bluish background emission in a
$H_2-O_2-N_2$ flame. He has ascribed this emission to the reaction

$$H + OH \rightarrow H_2O + h\nu \ . \qquad (i)$$

The emission intensity should then be proportional to the product [H][OH] of H-
and OH radical concentrations. However, it is very doubtful whether this reaction
really explains the radiation observed, because neither Dean and Stubblefield (1961),
working with H_2-O_2 flames, nor Zeegers (1966), working with H_2-air flames, have
been able to detect any background emission, whereas the value of [H][OH] in their
flames was at least 20% higher than [H][OH] in Padley's flame. From measurements
of the absolute intensity of background emissions Zeegers has concluded that the
spectral intensity of the background emission, if present at all in his H_2-air
flame (T = 2000 K), must be lower by a factor of 10^3 than the spectral emission
intensity in a shielded stoichiometric C_2H_2-air flame (T = 2300 K) at $\lambda \approx 4300$ Å.

Hydrogen flames containing halogen in appreciable amounts may show continua
in the visible part of the spectrum. For hydrogen-rich flames these continua have
been ascribed by Phillips and Sugden (1961) to the reaction

$$H + Z \rightarrow HZ + h\nu \ , \qquad (ii)$$

where Z stands for chlorine, bromine, or iodine. Addition of chlorine to hydrogen-
lean flames may give rise to a continuum according to

$$Cl(^2P_{\frac{1}{2}}) + Cl(^2P_{\frac{3}{2}}) \rightarrow Cl_2(^1\Sigma) + h\nu \ . \qquad (iii)$$

Chapt. IV   TYPES AND PROPERTIES OF NONSEEDED FLAMES

Hydrocarbon- and carbon monoxide-oxygen or -air flames show a quasi-continuum emission that extends throughout the visible till about 2100 Å, being strongest in the region 4500−4000 Å. Looking through the relevant literature, Alkemade and Zeegers (1971) have reached the conclusion that the quasi-continuum emission stems from excited CO_2 molecules, the formation of which has two origins: (i) association of CO and O, and (ii) collisions of ground-state CO_2 molecules with flame gas components.

According to Clyne and Thrush (1963) the chemiluminescent reaction proceeds through the 3-body associative excitation step: $CO + O + X \rightarrow CO_2(^3B_2) + X$ (see process no. 22 in Table II.1), followed by a radiationless, collisional transition to a higher lying excited state $CO_2(^1B_2)$. Radiative transitions from the latter state to the electronic ground state $CO_2(^1\Sigma_g^+)$ produces the chemiluminescence spectrum. In a recent study by Gaillard-Cusin and James (1977) the above reaction steps in the chemiluminscence of CO_2 has still been accepted. They have also found that a part of the chemiluminescence yield in the reaction zone of a $CO-O_2$ combustion at 4260 Å and at temperatures between 548 and 620 °C has to be ascribed to $O_2(^3\Sigma_u^+)$. Excited oxygen is supposed to be formed in the reaction $O + O + M \rightarrow O_2(^3\Sigma_u^+) + M$.

Measuring the absolute emission intensity of C_2H_2- and CO-air flames as a function of temperature and wavelength, Zeegers and Alkemade (1965b) and Zeegers (1966) were able to discriminate between the contributions of chemi-excitation by recombination of CO and O, and of collisional excitation of CO_2. The results of this work can be expressed empirically by

$$I_\lambda = a_\lambda [CO][O] + b_\lambda [CO_2] \exp[-hc/\lambda kT] , \qquad (IV.15)$$

where I_λ is spectral intensity in quanta per cm³, per second and per Å; $[CO_2]$ is the concentration of ground-state molecules. Penner et al. (1978) have used a similar approach to describe the equilibrium as well as non-equilibrium radiation observed in shock-tube studies of CH_4 oxidation. The coefficient a_λ includes the rate constant for the chemiluminescent reaction between CO and O. The product $b_\lambda \exp[-hc/\lambda kT]$ includes the rate constant for thermal collisional excitation of CO_2 from the ground-state. Both coefficients further include the fractional probability that de-activation of the excited state takes place by radiation in the spectral interval considered. From Eq. IV.15 and the values for a_λ and b_λ, presented graphically in Fig. IV.12, the quasi-continuum background emission of both C_2H_2- and CO flames can be evaluated within a factor of two (see Zeegers 1966).

On the basis of these results Zeegers (1966) concluded that collisional excitation may contribute significantly to the total background emission when chemical equilibrium is approached, at T equal to about 2100 K.

FLAME BACKGROUND SPECTRUM   Sect. IV.8

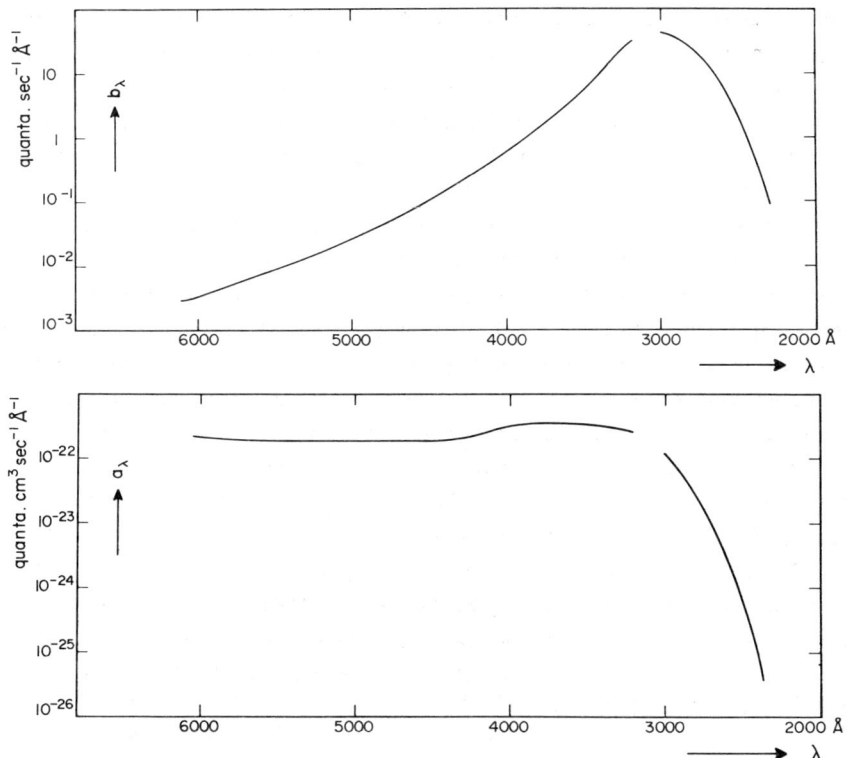

Fig. IV.12 The coefficients a_λ and b_λ occurring in Eq. IV.15 for the spectral intensity of the background radiation in C_2H_2- and CO flames are plotted as a function of λ. Near 3050 Å discrete OH bands appear and therefore no values of the coefficients could be determined here. (From Zeegers and Alkemade 1965b.)

The conclusion of Clyne and Thrush (1962) and of Kaskan (1959) that the intensity of the background emission is strictly linearly dependent on [CO][O] is, in general, not true. It will be true when [CO][O] exceeds largely the equilibrium product. It happens also to be true at chemical equilibrium at constant temperature, since the equilibrium ratio $[CO]_e [O]_e / [CO_2]_e$ is then constant.

Many flames in which oxides of nitrogen are present show a yellow-green quasi-continuum emission. This emission, which appears also when a small amount of nitric or nitrous oxide is added to hydrogen flames, is characterized by a short-

469

wavelength cut-off at 3970 Å (see Broida, Schiff and Sugden 1961), an intensity maximum at about 6500 Å (see Levitt 1965), and a gradual decrease toward longer wavelengths. The origin of this emission is undoubtedly a radiative transition between an excited NO_2 state and its ground-state. Two excited states seem to be involved: (i) an excited state that is formed in the collisions between NO and O, and (ii) a slightly lower-lying state from which the emission occurs. The crossing between these two states is supposed to occur rapidly, and it is, therefore, not rate-determining (see Clyne and Thrush 1962). Population of the radiative state may also occur by collisions of NO_2 with flame molecules (see Levitt 1965). This process is unimportant in equilibrated milieus at low temperatures, but it may contribute significantly at higher temperatures ($T > 2000$ K). Carrington and Polanyi (1972) have also come to the conclusion that two excited states of NO_2 are involved. However, in contrast to Levitt (1965), they have stated[†] that the emission intensity is proportional to [NO][O] under all conditions.

9. FLUCTUATIONS IN THE FLAME EMISSION

General. When the flame emission intensity is being measured, the meter usually shows irregular fluctuations around a certain mean value. (Even when the meter seems to stand still, fluctuations do nevertheless occur, but they are just too small to be perceived.) Fluctuations in the zeropoint (arising also with a closed photodetector, e.g., from dark-current noise) and in the signal itself may both make the meter indication unsteady. The fluctuations in the emission signal may be due to the following effects:

(i) the unavoidable shot effect in the photo-electron emission causing shot noise in the photocurrent (see Sect. III.5c);

(ii) random variations in the flame emission intensity caused, e.g., by fluctuations in the temperature, gas composition or metal supply, and by flame flicker;[*] this multiplicative noise component (see Sect. III.5d sub point 2) may, in turn, be induced by fluctuations in the flow of oxidant and fuel gases, by turbulence, draught and the statistical fluctuations in the number of aerosol particles supplied (called: shot effect of the mist droplets; see Alkemade 1954); it is characteristic for the flame;

[†] In their book Carrington and Polanyi (1972) say the emission intensity is proportional to [N][O]. This is obviously a misprint (compare their reaction 5.10).

[*] In the literature this kind of noise is sometimes called 'fluctuation noise' (as distinct from shot noise). We shall not use this term but prefer the expression: excess low-frequency noise (see below).

FLUCTUATIONS IN THE FLAME EMISSION   Sect. IV.9

(iii) instabilities in the optical system and random variations in the sensitivity of the photodetector and the amplifier + readout system.

In the following we shall disregard the fluctuations of type (iii) and focus our attention on the fluctuations in the flame emission (ii), in particular, in comparison with the shot noise of the photocurrent (i). Fluctuations in the flame conditions mentioned sub point (ii) will, of course, also affect the atomic absorption or atomic fluorescence and scattering signals when a metal vapour is introduced. However, there have been only a few experimental investigations of these effects and we shall not consider them here.

The emission of a flame fed by metal vapour consists of the flame background emission and the emission from the metal vapour proper, called simply: metal emission. The fluctuations in both emission components have many features in common and will therefore be treated together in this section. When the flame background fluctuations are dominant, they limit the *detectability* of weak metal emissions (see Sect. III.14f). When the fluctuations in the metal emission itself are dominant they limit the (*relative*) *precision* of the measurement. Fluctuations in both emission components play an important role in the evaluation of the analytical performance of a flame emission spectrometer. Fluctuations in the flame background or thermal metal emission may also limit the detectability and precision of analytical measurements in atomic absorption and fluorescence spectroscopy. In particular, the optimum choice of modulation frequency in a.c. measurements depends here on the noise spectrum of these emission fluctuations (see also Sects III.15 and 16). For similar reasons the emission fluctuations are also of interest in experimental absorption and fluorescence studies of metal vapours, for example, in the precise measurement of the efficiency of fluorescence (see Sect. III.16b).

Excess low-frequency noise and shot noise. There are some characteristic differences between the noise effects of type (ii) and the shot noise.

1.   Shot noise is imposed by the laws of statistics; its magnitude is determined by the number of independent elemental events occurring per unit of time (number of photo-electrons released per second; see Eqs III.23, 28 and 29). Noise of type (ii) is of a technical rather than a fundamental nature; its magnitude can be reduced, in principle, without limitation, e.g., by providing a more stable flame or a finer aerosol, or by excluding draught from the surrounding air.

2.   Shot noise has a 'white' spectrum; it is present equally strongly in the whole frequency range from zero up to 'infinity' (see Sect. III.5c and Eq. III.29). The spectrum of type (ii) noise increases, roughly speaking, with decreasing frequency and falls off to zero with increasing frequency (see, e.g., Fig. IV.13 below);

Chapt. IV   TYPES AND PROPERTIES OF NONSEEDED FLAMES

it usually dominates the shot noise in the low-frequency range and is therefore called: *excess low-frequency noise* or *e.l.f. noise* (see also Sect. III.14e).[†]

3. Because of its purely statistical character the root-mean-square (RMS) value of the shot noise increases proportionally to the square root of the mean photocurrent $\bar{i}_p$ (see Eq. III.28); the RMS value of the e.l.f. noise increases proportionally to $\bar{i}_p$ (see Eqs III.35 and 37).[*] In other words, the <u>fractional</u> RMS value of the shot noise decreases $\propto 1/\sqrt{\bar{i}_p}$ with increasing $\bar{i}_p$, whereas the <u>fractional</u> e.l.f. noise is independent of $\bar{i}_p$.

4. There may be a *statistical correlation* between the e.l.f. fluctuations in the flame emission arising at different moments or from different places and even between different components of the flame emission (background at different wavelengths and emission from different metals). Such correlations are essentially absent in the shot-noise component. The time-correlation of the e.l.f. noise is described by the autocorrelation function, which is directly related to the noise spectrum and which underlies the SNR calculations in Appendix A.7. When the flame-background noise and the metal-emission noise stem from a common noise source, e.g., temperature fluctuations, a (partial) statistical correlation between them is likely. As a consequence the RMS value of their combined noise effect will then differ from the square addition: $\sqrt{(RMS)_1^2 + (RMS)_2^2}$ of their separate RMS values numbered 1 and 2. In the case of a full, positive correlation we have

$$(RMS)_{comb} = (RMS)_1 + (RMS)_2 ; \qquad (IV.16)$$

in the case of a full, negative correlation we have

$$(RMS)_{comb} = |(RMS)_1 - (RMS)_2| . \qquad (IV.17)$$

Figure-of-merit. The 'quietness' of a given flame emission can best be characterized by the *fractional* RMS value of the associated e.l.f. noise, that is, by the RMS value divided by the mean value, both expressed in the same units.

[†] When we identify here and in the following, for didactic reasons, the noise spectrum of the flame emission with that of the e.l.f. noise in the photocurrent, it is tacitly assumed that no a.c. modulation is applied (modulation would shift the spectral noise components over a fixed frequency interval equal to f_{mod}).

[*] The significance of this statement should be well understood: the RMS e.l.f. noise increases <u>strictly</u> $\propto \bar{i}_p$ if we vary $\bar{i}_p$, under fixed flame conditions, by placing an attenuator in the light path or by changing the sensitivity of the detector. However, when we vary $\bar{i}_p$ by changing the metal solution concentration, the observation volume (see also sub point 4), the height of observation, etc., the RMS e.l.f. noise need not necessarily vary $\propto \bar{i}_p$. For example, the fractional fluctuations in the Na emission due to fluctuations in the salt transport are less at high concentrations, where self-absorption is strong, than at low concentrations.

FLUCTUATIONS IN THE FLAME EMISSION   Sect. IV.9

This ratio is independent of the magnitude of the photocurrent induced by the considered flame emission (see point 3 above). In order to make this ratio characteristic for the flame emission alone, i.e. independent of the detection system, we can standardize it by choosing $\Delta f = 1\,\text{Hz}$ and $f_0 = 1\,\text{Hz}$ in Eq. III.32. We then obtain from the latter equation for the standardized fractional e.l.f. noise the expression: $\sqrt{W_{\text{elf}}(1)}/\bar{i}_p$; here $W_{\text{elf}}(1)$ is the spectral noise power of the e.l.f. noise in i_p at frequency $f_0 = 1\,\text{Hz}$ (in the absence of a.c. modulation; see also footnote † on page 472). This expression is a rough estimate of the fractional noise due to fluctuations in the flame emission, as observed on a meter with a response time of 1 s, when we look at the meter for no longer than, say, 10 s.† The same fractional noise would be observed on the meter when we modulate the flame emission by a chopper and apply a.c. detection (see Sect. III.14f-4). When, however, in absorption or fluorescence measurements the background light source but not the flame emission is modulated at frequency f_{mod} and again a.c. detection is applied, the value $W_{\text{elf}}(f_{\text{mod}})$ should be inserted in the above expression. We may consider the reciprocal standardized fractional e.l.f. noise as a *figure-of-merit* for the characteristic flame emission noise in the frequency band under consideration (see Alkemade et al. 1979).

The reciprocal ratio of e.l.f. noise to shot noise is not a good figure-of-merit for the flame emission noise because this ratio depends on the quantum efficiency of the detector and on the attenuation of the flame radiation in the optical system. This is a consequence of the different ways in which e.l.f. noise and shot noise depend on $\bar{i}_p$ (see point 3 above). One can always make the shot noise component arbitrarily large compared to the e.l.f. noise component, for fixed flame conditions, by attenuating the radiant flux received by the detector or by lowering its quantum efficiency.

For a similar reason the *turnover frequency* f_t above which the spectral power of the shot noise becomes dominant over that of the e.l.f. noise (cf. also point 2 above) is not a good figure-of-merit either. For fixed flame conditions f_t can be made arbitrarily small simply by attenuating the radiant flux, etc.

When one compares flame spectrometers, any statements about the ratio of e.l.f. noise to shot noise, or about the turnover frequency, only make sense if the primary photocurrent emitted at the photocathode is specified, too.

---

† It is necessary to limit the duration of observation when the spectral noise power rises indefinitely for $f_0 \to 0$, as in the case of flicker noise (see also footnote † on page 390).

Chapt. IV   TYPES AND PROPERTIES OF NONSEEDED FLAMES

Because of the possible occurrence of statistical correlation effects in the e.l.f. noise (see point 4 above) one should judiciously specify to which part of the flame emission (both spectrally and spatially) the measured figure-of-merit relates. For example, when the observed homogeneous flame volume is halved, the mean photocurrent will be halved, too, if we disregard self-absorption. If the e.l.f. noises in the radiation from each half are fully correlated, the RMS value of the e.l.f. noise will also be halved if the observed flame volume is halved (see Eq. IV.16). The figure-of-merit then remains the same. However, if statistical correlation does not exist, the RMS value of the noise will be reduced by only a factor $\sqrt{2}$ upon halving the flame volume observed. The figure-of-merit will then be worsened by the same factor.

The existence or nonexistence of statistical correlations between the e.l.f. fluctuations in the Na emission from two distinct parts of a laminar C_3H_8-air flame with a Méker burner was investigated by Alkemade (1954, p.80; see also Alkemade et al. 1979) in the following manner. A diaphragm with two vertical slits 7.5 mm apart was placed in front of the flame. A synchronous meter (a.c. galvanometer with an indication period of 1 s; see Sect. III.6b) in combination with a light chopper was used to measure the RMS value of the noise components around 1 Hz in the radiations passing through one or the other slit and also in the radiation passing through both slits together. The contribution from shot noise was negligible. When the slits were positioned at equal height above the burner, the RMS value with both slits open appeared to equal, within 5%, the sum of the RMS values obtained with one or the other slit open. The RMS values found with each slit separately were made about equal to improve the statistical significance of this check. Considering Eq. IV.16 we must conclude that (nearly) full correlation exists. If there were no correlation, the RMS value with both slits open would have been about a factor $\sqrt{2}$ smaller than was found.

When in the above experiment the chopper was removed, the RMS value with both slits open appeared to equal, within 10%, the square addition of the two separate RMS values. Since the synchronous meter used was tuned at 50 Hz, there appeared therefore to be no correlation between the e.l.f. noise components around 50 Hz.

From the outcome of the above experiments we expect that the fractional e.l.f. noise or figure-of-merit of the flame noise will <u>not</u> change upon widening the observed flame part if we consider noise components around 1 Hz. It does <u>improve</u>, however, at frequencies around 50 Hz. The latter conclusion may be important in the application of fluorescence measurements with a modulated light source.

FLUCTUATIONS IN THE FLAME EMISSION   Sect. IV.9

Experiments. Noise spectra can be measured in different ways (see the literature cited in Sect. III.5c). At (very) low frequencies Fourier and autocorrelation techniques are appropriate. The Fourier transform of a digitized record of the noise signal in a restricted time domain is determined with the aid of a computer. From the Fourier transform an estimate of the noise spectrum is obtained simply. Distortions of the noise spectrum caused by the truncation and discrete sampling of the noise signal in the time domain should be attended to. The noise spectrum can also be determined indirectly by Fourier transformation of the measured autocorrelation function through the Wiener-Khintchine theorem (see Appendix A.7-1). Application of the 'Fast-Fourier-Transform' algorithm makes direct Fourier analysis more attractive in practice.

In the frequency range above about 3 Hz there are simple analog techniques available. They can be realized by using a series of fixed, frequency-selective filters and an a.c. power meter, or by using a lock-in meter with a continuously tunable response frequency which is set by a sine-wave generator producing the reference signal (see Sect. III.6b). The noise bandwidth, which varies with midfrequency of the filters, must be known in order to derive the spectral noise power, that is, the noise power per unit frequency bandwidth. However, if we want to determine only the ratio of the spectral noise power to that of pure shot noise, neither the noise bandwidth nor the amplifier characteristic, etc., need be known. We have then only to divide the flame noise power by the shot noise power measured with the aid of a stable incandescent light source at the same frequency and photocurrent. We can easily verify the occurrence of pure shot noise with the latter reference light source by checking the square-root dependence of the RMS noise on photocurrent (see point 2 above). If one determines the difference between the investigated noise and the shot noise, the secondary-emission noise of the photomultiplier drops out. The shot-noise spectrum can be calculated from theory (see Eq. III.29) when the primary photocurrent and the amplification factor are known. This procedure provides a simple possibility of determining a noise spectrum in absolute units.

Noise spectra have been reported for the emission of premixed and unpremixed, sheathed and unsheathed flames with and without metal vapour[†] (see Belyaev *et al.* 1968, Marinkovic and Vickers 1970, Mossotti and Abercrombie 1971, Alkemade *et al.* 1972, 1979, Hieftje and Bystroff 1975, and Talmi, Crosmun and Larson 1976). The total frequency range covered by these experiments together extends from about

---

[†] Noise spectra for the emission of other light sources including the hollow-cathode lamp, arc and tungsten lamp, have been reported by Belyaev *et al.* (1968), Lebedev and Dolidze (1969), Dolidze and Lebedev (1973), and Talmi, Crosmun and Larson (1976), and Bower and Ingle (1979).

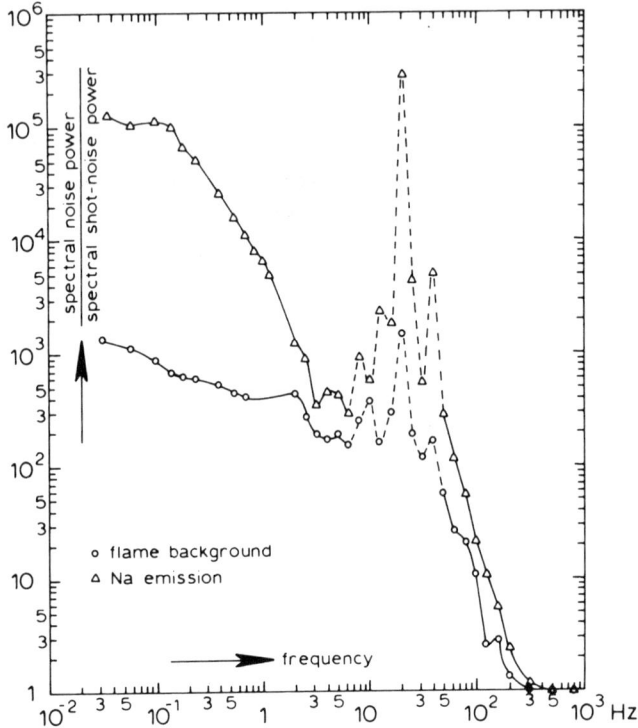

Fig. IV.13 The spectral noise power in the photocurrent of a photomultiplier tube divided by the spectral shot-noise power is shown as a function of frequency for the background and the Na-D emission in a premixed, laminar and sheathed C_2H_2-air flame at 5 cm above the Méker-burner top. The anode current was 2.8 and 68 nA, respectively; self-absorption was absent; an interference filter was used. Above 1 Hz a noise analyzer with fixed frequency-filters was used with constant frequency-resolution $\Delta f/f \simeq 0.2$. The precise position of the strong noise peaks was found to be 22 and 44 Hz when measured with a tunable lock-in analyzer having a better frequency-resolution. Below 1 Hz the noise spectrum was obtained by Fourier transformation of the digitized noise signal. (From Alkemade et al. 1979.)

FLUCTUATIONS IN THE FLAME EMISSION   Sect. IV.9

10^{-3} to 10^5 Hz. The noise spectra found exhibit generally an e.l.f. component rising more or less sharply with decreasing frequency and merging with the shot noise beyond a certain turnover frequency (depending on the measuring conditions; see above).[†] In addition, distinct peaks have been found in some noise spectra between roughly 10 and 100 Hz; they are probably associated with the presence of vortices or eddies in the boundary layers of the flame gases (see Alkemade *et al.* 1972, 1979, and Hieftje and Bystroff 1975). The disappearance of these peaks when the flame was shielded by a quartz tube supports this explanation (see Hieftje and Bystroff 1975). Acoustical resonances in the exit ports of the burner may bring about other noise peaks in the kilohertz range (see Talmi, Crosmun and Larson 1976).

An example of a noise spectrum measured over an extended frequency range with a premixed, laminar and sheathed C_2H_2-air flame, with and without Na vapour, is shown in Fig. IV.13.

[†] The appearance of a minimum in the excess flame noise spectrum as reported previously by Alkemade *et al.* (1972) has proved to be erroneous (see Alkemade, Hollander, Snippe and Zijlstra 1981).

CHAPTER V

Determination of Metal Concentration in the Flame

1. INTRODUCTION AND DEFINITIONS

The determination of the absolute or relative metal concentration in a flame for a given metal concentration in the sprayed solution is often the keystone for the quantitative interpretation of flame experiments. It also gives a clue to the calculation of detection limits in flame spectrometry as applied to the chemical analysis of metal solutions. The actual concentration of an element in a given flame depends on many instrumental factors such as rate of liquid aspiration, height of observation, temperature of spray chamber. Therefore it is practically impossible to derive its value from data found by other authors for similar flames and nebulizers. It has to be determined for each instrument and in each situation separately, if reliable values are to be obtained. This chapter is devoted to a rather detailed discussion of methods to determine absolute metal concentrations and concentration ratios in flames. Also attention will be paid to sources of error arising in these methods, which are sometimes underestimated in the literature.

The concept of metal concentration (= quantity of element per unit of volume) is not unambiguous. Metals are usually introduced in the form of a fine spray containing a metal salt as solute. The solute, however, may not be fully converted into the gaseous state owing to incomplete evaporation of the solvent or to incomplete volatilization of the desolvated aerosol particles. A discussion on incomplete desolvation and volatilization is given in Sect. III.9.

Further, in the gaseous state the element may be present in different forms, for example, as monoxides, free atoms, or singly charged ions. We define therefore: the *total metal concentration* denoted by suffix t, as the concentration of metallic element that is present in any form (combined or free, neutral or ionized, excited or unexcited) in the gaseous state; the *specific metal concentration* as

Chapt. V   DETERMINATION OF METAL CONCENTRATION IN THE FLAME

the concentration of a certain species of the element. We shall distinguish between these species by adding a suffix a for free neutral atoms, i for free ions, and m for molecules. The species X may also be denoted by adding X as a suffix. The suffix will be omitted whenever the meaning of the symbol is clear from the context. Concentrations of atoms or molecules in a specific energy level j ($j = 0, 1, 2, \ldots$; $j = 0$ for the ground state) are occasionally denoted by adding the suffix j.

We prefer to express the metal concentration in the flame as a (*number*) *density*, n, i.e. number of entities per unit volume, rather than as *partial pressure* p (see definition in Sect. IV.4b). The former is better adapted to the kinetic interpretation of many of the experimental results in which we are interested. The actual partial pressure of metal vapour is always so low (see also Sect. III.2h) that it never really contributes to the thermodynamics of the flame system as a whole. The relation between the partial pressure p (in atm) and the number density n (in cm^{-3}) in a gas at absolute temperature T (in K) is given through the ideal-gas-law as follows: $p = (nT/7.34) \times 10^{-21}$.[†]

In general, for a certain metal concentration, c, in the aspirated solution, n_t or n may vary inside a cylindrically symmetric flame with radial distance from the flame axis as well as with height above the burner. Often the homogeneity in a horizontal slab is sufficiently good to warrant the use of a unique value for n at a given height of observation. In other cases scanning methods may be applied to determine the variation of n in a horizontal cross section (see Sect. 2[*]).

Spectroscopic methods to determine the absolute concentration of a metal species or the ratio of metal concentrations will be dealt with in Sects 2 and 3, respectively. Some special but unusual methods will be mentioned briefly in Sect. 4, whereas the calculation of n_t from measured properties of the nebulizer and flame will be treated in Sect. 5. Section 6 will be devoted to metals that are present in the flame only as free, neutral atoms (known as *fully atomized elements*). These may serve as 'standards' in the determination of the *fraction atomized* $\beta_a \equiv n_a/n_t$ for other elements by a spectroscopic comparison method.

Each of the optical methods of determining absolute or relative n_a values can also be applied — in principle — to determine absolute or relative values of the oscillator strength f or transition probability A, if the theoretical expressions involved contain $(n_a f)$ or $(n_a A)$ as a product.

[†] See also Appendix A.1. Tables converting n into p, and vice versa, for a number of temperatures have been calculated by Boumans (1966).
[*] Section numbers quoted without a roman cipher refer to the same chapter.

2. SPECTROMETRIC DETERMINATION OF ABSOLUTE METAL CONCENTRATION

2a THE ABSOLUTE-INTENSITY METHOD

Case of negligible self-absorption. The absolute radiance B of a single, isolated spectral line in absence of self-absorption is given by (see Eq. II.281)

$$B = (h\nu_0/4\pi)(g_1/Q_X) \exp[-E_1/kT] An l , \qquad (V.1)$$

where A is the optical transition probability per second, h is the Planck constant, ν_0 is the frequency of line centre, g_1 is the statistical weight of the upper level with an excitation energy E_1, Q_X is the partition function of the species X emitting the line, kT is the Boltzmann constant times absolute flame temperature and is expressed in the same units as E_1, n represents here the absolute concentration of X in the gaseous state, while l is the thickness of the flame section with metal vapour along the axis of observation. Both T and n are assumed to be uniform in the flame section observed. The radiance B is expressed in radiant flux per sterad and per unit of flame surface projected perpendicular to the axis of observation and refers to the nonpolarized radiation of the line integrated over its total spectral width. If the absolute values of B and l are determined and all other parameters are known, n can be calculated. In case X is an atom, $n \equiv n_a$ and represents the concentration of free atoms in the ground and excited states together. If $Q_X \simeq g_0$ (g_0 = statistical weight of the ground state), n is virtually equal to the concentration of atoms in the ground state, n_0. The latter condition is often fulfilled for metal atoms and ions in flames, except, for example, for the transition elements and rare-earth elements where low-lying levels are present and Q_X may vary by about 10% when the flame temperature is raised from 2000 to 3000 K (see Sect. II.3b-2).

When T equals the 'true' translational temperature of the flame (see Sect. III.10b), Eq. V.1 only holds if the population of the excited state is in thermal equilibrium with the flame gases. However, this equation will also hold under conditions of nonthermal radiation (see Sect. VI.4), if we substitute for T the value measured by the line-reversal method employing the same line (that is, the 'excitation temperature'; see Sect. III.10c). When, however, the reversal temperature measured falls appreciably below the true flame temperature owing to radiative disequilibrium (see Sect. VI.4c), errors may arise by this substitution since the underpopulation of the excited state then depends on n. It should be recalled that the reversal temperature is usually measured with high metal concentrations giving sufficient absorption in the flame, whereas the application of Eq. V.1 is restricted to low concentrations where self-absorption is (virtually) absent.

Application of Eq. V.1 for the determination of n does not involve the

Chapt. V DETERMINATION OF METAL CONCENTRATION IN THE FLAME

spectral shape or width of the atomic line. We only have to assume here that the spectral bandwidth of the monochromator used exceeds the total spectral width of the emission line. On the other hand it requires the measurement of l and the absolute calibration of the line radiance (see Sect. III.14g). However, knowledge of all properties and dimensions of the optical system and photodetector used is not needed, if this calibration is performed in an appropriate way by comparison with a standard lamp with known spectral radiance at the wavelength of the spectral line (see Sect. III.14g). Of course, the angular aperture of the optical system and the area of the observed flame surface element have to be sufficiently small to ensure that l has a well-defined value and that a possible vertical temperature- or concentration gradient does not seriously interfere (compare Sect. III.14b).

Inhomogeneity of temperature and concentration along the line of observation may become apparent by scanning the radiance of the spectral line in a direction perpendicular to the flame axis (see Fig. V.1). If the distribution of T and n is homogeneous, then the measured radiance divided by the (varying) thickness of the coloured flame slab observed should be independent of the off-axis distance x (self-absorption being absent). If not, then the distribution of T and/or n is inhomogeneous.

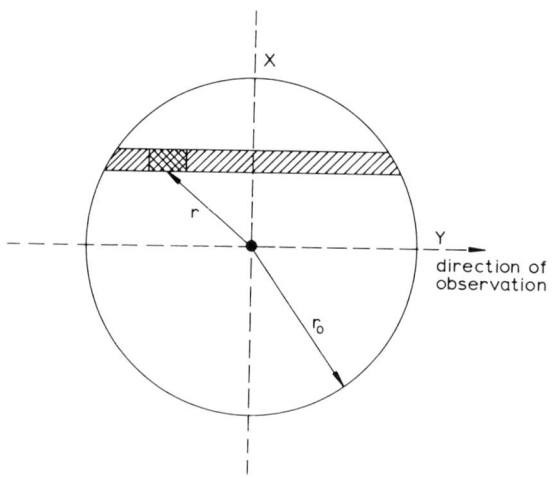

Fig. V.1 Horizontal cross section of cylindrical flame with vertical axis. Radius of flame column coloured by metal vapour is denoted by r_0. Hatched flame section is observed by the spectrometer. The distance of an arbitrary volume element within this section to the flame axis is denoted by r.

SPECTROMETRIC DETERMINATION OF ABSOLUTE METAL CONCENTRATION   Sect. V.2a

When the distribution of T in a horizontal cross section is uniform but $n(r)$ varies with distance r from the axis in a cylindrically symmetric flame (with radius r_0; see Fig. V.1), one can find $n(r)$ as follows. Consider $B(x)$ as a function of x by replacing in Eq. V.1 the product (nl) by

$$\int_0^{\sqrt{r_0^2 - x^2}} 2\, n(r)\, dy ,$$

with $r = \sqrt{(x^2 + y^2)}$. After substitution of r, as integration variable, for y one obtains from the modified Eq. V.1 *Abel's integral equation* (cf. Eq. III.55)

$$B(x) = \text{const. } 2 \int_x^{r_0} n(r)\, r (r^2 - x^2)^{-\frac{1}{2}}\, dr , \quad (V.2)$$

from which $n(r)$ can be solved if the function $B(x)$ is determined. The constant factor in Eq. V.2 comprises all factors of the right-hand side of Eq. V.1 except nl. The mathematical solution of $n(r)$ may be found by the same methods as discussed in Sect. III.10d sub point 4 in connection with the derivation of the radial temperature variation (see also Britske *et al.* 1973). The effect of the finite angular aperture of the optical system on the measurement of $n(r)$ has been discussed by Lapworth (1976). It was found that the angular aperture has no effect at all on the measurement of $n(0)$.

When the dependence of n on r is governed by a radial diffusion process, the function $n(r)$ can be found from $B(x)$ in a more straightforward way, as will be described in Chapt. X in connection with the determination of the diffusion constant.

When in a cylindrically symmetric flame both $T(r)$ and $n(r)$ vary with distance r as unknown functions, the measurement of $n(r)$ should preferably be made by one of the absorption methods discussed in the next sections. In these methods a variation of T hardly plays a role. However, if the function $T(r)$ is known, one can solve the product: $n(r) \exp[-E_1/kT(r)]$ as a function of r from the function $B(x)$ measured by the same inversion procedure as outlined in Sect. III.10d sub point 4. Then $n(r)$ is found by inserting the known function $T(r)$ in the exponent of the Boltzmann factor. It may be recalled that $T(r)$ can be derived by the two-line or the slope method independently of the possible variation of element concentration with distance r (see Sect. III.10d).

The use of a pencil-like laser beam tuned at an atomic resonance line has been suggested as a means for local sensing of physical parameters in a flame (see Omenetto 1975). When the population of the upper level of the line is saturated by the laser beam, the fluorescence observed at any point along the

Chapt. V   DETERMINATION OF METAL CONCENTRATION IN THE FLAME

beam is a direct measure of the atomic concentration at that point. By scanning the fluorescence measurements along the laser beam, the axial distribution can be found. This also applies to molecules (McDonald et al. 1979).

When the spectral line shows (*hyper-*)*fine-structure* and its spectral components are not resolved by the spectral apparatus, we should add up in Eq. V.1 the values of B for each component with appropriate g value (the other factors in this equation are in most cases virtually the same). Since in the case considered self-absorption is absent, the extent of possible intrinsic spectral overlap of these components does not really matter and the derivation of n from the total radiance is straightforward.

In principle, similar considerations apply for the determination of the absolute concentration of a *metal compound* from the radiance of its band spectrum (compare Eqs II.284 and 286). However, this presupposes that the term diagram of the molecular band and the appropriate values of g, Q_x, A, and E are known. The number of di-atomic metal compounds for which such data are available is still small but growing (see Smit 1966, and Kuznetsova et al. 1974). BaO is an example of a molecule with well-known transition probabilities, term diagram, etc. (see Johnson 1972, and Wentink and Spindler 1972).

The validity of Eq. V.1 is based on the *absence of self-absorption*, which holds good for nonresonance lines at practically all concentrations, but for resonance lines in the limit of low values of the product (nfl) only (f is the oscillator strength for absorption of the considered transition). A theoretical calculation from approximate data on line width and concentration suffices to check absence of self-absorption (see discussion of the curve-of-growth in Sect. II.5c-2).

Empirically, absence of (noticeable) self-absorption can be checked by measuring the absorption factor α with a background source emitting an identical resonance line. This can be realized by placing an auxiliary, identical flame, or, more simply, a mirror behind the flame being investigated. Alkemade (1954) has shown that for low α values ($\alpha < 20\%$) the relative loss in emission due to self-absorption is in fair approximation equal to $\frac{1}{2}\alpha$ for a-parameter values of order unity. To ensure absence of self-absorption one may also determine empirically the curve-of-growth and restrict the measurements to its initial, linear branch (compare Sect. 2d). Further, if the spectral apparatus allows resolution of the components of the resonance multiplet having different f-values, self-absorption is absent when their measured intensity ratios are equal to the corresponding f ratios (see also end of Sect. II.5c-3).

Case of strong self-absorption. When self-absorption is so strong that the square-root-law is applicable, the radiance of an atomic resonance line in a

SPECTROMETRIC DETERMINATION OF ABSOLUTE METAL CONCENTRATION   Sect. V.2a

homogeneous flame is given by

$$B = (h\nu_0^2/c)\ \exp[-E_1/kT]\left\{(g_1/Q_x)(\delta\nu_C+\delta\nu_N)/\pi\right\}^{\frac{1}{2}} (An\,l)^{\frac{1}{2}}\ ,\qquad (V.3)$$

which follows from Eq. II.299 when one substitutes there $\delta\lambda_L = (c/\nu_0^2)(\delta\nu_C + \delta\nu_N)$ and $\lambda_0 = c/\nu_0$ (compare Eqs II.208 and 243). In these equations $\delta\nu_C$ and $\delta\nu_N$ are the collisional and natural half-intensity widths of the line in Hz and all other symbols have the same meaning as in Eq. V.1. An absolute determination of the metal concentration from the measured radiance now requires the additional knowledge of $\delta\nu_C$; we may usually disregard $\delta\nu_N$ in comparison with $\delta\nu_C$ (see Sect. II.5b-2). The value of $\delta\nu_C$ depends on the line chosen, the composition and total pressure of the flame gas, and on T. Literature data on $\delta\nu_C$ are rather scarce, except for alkali and some alkaline-earth elements. The sum $\delta\nu_C + \delta\nu_N$ can also be calculated as $a\delta\nu_D/(\ln 2)^{\frac{1}{2}}$ (cf. Eq. II.246). The Doppler width $\delta\nu_D$ may easily be calculated from Eq. II.240 if T is known; the a-parameter must be found empirically or derived from literature data for the same line in a similar flame (see Tables VII.9 and 10 on pp.744 and 748). In the case of strong self-absorption, the uncertainty of the latter value may introduce an appreciable error into n, which is often difficult to estimate. We refer to the more detailed discussion in Sect. 2d on the uncertainty in the a-parameter. Usually one combines the determination of absolute metal concentration with that of the unknown a-parameter by one of the methods to be described in Sect. VII.3c.

It is assumed that in measuring B the bandwidth of the spectral apparatus is markedly larger than the spectral width of the line observed, which may be considerably broadened as a result of self-absorption (see Fig. II.14 on p.163).

In *inhomogeneous* flames where $n(r)$ varies with radial distance r, $n(r)$ can be solved from $B^2(x)$ by essentially the same procedure as outlined above for the case of negligible self-absorption. This can, however, be done only approximately, as Eq. V.3 may no longer be valid when x approaches the rim of the flame.

The occurrence of (*hyper-*)*fine-structure* of a self-absorbed line complicates the theoretical relation between B and n, especially in the case of partial intrinsic overlap of these components. We refer to Sect. 2d for a more detailed discussion.

In *conclusion*, the absolute-intensity method yields more reliable atomic metal concentrations if self-absorption can be ignored. In the presence of self-absorption the uncertainty in the a-parameter is likely to contribute markedly to the measurement error. When considerations of signal-to-noise ratio make it necessary to work at such high concentrations that self-absorption becomes noticeable, its effect can be corrected for by one of the methods discussed above for checking the absence of self-absorption. The absolute-intensity method is restricted by the

Chapt. V   DETERMINATION OF METAL CONCENTRATION IN THE FLAME

availability of standard lamps calibrated at the optical frequency of interest. Possible errors made in the absolute measurement of line radiance B are discussed in Sect. III.14g. A disadvantage of the absolute-intensity method as compared to absorption methods is that the flame temperature must be accurately known. For instance, an error of 25 K at $T \approx 2500$ K gives an error of about 10% in n for the Na-D doublet and even larger errors for lines with higher excitation energies. For some lines an uncertainty in the determination of n arises from appreciable errors in A or f, whereas a probable error of 5% in measured flame thickness l yields a relative error of equal magnitude in n. Since flames that are inhomogeneous with respect to n and T complicate the concentration measurement considerably, one should preferably use flames with a thick central zone fed by a homogeneous solution spray and surrounded by a metal-free flame mantle of about the same temperature (see Sect. III.2).

The determination of absolute metal concentration in flames from absolute line intensities has been described by several authors, for example, Alkemade (1954) and Gurvich and Ryabova (1964).

2b   THE INTEGRAL-ABSORPTION METHOD

Description. A metal resonance line can absorb radiation from a spectral continuum source such as a tungsten strip lamp or a xenon arc. Let B_ν^S be the spectral radiance at frequency ν of the source and $\Delta\nu_M$ the effective spectral bandwidth of the monochromator used, so that in the absence of flame absorption the radiant flux, Φ_0, received by the photodetector is given by $\Phi_0 = \text{const.} \, B_{\nu_0}^S \Delta\nu_M$ (see Eq. III.10 with λ being replaced by ν and $\Delta\nu_M \equiv \Delta\nu_{\text{eff}}$). The monochromator is assumed to be set at the central frequency ν_0 of the absorption line. It is also assumed that neither B_ν^S nor the sensitivity of the photodetector varies noticeably within a spectral range of order $\Delta\nu_M$. The radiant flux Φ_t received by the photodetector when part of the background radiation is absorbed by the metal line is then given by $\Phi_t = \text{const.} \, B_{\nu_0}^S \int \exp[-k(\nu)l] \, d\nu$ (compare Eqs II.262 and II.185) where $k(\nu)$ is the absorptivity of the line at frequency ν and l is the thickness of the homogeneous slab with metal vapour, while the integration extends over a spectral range $\Delta\nu_M$ around ν_0. It is assumed that the spatial extension of the beam from the background source is so small that l has a well-defined value and that $k(\nu)$, which is proportional to n (see Eq. II.185), is uniform along the beam. Later we shall discuss the more general case of a nonuniform $k(\nu)$.

The absorption factor α, that is, the absorbed radiant flux divided by the incident radiant flux from the continuum background source, is given by

$$\alpha \equiv (\Phi_0 - \Phi_t)/\Phi_0 = \text{const.} \, B_{\nu_0}^S \left\{ \Delta\nu_M - \int_{\Delta\nu_M} \exp[-k(\nu)l] \, d\nu \right\} \Big/ \text{const.} \, B_{\nu_0}^S \Delta\nu_M = A_t^{(\nu)}/\Delta\nu_M .$$

(V.4)

SPECTROMETRIC DETERMINATION OF ABSOLUTE METAL CONCENTRATION   Sect. V.2b

Here $A_t^{(\nu)}$ is the *integral absorption* defined by: $A_t^{(\nu)} \equiv \int_{\text{line}} \{1 - \exp[-k(\nu)l]\} d\nu$ where the integration extends over that spectral interval in which the line shows noticeable absorption. It has been assumed that this interval is less than $\Delta\nu_M$. When $\Delta\nu_M$ is smaller than the absorption line width, one finds the integral absorption by recording Φ_t as a function of wavelength-setting and determining the area under the absorption line profile in the record (see, for example, Kaskan 1965). The value of α is easily obtained from relative (linear) spectrometer readings with and without absorbing flame. Multiplication of α by $\Delta\nu_M$ (in Hz) yields $A_t^{(\nu)}$ (in Hz), from which n can be found (see Sect. II.5c-2). The absolute radiance of the lamp need not be known.

We shall distinguish here between: (a) the case of low metal concentrations where the maximum value of $k(\nu)l$, viz. $k(\nu_0)l$, is much less than unity, and (b) the case of high metal concentrations where $k(\nu_0)l \gg 1$. These cases correspond to low and high self-absorption, respectively, in the emission measurements.

In the limiting case of *low metal concentrations* $[k(\nu_0)l \ll 1]$ we have in good approximation from Eq. II.270 (while replacing herein A_t by $A_t^{(\nu)}$ according to Eq. II.268) and Eq. V.4

$$\alpha_{\text{low}} = 0.0265 \, (g_0/Q_X) \, \Delta\nu_M^{-1} \, fnl \, , \tag{V.5}$$

where g_0 is the statistical weight of the ground state, and the other factors have already been defined and are expressed in CGS units. Through this equation n follows directly from measured values of α_{low}, $\Delta\nu_M$ and l, and from tabulated values of g_0, Q_X and f. Neither T nor the absorption line width enter into this equation; thus only the determination of $\Delta\nu_M$ and l requires an absolute measurement.

A check on the validity of the linear relation between α_{low} and n in Eq. V.5, which implies: $k(\nu_0)l \ll 1$, can be obtained by making a dilution run and plotting α versus the metal solution concentration (see de Galan and Winefordner 1967). This check is valid only if in the considered concentration range n can be assumed to vary proportionally to the solution concentration; this condition will be further discussed in Sect. 2d. If the absorption line width $\delta\nu_{\text{abs}}$ and $\Delta\nu_M$ are both (approximately) known, the condition $k(\nu_0)l \ll 1$ underlying Eq. V.5 can also be checked by calculating the value of $\alpha\Delta\nu_M/\delta\nu_{\text{abs}}$. The latter quantity is a rough estimate for the peak absorption factor at the line centre and thus for $k(\nu_0)l$. With regard to other possible sources of error in the determination of n, which are likely to exceed 10%, the value $0.1-0.2$ may be considered as a safe upper limit for $k(\nu_0)l$.

A decrease in $k(\nu_0)l$ will usually be accompanied by an increase in the relative accidental error of the α value measured. So a compromise must be found

Chapt. V   DETERMINATION OF METAL CONCENTRATION IN THE FLAME

for the magnitude of $k(\nu_0)l$ if low systematic as well as low accidental errors in the determination of n through Eq. V.5 are wanted.

In the limit of *high metal concentrations*, i.e., $k(\nu_0)l \gg 1$, we have according to Eq. II.275 [while replacing here A_t by $A_t^{(\nu)}$ according to Eq. II.268, and $\delta\lambda_L$ by $(c/\nu_0^2)(\delta\nu_C + \delta\nu_N)$] and to Eq. V.4

$$\alpha_{high} = 0.230 \, (g_0/Q_X)^{\frac{1}{2}} \left\{(\delta\nu_C + \delta\nu_N)^{\frac{1}{2}}/\Delta\nu_M\right\} (fnl)^{\frac{1}{2}}. \qquad (V.6)$$

Comparing Eqs V.5 and V.6, we see that now in addition to the monochromator bandwidth $\Delta\nu_M$ the collisional half-intensity width $\delta\nu_C$ ($\gg \delta\nu_N$) of the absorption line appears. Here the same discussion applies as given in Sect. 2a with regard to the uncertainty in $\delta\nu_C$.

For a given flame and metal concentration n the measured absorption factor α is inversely proportional to the monochromator bandwidth $\Delta\nu_M$. The latter quantity may also affect the precision of the absorption measurements when the precision is limited by the noise associated with the detection of the unabsorbed radiation $\Phi_0 \propto B_{\nu_0}^S \Delta\nu_M$. For a quantitative discussion of the signal-to-noise ratio we refer to Sect. III.15d.

The validity of Eqs V.5 and V.6 requires that $\Delta\nu_M$ exceeds sufficiently the width of the metal absorption line. The latter increases roughly proportional to $n^{\frac{1}{2}}$ in the range of high n values where $k(\nu_0)l \gg 1$ and the square-root-law is applicable (see Sect. II.5c-2). So one has to be cautious whenever the actually measured value of α approaches unity for high metal concentrations since then the above mentioned condition for $\Delta\nu_M$ will certainly no longer be fulfilled.

Calculations by Kostkowski and Bass (1956) and Dalton (1965) have shown that the error introduced by the finite value of $\Delta\nu_M$ will be less than 3% if $\Delta\nu_M$ exceeds the half-intensity width of the absorption line by at least a factor of 4. This conclusion holds if the instrumental profile of the monochromator (see Sect. III.4b) and the spectral profile of the absorption line are both Gaussian. When the instrumental profile has a triangular form, the error introduced will be somewhat larger (see de Galan and Winefordner 1968).

In *inhomogeneous* flames with cylindrical symmetry, where $n(r)$ depends on the radial distance r from the axis, we can find $n(r)$ by scanning $\alpha(x)$, just as was done for $B(x)$ in Sect. 2a. In the case of high metal concentrations we may apply a similar squaring procedure as outlined in the same section in the case of strong self-absorption.

Inhomogeneity with respect to the flame temperature T will not interfere at all in cases where Eq. V.5 can be applied with low metal concentration. At higher concentrations where $\delta\nu_C$ becomes important according to Eq. V.6, a possible variation in T could interfere since $\delta\nu_C$ depends on it. This dependence is,

SPECTROMETRIC DETERMINATION OF ABSOLUTE METAL CONCENTRATION   Sect. V.2b

however, not very strong (see Sect. 2d), so that the interference of varying T along the optical axis may be ignored in first approximation.

The occurrence of (*hyper-*)*fine-structure* has a similar effect on the theoretical relations between α and n as on the relations between B and n discussed in Sect. 2a. This is not surprising since the thermal line-radiance and integral absorption are both related to n through the common factor A_t (compare the general Eq. II.295 for the line radiance with Eq. V.4).

Absorption measurements of molecular bands starting from the ground level or some low-lying (vibrational) level could, in principle, be used for the absolute determination of *metal-compound concentrations* (see Sect. II.5a-3). The number of di-atomic metal compounds, however, for which reliable data on transition probabilities and level diagram are available is still small but growing (see also Sect. 2a). A practical difficulty is that the absorption factors expected are much lower than those for most atomic resonance lines, even for molecular resonance bands (see Sect. II.5a-3). However, Fiorino, Kniseley and Fassel (1968) have found absorption bands of LaO, YO and ScO in the secondary combustion zone of slightly fuel-rich $C_2H_2-O_2$ flames. Overlap of bands or individual lines in the molecular spectrum may additionally complicate the determination of absolute concentrations by absorption (as well as emission) measurements.

For a discussion of the integral-absorption method we refer also to Frish (1970).

Comparison of the absolute-intensity and the integral-absorption method.
Although the basic theoretical equations underlying the determination of n from absolute-intensity and integral-absorption measurements are closely related, the experimental possibilities and procedures are quite different for the two methods. In contrast to emission measurements, absorption measurements in flames are restricted to optical transitions that start from the ground level or some low-lying level. An example of the latter case is the Ni 3414 Å line which has been measured in absorption (see McGee and Winefordner 1967), and the Tl 5350 Å line showing self-absorption (see Hinnov and Kohn 1957). Bleekrode (1968, 1968a) has measured the absorption of nonresonance Fe lines with lower levels up to 1 eV above the ground level in a low-pressure flame using a continuum source. The equations quoted in this section apply to the latter cases if we replace g_0 by the product of statistical weight and Boltzmann factor for the lower level of the line studied. Of course, this requires a knowledge of T, but because of the low excitation energy involved the value of T is not very critical.

An essential advantage of the absorption over the emission method is that the flame temperature does not play a critical role. This advantage is especially obvious when the flame is inhomogeneous with respect to temperature along the axis of observation.

Deviations from excitation equilibrium in the flame do not affect the results of absorption measurements for resonance lines. For this reason the absorption method is to be preferred when determining metal atom concentrations in cool flames or in the combustion zone. The method is also preferred when ultraviolet resonance lines are to be used that are only weakly excited in flames.

Since f and A are closely related through Eq. II.168 and always occur in the combination (fn) or (An) in the theoretical equations, any uncertainty in these values affects the n values determined by the absorption and emission method to the same extent. In general, this uncertainty may introduce an error in n of the order of 1% for the alkali first resonance lines, and of the order of 10% for most other lines.

It may be interesting to compare also the experimental techniques involved in the absolute-emission and integral-absorption method, respectively. When the former is applied, an accurately calibrated 'thermometer', for example a standard lamp and line-reversal equipment are required. Their careful operation requires some skill, otherwise unnoticed systematic errors may easily creep in. In the determination of the absolute line radiance the product of optical conductance and transmission factor of the optical apparatus times the sensitivity of the photoelectric measuring instrument must be known (see Eq. III.60). This product can be found with the aid of a calibrated tungsten lamp, if the spectral bandwidth of the apparatus is known (see Sect. III.14g). In the integral-absorption method this bandwidth is the only instrumental factor that must be known (see Eqs V.5 and V.6). So for both methods an absolute determination of the spectral bandwidth of the monochromator used is required. In the emission method, however, the choice of bandwidth is rather arbitrary and it may be convenient to choose it not too small to ensure that its value can be reliably determined and kept constant. In absorption measurements a comparatively large bandwidth is undesirable, since it would make great demands on the stability of the background source and of the optical equipment; in practice, therefore, smaller bandwidths are preferable.

The essential point of difference between the two experimental techniques is that absorption measurements are based on the *difference* between two signals of comparable magnitude. Consequently the stability of the whole apparatus must usually be much better than in the emission methods for obtaining the same statistical error in the final result. Special techniques to reduce the statistical error in absorption measurements and to eliminate the thermal flame emission have been discussed in Sect. III.15.

The range of n values that can be reliably determined by absolute-intensity measurements is usually much larger, since here the n value is not derived from

SPECTROMETRIC DETERMINATION OF ABSOLUTE METAL CONCENTRATION   Sect. V.2c

the difference of two relatively large photosignals. The detectability of low n values is here, however, ultimately restricted by the noise in the flame background emission or in the photo-electric measuring device. Usually the latter does not play a role in absorption measurements if the lamp emission is sufficiently strong (see Sects III.14 and 15 for a discussion of the signal-to-noise ratio).

Integral-absorption measurements with a continuum background source are coming more and more into use in *analytical atomic absorption spectrometry*, where one is not interested in absolute concentrations in the flame but in concentrations of metal solutions (see also near end of Sect. 2c). Integral-absorption measurements with a continuum background source often offer the advantage of better source stability and greater versatility in multi-element analysis than in the case when a spectral line source, e.g. a hollow-cathode lamp, is used (see also Sect. 2c). Detection limits (that is, the minimum detectable element concentration) are usually better in the latter case, however. The reader who requires further information is referred to the general textbooks on analytical atomic absorption spectroscopy listed in Chapt. I.

2c   THE PEAK-ABSORPTION METHOD

Description. The peak absorption of a single, isolated and unshifted resonance line with a symmetric spectral profile and central frequency ν_0 can be measured with a monochromatic background source. This source should emit a single spectral line at ν_0 with half-intensity width considerably less than the absorption line width $\delta\nu_{abs}$ in the flame. The *peak absorbance* is defined by: $A(\nu_0) \equiv -\log_{10}(\Phi_t/\Phi_0)$, where Φ_t is the radiant flux at ν_0 received by the photo-detector from the background lamp when absorption in the flame takes place and Φ_0 is the corresponding flux without absorption. The value of $A(\nu_0)$ follows from the corresponding spectrometer readings. For a homogeneous slab with metal vapour having a well-defined thickness l along the axis of observation, we find for an absorption line with effective width $\Delta\nu_{eff}$ (defined by Eq. II.180) by using Eqs II.264 [with $\kappa_m \equiv \kappa(\nu_0)$ and $A^{abs} \equiv A$] and II.187 while expressing n_0 by $(g_0/Q_X)n$

$$A(\nu_0) = 0.434\,\kappa(\nu_0)\,n_0 l = 1.15\times 10^{-2}\,(g_0/Q_X)\,fnl/\Delta\nu_{eff}, \qquad (V.7)$$

if we confine ourselves as usual to resonance lines and express all quantities in CGS units. In the case of pure Doppler broadening we find from this with the aid of Eq. II.241

$$A(\nu_0)_D = 1.08\times 10^{-2}\,(g_0/Q_X)\,fnl/\delta\nu_D. \qquad (V.7a)$$

In the case of pure collisional (and natural) broadening, resulting in a Lorentzian

Chapt. V   DETERMINATION OF METAL CONCENTRATION IN THE FLAME

line profile, we get with the aid of Eq. II.216

$$A(\nu_0)_C = 7.30 \times 10^{-3} (g_0/Q_X) fnl/\delta\nu_C . \quad (V.7b)$$

For an absorption line with an arbitrary Voigt profile we have to use Eq. II.251 for $\Delta\nu_{eff}$ in order to find $A(\nu_0)$ from Eq. V.7. Instead of using Eq. II.251, we can calculate $A(\nu_0)$ more simply but less accurately (maximum error about 10%) by substituting $(\delta\nu_D + \delta\nu_C)$ for $\delta\nu_D$ in Eq. V.7a if $0 \leqslant a \leqslant 2$, or by using Eq. V.7b if $a > 2$. The resulting approximate expressions will be used here in a discussion of the effect of $\delta\nu_C$ on the determination of n.

If $\delta\nu_C \ll \delta\nu_D$, Eq. V.7a enables us to derive n, once $A(\nu_0)$ and l are measured and $\delta\nu_D$ is calculated from Eq. II.240, the other factors being easily obtained from the literature. Only when $\delta\nu_C$ is comparable to, or larger than $\delta\nu_D$, may difficulties arise due to the uncertainty in $\delta\nu_C$; these are discussed in Sect. 2d. However, an uncertainty of 100% in $\delta\nu_C$ would result in an error of only a factor of 2 in n, if $\delta\nu_C \approx \delta\nu_D$. Anyway, if $\delta\nu_C$ is not much larger than $\delta\nu_D$, the peak-absorption measurements depend less critically on $\delta\nu_C$ than do the absolute-intensity or integral-absorption measurements when applied to high concentrations where the square-root-law holds; this follows from a comparison with Eqs V.3 and V.6. Parsons, McCarthy and Winefordner (1966) have computed values of $\delta\nu_D$ and $\delta\nu_C$ for a large number of atomic metal lines and for four flames of different composition and temperature. Because of the lack of experimental data only the order of magnitude of $\delta\nu_C$ could be calculated by starting from reasonable estimates about the maximum and minimum values expected for the cross section of collisional broadening. These calculations, however, give some idea about the applicability of Eqs V.7a and V.7b which hold strictly only for $\delta\nu_C \ll \delta\nu_D$ and $\delta\nu_C \gg \delta\nu_D$, respectively. If the condition $\delta\nu_C \ll \delta\nu_D$ is fulfilled, the peak-absorption method seems to be ideally suited for determining absolute values of n, since it involves only relative intensity measurements and one absolute measurement of l. The calculation of $\delta\nu_D$ requires only an approximate knowledge of the flame temperature, as it is proportional to $T^{\frac{1}{2}}$. A further advantage is that Eqs V.7 to V.7b hold strictly in an unlimited range of metal concentrations if the underlying conditions are fulfilled. Equations V.7 to V.7b are in fact a direct consequence of Beer's absorption law for monochromatic radiation.

A disadvantage of the peak-absorption method is that it is more strongly affected by (*hyper-*)*fine-structure splitting*, *line shift* and *line asymmetry* than the integral-absorption and absolute-intensity methods are. If the fine-structure components do not intrinsically overlap and are not resolved by the spectral apparatus, one could apply Beer's law to each component separately and easily derive the combined absorption effect. When these components have different f-values, as is the case for the Na-D doublet components with f-values equal to $\frac{1}{3}$ and $\frac{2}{3}$, respectively, the combined absorption effect can no longer be simply described by Beer's law and the proportionality between the measured peak absorbance and n is lost.

When the absorption line has h.f.s. components that show appreciable but not complete overlap in the flame, the theoretical relationship between peak absorbance and n is more complicated. A detailed analysis is then required and the intensity ratios of the spectral components in the lamp source, the peak-absorbance ratios of the components of the absorption line as well as their spectral line shape have to be taken into account (see Willis 1971). Winefordner and Vickers (1964) have calculated a correction factor of 0.97 to be added to the right-hand side of Eq. V.7a for the 5890 Å Na line showing two groups of h.f.s. components (due to nuclear spin) at a mutual distance equal to 0.45 times δv_D (at $T = 2700$ K) and with an intensity ratio of 3:5.

One can take advantage of isotopic line splitting for determining *isotopic abundances* in metal salt solutions. Zaidel and Korennoi (1961) successfully used a hollow-cathode discharge lamp containing a pure Li isotope as a monochromatic background source while spraying the Li sample to be analyzed into an C_2H_2-air flame (see also L'vov 1970).

A *spectral line shift* induced by collisions might be a source of error, since this collisional effect depending on the nature, pressure, and temperature of the environmental gas will differ in the lamp used as background source and in the flame. The shift δv_S of the centre of the 5890 Å Na line, induced by collisions with N_2- or CO_2 molecules in flames at 1 atm, may amount to about 0.5 δv_C (see Behmenburg 1964). If δv_D is not much larger than δv_C, this shift could noticeably affect the peak absorbance measured. If δv_C exceeds δv_D markedly, then according to L'vov (1961) the absorbance measured for the shifted line is given approximately by

$$A(v_0)_{C,S} = 7.30 \times 10^{-3} (g_0/Q_X) fnl \, \delta v_C / (\delta v_C^2 + 4\delta v_S^2), \tag{V.8}$$

where δv_S is the shift of the absorption line centre with respect to the (unshifted) centre of the lamp emission line. A shift δv_S equal to $0.5 \delta v_C$ as reported above would thus result in a decrease of absorbance by a factor of 2. Willis (1971) has shown that h.f.s. and line shift may affect n values of Na, Cu, Ag and Au derived from peak-absorption measurements by a factor of 1.1 to 2.1. For a further discussion of collisional line shift we refer to Chapt. VII.

So far we have assumed the *width of the lamp emission line* to be much smaller than that of the flame absorption line. Hollow-cathode lamps operated at relatively high discharge currents may emit lines that do not meet this condition and sometimes even show marked self-reversal (see Prugger 1969, Kirkbright and Sargent 1970, Bruce and Hannaford 1971, Wagenaar, Novotny and de Galan 1974, and Jansen, Hollander and Franken 1974; see also Sect. III.7c). The derivation of absolute concentrations from absorbance measurements would then require a detailed and cumbersome analysis of the actual shape of both the emission and absorption spectral line. Winefordner and Vickers (1964) calculated the error in n as a function of the ratio of emission line width to absorption line width in the case when n is determined from Eq. V.7a. This has been done for different values of $k(\nu_0)l$. It was found that for a Doppler-broadened absorption line with $k(\nu_0)l = 0.5$ the error in n amounts to 3, 10.5, 31, and 58% if the corresponding ratio of line widths equals 0.25, 0.50, 1.0, and 2.0, respectively. The occurrence of a relatively broad emission line in the hollow-cathode lamp would, in general, become manifest by a dependence of the measured n value on the discharge current of the lamp. In this case a plot of measured absorbance versus metal solution concentration would also deviate from a straight line through the origin (assuming that n is proportional to the metal solution concentration).

Wagenaar, Novotny and de Galan (1974), Wagenaar and de Galan (1975), L'vov, Katskov, Kruglikova and Polzik (1976), and Wagenaar (1976) have investigated both theoretically and experimentally the influence of the source line width, hyper-fine-structure and line shift on the absorption measurements.

The possibility of determining absolute atomic concentrations from absorption measurements was theoretically considered by Rann (1968) and L'vov, Katskov, Kruglikova and Polzik (1976) in the case when the source line is not small compared to the absorption line. The problem is considerably simplified when an auxiliary flame of identical composition and temperature is used as a source. The Doppler width and the a-parameter of the source line and the absorption line are then the same. Absolute calibration curves of absorbance versus atomic concentration can then be obtained by computer calculations. An experimental check has been made by using a-values from the literature and by calibrating the nebulization efficiency (see also Sect. V.5a).

In the ideal case where the lamp emits merely one resonance line and the flame's own thermal emission does not interfere, no monochromator would be required at all to perform peak-absorption measurements. In practice, however, the lamp may emit satellite lines of the metallic element and/or a gas-discharge spectrum, and the thermal metal or background emission of the flame may not be negligible. Interference by the latter emissions can be overcome by modulating the lamp emission and applying an a.c. measuring technique; the shot-noise contributions of these emissions, however, will remain (see Sect. III.14f). For these reasons some spectral

SPECTROMETRIC DETERMINATION OF ABSOLUTE METAL CONCENTRATION   Sect. V.2d

selecting device (filter or monochromator) is desirable, but the requirements for its bandwidth are much less stringent here than in the case of integral-absorption measurements with a continuum source.

For *inhomogeneous* flames having cylindrical symmetry $n(r)$ can be found as a function of radial distance r in an analogous way as discussed in Sects 2a and 2b.

Comparison of the peak-absorption and the integral-absorption method. The ratio of the sensitivities (percentage absorption per unit of metal concentration) of the two methods is of the order $\Delta\nu_M/\delta\nu_{abs}$ where $\Delta\nu_M$ is the monochromator bandwidth used with the integral-absorption method and $\delta\nu_{abs}$ is the half-intensity width of the absorption line. Since $\Delta\nu_M$ is usually much larger than $\delta\nu_{abs}$, the peak-absorption method has the advantage of greater sensitivity. With both methods the range of n values where reliable measurements can be performed is limited. Values of n giving absorption factors of order 10^{-3} or less would make considerable demands on the reliability of the measuring equipment (see also Sect. III.15d). Although ideally the peak absorbance rises proportionally to the atomic concentration however large, n values giving nearly 100% absorption are difficult to measure in practice. Even if the detection equipment were perfectly stable, considerable errors might arise from spurious emission lines of the lamp that are not absorbed in the flame, from h.f.s. components of the absorption line with different f-values, or from the fact that l is not equal for all rays from the background lamp that traverse the flame. These effects become increasingly critical at high absorption factors, and may invalidate the application of Eqs V.7 to V.7b. In integral-absorption measurements difficulties of this kind are much less serious. By enlarging the monochromator bandwidth one can simply arrange for the absorption to remain sufficiently below 100%.

Gallagher and Lapp (1967) have devised a method of determining atomic vapour concentrations from absorption measurements without requiring line-width data. Their method is based on the *combined* measurements of peak and integral line absorptions. The oscillator strength of the transition is required for the determination of absolute atomic concentrations. The method has been tested with the Cu resonance lines at 3248 and 3274 Å in a H_2-O_2 flame.

2d   CURVE-OF-GROWTH METHODS (IN EMISSION AND ABSORPTION)

Description. The (*theoretical*) *curve-of-growth* (COG) of a resonance line has been introduced in Sect. II.5c-2 as the (usually double-logarithmic) plot of $A_t^{(\nu)}(\ln 2)^{\frac{1}{2}}/\delta\nu_D$ versus $n_0 fl(\ln 2)^{\frac{1}{2}}/\pi\delta\nu_D$ for a homogeneous metal vapour, where $A_t^{(\nu)}$ is the 'integral absorption' (see Eqs II.267 and 268), $\delta\nu_D$ is the Doppler half-intensity width, and n_0 is the atomic ground-state concentration, while the other symbols are defined in Sect. 2a. The position and form of this curve are

495

Chapt. V   DETERMINATION OF METAL CONCENTRATION IN THE FLAME

fully determined by the parameter $a \equiv (\ln 2)^{\frac{1}{2}}(\delta\nu_C + \delta\nu_N)/\delta\nu_D$ where $\delta\nu_C$ and $\delta\nu_N$ are the collisional and natural half-intensity width of the line, respectively. In this section we shall ignore $\delta\nu_N$ as before.

The quantity A_t is proportional to the total line intensity, that is the spectral intensity integrated over the whole line width, if we assume a uniform temperature T (see Eq. II.295). A_t, by its very definition, is proportional to the radiant power absorbed by the line, over its whole spectral width, from a continuum background source. We now make the additional assumption that n_0 is proportional to the metal concentration c in the nebulized solution. Thus when we make a 'dilution run' and plot double-logarithmically the total intensity or integral absorption of the line, both expressed in arbitrary units, versus solution concentration c, an *experimental COG* is obtained that has the same shape as the theoretical COG with appropriate a-value. Superposition of the experimental COG on the corresponding theoretical curve by a shift parallel to the axes directly yields the absolute concentration n_0 corresponding to an arbitrary solution concentration c. The total atomic concentration, n, is then found from the Boltzmann relation: $n = (Q_X/g_0)n_0$ (for meaning of symbols see Sect. 2a). This holds if we assume the values of $l, f, \delta\nu_D$, and a to be known. The value of l can be measured simply; that of f can be derived from tables in the literature as it depends only on the optical transition involved (see also Appendix A.4), while $\delta\nu_D$ can be easily calculated through Eq. II.240.

The *a-parameter* depends on the line considered as well as on the composition, pressure and temperature of the flame. Its value must be found by experiment, since no satisfactory theory exists to predict a. Fortunately, a does not seem to depend very critically on the flame variables just mentioned if we restrict ourselves to normal flames diluted by N_2. It should vary proportionally to the total pressure at a given flame temperature and composition. The quantity $\delta\nu_C$ should vary $\propto T^{-0.7}$ according to theory if we can describe the interaction between the radiating atom and the colliding flame molecules by a simple van-der-Waals potential (for a fuller discussion see Sect. VII.2b-3 sub point 1). Since the Doppler width $\delta\nu_D$ varies $\propto T^{\frac{1}{2}}$, we expect a to vary $\propto T^{-1.2}$. But this dependence has not yet been firmly established by experiment. Thus, for an approximate knowledge of a we have to resort to observations on the same line in similar flames reported in the literature, while correcting for a possible difference in T. There are, however, only a few data available and these refer mainly to the first resonance lines of the alkali and alkaline-earth atoms (see Tables VII.9 and 10 on pp. 744 and 748). Moreover, these data still show appreciable spread, even when referring to similar flames. It is safe, anyway, to choose the error limits not too narrow in values taken from the literature.

SPECTROMETRIC DETERMINATION OF ABSOLUTE METAL CONCENTRATION   Sect. V.2d

When no reliable literature data are available for the flame type used, it is possible to determine the appropriate a-value by starting from a family of theoretical COG's with different a-parameters and selecting that curve whose shape best fits the experimental COG. This procedure, however, only works well if $a \leqslant 1$, since only then has the shape of the curve a pronounced dependence on a (see Fig. II.15 on p.170). In Sect. VII.3 we shall deal more generally with the experimental methods of determining this parameter. In this section we shall henceforth assume that a is known to a sufficient degree of accuracy.

Once a is known, it is not necessary to measure the full experimental COG. It is sufficient to measure its initial and final asymptotes only, corresponding to low and high solution concentrations where the emission (or integral absorption) is proportional to c and to $c^{\frac{1}{2}}$, respectively. From Eqs II.277 and 208 we find for the total atomic concentration n^* that corresponds to the point of *intersection of the initial and final asymptotes* in a double-logarithmic plot

$$n^* = 75.2\,(Q_x/g_0)\,\delta\nu_c/f\ell\,, \tag{V.9}$$

where all quantities are expressed in CGS units. One clearly sees from this equation that a given percentage error in $\delta\nu_c$ (or in a) causes an equally large percentage error in the concentration measurement. The flame temperature, however, does not enter explicitly into Eq. V.9; it does, of course, implicitly do so through $\delta\nu_c$ which depends on T. The f-value has to be known, of course, but this condition is shared by all spectroscopic methods of determining n.

Since n is supposedly proportional to the solution concentration c, we find the *conversion factor* K by which any c (in mol ℓ^{-1}) has to be multiplied in order to get the corresponding n value (in cm^{-3}) from $K = n^*/c^*$ where c^* corresponds to n^* (see also Sect. III.2h).

Application of the COG method is restricted to resonance lines, or lines with a lower level at a short distance above the ground level, with not too small an f-value. For these lines marked (self-)absorption is obtained for reasonably high metal concentrations. Ultraviolet resonance lines with weak thermal excitation in the flame should preferably be measured in absorption (see McGee and Winefordner 1967). When the two asymptotes of the experimental COG are used only, measurements should be extended far enough into the square-root-region in order to determine the position of the high-concentration asymptote with sufficient accuracy. When the full experimental COG is used, this demand may be relaxed somewhat, since the characteristic shape of the intermediate portion of this curve may be an extra help in finding the correct superposition on the theoretical COG. The use of this intermediate portion is especially recommended when, at high metal concentrations, complications arise that distort the shape of the curve in the square-root-region

Chapt. V   DETERMINATION OF METAL CONCENTRATION IN THE FLAME

(see below).
Classical theory, upon which the theoretical COG is usually based, does not allow for the effects of collisionally induced line shift and line asymmetry (see Chapt. VII). The consequences of these effects for the validity of the classical COG theory have experimentally been investigated (see Jansen 1976, and Jansen and Hollander 1977). The results of this investigation as well as the satisfactory agreement that is usually found between the experimental COG and classical theory, in particular in the square-root-region, gives confidence in its use in atomic flame spectroscopy.

Also the neglect of resonance broadening, which could make a depend on n at high n values, as explained in Sect. II.5b, will not introduce noticeable errors under the usual flame conditions.

($Hyper$-)$fine$-$structure$ of the line observed will, in general, interfere with the application of the COG method (see Sect. II.5c-2). For the first alkali resonance lines (except for Li) the doublet components virtually do not intrinsically overlap. When these components are not resolved by the spectral apparatus, the measured COG is called the $doublet$-COG. The value of n^*_{doub} corresponding to the point of intersection of the asymptotes of the theoretical doublet-COG follows from Sect. II.5c-2

$$n^*_{doub} = 75.2\,(Q_X/g_0)(\delta\nu_C/l)(f_1^{\frac{1}{2}}+f_2^{\frac{1}{2}})^2/(f_1+f_2)^2 \quad \text{(in CGS units)}, \quad (V.10)$$

where f_1 and f_2 are the oscillator strengths of the (nonoverlapping) doublet components.† It is easily seen by comparing Eqs V.9 and V.10 that neglect of the fine-structure of the Na-D doublet with $f_1 = \frac{2}{3}$ and $f_2 = \frac{1}{3}$ would cause an error of a factor 1.95 in the determination of the Na concentration.

A distortion of the COG might be expected for very high n values where the spectral profiles of the doublet components begin to overlap as a result of self-absorption broadening (see Sect. II.5c-3). This would manifest itself by a deviation from the square-root-dependence on n. Actually, however, such distortion has not been found by Behmenburg and Kohn (1964) for the 5890 Å Na line, even for n values that exceed n^* by a factor of about a thousand. Thus this disturbing effect may safely be ignored in flames for lines with similar or larger doublet splitting.

H.f.s. splitting due to nuclear spin or isotopic shift is more difficult to handle, if their spectral components show partial, but no full intrinsic overlap. In the calculation of their effect on the COG an overlap parameter can be defined

---

† It is noted that Eq. (7) in the paper of Winefordner, Vickers and Remington (1965) yields only an approximate value for n^*_{doub} in the case of fine-structure.

SPECTROMETRIC DETERMINATION OF ABSOLUTE METAL CONCENTRATION   Sect. V.2d

that depends on the ratio of the mutual distance of the h.f.s. components in the spectrum to the half-intensity width of each component. The latter is related to a, which is usually assumed to be the same for all h.f.s. components. From Behmenburg's (1964) calculations for the Na-D_2 line under usual flame conditions the neglect of h.f.s. makes n^* as calculated from Eq. V.9 too low by about 18%. For the 8521 Å Cs line, van Trigt, Hollander and Alkemade (1965) have estimated a corresponding deviation of about 30−35%.[†] From calculated data in the latter paper it is concluded for the Na- and K first resonance doublets that a neglect of doublet structure and h.f.s. together may lead to a calculated n^* value that is too low by a factor of 2.05−2.28, and 1.96−2.00, respectively. The spread in these factors results from some ambiguity in choosing the appropriate overlap parameter. In absence of h.f.s. both factors would equal 1.95.

The obvious *advantage* of the COG method of determining n is that only relative emission intensities or integral absorptions need to be measured if the a-parameter is known. This contrasts with the other spectroscopic methods discussed in Sects 2a and 2b. Of course, this advantage must be paid for by drawbacks in other respects. An essential condition for the application of the COG method is that n is strictly proportional to the solution concentration c over at least two, preferably three orders of magnitude. The factors that may cause a distortion of the experimental COG and may thus interfere with the determination of n will now be discussed.

Distortion of the experimental COG . In the range of low concentrations *partial ionization* of the metal vapour may cause the atomic metal concentration in the flame to rise faster than in proportion to the solution concentration c. This holds because the fraction of element ionized usually decreases with increasing metal concentration on account of the Saha equilibrium law (see Sect. IX.3b-1). At high concentrations the fraction ionized is usually negligible and the experimental COG remains unaffected. In the low-concentration range, however, the experimental COG of the atomic line is shifted downwards, and the initial slope of the curve in the double-logarithmic plot will exceed 45°. As a consequence the intersection point of the two asymptotes is displaced, which results in too low n values for a given solution concentration. Such a distortion of the experimental COG may be expected to occur for the alkali metals Cs, Rb, K, Na, and Li (in order of decreasing effect) at flame temperatures roughly above 2200 K. For the alkaline-earth metals disturbing ionization effects were found by Hofmann and Kohn (1961) at flame temperatures of 2500 K and higher (see Fig. V.2).

[†] The statement on page 824 in the latter paper that too high n values are found when neglecting h.f.s. is incorrect; rather too low n values are found.

Chapt. V DETERMINATION OF METAL CONCENTRATION IN THE FLAME

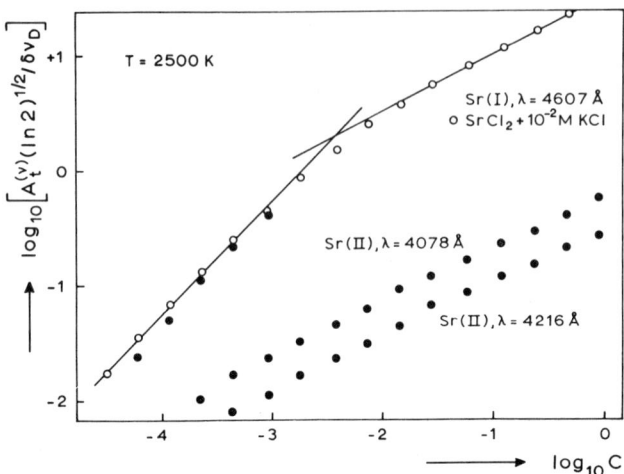

Fig. V.2 Intensity of atomic and ionic resonance lines of Sr in a shielded C_2H_2-air flame at 2500 K as a function of molar Sr concentration (c) in the solution (according to Hofmann and Kohn 1961). The intensity values are converted to absolute values of integral absorption $A_t^{(\nu)}$. This conversion was achieved by an additional absolute measurement of the integral absorption for the two lines (at high Sr concentration). Closed circles represent experimental points for a pure Sr solution. Open circles refer to measurements with K added as a de-ionizer. The position of the point of intersection of the two straight asymptotes of the atomic curve-of-growth yields the a-parameter and the atomic Sr concentration in the flame through Eq. V.9.

The simplest way of eliminating ionization effects is to de-ionize the element by adding to the solution a constant excess of a second, easily ionizable element. If the concentration of free electrons produced by this *ionization buffer* exceeds largely that produced by the original element alone, then the fraction ionized of the latter element is negligible or at least independent of its concentration according to Saha's law. Then the distorting effect of ionization on the shape of the experimental COG is eliminated. The atomic concentration found by the COG method then, of course, refers to a situation with the buffer element present.

Good results have been reported by Hofmann and Kohn (1961), who added a K salt to solutions containing Li, Na, Rb, Cs, Ca, Sr, or Ba sprayed into flames with $T = 2500$ and 2760 K (see also Fig. V.2). They have used Cs as a buffer element when K was to be measured. It may happen that even the highest buffer concentrations feasible are still inadequate for practically complete de-ionization.

This may be the case when Cs itself is the element to be de-ionized and another alkali element acts as a buffer. In that case one can try to estimate a correction factor by plotting the relative enhancement of atomic Cs emission or absorption as a function of the varying K concentration and by extrapolating to infinitely high K concentration. A numerical correction factor for incomplete de-ionization might also be calculated from the Saha equation at known flame temperature (see Sect. IX.3b-2). This correction factor depends on the absolute concentration of the element considered. One might start by inserting in the Saha equation the absolute atomic concentration derived from the uncorrected COG and then apply an iteration procedure. This method presupposes, however, that the Saha equilibrium has indeed been attained at the height of observation in the flame, which may not always be the case (see Chapt. IX).

The possible presence of free electrons produced by the flame gases themselves or the formation of negative ions could make such corrections uncertain.

Of course, additional electrical measurements of the free electron content in the flame could help to correct for the ionization effect (see Hofmann and Kohn 1961), but the apparatus required is rather elaborate and often not available.

Observations on the ionic line intensity could also provide useful information on the relative extent of ionization (compare Fig. V.2). For the alkaline-earth metals, which emit ionic lines, complete de-ionization, however, happens to be no great problem. The formation of metal compounds could strongly reduce the influence of ionization on the atomic metal concentration (see Sect. IX.3b-3). For this reason de-ionization of the alkaline-earth elements, which are present largely as (hydro-)oxides, is easily achieved (see Hofmann and Kohn 1961, and van der Hurk 1974).

An elegant method of eliminating the distortion of the COG by ionization can be applied for the alkaline-earth elements which emit molecular bands as well as atomic resonance lines (see van Trigt, Hollander and Alkemade 1965, and Jansen 1976). The molecular bands show practically no self-absorption. Their emission is thus proportional to the molecular concentration. On the other hand, ionization affects the concentration of metal compounds by the same factor as the atomic concentration. This holds because the ratio of the two concentrations is expected to be constant in chemical equilibrium on account of the mass-action-law (see Sect. VIII.1). Consequently, by plotting the atomic line emission versus the molecular band emission in arbitrary units for a series of solution concentrations, we obtain the true shape of the COG of the atomic resonance line without distortion by ionization.

Numerous papers have appeared describing applications of the ionization-

Chapt. V   DETERMINATION OF METAL CONCENTRATION IN THE FLAME

buffer method. In analytical flame spectroscopy, where a straight analytical curve is convenient, this method is often applied both in emission and in absorption measurements when hot flames are used. Application of this method in special investigations of the COG has been reported, among others, by Alkemade (1954), Hofmann and Kohn (1961), Behmenburg and Kohn (1964), Hollander (1964), and McGee and Winefordner (1967).

Formation of *metal compounds* is not expected to affect the shape of the atomic COG, since the fraction of dissociated element is usually independent of its concentration (see Sect. VIII.1).

In flames diluted with a noble gas underpopulation of the excited level can occur owing to *radiative disequilibrium* as explained in Sect. VI.4. Since for resonance lines the percentage underpopulation decreases with increasing atomic concentration, a distortion of the shape of the emission COG may result. The initial asymptote of the COG in emission experiences a downward shift, which results in a displacement of the point of intersection of the two asymptotes to higher abscissa values. The increase in the abscissa value of the intersection point may amount to a factor of three for Na in a H_2-O_2-Ar flame at 1 atm (see Hooymayers 1966, and Hooymayers and Alkemade 1966). This deviation was in accordance with theoretical predictions based on radiative disequilibrium.

Suprathermal *chemiluminescence* (see Sect. VI.3) is not expected to distort the emission COG. This holds if the chemiluminescent excitation rate is independent of the metal concentration, as is usually the case, and if it is uniform in a horizontal cross section of the flame. Distortions might, however, occur when the chemiluminescent spectral line shows an extra Doppler broadening (see Sect. VII.2b-9); but this is not likely outside the primary combustion zone.

Although the flame temperature does not explicitly play a role in the determination of n from relative COG measurements, the occurrence of a radial T variation can distort the COG when measured in emission. A fall of T with increasing distance from the flame axis may result in *self-reversal* of the resonance line observed (see Sect. II.5c-3). This causes an additional depression of the atomic line intensity in the high-concentration range, and makes itself manifest by a deviation from the square-root-dependence, the stronger the higher the metal concentration. No serious errors in the determination of n need result from this deviation, if at least the initial part of the square-root-asymptote is unaffected, so that it can be located unambiguously. But this depends on the radial temperature gradient.

Self-reversal does not interfere when the COG is measured in absorption. In emission experiments it can be precluded by shielding the central flame containing metal vapour by a flame mantle without metal vapour of about the same temperature (see Sect. III.2).

Of course, a disturbing effect of opposite sign may occur when the temperature increases towards the edge of the flame. This could happen in a fuel-rich flame when secondary combustion with oxygen from the surrounding air produces additional heat at the flame edge.

When the efficiency of the nebulizer decreases with increasing salt concentration in the solution or when complete volatilization of the desolvated aerosol particles is inhibited at high salt concentrations (see Sect. III.9), an additional downward curvature of the COG may again result. The position of the high-concentration asymptote may be affected by this *nebulization effect* and the square-root-dependence will not be observed, as in the case of self-reversal. Contrary to the self-reversal effect, this nebulization effect is observed both in emission and absorption measurements and will affect atomic resonance lines as well as molecular bands. In fact, it can be corrected for by plotting the resonance line intensity not as a function of solution concentration but as a function of the intensity of a molecular band emitted by the same element (see Hollander 1964, Hollander *et al.* 1970, and Jansen 1976). This method resembles that of correcting for ionization effects mentioned earlier in this section. A similar correction may be obtained by adding a constant amount of Ca to the metal solutions and watching the possible decrease of CaOH band emission with increasing metal concentrations (see Hooymayers 1966, and van der Hurk 1974).

Whenever a deviation from the square-root-law is observed, flame-duplication measurements may be helpful in establishing the cause (see Sect. 2e).

Finally, we summarize the *experimental requirements* for a reliable measurement of the COG (see also Alkemade 1954, Hollander 1964, and McGee and Winefordner 1967).

(i) A homogeneous distribution of temperature in the observed section should exist in emission measurements, in order to avoid self-reversal. A homogeneous distribution of metal vapour is not required however, since the only thing that matters is the integral of the concentration over the axis of the observed radiation beam ($= y$ axis).

In cylindrically symmetric flames where $n(r)$ depends on the radial distance r to the flame axis similar considerations as given in the preceding Sects 2a and 2b hold when applying the COG method. In flames where the temperature varies with r in an unknown way, the experimental COG should preferably be measured in absorption.

(ii) The product (nl) in a homogeneous flame or the integral $\int n \, dy$ in an inhomogeneous flame should be equal for all light rays that contribute to the optical signal measured. This condition may impose some restrictions on the angular

Chapt. V   DETERMINATION OF METAL CONCENTRATION IN THE FLAME

aperture and the stops used in the optical train.

(iii)   When doublet components are measured together, care should be exercised that the spectral apparatus transmits both components equally. If a difference in transmission factor exists, it should be accounted for in the comparison of the experimental COG with the theoretical doublet COG.

(iv)   Partial ionization in the range of low concentrations or a reduction in the efficiency of nebulization or fraction volatilized in the range of high solution concentration should be avoided or corrected for.

(v)   Good proportionality of the photodetector and electric measuring system over a large range of intensity ratios (2 or 3 orders) should exist when the COG is measured in emission. Methods of checking this proportionality are discussed in Sect. III.13. This requirement can be relaxed if optical attenuators are available with known attenuation factors.

In integral-absorption measurements the latter requirement is less stringent. But here special precautions in the experimental set-up are called for that permit the measurement of small absorption factors ranging from below 0.01 to somewhat above 0.1 (see McGee and Winefordner 1967; see also Sect. III.15d-1). The thermal emission of the flame itself, if disturbing, should be effectively eliminated or corrected for by subtraction of the readings. Care should be taken in the absorption measurements that the spectral bandwidth of the monochromator exceeds sufficiently the width of the absorption line, which increases with rising concentration when the line centre has become opaque (see also Sect. 2b). When the absorption approaches the limit of 100%, the usual COG theory is no longer applicable (see also Alkemade 1968); a larger monochromator bandwidth should then be chosen.

(vi)   During repeated measurements contamination between solutions with large differences in salt concentration should be avoided; this may be done by rinsing the nebulizer with distilled water between measurements.

Measurements of the emission COG yielding absolute metal concentrations in flames have frequently been reported. In some cases the COG obtained was also used to determine the appropriate α-parameter; more often this parameter was derived from literature data. Applications of the COG method in absorption are less frequent (see, e.g., McGee and Winefordner 1967, and Snelleman 1967).

2e   DUPLICATION METHODS

The total intensity or integral absorption of a resonance line (integrated over its full spectral width) depends on the product $(n_0 fl)$ in such a way that a doubling of metal ground-state concentration n_0 has essentially the same effect

SPECTROMETRIC DETERMINATION OF ABSOLUTE METAL CONCENTRATION   Sect. V.2e

as a doubling of oscillator strength f or coloured-flame thickness l. The factor by which the emission or absorption varies upon duplication of $(n_0 fl)$ can be read off from the curve-of-growth (COG). Inspection of Fig. II.15 on p.170 immediately shows that this factor depends on the absolute value of $(n_0 fl)$ and on the a-parameter in that range of $(n_0 fl)$ values where the COG changes from a linear to a square-root-dependence. Conversely, a measurement of the relative increase in emission or absorption caused by duplicating (or halving) either n_0, f, or l, can inform us about the absolute value of n_0 and thus of n, if the other relevant parameters are known. This constitutes the general basis of the duplication methods to be discussed in this section.

The relative enhancement of line emission is described by the *duplication factor* D defined by Eq. II.300. The theoretical curve relating D to $n_0 fl (\ln 2)^{\frac{1}{2}}/\pi \delta\nu_D$ for a resonance line is called a *duplication curve* (or *Gouy curve*) and is usually plotted double-logarithmically. Its position and shape depend on a, as discussed in Sect. II.5c-3. The value of D can be easily measured from the relative line intensities I_1 and I_2 found with $(n_0 fl)$ and $(2 n_0 fl)$, respectively, according to

$$D = (I_2 - I_1)/I_1 . \qquad (V.11)$$

Roughly speaking, D as a function of n_0 represents the first derivative of the COG with respect to n_0, and the information contained in the D-curve is closely related to that in the COG. The linear and square-root asymptotes of the COG for $n_0 \to 0$ and $n_0 \to \infty$, correspond to two horizontal asymptotes of the D-curve with ordinate values $D = 1$ and $D = 2^{\frac{1}{2}} - 1 = 0.415$, respectively (see Fig. II.18 on p.189). A minimum at finite n_0 value occurs in the D-curve when $a \leqslant 1$. The occurrence of this minimum offers a possibility of determining a (see Sect. VII.3c).

From the measured value of D, the absolute value of n_0 can be derived if a, f, l, and $\delta\nu_D$ are known. If a minimum occurs in the D-curve, an ambiguity might be introduced in the determination of n_0, but this is easily removed if D is also measured at a slightly higher or lower metal concentration. Obviously, reliable n_0 values can be found only when D deviates sufficiently from its asymptotic values 1 and 0.415. This restricts the range of measurable n_0 values to one or two orders, depending on the a-parameter.

Duplication factors could be measured by doubling or halving the solution concentration c, but this suffers essentially from the same limitations as the COG-method discussed in Sect. 2d. These limitations are connected with the requirements of proportionality between c and n or n_0, which may be violated by ionization or incomplete volatilization. These difficulties are avoided if we measure D not by duplicating c but by duplicating f or l, while keeping c constant.

Chapt. V   DETERMINATION OF METAL CONCENTRATION IN THE FLAME

This can be achieved in several ways.

Duplication of flame thickness l may be obtained effectively by placing an auxiliary, identical flame behind the original flame on the axis of the optical system. This auxiliary flame should be fed by the same metal solution nebulized by an identical nebulizer. We then measure the ratio of radiant flux received by the photodetector when the auxiliary flame is present and removed (or screened-off), respectively. This method of l-duplication, which has actually been applied by Gouy (1879) and extended by Bonner (1932) to a variable number of identical flames in a row, is not suitable for quantitative measurements, since two identical flames can hardly be realized and maintained in practice. It has a historical significance only. This difficulty is avoided if we place behind the original flame a (concave) mirror that reflects the radiation from a given spot in the flame back to the same spot. The D-factor is then derived from the relative increase in photosignal obtained with the mirror present and removed (or screened-off), respectively. This method has been tried by Gouy (1879) and later refined by Alkemade (1954), Hollander (1964), and van Trigt, Hollander and Alkemade (1965). Vanpee, Kineyko and Caruso (1970) used l-duplication to determine the absolute atomic Al concentration in a low-pressure C_2N_2-O_2 flame where a was practically zero. Huldt and Knall (1956) applied this method for a different purpose, viz. for checking the absence of Sr polymer compounds in flames; they compared the relative enhancement of the Sr emission signal as a result of doubling the Sr solution concentration with the enhancement resulting from doubling the effective flame thickness with the aid of a back mirror.

Because of reflection and other losses (which may amount to the order of 30%) the relative increase in line intensity measured in the presence of the back mirror does not directly yield the duplication factor. But these losses can be taken into account by normalizing the measured asymptotic D-factors to the theoretical value 1 or 0.415 for c going to zero or to infinity respectively.

Experimental duplication curves are shown in Fig. V.3 for Na in a shielded C_2H_2-air flame ($T = 2389$ K) and for Li, Na, K, and Cs in a shielded CO-air flame ($T = 1964$ K). Once a is known, comparison of the D-factor measured for any solution concentration (after normalization) with the corresponding theoretical D-curve in Fig. II.18 on p.189 yields n_0, the other quantities supposedly being known. When doublet lines are measured together (as was the case for Li, Na, and K, but not for Cs in Fig. V.3), the experimental curves should be compared with the theoretical doublet-duplication curves. The latter curves are simply derived from the duplication curves for a single line by taking into account the difference in f-values of the two lines, as was done for the doublet-COG in the preceding subsection.

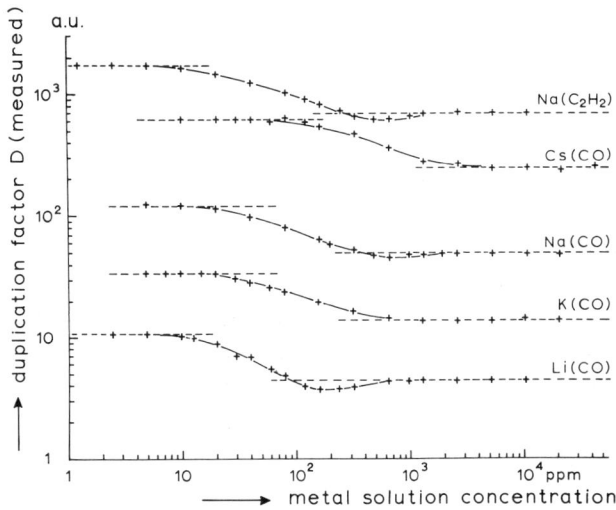

Fig. V.3 Experimental duplication curves for Na (at 5890/96 Å) in a shielded C_2H_2-air and CO-air flame, and for Li (at 6708 Å), K (at 7665/99 Å), and Cs (at 8521 Å) in a CO-air flame (according to van Trigt, Hollander and Alkemade 1965). The duplication factor is expressed in arbitrary units. The initial and final asymptotes corresponding to absolute values of $D = 1$ and $= 0.415$, respectively, are indicated by dotted horizontal lines.

The duplication method involves only relative intensity measurements and does not require n (or n_0) to be proportional to c. The method still works when for example ionization affects the relationship between n_0 and c (in Fig. V.3, incidentally, Cs was added in excess to de-ionize Na in the C_2H_2-flame, in order to check the theoretical D-curve). No great demands are made on the linearity of the measuring equipment, as the ratio of intensities to be measured varies by at most a factor of two.

When *self-reversal* occurs in the high-concentration range owing to a relatively cooler flame edge (see also Sect. 2d), the asymptotic value $D = 0.415$ will not be attained and the experimental D-factors will remain below the theoretical values. In fact, the behaviour of D at high concentrations is a sensitive criterion for the occurrence of self-reversal. This criterion is not affected by a possible deterioration in the efficiency of nebulization at high salt concentrations, which does distort the COG.

The D-factors measured in the low-concentration range may also help to correct line intensity measurements for the intensity loss due to the onset of self-absorption, as mentioned in Sect. 2a.

Chapt. V   DETERMINATION OF METAL CONCENTRATION IN THE FLAME

If the spectral apparatus used allows for sufficient spectral resolution, a measurement of the intensity ratio of two doublet resonance lines with f-values that differ by a factor of two may also be used to determine absolute metal concentrations. This method is in effect equivalent to a *duplication of the oscillator strength* f, and its theoretical treatment as well as practical advantages and limitations are similar to those discussed for l-duplication.

The first resonance doublets of the alkali metals, having well established f-ratios of 2:1, are suitable for this purpose. Li is an exception because of the appreciable overlap of its doublet components. When the transmission factor of the spectral apparatus or the sensitivity of the photodetector is markedly different for the two components, a normalization procedure similar to that outlined above in the case of an imperfect back mirror may be applied. In contrast to the mirror method, however, the additional assumption has now to be made that the a-parameter (and thus the collisional line broadening) is the same for the doublet lines. Behmenburg (1964) states that this assumption is justified, not only from theoretical considerations but also from observations on the COG made by Hofmann and Kohn (1961). The latter authors measured the intensity ratio of alkali-doublet components as a function of concentration. Application of this method further assumes that the metal concentration is not so high that the doublet components begin to overlap because of self-absorption broadening.

Measurement of the intensity ratio of doublet lines may also be used to correct the intensity of these lines for self-absorption loss. This application has already been reported by Heiërman (1937).

One may even think of using the intensity ratio of (alkali) doublet lines as a measure for the metal solution concentration for *analytical purposes* (see Schmidt 1963). An analytical curve relating this intensity ratio to the solution concentration has then to be constructed by nebulizing a series of reference solutions. This method, just as the atomic absorption method, is virtually independent of variations in flame temperature. A disadvantage, however, is that the former is only applicable in the limited concentration range in which the intensity ratio gradually changes from 2 to $\sqrt{2}$ with increasing self-absorption.

3. SPECTROMETRIC DETERMINATION OF CONCENTRATION RATIOS

Ratios of two metal concentrations can, of course, be found by measuring each concentration in absolute value by one of the methods described in Sect. 2, and by dividing the two results. More straightforward methods based on relative measurements of intensities or absorption factors will be described in this section.

SPECTROMETRIC DETERMINATION OF CONCENTRATION RATIOS   Sect. V.3a

3a   CONCENTRATION RATIOS OF ONE SPECIES

In some investigations one is interested in the ratio of concentrations of a given species in different situations in the flame, rather than in its absolute concentration. For example, a knowledge of the relative variation in atomic Li concentration with increasing height of observation may yield information on the decay of free H radicals with height (see Sect. IV.5b-2). Other examples are the measurement of the relative increase of atomic K concentration upon addition of Cs, caused by ionization suppression (see Sect. IX.3b-2), or the measurement of the relative concentration profile of metal vapour in flames in the determination of diffusion constants (see Chapt. X). In this section we discuss methods of performing such measurements by spectroscopic means.

3a-1   *INTENSITY MEASUREMENTS*

In a radially homogeneous flame and in the *absence of self-absorption*, we can find the ratio of atomic concentrations n' and n'' of a given metal species in two different situations from the corresponding ratio of line radiances B' and B'' by applying Eq. V.1

$$B'/B'' = \exp\left[-(E_1/k)(1/T' - 1/T'')\right] n' l' / n'' l'' . \qquad (V.12)$$

We have neglected here, as usual, any dependence of the partition function Q_X on temperature T. The ratio B'/B'' follows directly from relative readings of a linear spectrometer if the measuring conditions are kept the same. No absolute intensity calibration is required. If the temperature difference $\Delta T \equiv T' - T''$ is small compared to T', we can approximate $1/T' - 1/T''$ by $-\Delta T/T^2$ where $T \equiv \frac{1}{2}(T' + T'') \approx T'$. This substitution may be profitable as temperature differences can be measured to a better accuracy than absolute temperatures.

Molecular concentration ratios can be determined in a similar way, but here the temperature dependence of the vibrational (and rotational) parts of the partition function cannot in general be neglected (see also Sect. II.5c-3). Molecular transition probabilities do not need to be known.

If $n'(r)/n'(0)$ and $n''(r)/n''(0)$ depend on the radial distance r in the same way, we can find $n'(0)/n''(0)$ directly from the measured ratio B'/B'' without the radial distribution functions being known. When, in addition, the flame temperature depends on r, but in such a way that $T'(0) - T'(r) = T''(0) - T''(r)$, Eq. V.12 can still be used if the variation of $T'(r)$ with r and/or the value of $\Delta T \equiv T'(0) - T''(0)$ are so small that $\exp\left[E_1 \Delta T/kT(r)^2\right]$ is virtually constant.

Under conditions of *strong self-absorption* where Eq. V.3 can be used, we find the ratio n'/n'' from

Chapt. V DETERMINATION OF METAL CONCENTRATION IN THE FLAME

$$B'/B'' = \exp[-(E_1/k)(1/T' - 1/T'')](n'l'/n''l'')^{\frac{1}{2}} (\delta\nu'_C/\delta\nu''_C)^{\frac{1}{2}}. \quad (V.13)$$

Since $\delta\nu_C$ does not depend critically on T, we usually have $\delta\nu'_C/\delta\nu''_C \approx 1$, assuming that the flame gas composition is not markedly altered. Equation V.13 can then be handled in a similar way as Eq. V.12.

If the extent of self-absorption differs greatly in the two situations compared, the pertinent equations for B'/B'' can be found by a simple combination of Eqs V.1 and V.3. Then knowledge of the absolute $\delta\nu_C$ value is required. The determination of concentration ratios will, in general, be more complicated in such mixed cases.

For a further discussion of complications that may occur with the intensity method we refer to Sect. 2a.

3a-2 *ABSORPTION MEASUREMENTS*

The ratio of absorption factors α'/α'' in the case of *integral-absorption* measurements with a continuum background source may yield the corresponding ratio $n'l'/n''l''$ of a given species by application of Eq. V.5 and Eq. V.6 in the case of weak and strong absorptions, respectively. In this relative measurement neither the bandwidth $\Delta\nu_M$ of the monochromator, nor the f-value need to be known. Care should be taken, however, that $\Delta\nu_M$ exceeds sufficiently the spectral width of the absorption line used (see Sect. 2b). Since both the total line intensity and the integral absorption depend on (nl) through the same factor A_t (see Sect. 2b), we can refer to the preceding subsection 3a-1 for a further discussion.

The ratio of the *peak absorbances* of a given line directly yields in good approximation $(n'l'/n''l'')(\delta\nu''_D + \delta\nu''_C)/(\delta\nu'_D + \delta\nu'_C)$ according to Sect. 2c if we assume: $0 < a < 2$. For $a \geqslant 2$ the Doppler half-intensity width $\delta\nu_D$ should be dropped in this expression. The relation between peak absorbance and (nl) holds for any value of (nl), if we disregard certain complications that may arise in practice (see Sect. 2c). If $\delta\nu_D \gg \delta\nu_C$, the ratio of absorption line widths simply equals $(T''/T')^{\frac{1}{2}}$. If $\delta\nu_C \gg \delta\nu_D$, some uncertainty may be introduced because the variation of $\delta\nu_C$ with T may not be precisely known. Anyway, this variation will not be very important as long as T' deviates from T'' by no more than, say, 100 K.

In first approximation a collisionally induced shift of the flame absorption line with respect to the lamp emission line (see Sect. 2c) does not interfere with the determination of concentration ratios. This holds because this shift is independent of the metal concentration, while its dependence on T is not strong (see Sect. VII.2b-3). Also possible (hyper-)fine-structure effects are in first approximation not critical when comparing peak absorbances of the same metal line. For a further discussion we again refer to Sect. 2c.

In the *intermediate case* between peak- and integral-absorption measurements,

SPECTROMETRIC DETERMINATION OF CONCENTRATION RATIOS   Sect. V.3b

the source line and the absorption line have comparable spectral widths. Absolute concentration measurements are difficult in this case because there exists no simple theoretical relation between the absorption factor and n. Relative concentration measurements, however, are still easy to perform. At relatively low concentrations where $k(\nu_0)l \ll 1$, the absorption factor is simply proportional to (nl), regardless of the actual shape and width of the source line. The ratio $n'l'/n''l''$ is then found directly from the ratio of absorption factors, if in the situations compared the absorption line width is not markedly different.

At higher concentrations where $k(\nu_0)l$ is no longer small compared to unity, $n'l'/n''l''$ can still be simply derived from the ratio of absorption factors. The variation of absorption factor with varying (nl) can then be empirically calibrated by plotting this factor versus solution concentration c in a given situation, that is with fixed T and l. We only have to assume that c is proportional to n and that the half-intensity width of the absorption line is not significantly different in the two situations for which $n'l'/n''l''$ is to be determined.

The latter method has been applied in cases where no narrow-line background sources were available or where the source line width was unknown (see, e.g., Alkemade 1954, McEwan and Phillips 1966, and Zeegers 1966).

3b   CONCENTRATION RATIOS BETWEEN DIFFERENT SPECIES

It is often desired to determine the concentration ratio between two different metal species in a flame. For example, a measurement of the concentration ratio of Ca- and Na atoms may yield the absolute atomic Ca concentration, if we know by other means the absolute Na concentration. For simplicity's sake we shall assume throughout this section that both species are compared under identical conditions and that the flame volume observed has a uniform temperature. For a treatment of the more general case of varying temperature the reader is referred to the foregoing sections.

In some investigations one is not really interested in the actual ratio of concentrations of two species, but only in the factor by which this ratio changes when certain conditions in the flame are varied. For example, we may want to deduce the relative variation of H concentration with height by measuring the factor by which the ratio of Li- and Na concentrations varies with height (see Sect. IV.5). Or we may want to determine the dissociation energy of LiOH from a knowledge of the factor by which the concentration ratio of Li atoms and LiOH molecules changes when the temperature is varied (see Sect. VIII.5). We shall not present here special equations for such cases, since the reader will have no difficulty in deriving them by combining the general equations presented in this section.

Chapt. V   DETERMINATION OF METAL CONCENTRATION IN THE FLAME

3b-1   *INTENSITY MEASUREMENTS*

Denoting the quantities corresponding to the two species X' and X'' to be compared by a single and double prime, respectively, we have from Eq. V.1 for the ratio of two atomic line intensities

$$B'/B'' = (\nu_0' g_1' A' Q_{X''} / \nu_0'' g_1'' A'' Q_{X'}) \exp[-(E_1' - E_1'')/kT] (n'l'/n''l''). \quad (V.14)$$

The ratio of the atomic parameters involved in this equation, the difference of excitation energies, and the temperature must be known in order to deduce the ratio $n'l'/n''l''$ from B'/B''. The ratio of transition probabilities involved is often known to a better accuracy than the absolute values themselves. Also T will be less critical the smaller $(E_1' - E_1'')/kT$. We refer to Sect. III.14g for the determination of the ratio of line radiances from the corresponding readings of a linear spectrometer.

When one or both of the lines emitted by the species to be compared show strong self-absorption, Eq. V.3 should be applied as a starting-point and the collisional half-intensity width $\delta\nu_c$ of the line(s) will now play an important role, as discussed before.

3b-2   *ABSORPTION MEASUREMENTS*

Application of Eq. V.5 to relative *integral-absorption* measurements yields for the ratio of integral absorption factors of two lines of different metals at low concentrations

$$(\alpha'/\alpha'')_{low} = (g_0' Q_{X''} / g_0'' Q_{X'})(\Delta\nu_M''/\Delta\nu_M')(f'/f'')(n'l'/n''l''). \quad (V.15)$$

The ratio of monochromator bandwidths can be found directly by comparison of the dispersion values at the two wavelengths considered, if the slit width is kept the same and the bandwidth is determined by dispersion mainly (no slit-diffraction effects; see also Sect. III.14g). Neither the flame temperature nor the dependence on wavelength of the spectrometer response plays a role here. In this respect the absorption method obviously has an advantage over the intensity method discussed in the preceding subsection.

In the limit of high concentrations of one or both of the metals one should use Eq. V.6, which refers to the square-root-region. Just as in the corresponding case of strong self-absorption mentioned in the preceding subsection, the collisional half-intensity width $\delta\nu_c$ comes in and might introduce an additional uncertainty in the comparison.

In the case of *peak-absorption* measurements application of Eq. V.7 to both absorption lines yields for the ratio of peak absorbances

$$A(\nu_0')/A(\nu_0'') = (g_0' Q_{X''} / g_0'' Q_{X'})(f'/f'')(\Delta\nu_{eff}''/\Delta\nu_{eff}')(n'l'/n''l''). \quad (V.16)$$

SPECIAL METHODS OF DETERMINING ABSOLUTE METAL CONCENTRATIONS   Sect. V.4a

In the case of dominant Doppler or dominant collisional broadening we can replace $\Delta\nu_{eff}$ by $\delta\nu_D$ or $\delta\nu_C$ respectively (see Eqs V.7a and V.7b). If Doppler broadening dominates and Q_X is, virtually, equal to g_0 in both cases, only f'/f'' and the atomic masses need to be known in order to derive $n'l'/n''l''$ from the measured ratio of absorbances. The flame temperature which was assumed to be constant drops out. Such a situation would be favourable for the measurement of concentration ratios of different metal species. However, some additional complications, such as h.f.s. splitting or line shift could influence the results. We refer to the discussion in Sect. 2c. Use of the simple and straightforward Eq. V.16 often enables us to draw interesting quantitative conclusions from the findings in analytical flame atomic absorption spectrometry. In particular, the fractions atomized can be compared for two different metal salts from the measured ratio of peak absorbances.

4. SPECIAL METHODS OF DETERMINING ABSOLUTE METAL CONCENTRATIONS

Apart from the spectroscopic methods treated in Sect. 2 and the calculation of absolute metal concentrations from the nebulization efficiency to be discussed in the next section, some other experimental methods are feasible for measuring absolute metal concentrations. In principle, each observable process (such as ionization of metal atoms) that depends on the absolute metal concentration in the flame may serve for that purpose. Because of the very limited use of these special methods in practice (owing to uncertainties about the underlying assumptions or experimental complications), we shall mention some of them only briefly.

4a USE OF THE SAHA EQUATION FOR IONIZATION EQUILIBRIUM

The degree of ionization β_i of a metal vapour that is partly present as neutral atoms and partly as singly charged ions, is defined by: $\beta_i \equiv n_i/n_t$ where $n_t \equiv n_i + n_a$, and n_i and n_a denote the concentrations of metal ions and atoms, respectively. Assuming that ionization equilibrium exists, that no metal compounds are formed, and that $n_i = n_e$ (n_e being the free electron concentration) we can derive from the Saha equation a simple relation between β_i, n_t, and the ionization constant K_i (see Eq. IX.14). If the flame is assumed to be homogeneous, a measurement of β_i and of the flame temperature yields n_t, since K_i can be calculated through Eqs II.60a or 60b. Once n_t and β_i are known, n_i and n_a can be calculated directly.

A measurement of β_i can be performed through *de-ionization* of the element by adding a large excess of an easily ionizable element (see Sects V.2d and IX.3b-2). Using one of the methods described in Sect. 3a one determines the factor by which

513

Chapt. V   DETERMINATION OF METAL CONCENTRATION IN THE FLAME

n_a is raised in the limit of an infinitely large amount of added element (that is, fully suppressed ionization). The reciprocal value of this factor equals $1-\beta_i$. This procedure involves relative intensity or absorption measurements only.

A direct measurement of the free electron concentration by means of a calibrated microwave resonant cavity can also lead to an absolute value of the atomic concentration through the Saha equation. This method has been applied by Jensen and Padley (1966) to determine Cs concentrations in H_2 flames where natural ions are absent and Saha equilibrium is attained at 3 cm distance from the combustion zone. In a similar way, Fells and Harker (1967) have calibrated the supply of K to a C_3H_8-air flame by measuring the free electron concentration with two platinum probes. See Sect. IX.4 for these measuring methods.

Although these methods seem attractive because of their simple character, their application in practice may be hindered by the following *complications*:

(i)   The method based on the measurement of β_i works only for elements that are (partly) ionized; this presupposes a comparatively low ionization energy and/or high temperature, and a low metal concentration. Alkali elements in C_2H_2-air flames at low concentrations meet these demands.

(ii)   Ionization equilibrium is an essential condition. In some flames and for certain elements this equilibrium condition may not be fulfilled owing to relaxation effects (see Sect. IX.5a).

(iii) The occurrence of natural ionization (see Sect. IV.7) may interfere with the degree of ionization of the metal vapour. It does not seem easy to allow for this interference.

(iv)   Formation of molecules, such as LiOH and BaO, may also interfere with the ionization of metal atoms. Although these effects can easily be allowed for in theory (see Sect. IX.3b-3), they may introduce uncertain factors in the actual determination of atomic concentrations from measurements of the degree of ionization. Additional complications arise when molecular ions, such as $BaOH^+$, are also formed (see Sect. IX.3b-3). Under some flame conditions formation of negative ions, such as Cl^-, might also complicate the interpretation of ionization measurements.

(v)   The requirements for a precise knowledge of T and for a uniform distribution of T and n_t over the cross section are generally more stringent for ionization methods than for spectroscopic methods. This holds good, because firstly the dependence of K_i on temperature is more critical than that of the atomic line excitation (the ionization energy exceeds the excitation energy). Secondly, a nonuniform distribution of n_t entails a nonuniform distribution of β_i, which cannot easily be corrected for in the measurements.

CALCULATION OF ABSOLUTE METAL CONCENTRATION FROM MEASURED SALT SUPPLY   Sect. V.5a

It can be *concluded* that methods based on the Saha equation could involve considerable uncertainties in the absolute concentrations determined. Whenever they are to be used, additional experiments should be made in order to ensure that the underlying conditions are fulfilled. The ionization method is much more popular in arc diagnostics than in flame diagnostics (see Boumans 1966). For a more general and detailed discussion of the problems involved in ionization measurements, the reader is referred to Chapt. IX.

4b   OTHER METHODS

There is, at least in principle, a large variety of other special methods to determine absolute atomic metal concentrations. Optical methods that have often been used in the past to determine f-values of spectral lines, might conversely be used to determine absolute concentrations once f is known (see Mitchell and Zemansky 1961). We mention the method of *anomalous dispersion* (see Boumans 1966). *Saturated fluorescence spectroscopy* has been proposed for measuring atomic and molecular concentrations in flames (see Pasternack, Baronavski and McDonald 1978). However, saturation may change the chemical reaction rates and affect the atomic concentration (see Muller, Schofield and Steinberg 1980).

5.   CALCULATION OF ABSOLUTE METAL CONCENTRATION
     FROM MEASURED SALT SUPPLY

All metal present in any form in the flame is supplied by some special device, usually a pneumatic nebulizer, if we disregard the presence of metal impurities from the surrounding air, etc. Knowing such flame properties as the quantity of gas burnt per second and the amount of metal salt that is actually leaving the burner per second on its way to the flame, we would be able to calculate the metal content per cm^3 of flame gas. In this section we shall discuss the performance and reliability of such calculations. We shall first consider in some detail the usual case of a nebulizer with constant aspiration of metal salt solution. Next we discuss the possibility of calculating metal concentrations from the loss in weight of a metal salt bead placed in the flame. Finally we briefly mention some other, special methods.

5a   CALCULATION OF METAL CONCENTRATION FROM
     MEASURED NEBULIZATION EFFICIENCY

From the definition of the efficiency of nebulization ε_n (see Sect. III.8a) it follows that the number of moles of metallic element introduced per second into the flame is given by $10^{-3} \varepsilon_n F_1 c$, if F_1 is the aspiration rate of the solution (in $cm^3 s^{-1}$) and c is the solution concentration (in mol/ℓ). The volume flow

515

Chapt. V   DETERMINATION OF METAL CONCENTRATION IN THE FLAME

F_b of burnt gases at 1 atm (in cm³ s⁻¹) is given by: $F_b = F_u \zeta\,(T/298)$, where F_u is the volume flow of unburnt gases supplied to the burner at 298 K and at 1 atm (in cm³ s⁻¹) and ζ is the number of moles of burnt gases produced per mole of unburnt fuel-oxidant mixture. This expression for F_b holds if we disregard the contribution of water vapour from evaporated spray droplets and the dilution of the flame by turbulent mixing with secondary air. Both effects may be important with unpremixed turbulent flames (see below), while the second effect may also be important with thin flames produced by a slot-burner.

Assuming that the aerosol is uniformly distributed over the flame gases and that no other salt losses occur than those accounted for by ε_n, we simply calculate the total number density, n_t, of metal present in the gaseous state (as free atoms, ions or molecules) by combining the above expressions

$$n_t = 10^{-3}\,\varepsilon_n F_1\, c\, N_A\,(298/T)/F_u \zeta \quad \text{(in cm}^{-3}\text{)}. \tag{V.17}$$

Here N_A is Avogadro's number which equals 6.022×10^{23} mol⁻¹. When the desolvation or the volatilization of the aerosol particles is incomplete (see Sect. III.9), a correction factor should be added to Eq. V.17.

There is another approach for calculating n_t from the nebulization efficiency. When v_r is the rise-velocity of the burnt flame gases (in cm s⁻¹) and S_f is the cross-sectional area of the coloured flame (in cm²), and when a uniform flame without additional salt losses is assumed, we have

$$n_t = 10^{-3}\,\varepsilon_n F_1\, c\, N_A / v_r S_f \quad \text{(in cm}^{-3}\text{)}. \tag{V.18}$$

This equation expresses the *continuity of mass flow* for the metal considered, $(n_t v_r S_f)$ representing the flow of metal through a horizontal plane at the height of observation.

When in an inhomogeneous but cylindrically symmetric flame with outer radius r_0, $n_t(r)$ is a function of radial distance r from the flame axis, we calculate $n_t(0)$ from

$$n_t(0) = 10^{-3}\,\varepsilon_n F_1 c\, N_A \Big/ \Big[2\pi v_r \int_0^{r_0} \{n_t(r)/n_t(0)\}\, r\,\mathrm{d}r\Big], \tag{V.19}$$

if $n_t(r)/n_t(0)$ is known as a function of r and if v_r is constant over the cross section. The normalized function $n_t(r)/n_t(0)$ can be measured by scanning the atomic emission or absorption line of the metal considered in a direction perpendicular to the axis (see Sects III.12 and V.2a, and Fig. V.1). This holds if the atomic fraction of $n_t(r)$ is independent of r.

In order to calculate n_t from Eqs V.17 or V.18, the flame and nebulizer characteristics contained in F_u, ζ, T (or v_r and S_f), and ε_n, F_1, c should be

CALCULATION OF ABSOLUTE METAL CONCENTRATION FROM MEASURED SALT SUPPLY   Sect. V.5a

known. Methods of measuring F_u, v_r, S_f and ε_n are discussed in Sects III.2, 11, 12 and 8, respectively, while F_1 can be simply determined by measuring the time needed for aspirating a given volume of solution. The temperature may either be measured (see Sect. III.10) or estimated since its role is not very important. The expansion factor ζ connected with the overall combustion reaction can be calculated once T and the composition of the unburnt gas mixture are known (see also Sect. IV.4). The calculation of this factor is not very critical and possible minor deviations from chemical equilibrium may be ignored. For a premixed C_2H_2-air flame at 2500 K, ζ is about 1.1, while for an unpremixed C_2H_2-O_2 flame at 2850 K it is about 1.2 according to calculations made by Winefordner, Vickers and Remington (1965).

The reliability of calculations obtained through Eqs V.17 and V.18 may be affected by *uncertainties and complications* of different kinds.

Any uncertainty in ε_n (see Sect. III.8) affects the outcome of both equations to the same extent. The actual determination of ε_n may involve large errors because we do not only have to measure the aerosol losses inside the spray chamber, but also the losses in the gas conduits and the burner house. Besides, losses may occur owing to incomplete desolvation of the aerosol droplets in the flame, especially so for direct-injection burners (see Sect. III.9). Precipitation of metal salt on the top of the burner and loss of aerosol droplets that are hurled out of the flame with turbulent, unpremixed flames should also be accounted for.

In unshielded flames part of the gas mixture that is supplied at the rim of the burner outlet might largely fail to attain the flame temperature owing to turbulence or cooling at the border. This could entail an apparent loss of the flame gas and of the aerosol contained in it, which makes the outcome of Eq. V.18 too high. We note however that this '*stripping*' *effect* does not affect the outcome of Eq. V.17 if we assume that the aerosol is uniformly distributed over the unburnt gas mixture, including the part that is 'lost' owing to this stripping effect at the flame border.

It is also possible that higher up in laminar flames loss of metal content occurs because of *radial diffusion* of metal vapour out of the flame. Then both Eqs V.17 and V.18 would yield too high values for n_t.

The situation is different again when the coloured flame is surrounded by a flame mantle into which no aerosol is sprayed. Outward diffusion of metal vapour from the central flame into the flame mantle does not result then, properly speaking, in a loss of metal content, but only tends to increase the effective thickness of the coloured flame section at a given height. If this thickness is measured and the corresponding value of S_f is inserted in Eq. V.18, no large error would arise in the calculation of n_t. More precise results are obtained for a cylindrically symmetric flame by measuring the radial atom distribution (which is affected by diffusion) and applying Eq. V.19. Direct application of Eq. V.18, however, would yield too high values for n_t, although a correction procedure for the diffusion effect is also feasible in this case.

These diffusion effects will be the less serious the thicker the coloured flame zone and the smaller the height of observation. It is usually unimportant in flames having a thickness of about 1 cm or larger, and at a distance of a few centimetres above the burner (see, e.g., Alkemade 1954, and Hollander 1964). The diffusion effect, of course, also depends on the diffusion constant, which is specific for the kind of metal and flame gases (see Chapt. X).

In an actual premixed laminar and cylindrical C_2H_2-air flame with colourless flame mantle, fed by a chamber-type nebulizer, n_t was calculated to be 7.5×10^{14} cm^{-3} according to Eq. V.17, and 10×10^{14} cm^{-3} according to Eq. V.18, for a given Na solution (see calculations made by de Galan and Winefordner 1967 on flame and nebulizer data given by Hollander 1964). In this comparison, ε_n and F_1 drop out. The accidental error in the values compared may be about 25%, so that reasonable agreement is found for this type of flame with a coloured flame thickness of about 0.5 cm, up to a height of 8 cm. It should be noted that the ε_n value quoted by de Galan and Winefordner (1967) for Hollander's flame refers only to aerosol losses inside the spray chamber. Considerable additional losses in the long tubings to the burner, etc., are likely to occur here, so the calculated values of n_t have no absolute meaning.

Rann (1968) has used Eq. V.17 to calculate n_t for Cu in the median plane of a shielded premixed laminar C_2H_2-air flame of rectangular cross section. A correction was made for the horizontal migration of Cu atoms, using the concentration profile determined by absorption measurements. It was suggested that explosion of the salt particles entering the combustion zone could have caused this migration, in addition to thermal diffusion of the atoms. The estimated error in n_t was 10%.

Application of Eq. V.17 or V.18 to *unpremixed, turbulent flames* with direct-injection burners is more problematic because of strong secondary-air entrainment, incomplete evaporation of the spray droplets, and loss of spray droplets that are hurled out of the flame (see also Alkemade and Herrmann 1979). Neither T, n_t, nor F_b are here expected to be uniform in a horizontal cross section. Besides, the fraction of desolvated aerosol is expected to vary with varying height of observation because of progressive evaporation of the grosser droplets with increasing time spent in the flame (see Sect. III.9). Attempts have been made by Simon (1960), Püschel, Simon and Herrmann (1964) and McGee and Winefordner (1967) to calculate

CALCULATION OF ABSOLUTE METAL CONCENTRATION FROM MEASURED SALT SUPPLY   Sect. V.5b

absolute metal concentrations in unpremixed turbulent H_2-O_2 and $C_2H_2-O_2$ flames from the supply of metal salt per second to the burner. Comparison of these calculations with absolute spectroscopic measurements of the atomic metal concentration reveals that these calculations are likely to yield n_t values largely in excess of the true values. Such calculations should thus be applied with much caution to these flames, even when the radial distribution of the metal vapour is explicitly taken into account (see Simon 1960).

5b   CALIBRATION OF METAL SUPPLY WITH WEIGHED BEADS
OF SALT AND SOME OTHER METHODS

A weak spot in the application of Eq. V.18 is the uncertainty in the amount of salt that is actually supplied by a chamber-type nebulizer. Heiërman (1937) has described a method of introducing alkali vapour by placing a bead of fused alkali salt in the flame near its bottom. The measured loss of weight of the bead in the flame during a known time interval provides a basis for calculating the alkali concentration, in analogy to Eq. V.18. Belcher and Sugden (1950) have used essentially the same method for calibrating the salt supply from a chamber-type nebulizer. They have placed a small weighed bead of fused NaCl near the burner in the flame and integrated over time the radiant flux emitted from a horizontal flame slab through which all evaporated Na atoms had to pass, until the bead was completely evaporated. By comparing this flux with the integrated radiant flux received by the same photodetector in a known time interval while constantly nebulizing a given Na solution, they were able to calibrate the absolute supply of Na delivered by the nebulizer. From this calibration and the measured rise-velocity the Na concentration follows directly (compare Eq. V.18).

Although the calibration procedure by means of weighed salt beads circumvents certain difficulties connected with the measurement of ε_n, it is not often used nowadays. Its application may be hampered by the disturbance caused by the presence of the bead and its supports inside the flame, by time-varying evaporation from the bead (which requires integration measurements), by the strong inhomogeneity of the resulting metal distribution, and by the possibility of salt losses owing to precipitation on the burner top. The latter effect was eliminated by Heiërman (1937) with the aid of an additional blow pipe placed below the bead. Metal supports heated to flame temperature might, besides, cause interference owing to thermionic emission of electrons, whereas flame radicals might recombine at the wall of the support.

The amount of metal seeded as an aerosol to the flame can also be determined by *trapping methods*, for example, by passing the carrier gas leaving the burner outlet, with the flame off, up a packed column in order to absorb the seed in water (see Fells and Harker 1967). The concentration of the cation in the water

Chapt. V　　　　　　　　DETERMINATION OF METAL CONCENTRATION IN THE FLAME

may be found by flame spectrometry or titration. When the dispersion of the aerosol is too fine to be absorbed in this way, the amount of seed may be measured by trapping it in a Terylene fibrous material contained in a tube fastened to the burner outlet (see Fells and Harker 1967). The tube and fibre are weighed before and after trapping. From the difference in weight and the duration of trapping, the concentration of the element in the flame can be calculated, if the volume flow of the flame gases is known.

Cotton and Jenkins (1968) used an electrostatic precipitator to collect a sample of aerosol above the burner.

The *single-drop generator* developed by Hieftje and Malmstadt (1969) allows for a simple absolute calibration of the amount of metal salt introduced. With this technique single drops of known, uniform diameter are injected by a special device directly into the flame gas (see also Sect. III.3b). The amount of solute brought into the flame can be calculated from the drop diameter and the solution concentration. The completeness of desolvation and volatilization can be checked by spectroscopic observations above the point of injection (see Bastiaans and Hieftje 1974; see also Sect. III.9).

6. THE COMPARISON METHOD AND ELEMENTS SUITABLE AS FULLY ATOMIZED STANDARDS

General. In many cases one is interested in the *fraction atomized*[†] $\beta_a \equiv n_a/n_t$; the atomic concentration, n_a, can be determined from spectroscopic observations (see Sect. 2), and the total concentration, n_t, can be calculated from Eq. V.17 or V.18. In the latter calculations large errors may be involved (see Sect. 5).

There is, however, another way to find β_a, which is called the *comparison method*. It involves the comparison of the atomic concentration n_a of the investigated element with the corresponding n_a^* of a 'standard' element that is known to be present only as free, neutral atoms. If we assume that the efficiency of nebulization is independent of the kind of metal present in the solution and that the desolvated aerosol particles completely volatilize, n_t in the gas phase must be the same for **equimolar** concentrations of different metals in the sprayed solution. Under this condition a spectroscopic measurement of n_a as well as of n_a^* then yields directly the fraction atomized of the former element through the relations

$$\beta_a = n_a/n_t = n_a/n_a^*.$$

[†] The fraction atomized is here defined with respect to free *neutral* atoms only, and does not include free *ionized* atoms (compare IUPAC 1976a).

520

THE COMPARISON METHOD & ELEMENTS SUITABLE AS FULLY ATOMIZED STANDARDS   Sect. V.6

The uncertainties inherent in calculations of the efficiency of nebulization are circumvented in the comparison method. The latter method is based on a spectroscopic measurement of the ratio of two atomic concentrations. Both methods, however, require the complete volatilization of the desolvated aerosol particles and the absence of noticeable diffusion effects which could be markedly different for the two elements compared.

We now discuss the *general criteria* that must be satisfied by a *fully atomized standard element*, and some methods to check these criteria in practice.

(i) The standard element should emit or absorb an atomic spectral line that has a well-known oscillator strength, h.f.s. splitting and half-intensity width (unless spectrally integral measurements are applied under condition of low optical thickness; see Sect. 2). Moreover, when emission methods are used, the excitation energy should be known and the excitation should be in thermal equilibrium.

(ii) The standard element should be completely released from the dry aerosol particles that remain after evaporation of the solvent. Possibilities of checking this criterion were discussed in Sect. III.9. It is not necessary that all droplets evaporate completely if the fraction of spray droplets that are desolvated is the same when spraying the standard element or the element to be investigated.

(iii) In the gas phase the standard element should be present only in atomic form, that is its β_a should equal unity.

A general *check of these criteria* as a whole can be obtained by comparing the absolute atomic concentration n_a^* found by spectroscopic observations, with the value of n_t that follows from calculations of the efficiency of nebulization. If n_t is found to equal invariably n_a^* for a number of different standard elements and under different operating conditions of nebulizer and flame, this comparison may be considered as a good check. This holds notwithstanding nebulization calculations should, in general, be considered with some reserve.

The absence of molecule- or ion formation may be argued by theoretical considerations which are based on the composition and temperature of the flame used and for which a rough estimate of the absolute metal concentration may suffice. However, there might occur departures from (chemical) equilibrium that invalidate such calculations.

It is always recommended to supplement these theoretical considerations with direct or indirect experimental evidence, especially so when extreme conditions as to the temperature and composition of the flame occur. Ionization is most easily checked indirectly by looking for any enhancement of the atomic concentration upon adding to the solution a second, easily ionizable element. Also a supralinear curve-of-growth in the range of low metal concentrations may be indicative for the

Chapt. V   DETERMINATION OF METAL CONCENTRATION IN THE FLAME

occurrence of ionization, as was mentioned in Sect. 2d. When in the hotter flames ionization is feared, it can be suppressed through addition of a sufficiently large excess of an electron donor element, such as Cs. Direct evidence on the extent of ionization may also be obtained from electrical measurements, as will be explained in Sect. IX.4.

Formation of metal compounds cannot so easily be excluded, especially so when these molecules emit no radiation. We refer to Chapt. VIII for a discussion of the occurrence of metal compounds.

Internal consistency can be proven when two or more elements are expected to behave as fully atomized standards. Under various conditions spectroscopic measurements should then yield the same value of the absolute atomic concentration when equimolar solutions of these elements are nebulized.

When the experimental set-up is suited to burn flames with various, known unburnt gas compositions, there is another indirect method to check the absence of noticeable compound formation, which has been often used by Sugden and his collaborators. This method is based on the observation of a so-called *family-effect* in the behaviour of the atomic line intensity of the element considered. A 'family' of, for example, H_2-O_2-N_2 flames is characterized by a constant N_2/O_2 volume ratio in the unburnt gas mixture; its members have different H_2/O_2 volume ratios ranging occasionally from typically fuel-rich through stoichiometric to typically fuel-lean. The range of temperatures within a given family may partly overlap that of a proximate family. Coincident temperatures may even be found between members of the same family that correspond to a fuel-rich and a fuel-lean mixture, respectively. For convenience, the total volume flow, F_u, of unburnt gas mixture is often adjusted to the same value for each flame. In order to ensure an unambiguous interpretation of the experimental results as a function of T, the temperature and metal vapour should be homogeneously distributed over a horizontal cross section of the coloured flame column.

When the metal atoms are able to react with flame molecules or radicals to form compounds, β_a will have a value which depends on temperature as well as on flame composition. Consequently β_a will be different for two flames belonging to distinct families but having the same temperature. Absence of any such family-effect over a relatively wide range of temperatures and flame compositions can conversely be considered as a proof for the absence of compound formation. Below we shall give an example of such a proof for Na in hydrogen flames.

Under normal conditions Na and Ag can be used as fully atomized standard elements for which β_a is, virtually, equal to unity. The following theoretical arguments and experimental evidence from the literature can be brought forward to corroborate this statement.

THE COMPARISON METHOD & ELEMENTS SUITABLE AS FULLY ATOMIZED STANDARDS   Sect. V.6

Sodium as fully atomized standard element. James and Sugden (1955) and Jensen and Padley (1966) have concluded from theoretical calculations that the formation of NaOH molecules and Na^+ ions should be negligible in fuel-rich H_2 flames up to 2300 K. At $T \approx 2500$ K ionization of Na may become noticeable at dilute Na concentrations, that is, solution concentrations of about 10^{-3} mol/ℓ or less. It is easily suppressed by adding about 0.1 mol/ℓ CsCl. At this temperature the ratio of NaOH concentration to atomic Na concentration could amount to about 10%. These conclusions about the insignificance of NaOH formation may also hold for hydrocarbon flames and a fortiori for (dry) CO flames at comparable temperatures. In fuel-lean H_2 flames the formation of NaO_2 molecules may become significant at temperatures equal to or below about 2000 K (see Sect. VIII.3a). The extent of NaO_2 formation seems to decrease with decreasing height in the flame. James and Sugden (1955) have found the absolute atomic Na concentration as derived from the emission curve-of-growth to agree within 30% with calculations from the efficiency of nebulization for laminar, premixed H_2 flames with a chamber-type nebulizer. The α-parameter was here assumed to equal 1.0. In accordance with these conclusions they have not found a family-effect. Fig. V.4 shows indeed $\log_{10} I$ to depend solely on T^{-1} and not explicitly on the composition of the H_2-O_2-N_2 flames used, in a large range of temperatures (1650 – 2250 K). The nebulizer delivered a constant amount of Na per second and the total flow F_u of unburnt gas was the same in all flames. The I values plotted in Fig. V.4 have been corrected for the effect of variations in T^{-1} and ζ on the dilution of the metal vapour in the flame gases, as is described by Eq. V.17. Self-absorption was negligible. The straight line through the experimental points has a slope which corresponds with the excitation energy of the Na-D lines (2.1 eV), as should be expected. This straight line also proves the absence of suprathermal chemiluminescence at temperatures down to 1650 K. The authors have ascribed the slight 'hook' effect for near-stoichiometric flames (at the high-temperature end of the plot) to a dilution of the central flame by mixing with the mantle flame.

Following the above method, Gurvich and Ryabova (1964a) have also found that formation of Na compounds (NaOH, Na_2) and Na^+ ions in shielded, premixed H_2-O_2-N_2 flames at 1600 – 2200 K is negligible. Moreover, they have found the atomic Na concentration as derived from absolute intensity measurements to agree within 10% with their nebulization-efficiency calculations.

Hollander (1964) has checked the constancy within 5% of the atomic Na concentration as a function of height up to 8 cm in premixed, laminar CO-air and C_2H_2-air flames with thickness $l = 1.5$ cm. The flame, which was surrounded by a colourless mantle flame, was fed by a chamber-type nebulizer. Ionization was suppressed. He concludes there is complete vaporization and volatilization of the

523

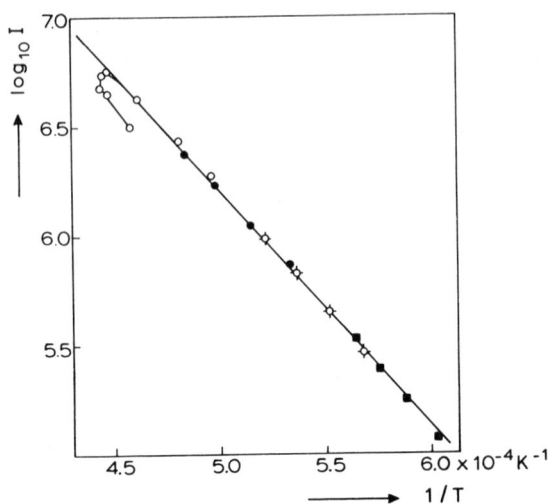

Fig. V.4 The relative intensity I of the Na-D lines is plotted on a logarithmic scale versus the reciprocal flame temperature T for four 'families' of H_2-O_2-N_2 flames, presented by ○, ●, -◊- and ■ (according to James and Sugden 1955). Each family has a constant pre-burnt N_2/O_2 ratio, while its members have varying H_2/O_2 ratios. The straight line through the experimental points is drawn with a slope corresponding with the excitation energy of the Na-D lines; this proves that the Na emission depends solely on T. Formation of molecular compounds can thus be excluded. An explanation of the slight 'hook' effect is given in the text.

wet and dry aerosol particles and no appreciable NaOH formation. Also diffusion seems not to affect the atomic Na concentration in these thick flames with rise-velocity of about 750 cm s^{-1}. In a CO-O_2-N_2 flame at 1964 K, where alkali hydroxide formation is a priori improbable, Hollander has found the same atomic alkali concentration within 10% when nebulizing equimolar concentrations of Li, Na, K, and Cs. These measurements, which were performed with the emission curve-of-growth method in absence of ionization, prove that β_a is most likely to be unity for all alkali metals in these flames.

De Galan and Winefordner (1967) have measured β_a by comparison of the absolute (low) atomic Na concentration, as derived from integral-absorption measurements, with the total concentration calculated from the nebulization efficiency. Additions of excess CsCl to the nebulized solution turned out to raise β_a from 0.5 to about unity, which points to partial ionization of Na at low concentrations

THE COMPARISON METHOD & ELEMENTS SUITABLE AS FULLY ATOMIZED STANDARDS   Sect. V.6

in the C_2H_2-air flame used.†

From integral-absorption measurements, de Galan and Samaey (1970) have concluded that Na is (nearly) completely atomized in fuel-rich flames of $C_2H_2 - N_2O$ ($T = 2950$ K), $H_2 - N_2O$ ($T = 2900$ K), C_2H_2-air ($T = 2450$ K), and H_2-air ($T = 2000$ K).

Willis (1971) has found that atomization of Na in a shielded stoichiometric, Méker-type C_2H_2-air flame ($T = 2320$ K) is essentially complete (see also below).

From equilibrium calculations reported by Rasmuson, Fassel and Kniseley (1973) the same conclusion can be drawn for Na in $C_2H_2 - N_2O$ flames with oxidant/fuel flow ratios between 1.95 and 2.8. They have calculated that at lower ratios appreciable formation of NaCN has to be expected, and at higher ratios formation of NaOH and NaO.

The formation of Na_2 molecules in flames can be excluded, because the shape of the experimental curve-of-growth for the Na-D lines is, in general, found to agree with theoretical predictions based on the assumption that n_a is proportional to the solution concentration (see also Sect. VIII.3).

Silver as fully atomized standard element. Silver is promising as a standard element because it has a high ionization energy of 7.6 eV (compared to 5.1 eV for Na) and because the only band emission observed in flames (see Bengtsson and Olson 1931) has been assigned to AgH with a dissociation energy of only 2.49 eV (see Mavrodineanu and Boiteux 1965). Moreover, the value of its oscillator strength is reasonably well-known. All f-values obtained for the Ag resonance line by various methods lie between 0.44 and 0.50 (see Penkin and Slavenas 1963, Lawrence, Link and King 1965, Moise 1966, Levin and Budick 1966, and Cunningham and Link 1967).

Willis (1971) has determined from absorption measurements with a continuum source the absolute atomic concentration of Na, Cu, Ag and Au in a shielded Méker-type air-acetylene flame. He found mutual agreement between the absolute atomic concentrations of Na, Ag and Au for equimolar solution concentrations, which points to full atomization of Ag.

Smyly *et al.* (1971) have made a special study of Ag as a fully atomized standard element. In twelve $H_2 - O_2 - Ar$ flames with various gas compositions and with temperatures ranging from 1795 to 2350 K they found that the absorption from a continuum source varied by less than about 5% for a given Ag solution concentration. Therefore they concluded that silver can be regarded as a good reference element for measuring free-atom fractions.

---

† From absorption measurements at the 3302 Å Na line de Galan and Winefordner (1967) calculated a value $\beta_a = 0.5$ for a Na concentration of about 10^{-2} mol ℓ^{-1}. In this calculation too high a value of the oscillator strength was used and the true value of β_a will be nearer to unity, as expected. This β_a value did not noticeably vary upon addition of Cs, which is to be explained by the low degree of Na ionization at this high Na concentration.

Chapt. V DETERMINATION OF METAL CONCENTRATION IN THE FLAME

Copper as fully atomized standard element. Copper has been investigated as a standard element because it has an ionization energy of 7.7 eV and it shows little tendency to form compounds in the flame. Moreover the residual formation of CuH, CuOH, CuO, or Cu halides can easily be checked by observing the corresponding molecular emission bands. Bulewicz and Sugden (1956a) in fact found that the intensity of the 3274 and 3247 Å Cu resonance lines as a function of T showed no family-effect in H_2-O_2-N_2 flames. They also found that the intensity is almost independent of the height above the burner in flames where T does not vary markedly with height. Hinnov and Kohn (1957) found $\beta_a = 0.82$ for Cu in C_2H_2-air flames from emission curve-of-growth measurements when assuming $\beta_a = 1$ for Na. De Galan and Winefordner (1967) concluded that Cu is an ideal standard element, which in their C_2H_2-air flame forms virtually no compounds or ions.

By borrowing a-values from the literature Rann (1968) concluded from his absorption measurements that Cu and Na are dissociated by only 37 and 40%, respectively, in an C_2H_2-air flame at $T = 2273$ K, whereas Ag is completely dissociated.[†] However, as Willis (1971) pointed out, Rann's interpretation of his Na- and Cu peak-absorption measurements was in error because he neglected hyper-fine-structure in the line sources (see also Hannaford and McDonald 1978). On the other hand, Willis (1971), using for the Cu 3247 Å resonance line the f-value of 0.32 measured by Moise (1966), concludes that under his experimental conditions the atomization of Cu is essentially complete. However, Bielski's (1975) critical survey of f-values measured by different authors employing various methods led to a best value $f = 0.43$, in accordance with the value measured later by Krellmann, Siefart and Weihreter (1975) employing the lifetime method. The question of whether Cu is fully atomized remains still open.

Rasmuson, Fassel and Kniseley (1973) found that the atomic Cu concentration in an C_2H_2-N_2O flame did not depend on flame gas composition. They concluded that atomization is complete because, if Cu compound formation had been significant, the degree of dissociation and thus the atomic concentration would have depended markedly on the flame gas composition.

Other elements. Also Au (see Willis 1971) and, in hot flames, Fe and Mg (see Rasmuson, Fassel and Kniseley 1973, de Galan and Winefordner 1967[*], and de Galan and Samaey 1970) may show practically complete atomization. These elements have not been investigated systematically as to their suitability as standard elements. The suitability of Tl has been investigated by several authors (see Hinnov

---

[†] For a critical discussion of Rann's conclusions see also de Galan (1969) and Rann (1969).

[*] When using a more recent f-value for the 2852 Å Mg line, we calculated for Mg: $\beta_a = 0.8$, instead of $\beta_a = 0.59$ as reported by these authors.

THE COMPARISON METHOD & ELEMENTS SUITABLE AS FULLY ATOMIZED STANDARDS   Sect. V.6

and Kohn 1957, and Bulewicz and Sugden 1958), but the uncertainty about the f-value of the 3775 Å Tl line does not allow a definitive conclusion to be drawn.

Conclusion. We can conclude that Na with negligible or suppressed ionization, and most probably Ag, can be used as fully atomized standard elements in normal flames, at least between 2000 and 2500 K. Their resonance lines at 5890/ 5896 Å and 3281/3383 Å, respectively, should be used. Comparison of the atomic concentrations of other elements with the atomic concentration of Na or Ag, for equimolar solution concentrations, may thus yield the fraction atomized in a much simpler and generally in a more accurate way than comparison with nebulization-efficiency calculations.

Possible systematic errors in the determination and comparison of atomic metal concentrations by spectroscopic methods, due to, for instance, errors in the f-values, may always be easily corrected for afterwards. Systematic errors in the determination of the efficiency of nebulization ε_n, which is strongly dependent on (unknown) instrumental factors, cannot so easily be checked and corrected for. The accuracy in β_a, as far as it depends on the standard element used in the comparison method, may be estimated to be of the order of 10% or perhaps somewhat better.

Thick, premixed, shielded and homogeneous flames and a short height of observation are recommended for application of the comparison method. Very high salt concentrations should be avoided on account of possibly incomplete volatilization. At temperatures below about 1500 K suprathermal chemiluminescence might, however, invalidate the comparison method. Excess amounts of chlorine in the flame should also be avoided, since a noticeable proportion of Na could then be bound as NaCl, as will be discussed in Sect. VIII.3c.

CHAPTER VI

Excitation and De-excitation of Metal Species †

1. INTRODUCTION

Survey and terminology. This chapter deals with the processes of (de-) excitation in relation to the optical radiation of metal vapours in flames.

By excitation, we mean any process that brings the atom or molecule from a lower state of internal excitation (often the ground state) to a higher state. We shall be concerned mostly with the (de-)population of electronic excited states from which optical transitions are possible in a wavelength range extending from the near-infrared to the ultraviolet part of the spectrum. We shall consider mainly metal atoms as radiating species, but there will be a few digressions relating to the excitation of metal compounds too.

(De-)excitation processes can be generally classified as *physical* or *chemical* processes (see also Sect. II.4h); these will be discussed in Sects 2 and 3, respectively.‡ The physical processes are here subdivided in *radiative* (Sect. 2c) and *collisional* processes (Sects 2a and 2b). *Stepwise radiative* (or *photo-*) *excitation* is a particular radiative process in which the atom is first raised to an intermediate level by photon absorption at wavelength λ_1, from which it is further excited to a higher level by photon absorption at (another) wavelength λ_2. The term *photolysis* will be used here for the chemical process no. 27 in Table II.1 on page 108, in which an excited atom A^* is produced by photodissociation of

---

† An abridged and older version of this chapter has been published by Alkemade and Zeegers (1971) in the book "Spectrochemical Methods of Analysis", edited by J.D. Winefordner. That version includes also some conclusions for the analytical application of flame spectroscopy.

‡ Section numbers without a roman cipher refer to this chapter.

Chapt. VI EXCITATION AND DE-EXCITATION OF METAL SPECIES

molecule AB. A *quenching*† process is here defined as a radiationless transition of an excited atom to another (usually lower) excitation level or ground level belonging to a different spectroscopic term [e.g. $Na(3^2P_{\frac{1}{2}} \to 3^2S_{\frac{1}{2}})$, $Cs(7^2D_{\frac{3}{2}} \to 8^2P_{\frac{1}{2}})$, $K(10^2P \to 11^2P)$ or $Hg(6^1P_1 \to 6^3P_0)$] This transition may be induced by a physical collision (*physical quenching*) or a chemical reaction (*chemiquenching*). When a collision induces a transition between different components of the same multiplet [e.g., $Na(3^2P_{\frac{3}{2}} \leftrightarrow 3^2P_{\frac{1}{2}})$ or $Tl(6^2P_{\frac{3}{2}} \leftrightarrow 6^2P_{\frac{1}{2}})$], we shall speak of an (*intramultiplet* or *fine-structure*) *mixing collision*.* When electronic excitation energy is transferred from one atom to another upon collision, the term (*electronic*) *excitation transfer* is used.‡ When photo-excitation is followed by emission of radiation at the same or a different wavelength, one speaks of *photoluminescence* or *fluorescence* (for short). Different kinds of fluorescence are distinguished in Sect. II.5a-1. When chemi-excitation is followed by emission of radiation, the term *chemiluminescence* is used.

In this chapter, the rate constants of transitions between Zeeman-substates of a given level (called M_J *mixing*) will not be considered. M_J mixing collisions tend to decrease the polarization and anisotropy of the angular distribution of the fluorescence radiation. Furthermore, they contribute to the atomic line broadening (see Chapt. VII). Depolarization of the $Cd(5^1P_1 - 5^1S_0)$ resonance fluorescence at 2288 Å in a H_2-air flame at or below 1 atm pressure has been studied by Chenevier and Lombardi (1972, 1974) (see also Sect.II.5c-4). The Na-D_1 fluorescence line, involving a $J=\frac{1}{2} \to J=\frac{1}{2}$ transition, should be unpolarized when excited by a linearly polarized radiation beam, as it was experimentally confirmed in flames at 1 atm by Jongerius, van der Bij, Hollander and Alkemade (1978). When the blue wing of the Na-D_2 line was excited by a linearly polarized laser beam about 5 Å off-resonance, the D_2 fluorescence emitted perpendicular to the polarization direction showed a degree of polarization of about 4% in a H_2-O_2-N_2 flame at 1 atm (see last reference). This polarization decreased with increasing

---

† This term was originally used, in a more restricted sense, in conjunction with fluorescence radiation. The extended definition refers to radiationless depopulation of excited states more generally.

* Note that the meaning of 'mixing' here is different from that in Quantum-mechanics ('mixing of quantum states'). The term 'mixing' collisions has also been used for collisional transitions between high-lying, neighbouring levels with same quantum number n but different quantum numbers l (see Gallagher, Edelstein and Hill 1977). In this book, however, these are called 'quenching' collisions. In the literature 'fine-structure mixing' is also called 'spin-orbit relaxation'.

‡ Krause (1975) uses this term in a broader sense, including also multiplet-mixing collisions with, e.g., noble gas atoms.

INTRODUCTION   Sect. VI.1

laser detuning.† In a H_2-O_2-Ar flame under similar measuring conditions the
degree of polarization was found to be zero. This outcome can be qualitatively
explained by the smaller quenching efficiency of this flame, which results in a
longer lifetime of the excited state. When the lifetime is longer, the number
of depolarizing collisions during this lifetime is increased (see also Pringsheim
1949, and Mitchell and Zemansky 1961). When the polarized laser beam was tuned
in the red wing of the Na-D_1 line, the fluorescence emitted by the other (D_2)
component, as a result of doublet-mixing collisions, was unpolarized, in accordance with theory (see Jongerius, van der Bij, Hollander and Alkemade 1978).*
No polarization was found in the fluorescence of other metal atoms in flames at
1 atm by Omenetto, Hart and Winefordner (1972), Lijnse, Zeegers and Alkemade
(1973), and Larkins and Willis (1974).

For a survey of more recent collisional depolarization experiments with
metal atoms in vapour cells as well as for a general outline of the theory and
the experimental methods, see Seiwert (1968), Massey, Burhop amd Gilbody (1971),
Krause (1972), Lijnse (1973b), Niewitecka, Skalinski and Krause (1974), Nikitin
(1975), and Niewitecka and Krause (1975) (see also Baylis 1979).

Curry et al. (1978) measured the rate at which the $M_J (\equiv M_S)$ substates of
the Na($^2S_{\frac{1}{2}}$) ground state are collisionally mixed (that is, the spin-relaxation
rate) in hydrocarbon flames at atmospheric pressure. They induced ground-state
spin polarization by optical pumping of the Na-D_1 line with the aid of a circularly polarized, c.w. dye-laser beam. When the sense of the circular polarization is chosen such that the selection rule $\Delta M_J = +1$ holds (cf. Sect. II.2c),
the population of the $M_J = -\frac{1}{2}$ ground substate is depleted. The Na atoms are
then trapped in the $M_J = +\frac{1}{2}$ ground substate, from where they cannot be excited
by the laser beam because the maximum M_J value of the $(^2P_{\frac{1}{2}})$ upper state is $+\frac{1}{2}$.
Consequently the laser-induced Na fluorescence intensity is also weakened. How-
ever, spin-exchange collisions partly restore this depletion and ensuing reduction of fluorescence intensity. This restoration can be enhanced by applying an
external d.c. magnetic field B in a suitable direction. When one plots the
fluorescence intensity as a function of B, a curve is obtained with a minimum

---

† At resonant excitation the degree of polarization appeared to be maximal but
lower than 10% (see Jongerius 1981).

* A full quantummechanical theory (in terms of generalized multipole-relaxation
matrices) for the effects of M_J- as well as multiplet mixing and quenching
collisions on the strength, polarization and anisotropy of the multiplet components of a fluorescence line excited by nonsaturating resonant or nearly resonant
light has been given by Nienhuis (1978a). The application of the general theory,
including Rayleigh scattering, to the ^2S-^2P transitions in alkali atoms has been
worked out by the same author.

Chapt. VI   EXCITATION AND DE-EXCITATION OF METAL SPECIES

at $B=0$ and a width that is a measure for the collisional spin-relaxation rate
(in analogy with the Hanle effect; see, e.g., Thorne 1974).

Quenching rate constants in flames are derived from the efficiency of resonance fluorescence (see Sects II.4a-4 and VI.2b); in its measurement one assumes the absence of any anisotropy in the fluorescence intensity, at least if self-absorption is negligible. Strong M_J mixing in flames at 1 atm pressure makes this assumption reasonable, even for polarized exciting beams (cf. above).

Processes will be considered that may occur in the gaseous phase in steady, chemical flames. Heterogeneous reactions involving the condensed phase (e.g., desolvated aerosol particles or soot) are insignificant for the (de-)excitation of free metal species and will not be considered here. Because of the absence of electrical accelerating fields and the low electron concentration, free electrons play only a subordinate role (see Sect. 2a-3). The excitation conditions are thus typically different from those in highly ionized plasmas like arcs, gas discharges or stellar atmospheres. Excitation and de-excitation in flames at 1 atm are caused predominantly by collisions and reactions between neutral atoms or molecules, and/or by photon emission and absorption.

Importance of (de-)excitation studies. The study of (de-)excitation processes and the measurement of their rate constants are of great importance in research concerning flames seeded by metal species. The excited-state populations and the resulting spectral radiations are, in thermal equilibrium, determined by the temperature and do not depend on the rate constants (see Sect. II.3). But in 'energetically open' systems like a flame, the existence of (local) thermal equilibrium is to be questioned. Whether an excited-state population is thermally equilibrated or only rate-controlled depends on the kinetics of the (de-)excitation processes and on the equilibration of the other degrees of freedom involved. After introducing the concept of thermal radiation, we shall consider quantitatively in Sect. 4 the causes of and conditions for significant deviations from thermal radiation. In particular, the influence of nonequilibrated flame-radical concentrations on the chemiluminescence in as well as above the combustion zone will be dealt with in Sect. 4. Experimental methods will be described in Sect. 4d to check the occurrence of (non-)thermal radiation of metal species. These considerations about thermal radiation are important for the application of optical methods of measuring flame temperatures and atomic densities. They are also of interest in connection with the rate of collisional metal ionization insofar as excited metal atoms are an intermediate step in the ionization process (see Chapt. IX).

Furthermore, explicit knowledge of quenching rates is required to determine the efficiency of fluorescence, which is an important parameter in analytical

INTRODUCTION   Sect. VI.1

atomic fluorescence spectroscopy (see Alkemade and Herrmann 1979). Quenching
collisions also contribute to the atomic line broadening (see Sect. VII.2b-5) and
also determine the degree of saturation obtained with a strong laser field tuned at
an atomic absorption line (see Sect. 2c). Finally, an insight into the (de-)excita-
tion processes enables us to predict how deviations from thermal excitation will
change with total pressure, composition or temperature of the flame.

The study of (de-)excitation processes is, more generally, of importance
in fundamental and applied physics and in allied fields such as astrophysics, the
physics of the upper atmosphere, plasma physics and chemical kinetics. The extensive kinetic data acquired about alkali atoms in different gas milieus may, for
example, be useful in the designing of optically pumped alkali lasers. In atomic
and molecular physics such studies have, in general, contributed to a better understanding of the interactions between (excited) atoms and molecules.

Advantages and disadvantages of flames studies. Rate constants for
(de-)excitation have been measured mainly in *bulk* systems (vapour cell, flame,
shock-wave) where the collision partners or reactants are usually distributed over
a broad range of initial velocities and internal states and where the collision
axis is isotropically distributed. Moreover, a given metal atom may interact with
several constituents of the flame gas, which is a multi-component system. This
complicates the interpretation of the experimental results in terms of specific
cross sections which describe the interaction as a function of relative velocity,
collision angle and quantum numbers of the initial and final states of a particular
collision partner. Furthermore, in the complicated theoretical analysis, simplifications must usually be made in order to obtain quantitative results. *Ad hoc*
assumptions are, moreover, often made because there is a lack of detailed information about the relevant interaction mechanism(s) and molecular properties. To test
the validity of these approximations and assumptions, we need experimental data
acquired over a wide range of temperatures or velocities, and relating to an
extended series of congenial collision partners.

Single- and crossed-molecular beam experiments are beginning to yield more
detailed information about collisional excitation processes (see Sects. 2a-1 and
2b-2).† The possibility of velocity-selection and initial-state preparation, and
of angular-distribution measurements give these methods a definite advantage

---

† Baylis (1977) has recently discussed the advantages of a single, well-collimated
thermal atomic beam, combined with fluorescence detection, for studying collisions between different beam particles with final-state selection.

533

Chapt. VI   EXCITATION AND DE-EXCITATION OF METAL SPECIES

over 'bulk' methods.[†] In other words, beam experiments enable us to 'probe' the interaction potentials more directly.

Not only has the understanding of physical and chemical (de-)excitation processes in flames largely profited from all these findings, but flame experiments have also, conversely, yielded new data for these processes (see this chapter). The rich variety of metallic species that can be introduced in an easy and controllable way into a flame, the controllable elevated temperature, and the absence of windows or walls[*] make the flame method a useful supplement to the other experimental methods. Besides, the flame is a milieu *par excellence* for the study of chemiluminescent reactions induced by free radicals that are not stable at lower temperatures. A disadvantage may be that the flame usually contains several components and that its total pressure cannot easily be varied.

2. PHYSICAL EXCITATION AND DE-EXCITATION PROCESSES

In this section, the physical processes leading to (de-)excitation of electronic states of metal species will be described and their role under flame conditions will be indicated. We successively discuss: (1) collisional (de-)excitation involving the conversion of translational energy without and with participation of internal energy of the collision partner, respectively (Sects 2a and 2b), and (2) radiative (de-)excitation (Sect. 2c). Our discussion will be restricted to atomic metal species, as but little is known about the much more complex

---

[†] In fluorescence quenching and mixing experiments in vapour cells one can, however, obtain a rough velocity-selection by scanning the exciting, narrow-band laser beam over the Doppler-broadened profile of the absorption line (see Apt and Pritchard 1976, 1979) or by varying the frequency of the exciting light beam in the application of photo-dissociation (see Sect. II.4a-4).
    Phillips *et al.* (1978) obtained relative differential cross sections for the mixing process: $Na(^3P_{\frac{1}{2}}) + Ar \to Na(^3P_{\frac{3}{2}}) + Ar$ in a crossed-beam experiment without mechanical selection of scattering angle by a movable Na detector. They used a single-frequency dye laser for exciting Na-beam atoms from the $3S_{\frac{1}{2}}(F=2)$ ground state to the $3P_{\frac{1}{2}}$ ($F=1,2$) states. A second, 'analysis' laser was directed along the relative-velocity axis of the incoming Na and Ar beams. The latter laser was tuned in such a way that only Na atoms that had been collisionally transferred to the $3P_{\frac{3}{2}}$ state under a certain scattering angle (in the centre-of-mass system) were excited to the 4D state. The population of the 4D state was therefore a relative measure for the inelastic differential cross section and was monitored by observing the stepwise line fluorescence of the $4P \to 3S$ transition at 3302 Å. The advantage of this method over conventional crossed-beam scattering experiments with excited alkali atoms is that it can discriminate between elastic and inelastic scattering channels.

[*] For example, Apt and Pritchard (1979) reported difficulties in stabilizing the Na vapour density in a cell with added CO_2 gas due to a chemical reaction between Na and CO_2, possibly on the cell wall.

PHYSICAL EXCITATION AND DE-EXCITATION PROCESSES   Sect. VI.2

(de-)excitation processes with metal compounds. For a survey of the older literature on the fluorescence of di- and poly-atomic species in the gaseous phase, we refer to Pringsheim (1949).†

We shall concentrate on those species that can act as collision partners in the flame region above the combustion zone and say very little about inelastic collisions of metal atoms with more complex molecules such as hydrocarbons.* Species such as C_2H_2 and C_2H may occur occasionally above the combustion zone of very carbon-rich C_2H_2 flames, but then only in minor concentrations ($\leqslant 1$ volume percent) (see Rasmuson, Fassel and Kniseley 1973).

The following treatment is exemplary rather than exhaustive. A selection of experimental cross sections for inelastic collisions ‡ that are of interest in the following discussions is collected in Table VI.1 at the end of Sect. 2b (see pages 603 ff.); dubious values are omitted. Theoretical explanations are given only briefly without mathematical details. There are still many controversial points in the theoretical analysis, but these will not be discussed within the limited scope of this chapter. Also the interpretation of the experimental results is not always unambiguous. This holds especially when the initial or final state of the metal atom undergoing a collisional transition is not well-defined. In work with metal vapour cells containing a gas admixture at low pressures, inelastic collisions between the metal atoms themselves might contribute to the observed transition rates. By keeping the metal vapour pressure sufficiently low one can avoid such complications. Also, in work with vapour cells at low pressures of the admixed gas, complications might arise due to a possible anisotropy and polarization of the observed fluorescence radiation ∅ or due to an ill-defined relative velocity-

---

† Wijchers *et al.* (1980) and Wijchers (1981) have utilized laser-induced fluorescence of YO bands in H_2 flames diluted by Ar or N_2 for finding inelastic collision rates of electronically excited YO molecules. Chan and Daily (1979) measured in a similar way quenching cross sections of OH radicals in a methane-air flame excited by a frequency-doubled pulsed dye laser. In both publications the steady-state populations of the excited levels are solved from a set of linear rate equations. Quenching cross sections of the BaCl $(C^2\Pi)$ state for He, Ar and N_2 were determined by Edelstein (1979) in a flow system with Ba vapour with the aid of time-resolved laser-induced fluorescence measurements; a single exponential decay curve was found, which directly yielded the quenching rate constant.

* See for quenching experiments with such species in vapour cells, e.g., Hrycyshyn and Krause (1970), Earl *et al.* (1972), and Siara and Krause (1973).

‡ In atmospheric-pressure flames the mean distance between two neighbouring particles is of the order of 100 Å. It appears that effective cross sections for inelastic collisions between a metal atom and another particle are, in general, much less that $(100 \text{ Å})^2$. This warrants the (often implicitly made) assumption that in usual flames such processes are truly bimolecular.

∅ One can minimize polarization effects by inserting a linear polarizer in the exciting or fluorescence beam with a suitably chosen polarization axis (see Deech, Pitre and Krause 1971, and Apt and Pritchard 1979).

Chapt. VI   EXCITATION AND DE-EXCITATION OF METAL SPECIES

distribution of the excited metal atoms produced, e.g., by photo-excitation with
the aid of a narrow-band laser beam or by photodissociation (see Czajkowski, Skardis
and Krause 1973a, Cuvellier et al. 1975, Gounand, Fournier and Berlande 1977, and
Apt and Pritchard 1979). In hot gas systems, e.g. flames, complications might be
caused by the occurrence of collisional ionization and recombination processes leading also to depopulation and population of excited levels (see van Dijk 1978, van
Dijk, Zeegers and Alkemade 1979, and van Dijk and Alkemade (1980a). The same might
hold for levels near the ionization limit in work with heated vapour cells or ovens
(see Cuvellier et al. 1975).

An extensive survey of the subject, including references to older literature, has been given by Massey, Burhop and Gilbody (1971). Gilmore, Bauer and
McGowan (1967) have tabulated rate constants known in 1967 for inelastic collisions
between atoms, molecules and electrons. A very extensive bibliography for chemical
kinetics and collision processes has been edited by Hochstim (1969). The latter
two publications also contain a classification of the processes considered. A
detailed and critical discussion of later experimental results and theoretical
interpretations has been presented by Lijnse (1972,1973), with emphasis on the
alkalis, and by Krause (1975). Alkali mixing experiments including higher-lying
doublets have been surveyed by Krause (1972, 1975).

2a   (DE-)EXCITATION OF ELECTRONIC STATES THROUGH CONVERSION
        OF TRANSLATIONAL ENERGY ONLY
        It is only by converting kinetic energy of relative motion into electronic
excitation energy that free electrons and, for example, noble gas atoms or hydrogen
radicals can excite metal atoms in flames. The excited levels of noble gas atoms
and hydrogen radicals lie too high above the ground level to give their role in
metal excitation any significance in flames (in contrast to oxygen radicals, which
have energy levels as low as ≈ 2 eV). Because of the principle of microscopic reversibility, the same conclusion also holds for the reverse de-excitation (quenching)
processes. There is a great difference in (de-)excitation mechanism and efficiency
between free electrons and free atoms as collision partners. We shall therefore
discuss them separately.

2a-1   *QUENCHING AND EXCITATION BY COLLISIONS WITH UNEXCITED ATOMS
         AND MOLECULES*
        Let M be a metal atom and Z a collision partner whose internal energy is
zero (or negligible) and not (significantly) changed in the collision. The metal
atom is supposed to have an energy level (M*) that lies some electronvolts above
the ground level (M). Excitation to this level from the ground level and
the reverse quenching process: $M + Z \rightleftharpoons M^* + Z$ thus involve the conversion of a

536

considerable amount of translational energy. The cross sections of such processes for atomic collision partners appear often to be small or negligible in comparison with the corresponding cross sections for molecular collision partners, which are usually of the order of the gas-kinetic cross sections or even higher (see Table VI.1). Cross sections for collisional transfer of metal atoms between neighbouring, higher excited levels, induced by atomic collision partners, are however not necessarily small (see Table VI.1).

Experiments relating to the lowest excited level. Careful *de-excitation measurements* using different methods (see Sect. II.4a-4) have now firmly established that the quenching cross sections of, e.g., the Na-D doublet for noble gas atoms are well below 1 $Å^2$. Lifetime measurements in an Na vapour bulb filled with a noble gas at 400 K have yielded an upper limit of 10^{-2} $Å^2$ for He...Xe (see Copley, Kibble and Krause 1967). Similar phase-shift measurements at 350 and 820 K have yielded an upper limit of 10^{-3} $Å^2$ for the quenching of the first resonance doublet of Cs by He (see Dodd, Enemark and Gallagher 1969). Previous experiments based on fluorescence-quenching measurements in vapour cells and flames have often led to too high cross sections because of certain experimental difficulties (see Sect. II.4a-4 and Lijnse 1972,1973). Fluorescence-quenching experiments with a vapour of boron produced by cathodic sputtering yielded, remarkably, cross sections as high as 5 $Å^2$ for the quenching of the B resonance lines at 2089 and 2497 Å by Xe at $T = 400$ K (see Hannaford and Lowe 1977). The quenching cross sections for Kr and Ar were found to be much smaller. Fluorescence-quenching measurements of the first resonance lines of the alkalis and Tl in flames diluted by a large excess of Ar and He have yielded only upper limits of the order of 1 $Å^2$ (see Jenkins 1966,1968, 1968b, 1968d). [†]

Quenching by He and Xe of the lowest metastable $^2D_{\frac{3}{2}}$ and $^2D_{\frac{5}{2}}$ levels of Bi, lying about 1.5 and 2 eV above the $^4S_{\frac{3}{2}}$ ground level, has been studied by Bevan and Husain (1976). The lifetime of these states produced by pulsed photolysis of Bi trimethyl and tri-ethyl molecules was determined by time-resolved measurements of the atomic absorption signals originating from these metastable levels. The rate constants found for He were of the order of 5×10^{-16} cm^3 molecule^{-1} s^{-1} (see Table VI.1), corresponding to effective cross sections of the order of 10^{-4} $Å^2$.

The small quenching cross sections for noble gases at thermal energies are corroborated by the following *excitation measurements* through the application of microscopic reversibility (see Sect. II.4b-4). From line-reversal measurements of the underpopulation of the Na-D doublet in shock-heated Ar gas at $T = 2750$ K, one

---

[†] The higher, tentative values obtained for Ar in similar flame experiments by Hooymayers and Alkemade (1966) should be discarded.

Chapt. VI   EXCITATION AND DE-EXCITATION OF METAL SPECIES

calculates through Eqs II.109, 111 and 131 a quenching cross section of the order of at most 10^{-3} $Å^2$ (see Tsuchiya 1964, and Tsuchiya and Kuratani 1964). Anderson, Aquilanti and Herschbach (1969) have measured the excitation of the first resonance doublet of K as a function of translational energy by injecting a beam of fast K atoms (produced by charge-exchange) into a chamber filled with noble gas and observing the resonance emission. They found threshold energies (in the centre-of-mass system) rising from 30 to about 100 eV for Xe, Kr, Ar, Ne and He in this order. The excitation cross section was estimated to be of the order of 1 $Å^2$ near threshold. Kempter, Kübler and Mecklenbrauck (1974) have done similar measurements for the separate K doublet lines with Ar, Kr and Xe. Düren, Kwan and Pauly (1974) found threshold energies of 59 and 69 eV for the excitation of $K(4^2P)$ and $K(5^2P)$, respectively, by Xe. Threshold energies for the excitation of $K(4^2P)$ by ground-state K-, Na- and Hg atoms were found to increase from 1.6 to 4.0 eV in this order, while the excitation cross section was about 3 $Å^2$ at 20 eV (see Kempter et $al.$ 1974). Relative differential cross sections for the excitation of the first K resonance doublet by the noble gas atoms have been measured by Zehnle et $al.$ (1978) in the energy range 40 - 800 eV.

Similar beam experiments have also been done for the excitation of the first Na- and first as well as second K doublets by di-atomic and poly-atomic molecules by Kempter et $al.$ (1970, 1971), Lacmann and Herschbach (1970), and Düren, Kwan and Pauly (1974). In contrast to the case of noble gases, the threshold in the centre-of-mass system appeared to be about equal to the alkali excitation energy for N_2 and O_2, and a few electronvolts higher for CO and CO_2. Since the molecular target gas was rotationally as well as vibrationally 'cold' ($T \approx 300$ K), excitation was effected by conversion of translational energy only. By applying the principle of microscopic reversibility, one deduces from these experiments a rough estimate of 0.2 - 1.0 $Å^2$ for the cross section of the reverse quenching process. Kempter, Koch and Schmidt (1974) determined similarly the ratio of the cross sections for the excitation of the $K(4^2P)$ doublet components by internally 'cold' N_2, O_2, CO_2 and NO_2 molecules as a function of translational energy. This ratio appeared to be, in general, independent of energy in the investigated range (up to 50 eV for some of these molecules) and to be determined solely by the statistical weight factors of the excited doublet levels. Microscopic reversibility then predicts that the cross sections for quenching by conversion to translational energy are the same for the two doublet levels. Practically equal quenching cross sections were also found by McGillis and Krause (1968) for these doublet levels in N_2 (see Table VI.1); in the latter experiment, however, the final internal states of the N_2 molecules could not be specified.

Differential cross sections for the excitation of $K(4^2P)$ by internally

PHYSICAL EXCITATION AND DE-EXCITATION PROCESSES   Sect. VI.2a

cold N_2 or CO molecules have been measured by Gersing, Pauly and Schädlich (1973), Loesch and Brieger (1977) and Martin et al. (1979). In the first two beam experiments a time-of-flight technique was applied to analyze the inelastically scattered K atoms. In the last experiment, the authors used the time-correlation between the photon emitted after excitation and the corresponding inelastically scattered atom. Loesch and Brieger concluded that at energies near threshold ($\simeq 2.4$ eV) strong vibrational excitation of N_2 occurred simultaneously with electronic excitation. Martin et al. concluded that ionic-curve crossing was involved in the excitation process (cf. Sect. VI.2b-2) and located the corresponding crossing point: $V(r_c = 2.77$ Å$) = 0.55$ eV for N_2 and $V(r_c = 2.48$ Å$) = 1.00$ eV for CO. They assumed that the internal degrees of the molecule were 'frozen' during the excitation process.

At thermal energies interconversion of electronic energy into translational energy alone is therefore effected much more readily by molecules than by noble gas atoms. On the other hand, the ability of molecules to quench electronic states appears to be significantly enhanced when (part of) the electronic energy is converted into internal molecular energy (see Sect. 2b).

Experiments relating to higher excited levels. So far we have mainly considered collisional transitions between the ground level and the lowest excited level or multiplet. When 'pumped' to a higher excited level, the atom may undergo various collisional as well as radiative transitions, either directly or in cascade, to lower excited levels in addition to the ground level. When sufficient thermal energy is available, endo-ergic transitions from the pumped level to adjacent higher levels may also occur. The number of (unknown) rate constants, $N(N-1)$, for collisional transitions increases rapidly with the number, N, of levels involved. When these rate constants are thermal and the temperature is known, application of the detailed balance relation, Eq. II.85, between the conjugated rate constants for each pair of levels halves this number. Alternatively, one may use this relation to check the mutual consistency of the separately measured, conjugated rate constants. However, this relation can be applied only when the rate constants are thermal.

When a fixed level is pumped from the ground level, e.g., by a tuned laser beam, the unimolecular collisional rate constants occur as coefficients in a set of N coupled differential equations. Each of these equations expresses the time-derivative, dn_k/dt, of the population n_k of an individual level ($k = 1, 2, ..., N$) in n_k and in the populations of all other levels from which collisional or radiative transitions to level k are possible.[†] In these linear expressions radiative

---
[†] When the set of N levels is 'closed', that is, when the total concentration of atoms in these levels is constant, these equations are not independent. By eliminating the population of one level one can obtain $N-1$ independent equations.

Chapt. VI   EXCITATION AND DE-EXCITATION OF METAL SPECIES

transitions to and from level k due to spontaneous emission as well as absorption are taken into account. The expression for the time-derivative of the pumped level contains a term that describes the pumping rate due to absorption of an external radiation beam, which is supposed to be tuned to this level only. When the atomic vapour is contained in a heated cell or oven, absorption of the continuous thermal emission from the walls might also induce radiative transitions between neighbouring levels in the infrared region, where thermal emission is usually strongest (see Pimbert 1972, and Cuvellier et al. 1975). The unimolecular thermal collisional rate constants can be expressed in the corresponding thermal effective cross sections, the density of the collision partner (we suppose only one species to be present) and the average relative velocity through Eq. II.133. Experimental information about these cross sections derives ultimately from observations on the level populations. The population of a level can be obtained by measuring the intensity of a fluorescence line originating from this level, if the optical transition probability involved is known (see Eq. II.279). We suppose that self-absorption is absent or can be corrected for (see Sect. II.5c-4). The population of a metastable level or the ground level can be obtained by absorption measurements on a spectral line that originates from that level.

In *kinetic* measurements of the time-resolved rise and/or decay of the level populations after short-pulse excitation one need only measure relative level populations. Such measurements may be done by the method of delayed coincidence, which, however, requires rather elaborate equipment. The populations measured as a function of time may be compared with the mathematical functions obtained by integrating the above-mentioned set of differential equations (see Gallagher, Cooke and Edelstein 1978). One may also integrate the latter functions multiplied by the appropriate Einstein coefficients in order to obtain expressions for the time-integrated fluorescence yields which can then be compared with experimental observations (see last reference). The unimolecular collisional rate constants are found by fitting the experimental data to these mathematical functions or expressions. The fitting procedure can be simplified when measurements are made at different gas pressures, p, and the results (e.g., decay rate, integrated fluorescence yield) are plotted as a function of p. This simplification is possible because at sufficiently low pressures some of the collisional transitions are eliminated because their rates become negligible compared to the competing radiative rates (which are usually known). Besides, (de-)population routes that involve two or more consecutive collisional transitions become negligible compared to single-collision routes when the pressure is sufficiently reduced.

When the pumping rate is constant or varies slowly compared to the lifetimes of the levels involved, the level populations are $(quasi-)stationary$, that is,

PHYSICAL EXCITATION AND DE-EXCITATION PROCESSES   Sect. VI.2a

the total population and depopulation rates for each level are balanced. The
mutually connected stationary level populations are determined by the above-
mentioned set of coupled differential equations where we have to put all time-
derivatives equal to zero. One can obtain information about the unimolecular
collisional rate constants by measuring the level populations as a function of time,
or the time-integrated or stationary level populations under long-pulse or d.c.
excitation. Under usual experimental conditions one may expect the relative population distribution over all levels to be independent of the total metal concentration, and the relative population distribution over the excited levels (that are
populated either directly or indirectly by the pumping process) to be independent
of the pumping rate. Then one need only measure relative distributions in order to
determine collisional rate constants. A measurement of the excited-level populations relative to the ground-level population is only useful if the pumping rate is
also known. If this rate is not known, only measurements of the excited-level
populations relative to each other are meaningful. This explains why in stationary
experiments one usually restricts oneself to the measurement of relative excited-
state populations through the corresponding relative fluorescence-line intensities.
To disentangle the various collisional transition rates one again measures the population ratios preferably as a function of pressure of the collision partner (cf.
above).

Since unimolecular collisional rate constants have dimension $[t]^{-1}$, one
needs an 'internal clock' to determine their absolute values (in s^{-1}) in
(quasi-)stationary experiments. The clock can be provided by the absolute pumping
rate per atom, if known. Usually this clock is provided by the optical transition
probabilities per second which occur together with the collisional rate constants
in the expressions for the stationary level populations. Our knowledge of absolute optical transition probabilities for higher levels of common metal atoms is
growing rapidly. One can also use one of the involved collisional rate constants
as a clock if its absolute value is known from a previous experiment.

In complicated systems (large N) the number of unknown collisional rate
constants involved may exceed the number of independent observables, even when
measurements are made as a function of gas pressure p over an extended range.
Optical selection rules, the limited spectral response of the photodetector, the
lack of suitable auxiliary light sources, and low signal-to-noise-ratios often restrict the number of fluorescence or absorption lines that can be used for measurements of level populations. One can augment the experimental data by repeating the
measurements with different pumped levels. For each pumping mode one obtains a
different set of kinetic equations, that may contain common collisional rate

Chapt. VI   EXCITATION AND DE-EXCITATION OF METAL SPECIES

constants (see Tudorache 1977). Whereas the spontaneous emission probabilities are usually known, the excitation rates due to absorption of continuous radiation from the cell enclosure may introduce additional unknowns into the rate equations. These are difficult to calculate unless one is dealing with black-body radiation, as in an oven at known temperature. Experimental information about the latter rates may be obtained, however, through additional measurements on the level populations in the absence of collision partners (see Cuvellier et $al.$ 1975). In order to simplify the analysis still further, one often makes extra, plausible assumptions. For example, one may assume a priori that a partial Boltzmann equilibrium is rapidly established for the relative populations of neighbouring levels whose energy separations are small compared to kT. However, whenever a partial Boltzmann equilibrium is established within a group of levels, it is in principle impossible to determine rate constants for collisional transitions between these levels (see Sect. II.4a-2). At best one can determine weighed average values of the rate constants for collisional transitions from this group of levels to other levels and vice versa. Furthermore, one may, a priori, disregard fine-structure splitting and assume that collisional rate constants for transitions to or from a multiplet term are the same for each component. When the energy separation between two levels is much larger than kT, it is often reasonable to assume that the rate of endo-ergic collisional transitions between them is negligible. Finally, one can always reduce the number of unknown collisional rate constants in the analysis of multi-level systems by simply restricting the pressure of the collision partner (cf. above). But this, of course, also restricts the scope of the experiment.

   For more detailed information about the various ways in which one can
   determine collisional rate constants involving higher atomic levels the reader
   is referred to, e.g., Pimbert (1972), Czajkowski and Krause (1974), Cuvellier
   et $al.$ (1975), Tudorache (1977), and Gallagher, Cooke and Edelstein (1978) for
   work with vapour cells with controllable pressure of admixed gas, and to van
   Dijk (1978), Allen et $al.$ (1979) and van Dijk, Zeegers and Alkemade (1979) for
   flames at 1 atm pressure.

   Self-quenching of higher-lying K and Cs states by ground-state K and Cs atoms, respectively, appears to be a very efficient process (see Table VI.1). In some cases quenching cross sections of the order of 100 to above 1000 $Å^2$ were found. The question whether we should speak here of 'quenching' or 'electronic excitation transfer' between similar collision partners remains open.
   When one deals with higher atomic levels, one often restricts oneself to the determination of an <u>overall</u> quenching cross section for the pumped level, which is a summation of the specific cross sections for all possible collisional

PHYSICAL EXCITATION AND DE-EXCITATION PROCESSES   Sect. VI.2a

transitions from that level. One has to make sure of the absence of repopulating collisions. This can be achieved, for example, by observing the time-decay of the investigated level population, following short-pulse pumping, over a period that is short compared to the actual lifetime(s) of the repopulating level(s) (see Gounand, Fournier and Berlande 1977). One is sometimes also content with determining the collisional rate constant for only one specific transition from the pumped level to another level. The experimental conditions should then be such that this transition dominates all other collisional population routes and that the known optical lifetime of the final level is much shorter than its collisional lifetime, if the latter is not (precisely) known.

Table VI.1 lists a selection of cross sections for the quenching of higher atomic alkali states by noble gas atoms. States with high principal quantum number n (so-called *Rydberg states*) can be pumped by a frequency-doubled laser beam (as in the case of the 12P,..., 22P states of Rb, investigated by Gounand, Fournier and Berlande 1977, or the 10P state of K, investigated by Gounand $et\ al.$ 1976) or by stepwise photo-excitation by means of two laser beams tuned to two consecutive atomic transitions (as in the case of the 5S and nD states of Na through the intermediate 3P state; see Gallagher, Edelstein and Hill 1977, and Gallagher, Cooke and Edelstein 1978). High-lying Na states can also be pumped by 2-photon absorption of a single laser beam whose frequency is tuned to half the excitation energy of the pumped level divided by h (see van Dijk 1978, and van Dijk, Zeegers and Alkemade 1979). Some of the listed cross sections relate to the overall quenching rate of the pumped level; others relate to a specific transition from the pumped level to another, neighbouring level. Some quenching cross sections of noble gas atoms for the higher alkali levels appear to be of the order of magnitude of 1-100 $Å^2$. The specific quenching cross sections for the $Cs(8^2P_{\frac{1}{2}} \rightarrow 7^2D_{\frac{3}{2}})$ transition by He and Ar are comparatively small ($\approx 1\ Å^2$), considering the small energy defect of 0.04 eV. On the other hand, cross sections of noble gas atoms of the order of $10^3\ Å^2$ were found for transitions between high-lying Na states with the same principal quantum number ($6 \leq n \leq 15$). Anyway, all these cross sections are several orders of magnitude larger than those for the quenching of the first alkali resonance doublets (see Table VI.1). This conclusion is corroborated by semiquantitative measurements of the rate constants for collisional transitions to the 3P level from a group of partially equilibrated higher levels adjacent to the pumped Na(3D) or (4D) level in an Ar-diluted H_2-flame at 1 atm (see van Dijk 1978, and van Dijk, Zeegers and Alkemade 1979; see also the pertinent discussion at the end of this subsection). In the case of the Rb(12P,...,22P) levels it appears that the overall quenching cross sections increase or reach a maximum when n is increased. For a given Rb level these cross sections increase in the order Ne-Ar-He. The cross sections of

Chapt. VI   EXCITATION AND DE-EXCITATION OF METAL SPECIES

He, Ne and Ar atoms for $Na(n^2D \to n, l)$ transitions with $l > 2$ appear to reach a maximum at $n \simeq 10$ and to decrease again for still higher n values (see Gallagher, Edelstein and Hill 1977). Gallagher and Cooke (1979) found a plateau in the cross section for collisional de-excitation of higher Na levels by He, Ar and Xe as a function of n.

A notable cross section of 5 Å² has been found for the overall quenching of excited $Ca(4^1P_1)$ atoms by Xe (see Ermisch 1966), whereas Kosasa, Maruyama and Urano (1977) found a lower, but nonnegligible value of the order of 0.5 Å² for the quenching of these atoms by Ar in a shock-tube experiment. The quenching process probably involves here a collisional transition to close-lying, lower excited Ca levels (see Seiwert 1968). The nonnegligible quenching cross sections of some noble gas atoms for, e.g., the $Hg(6^1D_2)$, $Cd(5^1P_1)$, $Cd(6^1D_2)$, $Pb(7^3P_1)$ and $Hg(6^1P_1)$ states are also noteworthy (see Table VI.1 and Granzow, Hoffman and Lichtin 1969). In all these cases lower-lying excited levels may be involved in the quenching transition.

Finally, Table VI.1 shows also the capability of Ar atoms transferring Hg and Cd atoms from excited singlet states to neighbouring triplet states.

Theory. A qualitative explanation for the low efficiency of noble gas atoms to quench or excite the first resonance doublets of alkali atoms can be given by considering the value of the Massey parameter ξ at thermal velocities (see Eq. II.142). If the collision partners approach each other in such a way that the vertical distance between the potential energy curves for $(M^* + Z)$ and $(M + Z)$ (cf. Fig. II.5) remains comparable to the atomic excitation energy, ξ is calculated to be of the order of 100. Since the collision partner is a 'heavy' particle, the duration of collision is long compared to the characteristic period of electronic oscillation ($\approx 10^{-15}$ s) in the atom to be excited. This leads to a large value for the Massey parameter.

A quenching collision: $M^* + Z \to M + Z$ could proceed more efficiently through one or more consecutive radiationless transitions at the pseudo-crossing point(s) of the potential energy curves associated with the temporary collision complex (see Sect. II.4e).[†] According to Laidler (1942, 1955), the quenching of excited $Na(^2P)$ atoms by ground-state $H(^2S)$ atoms, for example, might occur through a crossing transition from a singlet to a triplet state of the collision complex. However, according to Wigner's spin conservation rule, the probability of such a transition is low in this case.

Quenching through curve-crossings may, in general, be facilitated when the colliding atoms can form an intermediate ionic complex $(M^+ + Z^-)$ whose potential

---

[†] For the formal theory of interconversion of electronic and translational energy through curve-crossing see, e.g., Bates (1962), and Child (1979).

energy curve crosses the covalent (M^*+Z) as well as the covalent ($M+Z$) curve (see Fig. II.6). An ionic complex ($Na^+ + H^-$) with an attractive potential branch may be readily formed during the collision of Na^* and H. However, it appears to be unable to get rid of its excess vibrational energy so that it dissociates again to the initial, nonquenched state, i.e., it returns from branch III to branch I in Fig. II.6 (see Laidler 1942). Also calculations based on the assumption of an intermediate ($Na^+ + He^-$) complex gave only a low efficiency for the $Na(^2P)$ quenching by He (see Stamper 1965, and Fisher and Smith 1971), in accordance with experiments.

It is generally believed now that quenching of, e.g., excited $Na(^2P)$ and $Cs(^2P)$ atoms by ground-state (^1S) Ar- and He atoms, respectively, takes place as a result of potential-curve crossings that connect the initial and final states (see (Pringsheim 1949, Dodd, Enemark and Gallagher 1969, and Fisher and Smith 1971). An excited alkali atom in the ^2P state and an argon atom in the ^1S ground state can combine to a molecular $B^2\Sigma$ and an $A^2\Pi$ term (we neglect here fine-structure splitting). An alkali atom in the ^2S ground state combines with an argon (^1S) atom to a $X^2\Sigma$ term. Only the $A^2\Pi$ potential curve crosses the $X^2\Sigma$ curve, but this occurs in an energy region far above the alkali-excitation energy. Ionic potential curves with an attractive branch do not play a role here. Even if the internuclear distance r_c at the crossing point in Fig. II.5b were of the order of an atomic diameter and the transmission coefficient were near unity, the quenching rate constant would still be very low under thermal conditions, as a large activation energy E_A is involved. The excitation experiments with atomic alkali beams described above suggest that E_A may be of the order of 50 eV for alkali-noble gas collisions. This virtually prevents the occurrence of quenching at thermal energies, and at the same time leads to a very strong exponential dependence on $1/T$ of the excitation and quenching cross sections. Considerations similar to those presented above for $Na(^2P)$ quenching by Ar may also apply to the quenching of $Cr(^7P)$ by Ar (see Gaydon and Hurle 1963). Here too quenching seems to be achieved by the intersection of the repulsive potential curves of the ground-state $[Cr(^7S) + Ar]$ complex and the excited $[Cr(^7P) + Ar]$ complex.

The notably larger efficiency of noble gas atoms, in particular Xe, to quench the boron resonance line at 2497 Å may be explained as follows (see Hannaford and Lowe 1976). In contrast to the alkali resonance lines, the upper level of this boron line is a ^2S state, whereas the ground level is a ^2P state. A boron atom in the latter state can combine with a xenon atom to an $A^2\Sigma$ term; the corresponding, steeply rising potential-energy curve may cross the $B^2\Sigma$ curve originating from the excited boron atom in a thermally accessible energy region at 400 K.

Quenching of higher excited states to neighbouring excited states, however, may be facilitated by (successive) crossings of the corresponding potential-energy

Chapt. VI   EXCITATION AND DE-EXCITATION OF METAL SPECIES

curves with a manifold of other curves arising from different excited states. Also
an attempt has been made by Gounand, Fournier and Berlande (1977) to interpret semi-
quantitatively their above mentioned experimental cross sections for the quenching
of $Rb(n^2P)$ Rydberg states by noble gas atoms by classical arguments. It turns out
that the increase in quenching cross section with n is less rapidly than the increase in geometrical size ($\propto n^4$) of the excited atom. Since for Rydberg states
the valence electron is but loosely bound to the atomic core, one has sought for a
correlation between the quenching rate constant and the rate constant for elastic
scattering of a quasi-free electron having the same relative-velocity distribution
as the valence electron. A close correlation seems to exist (within a factor of
three) for the rate constants measured by Gallagher, Edelstein and Hill (1977) for
the collisional transfer of Na atoms in n^2D states ($6 \leq n \leq 15$) to neighbouring
n, l states ($l > 2$) by He (see Gounand, Fournier and Berlande 1977; see also Table
VI.1 on pages 603 ff.). Agreement was also found between these experiments and the
rate constants derived from quantummechanical calculations by Olson (1977a), using
asymptotic adiabatic potential-energy curves, for the collisional transfer of
$Na(n^2D \rightarrow n^2F)$ by noble gas atoms. However, Gounand, Fournier and Berlande (1977)
found no satisfactory agreement between the predictions of the simple elastic-
scattering theory and their own measurements on the $Rb(n^2P)$ quenching. They believe that a three-body approach including the core of the Rb atom is required.

Gounand et al. (1976) and Gounand, Fournier and Berlande (1977) have produced arguments to exclude the importance of ionic channels in the depopulation
of $K(10^2P)$ and $Rb(n^2P)$ states under their experimental conditions.

Specific calculations have been made for the quenching rate constant of
excited alkali atoms in collision with noble-gas atoms (see Stamper 1965, Nikitin
and Bykhovskii 1964, Nikitin 1966, 1967a, and Fisher and Smith 1971; see survey by
Lijnse 1972). Because of a lack of data on the atomic interactions, these theoretical calculations generally fail to make precise quantitative predictions. Moreover,
since the electronic transition involved in the quenching of the lowest excited
doublet takes place far up in the repulsive-potential region, the assumption of a
two-state model may not be adequate. Anyway, the present theory seems to explain,
at least qualitatively, the very low order of magnitude found experimentally under
thermal conditions.

The low efficiency of noble gas atoms to excite alkali atoms is not simply
a consequence of the fact that they have only translational energy. Molecules which
also have only translational energy in 'cold' beam experiments are much more efficient in exciting alkali atoms (see above). The difference is rather that for noble
gas atoms no suitable crossing points of the potential energy curves involved exist
in a thermally accessible range of internuclear distances. The scattering

experiments of Martin et al. (1979) with a fast atomic K beam excited by collisions in a 'cold' N_2 or CO gas indicated the importance of curve-crossings through an intermediate ionic complex (see above).

Beam experiments on the velocity- and angle dependence of scattering of metal atoms in the ground- or excited state by noble gas atoms can yield quantitative information about the potential-energy curves (see, e.g., Pritchard and Carter 1975, Carter et al. 1975, and van Deventer 1980). Of special interest are atomic-beam measurements of the differential cross section for the collisional excitation of ground-state alkali atoms by noble gas atoms (see earlier in this subsection). Potential-energy curves have been calculated by Baylis (1969), Pascale and Vandeplanque (1974), and Saxon, Olson and Liu (1977) for excited and unexcited Na atoms interacting with noble gas atoms (see also Baylis 1978).

Conclusions. In flames diluted with a noble gas, quenching of the lowest excited alkali states by noble gas atoms is negligible compared to the residual molecular quenching (see also Sect. VI.2b-2). This holds notwithstanding the expected strong relative increase of the noble gas quenching cross section with temperature. Conversely, noble gas atoms will make a negligible contribution to the collisional excitation of the first alkali resonance doublets in flames. Noble gas atoms, however, show a nonnegligible quenching efficiency for, e.g., Hg, Cd and Pb states for which transitions to lower-lying excited states are possible. Quenching cross sections of noble gas atoms for Rydberg states of alkali atoms may be quite large. In flames diluted by noble gases at 1 atm pressure a partial Boltzmann equilibrium may therefore be approached for the relative populations of the higher-lying excited states, whereas an infrathermal population may occur for the lowest excited state relative to that of the ground state (see also Sect. VI.4c-2 sub iv).

Van Dijk (1978) and van Dijk, Zeegers and Alkemade (1979) found by fluorescence intensity measurements a strong collisional coupling between neighbouring higher alkali levels in an Ar-diluted H_2 flame while pumping the 3D, 5S and 4D levels by a pulsed laser beam. In the first and third pumping modes a partial Boltzmann equilibrium was established for levels adjacent to the pumped level, whereas an inversion appeared to occur between the populations of the pumped level and some lower-lying excited levels. The average unimolecular rate constants for collisional transitions from the group of levels around the pumped level to the lower-lying 3P level were estimated to be about $3 \times 10^8 \, s^{-1}$. In this flame the lifetimes of the excited levels were sufficiently short compared to the duration of the laser pulse ($\approx 1 \, \mu s$) to ensure a quasi-stationary population of the Na levels. The circumstance that partial Boltzmann equilibrium may be readily obtained in flames at 1 atm and the difficulty of varying the gas pressure makes, however, flames less suitable for the measurement of collisional rate constants for transitions

Chapt. VI   EXCITATION AND DE-EXCITATION OF METAL SPECIES

between high-lying atomic levels.

Cross sections for self-quenching of alkali atoms in higher-lying states appeared to be two or three orders of magnitude larger than those for quenching by noble gas atoms. However, the relatively low concentration of alkali atoms in atmospheric-pressure flames (see Sect. III.2h) renders the contribution of self-quenching insignificant.

2a-2   *MIXING OF MULTIPLET COMPONENTS BY COLLISIONS WITH UNEXCITED ATOMS*
Experiments. Collisions with noble gas atoms are quite efficient to change the electron spin orientation with respect to the electron orbital angular momentum in excited Li-, Na- and K atoms at and above room temperature. An example of such a doublet-mixing process is

$$\text{Na}(^2P_{\frac{3}{2}}) + \text{Ar} \rightleftharpoons \text{Na}(^2P_{\frac{1}{2}}) + \text{Ar} \;(+\Delta E).\quad\quad\text{(i)}$$

Here ΔE is the positive electronic energy defect, which is considerably smaller than 1 eV in these instances and which is to be converted into or supplied from the translational energy of relative motion.

Mixing cross sections have in most cases been studied experimentally by the method of *stepwise line fluorescence*[†] (see Sect. II.5a-1; see also Krause 1966, 1975, and Drawin 1968). Filtered light of one of the Na-D lines, for example, is irradiated onto a cell containing Na vapour together with a noble gas at known, variable pressure and temperature. The ratio of emission intensities of the two fluorescent D lines is measured.[*] The vapour density is kept sufficiently low in order to avoid self-absorption and radiation trapping as well as the contribution of unexcited alkali atoms to the mixing process. This procedure is repeated while the other doublet component is irradiated onto the vapour cell. From the ratios thus obtained and the known optical lifetimes of the states concerned, the collisional mixing rates per excited alkali atom, or the mixing cross sections, are simple to calculate. The influence of quenching collisions on the intensity ratio of the fluorescent lines may safely be neglected, at least for the first resonance doublets, if molecular impurities are eliminated (see Sect. 2a-1). The thermal (phenomenological) effective cross sections σ_m and σ_{-m} for the forward exo-ergic and the reverse endo-ergic doublet mixing process should obey the detailed-balance relation

$$\sigma_m(T)/\sigma_{-m}(T) = \tfrac{1}{2}\exp\left[\Delta E/kT\right],\quad\quad\text{(VI.1)}$$

[†] Also called in the literature: 'sensitized fluorescence' or (more rarely): 'cross fluorescence'.

[*] See the footnote * on page 531 for a reference to the quantummechanical theory of the fluorescence intensities of alkali doublet components.

which follows from Eq. II.138. This relation provides a useful check for the consistency of the experimental results. The experimental method is liable to lead to systematic errors as a result of radiation trapping, chemical impurities, or spectral leakage of one collision-broadened doublet component through the interference filter used for selecting the other component (see Gallagher 1968). Several results obtained in earlier measurements had therefore to be discarded (see Lijnse 1972).

Relation VI.1 has been confirmed generally for the first and some higher resonance doublets of the alkali atoms in collision with noble gas species in vapour cells by, e.g., Siara (1972) (for a critical review see Lijnse 1972) as well as in flames by Lijnse (1973, 1974a).

The applicability of Eq. VI.1, which has been derived under equilibrium conditions, to the measured effective cross sections σ_m and σ_{-m} is not obvious. For example, when the exciting lamp radiation is strongly self-reversed and the absorption line shows Doppler broadening only, the velocity distribution of the photo-excited atoms will not be Maxwellian (see Gallagher 1968). This is because photo-excitation then takes place mainly in the wings of the absorption line profile. This might affect the measured value of σ_m, if σ_m depends markedly on relative velocity and if elastic collisions do not restore a Maxwellian distribution before mixing has a chance to take place.

Mixing of the first resonance doublet of K by a noble gas has been measured as a function of relative velocity in a crossed-beam experiment by Anderson et al. (1976) and by Cuvellier et al. (1979) (see also Table VI.1). There exists partial disagreement between these experiments as to the variation of cross section for He with relative velocity.

Mixing cross sections are tabulated in Table VI.1 for the lowest and some higher excited levels of the alkali atoms in collision with different kinds of noble gases, as measured in vapour cells (between 300 and 900 K) as well as in flames (at 1720 K). Whereas the mixing cross sections for the first Na- and K doublets, the second Rb doublet and the third Cs doublet are at least of the same order of magnitude as the gas-kinetic cross sections, they appear to be much smaller for the first Rb- and Cs doublets. The mixing cross sections depend only weakly on temperature with Na and K in the temperature range investigated, but show a strong, nearly exponential temperature dependence with the first resonance doublets of Rb and Cs (see Fig. VI.1A). Apt and Pritchard (1976) determined the velocity dependence of the cross sections for the mixing of the Na-D doublet by Ar and Xe by scanning a narrow-band laser beam over the Doppler-broadened profile of the Na-D lines in a vapour cell.

An upper limit of but 2×10^{-3} A^2 has been obtained for the mixing cross section of $Hg(6^3P_1 \rightarrow 6^3P_0)$ with Xe by Deech, Pitre and Krause (1971). A

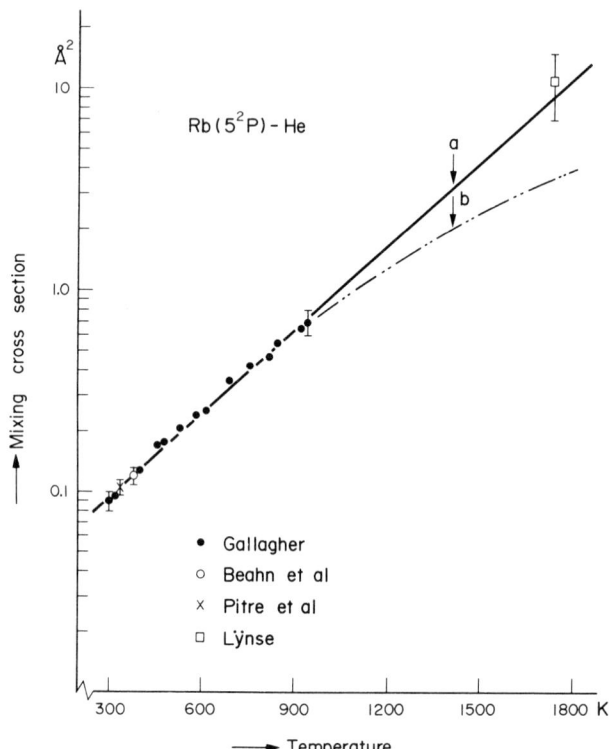

Fig. VI.1A shows the (phenomenological) mixing cross section for the $\text{Rb}\left(5^2\text{P}_{\frac{3}{2}} \to 5^2\text{P}_{\frac{1}{2}}\right)$ doublet with He atoms as a function of temperature. The experimental points are taken from Gallagher (1968), Beahn, Condell and Mandelberg (1966), Pitre, Rae and Krause (1966), and Lijnse (1974a). The experimental error is indicated by a vertical bar for the value obtained by Lijnse in a flame and for the outermost points obtained by vapour-cell measurements. Curve a shows the overall temperature dependence including the flame value. Curve b represents the extrapolation of the low-temperature data according to Gallagher's (1968) analysis. (Redrawn from Lijnse 1973, page 80.)

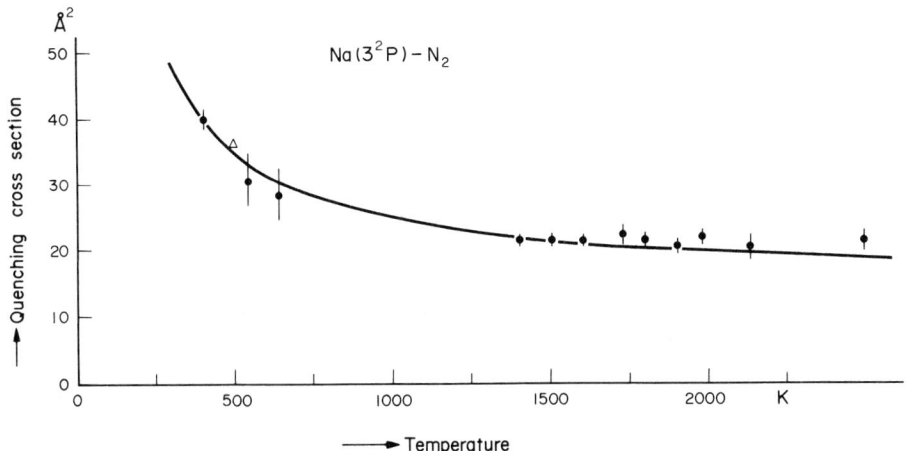

Fig. VI.1B shows the most reliable values of the (phenomenological) quenching cross section for the $Na(3^2P)$ doublet with N_2 molecules as a function of temperature, obtained from vapour-cell and flame measurements by Bästlein, Baumgartner and Brosa (1969), Hulpke, Paul and Paul (1964), Kibble, Copley and Krause (1967), Jenkins (1966, 1968), and Lijnse and Elsenaar (1972). The experimental error is indicated by a vertical bar. The drawn curve represents the theoretical temperature dependence that was fitted to the experimental points according to Lijnse (1973, 1974). However, recent direct lifetime measurements in flames by Ham and Hannaford (1979) have yielded cross section values at 1750 and 2100 K that lie systematically below the drawn curve (see also Table VI.1, pages 603-618). (Redrawn from Lijnse 1973, page 58.)

semiquantitative cross section value has been obtained for the mixing of the ground-state doublet of $Tl(6^2P_{\frac{3}{2}} \to 6^2P_{\frac{1}{2}})$ by He in a Tl vapour cell by Hsieh and Baird (1972) using the method of zero-field level crossing (Hanle effect). For the same process upper limits were found with Ar in a shock-tube by Naumann and Michel (1972) and with He, Ne and Ar in a cell with TlI vapour, dissociated by flash-photolysis, by Bellisio and Davidovits (1970). In the latter two experiments, the relaxation time for the population of the metastable $6^2P_{\frac{3}{2}}$ state of Tl was determined; the variation with time of this population was measured by monitoring an absorption line from this state. These measurements yielded very small values of the order of 10^{-3} $Å^2$ or less for the transition $J = \frac{3}{2} \to \frac{1}{2}$ induced by He and Ar. More exact cross sections of the same mixing process for noble gas atoms were obtained by Aleksandrov, Baranov and Kulyasov (1978). They again used atomic absorption to measure the collisional relaxation time of the metastable atoms, which were produced by a cascade transition from photo-excited $Tl(7^2S)$ atoms in a Tl vapour. In this way they avoided the competing quenching effect of volatile Tl compounds, as existing in the photodissociation experiments. Cross sections of 5×10^{-8}, 1.2×10^{-6} and 1.7×10^{-5} Å^2 were found for Ar, Kr and Xe, respectively, at 660 K. By a similar method as used by Bellisio and Davidovits (1970), Husain and other workers determined (upper limits of) rate constants for the exo-ergic mixing of the ground triplet states ($^3P_{0,1,2}$) of Sn and Pb by Ar, Kr and Xe at 300 K (for references see Foo *et al.* 1976). The energy separations of these triplet states range from about 0.2 to 1.0 eV; the mixing cross sections are far below 1 Å^2.

Theory. A qualitative theoretical interpretation of the results obtained may be presented by considering the Massey parameter ξ, given by Eq. II.142, for the transition involved. Estimating the range of interaction Δr to be of the order of 10^{-8} cm, and taking a value of 5×10^4 cm s^{-1} for the mean relative velocity $\bar{g}$ near room temperature, one calculates $\xi \simeq 1$ for $\Delta E = 3 \times 10^{-3}$ eV. For the first Na doublet ($\Delta E = 2 \times 10^{-3}$ eV) we thus expect ξ to be of the order of unity if ΔE at the transition point is about equal to the atomic doublet splitting. For the first K doublet ($\Delta E = 7 \times 10^{-3}$ eV) ξ is somewhat larger. For the first Rb doublet ($\Delta E = 3 \times 10^{-2}$ eV) and the first Cs doublet ($\Delta E = 7 \times 10^{-2}$ eV), ξ will be considerably larger (near-adiabatic collisions). This may explain the small mixing cross section for the latter two doublets as well as its pronounced dependence on T (through $\bar{g}$). Since the splittings of the higher resonance doublets of Rb and Cs are appreciably smaller, a much larger mixing cross section is expected here, again in accordance with the experiments (see above). Note that the mixing collisions for the Tl ground-state doublet ($\Delta E = 0.97$ eV) are also near-adiabatic and that the corresponding cross sections are very small.

The characteristic time, $\hbar/\Delta E$, occurring in the expression for ξ can be classically understood as the period of the spin-orbit precession. For near-adiabatic collisions, the collision duration is large compared to this period, and the total electronic angular momentum J may be considered as a good quantum number. There is thus a probability close to unity that J will be conserved during the collision and that no mixing between different J states will occur.

When for the exo-ergic doublet transfer, $\ln \sigma_m$ is plotted versus (a roughly defined) Massey parameter ξ, most experimental points fit fairly well the same, nearly straight curve with a negative slope (see Hooymayers and Alkemade 1969). Different values of ξ ranging from below 1 up to 10, and of σ_m ranging from about 10^2 to 10^{-5} Å^2, were obtained by these authors by assembling the data then available for the Na-, Rb-, and Cs doublets with different noble gases and different relative velocities. This fit is achieved by choosing a range of interaction Δr of about 1 Å and by putting ΔE equal to the doublet splitting at infinite separation. As pointed out by Dashevskaya, Nikitin and Reznikov

(1970) on theoretical grounds, this fit may have at most a semiquantitative meaning. In any case, it demonstrates the sensitive dependence of the mixing cross sections on the doublet splitting (see also Siara 1972 who measured and compared mixing cross sections of the 2nd and 3rd alkali resonance doublets; see also Krause 1966, 1975).

Considerable efforts have been expended on the calculation of the alkali-doublet mixing cross section for noble gas atoms.[†] Approximate analytical expressions have been derived by Nikitin and co-workers for the case of near-adiabatic mixing collisions (i.e., large ξ values). There is reasonable agreement with experiment as to absolute value and temperature dependence; Anderson et al. (1976) were able to fit the results of their K mixing experiments with He as a function of relative velocity to the semiclassical curve-crossing model of Nikitin (1965a). Mixing is brought about by transitions between three adiabatic quasi-molecular states. Two of these states ($A^2\Pi_{\frac{3}{2}}$ and $B^2\Sigma_{\frac{1}{2}}$) correlate with an alkali atom in the $^2P_{\frac{3}{2}}$ state at infinite separation, while the third one ($A^2\Pi_{\frac{1}{2}}$) correlates with the alkali $^2P_{\frac{1}{2}}$ state. The potential-energy curve for the last state does not cross the curves for the other states. It was concluded that exchange interaction was dominant over polarization interaction in this process.

Roueff and Launay (1977) determined by quantummechanical close-coupling calculations the cross section for mixing of the K(4P) states by He as a function of relative velocity. The results obtained by using the numerical interaction potentials of Pascale and Vandeplanque (1974) as well as those of Baylis (1969) appeared to be smaller by a factor of two than the values measured by Krause (1966) in vapour cells and by Lijnse (1974a) in flames. In the calculations the velocity-dependent cross sections were not averaged over a Maxwell distribution.

For nonadiabatic mixing collisions, numerical solutions of the basic quantummechanical equations have been computed for some special cases (see for a review Lijnse 1972, 1973). The absolute magnitude as well as relative velocity-dependence of the mixing cross sections of the Na-D doublet with Ar calculated by Pascale and Olson (1976) appeared to agree very well with the experiments of Apt and Pritchard (1976, 1979). However, the calculated absolute values were not in agreement with their experiments for Ne, Kr and Xe, whereas the relative velocity-dependence was in agreement for Xe. In the calculations use was made of the interaction potentials of Pascale and Vandeplanque (1974).

[†] See, e.g., Callaway and Bauer (1965), Nikitin (1965a, 1967b, 1975), Kumar and Callaway (1968), Reid and Dalgarno (1969, 1970), Dashevskaya, Nikitin and Reznikov (1970, 1970a), Nikitin and Reznikov (1971), Pascale and Stone (1976), Pascale and Olson (1976), Saxon, Olson and Liu (1977), Roueff and Launay (1977), and the reviews by Lijnse (1972), and Phaneuf (1973).

Chapt. VI. EXCITATION AND DE-EXCITATION OF METAL SPECIES

Reid and Dalgarno (1969, 1970) have presented a quantummechanical theory for calculating fine-structure mixing cross sections, which is an extension of earlier theoretical work on rotational energy transfer in collisions between a diatomic molecule and an atom. The mixing cross sections for alkali-noble gas systems can be calculated numerically when the $A^2\Pi$ and $B^2\Sigma$ potential-energy curves are known. Saxon, Olson and Liu (1977) applied this theory to the Na first-doublet mixing by Ar atoms while using their ab-initio calculations of the potential-energy curves concerned. After averaging the absolute mixing cross sections calculated as a function of collision energy over a Maxwell distribution, they found good agreement with the corresponding experimental cross sections listed in Table VI.1 at $T = 400$ K. Good agreement was also found between the dependence of the theoretical mixing cross sections on collision energy and that of the relative cross sections measured by Phillips, Glaser and Kleppner (1977) and of the absolute cross sections measured by Apt and Pritchard (1979) in the range of 0.01 - 0.2 eV. Similar results were obtained when the potential-energy curves calculated by Pascale and Vandeplanque (1974) were used instead of the above ab-initio calculated curves, although the well depth of the $A^2\Pi$ curve for Na(3P) + Ar was about twice as small as that calculated by Saxon, Olson and Liu (1977).

The relative variation of the mixing cross sections of the first K-, Rb- and Cs doublets plotted as a function of atomic number of the noble gas turned out to correlate closely with that of the elastic electron scattering cross sections (see Krause 1975). However, the absolute values for the mixing process were found to be one order of magnitude larger.

Cross sections for the Na-D doublet mixing ($J = \frac{1}{2} \to \frac{3}{2}$) induced by H atoms were calculated by Bender et $al.$ (1967) and Lewis, McNamara and Michels (1971), and were found to be about 50 Å^2 at 5000 K. They are not strongly dependent on temperature.

Theoretical mixing cross sections for the third Cs ^{2}P doublet with noble gases, in which crossings with potential-energy curves corresponding to lower-lying excited states were taken into account, were smaller by one or two orders of magnitude than the experimental values (see Nikitin and Reznikov 1971, and Pimbert 1972). The mixing cross sections calculated by Pascale and Stone (1976) for the second and third ^{2}P doublets of Rb and Cs in collision with He, Ar and Xe agreed with the experimental values within a factor of two, except for Cs-He. These authors applied a semiclassical method involving potential-energy curves.

The inefficiency of atomic species in mixing the Hg($6^3P_1 \leftrightarrow 6^3P_0$) states has been qualitatively argued by Laidler (1942a, 1955) upon consideration of the potential-energy curves.

In view of the theoretical models used in the calculations, it is

PHYSICAL EXCITATION AND DE-EXCITATION PROCESSES   Sect. VI.2a

interesting to note that Copley and Krause (1969a) provided evidence for the absence of a long-lived (K+Xe) complex with lifetime $\geqslant 10^{-9}$ s. They applied a delayed-coincidence method in measuring the stepwise line fluorescence of the first K doublet.

Self- and mutual mixing. A special case of mixing between doublet resonance states occurs in pure or mixed alkali vapour(s). Here *self-mixing* or *mutual mixing* is induced by collisions between the excited atoms and unexcited, identical or dissimilar alkali atoms (see Seiwert 1956, Pitre and Krause 1968, Hrycyshyn and Krause 1969, and Krause 1966, 1972, 1975). Examples are:

$$\text{Na}(^2P_{\frac{3}{2}}) + \text{Na}(^2S_{\frac{1}{2}}) \rightleftharpoons \text{Na}(^2P_{\frac{1}{2}}) + \text{Na}(^2S_{\frac{1}{2}}), \qquad \text{(ii)}$$

$$\text{K}(^2P_{\frac{3}{2}}) + \text{Rb}(^2S_{\frac{1}{2}}) \rightleftharpoons \text{K}(^2P_{\frac{1}{2}}) + \text{Rb}(^2S_{\frac{1}{2}}). \qquad \text{(iii)}$$

The difference in excitation energy is again made up by the translational energy of the colliding atoms. Mixing cross sections are found experimentally by the method of stepwise line fluorescence. The possible effect of the trapping of resonance radiation on the apparent lifetime of the states (see Sect. II.5c-3) should either be corrected for, or eliminated by choosing extremely low vapour densities. Low densities, however, make the measurement of self-mixing more difficult. Experimental values obtained in vapour cells are listed in Table VI.1, pages 603-618. The very large cross sections up to 500 Å^2 for self-mixing of Na and K, which have a relatively small energy defect, are noteworthy. The self-mixing cross sections plotted for different alkali doublets as a function of energy defect ΔE appear to vary $\propto (\Delta E)^{-1}$, at least in the case of (near-)adiabatic collisions, in accordance with Franck's Rule (see Pimbert 1972, and Krause 1966, 1972, 1975). The self-mixing cross sections for the first Na doublet obtained by Seiwert (1956) appear to be 3 times smaller than those listed in Table VI.1 (see for discussion Seiwert 1968 and Krause 1975).

The theoretical treatment of self-mixing for alkalis has been discussed, e.g., by Seiwert (1968), Dashevskaya, Voronin and Nikitin (1969), Dashevskaya and Nikitin (1969), Lue-Yung Chow (1969), Hrycyshyn and Krause (1969), Zembekov and Nikitin (1971), and Dashevskaya (1979) (see also the literature review in Lijnse 1973b). Resonant dipole-dipole interaction alone does not seem to explain the cross sections found; exchange interaction should be taken into account. In mutual-mixing collisions, the former type of interaction seems to be dominant (see Hrycyshyn and Krause 1969).

Self-mixing of the Tl ground-state doublet:

$$\text{Tl}(6^2P_{\frac{3}{2}}) + \text{Tl}(6^2P_{\frac{1}{2}}) \rightarrow \text{Tl}(6^2P_{\frac{1}{2}}) + \text{Tl}(6^2P_{\frac{1}{2}})$$

has been studied by Picket and Anderson (1969) who determined the lifetime of the metastable $(6^2P_{\frac{3}{2}})$ state with the aid of time-resolved atomic absorption measurements (cf. earlier in this subsection).

Conclusions. In flames diluted with a noble gas, collisions with noble gas atoms may make a noticeable contribution to the mixing of the first and higher resonance doublets of Na, K and Rb. Consequently, they may contribute to the diabatic collisional line-broadening, as they shorten the actual lifetime of the excited state. They likewise influence the degree of saturation attainable when the flame is irradiated by a strong laser beam tuned at one of the doublet components (see Sect. VI.2c-5). Mixing of the Tl ground-state doublet by noble gas atoms, however, is expected to be insignificant under flame conditions. The mixing cross section for the first resonance doublet of Rb is expected to vary markedly with flame temperature, in contrast to the corresponding cross sections for K and Na. Since kT exceeds the doublet splitting for all alkalis, the values of σ_m and σ_{-m} are comparable (see Eq. VI.1). Because of the low partial pressure of metal vapours in flames (see Sect. III.2h), self- or mutual-mixing collisions are insignificant, notwithstanding their large cross section values.

Although the (theoretical) Na-D doublet mixing cross section for H atoms is comparable to that for noble gases and di-atomic molecules, they play only a subordinate role in flames because of the low partial pressure of H radicals (see Sects IV.4 and 5).

2a-3 *QUENCHING AND EXCITATION BY COLLISIONS WITH FREE ELECTRONS*

Electrons are considerably more efficient than noble gas atoms in (de-)exciting metal atoms. This is in accordance with the rules for kinetic energy transfer in classical mechanics if we would consider the excitation process as a direct collision between the atomic electron and the collision partner. Microscopic effective cross sections, σ_e, for excitation of alkali emission lines by collisions with accelerated electrons have been measured as a function of electron energy E_e in electron-beam experiments. Just above threshold energy ($E_e \simeq E_{exc}$) this function generally shows a rather steep rise from zero, whereas it varies only slowly for higher E_e values. From this we can calculate the thermal rate constant, k_e, of the collisional *excitation* process:

$$M + e^- \rightarrow M^* + e^- \qquad (iv)$$

at temperature T by averaging over a Maxwell distribution of E_e. When on an energy scale of the order of kT, σ_e can be approximated by a constant value, σ_{thr}, for E_e above threshold, and by zero for E_e below threshold, one obtains from Eqs II.130 and 132b

$$k_e = \sigma_{thr} \sqrt{8kT/\pi m_e}\ (1 + E_{exc}/kT) \exp[-E_{exc}/kT] \ , \qquad (VI.2)$$

where m_e is the electron mass and the square-root factor is the mean electron velocity $\bar{v}_e$ (see Alkemade and Hooymayers 1966, and Riedmüller, Brederlow and Salvat 1968).

When σ_e rises smoothly with E_e above threshold and can be approximated by a linear function: $\sigma_e = (d\sigma_e/dE_e)_0 (E_e - E_{exc})$ in a range where E_e/kT is of the order of unity, we obtain the following approximate expression for k_e through Eq. II.129

$$k_e = (d\sigma_e/dE_e)_0 \, kT \sqrt{8kT/\pi m_e} \, (2 + E_{exc}/kT) \exp[-E_{exc}/kT]. \qquad (VI.3)$$

For $E_{exc} \gg kT$, this expression becomes identical to Eq. VI.2 when we replace σ_{thr} in the latter by the equivalent value: $(d\sigma_e/dE_e)_0 \, kT$. Values of σ_e for the first resonance lines of the alkalis and Ba and Ca are listed in Table VI.1. Excitation (and polarization) by electron impact near and above threshold has been studied for alkali lines by Zapesochnyi (1967), Korchevoi and Przonski (1967), Enemark and Gallagher (1972), Hafner and Kleinpoppen (1967), Stumpf, Becker and Schulz (1978), and Chen and Gallagher (1978). Gilmore, Bauer and McGowan (1967) have tabulated experimental rate constants for excitation of alkali atoms, noble gas atoms, H, Fe, Hg, Tl, and some molecules, by low-energy electrons. They also compare cross section values calculated by classical and quantummechanical methods. Classical results seem to agree generally with experiment within a factor of about 3 far above threshold, but may be largely in error near threshold. Quantummechanical close-coupling calculations for the Na-D lines are in excellent agreement with experiment near threshold (see Moores and Norcross 1972 and Enemark and Gallagher 1972).[†] For a comparison of experimental results and quantummechanical calculations for the alkaline-earth atoms see Robb (1975), and Damburg and Fabrikant (1975).

The pre-exponential factor in Eq. VI.2, which holds in the ideal case of a steeply rising excitation function, is seen to vary $\propto T^{-\frac{1}{2}}$ for $E_{exc} \gg kT$.[*] The corresponding factor in Eq. VI.3, which holds for a linear excitation function, varies $\propto T^{\frac{1}{2}}$ for $E_{exc} \gg kT$. A positive exponent of T also occurs in the empirical expression for the excitation rate of Cs up to $T = 20,000\,\text{K}$ given by Gilmore, Bauer and McGowan (1967). This agrees with the linear energy dependence of σ_e above threshold found by Chen and Gallagher (1978).

[†] For a more general discussion of inelastic electron-atom collisions we refer, e.g., to Bates and Estermann (1966, 1973), Drawin (1968), Bauer (1969a), Massey, Burhop and Gilbody (1969), Schulz (1973), and Vriens and Smeets (1980). An introduction to this subject has been given by Geltman (1969) and Bransden (1970).

[*] According to measurements by Enemark and Gallagher (1972), the excitation function for the Na-D lines e.g. rises rather sharply to about $20\,\text{Å}^2$ above threshold and levels off to about $30\,\text{Å}^2$ for E_e above 5 eV. A more smoothly rising excitation function was previously found by Zapesochnyi and Shimon (1965). Differences in electron-energy resolution might have played a part here.

Chapt. VI   EXCITATION AND DE-EXCITATION OF METAL SPECIES

For a steeply rising excitation function, the *quenching* rate constant k_{-e} for electron impact is derived from Eq. VI.2 through the detailed-balance relation Eq. II.136

$$k_{-e} = \sigma_{thr}(1 + E_{exc}/kT)(g_0/g^*)\bar{v}_e \,, \qquad (VI.4)$$

where g_0 and g^* are the statistical weights of M and M^*, respectively. Applying Eq. II.130, we get from this for the effective cross section σ_{-e} of quenching by electron impact

$$\sigma_{-e} \equiv k_{-e}/\bar{v}_e = \sigma_{thr}(1 + E_{exc}/kT)(g_0/g^*) \,. \qquad (VI.5)$$

For the first resonance doublet of Cs, we calculate, assuming $\sigma_{thr} \simeq 60 \text{ Å}^2$ (see Table VI.1): $\sigma_{-e} = 180 \text{ Å}^2$ at $T = 2500 \text{ K}$. The latter value markedly exceeds the cross sections for quenching by atomic and molecular partners (see Table VI.1). Nevertheless, in a flame at 1 atm pressure and $T \approx 2500 \text{ K}$ which is mainly composed of N_2 molecules the quenching rate of the Cs doublet for N_2 exceeds the rate for electron impact by about a factor of 120 for an electron concentration $[e^-] = 10^{13} \text{ cm}^{-3}$ corresponding to a partial pressure p_e of about 3×10^{-6} atm. This high electron concentration may be obtained by seeding the flame with Cs at a strong solution concentration. The ratio of quenching rates for N_2 and electrons, respectively, is determined by the corresponding ratio of quenching cross sections, multiplied by $(m_e/\mu)^{\frac{1}{2}}(p_{N_2}/p_e)$; here μ is the reduced mass of N_2 and Cs.

In a stoichiometric H_2-O_2 flame diluted by Ar at 1 atm, the overall quenching rate of the Cs doublet by molecular collisions may be about a factor of 10 lower than in a flame diluted with N_2. Even then, de-excitation by electron impact may become important only for electron concentrations of the order of 10^{14} cm^{-3} or higher.[†]

Expressing the overall rate, $k'[M^*]$, of molecular quenching in the efficiency of fluorescence Y and optical transition probability A through Eq. II.109a or 109b, we generally have for the ratio of de-excitation (or excitation) rates for electron and molecular impact, respectively, with the aid of Eq. VI.4 (see Alkemade and Hooymayers 1966)

$$\frac{\{(de\text{-})excit.\ rate\}_{el.}}{\{(de\text{-})excit.\ rate\}_{mol.}} = [e^-]\sigma_{thr}\bar{v}_e(g_0/g^*)(1 + E_{exc}/kT)\{Y/(1-Y)A\} \,. \qquad (VI.6)$$

This equation holds in the equilibrium case and is applicable under the same approximative condition which also underlies Eq. VI.2, namely that the electron excitation

---

[†] For a comparison of Cr excitation rates by collisions with electrons ($[e^-] \approx 10^{14} \text{ cm}^{-3}$) and with Ar atoms in shock-heated Ar gas at $T = 8000 \text{ K}$, see Shackleford and Penner (1966).

PHYSICAL EXCITATION AND DE-EXCITATION PROCESSES   Sect. VI.2a

cross section rises steeply to a constant value above threshold.
    When the excitation cross section grows linearly with E_e above threshold,
Eq. VI.3 should be used instead of Eq. VI.2. For $E_{\text{exc}} \gg kT$, Eqs VI.4, 5 and 6
are then again obtained if we replace there σ_{thr} by $(d\sigma_e/dE_e)_0 kT$. It thus follows that k_{-e} varies roughly $\propto T^{-\frac{1}{2}}$ and $\propto T^{\frac{1}{2}}$ in the case of a steeply and a slowly
rising excitation function, respectively.
    The above calculations are based on the assumption of equipartition of the
electron energy. This assumption may not be valid in the combustion zone where
'hot' electrons may be produced by the combustion reactions (see Sect. IV.7). Above
the combustion zone and at 1 atm pressure, equilibration of the electron energy by
collisions with flame molecules is expected to be established within a very short
time interval. In the case of elastic collisions and in a flame diluted with N_2
at 1 atm pressure, this time interval is of the order of $(m_{N_2}/m_e) \nu_e^{-1} \approx 10^{-7}$ s; here
m_{N_2} is the mass of a N_2 molecule and ν_e is the collision frequency for an electron in a gas of N_2 molecules. The actual equilibration time is expected to be
even considerably shorter owing to the contribution of inelastic collisions with
molecules. We should therefore expect an equilibrium population of the excited
state above the combustion zone if this state were (de-)populated by electron collisions only.
    We can calculate from Eq. VI.6 that under usual flame conditions excitation
process (iv) and the reverse quenching process are unimportant, at least above the
combustion zone.
    In the (de-)population of atomic states near the ionization limit, ion-
electron recombination in the presence of a third body X,

$$M^+ + e^- + X \rightarrow M^* + X , \qquad (v)$$

may compete with process (iv) and with molecular excitation processes in general.
Then the above considerations are no longer valid, and, in particular, the excited-
state population may be affected by possible deviations from Saha ionization equilibrium (see Sect. IX.3d). Alkemade and Hooymayers (1966) have estimated that in
ordinary flames at 1 atm pressure and $T \approx 2500$ K, (de-)excitation of the Na-D doublet by process (v), or its reverse, is at least four orders of magnitude less than
by molecular impact. The situation might be different, however, for levels with
$E_{\text{exc}} \approx 4$ eV or higher, as the relative importance of process (v) grows roughly
$\propto \exp[+E_{\text{exc}}/kT]$ with increasing excitation energy (see Alkemade and Hooymayers
1966).
    We can generally conclude from this subsection that even in flames seeded
with a large concentration of an easily ionizable element the relative density of
electrons will still be too low to play a part in the (de-)excitation of the lower
atomic levels. This holds at least in the flame region above the combustion zone.

Chapt. VI   EXCITATION AND DE-EXCITATION OF METAL SPECIES

2b   (DE-)EXCITATION OF ELECTRONIC STATES INVOLVING CONVERSION
     OF INTERNAL ENERGY OF COLLISION PARTNER

2b-1 *INTRODUCTION*

In this section we shall consider the (de-)excitation of a metal atom M through a collision with an atomic or molecular partner, Z, whose internal energy is also changed according to, e.g.,

$$M + Z' \rightleftharpoons M^* + Z''$$

or

$$M^* + Z' \rightleftharpoons M^{**} + Z'' \,.$$

In these examples M^{**}, M^*, and M denote a metal atom in two different excited states and the ground state, respectively, while Z' and Z'' denote a collision partner in two different energy levels (compare processes nos. 3-6 in Table II.1 on page 108). Energy balance requires that the total internal-energy defect in the collision process is made up by a change in the translational energy of relative motion. A special case is the multiplet-mixing process with M^* and M^{**} belonging to different states of the same multiplet.

There is a big difference between atoms and molecules as collision partner Z as far as behaviour and mechanism are concerned. The former have only electronic energy levels, which are rather widely spaced (apart from fine-structure splitting) in an energy range up to a few electronvolt. Molecules have, in addition, rotational and vibrational energy levels. The energy spacing of the rotational levels is small, even less than 10^{-2} eV for levels up to quantum number $J = 8$ in N_2. Lighter molecules such as H_2, however, have larger rotational-energy spacings, which are of the order of 0.1 eV for low J values. Usually the discrete vibrational energy levels have a spacing of the order of a few tenths of an electronvolt. Both the rotational and vibrational energy levels show an isotopic effect. The lowest electronic excitation levels of molecules that may occur as bulk species in flames are usually markedly higher than the lowest excited multiplets of metal atoms. The significance of electronically excited flame molecules in the (de-)excitation process is therefore limited. A typical exception is O_2 which has some electronic excitation levels below the $Na(3^2P)$ doublet (for other examples see Laidler 1955).

(De-)excitation by atomic collision partners involving electronic-excitation transfer will be treated in Sect. 2b-4. Transitions between atomic states belonging to different multiplets and induced by molecular collision partners will be discussed in Sect. 2b-2. Mixing collisions induced by molecular partners will be discussed separately in Sect. 2b-3. It should be realized, however, that in actual measurements, quenching and mixing processes are often concurrent.

With molecules, an additional difficulty may arise through the competition of chemical reactions with the physical (de-)excitation collisions. Chemical

560

quenching of excited states (through a dissociation reaction, process 24, or an exchange reaction, process 18; see Table II.1 on page 108) may contribute to the total quenching cross section measured for a certain partner. The contribution of a specific chemical quenching reaction in flames can sometimes be estimated from the known rate constant of the inverse chemiluminescent reaction by application of detailed-balance relations. We shall treat chemical excitation and chemiquenching in Sect. 3. In most cases chemiquenching appears to be less important in flames than physical quenching (see Sect. 3c-2).

The participation of various forms of internal molecular energy complicates a quantitative theoretical interpretation. It also complicates the relation — through microscopic reversibility — between the effective quenching and reverse excitation cross sections when the two are measured under different experimental conditions (see Sect. 2b-2). Additional information about the distribution of internal energy in the product molecule and about the velocity dependence is needed for an unambiguous interpretation. Experimental studies are therefore aimed at determining not only the absolute overall cross sections, but also the internal-energy distribution and the velocity- or temperature dependence (see below). Most attention has been paid to the role of the vibrational energy, both experimentally and theoretically. Rotational transitions have occasionally been taken into account in the explanation of resonance effects in alkali doublet mixing (see, e.g., McGillis and Krause 1967a, 1968, 1968a, and Siara and Krause 1973). Rotational transitions have also been taken into account in the theoretical analysis of the Cs- and Rb doublet mixing by deuterated methanes (see Baylis et $al.$ 1973, and Walentynowicz, Phaneuf, Baylis and Krause 1974).

2b-2   *QUENCHING AND EXCITATION BY COLLISIONS WITH MOLECULES*

General. There is ample experimental evidence that molecules induce transitions between electronic states with energy separation of the order of 1 eV much more readily than noble-gas atoms. The quenching cross sections for most molecules are comparable to gas-kinetic cross sections and sometimes much larger (see Table VI.1 on pages 603-618).

Absolute cross sections for de-excitation of the first and higher excited multiplets by molecules have been measured in vapour cells and flames by methods similar to those applied in the case of noble-gas atoms (see Sects VI.2a-1 and II.4a-4). Since molecules induce transitions between atomic multiplets with large energy separation much more readily than noble-gas atoms, the determination of rate constants of collisional transitions in multi-level systems is more complicated in the former case. We have to recur to the methods described in Sect. 2a-1 for disentangling the more numerous (de-)population routes when the quenching of higher-lying states by molecules is studied. In this respect vapour-cell experiments have

Chapt. VI   EXCITATION AND DE-EXCITATION OF METAL SPECIES

an advantage over flame experiments in that in a cell measurements can be readily made as a function of density of the collision partner (see Sect. 2a-1). Crossed-molecular beam experiments have again an advantage over these bulk experiments, as in the beam experiments the product-state distribution is not disturbed by secondary collisions. Moreover, beam experiments allow us to measure differential cross sections for excitation as well as quenching, from which information can be obtained more directly about the possible formation of a short-lived collision complex (see Silver, Blais and Kwei 1977, 1979). However, beam experiments often yield only relative (differential) cross sections. Further advantages of beam experiments over bulk experiments have already been mentioned in Sect. 1. For a survey of beam experiments see Kempter (1975), Hertel and Stoll (1977), and Hertel (1980).

When the first resonance doublet of an alkali atom is excited, the interpretation of quenching measurements with molecules may be complicated by the competition between quenching and doublet-mixing processes. In contrast, for noble-gas partners quenching can safely be disregarded (see Sect. 2a-1). When the resonance doublet is not spectrally resolved in the fluorescence measurement, some sort of weighed average value is obtained for the quenching rate constant of the doublet as a whole. The relation between this average value and the 'true' rate constants for the individual doublet states involves the mixing and optical rate constants as well as the ratio of the photo-excitation rates of the two doublet states (see Lijnse, Zeegers and Alkemade 1973). The general expression for the average unimolecular quenching rate constant k'_{I+II} is therefore complicated. However, it can be simplified considerably if the mixing rate constants are appreciably larger than the quenching and optical rate constants. One then obtains (see Lijnse, Zeegers and Alkemade 1973)[†]

$$k'_{I+II} \simeq \frac{A_{II}+A_{I}}{2(A_{II}+rA_{I})} (k'_{II}+rk'_{I}) , \qquad (VI.7)$$

where k'_{II}, k'_{I} and A_{II}, A_{I} are the quenching and optical rate constants (in s^{-1}) for the separate doublet states, respectively, and r is the ratio of mixing rate constants or cross sections, which is given by Eq. VI.1. For the first Cs doublet A_{I} and A_{II} differ from each other by about 10% because of the wavelength factor contained in Eq. II.162. For the first resonance doublets of the other alkalis this difference is less.

In addition, the 'true' or individual quenching rate constants (as well as the mixing rate constants) can — in principle — be determined in flames by measuring four independent fluorescence efficiencies (see Lijnse, Zeegers and Alkemade 1973). These fluorescence efficiencies are defined as the four ratios of the

---
[†] A similar expression was derived earlier by Jenkins (1968a) on the simplifying assumption that A_I equals A_{II}.

PHYSICAL EXCITATION AND DE-EXCITATION PROCESSES   Sect. VI.2b

fluorescence intensity of each doublet component to the absorbed primary-radiation intensity when either doublet component is excited separately.

Flames usually contain more than one quenching and mixing species as bulk constituent. In order to distinguish between the separate contributions of these species to the total rate constant measured, observations are made in a number of isothermal flames with widely varying quantitative composition. In this way one obtains a set of linear equations of the same kind as Eq. II.80, from which the unknown (bimolecular) quenching and mixing rate constants for each species can be solved. These rate constants can be expressed in phenomenological cross sections through Eq. II.130. A simplification is obtained when one of the bulk components is a noble gas whose quenching efficiency can be considered as negligible † (see Sect. 2a-1).

In vapour cells containing metal vapour and only one quenching agent a different procedure is followed to obtain the quenching rate constants for the individual doublet components. We assume here that the metal atom concentration is sufficiently low for allowing us to neglect inelastic collisions between the metal atoms themselves. One measures the ratio of resonance fluorescence intensity in the presence of quenching gas to that obtained in its absence, as a function of gas pressure. This is done for each doublet component separately. The constancy of metal atom concentration and exciting light intensity has to be monitored during these measurements. One also measures the ratio of fluorescence intensities of the doublet components as a function of pressure when either component is excited by the primary light source. From these measurements one derives the individual (bimolecular) quenching as well as the mixing rate constants. The rather complicated equations used in this derivation have been presented by McGillis and Krause (1967a, 1968a) (for a more general mathematical description see Tudorache 1977). If mixing collisions are unimportant in comparison with quenching collisions, the equations for the quenching rate constants reduce to the simple linear Stern-Volmer relation (see Eq. II.110).

Quenching cross sections for molecular species have mostly been determined from stationary fluorescence intensity measurements as described above. The average quenching cross section for the Na doublet in a flame with N_2 as diluent gas has first been derived by Boers, Alkemade and Smit (1956) from a measurement of the absolute fluorescence efficiency Y. Hooymayers, Jenkins and Lijnse, using in principle the same method, have later determined quenching cross sections in flames at

---

† One should be careful about neglecting *a priori* the contribution of noble-gas atoms when the investigated level lies close to other excited levels, as in the case of the $Sr(5^1P_1)$ level lying 0.2 eV above the (4^1D_2) level.

Chapt. VI   EXCITATION AND DE-EXCITATION OF METAL SPECIES

1 atm pressure for the first resonance doublets of the alkalis and for some other atomic resonance lines (see Table VI.1 on pages 603-618). Quenching cross sections for the individual alkali doublet components could not be successfully determined in flames because of the relatively strong mixing collisions (see Jenkins 1968a, and Lijnse, Zeegers and Alkemade 1973). McGillis and Krause (1967, 1967a, 1968, 1968a)[†] were the first to determine the individual quenching cross sections of the first K- and Cs doublet components from stationary fluorescence experiments in vapour cells.

Lifetime, phase-shift and photodissociation methods (see Sect. II.4a-4) have also been applied to measure quenching cross sections for molecules (see Table VI.1). Bellisio, Davidovits and Kindlmann (1968) were the first to measure quenching cross sections for the separate $Rb(5^2P)$ doublet components by a lifetime method in vapour cells. Ham and Hannaford (1979) and Hannaford (1979) have recently reported the first lifetime measurements in flames for the first excited resonance levels of the alkalis and a few other metal atoms. They used a tunable, pulsed, N_2-laser pumped dye laser as excitation source in combination with a storage oscilloscope. Their time resolution was about 2 ns, which prevented measurements from being done on the Na-D doublet in C_2H_2-air and C_2H_2-N_2O flames at 1 atm, in which the collisional + optical lifetime is too short because of strong quenching. H_2-O_2 flames diluted by Ar or by a mixture of Ar and N_2 proved to be most suitable for the determination of specific quenching cross sections for the Na-D doublet with H_2O, H_2, O_2 and N_2 as perturbers. The main advantage of the lifetime method over the usual stationary method is that fluorescence-intensity measurements are required on an arbitrary scale only. Moreover, no assumption need be made about the isotropy of the fluorescence radiation; this isotropy is an essential condition in the measurement of absolute fluorescence efficiencies. Furthermore, scattering of the exciting laser radiation by particles in the flame does not interfere with the lifetime measurements, which are done after the laser pulse has terminated.

Quenching by molecules has been studied mainly for the first alkali resonance doublets and the excited $Hg(6^3P_1)$ state. Table VI.1 on page 603 ff. shows the corresponding cross sections as well as those for some higher alkali states and for the excited states of a number of other elements that might be of interest in flame studies. Quenching cross sections that so far have been measured in vapour cells only might be of interest in the context of this book, too, when one wants to estimate fluorescence efficiencies of atomic metal lines in flames of known composition. The only uncertain factor in this estimation is the (usually weak) temperature-

---

[†] The figures quoted in the 1967 publication have been corrected in subsequent papers of the authors.

dependence of the quenching cross sections (see later in this section). Besides, the numerous quenching cross sections obtained by nonflame methods are useful to reveal general trends and to test theoretical models.

Older values are found in Pringsheim (1949) and in Massey, Burhop and Gilbody (1971); they have mostly not been included in Table VI.1. The reliability of some of the older fluorescence intensity measurements in vapour cells has been questioned (see Jenkins 1966, and Lijnse 1972; see also Sect. II.4a-4).

Experiments. The alkali- and Hg atoms will be dealt with separately.

Alkali atoms. The following trends are noted in the quenching cross sections of the *first resonance levels* (see also Lijnse 1973). For the quenching of the first resonance doublets by the di-atomic molecules N_2, O_2, and CO, the cross sections are roughly of the order of 10-100 Å^2, whereas they are of the order of 1-10 Å^2 for H_2, HD and D_2. For the tri-atomic molecules CO_2 and SO_2, values of the order of 100 Å^2 have been reported, but for H_2O the values were much smaller, namely of the order of 1-10 Å^2. Saturated organic molecules have very small quenching cross sections ($\leqslant 1$ Å^2).

The quenching cross sections found in vapour-cell experiments for the individual components of the first resonance doublet of K, Rb and Cs seem to be approximately equal, taking the possible error limits into account. This holds at least for N_2 as collision partner. We note, however, from Table VI.1 that a considerable systematic error might be present in the absolute magnitude of some of the values for $Rb(5^2P_{\frac{1}{2},\frac{3}{2}})$ with N_2 as measured by different authors using two different methods. Anyway, one may expect from the cell measurements that in flames too the quenching cross sections for the individual doublet components are approximately the same; this permits a simple interpretation of the flame results obtained for the doublet as a whole.

Quenching experiments have also been done for the *higher-lying energy levels*. Total quenching cross sections of N_2 for the second resonance doublets of Rb and Cs are roughly 2 to 3 times larger than for the first doublets, whereas they are roughly one order of magnitude larger in the case of H_2. We note, however, from the table that the vapour-cell measurements on the quenching of Rb(5P) obtained by various authors show a considerable spread. In contrast, the quenching cross section of N_2 for the second resonance doublet of Na seems to be roughly 2 times smaller according to Czajkowski, Krause and Skardis (1973), whereas it is virtually equally large according to Gallagher, Cooke and Edelstein (1978), who used a different experimental method. Remarkably, large values of roughly 100 Å^2 have been found for the quenching of the second $K(5^2P)$- and $Rb(6^2P)$ doublets by CH_4 (see Earl 1973 and Table VI.1). Note that the <u>total</u> cross section for quenching of the $Cs(8^2P_{\frac{1}{2},\frac{3}{2}})$ states by N_2, involving transitions to a number of lower states, appears to be very large, whereas the quenching cross section related to a specific collisional transition from the same state to the $7^2D_{\frac{3}{2}}$ state is markedly smaller. The same observation can be made when comparing the total quenching cross sections of Na(5S) and Na(4^2P) for N_2 with the specific quenching cross sections of these states as measured by Gallagher, Cooke and Edelstein (1978) and listed in Table VI.1. The total cross section for a high-lying state, which is the sum of all specific cross sections, will be large if many inelastic channels are open.

Czajkowski, Krause and Skardis (1973) also measured total quenching cross sections of 15 excited states of Na for N_2. A regular dependence of cross section on excitation energy was found within each S-, P-, and D series, with a minimum at the 7S-, 6P- and 5D states lying around 4.7 eV, and a sharp maximum at the 8S-, 7P- and 7D states lying around 4.8 eV. These authors measured the effect of adding N_2 gas to a vapour cell containing a mixture of Hg- and Na atoms; the Na atoms were excited by collisional transfer of electronic excitation energy from photo-excited Hg atoms. The stationary populations of a series of Na levels were derived from fluorescence-intensity measurements.

The above findings of Czajkowski, Krause and Skardis (1973) are in clear contrast to the experimental results obtained by Humphrey *et al.* (1978) for the S- and D states of Na using a lifetime method (see Table VI.1). Most absolute cross sections found by the two groups of authors deviate from each other by much more than the combined accidental error. Humphrey *et al.* have criticized the assumption of Czajkowski *et al.* that N_2 quenches the high-lying states through direct transitions to the ground state or some low-lying excited state. The experimental data of the former authors seem to indicate that quenching to energetically nearby lower states plays a dominant role. This indication might upset the interpretation of the excited-state populations in terms of quenching cross sections as given by Czajkowski *et al.*

The comparatively large probability of collisional transitions from a high-lying level to energetically neighbouring levels (including levels that lie within about kT above the initially excited level) may lead to a partial equilibrium population within this group of levels, if the perturber density is sufficiently high (cf. the pertinent discussion in Sect. 2a-1). Under these conditions one determines only a weighed average value of the cross section for quenching transitions from these levels to lower levels that lie outside the group; this holds even if only one level of the group is initially excited. The quenching cross sections listed in Table VI.1 for the Na(4^2D ... 8^2D) states with N_2 and measured by Humphrey *et al.* (1978) using a lifetime method in a vapour cell, relate in fact to such a group (called: 'manifold') of nearby levels. These levels all have the same principal quantum number n but different values of azimuthal quantum number $l \geqslant 2$. Whereas

PHYSICAL EXCITATION AND DE-EXCITATION PROCESSES   Sect. VI.2b

the quenching cross sections of the Na(nS) levels (which are well separated from other levels) measured by the same authors for N_2 show a tendency to increase slowly with increasing n, the cross sections of the manifold Na($n,l \geq 2$) levels for N_2 appear to decrease with increasing n. The authors have noted in this connection that the number of levels in the manifold is expected to increase with n and they have argued that this could lead to a reduction in effective quenching cross section that involves only transitions to lower levels outside the manifold.

Gallagher et al. (1977), using the same lifetime method, were able to measure the cross sections for collisional transitions from initially excited Na(n^2D) states to Na($n,l > 2$) states, induced by N_2 and CO, for $5 \leq n \leq 14$. These measurements, which had to be done at sufficiently low perturber densities, yielded very high cross sections of the order of 10^3 Å^2 which sharply increased when n is increased from 5 to 10 (see Table VI.1). This outcome indicates that collisional transitions from high-lying doublets to neighbouring doublets are indeed very efficient.

In Sect. 3c-2 we shall discuss the possible contribution of *chemiquenching* to the total cross sections measured for alkalis with H_2O and H_2.

The quenching cross sections of the first alkali resonance doublets for diatomics appear to decrease with increasing *relative velocity*.[†] This tendency has been confirmed by the dependence of quenching cross sections on temperature. For example, the variation of Na(3^2P) quenching cross sections for N_2 between 400 and 640 K measured in vapour cells is consistent with this tendency (see literature cited by Lijnse 1972, 1973, and Krause 1975). Also, the quenching cross sections measured in flames at about 2000 K for all alkalis with N_2 and for Na with H_2 are roughly 2 times smaller than those measured in vapour cells at about 400 K (see Fig. VI.1B on page 551). However, recent lifetime measurements by Ham and Hannaford (1979) of the quenching cross sections for Na(3^2P) with N_2 in flames yielded values that are about 20-30% lower than those obtained in similar flames by Jenkins and Lijnse who used the stationary fluorescence-quenching method (see Table VI.1 page 603 ff.). The results obtained by Ham and Hannaford indicate a more pronounced decrease of quenching cross section with increasing temperature. Also with the first resonance doublets of K, Rb and Cs a decrease of quenching cross section for N_2 with increasing temperature is noted (see Lijnse 1973, 1974). This holds at least if we tentatively discard the value found by Jenkins (1968c) for Cs in a flame at $T = 1400$ K, which is nearly twice as high as the value found by Lijnse, Zeegers and Alkemade (1973a) in a flame at $T \approx 1700$ K. No explanation of this discrepancy

---

[†] See Hanson (1955), Gatzke (1963), Brus (1970), Earl et al. (1972, 1973), and Barker and Weston (1973, 1976).

Chapt. VI   EXCITATION AND DE-EXCITATION OF METAL SPECIES

has been offered so far (for a discussion see the last mentioned paper). Remarkably, the quenching cross sections for the $Cs(7^2P)$ second resonance doublet components measured in a vapour cell with H_2 by Lukaszewicz (1974, 1975) appear to increase by a factor of 2 when the temperature is raised from 400 to 600 K.

Barker and Weston (1976), using the photodissociation method, have measured $Na(3^2P)$ quenching cross sections as functions of the kinetic energy, E_k, of relative motion. Combining their results with previous quenching measurements in vapour cells and flames as a function of temperature, they found that the phenomenological quenching cross section for N_2, H_2 and CO_2 can be described by: $\langle \sigma_q \rangle = A(1 - B/\hat{E}_k)$ where $\hat{E}_k$ is the most probable kinetic-energy value. The constants A and B depend on the quenching partner; $\hat{E}_k$ ranged from roughly 0.03 to 0.3 eV. The theoretical curve drawn through the experimental points in Fig. VI.1B, representing $\langle \sigma_q \rangle$ of $Na(3^2P)$ with N_2 as a function of temperature T, obeys the relationship: $\langle \sigma_q \rangle = A'(1 - B'/T)$ (see Lijnse 1973, 1974). Assuming that the true quenching cross section as a function of E_k, under mono-energetic conditions, is given by a similar relationship, then one finds that the conclusions of Barker and Weston (1976) and Lijnse (1974) are, at least qualitatively, consistent.

It should be noted, however, that the results of Barker and Weston point to a stronger T-dependence than was found in flames by the stationary fluorescence-quenching method. This stronger T-dependence is in better agreement with the flame values found by the lifetime method (see Ham and Hannaford 1979).

Alkali excitation rates have been directly measured in *crossed-beam experiments* with a hot molecular beam crossing a beam of alkali atoms (see Mentall, Krause and Fite 1967, 1967a, Kalff 1971, and Krause, Fricke and Fite 1972; for a general survey see Hertel and Stoll 1977). A beam of N_2-, H_2- or D_2 molecules effusing from a tungsten furnace with temperatures ranging from 2000 to about 3000 K crossed at right angle an alkali beam effusing from a relatively cool furnace containing metallic alkali. In the experiment by Krause, Fricke and Fite (1972) velocity selection of the molecular beam was applied. The intensity of the first resonance doublet of the collisionally excited alkali atoms was measured. In one case (see Kalff 1971) the K doublet components were measured separately. The confrontation of these beam experiments with the reverse quenching experiments will be discussed below.

Relative $Na(3^2P)$ quenching by NO, N_2, etc. has also been measured in a crossed-beam experiment as a function of recoil velocity and of scattering angle (Silver, Blais and Kwei 1977, 1979). Rost (1977), Hofmann (1977), and Hertel, Hofmann and Rost (1976, 1977) reported relative differential cross sections at fixed scattering angle for the quenching of $Na(3^2P)$ by $H_2, D_2, N_2, O_2, CO, CO_2, N_2O$ and C_2H_4

568

as a function of the velocity of the <u>scattered</u> Na atoms. Hofmann (1977) and Hertel, Hofmann and Rost (1977a) also investigated the dependence of the differential quenching cross section of Na(3^2P) with $N_2(v=0)$ on the polarization of the exciting laser beam, that is, on the alignment of the excited Na atoms (see also Hertel and Stoll 1976). They concluded from the dependence found that the quenching cross section is smaller when during the collision the system Na(3P) + N_2 is in the $A^2\Pi$ state than when it is in the $B^2\Sigma$ state (cf. Fig. VI.4 on page 588). For a survey we refer again to Hertel and Stoll (1977).

Hg atoms. Quenching of the Hg resonance radiation at 2537 Å ($6^3P_1 \rightarrow 6^1S_0$) by a large number of molecular species has been studied extensively (see Table VI.1).[†] It was often uncertain whether an intramultiplet mixing to the metastable 6^3P_0 state (lying 0.218 eV below the 6^3P_1 state) or a real quenching to the 6^1S_0 ground state was involved. Other investigations have shown that most molecules (e.g., N_2, H_2O and CO, but not CO_2) induce a transition to this metastable state rather than to the ground state (see Callear and Wood 1971, and Vikis, Torrie and Le Roy 1972; for a literature survey see Lijnse 1973). In contrast to the case of the alkalis, N_2 turns out to be a poor quencher of the Hg resonance radiation, whereas H_2 and D_2 are efficient quenchers. In the latter case, a dissociation reaction:

(i) $\quad Hg^* + H_2 \rightarrow Hg + 2H$

or an exchange reaction:

(ii) $\quad Hg^* + H_2 \rightarrow HgH + H$

may be involved, too. According to quantummechanical phase-space calculations by Yang, Paden and Hassell (1967), reaction (i) and the physical quenching process are dominant[*].

Preliminary measurements of the collisional excitation rate of $Hg(6^1S_0 \rightarrow 6^3P_1)$ by excited, metastable $CO(a^3\Pi)$ in a crossed-beam experiment have been reported by Van Itallie and Martin (1972). A time-of-flight analysis of the CO- and Hg resonance radiation revealed a decreasing excitation cross section when the kinetic energy was increased from 0.01 to 0.06 eV.

[†] See, for example, the experiments by Deech and Baylis (1971), Deech, Pitre and Krause (1971), and Michael and Suess (1974). The latter authors have compared their quenching cross sections with previous values from the literature while correcting for differences in the Einstein transition probability. For a survey of previous experiments see Lijnse (1973b) and Krause (1975).

[*] Callear and Hedges (1970) concluded from their HgH absorption experiments that reaction (ii) was the main quenching channel (see also Callear and McGurk 1972). The latter conclusion has been questioned, however, by Lee *et al.* (1973) and Jennings *et al.* (1973) who observed the presence of H atoms and vibrationally excited H_2 molecules as a function of time (see also below).

Chapt. VI   EXCITATION AND DE-EXCITATION OF METAL SPECIES

Crossed-beam experiments have also been employed to measure relative rate constants for the quenching processes: $Hg(6^3P_2) + NO(X^2\Pi) \rightarrow Hg(6^1S_0) + NO(A^2\Sigma^+)$ or $NO(B^2\Pi)$ (see Liu and Parson 1976). Krause and Datz (1976) obtained in a similar experiment an absolute value for the rate constant of the process that leads to NO in the $A^2\Sigma^+$ state.

We also note from Table VI.1 the high cross section found by Barrat-Rambosson and Kucal (1978) for collisional transitions from the $Hg(6^3D_1)$ triplet state to the neighbouring 6^1D_2 singlet state induced by N_2.

Other atoms. There are also reports of quenching experiments with other elements in vapour cells or flames (see Table VI.1). In some instances, it may be doubted whether the lower state of the quenching transition was the ground state or a lower lying excited state (see Lijnse 1973). Often the lower state was not specified in the experiment. Some striking results are the great efficiency of H_2O in quenching of the $Sr(5^1P_1)$ state when compared to the quenching of alkali states (see Hollander *et al.* 1972), the unusually large quenching cross section of O_2 for the same state (see Jansen 1976, and Jansen, Hollander and van Helvoort 1977[†]), and the very small quenching cross section of H_2 for $Tl(7^2S_{\frac{1}{2}})$ in flames (see Jenkins 1968d). The existence of a $Sr(4^1D_2)$ state lying close below the 5^1P_1 state should be noted (see earlier in this subsection). In contrast to the case of $Sr(5^1P)$, the quenching cross sections for the lowest metastable $Mg(3^3P)$ state with N_2, H_2 and CO are very small. These cross sections were measured by Blickensderfer, Breckenridge and Moore (1975) who monitored the decay of this state after pulsed-laser excitation by means of atomic absorption spectroscopy.

In some cases (for example, $Cu^* + O_2$; $Si^* + O_2$ or H_2 ; $Bi^* + O_2$ or H_2) *chemical reactions* between the excited metal atom M^* and the collision partner may participate in the quenching process.

The participation of internal molecular energy. The participation of internal (mostly vibrational and rotational) molecular energy in the (de-)excitation of metal atoms is evidenced by a large variety of experiments. The great (de-)excitation efficiency of molecules compared to noble-gas atoms has often been considered in the literature to be an indication that conversion of internal molecular energy plays a role. However, such a conclusion is not quite straightforward. This is demonstrated by the outcome of the excitation experiments with fast alkali beams impinging on a gas of 'internally cold' molecules (see Sect. 2a-1). The importance

---

[†] The values for CO_2, CO and O_2 published in the latter paper replace those reported earlier by Hollander *et al.* (1972, 1973).

of resonance effects in the interconversion of atomic excitation energy and internal molecular energy has often been over-estimated in the literature (for a critical discussion see Lijnse 1972, 1973, and for a review of later experiments including laser-excited atoms in an atomic beam Hertel and Stoll 1977, and Hertel 1980). We shall return to the latter problem later in this subsection. We review first the available (direct or indirect) experimental evidence for the participation of internal molecular energy. In most studies, only the conversion of vibrational energy is considered and the possible participation of rotational energy is neglected. The rotational energy levels are more closely spaced and their population shows a much shorter relaxation time. This hampers the experimental assessment of their role in the (de-)excitation process as distinct from the role played by the translational degree of freedom.

(i) Direct spectroscopic evidence as to the vibrational levels that are involved in the quenching of an electronic state, has been obtained from *infrared studies*. The vibrational bands emitted by polar quenching molecules such as CO, NO and HF, give us a clue to the distribution of vibrational energy directly after quenching. In a low-pressure flow system, several CO vibrational levels with quantum numbers up to $v = 9$ were found to be excited upon quenching of photo-excited Hg atoms.[†] This outcome implies that less than half of the electronic energy is converted into vibrational energy, the other part being converted into translational and/or rotational energy. It is essential that the experimental conditions are such that the observed vibrational level populations are, indeed, identical to those primarily produced in the quenching process.[*] It has been found that the majority of CO collisions with photo-excited $Hg(6^3P_1)$ atoms induces a transition to the 6^3P_0 state rather than to the ground state (see earlier in this subsection). So the observed vibrational-energy distribution results from (subsequent) quenching by CO of the metastable 6^3P_0 state rather than of the 6^3P_1 state (see Lijnse 1973). With NO, vibrational levels up to $v = 17$ were observed in emission. Thus at most two-thirds of the electronic excitation energy was here converted into vibrational energy.

[†] See Polanyi (1963), Karl and Polanyi (1963), Karl, Kruus and Polanyi (1967), Karl, Kruus, Polanyi and Smith (1967), Heydtmann, Polanyi and Taguchi (1971), and Fushiki and Tsuchiya (1973). It should be noted that the population of the $v = 0$ and 1 levels of CO could not be determined in the experiment.

[*] The influence of vibrational relaxation on the observed vibrational level population was eliminated by Fushiki and Tsuchiya (1973) who detected only the a.c. component of the CO luminescence intensity while modulating the Hg* photo-excitation rate at 4 kHz. They established that the vibrational-energy distribution measured previously by stationary methods was indeed markedly affected by vibrational relaxation.

Chapt. VI   EXCITATION AND DE-EXCITATION OF METAL SPECIES

In similar experiments Hassler and Polanyi (1967) observed vibrational excitation of CO up to $v=3$ as a result of Na(3^2P) quenching. Hsu and Lin (1976) measured the vibrational-energy distribution for the same process by observing infrared resonance absorption from different v levels employing a c.w. CO laser as background source. The latter authors eliminated the distorting effect of vibrational-vibrational relaxation by using a flash-lamp pumped dye laser to excite the Na vapour in a low-pressure vessel filled with CO- and Ar gas. They observed the time-resolved absorption signals over a period of 25 μs and extrapolated their magnitudes to the appearance time of strongest absorption, occurring shortly after the dye-laser initiation. Vibrational excitation up to the limit of the Na excitation energy, with a peak at $v=2$, was found.

From these observations it appears that several channels for internal energy transfer are operative in the molecular quenching of excited Hg- and Na atoms. Resonance effects in the electronic-vibrational energy transfer are obviously absent here. By applying microscopic reversibility, a similar negative conclusion can be drawn for the reverse excitation process. In the excitation process, an appreciable fraction of the excitation energy can therefore be supplied from the translational and/or rotational degrees of freedom.

(ii)   The observed emission of a series of K- and Na lines from different atomic levels that are populated by energy transfer from vibrationally excited N_2 molecules also demonstrates the role played by this degree of freedom.[†] Vibrationally excited N_2 was produced by a *microwave discharge* in a N_2 flow, while most observations were made in the after-glow. In this system, the translational and rotational 'temperatures' are of the order of room temperature. In contrast, the vibrational energy distribution may correspond to temperatures of a few thousand kelvin in the lower v range, whereas an excess population may be present for $v>10$ (Schmidt, Haug et al. 1974). Because the vibrational energy distribution is not well known, specific excitation cross sections can at present hardly be derived from these experiments.[*] Anyway, the appearance of a large series of atomic emission lines suggests that a close match of electronic and closest vibrational level is not a stringent condition for excitation.

[†] See Starr (1965), Starr and Shaw (1966), Milne (1967, 1970), Sadowski, Schiff and Chow (1972).

[*] The excitation (or transfer) cross sections derived also depend on the assumed quenching cross sections. Czajkowski, Krause and Skardis (1973) have criticized the quenching values adopted by Sadowski, Schiff and Chow (1972).

PHYSICAL EXCITATION AND DE-EXCITATION PROCESSES   Sect. VI.2b

In a way, the latter excitation experiments may be considered as a counterpart of the quenching experiments described sub (i). However, in addition to the fact that different collision partners were investigated in the two types of experiment, there is another distinction that impedes a direct comparison through microscopic reversibility. In the excitation experiments sub (ii), the kinetic temperature of the gas flow is low. Consequently, only vibrational levels with energies at least equal to the atomic excitation energy can contribute substantially to the observed atomic emissions. In the quenching experiments sub (i), also performed at low kinetic temperature, only vibrational levels with energies at most equal to the atomic excitation energy can be excited. Thus, the specific vibrational transitions occurring in the two experiments are different and cannot be directly connected through microscopic reversibility.[†] Therefore we must not conclude from the experiments sub (i) that molecules possessing vibrational energy only are incapable of exciting Hg atoms. The only conclusion that we can draw from this is the following. A CO molecule possessing vibrational energy up to half the required excitation energy plus enough translational and rotational energy to make up for the deficit excites the 6^3P state of Hg more efficiently than one possessing the required excitation energy in the form of vibrational energy only.

(iii)   *Super-elastic scattering of electrons* by vibrationally excited N_2 molecules has been observed in the presence of photo-excited Rb atoms by Burrow and Davidovits (1968). N_2 molecules were vibrationally excited up to $v = 5$ by quenching collisions with the excited Rb atoms.

(iv)   Lee et al. (1973) have applied *vacuum u.v. absorption spectroscopy* to observe H_2 molecules that were vibrationally excited by energy transfer from photo-excited alkali- and Hg atoms. Absorption bands arising from the $v = 1$ and 2 levels but not from higher v levels in the electronic ground state were detected. Time-resolved absorption measurements indicated that this was probably due to a vibrational relaxation effect occurring at H_2 pressures above several Torr.

Note that the possibility of detecting electronic-vibrational energy transfer in the above experiments is essentially connected with the comparatively long persistence of the vibrational energy (i.e., at not too high pressure; see also Sect. II.3c-2). Degradation of the vibrational energy to lower vibrational levels by inelastic collisions during the optical lifetime would have invalidated the experiments described sub (i). In order to maintain a sufficiently large vibrational excitation in the experiments sub (ii) and (iii), the conversion of

---

[†] For a somewhat different view see, however, Gilmore, Bauer and McGowan (1967, 1969).

Chapt. VI   EXCITATION AND DE-EXCITATION OF METAL SPECIES

vibrational to translational energy should be slow here too. In the experiments sub (iv), a degradation of the population of the higher levels to the $v=1$ and 2 levels could not be avoided. However, the molecules persist long enough in these lower levels to be detected. In general, the persistence of the vibrational energy will depend on the kind of molecule and on the experimental conditions (gas pressure, occurrence of wall-collisions). On the contrary, rotational energy is known to be much more readily degraded and converted into translational energy (see Sect. II.3c-2), which hampers the assessment of the initial rotational energy distribution in the above experiments.

The persistence of the vibrational energy also explains why Jennings, Braun and Broida (1973) were able to detect vibrationally excited di-atomic molecules (activated by energy transfer from one photo-excited alkali species) by observing the luminescence of a second alkali species having a lower excitation energy. The role played by the vibrational degree of freedom in the quenching as well as excitation of alkali atoms is here demonstrated in one and the same experiment.

(v) *Line-reversal measurements in shock-tubes* with Na vapour have shown that the vibrational degree of freedom of N_2 and CO is involved in the excitation of the first Na resonance doublet.[†] Shortly after passage of the shock-front, the line-reversal temperature (which is a measure for the relative population of the electronic state concerned; see Sect. II.5c-3) appeared to be systematically lower than the calculated equilibrium temperature. With increasing time, its value approaches asymptotically the equilibrium temperature. The corresponding relaxation time appeared to be the same as that for the vibrational temperature, which is known to lag behind the translational temperature. This lag is caused by the comparatively slow rate at which vibrational and translational energy are exchanged (see Sect. II.3c-2).

According to Callear (1965), the similarity of relaxation times does not necessarily imply that the line-reversal and vibrational temperatures are also identical. This would only be the case if practically the whole energy needed for excitation of the metal atom is supplied from the vibrational degree of freedom. If the translational or rotational energy also takes a substantial part in the excitation process, the value of the line-reversal temperature will lie somewhere between the translational and the vibrational temperatures (see Callear 1967). But the relaxation times for the approach of the reversal and vibrational temperatures to the final equilibrium value will then still be nearly the same. These considerations

---

[†] See Clouston, Gaydon and Glass (1958), Clouston, Gaydon and Hurle (1959), Gaydon and Hurle (1963), Hurle (1964), Tsuchiya (1964), Tsuchiya and Kuratani (1964), and Eremin, Kulikovsky and Naboko (1977). Some of the previous results obtained by Hurle have been corrected in Hurle (1964).

hold on the assumption that the lifetime of the electronic state involved in the line-reversal measurement is much shorter than the vibrational relaxation time. Lifetimes may be of the order of 10^{-7} to 10^{-9} s in a di-atomic gas at 1 atm (see Sect. 4c-1), whereas vibrational relaxation times are of the order of 10^{-4} s. From the above relaxation measurements one can therefore conclude only that vibrational energy transfer is, at least partially, involved in the excitation process. The measurements do not, however, exclude the possibility that other degrees of freedom may substantially contribute to the excitation as well.

(vi)   The participation of internal molecular energy can also be concluded from the *beam excitation experiments* described earlier in this subsection. The following points should be noted:

In the velocity-selected crossed-beam experiment, the deficiency in excitation energy must be made up by the internal molecular energy. A thermal vibrational energy distribution corresponding to the furnace temperature is expected for the beam molecules effusing from the furnace. An analysis of the velocity-selected experiments with N_2 and Na gave the approximate weight factors for the various possible vibrational transitions.

The absolute Na excitation rate constant measured as a function of N_2-furnace temperature T_1 (without velocity selection) is understandable only if conversion of vibrational energy is taken into account. The experimental results have been compared with theoretical expectations for two extreme cases (see Mentall, Krause and Fite 1967a). First, it was supposed that Na excitation takes place solely through conversion of the translational energy of relative motion. A much stronger increase in rate constant with increasing T_1 would be expected in this case than has in fact been found experimentally. This expectation is based on the fact that the distribution function for the energy of relative motion of N_2-Na is described by an effective temperature that is about twice as small as T_1. This, again, follows from the considerably lower Na furnace temperature and from the masses of N_2 and Na, which are of comparable magnitude. Moreover, the effective cross section for those molecules that have sufficient kinetic energy to excite Na would then have to be of the order of 10^5 Å^2 in order to explain the magnitude of the experimental rate constant. The true value, however, is expected to be at most of the order of $10-100 \text{ Å}^2$. These discrepancies are absent if it is assumed, in the other extreme case, that only the vibrational energy of N_2 is used for the Na excitation.

A confrontation of the absolute rate constants for Na excitation, measured as a function of T_1 by Krause, Fricke and Fite (1972), with those for the reverse quenching process, measured as a function of gas temperature, may yield information about the distribution of vibrational energy jumps in the quenching process (see

Chapt. VI   EXCITATION AND DE-EXCITATION OF METAL SPECIES

Lijnse 1973, 1974). In this confrontation, microscopic reversibility is applied to relate the cross sections for the specific excitation and quenching subprocesses. Certain ad-hoc assumptions had to be made, and the rotational energy was disregarded. It was found that a smooth distribution function with a maximum at about half the Na excitation energy agreed best with the experimental observations. A peaked resonant distribution function or a function peaked at small vibrational energy jumps was shown to be inconsistent with the experiments.

The latter conclusions about the detailed kinetics of the process could be drawn due to the fact that the kinetic energy in the beam experiments is not equilibrated with the internal molecular energy. Confrontation of excitation and quenching rate constants both measured under the same equilibrium conditions does not provide any new information, for these constants are invariably connected through detailed balance. Detailed-balance relations involve the temperature and the properties of the isolated atom, but not the kinetic parameters of the process (see also Sect. II.4a-2).

The quenching cross section estimated indirectly from the excitation experiments performed near threshold with a fast alkali beam impinging on a 'vibrationally cold' molecular gas (see Sect. 2a-1) turns out to be 1 or 2 orders of magnitude smaller than the directly measured value. This difference is again explained by the fact that in this excitation experiment internal molecular energy does not play a role, whereas it obviously does in the quenching experiment.

(vii)   More direct information about the vibrational energy distribution of diatomic molecules after quenching of $Na(3^2P)$ atoms has been obtained in a *crossed-beam quenching experiment* by Hertel, Hofmann and Rost (1976, 1977), Hofmann (1977), and Rost (1977) (see also Hertel and Stoll 1977). They measured the kinetic energy distribution of laser-excited Na atoms that were scattered at a fixed angle by a cold beam of N_2 molecules. A nonresonant electronic-to-vibrational energy transfer was established, e.g., with H_2, CO and N_2; the N_2 levels at $v = 3-4$ showed a peak population. A near-resonant behaviour was found, e.g., with O_2, CO_2 and N_2O.

In beam work Silver, Blais and Kwei (1977, 1979) found that in the quenching of Na(3P) by NO about 35% of the available electronic energy is converted into kinetic energy of relative motion whereas the rest is converted into vibrational energy. Some differences exist between their and Hertel's results for NO and O_2.

In the beam experiments by Liu and Parson (1976) on the quenching of $Hg(6^3P_2)$ by NO, described earlier in this subsection, product NO molecules were detected in two excited electronic states with $v = 0$ and 1. Krause and Datz (1976) found in their beam experiments on the same system an enhanced effective rotational temperature in the product $(A^2\Sigma; v = 0)$ state of NO; they noted that this state is nearly resonant to the excited $Hg(6^3P_2)$ state.

Further literature on the role played by the interconversion of vibrational and electronic energy (for example, in the Na emission of the aurora) is found in Starr and Shaw (1966), Mentall, Krause and Fite (1967a), and Lijnse (1972).

Resonance effects. Resonance effects have often been claimed in the literature to be an indication for the participation of a specific molecular transition in the (de-)excitation process. The experiments described above, however, show that vibrational-electronic resonance effects are the exception rather than the rule. Lijnse (1972, 1973) has critically reviewed the alleged experimental evidence for such effects and warned against too rash conclusions. In the literature conclusions were sometimes based on erroneous or incomplete experimental data. It will be shown later in this subsection that efficient physical (de-)excitation occurs often through potential-energy curve-crossing(s). From a theoretical point of view resonance effects are in this case not likely to occur.

Quenching experiments with isotopic molecular species like H_2, HD, D_2 are of special interest. The chemical nature of these species is the same, but the spacings between their vibrational energy levels and thus the matching of the nearest vibrational level to the atomic level are different. These isotopic species have frequently been studied as quenchers of the first (and some higher) alkali resonance doublets in vapour cells (see Table VI.1 and Fig. VI.2).[†] Bästlein, Baumgartner and Brosa (1969) found, within an experimental error of about 10%, no change in quenching cross section for $Na(3^2P)$ when $^{15}N_2$ was substituted for $^{14}N_2$ in their vapour cell experiments. The smallest (exo-ergic) vibrational-electronic energy defect, ΔE, amounted to +0.16 and +0.035 eV for the two isotopic molecules, respectively. (A positive defect means that the Na level lies above the nearest vibrational level.) They therefore concluded that a resonance process did not play an important role.

Following Hooymayers and Alkemade (1966), Lijnse (1973) has plotted phenomenological quenching cross sections of the first resonance doublets of Na and K for various di-atomic molecules versus the absolute energy defect $|\Delta E|$. This plot, which contains flame measurements at $T \approx 1600$ K as well as vapour cell measurements at $T \approx 400$ K, is reproduced in Fig. VI.2.

[†] See, e.g., McGillis and Krause (1968, 1968a), Copley and Krause (1969), Hrycyshyn and Krause (1970), Siara and Krause (1973), Bästlein, Baumgartner and Brosa (1969); see also Pringsheim (1949) for older quenching measurements of excited Hg- and Cd atoms with H_2 and D_2.

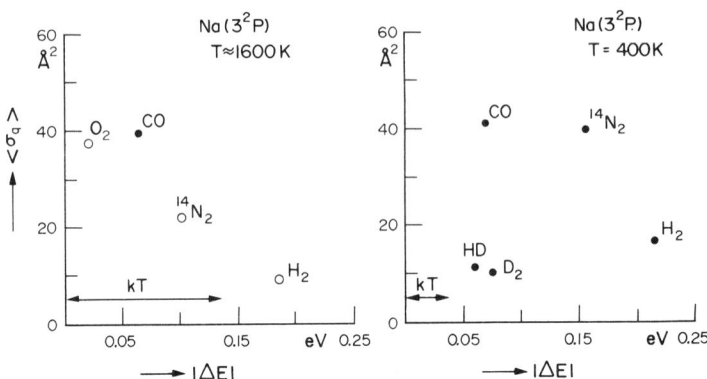

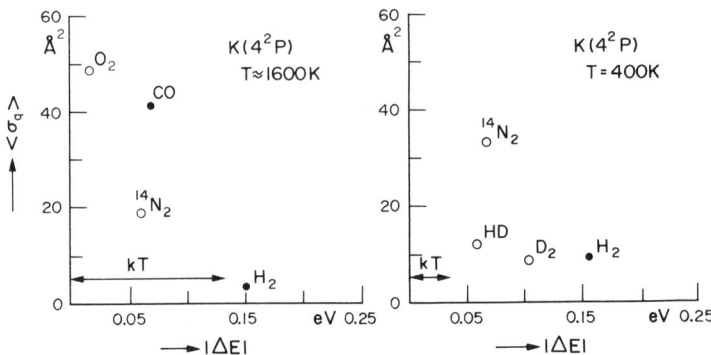

Fig. VI.2 Phenomenological quenching cross sections, $\langle \sigma_q \rangle$, of the Na(3^2P)- and K(4^2P) doublets, measured at $T \approx 1600$ and 400 K, are plotted as a function of absolute energy defect $|\Delta E|$ for various di-atomic molecules (according to Lijnse 1973, 1974). Open and closed circles refer to the case when the nearest attainable vibrational energy level lies above and under the (mean) doublet level, respectively. For a proper intercomparison, the cross sections denoted by the open circles should be multiplied by the appropriate Boltzmann factor to account for an activation energy equal to $|\Delta E|$. For $|\Delta E| = 0.1$ eV, this Boltzmann factor amounts to about 20 at $T = 400$ K, and about 2 at $T = 1600$ K. The value of kT at the two temperatures is indicated by arrows along the abscissa.

From these and similar plots for Rb and Cs, Lijnse (1973, 1974) concludes that there is no systematic overall dependence of quenching cross section on $|\Delta E|$.[†] This conclusion is supported by the results of the infrared studies and beam experiments mentioned previously in this subsection sub (i), (vi) and (vii). Vibrational-electronic resonance effects therefore do not seem to be important for the lowest alkali doublets with di-atomics. It should be noted, however, that the rotational energy levels have not been taken into consideration here.

The quenching cross sections of N_2 for the higher levels of Na [excited by electronic energy transfer from $Hg(6^3P_0)$], however, seem to suggest an electronic-vibrational resonance effect (see Czajkowski, Krause and Skardis 1973; see also Table VI.1). The 5S,..., 9S states were assumed to be quenched to the 3P- and 4P states, and their cross sections were plotted versus the energy defect with respect to the best matching N_2 vibrational level. The smooth curve obtained showed a resonant behaviour. We recall, however, that the quenching cross sections of the $Na(n,S)$ states found by Humphrey et al. (1978) deviate markedly from those found by Czajkowski et al. and do not point to a resonant behaviour (see Table VI.1 and the discussion earlier in this subsection). To enable us to draw a more definite conclusion, we need direct information about the states to which the excited atoms are quenched; such information could, for example, be obtained through luminescence measurements at various atomic lines.

Wade, Czajkowski and Krause (1978) also suggested the occurrence of an electronic-vibrational resonance effect in the quenching of the $Tl(6^2D_{\frac{5}{2}}; 7^2S; 8^2S)$ states by N_2 investigated by them in a vapour cell (see Table VI.1). There is, however, a systematic difference between the value for $Tl(6^2D_{\frac{5}{2}})$ found by them and that found by Gelbhaar and Hanle (1976) at comparable temperatures (see again Table VI.1).

A resonance effect has been found in the quenching of $Zn(4^1P_1)$ atoms by NO, yielding <u>electronically</u> excited $NO(A^2\Sigma^+)$ molecules. This was derived from an analysis of the NO fluorescence spectra measured in a flowing gas system at $T \approx 600$ K by Breckenridge, Blickensfelder, FitzPatrick (and Oba) (1976, 1979). The rate of production of $NO(A^2\Sigma^+)$ molecules in the near-resonant $v=1$ state appeared to exceed that in the $v=0$ state by one order of magnitude. The authors concluded that an efficient near-resonant dipole-dipole interaction is the dominant quenching mechanism in this system.

[†] This conclusion, based on more extensive experimental data, replaces the dissentient tentative conclusion drawn by Alkemade and Zeegers (1971) from a similar plot.

Chapt. VI   EXCITATION AND DE-EXCITATION OF METAL SPECIES

Conclusions from the experiments. So far we have mostly dealt with nonflame experiments. Now some conclusions will be drawn from the available experimental data, relating to the excitation and quenching of metal atoms by molecular species in flames.

In flames at 1 atm pressure consisting mainly of molecular components the quenching rate will be appreciably higher than the spontaneous emission rate. This holds even for the strongest optical transitions. A direct consequence of this statement and, in fact, one of the experimental foundations for it is that the fluorescence efficiency Y is small compared to unity. This is, of course, a disadvantage when one applies atomic fluorescence spectroscopy in chemical analysis (see Alkemade and Herrmann 1979). Furthermore, a small Y value suppresses the effect of suprathermal chemiluminescent excitation (see Sect. 3c-2), and it makes the attainment of saturation by a strong laser beam more difficult (see Sect. 2c-5).

In the above kind of flames the de-excitation and excitation of atomic states are controlled mainly by molecular collisions and not by radiative processes. We disregard here excitation caused by an external light source (see Sect. 2c) or chemiluminescent reactions (see Sect. 3). Since the internal energy of the flame molecules is rapidly equilibrated above the combustion zone, thermal radiation of the atomic metal vapour is then to be expected (see Sect. 4). The line-reversal temperature is then a true measure for the flame temperature. The thermal equilibrium population of the excited levels is also of importance for the close approach of the ionization rate constant to its thermal value (see Sect. IX.6).

Even in flames diluted by a noble gas (de-)excitation by the residual molecular components will largely dominate that by the noble gas atoms, at least for the first resonance levels. On the other hand, the residual molecular quenching rate here may no longer dominate the spontaneous emission rate (i.e., Y may become of order unity). This situation may also exist in low-pressure flames with a molecular diluent. However, the quenching rates of the higher levels for noble gas atoms may compete with those of molecules, at least in the case of alkali atoms (see Sect. 2a-1 and Table VI.1; see also van Dijk 1978, van Dijk, Zeegers and Alkemade 1979).

The actual lifetime of the first alkali doublet levels in flames diluted by N_2 at 1 atm is controlled mainly by quenching collisions and may be of the order of 10^{-9} s. This time is about 20 times shorter than the radiative lifetime. This is of importance for the attainment of a stationary level population during pulsed excitation (see Sect. 4c-1). This reduction in lifetime also causes 'quenching broadening' of the atomic line profile (see Sects II.5b-2 and VII.4d). In the general case, however, quenching broadening is not expected to be dominant over adiabatic collision broadening in flames. Finally, the depolarization of resonance fluorescence by M_J-mixing collisions may be partially offset by this considerable reduction in lifetime. For an explanation of the effect of lifetime reduction on the polarization we refer to Sect. 1.

Theory. A definitive quantitative theory for the quenching and excitation processes involving internal molecular energy does not yet exist. Nevertheless several theoretical models have been put forward to account, at least qualitatively, for some of the following features of the quenching cross sections measured:

(1) the absolute order of magnitude of the molecular quenching cross sections, which appears to exceed largely those found for noble gas atoms for transitions with large energy jumps;

(2) the dependence of the quenching cross sections on relative velocity or gas temperature, and the possible existence of an activation energy;

(3) the distribution of energy over the internal degrees of freedom of the molecule after quenching, and the possible role played by resonance effects;

(4) the competition with chemiquenching processes (to be discussed in Sect. 3c-2);

(5) the variation of the quenching cross sections within a series of similar metal atoms (e.g., the alkalis) for a given molecular quencher;

(6) the variation of the quenching cross sections for a given atomic state within a series of similar molecular quenchers (e.g., di-atomic molecules, isotopic species or saturated hydrocarbons);

(7) the variation of the quenching cross sections among different excited states for a given metal atom and molecular quencher.

An additional test of the theoretical model is provided by the excitation experiments, which can be related to the quenching experiments through microscopic reversibility.

Despite numerous efforts in recent years, relating mainly to Hg and the alkalis in collision with di-atomic molecules, a fully satisfying theory has not yet been found, as atom-molecule collisions are much more difficult to describe than atom-atom or atom-electron collisions because of the larger number of degrees of freedom involved. The course of a collision complex is often described by trajectories over multi-dimensional potential-energy surfaces between which multiple transitions may occur. These surfaces depend in general on the orientation of the molecule with respect to the atom-molecule axis. In some cases the possibility of a chemical reaction channel introduces yet another complication.

A fully quantummechanical treatment, such as exists for atom-atom collisions, would be very tedious and would require unduly long computer time for obtaining numerical results. Besides, lack of detailed or even approximate qualitative information on the interaction forces or potential-energy surfaces makes such treatment hardly meaningful. In practical cases, one has to resort to simplified semiclassical methods in which the relative motion of the collision partners is calculated classically. The transitions between different energy surfaces are thereby calculated using approximate quantummechanical formulas. In the general case, no ab initio calculations of the wave functions and energy surfaces of tne temporary collision complex are available. One often contents oneself with adopting trial functions based on qualitative considerations or suggested by analogy with related cases that are already known. Angle-averaged potential-energy functions are often introduced to simplify the calculations. As a consequence, rotational transitions in the quenching molecules are left out of account in most cases and one considers only electronic-vibrational transfer. The transition probabilities are often described by parameters still unknown which are matched afterwards in order to comply with the experimental data.

The applicability of semiclassical methods is limited. For low relative velocities the de Broglie wavelength may become comparable to the interaction range of the colliding particles. At high velocities, the 'effective' potential-energy surfaces experienced by the collision system may change. In most cases, therefore, one can expect predictions to be at best semiquantitative. Nevertheless, comparison of the theoretical results with experiments has often improved our insight into the dominant interaction mechanism. From this comparison we may also recognize which are the main atomic and molecular properties that are decisive for the magnitude of the (de-)excitation efficiency.

We shall now give a short description of some of the theoretical models that have been used to explain the quenching of excited Hg- and alkali atoms by di-atomic molecules.

Polanyi (1963, 1965) has constructed a purely *classical model* for calculating the highest fraction of electronic energy that can be converted into vibrational energy in the *quenching of mercury* by di-atomics. The quenching process is considered as a 'sudden' or 'impulsive' collision between the excited atom M^* and ground state molecule AB. A colinear configuration M^*-A-B is assumed. If the excitation energy is released instantaneously, it will be converted initially into translational energy of the motions of M and the nearest atom A only. In that case a force is acting between M and A during a very short time interval. Part of the translational energy gained by A will go into the relative motion of this atom with respect to B (i.e., vibration). The other part will go into the translational energy associated with the motion of the centre of gravity of AB. From the

PHYSICAL EXCITATION AND DE-EXCITATION PROCESSES   Sect. VI.2b

conservation law for linear momentum, it is simple to calculate the fraction of
excitation energy E_{exc} that can go into vibrational energy E_v ; this fraction is

$$E_v/E_{exc} = m_B m_M/(m_A+m_B)(m_M+m_A). \qquad (VI.8)$$

Here m_M, m_A, and m_B are the atomic masses of M, A, and B, respectively. In
other collision configurations (e.g., M*-B-A) and if one makes different assumptions
about the duration of the interaction force, the fraction E_v/E_{exc} will be different.

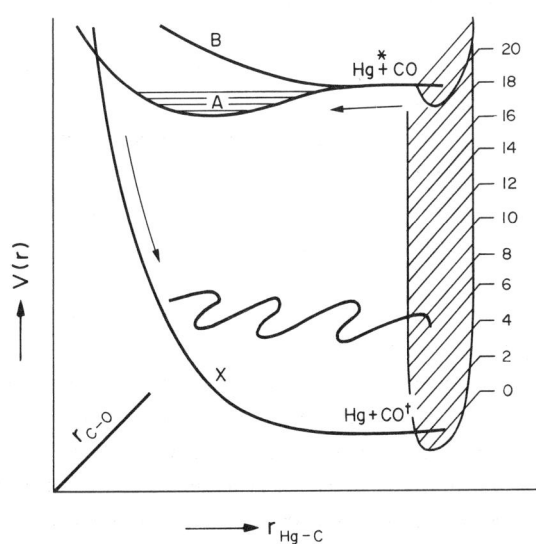

Fig. VI.3 Schematic presentation of the potential energy as a function of r_{Hg-C} (in the plane of the paper), and of r_{C-O} (perpendicular to the plane of the paper), illustrating the quenching of an excited Hg* atom by CO in the vibrational ground-state, according to Polanyi (1965). The course of the quenching process is indicated by arrows. The quantum numbers at the right-hand side of the figure denote the vibrational energy of CO in the final state (CO†).

Polanyi (1965) has used expression VI.8 to estimate an upper limit for the
fraction of Hg(6^3P) excitation energy that can be converted into vibrational energy
of CO. Fig. VI.3 schematically shows the potential energy as a function of the
distance r_{Hg-C} of the mercury atom from the neighbouring carbon atom, and of the
distance r_{C-O} of the carbon atom from the neighbouring oxygen atom. A colinear
approach of Hg* towards CO is assumed. Quenching proceeds by crossing-over from
the state $A^3\Pi$ of the complex molecule to the electronic ground state $X^1\Sigma$. Because
of the steepness of the curve belonging to the state X, we may regard the process
classically as an instantaneous release of repulsive energy between Hg and CO in
a colinear configuration. Application of Eq. VI.8 predicts that the highest fraction of electronic energy that is converted into vibrational energy of CO corresponds to the vibrational level $v=11$. There is a reasonable agreement with the
maximum experimental v value found from the infrared emission bands (see the above
discussion on the participation of internal molecular energy sub i).

The existence of a well in the potential curve for the A state (see Fig.

583

Chapt. VI   EXCITATION AND DE-EXCITATION OF METAL SPECIES

VI.3) favours the formation of a relatively long-lived collision complex. Arguments for the formation of a weakly bound complex between $Hg(6^3P)$ and CO or NO have been put forward by Karl, Kruus and Polanyi (1967) and Karl, Kruus, Polanyi and Smith (1967). The quenching cross section of CO appears to be of gas-kinetic order at about room temperature. So only a small or no activation energy is needed to reach the crossing point of the A- and X curves.

A similar 'impulsive' model has later been utilized for calculating *quantum-mechanically* the vibrational-energy distribution of CO after quenching of $Hg(6^3P_0)$ (see Gonzalez, Karl and Watson 1972, Levine and Bernstein 1972, and Wilson and Levine 1974). The CO molecule was conceived as a harmonic oscillator. The first mentioned authors took the angle dependence of the interaction explicitly into account and were thus able to calculate the rotational excitation of CO too. A mean rotational quantum number $\bar{R}=60$ was found. The vibrational levels with $v=0$ and 1 were calculated to be the most populated.

The distributions of vibrational and rotational energy in H_2 after quenching of $Hg(6^3P_1)$ have been calculated by phase-space theory by Yang, Paden and Hassell (1967). The population of the vibrational levels declines with increasing v.

Most theoretical work on *alkali quenching* by di-atomics has been done on the $Na-N_2$ system. Different theoretical models have been studied with varying success.

Dickens, Linnett and Sovers (1962) have examined the possible role of resonance effects in the quenching of Na(3P) by N_2. They assumed that a direct transition occurred between the initial covalent Π state of the $[Na(3P)-N_2]$ complex and the final covalent Σ state of $[Na(3S)-N_2]$. They assumed there was neither (pseudo-) crossing nor convergence of the corresponding potential curves. The calculated cross section was, however, many orders of magnitude smaller than the experimental values. This was a consequence of the very low probability of a multi-quantum jump in the vibrational energy. A jump $\Delta v \geqslant 1$ would be required to make the resonance defect ΔE sufficiently small.

In another model, a long-lived complex between the excited atom and the molecule was believed to exist and a direct transition through curve-crossing between the initial and final covalent states was assumed (see Nikitin 1965). However, it failed to explain the Na(3P) quenching by N_2 (see Bjerre 1968).

Several attempts to describe the alkali-quenching by di-atomics by the *ionic-curve crossing model* involving two successive electronic transitions into and out of an intermediate ionic, i.e. charge-transfer state (see Fig. II.6) have been more successful. We have, for example, in the case $Na-H_2$:

$$Na(3^2P) + H_2(^1\Sigma_g^+) \rightarrow (Na-H_2)^* \rightarrow (Na^+-H_2^-) \rightarrow (Na-H_2)^0 \rightarrow Na(3^2S) + H_2(^1\Sigma_g^+).$$

Two-dimensional potential surfaces of the electronic states involved have been proposed or calculated for Na and Li in collision with N_2 and H_2 by Magee and Ri (1941), Laidler (1942), Mori (1962), Bjerre and Nikitin (1967), Krauss (1968), Tully (1973), and Bottcher (1975). Krauss (1968) derived the potential surfaces for Li(2S) and Li(2P) with H_2 by the Hartree-Fock method. The potential surfaces calculated in this way were believed to have semi-quantitative accuracy. For various fixed H_2 internuclear distances, he found a strongly attractive curve intersecting the ground-state curve in a triangular H_2-Li configuration, but not in a colinear configuration.† No activation energy was required to reach the crossing point. The attractive curve could in fact be characterized as belonging to a charge-transfer state; the corresponding H_2^- ion state was correlated to the lowest-energy resonance state of a free H_2^- ion.* The marked dependence of the calculated curves on the H_2 internuclear distance suggested a strong coupling between the electronic and vibrational states. For similar calculations on Na-N_2 see Habitz (1980).

This intermediate ionic-curve crossing model is the current model for explaining the great efficiency of di-atomic molecules in quenching the first alkali resonance doublets. Because of the two-dimensional extension of the energy surfaces, a larger number of crossing possibilities may be expected here than in atom-atom collisions. Moreover, selection rules that forbid certain crossing transitions with atomic quenchers may be relaxed in the case of poly-atomic quenchers (see Gaydon and Wolfhard 1979).

The intermolecular distance at which crossing from the initial curve to the ionic curve may take place is expected to increase with decreasing $(E_{ion}-E_{aff})$, other things being kept constant (see Fig. II.6; see Jenkins 1968b, Bauer, Fisher and Gilmore 1969a, Fisher and Smith 1971, and Earl 1973). A small value of alkali ionization energy, E_{ion}, or a large (positive) value of molecular electron-affinity,‡ E_{aff}, is thus favourable for obtaining a large quenching cross section. Also the

---

† In the case of N_2 crossings were believed to exist in both configurations (see Mori 1962, and Krauss 1968); this circumstance would enhance the efficiency of alkali quenching by N_2.

* Negative molecular-ion states are known to play a role in the vibrational excitation by impact of low-energy electrons (see Schulz 1973).

‡ Note that N_2 and CO have a negative electron affinity of -1.9 and -1.75 eV, respectively, whereas O_2 has a positive affinity of about $+0.5$ eV (see values for CO and N_2 quoted by Können, Haring and de Vries 1975, and Bauer, Fisher and Gilmore 1969, and the values for O_2 reported by Gilmore 1965, Celotta et al. 1972, and Baede 1975). A free N_2^- ion is therefore unstable; it has an electron resonance state with a lifetime of about 2×10^{-15} s, thus markedly shorter than the vibrational period (see Ehrhardt and Willmann 1967). But N_2^- may become stable when it is found at a short distance from the alkali ion because of the Coulomb attraction (see Fisher and Smith 1971 and Rost 1977).

Chapt. VI   EXCITATION AND DE-EXCITATION OF METAL SPECIES

polarizability of the ionic complex plays a role, as it affects the shape of the ionic curve (see Fisher and Smith 1971).

A correlation between the excitation (or quenching) efficiency and the character of the lowest-energy vacant molecular orbital has been shown to exist by Anderson, Aquilanti and Herschbach (1969) (for a review see Lijnse 1973). This correlation may explain the typical differences found between N_2 and O_2 as quenchers on the one hand, and H_2 and Cl_2 on the other. It may also explain the inefficiency of H_2O in quenching the first alkali doublets.

The striking differences found in some cases between the quenching cross sections for different excited states of a given atom might be qualitatively interpreted if one considers the position of the ionic curve relative to that of the excited-state curves concerned in Fig. II.6. When the ionic curve lies relatively high, it might intersect the curve for the lowest excited state somewhere in its repulsive part. The quenching cross section may then be small owing to a high activation energy. The same ionic curve may intersect the curve belonging to a higher excited state at a larger intermolecular separation where no repulsion is felt. Such a situation might explain why H_2O (presumable electron affinity $\approx -3\,eV$) and CH_4 quench K(5P) much better than K(4P) (see Earl 1973, and Earl and Herm 1974) and why CH_4 quenches Rb(6P) much better than Rb(5P) (see Hrycyshyn and Krause 1970, and Siara and Krause 1973). On the other hand, if $(E_{ion}-E_{aff})$ is smaller than the atomic excitation energy E_{exc}, the whole ionic curve may lie below the excited-state curve and no ionic-curve crossing occurs for this state (unless v' is sufficiently high; see Fig. VI.4). This may explain the low efficiency of I_2 in quenching K(5P) compared to K(4P) (see Earl 1973).

Quantitative calculations concerning the quenching of the first alkali resonance doublets in which the vibrational structure of the quencher (N_2, CO, O_2) was explicitly taken into account, have been done by Bjerre and Nikitin (1967), Bjerre (1968), Bauer, Fisher and Gilmore (1969a), Fisher and Smith (1971, 1972), Meitlis and Fishman (1972), and Andreev (1972, 1973). These calculations yielded not only the total quenching cross section, but also the distribution of vibrational energy after quenching.

Bjerre and Nikitin calculated the pertinent two-dimensional potential surfaces and the classical trajectories for the (Na-N_2) complex. The transition probabilities at the (pseudo-)crossing points between the covalent and ionic surfaces were calculated by an extended, two-dimensional Landau-Zener formula (compare Eqs II.143-144). Here the intermolecular distance between Na and N_2 and the internuclear distance of N_2 acted as coordinates. This treatment failed to give accurate cross sections, and the calculated vibrational distribution was not in agreement with experiment (see the above experimental discussion sub vi). A critical evaluation of the calculations has been given by Andreev (1972) and Lijnse (1973a).

More satisfactory quantitative results have been obtained by the other authors mentioned above, who treated the course of the quenching collision formally as a 'diffusion' of the collision complex through a one-dimensional 'grid' or 'network' of effective potential-energy curves. This grid was constructed on the approximative assumption that the electronic interaction potential is independent of vibrational level. The effective potential curves for successive v numbers are then obtained by a displacement along the energy axis corresponding to the vibrational energy spacing of the di-atomic molecule. In this way, a set of parallel curves was obtained for each electronic state of the collision complex (see Fig. VI.4). The vibrational spacing of the unstable N_2^- ion being not known, it had to be estimated from the corresponding parameters of the iso-electronic NO and O_2^+ molecules (see Gilmore 1965, Bauer, Fisher and Gilmore 1969a, and Andreev 1972). The interaction potentials were averaged over the orientation of the molecule, so rotational excitation was not included. In contrast to the other authors, Andreev (1972, 1973) included also a weakly attractive van der Waals term in the covalent excited-state curve (see also the discussion below on the kinetic energy dependence).

The motion of the collision complex along these diabatic effective-potential curves is treated classically. The excited collision complex, $M^* + N_2(v = 0)$, may enter the 'grid' on the $A^2\Pi$ curve (the strongly repulsive $B^2\Sigma$ curve would not actually lead to an ionic-curve crossing at thermal energies). Transitions from the A curve are now possible at each of the crossing points with the set of ionic curves with $v' = 0, 1, 2, \ldots$. Because of the attractive shape of the ionic curve, the collision complex is drawn inwardly into the grid. Here it may make a number of subsequent transitions at any of the numerous crossing points between the sets of ground-state and ionic curves. After 'diffusing' through this grid, there is a certain probability that the collision complex will emerge from it on a ground-state curve with $v'' = 0, 1, 2, \ldots$. It may also return to the initial curve, in which case only an elastic collision has taken place. In the inelastic case, we are left with a quenched alkali atom and a N_2 molecule in vibrational state v''. The collision complex may follow a multitude of 'channels' that connect the initial state to the final state with different v'' numbers. One such quenching channel is indicated in Fig. VI.4 by arrows. The initial kinetic energy and the initial vibrational energy determine which channels are 'open'.

Because of the assumed independence of the vibrational and electronic motions, the transition probability at each crossing point can be calculated approximately through the Landau-Zener formula (see Eqs II.143-144) where the squared matrix element, $|V_{12}(r_c)|^2$, at the crossing point r_c, is written as a product of

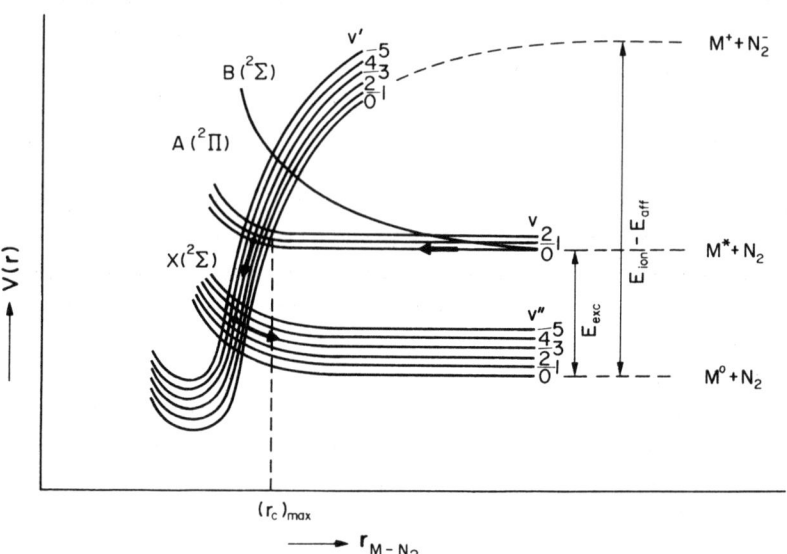

Fig. VI.4 Electronic-vibrational energy curves are obtained by adding the (constant) vibrational energy of the molecule to the electronic energy of the collision complex (M-N_2), with M = alkali atom. The resulting energy is plotted as a function of distance between M and centre of gravity of N_2 (schematic). The vibrational quantum numbers $v = 0, 1, 2, \ldots$ are indicated. The covalent excited-state curves, $A^2\Pi$ and $B^2\Sigma$, correlate to an excited atom $M(^2P)$ and ground-state $N_2(^1\Sigma)$ at infinite separation; E_{exc} is the atomic excitation energy. The covalent ground-state curve $X^2\Sigma$ correlates to a ground-state atom $M(^2S)$ and $N_2(^1\Sigma)$. Possible (small) attractive terms are disregarded in the covalent curves. The ionic $(M^+-N_2^-)$ curves represent a charge-transfer state with E_{ion} = ionization energy of $M(^2S)$ and E_{aff} = electron affinity of N_2. Because E_{aff} happens to be negative in this case, the $(M^+-N_2^-)$ state is not stable at larger separations.

Inelastic collisions occur through successive crossings between these sets of diabatic potential curves belonging to different electronic states. A possible channel for the quenching process: $M(^2P) + N_2(v = 0) \rightarrow M(^2S) + N_2(v'' = 3)$ is indicated by arrows. This channel involves only two successive crossings with the intermediate ionic state.

the squared electronic matrix element and a Franck-Condon factor. The latter factor describes the overlap of vibrational wave functions of the two states involved (see Sect. II.2d). In the case of vibrational transitions between N_2 and N_2^-, this factor has been estimated from known F.-C. factors for analogous transitions in NO by Bauer, Fisher and Gilmore (1969). The electronic matrix element is related to the resonance splitting (not shown in Fig. VI.4) of the adiabatic curves near the pseudo-crossing point (see Eq. II.141) and has, in principle, to be calculated by quantummechanical methods. However, it is usually estimated on a semi-empirical basis. The results are sensitive to the choice of potential-energy parameters, which are not well known (for a further discussion see Olson, Smith and Bauer 1971, Andreev 1972, and Anderson and Herschbach 1975 who deal with the dependence of the matrix element on the molecular orientation; see also Bottcher and Sukumar 1977).

If we assume that successive crossings occur independently, we can find by computer calculations the cross section for each outgoing channel at a given initial kinetic energy in the centre-of-mass system. The total quenching cross section is then obtained by summation. By averaging over a Maxwellian velocity distribution, the phenomenological quenching cross section is finally found as a function of translational temperature.

The *absolute* cross sections calculated have a semiquantitative meaning only. There are several uncertain parameters involved (such as the polarizability of N_2^-), and simplifications of different kinds have been made in the theoretical model. For most systems involving an alkali atom and a di-atomic molecule, the theoretical results can be reconciled with the quenching and excitation experiments by choosing reasonable values for the unknown parameters.

In addition to the limitations and uncertainties in the theoretical model that were already mentioned, the assumption of independent curve crossings and quasi-stationary vibrational populations (the collision time may be comparable to the oscillation period) may be questioned. This holds in particular for the quenching by O_2 which possesses two singlet states at 0.98 and 1.63 eV above the triplet ground state; a crowding of crossing points may therefore be expected here (see Fisher and Smith 1971, and Hertel, Hofmann and Rost 1977). One should also bear in mind the limited applicability of the first-order Landau-Zener theory, which was derived for a two-state model (for a critical discussion see Nikitin 1968a).

Bottcher and Sukumar (1977) have more recently compared the results from various theoretical models with quenching experiments involving alkali atoms in the first and higher-lying excited states and di-atomic molecules. The simplest model, called by them 'first-order theory', relates the quenching process to the scattering of the quasi-free valence electron by the molecule in a binary encounter. This model yields much too small total quenching cross sections for the first excited alkali states. The 'complex-formation model' in which partial charge transfer occurs, as described earlier in this subsection, yields quenching cross sections of about the same size as the experimental values, due to the formation of what they called a long-lived collision complex which allows the reactants to interact repeatedly until quenching takes place. The authors considered next a 'hybrid model' in which the total cross sections are derived from the complex-formation model but the branching ratios for the final population of the vibrational levels are taken from the first-order theory. Finally, a 'unified model' based on the standard quantummechanical theory of resonances yields for the Na(3P) state a total quenching cross section that is close to the value obtained in vapour cell experiments for N_2 but is too large by a factor of two for H_2.[†] This last model predicts by a simple formula the vibrational-branching ratios for Na(3P) in collision with N_2, H_2 and D_2, if these molecules are assumed to be initially in their vibrational ground state (see also below). The authors have also outlined the application of the first-order theory to the quenching of highly excited states by molecules. Finally, they have briefly discussed the dependence of the quenching on the rotational state of the molecule.

Recently Prunelé and Pascale (1979) using the first-order theory calculated cross sections for the quenching of Na(n^2D) states by N_2 that agreed satisfactorily as to magnitude and n-dependence for $n > 10$ with the experimental data of Gallagher, Olson, Cooke, Edelstein and Hill (1977).

An extension of the curve-crossing model of Bauer, Fisher and Gilmore (1969) has been made by Humphrey et $al.$ (1978) to calculate an upper limit on the quenching cross sections of the higher Na(n,S) states for N_2. The monotonous increase in the calculated quenching cross sections with n increasing from 5 to 9 was in qualitative agreement with the experimental results of Humphrey et $al.$ but not with those obtained by Czajkowski, Krause and Skardis (1973); the latter authors found a more resonant behaviour (see also Table VI.1).

Of particular interest are the conclusions one can draw from the theory about the <u>relative</u> dependence of (partial) quenching cross sections on kinetic energy or temperature, on initial and final vibrational energy, and on the kind of species involved (see points 2, 3, 5 and 6 at the beginning of this theoretical survey).

The weak decrease of quenching cross section with increasing $temperature$

[†] The authors criticized the previous calculation by Bjerre and Nikitin (1967) of the matrix elements describing the coupling of the intermediate-complex state with the initial and final states.

or *relative velocity* for the first alkali doublets (see Fig. VI.1B and the above experimental discussion) is semi-quantitatively explained if one assumes that the initial potential-energy curve is weakly attractive near the crossing point. An attractive curve is obtained when one includes a van der Waals term in the covalent excited-state curve (see Lijnse 1973, 1974, and Andreev 1973) or when one assumes that the quenching process is governed by the adiabatic rather than the diabatic potential-energy curves (see Barker and Weston 1973, 1976). Collision partners with impact parameters larger than the maximum ionic-crossing distance $(r_c)_{max}$ will be attracted to each other, so that the collision complex can still penetrate the potential-energy grid shown in Fig. VI.4. The lower the initial relative velocity of the collision partners, the larger the resulting enhancement of the quenching cross section. The relationship between this enhancement and the relative velocity depends on the location of the centrifugal barrier with respect to the crossing point (see Lijnse 1974). Since the course of the attractive curve is unknown, a quantitative prediction cannot be made. But the values of the potential energy at the crossing point needed to explain the alkali experiments with N_2 seem to be reasonable as to order of magnitude. However, the tentative values adopted by Lijnse (1973, 1974) on the one hand, and by Barker and Weston (1976) and Ham and Hannaford (1979) on the other, on the basis of their respective experimental temperature dependencies differ among each other by a factor of 2-3.[†]

Fisher and Smith (1971, 1972) calculated that the total quenching cross section of Na(3P) by $N_2(v)$ was practically independent of the *initial vibrational quantum number* up to $v = 3$. Higher initial vibrational states are not expected to play a noticeable role at flame temperatures. Only the relative distribution over the final vibrational quantum numbers changed markedly with changing initial quantum number.

Fisher and Smith (1971) calculated a rather broad distribution curve for the *final vibrational energy* of N_2 after quenching of the first alkali resonance doublets. For Na and N_2 with initial quantum number $v = 0$, this curve showed a pronounced maximum at about $v'' = 3$, whereas for resonant energy transfer a value $v'' = 7$ or 8 would be expected. For Na(3P) and K(4P) in collision with $N_2(v = 0)$ Andreev (1973) calculated a distribution function that was virtually flat, within a factor of two, from $v'' = 0$ up to the resonant v'' value. The distribution calculated by the former authors seems to be in better agreement with experiments. Hsu and Lin (1976) calculated the vibrational energy distribution of CO after quenching

---

[†] Recently a more elaborate calculation of the effect of long-range interatomic attraction on the total cross section for curve-crossing collisions has been presented by Chang and Pritchard (1979).

Chapt. VI   EXCITATION AND DE-EXCITATION OF METAL SPECIES

of Na(3P) on the basis of the ionic-curve crossing model as well as on that of the near-colinear impulsive model treated by Levine and Bernstein (1972) and Wilson and Levine (1974). The results obtained with both these models can be matched reasonably well to the experimental results of the first mentioned authors (see sub i earlier in this subsection).[†] This conclusion has also been reached by Hofmann (1977), Hertel, Hofmann and Rost (1977), and Hertel and Stoll (1977) for Na(3P) quenching by N_2. The unified quantummechanical theory of Bottcher and Sukumar (1977) predicted for the quenching of Na(3P) by N_2 a v''- distribution which peaks at $v'' \simeq 1-3$ (see also above). For H_2 a more pronounced peak at $v'' = 3$ is predicted by this theory.

The theoretical results of the ionic-curve crossing model depend strongly on the assumption that the motion of the particles can be described by purely classical laws (see Lijnse 1972). Although this assumption may not be absolutely correct, it can be concluded from the results obtained with this model that resonance effects are not important in the cases considered. This is in general agreement with existing experimental evidence (see above). The irrelevance of resonance effects is understandable, as the transitions take place in a range of intermolecular separations where the adiabatic potential-energy curves are strongly distorted by charge transfer. Hertel, Hofmann and Rost (1977) gave a qualitative explanation for the fact that quenching of Na(3P) by O_2, N_2O and CO_2 has a more resonant character (see this subsection sub vii); they used the same model while pointing to the larger number of degrees of freedom for tri-atomic quenchers and to the existence of several low-lying electronic states for O_2.

According to the recent qualitative model presented by Taylor (1979) a non-resonant transfer of electronic energy to vibrational energy in the quenching of an excited alkali atom A^* by a molecule X is to be expected whenever the geometries of X and X^- are alike (as in the case of, e.g., N_2, H_2, CO). When these geometries are greatly different (as in the case of CO_2 and N_2O), a resonant transfer is predicted. The question which case applies can be simply decided by molecular orbital theory without a knowledge of the potential-energy surfaces of the A-X system. In this model the quenching process is envisioned as the scattering of the slow valence electron of the excited alkali atom by the molecule, in the presence of the A^+ core.

In order to explain the experimental variation of total quenching cross section among *different alkali species* with N_2, Fisher and Smith (1971) had to

---

[†] However, the scattering experiments on the quenching of Na(3P) by CO in a crossed-molecular beam by Hofmann (1977) and Hertel, Hofmann and Rost (1977) showed a peak in the v''- distribution function at a larger v'' value than was measured by Hsu and Lin (1976) and calculated by Fisher and Smith (1970, 1972). Possible explanations for this discrepancy have been suggested by the first mentioned authors.

PHYSICAL EXCITATION AND DE-EXCITATION PROCESSES   Sect. VI.2b

assume different polarizabilities of the corresponding ionic collision complex. Since no independent measurements of the latter quantity exist, their interpretation could not be checked. It should be noted, however, that the polarizability of the ionic complex is mainly due to the polarizability of N_2^-. Large differences in this quantity for the ionic complex are therefore not to be expected among different alkali species.†

Additional information about the potential curves may be obtained from beam-scattering experiments with excited as well as unexcited alkali atoms. This may help considerably to improve the accuracy of the theoretical calculations of quenching cross sections.

2b-3   *MIXING OF MULTIPLET COMPONENTS BY COLLISIONS WITH MOLECULES*

Mixing of multiplet components by collisions with molecules is usually accompanied by transfer of internal molecular energy. The magnitude of the multiplet splitting determines whether herewith rotational energy is transferred only, or whether vibrational energy is transferred as well.

Experiments. It is often easier to measure mixing cross sections with molecular collision partners than with atomic partners because the former cross sections are usually considerably larger, at least with large multiplet splittings. However, complications may arise because, in the case of molecules, mixing and quenching may often concur at comparable rates.

Mixing cross sections have usually been determined by the method of stepwise line fluorescence (see Sects II.5a-1 and VI.2a-2). Most mixing cross sections have been measured in vapour cells at temperatures below about 1000 K. Doublet mixing cross sections are found by measuring the intensity ratio of the fluorescent doublet lines as a function of molecular gas pressure while either doublet component is excited separately (cf. Sect. 2b-2).* Mixing collisions affect this ratio to first order of gas pressure, whereas quenching collisions produce a second-order effect; one can therefore eliminate the latter by extrapolation to sufficiently low pressures (see McGillis and Krause 1967a, 1968a, and Apt and Pritchard 1979).

This procedure is particularly advantageous in measuring mixing cross

---

† Fisher and Smith (1971) found that the polarizability must have an unrealistically high value in order to explain the Cs quenching cross sections measured. However, if one takes into account the effect of an additional van der Waals term in the potential curve, a considerably smaller polarizability has to be assumed (Dr P.L. Lijnse, personal communication).

* See footnote ⋆ on p.531 for a reference to a quantummechanical theory of the fluorescence intensities of doublet components in the presence of collisions.

sections for the higher alkali doublets. In the case of the second resonance doublet of Rb, for example, more than 20 rate equations are needed to describe fully the population of the doublet levels at arbitrary gas pressure (see Siara 1972, and Siara and Krause 1973).

In flames at constant pressure, mixing cross sections for the first alkali resonance doublets can — in principle — be determined by measuring the four independent fluorescence efficiencies defined in Sect. 2b-2. The determination is considerably simplified when one assumes that the quenching rate constants of the doublet components are, virtually, similar. Each of these rate constants is then equal to the weighted average value of the quenching rate constant for the doublet as a whole, which can be simply determined in a separate experiment as described in Sect. 2b-2 (see Lijnse, Zeegers and Alkemade 1973, and van Calcar $et\ al.$ 1979). In this way mixing cross sections for K, Rb and Cs have been determined in flames (see Table VI.1), whereas a lower limit of about 5 $Å^2$ was obtained for the exo-ergic mixing of the Na-D doublet in a nearly stoichiometric H_2-O_2-Ar flame at $T = 1700$ K by van Calcar $et\ al.$ (1979) and van Calcar (1980). In a similar flame Omenetto $et\ al.$ (1979) obtained a value of (at least) 15 $Å^2$ for the same mixing cross section.

The detailed-balance relation Eq. VI.1 has been proved to hold generally for the first alkali doublets with molecular collision partners (for a review see Lijnse 1972). In flames, the validity of this relation has been checked directly within an error of 2% by measuring the ratio of only two fluorescence efficiencies without using explicitly the quenching and mixing cross sections themselves (see Lijnse, Zeegers and Alkemade 1973, 1973a, and Lijnse and Hornman 1974).

Possible sources of errors and experimental techniques to eliminate them have been discussed in Sects II.4a-4 and III.16. Errors due to spectral leakage of the filters used for selecting the doublet components (see Sect. VI.2a-2) are far less serious for molecular collision partners than for atomic partners.

Other methods of measuring mixing cross sections (such as photodissociation crossed-molecular beam and Hanle experiments) have been surveyed by Lijnse (1972, 1973) and Krause (1972, 1975).

A selection of experimental results that may be of interest in the present discussion is presented in Table VI.1 (for a more complete review of earlier experimental data see Lijnse 1973b and Krause 1975). The mixing cross sections obtained from flame measurements are also included in this table. Mixing cross sections for molecules have been studied less extensively than mixing cross sections for noble gas atoms. Mixing cross sections have been investigated also less extensively than quenching cross sections in the case of molecules. Mixing of the first resonance doublets of the alkalis and of the excited 6^3P triplet of mercury have been investigated most often. Most studies refer to nitrogen molecules and hydrogen molecules

PHYSICAL EXCITATION AND DE-EXCITATION PROCESSES   Sect. VI.2b

(together with their isotopes) as collision partners. The following trends are found in the experimental data obtained so far.

The cross sections for nonadiabatic mixing of the alkali doublets are of the same order of magnitude (20-200 $Å^2$) for nonpolar di-atomic molecules as for noble gas atoms. However, for adiabatic mixing collisions the cross section values for molecular species greatly exceed those for noble gas atoms. In the case of molecules the doublet mixing is much less dependent on the magnitude of the doublet splitting than in the case of atomic collision partners. Remarkably, the cross section for mixing of the ground doublet of $Tl(6^2P_{\frac{3}{2}} \to 6^2P_{\frac{1}{2}})$ at about 600 K is quite small for N_2, H_2, CO and CO_2, in contrast to that for NO and O_2 (see Bellisio and Davidovits 1970, Gedeon, Edelstein and Davidovits 1971, and Wiesenfeld 1973).[†] The last mentioned author explained the larger efficiency of O_2 by the near-resonance between the excited Tl state and the excited $O_2(^1\Delta_g)$ state. Furthermore we note that the cross sections for the exo-ergic mixing process found with the first alkali resonance doublets are larger than the corresponding quenching cross sections in the case of H_2, H_2O and saturated molecules like CH_4. This also applies to the first resonance doublets of Na and K in the case of N_2.

In discussing the temperature dependence of the mixing cross sections we shall disregard the trivial Boltzmann factor that occurs in the expression for the ratio of the endo-ergic to the exo-ergic mixing cross sections (see Eq. VI.1). (This Boltzmann factor explains why the σ_{12} values found in vapour cells for the first resonance doublets of Rb and Cs are less than the σ_{21} values, in contrast to the situation in the hotter flames; see Table VI.1.) When comparing the results obtained in vapour cells with those obtained in flames, a moderate increase of mixing cross section with increasing temperature is found for K(4P), Rb(5P) and Cs(6P) in collision with N_2 (see Table VI.1). The temperature dependence of the mixing cross sections of the first and second resonance doublets of Rb and Cs for various deuterated hydrogen- and hydrocarbon molecules is rather weak between 300 and 650 K (see Phaneuf 1973, Phaneuf, Skonieczny and Krause 1973, Walentynowicz, Phaneuf and Krause 1974, Lukaszewicz 1975, and Krause, Siara and Dubois 1975). The temperature dependence was found to be different for each of these species. Krause, Datz and Johnson (1973) found from crossed-molecular beam experiments that the mixing cross section for $Hg(6^3P_2 \to 6^3P_1)$ in the case of N_2 and NO decreased in general $\propto E^{-0.5}$ with increasing kinetic energy E, whereas they observed an additional, detailed structure in the energy dependence of this cross section in the case of H_2 and D_2 (see also Krause and Datz 1976, who determined an absolute cross section for the

---

[†] For a more recent absolute determination of the cross section for this mixing process with N_2 at $T = 660$ K see also Aleksandrov, Baranov and Kulyasov (1978).

mixing of these states in the case of N_2 by a beam experiment). A different energy dependence for this process was, however, found by Liu and Parson (1976) in crossed-molecular beam experiments with various molecules at higher energies. Wiesenfeld (1973) derived an activation energy of about $+0.8\,eV$ from the temperature dependence of the exo-ergic mixing cross section for $Tl\,(6^2P_{\frac{3}{2}} \rightarrow 6^2P_{\frac{1}{2}})$ with H_2 and O_2 and a negative activation energy of about $-1.6\,eV$ for CO.

The existence of resonance effects in the mixing of the first alkali resonance doublets has long been disputed. Of special interest in this discussion are experiments with various isotopic species of the collision partner (H_2, HD or D_2; $^{14}N_2$ or $^{15}N_2$) with a similar electronic structure but different rotational-vibrational energy levels. Because of the relatively small doublet splitting, resonances with vibrational transitions are not expected here. Resonances with rotational transitions are more likely, but so far no definite conclusions can be drawn from the available experimental data (see Krause 1972,1975, Siara 1972, and Lijnse 1973).

Because of the larger $Hg(6^3P)$ multiplet splitting, resonance with a single vibrational quantum jump might be feasible here for a number of di-atomic molecules. Such resonance effects, however, were not found for the mixing process $Hg(6^3P_1 \rightarrow 6^3P_0)$ on the basis of the older cross section data summarized by Mitchell and Zemansky (1961) and re-examined by Callear (1965, 1967) (see also Pringsheim 1949). In the search for resonance effects, these cross sections were corrected for an activation-energy factor whenever the vibrational excitation energy was larger than the separation of the Hg levels considered. More recent experiments by Deech and Baylis (1971), Deech, Pitre and Krause (1971), and Vikis, Torrie and Le Roy (1972) confirmed the above conclusion. Resonance effects also appeared to be unimportant for the mixing process $Hg(6^3P_2 \rightarrow 6^3P_1)$ investigated by Doemeny, Van Itallie and Martin (1969), Van Itallie, Doemeny and Martin (1972), and Krause, Datz and Johnson (1973) in crossed-molecular beam experiments.

The fact that the mixing process $Hg(6^3P_1 \rightarrow 6^3P_0)$ induced by CO is actually accompanied by excitation of the first vibrational level of CO was demonstrated directly by Callear and Wood (1971). These authors have analyzed the vibrational excitation of the product molecule by kinetic absorption spectroscopy in the vacuum ultraviolet. Mercury vapour was excited in this experiment to the initial 6^3P_1 state by a mercury flash lamp.

Theory. The theoretical interpretation of multiplet mixing by molecules is much less advanced than in the case of mixing by atoms or in the case of quenching. Because of the possible involvement of the internal molecular motions, the Massey parameter, so important for mixing by atoms, may not be a good criterion for judging the efficiency of the mixing process. Dickens, Linnet and Sovers (1962) have tried to treat the mixing process of $Hg(6^3P_1 \to 6^3P_0)$ as a resonance process in which the electronic energy released is converted into vibrational energy of a di-atomic molecule. The efficiency of this process should decline sharply with increasing energy defect for a single quantum jump of vibrational energy. Such resonance effects, however, have not been found experimentally (see above).

Attempts have also been made to treat the latter process theoretically on the basis of the crossing of potential-energy surfaces. Laidler (1955) has discussed qualitatively the probability of collisional mixing of the $Hg(6^3P_{0,1})$ states, while assuming that transitions between the initial and final states of the collision complex take place through an intermediate ionic $(Hg^+ + Z^-)$ state. Resonance effects were not found in this analysis. Bykhovskii and Nikitin (1964) have derived an analytical expression for the cross section of the above ionic-curve crossing process, taking into account the possibility of vibrational excitation of the diatomic molecule involved. According to their theory, the cross section for the collision transfer $Hg(6^3P_1 \to 6^3P_0)$ should exhibit a marked dependence on translational energy, but not on resonance defect. This theory, which involves several unknown parameters, agrees at least qualitatively with the experiments made by Krause, Datz and Johnson (1973) (see above). A curve-crossing process was also suggested by French and Lawley (1977) for the collisional mixing of the $Hg(6^3P_2)$ state with the $^3P_1$ and $^3P_0$ states by H_2 and HD.

Andreev and Voronin (1969) have suggested that Rb(5P)- and Cs(6P) mixing by N_2 also takes place via an intermediate ionic state. On this basis mixing cross sections have been calculated, but the approximations made were rather crude (see Lijnse 1972). Krause (1975) concludes from experimental data that the ionic-curve crossing model is applicable to adiabatic mixing collisions with N_2. Lijnse (1973) and Lijnse and Hornman (1974) believe that ionic-curve crossing alone does not fully explain the large cross sections found for the mixing of the Rb- and Cs doublets by N_2. Coupling between the atomic fine-structure transitions and the molecular rotation is expected to play an important role.

Starting from two previously derived adiabatic potential surfaces for the complex $Na(3P) + N_2$ Amaee and Bottcher (1978) have recently calculated the mixing cross section for the Na-D doublet as a function of the kinetic energy. They found good agreement with the experimental values listed in Table VI.1 at $T = 400$ K.

The smallness of the cross sections for $Tl(6^2P_{\frac{3}{2}} \to 6^2P_{\frac{1}{2}})$ mixing by N_2 at moderate temperatures (see above) might be explained by assuming that the potential-energy curve for the intermediate ionic-state crosses the initial potential-energy curve in the repulsive region (see Fisher and Smith 1971). This may lead to a high activation energy compared to kT.

A semiclassical theory in which rotational transitions are explicitly taken into account by the use of a nonspherical interaction potential has been advanced by Baylis *et al.* (1973) and by Walentynowicz, Phaneuf, Baylis and Krause (1974) for the doublet mixing in Cs and Rb by deuterated methanes (see also Phaneuf 1973). A similar theory was applied by Lukaszewicz (1975) to explain the temperature dependence of $Cs(7^2P)$ mixing by H_2.

The measured nonadiabatic mixing cross sections for Na in collision with several hydrocarbon molecules appeared to be in fair agreement with the theory developed by Callaway and Bauer (1965) for the mixing by noble gas atoms. This theory, which is based on a van der Waals interaction, predicts a dependence of the cross section on the polarizability, α, of the collision partner according to $\alpha^{\frac{2}{5}}$.

The theory for intramultiplet mixing of Tl, Hg, Pb and other heavy atoms by H_2 and HD has been examined by French and Lawley (1977). These authors demonstrated the importance of electronic and rotational-vibrational energy matching in the case of transitions that were allowed by quadrupole-quadrupole coupling.

For a survey of the older theoretical literature in this field we refer again to Lijnse (1972, 1973).

Conclusions. We shall consider the consequences of multiplet mixing by molecules under flame conditions. The phenomenological mixing cross sections in flames may differ from those measured at lower temperatures in vapour cells for two reasons. Firstly, the average translational energy in flames is higher than in vapour cells while the mixing cross sections may depend on the energy of relative motion. Secondly, they may differ because in the hotter flame the higher vibrational and rotational levels are more populated; the cross sections may depend on the internal energy of the molecule before collision. This dependence is, however, not well known at present.

When in atomic fluorescence spectroscopy alkali vapour in the flame is irradiated by only one of its doublet lines, the population ratio of the doublet levels will obey Boltzmann's law only if mixing is a much faster process than quenching (and radiation). The latter condition is reasonably well fulfilled for the first K- and Na doublets for most molecules that may occur as major constituents in the flame; it does not hold, however, for the first Cs doublet (see Table VI.1).[†] The fulfillment of this condition may be of importance when one applies nonresonance fluorescence by irradiating the alkali vapour with only one doublet line while observing the fluorescence emitted at the other doublet line. Anyway, the state of mixing of the doublet components is not expected to vary strongly with flame temperature since

---

[†] Note, however, that the only cross section value for the mixing of the Na-D doublet by CO_2 listed in Table VI.1 and measured by Apt and Pritchard (1979) is surprisingly low compared to the corresponding quenching cross sections listed.

PHYSICAL EXCITATION AND DE-EXCITATION PROCESSES   Sect. VI.2b

neither the mixing nor the quenching cross sections depend strongly on temperature. The Boltzmann factor occurring in Eq. VI.1 for the ratio of σ_m and σ_{-m} is not important in flames, as the alkali doublet splitting is in general less than kT.

In fuel-rich N_2-diluted flames the relative population of the $Tl(6^2P)$ doublet levels may deviate markedly from Boltzmann's ratio whenever Tl atoms are raised by a strong laser beam from one of these levels to a higher level. This deviation is expected because of the very small mixing cross sections of N_2, H_2 and CO for this doublet.

The mixing cross sections and the partial pressures of the molecules present in the flame determine whether, in laser-saturation experiments, the multiplet involved in the optical transition can be treated as a single level or not (see also Omenetto, Benetti, Hart, Winefordner and Alkemade 1973 and van Calcar et al. 1979).

Finally, mixing of alkali doublet levels by collisions with molecules in flames is not expected to have a marked effect on the collisional line broadening (see Sect. VII.2b-6).

2b-4   *ELECTRONIC-EXCITATION TRANSFER BETWEEN DISSIMILAR ATOMS UPON COLLISION*

Electronic excitation energy can be exchanged between two dissimilar atoms upon collision according to

$$A^* + B \rightleftharpoons A + B^*.$$

Here A and B denote the atoms in the ground state, and A^* and B^* denote excited atoms. The difference in excitation energy (with absolute value ΔE) must be made up by the translational energy of relative motion of the colliding atoms. The case of excitation-energy exchange between atoms of the same element has been treated in Sect. 2a-2 under the heading 'self-mixing'.

Excitation transfer between metal atoms is a priori expected to be unimportant in *flame spectroscopy* (see Winefordner and Mansfield 1967, and Alkemade 1968). Even if a metal salt solution of 1 mol l^{-1} is sprayed by a direct-injection burner into an unpremixed flame at 1 atm pressure, the partial pressure of the fully atomized metallic element is still less than 10^{-3} atm (see Sect. III.2h). Thus, when a sprayed solution contains, in addition to element A, another element B at a concentration as high as 1 mol l^{-1}, the chance that a given excited atom A^* will collide with a B atom instead of with an arbitrary flame molecule is less than 1:1000. Now the largest known cross sections for excitation transfer between two dissimilar metal atoms appear to be of the same order of magnitude as the quenching cross sections for most flame molecules (see Table VI.1). So quenching of A^* atoms by energy transfer to other metal atoms is unimportant under usual flame conditions. By application of detailed balance, a similar conclusion can be drawn for the excitation of A by excited metal atoms B^*. Only in a flame that is mainly composed

Chapt. VI   EXCITATION AND DE-EXCITATION OF METAL SPECIES

of poor quenchers, such as Ar, and for unusually large transfer cross sections, might an exchange of electronic energy between dissimilar metal atoms at high metal concentrations play a role. Up till now such transfer effects have not been found in flames.[†] The excitation energies of atomic flame constituents, like H or Ar (but not O), are generally so great that they can also be excluded as partners in the above exchange process under flame conditions. We shall restrict ourselves to only a short survey of electronic-excitation transfer processes.

The two types of transfer processes which have been most extensively studied are firstly:[*]

$$Hg(6^3P_1) + M \rightarrow Hg(6^1S_0) + M^*, \qquad (i)$$

where M is a metal atom (for example, Na, Cd, Tl, In, Zn); excited Cd- and Mg atoms have been occasionally used as donor atoms as well. The second type is:

$$A_1(^2P_J) + A_2(^2S_{\frac{1}{2}}) \rightleftharpoons A_1(^2S_{\frac{1}{2}}) + A_2(^2P_{J'}), \qquad (ii)$$

where A_1 and A_2 are two dissimilar alkali atoms in their ground state ($^2S_{\frac{1}{2}}$) or lowest excited ^2P state. The quantum numbers of total electronic angular momentum, J and J', may each have the value $\frac{1}{2}$ or $\frac{3}{2}$.

Other transfer processes, not discussed here, involve a pair of noble gas atoms like He and Ne; they play an important role in gaseous lasers pumped by a gas discharge.

Experiments. Excitation transfer between metal atoms can be studied experimentally by the *method of sensitized fluorescence* (for definition, see Sect. II.5a-1). In studying *type (i) processes*, a quartz bulb containing, for example, a mixture of Hg- and Na vapours, is irradiated by the 2537 Å resonance line of a mercury discharge lamp. Upon absorbing this radiation Hg atoms are raised from the ground level to the 6^3P_1 level lying at 4.86 eV. They may also be raised to this level (and to the 6^3P_0 level) by electron impact in an electric discharge through a mixture of He, Hg and Na (see Frish and Bochkova 1963). In addition to the Hg resonance radiation, several series of sensitized fluorescence lines of Na are

---

[†] In some papers on analytical flame spectroscopy, such effects have erroneously been held responsible for the observed enhancement of analyte emission in the presence of a concomitant element. For a critical discussion of these literature reports see Alkemade (1968) and Sect. VI.4e.

[*] For a survey of the earliest experiments and methods of measuring see Mitchell and Zemansky (1961) and Pringsheim (1949). Later experiments and theoretical interpretations have been included in the reviews by, e.g., Bates (1962a), Callear (1967), Seiwert (1968), Kraulinya (1968, 1969), and Massey, Burhop and Gilbody (1971), and numerical data have been compiled by Gilmore, Bauer and McGowan (1967). Experimental and theoretical work of more recent date has been reviewed by Krause (1972, 1975), Lijnse (1973b), and Elbel (1979).

PHYSICAL EXCITATION AND DE-EXCITATION PROCESSES   Sect. VI.2b

observed; these series arise from the radiative decay of high-lying S-, P- and D
levels (see Kraulinya 1964, Czajkowski, Skardis and Krause 1973a, and Czajkowski,
Krause and Skardis 1973; see also Table VI.1). The highest Na excitation levels
that contribute to the observed sensitized fluorescence lie about a tenth of an
electronvolt above the $Hg(6^3P_1)$ level. The high-lying Na levels are obviously
populated by transfer of electronic energy upon collision with an excited mercury
atom. The cross section for excitation transfer to a particular Na level can be
deduced from absolute intensity measurements of the fluorescence lines arising from
this level and from higher-lying Na levels (that contribute to the population of
this level by radiative cascade transitions) as well as of the Hg resonance line;
the Na- and Hg-vapour densities should be known (see Czajkowski, Skardis and
Krause 1973a). The largest transfer cross section (see Table VI.1) was found for
the $Na(9^2S)$ level, which lies 0.019 eV above the $Hg(6^3P_1)$ level.[†]

The cross sections for energy transfer from excited Hg atoms to various
Na levels showed a clear resonance dependence on ΔE, at least within each series
of S-, P- and D levels (see Czajkowski, Skardis and Krause 1973a; see Table VI.1).
The cross sections for excitation transfer from $Hg(6^3P_1)$ to $Cd(5^3P_1)$ (see
Czajkowski and Krause 1974)[*] and to various Tl states (see Wade, Czajkowski and
Krause 1978a) lie well on the same, empirical 'resonance curve' as established for
the Na states. A systematic resonance effect was also found in the electronic-
energy transfer from Cd- to Cs atoms for $\Delta E < 0.05$ eV (see Seiwert 1968).
Kraulinya et al. (1975) found that the transfer efficiency from excited $Cd(5^3P_1)$
atoms to Cs atoms was one order of magnitude greater for the levels of the D- and
F series than for those of the S series.

It was also found experimentally that the probability of energy transfer is
not influenced by the optical selection rules that hold for the transitions in the
separate atom (see Callear 1965, 1967). The transition from the Na ground level
to the 9^2S level, for example, is optically 'forbidden'. But the excitation of
this level is stronger than that of the $Na(8^2P)$ level (with $\Delta E = 0.01$ eV), for which
the transition would be 'allowed' (see Czajkowski, Skardis and Krause 1973a).

Cross sections for exo-ergic *transfer processes of type (ii)* have been determined for the systems $Rb^* \to Cs$ and $K^* \to Rb$ (see Krause 1972, 1975 for a survey;
see also Table VI.1). The method of measuring is analogous to that for mixing cross

---

[†] For a possible explanation of the large discrepancy between the earlier results of Kraulinya (1964) and the quoted results of Czajkowski and others, and of other (seeming) discrepancies occurring in Table VI.1 we refer to Czajkowski and Krause (1974), Krause (1975), and Wade, Czajkowski and Krause (1978a).

[*] Note, however, in Table VI.1 the discrepancy between the value found by Czajkowski and Krause (1974) and the one found by Morozov and Sosinskii (1973).

sections. The magnitude of the transfer cross sections appears to be systematically smaller than those for self-mixing at comparable ΔE value. Although there are some discrepancies between the cross sections measured by different authors, again a similar resonance effect as found for $Hg^* \to Na$ seems to be present (see Czajkowski, Skardis and Krause 1973a, and Krause 1975).

Theory. A satisfactory theoretical explanation of the experimental results, in particular of the resonant dependence on ΔE, has not yet been given. Besides, some experimental results still contradict each other. The theory for a quasi-resonance process which is based on a long-range resonance interaction (see Sect. II.4g) does not seem to provide an explanation for the effects described in this section. On the basis of a quasi-resonance process one would expect a much larger cross section for an optically allowed transition than for an optically forbidden transition at the same ΔE value (see Bates 1962).

Electronic-energy transfer is more likely to take place through a pseudo-crossing of potential-energy curves of the colliding atoms, involving short-range interaction (see Laidler 1955, Dashevskaya, Nikitin, Voronin and Zembekov 1970, and Nikitin 1975). Callear (1965) has pointed out that a small energy defect ΔE may be advantageous also when pseudo-crossing of potential-energy curves is involved. Pritchard and Apt (1975) found the transfer cross section of $Na(3P_{\frac{1}{2},\frac{3}{2}}) \to Rb(5P_{\frac{1}{2},\frac{3}{2}})$ to vary as the inverse of the relative velocity, which suggested a curve-crossing process. They applied velocity selection through the Doppler effect by exciting the Na vapour in a cell with a monochromatic dye laser whose frequency was tuned across the Doppler profile of the Na line. Quantitative comparison of the experimental results with theoretical calculations, however, is hampered by a lack of information about the course of the potential-energy curves.

Whenever large cross sections are found for energy transfer from excited Hg vapour to metal atoms M with large energy defect, the formation of (quasi-) molecules HgM and Hg_2 at comparatively high Hg pressures should be considered, too (see Seiwert 1968, and Kraulinya 1969).

PHYSICAL EXCITATION AND DE-EXCITATION PROCESSES   Sect. VI.2b

GENERAL NOTES AND LEGENDS TO TABLE VI.1

† The table lists a selection of measured absolute phenomenological cross sections of collisional processes that lead to *quenching* (i.e. intermultiplet transitions) and *mixing* (i.e. intramultiplet transitions; see also footnote † on page 530) of electronic states of metal atoms or to *electronic-excitation transfer* between metal atoms, and of microscopic cross sections relating to *excitation* of metal atoms by *electron impact*. Cross section values obtained in atomic-beam or shock-tube experiments are as a rule not included (for a recent review of atomic-beam excitation experiments see Hertel and Stoll 1977).

* The relevant quantum numbers of the excited metal atom(s) involved are specified; when the measurements refer to a multiplet as a whole, J numbers are not specified. Ground-state metal atoms are simply denoted by a superscript 0 (for their designation see Appendix A.4). The internal state of the collision partner is not specified unless it is also a metal atom; the collision partners are usually in their electronic ground state prior to collision. When the product state or compound of the metal atom is unknown or not specified in the literature, there is a question mark.

** Dubious values or values that have been replaced later by more reliable ones have not been included. The references listed are not exhaustive except in the case of flame experiments; the references are not selected on the basis of an evaluation of merits. No error limits are indicated, as unknown systematic errors often outweigh the accidental errors stated. Some literature values have been rounded off or averaged.

§ Temperature values (in K) are rounded off. When a non-Maxwellian, peaked distribution of relative velocities g occurred, the mean value $\bar{g}$ is stated. When measurements were made in a wide range of temperatures or mean velocities, their extreme values and the corresponding extreme cross section values are listed.

§§ The experimental system and method of measuring are indicated briefly. *Legends: fla* = metal-seeded flame; *ves* = vessel with metal vapour; *beam* = atomic or electron beam; *phdis* = photodissociation; *flq* = quenching of fluorescence; *slf* = stepwise line fluorescence (see definition in Sect. II.5a-1); *sens* = sensitized fluorescence (as defined in a restricted sense in Sect. II.5a-1); *lif* = lifetime measurement or kinetic measurement.

ACKNOWLEDGEMENT

The authors thank Dr P.L. Lijnse of the Fysisch Laboratorium, University, Utrecht, for his substantial contribution to the compilation and evaluation of the older literature data.

For the *SPECIAL NOTES* numbered (1) - (20) see end of table on page 618.

Chapt. VI EXCITATION AND DE-EXCITATION OF METAL SPECIES

Table VI.1
Cross Sections of Collisional (De-) Excitation Processes[+]

Process[*]	Cross section ($Å^2$)[**]	Temperature (K) [Relative velocity $\bar{g}$ (km/s)]	Method of determination	References[**]
Quenching				
$Li(2^2P) + N_2 \rightarrow Li^0 + N_2$	21	1400	fla;flq	Jenkins (1968b)
H_2	16			
CO	40			
CO_2	29			
H_2O	6			
Ar	≤1			
$Li(2^2P) + N_2 \rightarrow Li^0 + N_2$	26 → 19	$\bar{g} = 2.5 \rightarrow 4.6$	ves,phdis,flq	Lin, Weston (1976)
H_2	25 → 23	$\bar{g} = 3.5 \rightarrow 4.6$		
CO_2	40 → 26	$\bar{g} = 2.7 \rightarrow 4.2$		
$Na(3^2P) + N_2 \rightarrow Na^0 + N_2$	22 (i) / 17 (v)	(i) 1400→1800	(i)–(iv) fla;flq	(i) Jenkins (1966)
	21 (ii) / 14 (v)*	(ii) 1900	(v) fla;lif	(ii) Hooymayers, Lijnse (1969)
	9.0 (iii) / 8 (ii) 9.3–6.8 (iv)	(iii) ≈ 2100		(iii) Lijnse, Elsenaar (1972)
	6.5 (v)	(iv) 1500→2500		(iv) Lijnse, v.d. Maas (1973)
H_2	39 (i) / 34 (ii) 39 → 31 (iv)	(v) 1500→2500		Ham, Hannaford (1979);
	24 (v) / 20 (v)*			Hannaford (1979)
CO	38 (i) / 41 (ii)	1750		
CO_2	54 (i) / 50 (ii)	2100		
H_2O	1.5 (i) / 2.2 (ii)* 2.2 (iii)			
	<2.0 (v)			
Ar,He	<0.3 (i)			
$Na(3^2P) + {}^{14}N_2 \rightarrow Na^0 + {}^{14}N_2$	40 (i)	(i) 400	(i) ves;lif	(i) Kibble, Copley, Krause (1967)
${}^{15}N_2$	30 (iii)	(ii) 500	(ii) ves;lif	(ii) Hulpke, Paul, Paul (1964)
H_2	16 (i)	(iii) ≈570	(iii) ves;lif	(iii) Bästlein, Baumgartner, Brosa (1969)
HD	12 (i)	(iv) 900	(iv) ves;phdis;flq	(iv) Earl (1973); Earl, Herm (1974)
D_2	10 (i)	(v) 400	(v) ves;lif	(v) Copley, Kibble, Krause (1967)
CO	41 (iii)	(vi) ≈900	(vi) ves;phdis;flq	(vi) Edwards (1969)
I_2	200 (iv)	$\bar{g} = 1.5$	(vii) beam (20)	(vii) Speller, Staudenmayer, Kempter (1979)
CO_2	65 → 40 (iv)	$\bar{g} = 1.3 \rightarrow 2.2$		
H_2O	3 (iv)			
D_2O	<1 (iv)			
SO_2	130→80 (iv)			
CH_4	<1 (iv)	$\bar{g} = 1.2 \rightarrow 2.3$		
noble gas	<10^{-2} (v)(vii) <0.1–0.5 (vi)			

604

Table VI.1 (continued)

Process			Reference
$Na(3^2P) + N_2 \to Na^0 + N_2$	19→12	$\bar{\sigma} = 1.5 \to 2.3$	Barker, Weston (1976)
H_2	8.1	$\bar{\sigma} = 3.1$	
D_2	8.4→7.1	$\bar{\sigma} = 2.3 \to 2.8$	
CO	34→24	$\bar{\sigma} = 1.5 \to 2.3$	
CO_2	50→28	$\bar{\sigma} = 1.4 \to 2.3$	
$Na(3^2P) + N_2 \to Na^0 + N_2$	41 (i)		
$Na(4^2P)$ $Na(?)$	23 (i)		
$Na(5^2P)$	8 (i)		
$Na(6^2P)$	3 (i)		
$Na(7^2P)$	≈380 (i)		
$Na(5^2S)$	27 (i)	43 (iii)	ves;phdis;lif
$Na(6^2S)$	41 (i)	86 (iii) (iii)	[(i) Czajkowski, Krause, Skardis (1973)
$Na(7^2S)$	16 (i)	84 (iii)	(ii) Humphrey et al. (1978)
$Na(8^2S)$	95 (i)	82 (iii) [(i) ≈500	ves;sens;flq (iii) Gallagher, Cooke, Edelstein (1978)
$Na(9^2S)$	55 (i)	91 (iii) (ii) 420	ves;lif
$Na(4^2D)$	38 (i)	104 (iii) (iii) 420	ves;lif
$Na(5^2D)$	7 (i)	43 (iii)	
$Na(6^2D)$ (14)	10 (i)	36 (iii)	
$Na(7^2D)$	45 (i)	32 (iii)	
$Na(8^2D)$	25 (i)	18 (iii)	
$Na(5S) + N_2 \to Na(3^2D) + N_2$	10	420	ves;lif Gallagher, Cooke, Edelstein (1978)
$Na(4^2P) + N_2 \to Na(3^2D) + N_2$	19		
$Na(5S) + N_2 \to Na(4^2P) + N_2$	32		
$Na(6^2D) + He \to Na(6, l) + He$ $(l > 2)$	380 (i)	[(i) 425	ves;lif [(i) Gallagher, Edelstein, Hill (1977)
Ar	510 (i)	(ii) 435	(ii) Gallagher et al. (1977)
N_2	290 (ii)		
CO	720 (ii)		
$Na(10^2D) + He \to Na(10, l) + He$ $(l > 2)$	2.2×10^3 (i)	[(i) 425	ves;lif [(i) Gallagher, Edelstein, Hill (1977)
Ar	4×10^3 (i)	(ii) 435	(ii) Gallagher et al. (1977)
N_2	1.9×10^3 (ii)		
CO	1.7×10^3 (ii)		
$Na(15^2D) + He \to Na(15, l) + He$ $(l > 2)$	1.4×10^3	425	ves;lif Gallagher, Edelstein, Hill (1977)
Ar	3.7×10^3		

Chapt. VI   EXCITATION AND DE-EXCITATION OF METAL SPECIES

Table VI.1 (continued)

$K(4^2P) + N_2 \to K^0 + N_2$	18 (i)	19 (ii)	19 (iv)	1400	fla;flq	(i) Jenkins (1968a)
H_2	3.3 (i)	3.3 (ii)	3.4 (iv)	1500-2500		(ii) Lijnse, Hornman (1974)
O_2	49 (i)	63 (ii)	49 (iv)	1900		(iii) Hooymayers, Alkemade (1966)
CO	39 (i)	44 (iii)	44 (iv)	1900		(iv) Hooymayers, Lijnse (1969)
CO_2	67 (i)	66 (iii)	66 (iv)	≈ 2100		
H_2O	2.8 (i)	3.6 (ii)	2.6 (iv)*			
Ar	<0.6 (i)					
He (5)	<0.2 (i)					
$K(4^2P) + N_2 \to K^0 + N_2$	34 (i)			400	ves;llf	(i) Copley, Krause (1969)
H_2	9.4 (i)	3 (ii)		900	ves;phdis;flq	(ii) Earl (1973); Earl, Herm (1974)
HD	12 (i)					
D_2	8.0 (i)	3 (ii)				
I_2	215 (i)					
H_2O	2 (ii)					
SO_2	130→80 (ii)					
CH_4	<1 (ii)					
$K(4^2P) + He \to K^0 + He$	<0.04 (i)	$<3\times10^{-3}$ (ii)		$\bar{g} = 1.0$	(1)	
Ar	<0.07 (i)	$<3\times10^{-3}$ (ii)		$\bar{g} = 0.85 \to 1.5$		
$K(4^2P_{\frac{1}{2}}) + N_2 \to K^0 + N_2$	35			≈ 900	ves;phdis;flq (1)	(i) Edwards (1969)
H_2	7				beam (20)	(ii) Speller, Staudenmayer, Kempter (1979)
HD	11					
D_2	2					
$K(4^2P_{\frac{3}{2}}) + N_2 \to K^0 + N_2$	39			370	ves;flq	(1) McGillis, Krause (1968)
H_2	4					
HD	14					
D_2	1					
$K(5^2P) + N_2 \to K(?) + N_2$	31 (i)(2)	10 (ii)(2)		(i) 1900	fla;flq	(1) McGillis, Krause (1968)
H_2 (5)	40 (i)(2)	2 (ii)(2)		(ii) 900	ves;phdis;flq	
H_2O (5)	7 (i)(2)	15 (ii)(2)				
$K(10^2P) + K^0 \to K(9^2D, 11S) + K^0$	800			470	ves;slf	(i) Hooymayers, Alkemade (1966)
$K(10^2P) + K^0 \to K(?) + K(?)$	6×10^3			470	ves;flq	(ii) Earl (1973); Earl, Herm (1974)
$K(10^2P) + He \to K(?) + He$	18			470	ves;flq	Gounand et al. (1976)
Ar	8					Gounand et al. (1976)
Xe	40					Gounand et al. (1976)

606

Table VI.1 (continued)

$Rb(5^2P) + N_2 \to Rb^0 + N_2$					
H_2	20 (i)	19 (ii)	25 (v)	(i) 1400	(i) fla;flq
O_2	1.9 (i)	2.1 (ii)	3.6 (v)	(ii) ≈1700	(ii) fla;flq
CO	83 (i)	90 (ii)	83 (v)	(iii) ≈1700	(iii) fla;flq
CO_2	38 (i)			(iv) ≈ 900	(iv) ves;phdis;flq
H_2O	80 (i)	74 (iii)		(v) 1900	(v) fla;flq
Ar	4.0 (i)	3.9 (ii)	4.0 (v)*	(vi)* ≈2100	(vi) beam (20)
He	<1 (i)	<0.14 (iv)	<2×10⁻³ (vi)		
Xe	<0.4 (i)	<2×10⁻³ (v)			
	<0.21 (iv)	<2×10⁻³ (vi)			
$Rb(5^2P_{1/2}) + N_2 \to Rb^0 + N_2$					
H_2	37 (i)	80 (ii)	58 (iii)	(i) 300	(i) ves;lif
HD	6 (iii)			(ii) 300	(ii) ves;lif
D_2	6 (iii)			(iii) 340	(iii) ves;flq
CH_4	3 (iii)				
	<1 (iii)				
$Rb(5^2P_{3/2}) + N_2 \to Rb^0 + N_2$					
H_2	36 (i)	80 (ii)	43 (iii)	ditto	ditto
HD	3 (iii)				
D_2	5 (iii)				
$Rb(6^2P_{1/2}) + N_2 \to Rb(?) + N_2$					
H_2		128		430	ves;flq
HD		36			
D_2		47			
CH_4		28			
		129			
$Rb(6^2P_{3/2}) + N_2 \to Rb(?) + N_2$					
H_2		126		430	ves;flq
HD		31			
D_2		38			
CH_4		21			
		114			
$Rb(12^2P) + Ar \to Rb(?) + Ar$					
Ne		15		460	ves;lif
He		5			
		38			
$Rb(22^2P) + Ar \to Rb(?) + Ar$					
Ne		30		460	ves;lif
He		13			
		60			

References:
- (i) Jenkins (1968c) (16)
- (ii) Lijnse, Zeegers, Alkemade (1973)
- (iii) Lijnse, P.L. (unpubl.)
- (iv) Edwards (1969)
- (v) Hooymayers, Nienhuis (1968)
- (vi) Speller, Staudenmayer, Kempter (1979)
- (i) Bellisio, Davidovits, Kindlman (1968)
- (ii) Bulos, Happer (1969)
- (iii) Hrycyshyn, Krause (1970)

ditto

Siara, Krause (1973)

Siara, Krause (1973)

Gounand, Fournier, Berlande (1977)

Gounand, Fournier, Berlande (1977)

Table VI.1 (continued)

Reaction	Partner	Value	Value	Value(s) (K)	Method	Reference
Cs(6²P) + N₂ → Cs⁰ + N₂	N₂	66 (i)		(i) 1400	(i)–(iii) fla;flq	(i) Jenkins (1968c) [16]
	H₂	4.7 (i)	35 (ii)	(iii) ≈1700	(iv) fla;flq	(ii) Lijnse, Zeegers, Alkemade (1973a)
	O₂	134 (ii)	5.0 (ii)	(iii) 1700		(iii) Lijnse, P.L. (unpubl.)
	CO₂	86 (ii)		(iv) 1750		(iv) Hannaford (1979)
	H₂O	15 (i)				
	Ar	<3 (i)				
	He	<1 (i)				
Cs(6²P) + He → Cs⁰ + He		<10⁻³ (i)	9.0 (ii)	(i) 350–800	(i) ves;llf	(i) Dodd, Enemark, Gallagher (1969)
			<6 (iv)	(ii) ≈900	(ii) ves;phdis;flq	(ii) Edwards (1969)
Cs(6²P½) + N₂ → Cs⁰ + N₂				(15)		Antonov, Korchevoi (1976)
	H₂	86 (i)				
	HD	7 (i)	60			
	D₂	4 (i)				
		8 (i)				
Cs(6²P_{3/2}) + N₂ → Cs⁰ + N₂				(i) 315	ves;flq	(i) McGillis, Krause (1968a) [16]
	H₂	75 (i)	10 (ii)	(ii) 420		(ii) Lukaszewicz (1975)
	HD	5 (i)				
	D₂	3 (i)				
		7 (i)				
Cs(5²D) + N₂ → Cs(?) + N₂			7 (ii)	(i) 315	ves;flq	(i) McGillis, Krause (1968a) [16]
				(ii) 420		(ii) Lukaszewicz (1975)
Cs(7²P½) + N₂ → Cs(?) + N₂		≈160 (i)		(15)		Antonov, Korchevoi (1976)
		110→210 (ii)	66			
Cs(7²P_{3/2}) + N₂ → Cs(?) + N₂		≈160 (i)		(i) 400–650	ves;flq	(i) Stara, Dubois, Krause (1974)
		46→100 (ii)		(ii) 400→600		(ii) Lukaszewicz (1975)
Cs(7²P½) + Ar → Cs(6D_{3/2}) + Ar; Cs(7²P_{3/2})		≤2×10⁻³ (i)		(i) 400–650	ves;flq	(i) Stara, Dubois, Krause (1974)
		4×10⁻³ (i)		(ii) 400→600		(ii) Lukaszewicz (1975)
Cs(7²D) + He → Cs(?) + He	Ar	2 (i)		450	ves;slf	Cuvellier et al. (1975)
	Xe	0.8 (i)				
	H₂	3.8 (i)				
		80 (ii)				
Cs(8S) + He → Cs(4²F) + He	Ar	8 (i)	(?)	450	ves;flq	Dmitrieva et al. (1977)
	Xe	12 (i)	(?)			
	H₂	19 (i)	(?)			
	Cs⁰	850 (ii)		525	(i) ves;slf	(ii) Krause (1975)

Table VI.1 (continued)

Reaction			Reference
$Cs(8^2P_{\frac{1}{2}}) + N_2 \to Cs(7^2D_{\frac{3}{2}}) + N_2$ H_2 He Ar Cs^0	(i) 1.8 (i) 4.7 (i) 1.4 (i) 0.3 (ii) 21	(i) 450 (ii) 420	ves;slf	(i) Rocchiccioli (1972) (ii) Pimbert (1972)
$Cs(8^2P_{\frac{1}{2}}) + N_2 \to Cs(7) + N_2$ $Cs(8^2P_{\frac{3}{2}})$	135 135	450	ves;flq	Dmitrieva et al. (1977)
$Cs(8^2P_{\frac{3}{2}}) + Cs^0 \to Cs(8S) + Cs^0$	120	525	ves;llf	Krause (1975)
$Mg(3^3P) + N_2 \to Mg(3^1S) + N_2$ H_2 CO He	12×10^{-3} 10×10^{-3} 28×10^{-3} $<1 \times 10^{-6}$	800	ves;llf	Blickensderfer, Breckenridge, Moore (1975)
$Sr(5^1P) + N_2 \to Sr(?) + N_2$ H_2 O_2 CO CO_2 H_2O Ar	(i) 16 (ii) 22 (i) 150 (i) 50 (i) 30 (ii) 67 (ii) <1	(i) 2700 (ii) ≈1950	fla;flq	(i) Jansen, Hollander, v. Helvoort (1977) (ii) Hollander et al. (1972)
$Cu(4^2P_{\frac{1}{2}}) + N_2 \to Cu^0 + N_2$ H_2 CO_2	14 22 50	—	ves;flq	Bleekrode, v. Benthem (1969)
$Cu(4^2P_{\frac{3}{2}}) + N_2 \to Cu^0 + N_2$ H_2 CO_2	19 23 36	—	ves;flq	Bleekrode, v. Benthem (1969)
$Cu(4^2D_{\frac{5}{2}}) + O_2 \to ?$ (5,13)	$k = 3.5 \times 10^{-12}$ (11,12) 1.1×10^{-12}	610 560	ves;phdis;llf	Trainor (1976)
$Tl(7^2S) + N_2 \to Tl(6^2P) + N_2$ H_2 O_2 CO CO_2 H_2O Ar He	20 0.1 41 42 102 5.5 <0.3 <0.3	1400	fla;flq	Jenkins (1968d)

Table VI.1 (continued)

$Tl(6^2D_{\frac{3}{2}}) + N_2 \to Tl(?) + N_2$	94 (i)	40 (ii)	[(i) 700; (ii) ≈750]	ves;flq	[(i) Gelbhaar, Hanle (1976); (ii) Wade, Czajkowski, Krause (1978)]
H_2	43 (i)				
$Tl(8^2S) + N_2 \to Tl(?) + N_2$		110	≈750	ves;flq	Wade, Czajkowski, Krause (1978)
$Tl(7^2S)$		10			
$Pb(7^3P_1) + N_2 \to Pb(?) + N_2$	18 (i)	18 (ii)	[(i) 1400; (ii) 900]	[(i) fla;flq; (ii) ves;flq]	[(i) Jenkins (1969); (ii) Mandl, Hao-Lin Chen (1976)]
H_2	1.2 (i)	<0.05 (ii)			
O_2	47 (i)				
CO	41 (i)	27 (ii)			
CO_2	90 (i)	63 (ii)			
H_2O	25 (i)				
Ar	<5 (i)	<0.3 (ii)			
He	<2 (i)				
$Hg(6^3P_0) + N_2 \to Hg^0 + N_2$	<1×10⁻³		300	ves;lif	Horiguchi, Tsuchiya (1977)
H_2	6.0				
O_2	35				
CO	1.8				
CO_2	0.1				
$Hg(6^3P_1) + N_2 \to Hg(?) + N_2$	3.0 (i)⁽³⁾	0.7 (ii)	[(i) 1400; (ii) 300; (iii) –]	[(i) fla;flq; (ii) ves;lif; (iii) ves;flq]	[(i) Jenkins, D.R. (unpubl.)⁽⁴⁾; (ii) Deech, Pitre, Krause (1971); (iii) Gleditsch, Michael (1975); Michael, Suess (1974)]
	16 (i)⁽⁵⁾	25 (ii)			
O_2		60 (ii)			
CO	16 (i)	22 (ii)			
CO_2	20 (i)	10 (ii)			
$Hg(6^1D_2) + N_2 \to Hg(6^3D_2) + N_2$		27	300	ves;slf	Lecler, Laniepce (1976)
Ar		12			
$Hg(6^3D_1) + N_2 \to Hg(6^1D_2) + N_2$		100 ⁽¹⁸⁾	–	ves;slf	Barrat-Rambosson, Kucal (1978)
$Cd(5^3P_1) + N_2 \to Cd(5^1S) + N_2$⁽³⁾		1.7	≈500	ves;flq	Czajkowski, Krause (1974)
$Cd(5^1P_1) + N_2 \to Cd(?) + N_2$⁽¹⁷⁾		50	≈550	ves;lif	Morten et al. (1974)
H_2		11			
CO		140			
CO_2		150			
Ar		1.1			
He		0.3			

Table VI.1 (continued)

Reaction	k	T	Method	Reference
$Cd(5^1P_1) + N_2 \to Cd(5^3P_J) + N_2$ (17)	12	720	ves; slf	Morozov, Sosinskii (1973)
$Cd(6^1D_2) + Ar \to Cd(6^3D_2) + Ar$ / $Cd(7^3P_2)$	1.7 / 9	500	ves; slf	Gordeev, Shevtsov (1977)
$Zn(4^3P_1) + N_2 \to Zn(4^1S) + N_2$ (19)	0.19 (i) 0.37 (ii)	(i) ≈ 600	ves; flq	(i) Czajkowski, Krause (1976)
H_2 (5)	14 (ii)	(ii) 575		(ii) Yamamoto et al. (1976, 1980)
$Bi(6^2D_{3/2}) + N_2 \to ?$	$< 2 \times 10^{-15}$	300	ves; phdis; lif	Bevan, Husain (1976)
H_2	8×10^{-14}			
CO	7.5×10^{-15}			
CO_2	$< 4 \times 10^{-15}$			
He	5×10^{-16}			
$Bi(6^2D_{3/2}) + N_2 \to ?$ (5,13) $k =$	$< 5 \times 10^{-15}$	300	ves; phdis; lif	Bevan, Husain (1976)
H_2	5.5×10^{-12}			
CO	5.4×10^{-13}			
CO_2	2.1×10^{-13}			
He	4×10^{-16}			
$Bi(6^2D_{5/2}) + N_2 \to ?$ (5) $k =$	$0 \to 1.4 \times 10^{-15}$	300→550	ves; phdis; lif	Trainor (1977)
H_2	$7.3 \times 10^{-15} \to 2.9 \times 10^{-14}$			
O_2	$3.7 \times 10^{-13} \to 3.7 \times 10^{-12}$			
CO	$0 \to 1.3 \times 10^{-13}$			
CO_2	$0 \to 3.0 \times 10^{-15}$			
$Bi(6^2D_{5/2}) + N_2 \to ?$ (5,13) $k =$	$0 \to 2.8 \times 10^{-14}$	300→550	ves; phdis; lif	Trainor (1977)
H_2	$1.0 \times 10^{-11} \to 2.2 \times 10^{-11}$			
O_2	$2.3 \times 10^{-11} \to 2.5 \times 10^{-11}$			
CO	$4.7 \times 10^{-13} \to 1.8 \times 10^{-12}$			
CO_2	$2.4 \times 10^{-14} \to 9.2 \times 10^{-14}$			
$Si(3^1D_2) + H_2 \to ?$ (5) $k =$	8×10^{-11}	300	ves; phdis; lif	Husain, Norris (1978)
O_2	2.3×10^{-11}			
He	$\leqslant 10^{-15}$			
$Si(3^1S_0) + H_2 \to ?$ (5) $k =$	$\leqslant 1 \times 10^{-14}$	300	ves; phdis; lif	Husain, Norris (1978)
O_2	1.5×10^{-11}			
He	$\leqslant 1.3 \times 10^{-15}$			
$B(2s^23s; {}^2S) + Xe \to B^0 + Xe$	4	400	ves; flq	Hannaford, Lowe (1977)
$B(2s2p^2; {}^2D) + Xe \to B^0 + Xe$	5	400	ves; flq	Hannaford, Lowe (1977)

Table VI.1 (continued)

	$\langle\sigma_{21}\rangle(\text{Å}^2)$ (6)**	Mixing $\langle\sigma_{12}\rangle(\text{Å}^2)$ (6)**			
$Na(3^2P_{\frac{3}{2}}) + N_2 \rightleftharpoons Na(3^2P_{\frac{1}{2}}) + N_2$	76 (i)	144 (i) $\Big\{$ 152 (iv) $k = (12 \to 17) \times 10^{-10}$ (iv)* (11)			[(i) Stupavsky,Krause (1968) (ii) Pitre, Krause (1967)]
H_2	42 (i)	80 (i)			
HD	44 (i)	84 (i)			
D_2	52 (i)	98 (i)			
CO_2		36 (iv)			
CH_4	77 (iii)	148 (iii)			
He	45 (ii)	86 (ii) 107 (v)			
Ne	35 (ii)	$\Big\{$ 67 (ii) 78 (v) 78 (vi) 69 (vi)	$\Big\}$	[(i)-(iii)] ≃ 400 [(v),(vi)] $\bar{\sigma} \simeq 0.8$ (iv) $\bar{\sigma} \simeq 0.8$ (iv)* $\bar{\sigma} = 0.3 \to 1.4$	ves; slf	[(iii) Stupavsky,Krause (1969) (iv) Apt, Pritchard (1979) (v) Gay, Schneider (1976) (vi) Jordan, Franken (1966)]
Ar	56 (ii) 72 (iv)	110 (ii) 129 (iv) 130 (v) 124 (iv)* (11) $k = (10 \to 17) \times 10^{-10}$			
Kr	44 (ii)	$\Big\{$ 85 (ii) 114 (iv) 112 (v) 131 (vi)			
Xe	46 (ii) 54 (iv)	$\Big\{$ 90 (ii) 100 (iv) 105 (v) 119 (vi)			
$Na(3^2P_{\frac{3}{2}}) + Na^0 \rightleftharpoons Na(3^2P_{\frac{1}{2}}) + Na^0$	283 (i)	532 (i)	550	ves; slf	[(i) Pitre, Krause (1968) (ii) Seiwert (1956)]
K^0	45 (ii)	65 (ii)			
$K(4^2P_{\frac{3}{2}}) + N_2 \rightleftharpoons K(4^2P_{\frac{1}{2}}) + N_2$	100 (i)	190 (i)			[(i) Lijnse,Hornman (1974) (ii) Lijnse, P.L. (unpubl.) (iii) Lijnse (1974a)]
O_2	40 (i)	75 (i)			
CO_2	335 (ii)	640 (ii)	1720	fla; slf	
H_2O	50 (i)	100 (i)			
He	55 (iii)	105 (iii)			
Ar	15 (iii)	30 (iii)			
$K(4^2P_{\frac{3}{2}}) + N_2 \rightleftharpoons K(4^2P_{\frac{1}{2}}) + N_2$	66 (i)	100 (i)			[(i) McGillis,Krause (1968) (ii) Chapman, Krause (1966)]
H_2	53 (i)	76 (i)			
HD	49 (i)	74 (i)	370	ves; slf	
D_2	50 (i)	72 (i)			
He	41 (ii)	59 (ii)			
Ar	22 (ii)	37 (ii)			

Table VI.1 (continued)

Process				Reference
$K(4^2P_{3/2}) + K^0 \rightleftharpoons K(4^2P_{1/2}) + K^0$	250 (i)	370 (i)		ves; slf	(i) Chapman, Krause (1966)
Rb^0	175 (ii)	260 (ii)			(ii) Hrycyshyn, Krause (1969)
$K(4^2P_{3/2}) + He \rightleftharpoons K(4^2P_{1/2}) + He$		$\begin{bmatrix} 0 \rightarrow 200 \text{ (i)} \\ 80 \rightarrow 130 \text{ (ii)} \end{bmatrix}$ (?) (?)	370	beam; slf	Anderson et al. (1976); Cuvellier et al. (1979)
$Rb(5^2P_{3/2}) + N_2 \rightleftharpoons Rb(5^2P_{1/2}) + N_2$	60 (i)	99 (i)	$\begin{bmatrix} (i) \bar{v} = 1.0 \rightarrow 3.5 \\ (ii) \bar{v} = 2.0 \rightarrow 4 \end{bmatrix}$		
H_2	>30 (i)	>50 (i)			(i) Lijnse, Zeegers, Alkemade (1973)
O_2	40 (i)	66 (i)		fla; slf	(ii) Lijnse, P.L. (unpubl.)
CO_2	69 (ii)	112 (ii)	1720		(iii) Lijnse (1974a)
H_2O	73 (i)	120 (i)			
He	11 (iii)	18 (iii)			
$Rb(5^2P_{3/2}) + N_2 \rightleftharpoons Rb(5^2P_{1/2}) + N_2$	23	16			
H_2	15	11			
HD	25	18	340	ves; slf	Hrycyshyn, Krause (1970)
D_2	30	22			
CH_4	42	30			
$Rb(5^2P_{3/2}) + He \rightleftharpoons Rb(5^2P_{1/2}) + He$	$0.1 \rightarrow 0.7$	$0.06 \rightarrow 1$	$300 \rightarrow 900$	ves; slf	Gallagher (1968)
Ar	$8 \times 10^{-4} \rightarrow 8 \times 10^{-3}$	$5 \times 10^{-4} \rightarrow 1 \times 10^{-2}$			
$Rb(5^2P_{3/2}) + Rb^0 \rightleftharpoons Rb(5^2P_{1/2}) + Rb^0$	68	53	360	ves; slf	Rae, Krause (1965)
$Rb(6^2P_{3/2}) + N_2 \rightleftharpoons Rb(6^2P_{1/2}) + N_2$	70 (i)	107 (i)			
H_2	26 (i)	41 (i)			
HD	27 (i)	42 (i)			(i) Siara, Krause (1973)
D_2	27 (i)	42 (i)	430	ves; slf	(ii) Siara, Hrycyshyn, Krause (1972)
CH_4	24 (i)	38 (i)			
He	19 (ii)	29 (ii)			
Ar	15 (ii)	24 (ii)			
$Rb(6^2P_{3/2}) + Rb^0 \rightleftharpoons Rb(6^2P_{1/2}) + Rb^0$	163	245	430	ves; slf	Pace, Atkinson (1974)
$Cs(6^2P_{3/2}) + N_2 \rightleftharpoons Cs(6^2P_{1/2}) + N_2$	55	70			
O_2	<90	<115	1720	fla; slf	Lijnse, Zeegers, Alkemade (1973a)
H_2O	200	250			

Chapt. VI   EXCITATION AND DE-EXCITATION OF METAL SPECIES

Table VI.1 (continued)

Reaction				Reference
$Cs(6^2P_{3/2}) + N_2 \rightleftarrows Cs(6^2P_{1/2}) + N_2$	25 (i)	4.7 (i)		[(i) McGillis, Krause (1968a)
	{44 (i)	{6.7 (ii)		(ii) Lukaszewicz (1975)
	{30 (ii)	{9 (ii)		(iii) McGillis, Krause (1969)]
H_2				
D_2	28 (i)	4.2 (i)		
HD	32 (i)	4.8 (i)		
CH_4	20→18 (iii)	40 (iii)	(i) 315 (ii) 420 (iii) 300→450	ves; slf
$Cs(6^2P_{3/2}) + He \rightleftarrows Cs(6^2P_{1/2}) + He$	$3\times10^{-4} \to 4\times10^{-3}$	$4\times10^{-5} \to 3\times10^{-3}$	300→900	ves; slf	Gallagher (1968)
$Cs(6^2P_{3/2}) + Ar \rightleftarrows Cs(6^2P_{1/2}) + Ar$	5.2×10^{-5}	1.6×10^{-5}	300	ves; slf	Czajkowski, McGillis, Krause (1966)
$Cs(6^2P_{3/2}) + Cs^0 \rightleftarrows Cs(6^2P_{1/2}) + Cs^0$	31	6.4	350	ves; slf	Czajkowski, Krause (1965)
$Cs(7^2P_{3/2}) + N_2 \to Cs(7^2P_{1/2}) + N_2$	18→34 (i)	19→45 (i)	(i) 390→540	ves; slf	[(i) {Siara, Dubois, Krause (1974); Krause, Siara, Dubois (1975)}
	{22→25 (ii)	≈20 (ii)	(ii) 410→500		(ii) Lukaszewicz (1975)
	{20→16 (iii)	12 (iii)	(iii) 450		(iii) Cuvellier et al. (1973)]
H_2					
He	11				
Ar	0.10	0.12			
$Cs(7^2P_{3/2}) + Cs^0 \rightleftarrows Cs(7^2P_{1/2}) + Cs^0$	107	121	440	ves; slf	Pace, Atkinson (1974a)
$Cs(8^2P_{3/2}) + N_2 \to Cs(8^2P_{1/2}) + N_2$	$<2\times10^{-3}$ (i) 5×10^{-4} (ii)	55 (i) 45 (iii)	(i) 450	ves; slf	[(i) Rocchiccioli (1972)
H_2	0.04 (i)	80 (i)	(ii) 420-620		(ii) Pimbert, Rocchiccioli, Cuvellier, Pascale (1970)
O_2	28 (i)	34 (i)	(iii) 450		(iii) Dmitrieva et al. (1977)]
He	0.01 (i)	5.5 (ii)			
Ar	$<2\times10^{-3}$ (i)				
CO	$<2\times10^{-3}$ (i)				
CO_2					
(He, Ne, Ar)					
$Tl(6^2P_{1/2}) + N_2 \to Tl(6^2P_{3/2}) + N_2$	5.4 (i) 4.5 (ii)		≈600	[(i) ves; phdis; llf (ii) ves; llf]	[(i) Bellisio, Davidovits (1970) (ii) Aleksandrov, Baranov, Kulyasov (1978)]
$Tl(6^2P_{1/2}) + Tl(6^2P_{1/2}) \to Tl(6^2P_{3/2}) + Tl(6^2P_{1/2})$ H_2		71 87	[(i) ≈900 (ii) 660]	ves; slf	[(i) Picket, Anderson (1969) (ii) Aleksandrov, Baranov, Kulyasov (1978)]
$Tl(6^2D_{5/2}) + N_2 \to Tl(6^2D_{3/2}) + N_2$ H_2			700	ves; slf	Gelbhaar, Hanle (1976)

Table VI.1 (continued)

Reaction	k	T (K)	Method	Reference
$Hg(6^3P_1) + Xe \rightarrow Hg(6^3P_0) + Xe$	$<2\times10^{-3}$	300	ves;lif	Deech, Pitre, Krause (1971)
$Hg(6^3P_1) + N_2 \rightarrow Hg(6^3P_0) + N_2$	0.77	300	ves;lif	Horiguchi, Tsuchiya (1977)
$\quad H_2$	<0.3			
$\quad O_2$	<6			
$\quad CO$	22			
$\quad CO_2$	0.19			
$Hg(6^3P_2) + N_2 \rightarrow Hg(6^3P_1) + N_2$	19	(300)	beam	Krause, Datz (1976)
$Hg(6^3D_2) + N_2 \rightarrow Hg(6^3D_1) + N_2$	22 (i)	(i) 300	ves;slf	(i) Lecler, Laniepce (1976)
$\quad Ar$	6.0 (i)	(ii) —		(ii) Barrat-Rambosson, Kucal (1978)
$Hg(6^3D_2) + N_2 \rightarrow Hg(6^3D_3) + N_2$	42	300	ves;slf	Lecler, Laniepce (1976)
$\quad Ar$	23			
$Hg(6^3D_1) + N_2 \rightarrow Hg(6^3D_3) + N_2$	33	—	ves;slf	Barrat-Rambosson, Kucal (1978)
$Sn(5^3P_1) + N_2 \rightarrow Sn(5^3P_0) + N_2$	$k < 3\times10^{-13}$ (ii)	300	ves;phdis;lif	(i) Foo, Wiesenfeld, Husain (1975)
$\quad H_2$	$<2\times10^{-12}$ (i)			(ii) Foo et al. (1976)
$\quad O_2$	8×10^{-11} (i)			
$\quad CO$	1.7×10^{-12} (i)			
$\quad CO_2$	3×10^{-13} (ii)			
$\quad Ar$	$<5\times10^{-16}$ (i)			
$Pb(6^3P_1) + N_2 \rightarrow Pb(6^3P_0) + N_2$	$k=1.7\times10^{-15}$ (ii)	300	ves;phdis;lif	(i) Husain, Littler (1974)
$\quad H_2$	3×10^{-15} (i)			(ii) Ewing, Trainor, Yatsiv (1974)
$\quad O_2$	4.5×10^{-11} (i)			
$\quad CO$	2×10^{-12} (i)			
$\quad Ar$	$<2\times10^{-16}$ (i)			
$Pb(7^3P_1) + N_2 \rightarrow Pb(7^3P_0) + N_2$	0.20	900	ves;slf	Bolshov, Zybin, Koloshnikov, (1979)
$\quad H_2$	1.5×10^{-3}			
$\quad H_2O$	1.5			
$\quad Ar$	$<2\times10^{-4}$			

Table VI.1 (continued)

Electronic-excitation transfer		(ΔE; eV)	(column headings as on page 604)		
$K(4^2P_{\frac{1}{2}}) + Rb^0 \to K^0 + Rb(5^2P_{\frac{3}{2}})$		40 (i) 5.3 (ii)			
	$Rb(5^2P_{\frac{1}{2}})$	2.7 (i) 2.3 (ii)			
$K(4^2P_{\frac{3}{2}})$	$Rb(5^2P_{\frac{3}{2}})$	27 (i) 5.5 (ii)			
	$Rb(5^2P_{\frac{1}{2}})$	1.9 (i) 2.5 (ii)			
$Rb(5^2P_{\frac{1}{2}}) + Cs^0 \to Rb^0 + Cs(6^2P_{\frac{3}{2}})$		1.5	≈ 400	ves ; sens	(i) Hrycyshyn, Krause (1969a)
	$Cs(6^2P_{\frac{1}{2}})$	0.5			(ii) Stacey, Zare (1970)
$Rb(5^2P_{\frac{3}{2}})$	$Cs(6^2P_{\frac{3}{2}})$	0.9	≈ 400	ves ; sens	Czajkowski, McGillis, Krause (1966)
	$Cs(6^2P_{\frac{1}{2}})$	0.3			
$Rb(12^2P) + Rb^0 \to Rb(?) + Rb(?)$		$\approx .3 \times 10^3$	460	ves ; lif	Gounand, Fournier, Berlande (1977)
$Rb(22^2P)$		$.6 \times 10^4$			
$Hg(6^3P_1) + Na^0 \to Hg^0 + Na(10^2S)$		3.4 (-0.07)	500	ves ; sens	Czajkowski, Skardis, Krause (1973a)
	$Na(9^2S)$	38 (0.02)			
	$Na(8^2S)$	4.9 (-0.05)			
	$Na(9^2P)$	8.1 (0.06)			
	$Na(8^2P)$	9.4 (-0.01)			
	$Na(7^2P)$	2.9 (-0.10)			
	$Na(9^2D)$	8.1 (0.09)			
	$Na(8^2D)$	31 (0.04)			
	$Na(7^2D)$	4.5 (-0.03)			
$Hg(6^3P_1) + Tl^0 \to Hg^0 + Tl(8^2S)$		3.0	700 - 800	ves ; sens	Wade, Czajkowski, Krause (1978a)
	$Tl(6^2D_{\frac{3}{2}})$	0.3			
	$Tl(7^2S)$	0.03			
$Hg(6^3P_1) + Tl^0 \to Hg^0 + Tl(8^2S)$		5.8	1120	ves ; sens	Kraulinya, Lezdin (1977)
	$Tl(6^2D_{\frac{3}{2}})$	9.2			
	$Tl(7^2S)$	≤ 0.1			
$Hg(6^3P_1) + Cd^0 \to Hg^0 + Cd(5^3P_1)$		4.3×10^{-2} (i)	≈ 500 (i)	ves ; sens	(i) Czajkowski, Krause (1974)
		1.0 (ii)	≈ 800 (ii)		(ii) Morozov, Sosinskii (1973)
		0.34 (iii)	530 (iii)		(iii) Chéron (1975)
$Hg(6^3P_1) + Zn^0 \to Hg^0 + Zn(4^3P_1)$		0.02	≈ 800	ves ; sens	Morozov, Sosinskii (1973)

Table VI.1 (continued)

	Excitation by electron impact [10]		
		Kinetic energy (eV) of relative motion	
$Li^0 + e^- \to Li(2^2P) + e^-$	$0 \to 20$	$\approx 1.9 \to 6$	beam	Aleksakhin, Zapesochnyi (1967)
$Na^0 + e^- \to Na(3^2P) + e^-$	(i) $0 \to 20$ (ii) $0 \to 30$	(i) $\approx 2.1 \to 5$ (ii) $\approx 2.1 \to 5$	beam	(i) Zapesochnyi, Shimon (1965) (ii) Enemark, Gallagher (1972)
$K^0 + e^- \to K(4^2P) + e^-$	(i) $0 \to 30$ (ii)	$\approx 1.6 \to 2.0$	beam	(i) Chen, Gallagher (1978) (ii) Korchevoi, Przonski (1967)
$Rb^0 + e^- \to Rb(5^2P) + e^-$	(i) $0 \to 40$ (ii) $0 \to 80$	$\approx 1.6 \to 2.0$	beam	ditto
$Cs^0 + e^- \to Cs(6^2P) + e^-$	(i) $0 \to 40$ (ii) $0 \to 60$	$\approx 1.4 \to 1.8$	beam	ditto
$Mg^0 + e^- \to Mg(3^1P) + e^-$	$0 \to 3.5$	$\approx 4.3 \to 5.0$	beam	Leep, Gallagher (1975)
$Ca^0 + e^- \to Ca(4^1P) + e^-$	(i) $0 \to 25$ (ii) $0 \to 12$	(i) $\approx 2.9 \to 10$ (ii) $\approx 2.9 \to 9$	beam	(i) Ehlers, Gallagher (1973) (ii) Garga et al. (1974)
$Ba^0 + e^- \to Ba(6^1P) + e^-$	$0 \to 21$	$2.2 \to 6.0$	beam	Aleksakhin et al. (1975)

Chapt. VI   EXCITATION AND DE-EXCITATION OF METAL SPECIES

SPECIAL NOTES TO TABLE VI.1

(1)   Measurements made under thermalized conditions.

(2)   Recalculated from original data with the aid of the more recent value for the optical lifetime: $\tau_r = 8 \times 10^{-7}$ s reported by Wiese, Smith and Miles (1969).

(3)   For Hg, a mixing transition to the $^3P_0$ state rather than a quenching transition to the $^1S_0$ ground state may occur in the case of N_2 and CO; in the case of CO_2 direct quenching to the ground state is predominant (see Vikis, Torrie and Le Roy 1972). For $Cd + N_2$, mixing collisions involving the other ^3P multiplet states may also occur.

(4)   Personal communication in Krause (1975).

(5)   Chemical quenching may contribute noticeably to (some of) the quenching cross section(s) listed in this group (see Sect. VI.3c-2).

(6)   The phenomenological cross sections for the exo-ergic mixing process are denoted by $\langle \sigma_{21} \rangle$, those of the reverse, endo-ergic process by $\langle \sigma_{12} \rangle$.

(7)   The relative total cross sections obtained in this crossed-beam experiment were reported to be normalized with respect to the absolute phenomenological mixing cross section measured in a vapour vessel by Krause (1966) (i), and Chapman and Krause (1966) (ii), respectively.

(8)   The energy discrepancies, ΔE, between the excitation levels of donor and acceptor atom are also given.

(9)   Cross sections were calculated from the rate constants reported.

(10) For an extensive bibliography see Kieffer (1976), and Beaty and Gallagher (1976); for simple analytical formulas see Vriens and Smeets (1980).

(11) Only the reported bimolecular rate constants k $(= \sigma \bar{g})$, expressed in $cm^3 s^{-1}$, are listed.

(12) The first mentioned value refers to photodissociation experiments with CuI; the second value refers to similar experiments with CuCl.

(13) Mixing processes might be involved.

(14) According to Humphrey *et al.* (1978) the initially populated D states are rapidly mixed with the close-lying, higher l-states so that the quenching cross sections measured refer to a manifold of $(l \geq 2)$-states.

(15) This measurement was performed by observing the quenching of the emission from the considered state after excitation in a gas-discharge. The temperature of the thermostat was reported to correspond to a saturated Cs vapour density of 1.2×10^{14} atoms per cm^3.

(16) Recalculated from the original data while using the transition probability reported by Link (1966). The latter value was also used by Lijnse, Zeegers and Alkemade (1973, 1973a).

(17) For recent studies of the quenching of $Cd(5^1P_1)$ by molecules at $T \simeq 450\,K$ see Breckenridge and FitzPatrick (1976), Breckenridge and Renlund (1978), and Breckenridge, Donovan and Kim Malmin (1979).

(18) Umemoto, Tsunashima and Sato (1979) and Breckenridge and Renlund (1979) obtained much lower values of about 0.03 $Å^2$ at $T \simeq 500\,K$.

(19) Breckenridge and Renlund (1979a) have measured quenching cross sections of 19, 12 and 30 $Å^2$ for N_2, H_2 and CO at $T = 625\,K$.

(20) Excitation experiments combined with microreversibility.

PHYSICAL EXCITATION AND DE-EXCITATION PROCESSES   Sect. VI.2c

2c   RADIATIVE (DE-)EXCITATION OF ELECTRONIC STATES

Excitation and de-excitation of an electronic state can occur also through the absorption and emission, respectively, of a photon at a discrete frequency $\nu_0 = c/\lambda_0$, according to the process: $M + h\nu_0 \rightleftharpoons M^*$. For simplicity, we consider here the case where M represents a metal atom in the ground state (in this subsection we have dropped the superscript o), while there is only one excited state, denoted by M^*. The probability of a spontaneous downward optical transition (per second and per excited atom) depends on neither the number of atoms present nor on the flame conditions or radiant density. On the other hand, the radiative excitation rate per second and per atom in the ground state as well as the induced-emission rate do depend on the spectral radiant density ρ_λ at $\lambda \approx \lambda_0$. This introduces a complication which has no counterpart in the discussion of collisional excitation rates. For, since emission by the excited atoms contributes to the radiant density in the flame, we expect ρ_λ to depend on $[M^*]$. But $[M^*]$, in turn, will depend markedly on ρ_λ when the radiative excitation rate is at least of comparable order to the collisional excitation rate. This leads, as we shall see presently, to a set of coupled equations which can be solved, in general, by an iteration procedure.

Two distinct cases will be considered. In the first, the radiation field produced by the atomic vapour in the flame itself is the only source of radiative excitation (the coloured flame is the only source of radiation). In the second case, the radiation field is produced, for the main part, by an external light source having spectral components at $\lambda \approx \lambda_0$. The latter situation applies in flame fluorescence spectroscopy. Only steady-state solutions will be considered here.

2c-1   *GENERAL EXPRESSIONS FOR THE RADIATIVE (DE-)EXCITATION RATES*

Let k_1 be the radiative excitation-rate constant (expressed per second and per atom M in the ground state), so that the rate, v_1, of *radiative excitation* (i.e., the number of upward transitions per second and per unit volume) is given by

$$v_1 = k_1 [M] . \qquad (VI.9)$$

Using Eq. II.173 for the absorbed radiant energy and writing: $\Phi(\lambda) = \Phi_\lambda d\lambda = \rho_\lambda c O d\lambda$ (cf. Eq. II.175) we find for v_1, after dividing by $O dx$ and dt,

$$v_1 = \int (\rho_\lambda / h\nu_0) c k(\lambda) d\lambda , \qquad (VI.10)$$

where the integration extends over the relatively small spectral interval in which the absorptivity $k(\lambda)$ differs noticeably from zero. The spectral volume density of the radiation field is denoted by ρ_λ (energy per unit volume and per unit wavelength interval), while c is the velocity of light. Substituting $h\nu_0 = hc/\lambda_0$ and

619

Chapt. VI EXCITATION AND DE-EXCITATION OF METAL SPECIES

defining the weighted average of wavelength-dependent ρ_λ by [†]

$$\bar{\rho}_\lambda \equiv \int \rho_\lambda(\lambda) k(\lambda) d\lambda \bigg/ \int k(\lambda) d\lambda , \qquad (VI.11)$$

we find for v_1

$$v_1 = (\lambda_0/h) \bar{\rho}_\lambda \int k(\lambda) d\lambda .$$

Using Eq. II.183 with $n_p \hat{=} [M]$, we find from this for the rate constant k_1 of radiative excitation

$$k_1 \equiv v_1 / [M] = (\pi e^2 / m_e hc^2) \lambda_0^3 f \bar{\rho}_\lambda . \qquad (VI.12a)$$

The oscillator strength f for absorption is related to the spontaneous transition probability A by Eq. II.168 so that Eq. VI.12a can also be written, with g^* and g being the statistical weight of the excited and ground state, respectively, in the form

$$k_1 = (\lambda_0^5/c_3)(g^*/g) A \bar{\rho}_\lambda \quad \text{with} \quad c_3 \equiv 8\pi hc . \qquad (VI.12b)$$

It follows from Eq. VI.12a that the ratio of k_1 (or v_1) to its equilibrium value $(k_1)_e$ [or $(v_1)_e$] is given by [*]

$$k_1/(k_1)_e = v_1/(v_1)_e = \bar{\rho}_\lambda / \rho_{\lambda_0}^b . \qquad (VI.13)$$

Here $\rho_{\lambda_0}^b$ is the spectral radiant density at λ_0 inside a black body at flame temperature (see Eq. II.65a). When $\bar{\rho}_\lambda \neq \rho_{\lambda_0}^b$, *radiative disequilibrium* will exist, that is, the radiative excitation and de-excitation rates are not balanced (see Sect. VI.4c-2).

The rate v_{-1} of *radiative de-excitation* (i.e., the number of downward transitions per second and per unit volume) is the sum of the spontaneous-emission rate and the induced-emission rate

$$v_{-1} = A[M^*] + ([M^*]g/[M]g^*) v_1 . \qquad (VI.14)$$

[†] For a monochromatic radiation field having a spectral width very small compared with the atomic line width and tuned at λ_0, we have from Eq. VI.11:

$$\bar{\rho}_\lambda = \rho k(\lambda_0) / \int k(\lambda) d\lambda \equiv \rho/\Delta\lambda_{eff} ,$$

where ρ is the integrated radiant density and $\Delta\lambda_{eff}$ the effective atomic line width (cf. Eq. II.210). When the monochromatic field has an arbitrary wavelength λ, we have more generally: $\bar{\rho}_\lambda = \rho S_\lambda(\lambda)$ where $S_\lambda(\lambda)$ is the normalized spectral distribution of the atomic line (see Eq. II.207). For a (quasi-)continuous radiation field we have: $\bar{\rho}_\lambda = \rho_{\lambda_0}$, where ρ_{λ_0} is the spectral radiant density at the line centre.

[*] Greig (1965) has expressed the right-hand side of Eq. VI.13 in an effective emission factor of the radiating gas (no external light source present). The equation given for the effective emission factor erroneously contains the bandwidth of the monochromator. The values of $k_1/(k_1)_e$ calculated in Greig's paper are consequently considerably smaller than those found from Eq. VI.13.

PHYSICAL EXCITATION AND DE-EXCITATION PROCESSES   Sect. VI.2c

Here we have made use of Eq. II.158 for the relation between the absorption rate v_1 and the induced-emission rate.

Rather than these absolute rate expressions, the ratio of the radiative (de-)excitation to the collisional (de-)excitation rates will be of interest in flames. Let k'_2 and k'_{-2} be the unimolecular rate constants (in s^{-1}) for collisional excitation and quenching, respectively (see Sect. II.4a-1). The ratio of radiative to collisional de-excitation rates is then, when we neglect induced emission, found to be

$$v_{-1}/v_{-2} = A[M^*]/k'_{-2}[M^*] = A/k'_{-2} \ . \qquad (VI.15)$$

By using Eq. II.306 with $k' \hat{=} k'_{-2}$, this ratio can be expressed in terms of the fluorescence efficiency Y of resonance fluorescence, which can be directly found by experiment, according to

$$v_{-1}/v_{-2} = Y/(1-Y) \ . \qquad (VI.16)$$

In flames of 1 atm pressure and with nitrogen as diluent gas, the ratio $Y/(1-Y)$ is of the order 0.01 - 0.1 for the alkali resonance lines. In stoichiometric H_2-O_2-Ar flames, however, this ratio may exceed unity (see Sect. VI.2b).

When the molecular collision partners in the collisional (de-)excitation process are themselves in physical and chemical equilibrium, we have $k'_{-2} = (k'_{-2})_e$ and $k'_2 = (k'_2)_e$. Then it follows, through Eqs VI.15 and 16, that

$$v_{-1}/v_{-2} = (v_{-1})_e/(v_{-2})_e = Y/(1-Y) \qquad (VI.17)$$

irrespective of any deviation of $[M^*]$ from its Boltzmann equilibrium value $[M^*]_e$. Because of the principle of detailed balance, we also have:

$$(v_1)_e/(v_2)_e = (v_{-1})_e/(v_{-2})_e \ .$$

Combining this with Eq. VI.17, we obtain

$$(v_1)_e/(v_2)_e = (v_{-1})_e/(v_{-2})_e = Y/(1-Y) \ . \qquad (VI.18)$$

From this equation and Eq. VI.13, we find for the ratio of excitation rates $v_1/(v_2)_e$, or $k_1/(k_2)_e$, in the case of radiative disequilibrium,

$$v_1/(v_2)_e = \{Y/(1-Y)\} \bar{\rho}_\lambda/\rho_{\lambda 0}^b \ . \qquad (VI.19)$$

When the radiation field is produced by photon emission from the excited atoms in the flame itself, $\bar{\rho}_\lambda/\rho_{\lambda 0}^b$ can be, at most, equal to unity, if we suppose a uniform temperature distribution and equilibration of the flame molecules. This limit will be approached inside the flame in the range of high metal concentrations. A volume element in the interior is then surrounded by a hot vapour that is at

Chapt. VI   EXCITATION AND DE-EXCITATION OF METAL SPECIES

flame temperature and is optically thick at $\lambda \approx \lambda_0$ † and thus acts effectively as a black body (cf. Sect. II.3b-5). A volume element at the flame border practically receives black body radiation at $\lambda \approx \lambda_0$ from one hemisphere only. It receives no radiation at all from the outer hemisphere, so that $\bar{\rho}_\lambda$ is virtually equal to $\tfrac{1}{2}\rho_{\lambda_0}^b$ here. Consequently, we must expect that at the very flame border v_1 will be, at most, about $\tfrac{1}{2}(v_1)_e$ in the limit of high metal concentrations (see Eq. VI.13). On the other hand, in the limit of very low metal concentrations, $\bar{\rho}_\lambda$ will approach zero everywhere, and the relative contribution of radiative excitation will also vanish.

2c-2   *GENERAL DEPENDENCE OF RADIANT DENSITY ON EXCITED-STATE POPULATION (FLAME AS THE ONLY SOURCE OF RADIATION)*

At the end of the preceding subsection, the magnitude of the spectral radiant density was considered in some borderline cases only. Here we shall derive a more general expression that holds for arbitrary atom densities and at any place in a flame with assumed uniform distribution of temperature and metal content. In order to avoid uninteresting mathematical complications, we shall consider a simplified model of the flame, consisting of an infinite slab of gas with two parallel, plane boundaries at mutual distance d (see Fig. VI.5). The space-dependent quantities ρ_λ and $[M^*]$ are then only functions of the distance x to one of the flame boundaries. The spectral radiance B_λ inside the flame is, moreover, a function of the angle θ between the direction considered and the x axis. All these functions have a plane of symmetry at $x = \tfrac{1}{2}d$.

In the absence of an external (laser) source we can safely ignore induced emission in the following derivation. We then have to use consistently Wien's law for the black-body radiance (see Eq. II.71a and the pertinent discussion in Sect. II.5c-3).

We now have the relation (following Hooymayers and Alkemade 1966a)

$$\rho_\lambda(x) = c^{-1} \int B_\lambda(x,\theta)\, d\Omega, \tag{VI.20}$$

where $d\Omega$ represents an infinitesimal solid angle and the integration extends over all space directions. The integrand can be calculated if the spatial distribution of the excited-atom density in the flame is known. Consider the shaded volume in Fig. VI.5 which is part of an infinitesimal cone with solid angle $d\Omega$ and direction θ. The distance of this element to the flame surface, measured along the axis of this cone, is denoted by r', while its normal distance to the flame surface is

---

† In this context a cylindrical flame with metal vapour is *optically thick* for the spectral line considered if its radius exceeds $k(\lambda)^{-1}$ at any λ for which $k(\lambda)$ is larger than, say, 1% of its peak value $k(\lambda_0)$.

PHYSICAL EXCITATION AND DE-EXCITATION PROCESSES Sect. VI.2c

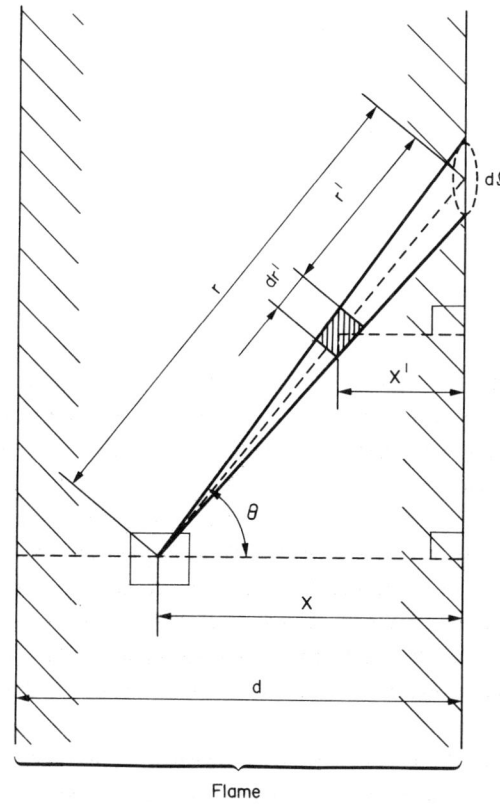

Fig. VI.5 Cross section of a slab of flame gas with plane, parallel boundaries at mutual distance d. The distance from the considered flame element to the right-hand boundary is denoted by x.
(From Alkemade and Zeegers 1971.)

denoted by x'. The variable r', for given value of θ, ranges from zero to $r = x/\cos\theta$. The width of this volume element, measured along the cone axis, is dr'. The contribution of the emission from this infinitesimal element to $B_\lambda(x,\theta)d\Omega$ at the top of the cone is given by:

$$k(\lambda)\,dr'\,P(x')\,B_\lambda^b\,d\Omega\,\exp\left[-(r-r')\,k(\lambda)\right].$$

Here use is made of Eq. II.294 in combination with the asymptotic relation: $\alpha(\lambda) \simeq k(\lambda)\,dr'$ (cf. Eq. II.261). The population factor $P(x')$ equals the ratio of the actual excited-state population to its equilibrium value. This factor is added in order to take into account the effect of the underpopulation of the excited state at position x' (due to radiative disequilibrium) on the emission intensity. [Radiative disequilibrium is here not expected to affect noticeably the ground-state population or $k(\lambda)$, which is proportional to it.] The relative spectral distributions of the same line seen in emission and in absorption are considered

Chapt. VI   EXCITATION AND DE-EXCITATION OF METAL SPECIES

here to be identical. For thermal radiators this is a consequence of Kirchhoff's law (see Eq. II.293). The applicability of this assumption under flame conditions has been discussed by Hooymayers (1966) and Hooymayers and Alkemade (1966a) and experimentally tested by Jansen (1976) and Jansen, Hollander and Alkemade (1977b). The exponential factor describes the fraction of radiant energy that is transmitted through a flame column with uniform absorptivity $k(\lambda)$ and length $(r-r')$ (cf. Eqs II.257 and 185). This is the distance that separates the radiating volume element dr' from the top of the cone, where ρ_λ is to be calculated.

By integration over r', we find from the above expression

$$B_\lambda(x,\theta)d\Omega = B_{\lambda_0}^b d\Omega \int_0^r P(r'\cos\theta)\exp\left[-(r-r')k(\lambda)\right]k(\lambda)\,dr',$$

or, by putting $k(\lambda)r' \equiv \tau'$,

$$B_\lambda(x,\theta)d\Omega = B_{\lambda_0}^b d\Omega \exp[-k(\lambda)r] \int_0^{k(\lambda)r} P\{\tau'\cos\theta/k(\lambda)\}\exp[\tau']\,d\tau'. \quad (VI.21)$$

Combining Eqs VI.11, 20 and 21, and using Eq. II.66, we finally obtain

$$\bar{\rho}_\lambda(x) = \frac{(\rho_{\lambda_0}^b/4\pi)\int k(\lambda)\,d\lambda \left\{\int d\Omega \exp[-k(\lambda)r] \int_0^{k(\lambda)r} P\{\tau'\cos\theta/k(\lambda)\}\exp[\tau']d\tau'\right\}}{\int k(\lambda)\,d\lambda} \quad (VI.22)$$

This expression holds for an infinite slab of flame gas with plane, parallel boundaries. It may also be applied, in good approximation, for a cylindrical flame of finite length if the flame is optically thick for the spectral line considered.

It is noted that Eq. VI.21 is in fact an integration of the equation for radiative transfer, as was discussed in Sect. II.5c-3 (see also Hearn 1963, and Richter 1968).

2c-3 *GENERAL DEPENDENCE OF EXCITED-STATE POPULATION ON RADIANT DENSITY AND ON FLUORESCENCE EFFICIENCY*

Equation VI.22 relates the weighted average of the spectral volume density, $\bar{\rho}_\lambda(x)$, at a given place x to the excited-state population described by the population factor $P(x')$ at any other place x' in the flame. Now we shall derive an independent expression that relates, conversely, the excited-state population at a given place to the value of $\bar{\rho}_\lambda$ at the same place and to the fluorescence efficiency Y. For generality we shall include induced emission too, which may be important when the flame is irradiated by a laser. Homogeneous line broadening is assumed throughout in the following discussion (see also Sect. II.5a-2).[†]

[†] The more complicated inhomogeneous-broadening case has been analyzed by Piepmeier (1972) for an atomic vapour in a flame irradiated by a monochromatic laser beam. An analysis of the intermediate case has been given by Piepmeier (1972a).

PHYSICAL EXCITATION AND DE-EXCITATION PROCESSES   Sect. VI.2c

In the steady state, the total excitation and de-excitation rates must be balanced at each point: $v_1 + v_2 = v_{-1} + v_{-2}$; here the index 1 refers to radiative transitions and 2 to collisional transitions. We have from Eqs VI.9 and 12b:

$$v_1 = [M](g^*/g) A \bar{\rho}_\lambda \lambda_0^5 / c_3 ,$$

where $c_3 \equiv 8\pi hc$, and from Eq. VI.14:

$$v_{-1} = A[M^*] + ([M^*]g/[M]g^*) v_1 .$$

The collisional rates can be expressed in the respective unimolecular rate constants k'_2 and k'_{-2} by: $v_2 = k'_2[M]$ and $v_{-2} = k'_{-2}[M^*]$. Inserting these expressions into the above balance equation we get after rearrangement of terms

$$[M]\{1 - ([M^*]g/[M]g^*)\}(g^*/g) A\bar{\rho}_\lambda \lambda_0^5 / c_3 + [M]k'_2 = [M^*](A + k'_{-2}) . \quad (VI.23)$$

We can transform the latter equation into

$$([M^*]g/[M]g^*)(A + k'_{-2} + A\bar{\rho}_\lambda \lambda_0^5/c_3) = (A\bar{\rho}_\lambda \lambda_0^5/c_3) + k'_2 g/g^* . \quad (VI.24)$$

Using $k'_2 = (k'_2)_e$ and $k'_{-2} = (k'_{-2})_e$ (see Sect. VI.2c-1) we can write:

$$k'_2 = (k'_{-2} g^*/g) \exp[-E/kT]$$

with $E = hc/\lambda_0$ (cf. Eq. II.85). The quenching rate constant k'_{-2} can be expressed in Y and A through: $k'_{-2} = A(1-Y)/Y$, while we have for the sum: $A + k'_{-2} = A/Y$ (cf. Eq. II.306). Inserting these substitutions into Eq. VI.24 we finally get:

$$\frac{[M^*]g}{[M]g^*} = \frac{c_3 \exp[-E/kT](1-Y)/Y\lambda_0^5 + \bar{\rho}_\lambda}{c_3/Y\lambda_0^5 + \bar{\rho}_\lambda} , \quad (VI.25)$$

with $c_3 \equiv 8\pi hc$. Essentially the same result has been obtained by Omenetto, Benetti, Hart, Winefordner and Alkemade (1973). It is easily verified from the latter equation that:

(i)   in thermodynamic equilibrium when $\bar{\rho}_\lambda$ conforms to the exact Planck law: $\rho_{\lambda 0}^b = (c_3/\lambda_0^5) \{\exp[hc/\lambda_0 kT] - 1\}^{-1}$ at flame temperature, the right-hand side of the equation reduces to: $\exp[-E/kT]$ in accordance with Boltzmann's law (see Eq. II.33b); note that this result is strictly independent of Y;

(ii)   in the absence of external radiation and for negligible thermal emission (i.e. low atom concentration) we have: $\bar{\rho}_\lambda \simeq 0$; $[M^*]/[M]$ is found to be $(1-Y)$ times the Boltzmann ratio and the atomic line intensity is thus infrathermal (see also Sect. VI.4c-2);

(iii) when $\bar{\rho}_\lambda$ is made very large, e.g., by the use of a strong laser beam tuned at λ_0, the ratio $[M^*]g/[M]g^*$ tends to unity, i.e. the optical transition becomes saturated (see Sect. II.5a-2); the net absorptivity then approaches zero (see Eq. II.191) and the excited-atom density becomes independent of $\bar{\rho}_\lambda$ and equal to $[M]_t g^*/(g+g^*)$ where $[M]_t \equiv [M]+[M^*]$; the density of ground-state atoms is then reduced to $[M]_t g/(g+g^*)$; for $g=g^*$, both $[M^*]$ and $[M]$ become equal to $\frac{1}{2}[M]_t$ (see Sect. VI.2c-5).

Even for conventional light sources as used in flame atomic fluorescence spectroscopy, $\bar{\rho}_\lambda$ may often be the dominant term in the numerator of the expression in Eq. VI.25. This is, in particular, true for ultraviolet resonance lines and/or in cool flames where the exponential factor will be very small. However, $\bar{\rho}_\lambda$ will be negligibly small when compared to the term $c_3/Y\lambda_0^5$ occurring in the denominator. This is tantamount to saying that induced emission has a negligible effect on the population of the excited state. It should be remembered that c_3/λ_0^5 exceeds the Planck spectral radiant density by a factor $\exp[E/kT]$ (see Eq. II.69), which amounts, e.g., to 10^4 for the Na-D doublet at $T \approx 2500\,\mathrm{K}$. The fluorescence efficiency ranges roughly from 0.01 to 0.5.

Disregarding $\bar{\rho}_\lambda$ in the denominator we find from Eq. VI.25 for the population factor

$$P \equiv [M^*]/[M^*]_e = 1-Y+Y(\bar{\rho}_\lambda/\rho_{\lambda_0}^b) . \qquad (VI.26)$$

In thermodynamic equilibrium, i.e. when $\bar{\rho}_\lambda = \rho_{\lambda_0}^b$, we do indeed find $P=1$, irrespective of the value of Y. For a strongly self-absorbed resonance line, this condition is approximately fulfilled in the interior of the flame, whereas $\bar{\rho}_\lambda = \frac{1}{2}\rho_{\lambda_0}^b$ at the flame border (see Sect. VI.2c-1). However, when $Y \ll 1$ and $\bar{\rho}_\lambda \leqslant \rho_{\lambda_0}^b$, P is practically equal to unity; under these conditions radiative (de-)excitation processes are obviously unimportant compared to collisional processes.[†] When these conditions are not fulfilled, a spatial variation of $\bar{\rho}_\lambda$ will induce also a spatial variation of P, even when the flame temperature is uniform (see Sect. VI.4c-2.sub iv).

Equation VI.26 is usually derived in a somewhat different form in the theory of imprisonment of resonance radiation (see Holstein 1947, 1951, and Richter 1968; see also Sect. II.5c-3).

2c-4 *APPROXIMATE SOLUTION WHEN THE FLAME IS THE ONLY SOURCE OF RADIATIVE EXCITATION*

When the radiation field is produced solely by photon emission from excited atoms inside the flame, calculation of $\bar{\rho}_\lambda$ by means of Eq. VI.22 requires a knowledge of the population factor P throughout the flame. The latter factor, however, is again connected to $\bar{\rho}_\lambda$ through Eq. VI.26. Solutions can be obtained in such

---

[†] In plasma physics one speaks then of a *collision-dominated* plasma (see Richter 1968).

PHYSICAL EXCITATION AND DE-EXCITATION PROCESSES   Sect. VI.2c

cases by an iteration procedure, carried out either by deriving closed analytical expressions in successive orders of approximation (see, e.g., Hooymayers 1966 and Hooymayers and Alkemade 1966a) or numerically by computer calculations. Here we shall derive, for a simple case, closed expressions for $\bar{\rho}_\lambda$ and P that hold to the first order in the atomic density [M], that is, for relatively low densities (compare also Hooymayers and Lijnse 1969).

In order to obtain $\bar{\rho}_\lambda$ in the first order in [M], we substitute in Eq. VI.22 for P its zero-th order value, $1-Y$, which follows from Eq. VI.26 for $\bar{\rho}_\lambda = 0$. Since this approximated P value is independent of position, we can apply Eq. VI.22 to flames with arbitrary boundaries (but still with uniform metal densities). The integral over τ' in Eq. VI.22, multiplied by $\exp[-k(\lambda)r]$, then yields $k(\lambda)r$ in the first order in [M] which is proportional to $k(\lambda)$ (see Eq. II.177). The approximation made here is equivalent to neglecting any loss in radiant energy due to self-absorption. Then Eq. VI.22 yields for $\bar{\rho}_\lambda(x)$

$$[\bar{\rho}_\lambda(x)]_{\mathrm{appr}} = \frac{\rho_{\lambda 0}^b (1-Y)\{\int k^2(\lambda)\,d\lambda / \int k(\lambda)\,d\lambda\} \int r\,d\Omega}{4\pi}, \quad (\mathrm{VI.27})$$

where x now stands for an arbitrary set of spatial coordinates. Defining the weighted average value of the absorptivity by $\overline{k(\lambda)} \equiv \int k^2(\lambda)\,d\lambda / \int k(\lambda)\,d\lambda$, and the average distance $\bar{r}(x)$ of the considered position x to the flame surface by $\bar{r}(x) \equiv \int r\,d\Omega/4\pi$ (see Fig. VI.5), we finally obtain

$$[\bar{\rho}_\lambda(x)]_{\mathrm{appr}} = \rho_{\lambda 0}^b (1-Y)\,\overline{k(\lambda)}\,\bar{r}(x). \quad (\mathrm{VI.28})$$

From Eqs VI.26 and 28 we immediately obtain the population factor to the first order in [M]

$$[P(x)]_{\mathrm{appr}} = (1-Y)\{1 + Y\overline{k(\lambda)}\,\bar{r}(x)\}. \quad (\mathrm{VI.29})$$

It follows upon closer inspection that $\bar{r}(x)$ is approximately constant, within 10% accuracy, and equal to about $1.6R$ for a relatively long cylindrical flame with radius R (see Hooymayers and Lijnse 1969). For a homogeneous cylindrical flame $[P(x)]_{\mathrm{appr}}$ is thus virtually independent of position.

In fact, $[1 - \overline{k(\lambda)}\,\bar{r}(x)]$ is a first-order expression for the probability, p, for a photon to escape from the flame without being re-absorbed. In Sect. II.5c-3, p was introduced in order to define an effective optical lifetime: $(\tau_r)_{\mathrm{eff}} = (pA)^{-1}$. Interpreting $A_{\mathrm{eff}} \equiv Ap$ as an effective optical transition probability, we can define an effective fluorescence efficiency Y_{eff} by inserting A_{eff}, instead of A, into Eq. II.306. We then find that the right-hand side of Eq. VI.29 for $P(x)$ becomes: $(1-Y_{\mathrm{eff}})$, to first order in $\overline{k(\lambda)}$ or in [M].

Chapt. VI   EXCITATION AND DE-EXCITATION OF METAL SPECIES

This example illustrates how the concept of effective optical transition probability or lifetime, borrowed from the theory of radiative transfer (see Sect. II.5c-3), may be used to describe the contribution of the radiative excitation process to the excited-state population.

Instead of expressing $\overline{k(\lambda)}$ theoretically in the parameters of the atomic line, one can also find its value experimentally from (self-)absorption measurements. At sufficiently low metal concentrations, the fraction of radiation lost due to self-absorption equals $\frac{1}{2}l\overline{k(\lambda)}$ in first-order approximation (see Eqs II.274 and 295). Here l is the thickness of the flame along the line of observation. This fractional loss, which is proportional to [M], can be found experimentally from the initial deviation of the curve-of-growth from its low-density asymptote. The value of $l\overline{k(\lambda)}$ can also be determined by the method of flame duplication (see Sect. II.5c-3). Another method of determining $l\overline{k(\lambda)}$ is based on the apparent decrease in the efficiency of resonance fluorescence, caused by self-absorption, when the metal concentration in the flame is enhanced (see Hooymayers and Lijnse 1969).

Some consequences of the above analysis for the total radiance and the spectral shape of a self-absorbed emission line, as well as for the defect in line-reversal temperature, will be considered in Sect. VI.4c.

2c-5   *APPROXIMATE SOLUTION FOR A FLAME IRRADIATED BY AN EXTERNAL LIGHT SOURCE*

Case of conventional light sources (no saturation). We suppose that the flame is irradiated by a conventional light source. The radiation field at $\lambda \approx \lambda_0$ inside the flame is then composed of photons emitted by the light source as well as by the atomic vapour in the flame. The latter contribution to the radiation field depends on the population of the excited state (see Eq. VI.22), which, in turn, may depend on the total radiant density (see Eq. VI.26). It is not easy to solve from these equations the position-dependent radiant density and excited-state population in the general case. In the limiting case: $Y \ll 1$ or $[M] \to 0$, however, the solution is again simple (see also Alkemade and Zeegers 1971).

We shall here derive an expression for the population factor which holds for arbitrary Y value in the limiting case when $[M] \to 0$. Then ρ_λ is practically everywhere equal to the spectral density of the unweakened external light beam, which we shall suppose to be uniform throughout the illuminated flame part. Two extreme cases will here be considered in regard to the spectral distribution of ρ_λ.

(i)   Case of a light source emitting a *spectral (quasi-)continuum* with constant ρ_λ equal to ρ_{λ_0} for all wavelengths $\lambda \approx \lambda_0$ where $k(\lambda)$ differs noticeably from zero. Substituting in Eq. VI.26 $\overline{\rho_\lambda} = \rho_{\lambda_0}$ (compare also footnote on page 620) we find

$$P = 1 - Y + Y(\rho_{\lambda_0}/\rho_{\lambda_0}^b) \ . \tag{VI.30}$$

PHYSICAL EXCITATION AND DE-EXCITATION PROCESSES   Sect. VI.2c

In the absence of the external light beam, we have: $P_0 = 1-Y$. The relative increase in the excited-state population, induced by the external light beam, is thus given by the ratio

$$(P-P_0)/P_0 = \{Y/(1-Y)\}(\rho_{\lambda_0}/\rho_{\lambda_0}^b) \,. \tag{VI.31}$$

This ratio also equals the intensity ratio of the fluorescent emission and the thermal flame emission, if we suppose that the whole flame volume observed is uniformly illuminated by the light beam. When the spectral radiance of the continuum source at $\lambda \approx \lambda_0$ is equivalent to that of a black body radiator with temperature T_r, the maximum obtainable value of ρ_{λ_0} in the flame is $\rho_{\lambda_0}^b(T_r)(\Omega/4\pi)$; here $\rho_{\lambda_0}^b(T_r)$ is the spectral radiant density inside a black body at temperature T_r, and Ω is the solid angle subtended by the exciting light beam in the flame. For given values of Ω and radiance temperature T_r, the maximum gain of the fluorescent emission compared to the thermal flame emission is then, according to Eq. VI.31,

$$\{(P-P_0)/P_0\}_{\max} = \{Y/(1-Y)\}\{\rho_{\lambda_0}^b(T_r)/\rho_{\lambda_0}^b(T_f)\}(\Omega/4\pi) \,. \tag{VI.32}$$

For a clear distinction, the spectral black body radiant density at flame temperature T_f and at $\lambda = \lambda_0$ is here denoted by $\rho_{\lambda_0}^b(T_f)$. When, for instance, $Y = \frac{1}{2}$ and $\Omega = 1\,\text{sr}$, the maximum gain factor at $\lambda_0 = 3000\,\text{Å}$ is 2.2×10^6 for $T_r = 7000\,\text{K}$ and $T_f = 2000\,\text{K}$. It is interesting to note that Y and λ_0 are the only atomic parameters contained in Eq. VI.32. This equation, which holds for resonance lines in the limit of vanishing metal concentrations only, may be applied in the discussion of detection limits obtainable by the method of resonance fluorescence.

(ii) Case of a light source emitting a relatively *narrow atomic line*. We suppose that the spectral width of the source line is markedly smaller than the width of the same resonance line in the flame. According to the definition of $\bar{\rho}_\lambda$, we now have (compare also footnote † on page 620): $\bar{\rho}_\lambda = \rho/\Delta\lambda_{\text{eff}}$ where $\rho \equiv \int \rho_\lambda \, d\lambda$ is the integrated spectral volume density (energy per unit volume) of the exciting light beam at the place considered and $k(\lambda_0)\Delta\lambda_{\text{eff}} \equiv \int k(\lambda)\,d\lambda$. We obtain for the relative increase in excited-state population induced by the external light beam through Eq. VI.26

$$(P-P_0)/P_0 = \{Y/(1-Y)\}\rho/(\rho_{\lambda_0}^b \Delta\lambda_{\text{eff}}) \,. \tag{VI.33}$$

Here P_0 is again defined as the population factor in the absence of the light beam. This equation holds only for resonance lines in the limit of vanishingly small metal concentration. The only parameters of the atomic line contained in Eq. VI.33 are λ_0, $\Delta\lambda_{\text{eff}}$, and Y. Supposing $Y = \frac{1}{2}$, $\lambda_0 = 2300\,\text{Å}$ and $\Delta\lambda_{\text{eff}} = 0.02\,\text{Å}$, the gain factor of the fluorescent emission is calculated from Eq. VI.33 to be 0.73×10^8, if $\rho = 3 \times 10^{-7}\,\text{erg cm}^{-3}$, for a flame temperature of $2000\,\text{K}$.

629

Chapt. VI   EXCITATION AND DE-EXCITATION OF METAL SPECIES

The integrated spectral radiance of a cadmium electrodeless discharge lamp is about 10^4 erg s^{-1} cm^{-2} sr^{-1} at 2290 Å (see Mansfield *et al.* 1968). When this source is imaged under a solid angle of 1 sr in the flame, the value of ρ assumed in the above calculation can be realized. Normal hollow-cathode lamps, however, usually have a radiance that is one or two orders of magnitude less than that of the cadmium lamp just mentioned (see Prugger 1969). According to Eq. VI.32, the same gain factor will be attained when a continuum light source (e.g., a high-pressure xenon lamp XBO 450) is used having a radiance temperature T_r = 5900 K, under otherwise identical conditions (see also Zeegers, Smith and Winefordner 1968). The Osram cadmium vapour lamp has been reported to yield a radiance of about 10^5 erg s^{-1} cm^{-2} sr^{-1} (see Prugger 1969), corresponding roughly to a radiance temperature $T_r \approx$ 7000 K (for an estimated value of $\Delta\lambda \approx$ 0.03 Å). For lamp intensity data see also Fig. III.15.

Case of laser sources (with saturation).† In the case of intense laser beams causing (partial) saturation of the optical transition, we must have recourse to the exact Eq. VI.25 which includes induced-emission transitions. As before, we shall disregard the thermal excitation, so that Eq. VI.25 can be simplified to

$$\frac{[M^*]g}{[M]g^*} = \frac{\bar{\rho}_\lambda}{c_3/Y\lambda_0^5 + \bar{\rho}_\lambda} , \qquad (VI.34a)$$

where $\bar{\rho}_\lambda$ is solely due to the laser radiation. Introducing the parameter $z \equiv \bar{\rho}_\lambda Y\lambda_0^5/c_3$ (with dimension unity) as a measure for the degree of saturation, we can re-write the latter equation as

$$\frac{[M^*]g}{[M]g^*} = \frac{z}{1+z} . \qquad (VI.34b)$$

For a broad-band laser beam which can be regarded as a quasi-continuum with respect to the relatively narrow atomic line, we have (see footnote † on page 620): $\bar{\rho}_\lambda = \rho_{\lambda 0}$ and $z = \rho_{\lambda 0} Y\lambda_0^5/c_3$. For a monochromatic laser beam with integrated radiant density ρ at discrete wavelength λ we have: $\bar{\rho}_\lambda = \rho S_\lambda(\lambda)$ (see Daily 1979) and thus: $z = \rho S_\lambda Y\lambda_0^5/c_3$; $S_\lambda(\lambda)$ is here the normalized spectral distribution function of the atomic line (see Eq. II.207).

Saturation sets in when z becomes comparable to unity. For $z \ll 1$, the excited-state population and thus the fluorescence intensity grow proportionally with the laser intensity. For $z \gg 1$, full saturation is obtained, that is:

---

† When very short laser pulses are applied, the steady-state solutions presented here are no longer valid. See Daily (1979) for a critical discussion of the applicability of the conventional rate equations for photo-excitation in flames and for a description of quantummechanical transient effects.

PHYSICAL EXCITATION AND DE-EXCITATION PROCESSES   Sect. VI.2c

$[M^*]g/[M]g^* \simeq 1$; the excited-state population has then reached a constant value (plateau). For a 2-level atom we have: $[M]+[M^*]=[M]_t$ where $[M]_t$ is the total atomic density. It follows from Eq. VI.34b† that

$$\frac{[M^*]}{[M]_t} = \frac{g^*z}{g + z(g+g^*)} \tag{VI.35a}$$

and

$$\frac{[M]}{[M]_t} = \frac{g(1+z)}{g + z(g+g^*)}, \tag{VI.35b}$$

whereas

$$\frac{[M^*]g/g^*}{[M] - [M^*]g/g^*} = z. \tag{VI.36}$$

At full saturation we thus have for the plateau value: $[M^*] \to [M]_t g^*/(g+g^*)$, while $[M] \to [M]_t g/(g+g^*)$. For $g=g^*$, both $[M^*]$ and $[M]$ tend to $\frac{1}{2}[M]_t$ when $z \to \infty$.

The net absorptivity $k'(\lambda)$ is proportional to: $\{[M]-[M^*](g/g^*)\}$ (see Eq. II.190) which can be expressed by making use of the above equations as

$$k'(\lambda) \propto [M]-[M^*](g/g^*) = \frac{[M]_t}{1 + z(g+g^*)/g}. \tag{VI.37}$$

We have assumed here that the conditions underlying the validity of Eq. II.190 are fulfilled in flames (see Sect. II.5a-2). For $z \to \infty$, $k'(\lambda)$ tends to zero as $1/z$; the product of absorptivity and radiant power, i.e. the <u>absolute</u> absorbed radiant power, then becomes constant. This causes the bleaching effect mentioned already in Sect. II.5a-2. Since in the absence of the laser beam $[M] \simeq [M]_t$, we conclude from Eq. VI.37 that $k'(\lambda)$ is reduced to <u>half</u> the absorptivity $k(\lambda)$ that exists without laser radiation, when $z = g/(g+g^*)$. For this z value the excited-state population is also <u>half</u> the saturated value. The value of $\bar{\rho}_\lambda$ at which $z = g/(g+g^*)$ is now defined as the *saturation parameter* $\bar{\rho}_\lambda^s$ (cf. Omenetto, Benetti, Hart, Winefordner and Alkemade 1973, and Omenetto, Winefordner and Alkemade 1975*). From the above definition of z we thus find

$$\bar{\rho}_\lambda^s = gc_3/Y\lambda_0^5(g+g^*) = 8\pi hcg/Y\lambda_0^5(g+g^*) \tag{VI.38a}$$

and

$$z = (\bar{\rho}_\lambda/\bar{\rho}_\lambda^s)g/(g+g^*). \tag{VI.39a}$$

---

† After some algebra Eq. VI.35a appears to be identical to Eq. (27) in Piepmeier (1972a) when use is made of the relation between the Einstein coefficients and the definition of Y. Unlike the latter author, we believe that this equation is valid for <u>any</u> z as long as the case of homogeneous broadening applies.

* A misprint is found in Eq. (12) of this paper, where $B_{\nu_0}^s$ should read: $E_{\nu_0}^s$.

Chapt. VI EXCITATION AND DE-EXCITATION OF METAL SPECIES

Converting the spectral volume density ρ_λ into spectral irradiance E_λ (see Appendix A.5), we can also define the saturation parameter $\bar{E}_\lambda^s$

$$\bar{E}_\lambda^s (\equiv \bar{\rho}_\lambda^s c) = 8\pi hc^2 g / Y \lambda_0^5 (g + g^*) = 1.50 \times 10^{-4} g / Y \lambda_0^5 (g + g^*) , \qquad (VI.38b)$$

where $\bar{E}_\lambda^s$ and λ_0 are expressed in CGS units. Instead of Eq. VI.39a we can then also write

$$z = (\bar{E}_\lambda / \bar{E}_\lambda^s) g / (g + g^*) . \qquad (VI.39b)$$

The practical realization of (near-)saturation with atomic resonance lines in flames by using available tunable dye lasers can be assessed as follows.

We consider the $3^2S \rightarrow 3^2P$ transition of Na in a H_2-O_2-Ar flame. In such a flame mixing collisions between the $3^2P_{\frac{1}{2}}$ and $3^2P_{\frac{3}{2}}$ states are more effective than quenching collisions of either state to the ground state $3^2S_{\frac{1}{2}}$. Under this condition the $3^2P_{\frac{1}{2}}$ and $3^2P_{\frac{3}{2}}$ states are approximately populated in the ratio, $g_1/g_2 = \frac{2}{4}$, of their statistical weights, as the difference in their excitation energies is small compared to kT_f. The population of the 3P doublet as a whole is then given by Eqs VI.35a and 39a (or 39b) if we substitute the sum $(g_1 + g_2)$ for g^* (see, e.g., van Calcar et al. 1979). When only the $P_{\frac{3}{2}}$ component is directly excited by the laser, we have to replace g in Eq. VI.38a (or 38b) by: $g(g_1 + g_2)/g_2$; when only the other component is directly excited, a replacement by: $g(g_1 + g_2)/g_1$ is appropriate. With the reasonable value of $Y \simeq 0.5$, we calculate from the modified Eq. VI.38b for the saturation parameter: $\bar{E}_\lambda^s \simeq 1.6 \times 10^{17}$ erg s^{-1} cm^{-3} for excitation at $\lambda = 5890$ Å.

For a commercial flash-lamp pumped dye laser (Zeiss) typical values for the instrumental parameters are: peak output power $\simeq 10$ kW, pulse duration $\simeq 1$ μs, spectral bandwidth $\simeq 150$ mÅ and beam cross section $\simeq 7 \times 10^{-2}$ cm^2 (cf. Table III.3). The spectral irradiance obtainable with this broad-band laser is then: $E_\lambda \simeq 9.5 \times 10^{20}$ erg s^{-1} cm^{-3}. By applying the modified Eq. VI.39b we find in the above example: $z = 1.5 \times 10^3$, which means almost full saturation in the stationary state.

Typical instrumental values for a c.w. dye laser (Coherent Radiation Model 590) are: output power $\simeq 400$ mW, spectral bandwidth $\simeq 60$ mÅ and beam cross section $\simeq 1.0 \times 10^{-2}$ cm^2, which yield: $\bar{E}_\lambda \simeq 6.7 \times 10^{17}$ erg s^{-1} cm^{-3}.[†] By applying the modified Eqs VI.39b and 34b, we find: $z = 1.1$ and for the degree of saturation: $[M^*]g/[M]g^* = 0.52$. So, when Na in a H_2-O_2-Ar flame is irradiated with a c.w. laser, saturation about starts to set in; the excited-state population no longer grows proportionally to laser intensity, but a constant value has not yet been reached.

---

† This value is only approximate, as the laser bandwidth is here no longer large compared with the absorption line width.

It should be realized that the theoretical treatment presented in this section was restricted to the simplest case of a 2-level atom radiating a single, isolated resonance line in a flame with homogeneous distribution of temperature and atomic density. The treatment must be considerably more complex when the atom possesses several excitation levels between which optical and/or collisional transitions can occur.[†] These levels may belong to different terms, as in the case of the Sr atom having several singlet and triplet terms in the vicinity of the first resonance level (see Jansen 1976, and Jansen, Hollander and van Helvoort 1977 for a discussion of their effect on the measured resonance-fluorescence efficiency). These levels may also be components of the same multiplet showing fine-structure splitting, as in the case of the excited alkali P doublets. The following authors have discussed theoretically the effect of intramultiplet-mixing collisions on the atomic fluorescence in flames without saturation: McCarthy, Parsons and Winefordner (1967)[*], Jenkins (1968a), Finn and Jefferies (1968, 1968a), Hooymayers and Lijnse (1969), Omenetto and Rossi (1970), and Lijnse, Zeegers and Alkemade (1973); other authors have included saturation in their discussion: Piepmeier (1972), Kuhl, Neumann and Kriese (1973), Alkemade (1977), Daily and Chan (1978), van Calcar et al. (1979), Lucht and Laurendeau (1979), and Omenetto and Winefordner (1979a). The occurrence of stepwise line fluorescence as a result of collisional doublet mixing was already discussed in Sect. VI.2a-2.

Complex situations also arise when the atomic density, flame temperature, fluorescence efficiency or beam intensity vary with position in the flame, or when suprathermal chemiluminescent line emission occurs. Extra complications may arise when (partial) saturation occurs while the irradiance is not uniform across the laser beam or not constant during the laser pulse (see Sect. II.5c-4).

[†] A general theoretical treatment of a 3-level atom under pulsed-laser conditions has been presented by Bolshov et al. (1977). In this model conventional rate equations were used to calculate the time-behaviour of the level populations, with the neglect of quantummechanical transient effects (cf. Daily 1979). The theoretical equations were applied to the interpretation of nonresonance fluorescence experiments with a nonflame Pb atomizer. In particular, the effect of accumulation of Pb atoms in the metastable $^3P_{1,2}$ levels during pulse excitation was studied.

[*] Since only downward collisional transitions between excited states were considered, the results obtained are incomplete.

Chapt. VI   EXCITATION AND DE-EXCITATION OF METAL SPECIES

3. CHEMICAL EXCITATION AND DE-EXCITATION PROCESSES
(CHEMILUMINESCENCE)

3a   INTRODUCTION

In this section, the processes in which chemical energy is converted into excitation energy (chemi-excitation) and the reverse processes (chemiquenching) will be discussed. In contrast to the preceding section, here the formation of excited species will be emphasized rather than the quenching thereof. This approach is chosen because the interconversion of chemical energy and excitation energy is most easily studied by considering the formation of excited species resulting in chemiluminescence. Of course, by applying the principle of microreversibility, conclusions concerning the reverse processes can (and will) be drawn.

Chemiluminescent processes can be studied only in flame parts where a departure from chemical equilibrium is known to exist, that is, in the inner zone and in the lower part of the interzonal region (see Sect. IV.5). In these flame parts, the concentrations of some of the species that take part in the excitation reactions are higher than their equilibrium values. Therefore, the rate of excitation reactions and consequently the concentration of the excited species are found too high when compared with these quantities under equilibrium conditions. As a result the measured emission may appear to be suprathermal (see Sect. VI.4c). In the literature this phenomenon is usually referred to as chemiluminescence. However, the same processes occur in equilibrium as well as outside equilibrium; only the rates of these processes may be different in the two cases. We prefer, therefore, to use the term *chemiluminescence* for any process whereby species are excited to produce light emission as a result of a chemical reaction in which part of the chemical energy is directly converted into electronic excitation energy. Where these processes lead to suprathermal emission, we shall refer to this emission as *suprathermal chemiluminescence*. Processes in which electronic excitation energy is converted into chemical energy will be called *chemiquenching* whether or not these processes occur under equilibrium conditions. Experimental methods to test the possible occurrence of suprathermal emission (i.e. deviation from Boltzmann equilibrium) will be considered in Sect. VI.4d.

Chemiluminescence of metal additives is discussed in two parts: the so-called *hard* chemiluminescence of metal lines with excitation energy E_{exc} up to 9 eV in Sect. 3b, and the so-called *soft* chemiluminescence of metal lines with $E_{exc} < 5$ eV in Sect. 3c. Infrared chemiluminescent emission from rotational and vibrational levels is not discussed.

CHEMICAL EXCITATION AND DE-EXCITATION PROCESSES (CHEMILUMINESCENCE)   Sect. VI.3b

3b   HARD CHEMILUMINESCENCE

In the combustion zone of hydrocarbon flames and in very fuel-rich acetylene-air or oxygen flames (C/O > 1) the emission of a great number of metals with excitation levels up to 9 eV has been observed, for example, Zn at 2516 Å with $E_{exc} = 9.0$ eV (see Gilbert 1963). It is well known that there exist in these cases a relatively large number of very reactive, short-lived species (C, CH, C_2, C_2O, CHO^+ ; see Hand and Kistiakowski 1962, Becker and Bayes 1968, Arrington et al. 1965, 1965a, and Gaydon 1974). The energy released in the fast reaction of these species with oxygen, oxides, or electrons is often high enough to account for the observed excess population of high-lying atomic levels. In particular, the high energy of association of C and O to CO (11.1 eV) is of interest.

The occurrence of very reactive, reducing species in appreciable amounts may also lead to a marked depletion of the concentration of metal oxides with high dissociation energy, and consequently to an enhancement of the free-atom population. However, this fact does not necessarily imply that the population of an excited state deviates from the Boltzmann law at flame temperature. A number of elements show an atomic emission enhancement in fuel-rich flames probably only as a result of the increased free-atom population. This may apply especially for the emission from low-lying excited levels $(E_{exc} < 3$ eV$)$ of elements with high dissociation energies: La, Ce, Pr, Nd, Sm, Eu, Tb, Dy, Ho, Er, Tm, Yb, Lu, Sc, Y, V, Nb, Ti, Mo, and Re (see Fassel, Curry and Kniseley 1962, and Fassel, Myers and Kniseley 1963). A discussion of the effect of reducing conditions in the flame upon the fraction of atomized element will be given in Chapt. VIII. A clear distinction must be made between the two possible origins of the observed emission enhancements when discussing flame spectrometric results obtained in fuel-rich flames.

An extensive survey of elements that may show strong suprathermal chemiluminescent excitation under the conditions mentioned has been given by Gilbert (1963). For the explanation of this phenomenon a great variety of mechanisms has been proposed. We shall discuss them separately for the cases when metal atoms (Subsection 3b-1), metal ions (Subsection 3b-2) and metal compounds (Subsection 3b-3) are involved as reactants.

3b-1   *CHEMI-EXCITATION REACTIONS IN WHICH ONLY ATOMIC METAL SPECIES ARE INVOLVED*

According to Broida and Shuler (1957), metal atoms are not excited to high levels directly as a third partner in a termolecular hard chemi-excitation reaction, since termolecular collisions are comparatively rare. More probably the excited metals are formed in a bimolecular two-step process (see Table II.1, reaction no. 42); initially a precursor, that is, a flame molecule with high internal energy, is

Chapt. VI   EXCITATION AND DE-EXCITATION OF METAL SPECIES

formed and then excitation of the metal atom follows by energy exchange between precursor and metal atom. Flame molecules that are formed in highly exo-ergic reactions and that have sufficiently high-lying excited levels to store an appreciable fraction of the reaction energy released without disintegrating may serve as precursors. Only CO fulfils these requirements when excitation energies up to 9 eV are considered.† Excited CO* may be formed by the reactions

$CH + O \rightarrow CO^* + H$ ($\Delta H = -7.6$ eV)   (see Broida and Shuler 1957)   (i)
$CH + OH \rightarrow CO^* + H_2$ ($\Delta H = -7.7$ eV)   (ii)
$C_2O + O \rightarrow CO^* + CO$ ($\Delta H = -8.85$ eV)   (see Becker and Bayes 1968).   (iii)

Here CO* stands for CO in one of the excited electronic states $A^1\Pi$, $d^3\Delta$, and $e^3\Sigma$. The reaction energy listed, ΔH, is calculated for the products in their ground states; it is, therefore, the maximum energy that can be given to the internal degrees of freedom of CO. Subsequently, CO* may excite a metal atom according to

$$CO^* + M \rightarrow CO + M^*.$$ (iv)

Of course, CO may also transfer its internal energy upon collision to another flame molecule which thereupon may excite the metal atom by a collision as in reaction (iv) However, only CO and N_2, if present, will be effective as intermediate carriers in transferring energies up to 9 eV, because the dissociation energy of other flame species, H_2, CO_2, and H_2O, is too low.* Consequently a collision of CO* with one of the latter molecules may result in dissociation of these molecules or in a redistribution of the internal energy of CO* over the various degrees of freedom of both collision partners. This latter process will decrease rapidly the number of molecules having sufficient internal energy to excite metal atoms to high excitation levels. It may be noted here that only reaction (iii) yields enough energy for being a step in the chemi-excitation mechanism of Zn levels with E_{exc} equal to 7.8, 8.5, and 9.0 eV, and of the Cd level with E_{exc} equal to 8.1 eV. Chemiluminescent emission of these levels has been reported by Gilbert (1963), Buell (1963),

---

† Cummings and Hutton (1966) have suggested that OH may serve as an energy carrier for the excitation of iron to levels up to 7.55 eV. These authors have supposed that OH* is formed by: $CH + O_2 \rightarrow CO + OH^*$, which is exo-ergic by 6.95 eV. However, it is not known whether OH can store such a large amount of energy. The dissociation energies of OH are 4.4 and 6.4 eV for dissociation into ground-state atoms, $O(^3P)$ and $H(^2S)$, and into $O(^1D)$ and $H(^2S)$, respectively. Both dissociation energies are substantially lower than the excitation energy of the highest Fe level (7.55 eV) for which chemi-excitation has been established.

* It has been found by Alder, Thompson and West (1970) that chemiluminescent emission is more intense in Ar-diluted flames than in flames with N_2. This may indicate that chemi-excitation occurs mainly without intermediate energy carriers. However, the greater intensity in Ar-diluted flames can also be explained by the smaller quenching efficiency of Ar in collision with M*.

CHEMICAL EXCITATION AND DE-EXCITATION PROCESSES (CHEMILUMINESCENCE)   Sect. VI.3b

and Alder, Thompson and West (1970) in the primary combustion zone of fuel-rich $C_2H_2-O_2$ flames.

In addition to the above processes, Sugden (1962) has proposed the reaction

$$CH + O + M \rightarrow CHO + M^* \quad (\Delta H = -8.9 \text{ eV}) \quad \text{(v)}$$

(cf. Table II.1, reaction no. 23). Here ΔH was calculated for the products in their ground state. The following discussion holds also when reaction (v) occurs in actual fact as a two-step process, that is, when first a highly excited flame molecule, X^*, is formed that then transfers its energy to M (cf. two-step process no. 41 in Table II.1). Also, in the case when reaction (v) results in: $CO + H + M^*$ or in: $CO + H + X^*$ (that is, when the reaction is assumed to occur as a two-step process; see Broida and Shuler 1957), the final conclusion of the following discussion remains unchanged.

As mentioned above, a termolecular reaction is usually improbable, at least when compared to excitation by (a set of) bimolecular reactions. This can be shown very easily as follows.

A competing set of bimolecular reactions consists of reaction (i) and reaction (iv), with rate constants k_i and k_{iv}, respectively, and an inelastic collision in which CO^* is de-activated by a flame species X other than the metal atom M, with rate constant k_{vi}, according to

$$CO^* + X \rightarrow CO + X. \quad \text{(vi)}$$

The products of this inelastic collision may be either excited X molecules or dissociation fragments of X. It is supposed that the products of collision process (vi) no longer have sufficient internal energy to excite M. The ratio of the rate of M^* production by the termolecular reaction, v_{ter}, to that of the set of bimolecular reactions, v_{bi}, is given by

$$\frac{v_{ter}}{v_{bi}} = \frac{k_{vi} k_v}{k_{iv} k_i} \frac{[CH][O][M]}{[CH][O][M]/[X]} = \left(\frac{k_v}{k_i}\right)\left(\frac{k_{vi}}{k_{iv}}\right)[X]. \quad \text{(VI.40)}$$

Equation VI.40 holds for the realistic situation that [M] ≪ [X]. At atmospheric pressure gas-kinetic three-body collisions are about 10^{-3} as frequent as two-body ones (see Sect. II.4b-1). If the efficiencies of reactions (v) and (i) are the same (that is, if the probabilities that M^* or CO^* is actually formed when a termolecular or a bimolecular collision, respectively, takes place are equal and if the rate constants k_{iv} and k_{vi} are of the same order, then v_{ter}/v_{bi} equals roughly 10^{-3}. The actual ratio may be different by 1 or 2 orders of magnitude if the reactions compared have different efficiencies.

It is clear from Eq. VI.40 that, when the pressure decreases, the contribution of termolecular excitation mechanisms decreases still further with respect to that of bimolecular reactions. In conclusion it can be said that the contribution of termolecular reactions to the population of excitation levels is likely to be small when this population can also occur by (a set of) bimolecular reactions.

Working in hydrogen-oxygen flames with methane as an additive, Blades (1971) has found the chemiluminescence from excited levels up to 6.8 eV of several atoms (Cr, Mo, Sn, Sb and As) to be first-order in methane concentration. Therefore, in the chemi-excitation mechanism a single-carbon species or one of its derivatives should occur. Blades favours the bimolecular association reaction[†]

$$CH + O \rightarrow CHO^* \qquad (vii)$$

and a subsequent energy-transfer reaction

$$CHO^* + M \rightarrow CHO + M^*. \qquad (viii)$$

However, according to Cummings and Hutton (1966) the actual lifetime of the unstable CHO^* will be of the order of the period of one vibration (10^{-13} s), and thus chemi-excitation of M by reactions (vii) and (viii) proceeds virtually as the termolecular reaction (v), which was already discussed. Furthermore, Fontijn (1966) has found that CH is definitely not the precursor of the Vaidya band emitter, which is, with a fair degree of certainty, CHO^* (see Vaidya 1964). In conclusion, it is not very likely that CHO^* plays a part in the hard chemiluminescence of metals.

3b-2 *CHEMI-EXCITATION REACTIONS INVOLVING METAL IONS*

In the combustion zone of hydrocarbon flames a large number of ionized species have been detected (see Sect. IV.7). On the basis of the existence of H_3O^+ in rather large quantities in and above the combustion zone, Calcote (1962) and Gilbert (1963) have proposed the reaction

$$H_3O^+ + M \rightarrow M^+ + H_2O + H \qquad (ix)$$

followed by the recombination process leading to excited neutral M

$$M^+ + e^- + X \rightarrow M^* + X. \qquad (x)$$

At the time that these reactions were proposed, there existed a discrepancy between values reported in the literature for the heat of formation of H_3O^+, $H_{fo}(H_3O^+)$. Using $H_{fo}(H_3O^+) = 8.5$ eV Gilbert (1963) calculated the available reaction energy to be 8.7 eV, which is high enough to ionize most metals. However, according to the JANAF tables (1971), the estimate of $H_{fo}(H_3O^+) = 6.02$ eV is very likely the better value. Then the available energy of reaction (ix), 6.2 eV, is high enough

---

[†] Blades wrongly suggests that this reaction has been proposed by Sugden (1962). However, Sugden has discussed the termolecular reaction (v).

only for the ionization of the alkali and alkaline-earth metals, the lanthanides, aluminium, gallium and thallium. Besides, reaction (x), being termolecular, is a priori not very probable. Altogether, it seems very unlikely that ions play a part in hard chemiluminescent excitations.

3b-3   *CHEMI-EXCITATION REACTIONS INVOLVING METAL COMPOUNDS*

According to Gilbert (1963), it may be possible that excited metal atoms are formed by the sequence of reactions

$$H_3O^+ + MO \rightarrow MOH^+ + H_2O \qquad (xi)$$

$$MOH^+ + e^- \rightarrow M^* + OH \qquad (xii)$$

with overall $\Delta H = -8.1 + D(MO) + E_{exc}(M)$ eV, using the now adopted value of $H_{fo}(H_3O^+) = 6.02$ eV. Of course, only metals that form hydroxides may be excited in this way. It is hard to figure out which metals can in fact be excited, because of lack of information about the dissociation energies and, especially, the ionization energies involved. In view of the now adopted lower value of $H_{fo}(H_3O^+)$, reaction (xi) is probably less general than Gilbert suggests.

Instead of reactions (xi) and (xii), their overall reaction has been proposed as a single chemi-excitation step

$$H_3O^+ + MO + e^- \rightarrow M^* + H_2O + OH \qquad (xiii)$$

with $\Delta H = -8.1 + D(MO) + E_{exc}(M)$ eV. As pointed out, significant contribution to the excitation of M by a termolecular reaction is unlikely when bimolecular reactions are also possible. Besides, the reaction energy of reaction (xiii) is only high enough to account for the emission from the highest excited levels that have been observed for bismuth, tellurium, cobalt, nickel, and the alkaline-earth metals, except magnesium and beryllium.

In flames with a high atomic carbon content chemi-excitation might occur by the reaction first proposed by Sternberg, Gallaway and Lones (1961) (see also Gilbert 1963)

$$C + MO \rightarrow CO + M^* \qquad (xiv)$$

with $\Delta H = -11.1 + D(MO) + E_{exc}(M)$ eV. Although this reaction is energetically more favourable than reaction (xiii), its energy is still not high enough to explain the chemi-excitation of a great number of atomic lines. In some cases this reaction is spin-forbidden by Wigner's spin conservation rule (see Sect. II.4f); for example, excitation of Be($^1P_1$), although energetically possible, is spin-forbidden, the ground states of C, CO, and BeO being a triplet, a singlet, and a singlet, respectively. It should be pointed out that Wigner's rule cannot always be applied as a criterion, since the designation of the electronic ground state of a number of oxides is not known unambiguously.

Another kind of chemi-excitation reaction has also been suggested by Sternberg, Gallaway and Lones (1961)

$$CO^* + MO \to CO_2 + M^* \qquad (xv)$$

with $\Delta H = -5.1 - E_{exc}(CO^*) + D(MO) + E_{exc}(M)$ eV. According to reactions (i) and (iii), the excitation energy of CO^* may range from 7.6 to 8.9 eV. Reaction (xv) is then energetically feasible for all elements of which chemiluminescence has been observed. This reaction is also less severely limited by the spin conservation rule, since CO^* may be present in singlet as well as in triplet states. Alder, Thompson and West (1970, 1972) have found that elements forming refractory oxides (Ge, Mo, Sn and V) require more fuel-rich flames for showing chemiluminescence than elements (such as I, Cd and As) with lower $D(MO)$ values. They concluded therefore that breakdown of the monoxide is a necessary preliminary step before chemi-excitation can occur and consequently they consider reaction (xv) to be unlikely.

The feasibility of the reactions suggested is discussed here mainly in terms of reaction energy. With the current limited knowledge about the gas composition in the combustion zone and in very fuel-rich flames, it is impossible to predict the absolute rate of production of excited atoms by the respective reactions. There are estimates of the concentration of CH and C_2 in the ground and the excited state (see Bleekrode 1966), but all other quantities, for example [O], [CO^*] and, in particular, the reaction rate constants, are unknown. Therefore, it is difficult to estimate to what extent each reaction may contribute to the chemi-excitation. Nevertheless there are reasons to expect that reaction (iv) may play an important part. Energetically this reaction is feasible for all elements that show hard chemiluminescence. Electronically excited CO^* may be found in singlet as well as in triplet states, so that there are more possibilities to obey Wigner's spin rule. The simultaneous occurrence of maximum chemiluminescent intensity and a peak in the CH emission as well as in the electron concentration, observed by Bulewicz and Padley (1961), is not incompatible with our suggestion that chemiluminescence is related primarily to [CO^*] (see Bowman and Seery 1968).

The suggestion by Alkemade and Zeegers (1971) has been confirmed experimentally to some extent by the work of Alder, Thompson and West (1970, 1972). They have observed in acetylene-air and acetylene-oxygen-argon flames maximum CO emission (fourth positive band) and maximum chemiluminescent intensity of atomic levels with $E_{exc} > 5$ eV to occur at the same fuel-oxidant ratio.

It is essential for the occurrence of hard chemiluminescence that there be present in the flame gas fragments of hydrocarbons with two or more C atoms. These fragments may be found not only in the combustion zone of hydrocarbon flames and in the central region of very fuel-rich hydrocarbon flames, but also in fuel-rich H_2

CHEMICAL EXCITATION AND DE-EXCITATION PROCESSES (CHEMILUMINESCENCE)   Sect. VI.3c

flames in which organic solvents are sprayed. Under these conditions there is no evidence that chemical equilibrium exists for the carbonaceous species. On the contrary, all available information points to the existence of strong deviations from chemical and Boltzmann equilibrium. Therefore, we may expect that this type of chemiluminescence is suprathermal in character.

The intensity of suprathermal chemiluminescent emission may be much higher than the intensity of the thermal emission of the same atomic line. For instance, the analytical detection limit (see definition in Sect. III.14f-1) of cadmium (2288 Å) in an unpremixed hydrogen-air flame is decreased from 50 ppm to 0.3 ppm upon addition of isopropanol (see Gilbert 1963). The temperature in both flames is about equal. For cadmium no enhancement of the free-atom content is expected to result from the addition of isopropanol because cadmium does not form oxides in an appreciable amount. Therefore the observed emission enhancement (decrease of detection limit) is caused by suprathermal chemiluminescence.

Another example has been given by Gaydon and Wolfhard (1979). Using the line-reversal method, these authors have shown that the population of excited Fe states in the combustion zone of a low-pressure C_2H_2-air flame is far in excess over that at equilibrium. Mavrodineanu and Boiteux (1965) have calculated from the reversal temperatures found by Gaydon and Wolfhard that the ratio $[Fe^*]/[Fe^*]_e$ was about 1.7×10^5 for the Fe state ($E_{exc} = 4.991\,eV$) that gives rise to the radiation at $\lambda = 2483.3$ Å.

An extensive survey of the elements that may be detected in flame emission spectroscopy by hard chemiluminescence and of the optimum flame conditions for each element is given by Gilbert (1963) and Buell (1963).

3c   SOFT CHEMILUMINESCENCE
3c-1 *QUALITATIVE CONSIDERATIONS*

The chemi-excitation reactions discussed in the previous section occur only under nonequilibrium conditions. The very high excitation energy involved was ultimately released by the irreversible oxidation of highly endo-ergic species. The occurrence of this type of chemi-excitation reaction is restricted to the combustion zone or to very fuel-rich, luminous flames. In the combustion zone H, OH, and O radicals are formed in excess of their equilibrium values (see Bulewicz, James and Sugden 1956, and Kaskan 1959; see also Sect. IV.5). The establishment of chemical equilibrium for these radicals is governed by relatively slow three-body recombination reactions (see Zeegers 1966 and Bulewicz, James and Sugden 1956). The relaxation toward equilibrium occurs within 1—10 ms. Depending on the kind of third body involved, part of the energy may be converted into translational, ionization, or internal energy of the third body. This means, for instance, that as long as the

Chapt. VI   EXCITATION AND DE-EXCITATION OF METAL SPECIES

radical concentrations exceed their equilibrium values, the population of an excited state of the third body may be higher than its thermal value. Since the recombination energy of all these reactions ranges typically to about 5.5 eV, only excited states that lie at most 5.5 eV above the ground state may show this type of chemiluminescence.

When in chemical equilibrium chemiluminescent reactions are — in addition to inelastic collisions and absorption of radiation — one of the ways in which the thermal population of an excited state is maintained, the population of this state depends only on temperature according to the Boltzmann formula. This formula holds irrespective of the kind of prevailing excitation mechanisms. On the other hand, this means that chemiluminescent reactions can be studied only in those parts of the flame where a known departure from chemical equilibrium exists. In fact, only the suprathermal chemiluminescence can be studied, but the results derived from such investigations can be applied to calculate chemi-excitation rates under thermal conditions too.

Chemi-excitation of metal atoms or flame species such as OH may occur by one or more of the following overall reactions,[†] which might proceed in reality by two steps, such as reaction no. 41 in Table II.1

$$H + OH + M \rightarrow H_2O + M^* \quad (\Delta H = -5.2 \text{ eV}) \quad \text{(xvi)}$$

$$H + H + M \rightarrow H_2 + M^* \quad (\Delta H = -4.5 \text{ eV}) \quad \text{(xvii)}$$

$$CO + O + M \rightarrow CO_2 + M^* \quad (\Delta H = -5.5 \text{ eV}) \quad \text{(xviii)}$$

$$O + O + M \rightarrow O_2 + M^* \quad (\Delta H = -5.2 \text{ eV}) \; . \quad \text{(xix)}$$

Sugden (1962) and his co-workers have shown that reactions (xvi) and (xvii) may lead to suprathermal excitation of a number of metals in stoichiometric and fuel-rich H_2-N_2-O_2 flames at 1 atm. Carabetta and Kaskan (1967) favour reaction (xix) under fuel-lean conditions. Reaction (xviii) has been suggested by Hollander (1964) to explain the observed suprathermal emission of the blue K doublet in CO-air flames, and Zeegers (1966) has found some evidence that this reaction may contribute to the suprathermal excitation of the same K doublet and of OH in C_2H_2-air flames. In most flames more than one of the above reactions may be significant. Which reaction contributes predominantly to the chemi-excitation of a particular atom, under certain flame conditions, is determined by the reaction rate constant and the concentration of the relevant reactants. For instance, Padley and Sugden (1959) have reported that for fuel-rich H_2 flames reaction (xvii) was predominant for the chemi-excitation of Tl ($E_{\text{exc}} = 3.28$ eV), whereas McEwan and Phillips (1965) have

---

[†] See Page (1973) for a survey, and Golde and Thrush (1975) for an introduction.

CHEMICAL EXCITATION AND DE-EXCITATION PROCESSES (CHEMILUMINESCENCE)   Sect. VI.3c

found reaction (xvi) to be the predominant one when the H_2 flame was oxygen-rich. Both pairs of authors consider reaction (xvi) to be the main reaction for the chemi-excitation of Pb ($E_{exc} = 4.33$ eV), irrespective of the different gas compositions of the flames in which they performed their experiments. Phillips and Sugden (1961) have found for the ratio k_{xvi}/k_{xvii} the values 2.2 and about 100 for thallium and lead, respectively. These results show that reaction (xvii) may be the main chemi-excitation reaction for thallium if the flame composition is chosen so that [H] is at least a small factor larger than [OH]. For the chemi-excitation of lead, reaction (xvi) will always be predominant because it is practically impossible to make flames in which the ratio [H]/[OH] is so high that the effect of the much greater efficiency of reaction (xvi) (i.e., $k_{xvi} \gg k_{xvii}$) is compensated for.

3c-2   *QUANTITATIVE CONSIDERATIONS*

Rate constants of chemi-excitation reactions. For a chosen fuel/oxidant ratio, that is, for given radical concentrations, the relative contributions from the various chemi-excitation reactions listed above are determined only by their reaction rate constants. These constants may be found by measuring the chemiluminescent emission as a function of known excess radical concentrations in a number of flames with sufficiently different flame gas compositions. Padley and Sugden (1959) have tried to achieve this goal by measuring the peak in the intensity of an atomic line as a function of height in several 'flame families'.[†] Usually this peak occurs very near the combustion zone. They have determined the radical concentrations at the point where the peak occurs by extrapolating the radical concentrations measured at points further downstream from the combustion zone (see also Sect. IV.5). There is some discussion as to whether this extrapolation procedure is valid for points very near the combustion zone. In these measurements the influence of the varying flame gas composition on the efficiency of fluorescence Y should be taken into account. Moreover, the emission intensity is unambiguously related to the chemi-excitation rate only if it largely exceeds the thermal intensity (see Sect. VI.4c-3). However, also when the thermal emission is not comparatively small, the chemi-excitation rate constant can be determined by measuring the difference between the actual and the thermal emission intensity, the excess radical concentrations, and the efficiency of fluorescence, Y, of the excited state considered (see Zeegers and Alkemade 1965 and Zeegers 1966).

In order to calculate the stationary excited-state population and the suprathermal chemiluminescence intensity, consider the following set of processes, the

---

[†] A 'flame family' is a set of flames each having the same fuel/oxidant ratio, but a different oxidant/diluent gas ratio.

first of which may represent any of the reactions (xvi) - (xix)

$$X + Y + M \rightleftharpoons XY + M^* \qquad (xx)$$

$$M + Z \rightleftharpoons M^* + Z \qquad (xxi)$$

$$M^* \xrightarrow{A} M + h\nu \; . \qquad (xxii)$$

Here M denotes a metal atom, X and Y stand for flame radicals, and XY and Z represent stable flame molecules. The forward and backward steps of each reaction need not necessarily be in balance. The concentrations of X and Y may be higher than their equilibrium values. It is well known that everywhere in the central region of the flame the concentrations of the major stable species, for example, $XY = H_2O$ and $Z = N_2$, are nearly equilibrated (see Sect. IV.5). This equilibration is practically independent of the radical disequilibrium, because radicals are only minor constituents. Their total concentration amounts, at most, to about 3% of the flame gas. We assume, therefore, that [XY] equals $[XY]_e$ and that [Z] equals $[Z]_e$ whether or not the radical concentrations are equal to their equilibrium values. Consequently the rate of the forward reaction (xx) will exceed the rate of the reverse reaction.

The number of photons emitted per second and per cubic centimetre, in the absence of self-absorption, is given by

$$I = Y[M](k_{xx}[X][Y] + k_{xxi}[Z]_e \; , \qquad (VI.41)$$

where $Y \equiv A/(k_{-xx}[XY] + k_{-xxi}[Z]_e + A)$, and k_{-xx} refers to the backward reaction (xx), etc. In equilibrium we would have $I = I_e$ given by

$$I_e = Y[M](k_{xx}[X]_e[Y]_e + k_{xxi}[Z]_e) \qquad (VI.42)$$

if the values of Y, [M], k_{xx} and k_{xxi} are assumed to be practically independent of the excess concentrations of X and Y, which are but minor species. The excess emission per ground-state M atom follows from Eqs VI.41 and 42

$$(I - I_e)/[M] = Yk_{xx}([X][Y] - [X]_e[Y]_e) \; . \qquad (VI.43)$$

If [M] is constant and if the temperature at each point in the flame is known, the excess chemiluminescent emission per M atom (i.e., the left-hand side of Eq. VI.43) may be found experimentally as follows. First, the thermal emission per M atom (i.e., $I_e/[M]$), calculated in absolute measure (from Eq. II.281), is plotted on a logarithmic scale as a function of height in the flame on a linear scale. Second, the experimental intensity values I_{exp}, expressed in arbitrary units, are inserted in the same graph. The latter plot is now shifted along the logarithmic intensity axis in such a position that the ratio $(I_{exp} - I_e)_1/(I_{exp} - I_e)_2$ at two extreme heights above the combustion zone fits in with the ratio predicted by Eq. VI.43 from the known radical concentrations at these two heights. Then $(I - I_e)/[M]$ is

CHEMICAL EXCITATION AND DE-EXCITATION PROCESSES (CHEMILUMINESCENCE)   Sect. VI.3c

found in absolute measure as a function of height. Finally a plot is constructed that represents $(I-I_e)/[M]$ on an absolute scale as a function of $([X][Y]-[X]_e[Y]_e)$. Such a graph yields a value for Yk_{xx}. Equilibrium concentrations may be found by calculation, and actual concentrations by experiment (see Sects IV.4 and 5).

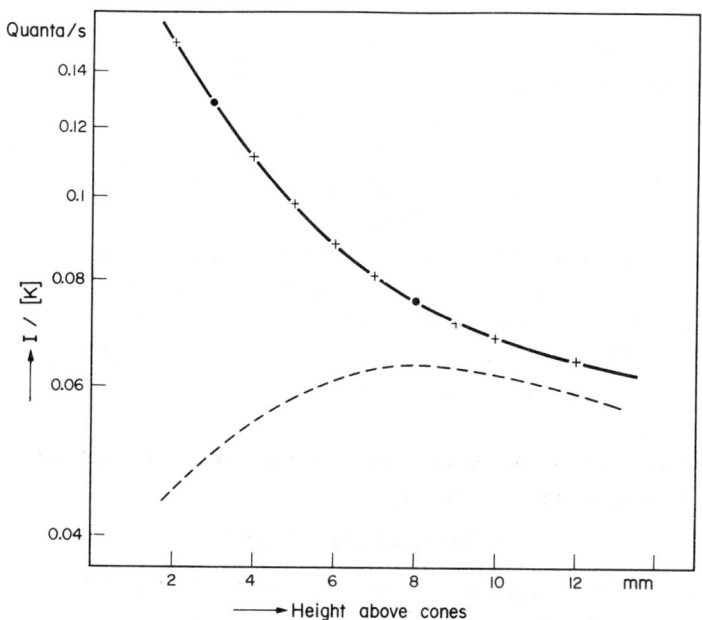

Fig. VI.6 Dependence of the absolute emission intensity per K atom at 4044/47 Å on height above the combustion zone in a H_2-O_2-N_2 flame (T_{max} = 1980 K). The dashed line represents the thermal curve as calculated from the known local temperature by applying Eq. II.281. The position of the experimental curve with respect to the absolute intensity scale was chosen in such a way that the difference in intensity between the two curves agreed with Eq. VI.43 at two distinct heights as indicated by closed circles. (From Zeegers 1966.)

Zeegers and Alkemade (1965) and Zeegers (1966) have employed this adaptation procedure to study the chemi-excitation of $K(5^2P)$ and $OH(^2\Sigma, v'=0)$. In their flames only reaction (xvi) appeared to contribute to the chemi-excitation. An illustration of the adaptation procedure is given in Fig. VI.6; Fig. VI.7 presents the final graphs in the case of the chemi-excitation of potassium in a nearly stoichiometric H_2-air flame and H_2-O_2-Ar flame. The lines drawn in Fig. VI.7 have by definition the correct slope, since these straight lines connect the two points where the experimental values (in Fig. VI.6) were adapted to the thermal emission curve according to Eq. VI.43. The internal consistency of the whole procedure follows from the fact that the other experimental points fit in well with these straight

645

Chapt. VI   EXCITATION AND DE-EXCITATION OF METAL SPECIES

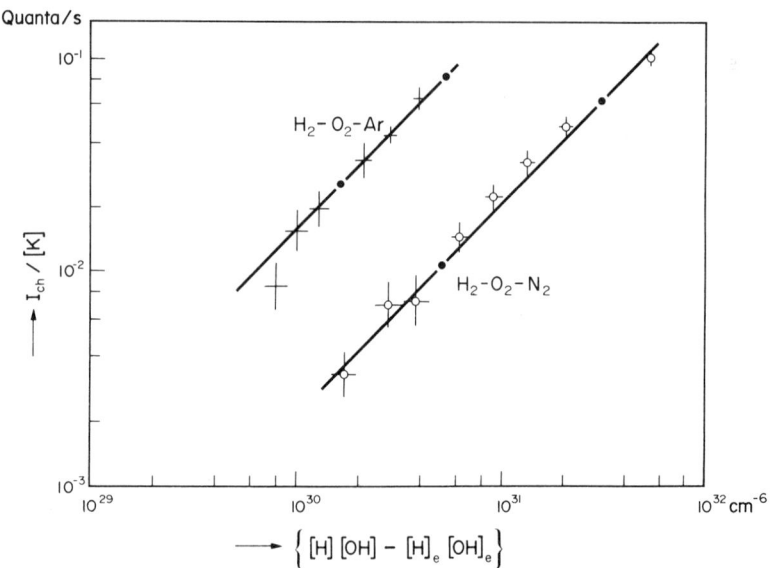

Fig. VI.7 Dependence of suprathermal emission intensity of K at 4044/47 Å on the product of radical concentrations for two hydrogen flames with nitrogen and argon, respectively, as diluent gases. Closed circles present points used in the adaptation procedure to find the suprathermal emission as an absolute value.   (From Zeegers 1966.)

lines. Graphs like the ones in Fig. VI.7 yield the product Yk. In combining results from measurements in different flames, one must bear in mind that Y may vary strongly with flame gas composition. This point is clearly emphasized in Fig. VI.7. Here the Yk_{xvi} values in the two flames differ by a factor of 6.5, which is caused mainly by the difference of a factor 6 between both Y values, Y being 4.3×10^{-3} and 7×10^{-4} in the argon and the air flame, respectively. For each flame it is reasonable to assume Y to be independent of height, because the main gas composition will not vary markedly with height.

In general, the situation may be more complicated, since more than one reaction of the set of reactions (xvi)-(xix) may contribute to the suprathermal chemiluminescence. To describe all these contributions theoretically, we have to multiply the term ([H][OH] - [H]$_e$[OH]$_e$) in Eq. VI.43 by a factor that contains the ratios of the relevant concentrations and rate constants. For most flames this factor will be constant within 10% at the various heights of measurement (see, e.g., Zeegers 1966). Then a straight line such as that shown in Fig. VI.7 should again be obtained.

The relative contribution of each possible radical reaction will depend on

CHEMICAL EXCITATION AND DE-EXCITATION PROCESSES (CHEMILUMINESCENCE)   Sect. VI.3c

flame gas composition and temperature. Therefore, combined measurements in flames with different compositions may yield the rate constant of each chemi-excitation separately.

Rate constants of chemi-excitation reactions involving K and Na atoms have been listed by Jensen and Jones (1978).

Rate constants of chemiquenching reactions. When the chemi-excitation rate constant of, for instance, reaction (xx) has been found, one may derive from it the contribution of *dissociative quenching* of the considered state by the reversed reaction (xx), to the total quenching by flame molecules. When Eq. II.86 is applied, the chemiquenching rate constant is given by

$$k_{-xx} = \frac{k_{xx} K_d (g_0/g^*)}{\exp[-E_{exc}/kT]} \quad , \tag{VI.44}$$

in which K_d stands for the dissociation constant of $XY \rightleftharpoons X + Y$. By applying Eqs II.109b and VI.44, one finds for the ratio of unimolecular quenching rate constants (if $Y \ll 1$)

$$\frac{k_{-xx}[XY]}{\text{total quenching}} = \frac{Y k_{xx} K_d}{A} \times \frac{g_0 [XY]}{g^* \exp[-E_{exc}/kT]} \quad . \tag{VI.45}$$

Since $Y k_{xx}$ follows from Eq. VI.43, Y need not be known.[†] The ratio of chemiquenching by dissociation of H_2O to total quenching is found to be about 0.1 and 0.2 for $K(5^2P)$ and $OH(^2\Sigma, v'=0)$, respectively, in H_2-air flames with $T \approx 2000\,K$ (see Zeegers 1966).

If the specific quenching cross section of XY is known, the ratio of physical to chemical quenching by XY may also be calculated. Specific quenching cross sections may be found from fluorescence measurements, as described in Sect. VI.2. In these measurements physical and chemical quenching cannot be distinguished from one another; only the sum effect is measured. Therefore, it is clear that fluorescence measurements yield only an upper limit for k_{-xx} and, by applying Eq. VI.44, an upper limit for k_{xx} (see Jenkins 1968).

Using the specific (physical + chemical) quenching cross section found by Hooymayers (1966) for $K(5^2P)$ and applying Eq. VI.44 Zeegers (1966) calculated that dissociative quenching of this state by H_2O could be 1.5 times more efficient than physical quenching by H_2O in flames at about 2000 K. Because some activation energy is involved in the former process, dissociative quenching will be less

---

[†] For arbitrary Y, A in the denominator of Eq. VI.45 must be replaced by $A(1-Y)$; so for large values of Y the ratio cannot be determined without knowledge of Y.

Chapt. VI EXCITATION AND DE-EXCITATION OF METAL SPECIES

important in vapour cell experiments, where the temperature is much lower.

Similar calculations are feasible for other atomic lines for which the quenching cross section of H_2O at flame temperature (see Table VI.1) and the chemiluminescence rate constant (see below) are known.

When XY would quench the excited M^* predominantly by dissociation, that is, by the reverse reaction (xx), one expects from Eq. VI.44 that the quenching cross section will increase exponentially with increasing temperature. The temperature dependent factor in Eq. VI.44 is $\exp\left[-(D_0 - E_{exc})/kT\right]$, where the dissociation energy of XY, D_0, usually exceeds E_{exc} by some electronvolts. For instance, let us consider the quenching of $Na(3^2P)$ by O_2, which may proceed by inelastic collisions as well as by the reversed reaction (xix) with rate constant k_{-xix}. By applying Eq. VI.44 and using $k_{xix} \simeq 1 \times 10^{-29}$ cm^6 s^{-1}, as reported by Carabetta and Kaskan (1967), one finds $k_{-xix} = 4 \times 10^{-12}$ cm^3 s^{-1} at $T = 1950$ K. When this result is combined with the total quenching rate constant, k_q, of O_2 for the same Na doublet, found from Lijnse and van der Maas' (1973) fluorescence experiments: $k_q = \bar{g}_v \sigma_q = 7 \times 10^{-10}$ cm^3 s^{-1} (see Eq. II.130 and Table VI.1) at $T = 1950$ K, it follows that only about 1% of the quenching collisions results in dissociation of O_2. No strong T-dependence of the total quenching cross section is then expected, since the physical quenching rate is not supposed to vary strongly with temperature (see Sect. VI.2b-2). However, if the temperature is increased to about 2700 K, then k_{-xix} is estimated from Eq. VI.44 to attain the same order of magnitude as the physical quenching rate constant. Therefore, for temperatures higher than 2700 K one has to expect that the temperature dependence of the quenching rate of $Na(3^2P)$ by O_2, as determined, for instance, from fluorescence studies, will be given by $10^{-15120/T}$ (T in K) approximately.

Chemiquenching of excited alkali atoms by H_2O might also proceed through an *exchange reaction* of the type

$$M^* + H_2O \rightarrow MOH + H. \qquad (xxiii)$$

This process is exo-ergic for the first resonance doublets of all alkali atoms. Jenkins (1968, 1968b) has tried to explain qualitatively the trend in quenching cross sections found in flames for Li-Na-K by the corresponding decrease in absolute enthalpy change of the above reaction ($\Delta H = -1.1$, -0.3 and -0.1 eV for Li, Na and K). He assumed the quenching reaction to take place via crossing of potential-energy surfaces (see Sect. VI.2b-2).

Earl and Herm (1974) have suggested that in vapour cell experiments quenching of excited $K(5^2P)$ could also proceed through the above exchange reaction (xxiii).

CHEMICAL EXCITATION AND DE-EXCITATION PROCESSES (CHEMILUMINESCENCE)   Sect. VI.3c

Another exchange reaction has been suggested by Jenkins (1968b) for the quenching of the first resonance doublet of Li by H_2, according to: $Li^* + H_2 \rightarrow LiH + H$. The positive enthalpy change of the reaction is but small (0.14 eV); it becomes progressively larger for Na (0.37 eV) to Cs (1.3 eV), which makes this quenching reaction less probable for the other alkalis under flame conditions.

Similar dissociative and exchange reactions were also considered for the quenching of excited Hg atoms by H_2 (see Sect. VI.2b-2).

Bevan and Husain (1975, 1976) discussed the results of their quenching measurements with excited $Sb(5^2D_{\frac{3}{2},\frac{5}{2}})$ and $Bi(6^2D_{\frac{3}{2},\frac{5}{2}})$ atoms in collision with molecules such as H_2, O_2, CO_2 (see Table VI.1) in terms of correlation diagrams connecting, e.g., the initial $(Bi^* + O_2)$ state with various $(BiO + O)$ states (see also Donovan and Husain 1970 and Husain 1977). Whether the collision complex will again end up in a free, but de-excited metal atom (physical quenching) or in a metal compound (chemical quenching) still depends on the probability of subsequent diabatic transitions during the collision. They were able to explain qualitatively on this basis the difference in quenching cross sections measured for the doublet components separately. For these 'heavy' atoms the doublet splitting is relatively large, so strong spin-orbit or (J,Ω)-coupling was assumed in the correlation diagrams. Similar studies have been made by Trainor (1976, 1977) for the chemiquenching of excited Bi-, Cu- and Si atoms by O_2 or H_2, and by Breckenridge and Renlund (1978a) for the chemiquenching of excited $Cd(^3P_J)$ and $Cd(^1P_1)$ atoms by isotopic hydrogen molecules.

Discussion of some experimental rate constants. At present only a few rate constants of chemi-excitation reactions have been determined, for example, k_{xvi} for $K(5^2P)$ and $OH(^2\Sigma, v' = 0)$ by Zeegers (1966) and Zeegers and Alkemade (1965), k_{xvi} and k_{xvii} for $Na(3^2P)$, $Tl(7^2S)$, and $Pb(7^3P)$ by Jenkins (1968), Padley and Sugden (1959), and Phillips and Sugden (1961); and k_{xiv} for $Na(3^2P)$ by Carabetta and Kaskan (1967). The chemiluminescent character of many more atomic emission lines has been established, but lack of knowledge about the fluorescence efficiency involved has made it impossible to evaluate rate constants.

For some termolecular chemi-excitation rate constants, values in the range from 10^{-29} to 10^{-30} cm^6 s^{-1} have been reported. These values are surprisingly high for termolecular processes, and they have evoked speculation about the detailed reaction mechanism involved. Alkemade (1963) has suggested the occurrence of a two-step mechanism (see Sect. II.4h) and has compared semiquantitatively the two-step chemi-excitation rate with the rate of excitation through a direct process in terms of a persistence factor. The lower the energy of the excited state concerned, the higher the persistence factor of vibrational energy. This might explain the

649

Chapt. VI   EXCITATION AND DE-EXCITATION OF METAL SPECIES

observation by Phillips and Sugden (1961) that k_{xvii} is higher for Tl (E_{exc} = 3.3 eV) than for Pb (E_{exc} = 4.35 eV). One has to be careful in reaching such a conclusion, as was pointed out by Zeegers (1966). He has found a surprisingly high value of 3.5×10^{-30} cm^6 s^{-1} for k_{xvi} in the case of K(5^2P). If a two-step process plays really an important part, one has to expect that the overall rate constant is higher in flames diluted with N_2 than in flames diluted with argon, since the latter has only translational degrees of freedom to store the energy to be carried over to the metal atom in the second step. It is well known that the persistence factor for excess translational energy is practically unity (see Sect. II.3c-2). However, Zeegers has found the same value for the rate constant in flames with nitrogen and with argon as diluent gas, so that a two-step process must be ruled out.

The difference between the K- and the OH chemi-excitation rate constants found by Zeegers (1966) may be understood by considering that in reaction (xvi) OH may be excited to each of the many vibrational-rotational levels of the electronic ground state and of the $^2\Sigma$ state with $v' = 0, 1, 2$ and 3. Only a fraction of the total number of exciting reactions will produce OH in the observed ($^2\Sigma, v' = 0$) state. Such an explanation may also hold for the observed difference between thallium and lead, since the latter element has many more levels to which it may be excited.

3c-3   *REACTIONS INVOLVING A METAL COMPOUND*

Phillips and Sugden (1961) have suggested another type of two-step process

$$X + M + Z \rightarrow MX^* + Z \qquad \text{(xxiv)}$$

$$MX^* + Y \rightarrow M^* + XY . \qquad \text{(xxv)}$$

Here an intermediate compound between a metal atom M and a flame radical X is formed. Subsequently, this compound yields the excited atom upon reaction with another flame radical Y. It is very difficult to say whether such a mechanism may occur for elements of which stable excited intermediates, that is, MX*, are unknown. For elements that form such stable intermediates, for example CuH*, the actual occurrence of this mechanism may be checked by measuring the chemiluminescent intensity of the band emission as well as of the atomic line emission. The band emission is expected to vary linearly with radical concentration, whereas for the line emission a quadratic dependence on radical concentration is anticipated. There is no doubt that reactions such as (xxiv) may play a part in the excitation of a metal compound, but whether such a step occurs in the excitation of an atom has never been proved.

Reid and Sugden (1962) have ascribed the chemiluminescence of CuOH and MnOH to reaction (xxiv). In this study they have also investigated the chemi-

CHEMICAL EXCITATION AND DE-EXCITATION PROCESSES (CHEMILUMINESCENCE)   Sect. VI.3c

excitation of the alkaline-earth hydroxides. From the rather strong band emissions observed, they have concluded that the excited states involved are populated suprathermally. This should be brought about by

$$MO + H + Z \rightarrow MOH^* + Z \;. \qquad (xxvi)$$

However, on the basis of revised dissociation energies of the alkaline-earth hydroxides, Cotton and Jenkins (1968) have suggested that the rather strong emission of the hydroxide bands may be due, not to suprathermal chemiluminescence, but to the large proportion of hydroxides present.

The reverse of the quenching exchange reaction (xxiii), $MOH + H \rightarrow M^* + H_2O$, might be an alternative route for chemi-excitation of alkali atoms in flames, besides the termolecular recombination reactions (xvi - xix). This bimolecular excitation reaction is endo-ergic by 1.1 eV for Li and by a lesser amount for the other alkalis. The probability of this reaction for Li is favoured by the fact that this element occurs mainly as hydroxide in flames (see Sect. VIII.3b). The actual significance of this reaction in flames is difficult to assess because it need not lead to suprathermal emission in the presence of excess H radicals, in contrast to the recombination reactions (xvi) and (xvii). This is connected with the expected partial equilibrium between the reactants and products of this bimolecular reaction (see end of Sect. VI.4c-3 for further discussion).

Metal compounds may also be involved in chemical association reactions that produce a continuum radiation according to process no. 26 in Table II.1. Such continua have been observed in hydrogen- and acetylene-air flames containing alkali metals. The intensity of these continua is highest in the green-blue part of the spectrum and decreases gradually down to about 3100 Å, except for the lithium continuum, which extends to about 2800 Å (see James and Sugden 1958). These continua are now generally ascribed to the chemiluminescent reaction

$$M + OH \rightarrow MOH + h\nu \qquad (xxvii)$$

(M being an alkali metal). Accordingly, the spectral intensity of the continuum radiation is given by

$$I_\lambda = k_\lambda [M][OH] \;. \qquad (VI.46)$$

This dependence of I_λ on [M] and [OH] has been confirmed experimentally. James and Sugden (1958) have found that the reaction rate constant k_λ varies but weakly with T. The dependence of k_λ on wavelength has not been investigated extensively, although such a study may be of interest in relation to the shape of the potential curves of the complex involved. Only the spectral intensity of the potassium continuum has been investigated as a function of λ to some extent by Alkemade (1959). The main result of this study is the disclosure of a short wavelength cut-off at

651

3150 ± 100 Å corresponding to a maximum energy released of 3.9 ± 0.1 eV. This is only slightly higher than the dissociation energy of KOH, which is 3.75 eV. In order to draw conclusions concerning the shape of the potential curve of the activated complex formed in the association of K and OH the dependence of cut-off wavelength on temperature must be studied.

Conclusion. Soft chemiluminescence may cause a relative overpopulation as high as 10^2 in low-temperature flames of 1500-1800 K. This effect is of no practical use in analytical chemistry, since an emission enhancement of the same order can be obtained by choosing a flame with a temperature of 2500 K or higher. The occurrence of soft chemiluminescence may cause serious errors when deriving flame temperatures or ground-state metal concentrations from intensity measurements. This subject will be considered in Sect. VI.4c.

4. NONTHERMAL RADIATION OF ATOMIC LINES

In the previous sections, we were concerned mainly with the processes of (de-)excitation themselves and their relation to the population of the excited state. Here we shall discuss the effect they may have on the intensity of the outgoing radiation as received, for example, by a spectrometer. Ultimately, it is this outgoing radiation that is measured to extract from it information on the flame temperature, metal density, flame radicals, and so on, or that is utilized in analytical flame spectroscopy. In particular we shall here pay attention to the deviations from thermal emission intensity for atomic metal lines. Fluorescence radiation, which is an obvious case of nonthermal radiation, has been discussed extensively in Sects II.5c-4 and VI.2c-5 and will not be considered again. Induced emission will also be disregarded here (see Sect. II.5a).

4a ON THE CONCEPT OF 'THERMAL RADIATION'

According to the definition given in Sect. II.5c-3, the radiation of an atomic line from an isothermal region is called *thermal* if the line radiance as well as the spectral line profile conform to Maxwell-Boltzmann equilibrium. In particular, the <u>relative</u> populations of the involved upper and lower levels with respect to the ground level should conform to the Boltzmann distribution law (see Eq. II.30). The line-reversal temperature, measured with a thermal emission line, then equals the true temperature of the flame region observed (see Sect. II.5c-3).

When the temperature varies along the line of observation, we might still speak of *composite thermal radiation* if the Maxwell-Boltzmann distribution laws are <u>locally</u> obeyed by the radiating atoms at the prevailing temperature. Neither the total line radiance nor the emission factor can then be described by a simple expression, but they depend explicitly on the spatial distribution of temperature and of

NONTHERMAL RADIATION OF ATOMIC LINES   Sect. VI.4a

atomic density. The line-reversal temperature measured will lie somewhere between the maximum and minimum temperature values in the observed flame region. In general, it will vary with the varying supply of metal atoms to the flame and with varying radial distribution of the atoms inside the flame (see Sect. IV.6b). When the coloured flame border is cooler than the interior parts, a relative minimum or dip may appear at the centre of a resonance emission line at high atomic densities (self-reversal; see Sect. II.5c-3). Because of these complications, we shall restrict the discussion to the case when the flame temperature is uniform along the line of observation, whereas its value may still depend on the height of the observed flame section above the burner.

Some annotations are useful in regard to the concept of thermal radiation.
(i)   In order to be meaningful, this concept requires the possibility of defining unambiguously the 'true' flame temperature. When not all degrees of freedom in the flame gas are equilibrated as to their energy distributions, we may most conveniently define the flame temperature as the translational temperature of the bulk flame constituents (see Sect. III.10a).

(ii) It should be realized that 'thermal radiation' is essentially an approximative concept, when applied to open flames. The very absence of an equally hot enclosure around the flame means that the spectral volume density of the radiation field inside the flame is lower than its thermodynamic equilibrium value as determined by the Planck law. The relative population of the excited state will then not conform exactly to the Boltzmann law. However, the Boltzmann law is virtually obeyed in the case of strong self-absorption and, more generally, if the rate of radiative de-excitation is unimportant compared to that of the other de-excitation processes (that is, if $Y \ll 1$; see Sects VI.2c-1 and 2c-3). The higher the total gas pressure the better this condition is fulfilled. The outgoing radiation can then be called 'thermal' in an approximative sense. Also, the limited accuracy obtained in the practical determination of the true flame temperature allows some margin in the applicability of this concept (cf. Sect. III.10).

A deviation of the radiation field from Planck's law might, in principle, also have an effect on the spectral distribution of the spontaneously emitted photons, if there exists some degree of coherence between the re-absorbed and subsequently re-emitted photons (see Sect. II.5a-1). The spectral distribution of the emitted photons may then be influenced by the spectral distribution and the intensity of the radiation field present. This holds if the radiative excitation rate (by re-absorption) is at least comparable to the collisional excitation rate and if the coherence is not destroyed by collisional and Doppler broadening. In flames at 1 atm pressure we may neglect coherence effects (see Sect. II.5a-1).

Chapt. VI   EXCITATION AND DE-EXCITATION OF METAL SPECIES

(iii)   In the definition of thermal radiation, only the population of the excited levels relative to that of the ground level is essential. Deviations from dissociation or ionization equilibrium are thus irrelevant in regard to the existence of thermal radiation if they affect only the total free atom density, not the relative distribution over the atomic states. Under this acceptable condition, the line-reversal temperature is independent of such deviations.

(iv)   It should be emphasized that the adjective 'thermal' applies only to the nature of the outgoing radiation, and does not specify the kind of underlying excitation process. Thermal radiation is generally found when detailed balance exists for the dominant excitation and de-excitation processes, whatever they may be. The question of whether (de-)excitation is achieved mainly by physical collisional processes or by chemical reactions is irrelevant as long as each single process is balanced by its reverse. Under this condition, an equilibrium population of the excited state is maintained independently of the actual excitation process (see Sect. II.4a-2). Chemiluminescent reactions do not lead to nonthermal radiation if the reaction partners are in chemical and physical equilibrium. The excitation rate of the chemiluminescent reaction is then balanced by the quenching rate of the inverse chemical reaction (see also Sect. VI.3c). It is misleading to consider 'thermal radiation' and 'chemiluminescence' as intrinsically opposite concepts. Line-reversal temperatures measured under conditions of dominant chemiluminescent excitation need not necessarily deviate from the true flame temperature.

Even when a specific chemiluminescent reaction leads to the formation of excited atoms with high velocities, the Doppler broadening of the emission line will not be affected under equilibrium conditions. Again, this holds because the principle of detailed balance requires that excited atoms with high velocities are also more readily quenched by the reverse chemical reaction.

It is not always fully realized in the literature that conclusions about the prevailing type of (de-)excitation process cannot be drawn from the mere existence of thermal radiation.

4b   RELATIONSHIP BETWEEN OUTGOING RADIATION INTENSITY AND EXCITED-STATE POPULATION

In Sect. II.5c-3 we have derived, by an application of Kirchhoff's law, a simple, general relationship between the thermal radiation intensity, the spectral intensity of a black body and the integral absorption, A_t, taken along the line of observation. When A_t, the central wavelength and the flame temperature are given, one needs no further information about the light source or the line to calculate the thermal line radiance (see Eq. II.295). Furthermore, the spectral profile of the outgoing thermal radiation is identical to that of the absorption factor $\alpha(\lambda)$

(see Eq. II.294).

In the calculation of the nonthermal radiation intensity of a resonance line one has in general to know explicitly the excited-state population $n_1(y)$ and the ground-state population $n_0(y)$ as functions of the spatial coordinate, y, measured along the line of observation, as well as the normalized spectral distribution, $S_\lambda(\lambda)$ (see Sect. II.5b-1), of the spontaneously emitted photons. It is noted that in absence of equilibrium the spectral profile of $S_\lambda(\lambda)$ might deviate from that of the absorptivity; the latter is given by: $k(\lambda,y) = \kappa(\lambda)\,n_0(y)$ (see Eq. II.185).

We now derive an expression for the nonthermal spectral radiance B_λ at arbitrary λ, from which the total line radiance B can be obtained by an integration over λ (see also Alkemade 1963, and Omenetto, Winefordner and Alkemade 1975). We make the simplifying assumption that neither $S_\lambda(\lambda)$ nor $\kappa(\lambda)$ depend on the location in the flame. This assumption is a realistic one if we assume that the flame temperature T and gas composition are uniform and that any coherence between re-absorbed and subsequently re-emitted photons is absent (see Sect. 4a). The spectral radiance B_λ of the outgoing radiation will be calculated by first considering the value $dB_\lambda'(y)$ of an infinitesimal segment with thickness dy and at distance y from the flame border as measured along the line of observation. Because this segment is infinitesimally thin, we can use Eqs II.279 and 283, which hold in the absence of self-absorption, for calculating $dB_\lambda'(y)$. Replacing in Eq. II.279 $\int n_1(y)\,dy$ by $n_1(y)\,dy$ and using Eq. II.283 we find

$$dB_\lambda'(y) = (Ahc/4\pi\lambda_0)S_\lambda n_1(y)\,dy, \qquad (VI.47)$$

where A is the transition probability for spontaneous emission. Since the radiation from this segment has yet to travel over a distance y to the flame border, $dB_\lambda'(y)$ must be multiplied by the corresponding transmission factor $\tau(\lambda,y)$ to obtain the contribution, $dB_\lambda(y)$, of this segment to the spectral radiance of the outgoing radiation. Using Beer's law, Eq. II.256, we get

$$dB_\lambda(y) = dB_\lambda'(y)\,\exp\left[-\int_0^y \kappa(\lambda)\,n_0(y')\,dy'\right]. \qquad (VI.48)$$

Combination of Eqs VI.47 and 48 and integration over y between the limits 0 and l (= depth of flame along line of observation) yields

$$B_\lambda = (Ahc/4\pi\lambda_0)\,S_\lambda \int_0^l n_1(y)\,\exp\left[-\int_0^y \kappa(\lambda)\,n_0(y')\,dy'\right]dy. \qquad (VI.49)$$

Introducing the function:

$$u(y) \equiv \int_0^y \kappa(\lambda)\,n_0(y')\,dy',$$

Chapt. VI   EXCITATION AND DE-EXCITATION OF METAL SPECIES

with $du/dy = \kappa(\lambda) n_0(y)$, we can formally rewrite the latter equation as

$$B_\lambda = (Ahc/4\pi\lambda_0) \left\{S_\lambda/\kappa(\lambda)\right\} \int_0^l \left\{n_1(y)/n_0(y)\right\} \exp[-u] (du/dy) dy . \quad (VI.50)$$

The deviation from thermal equilibrium can be characterized by the *population factor* P which is defined as the ratio of the actual excited-state population to its Boltzmann equilibrium value. In our notation we thus have

$$P(y) \equiv n_1(y)/\{n_1(y)\}_e . \quad (VI.51)$$

Inserting $P(y)$ into Eq. VI.50 we get

$$B_\lambda = (Ahc/4\pi\lambda_0) \left\{S_\lambda/\kappa(\lambda)\right\} \left\{n_1(y)/n_0(y)\right\}_e \int_0^l P(y) \exp[-u] (du/dy) dy , \quad (VI.52)$$

where we have used the fact that the equilibrium ratio of n_1 and n_0 is independent of y in an isothermal section, while assuming that n_0 is, virtually, the same as in the case of Boltzmann equilibrium (because $n_1 \ll n_0$). Applying Boltzmann's law for the equilibrium ratio we get from the latter equation

$$B_\lambda = (Ahc/4\pi\lambda_0) \left\{S_\lambda/\kappa(\lambda)\right\} (g_1/g_0) \exp[-hc/\lambda_0 kT] \int_0^l P(y) \exp[-u](du/dy) dy .$$
$$(VI.53)$$

If we make the additional assumption that the deviation from Boltzmann equilibrium is independent of position, i.e. $P = $ constant, we can take P outside the integral. Realizing further that:

$$\int_0^l \exp[-u] (du/dy) dy = \int_0^{u(l)} \exp[-u] du = 1 - \exp[-u(l)] \equiv \alpha(\lambda) ,$$

we can reduce Eq. VI.53 to

$$B_\lambda = P\alpha(\lambda)(Ahc/4\pi\lambda_0) \left\{S_\lambda/\kappa(\lambda)\right\} (g_1/g_0) \exp[-hc/\lambda_0 kT] , \quad (VI.54)$$

where $\alpha(\lambda)$ is the absorption factor at λ along the line of observation.

In thermal equilibrium we have $P \equiv 1$, while Eq. II.178 together with Eq. II.209 can be used for the relation between AS_λ and $\kappa(\lambda)$. We then find from Eq. VI.54 for the thermal spectral radiance

$$(B_\lambda)_e = \alpha(\lambda)(2hc^2/\lambda_0^5) \exp[-hc/\lambda_0 kT] , \quad (VI.55)$$

which is indeed identical to Eq. II.294 if we use Wien's law for the black-body spectral radiance. Similarly the total thermal line radiance B_e, obtained by replacing $\alpha(\lambda)$ by the integral absorption A_t, is found to agree with Eq. II.295.

Wien's approximate law is here appropriate because we have disregarded induced emission. If we had taken induced emission into account, the exact Planck expression would have to be used, while $\alpha(\lambda)$ would have been replaced by the <u>net</u> absorption factor, $\alpha'(\lambda)$, defined in Sect. II.5c-2 (see Omenetto, Winefordner and Alkemade 1975).

The above, direct derivation of $(B_\lambda)_e$ shows more clearly than the derivation through Kirchhoff's law in Sect. II.5c-3 which premises are essential for obtaining thermal radiation. The usage of Eq. II.178 for the proportional relationship between $S_\lambda(\lambda)$ and $k(\lambda)$, and of the Boltzmann distribution law are seen to be essential, as well as the assumption of a uniform temperature.

When P is uniform but deviates from unity, while the spectral distributions of $S_\lambda(\lambda)$ and $\kappa(\lambda)$ still conform to equilibrium, we find from Eqs VI.54 and 55, using Eq. II.294,

$$B_\lambda = P(B_\lambda)_e = P\alpha(\lambda) B^{\mathbf{b}}_{\lambda_0}(T) , \qquad (\text{VI.56a})$$

$$B = PB_e = PA_t B^{\mathbf{b}}_{\lambda_0}(T) . \qquad (\text{VI.56b})$$

We can express P in terms of the excitation temperature, T_{exc}, of the considered emission line through Eq. II.114; T_{exc} equals the flame temperature in the case of a thermal emission line.

We can, finally, express B_λ and B also in terms of n_1 by starting from Eq. VI.50, while assuming that (n_1/n_0) is independent of y and that the spectral profile functions conform to equilibrium. The result is

$$B_\lambda = \alpha(\lambda)(2hc^2/\lambda_0^5)(g_0/g_1)(n_1/n_0) , \qquad (\text{VI.57a})$$

$$B = A_t(2hc^2/\lambda_0^5)(g_0/g_1)(n_1/n_0) . \qquad (\text{VI.57b})$$

We note that in the expression of Eq. VI.47 no extra term occurs that describes the contribution of photons being re-emitted within the considered flame segment after absorption of photons received from the outside. The contribution of these re-emitted photons is already implicitly accounted for by taking into account the contribution of radiative excitation to the excited-state population, $n_1(y)$, as described in Sect. VI.2c-3. These re-emitted photons must thus not be counted twice. This treatment is made possible by the absence of coherence between absorbed and re-emitted photons as discussed before.

4c CAUSES AND CONDITIONS OF NONTHERMAL RADIATION

Nonthermal radiation arises ultimately from deviations of the (de-)excitation rates from their thermodynamic equilibrium values. Following the same classification of excitation processes as used in Sects VI.2 and 3, we shall

Chapt. VI   EXCITATION AND DE-EXCITATION OF METAL SPECIES

describe here under which conditions and to what extent nonthermal radiation can be expected in flames.

4c-1   *NONTHERMAL RADIATION DUE TO A RELAXATION IN THE EXCITED-STATE POPULATION*

Even when, immediately after leaving the combustion zone, the flame molecules and radicals have reached a state of thermal equilibrium and the temperature is uniform, deviations from thermal metal radiation might, in principle, be caused by a *slow relaxation* in the attainment of a *stationary excited-state population*. When the excited state is depopulated by spontaneous emission and by quenching collisions with molecules Z (quenching rate constant = k_q), the relaxation time τ for the establishment of a stationary population is given by

$$\tau^{-1} = A + k_q[Z] \qquad (VI.58)$$

(compare with Eq. II.106 and the subsequent discussion of experimental relaxation methods). When radiation imprisonment is important, A should here be replaced by the reciprocal of the effective radiative lifetime τ_{eff} (see Sect. II.5c-3). In usual flames at 1 atm, the quenching term will be dominant and τ is expected to be less than 10^{-7} s. The flame gases travel over a vertical distance of 0.1 mm in roughly 10^{-5} s. Thus a stationary excited-state population will be attained practically immediately. Even when there is an appreciable vertical temperature gradient (say, 100 K mm^{-1}), the excited-state population will follow the variation in temperature very closely. Therefore, relaxation in the attainment of a stationary population in the excited state is not expected to be an important cause of nonthermal radiation, at least in the flame region outside the combustion zone (see also Gaydon and Wolfhard 1979).

The above relaxation time τ is independent of the <u>excitation</u> rate constant. For a given quenching rate per excited atom $k_q[Z]$, and temperature, the rate constant of the reverse collisional excitation process drops exponentially to zero when the excitation energy E_{exc} goes to infinity (cf. Eq. II.122 with $E_A \simeq E_{exc}$). For a sufficiently high value of E_{exc}, the probability for a given, <u>single</u> ground-state atom M to undergo at least one exciting collision during its residence time in the flame, may even become markedly less than unity. Nevertheless, [M*]/[M] will still attain its stationary value within the same (short) relaxation time predicted by Eq. VI.58. A decreased excitation rate constant has here only the effect that the stationary value of [M*]/[M] is decreased in the same proportion, too. This stationary value corresponds to a statistical <u>average</u> over a <u>large number</u> of atoms.

Relaxation effects in the attainment of a stationary excited-state population may, however, show up when the atomic vapour is irradiated by a very short laser pulse with pulse duration typically in the range of nanoseconds (cf. footnote † on page 630).

4c-2 *NONTHERMAL RADIATION CAUSED BY NONEQUILIBRATED PHYSICAL
       (DE-)EXCITATION PROCESSES*

(i)   (De-)excitation processes in which the electronic excitation energy is fully converted into *translational energy*, and vice versa, will not lead to nonthermal radiation, as the translational degrees of freedom are readily equilibrated in flames. If these processes are the dominant ones, the occupation of the excited state is expected to conform to its equilibrium value at the translational temperature of the flame.

This conclusion applies also to the case in which the atoms are excited by collisions with free electrons, insofar again as the electron velocity distribution conforms to the Maxwell law. The free-electron concentration, which may vary between wide limits, is, as such, irrelevant here. Even if the electron concentration deviates from its Saha equilibrium value, thermal radiation will still exist in this case. Only when excited neutral atoms are directly formed by the recombination of the parent ion and an electron (process no. 12 in Table II.1), will deviations from Saha equilibrium induce nonthermal radiation. In actual flames, the latter case may be important only for excited levels near the ionization limit (cf. Sect. IX.6). In the combustion zone of hydrocarbon flames, 'hot' electrons may be found with an excess of translational energy in significant concentrations (see Sect. IV.7). They may induce suprathermal radiation of metal lines. Greig (1965) has discussed the conditions under which the Na line-reversal temperature may markedly deviate from the flame gas temperature when the effective electron temperature exceeds the latter by a factor of 10.

(ii)   (De-)excitation processes in which the electronic excitation energy is (partially) converted into *internal molecular energy* are not likely to cause deviations from thermal metal radiation in the burnt flame gases at 1 atm. Relaxation effects in the equilibration of the vibrational degrees of freedom are much less pronounced here than in shock waves because of the difference in gas velocity and thus in residence time. In the combustion zone, certain combustion products may be found in a high state of vibrational or electronic excitation and may induce suprathermal metal emission upon collision. For example, electronically excited CO^* formed here in excess concentrations may have this kind of effect (see Sect. VI.3b).

In low-pressure flames, the equilibration of the vibrational degree of freedom is expected to proceed at a rate that is lowered in proportion to the pressure. The resulting persistence of an underpopulation of the vibrational levels above the combustion zone could then be a cause of infrathermal metal radiation.

Chapt. VI   EXCITATION AND DE-EXCITATION OF METAL SPECIES

It holds generally, for excitation by physical collisions, that deviations in the concentration of the collision partner from chemical equilibrium do not affect the equilibrium population of the excited metal state. The only condition is that the energy distribution of the partner conforms to the Maxwell-Boltzmann law.

(iii)   When excitation proceeds by the *exchange of electronic excitation energy* between metal atoms upon collision, no deviation from thermal radiation is expected if the relative population of the excited level in the donor atom conforms to the Boltzmann law. This is a consequence of the principle of detailed balance. Deviations from ionization or dissociation equilibrium for the donor atoms will not per se upset the thermal radiation of the other partner. Also, it was argued in Sect. VI.2b-4 that under flame conditions the exchange of energy between metal atoms is usually unimportant compared to the other excitation processes.

(iv)   *Infrathermal radiation* may arise from *radiative disequilibrium*†, that is, when the rate of photon emission is not balanced by an equal rate of photon absorption. For resonance lines, the deviation from the thermal intensity is stronger, the larger the fluorescence efficiency Y and the lower the metal concentration [M]. This is seen from Eq. VI.29, which describes the population factor P to the first order in the metal concentration. In the limit of very low metal concentration, the relative deviation $(= 1-P)$ equals Y. In most flames and for the alkali resonance lines, Y amounts to but 1-10%. It may be considerably larger in low-pressure flames, where the collisional rate is low, and in flames diluted by a noble gas, whose quenching efficiency is very small. In a stoichiometric H_2-O_2-Ar flame at 1 atm, Y values have been reported by Jenkins (1966) and Hooymayers and Lijnse (1969) to equal about 0.6 for the Na-D doublet.* However, in the limit of $[M] \to \infty$, the total line radiance approaches again its thermal equilibrium value.

The total line radiance for intermediate concentrations is calculated by first solving $P(x)$, as a function of position x, from Eqs VI.22 and 26. Once this function is known, the outgoing line radiance at uniform flame temperature T is obtained by solving B_λ from Eq. VI.53 and integrating $B_\lambda(\lambda)$ over wavelength. This may lead to complicated expressions, even in the case where the spectral distributions of the photon emission and photon absorption process are identical and the metal concentration is uniform. Asymptotic expressions for the intensity of a resonance line which hold in the range of high metal concentrations have been given

---

† Also called *radiation depletion* by Gaydon and Wolfhard (1979).

* The statement that radiative disequilibrium would become important at very low pressures only (see Schuler 1953) is thus not fully justified.

by Hooymayers (1966). In this range, the relative deviation of the line intensity from its equilibrium value decreases $\propto [M]^{-\frac{1}{2}}$, when only collisional and natural line broadening exist. This asymptotic behaviour is expected to apply also to resonance lines in flames at 1 atm where collisional and Doppler broadening are of comparable order of magnitude. This has been concluded theoretically and confirmed experimentally for the Na-D doublet in a H_2-O_2-Ar flame at 1 atm by Hooymayers (1966), Hooymayers and Alkemade (1966), and Hooymayers and Lijnse (1969).

A qualitative insight into this asymptotic behaviour $\propto [M]^{-\frac{1}{2}}$ may be gained by considering the spectral profile of the outgoing line radiation as a function of [M] or $k(\lambda_0)l$ (see Fig. VI.8). In the range of high atomic densities, the spectral

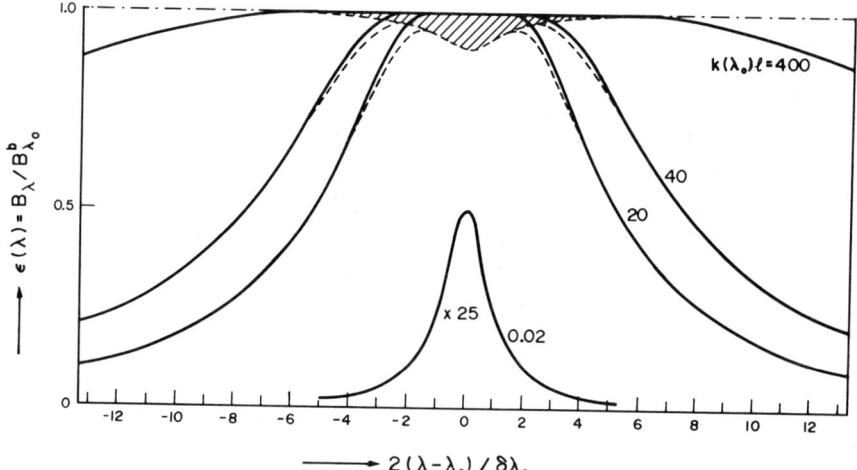

Fig. VI.8 Calculated spectral profiles of self-absorbed resonance line (solid curves) are shown for various values of $k(\lambda_0)l$, where $k(\lambda_0)$ is the absorptivity at line centre and l is the flame depth. The absorptivity is strictly proportional to the atomic density in the flame. The emission factor ε [defined by the ratio of spectral radiance $B_\lambda(\lambda)$ of the line to the spectral radiance $B_\lambda^b(\lambda_0)$ of a black body] is plotted as a function of normalized wavelength difference for the case of pure collisional broadening with half-intensity width $\delta\lambda_c$. The ordinate scale for the lowest value of $k(\lambda_0)l$ is enlarged by a factor of 25; the corresponding profile is essentially described by the function $k(\lambda)l$.

The dashed curves describe the distortion of the spectral profile at high atomic densities, caused by radiative disequilibrium near the flame border (see Sect. 4c-2 sub iv). The extent of this distortion depends on the efficiency of resonance fluorescence, which was here assumed to be 0.32. The shaded area represents the shortage in line radiance for the curve with the highest value of $k(\lambda_0)l$.

(The curves were redrawn from Figure 4 in Hooymayers 1966.)

profile of the self-absorbed line radiation shows a dip around the line centre (see dashed curve in Fig. VI.8). This dip arises because most photons at the line centre λ_0, escaping the flame, are produced in a boundary layer of thickness $\approx k(\lambda_0)^{-1}$. In this layer, the spectral radiant density is lower than in the interior parts of the flame (see Sect. VI.2c-1). As a consequence, the radiative excitation rate and the excited-state population are here below their thermodynamic equilibrium values. The closer the fluorescence efficiency Y is to unity (see Eq. VI.26), the larger this deviation of the excited-state population. Photons at the wings of the $k(\lambda)$ profile can travel over a larger distance without being re-absorbed. Most of the outgoing radiation at these wavelengths will thus be produced in the interior parts, where Boltzmann equilibrium is closely approached if strong self-absorption prevails (see Sect. VI.2c-1).

The depth of the dip is seen not to change greatly with [M], once [M] has become large enough to make the flame optically thick for $\lambda \simeq \lambda_0$ (see Fig. VI.8). Its width is of the order of that of the $k(\lambda)$ profile represented by the lowest curve in the figure and is practically independent of [M]. The total shortage in line radiance, corresponding to the shaded area in the figure, is thus approximately independent of [M]. On the other hand, the line radiance itself grows approximately $\propto$ [M]$^{\frac{1}{2}}$ (see Sect. II.5c-3). The relative deviation from the thermal line radiance is thus expected to vary $\propto$ [M]$^{-\frac{1}{2}}$.

The central dip in the line profile discussed here should be clearly distinguished from the well-known dip due to self-reversal (see Sect. II.5c-3). The latter occurs in flames with a relatively cool outer layer. The dip discussed above, however, is calculated for a strictly uniform flame temperature; a similar dip may also be found in low-pressure discharge sources where Y is close to unity, under uniform collisional excitation conditions (see computer calculations by Falk, Becker-Ross and Schiller 1971).

The total radiance of an infrathermal, slightly self-absorbed resonance line can be calculated, to second order in [M], from $B = PB_{\lambda_0}^b A_t$ (see Eq. VI.56b). Using the approximate expressions for P (Eq. VI.29) and for A_t (Eq. II.274) we obtain, while using $\kappa(\lambda)n_p = k(\lambda)$,

$$B \simeq (1-Y)\{1 + Y\overline{k(\lambda)}\,\bar{r}\} B_{\lambda_0}^b \int lk(\lambda)\,d\lambda\,\{1 - \tfrac{1}{2}\overline{lk(\lambda)}\}\ .$$

Re-arranging terms yields, to second order in [M],

$$B \simeq \{B_{\lambda_0}^b \int lk(\lambda)\,d\lambda\,\}\,(1-Y)\{1 + \overline{k(\lambda)}(Y\bar{r} - \tfrac{1}{2}l)\}\ .$$

When a cylindrical flame is viewed in a direction perpendicular to the flame axis, we have $\bar{r} \simeq 0.8\,l$ (see Sect. VI.2c-4). Using the general expression for $\int k(\lambda)\,d\lambda$ given by Eq. II.183 we finally obtain

$$B \propto B_{\lambda_0}^b\, l\, [M](1-Y)\{1 + \overline{k(\lambda)}\, l\,(0.8\,Y - \tfrac{1}{2})\}\ . \tag{VI.59}$$

The constant of proportionality follows from the constant factors occurring on the right-hand side of Eq. II.183. Realizing that $\overline{k(\lambda)}$ is proportional to [M], we see that the convex curvature in the intensity-density graph, due to self-absorption and described by the negative term $-\frac{1}{2}\overline{k(\lambda)}l$, is opposed by the effect of radiative disequilibrium described by the positive term $0.8Y\overline{k(\lambda)}l$. At strong radiative disequilibrium $(Y > 0.6)$, even an initial concave curvature may be found in this graph, as long as this second-order approximation is still valid.

As a further consequence of infrathermal radiation, the line-reversal temperature will be lower than the true flame temperature T_f.[†] The deviation ΔT is maximum for low atomic densities and is approximately equal to $(\lambda_0 k T_f^2/hc) \ln(1-Y)^{-1}$ (see Hooymayers and Alkemade 1966a). For the Na-D lines in a H_2-O_2-Ar flame $(T_f \approx 2300 \text{ K})$ at 1 atm, the maximum temperature defect is calculated to be 170 ± 8 K in agreement with the experimental value of 155 ± 15 K (see Hooymayers and Lijnse 1969). The true flame temperature is approached asymptotically when the metal concentration grows to infinity (and self-reversal is avoided). The true flame temperature can be found by extrapolation for [M] → ∞, while making use of the above asymptotic dependence $\propto [M]^{-\frac{1}{2}}$ for the relative deviation in total line radiance (see Hooymayers and Lijnse 1969). Computer calculations of the line-reversal temperature as a function of the Na concentration in a cylindrical low-pressure flame have been made by Schulz and Gutjahr (1968).[*]

The problem of radiative disequilibrium for doublet lines may be complicated by collisional mixing. The fractional underpopulations of the doublet levels are interconnected by mixing collisions, and they are unequal in the general case. Approximate expressions for the radiance of the doublet lines have been derived by Hooymayers and Lijnse (1969) for the Na-D doublet in flames at 1 atm. Detailed computer calculations have been given for the population factor of each doublet level as a function of position by Finn and Jefferies (1968, 1968a) for Y values close to unity. These Y values are not realistic in flames at 1 atm pressure.

4c-3 *SUPRATHERMAL RADIATION CAUSED BY NONEQUILIBRATED CHEMILUMINESCENT REACTIONS*

Nonthermal emission of a quite different nature arises when reactive species in excess concentrations or with excess internal energy give rise to nonequilibrated

---

[†] The reversal temperature is, however, not a measure for the value of the population factor as averaged over the flame cross section, unless the flame is optically thin $[k(\lambda_0)l \ll 1]$. In a moderately thick flame, the radiation near the line centre stems mainly from a thin boundary layer where the deviation of the population factor from the Boltzmann value is the largest (see Fig. VI.8).

[*] Some misprints are found in their equations nos. 3 and 6. These calculations were based on a rather crude approximation without iteration steps. This approximation may be acceptable in the high-concentration range, but may lead to a serious error at low concentrations because Y is close to unity in the low-pressure flame.

Chapt. VI   EXCITATION AND DE-EXCITATION OF METAL SPECIES

chemiluminescent reactions. Examples were given in Sect. 3 for the combustion zone as well as for the adjacent region in the burnt flame gases. Here we shall present quantitative expressions for a particular type of chemiluminescent reaction, in order to elucidate the general influence of certain factors on the deviation from thermal radiation.

Case of (induced) chemiluminescence through 3-body association reaction. Consider the case in which metal atoms M are excited only by inelastic collisions with molecules Z and by the 3-body association reaction: $M + X + Y \to M^* + XY$ (see processes no. 3 and 23 in Table II.1). An example of such a chemiluminescent reaction where X and Y are H and OH radicals, respectively, was discussed in Sect. VI.3c. In the following the de-excitation rate by spontaneous emission and radiative disequilibrium effects will be disregarded; de-excitation is assumed to proceed only through the above processes in reverse. Because of the rapid establishment of a stationary excited-state population (see Sect. 4c-1), we find, using Eq. II.108 and assuming that all rate constants as well as [Z] conform to equilibrium, for the ratio $P \equiv [M^*]/[M^*]_e$

$$P = \frac{1 + W[X][Y]/[X]_e[Y]_e}{1 + W[XY]/[XY]_e} . \qquad (VI.60)$$

Here W is defined by

$$W \equiv k_3[X]_e[Y]_e / k_2[Z]_e , \qquad (VI.61)$$

where k_2 and k_3 are as defined in Sect. II.4a-3. The ratio W is a measure for the importance of the chemiluminescent excitation relative to the collisional excitation, in the case of equilibrium. Because of detailed balance, W also equals the ratio of the corresponding quenching rates in equilibrium.

The assumption that $[Z] = [Z]_e$ is reasonable, since Z is usually a bulk flame molecule such as N_2 or CO_2, whose concentration is affected but little by the deviation of [X] and [Y] from chemical equilibrium. It is noted that [Z] refers to the total concentration of Z in all states of excitation, including the ground state; the rate constants k_2 and k_{-2} are defined accordingly. The assumption that the rate constants also equal their equilibrium values is reasonable for the flame region above the combustion zone, where physical equilibrium is rapidly established. Inside the combustion zone, the flame species taking part in a chemiluminescent reaction may have excess internal energy; the corresponding rate constants may be affected by that.

At first sight, the neglect of radiative excitation might not seem justified when the suprathermal emission due to chemiluminescence is so strong that the spectral radiant density in the flame at the atomic line centre exceeds the

Planck equilibrium value. This situation might occur in the combustion zone, in particular for ultraviolet lines, for which the Planck value is comparatively small. However, as long as the fractional number of photons produced per chemiluminescent reaction is small compared to unity, the radiative excitation rate by re-absorption of these photons can certainly be neglected relative to the chemiluminescent rate. A small photon yield is expected beforehand as collisional quenching usually dominates over photon emission.

When the deviation of P from unity, resulting from suprathermal chemiluminescence, is uniform in the observed flame part, the line radiance B is given by Eq. VI.56b. For resonance lines with self-absorption, the relative variation of B with atomic density or flame depth is then described by the same factor A_t as in the case of thermal radiation.

Inspection of the above equations reveals that an overpopulation of the excited state ($P>1$) occurs if $[X][Y]/[X]_e[Y]_e$ exceeds the ratio $[XY]/[XY]_e$, that is, for $[X][Y]/[XY] > K_d$. Here K_d is the dissociation constant of: $XY \rightleftharpoons X+Y$ at equilibrium. Such deviation from chemical equilibrium may exist for the dissociation of H_2O into $H+OH$ (see Sect. IV.5). This deviation and the associated suprathermal metal emission usually diminish with increasing height above the combustion zone. Above this zone, we usually have: $[XY] \simeq [XY]_e$. The suprathermal emission intensity is then related only to $[X][Y]/[X]_e[Y]_e$ and W. A small value of W entails that the influence of chemical disequilibrium on P is only weak because of the relatively much stronger collisional excitation rate, unless $[X][Y]$ is considerably larger than $[X]_e[Y]_e$. For $W \ll 1$ and $[XY] \simeq [XY]_e$, we may write, instead of Eq. VI.60,

$$P \simeq 1 + W[X][Y]/[X]_e[Y]_e . \tag{VI.62}$$

In general, the presence of reactants (X and Y) in excess concentrations is not a sufficient condition for suprathermal radiation. The possible excess concentration of the reaction product (XY) must also be considered. Of course, an excess of both X and XY can occur only if there exists another compound of X whose concentration in the flame is below the equilibrium value.

Factors influencing suprathermal chemiluminescence. We shall now discuss separately the influence of some specific factors on the suprathermal chemiluminescence.

(i) When the *flame pressure* is reduced, the approach to dissociation equilibrium for: $XY \rightleftharpoons X+Y$ is generally slower, and suprathermal radiation may persist over a much longer distance from the combustion zone (see Sugden 1962).

(ii) The dependence of P on the *excitation energy* E_{exc}, for given excess concentrations of X and Y, may be judged from Eqs VI.61 and 62. If the energy released by recombination of X and Y is larger than E_{exc}, the chemical excitation rate constant k_3 is not expected to depend strongly on E_{exc}. On the other hand, the collisional excitation process includes an energy of activation which is approximately equal to E_{exc}. The corresponding rate constant k_2 is described by Eq. II.124 with $E_{\text{Arrh}} \approx E_{\text{exc}}$. The factor W occurring in Eq. VI.62 is then expected to increase exponentially with increasing E_{exc}. This explains why suprathermal radiation becomes more conspicuous for ultraviolet metal lines than for visible lines if $h\nu = E_{\text{exc}}$ (cf. Fig. VI.9 in the following subsection).

(iii) The influence of the *temperature* on the overpopulation may be judged by similar considerations as sub (ii). It can be assumed that the excess concentrations at which X and Y are formed in the combustion zone are not influenced strongly by the final flame temperature. The rate of their decay by termolecular recombination downstream from the combustion zone will also not depend critically on T. According to Eq. VI.62 the dependence of P on T is then governed mainly by the temperature dependence of $W/[X]_e[Y]_e = k_3/k_2[Z]_e$. The chemiluminescent rate constant k_3 will not vary strongly with T as long as the recombination energy of X+Y exceeds E_{exc} (compare the above). In contrast, k_2 depends exponentially on T^{-1} according to Eq. II.124. The concentration of Z will depend but weakly on T, since Z is supposed to be a stable flame species occurring in bulk concentration. Consequently, the second term in the right-hand side of Eq. VI.62 varies with temperature mainly through the exponential T^{-1}-dependence of k_2^{-1} contained in W (see Eq. VI.61). Thus suprathermal chemiluminescent emission becomes more pronounced in cooler flames, all other factors being kept the same. For example, McEwan and Phillips (1965) found noticeable suprathermal emission for the first resonance doublet of lithium in cool hydrogen flames $(T<1600\,\text{K})$ only.

Sugden (1965) has estimated the rates of chemiluminescent excitation and of collisional excitation, for different values of flame temperature and excitation energy (see Table VI.2). Referring to typical conditions in H_2 flames at 1 atm, he assumed (roughly):

$$[Z] \approx 10^{19}\,\text{cm}^{-3}, \quad [X] \approx [Y] \approx 10^{17}\,\text{cm}^{-3},$$

$$k_2 \approx 10^{-12} \exp[-E_{\text{exc}}/kT]\,\text{cm}^3\,\text{s}^{-1}, \text{ and } k_3 \approx 10^{-32}\,\text{cm}^6\,\text{s}^{-1}.$$

Table VI.2 clearly shows the strong effect of T on the ratio of the two rates. This effect is stronger, the larger E_{exc} (for resonance lines, an excitation energy of 1.7 eV and 3.5 eV corresponds to $\lambda \approx 7000$ and $3500\,\text{Å}$, respectively).

The ratio, $k_3[X][Y]/k_2[Z]$, of the excitation rates considered in Table VI.2 is identical to the term $W[X][Y]/[X]_e[Y]_e$ occurring on the right-hand side of Eq. VI.60. The value of this term is indicative of suprathermal emission

Table VI.2 [†]
Estimated Rate Constants (in s⁻¹) for Collisional Excitation
($k_2[Z]$) and Chemiluminescent Excitation by Radical Recombination ($k_3[X][Y]$) in Flames at 1 atm

T, K	$k_2[Z]$ (E_{exc} = 1.7 eV)	$k_2[Z]$ (E_{exc} = 3.5 eV)	$k_3[X][Y]$
1000	10^{-1}	10^{-9}	10^2
2000	10^3	10^{-1}	10^2
4000	10^7	10^7	10^2

if $W \ll 1$ and [XY] ≃ [XY]$_e$ (see Eq. VI.62). However, when $W \gg 1$ or when [X][Y] / [X]$_e$[Y]$_e$ ≫ 1/W, it is only the deviation of [X][Y] / [X]$_e$[Y]$_e$ from unity that is a measure for the overpopulation of the excited state. In the latter case the ratio of the excitation rates listed in Table VI.2 as such is not a criterion for the existence of (non-)thermal radiation (see Alkemade 1965).

(iv) The effect of chemical disequilibrium on P depends on k_2 through the factor W occurring in Eq. VI.60 or 62. Thus the overpopulation of the excited state, due to chemical disequilibrium, depends in an indirect way on the average *excitation efficiency* of the *bulk flame species*. Suprathermal chemiluminescence will therefore be more pronounced in flames diluted with a noble gas, for example, than in similar flames diluted with nitrogen, because noble gas atoms are known to (de-)excite atomic states very poorly (see Sect. VI.2b). A value $W = 0.12$ has been measured by Zeegers (1966) in a H_2-O_2-Ar flame at 1 atm and 1950 K for the case of chemiluminescent excitation of the blue K doublet by recombination of H and OH. When k_2 and thus k_{-2} are so small that the collisional quenching rate per excited M* atom is comparable to, or less than, the spontaneous emission rate, the above analysis no longer holds. Eq. VI.60 should then be extended by including a term describing the rate of radiative (de-)excitation (cf. Sect. VI.3c-2).

Case of (true) chemiluminescence through 2-body exchange reaction. So far we have considered, as an example, suprathermal chemiluminescence caused by a termolecular radical association reaction. Whenever the concentrations of the radicals exceed their equilibrium values, suprathermal emission results. This conclusion need not be necessarily true when ('true') chemi-excitation takes place by a bimolecular exchange reaction between a flame radical and a metal compound

---

[†] Derived from Sugden (1965).

Chapt. VI   EXCITATION AND DE-EXCITATION OF METAL SPECIES

(see reaction no. 19 in Table II.1). Consider the following pair of exchange
reactions which lead to the formation of an excited and of a ground-state atom,
respectively (see also Sects VI.3c-2 and VIII.4a)

$$MOH + H \rightleftharpoons M^* + H_2O \qquad (i)$$

$$MOH + H \rightleftharpoons M + H_2O \ . \qquad (ii)$$

Just because these reactions are bimolecular, we can suppose that a partial equilibrium is readily established for each of these reactions, with equilibrium constants
K_i and K_{ii} (see Sect. II.4a-2). It is easily seen that now: $[M^*]/[M] = K_i/K_{ii}$,
which is equal to the ratio expected under full chemical equilibrium conditions,
$[M^*]_e/[M]_e$. If such additional partial equilibria are established for the
ground-state and excited-state species simultaneously, the emission will be thermal,
independently of any deviation of [H] from equilibrium. This means that the ratio
$[M^*]/[M]$ will obey Boltzmann's law, although $[M^*]$ and $[M]$ may not have attained
chemical equilibrium values. Also, when M is excited by inelastic collisions in
addition to reaction (i), the ratio $[M^*]/[M]$ will be thermal. Even when, in addition to reactions (i) and (ii), a nonequilibrated, slow reaction like (xx) in Sect.
VI.3c-2 contributes to the excitation of M, the ratio will be thermal. The only
condition is that the production rate of reaction (xx) be slow enough compared to
that of (i), so that (i) is equilibrated irrespective of the occurrence of (xx).

Extra Doppler broadening. Chemiluminescence reactions may also induce
extra Doppler broadening of the line. This may occur when excess chemical energy
is available for partitioning as translation energy between the reaction products
including the excited atom. An essential condition for extra Doppler broadening is
that the chemiluminescent excitation rate is not balanced by the rate of the reverse
chemiquenching reaction (see Sect. II.5b-2). We refer to Sect. VII.2b-9 for a further discussion. It is only concluded here that Eqs VI.56a and 56b then no longer
describe the suprathermal chemiluminescent (spectral) radiance as a function of
atomic density and wavelength. In particular the loss of outgoing emission intensity due to self-absorption is then no longer described by the curve-of-growth
theory.

4d   EXPERIMENTAL METHODS OF EVALUATING NONTHERMAL RADIATION
      Some methods will be described which enable us to evaluate experimentally
the deviation from thermal radiation. We shall restrict ourselves to the case in
which the temperature is uniform throughout the flame region viewed by the optical
system.

(i) The most direct method giving evidence of any deviation from thermal radiation is the measurement of the excitation temperature T_{exc} by the line-reversal method (see Sects II.5c-3 and III.10c). When T_{exc} deviates from the known true flame temperature T_f, nonthermal radiation exists for the resonance line considered. This can be concluded a fortiori when the suprathermal radiation appears to be so strong that no line-reversal can be obtained at all, when a continuum background source whose radiance temperature exceeds markedly the maximum theoretical flame temperature is used. In the combustion zone of an C_2H_2-air flame at low pressure, there have even been observed nonresonance lines of lead that still appear as emission lines against a carbon arc as background source (see Gaydon and Wolfhard 1979).

The application of this method presupposes that T_f is known. Its value can be determined by applying the method of line-reversal to a resonance line with such low excitation energy that its radiation is unlikely to be affected by chemiluminescence (see Sect. 4c-3 sub ii). Additional evidence for the latter supposition may be gained by checking the mutual agreement between the line-reversal temperatures obtained for a number of resonance lines with low excitation energies. To ensure reliability, it should also be verified that the flame temperature thus measured is independent of the atomic density (cf. Sect. 4c-2 sub iv). The flame temperature can also be found by measuring the rotational or vibrational temperature of molecular flame bands (see Sect. II.5c-3), if rotational or vibrational equilibrium exists. The flame temperature may also be obtained independently by other methods described in Sect. III.10.

The method of line-reversal yields essentially the same result as that obtained by a combined measurement of the line radiance and the atomic density [M] through Eq. II.281 in the absence of self-absorption. This method, involving absolute measurements, may be followed for nonresonance lines for which line-reversal cannot practically be applied or for suprathermal resonance lines with very high excitation temperatures. The difficulty is, however, that [M] must be determined. For resonance lines, [M] can be found from absorption measurements (see Sects V.2b, 2c and 2d). Since the oscillator strength for absorption f is related to the transition probability for emission A through Eq. II.163, the value of T_{exc} found is independent of the adopted values of A and f. When nonresonance lines are investigated, [M] can be calibrated by absorption measurements with a resonance line of the same atom. It is also possible to calibrate [M] by the comparison method (see Sect. V.6), if the absolute density of the standard element can be found by optical methods and if M is known to be fully atomized.

Of course, [M] may also be eliminated by measuring only the intensity ratio of the investigated nonthermal line and a thermal reference line of the same atom. When T_f is known, application of Eq. II.281 provides a check on the nonthermal character of the line in question.

The line-reversal method has been applied, for example, to prove the existence of suprathermal radiation of metal lines in the combustion zone of an C_2H_2-air flame at reduced pressure (see Gaydon and Wolfhard 1979). The reversal temperature for the 5890/96 Å Na doublet appeared to be close to the final flame temperature ($\approx$ 2000 K), but for a series of ultraviolet Fe and Pb resonance lines the reversal temperature increased monotonously up to 3450 K with decreasing wavelength. De Galan and Winefordner (1967a) have measured excitation temperatures by the line-reversal method for a number of metal lines in unpremixed, turbulent H_2-O_2 and H_2-air flames at 1 atm (see Fig. VI.9). These flames are typical of those used in analytical flame spectrometry. Measurements were made at a distance of some centimetres above the base of the flame. The mutual consistency found in the H_2-O_2 flame proves that thermal radiation occurs here. The systematic increase of T_{exc} with increasing energy E_{exc} in the other flame points to suprathermal radiation of metal lines with E_{exc} exceeding 22000 cm^{-1} ($\hat{=}$ 2.7 eV). According to de Galan and Winefordner, this is probably due to the persistence of radicals in excess concentrations over a large height interval (cf. Sect. 4c-3).

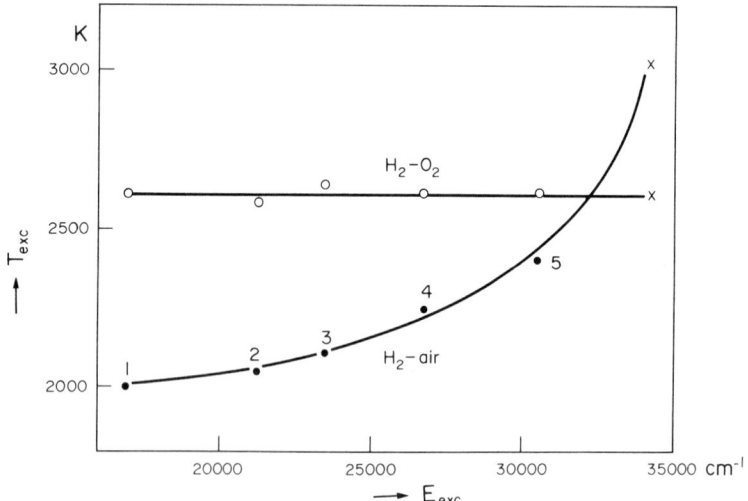

Fig. VI.9 The excitation temperature T_{exc} (measured by the method of line reversal) is plotted as a function of the excitation energy E_{exc} of the emission line observed, for an unpremixed H_2-O_2 and H_2-air flame, respectively. Legend: (1) Na 5890 Å; (2) Sr 4607 Å; (3) Ca 4227 Å; (4) Fe 3720 Å; (5) Cu 3247 Å; X is the Fe-excitation temperature derived from a plot of $\ln(A/I)$ versus E_{exc} for Fe lines with $E_{exc} \approx 34000$ cm^{-1} (see text sub ii). $E_{exc} = 10000$ cm^{-1} corresponds to 1.24 eV.
(Reprinted from de Galan and Winefordner 1967a, p.711.)

NONTHERMAL RADIATION OF ATOMIC LINES   Sect. VI.4d

Behmenburg, Kohn and Mailänder (1964) have found a higher reversal temperature for the 4227 Å Ca line than for the 5890 Å Na line in a premixed C_2H_2-O_2-N_2 flame at 1 atm and $T = 2500$ K slightly above the combustion zone. In a flame diluted with He instead of N_2, the difference was more pronounced. Suprathermal chemiluminescent excitation probably plays a role with the Ca line having the larger excitation energy.

When the reversal temperature increases systematically with metal concentration, while approaching constant asymptotic values for low and high concentrations, respectively, infrathermal radiation due to radiative disequilibrium probably exists (see Sect. 4c-2 sub iv). This conclusion can be drawn only if the temperature is really uniform in the coloured flame region under observation.

(ii) When for a set of non-selfabsorbed lines of the same element, with known values of E_{exc}, transition probability A and statistical weight g^* of the upper level, relative intensities I are measured, a plot of $\ln(Ag^*/I\lambda)$ versus E_{exc} should yield a straight line in the case of thermal radiation (cf. Eq. II.280). The slope of this line is determined by $(kT_f)^{-1}$ (so-called slope method of temperature measurement; see Sect. III.10c-2). Deviations from a straight line or the appearance of a different slope point to nonthermal radiation. When the lines are closely grouped in the spectrum, the proportionality constant relating the intensity and the meter reading may be assumed to be independent of wavelength. Note that in Fig. VI.9 the Fe-excitation temperature determined from the slope of such a plot agrees well with the true temperature for the H_2-O_2 flame as determined by line-reversal measurements. It does not agree, however, for the unpremixed H_2-air flame, where suprathermal radiation occurs for lines with $E_{exc} \geqslant 3$ eV. Broida and Shuler (1957) have proved by this method the nonthermal excitation of Fe lines in the combustion zone of an C_2H_2-O_2 flame. Gilbert (1963, 1966) has shown in the same way the probable existence of suprathermal radiation for a series of ultraviolet Sn lines in an incandescent unpremixed C_2H_2-O_2 flame.

(iii) The question of thermal radiation in the region above the combustion zone may also be solved by relative intensity measurements at different, known flame temperatures. If E_{exc} is known and $[M]l$ remains constant, it can easily be checked whether or not the relative intensity varies with flame temperature in accordance with Eq. II.281. The transition probability need not be known. When $[M]l$ varies, its influence on the line intensity can be corrected for by additional measurements of the relative variation in the absorption factor. Additional measurements of the relative variation in intensity of another line of the same metal which is known to be thermally excited may also be useful in this respect. When the investigated line is self-absorbed, the influence of a known variation in $[M]l$ on the intensity can be derived from the experimental curve-of-growth.

Chapt. VI   EXCITATION AND DE-EXCITATION OF METAL SPECIES

A special application of this method is the measurement of the relative variation in intensity with height in the flame, reported by Padley and Sugden (1959) and Alkemade (1963). When the variation of T_f with height is known and the variation in [M]l can be corrected for, the presence of nonthermal radiation in this height interval is easily checked in that way. When strong suprathermal excitation occurs because of the presence of excess flame radicals, the variation of intensity with height reflects the decay of these radicals with height (see Sect. 4c-3 and Eq. VI.62) rather than the temperature variation. Essentially the same method was followed by Vickers, Cottrell and Breaky (1970) to check nonthermal radiation of Cu, Sn and OH in a turbulent H_2-air flame with slot burner when alcohol was added to the sprayed solution. These authors measured the ratio of the relative line intensity to the absorption factor of the same line with a continuous background source, as a function of height.

(iv)   The existence of nonthermal radiation above the combustion zone may also be proved by measuring the relative variation $\Delta I/I$, caused by a comparatively small, unknown variation ΔT_f, for a set of different lines of the same atom (see Alkemade and Zeegers 1971). When the variation in temperature T_f occurs in combination with a variation in [M]l, we have, in good approximation, for lines with negligible self-asborption from Eq. II.281

$$\Delta I/I = (E_{exc}/kT_f)(\Delta T_f/T_f) + \Delta([M]l)/[M]l \; . \qquad (VI.63)$$

A plot of $\Delta I/I$ versus E_{exc} should yield a straight line if all spectral lines are thermally excited. Neither the transition probabilities nor ΔT_f need be known in this test. As an additional result, we find the relative variation $\Delta([M]l)/[M]l$ from the point of intersection of the line with the ordinate axis.

(v)   When the relative intensity of a moderately self-absorbed line varies more strongly with the metal concentration in the solution than the integral absorption A_t, infrathermal radiation caused by radiative disequilibrium exists (see Sect. 4c-2 sub iv). The concentration dependence of A_t can be determined from integral absorption measurements or from the theoretical curve-of-growth (see Sect. II.5c-2). There are, however, many other causes that may distort the relationship between A_t and the solution concentration (see Sect. II.5c-2). Besides, in nonuniform flames self-reversal may also affect the resonance line intensity at high concentrations (see Sect. II.5c-3).

On the other hand, the observation that the emission intensity follows A_t as a function of concentration does not exclude the possibility of nonthermal chemiluminescent radiation (see the occurrence of A_t as a factor in Eq. VI.56b).

Whenever a deviation from Boltzmann equilibrium is suspected for the upper level of an atomic emission line, it is advisable not to use this line in temperature measurements by the line-reversal or slope method (see Sect. III.10). Temperature measurements by atomic absorption or atomic fluorescence methods (see Sect. III.10) are then more reliable; they are based on the relative population of low-lying excited levels, which are much less affected by chemiluminescence (see Sect. 4c-3 sub ii).

4e   SOME CONCLUSIONS FOR THE ANALYTICAL APPLICATION OF
      FLAME EMISSION SPECTROSCOPY

The theoretical question of thermal or nonthermal radiation is not directly relevant in the analytical application of flame spectroscopy. It suffices that the intensity for a given analyte concentration in the solution is reproducible, because reference solutions with known analyte concentrations are used for calibration.

Suprathermal radiation has been applied to improve the detection limit not only in flame atomic fluorescence spectroscopy but also in *flame chemiluminescence spectroscopy*. The former application has been briefly dealt with in Sects II.5c-4 and III.16. Here we shall summarize some conclusions from the foregoing subsections with regard to the latter application only. For a more extensive discussion and literature survey we refer to Herrmann and Alkemade (1963), Mavrodineanu and Boiteux (1965), Alkemade (1969), Gilbert (1963, 1970), and Alkemade and Herrmann (1979). Chemiluminescence spectroscopy includes atomic lines as well as molecular bands, and is utilized for the analysis of metallic as well as nonmetallic elements. Here we shall restrict ourselves to the chemiluminescence of atomic metal lines.

(i)   The *shape of the analytical curve* for self-absorbed resonance lines showing suprathermal chemiluminescence need not necessarily deviate from that found under thermal excitation conditions. This is seen from Eq. VI.56b which directly expresses the proportionality between the nonthermal and the thermal radiance if the excitation conditions are at least uniform in the flame region under observation. Such conditions are not expected, however, when the chemiluminescent reactions involve short-lived radicals that are typical for the combustion zone. The concentration of these radicals may vary strongly with radial distance from the flame axis.

However, when the nonequilibrated chemical excitation reaction leads to an extra Doppler broadening of the emission line (see Sect. 4c-3), the spectral distribution of the photon emission process will be broader than that of the photon absorption process. The loss of radiation due to self-absorption will be somewhat reduced by that difference.

(ii)   Flame chemiluminescence spectroscopy is especially exploited with the goal of obtaining better *analytical detection limits* (see definition in Sect. III.14f-1). The gain in detection limit depends for a great deal on the increase in atomic line intensity relative to that of the flame background fluctuations. Hard chemiluminescence of metal additives (see Sect. 3b) can be brought about by choosing a very fuel-rich hydrocarbon flame or by using an alcoholic solvent. But the flame background intensity will generally be increased, too (see Gilbert 1963).

The gain in atomic line intensity as compared to thermal intensity can hardly be computed quantitatively by theory alone. The pertinent equations given in Sect. 4c-3 are applicable only to a special type of soft chemiluminescence and under idealized flame conditions. The few available values of the absolute rate constants and the expected excess H, OH and O radical concentrations render soft chemiluminescence of only limited practical use in analytical flame spectroscopy (see also end of Sect. 3c). The rate constants involved in hard chemiluminescence are unknown in most cases. Besides, the physical and chemical conditions in the primary combustion zone, where hard chemiluminescence naturally occurs, are usually ill-defined.

Nevertheless, the equations and discussions in Sects 3 and 4 may be helpful in giving us at least a qualitative insight into the general conditions favourable for the appearance of strong, suprathermal atomic lines in practice. From the discussion in Sect. 4c-3 on the influence of the excitation energy and flame temperature, one expects the effect of suprathermal chemiluminescence to be most conspicuous for u.v. lines and in cool flames, which generally also exhibit a lower background. Also, it appears from the same subsection that suprathermal chemiluminescence is more pronounced in flames diluted with a noble gas than with a molecular gas.

From the discussion on the antagonistic effect of the chemical excitation and the reverse chemiquenching reaction in Sect. 4c-3, we conclude that a large concentration of reactant flame radicals is not a sufficient condition for the occurrence of suprathermal radiation. The possible excess concentration of the flame species that occurs as a product in the reaction considered should be taken into account, too. Only if the reactant species are present in a larger excess over chemical equilibrium than the product species can suprathermal chemiluminescence be expected. The latter condition is not in all cases a sufficient one either, as we showed at the end of Sect. 4c-3 for a special pair of simultaneous, partially balanced exchange reactions.

(iii) The possibility of *excitation interference* between the analysis element and a concomitant metallic element in emission flame spectroscopy has been debated in the past (see Johnson and Schrenk 1964, Slavin 1964, West 1967, and Alkemade 1968). If the concomitant atoms are suprathermally excited in some way, they could, in principle, cause an overpopulation of the excited level of the analyte atoms upon collision. The analyte emission would then be enhanced by the presence of the

NONTHERMAL RADIATION OF ATOMIC LINES   Sect. VI.4e

concomitant in the solution and an interference would occur. When the relative populations of the excited levels of both analyte and concomitant atoms obey Boltzmann's distribution law, such interference is, however, excluded by the principle of detailed balance (see Sect. 4c-2 sub iii). However, even when this condition is not fulfilled, there will be hardly any disturbing effect on the analysis element, because a collision between two metal atoms in the flame is but a comparatively rare event (see Sect. 2b-4).

There still remain some unexplained observations of anomalous intensities of atomic metal lines in analytical flame spectroscopy (see, e.g., Alkemade and Herrmann 1979). Attempts to explain them often fail because the flame conditions under which they were found are complicated, ill-defined and far from uniform. This applies especially to nonshielded unpremixed flames. Also the experimental factors on which these anomalous intensities may depend are often incompletely investigated. The first thing to decide is whether the high atomic line intensity is caused by an anomalous atomization (see Chapt. VIII) or by a relative overpopulation of the excited level (see Sect. 4d). When the observations are made in fuel-rich flames with excess radicals, abnormal reduction of stable metal (hydro-)oxides and suprathermal chemiluminescence may concur and complicate matters.

CHAPTER VII

Broadening and Shift of Atomic Metal Lines

Prof. Alkemade

1. INTRODUCTION

The importance of collisional line-broadening processes in flames has already been outlined in Sect. II.5b. In this chapter we focus on the possibility of using these processes to study the interaction between the radiating (or absorbing) atom and the surrounding flame gas molecules (perturbers). We will therefore treat quite extensively semi-classical and quantummechanical line-broadening theories, discuss experimental methods for the derivation of line profiles and try to make a confrontation between theoretical models and experimental results.

In the theoretical discussion we will primarily consider the spectral profile of isolated, single atomic lines and ignore degeneracy of the atomic levels and (hyper-)fine-structure splitting. The effect of degeneracy (M_J-mixing) and of intramultiplet mixing collisions (J-mixing) will be briefly discussed in Sect. 2b-6.[†] The effect of partially overlapping (hyper-)fine-structure components on the integrated profile of optically thick lines as well as on the line profile itself will be briefly mentioned in Sect. 2b-7. Isolated line of molecular bands can be treated similarly and are not discussed separately. For the case of partially overlapping lines of molecular bands the reader is referred to Penner (1959), and Arnold, Whiting and Lyle (1969).

Furthermore we confine ourselves in the theoretical discussion of line-broadening phenomena to optically thin media (no self-absorption or concentration broadening; see also Sect. II.5b). We assume, if not explicitly stated otherwise, that Kirchhoff's law applies, i.e. that the absorption and emission line profiles

---

[†] Section numbers without a roman cipher refer to the same chapter.

are identical, as it should be in strict thermodynamic equilibrium (see also Sect. II.5c-3; see Nienhuis 1973).

In Section 2 a survey will be given of theories for collision broadening due to foreign, neutral perturbers. This limitation is based on the fact, that under usual flame conditions Stark effect and resonance broadening can almost always be neglected (see also Sect. II.5b-2); we will justify this neglect quantitatively in Sect. 2a. Emphasis is laid on an account of basic assumptions underlying the various theories, i.e. model interaction potentials, boundary conditions, etc., rather than on extensive, detailed derivations of formulas. The reader who is interested in these derivations is referred to books by Mitchell and Zemansky (1961), Breene (1961), Unsöld (1968), Sobel'man (1972), and to survey articles by Sobel'man (1957), Margenau and Lewis (1959), Baranger (1962), Griem (1964), Traving (1968), Hindmarsh and Farr (1972), Schuller and Behmenburg (1974), and Behmenburg (1979). As far as feasible the ranges of validity, possible overlapping of the theories, and boundary conditions of the various theories are discussed in Sect. 2b-2. It is attempted to demonstrate the links but also the difference between (semi-)classical and quantummechanical theories, and more specifically the way in which (semi-)classical theories follow from the quantummechanical ones and which assumptions and simplifications have to be made (see Sect. 2b-4). The computability of line profiles from the various theories and models is also discussed throughout Sect. 2b. In more detail we will treat the theories pertaining under the usual flame conditions and give closed expressions for (parts of) the line profile under specified conditions of kind and pressure of perturbers.

Realistic line profiles are the result of different, simultaneously acting broadening processes, which may be mutually dependent. We will discuss the validity and relevance of the convolution of line profiles, and the conditions necessary for its application and justification in Sect. 2b-5. Necessarily the idealized concept of *Voigt profiles* (see Sect. II.5b-3) will be confronted with more realistic approaches. Further the influence of intramultiplet- or J-mixing collisions and of degeneracy of (one of) the atomic levels involved in the optical transition (leading to M_J-mixing or re-orientation collisions) will be discussed in Sect. 2b-6.

In Section 3 experimental methods for the study of line profiles will be discussed. In the discussion of the experimental results (a-parameters, phenomenological cross sections, interaction potentials, etc.) in Section 4 we will mainly be concerned with flame experiments, but for comparison also data from nonflame milieus (absorption cells, etc.) will be cited. The latter data will appear to be of much relevance in the confrontation of theory and experiment (see Sect. 4b).

2. THEORETICAL

2a FORMULATION OF THE PROBLEM

The absorptivity of a spectral line is given by Eq. II.177

$$k(\nu) = (\pi e^2/m_e c)\, n_0\, S_\nu(\nu)\, f \qquad \text{(VII.1)}$$

where f=oscillator strength of the spectral line considered, n_0=number density of absorbing atoms (assumed to be in the ground state, for example), $S_\nu(\nu)$ is the line profile function, and the other symbols have their usual meaning. The profile function is assumed to be normalized (see Eq. II.206), that is

$$\int_{\text{line}} S_\nu(\nu)\, d\nu \equiv 1. \qquad \text{(VII.2)}$$

In the following we are concerned only with the calculation of the profile function S_ν. We confine ourselves, if not otherwise stated, to:

(i) Isolated atomic lines without h.f.s.
(ii) Cases in which Kirchhoff's law (see Eq. II.293) is obeyed[†].
(iii) Optically thin media (see Sect. II.5b-1) at a uniform temperature and with a homogeneous density of absorbers (or radiators) and perturbers.
(iv) Nondegenerate atomic states (two-state model).

Three major causes of line broadening exist in flames (see also Sect. II.5b):[*]

(1) Interaction of the atom with the radiation field, resulting in natural line broadening.
(2) Thermal motion of the absorbing (or radiating) atom, resulting in Doppler broadening.
(3) Interaction of the absorbing (or radiating) atom with surrounding particles, resulting in collision broadening.

[†] Experimental evidence for the identity of emission and absorption line profiles in flames has been produced by Jansen, Hollander and Alkemade (1977b) for the case of Ba and Sr resonance lines in C_2H_2-air flames at 1 atm.
 A case in which Kirchhoff's law is not obeyed is found in Sect. 2b-9. Extra Doppler broadening of nonthermal emission lines (due to chemi-excitation reactions) will be discussed there.

[*] In strong radiation fields the radiative lifetime of the atom in the upper and lower state of the optical transition may be shortened. As discussed in Sect. II.5b-1 we can neglect this extra broadening effect if conventional light sources are used in the absorption experiments.

Chapt. VII   BROADENING AND SHIFT OF ATOMIC METAL LINES

The broadening causes (1) and (2) were already discussed extensively in Sect. II.5b-2. In Sect. 2b we focus on collision broadening, which was introduced in Sect. II.5b-2, too; in Sect. 2b-5 we discuss the calculation of the total line profile by semiclassical and quantummechanical theories in which the contributions from collision broadening and the other relevant broadening processes are simultaneously taken into account.

For the *collision broadening*, characterized according to the type of interaction and/or the kind of collision partner, we obtain the following general classification:

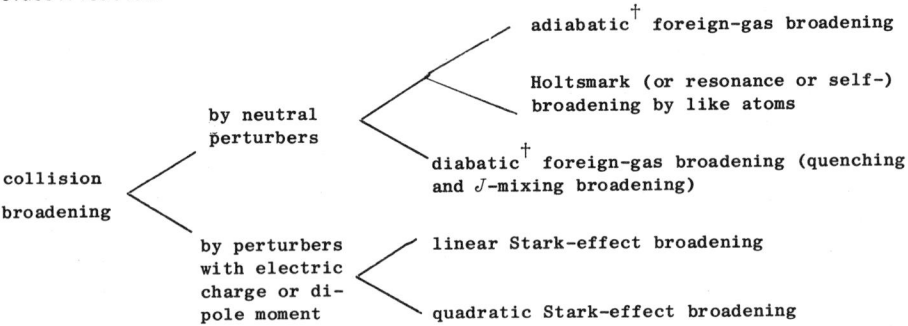

Some of these broadening causes have negligible influence on the resulting line profile under normal flame conditions. More specifically (see also Sect. II.5b-2) Holtsmark broadening and broadening by the (linear and quadratic) Stark effect can practically be neglected.

In the justification of the neglect of *Stark-* and *resonance broadening* under usual flame conditions it is assumed that the impact theory yields an adequate description of the adiabatic line profiles in flames. It should be recalled that the *linear Stark effect* occurs mainly for atomic hydrogen lines (see Sect. II.2a-2) and is not relevant in flames. Furthermore we take the line half-intensity width $\delta\nu$ (as defined in Sect. II.5b-1), for a certain collisional broadening process as a measure for its contribution to the total line profile. The $\delta\nu$ value for Stark- and resonance broadening can be found either theoretically from values of the force constant C_3 (see Table VII.1) or from experimental cross section values.

Let us first consider the neglect of *resonance broadening* in flames. Behmenburg (1964) has calculated that at the highest possible metal concentrations in his C_2H_2-O_2-inert gas flame at $T = 2000$ K, the maximum $\delta\nu_{res}$ value is in the order of 2×10^7 s^{-1}, which is much smaller than the value of 3×10^9 s^{-1} of the half-intensity width, $\delta\nu_{c''}$, due to adiabatic foreign-gas broadening. In this

---

† For the definition of adiabatic and diabatic collision broadening see Sect. II.5b-2 and Sect. 2b-3.

680

THEORETICAL   Sect. VII.2a

estimate the force constant C_3 was calculated according to: $C_3 = e^2 f/8\pi^2 \nu m$ (m = atomic mass). De Galan and Wagenaar (1971) using a similar method of estimating found a value of $\delta\nu_{res} = 2\times 10^8$ s^{-1} at an unrealistically high atomic density of 3×10^{15} cm^{-3}. At more realistic densities they expect $\delta\nu_{res} \leqslant 2\times 10^7$ s^{-1}, in agreement with Behmenburg's estimate.

One may also insert experimental cross section values for resonance broadening in Eq. II.231 in order to justify the neglect of resonance broadening in flames.

As to the neglect of *quadratic Stark broadening*, one has only estimated values of the force constant C_4. For interactions with electrons the C_4 value is less than 10^{-15} cm^4 s^{-1} (see de Galan and Wagenaar 1971) and one finds $\delta\nu_{Stark} \leqslant 2\times 10^4$ s^{-1} for an electron density of 10^{11} cm^{-3} in Behmenburg's C_2H_2 flames for the Na-D doublet. This estimate is also reasonably valid for other metal lines (see de Galan and Wagenaar 1971).

Stark broadening due to the interaction with permanent dipoles like H_2O may be more important. Behmenburg (1964) has calculated $\delta\nu_{Stark}$ in a flame with a H_2O density of 2×10^{17} cm^{-3}, being a representative value under flame conditions. For the Na-D doublet he found that the contribution of this Stark broadening to the half-intensity width was less that 1% and thus negligible. It should be noted that the Stark effect for this doublet is rather small, but that for most metal lines, for example the alkaline-earth lines, its magnitude is not known at all.

In semi-classical (SC) and also in some quantummechanical (QM) theories the individual line profiles are calculated first and most commonly the *total line profile* is expressed as a convolution of the individual profiles. Convolution is, generally speaking, allowed only if the broadening mechanisms are not interrelated, so that the total correlation function is the product of the individual correlation functions involved (see Sect. II.5b-3). But when there is, for instance, a correlation between the motion of the particles and the phase perturbation, as may be the case for Doppler-broadened adiabatic profiles, convolution is strictly speaking not allowed.

Dicke-narrowing (see Sect. II.5b-2) illustrates the influence of collisions on the Doppler profile. In this case Smith *et al.* (1971) have shown by QM theory that convolution of the narrowed Doppler profile with the collisional profile may not be permitted. The case of Dicke-narrowing can also be treated semi-classically (see Sobel'man 1957, Galatry 1961, Rautian and Sobel'man 1967, Gersten and Foley 1968, Berman and Lamb 1971, and Zaida 1972; see also Sect. 2b-5). McCartan and Lwin (1977) have reported absorption-cell measurements on the Li resonance line in a Xe milieu. They found Li core profiles significantly deviating from the Voigt shape; the experimental profiles could adequately be described by a theory from Ward, Cooper and Smith (1974), which took Dicke-narrowing into account.

Chapt. VII   BROADENING AND SHIFT OF ATOMIC METAL LINES

2b   COLLISION BROADENING

2b-1 *GENERAL CONSIDERATIONS*

Collision broadening involves the interaction of the absorbing or radiating atom or molecule (further to be referred to as '*atom*') with surrounding particles (further simply called '*perturbers*') which causes a frequency-perturbation and hence a phase shift $\eta(t+s,t)$ in the time interval $(t,t+s)$ (see Sect. II.5b-2). This phase shift is a real number when the collisions are adiabatic. Diabatic collisions, which cause the amplitude of the oscillator (in the classical model of the perturbed atom) to change, may be included in the imaginary part of a complex phase shift.

The phase shift is in general the result of interactions with many perturbers. Especially when different perturbers act simultaneously, we have to consider mutual interference of their perturbations. This interference is dependent on the type of interactions involved. The case of adiabatic broadening by neutral foreign-gas particles is relatively simple; here the perturbations and hence the phase shifts add like scalars. In the linear Stark effect the perturbations depend linearly on the field and add like vectors. The quadratic Stark effect depends on the square of the vector sum of the perturbing fields. For Holtsmark broadening, for instance, there is no simple scheme of superposition.

In all *quasi-static* (or *statistic*) *theories* of collision broadening, in which the motion of the perturber can be neglected, this superposition problem arises. It can only be avoided when solely the interaction with the nearest perturber is taken into account; this leads to the *nearest-neighbour approximation* (binary collision concept).

The superposition problem is also circumvented in the *impact approximation*. Here we assume that the moving perturbers act one after another in the time sequence of their closest approach to the radiating atom (binary collision concept).

As we will show in the discussion of these theories, the quasi-static and impact theories have a limited validity region as to perturber density and wavelength range of the line profile. There are also semi-classical and quantummechanical theories that do not have these restrictions and are therefore called *unified theories*, *general-pressures theories*, or *generalized theories of pressure broadening* (see, e.g., Lindholm 1945, Anderson and Talman 1955, Futrelle 1972, Hindmarsh and Farr 1972, and Lee 1974). These theories describe, in principle, adequately the entire line profile (including the far line wings) at any perturber density. Impact as well as quasi-static theories can thus be considered as limiting cases of this more universal concept of collision broadening. Quite often the complexity of the formal expressions derived in the unified theories makes it necessary to consider only these limiting cases, which are (sometimes) computable. The existing line-

THEORETICAL
Sect. VII.2b

broadening theories divide mainly into two categories:

(i) General *quantummechanical* (*QM*) *theories*, from which it is very hard to calculate actually the entire line profile (see Baranger 1958, 1958a, Fano 1963, Smith *et al.* 1971, Chappell *et al.* 1971, Futrelle 1972, and Nienhuis 1973).

(ii) Simpler *semi-classical* (*SC*) *theories*, which can be used for numerical calculations of the line profile (see, e.g., Lindholm 1942, 1945, Anderson 1949, Anderson and Talman 1955, Margenau and Lewis 1959, Fox and Jacobson 1969, Hindmarsh and Farr 1972, and Schuller and Behmenburg 1974).

Semi-classical theories of collision broadening start from the concept that the relative motion can be treated classically whereas the internal states of the particles are described quantummechanically. In the QM theories the atom plus all relevant perturbers may be considered as a single but very complex system, and the profile function is derived from the QM expression for the power of the radiation with angular frequency ω_{if} which is emitted by this system in the transition from an initial state (i) to a final state (f).

The QM theories are evidently the more general ones. An important problem is how the SC theory is found as a limiting case of the general QM theory and under which conditions this limiting case is realized (see Sect. 2b-4).

2b-2 *SURVEY OF COLLISION BROADENING THEORIES*

A survey of collisional line-broadening theories is presented in Fig. VII.1. In this diagram the mutual connections between the various theories and their simplifications, basic assumptions and limitations can easily be recognized. The detailed discussion of these theories can be found in the following sections. We make the following general assumptions:

(i) Only collision broadening due to neutral, foreign perturbers is considered.

(ii) The interaction potentials involved are spherically symmetric.

(iii) In most cases degeneracy of the atomic states is ignored; a two-state model is assumed. M_J-mixing occurring with degenerate states will be discussed separately in Sect. 2b-6.

2b-3 *SEMI-CLASSICAL (SC) THEORIES OF COLLISION BROADENING*

From a formal standpoint it seems logical and straightforward to start the discussion with the QM theories since they are the most general ones. For practical reasons, however, it is more convenient to start with the (older) SC theories; they give more direct insight in the physical background, are simpler and yield

683

Chapt. VII   BROADENING AND SHIFT OF ATOMIC METAL LINES

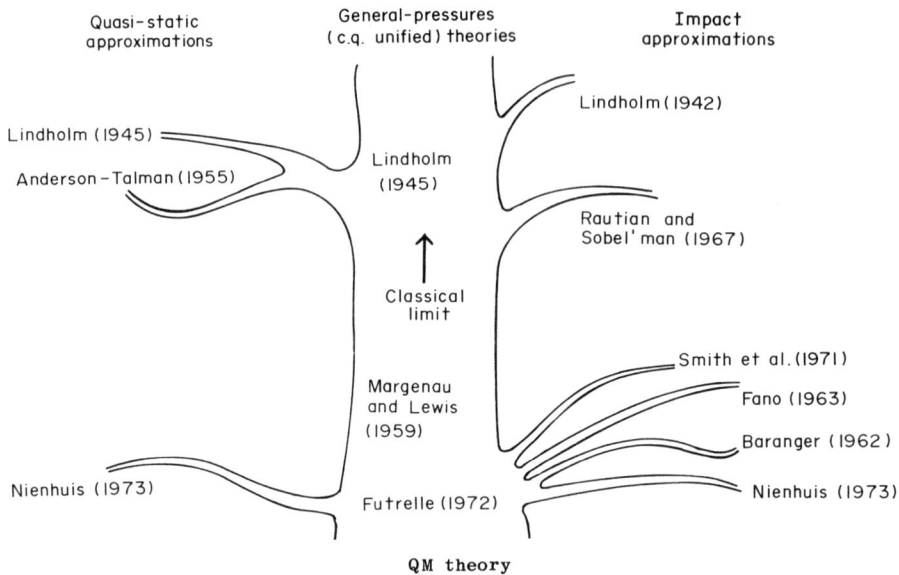

Fig. VII.1 Schematic survey of some representative collisional line-broadening theories ('tree-model') and their authors. The roots of the 'tree' represent first principles; classical limits are found by climbing the trunk of the 'tree'. Branches to the left and the right, respectively, represent QS and impact approximations.

generally speaking computable expressions for the line profile. Besides, in taking this sequence we are able first to discuss the various approximations: impact theory and quasi-static theory, and then the general-pressures theory. Finally, since the SC theories will be found as (classical) limiting cases of the more general QM theories, the sequence chosen makes a comparison of the results from SC and QM theories more readily feasible.

In the SC theories *diabatic* (inelastic) collisions are, generally speaking, not considered. We will describe this type of line broadening by making use of the similarity between diabatic and natural broadening, as was discussed in Sect. II.5b-2. We will return to this subject in Sect. 4a.

Under flame conditions the collisions are usually *adiabatic*. In the SC picture the atom emits wave trains which are subject to perturbations caused by collisions. As a result of these collisions, time-dependent frequency-perturbations $\Delta\nu(t)$ and, with them, real phase-perturbations: $\eta = 2\pi \int \Delta\nu(t)\,dt$ are produced. As

THEORETICAL   Sect. VII.2b

a general criterion for adiabatic collision broadening with nondegenerate states one may state: the quantum state of the atom is maintained throughout a collision except for its alteration through the radiation process (see Sect. II.5b-2). The line profile is in these theories usually derived by calculating the autocorrelation function, $\phi(s)$, of the complex amplitude $A(t)$ of the oscillation (see Sect. II.5b-2 and Eqs II.233 and 234); the Fourier transform of $\phi(s)$ is the profile function, $S_\nu(\nu)$, sought (see Traving 1968).

2b-3.1 *Impact Approximation*

In this theory (here only considered for adiabatic collisions) the atom is assumed to be most of the time without any phase-perturbation and during the short duration of collision to interact only with the one colliding particle (i.e. 'momentaneous collision' or 'sudden impact'). The perturbations are mutually independent. The collision duration τ_{coll} is small compared to the time, τ_{ic}, between two consecutive collisions (see, e.g., Margenau and Lewis 1959, and Jefferies 1968); the interaction of the atom with more than one perturber at a time may then be neglected. The latter condition is met (see below) if the density of perturbers is low enough.

More explicitly, we make the following 'assumptions about collisions':
(i)   Only those perturbers, j, that pass the atom with an impact parameter, b, less than a critical value b_0 in the time interval $(t,t+s)$ contribute to $\eta(t+s,t)$ (for the definition of impact parameter see Sect. II.4e).

(ii)   The value of $\eta_j(t+s, t)$ is replaced by the phase shift $\eta(b)$ due to a completed collision with impact parameter b; of course, $\eta(b)$ depends also on the relative velocity v.

With these assumptions and neglecting the velocity distribution, we find a dispersion profile (as is the usual profile in the impact approximation; see also Sect. II.5b-2 and Eq. II.228), the maximum of which is shifted with respect to the undisturbed frequency ν_0.

In order to calculate the profile function, and more specifically the line half-intensity width $\delta\nu$ and the line shift $\delta\nu_s$, further assumptions have to be made concerning the type of interaction and the path and motion of the perturbers.

The *Lindholm-Foley theory* (see Lindholm 1942, and Foley 1946; see also Margenau and Landwehr 1958, and Sobel'man 1957, 1972) assumes a straight perturber path and mono-energetic perturbers, i.e. the collision partners are assumed to have the same relative velocity being equal to the mean value: $\bar{v} = (8kT/\pi\mu)^{\frac{1}{2}}$ (see Eq. II.29). The interaction law is an inverse-power law: $V(r) = C_p/r^p$ (with $p > 0$ and r is the interatomic distance); $V(r)$ is essentially a <u>difference</u> potential, i.e. the difference between the interaction potentials pertaining to the upper and

685

the lower state of the radiating atom, respectively. Mostly the interaction potentials applied in the collision broadening theories are spherically symmetric. This type of potential gives a satisfactory description for the interaction of atoms with noble gases, but not for molecular perturbers.

Table VII.1

Line Broadening and Shift according to Lindholm-Foley Collision Theory

quantity $p =$	3	4	6
$\delta\nu$	$2\pi^2 C_3 n_p$	$6.18\, C_4^{\frac{2}{3}} \bar{v}^{\frac{1}{3}} n_p$	$2.71\, C_6^{\frac{2}{5}} \bar{v}^{\frac{3}{5}} n_p$
$\delta\nu_s$	0	$5.31\, C_4^{\frac{2}{3}} \bar{v}^{\frac{1}{3}} n_p$	$0.98\, C_6^{\frac{2}{5}} \bar{v}^{\frac{3}{5}} n_p$
$\delta\nu/\delta\nu_s$	∞	1.16	2.76

In Table VII.1 the expressions for $\delta\nu$ and $\delta\nu_s$ are shown for the cases of $p=3$, 4 and 6 (see Traving 1968). Here $p=3$ describes resonance broadening, $p=4$ relates to the quadratic Stark effect due to electrons, and $p=6$ to foreign-gas broadening with a van der Waals interaction potential. A characteristic of the impact approximation is recognized in the linear dependence of $\delta\nu$ and $\delta\nu_s$ on the perturber density n_p. The force constants C_3, C_4 and C_6 of the difference interaction potentials can be calculated from theory (see Slater and Kirkwood 1931, Traving 1968, Unsöld 1968, and Davison 1968).

Broadening is mostly due to strong collisions with a small impact parameter, b, whereas the shift of the line is caused by the relatively larger number of weak collisions, i.e. collisions with a large b value. In order to scale the b value we introduce the *Weisskopf radius*, b_w, being that particular impact parameter for which the phase perturbance $\eta(b_w) \equiv 1$ (see Traving 1968, and Behmenburg 1970). For the above one-term potentials one gets: $b_w = (\alpha_p C_p / \bar{v})^{1/(p-1)}$ with

$$\alpha = \sqrt{\pi}\, \Gamma\{\tfrac{1}{2}(p-1)\}/\Gamma(\tfrac{1}{2}),$$

where Γ stands for the gamma-function. It is simple to see from the mathematical derivation of S_ν (see, e.g., Traving 1968) that the broadening is mostly due to collisions with $b \leqslant b_w$, whereas the shift of the line is caused by collisions with $b \geqslant b_w$. For two- and more-term potentials a closed expression for the Weisskopf radius cannot be given.

In order to clarify the role of the two basic assumptions made in the impact theory we consider these assumptions in more detail. The second assumption, called the *Markov* or *completed-collision assumption*, states that the time, s, over which

THEORETICAL   Sect. VII.2b

the phase-shift is considerd has to be much greater than the average collision duration: $s \gg \tau_{coll}$. This assumption yields the *frequency-criterion* (see below). The profile function S_ν calculated on the basis of this assumption has only real significance if the *impact approximation* is valid, too; the latter states that $\tau_{coll} \ll \tau_{ic}$ (i.e. sudden impacts, momentaneous collisions) and is essentially the *density-criterion*. In flames at pressures up to 1 atm the impact assumption appears to describe adequately the core of the line profile.

The range of application of the impact theory can be derived from the above two assumptions:

(i) The completed-collision assumption implies that $s_0 \gg b_w/\bar{v}$. The consequence of this is that the impact theory determines the inner part of the line profile only ('line-core'). The frequency-limit is given by: $\Delta\omega_H (\equiv 2\pi\Delta\nu_H) = \bar{v}/b_w$ (see Holstein 1950). The wings of the line, where $\Delta\omega(\equiv |\omega-\omega_0|) \gg \delta\omega$, can be found from quasi-static theory (see Sect. 2b-3.2).

(ii) The impact assumption implies that the phase changes due to different perturbers in the interval $(t, t+s)$ are mutually independent. This means that, except for scalar additive perturbations, it must be very improbable that two or more collisions occur at the same time. This imposes an upper limit on the perturber density: $n_p b_w^3 \ll 1$. It can be shown that, if this condition is fulfilled, S_ν is, virtually, zero for $\Delta\omega > \Delta\omega_H$.

The expressions for half-intensity width $\delta\nu_{c''}$ and shift $\delta\nu_s$ for adiabatic collision broadening on the basis of a van der Waals interaction potential (case $p = 6$ in Table VII.1) will be compared with the corresponding expressions when another (mostly a Lennard-Jones) potential or a hard-sphere model is chosen. In general, the various modifications of the theory treated in the literature assume a spherically symmetric interaction potential and lead to a dispersion (or Lorentzian) line profile as described by Eq. II.228. In the line wings, i.e. at $\Delta\nu (\equiv |\nu-\nu_0|) \gg \delta\nu$, the profile function thus follows a $(\Delta\nu)^{-2}$ law.

Behmenburg (1964, 1968, 1970) has calculated $\delta\nu_{c''}$ and $\delta\nu_s$ for a Lennard-Jones (L.-J.) potential: $V(r) = h(-C_6 r^{-6} + C_{12} r^{-12})$ where C_6 and C_{12} are positive. The results are

$$\delta\nu_{c''} = f_w(\alpha) \bar{v}^{\frac{3}{5}} C_6^{\frac{2}{5}} n_p \ , \qquad (VII.3)$$

$$\delta\nu_s = f_s(\alpha) \bar{v}^{\frac{3}{5}} C_6^{\frac{2}{5}} n_p \ , \qquad (VII.4)$$

with

$$\alpha \equiv \frac{\pi}{\bar{v}} (c_6 C_6)^{\frac{11}{6}} / (c_{12} C_{12})^{\frac{5}{6}} \ , \qquad (VII.5)$$

where c_6 and c_{12} equal $3\pi/8$ and $63\pi/256$, respectively. Behmenburg (1964) calculated the functions $f_w(\alpha)$, $f_s(\alpha)$ and $\xi(\alpha) \equiv \delta\nu_{c''}/\delta\nu_s = f_w(\alpha)/f_s(\alpha)$ and presented them in graphical form. Numerical values of similar functions have independently been reported by Hindmarsh, Petford and Smith (1967). A comparison of Eqs VII.3 and VII.4 with the analogous expressions in Table VII.1 shows that, while for a van der Waals potential the ratio $\xi \equiv \delta\nu_{c''}/\delta\nu_s$ has a constant value of 2.76, $\xi(\alpha)$ for a L.-J. potential depends critically on α, and thus on C_6 and C_{12}. With a L.-J. potential usually red, but also blue line shifts may be found, whereas with a van der Waals potential only red shifts[†] are found. Blue shifts have been found experimentally for Hg- and alkali lines with helium as a perturber (see Behmenberg 1964, 1968).

Schuller (1962) has calculated $\delta\nu_{c''}$ and $\delta\nu_s$ under the assumption that atoms and perturbers were *hard spheres* that attract each other according to a van der Waals potential. When the sum of the radii of atom and perturber $(r_A + r_B \equiv D)$ was small compared to the Weisskopf radius b_w, he found expressions for $\delta\nu_{c''}$ and $\delta\nu_s$ only slightly differing from the ones cited in Table VII.1 for the case $p=6$. When, however, $D \gg b_w$, the dependence of $\delta\nu_{c''}$ and $\delta\nu_s$ on $\bar{v}$ and C_6 changes appreciably and the ratio ξ is no longer constant but heavily dependent on $\bar{v}$, C_6 and D.

Herman and Coulaud (1970) have calculated the influence of a curved (= hyperbolic) path on the $\delta\nu_{c''}$ and $\delta\nu_s$ functions for a Lennard-Jones potential. They assumed that the Lennard-Jones potential should include a centrifugal term $(\propto r^{-2})$ and that the r^{-12} term should be replaced by a r^{-10} term. A typical difference with the above impact theories is that here the distance of closest approach, r_{min}, is in general not equal to the collision parameter b. The expressions for $\delta\nu_{c''}$ and $\delta\nu_s$ found from their theoretical approach have essentially the same form as Eqs VII.3 and VII.4. The difference enters in the value of $f_w(\alpha)$ and $f_s(\alpha)$: for α values up to 1.0 the differences with the values from Behmenburg (1964) and Hindmarsh, Petford and Smith (1967) are at most 3%. This suggests that under usual flame conditions, where $\alpha < 1$, the straight-path assumption is valid.

Up till now there is little evidence whether the adiabatic (foreign-gas broadening) line profiles as to be expected in flames at 1 atm differ significantly from the ones predicted by the impact theory, at least as far as the line core

---

[†] This holds at least if the usual assumption is true that the C_6'' coefficient in the potential of the upper state is larger than the C_6' coefficient in the lower-state potential. Only then the above C_6 coefficient in the <u>difference</u> potential is positive. The blue wing arises from the repulsive part of the difference potential if $C_{12}'' > C_{12}'$ and thus $C_{12} > 0$. Lindholm (1945) using a van der Waals potential, however, predicted a $(\Delta\nu)^{-7/3}$ law for the blue wing.

THEORETICAL   Sect. VII.2b

($\delta\nu_{C''}$ and $\delta\nu_{S}$) is concerned. The evidence available is based on the intercomparison of line profiles calculated (under assumption of a certain interaction potential) with the impact-approximation and the general-pressures theory (see Sect. 2b-3.3) and on the comparison with experimental profiles. In the latter comparison, the experimental profile, especially the line core, is curve-fitted to a convolution product of Doppler, adiabatic, natural and quenching profiles, resulting in the total theoretical line profile. It will be clear that any evidence obtained in this way is laden with additional ad-hoc assumptions and is not very convincing, the less so since the deviations from the theoretical shape to be expected in the line core are in the order of a few percent (see below; see also Sects 2b-3.3 and 2b-5). Similar evidence can be derived from the curve-fitting of experimental line profiles to *Voigt* profiles (see Sect. 2b-5). The good agreement found between the a-parameters (defined by Eqs II.246 and 243) derived from integral-absorption measurements and interferometric line-profile measurements yields another piece of evidence (see, e.g., Hollander, Jansen, Plaat and Alkemade 1970, Jansen and Hollander 1973, and Jansen, Hollander and Alkemade 1977a).

The usual simplifications made in flame work and based on the assumed validity of the impact approximation are:

(i)   The collisional half-intensity width $\delta\nu_{C''}^{(j)}$ due to one kind of perturbers is proportional to the number density of perturbers $n_{p}^{(j)}$ and relative velocity $\bar{v}_j$. We define the *adiabatic-broadening cross section* $\sigma_{C''}^{(j)}$ according to (cf. Eq. II.230)

$$\delta\nu_{C''}^{(j)} = \sigma_{C''}^{(j)} \bar{v}_j n_{p}^{(j)}/\pi . \qquad (VII.6)$$

Similarly we define

$$|\delta\nu_{S}^{(j)}| = \sigma_{S}^{(j)} \bar{v}_j n_{p}^{(j)}/2\pi . \qquad (VII.7)$$

(ii)   The collisional half-intensity width $\delta\nu_{C''}$ due to the combined action of different perturbers of kind j is expressed as

$$\delta\nu_{C''} = \sum_{j} \sigma_{C''}^{(j)} \bar{v}_j n_{p}^{(j)}/\pi . \qquad (VII.8)$$

For the line shift $\delta\nu_S$ one has accordingly (if the sign of $\delta\nu_S^{(j)}$ is the same for all j's)

$$|\delta\nu_S| = \sum_{j} \sigma_{S}^{(j)} \bar{v}_j n_{p}^{(j)}/2\pi . \qquad (VII.8a)$$

Equations VII.8 and 8a imply mutual independence of perturbations due to different perturbers.

The experimental $\sigma_{C''}^{(j)}$ values contain, in principle, information about $V(r)$, albeit that an additional knowledge of $\delta\nu_S$ is necessary when $V(r)$ is not a simple van der Waals potential (see Eqs VII.3 and VII.4). The dependence of $\sigma_{C''}^{(j)}$ on temperature is critically dependent on the (assumed) interaction potential (see Eqs VII.3 and 6). For a van der Waals potential we find: $\sigma_{C''}^{(j)} \propto T^{-0.2}$ from Table VII.1 and Eq. VII.6. For a L.-J. potential the temperature-dependence is much more complex and can only be found when α is (approximately) known (see Eqs VII.3 and 5).

2b-3.2 *Quasi-static (QS) Theories of Adiabatic Line Broadening*

Only for adiabatic line broadening due to foreign perturbers a QS theory is available.

In the QS theory we suppose that a strong perturbation occurs, resulting in a large frequency shift. It is then sufficient to consider the phase shift over a time interval s that is short compared to τ_{coll} (the opposite case occurs in the impact approximations; see Sect. 2b-3.1). We may then analyze the problem as if the perturbers were at rest, so that the phase shift is linear in s. A certain spatial configuration of the perturbers produces a frequency shift, $\nu-\nu_0$, that is equal to the corresponding perturbation in the difference potential divided by h. The frequency distribution of the radiation emitted by a statistical ensemble of atoms follows then directly from the distribution of the probability for stationary perturbers to produce a given interaction with the atom.

The QS theories are expected to describe correctly the line *wings*; this conclusion is based on the interpretation of the condition: $s \ll \tau_{coll}$ in a similar way as was done in Sect. 2b-3.1. In analogy with the derivation of the density-criterion in the impact approximation, one could now expect the QS theories to be valid for high perturber densities, i.e. certainly at higher pressures than 1 atm. In general this statement is true, but the $\Delta\nu$- and n_p-criteria are mutually connected.

At high perturber densities the simultaneous interaction with many particles plays a role, thus superposition problems have to be faced. There are two different ways to overcome these problems:

(I) The perturbation is thought to be due only to the nearest neighbour (binary collision concept). The theory describes the more adequately the line profile nearer to the line centre the higher the perturber density is, or stated in the opposite way: at lower perturber densities $\Delta\nu$ has to be larger in order to guarantee the validity of the *nearest neighbour approximation*. Large $\Delta\nu$ values mean close encounters and for large $\Delta\nu$ the probability that only one perturber interacts is thus relatively high. On the other hand, at high perturber densities the justification of the nearest-neighbour approximation requires large values of $\Delta\nu$, corresponding to a small nearest-neighbour distance (see Behmenburg 1970, and Hindmarsh and Farr 1972).

THEORETICAL   Sect. VII.2b

(II)   The perturbations and hence the phase shifts due to different perturbers
are assumed to add like scalars. This is the case when the atom is surrounded by
neutral atoms of a foreign gas, leading to a van der Waals or L.-J. interaction.

Various QS theories have been worked out for different interaction potentials (see Fig. VII.1). Instead of giving a detailed description, we rather summarize the basic characteristics of the QS theories:

(i)   The QS theories give a correct description for strong but slowly varying perturbations.

(ii)   The QS theories assume low relative perturber velocity and disregard the time-dependence of the perturbation: actually the perturbers are thought to be at rest with respect to the atom. This is identical with the statement that QS theory describes the small-time (or large $-\Delta\nu$) limit ($s \ll \tau_{coll}$). The time is so short that the perturbers have not moved and the interaction potential has not changed in this time.

(iii) The probability distribution for stationary perturbers that produce a given interaction with the atom is determined by their spatial distribution, which is assumed to be random. This yields the QS line profile sought, if $V(r)$ is known.

(iv)   Superposition problems can be avoided either by using the binary collision concept, or by using $V(r)$ functions that yield perturbations adding like scalars.

In order to illustrate the differences and common features of types I and II of QS theories we give in Table VII.2 the line profiles S_ν calculated for a van der Waals potential, which gives rise to a red wing only (see above).

Table VII.2
Typical QS Expressions for the Line-profile Function S_ν
(in the Case of a van der Waals Potential)

QS theory (nearest-neighbour approximation) (Kuhn 1934; see also Kuhn 1969)	QS theory (scalar additivity) (Margenau 1935)
$S_\nu = C_6^{\frac{1}{2}} n_p \dfrac{1}{3(\nu_0-\nu)^{\frac{3}{2}}} \times \exp\left[-\dfrac{4\pi}{3} \dfrac{n_p C_6^{\frac{1}{2}}}{(\nu_0-\nu)^{\frac{1}{2}}}\right]$ (VII.9)	$S_\nu = C_6^{\frac{1}{2}} n_p \dfrac{1}{3(\nu_0-\nu)^{\frac{3}{2}}} \times \exp\left[-\dfrac{4\pi^2}{9} \dfrac{C_6 n_p^2}{(\nu_0-\nu)}\right]$ (VII.10)
No explicit expressions for $\delta\nu_s$ and $\delta\nu_{c''}$ have been reported	$\delta\nu_{c''} = 0.411\,\pi^2\,C_6\,n_p^2$ $\delta\nu_s = 0.148\,\pi^2\,C_6\,n_p^2$

Both S_ν expressions have the same asymptotic behaviour for $\nu_0 - \nu \gg C_6 n_p^2$ indicating that the strong perturbations are always due to one single perturber which is very close to the atom. (There is a difference between the exponential factors in both S_ν expressions.)

From the expressions for $\delta\nu_S$ and $\delta\nu_{C''}$ according to Margenau's theory we conclude:

(i) Shift and half-intensity width are proportional to n_p^2, this in contrast to the impact approximation where $\delta\nu_S$ and $\delta\nu_{C''}$ are proportional to n_p (see Eqs VII.5 and 7).

(ii) The ratio $\delta\nu_{C''}/\delta\nu_S$ in the QS theory is equal to 2.78, in good agreement with the value of 2.76 from the impact approximation (see Table VII.1).

For a L.-J. potential one gets in the line wings ($|\nu - \nu_0| \gg \delta\nu$) according to Behmenburg (1970) (cf. Eqs VII.9 and 10) at the red side: $S_\nu \propto (\Delta\nu)^{-3/2}$ and at the blue side: $S_\nu \propto (\Delta\nu)^{-5/4}$. It turns out that these frequency-dependencies are the same as those for a pure van der Waals potential (red wing) and a pure repulsive potential (blue wing), respectively.

As to the practical application of the QS theories to line broadening in flames, we note that, in contrast to the impact approximation, there are no clear opinions in the literature about the $\Delta\nu$- and n_p regions where the QS theory is valid. Some considerations on this subject are found in Margenau and Lewis (1959), Behmenburg (1970), and Hindmarsh and Farr (1972). Since in usual flames the perturber pressure is not higher than 1 atm, one may expect that the line core and very near wings should be described by the impact approximation; the only remaining question is at which $\Delta\nu$ values the QS theory starts to describe the (far) wings adequately. This means that QS expressions for $\delta\nu_S$ and $\delta\nu_{C''}$ have no significance for flame work. Some experimental evidence about the validity region of QS theory can be derived from line wing measurements in absorption cells (see Sect. 4).

The question remains of how the frequency- and density ranges where the impact- and QS approximations are valid, are linked. In particular, the line profile in the intermediate frequency range where a $(\Delta\nu)^{-2}$ dependence (as predicted by impact theory) merges into a $(\Delta\nu)^{-3/2}$ or $(\Delta\nu)^{-5/4}$ dependence has not yet been derived. An answer to these questions must come from the unified theories.

2b-3.3 *General-Pressures Theories of Adiabatic Line Broadening*

As stated earlier the *general-pressures theories* describe in principle the entire line profile at any perturber number density. In QM theories one generally calls such theories *unified theories*: the term general-pressures theory more or less belongs to SC theories (see Hindmarsh and Farr 1972, Futrelle 1972, and

THEORETICAL
Sect. VII.2b

Nienhuis 1973). Two SC theories will be discussed here: the theory for general pressures of Lindholm (1945) and that of Anderson and Talman (1955). The basic assumptions of both theories are practically identical; the mathematical treatment is different: Lindholm applies Fourier analysis of the perturbed wave trains for the derivation of the line profile functions, whereas Anderson and Talman used the autocorrelation function method. The assumptions underlying these theories are the following:

(i) The perturbers follow classical, straight paths with constant relative velocity, $\bar{v}$.

(ii) The perturbers have a random distribution in space and are mutually independent, i.e. have no interaction.

(iii) As opposed to the impact approximation, the collision duration τ_{coll} is no longer small compared to the time between two consecutive collisions.

(iv) The frequency-perturbations are additive.

(v) In Lindholm's theory the (angular) frequency-perturbation $\Delta\omega(t)$ is assumed to be constant during τ_{coll}, i.e. the phase-perturbation

$$\eta(t) = \int_0^t \Delta\omega(t')\,dt'$$

varies linearly with t in the interval $0 < t < \tau_{coll}$. The collision duration is further assumed to be proportional to the impact parameter b: $\tau_{coll} = 2\pi\kappa b/\bar{v}$. The factor κ can be found in the case of a van der Waals potential by demanding that the line profile in the wings calculated through the general-pressures theory agrees with the result from the QS theory. With a L.-J. potential, κ is fixed by adapting the theory to the experiment (see Behmenburg 1968, 1970). In the theory of Anderson and Talman (1955, 1956) the time-dependent frequency-perturbation $\Delta\omega(t)$ is taken to be proportional to an inverse power of the distance to the perturber.

The profile functions S_ν have been calculated for van der Waals and L.-J. interaction potentials by several authors. Because of the complicated form of the profile functions we will not give them explicitly. The reader may find them in the literature cited below.

Lindholm (1945) reported the profile function in closed form for a van der Waals potential. No numerical calculations were done by this author. Anderson and Talman (1955, 1956) have given the profile function also for a van der Waals potential.

Behmenburg (1968, 1970) has numerically calculated the S_ν function for a L.-J. potential, following Lindholm's theory for the system Hg—Ar ($^3P_1 \to {}^1S_0$ transition of Hg at 2537 Å). Jansen, Hollander and Behmenburg have performed a

693

similar calculation[†] for Sr—Ar ($^1P_1 \to {}^1S_0$ transition of Sr at 4607 Å).

Takeo (1970) calculated numerically the profile according to Anderson-Talman's theory for the system Cs—Xe, using a L.-J. potential. It should be noted that the numerical calculations for a L.-J. potential are much more complicated than for a van der Waals potential. The calculations according to Lindholm's theory are very much complicated by the necessity of substituting a reasonable value for the factor of proportionality κ (see above). The Anderson-Talman theory is less difficult to handle in this respect and more simply accessible for computer calculations. Computer calculations of the line profile function for $\Delta\nu$ values in the order of 100 to 150 times the half-intensity width or more, are very hard and practically impossible.

The typical difference between the line profile from general-pressures theory and that from the impact theory is the possible occurrence of asymmetry in the line core (the line wings are asymmetric in a similar way as according to the QS theory). The line core asymmetry, usually expressed in terms of the asymmetry ratio, depends on $V(r)$ and n_p. As an example we mention the calculated *asymmetry ratio*[*] of 1.02 for the Sr line at 4607 Å perturbed by Ar at 2500 K and 0.8 atm (see Jansen 1976).

As we will discuss in Sect. 4a the shape of the line core and near wings calculated from general-pressures theory is under usual flame conditions not significantly different from that derived from impact theory. This can be expected as the impact approximation derives these shapes from the general-pressures theory in the limit of low perturber density and small $\Delta\nu$.

Apart from the distortion due to asymmetry, the other characteristics: $\delta\nu_s$, $\delta\nu_{c''}$, and the asymptotic $(\Delta\nu)^{-2}$ behaviour of the profile, are found to be the same from both theories. The new information one expects to get from general-pressures theory is the line profile in the ν-region where neither impact nor QS theory is actually valid.

In the '*semi-QS*' *theory* developed by Fox and Jacobson (1969) (see also Jacobson 1971, and Atakan and Jacobson 1972) the perturbers are assumed to move with variable velocities along curved classical paths determined by the potential of the initial state of the atom. So the classical paths are determined by the upper-state potential V_u ($\equiv V'$) in the case of emission, and by the lower-state potential, V_l ($\equiv V''$) in the case of absorption line profiles. Likewise the

---

[†] Jansen, Hollander and Behmenburg succeeded in calculating the line profile for the system Sr—Ar up to $\Delta\nu = 2$ Å, i.e. 50 times the half-intensity width (unpublished work).

[*] The asymmetry ratio of the line core is defined as the ratio of the half of the half-intensity width of the red part of the core to the corresponding width of the other part.

THEORETICAL   Sect. VII.2b

possible initial configurations are weighted by a Boltzmann factor including the initial potential function. These assumptions cause the emission and absorption line profiles to be different, thus violating Kirchhoff's law. For a further discussion the reader is referred to Sect. 2b-4.

2b-4   *QUANTUMMECHANICAL (QM) THEORIES OF COLLISION BROADENING*

In this section we discuss some QM theories, demonstrating their qualities and possibilities, advantages and drawbacks, as compared to the SC ones. In particular the computability of the expressions derived is of interest, but also the connections between QM and SC theories. Since we are mainly interested in line broadening processes in flames, we leave also out any discussions of resonance and Stark broadening (see Sect. 2b-3.1); we focus again on foreign, neutral-gas broadening, i.e. adiabatic collision broadening and quenching broadening, excluding J-mixing and M_J-mixing collisions (see Sect. 2b-6).

As shown in Fig. VII.1, there are various QM theories for collision broadening; most of them lead to expressions from which the entire line profile is hard to calculate. Detailed discussions of these theories are found in various handbooks and (survey) papers cited in Sect. 2b-1. In this section we focus more closely on three QM theories: Smith, Cooper, Chappell and Dillon (1971) (see also Chappell, Cooper, Smith and Dillon 1971), Futrelle (1972), and Nienhuis (1973). (Baranger 1958, 1958a, 1958b was the first author who produced an impact QM theory including diabatic, adiabatic, J- and M_J-mixing collisions.) The impact theory of Smith *et al.* yields a line profile due to the simultaneous action of Doppler effect and adiabatic collision broadening. Futrelle developed a unified theory of adiabatic collision broadening (with Doppler effect) including impact and QS theory as limiting cases; also the effect of the duration of collision is taken into account. Nienhuis' theory includes in principle Doppler effect, adiabatic collision and quenching broadening; in the impact- as well as in the QS approximation line profile functions are reported. All three theories are in principle capable of including J- and M_J-transitions.

First we discuss **Nienhuis'** theory (see Nienhuis 1973) for the case of simultaneous Doppler effect and adiabatic broadening; an isolated atomic line is studied (under optically thin conditions and without self-broadening). The assumptions made are:

(i)   Two nondegenerate internal states of the atom, $|u\rangle$ and $|1\rangle$, with internal energies V_u and V_1 are considered (two-state approximation).

(ii) No collisional transitions occur between $|u\rangle$ and $|1\rangle$.

(iii) Atoms are not kept stationary (as is usually assumed in older QM theories),

695

Chapt. VII BROADENING AND SHIFT OF ATOMIC METAL LINES

i.e. the Doppler effect is accounted for.

The absorption (and emission) line-profile functions are expressed as Fourier transforms of the correlation function $\Phi(t)$ of the system consisting of a single atom in a thermal bath of perturbers at density n_p. The expression for $k(\nu)$ [$\propto \alpha(\nu)$] and $\varepsilon(\nu)$ obey Kirchhoff's law (Eq. II.293), as is explicitly shown in Nienhuis' paper.

For example, the absorption line-profile function $k(\omega)$ reads as follows ($\omega \equiv 2\pi\nu$) :

$$k(\omega) = n_p \frac{2\pi\omega}{3\hbar c}\left(1-\exp[-\hbar\omega/kT]\right) \int_{-\infty}^{\infty} dt \, \exp[i\omega t] \, \Phi(t) , \qquad (VII.11)$$

where

$$\Phi(t) \equiv \text{Tr} \, \exp[-H/kT]\exp[iHt/\hbar]\exp[-i\mathbf{k}\cdot\mathbf{R}]\boldsymbol{\mu}\cdot$$
$$\cdot \exp[-iHt/\hbar]\,\boldsymbol{\mu}\exp[i\mathbf{k}\cdot\mathbf{R}]/Z(T). \qquad (VII.12)$$

Here $\boldsymbol{\mu}$ is the electric dipole moment operator of the atom; $\mathbf{R}$ is the position operator of the atom; H is the Hamiltonian of the system; $\mathbf{k}$ is the wavevector of the incident light beam, and Z is the canonical partition function. The Doppler effect is accounted for by the factor $\exp[\pm i\mathbf{k}\cdot\mathbf{R}]$. (The dots in the expression for $\Phi(t)$ denote the inner product of two vectors, whereas Tr is an abbreviation of 'Trace'). The two-state Hamiltonian can be expressed as: $H = K + H_0 + V$, where K is the kinetic energy; H_0 is the undisturbed internal energy of the atom; and V denotes the interaction of all particles. The Hamiltonian in matrix form does not contain off-diagonal elements, since diabatic effects are not considered (see assumption ii). Note that the Hamiltonian of the radiation field and its interaction with the atom are not included in H.

In order to obtain computable expressions, first the *classical limit* of $\Phi(t)$ is considered by taking $\hbar \to 0$. As shown by Nienhuis (1973) one may distinguish two cases:

(a) $\hbar \to 0$, $t \to 0$ while $\frac{\hbar}{t}$ = constant. This limit leads to a QS theory, since the *small-time limit* is considered (see Sect. 2b-3.2). The resulting $k(\omega)$ expression for this case is the same one as found from the QS theories of Kuhn (1934) and Margenau (1935). The result is different from that of Fox and Jacobson (1969) (see discussion in Sect. 2b-3.2). The Doppler contribution drops out in a 'natural way' as should be the case in the QS limit where perturbers are assumed to be, virtually, at rest.

(b) $\hbar \to 0$ without simultaneously taking the small-time limit as in case (a). In other words we now try to find a classical expression that is valid for large times

THEORETICAL   Sect. VII.2b

and can ultimately yield the *impact approximation*. This can be done by letting
$\hbar \to 0$ together with $\Delta V \ (\equiv V_u - V_1) \to 0$ while $\frac{\Delta V}{\hbar} =$ constant.

The result is a generalization of Anderson-Talman's theory (see Sect. 2b-3.3), but the Doppler effect is still included here. The (classical) path of the perturber is here determined by the <u>average</u> potential $V_0 \equiv \frac{1}{2}(V_u + V_1)$.

From case (b) one gets computable expressions for $k(\omega)$ (or $\varepsilon(\omega)$), when a number of simplifying assumptions are made:

(i) The perturbers are assumed to be statistically independent; the perturber-atom interaction is pairwise additive.

(ii) The binary collision concept is adopted (see also Sect. 2b-3.1), i.e. it is assumed that at each moment not more than one perturber interacts with the atom, and that the successive binary collisions are completely uncorrelated. In this way the theory is typically applicable to the low-density case. (The recoil contribution to the Doppler effect as well as the Doppler effect itself are formally still included in the profile functions, but the former effect has to be neglected; see (iii) and Ward, Cooper and Smith 1974a.)

(iii) The final simplification is produced by considering the *impact limit*: the collisions are assumed to be momentaneous (i.e. they occur precisely at the moment of closest approach). This means $\tau_{coll} \ll \tau_{ic}$ (see Sect. 2b-3.1 Furthermore the completed-collision assumption is made as usual: $s \gg \tau_{coll}$.

The classical collision trajectory is here approximated by a straight path that has different directions after each 'collision' but retains the same absolute value of the relative velocity. The deflection angle in the centre-of-mass system is determined by the average interaction potential V_0 (see above). It cannot be proved that under usual flame conditions the recoil Doppler effect is negligible (see Nienhuis 1973). However, in order to obtain a <u>computable</u> line profile function the recoil Doppler effect is neglected.

If the recoil effect is neglected, we find a line profile that is essentially an average of a set of dispersion profiles, one for each initial momentum **P** of the atom with mass M; one gets

$$k(\omega) \propto \int d\mathbf{P} \ \exp[-\mathbf{P}^2/2MkT] \ \frac{\delta\omega/\pi}{\delta\omega^2 + (\omega - \omega_0 - \mathbf{k}\cdot\mathbf{P}/M - \delta\omega_S)^2} \ . \tag{VII.13}$$

The half-intensity width $\delta\omega$ and the shift $\delta\omega_S$ are functions of $P \equiv |\mathbf{P}|$, and are found to be

$$\delta\omega = 2\pi n_p (2\pi kT)^{-\frac{3}{2}} \int d\mathbf{p} \ \exp[-\mathbf{p}^2/2mkT] \ v_r \int_0^\infty b \, db (1 - \cos \eta_0) \tag{VII.14}$$

Chapt. VII   BROADENING AND SHIFT OF ATOMIC METAL LINES

$$\delta\omega_S = 2\pi n_p (2\pi kT)^{-\frac{3}{2}} \int d\mathbf{p} \, \exp[-\mathbf{p}^2/2mkT] \, v_r \int_0^\infty b\,db \, \sin\eta_0 \, . \qquad \text{(VII.15)}$$

Here $\mathbf{p}, m$ are initial momentum and mass of the perturber, respectively. The phase shift η_0 is a function of v_r and b: the relative velocity v_r is defined as $\left|\frac{\mathbf{p}}{m} - \frac{\mathbf{P}}{M}\right|$. The profile function, as well as the $\delta\omega$- and $\delta\omega_S$ expressions are now computable.

In the $k(\omega)$ expression of Eq. VII.13 the pure Doppler shift, represented by $\mathbf{k} \cdot \mathbf{P}/M$, is present. Only when this term and the $|\mathbf{P}|$-dependence of $\delta\omega$ and $\delta\omega_S$ are neglected, we find from Eq. VII.13 the Voigt profile, being the convolution of a pure Doppler and a pure adiabatic (dispersion) profile. If we do not neglect this $|\mathbf{P}|$ dependence, even if the Doppler effect is neglected, we do not find a pure dispersion profile in the impact limit!

This rather detailed discussion of a typical QM theory of line broadening has demonstrated the serious problems one has to face when trying to perform an actual line profile calculation. It may be clear from this discussion that QM theories can relatively simply include all kinds of simultaneously acting line broadening processes (see also below), but that they have to be simplified to such a point that they are only slightly more general than the SC ones, when one is interested in feasible numerical line-profile calculations. Furthermore it is hard to say whether certain refinements that are not present in SC theories but can be made in QM theory, actually lead to real improvements of the results relevant under usual flame conditions. Definite advantages of QM theory are the facts that adiabatic and diabatic collisions can simultaneously be treated and that the effect of Doppler broadening on the resulting line profile can be accounted for in a 'natural' way. These advantages are more closely looked at in Sect. 2b-5.

Futrelle (1972) discussed a unified QM theory, i.e. his theory includes impact- and QS approximations. (A similar theory is also described by Lee 1974, but the latter one seems to be less general and not easily accessible for numerical calculations.) Futrelle also incorporated effects of the duration of the collisions, which link the QS- and impact regimes. The basic assumptions of his theory, developed for the case of isolated atomic lines with neutral, foreign-gas perturbers, at low atomic density (no resonance-broadening), are:

(i)   Two internal atomic states are considered $|u\rangle$ and $|1\rangle$.

(ii)   Structureless perturbers; perturbers act independently.

(iii) $V_u(r)$ and $V_1(r)$ are spherical, pair-wise additive potentials.

(iv)   No collisional transitions between $|u\rangle$ and $|1\rangle$ are allowed, i.e. only adiabatic collision broadening is considered.

THEORETICAL   Sect. VII.2b

(v)   The atoms are kept stationary, i.e. no Doppler effect is taken into account. Only the $k(\omega)$ expression is derived, but it is claimed that detailed balance is obeyed, i.e. Kirchhoff's law is obeyed (see above).

The impact theory follows from a long-time analysis of the correlation function $\Phi(t)$. The QS limit is found by studying the short-time behaviour of $\Phi(t)$. Unified methods for calculating the entire line shape are suggested by using interpolation methods in the intermediate region between long- and short-time behaviour of $\Phi(t)$. It is expected, but not yet demonstrated by the author, that this procedure leads to a computable expression for the entire line profile.

The third QM theory from Smith, Cooper, Chappell and Dillon (1971) considers again isolated atomic lines with foreign-gas perturbers. Resonance broadening is excluded by assuming no interaction between like atoms. The authors do not confine themselves to two nondegenerate atomic levels (as was done by Nienhuis 1973 and Futrelle 1972), thus M_J-mixing collisions are included. Also diabatic collisions are not *a priori* excluded.

The perturbers are taken to be statistically independent particles; thus the perturbers can influence the spectral emission from the atom only through their direct interaction with the atom. The perturber-atom interaction is pair-wise additive. Since the atoms are not kept stationary, the Doppler effect is included. Furthermore the particles do not follow classical paths.

In order to obtain expressions for the line profile due to the simultaneous action of Doppler effect, adiabatic collision broadening (and, in principle, diabatic broadening), first the impact approximation is derived from the general formalism.

In this stage of development of their theory, the authors demonstrate that the Dicke-narrowing as calculated by Rautian and Sobel'man (1967) from SC theory (see Traving 1968), can also be found from their theory. Their results are just a QM generalization of the Rautian and Sobel'man results.

The influence of diabatic collisions on the internal state of the atom can be removed by assuming that the interaction potential has no off-diagonal terms (see above). The authors make explicitly in their further calculations the no-quenching assumption, so only elastic collisions are accounted for.

At the point reached here one can show that the usual Voigt profile is recovered from the above results. For that purpose it is necessary to neglect recoil effects, i.e. collisions have no influence on the atomic trajectory.

The line profile function including Doppler and adiabatic broadening, obtained in the impact approximation, can be calculated numerically without much difficulty, especially so when lower-state interactions can be neglected.

Chapt. VII BROADENING AND SHIFT OF ATOMIC METAL LINES

2b-5 *THE LINE PROFILE DUE TO THE SIMULTANEOUS ACTION OF NATURAL, DOPPLER, ADIABATIC AND DIABATIC (FOREIGN GAS) COLLISION BROADENING*

The problem to be discussed here is how one may derive the line profile function S_ν when the atoms are subject to different, simultaneously acting broadening processes. We confine ourselves to natural, Doppler, adiabatic collision (excluding M_J-mixing) and diabatic collision (quenching) broadening (excluding J-mixing; see Sect. 2b-6). One may get the resulting line profile in the following ways:

(i) The simplest way is to use the Voigt concept (see also Sect. II.5b-3), i.e. natural, adiabatic and quenching profiles are assumed to be dispersion profiles. The resulting profile is then a convolution product of Gaussian and dispersion profiles, based on the assumed mutual independence of these broadening effects.

(ii) By convolution of the individual profiles due to natural, Doppler, adiabatic and quenching broadening. The adiabatic profile follows from SC theory and is here not necessarily a dispersion profile. The broadening mechanisms are again taken mutually independent.

(iii) In some cases convolution of Doppler and adiabatic collision profiles is not allowed when collisions influence the Doppler profile (Dicke-narrowing). Convolution is still applied, but in a different way as compared to case (ii).

By using the QM theory the convolution problem seems to be circumvented (see Sect. 2b-4), but the formal computability of the resulting line profile is then the big problem, i.e. also here one has to convolute individual profiles.

Ad (i). This Voigt concept is an extension of the original Voigt concept (see Mitchell and Zemansky 1961) in which only natural and Doppler broadening were taken into account. The mutual independence of these broadening processes is generally accepted. Under flame conditions natural broadening is usually negligible compared to Doppler and adiabatic broadening, and can be deleted, but quenching broadening might play a significant role. Thus the addition of a quenching profile may be necessary, but then even more the statistical independence of the individual broadening processes is questionable. In the Voigt concept this independence and thus the validity of convolution is simply taken for granted.

Ad (ii). The criterion for application of convolution is again the independence of the broadening processes. Under usual flame conditions (see also Sect. II.5b-2) Doppler and adiabatic collision broadening are considered as statistically independent. Again there is no problem with natural broadening in this respect, but with quenching broadening the same sort of complications may arise as with

THEORETICAL   Sect. VII.2b

adiabatic broadening. Furthermore the question of mutual independence of adiabatic and quenching broadening is unanswered in the theory. So one can only proceed with assumed statistical independence; there is no definite evidence pro or contra.

The total line profile is calculated by adding up the half-intensity widths of the dispersion profiles from natural and quenching broadening. The resulting dispersion profile is convoluted into the Doppler profile and this result is, again, folded with the adiabatic profile. This latter profile may differ from a dispersion function and is at present usually calculated from general-pressures theory (according to Jansen, Hollander and Behmenburg, unpublished work).

Ad *(iii)*. When Doppler and adiabatic collision broadening are not taken to be independent (see Galatry 1961, Rautian and Sobel'man 1967, Gersten and Foley 1968, Berman and Lamb 1971, and Zaidi 1972), the only way in SC theory for calculating the total line profile is to fold the modified (= Dicke-narrowed) Doppler profile into the adiabatic profile (and possibly the quenching profile). This means procedure (ii) with the modified Doppler profile instead of the undistorted one. It is believed that this method is not correct (see Smith *et al.* 1971), but in the framework of SC theory no other possibility exists so far.

It is recalled that QM theories can in principle describe the total line profile due to simultaneous action of the various broadening processes in a formally adequate manner. Diabatic and adiabatic collisions can both be incorporated by choosing an interaction potential that accounts for both effects. [In the matrix form of $V(r)$ the inclusion of diabatic effect means considering also off-diagonal terms; see Nienhuis 1973.] Adiabatic effects are considered in several QM theories (see Baranger 1958b, Fano 1963, and Nienhuis 1973).

For the line profile resulting from simultaneously acting Doppler, adiabatic and quenching broadening Nienhuis (1973) has presented a closed expression. His expressions for $\delta\lambda$ and $\delta\lambda_s$ (in the impact approximation) can be computed numerically; recoil Doppler effect is neglected. A factor is present that accounts for diabatic collisions; this factor reflects the reduction in the duration of the radiation process due to quenching. This factor can be derived from the interaction potential and numerical calculations then become feasible. In most cases, however, the interaction potential is not known and but little can be derived about the quenching broadening, even when the quenching cross section is known from experiment (see Sect. VI.2; see also Sect. 4).

Summarizing one may conclude that there are two ways of calculating total line profiles. The SC theory leads to computable expressions but, in general, questionable assumptions have to be made. In QM theory one can formally correctly describe the total line profile, but the formulas are not amenable to numerical calculations unless one makes similar approximations as in SC theory.

Chapt. VII   BROADENING AND SHIFT OF ATOMIC METAL LINES

2b-6   *INFLUENCE OF J- AND M_J-MIXING COLLISIONS ON THE LINE PROFILE*

In the theoretical considerations presented thus far we assumed the upper and lower state of transition involved to be nondegenerate, which is only true, in the absence of external fields, for states with zero angular momentum. This is not very realistic, since an electric dipole transition between states with $J=0$ is forbidden (see Sect. II.2c). It has been shown that the theory remains valid for degenerate states provided that the magnetic substates are affected by perturbers in the same way; this is the case for a state with $J=\frac{1}{2}$ perturbed by particles with $J=0$ (see Baranger 1958). Only a qualitative description of the effect of (diabatic) J-mixing collisions (transitions between fine-structure components) and of M_J-mixing collisions (transitions between magnetic substates) will be given. We shall here neglect collisional transfer between the upper and the lower terms.

A collision that causes a transition between two magnetic substates brings about a change of the orientation of the atom (see Sect. II.5b-2). In case of J-mixing collisions the lifetime of the radiating atom is shortened, which causes an extra line broadening in analogy with the effect of quenching collisions (see also Sect. II.5b-2) if the f.s. splitting is comparatively large.

In the discussion of line broadening processes in flames the effect of degenerate atomic states has not yet been taken into account so far. In the literature the question of degenerate states is considered by some authors who reported line broadening measurements in absorption cells, for example, Ch'en and Garrett (1966, 1966a and 1967), Ch'en, Gilbert and Tan (1969, 1969a), McCartan and Hindmarsh (1969), Rostas and Lemaire (1971), and Butaux, Schuller and Lennuier (1972) The measurements refer to noble-gas perturbers and the lines investigated are usually Hg 2537 Å ($^3P_1 \to {}^1S_0$) and the first and second resonance lines of the alkali atoms.

In Table VII.3 we have collected theories, authors, computability, etc., of J-mixing broadening. For completeness' sake also the treatment of quenching collisions is listed. The perturbers are assumed to have zero angular momentum (noble-gas atoms). We will briefly outline how the effect of J- and M_J-mixing collisions on the line shape is accounted for in SC and QM theories.

The relatively simplest cases in theory are the $^1P_1 \to {}^1S_0$ transitions in the alkaline-earth atoms and the 2537 Å resonance line of Hg ($^3P_1 \to {}^1S_0$ transition). The upper level here is three-fold degenerate, i.e. $M_J = 0, \pm 1$. In this case there is not one single interaction potential for the upper state, but two different ones, viz. a V_σ-potential for $M_J = 0$ and a V_π-potential for $M_J = \pm 1$. The interaction potential in the upper state consists thus of a sort of combination of two different potentials, V_σ and V_π; the effective upper-state potential has an anisotropic form. Butaux, Schuller and Lennuier (1972) have shown that this effective potential

Table VII.3
Survey of Theoretical Work on Quenching, M_J- and J-Mixing Effects on Atomic Line Profiles

Authors	Type of theory[+]	Quenching broadening	J-mixing	M_J-mixing	Formal theory/ computable expressions	Limitations[+]
Anderson (1949)	SC	Yes	Yes	Yes	computable, but many approximations	impact
Baranger (1958, 1958a)	QM	No	No	Yes	formal	impact
Baranger (1958b)	SC	Yes	Yes	Yes	formal	impact
Fano (1963)	QM	Yes	Yes	Yes	formal	impact
Schuller (1971); Schuller and Oksengorn (1969; 1969a)	SC	No	No; No; Yes	Yes	computable	impact
Nienhuis (1973)	QM	Yes	No	No	computable	impact and QS limits
Nienhuis (1974)	QM	No	Yes	Yes	computable	impact and QS limits

[+] SC and QM mean 'semi-classical' and 'quantummechanical', respectively. QS means 'quasi-static'.

can be determined correctly from the experimental data on the broadening and shift of the 2537 Å Hg line when this anisotropy is accounted for. The method makes use of approximative calculations in which the potential of the upper state is expressed in terms of V_σ and V_π potentials for which approximative force constants are reported in the literature. By confronting the 'effective' upper-state potential with the experimental results, a best fit is obtained and thus the best fitting force constants of this potential are derived.

The alkali resonance doublets have upper states $^2P_{\frac{1}{2}}$ and $^2P_{\frac{3}{2}}$ and a lower state $^2S_{\frac{1}{2}}$. Schuller and Oksengorn (1969) have calculated the J-dependence of the broadening and shift of these doublets in the impact approximation using a van der Waals potential. They showed that the $^2P_{\frac{1}{2}}$ component behaves as in Lindholm's theory with isotropic potentials. For the $^2P_{\frac{3}{2}}$ component there is a difference due to the anisotropy of the excited-state interaction. Schuller (1971) has also discussed the more general case of a L.-J. interaction potential. In both papers the influence of transitions induced by diabatic collisions between the fine-structure levels (J-mixing collisions) was not taken into account. In a subsequent paper Schuller and Oksengorn (1969a) have also included the effect of J-mixing. The authors showed that this effect is measurable only in lithium. So with noble-gas perturbers, M_J-mixing collisions do influence the profiles of alkali doublet lines much more than J-mixing collisions do. Also here it can be remarked that knowledge of J- and M_J-mixing cross sections does not contribute to our knowledge of the interaction potentials sought.

Nienhuis (1974) has discussed the broadening and shift of alkali fine-structure components in the impact- and the QS limit of QM theory. He includes J- and M_J-mixing collisions in a formal theory. Also his resulting expressions are only numerically computable when interaction potentials can be substituted.

2b-7 *(INTEGRATED) LINE PROFILES WITH H.F.S.*

We first discuss the influence of hyper-fine-structure on the *integrated profile* of an optically thick line by considering curves-of-growth (and duplication curves; see Sects II.5c-2 and 5c-3). Several authors have derived expressions for the integral absorption, A_t, in the case of a line with h.f.s. (e.g. Matossi, Mayer, and Rauscher 1949, Preobrazhenskii 1963, Rukosueva 1964, Behmenburg and Kohn 1964, van Trigt, Hollander and Alkemade 1965, and Gallagher and Lapp 1967). In these expressions, partially overlapping lines with equal half-intensity widths and a-parameters were considered. The overlap parameter γ (see Sect. II.5c-2) depends on $\delta\nu$ and thus on the a-parameter.

The combined curve-of-growth or duplication curve for an arbitrary a-value can be computed numerically when relative position and relative strength of the

THEORETICAL   Sect. VII.2b

h.f.s. lines are known from literature. The combined duplication curve is derived from the combined curve-of-growth in the usual manner (see Sect. II.5c-3).

More specifically there are a number of calculations on the effect of h.f.s. on the ordinate value of the intersection point of the asymptotes in the theoretical curve-of-growth. It is recalled from Eq. II.276 that the integral absorption, A_t^*, at the intersection point for a single line is given by

$$A_t^* = 2\delta_L \equiv 2a\delta\nu_D/\sqrt{\ln 2} \,, \qquad (VII.16)$$

where we have made use of Eq. II.246. For two nonoverlapping line components that are measured together and that have oscillator strengths f_1 and f_2, and the same a-parameter, we can derive (cf. Sect. II.5c-2)

$$A_t^* = \left\{(\sqrt{f_1} + \sqrt{f_2})^2 / (f_1 + f_2)\right\} 2a\delta\nu_D/\sqrt{\ln 2} \,. \qquad (VII.17)$$

The value of A_t^* for two partially overlapping line components lies in between the values obtained from Eqs VII.16 and 17. Values of the integral absorption in the intersection point of the asymptotes in the case of partially overlapping h.f.s. components have been reported for the Na-D_2 line by Behmenburg and Kohn (1964) and for the first resonance doublets of Li, Na, K, Cs by van Trigt, Hollander and Alkemade (1965). Gallagher and Lapp (1967) have calculated the effect of h.f.s. on the integral absorption for the u.v. resonance lines of Cu.

The experimental derivation of a-parameters from integral absorption measurements with h.f.s. is only feasible by an iteration procedure, since the experimental a-parameter depends on the assumed γ-parameter, which in turn depends on the a-parameter sought (see van Trigt, Hollander and Alkemade 1965).

In calculations of the influence of h.f.s. on the *spectral profile* of an optically thin line, one usually assumes that the h.f.s. lines have the same individual profile. In the literature one has only considered the case of Voigt line shapes (see, e.g., Schulz and Stopp 1967, 1969, and Wagenaar, Pickford and de Galan 1974). The knowledge of the relative position and intensity of the h.f.s. lines plus a known or assumed a-parameter (and known $\delta\nu_D$) leads to a value of the γ-parameter and thus to a combined line profile with h.f.s. By varying the trial-value of an unknown a-parameter one may derive the true, individual line profile from the experimental profile by a curve-fitting procedure (see Wagenaar, Pickford and de Galan 1974).

2b-8   *THE OCCURRENCE OF SATELLITE BANDS*

For more than 40 years there have been reports of narrow diffuse bands produced near the atomic lines of various metals by collisions with foreign perturbers.

The earliest reference was presented by Kunze (1931). Most observations have been made on resonance lines (alkali, mercury) in absorption (vapour cells with temperatures roughly between 350 and 900 K) and have shown that 'satellites' occur both on the short-wavelength side (blue satellites) as well as on the long-wavelength side (red satellites) of these lines. We do not intend to be complete when mentioning the following theoretical papers and experimental investigations in nonflame milieus: Ch'en and Takeo (1957), Ch'en and Wilson (1961), Granier and Granier (1973), Ch'en and Garrett (1966a, 1967), Kielkopf and Gwinn (1968), Granier (1969), Hindmarsh and Farr (1969, 1972), Szudy (1970), Mahan (1972), Atakan and Jacobson (1973), Ch'en and Phelps (1973), and Sando (1974).

Satellite bands are induced by the noble gases as well as by perturbers such as H_2, D_2, N_2 and many hydrocarbons. The experimental results on alkali resonance lines in vapour cells with inert-gas perturbers are in summary:

(i)   On the long-wavelength side of the resonance lines in the presence of Xe and Kr a comparatively strong satellite appears. With the lighter noble gases and N_2 this satellite is absent (see McCartan, Farr and Hindmarsh 1974, Ottinger *et al.* 1975, and McCartan and Farr 1976). The satellite becomes as intense as the line itself at high perturber pressures.

(ii) Also a series of weak satellites is found in the red wings, some of which appear to converge and some of which do not.

(iii) A blue satellite appears on the $^2P_{\frac{3}{2}}$ component in each noble gas. Structure has been observed on this blue satellite and frequently two satellites are reported. Similar observations have been reported about the satellites on the wings of the Hg 2537 Å line by Granier and Granier (1973) (see also Losen and Behmenburg 1973). The number of both blue and red satellites depends on both the radiating and perturbing atoms. There appeared, however, an essential difference between red and blue satellites: blue satellites are observed for all perturbing gases provided the pressure is high enough; their intensity increases with the increase in pressure. The red satellites are only produced by heavier gases (not by He and Ne, which give rise to strong blue satellites) and disappear entirely at higher pressure; see references under (i).

In flames, only one observation of satellite bands has been reported by Hollander, Jansen and Alkemade (1977). A blue satellite was found on the wing of the Sr resonance line in a $CO-N_2O$ flame. Red satellites were found on the $Na-D_1$ wings in H_2-O_2-Ar and $H_2-O_2-N_2$ flames.

Different approaches have been made in order to explain the origin of satellites. The two most frequently used explanations are the following:

THEORETICAL   Sect. VII.2b

The satellites are due to the existence of vibrational levels of the temporary van der Waals molecule formed by radiator and perturber. Red and/or blue satellites arise from transitions between bound states of this quasi-molecule (see Michels, de Kluiver and ten Seldam 1959, Klein and Margenau 1959, and Mahan and Lapp 1969).

Another explanation, which is more widely used, ignores the existence of vibrational levels. It concerns only the red satellites and is based on the quasi-static theory of pressure broadening. In this model a satellite band results from any minimum or maximum in the difference potential $V(r)$ (i.e. the points where upper- and lower-state potential curves are parallel; see, e.g., Kielkopf and Gwinn 1968, Hindmarsh and Farr 1969, 1972, and Mahan 1972).

Atakan and Jacobson (1973) have discussed the theoretical models of satellites, and they critically compared their theoretical explanation with experimental data from Ch'en and coworkers (see literature cited above). Their explanation is essentially that red satellites originate from an extremum in $V(r)$ at fairly large values of r, whereas blue satellites result from transitions to a bound state of the quasi-molecule formed at smaller values of r.

When following the calculations of Hindmarsh and Farr (1969, 1972) or Kielkopf and Gwinn (1968) and Kielkopf, Davis and Gwinn (1970) for a L.-J. interaction potential, one finds a definite relation between the position of the (one) red satellite band and the (positive) force constants of the difference potential, C_6 and C_{12}. Writing for the difference potential: $V(r) \equiv h\Delta\nu$, one obtains for $\Delta\nu$ the following expression:

$$\Delta\nu = \frac{V(r)}{h} = \frac{C_{12}}{r^{12}} - \frac{C_6}{r^6} = 4(\varepsilon/h)\left\{\left(\frac{R}{r}\right)^{12} - \left(\frac{R}{r}\right)^6\right\} \qquad (VII.18)$$

where

$$\varepsilon \equiv \frac{C_6^2 h}{4C_{12}} \quad \text{and} \quad R \equiv \sqrt[6]{C_{12}/C_6} \,.$$

One finds a red satellite at a frequency distance: $\Delta\nu = -\frac{\varepsilon}{h}$. The physical meaning of ε is the depth of the difference-potential well, which is located at $r = r_m$ (compare Fig. X.1b on page 912): at $r = R$ the potential is zero and we have the relation: $r_m = \sqrt[6]{2}R$. In principle it appears feasible to derive a relation between the C_6 and C_{12} values from the position of the red satellite.

Mahan (1972) has extended this theory of red satellites by using an arbitrary difference potential. When such a difference potential has a minimum with the same values of r_m, ε and curvature K_m as above, identical results are obtained for the position of the red satellites. The relative intensity of the satellite band is $\propto r_m^2 K_m^{-\frac{1}{2}}$. This theory predicts that satellite-band <u>positions</u> are <u>independent</u> of temperature and perturber density, in agreement with experimental results.

707

Chapt. VII   BROADENING AND SHIFT OF ATOMIC METAL LINES

2b-9 *EXTRA DOPPLER BROADENING OF NONTHERMAL EMISSION LINES*

So far we have assumed in our discussion of line broadening processes that the relative spectral distribution of the photons emitted by the excited atoms was the same as under thermal-radiation conditions, and that emission and absorption line profiles were identical. However, in the general case of nonthermal radiation this assumption may not be valid: extra Doppler broadening may occur and may cause both profiles to be different when the translational-energy distribution of the excited or absorbing atoms differs from Maxwell's distribution law.

In chemiluminescent reactions, the law of conservation of total linear momentum must be obeyed. Suppose that the excess reaction energy left after excitation of the metal atom M is converted mainly into translational energy of the reaction products. In the case of the two-body exchange reaction: $MX + Y \to M^* + XY$ (see process no. 19 in Table II.1 on p.108), the conservation laws of energy and linear momentum determine which fraction of the excess energy ΔE is imparted to each of the reaction products, M^* and XY. The metal atom thus excited will accordingly fly away with an extra velocity. The spectral line emitted by this particular class of excited atoms may exhibit an extra Doppler broadening when photon emission occurs before thermalization of the velocity through collisions; the spectral Doppler profile is no longer a Gaussian function. It was argued in Sect. II.5b-2, on the basis of detailed balance, that in thermal equilibrium the velocity-distribution of the whole ensemble of excited atoms is not affected by the occurrence of a particular excitation process at large rate, so that the Doppler broadening conforms to equilibrium. However, when a suprathermal chemical excitation reaction (cf. Sect. VI.4c) is not balanced by the reverse chemi-quenching reaction, an extra Doppler broadening might result.

If excited metal atoms are produced mainly by the above binary reaction, and if the reverse quenching reaction proceeds at a negligible rate, the Doppler width $\delta\lambda_D$ exceeds its equilibrium value $(\delta\lambda_D)_e$ by the factor (see Alkemade 1963)

$$\frac{\delta\lambda_D}{(\delta\lambda_D)_e} = \left(1 + \frac{m_{XY}}{m_{XY}+m_M} \cdot \frac{\Delta E}{\tfrac{3}{2}kT}\right)^{\tfrac{1}{2}}. \qquad (VII.19)$$

Here m_{XY} and m_M are the mass of the reaction products XY and M^*, respectively. This equation is valid if the velocity of M^* is not changed before photon emission (see also below). For reactions of the type: $MO + C \to M^* + CO$, one calculates the following values of this factor in flames: 1.3 for the 2863 Å Sn line, 1.5 for the 3068 Å Bi line, 1.5 for the 2311 Å Sb line, and 1.7 for the 2366 Å Cr line.

This deviation may have consequences for the spectral profile of the emission line and, in particular, for the radiation loss due to self-absorption of resonance lines. The effect of this deviation becomes more conspicuous in low-pressure flames, where collision broadening is small compared with Doppler

THEORETICAL   Sect. VII.2b

broadening. The relative variation of the resonance line intensity with metal concentration can then no longer be described by the factor A_t (see Eq. II.295). The shape of the intensity-concentration curve deviates from that of the curve-of-growth for the thermal resonance line in the range where self-absorption is noticeable. It is noted that the Doppler broadening of the same line seen in absorption is still thermal, since the velocity distribution of the ground-state atoms will not be significantly affected.

In fact the theoretical analysis is more complex than suggested above. When one of the reaction products is a di- or poly-atomic molecule, part of the excess energy may also be distributed over the internal degrees of freedom. This would reduce the extra Doppler effect. When, before radiating, the excited metal atom undergoes on the average a few elastic collisions with the surrounding inert gas molecules, its velocity distribution will be thermalized. No extra Doppler broadening is then to be expected. However, when the quenching efficiency of these molecules in a gas-kinetic collision is significant, most of the (few) emitted photons will be produced by M^* atoms that have not undergone a collision. Effective thermalization and consequently normal Doppler broadening of suprathermal chemiluminescent lines are thus expected with Ar as diluent gas, but not with N_2 (see Sect. VI.2).

In a similar way, extra Doppler broadening may also arise when M atoms are excited by transfer of electronic excitation energy according to: $M^0 + A^* \to M^* + A^0$, the excitation energy of A^* being larger than that of M^* (see process no. 6 in Table II.1 on p.108). In the case of sensitized fluorescence of Tl by photo-excited Hg (see Sect. II.5a-1), the extra Doppler broadening has been tested experimentally in vapour cells (see Mitchell and Zemansky 1961). The magnitude of the effect is described by a similar equation as Eq. VII.19.

Kraulinya, Liepa and Skudra (1976) described the measurement of thallium spectral line profiles in the sensitized fluorescence of a mixture of Hg- and Tl vapours in cells. Extra Doppler broadening was expected, and found experimentally, according to the reaction:

$$Hg(6^3P_1) + Tl(6^2P_{\frac{1}{2}}) \to Hg(6^1S_0) + Tl^* + \Delta E .$$

When Na^* is produced by photo-dissociation of NaI vapour with a photon energy above threshold (see Sect. VI.1 and process no. 27 in Table II.1), the excess energy is again transformed into relative translational energy of the dissociation products. An extra Doppler broadening of the Na line results, which increases with increasing photon energy (see Mitchell and Zemansky 1961, and Zare 1964). The extra Doppler broadening was found experimentally to be somewhat smaller than predicted by theory (see Hanson 1967). The latter two excitation effects are, however, not believed to be relevant in flames (see also Sect. VI.4c-2 under iii).

Collision broadening is not expected to be affected by the occurrence of nonequilibrated chemiluminescent reactions. A slight deviation might perhaps occur if the collisional half-intensity width were to depend on the velocity of the radiating atom (see Mizushima 1967). The extra mean velocity of the excited atoms produced by suprathermal chemiluminescent reactions might then induce a change in the collisional half-intensity width as well (see also Alkemade 1963).

In conclusion we may state that extra Doppler broadening of suprathermal chemiluminescent lines might be expected to occur under special flame conditions, but has not been investigated so far in flame spectrometry.

3. EXPERIMENTAL

We confine ourselves mainly to methods that have been applied to experiments in flames. In Subsections 3a and 3b we discuss various experimental methods of determining (parts of) spectral line profiles including line shifts, in emission as well as in absorption. In Subsection 3c we discuss methods of obtaining information about the half-intensity width from measurements of integrated line intensities (integrated over the spectral line profile). Spectral profiles of fluorescence lines can in principle be measured by similar methods as emission line profiles. Only when pulsed light sources are used for inducing fluorescence, may special methods be needed. Photographical measuring methods are not mentioned.

3a   DETERMINATION OF EMISSION LINE PROFILES IN FLAMES
3a-1   *SCANNING INTERFEROMETER*

Spectral profiles can be determined by means of a pressure- or piezo-electrically scanned Fabry-Pérot interferometer. Its most important properties are the spectral resolution ($\lambda/\Delta\lambda$) and the free-spectral-range (FSR), which are interconnected and can be varied by changing the distance of the interferometer plates. For details we refer, e.g., to Jansen, Hollander and Alkemade (1977) (pressure-scanning) and to Wagenaar and de Galan (1973) and Wagenaar (1976) (piezo-electric scanning). The cores of atomic line profiles in flames have been measured with these techniques with a spectral resolution of about $(1-2) \times 10^6$.

As an illustration we describe here some features of the pressure-scanned Fabry-Pérot interferometers that have been used for line-broadening work in flames by Jansen, Hollander and Franken (1974) and Jansen, Hollander and Alkemade (1977). For measurements of the Sr 4607.33 Å line profile ($^1P_1 \rightarrow {}^1S_0$ transition) a plate separation of 1 cm was used, while the finesse was about 35. From this one gets an

EXPERIMENTAL Sect. VII.3a

FSR of 108.3 mÅ and a $\Delta\lambda$ of about 3.0 ± 0.5 mÅ, or $\lambda/\Delta\lambda \approx 1.5 \times 10^6$. The apparatus half-intensity width $\Delta\lambda$ consisted of a Lorentzian part $\Delta\lambda_{Lor} \simeq 1.6$ mÅ and a Gaussian part $\Delta\lambda_{Gauss} \simeq 2.4$ mÅ; the total $\Delta\lambda$ of the resulting Voigt profile follows from Eq. VII.21 (for further details see also Kirkbright and Troccoli 1973). When one chooses a plate separation of 0.1 cm (in practice a minimum value) one gets in our example: FSR $\simeq 1.1$ Å and $\Delta\lambda \simeq 0.03$ Å; thus $\lambda/\Delta\lambda \simeq 1.5 \times 10^5$. This resolution can also be expected with a grating scanning monochromator, but then a considerably larger FSR is of course available.

The line shift is usually derived by measuring the line profiles from flame and hollow-cathode lamp simultaneously (see Wagenaar and de Galan 1973) or alternately (see Jansen, Hollander and Alkemade 1977). The HCL at sufficiently low current has an unshifted Doppler line profile (see Wagenaar and de Galan 1973 and Jansen, Hollander and Franken 1974).

Three types of corrections have to be considered (see, e.g., Jansen, Hollander and Alkemade 1977a):

(1) One has to correct for the known apparatus profile by deconvolution. For the measurement of flame line profiles this correction is only of minor importance ($\delta\lambda_{line} \geqslant 30$ mÅ as compared to $\Delta\lambda_{app} \approx 3$ mÅ).

(2) The limited free-spectral-range at $\lambda/\Delta\lambda \approx 1.5 \times 10^6$ may cause partially overlapping orders, especially when self-absorption causes extra broadening of the line profiles. One may carry out corrections for this effect when the line profile can be represented by a Voigt profile (with known a-parameter).

(3) Under optically thin conditions the emission signal may be weak and the signal-to-noise ratio may be unfavourable. One then has to rely on measurements of line profiles at higher densities, that are extra broadened due to self-absorption (see also sub 2). The correction procedure in this case makes use of an assumed Voigt shape of the optically thin profile. One measures line profiles at different atomic densities and extrapolates, often with the help of a computer, towards optically thin conditions. The a-parameter has to be known.

When line profiles can be measured at sufficiently low atomic densities, the extrapolation towards zero density can be worked out simply 'by hand' by making a 'Beer-plot' (cf. Subsection 3b-2) at each frequency without knowing the a-parameter (see also Kirkbright, Troccoli and Vetter 1973, and Alger, Kirkbright and Troccoli 1973).

The correction for self-absorption and that for partially overlapping interferometer orders can be carried out simultaneously with a computer when the

Chapter VII   BROADENING AND SHIFT OF ATOMIC METAL LINES

a-parameter of the (assumed) Voigt profile is known.

3a-2   *SCANNING MONOCHROMATOR*

With (grating) monochromators the entire line profile can be scanned, but usually with much lower resolution than with interferometers (see Table III.2 on p.246). The correction for the apparatus profile is the major problem here (see, e.g., Carrington 1959). The applicability of these instruments to line profile measurements is therefore mainly restricted to line wing measurements (starting, say, from about 0.5−2 Å from the line centre) and observation of satellite bands. For these measurements a resolution of the order of 10^5 is generally sufficient (see, e.g., Hedges, Drummond and Gallagher 1972, Gallagher 1975, and Hollander, Jansen and Alkemade 1977).

The question of overlapping orders does usually not occur with monochromators. (Self-absorption effects can usually be neglected in the line wings; with regard to line core measurements, the corrections are in principle similar to those with interferometers.) Shift measurements are not feasible in view of too low spectral resolution.

3b   DETERMINATION OF ABSORPTION LINE PROFILES IN FLAMES

Absorption line profiles can be measured with atomic resonance lines that show sufficient absorption. In Table VII.4 we have collected relevant information about the measuring techniques used with flames.

3b-1   *SCANNING INTERFEROMETER*

Basically this method proceeds in a similar way to emission interferometry. The same corrections and correction procedures apply here (see Subsection 3a-1).

A spectral continuum light source is used as background source. Two applications to absorption line profiles in flames have been reported, viz. Kirkbright and Troccoli (1973), and Wagenaar, Pickford and de Galan (1974).

The main problems arose here from the relatively wide band pass of the preselecting monochromator. In emission measurements order-overlap is usually not a great problem even when using a fairly large band pass; this is because the atomic emission lines in a flame spectrum are usually sufficiently far apart. In absorption measurements with a continuum source, however, the monochromator will transmit a large number of interferometer orders (each corresponding to a different wavelength region), only one of which will be absorbed in the atomic line profile considered; as a result the absorption signals are relatively weak ('poor contrast') and the SNR is rather poor. The latter disadvantage can be overcome by the use of electronic averaging techniques (see Kirkbright and Troccoli 1973) and high absorber concentrations.

Table VII.4

Methods for Determining Absorption Line Profiles

Method/instrument	Part of profile measured	Spectral resolution obtainable ($\lambda/\Delta\lambda$)	Literature
Interferometer (1) pressure-scanning (2) piezo-electric scanning	line core	$(1-2) \times 10^6$	(1) Jansen, Hollander and Alkemade (1977); Kirkbright and Troccoli (1973); (2) {Wagenaar and de Galan (1973); Wagenaar, Pickford and de Galan (1974)
Zeeman scanning	line core (± 0.20 Å at 20,000 G)	$(5-25) \times 10^5$ (depending on $\delta\lambda$ of source line)	Hollander and Broida (1967) and (1969); Strumia (1967); Joli, Strumia and Moretti (1971); Jansen, Hollander and Alkemade (1977b)
Scanning grating monochromator	entire profile	5×10^4 per order[+]	Losen and Behmenburg (1973); Carrington and Gallagher (1974)
Tunable dye laser	entire profile	for pulsed laser: $> 2 \times 10^5$ for c.w. laser: $> 2 \times 10^6$	Kuhl, Marowsky and Torge (1972); van Dijk (1977); Walther (1974); Jongerius, Hollander and Alkemade (1978)

[+] A much higher resolution can be obtained by the application of a spectrograph with scanning slit (see Subsection 3b-3).

De Galan and co-workers improved the 'contrast' of this method by using a system proposed by Jaffe, Rank and Wiggins (1955). Here both the interferometer and the monochromator are scanned, in exact synchronization, in such a way that one and the same order of the interferometer is observed while it covers a wavelength region determined only by the properties of the interferometer (see Wagenaar, Pickford and de Galan 1974).

Line shifts can be derived by using a reference absorption cell producing a sufficiently narrow, unshifted (Doppler) line of the same atom.

3b-2 *ZEEMAN SCANNING*

A narrow-line source (viz. an EDL; see Sect. III.7c) is placed in a variable magnetic field and the light beam emerging from it, viewed along the magnetic-field direction, is used for the scanning of the absorption line profile. Due to the Zeeman effect at least two σ-components with opposite circular polarization directions are present (with the magnetic field switched on) in the background line source. With a quarter-wave plate + polarizer (for example a Glan Thomson prism) either σ-component can be isolated and used for the scanning of the absorption line to higher and lower wavelength values. The absorption factors are measured as a function of magnetic field strength and, after conversion, as a function of $\Delta\lambda$. Collisional line shifts are derived from the difference in wavelength of the background line (which is supposedly not shifted by collisions) at zero field and of the peak of the shifted absorption line. This method is discussed in detail in the literature (see, e.g., Hollander and Broida 1967, 1969, Hollander, Jansen, Plaat and Alkemade 1970, and Jansen, Hollander and Alkemade 1977b). With magnetic fields up to 20,000 G the scanning range is roughly ± 0.2 Å, i.e. of the order of about (5-7) times the absorption line half-intensity width in flames.

Some aspects of this measuring method have to be considered more closely (see also Sect. III.15):

(i) With a double-beam set-up one can measure a 0.1% absorption signal at an $SNR \approx 10$. Sampling and signal-averaging techniques have then to be used. It is much easier to measure line profiles at higher concentrations, which may, however, be extra broadened by concentration-broadening (see Sect. II.5c-2 and Fig. II.14 on p.163); the absorption factors are then also higher. At least in the line core a correction for concentration-broadening can be applied in order to obtain the desired, optically thin line profile, i.e. the $k(\lambda)$ profile.

This correction for concentration-broadening can be carried out simply by measuring $\alpha(\lambda)$ at fixed $\Delta\lambda$ position as a function of absorber concentration and by making use of Beer's law, Eq. II.256. When $\ln(1-\alpha)$ is plotted versus absorber concentration, one obtains a value for $k(\lambda)$ per unit concentration from the slope

EXPERIMENTAL   Sect. VII.3b

of the straight line. (A straight line is expected from Beer's law if the background line source is sufficiently 'monochromatic' and the absorber concentrations not too high.) One repeats this procedure for different $\Delta\lambda$ positions of the background line in order to obtain $k(\lambda)$ as a function of λ. Signal-to-noise problems are avoided by increasing the range of absorber concentrations n_0 whenever $k(\lambda)$ gets too low in near-wing measurements.

(ii) The transitions considered should be simple ones. The atomic line studied should have a normal Zeeman effect (as in the $^1P_1 \leftrightarrow {}^1S_0$ transitions of the alkaline-earth resonance lines) and h.f.s. should be absent.

(iii) The resolution obtained in the Zeeman scanning method is entirely determined by the background-line profile. So far EDL's have been used as background sources; their narrowest line profile corresponds to a Doppler profile at about 400 K (see Jansen, Hollander and Franken 1974). This means that the resolution obtained, for instance, for the alkaline-earth resonance lines amounts to roughly $(5-8) \times 10^5$.

The experimental absorption line profile has to be deconvoluted for the background line profile. Only if the background line has a Doppler profile can this convolution be carried out relatively simply. The $\delta\lambda$ and $\delta\lambda_s$ of the absorption line profile can also be derived when the background line profile differs from a Doppler shape, but then the background profile has to be known from interferometric measurements (see Jansen, Hollander and Franken 1974).

(iv) The solid angle under which the background source is seen by the detector is limited by the small hole in the drilled-through pole piece of the magnet, which is of the order of 6 mm. The presence of a $\frac{1}{4}\lambda$-plate, a polarizer, mirrors, etc., markedly reduce the light output of the background source. When at the same time a high resolution is required (i.e. a narrow background line profile), the radiance of the source is certainly not at its maximum; a compromise has to be found between a narrow line (no self-absorption) and a high radiance of the source.

3b-3 SCANNING MONOCHROMATOR

When a spectral continuum source is used as a background source, the absorption line profile can be determined with a scanning monochromator. The resolution obtained is the same as in the corresponding emission measurements (see Sect. 3a-2). Shift measurements are therefore also not feasible here. In practice this method is only applicable to the line wings, because in line core measurements correction for the apparatus profile becomes usually too large. The precision of line wing measurements may be limited by too low absorption factors, even at high atomic concentrations. The use of a double-beam technique is then recommended (see Sect. III.15).

Instead of a scanning monochromator, one may also use a high-resolution spectrometer with scanning slits. Smith (1975) used a 12 m focal-length spectrograph for his Ca absorption line core measurements; his background source was an Xe arc. McCartan and Farr (1976) used for their line core and wing measurements a 4.5 m focal-length instrument with a half-intensity width of 5.0 mÅ for the Gaussian component and of 5.0 mÅ for the Lorentzian one; the combined half-intensity width was about 8 mÅ.

3b-4   *TUNABLE DYE LASER*

When a narrow-band tunable dye laser (see for specification Tables III.3 and III.4 on pp.297 and 301) is used as background source, the absorption line core profile can readily be measured with satisfactory resolution and good accuracy by detuning the laser wavelength. For line-wing measurements, where the absorption is weak, a double-beam method is recommended to eliminate fluctuations in the laser power (see, e.g., van Dijk 1977); but then the laser bandwidth need not be small compared to the absorption line width. When measurements are made in the line core, the laser power should be kept sufficiently low to avoid saturation broadening (see Sect. II.5b-1).

The resolution that was claimed by, e.g., Kuhl, Marowsky and Torge (1972) for a pulsed laser amounted to more than 10^6; they measured the core of the Na-D lines in a Na-vapour cell. For the line wings this scanning procedure proved rather cumbersome in view of the wavelength-calibration of the laser source. Reproducible wavelength-tuning and adjustment is largely dependent on the mechanical properties of the optical parts of the laser. Furthermore, the bandwidth of the laser also depends strongly on the mechanical properties of the laser cavity; stabilization of the laser wavelength and its width has been achieved in the case of a c.w. laser by, e.g., Walther and Hall (1970) and Walther (1974). These authors claimed a half-intensity width with their c.w. argon-ion pumped dye laser of less than 2.5 mÅ at 5900 Å; their wavelength-adjustment is better than this figure.

For pulsed dye lasers the above claimed half-intensity widths of less than 5 mÅ at 5800 Å are hard to realize; more likely are values in the 30-50 mÅ range, depending on the output power. In a later paper Kuhl, Neumann and Kriese (1973) have reported a more realistic $\Delta\lambda$ value of about 33 mÅ at 6000 Å (see also Sect. III.7).

When the absorption is relatively weak (especially in the far line wings) and/or the laser output power is not very stable, one can better measure the *fluorescence-excitation function* instead of measuring directly the absorption line profile. The fluorescence-excitation function describes the dependence of the fluorescence intensity (integrated over the fluorescence line profile) on the wavelength of the tunable dye laser that is used for inducing the fluorescence (cf.

Fig. II.11 on p.141). If the laser bandwidth is sufficiently narrow and the laser power sufficiently low to preclude saturation and if no self-absorption of the fluorescence radiation occurs, the fluorescence-excitation function reflects directly the absorption line profile. This technique should be well distinguished from the spectral scanning of fluorescence line profiles as applied in vapour cells by, e.g., Hedges, Drummond and Gallagher (1972), Gallagher (1975), Ottinger et al. (1975), and York, Scheps and Gallagher (1975). Since the bandwidth of c.w. dye lasers is much smaller than for pulsed dye lasers, the former are preferred in the study of core excitation profiles. We note the following papers:

Kuhl, Neumann and Kriese (1973) measured the Na-D_2 core profile in a flame with a flashlamp-pumped tunable dye laser. Van Dijk (1977) did measurements in H_2-O_2-Ar flames with the same type of laser but with larger bandwidth ($\Delta\lambda \simeq 150$ mÅ) in the Na-D line wings up to ± 250 Å from line centre. Jongerius, Hollander and Alkemade (1978) used a c.w. tunable dye laser for measuring line wings up to ± 250 Å from line centre of the Na-D lines in C_2H_2-air and H_2-O_2-Ar- or N_2-diluted flames. They used a laser bandwidth of 100 mÅ for the wing measurements which covered a $\Delta\lambda$ range from about 200 mÅ to about 250 Å. Recently Jongerius et al. (1981) measured with the same laser and measuring method the core profiles of Na-D_1 and Na-D_2 lines in C_2H_2-air and H_2-O_2-Ar- or N_2-diluted flames. They succeeded in reducing their laser bandwidth to less than 3 mÅ and in obtaining a scanning range of ± 350 mÅ from line centre. Line shifts were derived by making use of a Na vapour cell (without perturber gas and at about 400 K) as a reference source. The authors also succeeded in extending their previously reported $\Delta\lambda$ range to 300 Å and 400 Å for the outer blue and red wing, respectively.

The applicability of the fluorescence technique is based on the complete spectral redistribution of the fluorescence radiation. In other words, the laser detuning should have no influence on the fraction of absorbed laser power that is re-emitted and the spectral distribution of the fluorescence radiation should remain centred around the atomic line centre independently of detuning. There has been found a contribution from Rayleigh scattering in flames by the above authors, but its relative magnitude is negligible (see also Sect. II.5a-1). When complete doublet-mixing occurs in the flame, it is not essential whether one doublet component or both components are detected in fluorescence (for a discussion see van Calcar et al. 1979). However, the laser detuning should not go so far that other optical transitions are induced in the atomic vapour; this did, however, not occur in the measurements reported above.

Jongerius, Hollander and Alkemade (1978) have checked that the core and wings of the Na-D lines measured, in the same C_2H_2-air flame at 1 atm pressure, by the absorption and the fluorescence techniques were indeed identical.

Chapt. VII   BROADENING AND SHIFT OF ATOMIC METAL LINES

3c   DETERMINATION OF LINE WIDTHS FROM INTEGRATED LINE INTENSITIES

General. The methods discussed here deal with the determination of the half-intensity width and thus of the a-parameter (see Eq. II.246) from measurements of integrated emission and/or absorption line intensities or from a combination of these. These methods generally involve curves-of-growth (COG) and/or duplication curves (for definition see Sect. II.5c). The 'combined measurement' of integrated absorption and corresponding peak absorption may also be used. The measurement of integrated fluorescence intensity as a function of concentration (see Sect. II.5c-4) may be considered as a pendant of the usual curve-of-growth method; the interpretation is, however, more intricate.

The assumptions underlying the applicability of most of the measuring methods discussed here are: (1) Kirchhoff's law is valid, i.e. emission- and absorption line profiles are identical; (2) the spectral lines studied have Voigt profiles; and (3) the lines should be resonance lines showing strong absorption at high concentrations. From the integrated-intensity measurements no information about the line shift can be derived; only a-parameters and $\delta\lambda$ values can be found. We shall not discuss the influence of h.f.s. but refer to Sect. 2b-7 and the literature cited therein. In Table VII.5 the measuring methods and literature are summarized.

Table VII.5

Methods of Determining Line Profile Characteristics
from Integrated Intensities

No.	Method	Line profile(s) involved	References
1	Emission COG (relative)	Emission + self-absorption	1, 2, 3
2	Absorption COG (relative)	Absorption	4
3	Duplication curve	Emission + self-absorption	1, 2, 3
4	Emission COG + absolute integral absorption	Emission + self-absorption	5, 6
5	Absorption COG (absolute)	Absorption	7, 8
6	Emission COG + duplication curve	Emission + self-absorption	3
7a	Fluorescence COG with background line source	(Self-)absorption + emission	9, 10
7b	Fluorescence COG with background continuum source	(Self-)absorption + emission	9, 10
8	Integral + peak absorption	Absorption	11

LEGENDS TO TABLE VII.5

1 = Alkemade (1954); 2 = Hollander (1964); 3 = van Trigt, Hollander and Alkemade (1965); 4 = Mitchell and Zemansky (1961); 5 = Hinnov (1957); 6 = Hofmann and Kohn (1961); 7 = de Galan, McGee and Winefordner (1967); 8 = McGee and Winefordner (1967); 9 = Hooymayers and Alkemade (1967); 10 = Zeegers, Smith and Winefordner (1968); 11 = Gallagher and Lapp (1967).

3c-1 *EMISSION CURVE-OF-GROWTH*

The relative emission curve-of-growth is determined by measuring the integrated line intensity in arbitrary units as a function of metal concentration in the flame (see Sect. V.2d). Comparison of the experimental curve with a set of theoretical curves with a as parameter may yield the a-parameter and therefrom $\delta\lambda$ if $\delta\lambda_D$ is known (see Sect. II.5b-3). This comparison is realized by shifting the experimental curve, plotted on double-logarithmic scales, along the ordinate and abscissa axis of the set of theoretical curves. Apart from the practical problem that the shape of the curves-of-growth may not critically depend on the a-parameter, the method relies on the proportionality of the sprayer performance, on the complete volatilization and on the absence of ionization effects. Methods of avoiding such complications or of correcting them have been described in Sect. V.2d.

This method for deriving a-parameters is not very accurate when $a > 1$ (see, e.g., Hollander 1964).

3c-2 *ABSORPTION CURVE-OF-GROWTH (RELATIVE)*

Here a spectral continuum source is used as background and the absorption curve-of-growth is measured (see Sect. III.15). For its measurement the same remarks apply as made in Subsection 3c-1. In addition, special problems arise when at low metal concentrations the absorption factor becomes too small to be measured precisely. On the other hand one should avoid that at very high concentrations the width of the absorption line profile does compare with the monochromator bandwidth, owing to concentration-broadening (see Sect. II.5c-2).

3c-3 *DUPLICATION CURVE*

In the methods utilizing the comparison between theoretical and experimental curves-of-growth a special problem is the similarity in the shape of COG's especially for $a > 1$ (see Sect. II.5c-2 and Fig. II.15 on p.170). The shape of the duplication curves (see Sect. II.5c-3 and Fig. II.18 on p.189) is more sensitive to the a-parameter than the shape of the COG for any a. For $a \leqslant 1$ the duplication curves show a minimum value that is critically dependent on the a-parameter.

Another advantage of the use of duplication curves is that these curves, in contrast to COG's, are less affected by nonlinearities between metal solution concentration and atomic density in the flame; in particular the position of the low- and high-density asymptotes and the value of the minimum are not affected by ionization effects, nonlinearity of the sprayer performance, etc. When use is made of the entire experimental curve in the comparison with a set of theoretical curves with different a-parameters, one can correct for nonlinearities in the manner described in Sect. V.2d.

The duplication-curve technique in its present form was introduced by Alkemade (1954) (see also Hübner 1933). This method of deriving a-parameters is often more reliable than the curve-of-growth method, especially for $a<1$, but the accurate measurement of the duplication factor requires special precautions (see Hollander 1964). In practice the required accuracy of $\leq 0.5\%$ is hardly obtainable.

3c-4 *EMISSION CURVE-OF-GROWTH PLUS ABSOLUTE INTEGRAL ABSORPTION*

One of the big problems of the relative methods is the fitting to theoretical curves. Hinnov and Kohn (1957) introduced a measuring method that overcomes most of the difficulties involved in this fitting. After determining the low- and high-density asymptotes of the relative experimental COG in emission, they also determined the absolute integral absorption, A_t, for a number of metal concentrations situated on the high-density asymptote (see Sect. III.15d-1). They thus obtained the absolute value of the integral absorption, A_t^*, at the intersection of the asymptotes, from which $\delta\lambda_L$ or a can be directly found through Eq. II.276 or Eq. VII.16. The experimental procedure has been outlined by Hinnov and Kohn (1957), Hofmann and Kohn (1961), and Behmenburg and Kohn (1964), and is referred to briefly in Sect. III.15.

The remaining experimental difficulties are the same ones as discussed in Subsection 3c-1.

3c-5 *ABSORPTION CURVE-OF-GROWTH (ABSOLUTE)*

Winefordner and co-workers (see, e.g., de Galan, McGee and Winefordner 1967, and McGee and Winefordner 1967) measured absorption COG's on an absolute absorption scale by making use of a continuous background source and a monochromator with calibrated spectral bandwidth or instrumental profile (see Sect. III.15d-1). In this way they determined A_t^* at the intersection point of the asymptotes, from which a followed (see Subsection 3c-4). In the application of this method the same difficulties may arise as mentioned in Subsection 3c-2. Besides, the absolute calibration requires an accurate measurement of the instrumental profile.

3c-6 *EMISSION CURVE-OF-GROWTH PLUS DUPLICATION CURVE*

Here the relative emission COG and duplication curve are measured under identical experimental conditions. Corrections for the distortion of the

EXPERIMENTAL   Sect. VII.3c

concentration scale are made as mentioned in Subsection 3c-1. In this combinatory method, introduced by van Trigt, Hollander and Alkemade (1965), the a-parameter is found from the intersection point of two trial curves, one representing the theoretical connection between the (unknown) a-parameter and the (unknown) atomic density for a <u>given</u> (known) D_i value. The other one representing in a similar way the connection between a and density for a <u>given</u> (known) y_i value. Here D_i is the duplication factor measured at solution concentration c_i, and y_i is the ratio of the integral absorption, $(A_t)_i$, measured at the same solution concentration, to the value A^* at the intersection point of the two asymptotes of the COG. (For a detailed discussion of this 'combinatory method' see Hollander 1964, van Trigt, Hollander and Alkemade 1965, and Hollander, Jansen, Plaat and Alkemade 1970.)

There are two main advantages of this method: (a) only relative measurements are needed (and a set of theoretical COG's and duplication curves), and (b) the experimental duplication curve provides a means of establishing unambiguously which concentration values are situated in the ranges of the low- and high-density asymptotes of the COG. This means that we have an additional indication for the proper determination of the position of the asymptotes. Moreover, we can check the internal consistency of the combinatory method by varying the value, c_i, of the solution concentration.

The combinatory method is very hard to apply for $a \geqslant 2$, unless D- and y_i values can be measured with accuracies of better than 0.3—0.5% (see Jansen 1976).

3c-7   *FLUORESCENCE 'CURVE-OF-GROWTH'*

As discussed in some detail by Hooymayers (1968), the fluorescence 'curve-of-growth' (integrated fluorescence intensity as a function of metal concentration) also contains information about the a-parameter (see Sect. II.5c-4). There are two kinds of growth-curves, depending on whether one uses a narrow-line source or a continuum source for irradiating the flame. Examples of theoretical fluorescence 'curves-of-growth' are given in Fig. II.20 on p.197.

The typical difference between the two kinds of curves is the slope of the high-density asymptotes: with a continuum source one has a horizontal asymptote, with a line source an asymptote with negative slope with arctan $= -\frac{1}{2}$. The shape of the fluorescence curve depends on the a-parameter, but also on the dimensions of the flame region irradiated by the primary source, etc. (see Zeegers, Smith and Winefordner 1968).

When the relevant dimensions of the flame and the irradiating beam are known, a comparison of experimental curves with a set of theoretical ones yields the a-parameter. Since the fluorescence growth-curves are subject to the same sort of experimental complications (ionization, nonlinearity of sprayer performance, etc.) and, furthermore, since their shape depends sensitively on the geometrical dimensions, the use of these curves is even more disputable than that of the normal COG's. The fluorescence growth-curves have, however, a real advantage over the absorption COG's, viz. they are very simple and accurately measurable especially for low metal concentrations (see also Zeegers, Smith and Winefordner 1968).

3c-8 *INTEGRAL ABSORPTION PLUS PEAK ABSORPTION*

With this method the absolute integral absorption is determined for a solution concentration at which the spectral line is optically thin. Furthermore, under the same experimental conditions one determines the peak absorption factor with a narrow-line source (see Sect. III.15d-2). When it is assumed that (a) the line profile measured has a Voigt shape, (b) the line shift is known or can be determined, and (c) corrections for the effect of concentration-broadening and h.f.s. can be made, one obtains from the measured experimental data the $\delta\lambda$ value and thus the a-parameter. This method was introduced by Lapp and co-workers (see, e.g., Gallagher and Lapp 1967). It is essentially based on a determination of the effective width, $\Delta\lambda_{eff}$, of an optically thin spectral line by taking the ratio of the asymptotic expressions in Eqs II.270 and II.265 for the integral-absorption and peak-absorption factor, respectively.

The values of integral and peak absorption under optically-thin conditions can be found by extrapolating the absorption values found as a function of metal concentration towards zero concentration. Because the line shift has to be known, one has to carry out an additional and rather elaborate $\delta\lambda_s$ measurement. Summarizing one can say that the method has its attractions for lines where the shift is known beforehand.

4. RESULTS FROM FLAME EXPERIMENTS IN COMPARISON WITH NON-FLAME RESULTS AND WITH THEORY

In this section we discuss primarily results of flame experiments, but supplement these data occasionally with results of nonflame experiments. The experimental results, in turn, are compared with theory.

All flame results reported here are interpreted on the basis of the simple two-state model (see Sect. VII.2b-1) in which J- and M_J-mixing collisions are not taken into account. The results of flame experiments generally refer to multicomponent systems where one component is dominant. It is, in principle, possible to derive from these experiments collision-broadening cross sections that are specific for an arbitrary (major) component, by varying the quantitative flame gas composition (cf. Sect. II.4a-1). This has been done by Jongerius (1981).

FLAME RESULTS COMPARED WITH OTHER RESULTS AND THEORY   Sect. VII.4a

Line profile studies in flames are often aimed at deriving only auxiliary information for use in the interpretation of other flame studies (e.g., pertaining to the curve-of-growth) or in analytical flame atomic absorption or fluorescence spectroscopy. In some instances, however, line profile data obtained in flames under well-defined conditions and with different gas compositions have proved to be of more fundamental interest. Most of these data deal with line-core measurements but some pertain also to line wings and satellite bands.

In flames the question of statistical independence of Doppler and collision broadening, as well as that of adiabatic and diabatic (mostly quenching) broadening, can only be studied indirectly, as we shall show. In vapour cells the influence of Doppler broadening is reduced and quenching broadening does not occur if the perturber is a noble gas.

Experimental flame and nonflame results are compared with theoretical models in Sects 4a and 4b, respectively. The derivation of force constants of difference interaction-potentials from experiments is treated in Sect. 4c. The derivation of a-parameters and collision-broadening cross sections is discussed in Sect. 4d, with a short excursion to a-parameters measured under less known flame conditions (Sect. 4e). Finally we present results of the influence of hyper-fine-structure on the line profiles measured in Sect. 4f.

4a   COMPARISON OF LINE PROFILES MEASURED IN
FLAMES WITH THEORETICAL MODELS

We shall first briefly discuss the expected effects of J-mixing and quenching collisions on the line broadening measured in flames. Next we shall compare line profile measurements with theoretical predictions based on an assumed difference interaction-potential, and with theoretical Voigt profiles (directly via curve-fitting and indirectly through COG measurements). We shall go on to discuss the statistical independence of Doppler and collision broadening in the light of the flame experiments, and, finally, the question of the validity of Kirchhoff's law.

Broadening effect of J-mixing and quenching collisions in flames. In flames no measurements have been reported on the influence of M_J- and J-mixing collisions. Jansen (1976) has only estimated the influence of J-mixing collisions on the Na-D line profiles in H_2-O_2-Ar and H_2-O_2-N_2 flames. Jansen used experimental values of J-mixing cross sections for the Na-D doublet extrapolated to flame temperatures,[†] and estimated a half-intensity width due to the reduction in

---

[†] According to Dr P.L. Lijnse (personal communication); see also Table VI.1 on p.604 ff.

lifetime by J-mixing, that was less than (5–10)% of the adiabatic collisional half-intensity width $\delta\nu_{c''}$ for both D lines.[†]

The influence of quenching broadening was estimated by Behmenburg (1964) for the Na-D_2 line in C_2H_2-O_2-N_2 flames and by Jansen, Hollander and Alkemade (1977) for the Sr 4607.33 Å line in H_2-O_2-N_2, H_2-O_2-Ar and CO-N_2O flames. These estimates were essentially based on Eqs II.225 and II.227. Behmenburg (1964) made corrections for the influence of quenching collisions by subtracting an estimated value of the corresponding cross section $\sigma_{c'}$ from the total collisional-broadening cross section σ_c measured; the $\sigma_{c'}$ value was derived from the independently measured efficiency of fluorescence Y (see Eq. VII.23). This procedure has also been reported by van Trigt, Hollander and Alkemade (1965) for the alkali resonance lines in CO-O_2-N_2 and C_2H_2-O_2-N_2 flames (see also Table VII.9 in Sect. 4d).

Jansen, Hollander and Alkemade (1977) have made a systematic attempt to tackle the problem of the influence of quenching collisions experimentally (see also Jansen, Hollander and van Helvoort 1977). For the Sr 4607.33 Å line, resonance fluorescence measurements were done in H_2-O_2-Ar, H_2-O_2-N_2 and CO-N_2O flames in order to find the flame conditions with the lowest Y value, i.e. the largest influence of quenching to the line profile (see Eq. II.227). In the N_2-diluted H_2 flames, Y values as low as 0.050 ($\pm$ 0.005) could be realized, whereas in the CO flames the minimum Y value was found to be 0.093 ($\pm$ 0.002). Translating these results into terms of quenching half-intensity widths, one readily sees that the contribution of the quenching broadening is at most 15%.

In Table VII.6 we have collected the $\delta\lambda_s$, $\delta\lambda$, $\delta\lambda_{c'}$ and $\delta\lambda_{c''}$ values found for this Sr line in H_2-O_2-Ar flames for illustration, where $\delta\lambda$ and $\delta\lambda_{c'}$ are the total and the quenching half-intensity width, respectively.

The conclusion from the above (sparse) estimates may be that we can practically neglect the influence of quenching and J-mixing collisions on line profiles measured in flames and that one is allowed to treat here the collisional broadening as if it were <u>solely</u> due to adiabatic collisions.

[†] Jongerius et al. (1981) have recently found definite differences between the $\delta\lambda_{c''}$ values of the Na-D doublet lines in H_2-O_2-Ar flames. The ratio of the specific values for Ar perturbers,

$$\frac{\delta\lambda_{c''}(D_1)}{\delta\lambda_{c''}(D_2)},$$

was found to be 1.15 $\pm$ 0.06. For N_2 perturbers this ratio amounted to merely 1.03 $\pm$ 0.05.

FLAME RESULTS COMPARED WITH OTHER RESULTS AND THEORY   Sect. VII.4a

Table VII.6

$\delta\lambda$-, $\delta\lambda_s$-, $\delta\lambda_{c'}$- and $\delta\lambda_{c''}$ Values (in mÅ) for the Sr First
Resonance Line Measured in H_2-O_2-Ar Flames*

Flame No.	Temperature (K)	$\delta\lambda$	$\delta\lambda_s$	$\delta\lambda_{c'}$ †	$\delta\lambda_{c''}$
1	1765	32 ± 2	8.7 ± 0.8	1.1 ± 0.1	24.1 ± 2.3
2	2000	35 ± 2	9.1 ± 0.6	1.5 ± 0.2	26.3 ± 2.2
3	2365	38 ± 2	7.9 ± 0.7	3.4 ± 0.4	26.9 ± 2.2

* According to Jansen, Hollander and Alkemade (1977).
† $\delta\lambda_{c'}$ values were estimated by using Eq. VII.22 and Y values reported by Hollander et al. (1972).

Line-core and very near line-wing profiles. As to the line core and very near wings † (i.e., $|\Delta\lambda| \leqslant 0.3$ Å) we mention the line profile measurements of:

(i)   Behmenburg (1964) on the Na 5890 Å line in C_2H_2-O_2 flames diluted with N_2, He, Ar and CO_2 at 1 atm, with a Fabry-Pérot interferometer;

(ii)   Hollander and Broida (1967, 1969) on Zn 3075.9 Å and OH 3072.009/063 Å lines in H_2-O_2-N_2 and C_2H_2-O_2-N_2 flames at 1 atm and at pressures down to 50 Torr with the aid of Zeeman scanning;

(iii) Hollander, Jansen, Plaat and Alkemade (1970) on the Ca, Sr and Ba resonance lines in C_2H_2-air flames at 1 atm, also with Zeeman scanning;

(iv)   Wagenaar and de Galan (1973) on resonance lines of Al, Ca, Ca^+, Cr, Ga, In, K and Mn in C_2H_2-N_2O flames at 1 atm, with a Fabry-Pérot interferomenter (see also Wagenaar, Pickford and de Galan 1974, and Wagenaar 1976);

(v)   Vasileva et al. (1975) who measured profiles of the Na and K first resonance lines in a CH_4-air flame by emission interferometry;

(vi)   Jansen, Hollander and Alkemade (1977, 1977b) on Sr 4607 Å line profiles in H_2-O_2-Ar (see Fig. VII.2) and C_2H_2-air flames at 1 atm, and on Ba 5536 Å and Sr 4607 Å in C_2H_2-air flames at 1 atm;

(vii) Jongerius et al. (1981) on the Na-D_1 and D_2 line profiles in C_2H_2-air, H_2-O_2-Ar and H_2-O_2-N_2 flames at 1 atm.

---

† In the following we distinguish between: (i) core + (very) near wings where collisions are expected from theory to produce a dispersion line profile, (ii) far wings where the QS approximation is valid, and (iii) the intermediate range of $\Delta\lambda$ values.

Chapt. VII   BROADENING AND SHIFT OF ATOMIC METAL LINES

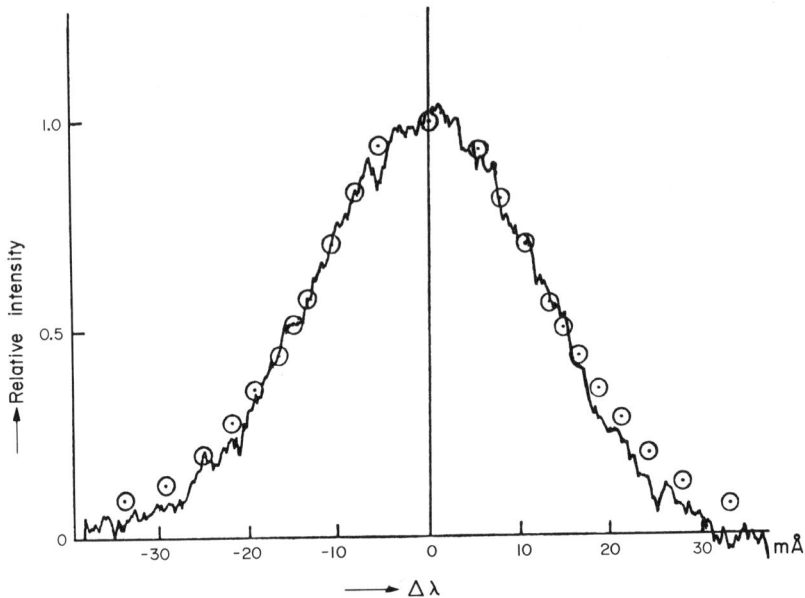

Fig. VII.2 An interferometric scan is shown of an optically thin spectral profile of the 4607.33 Å Sr emission line in a H_2-O_2-Ar flame at 1765 K and 1 atm. The corresponding theoretical Voigt profile with a-parameter equal to 1.4 is plotted by ⊙. The position of 0 mÅ coincides with the maximum of the measured line profile (shift not indicated). (From Jansen, Hollander and Alkemade 1977.)

The conclusions that can be drawn from these experiments are the following:
— The line profiles are practically symmetric and have a Voigt shape; the authors cited have proved this by curve-fitting.[†] De Galan and co-workers have also shown that in the case of hyper-fine-structure their experimental profiles can always be fitted to a Voigt profile when one makes proper allowance for the h.f.s. effect. Jongerius et al. (1981) recently reported the same result for the Na-D doublet line profiles in H_2 and C_2H_2 flames. After correcting for hyper-fine-structure all their experimental profiles fitted within the experimental scatter to Voigt profiles. This fit was feasible for $\Delta\lambda$-values of about ± 350 mÅ. One may wonder if the slight systematic deviation of the experimental Sr profile from the theoretical Voigt profile as demonstrated in Fig. VII.2 is also due to h.f.s. influence, which was not corrected for.

[†] The asymmetry ratio previously reported by Hollander, Jansen, Plaat and Alkemade (1970) for the Ba resonance line profile has been proved by Jansen, Hollander and Alkemade(1977b) to be an experimental error.

FLAME RESULTS COMPARED WITH OTHER RESULTS AND THEORY   Sect. VII.4a

— All flame line profiles show a definite red shift. In absorption cell experiments one usually finds a blue shift with He as perturber; for the other noble gases one has found only red shifts (see, e.g., Behmenburg 1968). The experimental value of $\xi(\alpha) \equiv - \delta\lambda_c\prime\prime/\delta\lambda_s$; see Sect. 2b-3.1) is in all cases different from the value of 2.76 that is predicted in the case of pure van der Waals interaction; consequently interaction potentials with inclusion of repulsive terms are more realistic for describing flame experiments.

— In the very near wings the line profiles show indeed a dispersion form: $S_\nu \propto \Delta\nu^{-2}$ (see Hollander, Jansen, Plaat and Alkemade 1970).

Recently laser-induced fluorescence experiments on Na-D line profiles in H_2-O_2-Ar flames were carried out by van Dijk (1977) and in H_2-O_2-Ar or -N_2 and C_2H_2-air flames by Jongerius, Hollander and Alkemade (1978). From the former experiments a $\Delta\nu^{-2}$-behaviour was found for the red Na-D_1 line wing in the range $1.2 < \Delta\lambda < 5$ Å and for the blue D_2 wing a $\Delta\nu^{-(2.2\pm0.1)}$ behaviour from 1 to 5 Å. The measurements by Jongerius, Hollander and Alkemade (1978) were performed with higher spectral resolution than van Dijk's measurements and could thus be extended closer to the line core; the results revealed a $\Delta\nu^{-2}$-behaviour for the red D_1 wing between 0.3 and 7 Å, and similar for the red D_2 wing and the blue D_1 wing between 0.3 and 3 Å, whereas for the blue D_2 wing a $\Delta\nu^{-(2.19\pm0.05)}$ behaviour was found between 0.3 and 7 Å (see Figs VII.3a and 3b). An explanation for this anomalous blue D_2 wing behaviour, especially in the (very) near line wings ($0.3 < \Delta\lambda < 3$Å), cannot yet be given. Jongerius et al. (1981) confirmed experimentally the above anomaly of the blue D_1 wing down to about 0.3 Å from line centre; they recovered a dispersion profile in the cores of both D-lines (i.e. for $\Delta\lambda < 0.3$ Å) in all H_2- and C_2H_2 flames investigated.

The above results lead us to the view that adiabatic collision broadening in flames up to 1 atm in line core and (very) near wings can satisfactorily be described by the classical impact theory, which predicts a dispersion profile (see Sect. 2b-3.1).

The applicability here of the impact theory, as suggested by the experimental results, depends on the $\Delta\nu$- and n_p-criteria discussed in Sect. 2b-3.1. As to the n_p-*criterion* we have the condition: $n_p b_w^3 \ll 1$. For van der Waals interaction, b_w has a unique value and follows from:

$$b_w = \left(\frac{3\pi}{8} C_6/\bar{v}\right)^{1/5} ;$$

for purely repulsive interaction, one has:

$$b_w = \left(\frac{63\pi}{256} C_{12}/\bar{v}\right)^{1/11} .$$

Taking, for example, the b_w values for the Na 5890 Å line in N_2, He, Ar and CO_2

BROADENING AND SHIFT OF ATOMIC METAL LINES

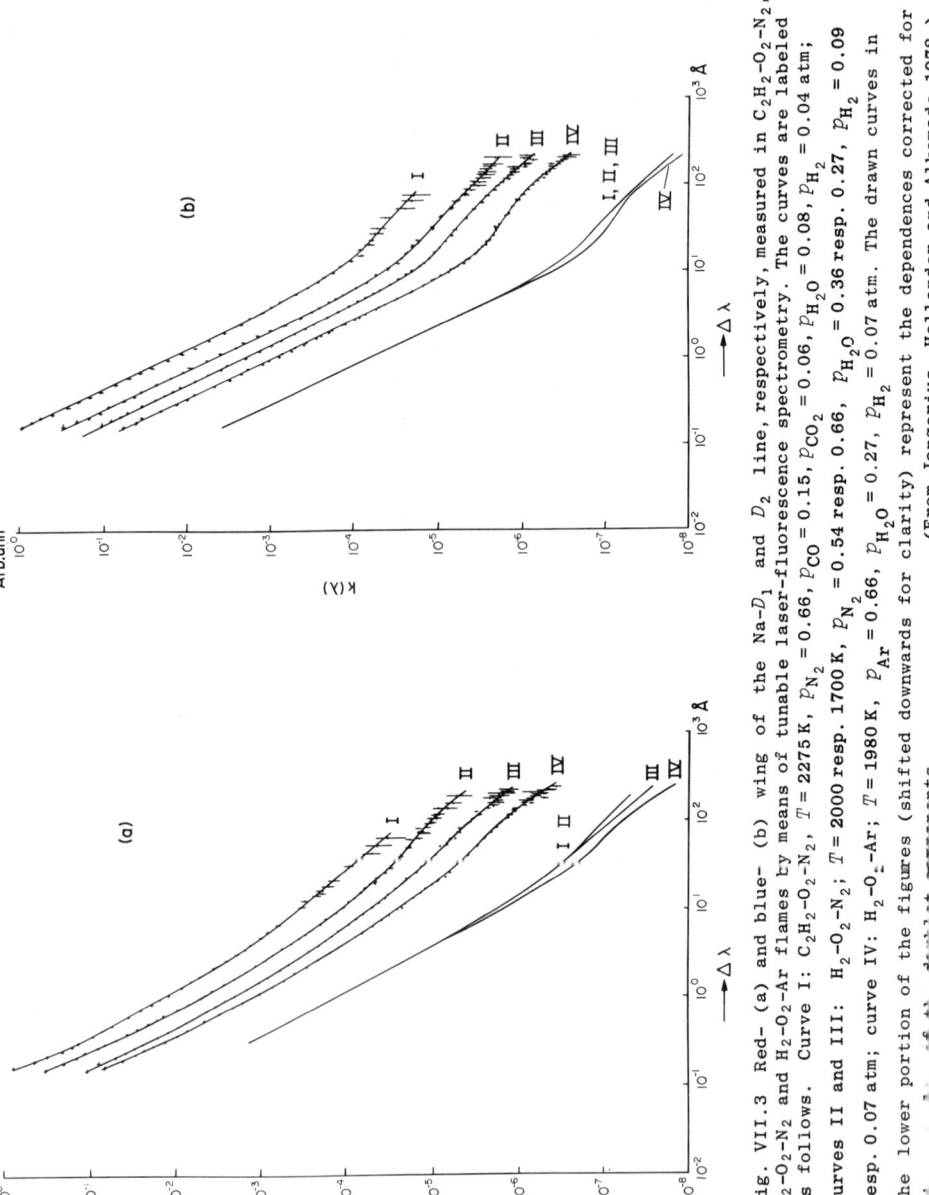

Fig. VII.3 Red- (a) and blue- (b) wing of the Na-D_1 and D_2 line, respectively, measured in $C_2H_2-O_2-N_2$, $H_2-O_2-N_2$ and H_2-O_2-Ar flames by means of tunable laser-fluorescence spectrometry. The curves are labeled as follows. Curve I: $C_2H_2-O_2-N_2$, $T = 2275$ K, $p_{N_2} = 0.66$, $p_{CO} = 0.15$, $p_{CO_2} = 0.06$, $p_{H_2O} = 0.08$, $p_{H_2} = 0.04$ atm; curves II and III: $H_2-O_2-N_2$; $T = 2000$ resp. 1700 K, $p_{N_2} = 0.54$ resp. 0.66, $p_{H_2O} = 0.36$ resp. 0.27, $p_{H_2} = 0.09$ resp. 0.07 atm; curve IV: H_2-O_2-Ar; $T = 1980$ K, $p_{Ar} = 0.66$, $p_{H_2O} = 0.27$, $p_{H_2} = 0.07$ atm. The drawn curves in the lower portion of the figures (shifted downwards for clarity) represent the dependences corrected for

(From Jongerius, Hollander and Alkemade 1978.)

from Behmenburg's (1964) flame work to be 4.6, 2.6, 4.6 and 5.2 Å, respectively, we see that at atmospheric pressure (i.e., $n_p \simeq 3 \times 10^{18}$ cm^{-3} at $T \simeq 2300$ K) this inequality is fully satisfied. Behmenburg's calculations referred to van der Waals interaction; as pointed out by Behmenburg (1970) in a later paper, in the case of L.-J. interaction b_w has no unique value.

As to the $\Delta \nu$-*criterion* we present some data for the Sr 4607 Å line perturbed by Ar (at $T = 2265$ K); a similar estimate has been worked out by Hindmarsh and Farr (1972). For van der Waals interaction, this criterion requires:

$$(\omega_0 - \omega) \ll \left(\frac{\bar{v}}{\sqrt{\pi}}\right)^{6/5} \left(\frac{1}{C_6}\right)^{1/5}.$$

Since C_6 is positive, the criterion refers to the red wing; for L.-J. interaction, the red-wing criterion is the same. The blue-wing criterion in the case of L.-J. interaction is the same as for purely repulsive interaction, and is determined by C_{12} according to

$$(\omega - \omega_0) \ll \left(\frac{\bar{v}}{\sqrt{\pi}}\right)^{12/11} \left(\frac{1}{C_{12}}\right)^{1/11}$$

(see also above).

When taking in our example of Sr—Ar as representative values: $C_6 \approx 10^{-32}$ cm^6 s^{-1}, $C_{12} \approx 10^{-77}$ cm^{12} s^{-1} and $\bar{v} \approx 10^5$ cm s^{-1} (corresponding to $T \simeq 2265$ K), we get for the red-wing boundary: $\omega_0 - \omega \ll 1.5 \times 10^{12}$ s^{-1}, i.e. $\Delta\lambda_{red} \ll 1.5$ Å, and for the blue wing: $\Delta\lambda_{blue} \ll 3$ Å. Since in practice dispersion profiles for adiabatic broadening have been found up to at least 1—2 Å from line centre, we conclude that the $\Delta\nu$-*criterion* is satisfied.

The validity of the impact theory for adiabatic line broadening under flame conditions is also supported by computer calculations on the adiabatic profile of the Sr resonance line in Ar- and N$_2$-diluted flames at 1 atm and with 2000 K $< T <$ 2500 K on the basis of Lindholm's general-pressures theory (see Jansen, Hollander and Behmenburg 1974, and Jansen 1976). The computed Sr line profiles with assumed L.-J. interaction potentials as derived from work by Farr and Hindmarsh (1971) and from Hollander, Jansen, Plaat and Alkemade (1970), respectively, showed asymmetry ratios of at most 1.02. The collisional line profiles proved, within a few percent, to be equal to the simple dispersion profiles following from impact theory for $|\Delta\lambda| < 2$ Å.

Far-wing profiles. The following far-wing experiments were carried out in flames:

(i) Line wings of Na-D doublet and K first resonance doublet were measured in emission in a CH$_4$-air flame jet at 2050 K by Vasileva, Deputatova and Nefedov (1975).

Chapt. VII   BROADENING AND SHIFT OF ATOMIC METAL LINES

(ii)   Wing measurements of Sr 4607.33 Å line in CO$-N_2$O flame ($T \approx 2800$ K) and of
        Na-D doublet lines in H_2-O_2-Ar or $-N_2$ flames ($T \approx 2000$ K) were made by
        Hollander, Jansen and Alkemade (1977).

(iii)   Line wings of the Na-D doublet in H_2-O_2- Ar flame ($T \simeq 1800$ K) were measured by van Dijk (1977).

(iv)   Line wings of the Na-D doublet in C_2H_2-air, H_2-O_2-Ar or $-N_2$ flames at
        $T \simeq 2300$ K and $\simeq 2000$ K were carried out by Jongerius, Hollander and
        Alkemade (1978) and over a larger $\Delta\lambda$ range by Jongerius (1981) and
        Jongerius et al. (000) in an extended set of H_2- and C_2H_2 flames.

        The line profiles under (ii) are presented in Figs VII.4 and 5, those
under (iv) are included in Fig. VII.3 (see above).

        As can be seen from the Figs VII.3 to 5, it is very difficult to decide
where a particular (constant) slope begins or ends. The choice of the $\Delta\lambda$
range also influences the value of the slopes cited. Therefore a direct comparison of the curves reveals more about the (in)consistencies between the results
of the various authors than a mere comparison of numerical slope data.

        Figure VII.3 also shows corrected ('true') wavelength dependences for the
Na lines; in the correction for spectral overlap of the doublet components we
assumed equal wavelength-dependences in the wings of the two components (see
Hollander, Jansen and Alkemade 1977).

        When comparing the experimental line wing results from Vasileva, Deputatova
and Nefedov (1975) with those from Hollander, Jansen and Alkemade (1977) as presented in Figs VII.4 and 5 and from van Dijk (1977) and Jongerius, Hollander and
Alkemade (1978) as presented in Fig. VII.3, one may draw some tentative conclusions:

        — The Na blue and red wings from Jongerius, Hollander and Alkemade and from
van Dijk, in the wavelength range up to about 10 Å from line centre, are in good
mutual agreement; this holds for Ar- as well as for N_2-diluted flames. The blue
Na-D_2 line wing from Vasileva et al. is in good agreement with these blue D_2-line
wing results, but their red D_1 line wing appears to differ significantly.

        — The (far) blue wings of the Na and K lines for $|\Delta\lambda| > 20$ Å are in good
mutual agreement; the same holds roughly for the (far) red wings of the Na lines
in N_2-diluted flames from Vasileva et al. and Jongerius et al., and to a lesser
extent for the far red wing of the K lines from Vasileva et al.

        — The Sr line wing for 1 Å $< \Delta\lambda <$ 10 Å differs significantly from the (-2)law
found experimentally for the Na and K lines.

        — The Na-line results in the H_2-O_2- Ar flames from van Dijk and Jongerius
et al. are in good agreement for the whole $\Delta\lambda$ range observed, i.e. including the
intermediate range.

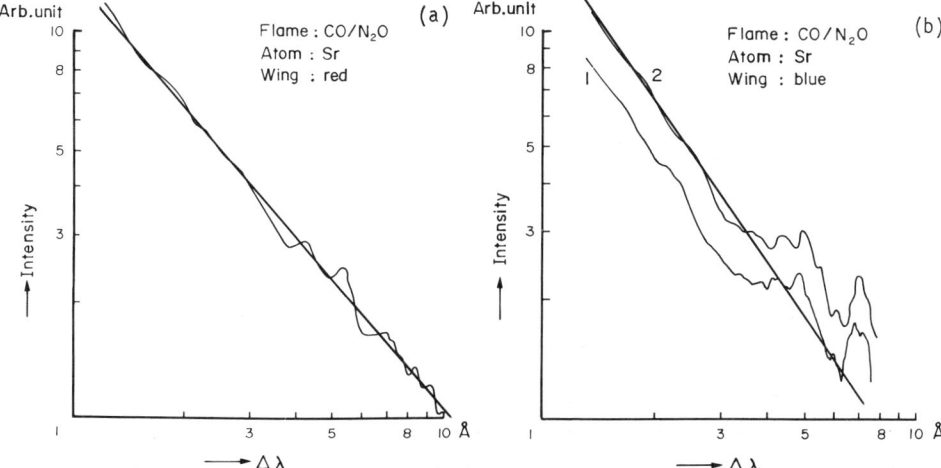

Fig. VII.4 Double-logarithmic plot of the red- (a) and the blue- (b) wings of the 4607.33 Å Sr line as a function of wavelength separation from line centre, $\Delta\lambda$, measured in emission by means of a scanning monochromator. (From Hollander, Jansen and Alkemade 1977.)

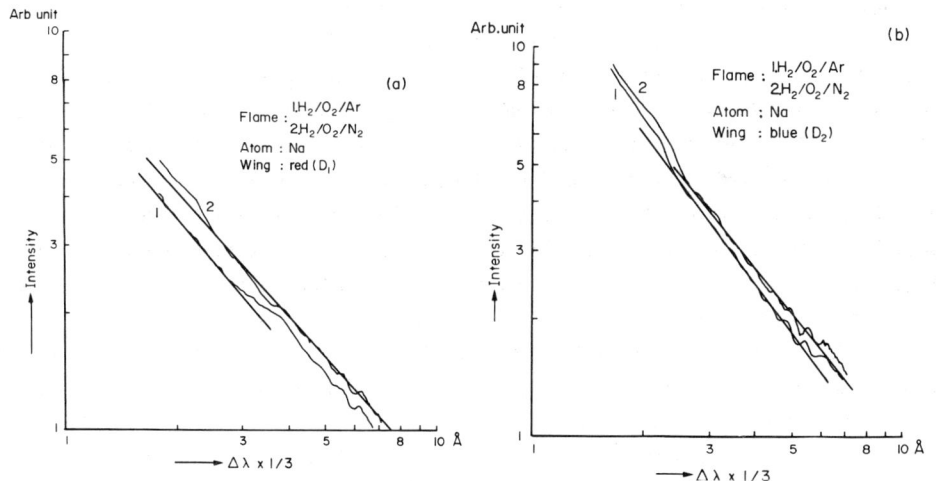

Fig. VII.5 Double-logarithmic plot of the red- (a) and the blue- (b) wings of the Na-D_1 and D_2 emission line, respectively, as a function of $\Delta\lambda$. (From Hollander, Jansen and Alkemade 1977.)

— The blue and red Na line wings in the H_2-O_2- Ar and $-N_2$ flames from Hollander *et al.* appear to differ significantly but unexpectedly from the corresponding results reported by van Dijk and by Jongerius *et al.*; there is good agreement, however, of the former results with the Na results from Vasileva *et al.*

— The blue and red Na wing dependences obtained by van Dijk and Jongerius *et al.* in Ar-diluted flames on the one hand, and N_2-diluted ones on the other show significant differences, especially in the intermediate range. These differences may be explained by the difference in dominant perturber species.

Confrontation with theory. We now attempt to confront the preceding linewing results (see also Figs VII.3,4 and 5) with theoretical predictions:

— If one assumes a L.-J. difference potential, one may expect for the far wings a $(-3/2)$law for the red, and a $(-5/4)$law for the blue wing. In the literature L.-J. interactions have been reported for the systems Na—Ar, Na—N_2, K—Ar and Sr—Ar (see Hindmarsh and Farr 1972, and Pascale and Vandeplanque 1974). The wavelength-dependence found experimentally for the far blue and red wings of the Na lines in Ar-diluted flames is in good agreement with theory.

— Both Na line wings in all N_2-diluted CH_4, C_2H_2 and H_2 flames presumably do not reach the far-wing behaviour predicted by QS theory.

— The Na blue and red wing behaviour in the intermediate range, found in N_2-diluted flames by Jongerius *et al.*, differ significantly from those found in the Ar-diluted flames by Jongerius *et al.* and by van Dijk. One is tempted to assume that, since Na-Ar and Na-N_2 systems have different interaction potentials, the intermediate ranges will, as a result, behave differently. A straightforward theoretical explanation cannot be given, however.

— When comparing the Na line wing behaviour in the three N_2-diluted flames investigated by Jongerius *et al.*, one may tentatively conclude that differences in flame gas composition do not have a measurable influence on the blue-wing profiles, i.e. all major molecular constituents (N_2, H_2O, CO, CO_2) affect the blue wing in about the same way (see also Jongerius 1981, and Jongerius *et al.* 000).

Conclusions about line wings. The differences in intermediate and far blue line wings of Na between Ar- and N_2-diluted flames are probably indicative not only of specific differences in interaction potential for the Na—Ar and Na—N_2 systems, but of differences in interaction with Na between Ar and molecular perturbers in general. For the red wing a similar conclusion cannot yet be drawn.

The blue Na—D_2 and $-D_1$ wings in the range $0.3 < \Delta\lambda < 7$Å and $0.3 < \Delta\lambda < 3$Å respectively appear to deviate from the expected (-2)law and so do the red Na wings in the corresponding λ-ranges. Since both blue and both red Na wings show roughly an equal, but asymmetric deviation from the (-2)law (see Jongerius 1981), the expected

FLAME RESULTS COMPARED WITH OTHER RESULTS AND THEORY   Sect. VII.4a

effect on the COG will therefore be negligible. A qualitative explanation for these deviations has been presented by Jongerius (1981).

The Sr line wing results are presumably still in the intermediate range where the QS theory does not yet apply (compare Sr results with Na results). Since no knowledge of the difference interaction potentials for Sr and the major flame gas species is available, one cannot check the $\Delta\nu$-criterion in order to justify this assumption.

Satellites. Under flame conditions satellite-like structures appeared to be present (see Figs VII.4 and 5):

— On the blue Sr line wing there is a hump with superimposed undulations extending from about 4 to 6 Å from line centre. The hump itself may be the result of an interaction of Sr with some perturber in the $CO-N_2O$ flame. If so, this hump is a satellite. Since the difference interaction potentials are not known, no comparison of the experimental λ position with theory could be made.

— On the red Sr line wing undulating structures only but no satellite seem to be present. According to QS theory one would expect, under the flame conditions where the perturber pressure is relatively low (1 atm), to find a satellite on the red wing rather than on the blue wing (see, e.g., Hindmarsh and Farr 1972).

— On the blue wings of the Na line in the H_2 flames with Ar- or N_2-diluent gas, similar undulations are present in the range from 16 to 20 Å, but there are no satellites.

— On the red Na line wings undulations are also found in both flames in the range from 16 to 20 Å but they are less alike. In both flames at about 12 Å from line centre a structure can be observed that resembles a 'smoothed-out' satellite. The λ position of the red satellite could be calculated via Eq. VII.18 for the system Na—Ar, by using the interaction potentials as calculated by Pascale and Vandeplanque (1974). This theoretical λ position was found to be about 2 Å from line centre, in contrast to the positions found experimentally. No satellites were found in the H_2-O_2-Ar flame in the range from 0.5 to 3.5 Å (see Hollander, Jansen and Alkemade 1977).

Adequacy of the Voigt model. The model of Voigt profiles is widely accepted as a good description of the core and (very-)near wings of line profiles in flames. As was discussed earlier (see also Sect. II.5c-2), it is the basis of the curve-of-growth calculations. Evidence for the validity of this model can be derived from the following three classes of experimental tests:

(i)   Curve-fitting of experimental profiles with Voigt profiles for the line core and additional observations on the near line wings (see Fig. VII.3; see also van Dijk 1977).

Chapt. VII   BROADENING AND SHIFT OF ATOMIC METAL LINES

(ii) Comparison of $\delta\nu$ values from scanned line profiles with $\delta\nu$ values derived indirectly from COG measurements under the same experimental conditions.

(iii) Comparison of experimental COG's with those calculated from the Voigt model as well as from a 'more realistic' theory of line broadening.

Ad (i). From the experimental results presented earlier in this section (see, e.g., Figs VII.2 and 3) one may conclude that the line core and (very-)near wings of atomic line profiles in flames are within the experimental scatter adequately described by the Voigt model.

Ad (ii). In Table VII.7 we have collected $\delta\nu$ values calculated from a-parameter values derived from COG measurements; these are compared with $\delta\nu$ values derived from line-scanning experiments made under identical experimental conditions.

The data from Table VII.7 demonstrate that both methods of determining $\delta\nu$ yield consistent values within the experimental error. This outcome proves directly the usefulness and indirectly the validity of the Voigt model for line core profiles under flame conditions (see Jansen and Hollander 1977).

Ad (iii). An indirect way of checking the validity of the Voigt model is via the COG. Experimental COG's in flames show a $\sqrt{n_0}$-dependence at high n_0 values (with n_0 = atomic density in the ground state; see Sect. II.5c-2) as expected from the Voigt model. As has been shown, however, the (far) wings may differ from the $\Delta\nu^{-2}$-dependence as predicted by the Voigt model. In order to check the influence of this deviation, Jansen and Hollander (1977) have constructed line profiles where the line core and very near wings had a Voigt shape but the line wings had a frequency-dependence according to QS theory for van der Waals or L.-J. potentials. These authors also assumed various transition points in the wings where the $\Delta\nu^{-2}$-dependence of the line core changed over to a different frequency-dependence. Using these artificial profiles they calculated COG's; it was concluded that possible deviations from a Voigt profile in the wings of the line will not noticeably influence the position and slope of the high-density asymptote of the COG as measured in flames. No conclusions concerning such deviations can therefore be drawn from COG's measured in flames. Furthermore, the values of a-parameter and n_0 derived via experimental COG's (see Sect. II.5c-2) will hardly (< 5%) be affected by the possible deviations of the line wings from a Voigt profile.

In conclusion one may say that the line core and near wings (0.1 < $\Delta\lambda$ < 2 Å) are, within the experimental scatter, adequately described by the Voigt formula; more specifically this means that $\delta\lambda$ values calculated from a-parameters obtained via COG- and duplication-curve measurements are consistent with $\delta\lambda$ values directly measured by scanning the line profile. Scanning measurements, however, yield also $\delta\lambda_s$, which may provide us with extra information about the difference potential $V(r)$

FLAME RESULTS COMPARED WITH OTHER RESULTS AND THEORY   Sect. VII.4a

Table VII.7

A Comparison between Half-intensity Widths, $\delta\nu$, Derived from COG Measurements and Scan Data

Metal	Wavelength (Å)	Flame	Temperature (K)	$\delta\nu$ (s^{-1}) COG meas.	$\delta\nu$ (s^{-1}) Scan data	Measuring method[+]	Ref.
Ca	4227.7	C_2H_2/air	2275	4.6×10^9	4.9×10^9	ZSC	1
Sr	4607.3	C_2H_2/air	2275	4.7×10^9	4.9×10^9	ZSC	1
Ba	5535.5	C_2H_2/air	2275	4.4×10^9	4.2×10^9	ZSC	1
Sr	4607.3	H_2/O_2/Ar	1765	4.2×10^9	4.5×10^9	FPI	2
Sr	4607.3	C_2H_2/air	2275	4.9×10^9	4.7×10^9	FPI	3
Sr	4607.3	CO/N_2O	2800	5.1×10^9	5.0×10^9	FPI	3
Na	5890	$C_2H_2/O_2/CO_2$	2409	6.1×10^9	6.2×10^9	FPI	4,5 *

[+] ZSC = Zeeman scanning (absorption); FPI = Fabry-Pérot interferometer (emission).

* The $\delta\nu$ values for Na 5890 Å line in C_2H_2/O_2 flames (T = 2409 K) diluted by N_2, He and Ar respectively, obtained from curve-of-growth (Ref. 4) and Fabry-Pérot interferometer measurements (Ref. 5) showed a systematic deviation of 18, 25 and 15%, respectively.

REFERENCES

1 = Hollander, Jansen, Plaat and Alkemade (1970)

2 = Jansen, Hollander and Alkemade (1977)

3 = Jansen and Hollander (1973)

4 = Behmenburg and Kohn (1964)

5 = Behmenburg (1964)

Chapt. VII   BROADENING AND SHIFT OF ATOMIC METAL LINES

(see Sect. 2b-3.1).

The agreement found between the line profiles up to $\Delta\lambda \approx 2$ Å and a Voigt profile and the applicability of the Voigt model for the curve-of-growth in flames up to atomic densities of roughly $(5-10) \times 10^{13}$ cm^{-3} are consistent with the following theoretical assumptions:

(i)   Resonance broadening is negligible.

(ii)   No Dicke-narrowing of the Doppler profile occurs (see also Eq. II.242).

(iii) Doppler- and collision-broadening are statistically independent (see also Sect. 2b-3.1).

(iv)   Collision broadening produces a dispersion profile in the core and very near wings. Lindholm's general-pressures theory predicts that the adiabatic line profile in flames differs from a dispersion form by only 1—2% in the near wings. The flame results are thus consistent with the impact approximation of the semi-classical theory (see Sect. 2b-3.1).

It should be realized, however, that the flame results obtained so far prove nothing much about other assumptions that are usually made (implicitly) in applying the theory of line broadening to flame experiments. Examples of such assumptions are: the validity of the procedure of correcting for quenching broadening, the assumed absence of a line shift due to this broadening effect, or the applicability of an atomic model involving only two, nondegenerate states.

Validity of Kirchhoff's law. In flames it is possible and fairly simple to study simultaneously thermal-emission and absorption line profiles (incidentally, that is one of the real advantages of flames in the study of line broadening). Jansen, Hollander and Alkemade (1977b) compared the line core data of the Sr and Ba resonance lines in C_2H_2-air flames in emission via interferometry and in absorption via Zeeman scanning. They concluded that absorption and emission half-intensity widths are identical within the experimental scatter of 5%; the shifts are the same within 20%. The scanned line profiles of Ba and Sr measured by these authors show that the thermal-emission and absorption line profiles coincide within the experimental scatter up to $|\lambda-\lambda_0| \simeq \delta\lambda$ (see Fig. II.16); it seems therefore that Kirchhoff's law is obeyed in the flame region above the primary combustion zone at 1 atm pressure.

4b   COMPARISON OF SOME NONFLAME LINE-BROADENING
      EXPERIMENTS WITH THEORETICAL MODELS

The rather indirect evidence from flame experiments for the validity of the impact approximation for describing the *line core* can be supported by more direct

FLAME RESULTS COMPARED WITH OTHER RESULTS AND THEORY   Sect. VII.4b

evidence from nonflame absorption experiments in *metal vapour cells* filled with a
one-perturber gas, and in other one-perturber milieus. We briefly mention here
some of these other experiments.

Smith (1967, 1972) has investigated the *broadening* and *shift* of the Ca
intercombination line at 6437 Å and of the Ca resonance line at 4227 Å in a pure
He, Ne, Ar, Kr and Xe milieu; the temperatures ranged from about 700 to 900 K, and
the perturber pressures varied between 10^{-2} and 1 atm. The author studied the
absorption line profiles under high resolution (see also Smith 1975). Similar measurements on the Sr 4607 Å line were reported by Farr and Hindmarsh (1971). The
line profiles corrected for Doppler broadening were symmetric dispersion profiles
at pressures up to 1 atm for He and Ne; for Ar, Kr and Xe a breakdown of the
impact theory was established at 1 atm. These results are in good agreement with
the line profile calculations for Sr—Ar discussed earlier in this chapter. Voigt
profiles were also found by Butaux, Schuller and Lennuier (1972) for the Hg 2537 Å
line in noble gases.

Behmenburg (1968, 1970) has reviewed a large number of experiments by S.Y.
Ch'en and co-workers on *shift, half-intensity width* and *asymmetry ratio* of alkali
resonance lines in noble-gas milieus up to $n_p \approx 3 \times 10^{21}$ cm^{-3} ($T \approx 1000$ K). The plots
of $\delta\nu_s$ and $\delta\nu_{c''}$ against n_p show a linear part for perturber pressures up to 1 atm,
indicating that the impact approximation still holds here (cf. Eqs VII.3 and 4),
while the asymmetry ratio remains virtually 1.0, as would be expected. Behmenburg
has attempted to fit the experimental points to curves calculated on the basis of
Lindholm's general-pressures theory. A reasonable agreement was found for perturber pressures > 1 atm.

As to the *line wings* we mention a few representative results.

— The 2537 Å line wings of Hg in an Ar or Ne milieu were investigated in
absorption by Behmenburg (1970), and Losen and Behmenburg (1973). For $6.4 < \Delta\lambda < 16$ Å
a (−3/2)law was found in the red wings; the expected (−5/4)law in the far blue
wings was not observed.

— Ch'en and Phelps (1973) observed the blue and infrared Cs resonance lines
in an Ar milieu (400 < T < 600 K) at perturber densities up to 10^{19} cm^{-3}. They
found a (−3/2)law for the red wings of both lines; for the blue wings they observed an unexpected (−7/3)law.

— Gallagher (1975) measured the wings of the 6708 Å Li resonance line in
noble-gas milieus at 600−1800 Torr ($T \approx 400$ K). The author observed wavelength-
dependences in both wings for $1 < \Delta\lambda < 15$ Å that were very close to a (−2)law for
He, Ne and Ar, but an unexpectedly steeper slope of about −2.4 for Kr and Xe.

— Ottinger *et al.* (1975) found for the first resonance-doublet lines of Rb in Ar a slope of about -1.5 for $1 < \Delta\lambda < 5$ Å in the red wing, and of -2.8 for $0.8 < \Delta\lambda < 4.5$ Å in the blue wing.

— McCartan and Farr (1976) observed the red Na-D_1 line wing in N_2, Ar, Ne, Xe, He and Kr milieus at 0.3 atm. The $\Delta\lambda$ range considered was from 0.1 to 36 Å. For N_2 they found a (-2)law for $0.1 < \Delta\lambda < 0.7$ Å, and a $(-3/2)$law from 0.7 Å to 36 Å. For all other perturber milieus a (-2)law was found throughout the whole $\Delta\lambda$ range.

From the above results no definite conclusions can be drawn when these results are compared with the results in Ar- or N_2-diluted flames (see Sect. 4a), or when they are compared with QS theory.

A large number of nonflame results on *satellite bands* have been reported in the seventies. We mention a few: Granier (1969), and Granier and Granier (1973) investigated satellites in the wings of (forbidden) alkali lines at pressures $\geqslant 1$ atm. Behmenburg (1970), and Losen and Behmenburg (1973) found blue satellites on the Hg 2537 Å line wing in an Ar milieu at $\Delta\lambda = 4.6$ and 10.8 Å. Kielkopf *et al.* (1970, 1968) have compared the spectral intensity profiles of Cs principal-series lines and the associated red satellite bands at Xe pressures up to 560 Torr with theory on the basis of a L.-J. interaction. McCartan, Farr and Hindmarsh (1974) studied the red satellite of the 4047 Å K line perturbed by Kr. McCartan and Farr (1976) found a red satellite on the Na-D_1 line wing in Xe- and Kr milieus at 12.2 and 6.25 Å, respectively, from the line centre.

For most of the above results a satisfactory theoretical explanation could be given for the λ position and intensity ratio of the satellite bands to the parent line. This is in contrast to the satellite bands measured in flames (see Sect. 4a).

Summarizing we conclude that the nonflame results for line profiles point to the following general picture:

— The adiabatic line profile in core and near wings is adequately described by the impact approximation at perturber pressures $\leqslant 1$ atm.

— When Doppler broadening cannot be neglected, the core profiles have a Voigt shape.

— The far line wings can be described reasonably well by QS theory.

— The same holds for the occurrence of satellite bands, in contrast to the flame results.

— The intermediate λ range is experimentally not clearly established (see also Sect. 4a).

4c   DERIVATION OF FORCE CONSTANTS OF (DIFFERENCE)
     INTERACTION POTENTIALS FROM LINE PROFILES

In principle the force constants occurring in model expressions for the difference, $V(r)$, of the interaction potentials pertaining to the lower and upper states of the optical transition can be derived from the characteristics of the spectral line broadening due to collisions. The derivation of $V(r)$ from line profile measurements in flames suffers from the difficulty that perturbers of various kinds are present here. This difficulty is less serious in flames where one species (e.g., N_2 or Ar) is dominant. Metal-vapour cells containing only one perturber species have a definite advantage in this respect. But here, as in flames, the derivation of interaction potentials from line profile measurements is subject to the following uncertainties and limitations:

(i)   In general, line profile data yield *difference* potentials (see Sect. 2b.3-2); we need additional data from, for instance, elastic scattering experiments with ground-state atoms in order to derive the force constants of the separate potentials pertaining to the ground state and the excited state.

(ii) In order to derive $V(r)$ one has to know which model describes the adiabatic collision broadening adequately, i.e., which theory is correct for which part of the line profile (impact, quasi-static, or general-pressures theory).

(iii) Beforehand, one has to decide on which profile characteristics are considered in the derivation of $V(r)$. When only line core data are used in the framework of impact theory, it suffices to know $\delta\nu_{c''}$ for deriving C_6 of the difference potential in the case of a van der Waals interaction (see Table VII.1); in the case of a L.-J. interaction one has to know $\delta\nu_{c''}$ as well as $\delta\nu_s$ (see Eqs VII.3, 4 and 5). When general-pressures theories are used, additional knowledge of asymmetry ratios and line shape are required to derive $V(r)$.

(iv) The experimental adiabatic profile can always be curve-fitted to a theoretical one using a certain model by introducing more terms into the expression for $V(r)$ (see also below).

(v)   The assumption of spherically symmetric $V(r)$ functions is only valid for structureless perturbers (i.e. noble gases) and for nondegenerate atomic states. (It should be noted that $J=\frac{1}{2} \rightarrow J=\frac{1}{2}$ transitions may be treated as if the states were nondegenerate; see also Sect. 4f.) In practice, however, this assumption is not correct for most of the experimental situations.

In the following we use the simplification of nondegeneracy of atomic states and structureless perturbers. Furthermore the impact approximation is applied for L.-J. interactions and only occasionally for van der Waals interactions. Where the experimental values of $\xi(\alpha)$ were found to be different from the value 2.76 for a van der Waals interaction, only a L.-J. potential was adopted (see Table VII.1). Finally, adiabatic and quenching broadening processes are assumed to be statistically independent and to have dispersion profiles, viz. $\delta\nu_{C'} + \delta\nu_{C''} = \delta\nu_L$ (see Eq. VII.22; natural line broadening is neglected). The C_6- and C_{12} values thus derived from $\delta\nu_{C''}$ and $\delta\nu_S$ values measured in flames through Eqs VII.3, 4 and 5 are presented in Table VII.8.

Table VII.8
Force Constants of van der Waals or Lennard-Jones Difference Potential Derived from Flame Experiments

Metal	Wavelength (Å)	Perturber (dominant)	$C_6 \times 10^{32}$ (cm^6 s^{-1}) *	$C_{12} \times 10^{76}$ (cm^{12} s^{-1}) *	Ref.
Na	5890	He	2.7	-	1
Na	5890	He	5.0 †	1.0 †	1
Na	5890	N_2	3.7	-	1
Na	5890	Ar	6.1	-	1
Na	5890	CO_2	2.1	-	1
Ca	4227	N_2	20	38	2
Sr	4607	N_2	54	3.7×10^2	2
Sr	4607	Ar	2.2 ± 0.5 ≠	10 ± 5 ≠	3
Ba	5536	N_2	43	6.0×10^2	2

† The adiabatic line shift was obtained from the total line shift measured by assuming that according to Lindholm (1945) quenching collisions cause a shift $\delta\nu_S^{qu} = 0.36\, \delta\nu_{C'}$.

≠ In H_2-O_2-Ar flames of different quantitative gas composition the $\delta\nu_{C''}$ and $\delta\nu_S$ values due to the simultaneous action of Ar and H_2O perturbers were measured. Making use of Eqs VII.8 and 8a the specific $\delta\nu_{C''}$ and $\delta\nu_S$ values for Ar perturbers were found; from this the listed C_6- and C_{12} values were calculated.

* When no C_{12} value is listed, a van der Waals potential was assumed.

REFERENCES
1 = Behmenburg (1964);
2 = Hollander, Jansen, Plaat and Alkemade (1970);
3 = Jansen, Hollander and Alkemade (1977).

FLAME RESULTS COMPARED WITH OTHER RESULTS AND THEORY   Sect. VII.4c

The flame results in Table VII.8 can be compared with corresponding non-flame results practically only for Sr—Ar and Na—Ar (see below). The C_6- and C_{12} values from vapour-cell experiments reported in the literature have been derived for assumed van der Waals, L.-J. and (6,8,12) interaction potentials[†]. The (6,8,12) potential introduced by Hindmarsh and co-workers (see Smith 1967, and Hindmarsh and Farr 1972) is a refined version of the (6,12) L.-J. potential and appears to yield C_6 values in better agreement with values calculated through London's formula (see Eqs VII.20 and 20a).

Comparison of the flame results in Table VII.8 as well as of those from absorption cells may be made with theoretical C_6 values calculated from London's formula (see Unsöld 1968) which reads for the upper state (denoted by a single prime)

$$C_6' = \alpha_0 e^2 \langle r^2 \rangle \qquad \text{(VII.20)}$$

with

$$\langle r^2 \rangle = \tfrac{1}{2} a_0^2 n^{*2} \{5n^{*2} + 1 - 3l(l+1)\} . \qquad \text{(VII.20a)}$$

Here α_0 = dipole polarizability of perturber; e = electronic charge; a_0 = radius of first Bohr orbit; n^* = effective principal quantum number and l = orbital angular-momentum quantum number. A similar expression holds for the lower state and yields C_6''.

Comparison with 'expected' C_{12} values is not fruitful, since the empirical relation found by Hindmarsh and Farr (1972) refers to too limited a selection of experimental C_{12} values.

There are only very few data from absorption-cell measurements that refer to the same atom-perturber combinations as have been studied in flames. We report here the cell data on the Sr 4607 Å line perturbed by Ar obtained by Penkin and Shabanova (1968). They derived a C_6 value of 1.63×10^{-31} cm^6 s^{-1} when interpreting their measurements on the basis of a van der Waals interaction, whereas they obtained with a L.-J. potential: $C_6 = 2.6 \times 10^{-32}$ cm^6 s^{-1} and $C_{12} = 0.2 \times 10^{-75}$ cm^{12} s^{-1}. Farr and Hindmarsh derived force constants from their line broadening experiments on the Sr 4607 Å line in an Ar milieu: $C_6 = (6.5 \pm 0.6) \times 10^{-32}$ cm^6 s^{-1} and $C_{12} = (1.9 \pm 0.4) \times 10^{-75}$ cm^{12} s^{-1}.

An inspection of the corresponding data from Table VII.8 shows that the flame data on C_6 are consistent with those from Penkin and Shabanova, whereas the C_{12} values are closer to those from Hindmarsh and Farr.

[†] Recently Kielkopf (1976) reported on semi-empirical (6,8,10) potentials for alkali collision broadening in noble gases; experimental $\delta\nu_{c''}/n_p$- and $\delta\nu_s/n_p$ values (at $T = 500$ K) were in good agreement (within 20%) with the values calculated on the basis of these potentials. Kielkopf used for these calculations the general-pressures theory from Margenau and Lewis (1959).

Theoretical C_6 values are available for the Na 5890 Å line perturbed by N_2 and Ar (calculated by Behmenburg 1964 from Eq. VII.20) and for the Sr 4607 Å line in an Ar milieu (calculated by Behmenburg and Kohn 1964; Penkin and Shabanova 1968, and Farr and Hindmarsh 1971, respectively).

Recently Exton and Snow (1978) have reported on the derivation of upper-state potentials from measured line profiles (including satellite bands) of Cs lines perturbed by Xe and neopentane in an absorption cell at perturber pressures under 0.5 atm ($T \leqslant 500$ K). Starting with a constructed difference potential (built up out of two Morse potentials), they applied an iterative method in which they made use of the shape and λ position of satellite bands and the line wing profiles of the Cs lines measured. The best-fit difference potential thus found was combined with a ground-state potential derived from scattering experiments to derive the upper-state potential.

Surveying the data presented we may draw the following tentative conclusions:

— Flame data are scarce and refer mainly to N_2 and Ar perturbers; cell data refer mainly to noble gases, so that a mutual comparison can only be made for Ar.

— The available theoretical and experimental data for the Na 5890 Å line in inert-gas milieus show a good agreement for N_2 perturbers, but are inconsistent for Ar perturbers.

— The flame data on the Sr—Ar system show that the C_6 values from flame experiment are consistent with cell data from Penkin and Shabanova (1968) and close to the theoretically predicted value from Eq. VII.20. Furthermore, the C_{12} values from flame experiments are close to the 'cell values' from Hindmarsh and Farr 1972); theoretical C_{12} values are not available.

— It is still uncertain whether spherically-symmetric potentials satisfactorily describe the interaction with molecular perturbers.

4d α- PARAMETERS AND (ADIABATIC) LINE-BROADENING CROSS SECTIONS DERIVED FROM FLAME EXPERIMENTS

In Table VII.9 we collected a selected number of experimental values of α-parameters, effective adiabatic-broadening cross sections $(\sigma_{C''})_{eff}$, and effective total collision broadening cross sections $(\sigma_C)_{eff}$. Only in one case specific cross sections have been determined from flame experiments (see footnote $\neq$ to the table). We have listed only those values that were determined in flames the temperature and quantitative gas compositions of which were sufficiently uniform and known.

The data in Table VII.9 were derived by using various experimental techniques, line scanning as well as COG techniques. It should be recalled that both techniques confirmed the applicability of the Voigt model for the line core (see

FLAME RESULTS COMPARED WITH OTHER RESULTS AND THEORY   Sect. VII.4d

LEGENDS TO TABLE VII.9

† Only experiments in flames with well-defined temperature and composition are considered.

≠ Specific adiabatic cross sections found by these authors were: $\sigma_{C''}^{(Ar)}/\pi = 54 \pm 4 \text{ Å}^2$ and $\sigma_{C''}^{(H_2O)}/\pi = 105 \pm 5 \text{ Å}^2$, respectively, at $T = 2000 \text{ K}$.

* For the Na 5896 Å and the K 7699 Å line (at $T = 2500 \text{ K}$) Hofmann and Kohn (1961) have calculated from theory: $\sigma_{C''}^{(N_2)}/\pi = 54.3 \text{ Å}^2$ and 65.0 Å^2, respectively. Jongerius et al. (1981) have derived the following specific adiabatic cross sections for the Na-D lines: $\sigma_{C''}^{(Ar)}/\pi = 71 \pm 2 \text{ Å}^2$ and $64.5 \pm 2 \text{ Å}^2$ for Na-D_1 and D_2 line, respectively; $\sigma_{C''}^{(N_2)}/\pi = 60 \pm 2 \text{ Å}^2$ and $59 \pm 2 \text{ Å}^2$; $\sigma_{C''}^{(H_2O)}/\pi = 63 \pm 6 \text{ Å}^2$ and $69 \pm 6 \text{ Å}^2$; $\sigma_{C''}^{(H_2)}/\pi = 35 \pm 6 \text{ Å}^2$ and $24 \pm 5 \text{ Å}^2$.

** In most literature reports on flame experiments the value of σ/π instead of σ (as defined in Eq. VII.6) is communicated.

REFERENCES

1 = Hinnov and Kohn (1957); 2 = Hofmann and Kohn (1961); 3 = van Trigt, Hollander and Alkemade (1965); 4 = Hayhurst (personal communication); 5 = Behmenburg and Kohn (1964); 6 = Behmenburg (1964); 7 = van der Held (1932); 8 = Alkemade (1954); 9 = Hollander, Jansen, Plaat and Alkemade (1970); 10 = Jansen, Hollander and Alkemade (1977); 11 = Hollander and Broida (1967); 12 = Hollander and Broida (1969).

NOTES

— Corrections for h.f.s. were carried out by the authors of Refs (3, 5, 6, 9, and 10).

— The a-parameter and $(\sigma_C)_{eff}/\pi$ values derived by Hinnov and Kohn (1957) for lines of Mn, Fe, Co, Ni, Cu, Ag, Tl and Cr are not listed. We also left out the data for the second resonance lines of Na and K derived by these authors. Their values for the first resonance lines of Na, K, Rb and Cs are overrated (see Ref. 2).

— From C_6- and C_{12} values for Sr—Ar measured in a vapour cell by Farr and Hindmarsh (1971) a $\sigma_{C''}^{(Ar)}/\pi$ value of 44 Å^2 for the 4607 Å line of Sr at $T = 2265 \text{ K}$ can be derived; this value is about 20% lower than the above flame result.

— Recently Lovett and Parsons (1977) have reported calculated $\sigma_{C''}$ and a-parameter values for the resonance lines of alkali and alkaline-earth metals in a $C_2H_2/O_2/N_2$ flame at 2500 K. London's formula was used (see Eqs VII.20 and 20a). In their comparison with experimental results the authors took the $\sigma_{C''}^{(N_2)}$ value for the Na-D line from Hofmann and Kohn (1961) as a reference for the subsequent calculations of $\sigma_{C''}$ values for other perturbers. They conclude that experimental and calculated values agree within ± 20%. Unfortunately they have deleted some representative experimental data for the alkali resonance lines, so that we conclude that the above agreement may be too optimistic.

Table VII.9
Experimental α-Parameters and Collision Broadening Cross Sections (in $Å^2$) from Flame Experiments†

Metal	Wavelength (Å)	Flame	Temperature (K)	Main perturber	α-Parameter	$(\sigma_{C''})_{eff}/\pi$ **	$(\sigma_{C'})_{eff}/\pi$ **	Ref.
Li	6708	C_2H_2/air	2500	N_2	0.55	–	43	1
	6708	C_2H_2/air	2500	N_2	0.58	–	46.5	2
	6708	CO/air	1964	N_2	0.29 ± 0.02	–	18	3
	6708	$H_2/O_2/N_2$	≈2000	N_2	0.3	–	19	4
Na	5890	C_2H_2/air	2500	N_2	0.71	47	53.5	5
	5890	C_2H_2/O_2/He	2463	He	0.74	–	108	5
	5890	C_2H_2/O_2/Ar	2493	Ar	0.77	–	54	5
	5890	$C_2H_2/O_2/CO_2$	2505	CO_2	0.82	–	56.5	5
	5890	C_2H_2/air	2500	N_2	0.83	58.5	65	6
	5890	C_2H_2/O_2/He	2463	He	0.95	–	138	6
	5890	C_2H_2/O_2/Ar	2493	Ar	0.88	–	62	6
	5890	$C_2H_2/O_2/CO_2$	2505	CO_2	0.82	–	54	6
	5890	$C_2H_2/O_2/N_2$	2760	N_2	0.79	–	65.9	6
	5890	C_2H_2/air	2500	N_2	0.79	53	59.7	2
	5890	C_2H_2/air	2500	N_2	0.78	52 *	58.9	2
	5896	city gas/air	2080	N_2	1.0	–	33	2
	5890/96	C_3H_8/air	2180	N_2	0.50 ± 0.02⁵	–	34	7
	5890/96	CO/air	1964	N_2	0.45 ± 0.03	20	27	8
	5890/96	CO/air	2450	N_2	0.33 ± 0.02	18	25	3
	5890/96	C_2H_2/air	2389	N_2	0.41 ± 0.02	25	30	3
K	7665	C_2H_2/air	2500	N_2	1.1	56	60.4	2
	7699	C_2H_2/air	2500	N_2	1.1	54 *	57.7	2
	7665/99	CO/air	1964	N_2	0.78 ± 0.04	28	31	3

Table VII.9 (cont.)

Rb	7801	C_2H_2/air	2500	N_2	2.06	–	79.3	2
	7948	C_2H_2/air	2500	N_2	1.94	–	73.3	2
Cs	8521	C_2H_2/air	2500	N_2	3.1	–	91.2	2
	8944	C_2H_2/air	2500	N_2	3.1	–	86.9	2
	8521/8944	CO/air	1964	N_2	2.0 ± 0.1	–	47	3
Ca	4227	C_2H_2/air	2800	N_2	0.61	–	56.6	2
	4227	CO/air	2450	N_2	0.41 ± 0.03	–	37	3
	4227	C_2H_2/O_2	2275	N_2	0.43 ± 0.02	–	36.5	9
	4227	$C_2H_2/O_2/N_2$	2760	N_2	0.57	–	57.3	2
Sr	4607	$C_2H_2/O_2/N_2$	2760	N_2	0.91	–	64.8	2
	4607	C_2H_2/air	2800	N_2	1.04	–	66.0	2
	4607	C_2H_2/air	2500	N_2	1.02	–	65	5
	4607	C_2H_2/O_2/He	2463	He	1.10	–	(380)	5
	4607	C_2H_2/O_2/Ar	2493	Ar	1.18	–	83	5
	4607	$C_2H_2/O_2/CO_2$	2505	CO	1.24	–	67	5
	4607	CO/air	2480	N_2	0.85 ± 0.04	–	53	3
	4607	C_2H_2/O_2	2275	N_2	1.25 ± 0.03	–	73.5	9
	4607	H_2/O_2, Ar	2000	Ar+H_2O	1.1 ± 0.1	‡	–	10
Ba	5535.6	C_2H_2/air	2800	N_2	0.83	–	37.3	2
	5535.6	$C_2H_2/O_2/N_2$	2760	N_2	0.84	–	41.4	2
	5535.6	C_2H_2/air	2275	N_2	2.02 ± 0.04	–	82	9
OH	3072.009/063	$H_2/O_2/N_2$	2050	N_2	0.6 ± 0.1	–	70 ± 10	11,12
	3072.009/063	$C_2H_2/O_2/N_2$	2290	N_2	0.45 ± 0.09	–	60 ± 10	11,12
	3072.009/063	$C_2H_2/O_2/N_2$	2180	N_2	0.6 ± 0.1	–	80 ± 20	11,12
Zn	3075.9	$C_2H_2/O_2/N_2$	2300	N_2	1.93 ± 0.12	–	180 ± 20	12

Chapt. VII   BROADENING AND SHIFT OF ATOMIC METAL LINES

Sect. 4a). This justifies the use of the following equations for deriving cross sections (for the meaning of symbols and derivations see Sect. II.5b-3):

$$\delta\nu = \tfrac{1}{2}\delta\nu_L + \sqrt{(\tfrac{1}{2}\delta\nu_L)^2 + \delta\nu_D^2} \ , \qquad (VII.21)$$

$$\delta\nu_L = \delta\nu_N + \delta\nu_{C'} + \delta\nu_{C''} \ , \qquad (VII.22)$$

$$\delta\nu_N + \delta\nu_{C'} = A/2\pi + A(1-Y)/2\pi Y = A/2\pi Y \ , \qquad (VII.23)$$

$$\delta\nu_{C''}^{(j)} = n_p^{(j)} \sigma_{C''}^{(j)} \bar{v}_j \ , \qquad (VII.24)$$

$$\delta\nu_{C''} = \sum_j n_p^{(j)} \sigma_{C''}^{(j)} \bar{v}_j \ , \qquad (VII.25)$$

with the a-parameter being defined by

$$a \equiv (\delta\nu_L / \delta\nu_D)(\ln 2)^{\tfrac{1}{2}} \ . \qquad (VII.26)$$

The subscripts L, C', C'', D and N refer to the total Lorentz broadening, diabatic (or quenching) collision broadening, adiabatic collision broadening, Doppler- and natural broadening, respectively. The efficiency of resonance fluorescence is denoted by Y and the spontaneous transition probability by A. The Doppler half-intensity width $\delta\nu_D$ follows from Eq. II.240. The superscript (j) refers to a specific perturber labeled j. For $\delta\nu_{C'}^{(j)}$ and $\delta\nu_{C'}$ analogous equations hold as Eqs VII.24 and 25 for $\delta\nu_{C''}^{(j)}$ and $\delta\nu_{C''}$.

When one perturber is dominant, we have calculated an *effective* adiabatic collision broadening cross section, $(\sigma_{C''})_{\text{eff}}$, from Eq. VII.25 by assuming that all other perturbers have the same $\sigma_{C''}$ value as the main perturber, while only allowing for the differences in density $n_p^{(j)}$ and mean relative velocity $\bar{v}_j$ as given by Eq. II.29. This effective cross section may be regarded as an estimate for the corresponding specific $\sigma_{C''}^{(j)}$ value of the dominant perturber. In a similar way we have defined an effective total collision broadening cross section, $(\sigma_C)_{\text{eff}}$, relating to adiabatic plus quenching collisions. The relation between the specific total collision broadening cross section and $\sigma_{C''}^{(j)}$ and $\sigma_{C'}^{(j)}$ is given by

$$\sigma_C^{(j)} = \sigma_{C'}^{(j)} + \sigma_{C''}^{(j)} \ . \qquad (VII.27)$$

Specific cross sections $\sigma_C^{(j)}$ and $\sigma_{C''}^{(j)}$ have occasionally been derived from flame experiments (through Eqs VII.21, 22, 25 and 27) by measuring $\delta\nu$ and thus $\delta\nu_C$, etc., in a set of flames with different but known gas compositions.

When considering the data shown in Table VII.9 one may draw the following (tentative) conclusions:

FLAME RESULTS COMPARED WITH OTHER RESULTS AND THEORY   Sect. VII.4e

(i)   The $(\sigma_C)_{eff}$ values reported in the literature for a given line and flame
      type show rather a large spread, especially for the alkali first resonance
      lines, where yet unexplained differences of a factor 2 to 3 exist.

(ii) For the alkaline-earth first resonance lines the $(\sigma_C)_{eff}$ values show a
      better mutual agreement; for Sr the results are fairly consistent.

(iii) It seems that the broadening cross sections for different alkalis increase
      with increasing mass of the atom; but among the alkaline-earth metals Ba
      shows an anomalous behaviour if we consider the results from Ref. (2), but
      not for those from Ref. (9).

(iv) Comparison of experimental $(\sigma_{C''})_{eff}$ values for the Na and K first reson-
      ance lines with theoretical predictions does not lead to a definite
      conclusion (see also van Trigt, Hollander and Alkemade 1965). For Sr some
      theoretical $\sigma_{C''}^{(j)}$ values are available for comparison with the experimental
      ones, but the underlying potential parameters are rather uncertain (see
      Sect. 4c).

4e   EXPERIMENTAL α-PARAMETERS OBTAINED IN
      ANALYTICALLY USEFUL FLAMES

   In Table VII.10 we present α-parameter values reported in the literature for
flames that are useful in analytical spectroscopy but whose temperature and quantitative gas composition may not be well-defined and/or uniform. The interpretation
of these α-values in terms of adiabatic and diabatic interactions may therefore not
be meaningful. These α-parameter values and the total line widths following from
them through Eqs II.246 and 250 may be used in the evaluation of the suitability
of the investigated lines and flames for analytical atomic absorption and fluorescence spectroscopy. They may be used in the calculation of the effect of hyper-
fine-structure or of the spectral width of the source line on the sensitivity, the
shape of the analytical curve, etc. In such calculations also the line shift in
the flame may play a significant role; line shifts are not listed in the table,
but such data are found, e.g., in references no. 1 and 3 cited in the table.

   One should be cautious in transferring the α-parameters listed to other
flames with the same combustible mixture that do not reproduce the experimental
conditions sufficiently well. Variations in temperature of limited extent are
perhaps not so critical, but variations in flame gas composition do have an
influence.

   We present the data in Table VII.10 without further discussion, mainly for
the use by the analytical flame spectroscopist.

Chapt. VII　　　　　　　　　　　　BROADENING AND SHIFT OF ATOMIC METAL LINES

Table VII.10

α-Parameters Measured in Analytically Useful Flames

Metal	Wavelength (Å)	Flame	Temperature (K)	α-Parameter	Exp. method *	Ref.
Ag	3281	C_2H_2/air	2400	0.73 ± 0.02 †	FPI(E)	1
Ag	3281	C_2H_2/air	≈2800	0.47	IPA	2
Ag	3383	C_2H_2/air	2480	0.56 ± 0.02 †	FPI(E)	1
Ag	3383	C_2H_2/air	≈2500	0.51	IPA	2
Al	3944	C_2H_2/N_2O	≈3000	0.81 ± 0.02 †	FPI(E)	3
Al	3961	C_2H_2/N_2O	≈3000	0.89 ± 0.03 †	FPI(E)	3
Ca	4227	C_2H_2/air	2450	0.73 ± 0.03 †	FPI(E)	1
Ca	4227	C_2H_2/air	2450	0.66	FPI(A)	4
Ca	4227	C_2H_2/air	2450	0.84 ± 0.04 †	FPI(A)	3
Ca	4227	C_2H_2/N_2O	≈3000	0.70 ± 0.02 †	FPI(A)	3
Ca †	3933	C_2H_2/N_2O	≈3000	0.48 ± 0.02	FPI(A)	3
Ca †	3933	C_2H_2/air	2450	0.57 ± 0.02	FPI(A)	3
Cu	3251	C_2H_2/air	2450	0.56 ± 0.02	FPI(A)	1
Cu	3251	C_2H_2/air	≈2800	0.35	IPA	2
Cu	3274	C_2H_2/air	2450	0.57 ± 0.03	FPI(E)	1
Cu	3274	C_2H_2/air	≈2500	0.345	IPA	2
Ga	4033	C_2H_2/N_2O	≈3000	1.12 ± 0.03	FPI(E)	3
Ga	4172	C_2H_2/N_2O	≈3000	1.14 ± 0.03	FPI(E)	3
In	4102	C_2H_2/N_2O	≈3000	1.40 ± 0.05	FPI(E)	3
K	4044	C_2H_2/N_2O	≈3000	1.08 ± 0.05	FPI(E)	3
K	4047	C_2H_2/N_2O	≈3000	1.51 ± 0.05	FPI(E)	3
Mn	4031	C_2H_2/N_2O	≈3000	1.60 ± 0.05	FPI(E)	3
Mn	4033	C_2H_2/N_2O	≈3000	1.10 ± 0.05	FPI(E)	3
Mn	4034	C_2H_2/N_2O	≈3000	0.88 ± 0.05	FPI(E)	3

* See Sect. 3; the abbreviations used have the following meaning:
　FPI(E): emission line scanning with Fabry-Pérot interferometer;
　FPI(A): absorption line scanning with Fabry-Pérot interferometer;
　IPA　 : measurement of integral and peak absorption using a wide band-pass monochromator.

† Hyper-fine-structure accounted for.

REFERENCES　　　1 = Wagenaar, Pickford and de Galan (1974);　2 = L'vov (1972);
　　　　　　　　3 = Wagenaar and de Galan (1973);　4 = Kirkbright and Troccoli (1973).

4f OBSERVATIONS OF THE EFFECT OF HYPER-FINE-STRUCTURE ON LINE PROFILE AND CURVE-OF-GROWTH IN FLAMES

H.f.s. splittings may be of the same order of magnitude as the width of atomic lines in flames. The derivation of a-parameters or collisional line widths from line-scanning and COG measurements may be complicated by the possible occurrence of h.f.s. One has to apply correction procedures in which use is made of the known h.f.s. splitting, the known intensity ratio(s) of the h.f.s. components, and of the reasonable assumption that collision broadening affects the individual components in the same way.

In the construction of COG's for lines with a h.f.s. one assumes again that the profile of each component can be described by a Voigt function with the same a-parameter. If the a-parameter is unknown, one has to apply an iteration procedure for finding a from the experiments after starting with a probable estimate of a.

Behmenburg and Kohn (1964) and van Trigt, Hollander and Alkemade (1965) calculated theoretical COG's and duplication curves while allowing for the effect of h.f.s. As discussed in Sect. 2b-6, the effect of h.f.s. on the ordinate value, A_t^*, of the intersection point of the two asymptotes is of special interest. The effect of h.f.s. on A_t^* found with the Na 5890 Å line in a C_2H_2-air flame ($T \simeq 2500$ K), the Li 6708 Å line, the 8521 Å Cs line and the resonance doublets of Na and K in a $CO-O_2-N_2$ flame ($T \simeq 2000$ K) amounted to 15-20, 5-8, 30-35, 6-15 and 1-3%, respectively, depending on the quantitative flame gas composition.

Willis (1971) accounted for the effect of h.f.s. on the integral absorption A_t measured in flames with a continuous background source. Gallagher and Lapp (1967) did the same in their absorption experiments with Cu resonance lines (see also Sect. 2b-6). Rice and Ragone (1965) corrected for the influence of h.f.s. on the position of the high-density asymptotes of the 3067 Å Bi line in flames. It should be remarked that h.f.s. affects the position (not the slope) of the high-density asymptote, but does not affect the position and slope of the low-density one.

In the derivation of line parameters from scanned profiles and in the curve-fitting of these profiles to theoretical models the influence of h.f.s. may complicate matters. Hollander, Jansen, Plaat and Alkemade (1970), studying the absorption profiles of Ca, Sr and Ba resonance lines in C_2H_2-air flames ($T \simeq 2275$ K), did not find a distortion due to h.f.s., in accordance with expectations based on the known h.f.s. splittings and line widths. Wagenaar and de Galan (1973) and Wagenaar, Pickford and de Galan (1974) reported absorption line profiles measured with a Fabry-Pérot interferometer in $C_2H_2-O_2-N_2$ flames at $T = 2450$ and 3000 K. They determined the Voigt profile for the resonance lines of, inter alia, Al, Ca, Cr, Ga, In, K, Mn, Ag and Cu after removing possible distortion by h.f.s. by an interation of the curve-fitting procedure. Jongerius *et al.* (1981) corrected their Na-D line profiles for the influence of h.f.s. They recovered a Voigt profile for both D-lines in four different H_2-O_2-N_2 flames, four different H_2-O_2-Ar flames and one C_2H_2-air flame.

CHAPTER VIII

Formation of Metal Compounds

1. INTRODUCTION

Metal atoms can react with other species in the flame forming mainly simple, stable molecules or radicals such as CaOH, Ca(OH)$_2$, CaO, CaCl, or CaCl$_2$. These other species may be combustion products or they may be introduced as an additive into the flame, e.g. chlorine. More complicated molecules, such as KHVO$_3$, or molecules which contain two or more metal atoms, e.g. Na$_2$, K$_2$VO$_3$, do not usually occur when metals are introduced by spraying salt solutions because of the relatively low metal concentration in the flame. Only when metal vapour is introduced in amounts larger than 0.5 vol. %, may these polymer compounds be expected, as has been ascertained by mass spectrometry. Large amounts of metal atoms can be introduced in the form of vapours, e.g. KI, VCl$_4$, WF$_6$ (see Farber and Srivastava 1973, 1973a). Only the formation of simple molecules will be considered here.

The relative extent of molecule formation is often characterized by $\phi \equiv [MX]/[M]$, where [M] and [MX] denote the concentration or (number) density of free metal atoms M and of molecules MX, respectively. We shall call ϕ the *association factor* of the considered metal compound; when several compounds, MX, MY, ..., are formed, we obtain the *total association factor* by summing [MX] + [MY] + ..., etc. Sometimes the *degree of dissociation* $\beta_d \equiv [M]/([M]+[MX]) = (1+\phi)^{-1}$ is also used as a measure for molecule formation (see Eq. II.45).

If chemical equilibrium between M, MX, and X prevails in the flame, ϕ can be found from the *mass-action-law* (see Eq. II.43)

$$\phi = [X]/K_d. \qquad (\text{VIII.1})$$

Here K_d is the *dissociation constant* at flame temperature T for the reaction: MX ⇌ M + X. The dissociation constant depends exponentially on T^{-1} and on the

Chapt. VIII FORMATION OF METAL COMPOUNDS

dissociation energy D_0 (see Eq. II.44). Even small differences in D_0 of the order of 0.2 eV entail large relative variations in K_d under flame conditions.

In thermodynamic equilibrium, Eq. VIII.1 holds independently of the reaction path through which MX is actually formed. However, if the concentrations of some of the partners in the dominant formation reaction deviate from equilibrium, ϕ will in general be affected. Knowledge of the dominant reaction path is then required to predict the effect of these deviations on ϕ. This will be discussed in detail in Sect. 4.[†]

The concentration of X in Eq. VIII.1 is usually much larger than the concentration of MX. When X is a flame constituent, its concentration is related through (partial) chemical equilibria to the concentrations of other flame constituents. One may compare the concentrations of the flame constituents listed in Sect. IV.4 with the highest value of about 10^{15} cm^{-3} that is reported in Table III.1 on page 235 for the total metal concentration when spraying metal salt solutions of 1 mol/l. Then the value of [X] is not expected to be affected by the formation of MX molecules and it can be calculated from the composition of the unburnt gas mixture and T, or from the total halogen supply in the case of halide formation. Because ϕ is then independent of the metal concentration, the absolute value of the latter need not be known in the study of atom/molecule equilibria for metallic elements. This greatly simplifies the interpretation of the experimental results. Only introduction of metals in large amounts may affect the concentrations of flame constituents or additives (see Sect. IV.5c).

Some *consequences of metal compound formation* are:

(i)   The concentration of free atoms decreases. Atomic line intensities are weakened, whereas molecular bands may appear in the flame spectrum. The atomic line intensity is no longer simply related to T through the Boltzmann factor only. The dependence of the metal dissociation on T has to be taken into account, too.

(ii) Relaxation effects in the establishment of chemical equilibrium between the flame gas components may indirectly affect the atomic metal concentration as a function of height.

(iii) Formation of molecules also buffers the extent of ionization of the metal atoms, whereas the formation of molecular ions, such as BaOH$^+$, may supply additional free electrons in the flame.

(iv) In some cases chemical reactions may produce directly molecules in an excited state. Deviations from chemical equilibrium might then lead to suprathermal chemiluminescent band emission.

[†] Section numbers without a roman cipher refer to this chapter.

INTRODUCTION   Sect. VIII.1

(v)   Diffusion of the metal vapour will also depend on the extent of molecule
formation, since the diffusion constants of the metal compound and the free atom
are, in general, different.

In *analytical flame spectrometry*, compound formation deteriorates the detection limits of elements that are determined by their atomic emission, absorption or fluorescence. It is often advisable to use highly reducing flame conditions to reduce the influence of oxide formation. Such conditions prevail in fuel-rich flames and close to the tip of the inner combustion zone. Also hotter flames can often help to overcome the limitations set by compound formation.

Formation of molecules that emit spectral bands in the flame may be used in the spectrochemical analysis of elements that do not radiate or absorb in atomic form. Halogen concentrations can be determined by measuring the intensity of the SrF bands at 5780 Å and of the InCl-, InBr-, and InI bands emitted at 3600, 3760 and 4100 Å, respectively (see Gilbert 1966, 1970, Dagnall, Thompson and West 1969, Herrmann 1974, Gutsche and Herrmann 1974). Measurements of CN-, CH-, NH-, NO-, HPO-, and S_2 bands, emitted in very cool $(T \approx 750 \text{ K})$ nitrogen-hydrogen diffusion flames, have been used for identification and quantitative determination of organic compounds (see Dagnall, Smith, Thompson and West 1969, Dagnall, Thompson and West 1967, and Gilbert 1970). In similar cool flames, Sn has been determined by measuring the SnH band intensity at 6095 Å (see Dagnall, Thompson and West 1968).

Spectral interference effects can occur when molecular bands emitted by a concomitant element overlap the spectral line used for analysis. The atomic absorption method is less disturbed by such spectral interference effects, as molecular bands of metal compounds normally show negligible absorption.

If Φ is independent of the total metal concentration, the shape of the analytical curves for the atomic analysis line will not be distorted by molecule formation.

For a further discussion of the effects of compound formation in analytical flame spectroscopy we refer to the textbooks cited in Chapt. I.

If the flame temperature and gas composition are known and a reliable value of the dissociation energy D_0 is available, the equilibrium value of ϕ can be calculated for a given molecular species. In Sect. 3 we shall present the results of such calculations for a number of molecules under different conditions. This is part of a more general survey of the molecular metal species to be expected in flames. The D_0 values adopted in these calculations will be given in tabular form in Sect. 2. For many molecules the D_0 values reported in the literature show considerable spread among each other. For a better understanding of the possible origins of this spread a brief introduction to some common methods of determining

Chapt. VIII   FORMATION OF METAL COMPOUNDS

D_0 will also be given in Sect. 2. In Sect. 5 we shall deal in more detail with the specific flame spectrometric methods of determining D_0. First, however, we shall consider in Sect. 4 the formation of metal compounds under nonequilibrium conditions. In particular, we shall discuss how information can be obtained about the specific reaction path through which the molecules are actually formed.

2. METHODS OF DETERMINING DISSOCIATION ENERGIES (D_0) AND VALUES REPORTED

The dissociation energy D_0 is a cornerstone in the calculation of the degree of dissociation of a metal compound at given temperature. Our knowledge of D_0 values is growing rapidly. However, the values obtained by different authors, under different measuring conditions, and by different methods, are often conflicting. An exhaustive discussion of the different methods used to determine dissociation energies and a critical evaluation of the results obtained are outside the scope of this book. The reader is referred to more authoritative works such as Cottrell (1958), Herzberg (1950, 1966), Lewis and von Elbe (1961), Vedeneyev et al. (1966), Schofield (1967), Walker and Straw (1967), Gaydon (1968), Brewer and Rosenblatt (1969), and Kondratiev (1974).

2a   METHODS OF DETERMINING D_0

The dissociation energy D_0 is the energy required for the decomposition of an isolated molecule XY in its lowest (real) energy level into neutral fragments X and Y that are in their ground states and have zero kinetic energy of relative motion (see Sect. II.3b-3; see also Cottrell 1958, and Gaydon 1968). D_0 for a single molecule is often expressed in eV/molecule, and D_0 for one mole of dissociating gas in kJ/mol (or in kcal/mol in the older literature; see also Appendix A.1).

D_0 is usually determined in an indirect way from dissociation measurements, which sometimes involve the solid or liquid phase. D_0 can also be calculated from the molecular structure, which is usually determined by optical methods. Then, auxiliary data and theoretical considerations are often required. Uncertainties in the final result stem mainly from doubts about the assumptions and data used.

Rather special methods such as the direct calorimetric measurement of the heat liberated by the exo-ergic recombination of atoms and radicals, or the transpiration method, will not be discussed. The methods discussed here can be divided into two broad classes:

I.   *molecular methods* where the structure and characteristics of an <u>individual</u> molecule are determined and D_0 is calculated therefrom;

METHODS OF DETERMINING DISSOCIATION ENERGIES (D_0) AND VALUES REPORTED   Sect. VIII.2a

II. *thermochemical methods* in which the heat, ΔH_T, required to dissociate one mole of a bulk substance at temperature T, is measured and subsequently converted into D_0 (see Sect. II.3b-3 for their relation).

2a-1 *I. MOLECULAR METHODS*

In these methods the temperature does not appear as a parameter and might only be important in so far as it controls the production of the investigated species in the gaseous phase or in a certain internal state. D_0 is derived (a) from molecular emission or absorption spectra, i.e. *spectroscopic methods*, or (b) from measurements of the threshold energy that is required for the appearance of fragments or reaction products in a certain state by electron impact, photon absorption or chemical reaction, i.e. *appearance-threshold methods*.

Ia. *Spectroscopic methods* yield data on the molecular structure. The optical quantities involved can usually be measured precisely. However, large systematic errors in D_0 may arise because of misinterpretation of the experimental data. These errors may be introduced by wrong assumptions about the energy-level diagram of the molecule or about the quantum states of the fragments into which the molecule is allowed to dissociate. Also, uncertain extrapolation procedures may often lead to appreciable systematic errors as shown below. It is evident that spectroscopic methods only work if the assignment of the molecular spectra studied is unambiguous. Di-atomic molecules are therefore the most suitable species for application of these methods.

When photodissociation occurs, the absorption spectrum of a molecular gas shows a continuum (see Sect. II.2e). A determination of the *long-wavelength limit of the absorption continuum* yields D_0 or at least an upper limit for it. The applicability of these methods relies heavily on the position and shapes of the molecular potential-energy curves involved in the optical transition. By way of illustration we refer to Fig. VIII.1a where the positions and shapes of the curves are such that an absorption transition can take place from the ground-level to an electronic excitation level. In this example, the long-wavelength limit at which the spectral continuum merges into a band structure, yields the energy of dissociation from the ground-state to fragments one of which is in an excited state. In order to find D_0 the energy of this state has to be subtracted from the energy corresponding to the long-wavelength limit. When a region of continuum absorption does not merge into a band spectrum at a definite convergence limit, only an upper limit of D_0 can be found.

An accurate value for D_0 can be obtained from the *convergence limit of the vibrational bands* for a given electronic state if all vibrational transitions up to this limit can be observed. The existence of a convergence limit is essentially related to the anharmonicity of the molecular vibrations, as explained in Sect. II.2b.

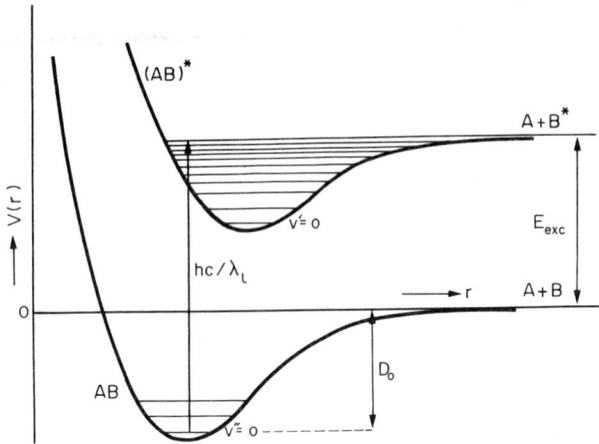

Fig. VIII.1a Potential curves are schematically drawn for ground-state molecule AB and stable, excited molecule $(AB)^*$ dissociating into ground-state atom A and excited atom B^* (excitation energy = E_{exc}). The long-wavelength limit for the absorption continuum from the lowest vibrational level $(v'' = 0)$ of the ground-state is denoted by λ_1. The dissociation energy D_0 is determined from: $D_0 = hc/\lambda_1 - E_{exc}$.

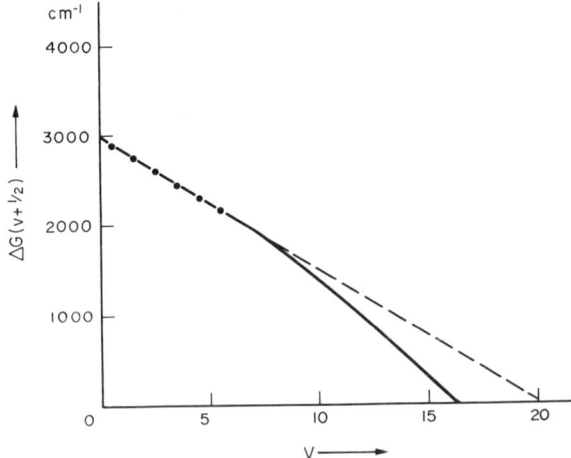

Fig. VIII.1b The difference $\Delta G(v+\tfrac{1}{2}) \equiv G(v+1) - G(v)$ between two successive vibrational energy terms is plotted versus vibrational quantum number v (schematic). The dashed line is obtained by linear Birge-Sponer extrapolation from measured points (filled circles). The solid line depicts the actual function. Integration of the linear function yields too high a D_0 value.

METHODS OF DETERMINING DISSOCIATION ENERGIES (D_0) AND VALUES REPORTED   Sect. VIII.2a

Usually only transitions within a limited range of vibrational levels can be observed and the convergence limit must be determined approximately by an extrapolation procedure. Including only the first two terms in the series expansion of the vibrational term $G(v)$ as a function of vibrational quantum number v, we find from Eq. II.13 that the separation, $\Delta G(v+\tfrac{1}{2})$, between two successive terms is a linear function of v, given by

$$\Delta G(v+\tfrac{1}{2}) \equiv G(v+1) - G(v) = \omega_e - 2x_e\omega_e(v+1) \ . \qquad \text{(VIII.2)}$$

This equation is exactly valid if the potential curve can be represented by a Morse function. Plotting a limited number of experimental $\Delta G(v+\tfrac{1}{2})$ values as a function of known v ($v = 0, 1, 2, ...$) we can determine the highest v value just below the dissociation limit by extrapolating this plot linearly towards $\Delta G = 0$. Adding up all positive ΔG's obtained from this plot, we find the dissociation energy D_0^* (expressed in wavenumber) counted from the lowest energy level having $v = 0$.

This extrapolation procedure is known as the *linear Birge-Sponer extrapolation* and is shown in graphical form in Fig. VIII.1b. Often this method yields too high D_0 values, owing to a more rapid convergence of ΔG to zero than is predicted by a linear function. This is also illustrated in Fig. VIII.1b where the full curve intersects the v axis at a lower value than the broken line. More refined methods of extrapolation which yield more accurate results are sometimes applied. Generally, the precision of the D_0 values obtained is estimated to be of the order of 10-20%. The method has found wide application in determining D_0 for di-atomic molecules, as is seen in Tables VIII.1 and 3 of Sect. 2b. Systematic errors in the interpretation of the optical transitions may be considerable (see Gaydon 1968).

Another procedure of finding dissociation energies of di-atomic molecules is based on the assumption of a suitable form of potential function, in which D_e occurs as a parameter (see Fig. II.2 on page 23). One obtains D_e by fitting the calculated potential-energy values in the turning points to the experimental $G(v)$ values. This curve-fitting method has recently been employed for the ground-state of BeO by Murthy and Prahllad (1978), of MgO by Murthy and Bagare (1978), and of AlO by Murthy, Bagare and Murthy (1978), yielding $D_e = 5.15 \pm 0.05$, 3.35 ± 0.4 and 4.15 ± 0.05 eV corresponding to $D_0 = 5.06$, 3.30 and 4.09 eV, respectively. The low value found for MgO supports the view that its ground-state is a $^1\Sigma$ state.

Predissociation which results into disappearance or diffuseness of the band structure at a certain place in the absorption spectrum (see Sect. II.2d), can sometimes be used to find a precise D_0 value. In all cases, predissociation yields at least an upper limit for D_0. Interpretation of the results is difficult, especially when the nature of the predissociating state is uncertain.

Chapt. VIII   FORMATION OF METAL COMPOUNDS

Ib. Impact of electrons with known energy on a molecule can result in its dissociation and the subsequent ionization of one of the fragments formed. The ionized fragment can be detected by electric or mass-spectrometric means. The *appearance potential*, i.e. the minimum electron energy required for the appearance of the ionized fragment, is measured. From the appearance potential and the known ionization energy of the fragment, D_0 can be calculated. This method is of special advantage for poly-atomic molecules.

In the *photo-ionization method* the dissociation of the gas molecules and subsequent ionization of the fragments is achieved by incident photons of sufficiently high energy. A monochromator is used to determine the minimum photon energy required for the appearance of the ionized fragment.

In the *photodissociation method* the minimum photon energy is determined, that is required to produce a certain neutral fragment (process no. 25 in Table II.1 on pages 108-9). The appearance of an atomic fragment may be detected by observing a specific chemical reaction that is induced by this fragment in the system. The appearance of an excited atomic fragment may also be detected by observing its luminescence. A well-known example of the latter variant is the process (no. 27 in Table II.1)

$$NaI + h\nu_c \to Na^* + I$$

followed by (no. 7 in Table II.1)

$$Na^* \to Na + h\nu_d ,$$

where $h\nu_c$ is a photon from a spectral continuum source and $h\nu_d$ the observed re-emitted photon of discrete frequency. The products of the photodissociation process may also be detected by a mass spectrometer.

A variant of this method, called *photofragmentation spectroscopy*, is based on the photo-excitation of a di-atomic molecule from its electronic ground-state to an excited repulsive state from which it dissociates into ground-state atoms (see Busch and Wilson 1972, and van Veen, de Vries and de Vries 1979, 1979a). The dissociation energy can be obtained from the energy balance: $D_0 = E_i + h\nu_d - E_k$, where E_i = the internal energy of the molecule in its initial rotational-vibrational state, $h\nu_d$ = discrete photon energy and E_k = kinetic energy of relative motion of the atoms. The last term can be derived from the velocity of one of the atoms, which is measured by a time-of-flight technique. This method has yielded accurate D_0 values of some alkali halides (see Table VIII.4 on page 769).

Chemical reactions have been produced in (*crossed*) *beam experiments* in order to determine D_0 of di-atomic molecules. A beam of metal atoms either intersects a beam of reactants, or traverses a scattering chamber filled with reactants. For example, a metal atom M and an oxidizer RO may react to form an electronically excited molecule MO^*, according to

METHODS OF DETERMINING DISSOCIATION ENERGIES (D_0) AND VALUES REPORTED   Sect. VIII.2a

$$M + RO \rightarrow MO^* + R .$$

Here RO may be: O_2, NO_2 etc. Of course the reaction between M and RO should be sufficiently exo-ergic. D_0(MO) follows from the energy balance

$$D_0(MO) = D_0(RO) + E_{exc}(MO^*) + [E(R) - E(RO) - E(M)] + \left(E''_{tr} - E'_{tr}\right) , \quad (VIII.3)$$

in which $E_{exc}(MO^*)$ is the excitation energy of MO^*, E is the internal energy, and E'_{tr} and E''_{tr} are the translational energies of relative motion before and after the reaction, respectively. The chemiluminescence spectrum emitted by MO^* is recorded; the photon energy corresponding with the *short-wavelength limit* of this spectrum yields the highest excitation energy, E_{max}, of MO^* if the optical transition is to the ground-state. If this transition is to an excited state of MO, then the excitation energy of the latter should be added to the photon energy to obtain E_{max}. From E_{max} and an estimate of the (spread in) internal and relative translational energy terms one obtains at least a lower limit for D_0. If there is an energy barrier in the incoming branch of the reaction path, only those atoms which have enough energy to surmount the barrier will be able to react (see Thrush 1973). Depending on the shape of the potential-energy surface, this activation energy should be available as additional energy of relative motion, or as extra vibrational energy (see Polanyi 1972).

Beam experiments have been carried out to determine D_0(AlO) from the reaction between Al + O_3 (see Gole and Zare 1972) and D_0(BaO) from Ba + NO_2 (see Jonah, Zare and Ottinger 1972), D_0(SmF) and D_0 (SmCl) from the reactions Sm + (F_2,Cl_2), D_0(YbO) and D_0(YbF) from the reactions Yb + (O_3, F_2) (see Yokozeki and Menzinger 1976) and D_0(YbO) from the reaction Yb + O_2 (see Cosmovici *et al.* 1977). In all cases a lower limit of these dissociation energies has been obtained.

A supersonic, nearly mono-energetic O_2 beam with widely variable mean translational energy has been crossed with a thermal atomic alkaline-earth beam (see Batalli-Cosmovici and Michel 1972). The dissociation energy of the metal oxide, produced in the endo-ergic reaction: $R + O_2 \rightarrow RO + O$, could be determined from the threshold energy of the O_2 beam for formation of the metal oxide. Appearance of the metal oxide is detected mass-spectrometrically.

The value of 4.9 eV found by the above authors for D_0(SrO) is rather high when compared with values obtained by other methods (see Table VIII.1 on pages 766-767).

In fact, beam methods yield the difference between the dissociation energies of a reactant and a product molecule, ΔD_0. The spread in the estimates of E and E_{tr} (see Eq. VIII.3) should be small in order to obtain a reliable value of

Chapt. VIII   FORMATION OF METAL COMPOUNDS

D_0(MO). In the beam experiments done so far the atomic species M are in their ground-state as are the R atoms if RO happens to be a di-atomic molecule.

2a-2   *II. THERMOCHEMICAL METHODS*

These methods are generally characterized by the appearance, in some manner or other, of the temperature as an essential parameter or variable in the measurements or calculations. The following particular methods may be reckoned among this broad class: (a) *calorimetric methods*, (b) *thermal-equilibrium methods*, and (c) *chemical-kinetics methods*.

IIa. Calorimetric measurements in conjunction with a *thermochemical cycle* can often lead to the calculation of the dissociation energy at 0 K. As an example we consider the relation of D_0(MX) for a metal halide MX to its heat of formation ΔH_{f_0} from solid metal (M)$_s$ and molecular halogen gas $(X_2)_g$, which is given by the following cycle:

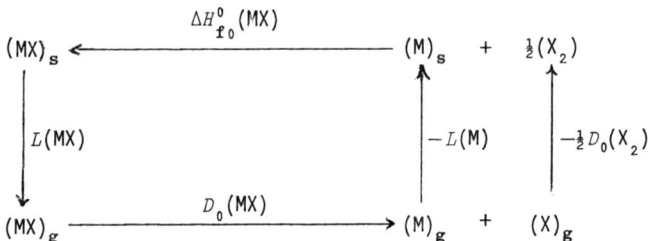

so that

$$D_0(\text{MX}) = -L(\text{MX}) - \Delta H_{f_0}^0(\text{MX}) + L(\text{M}) + \tfrac{1}{2} D_0(X_2).$$

The subscript s refers to the solid state and the subscript g refers to the gaseous state. In this presentation it is assumed that all thermochemical quantities that are dependent on temperature have been reduced to 0 K, i.e. the temperature at which D_0 is defined. Of the required thermodynamical values, the heat of sublimation L of the compound is usually the least certain.

IIb. Thermal-equilibrium methods are based on the relation between the dissociation energy and the dissociation constant K_d. For dissociation equilibria in the gaseous state: $XY \rightleftharpoons X + Y$, where all constituents obey the ideal-gas law, the dissociation constant $K_d = \lceil X \rceil \lceil Y \rceil / \lceil XY \rceil$ at temperature T is related to the change, ΔG_T^0, in Gibbs free energy through Eqs II.46 and 51. ΔG_T^0, in turn, can be converted into D_0 $(=\Delta H_0^0)$ by using specific-heat data (see Eqs II.49, 55 and 56 with $\vartheta = 0$). We can also derive D_0 from K_d by using Eq. II.44 and calculating partition functions from molecular and atomic constants, as indicated in Sect. II.3b-2. Thus if T is known and sufficient additional data are available, D_0 can be derived

METHODS OF DETERMINING DISSOCIATION ENERGIES (D_0) AND VALUES REPORTED   Sect. VIII.2a

from an experimental absolute value of K_d obtained by determining the concentration ratio of two species, $[X]/[XY]$, and the absolute concentration of the third one, $[Y]$. This method is known as the *third-law method*, since the third law of Thermodynamics is involved here.

Absolute concentration measurements may sometimes be difficult (see below). Such difficulties can be avoided when isomolecular reactions, e.g.: $MO + O \rightleftharpoons M + O_2$, are investigated instead of the direct dissociation reaction: $MO \rightleftharpoons M + O$. Then only the ratios $[M]/[MO]$ and $[O_2]/[O]$ need to be known to obtain the pertinent equilibrium constant and thus the isomolecular reaction energy. This reaction energy together with the known dissociation energy of $O_2 \to 2O$ yields the dissociation energy of $MO \to M + O$ (see Drowart, Exsteen and Verhaegen 1964, and Farber and Srivastava 1976).

A second thermal-equilibrium method of determining D_0 involves only the relative variation of K_d with temperature T. According to Eq. II.54a or 54b, $d(\ln K_d)/dT$ or $d(\ln K_d)/dT^{-1}$ is simply related to the heat of dissociation ΔH_T^0 at constant standard pressure and at temperature T. From ΔH_T^0 we can derive D_0 ($\equiv \Delta H_0^0$) by using Eq. II.55 (with $\theta = 0$) and data on the specific heats as a function of T. This method is called the *second-law method*. We can also apply this method by using the relation between $d(\ln K_d)/dT$ and D_0 that follows from Eq. II.44, and by deriving the relative T-dependence of the involved partition functions from molecular and atomic constants. The applicability of this method is favoured by the fact that ΔH_T^0 usually varies but little with T as long as only the gaseous state is involved. Its accuracy relies, however, heavily on the extension of the range of temperatures where K_d is measured. An accurate determination of temperature differences is required here. In the literature the results of the latter method are usually considered to be less accurate than those obtained by the third-law method. On the other hand the second-law method is not affected by uncertainties in some of the molecular constants that are used in the third-law method. These constants drop out when only the <u>relative</u> variation of K_d with temperature is considered. A more detailed discussion is found in Sect. 5a.

Both thermal-equilibrium methods may be applied in many varieties as regards the way in which the equilibrium conditions are realized and the concentrations of the involved species are measured. Di-atomic compounds often require high-temperature sources to produce measurable concentrations of their dissociation products. Often a Knudsen cell, which can be heated up to 2700 K, is used (see Babeliowsky 1962). It is made of tantalum or tungsten and contains the compound to be dissociated (for example, metal oxide) in the condensed phase. The gas mixture effusing from a small hole in the wall of this cell is usually analysed quantitatively by a mass spectrometer.

Chapt. VIII   FORMATION OF METAL COMPOUNDS

The great advantage of *mass spectrometry* is that this technique enables a qualitative as well as quantitative analysis of the different components in complex mixtures. This detection method in conjunction with high-temperature vapour sources has provided many new data on the dissociation energies of refractory di- and polyatomic molecules. However, the calibration of the mass spectrometer in terms of absolute concentrations or concentration ratios of different species is often difficult. Besides, dissociation of the molecules upon electron impact in the ion chamber of the spectrometer might affect the concentration ratios to be analyzed. It has been reported that approximately 50% of the BaO^+ ions dissociate in the ion chamber (see Inghram, Chupka and Porter 1955). This could lead to too low values of the dissociation energy. It is possible to correct for such disturbing effects (see Medvedev 1961). For a further discussion we also refer to Drowart and Goldfinger (1962).

Also the measured *rate of effusion* can inform us about the partial pressure of the components in the dissociating gas contained in the cell. The total pressure inside the cell and the molar mass of all components concerned have to be known. A molecular beam effusing from a hole in a furnace may also be analyzed by deflection of its components in a nonuniform magnetic field if these components happen to have different electromagnetic moments. This method has been used to determine D_0 values for the dimers of alkali metals, such as Cs_2.

Measurement of the *total pressure* for a given mass of dissociating gas contained in a closed vessel with known volume simply yields the degree of dissociation of the gas. This holds if the ideal-gas law can be applied and the dissociation products are known. A value of the equilibrium constant is found only if all gaseous constituents can be assumed to be in chemical equilibrium with each other at the temperature of the cell. The cell temperature can be determined by optical pyrometry.

Explosion-, *flame-*, *detonation-*, *and shock-wave methods* where the final pressure, flame temperature, and detonation or shock-wave velocity, respectively, are measured, have been used to determine D_0 values of molecular gases. The physical quantity measured is simply related to the degree of dissociation of the gas if thermal equilibrium has been attained adiabatically. The high temperatures reached are favourable for appreciable dissociation of gaseous compounds with strong bonds.

A particular method of more direct interest to the subject of this book is the *flame-spectrometric equilibrium method*, which will be treated more extensively in Sect. 5. It is based on the same principles as outlined above for the general thermal-equilibrium methods. The hot equilibrium system is provided by the flame gases themselves (acting as a heat bath), after they have left the combustion zone. The high temperature makes this method also suitable for the determination of

METHODS OF DETERMINING DISSOCIATION ENERGIES (D_0) AND VALUES REPORTED   Sect. VIII.2a

comparatively high D_0 values. This holds the more so since even small traces of free metal atoms may be detected spectrometrically with great accuracy. The main difficulties of the flame-spectrometric method are that often other molecular compounds of the same metal are formed, too, and that deviations from chemical equilibrium can occur.

Thermal-equilibrium methods are especially suited for the determination of dissociation energies of di-atomic molecules that have simple and stable dissociation products.

IIc. Chemical-kinetics methods are often well suited for poly-atomic and in particular organic molecules. Here it is assumed that the activation energy for the recombination reaction of the dissociation products is zero. Thus the activation energy, E_A, for the dissociation reaction is equal to the reaction energy. The precise relationship of the experimental activation energy to the dissociation energy can be found in Cottrell (1958). The unimolecular rate constant, k_d, for the dissociation reaction: $XY \to X+Y$ is defined by: $k_d \equiv -d[\ln[XY]]/dt$ and depends on E_A and T according to the Arrhenius expression (cf. Eq. II.124; the slight difference between E_A and E_{Arrh}, according to Eq. II.126, is here disregarded)

$$k_d = k_d^0 \exp[-E_A/kT]. \qquad (VIII.4)$$

A plot of $\ln k_d$ versus T^{-1} then yields E_A (cf. Eq. II.125a). The rate constant k_d can be determined by measuring the rate of production of the dissociation products with the aid of a mass spectrometer. As in all bulk methods, the temperature is an important parameter in the kinetic method and has to be known and controlled accurately.

Purely *theoretical calculations* of the stability of molecular compounds are still too uncertain to predict accurate D_0 values even for the simplest di-atomic molecules. Semi-empirical calculations, which are partially based on experimental data, have been proven to be of greater value (see Vedeneyev et al. 1966).

Finally, we mention the possibility of obtaining *estimates* of D_0 by means of interpolation for a series of molecules with similar electronic ground-state configurations. A comparison of dissociation energies for adjacent iso-electronic pairs of molecules has been given by Wilkinson (1963). General systematic trends in the D_0 values for a series of alkaline-earth halides and hydroxides have been considered by Ryabova and Gurvich (1965). Such estimates may be useful in deciding between controversial experimental values obtained by different methods. For a review we refer to Thorn (1966).

Chapt. VIII	FORMATION OF METAL COMPOUNDS

2b   REPORTED VALUES OF D_0

In Tables VIII.1, 2, 3, and 4 we have listed reported D_0 values for some typical metal monoxides, hydroxides, hydrides, and halides, respectively. The selection of molecules listed is rather arbitrary but most molecular species incorporated in the tables are believed to be of interest in flame studies.

The D_0 values listed in Table VIII.1 for the monoxides are specified according to the method of determination applied. Distinction is made between spectroscopic methods (Sp), thermochemical methods in bulk systems excluding flames (Th), and flame-spectrometric equilibrium methods (Fl). Recommended values based on a comparison of results obtained by different methods and authors are listed in a separate column, together with their reference. The last column contains the D_0 values adopted in Sect. 3 in our calculation of association factors. Since a critical discussion of measured and recommended values reported by different authors is outside the scope of this book, these adopted values are not intended to replace other recommended values or to reject certain measured values. When large discrepancies exist between spectroscopic values and values obtained by other methods, we have in general attached a greater weight to the latter ones. With some metal oxides, the D_0 value determined by the third-law method may be uncertain because of doubtful electronic partition functions of the molecule. In these cases, the listed D_0 value refers specifically to the value of the partition function assumed in the table. These assumptions should not be considered as a recommendation.

In the other tables, the D_0 values obtained by different methods are not listed in separate columns. The available literature data are usually less extensive here than for the monoxides. Some indication of the method(s) used is given in a separate column, where appropriate.

For all D_0 values listed, references to the appropriate literature are given. Reference is made to original papers as well as to review papers or compilations. Values cited in different references cannot always be considered as independent, since they often refer back to a common source. The reliability of a value listed is thus not connected to the number of references attached. No exhaustive compilation of references was intended. The references given only serve as a first guide to the reader who wants to check the origin of the listed values. For more complete information the reader is advised to consult the books included in the list of references.

Improbably low D_0 values, such as those obtained for the alkaline-earth monohalides by spectroscopic methods, are not included.

All values refer to the mean values reported in the quoted references, rounded off to 0.1 eV. No error limits have been added, as the discrepancies between the values found by different authors often exceed significantly the combined

METHODS OF DETERMINING DISSOCIATION ENERGIES (D_0) AND VALUES REPORTED Sect. VIII.2b

REFERENCES in Tables VIII.1,2,3,4

1. Gaydon (1968)
2. Brewer and Rosenblatt (1969)
3. Beckel, Shafi and Engelke (1971)
4. Vedeneyev *et al.* (1966)
5. Wilkinson (1963)
6. Appelblad and Lagerqvist (1974)
7. McDonald and Innes (1969)
8. Zare (1974)
9. Panchenkov, Gusarov and Gorokssov (1973)
10. Ottinger and Zare (1970)
11. Batalli-Cosmovici and Michel (1972)
12. Burns (1966)
13. Drowart *et al.* (1960)
14. Farber and Srivastava (1976)
15. Hildenbrand (1973)
16. Gurvich and Veits (1958)
17. Newman and Page (1971)
18. Jensen and Jones (1972)
19. Fontijn (1977)
20. Frank and Krauss (1974)
21. Uy and Drowart (1970)
22. Drummond and Barrow (1951)
23. Inghram, Chupka and Berkowitz (1957)
24. Stafford and Berkowitz (1964)
25. Colin, Goldfinger and Jeunehomme (1964)
26. Farber and Srivastava (1975)
27. Veits and Gurvich (1956)
28. Lagerqvist and Huldt (1954a)
29. Kalff and Alkemade (1973)
30. Ryabova and Gurvich (1965)
31. Uy and Drowart (1969)
32. Gilbert (1963)
33. Drowart, Exsteen and Verhaegen (1964)
34. Brewer and Searcy (1956)
35. Zeegers, Townsend and Winefordner (1969)
36. Lagerqvist and Huldt (1953)

37. Smoes *et al.* (1972)
38. Reid and Sugden (1962)
39. Balducci *et al.* (1971)
40. Hildenbrand (1975)
41. Jensen and Jones (1973)
42. Burns *et al.* (1963)
43. Coppens, Smoes and Drowart (1967)
44. Rauh and Ackermann (1975)
45. Kniseley, Butler and Fassel (1975)
46. Hildenbrand (1972)
47. Richards, Verhaegen and Moser (1966)
48. Alexander, Ogden and Levy (1963)
49. Brewer and Porter (1954)
50. Huldt and Lagerqvist (1950)
51. Bulewicz and Sugden (1959)
52. Padley and Sugden (1959)
53. Huldt and Lagerqvist (1951)
54. de Maria *et al.* (1960)
55. Hildenbrand and Murad (1970)
56. Grimley, Burns and Inghram (1961)
57. Huldt and Lagerqvist (1954)
58. Drowart, Colin and Exsteen (1965)
59. Friswell and Jenkins (1972)
60. Bulewicz (1958)
61. Colin, Drowart and Verhaegen (1965)
62. Behmenburg and Kohn (1964)
63. Porter, Chupka and Inghram (1955)
64. Balducci *et al.* (1972)
65. Hampson and Gilles (1971)
66. Drowart, Coppens and Smoes (1969)
67. Liu and Wahlbeck (1975)
68. Steiger and Cater (1975)
69. Murad and Hildenbrand (1975)
70. Darwent (1970)
71. Schofield (1967)
72. JANAF (1971)

73. Dougherty *et al.* (1971)
74. Ryabova, Khitrov and Gurvich (1972)
75. Cotton and Jenkins (1968)
76. Sugden and Schofield (1966)
77. Gorokhov, Gusarov and Panchevkov (1970)
78. Cotton and Jenkins (1969)
79. Kelly and Padley (1971)
80. Jensen (1970)
81. Bulewicz and Sugden (1956)
82. Gurvich, Novikov and Ryabova (1965)
83. Bulewicz and Sugden (1958)
84. Gurvich and Ryabova (1964a)
85. Jenkins and Sugden (1969)
86. Zeegers and Alkemade (1970)
87. McEwan and Philips (1967)
88. Gurvich *et al.* (1971)
89. Bulewicz and Sugden (1956a)
90. Neuhaus (1959)
91. Blue *et al.* (1963)
92. Ryabova and Gurvich (1965a)
93. Hildenbrand (1970)
94. Zmbov (1969)
95. Khitrov, Ryabova and Gurvich (1973)
96. Bulewicz, Philips and Sugden (1961)
97. Brewer, Somayajulu and Brackett (1963)
98. Schofield and Sugden (1971)
99. Brewer and Brackett (1961)
100. Guido, Gigli and Balducci (1972)
101. Hildenbrand (1968)
102. Ham (1974)
103. Ryabova, Gurvich and Khitrov (1971)
104. Murthy, Bagare and Murthy (1978)
105. Murthy and Bagare (1978)
106. Murthy and Prahllad (1978)
107. Frank and Krauss (1976)
108. van Veen, de Vries and de Vries (1979)
109. Balfour and Lindgren (1978)

Chapt. VIII   FORMATION OF METAL COMPOUNDS

Table VIII.1
Dissociation Energies of Metal Monoxides (in eV)

Molecule	Spectroscopic		Thermochemical		Flame equilibrium		Recommended D_0					Value adopted in Fig. VIII.2
	D_0^+	Ref.	D_0	Ref.	D_0	Ref.	Ref. 1	Ref. 2	Ref. 70	Ref. 71	Ref. 72	
AgO	2.5	1, 2					2	2				5.5
AlO	4.1; 5.2‡	1, 104; 8	5.0 – 5.2	12 – 15, 107	5.1 – 6.0	16 – 20	4.6	5.0	5.0		5.0	8.0
BO	7.4 – 9.3	1, 2, 3	8.3	21			8.0	8.3	8.1		8.3	8.0
BaO ⌀	6.7; 5.7‡	3, 9, 10	5.6 – 5.9	22 – 26	5.6 – 5.9	27 – 30	5.75	5.7	5.8	5.4↑		5.0↑
BeO ⌀	4.6; 5.1; 5.7	1, 4; 106; 3	4.6	4			4.6	4.2	4.6	4.4	4.6	5.0
BiO	3.4	1, 3	3.5	31	<3.8	32	3.1	3.7				4.0
CaO ⌀			3.6 – 5.0	22, 25, 33, 34	4.0 – 5.0	28 – 30, 32	4.3	3.6	4.8	3.8*		4.0↑
CrO	3.8; 5.3	1; 5	4.4; 4.7	4; 2	4.9; 5.3	35; 36		4.7	4.4			
CuO	2.8	6	2.8	37	<4.5	38		3.5			3.5	3.5
FeO ⌀	4.3; 5.0	5; 1	4.2	39, 40	4.2 – 4.4	35, 36, 41	4.3	4.1	2.5		4.2	4.3
GaO	2.8 – 3.0	1, 2	3.9	12	5.0	16	3.0	2.9	3.7			
InO	1.1 – 4.1	1, 2, 5	3.3	42	4.5	16	3.3	<3.3	8.1			
LaO	7 – 9.1	1, 3	8.2	43, 44	8.2	45	8.2	8.1	3.4		3.3	8.0
LiO ⌂			3.4; 3.5	4; 46	3.5	73	3.5	3.3	3.9	3.6*	4.0	3.5
MgO ⌀	1.7; 3.3	1; 105	3.5 – 4.8	47 – 49	4.0 – 5.2	35, 50, 51	4.1	3.4				4.5
MnO	4.5 – 4.8	1, 2			4.0 – 4.2	35, 52, 53	4.1	4.1				4.0

766

METHODS OF DETERMINING DISSOCIATION ENERGIES (D_0) AND VALUES REPORTED   Sect. VIII.2b

Species				Refs							
MoO	3.9 – 4.3					5.0				5.3	5.0
NaO†	6.0 ; 7.4						3.1			2.8	3.1
NiO	7.8 – 8.8						3.8				4.1
PbO	5.2 – 5.6		3.8 ; 4.0	54	3.9	3.9	3.8			8.2(?)	4.3
ScO		1	7.0	55		7.0	6.7				
SiO		5 ; 1	8.2	56			8.3				6.5
SnO		1, 3, 5	5.5	58 ; 4	4.0 ; 4.3	5.4	5.5				7.2
SrO‡	4.9†	1, 2, 5	3.7 – 4.9	43	7.1		4.0	5.6	4.1*		5.5
TiO	6.9	11	6.2 – 6.9	21		7.2	6.8	4.8		7.3	4.1ϕ
TlO		1		61			<3.2				
UO			7.8	22, 25, 33, 63	4.1 – 4.8	7.8	7.8				
VO	6.4	1, 3	6.4	64 – 67	<3.9	6.4	6.6				6.4
ZnO			2.8	54, 68		2.8	<2.8				
ZrO	7.8 ; 8.0	5 ; 1	7.8	4, 44, 69		7.8	7.8			7.0	

† Spectroscopic data are taken from linear Birge-Sponer extrapolations, except where stated otherwise.

‡ Data from appearance-threshold methods (with optical or mass-spectrometric detection).

† D_0 values converted for assumed $(Q_e)_{MO} = 4$.

‡ D_0 values converted for assumed $(Q_e)_{MO} = 1$, except where stated otherwise.

ϕ D_0 values converted for assumed $(Q_e)_{MO} = 6$.

* D_0 values converted for assumed $(Q_e)_{MO} = 3$.

Chapt. VIII   FORMATION OF METAL COMPOUNDS

Table VIII.3
Dissociation Energies of Metal Monohydrides (in eV)

Molecule	D_0	Method[+]	Ref.[*]
AgH	2.3 - 2.5	Sp, Th	1
AlH	2.9	Sp	1
AuH	3.1 - 3.7	Sp, Th	1
CuH	2.9	Sp, Th	1, 89
GaH	2.7 - 2.9	Sp	1, 90
InH	2.5	Sp	1
MgH	1.3	Sp	109
SiH	3.2	Sp	1
TlH	1.9; 2.0	Fl; Sp	83; 1

[+] Methods: Sp is spectroscopic;
            Th is thermochemical (nonflame);
            Fl is flame equilibrium.
[*] References: see Table VIII.1, page 765

Table VIII.2
Dissociation Energies of Metal Hydroxides (in eV)
(MOH → M + OH; M(OH)$_2$ → M + 2OH)

Molecule	D_0 (eV)	Method[+]	Ref.[*]	D_0 adopted in Fig. VIII.3
BaOH	4.6 4.7 - 4.9	Th Fl	24 29, 74, 75	4.7
Ba(OH)$_2$	9.0 9.0; 10.0	Th Fl	24 75; 76	
CaOH	4.1 - 4.5	Fl	29, 74, 75	4.3
Ca(OH)$_2$	8.8; 9.4	Fl	75; 76	
CsOH	3.7 3.9 - 4.0	E.I. Fl	77 78 - 80	4.0
CuOH	2.6	Fl	81	
GaOH	4.4 - 4.5	Fl	79, 82, 83	
InOH	3.7; 3.9	Fl	83; 82	
KOH	3.5 3.5 - 3.9	E.I. Fl	77 78 - 80	3.5
LiOH	4.5 4.4 - 4.6[#]	Th Fl	4 78-83, 84-87	4.52
MgOH	2.4 (3.5 - 4.0)	Fl estimated	51 30	
MnOH	3.2	Fl	52	3.3
NaOH	3.3 - 3.5	Fl	78 - 80	3.3
RbOH	3.6 - 3.8	Fl	78 - 80	3.6
SrOH	4.1 - 4.4	Fl	29, 75, 88	4.1
Sr(OH)$_2$	8.8; 9.3	Fl	75; 76	
TlOH	<3.0	Fl	16	

[+] Methods: Th is thermochemical (nonflame);
             Fl is flame-equilibrium;
             E.I. is electron impact.
[*] References: see Table VIII.1, page 765
[#] See also Table VIII.5, page 804

METHODS OF DETERMINING DISSOCIATION ENERGIES (D_0) AND VALUES REPORTED Sect. VIII.2b

Table VIII.4
Dissociation Energies of Metal Halides (in eV)*
$(MX \rightarrow M + M; \ MX_2 \rightarrow M + 2X)$

Metal	F D_0	F Ref.(Meth.)†,‡	Cl D_0	Cl Ref.(Meth.)	Br D_0	Br Ref.(Meth.)	I D_0	I Ref.(Meth.)	F_2 D_0	F_2 Ref.(Meth.)	Cl_2 D_0	Cl_2 Ref.(Meth.)
Ba	5.8 / 6.2	91(Th) / 92(Fl)	4.5 / 4.6	93,94(Th) / 74(Fl)	3.8; 4.4	95; 96(Fl)	*		12.1 / 12.1	97(Th) / 92(Fl)	9.4; 10.2 / 9.4; 9.6	93; 97(Th) / 98; 74(Fl)
Ca	5.4 / 5.9	91(Th) / 92(Fl)	4.1 / 4.1	93,94(Th) / 74(Fl)	3.3	95(Fl)	*		11.6 / 11.5	97(Th) / 92(Fl)	9.1; 10.2 / 9.3	93; 97(Th) / 74, 98(Fl)
Cs	5.0 – 5.2 / 5.3	4, 99(Th) / 96(Fl)	4.4 – 4.6 / 4.6	4, 99(Th) / 96(Fl)	4.2 / 4.3	99(Th) / 96(Fl)	3.6 / 3.6	99(Th) / 96(Fl)				
Cu	3.1*	1(Sp)	3.2 / 3.9	1(Sp) / 100(Th)	3.5 / 3.4	1(Sp) / 1(Th)	3.0	1(Sp)			9.8	100(Th)
Ga	6.0; 6.2	1; 4(Sp)	3.7; 4.9* / 4.9	1; 4(Sp) / 96(Fl)	4.3 / 4.4	4(Sp) / 96(Fl)	3.5 / 3.9	4(Sp) / 96(Fl)				
In	5.5 / 5.4	4(Sp) / 96(Fl)	4.5	4(Sp)	4.0	96(Fl)	3.4	96(Fl)				
K	5.1 / 5.1	96(Fl) / 99(Th)	4.4 / 4.4 / 4.3	4(Sp) / 99(Th) / 96(Fl)	3.9 / 3.90 / 3.9	96(Fl) / 108(Ph) / 99(Th)	3.33 / 3.3 / 3.3 / 3.4	108(Ph) / 4(Sp) / 94(Fl) / 96(Fl)				
Li	6.0 / 5.9	99(Th) / 96(Fl)	4.9 / 4.8; 5.1	99(Th) / 99; 16(Fl)	4.3 / 4.4	99(Th) / 96(Fl)	3.7 / 3.6	99(Th) / 96(Fl)				
Mg	4.6; 5.2 / 4.5; 4.7	1; 4(Sp) / 91; 101(Th)	3.5; 3.9 / 3.2; 3.4	1; 4(Sp) / 93; 102(Th)	3.2	1,4(Sp)						
Mn	3.9; 5.2	4; 1(Sp)	3.3 / 3.7	4(Sp) / 96(Fl)	2.9; 3.2 / 3.2	4; 1(Sp) / 96(Fl)	2.9	96(Fl)				
Na	4.9 / 5.3 / 5.2	99(Th) / 102(Sp) / 96(Fl)	4.23 / 4.26 / 4.23	99(Th) / 16(Fl) / 96(Fl)	3.8 / 3.8 / 3.77	99(Th) / 108(Ph) / 99(Th)	3.2 / 3.1 / 3.18	99(Th) / 96(Fl) / 108(Ph)				
Rb	5.0 / 5.2	99(Th) / 96(Fl)	4.4 / 4.4	99(Th) / 96(Fl)	3.9 / 4.0	99(Th) / 96(Fl)	3.3 / 3.5	99(Th) / 96(Fl)				
Sr	5.4 / 5.7	91(Th) / 30(Fl)	4.2 / 4.1	93(Th) / 88(Fl)	3.4	95(Fl)	*		11.6 / 11.5	97(Th) / 92(Fl)	9.0 / 9.1	93(Th) / 98, 103(Fl)
Tl	4.8 / 4.8	1, 4(Sp) / 96(Fl)	3.9 / 3.8	1(Sp) / 96(Fl)	2.9 / 3.4	1(Sp) / 96(Fl)	2.9	96(Fl)				

† Methods: Sp is spectroscopic; Th is thermochemical (nonflame); Fl is flame equilibrium; Ph is photofragmentation spectroscopy.
‡ References: see Table VIII.1, page 765.
* The dissociation energies of metal-monohalides summarized by Kondratiev (1974) are consistent within about 0.2 eV with most of the (mean) values listed in the table, except for CuF, CaI and SrI. For BaI, CaI and SrI this author quotes a dissociation energy of 2.8, 2.9 and 2.7 eV, respectively.

Chapt. VIII   FORMATION OF METAL COMPOUNDS

error intervals reported by the individual authors. A critical examination of the error limits would require a more thorough analysis of the methods used and the results obtained than can be afforded here. Whenever different mean D_0 values have been obtained by the same class of methods, only the total range of these values is indicated in the tables. No weight factors have been assigned to the individual results in these cases, although certain values might well be in error.

We have not attempted to attach a probable error interval to the final D_0 value adopted for the monoxides in Table VIII.1. The reliability of these values might be guessed, in first approximation, from the spread in values quoted from different sources.

Common systematic errors might occur in D_0 for some classes of compounds if in the calculation of D_0 from experimental observations common uncertain data have been used. This situation arises for the alkali hydroxides, where a value of D_0 for the dissociation of H_2O into $H+OH$ is used. Any new 'adjustment' of this value would shift all alkali hydroxide values by the same amount.

As an exception, we have mentioned in Table VIII.4 more precise values of D_0 for NaCl, yielding a mean value of 4.24 ± 0.01 eV. This mean value in particular is firmly established now and plays a role in the diagnostics of flame radicals by the NaCl-method, as is discussed in Sect. IV.5b-2. The same applies, for similar reasons, to the individual and mean D_0 values listed for LiOH in Table VIII.5; if LiOH is assumed to be a linear molecule, the weighted average value of D_0 is 4.52 ± 0.05 eV.

3. KIND AND ABUNDANCE OF METAL COMPOUNDS

Gaseous metal compounds occurring above the primary combustion zone of flames are usually of the simple di-atomic or tri-atomic type. Some species among them, such as CaOH, have free valencies and are not stable at ordinary temperatures. Under certain conditions more complex molecules such as $Ca(OH)_2$ may be found in noticeable concentrations. Generally, complicated metal salts, say Na_2SO_4, or organo-metallic compounds are not stable in the flame and will be quickly decomposed or burnt to smaller fragments. However, the persistence of some salts has been observed mass-spectrometrically, e.g., K_2VO_3, K_2CrO_4, K_2MoO_4, K_2WO_4, if a large amount of metal salt(s) is introduced into the flame in the form of a vapour (see Farber and Srivastava 1973, 1973a). Equilibrium calculations indicate that, e.g., H_2WO_4, WO_3 and WO_2 may occur in large fractions in fuel-lean $C_2H_2-N_2O$ flames (see Rasmuson, Fassel and Kniseley 1973, 1976). The occurrence of dimers such as Na_2 or Sr_2O_2 is most unlikely because their dissociation energy and the metal concentration in the flame are relatively low, even when high solution

KIND AND ABUNDANCE OF METAL COMPOUNDS   Sect. VIII.3a

concentrations are used. This has been corroborated by experiments in which the
emissions of molecular bands as well as atomic lines were studied as a function of
solution concentration up to high concentrations (see, for example, Lagerqvist and
Huldt 1954, 1955, James and Sugden 1955a, Huldt and Knall 1956, Hollander 1964,
Sugden and Schofield 1966; see also Sects V.2e and V.6).

In this section we present a survey of the metal compounds that may be
found in flames into which metal salt solutions are sprayed. The influence of the
flame conditions on the formation of some important species will be shown quantitatively by means of diagrams. Experimental evidence on the kind and extent of
compounds formation we shall be described. We shall discuss in succession the
metal oxides, hydroxides, hydrides, and halides.

3a   METAL OXIDES

Monoxides are the most common di-atomic metal compounds that are found in
flames burning with air, pure oxygen, nitrous oxide, and so on. In particular,
elements such as La, U, Sn, Ti will tend to form stable refractory monoxides.
Alkali elements and Cu, Ag and Au are virtually free from oxide formation (see
Sect. 4b for some exceptions).

The extent of monoxide formation in equilibrium at temperature T can be
theoretically evaluated by applying the mass-action-law, Eq. VIII.1. In this equation the equilibrium concentration $[O]_e$ of free oxygen atoms should be substituted
for $[X]$. The dissociation constant follows from Eq. II.44 if the dissociation
energy D_0 and all relevant molecular and atomic constants are known. Using the D_0
values listed in the last column of Table VIII.1 and the molecular constants tabulated by Herzberg (1950) and Cottrell (1958), we have calculated the reciprocal
association factor $\phi^{-1} \equiv [M]/[MO]$ as a function of T, with $[O]_e$ as parameter.
The results obtained are shown in Fig. VIII.2 for a number of elements. The electronic partition functions of the metal atoms were derived from de Galan, Smith and
Winefordner (1968). Whenever the designation of the molecular ground state was
doubtful, the electronic partition function of the molecule was chosen in consistency with the value adopted in Table VIII.1. The ϕ values do not have a high
accuracy, since the error in the adopted D_0 value might be considerable. An error
of 0.5 eV in D_0 would correspond to an error of one order of magnitude in ϕ at
$T = 2500$ K. A minimal error of about a factor two should therefore be attached to
all values of $[M]/[MO]$ plotted in Fig. VIII.2. The data in Fig. VIII.2 are representative for the fraction atomized β_a only if the oxide is the dominant compound.
For instance, the value of $[M]/[MO]$ for β_a appears to be one order of magnitude
smaller than that for Al, but the detection limit of Al in analytical atomic flame
spectroscopy is markedly worse (see de Galan and Winefordner 1967). This might be

771

Chapt. VIII FORMATION OF METAL COMPOUNDS

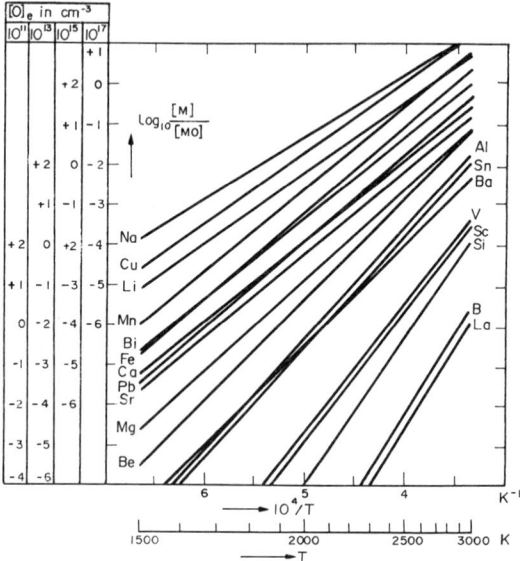

Fig. VIII.2 Diagram showing concentration ratio $[M]/[MO]$ between free metal atoms and monoxides, calculated as a function of temperature, with the equilibrium concentration $[O]_e$ of free oxygen atoms as parameter. The ordinate scale depends on $[O]_e$ as indicated. Dissociation energies D_0 adopted in the calculations are given in Table VIII.1. When one adopts the mean recommended values $D_0 = 5.5$, 4.5 and 8.2 eV listed in this table for BaO, BeO and SiO, respectively, one obtains for $\log_{10}[M]/[MO]$ at $[O]_e = 10^{17}$ cm^{-3}:

	Ba	Be	Si
$T = 2000$	− 7.72	− 4.12	− 12.5
$T = 3000$	− 3.17	− 0.37	− 5.62

KIND AND ABUNDANCE OF METAL COMPOUNDS   Sect. VIII.3a

explained by the presence of other Al compound(s) [probably $Al(OH)_2$, see Newman
and Page 1971 and Jensen and Jones 1972; or AlOH, see Farber and Srivastava$^+$].
Also the low volatility of aluminium oxide particles may play a part.

Inspection of Fig. VIII.2 together with Table VIII.1 reveals that in general
metal oxides with a dissociation energy below 4 eV are substantially dissociated in
normal flames with T ranging from 2000 - 2500 K, $[O]_e$ being of the order of 10^{15}
cm^{-3}. The oxides of Na and Ag which were introduced in Sect. V.6 as fully atomized standard elements, clearly belong to this class. Metal oxides with D_0 exceeding 6 eV are hardly dissociated in normal flames with T below 2500 K. It is only
in fuel-rich, hot C_2H_2-N_2O flames that a noticeable fraction of these elements may
exist in atomic form (see Rasmuson, Fassel and Kniseley 1973, 1976).

Even in the interconal zone of fuel-rich, premixed oxy-acetylene flames the
degree of dissociation for the monoxides of B, Hf, Ta and Th (with D_0 values
around 8 eV) seems not to be sufficient for good detectability in atomic absorption
analysis (see Fassel and Golightly 1967). Better conditions prevail in the fuel-
rich, hot C_2H_2-N_2O flame (except for Th; see textbooks cited in Chapt. I).

We note that the ϕ values for MnO and CaO differ by one order of magnitude, although for both molecules a same value of D_0 = 4.0 eV is used. This illustrates that not only D_0, but also other molecular and atomic parameters can play
an important role.

Obviously the *flame temperature* is an important parameter in controlling
the degree of dissociation in equilibrium. A temperature variation of a factor two
may change ϕ by several orders of magnitude, as is seen from Fig. VIII.2. The
larger D_0 the stronger the temperature influence. However, the temperature influences not only the dissociation constant of the metal oxide, but also that of
O_2, CO_2 and H_2O, and consequently a rise in T may affect ϕ through a variation
of both K_d and $[O]_e$.

Let us consider, for example, a family of oxygen-rich flames with different
temperatures but essentially constant $[O_2]$ values. In a plot of $\log([M]/[MO])$
versus the reciprocal temperature of these flames, the slope of the curves obtained
would correspond to $[D_0(MO) - \tfrac{1}{2} D_0(O_2)]$ rather than to $D_0(MO)$. This holds because
the increase of $[O]$ in the hotter flames partly compensates for the increase of the
dissociation constant of MO with temperature. We note that $\tfrac{1}{2} D_0(O_2)$ = 2.56 eV (see
JANAF 1971), which is less than $D_0(MO)$ for most metals considered here.

$^+$ Wittenberg, Haun and Parsons (1979) have recently calculated the equilibrium
fractions of 10 Al species [including ions but not $Al(OH)_2$] in C_2H_2-N_2O flames
at varying fuel/oxidant ratios. They applied the method of minimization of the
free enthalpy (cf. Sect. IV.4b).

For similar reasons a reduction of T down to 1500 K did not result in the formation of an appreciable amount of NaO in the oxygen-rich flames studied by McEwan and Phillips (1966). Using the equilibrium O_2 concentration calculated by these authors and taking into account the temperature dependence of the dissociation constant for $O_2 \rightleftharpoons 2O$ we calculated $[\text{NaO}]/[\text{Na}]$ for these flames. It appears that this ratio remains smaller than unity when T is varied from 2000 to 1500 K while the partial pressure of O_2 in the flame is kept equal to 0.1 atm.

The influence of temperature on the metal oxide dissociation and possibly also on the volatilization of the solid particles explains the great importance of premixed, hot $C_2H_2-N_2O$ flames in analytical flame spectroscopy (see Rasmuson, Fassel and Kniseley 1973, 1976). Besides, the reducing conditions in these flames are a favourable factor (see below).

The *atomic oxygen concentration* $[O]_e$ may vary by about six orders of magnitude among flames of practical interest, not only because of varying flame temperature but also because of varying fuel-to-oxidant ratio (see Figs IV.2a-f on pages 429-31). Fuel-rich or even incandescent hydrocarbon-O_2 flames are often recommended in flame spectroscopy since they yield a sufficiently large proportion of free metal atoms. The strong dissociation in these flames and possibly also chemiluminescent excitation largely improve the detection limits in emission measurements of elements such as Sn and Mo (see Gilbert 1963). These flames are also suitable for the study of atomic spectra of rare-earth elements (see Fassel, Curry and Kniseley 1962). Fuel-rich premixed $C_2H_2-O_2$ and $C_2H_2-N_2O$ flames have been applied to enhance the dissociation of stable oxides of refractory elements (see Kniseley, D'Silva and Fassel 1963, de Galan and Samaey 1970, and Rasmuson, Fassel and Kniseley 1973, 1976).

The favourable reducing conditions in the latter flames exist especially near the primary combustion zone. The atomic metal concentration often appears to drop sharply with increasing distance from this zone. Then the detectability of the element concerned is sensitively dependent on the place of observation in the flame. Special optical means are sometimes provided to select the desired area of observation.

From a theoretical point of view it is not quite clear whether the increase in extent of oxide formation with increasing height in these flames is actually due to a shift in chemical equilibrium. This shift could be caused by a rise in free oxygen concentration or a fall in temperature when the flame gases move upwards. An alternative explanation could be that dissociation equilibrium is not attained at all shortly above the combustion zone. Free metal atoms and flame radicals may be formed in large excess in the combustion zone and could persist for some time because of slow recombination. The actual chemical processes in and near the combustion zone of these flames are still uncertain. Nevertheless there is a strong evidence that even stable metal oxides are often strongly dissociated in and near this zone. Probably the reducing effect of radicals like C, C_2, CN in excess concentrations may play an important part (see Sect. 4b).

If $[O]$ exceeds $[O]_e$, as is often the case shortly above the combustion zone, we cannot derive the actual ϕ value by simply substituting $[O]$ for $[O]_e$ in Fig. VIII.2. This would only give the right answer if the formation reaction

$$M + O (+X) \rightarrow MO (+X)$$

is in partial equilibrium with the reverse dissociation reaction. However, if MO is predominantly formed by other reaction paths including other flame constituents, no direct relation between the metal dissociation and excess $[O]$ need exist (see Sect. 4).

A number of authors have studied the degree of dissociation of metal oxides in different flames and from different points of view. In some studies the determination of dissociation energies was the primary object (see also Sect. 5). In other papers dissociation was considered mainly as a correction for other effects or in relation to analytical applications (see de Galan and Samaey 1970, Koirtyohann and Pickett 1971, and Rasmuson, Fassel and Kniseley 1973, 1976).

Certainly not all observations reported are satisfactorily explained. For example, the remarkable, abnormal dissociation of SnO in H_2-air flames observed by Slavin (1966) still awaits further elucidation (see also Rubeška 1973).

Alkali superoxides such as NaO_2 have been found in cool oxygen-rich H_2 flames at $T \leqslant 2000$ K by Kaskan (1965), McEwan and Phillips (1967), and McEwan and Phillips (1966). Equilibrium concentrations seem to be attained within 1 ms only for NaO_2 at $T \geqslant 2000$ K. At a free-O_2 pressure of 0.05 atm in the flame and at $T = 2000$ K, the equilibrium ratio $[Na]/[NaO_2]$ is about unity. Under those conditions Na cannot be regarded as a fully atomized reference element (see also Sect. V.6). The stability of KO_2 and CsO_2 is greater than that of NaO_2, but LiO_2 appears to be less stable.

3b METAL HYDROXIDES AND HYDRIDES

Hydroxides are the most important poly-atomic metal compounds found in flames fed by hydrogen or hydrocarbons. Even in 'dry' CO flames where the gases are purified and dried, and dry Cu salt is directly evaporated into the flame without nebulization, some CuOH band emission still persists (see Hollander 1964).

Hydroxide formation in flames has been studied for a large number of metals. Among them the alkali and alkaline-earth metals have attracted particular attention. The hydroxides of the former group of elements do not emit visible or u.v. bands; the monohydroxides of the latter group and also CuOH do emit useful bands.

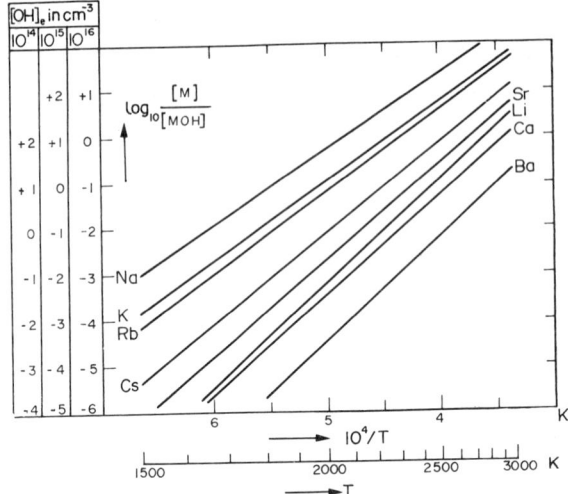

Fig. VIII.3 Diagram showing concentration ratio [M]/[MOH] between free metal atoms and monohydroxides, calculated as a function of temperature, with the equilibrium concentration $[OH]_e$ of free hydroxyl radicals as parameter. The ordinate scale depends on $[OH]_e$ as indicated. Dissociation energies adopted in the calculation are given in Table VIII.2.

Calculated equilibrium ratios $\phi^{-1} \equiv [M]/[MOH]$ for the monohydroxides of Li, Na, K, Rb, Cs, Ca, Sr, and Ba are shown in Fig. VIII.3 as a function of T and for various assumed equilibrium concentrations $[OH]_e$. Use was made of the dissociation energies listed in the last column of Table VIII.2. The results are plotted in Fig. VIII.3 for $[OH]_e = 10^{14}, 10^{15}$, and 10^{16} cm^{-3}, respectively. ϕ^{-1} as a function of $[OH]_e$ and T was calculated from the quotient of the equilibrium constant K_1 for reaction VIII.12 (see Sect. 4a) and that of the dissociation reaction: $H_2O \rightleftharpoons H + OH$. The value of K_1 was derived from the statistical-mechanical expression in Eq. II.44 with the help of the molecular constants quoted by Kelly and Padley (1971) for the alkali hydroxides, and those quoted by Gurvich, Ryabova, Khitrov and Starovoitov (1971) and by Ryabova, Khitrov and Gurvich (1972) for the alkaline-earth hydroxides. The equilibrium constant for the H_2O dissociation reaction was calculated from thermodynamic data given by Gaydon and Wolfhard (1970).

The discussion of the influence of T and $[OH]_e$ on the dissociation of the hydroxides in analogous to that given above for the dissociation of oxides. The range of variation of $[OH]_e$ among flames of practical interest amounts to three orders of magnitude and is thus much smaller than that for $[O]_e$ (see Figs IV.2a-f on pp.429-31). An increase of T at constant fuel-to-oxidant ratio affects not only the metal dissociation constant, but also $[OH]_e$. Therefore NaOH and KOH have been found to be relatively more important in the hotter, fuel-rich H_2 flames than in the cooler ones (see James and Sugden 1955, and Jensen and Padley 1966).

If H, OH, and O are present in excess over their equilibrium concentration, the plots in Fig. VIII.3 cannot be used to derive ϕ values. The actual hydroxide formation and dissociation reactions must then be considered (see Sect. 4). The *alkali monohydroxides* are the only important molecular compound of this group of elements in hydrogen-containing flames if we disregard halide formation.[†] Their association factors ϕ can be determined indirectly by the comparison method as explained in Sects V.3 and V.6, or can be calculated if the flame gas composition and temperature are known. Equilibrium constants have been determined by Jensen and Padley (1966), Cotton and Jenkins (1969), Zeegers and Alkemade (1970), and Kelly and Padley (1971). In all cases Li is found to yield the highest ϕ values, which may often be of the order 10. The order of stability found for the alkali hydroxides generally agrees with that expected from Fig. VIII.3. Apparent exceptions to this rule were occasionally found for K, Cs, and Rb (see, for example, Hofmann, Kohn and Schneider 1961) but might be explained by secondary effects, such as ionization.

The influence of excess flame radicals on the dissociation of alkali hydroxides can be evaluated by anticipating from Sect. 4b the actual dominant formation reaction in H_2- and hydrocarbon flames: $M + H_2O \rightleftharpoons MOH + H$. This reaction is balanced, and thus a partial equilibrium is set up between $[M]/[MOH]$ and $[H]/[H_2O]$. Since H_2O is a major species in these flames, its concentration is practically not affected by the deviations of the radical concentrations from their equilibrium values. Therefore, $[M]/[MOH]$ will be shifted to larger values if $[H]$ exceeds $[H]_e$. Usually $[H]$ decays to $[H]_e$ within a few milliseconds and so does $[M]/[MOH]$. The corresponding decrease of atomic line emission or absorption has frequently been observed for Li.

Direct evidence about the occurrence of *alkaline-earth monohydroxides* has been obtained from observations of the corresponding spectral bands[*][‡] (see van der Hurk, Hollander and Alkemade 1973). When spraying Mg solutions into H_2 flames, Bulewicz and Sugden (1959) have found a pronounced band system extending from 3700 to 4000 Å and have assumed the emitter to be MgOH. Indirect determination of the MOH content by the comparison method is difficult, because alkaline-earth metals also form oxides and dihydroxides. It is only in cases where MOH is the dominant compound that the comparison method may yield reliable results. The equilibrium MOH content may be calculated if the contribution of oxides and dihydroxides to

---

[†] In dry, O_2-rich $CO-O_2-N_2$ flames, however, appreciable amounts of LiO are also formed (see Dougherty, McEwan and Phillips 1973).

[*] Flame spectra of BaO + BaOH, CaOH and SrOH bands in the visible region are presented in Appendix A.8.

[‡] The conclusion drawn by Reid and Sugden (1962) about the suprathermal character of the alkaline-earth monohydroxide emission bands was based on the assumption of too low a value for the dissociation energies (see Cotton and Jenkins 1968).

Chapt. VIII FORMATION OF METAL COMPOUNDS

the total compound formation is either small or is calculated too (see Cotton and Jenkins 1968, Gurvich, Ryabova, Khitrov and Starovoitov 1971, Ryabova, Khitrov and Gurvich 1972, Kalff and Alkemade 1973, and van der Hurk, Hollander and Alkemade 1973).

As in the case of alkali hydroxides, the actual, balanced formation reaction has to be considered when discussing the effect of excess flame radicals on the dissociation of alkaline-earth monohydroxides. In Sect. 4b we shall see that the same type of reaction: $M + H_2O \rightleftharpoons MOH + H$ is operative here. A similar discussion as given above for LiOH applies here. There may be, however, one important difference. If MO occurs as a major species, then it buffers the variation of [M] with height that is brought about by the dependence of [MOH] on the varying excess H concentration.

In most flames containing hydrogen, Ca- and Sr monohydroxides are more abundant than their oxides. Even in $CO-N_2O$ flames with a low H_2O content, [CaOH] is higher than [CaO], and [SrOH] is about equal to [SrO] (see Kalff and Alkemade 1973). BaOH is usually less abundant than BaO in almost all flames. We refer to van der Hurk (1974) for a general survey of equilibrium concentrations of the (hydro-)oxides of Ca, Sr and Ba in various flames.

The stability of *Tl-*, *In-*, *and Ga hydroxides* increases in this order. From spectrometric measurements of the atomic line intensities of these elements, made as a function of T and flame composition by Bulewicz and Sugden (1958) and Kelly and Padley (1971), it was concluded that compound formation of Tl is negligible and that of Ga is strong; In takes an intermediate place in H_2 flames. No useful bands of these compounds are found in the flame spectrum, but these authors assumed that hydroxides were formed. Gurvich and Veits (1958), who did the same kind of experiments, interpreted their results under the assumption of In- and Ga-oxide formation. However, their value for D_0(GaO), i.e. 5.0 ± 0.13 eV, is not consistent with the value (3.92 ± 0.15 eV) determined mass-spectrometrically by Burns (1966) in a hydrogen-free milieu. Furthermore, evaluating the data reported by Kelly and Padley (1971) under the assumption of GaO formation, we calculated D_0(GaO) to be equal to about 8.7 eV, which is unreasonably high. Therefore, it is very likely that in fact GaOH is formed. By analogy, formation of InOH is more probable than InO. Formation of InOH and GaOH is governed by the same reaction as that for the alkali hydroxides (see Sect. 4b). Thus similar effects of excess flame radicals on the degree of dissociation are to be expected here.

Hydroxides of other elements such as Cu, Mn and Sn have been assumed to persist in hydrogen flames. These assumptions have not been satisfactorily proven so far.

The copper emission bands at 5350-5550 Å and at 6150-6250 Å have been ascribed to thermal emission of CuOH (see Bulewicz and Sugden 1956). This suggests that at least some CuOH may persist in H_2 flames. However, Reid and Sugden (1962) have ascribed these emission bands to chemically excited CuOH*, being formed as an intermediate in a chain of nonequilibrated reactions which lead ultimately to Cu and CuO (see also Sugden 1962). In their opinion, persistence of ground-state CuOH is not necessary to explain CuOH emission in H_2 flames. Both studies reveal that at most 1% of the Cu atoms, if any, is bound to CuOH.

When spraying Mn solutions into H_2 flames, Padley and Sugden (1959) have found a system of diffuse bands extending from 3500 to 4200 Å. They have ascribed these bands to thermally excited MnOH, suggesting persistence of ground-state MnOH. Reid and Sugden (1962) have ascribed these bands to chemi-excited MnOH* formed as an intermediate. Again, persistence of ground-state MnOH is not conclusively proven. The ratio [Mn]/[MnOH] is at least 100.

Bulewicz and Padley (1971) have studied the flame chemistry of Sn and attributed the diffuse bands in the region 4700-5100 Å to chemically excited SnOH*. They concluded that ground-state SnOH did not persist in their hydrogen-rich flames. They did observe that SnO was the predominant Sn compound.

In H_2 flames *alkaline-earth dihydroxides* may become dominant over the oxides and monohydroxides if the H_2 flow rate is chosen about four times as large as the O_2 flow rate. Only for Ba the oxide concentration will then still be of comparable magnitude as the dihydroxide concentration (see Sugden and Schofield 1966, and Cotton and Jenkins 1968). From the formation reaction of $M(OH)_2$ (see Sect. 4b) the atomic metal concentration is expected to decrease with increasing height, when the concentrations of the excess flame radicals decay with increasing height. This behaviour has been observed by Sugden and Schofield (1966) for Ca, Sr, and Ba in H_2 flames where dihydroxides were the dominant species.

Metal hydrides, which usually have low dissociation energies (see Table VIII.3 on page 768), are in general unimportant. The occurrence of AgH, AuH, CuH, PdH, PtH, SnH, and TlH can, however, be detected by their emission bands. The spectra of PtH and PdH in C_2H_2-air flames disappear at short distance above the combustion zone, probably because these molecules are completely dissociated in full equilibrium (see Mavrodineanu 1965, and Bulewicz and Padley 1971). Bulewicz and Sugden (1958) have studied the variation of CuH- and TlH-band intensities with temperature and distance from the combustion zone. Their conclusions about the formation reaction of these hydrides will be dealt with in Sect. 4b.

3c   METAL HALIDES

Metal halides may occur if halogen gas or a halogen compound is supplied to the flame (see Sect. III.3b) or if a special hydrogen-halogen flame is used.

Chapt. VIII   FORMATION OF METAL COMPOUNDS

Usually the halogen content in the flame is relatively small, say about 1% or less, so that the flame conditions will hardly be altered by the supply of halogen. An exception should perhaps be made for oxygen-rich H_2 flames with their low atomic H concentration, where addition of 0.1% of chlorine could disturb the flame appreciably (see McEwan and Phillips 1965). Halogen (X) is present in the flame mainly as free atoms X and as HX. X_2 molecules and metal halides occur usually in minor proportions.

The occurrence of *metal monohalides* MX can be indirectly concluded from the linear decrease of atomic metal emission with increasing total concentration $[X]_t$ of supplied halogen. For alkaline-earth metals the formation of MX can be concluded directly from the appearance of halide bands in the flame spectrum. Data of band spectra can be found in Rosen (1970). Dissociation energies of some important metal halides are listed in Table VIII.4 on page 769). D_0 values for Cr-, Mn-, and Ni halides have been given by Bulewicz, Phillips and Sugden (1961). In the table, certain regular trends in the stability of the monohalides for a given metal or for a given halogen are readily recognized. For example, all K halides are systematically more stable than the corresponding Na halides, but less stable than the Li halides. Further, all alkali fluorides, chlorides, bromides, and iodides show a systematically decreasing stability in this order.

The significance of the state of halide dissociation will be shown for NaCl. In Fig. VIII.4 the measured ratio I/I' of the Na resonance line intensities in the absence and in the presence of halogen, respectively, is shown as a function of total halogen concentration $[X]_t$. A H_2 flame with $T = 2215$ K was used. Since self-absorption was negligible, the reciprocal ratio I'/I is proportional to the reduction in atomic Na concentration owing to halide formation.

It is seen that a reduction $I'/I = 0.5$, corresponding to $[Na]/[NaCl] = 1$, is obtained for $[Cl]_t \approx 5 \times 10^{-3}$ atm. On the other hand, using the molecular constants of Bulewicz, Phillips and Sugden (1961), one calculates from Eqs II.43 and 44 that $[Na]/[NaCl] = 1$ for atomic $[Cl] = 2.9 \times 10^{-5}$ atm. This value is considerably lower than the total chlorine concentration obtained from Fig. VIII.4. This simply shows that a large proportion of chlorine in H_2 flames is tied up as HCl, which buffers the Na depression by halogen. This strong reduction of free halogen atoms in H_2- and hydrocarbon flames owing to the formation of halogen hydride also explains why the alkali emission in analytical flame spectroscopy is hardly depressed when 1 mol/l HCl is added to the sprayed solution (see Alkemade 1954, Kropp 1960, Püschel 1962, and Grove, Scott and Jones 1965). Other halogens in the solution are not expected to influence the alkali emission either. Direct evidence of the buffering effect of HCl has been given by Hollander (1964).[†] He found that

---

[†] Mandelstam (1939) has earlier explained the absence of any influence of a 1% NaF solution on the Rb emission by the formation of HF molecules in the flame.

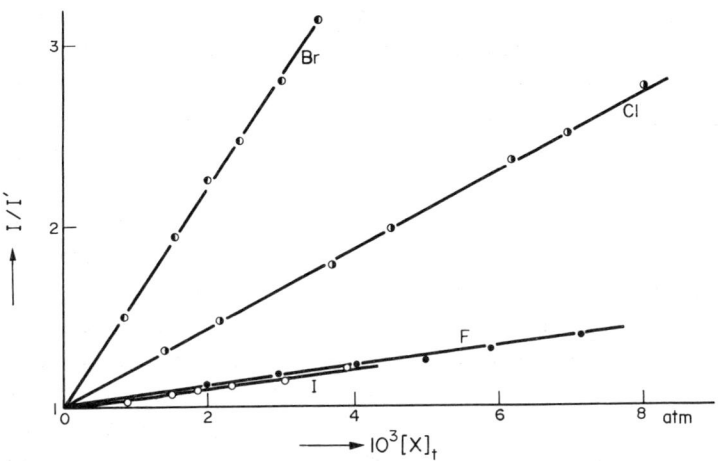

Fig. VIII.4 The ratio I/I' of the Na line intensities measured in the absence and in the presence of halogen, respectively, is shown as a function of total halogen concentration $[X]_t$ in the flame. Observations were made at 3 cm distance from the combustion zone of a H_2-O_2-N_2 flame by Bulewicz, Phillips and Sugden (1961).

the strength of CuCl bands was significantly smaller in a C_3H_8-air flame than in a CO-air flame under comparable conditions.

This *buffering effect of halogen hydride*, HX, depends on the stability of HX and will thus be different for different halogens. The stability of hydrogen halides can be estimated from the dissociation energy, which amounts to 5.8 eV for HF, 4.43 eV for HCl, 3.75 eV for HBr, and 3.1 eV for HI.

Differences in the buffering effect of HX also explain why the slopes of the straight lines in Fig. VIII.4, corresponding to different halogens, are not in the same order as the stabilities of the metal halides. The steepest slope would be expected with fluorine as it gives the most stable metal halide. But this is compensated for by the strong tendency of fluorine to be bound as HF.

Let us assume that full chemical equilibrium exists and that the halide is the only metal compound formed. From the mass-action-law we then have

$$[M][X] / [MX] = K_3 \qquad (VIII.5)$$

and

$$[H]_e [X]/[HX] = K_4 , \qquad (VIII.6)$$

where $[H]_e$ denotes the equilibrium concentration.

Using the mass-balance equations

$$[M]_t = [M] + [MX] = [M]_0 , \qquad (VIII.7)$$

Chapt. VIII   FORMATION OF METAL COMPOUNDS

where $[M]_0$ is the metal concentration in the absence of halogen, and

$$[X]_t = [X] + [HX] + 2[X_2] + [MX] \; , \qquad (VIII.8)$$

and neglecting the last two terms in the right-hand side of Eq. VIII.8 we find

$$[M]_0 / [M] = 1 + ([X]_t/K_3)/(1 + [H]_e/K_4) \; . \qquad (VIII.9)$$

In fact $[M]_0/[M]$ equals I/I', which is plotted as a function of $[X]_t$ in Fig. VIII.4. The slope of this plot is thus found to depend on $[H]_e/K_4$ as a consequence of HX formation. The factor $(1 + [H]_e/K_4)^{-1}$ in Eq. VIII.9 equals the ratio of free atomic halogen concentration $[X]$ to total halogen concentration $[X]_t$. In a fuel-rich H_2 flame at $T \approx 2200$ K, $[H]_e/K_4$ is of the order of magnitude 10^{+4} for F, 10^{+2} for Cl, 1 for Br, and 10^{-2} for I. This result explains why the slope of the curves for F and Cl in Fig. VIII.4 are strongly depressed with respect to those for Br and I.

Excess H radicals influence the degree of metal halide dissociation in a way that is determined by the actual formation reaction of MX and HX discussed in Sect. 4b.

Formation of other metal compounds may also have a buffering effect on the removal of free metal atoms by halide formation. This *buffering effect due to metal compound formation* can be evaluated as follows. When the metal forms also compounds with flame constituents, we should replace Eq. VIII.7 by

$$[M]_t = [M] + [MZ] + [MX] = [M]_0 + [MZ]_0 \; , \qquad (VIII.10)$$

where MZ is either an oxide, hydroxide or hydride. According to the mass-action-law, a constant ratio $[MZ]/[M] = [MZ]_0/[M]_0$ ($\equiv \phi'$) is maintained independently of the addition of halogen, since the concentration of the flame partner Z is independent of it. We then simply find, instead of Eq. VIII.9,

$$[M]_0/[M] = 1 + ([X]_t/K_3)(1 + [H]_e/K_4)^{-1}(1 + \phi')^{-1} \; . \qquad (VIII.11)$$

The slope of I/I' ($= [M]_0/[M]$) as a function of $[X]_t$ is now reduced by a factor $(1 + \phi')$. This factor may readily be of the order of 10 for LiOH in H_2 flames (see also Fig. VIII.3 on page 776). Consequently the atomic Li concentration is markedly less affected by halogen than the atomic Na concentration, although the halides of the former element are more stable (see Bulewicz, Phillips and Sugden 1961).

This buffering effect can be easily visualized if one realizes that free metal atoms removed by halide formation must partly be supplemented by further dissociation of, say, metal hydroxide molecules. For, in this way a <u>constant</u> ratio of metal atoms to hydroxide molecules is maintained. Most of the metal that is bound as halide is thus ultimately supplied by dissociation of metal hydroxide, if the

FORMATION REACTIONS OF METAL COMPOUNDS   Sect. VIII.4

latter is a dominant species.

A similar buffering action as depicted above exists when the metal atoms are partly ionized. It has been shown that this case can be handled by adding a factor $(1+[M^+]/[M])$ to the dissociation constant K_3 in the final formula (see Alkemade 1954). The formation of Cl^- ions could, however, interfere with the metal ionization and complicate matters.

Alkaline-earth hydroxychlorides (MOHCl) are supposed to be formed in fuel-rich H_2-O_2-N_2 flames (see Gurvich, Ryabova, Khitrov and Starovoitov 1971, Ryabova, Gurvich and Khitrov 1971, and Ryabova, Khitrov and Gurvich 1972). In studying the dissociation energy of metal chlorides using the method outlined above for alkaline halides, they have found dissociation energies which are higher than mass-spectrometric values. This discrepancy can be explained if alkaline-earth metals form at least one more compound containing one Cl atom. Such compounds may be MOHCl, MHCl and MOCl, of which MOHCl was supposed to be the more probable one (see Ryabova, Khitrov and Gurvich 1972). A plot of I/I' as a function of $[Cl]_t$ (compare Fig. VIII.4 on page 781) is still a straight line. Assuming that MOHCl is formed, then the slope of the plot depends not only on $D_0(MCl)$, but on $D_0(MOH-Cl)$ as well. In fact the last term in Eq. VIII.9 has to be multiplied by a factor $(1+[MOHCl]/[MCl])$.

From data reported by Gurvich, Ryabova, Khitrov and Starovoitov (1971) and by Ryabova, Gurvich and Khitrov (1971) we calculated the ratio $[SrOHCl]/[SrCl]$ to range from 10 to 60 for some fuel-rich H_2-O_2-N_2 flames with temperatures ranging from 1860 to 2210 K.

Alkaline-earth dihalides may be formed at high halogen concentrations, say at $[Cl]_t \geqslant 10^{-2}$ atm, in H_2 flames. If a plot of I/I' versus $[X]_t$ has a parabolic form in the range of high $[X]_t$ values, instead of the linear form shown in Fig. VIII.4, the formation of compounds like CaF_2 or $BaCl_2$ is likely. Measurements of this kind have been made by Ryabova, Gurvich and Khitrov (1971), and by Ryabova, Khitrov and Gurvich (1972) to determine the dissociation energies of alkaline-earth hydroxyhalides in the linear portion of the plot, and those of the dihalides in the parabolic portion.

Zhitkevich *et al.* (1963) have found that upon further increasing the halogen supply to the flame, the band intensities of the alkaline-earth monohalides decreased again after an initial rise. This has been explained as due to formation of dihalides at the cost of the monohalides at high halogen concentrations.

4. FORMATION REACTIONS OF METAL COMPOUNDS

Here we discuss the reactions through which the molecules described in Sect. 3 are actually formed. In particular, the rate constants of these reactions are of

Chapt. VIII   FORMATION OF METAL COMPOUNDS

interest because the establishment of dissociation equilibrium depends on them. Our discussion will be restricted to flames burning at 1 atm pressure, but the considerations given below can easily be extended to flames at lower pressure. Also, our discussion will be restricted mainly to reactions occurring above the combustion zone, since much less is known with certainty about reactions of metal species in this zone.

Knowledge of the actual formation reactions enables us to predict to what extent deviations from chemical flame equilibrium will affect the dissociation of metal compounds. This information is necessary to determine dissociation energies from the observed degree of dissociation, as will be described in Sect. 5. Conversely, knowledge of the specific formation reactions enables us to determine excess concentrations of flame radicals from the observed degree of dissociation, as was shown in Sect. IV.5. Of course, a study of metal reactions is also of interest as a subject in itself, as it enlarges our knowledge of the reaction kinetics of simple atoms and molecules at elevated temperatures.

We shall discuss in Sect. 4a the general kinetic aspects of the formation and dissociation of metal compounds and the general methods used to obtain experimental evidence about these reactions. This will be followed in Sect. 4b by a more detailed survey of the actual formation reactions and partial equilibria that have been found for some metal oxides, hydroxides, hydrides, and halides.

A recent tabulation of rate constants as a function of temperature for reactions that are of interest in studies of metal vapours in flames has been presented and commented upon by Jensen and Jones (1978).

4a   GENERAL CONSIDERATIONS OF METAL COMPOUND FORMATION

4a-1   *GENERAL KINETIC CONSIDERATIONS*

In thermodynamic equilibrium the actual reactions and reaction rates are essentially irrelevant. In this situation no information whatsoever about formation reactions can be gained from observed association factors. The degree of dissociation is, then, fully determined by the laws of chemical equilibrium as explained in Sect. II.3b-3. Observations of the variation of atomic line or molecular band intensities as a function of temperature and flame composition can, then, inform us only on the kind of molecule formed or the emitter of the observed bands, and on the dissociation energy. Knowledge of formation reactions, however, can be obtained from such observations only if *deviations from chemical equilibrium* exist (see Sect. II.4a-2). Two kinds of deviations can be distinguished:

(i)   some of the flame constituents, usually free radicals, that are involved in the metal reactions are not present in equilibrium concentrations;

FORMATION REACTIONS OF METAL COMPOUNDS   Sect. VIII.4a

(ii)   the reactions through which metal compounds are formed or dissociated are
too slow to set up an equilibrium distribution between metal atoms and metal compounds.

Two *main types of formation reactions* (and reverse dissociation reactions)
can be discerned (see Sugden 1956). Firstly we have the *bimolecular exchange reactions*, which for hydroxides read

$$M + H_2O \underset{k_{-1}}{\overset{k_1}{\rightleftharpoons}} MOH + H \quad \text{with} \quad K_1 \equiv k_1/k_{-1} \,. \qquad \text{(VIII.12)}$$

Secondly we have the *ternary recombination reactions*, which for hydroxides read

$$M + OH + Q \underset{k_{-2}}{\overset{k_2}{\rightleftharpoons}} MOH + Q \quad \text{with} \quad K_2 \equiv k_2/k_{-2} \,, \qquad \text{(VIII.13)}$$

where Q is a flame molecule acting as a 'third body' that carries off part of the
recombination energy (see also Sect. II.4h).

Theoretically, the criterion for reaction VIII.13 to become dominant over
reaction VIII.12 is given by: $k_{-2}[Q] \gg k_{-1}[H]$, on the assumption that the dominant
reaction is practically balanced, and conversely (see Sugden 1956[+]). The concentration ratio $[H]/[Q]$ may be estimated to be of the order of 10^{-3} in usual H_2
flames, since Q is a major flame constituent, say H_2O or N_2. The rate constants
(k_{-1} and k_{-2}) can be expressed by their frequency factors (A), and activation energies (E_A) according to: $k = A \exp[-E_A/kT]$ (cf. Eqs II.124 and 126). For reactions
with H atoms, A_{-1} may range from 10^{-12} to 10^{-10} cm^3 molecule^{-1} s^{-1}, corresponding
to a steric factor ranging from 10^{-2} to 1. The factor A_{-2} may be estimated to be
10^{-10} cm^3 molecule^{-1} s^{-1}, if no additional degrees of freedom of internal energy
are assumed to contribute to the activation of Q in the dissociating reaction (see
Sect. II.4b-2). If Q is a poly-atomic molecule, say H_2O, several additional degrees of freedom could be effective and might enhance A_{-2} by perhaps a factor 10^3.
Since the energy of dissociation of H_2O into $H + OH$ (5.1 eV; see JANAF 1971) exceeds
the dissociation energy of most metal hydroxides, the activation energy $(E_A)_{-1}$, if
any, may not be very important. Assuming that practically no activation energy is
required for the forward, recombination reaction VIII.13, we know that $(E_A)_{-2}$ equals
about D_0(MOH). The corresponding exponential factor in k_{-2} will thus be very
small. Sugden now concludes that the highest value of $(E_A)_{-2}$ [$\approx D_0$(MOH)] at which
the ternary recombination reaction could possibly dominate over the binary exchange
reaction is about 3 eV. In later papers, however, Reid and Sugden (1962) and Sugden

---

[+] Note that Sugden (1956) defines k_{-2} as a monomolecular rate constant, whereas it
is defined here as a bimolecular rate constant.

(1972) consider this criterion as improbably high and not to be a good measure for the dominance of reaction VIII.13. Anyway, for D_0 (MOH) values above 3 eV the dominance of reaction VIII.12 seems to be well established experimentally.

Similar considerations can be applied to the analogous formation reactions for metal hydrides

$$M + H_2 \rightleftharpoons MH + H \qquad (VIII.14)$$

or

$$M + H + Q \rightleftharpoons MH + Q . \qquad (VIII.15)$$

Formation of metal oxides may proceed through similar reaction pairs, viz. the exchange reactions

$$M + OH \rightleftharpoons MO + H \qquad (VIII.16)$$

or

$$M + H_2O \rightleftharpoons MO + H_2 \qquad (VIII.17)$$

or

$$M + CO_2 \rightleftharpoons MO + CO , \qquad (VIII.18)$$

on the one hand, and the recombination reaction

$$M + O + Q \rightleftharpoons MO + Q , \qquad (VIII.19)$$

on the other hand. We note that in the reverse reactions VIII.17 and 18 MO reacts with a molecule of a bulk flame constituent and not with an atomic radical. This could markedly increase the pertinent rate constant, according to the above theoretical considerations.

Violation of Wigner's spin conservation rule, as explained in Sect. II.4f, might be another factor that prevents a reaction from proceeding at a sufficiently high rate. It is, however, very difficult to handle this rule quantitatively. An example will be given in Sect. 4b.

So far we have compared the rates of two typically different reactions relative to each other. We still have to consider whether sufficient time is available for *equilibration of the dominant reaction*. We note that the state of dissociation in which the metal vapour leaves the primary combustion zone will often be quite different from the equilibrium state that corresponds to the flame temperature. Since a vertical distance of, say, 1 cm in the flame corresponds to about 1 ms travelling time, the reaction will be (nearly) equilibrated only if the reaction rates are at least of the order of 10^3 s^{-1}.

For reaction VIII.12 the rate of dissociation per MOH molecule is given by $k_{-1}[H]$ which ranges from 10^6 to 10^4 s^{-1} for the median values of k_{-1} and [H] considered above. Calculating the rate of dissociation per MOH molecule for reaction VIII.13 in the unfavourable case $[(E_A)_{-2} = 3\,\text{eV};\ A_{-2} = 10^{-10}\ \text{cm}^3\,\text{molecule}^{-1}\,\text{s}^{-1}]$,

FORMATION REACTIONS OF METAL COMPOUNDS   Sect. VIII.4a

we find: $k_{-2}[Q] \approx 15 \, s^{-1}$ at $T = 2000$ K. If the latter reaction is to become dominant over the former reaction, $k_{-2}[Q]$ has to be larger by at least, say, two orders of magnitude. In conclusion, one can say that the rate of reaction VIII.12 is always high enough to be balanced within 1 ms, and that the rate of reaction VIII.13 is sufficiently high to be balanced within 1 ms under conditions where it predominates over reaction VIII.12. An exception will be dealt with below.

Quantitative measurements of the rate of oxide formation by O_2 have been made by Fontijn, Kurzius and Houghton (1973) and Fontijn, Felder and Houghton (1975, 1977) in a fast gas-flow reactor for Fe and Al and by Kashireninov, Kuznetsov and Manelis (1977) in a diluted, low-pressure diffusion flame for the alkaline-earth atoms (for a survey see Fontijn and Felder 1979). The oxidation is believed to proceed here through the reaction: $M + O_2 \rightarrow MO + O$. Extrapolating the rate constant found for FeO at 1600 K to a temperature of 2500 K and assuming O_2 to be present in the flame at a partial pressure of 0.01 atm, one estimates equilibration to be attained within 10 μs. The presence of excess O radicals would shift the (partial) equilibrium to the left-hand side of the above reaction.

4a-2   *DETERMINATION OF DOMINANT REACTION PATH FROM VARIATION OF ASSOCIATION FACTOR WITH HEIGHT*

If the dominant reaction is fast enough to be balanced at any height, the local value of ϕ will be related to the local concentrations of the flame constituents involved, through the corresponding equilibrium constant. For example, if the exchange reaction VIII.12 is dominant and balanced, we have

$$\phi = K_1[H_2O] / [H] \,, \qquad (VIII.20)$$

whereas in the case when the reaction VIII.13 is dominant and balanced

$$\phi = K_2[OH] \,. \qquad (VIII.21)$$

These relations should also hold when H and OH occur in excess over their equilibrium concentrations. Then, any variation of excess [H] or [OH] with height in the flame will be accompanied by a variation of ϕ with height as predicted by either Eq. VIII.20 or 21. If the former equation applies, a decrease of excess radicals will manifest itself by an increase of ϕ, whereas in the other case ϕ will decrease. It is assumed for simplicity's sake that T and $[H_2O]$ are, virtually, constant. Under these conditions a relative measurement of atomic metal concentration as a function of height provides a basis for deciding which kind of reaction is predominant; it is assumed here that the total metal concentration (atoms plus molecules) is independent of height. The latter measurement also allows a determination of the extent to which [H] and [OH] deviate from their equilibrium values (see Sect. IV.5).

Reaction VIII.12 in the forward direction is endo-ergic by about 1 eV when M is a Li atom in the ground-state. When Li is in its first excited state, this reaction becomes exo-ergic. One may thus expect that the rate of LiOH formation is increased when an appreciable fraction of the Li atoms is brought into the excited state by a nearly saturating laser beam. Muller, Schofield and Steinberg (1978) have indeed found an enhanced depletion of Li atoms in a fuel-rich H_2-O_2-N_2 flame at 1 atm during a strong laser pulse $(\approx 2\,\mu s)$ tuned at the resonance line. A similar effect was found for laser-excited Na atoms near saturation (see Muller, Schofield and Steinberg 1978a). The decay of the saturated alkali fluorescence with increasing height of observation appeared to agree quantitatively with model calculations (Muller, Schofield and Steinberg 1980).

As long as these radical deviations do occur, Eqs VIII.20 and 21 cannot be satisfied at the same time. The equilibration set up between the partners of the balanced dominant reaction is therefore only a <u>partial</u> equilibrium (see Sect. II.4a-2). Only if the concentrations of all flame species equal their equilibrium values, can Eqs VIII.20 and 21 be satisfied at the same time. This comes to the equilibration of the water dissociation reaction: $H_2O \rightleftharpoons H + OH$.

The concept of partial equilibrium at flame temperature T makes sense only if the values of the equilibrium constants, such as K_1 and K_2, correspond to T and are not affected by the considered deviations from full chemical equilibrium. These equilibrium constants are determined mainly by the physical equilibrium conditions in the flame. This includes the equipartition of energy over the internal and external degrees of freedom of the reacting partners, as was explained in Sect. II.4a-2. The presence of excess flame radicals, which are a minor constituent of the flame, is not expected to upset noticeably this physical equilibrium.

The existence of certain partial equilibria between flame radicals and other flame constituents links together effectively several formation reactions for metal compounds. Because of the partial equilibrium

$$OH + H_2 \rightleftharpoons H_2O + H \qquad (VIII.22)$$

reactions VIII.16 and 17 lead to the same ϕ value. This holds irrespective of the extent to which flame radical concentrations deviate from equilibrium. Similarly, because of the well-known water-gas equilibrium

$$H_2 + CO_2 \rightleftharpoons H_2O + CO, \qquad (VIII.23)$$

reaction VIII.17 for the formation of metal oxides is indiscernible from reaction VIII.18.

FORMATION REACTIONS OF METAL COMPOUNDS   Sect. VIII.4a

4a-3 *DETERMINATION OF DOMINANT REACTION PATH FROM DEPENDENCE OF ASSOCIATION FACTOR ON T AND FLAME COMPOSITION*

If excess flame radicals occur, dominance of either reaction VIII.12 or 13 can also be deduced by measuring ϕ for a set of flame families (see Sect. IV.5c) with different but known composition and temperature. If reaction VIII.12 is predominant and balanced, Eq. VIII.20 must hold for any flame, and a plot of $\log_{10}(\phi[H]/[H_2O])$ versus T^{-1} should yield a nearly straight line. This holds because the equilibrium constant K_1 can usually be described in good approximation as a function of T by: $K_1 \propto T^n \exp[-\Delta E_0^0/kT]$. The magnitude of n is usually small, of order unity, so that $\log_{10} K_1$ is a linear function of T^{-1} in a restricted temperature range. If, on the contrary, reaction VIII.13 is predominant and balanced, a plot of $\log_{10}(\phi/[OH])$ versus T^{-1} would yield a nearly straight line. In the case of full chemical equilibrium, of course, both plots must give straight lines.

Figures VIII.5 and 6 show such plots for LiOH and CuH, respectively, measured for a set of H_2-O_2-N_2 flames containing different but known excess concentrations of H and OH. The plotted functions have been measured by Bulewicz, James and Sugden (1956) and by Bulewicz and Sugden (1956a), respectively, and have been further discussed by Sugden (1956). One clearly sees from the straight lines obtained in Figs VIII.5a and 6b that LiOH is formed by a binary exchange reaction of the type VIII.12, and that CuH is formed by a ternary recombination reaction of the type VIII.15. Indeed, D_0(LiOH), being 4.5 eV (see Table VIII.2 on page 768), lies well above the value of 3 eV which was estimated by Sugden (1956) from kinetic considerations as an upper limit for predominance of the recombination reaction. On the other hand, D_0(CuH), being about 2.8 eV, lies below this limit.[†] From the slope of the straight line the heat of reaction and the dissociation energy of the molecule can be derived. Experimental methods of measuring ϕ will be dealt with in Sect. 5. Kind and concentration of the third partner Q occurring in the ternary recombination reaction do not explicitly play a role, as long as this reaction is sufficiently rapid to be balanced.

[†] Discussing metal hydride formation Sugden (1972) comes to the general conclusion that binary exchange reactions of the type VIII.14 are completely dominant for dissociation energies of MH above about 2.6 eV, whereas recombination reactions of the type VIII.15 predominate below about 1.8 eV. It is not clear how the experimental results presented in Fig. VIII.6 fit into this general conclusion.

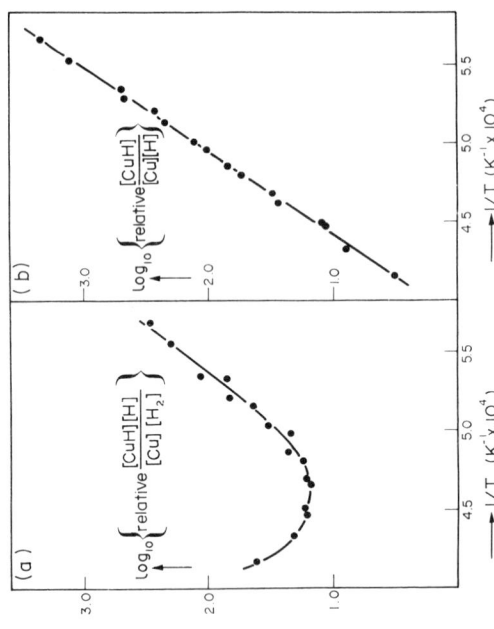

Fig. VIII.5 Plot of $\log_{10}\{[\text{LiOH}][\text{H}]/[\text{Li}][\text{H}_2\text{O}]\}$ versus T^{-1} (see a) and of $\log_{10}\{[\text{LiOH}]/[\text{Li}][\text{OH}]\}$ versus T^{-1} (see b) for a set of $\text{H}_2\text{-O}_2\text{-N}_2$ flames with different compositions and temperatures T, according to Sugden (1956). The straight line obtained in (a) proves the dominance of the balanced reaction $\text{Li} + \text{H}_2\text{O} \rightleftarrows \text{LiOH} + \text{H}$ in controlling the $[\text{LiOH}]/[\text{Li}]$ ratio. The concentration units were not specified in the right-hand figure.

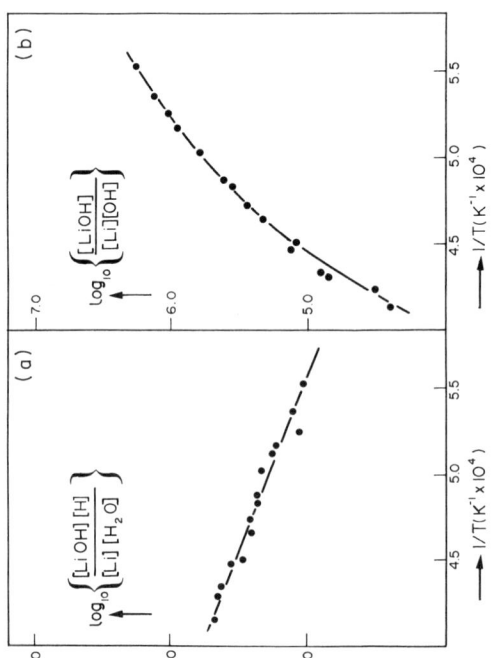

Fig. VIII.6 Plot of $\log_{10}\{\text{relative } [\text{CuH}][\text{H}]/[\text{Cu}][\text{H}_2]\}$ versus T^{-1} (see a) and of $\log_{10}\{\text{relative }[\text{CuH}]/[\text{Cu}][\text{H}]\}$ versus T^{-1} (see b) for a set of $\text{H}_2\text{-O}_2\text{-N}_2$ flames with different compositions and temperatures T, according to Sugden (1956). The straight line obtained in (b) proves the dominance of the balanced recombination reaction $\text{Cu} + \text{H} + \text{Q} \rightleftarrows \text{CuH} + \text{Q}$ in controlling the $[\text{CuH}]/[\text{Cu}]$ ratio.

FORMATION REACTIONS OF METAL COMPOUNDS   Sect. VIII.4b

4b   SURVEY OF FORMATION REACTIONS FOR SPECIFIC METAL COMPOUNDS
4b-1   *METAL OXIDES*

Our knowledge of specific formation reactions for metal monoxides in flames is rather limited.

The formation of *alkaline-earth oxides* in H_2 flames has been discussed by several authors (see, for instance, Sugden and Wheeler 1955, Reid and Sugden 1962, Sugden 1962, Gurvich and Ryabova 1965, and Sugden and Schofield 1966). It is generally believed that these molecules are formed and dissociated by the balanced binary exchange reactions VIII.16, involving flame radicals, and (or) VIII.17, involving only stable flame molecules. We note that the enthalpy change of these reactions for the alkaline-earth metals is, in absolute value, lower than about 1 eV. Reactions VIII.16 and 17 cannot be distinguished because of the partial equilibrium VIII.22. So, even if reaction VIII.16 were dominant, deviations of the radical concentrations from equilibrium would not upset the equilibrium dissociation of the metal oxide unless the concentrations of the bulk constituents H_2 and H_2O were noticeably disturbed by these deviations. However, if the recombination reaction VIII.19 were dominant, the occurrence of excess O radicals would obviously lead to too high ϕ values.

Although being indistinguishable from reaction VIII.17, the exchange reaction VIII.16 is a priori expected to be the less important one, because the concentrations of H and OH are usually lower than those of H_2 and H_2O, which occur in the former reaction.

In CO flames with little or no moisture, reactions VIII.16 and 17 could not be very effective. The balanced reaction VIII.18, involving CO_2 and CO, has been advanced as the dominant one in these flames by Alkemade, Hollander and Kalff (1965), Kalff (1971), and Kalff and Alkemade (1973).

The enthalpy change of reaction VIII.18 amounts to about +1.7, +1.4 and +0.15 eV for Ca, Sr, and Ba, respectively. No high activation energy is expected and thus from an energetic point of view this reaction may be easily balanced at any height in the flame. If MO is supposed to have a singlet ground-state (see Kaufman, Wharton and Klemperer 1965), no difficulties arise with respect to the spin-conservation rule (see Sect. II.4f), since all other partners have singlet ground-states too. If, however, MO has a (pure) triplet ground-state, then the spin rule is violated when all partners are in their electronic ground-states. The spin of the electronic ground-state of MO and the possible presence of other close-lying electronic states has been a subject of controversial discussions since Lagerqvist and Huldt (1954a). We refer to Gaydon (1968) and van der Hurk, Hollander and Alkemade (1975) for a review of these discussions.

In hydrocarbon flames both exchange reactions VIII.17 and 18 may be

Chapt. VIII   FORMATION OF METAL COMPOUNDS

operative. They are indistinguishable, since H_2 and H_2O on the one hand and CO and CO_2 on the other hand are linked together by the balanced reaction VIII.23.

In the *primary combustion zone* of fuel-rich hydrocarbon flames, free C atoms might contribute to the excess dissociation of oxides such as BeO, SnO, and VO according to Gibson, Grossman and Cooke (1963), Amos and Thomas (1965), and Gilbert (1966). The proposed reaction is

$$MO + C \rightarrow M + CO . \qquad (VIII.24)$$

This reaction may play also a role in the suprathermal chemiluminescence of metal atoms in this zone as was discussed in Sect. VI.3.

Kirkbright, Peters and West (1967) and Kirkbright, Semb and West (1968) have postulated that the excess dissociation of stable oxides in the interconal zone of fuel-rich C_2H_2 flames may proceed through reactions of the type

$$MO + NH \rightarrow M + NO + H \quad \text{or} \quad \rightarrow M + OH + N \qquad (VIII.25)$$

and

$$MO + CN \rightarrow M + CO + N . \qquad (VIII.26)$$

These postulates are based on their observations of strong emission of CN and NH bands in the interconal zone, from which they have concluded that these radicals are present in high concentration. However, strong emission does not necessarily mean a high concentration. Rasmuson, Fassel and Kniseley (1973, 1976) calculated that other species like HCN, H, and C_2H occur in larger equilibrium concentrations than CN and NH. Their conclusion is that it is highly speculatory to identify CN and NH, or possibly atomic carbon, as being primarily responsible for the reduction of metal oxides.

Coker and Ottaway (1971) and Coker, Ottaway and Pradhan (1971) have suggested that metal atoms in the primary combustion zone of hydrocarbon flames may be formed according to

$$MO + CX \rightarrow M + CO + X , \qquad (VIII.27)$$

where CX may stand for C_2, CH, CN, etc. No experimental evidence for the occurrence of these reactions is given.

Kaskan (1965) does not rule out the possibility that in fuel-lean H_2-O_2-N_2 flames the observed alkali-atom decay with height is consistent with the nonbalanced formation of *alkali monoxides* through

$$M + O_2 \rightarrow MO + O . \qquad (VIII.28)$$

However, he prefers the recombination reaction VIII.29 (see below) to interpret his experimental results.

In dry CO flames Li may form LiO by the balanced reaction VIII.18 if the

792

FORMATION REACTIONS OF METAL COMPOUNDS   Sect. VIII.4b

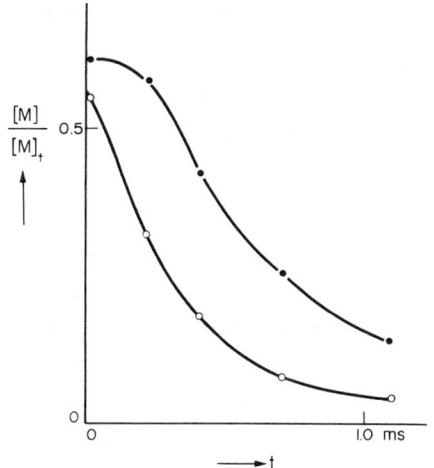

Fig. VIII.7 Plot of atomic sodium (●) and potassium (o) concentration [M] divided by total alkali concentration $[M]_t$, as a function of rise time, t, above the combustion zone of a H_2-N_2-O_2 flame at 1 atm pressure, as measured by McEwan and Phillips (1966). Unburnt gas composition: $H_2/N_2/O_2 = 1.5 : 6:1$; $T = 1490$ K. The decay of atomic alkali concentration is explained by a slow unbalanced recombination reaction between alkali atoms and O_2 molecules, resulting in the formation of alkali superoxides.

flame is fuel-rich, and by the reaction VIII.28 if the flame is fuel-lean (see Dougherty, McEwan and Phillips 1973).

Alkali superoxides can be formed by the slow recombination reaction

$$M + O_2 + Q \rightarrow MO_2 + Q .$$ (VIII.29)

This reaction has been studied by Kaskan (1965), McEwan and Phillips (1966), and Carabetta and Kaskan (1968) in oxygen-rich, cool H_2-O_2-N_2 flames at $T \approx 1500$-2000 K, and by Fontijn, Kurzius and Houghton (1973) in a hot, fast gas-flow reactor. In such flames the atomic alkali concentration appeared to decrease exponentially with increasing height as is shown in Fig. VIII.7. The recombination of alkali atoms with O_2, which is present in large concentration in these flames, does not go to completion within 1 ms at 1 atm pressure, if $T < 2000$ K. From the slope of the curve the termolecular-recombination rate constant can be derived if $[O_2]$ is assumed to be constant and equal to its calculated equilibrium value. From the curves depicted in Fig. VIII.7 one finds rate constants equal to about $3.1 \times$ and 2.2×10^{-33} cm^6 molecule^{-2} s^{-1} for K and Na, respectively. The variation of these constants with temperature has proved to be small from observations in similar flames with different temperatures.

4b-2   *METAL HYDROXIDES*

Formation of *alkali hydroxides* in H_2- and hydrocarbon flames is generally considered to proceed through the balanced binary exchange reaction VIII.12 (see James and Sugden 1955, Sugden 1962, Jensen and Padley 1966, Cotton and Jenkins 1969, and Kelly and Padley 1971). We note from Table VIII.2 that the dissociation energies

793

Chapt. VIII   FORMATION OF METAL COMPOUNDS

of all alkali hydroxides well exceed the critical value 3 eV above which the predominance of reaction VIII.13 becomes quite improbable. The predominance of the exchange reaction VIII.12 for LiOH is clearly shown by the experimental plot in Fig. VIII.5, which refers to H_2 flames with excess radical concentrations.

Formation of *alkaline-earth monohydroxides* is also believed to proceed through the balanced exchange reaction VIII.12, at least in fuel-rich H_2 flames (see Gurvich and Ryabova 1965, Ryabova and Gurvich 1965, Sugden and Schofield 1966, Gurvich, Ryabova, Khitrov and Starovoitov 1971, and Ryabova, Khitrov and Gurvich 1972).

Formation of *Ga- and In hydroxides* having D_0 values well above 3 eV has also been found to proceed through the balanced exchange reaction VIII.12 (see Bulewicz, James and Sugden 1956, Bulewicz and Sugden 1958, and Kelly and Padley 1971). Similar linear plots as shown for LiOH in Fig. VIII.5a on page 790 were obtained for these elements. Also, the ratio of the atomic In concentration found at different heights in the flame with different excess H concentrations proved to be consistent with the balanced reaction VIII.12.

The formation of *alkaline-earth dihydroxides* is assumed by Sugden and Schofield (1966) to proceed through the balanced exchange reaction

$$MOH + H_2O \rightleftharpoons M(OH)_2 + H \quad \text{with} \quad K_5 = [M(OH)_2][H] / [MOH][H_2O] \, . \quad (VIII.30)$$

Combination of this reaction with the balanced reaction VIII.12, leads to the overall balance (cf. process no. 32 in Table II.1 on pages 108-9)

$$M + 2H_2O \rightleftharpoons M(OH)_2 + 2H \, . \quad (VIII.31)$$

The simultaneous occurrence of the balanced reactions VIII.30 for $M(OH)_2$ and VIII.12 for MOH has been experimentally checked by Cotton and Jenkins (1968). They measured $\phi \equiv ([M]_t - [M])/[M]$ as a function of $[H]$ for a 'family' of fuel-rich, isothermal H_2-O_2-N_2 flames with constant $[H_2O]$. In these flames [MO] proved to be negligible, while ionization was absent. These measurements supplement the earlier experiments made by Sugden and Schofield (1966) in similar flames, from which they have concluded that $M(OH)_2$ was dominant over MOH.

4b-3 *METAL HYDRIDES*

The formation of *Cu- and Tl hydrides* with their low dissociation energies appears to proceed through the balanced ternary recombination reaction VIII.15 (see Bulewicz and Sugden 1956a, 1958). This has been proved both for CuH and TlH by 'family' plots of the kind shown in Fig. VIII.6. In accordance with the predominance of this reaction, it has also been found that the decrease of [CuH] with increasing height is similar to that of the excess H concentration. It is believed that also AuH and AgH, with dissociation energies comparable to that of CuH, are

FORMATION REACTIONS OF METAL COMPOUNDS   Sect. VIII.4b

formed by recombination reactions of the same kind (see Mavrodineanu and Boiteux 1965).

4b-4 *METAL HALIDES*

It is generally assumed that the *alkali monohalides*, MX, having comparatively high D_0 values, are formed in H_2- and hydrocarbon flames by the balanced binary exchange reaction (see Bulewicz, Phillips and Sugden 1961)

$$M + HX \rightleftharpoons MX + H \quad \text{with} \quad K_6 = [MX][H]/[M][HX] . \quad \text{(VIII.32)}$$

In turn, HX is formed by the balanced reaction

$$H_2 + X \rightleftharpoons HX + H \quad \text{with} \quad K_7 = [HX][H]/[H_2][X] . \quad \text{(VIII.33)}$$

Combining the equilibrium relations VIII.32 and 33 and allowing for the possible formation of other metal compounds MZ, with $\phi' \equiv [MZ]/[M]$, we find according to Bulewicz, Phillips and Sugden (1961)

$$[M]_0/[M] = 1 + [X]_t K_6 \left[[H](1+\phi') \left\{ 1 + ([H]/K_7 [H_2]) \right\} \right]^{-1} . \quad \text{(VIII.34)}$$

In this equation [H] denotes the actual H concentration, while $[M]_0$ and [M] are the atomic metal concentrations in the absence and in the presence of halogen, respectively. $[X]_t$ is the total concentration of halogen and halogen compounds present in the flame, which is assumed to be much larger than [MX]. This equation describes the effect of (excess) H radicals on the halide dissociation for a given value of $[X]_t$. It should be realized that ϕ' may also be a function of [H], as is seen from Eq. VIII.20 which holds if MOH formation proceeds through reaction VIII.12.

In full equilibrium where $[H] = [H]_e$, Eq. VIII.34 reduces to Eq. VIII.11 when K_6 and K_7 are expressed in K_3 and K_4 given by Eqs VIII.5 and 6. Then the actual formation reactions of MX and HX are no longer relevant. However, in the nonequilibrium case with $[H] > [H]_e$ Eq. VIII.34 should be used to describe the actual behaviour of $[M]_0/[M]$, and not Eq. VIII.11. This conclusion still holds if [HX] is negligibly small relative to $[X]_t$, as long as the metal halides are actually formed by reaction VIII.32 and both reactions VIII.32 and 33 are balanced.

Equation VIII.34 has been checked experimentally by Bulewicz, Phillips and Sugden (1961) for alkali metals and all halogens in H_2 flames with varying H concentration.

In CO flames with little or no moisture, the concentration of HX might be too low for balancing the reactions VIII.32 and 33. Direct recombination of metal atoms and halogen atoms could, then, be more significant than the exchange reaction VIII.32. Since the rate of such ternary recombination reactions is expected to be fairly low, a lag in the establishment of halide dissociation equilibrium might

Chapt. VIII   FORMATION OF METAL COMPOUNDS

occur in these flames. It has been found by Zeegers[†] that the ratio of association factors $\phi \equiv [MCl]/[M]$ for K and Na gradually approaches the equilibrium ratio over a height interval of 3 cm above the combustion zone. In this ratio the halogen content in the flame drops out and need not be known. This outcome might indicate that the establishment of full dissociation equilibrium for alkali chlorides has a relaxation time of the order of 6 ms in these flames.

The occurrence of *alkaline-earth halides*, *dihalides* and *hydroxyhalides* has been studied only under flame-equilibrium conditions by Ryabova, Gurvich and Khitrov 1971, and Ryabova, Khitrov and Gurvich 1972). The actual formation reactions are still unknown.

5. FLAME SPECTROMETRIC DETERMINATION OF DISSOCIATION ENERGIES

The *flame-spectrometric equilibrium method* has been successfully applied to metal compounds that are not readily treated by other thermochemical methods. It can be applied to molecules that are not stable at room temperature or that require a high temperature to be dissociated. *Advantages* of the flame method are:

(i)   measurements are done in situ, so that the equilibrium state of the system is not disturbed by the measuring procedure;

(ii)   flames can be produced with homogeneous composition and temperature which can be controlled over a wide range; the bulk composition can be calculated; the temperature can be measured accurately;

(iii)   measurement of (relative) metal concentrations is easy and its results are accurate.

Disadvantages are:

(i)   during the short residence time (say, 3 ms) of the metal vapour in the flame the flame gases may not be fully equilibrated; especially radical concentrations may deviate from their equilibrium value (see Sect. IV.5);

(ii)   sometimes it is difficult to distinguish between the different compounds that metals may form with flame constituents or additives; mass-spectrometric methods may be of help here.

In Sect. 5a we shall treat the theoretical and experimental basis of the flame spectrometric method, while in Sect. 5b some typical applications will be described. Further applications are found in the pertinent references quoted in

---

[†] Unpublished work (internal report V 2985, 1969, of Physisch Laboratorium der Universiteit, Utrecht).

FLAME SPECTROMETRIC DETERMINATION OF DISSOCIATION ENERGIES   Sect. VIII.5a

Tables VIII.1-4 on pp.766-767. A critical comparison of results obtained by the flame method with those obtained by other methods is beyond the scope of this book. The interested reader is referred, for example, to Cottrell (1958), Drowart, Exsteen and Verhaegen (1964), Schick (1966), Sugden and Schofield (1966), Schofield (1967), and Gaydon (1968).

5a   FLAME-SPECTROMETRIC EQUILIBRIUM METHOD

The *theoretical basis* of this method is the general relation between the dissociation energy D_0 and the dissociation constant K_d for the equilibrated dissociation reaction:

$$MX \rightleftharpoons M + X. \qquad (VIII.35)$$

The relation between K_d (in cm^{-3}) and D_0 (in eV) reads according to Eq. II.44

$$\log_{10} K_d = 20.274 + \tfrac{3}{2}\log_{10}\left\{(M_r)_M (M_r)_X/(M_r)_{MX}\right\} + \log_{10}(Q_M Q_X/Q_{MX})$$
$$+ \log_{10}(s_{MX}/s_M s_X) + \tfrac{3}{2}\log_{10} T - 5040\, D_0/T. \qquad (VIII.36)$$

Here M_r is the relative atomic or molecular mass ($M_r \equiv 12$ for ^{12}C atoms), Q is the internal partition function, and T is the absolute temperature in K. This relation involving molecular and atomic constants follows from Statistical Mechanics. It is equivalent to Eq. II.51 which involves thermodynamic functions. For metal compounds the required molecular constants are often more easily accessible by experiment or can be estimated with more confidence than the thermodynamic functions.

In the special case when MX is a heteronuclear di-atomic molecule, Eq. VIII.36 usually reduces to the approximated relation

$$\log_{10} K_d = 20.432 + \tfrac{3}{2}\log_{10}\left\{(M_r)_M (M_r)_X/(M_r)_{MX}\right\} + \log_{10}(g_0)_M (g_0)_X/(Q_e)_{MX}$$
$$+ \log_{10} B + \log_{10}\left(1 - 10^{-0.625\, \omega_e/T}\right) + \tfrac{1}{2}\log_{10} T - 5040\, D_0/T. \qquad (VIII.37)$$

Here g_0 is the statistical weight of the atom in its ground-state, $(Q_e)_{MX}$ is the electronic partition function of the molecule, B is the rotational constant of MX in cm^{-1}, and ω_e is the fundamental vibrational frequency in cm^{-1} (see Eqs II.38 and 40 and the pertinent discussion in Sect. II.3b-2).

The following assumptions on which Eq. VIII.37 is based (see also Gaydon 1968) are reasonably well fulfilled in flames:

(i)   validity of the ideal-gas law for all components;

(ii) neglect of the contribution of excited levels to the atomic partition functions; for some metal atoms and for Cl and O having low-lying excited levels, g_0 should be replaced, at $T > 2000$ K, by the appropriate partition function tabulated by de Galan, Smith and Winefordner (1968);

(iii) classical treatment of the rotational partition function of the molecule;

(iv) neglect of the interaction between molecular rotation and vibration, as well as neglect of the anharmonicity of the molecular vibration.

As was explained in Sect. 2, D_0 can either be found by an <u>absolute</u> determination of K_d at known temperature T (*third-law method*), or by determining $d(\log_{10} K_d)/dT^{-1}$ (*second-law method*). In the first variant, all constants and functions entering into Eq. VIII.36 (or 37) have to be known. For most di-atomic molecules, the constants $(Q_e)_{MX}$, B, and ω_e may be obtained with fair accuracy from spectroscopic band analysis (Huber and Herzberg 1979). In the second variant, B and, virtually, also $(Q_e)_{MX}$ drop out for di-atomic molecules, whereas the T-dependence of the vibrational term will usually be of secondary importance. Then a plot[†] of $\log_{10} K_d$ versus T^{-1} in a restricted range of temperatures should yield a nearly straight line, the slope of which is simply related to D_0. Only <u>relative</u> values of K_d are required here. Similarly a measurement of the ratio $K_d(T_1)/K_d(T_2)$ at two known temperatures can yield a value of D_0. The accuracy of the second-law method depends on the accuracy of the temperature and on the extension of the T-range covered in the experiment.

If the dissociation of a metal compound, MX, is not equilibrated, D_0(MX) can often be evaluated from the reaction energy of a *balanced binary exchange reaction* of the type: $M + XY \rightleftharpoons MX + Y$ with $K = [MX][Y] / [M][XY]$. If Y and M represent atoms while MX and XY are linear or nonlinear poly-atomic molecules, as is the case with reactions of the type VIII.12, we have for the equilibrium constant K from the general Eqs II.42 and 58

$$\log_{10} K = \tfrac{3}{2} \log_{10} \left\{ (M_r)_{MX} (M_r)_Y/(M_r)_M (M_r)_{XY} \right\} + \log_{10}(Q_e)_Y/(Q_e)_M + \log_{10}(Q_e)_{MX}/(Q_e)_{XY}$$

$$+ \log_{10}(Q_v)_{MX}/(Q_v)_{XY} + \log_{10}(Q_r)_{MX}/(Q_r)_{XY} + \log_{10}(s_{XY}/s_{MX}) - 5040 \Delta U_0^0/T \,.$$
(VIII.38)

ΔU_0^0 is the reaction energy (in eV) per molecule MX at 0 K and in the ideal-gas state. If MX and XY are di-atomic molecules, Q_v reduces to: $(1-10^{-0.625\omega_e/T})^{-1}$. If we are dealing with a poly-atomic molecule, Q_v is a product of such factors for the various normal modes of vibration. The term $\log_{10}(Q_r)_{MX}/(Q_r)_{XY}$ can be written as: $\log_{10}(B_{XY}/B_{MX})$ when MX as well as XY are di-atomic or linear poly-atomic molecules; it can be written as: $\tfrac{1}{2}\log_{10}(I_A I_B I_C)_{MX}/(I_A I_B I_C)_{XY}$ when MX and XY are both nonlinear poly-atomic molecules; and as:

$$\log_{10} I_{MX} - \tfrac{1}{2} \log_{10}(I_A I_B I_C)_{XY} - \tfrac{1}{2} \log_{10} T - 19.445$$

[†] Although $\log_{10} K_d$ is commonly plotted versus T^{-1}, it is more correct to plot $\log_{10}(K_d T^{-\tfrac{1}{2}})$ versus T^{-1} (see Eq. VIII.37) for heteronuclear di-atomic molecules.

FLAME SPECTROMETRIC DETERMINATION OF DISSOCIATION ENERGIES   Sect. VIII.5a

when MX is a linear and XY is a nonlinear poly-atomic molecule. Herein $I_A I_B I_C$ (in $g^3 cm^6$) is the product of the principal moments of inertia of the nonlinear molecule and I (in $g\,cm^2$) is the corresponding moment of the linear molecule. For a derivation of the above expressions see Sect. II.3b-2.

If the energy, $D_0(XY)$, of dissociation of XY into $X+Y$ is known, we find $D_0(MX)$ from

$$D_0(MX) = D_0(XY) - \Delta U_0^0 . \qquad (VIII.39)$$

When Y is a flame radical that is present in excess concentration, the actual value of [Y] (and of [MX]/[M]) and not its equilibrium value should be inserted in order to determine K. Again, either a third-law- or a second-law method can be applied to determine ΔU_0^0. The second-law method is to be preferred when uncertainties exist as regards the electronic partition functions and the geometry of the molecules, in particular when they are poly-atomic.

The determination of D_0 from the equilibrated dissociation reaction: $MX \rightleftharpoons M+X$ requires knowledge of the concentration of X. Any uncertainty in the assumed value of [X] affects the D_0 value found. For metal oxides, such uncertainties may arise due to infusion of secondary air, which can hardly be accounted for. With metal halides, the absolute calibration of the supply of halogen could involve errors. If [X] is unknown, the difference of dissociation energies of two metal compounds that contain the same partner X can still be determined from the ratio of their association factors ϕ. The only requirement is that the ϕ values are measured under identical flame conditions. Taking the ratio of ϕ values comes down to subtracting both sides of Eq. VIII.36 or 37 written down for each metal compound separately. Differences in D_0 for the alkaline-earth oxides have thus been determined by Hollander (1964) and Pungor (1967).

Tabulations of equilibrium constants of reactions involving metal species, as a function of temperature, are found in JANAF (1971, 1974, 1975), and Jensen and Jones (1978).

The *experimental determination of equilibrium constants* requires a measurement of the association factor ϕ ($\equiv$ [MX]/[M]) and knowledge of the concentration (ratio) of the other reaction partner(s). The latter concentration (ratio) can be calculated if it conforms to chemical equilibrium. Experimental determination of the actual radical concentration(s) (see Sect. IV.5) is required when the degree of dissociation of the metal compound is affected by radical disequilibrium (see Sect. 4b).

If the actual formation reaction of MX does not involve radical X, but only major flame constituents that occur practically in equilibrium concentrations, we must insert the calculated equilibrium value $[X]_e$ in Eq. VIII.1, even if the actual value [X] deviates from $[X]_e$. For instance, in CO flames alkaline-earth

Chapt. VIII   FORMATION OF METAL COMPOUNDS

oxides are not formed by direct recombination of M- and O atoms, but by the exchange reaction VIII.18 involving CO and CO_2. The concentration of these major flame constituents is practically not affected by radical disequilibrium so that: $[CO] \simeq [CO]_e$ and $[CO_2] \simeq [CO_2]_e$. Then we have for reaction VIII.18, with equilibrium constant K_{18}, the relation: $[MO]/[M] = K_{18}[CO_2]_e/[CO]_e$. For the dissociation of CO_2 we have the equilibrium relation: $K_d(CO_2) = [CO]_e[O]_e/[CO_2]_e$. Combining both expressions we thus find: $[MO]/[M] = \{K_{18}/K_d(CO_2)\}[O]_e = [O]_e/K_d(MO)$.

The *experimental determination of absolute* ϕ *values* can be performed in several ways:

(i)   The absolute concentration of free metal atoms [M] and the absolute total metal concentration $[M]_t$ are determined by one of the methods described in Sects V.2 and 5. The ratio $[M]/[M]_t$ can also be determined directly and often more accurately by the comparison method treated in Sect. V.6. From the balance equation: $[M]_t = [M] + [MX] +$ [other metal species], the ratio [MX]/[M] follows if the concentrations of the other species relative to [M] can be either neglected or accounted for by theoretical calculations. This method works only well if ϕ is markedly larger than unity and if the other metal species are not predominant.

(ii)   When the partner X that combines with the metal is supplied separately to the flame, as is the case for the halogens, ϕ can be found from

$$\phi = ([M]_t/[M])(1 - [M]/[M]_0) \, . \tag{VIII.40}$$

Here [M] and $[M]_0$ are the atomic metal concentrations that are found with and without supply of X. This relation holds independently of the occurrence of other compounds between metal atoms and flame constituents if the concentrations of these compounds relative to [M] are independent of the supply of X, as is usually the case. The ratio $[M]_t/[M]$ in Eq. VIII.40 can be found as indicated in (i). The ratio $[M]/[M]_0$ is simply derived from the relative decrease in atomic line intensity or absorption that is brought about by supplying X (see also Sect. V.3a). This method works well for both small and large ϕ values.

Equation VIII.40 is simply proved by considering the metal balance equation in the absence and presence of MX formation, respectively. We have
$$[M]_t = [M]_0(1+\phi') \quad \text{and} \quad [M]_t = [M](1+\phi') + [MX] \, .$$
Here ϕ' is the constant concentration ratio of the other metal compounds relative to the free metal atoms. Eliminating $(1+\phi')$ from the two equations and dividing by [M], one immediately finds Eq. VIII.40. This equation has been derived by Gurvich and Ryabova (1964) in their study of BaCl formation.

(iii) ϕ may also be found from the deviation from a straight line in the plot of $\ln I$ versus T^{-1} for flames with different composition and temperature (compare

FLAME SPECTROMETRIC DETERMINATION OF DISSOCIATION ENERGIES   Sect. VIII.5a

Fig. V.4 on page 524). Here I is the relative intensity of the atomic metal line, corrected for any relative variation of $[M]_t$ and of flame depth l with T. The relative variation of $[M]_t$ and l with T can be determined by measuring the corresponding variation for a fully atomized element (see Sect. V.6). In the temperature range where this plot is a straight line with a slope corresponding to the atomic excitation energy, $[M]$ equals $[M]_t$ and compound formation is likely to be absent (see Fig. V.4). A deviation from the extrapolated straight line in an adjacent temperature range then indicates that compound formation occurs here. Ionization effects should, of course, be eliminated. The extent of this deviation immediately yields ϕ if only one compound is formed. This method works well also when ϕ is rather small. It requires, however, a rather careful selection of a set of flames with different compositions and temperatures and constant delivery of metal salt.

Because of lack of absolute oscillator strengths, absolute intensity or absorption measurements of molecular bands have not yet been used in determining absolute ϕ values (see also Sects V.2a and 2b). The *relative variation of* ϕ with temperature, as used in the second-law method, can, of course, be derived from absolute ϕ measurements in flames with different temperatures. It can also be found more directly by one of the following methods:

(i)   If ϕ is sufficiently large to ensure that $[MX] \approx [M]_t$, its relative variation can be determined by measuring the relative variation of both $[M]$ and $[M]_t$ (see also Sects V.3, 5 and 6). It may happen that $[M]_t$ is practically constant in flames with different temperatures or that the variation of $[M]_t$ with temperature T, connected with the factor $(298/T)/\eta$ in Eq. V.17, can be easily accounted for by calculation. Then only the relative variation of $[M]$ need be measured by one of the methods described in Sect. V.3a.

(ii)   If the molecule studied emits a spectral band and the relative dependence of band excitation on T is known, the variation of ϕ with T can be found from the variation in band-to-line intensity ratio. The excitation energy of the atomic line is usually well-known. Self-absorption should be avoided or corrected for as was described in Sect. V.2a. The method works equally well for high and low ϕ values and is, moreover, independent of the possible formation of other metal compounds or metal ions. It is also independent of any variation in total metal content or flame dimensions. This method was used to determine the dissociation energy of AlO (see Newman and Page 1970, 1971) and of the monohalides and monohydroxides of alkaline-earth metals (see Gurvich, Ryabova, Khitrov and Starovoitov 1971, and Ryabova, Khitrov and Gurvich 1972). If $\phi \ll 1$ to the effect that $[M] \approx [M]_t$, only the variation of band intensity with T need be measured if possible variations of $[M]_t$ with T are corrected for. Of course, ionization effects have either to be suppressed or corrected for. In this way Bulewicz and Sugden (1956) have

Chapt. VIII   FORMATION OF METAL COMPOUNDS

determined D_0(CuH).

The emitter of the band observed should, of course, be identified and the emission of the band should not be contaminated by spectral overlap of other metal bands or continua. For instance, in the C_2H_2-air flame the bands in the barium spectrum at 4870 and 5120 Å are due to BaO as well as BaOH and cannot be used to determine dissociation energies.

Difficulties in the application of this method arise when deviations from thermal band excitation occur because of suprathermal chemiluminescence (see Sect. VI.3c-3).

Another difficulty will usually be our lack of knowledge about the temperature dependence of the thermal band excitation (see, for example, van der Hurk, Hollander and Alkemade 1974, 1975).

5b   SOME TYPICAL APPLICATIONS OF THE FLAME-SPECTROMETRIC
      EQUILIBRIUM METHOD

In this section we will describe actual applications of flame spectrometric methods to the determination of D_0 for some typical metal compounds. LiOH is chosen as a representative of metals that usually form only one compound with flame constituents, and BaCl as an example of compounds formed between a metal and an additive. Next we shall treat methods to obtain D_0 values when two different compounds of the same metal with flame constituents may occur in comparable concentrations. Determination of D_0(SrOH) by measuring the band-to-line intensity ratio as a function of temperature will be the first example. This method yields results that are independent of the abundance of other Sr compounds. Next the simultaneous determination of $D_0\{Ca(OH)_2\}$ and D_0(CaOH), and of D_0(BaOH) and D_0(BaO) in flames where either hydroxides and dihydroxides, or hydroxides and oxides occur in comparable concentrations will be discussed. These examples may illustrate the difficulties and ambiguities that may arise with flame spectrometric methods and suggest some possible ways to circumvent them. With proper precautions flame spectrometric methods will be found to yield not only precise but also accurate D_0 values.

5b-1   *DETERMINATION OF* D_0(LiOH)

LiOH is a typical example of a molecule whose degree of dissociation is affected by the possible occurrence of excess flame radicals. Under the latter condition, the dissociation reaction: LiOH ⇌ Li + OH is not equilibrated, but partial equilibrium is always maintained for the exchange reaction: LiOH + H ⇌ Li + H$_2$O, as was discussed in Sect. 4b. D_0(LiOH) can be derived from the measured equilibrium constant for the latter reaction, $K_1 = \phi[H] / [H_2O]$, by applying Eqs VIII.38 and 39; under equilibrium conditions D_0(LiOH) can also be derived directly from $K_2 = \phi/[OH]_e$ by applying Eq. VIII.36.

FLAME SPECTROMETRIC DETERMINATION OF DISSOCIATION ENERGIES   Sect. VIII.5b

Using the Na-comparison method Jensen and Padley (1966) have measured an absolute ϕ value in a H_2-O_2-N_2 flame at 2475 K at a sufficiently large height to ensure that [H] was equal to its equilibrium value. Slight corrections for the formation of Na^+ ions and NaOH molecules were made. Ionization of Li was insignificant. The equilibrium constant K_1 was calculated from ϕ and $[H]/[H_2O]$. A value of ΔU_0^0 for the corresponding exchange reaction was obtained through Eq. VIII.38 (third-law method). Inserting in Eq. VIII.39 the value $D_0(H_2O) = 5.10$ eV, the authors have found a value $D_0 = 4.40 \pm 0.10$ eV for LiOH → Li + OH. Data required to calculate the moments of inertia and vibrational partition functions were taken or estimated from JANAF (1960). Using newer estimates of the molecular parameters (JANAF 1966; see also Zeegers and Alkemade 1970), we recalculated from their experimental data $D_0(LiOH) = 4.50 \pm 0.10$ eV.

Zeegers (1966) has obtained a value for $D_0(LiOH)$ by applying the second-law method to the relative variation of K_1 with T in C_2H_2-air flames. At the height of observation, [H] was somewhat larger than its equilibrium value and the ratio $[H]/[H]_e$ was determined experimentally from the decay of atomic Li concentration with height (see Sect. IV.5). For $[H_2O]$ the calculated equilibrium concentration was substituted. ϕ values were again obtained by the Na-comparison method, while excess Cs salt was added to suppress the Na ionization. NaOH-formation was calculated to be insignificant. A plot of $\log K_1$ versus T^{-1} in a temperature range from 2000 to 2400 K yielded a straight line (compare Fig. VIII.5a on page 790 where a similar plot is shown). Combining the reaction enthalpy $\Delta H^0_{2200} = -0.63$ eV as derived from the slope of this plot for a median value of $T = 2200$ K, with the reaction enthalpy $\Delta H^0_{2200} = +5.35$ eV [†] for $H_2O \to H + OH$, one finds $\Delta H^0_{2200} = +4.72$ eV for LiOH → Li + OH. Using the specific heats of LiOH, Li, and OH in the gaseous state as a function of T between 0 and 2200 K, one calculates $\Delta H^0_0 (\equiv D_0) = 4.60 \pm 0.10$ eV for LiOH → Li + OH.

The experimental data reported by Zeegers (1966) also allow a determination of $D_0(LiOH)$ from an absolute value of K_1 at known T by the third-law method. Applying Eqs VIII.38 and 39, a value of $D_0(LiOH) = 4.65 \pm 0.10$ eV is obtained (see also Zeegers and Alkemade 1970). This value agrees very well with the value obtained from Zeegers' measurements by the second-law method. Thermochemical data and molecular constants were taken from JANAF (1966).

Accurate knowledge of K_1 at given temperature is essential for the determination of [H] (see Sect. IV.5b-2). K_1 may be calculated from $D_0(LiOH)$ and the molecular parameters by using Eqs VIII.38 and 39. The molecular structure of LiOH

[†] The latter value replaces the somewhat larger value used in Zeegers' thesis (1969).

Chapt. VIII FORMATION OF METAL COMPOUNDS

Table VIII.5

D_0(LiOH → Li + OH) (in kcal/mol[†]) Calculated for a Linear and a Bent Molecule from Measurements of Equilibrium Dissociation

Sources of experimental data used	linear molecule*		bent molecule*	
	3rd law	2nd law	3rd law	2nd law
Gurvich and Ryabova (1964)	104.8 ± 2	106.6 ± 2	105.5 ± 2	103.6 ± 2
McEwan and Phillips (1965, 1967)	102.0 ± 2		103.0 ± 2	
Zeegers (1966)	105.8 ± 2	108.8 ± 2	107.0 ± 2	105.4 ± 2
Jensen and Padley (1966)	102.3 ± 2		104.1 ± 2	
Cotton and Jenkins (1969)	103.2 ± 1		104.6 ± 1	
Kelly and Padley (1971)	103.0 ± 2	103.4 ± 2	104.1 ± 2	99.9 ± 2
Weighted mean value	103.4 ± 0.7	106.3 ± 1.2	104.7 ± 0.7	103.0 ± 1.2

[†] See Appendix A.1 for conversion factors of other units.
* See table below for assumed molecule parameters.

Molecule Parameters Used in Calculation of D_0 (Li-OH)

	Li-O (Å)	O-H (Å)	angle	$I_A I_B I_C$ (g³ cm⁶)	I (g cm²)	ω_1 (cm⁻¹)	ω_2 (cm⁻¹)	ω_3 (cm⁻¹)
linear [‡]	1.52	0.96	180°	–	2.2×10^{-39}	665	350 [§]	3600
bent [φ]	1.60	0.96	110°	6.156×10^{-118}	–	1000	1300	3700

[‡] According to Kelly and Padley (1971).
[φ] According to JANAF (1971).
[§] Twofold degenerate.

FLAME SPECTROMETRIC DETERMINATION OF DISSOCIATION ENERGIES   Sect. VIII.5b

is, however, uncertain. In JANAF (1966) a bent configuration is adopted; bond lengths are estimated from measured data of LiF and H_2O; vibration frequencies are estimated from measured values for Li_2O, and H_2O, D_2O and T_2O. Kelly and Padley (1971) have argued in favour of a linear configuration by analogy with the results for NaOH, RbOH and CsOH, obtained by Acquista, Abramowitz and Lide (1968) and Acquista and Abramowitz (1969),which show these molecules to be linear. Kelly and Padley (1971) have estimated bond lengths and vibration frequencies by extrapolating the respective results for NaOH, RbOH and CsOH.

We evaluated the most probable value of D_0(LiOH) from experimentally determined K_1 values or from experimental data on ϕ, flame composition and temperature found in the literature, by using second-law- as well as third-law methods. Table VIII.5 presents our results obtained using either the molecular parameters of JANAF (1971) or those of Kelly and Padley (1971). Since the value of the molecular constants has a different impact on the results of second-law- and third-law calculations, one can compare these results to find out which set of molecular parameters is correct. However, due to the spread in experimental data, the results in Table VIII.5 do not justify a conclusion as regards the molecular structure. For practical purposes, each set of parameters may be used to calculate K_1 if one uses consistently the D_0(LiOH) value derived with the same set.

5b-2   *DETERMINATION OF D_0(BaCl)*

As another typical example, we describe the flame spectrometric determination of D_0 for BaCl, reported by Gurvich and Ryabova (1964), Ryabova and Gurvich (1965), and Ryabova, Khitrov and Gurvich (1972). These authors have used doubly shielded H_2-O_2-N_2 flames, burning on a Méker burner, with known temperature and composition. Calibrated amounts of chlorine (up to 1 volume percent) were supplied in the form of CCl_4 vapour.

D_0(BaCl) has been determined by applying the second-law- and third-law method to the partial equilibrium: $Ba + HCl \rightleftharpoons BaCl + H$ with equilibrium constant K_6 (see reaction VIII.32 in Sect. 4b). Absolute values of K_6 were determined in two different ways, by using either Eq. VIII.40 together with calculated values of [HCl] and [H], or Eq. VIII.34. In the first variant, [Ba] was determined from absolute intensity measurements of the Ba resonance line, and $[Ba]_t$ was derived from calculations based on the nebulization efficiency (see Sect. V.5a). The equilibrium values of [HCl] and [H] were calculated through Eq. VIII.33 from the known flame composition and total concentration $[Cl]_t$. In the second variant, only the ratio $[Ba]_0/[Ba]$ of atomic Ba concentrations in the absence and presence of chlorine, respectively, had to be measured. A knowledge of the proportion ϕ' of other Ba compounds relative to Ba atoms is required, however. The slope of the plot

Chapt. VIII FORMATION OF METAL COMPOUNDS

$[M]_0/[M]$ versus $[Cl]_t$ then directly yields K_6, since $[H]$, $[H_2]$ and K_7 are also known (see Eq. VIII.34).

For flame temperatures lower than 2000 K and for values of $[Cl]_t$ up to 10^{-2} atm, Gurvich and Ryabova (1964) reported this plot to be linear, from which they concluded (see Sect. 4b) the absence of noticeable dihalide formation under these conditions. However, neither this conclusion nor the result obtained for $D_0(BaCl)$ have been confirmed by more recent experiments done by Ryabova, Khitrov and Gurvich (1972). For the same flames, the latter authors now report a linear dependence of $[Ba]/[Ba]_t$ on $[Cl]_t$ for $[Cl]_t < 5 \times 10^{-4}$ atm. For $[Cl] > 5 \times 10^{-4}$ atm, a quadratic dependence has been found, from which they have derived $D_0(BaCl_2)$. Moreover, a linear dependence only proves the absence of Ba compounds with more than one Cl atom, e.g. $BaCl_2$, but compounds like BaOHCl, BaOCl, BaOHOCl and BaHCl may still persist (see Schofield and Sugden 1971). $D_0(BaCl)$ values determined without taking into account a possible formation of other Ba compounds with one Cl atom are incorrect. The same conclusion holds for similar determinations of $D_0(CaCl)$ and $D_0(SrCl)$.

Ryabova, Khitrov and Gurvich (1972) have determined $D_0(BaCl)$ by measuring the ratio of relative Ba line intensity to relative BaCl band intensity at $\lambda = 5240$ Å (see Sect. 5a) as a function of temperature. This method yields a value of $D_0(BaCl)$ that is not affected by a possible occurrence of barium compounds other than BaCl.[†] However, D_0 values determined in this way may have a serious systematic error due to a possible lack of knowledge of the excitation energy of the molecular band. Errors of this kind have been reported for alkaline-earth oxides and hydroxides (see below in this section).

Dissociation energies for several monohalides have been determined by the above mentioned authors by one of the three methods discussed. Data necessary for the conversion of measured quantities (K_6 or ΔH_T) into D_0 values have been taken from Gurvich, Khackkuruzov and Medvedev et al. (1962) and Ryabova and Gurvich (1965). The results are found in Table VIII.4 on page 769. A probable error of 2% for the chlorides and of 6% for the fluorides has been reported.

5b-3 *DETERMINATION OF* $D_0(SrOH)$

If an element forms more than one compound, then a measurement of the band-to-line intensity ratio as a function of temperature is the only spectroscopic way to determine a D_0 value that is independent of assumptions concerning the abundance

---

[†] See also Newman and Page (1971) who have determined $D_0(AlO)$ by the same method in circumstances where other aluminium compounds may be expected.

FLAME SPECTROMETRIC DETERMINATION OF DISSOCIATION ENERGIES   Sect. VIII.5b

of the other compounds. However, this method is dependent on a knowledge of structure and excitation energy of the atoms and molecules involved. As an example we discuss the determination of $D_0(\text{SrOH})$.

The SrOH band to Sr line intensity ratio follows from Eqs II.286 and 281

$$(I_{\text{band}}/I_{\text{line}}) = C_1 ([\text{SrOH}]/[\text{Sr}]) \left\{ (Q_{r,u})_{\text{SrOH}} \, Q_{\text{Sr}}/Q_{\text{SrOH}} \right\} \times \exp[-(E_{\text{band}} - E_{\text{line}})/kT] ,$$
(VIII.41)

in which C_1 includes all factors that are independent of temperature, $(Q_{r,u})_{\text{SrOH}}$ is the rotational partition function of the upper state of SrOH, Q_{Sr} and Q_{SrOH} are the internal partition functions of Sr and SrOH, respectively, E_{band} is the sum of electronic and vibrational energy in the upper state of SrOH, and E_{line} is the excitation energy of Sr. The ratio [SrOH]/[Sr] is determined by the balanced reaction: $\text{Sr} + \text{H}_2\text{O} \rightleftharpoons \text{SrOH} + \text{H}$ with

$$K_{\text{Sr}} = [\text{SrOH}][\text{H}]/[\text{Sr}][\text{H}_2\text{O}] ,$$
(VIII.42)

which holds also in case of radical disequilibrium. The actual ratio $[\text{H}]/[\text{H}_2\text{O}]$ can be found with the LiOH-method (see Sect. IV.5b-2) from

$$[\text{H}]/[\text{H}_2\text{O}] = K_{\text{Li}}/\phi_{\text{Li}} ,$$
(VIII.43)

where $K_{\text{Li}} = [\text{LiOH}][\text{H}]/[\text{Li}][\text{H}_2\text{O}]$ and $\phi_{\text{Li}} = [\text{LiOH}]/[\text{Li}]$. Thus

$$[\text{SrOH}]/[\text{Sr}] = \phi_{\text{Li}} K_{\text{Sr}}/K_{\text{Li}} .$$
(VIII.44)

Inserting Eq. VIII.44 into VIII.41 and using for K_{Sr} and K_{Li} statistical-mechanical expressions (see Eq. VIII.38), one gets

$$I_{\text{band}}/I_{\text{line}} = C_2 \left\{ Q_{\text{Li}} (Q_{r,u})_{\text{SrOH}} / (Q_e Q_v Q_r)_{\text{LiOH}} \right\} \phi_{\text{Li}}$$
$$\times \exp\left[-\left\{ E_{\text{band}} - E_{\text{line}} + D_0(\text{LiOH}) - D_0(\text{SrOH}) \right\}/kT \right] ,$$

where E and D_0 are expressed in the same energy unit as kT. The latter expression can be rewritten as

$$\ln \left\{ (I_{\text{band}}/I_{\text{line}}) (Q_v Q_r)_{\text{LiOH}} / (Q_{r,u})_{\text{SrOH}} \phi_{\text{Li}} \right\} =$$
$$C_3 - \left\{ E_{\text{band}} - E_{\text{line}} + D_0(\text{LiOH}) - D_0(\text{SrOH}) \right\}/kT .$$
(VIII.45)

Again all temperature-independent factors are included in C_3. On the left-hand side of Eq. VIII.45, the factors $I_{\text{band}}/I_{\text{line}}$ and ϕ_{Li} can be measured and the factor $(Q_v Q_r)_{\text{LiOH}}/(Q_{r,u})_{\text{SrOH}}$ can be calculated if the molecular parameters of LiOH and SrOH are known. The ratio $(Q_r)_{\text{LiOH}}/(Q_{r,u})_{\text{SrOH}}$ is independent of temperature when the LiOH molecule has the same configuration as SrOH (both bent or both linear); then this ratio can be included in C_3 and only the T-dependence of $(Q_v)_{\text{LiOH}}$ has to be calculated. When the configurations are different, the temperature dependence of this ratio should be taken into account (see van der Hurk, Hollander and Alkemade 1974).

Plotting the left-hand side of Eq. VIII.45 versus reciprocal temperature one should obtain a straight line, the slope of which yields the sum:

$$E_{\text{band}} - E_{\text{line}} + D_0(\text{LiOH}) - D_0(\text{SrOH}).$$

The terms E_{line} and $D_0(\text{LiOH})$ are known, but the excitation energy of the band is usually not well-known; in any case one can take $h\nu_{\text{band}}$ as a lower limit for E_{band}. When $D_0(\text{SrOH})$ is to be determined from the slope of the above plot, any error in E_{band} shows up in D_0.

Gurvich, Ryabova, Khitrov and Starovoitov (1971) have used this method to determine $D_0(\text{SrOH})$, while assuming a rather questionable value for E_{band}. Their $D_0(\text{SrOH})$ value (4.04 ± 0.13 eV) is probably systematically too low, but it is impossible to estimate how much, because they do not report the precise E_{band} value used.

Van der Hurk, Hollander and Alkemade (1974) have determined the excitation energy of the SrOH band at 6060 Å using Kalff's (1971) value of $D_0(\text{SrOH})$, which has been determined without making any assumption concerning E_{band}. They have discussed in detail the possible systematic errors and the influence of an assumed molecular configuration upon the final result. Possible systematic and accidental errors are reported to be 0.2 and 0.1 eV, respectively. Although they wanted to determine E_{band} values, their discussion of the method also applies when used to determine D_0 values.

5b-4 *SIMULTANEOUS DETERMINATION OF $D_0(\text{CaOH})$ AND $D_0\{\text{Ca(OH)}_2\}$*

Alkaline-earth atoms may react with flame constituents to form oxides, hydroxides and dihydroxides. In general, one derives from the combined balanced formation reactions VIII.12, 31 and 17, with equilibrium constants K_1, K_{31} and K_{17}, respectively,

$$\phi = K_1[\text{H}_2\text{O}]/[\text{H}] + K_{31}[\text{H}_2\text{O}]^2/[\text{H}]^2 + K_{17}[\text{H}_2\text{O}]/[\text{H}_2]. \quad (\text{VIII.46})$$

Here the first term of the right-hand side accounts for MOH, the second for M(OH)_2, and the third for MO. Equation VIII.46 also holds under conditions of radical disequilibrium, because the formation reactions are balanced (see Cotton and Jenkins 1968, and Sect. 4). If flame conditions are chosen so that oxide formation is negligible, then the first two terms of the right-hand side of Eq. VIII.46 remain to be considered. Under this condition Cotton and Jenkins (1968) have been able to determine $D_0(\text{CaOH})$ and $D_0\{\text{Ca(OH)}_2\}$. They have used three sets of fuel-rich

H_2-O_2-N_2 flames, each set consisting of six flames. Unburnt gas flows have been chosen so that the temperatures and the H_2O concentrations in each flame within a set were the same. In the interconal region of each flame the (excess) H concentration and ϕ have been measured as a function of distance from the combustion zone.

Relative [H] values have been measured by the LiOH-method and have been converted into absolute values by applying this method, under identical experimental conditions, also in a hot flame where [H] has reached its known equilibrium value. Relative [Li] and [Ca] have been measured by a double-beam atomic absorption method, using hollow-cathode lamps (see also Sect. III.15). The absolute [Ca] has been obtained by the peak-absorption method (see Sect. V.2c). The total amount of Ca introduced has been measured by collecting a sample of mist leaving the burner (with extinguished flame) with an electrostatic precipitator. The precipitate was solved and the concentration of the Ca solution thus obtained was measured by atomic absorption analysis.

For a set of isothermal flames with constant $[H_2O]$ extrapolation of plots of ϕ against $1/[H]$ showed that $\phi \to 0$ for $1/[H] \to 0$, which proved that CaO formation, which is independent of [H], is negligible in these fuel-rich flames (see Fig. VIII.8). From the plots, the derivative $d\phi/d[H]^{-1}$ has been determined which showed a linear dependence on $[H]^{-1}$ (cf. Eq. VIII.46; see Fig. VIII.9). Then K_1 and K_{31} followed from the intercept and slope of this linear graph. From the equilibrium constants thus obtained, $D_0(CaOH)$ and $D_0\{Ca(OH)_2\}$ have been derived through third-law calculations. The results for the mono- and di-hydroxides of Ca, Sr and Ba are included in Table VIII.2 on page 767.

5b-5   SIMULTANEOUS DETERMINATION OF D_0(BaO) AND D_0(BaOH)

In principle, the dissociation energy of a compound can be determined by measuring absolute ϕ values if that compound is the only or at least the dominant species, or by measuring the band-to-line intensity ratio as a function of temperature if the excitation energy of the band is known. For situations in which these conditions are not fulfilled Kalff (1971) and Kalff and Alkemade (1973) have developed a procedure which results in the simultaneous assignment of most probable values to D_0(BaOH) and D_0(BaO). They have found that the contribution of $[Ba(OH)_2]/[Ba]$ to ϕ is insignificant in their hot CO-N_2O flames with temperatures ranging from 2650 to 2850 K and with low water content.

The negligible contribution of $[Ba(OH)_2]/[Ba]$ to ϕ in hot CO-N_2O flames, in contrast to the situation in cool and fuel-rich H_2-O_2-N_2 flames, can be understood by inspection of Eq. VIII.46. Equilibrium constants K_1, K_{31} and K_{17} are only weakly dependent on temperature, because the overall reaction energy of the

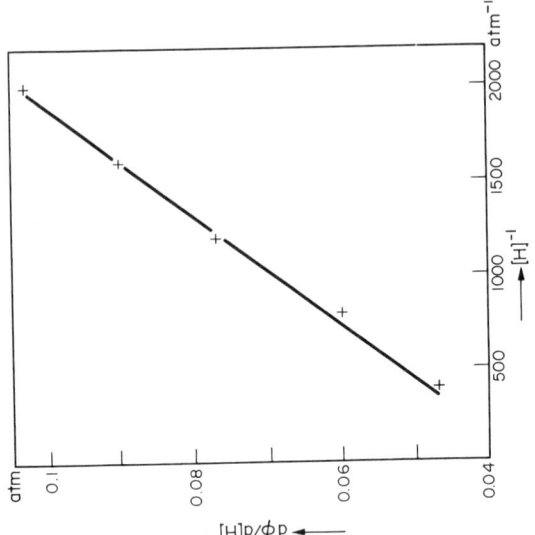

Fig. VIII.9 The derivative, $d\phi/d[H]^{-1}$, determined from the drawn curves for $T = 1570$ K in Fig. VIII.8, is plotted versus $1/[H]$.

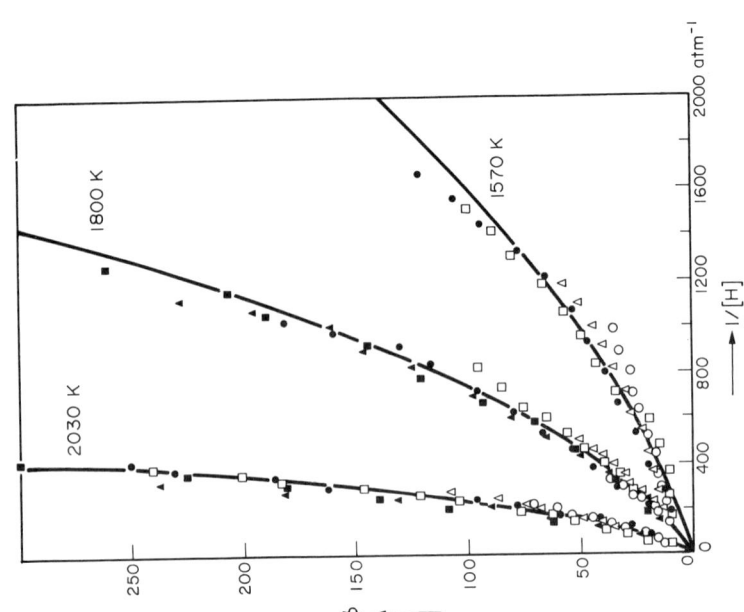

Fig. VIII.8 The association factor ϕ is plotted versus $1/[H]$ for calcium at three temperatures, as measured by Cotton and Jenkins (1968). For each temperature six different unburnt-gas compositions were chosen; in each flame, measurements were performed at various heights above the primary combustion zone, i.e. at various values of [H]. Solid curves are least-square fits to Eq. VIII.46 (without the last term of the right-hand side) of the experimental ϕ values.

FLAME SPECTROMETRIC DETERMINATION OF DISSOCIATION ENERGIES   Sect. VIII.5b

exchange reactions VIII.12, 31 and 17 for Ba is small. The ratio $[H_2O]/[H_2]$, which is related to $[CO_2]/[CO]$ by a balanced reaction, is also but little affected by temperature variations for given fuel/oxidant ratio. On the other hand, $[H_2O]/[H]$ strongly decreases with increasing temperature, so that $[BaOH]/[Ba]$ (first term of right-hand side of Eq. VIII.46) and even more $[Ba(OH)_2]/[Ba]$ will markedly decrease with increasing temperature.

Total ϕ values have been measured in six $CO-N_2O$ flames using the Na-comparison method (see Sects V.3b and 6). Special attention was paid to the suppression of or correction for ionization of Ba and Na. In each flame $[H_2O]/[H]$ has been measured. Measured ϕ values have been corrected for minor $Ba(OH)_2$ formation, if necessary. Putting the measured total ϕ-factor equal to either $[BaOH]/[Ba]$ or $[BaO]/[Ba]$, one finds an upper limit for $D_0(BaOH)$ and for $D_0(BaO)$, respectively. Then a trial value of $D_0(BaOH)$ lying below this upper limit was

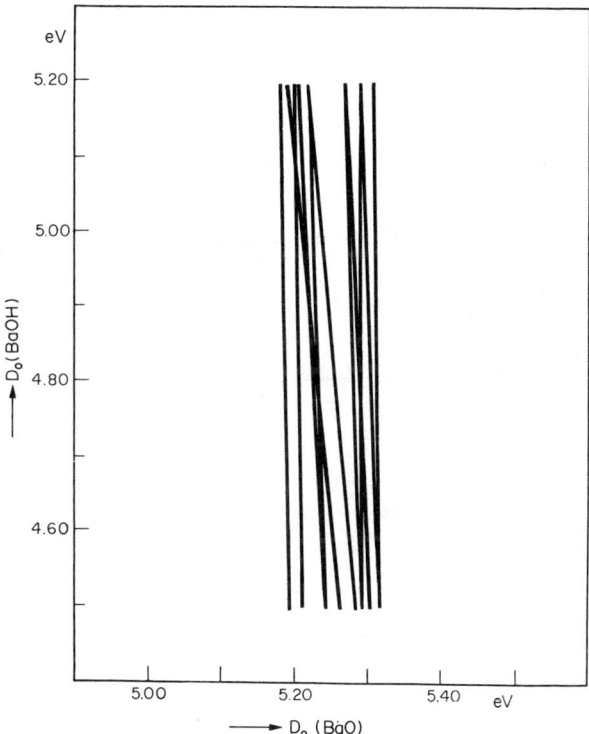

Fig. VIII.10 Graphical representation of the associated pairs of trial values for $D_0(BaO)$ and $D_0(BaOH)$, as calculated from measured total association factors in different flames by Kalff (1971) and Kalff and Alkemade (1973).

Chapt. VIII   FORMATION OF METAL COMPOUNDS

chosen, which yielded a calculated trial value for [BaOH]/[Ba]. By subtracting this trial value of [BaOH]/[Ba] from the measured ϕ, [BaO]/[Ba] and thus a value for $D_0(\text{BaO})$ were found. By starting from different trial values of $D_0(\text{BaOH})$, different pairs of associated $D_0(\text{BaOH})$ and $D_0(\text{BaO})$ values were obtained and these were plotted in a diagram (see Fig. VIII.10).

Similar curves were constructed for each total ϕ value measured under different flame conditions. Ideally, all these curves should intersect in one point, corresponding to the 'true' pair of $D_0(\text{BaOH})$ and $D_0(\text{BaO})$ values. Because of experimental errors no sharp intersection point is actually found, and only 'most probable' values for both dissociation energies together could be derived. From an estimate of the experimental errors, the uncertainty in the most probable D_0 values was calculated. This uncertainty depends largely on the steepness of the curves shown in Fig. VIII.10; it is relatively small for $D_0(\text{BaO})$, but large for $D_0(\text{BaOH})$. This outcome is a consequence of the dominance of BaO over BaOH in the set of flames investigated. Using the above method, Kalff (1971) and Kalff and Alkemade (1973) have determined most probable values for the dissociation energy of CaOH, SrO, SrOH and BaO. The latter results are included in Tables VIII.1 and 2 on pages 766-67, 768.

5b-6 *CONCLUSION*

Through the examples in Sect. 5b we believe to have shown that reasonably reliable D_0 values for metal compounds of different kinds may be obtained by flame spectrometric methods. Combination of second-law- and third-law determinations is recommended. Possible systematic errors in molecular and atomic parameters (partition functions, oscillator strengths, and so on) can then be checked by intercomparison. Systematic errors in flame temperature need not be serious when the Na-line reversal method is applied carefully in shielded flames with a homogeneous temperature distribution in the flame volume that contains the metal vapour. In some cases suprathermal chemiluminescence and infrathermal effects (the latter especially in Ar-diluted flames) could invalidate the application of intensity measurements. Absorption measurements are then recommended. In measuring ϕ values, the Na- or Ag-comparison method is preferred to calculations based on the nebulization efficiency, but possible disturbing effects with the former method, as described in Sect. V.6, should not be overlooked. In general, confidence about the results obtained can be much enhanced by repeating measurements under different flame conditions (by altering height of observation, flame temperature, flame composition, and so on) and by applying different methods.

CHAPTER IX

Ionization in Seeded Flames

1. INTRODUCTION

Electric conductivity measurements (see Wilson 1912, 1931) and optical observations on the depletion of neutral alkali atoms by Huldt (1948) and Smit and Vendrik (1948) long ago demonstrated quantitatively the occurrence of ionization in metal-seeded flames. The number density of charge carriers may be as high as 10^{13} cm^{-3} when concentrated alkali solutions are sprayed into a flame of, say, 2500 K at 1 atm pressure. This value exceeds markedly the maximum density of natural flame ions present in or close to the combustion zone of hydrocarbon flames (see Sect. IV.7). In contrast to these natural ions, the state of ionization of alkali atoms in the burnt gas of the hotter flames conforms reasonably well to Saha equilibrium (see Eq. II.59).

Even when the electron density is as high as 10^{13} cm^{-3}, the partial electron pressure is still less than 10^{-5} atm. The flame is thus essentially a weakly ionized plasma and the total energy consumed in the ionization is negligible in comparison to the heat content of the flame gas.

A flame with an electron density above about 10^8 cm^{-3} may be termed a *plasma* if we define the latter as a system in which collective motions of free charge carriers are in principle possible (see, e.g., Thorne 1974). A condition is that the *Debije shielding distance* ρ_D (see Sect. II.3b-4) is much less than the dimensions of the system. For electron concentrations above 10^8 cm^{-3} and at usual flame temperatures ρ_D is, indeed, about 10^{-2} cm or lower (see Eq. II.63).

The flame as a whole is electrically neutral. Since the Debije shielding distance is very small, *charge neutrality* is also preserved locally (see Thorne 1974); that is, the density of the free electrons plus singly charged negative ions balances everywhere the density of the singly charged positive ions.

813

Chapt. IX										IONIZATION IN SEEDED FLAMES

In contrast to the situation in highly ionized plasmas, collisions between electrons (or ions) and neutral flame particles dominate over electron-electron or electron-ion collisions. The electric conductivity and the attenuation of electromagnetic waves is thus governed by the exchange of momentum between charged species and neutral flame particles upon collision (see Sect. 3a[†]). This exchange is quantitatively described by the *collision frequency* ν (in s^{-1}), which is the probability per second of a collision between a given electron or ion and an arbitrary flame particle.

Flames are a suitable gaseous milieu for studying ionization of additives. *Advantages* of flames in this respect are:

1. the flame is a 'thermal bath' whose bulk properties are not (much) disturbed by the introduction of an additive in trace amounts; besides, flames fed by metal salt solutions are simple to construct;

2. the natural ionization in pure H_2 and CO flames is quite negligible compared to that of metallic additives (see Sect. IV.7);

3. the occurrence of excess natural ions in hydrocarbon flames or in H_2 flames with a trace of hydrocarbon fuel enables us to study their interactions with metal ionization (see Sect. 2b);

4. the ions produced are singly charged (see Sect. 2), which simplifies the analysis; some of them, such as $SrOH^+$, HWO_4^-, WO_2^- or BO^- (see Jensen and Miller 1969, 1970, and Miller 1972, 1978) are rather unusual species, which may be difficult to produce and to investigate in other milieus;

5. flames are by their very nature especially suited to study chemi-ionization processes; in general, ionization in flames is mainly brought about by heavy-particle collisions and chemical reactions (see Sect. 2b); photoionization and photodetachment play no role because of the low radiant-energy density (at least in the absence of a strong external radiation beam);

6. additives can be supplied in reproducible and easily controllable concentrations;

7. the temperature, bulk composition and total pressure of the flame can be varied over a wide range;

8. a variety of optical and electric techniques is available for ionization measurements, at least for those in the interzonal region (see Sect. 4).

[†] Section numbers without a roman cipher refer to the same chapter.

INTRODUCTION   Sect. IX.1

One of the *disadvantages* of the use of a flame is that the energy available for ionization is in the thermal range and thus limited, if we disregard certain exo-ergic chemi-ionization reactions. This energy-restriction may be relaxed when a strong laser beam is used to saturate an excitation level, especially so when this level lies near the ionization continuum. Furthermore, as in any bulk system, the rate constants determined are phenomenological ones; that is, they are a weighted average over the distributions of the thermal relative velocity and the internal energy. Even when in a flame a single excitation level is saturated by a tuned laser beam, the (enhanced) ionization rate constant is a weighted average over a group of neighbouring excitation levels. This holds because population redistribution by collisions is faster than collisional ionization. Here, crossed-beam experiments with velocity- and internal-state selection have a definite advantage (for a recent survey see Wexler and Parks 1979); but they are more difficult to perform. Finally, flames are multi-component systems; specific ionization rate constants or cross sections can be obtained only indirectly by varying the flame gas composition in a known manner and solving a set of equations with several unknowns (cf. Sect. II.4a-1). Also in this respect do crossed-beam experiments as well as shock-tube experiments (see, e.g., Shackleford and Penner 1966) have a definite advantage.

The *main aims* of ionization experiments in flames with additives have been the establishment or the measurement of:

1. the presence or absence of Saha equilibrium;

2. ionization energies and electron affinities (see Sect. 5b);

3. the kinds of ions formed;

4. the dominant ionization and neutralization processes, physical as well as chemical, for various species and under different conditions (see Sect. 2);

5. the rate constants of ionization, neutralization and/or ionic re-arrangement processes (see Sects 3c and 5a); a necessary condition for the realization of objectives 4 and 5 is that some deviation from Saha equilibrium must be present; in flames such deviations may occur naturally (see Sect. 3d);

6. activation energies for the forward ionization process from ionization rate constants measured as a function of temperature (see Sects 5a and 6);

7. the electron mobility or electron collision frequency at elevated temperatures (see Sects 4 and 5b).

Ionization studies in flames (with and without additives) have been greatly stimulated by their significance in combustion research and lately also by

Chapt. IX   IONIZATION IN SEEDED FLAMES

theoretical interests (see Sect. 6). Above all, such studies have been undertaken with a view to potential applications in MHD power generation, combustion control, space flight (interference of rocket exhaust with the transmission of radio waves), flame detectors in gas-chromatography and in warning or switching devices (see Lawton and Weinberg 1969). Analytical flame spectroscopy has profited especially from the findings of ionization studies in flames with metallic additives (see Sect. 6) and has also prompted several investigations in this field.

In this chapter we shall restrict ourselves to ionization phenomena occurring in the homogeneous gas phase in the interzonal region of stationary laboratory flames seeded by metallic species or halogens. Strong electric fields are supposed to be absent (see for an extensive treatment of field effects Lawton and Weinberg 1969). In Sect. 2 we introduce the terminology used and survey the kinds of ionization processes and ionic species expected in flames. Section 3 presents various theoretical relationships that will be used in subsequent sections, and a theoretical discussion of the important question of thermal or nonthermal ionization. The main experimental techniques applied with flames are briefly described in Sect. 4. Some typical experiments that have been performed under various flame conditions in order to determine ionization rate constants are dealt with in some detail in Sect. 5 and results obtained are given; other ionization experiments in flames are only mentioned briefly. Finally, Sect. 6 contains a discussion of the collisional-ionization rate constants found with the alkali atoms in flames; here conclusions are also drawn from the preceding sections as regards the significance of ionization effects in other flame studies and in analytical flame spectroscopy.

The reader who wishes further information is referred especially to the authoritative and well documented book by Lawton and Weinberg (1969), which is concerned with ionization in combustion in all its aspects and which contains a short section on ionization of metallic additives. Also recommended for further reading are chapters in the books by Cambel, Duclos and Anderson (1963), von Engel (1964), Mavrodineanu and Boiteux (1965), Miller (1968), and Gaydon and Wolfhard (1979). Reviews which (partly) cover the subject matter of this chapter are: Jensen and Miller (1969), Feugier (1970), Sugden (1971), Fontijn (1972, 1974), Jensen and Travers (1973), Page (1973), Hayhurst (1974), Miller (1976, 1978). Ionization in flames has, more briefly, been dealt with also in most of the textbooks on analytical flame spectroscopy cited in Chapt. I.

2. SURVEY OF IONIZATION PROCESSES

Ionization processes involving additives to the flame in the homogeneous gas phase that have been postulated in the current literature will be reviewed and illustrated by examples. Thermionic emission of solid particles (see Lawton and Weinberg 1969, and Page and Woolley 1974) and processes that take place during or after the beam formation in ion mass spectrometers (see Hayhurst and Kittelson 1974, and Burdett and Hayhurst 1979) will not be considered. The discussion in this section is mainly qualitative; we refer to Sects 3, 5 and 6 for quantitative theoretical relationships and experimental results. The reader who wants more detailed information on the subject of this section and on the relevant literature sources will find this in the following reviews or books: Berry (1963), Calcote and Jensen (1966), Lawton and Weinberg (1969), Feugier (1970), Sugden (1971), Fontijn (1972, 1974) and Hayhurst (1974), and in the relevant chapters in some books cited in Sect. 1. It should be pointed out, however, that the rapid growth of our knowledge in this field makes some of the older reviews incomplete and even partly obsolete.

2a CLASSIFICATION AND EXAMPLES OF IONIZATION PROCESSES

The terminology used in the literature for classifying the various processes involving ions is not always consistent. In particular the use of the term thermal ionization is confusing. In order to avoid ambiguities we shall adopt the following terminology which is an extension of the general terminology used in Table II.1 on page 108.

Any process (physical collision, chemical reaction, radiation) by means of which the number of (singly charged) positive or negative ions (not electrons) is increased is called an *ionization process*.[†] When the number of ions is reduced, we speak of a *neutralization process*. An *ion-recombination* process in which a positive ion recombines with a free electron or negative ion is a special type of neutralization process (see examples i, ii, iv, vi-viii and x in Table IX.1 on page 823; the examples iii and ix are a different type of neutralization process). When the number of (positive or negative) ions remains unchanged, we speak of an *ionic-rearrangement process* (see examples v, xi-xv). When the latter process takes place predominantly in one direction, as is the case when the system is far out of equilibrium, the ion produced is called a *secondary ion*. Because of the principle of microscopic reversibility (see Sect. II.4b-4) each possible ionization process corresponds to a possible, reverse neutralization process. In nonequilibrium

---

[†] The term ionization process is also used in a broader sense and covers any process in which ions are involved (cf. the title of this section).

Chapt. IX   IONIZATION IN SEEDED FLAMES

conditions the rates of the two processes may or may not be equal; when they are
(nearly) equal there is a state of partial equilibrium (see Sect. II.3c-3).

The processes mentioned above are accompanied by a transition of an electron
from a bound state to a continuum state (*detachment*) or vice versa (*attachment*), or
from a bound state in one species to a bound state in another species (*transfer*).
Note that in example (xi) of Table IX.1 it would be more natural to say that an
electron <u>defect</u> is transferred from one species to another (each of which contains
the same metallic element). A similar example, not important in flames, is:
$Cs_2^+ + X \rightleftharpoons Cs^+ + Cs + X$. We shall, however, classify these cases also as an electron
transfer process and treat them in the same way as for example: $Cl_2^- + X \rightleftharpoons Cl^- + Cl + X$
(not important in flames).

All the above processes can be divided into *physical* processes and *chemical*
processes (or *reactions*). In a *chemi-ionization* process ionization is accompanied
by the formation of new chemical bonds (cf. chemiluminescence in Sect. VI.1; see
also Sugden 1965 and Fontijn 1972). This may result in *association, dissociation*
or *exchange* of atoms or groups of atoms. Note that chemical dissociation may be
accompanied by ion-recombination (see example vi in Table IX.1).

Physical processes can be subdivided into *collisional* and *radiative* processes.† The latter class might be important in flames only if an intense laser beam
is used to induce ionization by a multiphoton process (for a relative estimate of
radiative versus collisional ionization rates see Berry 1963, and for a general
bibliography for radiative processes Kieffer 1976, and Beaty and Gallagher 1976).
In a collisional ionization process the translational energy of relative motion of
the partners as well as their internal energies may contribute to the total energy
balance. The well-known Penning process (important in gas discharges but not in
flames; see Fontijn 1972) is a special type of collisional ionization. When a
species is first brought to an excited state and then becomes ionized, one encounters the terms *cumulative, step-up* or *ladder ionization* (see Sect. 6).

A special case of step-up ionization (see process no. 45 in Table II.1) is
observed when an appreciable fraction of the metal atoms is first brought to an
excited state (with energy E_{exc}) by resonant absorption of a strong laser beam.
Since the activation energy for collisional ionization from this state is lower
than from the ground-state by an amount equal to E_{exc}, the ionization rate is
considerably enhanced. This enhancement has been detected by electric-probe measurements in flames with Na atoms excited by a pulsed or c.w. laser beam tuned at one

---

† The rapid decay of an auto-ionizing state of a single species to an ion and a
free electron will not be considered as a separate, monomolecular process but
rather as an intermediate stage in a collisional, radiative or chemical ionization process (see Sect. II.4h).

SURVEY OF IONIZATION PROCESSES   Sect. IX.2a

of the Na-D lines (see Green et al. 1976, Alkemade 1977, van Dijk 1978, and van
Dijk and Alkemade 1980a,b; see also Note 14 to Table IX.1). This *opto-galvanic
effect*, also called *laser-enhanced ionization* (LEI), has been proposed by Green
et al. (1976, 1976a) as a tool for analytic determinations of a large number of elements in flames (see also Turk et al. 1978, 1979). Element-specificity is obtained
by tuning the dye laser to one of the optical transitions of the analyte atom.
Advantages are the reduction of the analytic limit of detection[†] and the elimination of spectral interferences. On the other hand, new types of ionization interference may play a role here. As expected, the collisional ionization rate is
enhanced when higher-lying states are excited from the ground-state by two-photon
absorption or by a frequency-doubled laser beam, or when they are excited from a
thermally populated lower excitation level by one-photon absorption.

Since the population of the laser-excited state is redistributed by thermal
collisions over a multitude of other excited states before ionization takes place
(see van Dijk 1978, and van Dijk, Zeegers and Alkemade 1979), we are here dealing
with a multi-step process. This complicates the interpretation.

In the literature the term thermal ionization is often used as a counterpart to chemi-ionization, and is in fact used as a synonym for collisional ionization. This use of the term is misleading as it could suggest that chemi-ionization
per se leads to deviations from thermal equilibrium. An analogous misunderstanding is found in the literature with respect to the pair of terms thermal radiation
and chemiluminescence (see Sect. VI.4a). In accordance with our definition of
thermal radiation in Sect. VI.4a, we shall define thermal ionization as relating
to the state of a component in the gas and not to a particular process.[*] *Thermal
ionization of a given species* (X) exists if the concentrations of X^+ or X^-, X and
the free electrons obey the Saha law at the temperature of the flame when the latter
can be meaningfully defined (see Sect. II.3c-3). This definition does not require
that the electron concentration should also be in equilibrium with all other kinds
of ions, e.g., the natural flame ions, or that the atoms involved should be in
chemical equilibrium with their molecular compounds. Strictly speaking, when such
deviations from thermodynamic equilibrium occur, the concept of thermal ionization
can be applied in an approximate sense only (cf. the discussion on thermal radiation in Sect. VI.4a sub ii); but we shall ignore the latter theoretical subtlety.

[†] For a critical comparison with the detection limits obtainable by other laser-
assisted optical or electric techniques see Alkemade (1981).

[*] In this respect our definition of thermal ionization conforms to that given by
Sugden (1965). The latter author, however, uses this term in a broader sense
than we shall do, and he also includes the case of a steady-state balance that
is shifted by the presence of excess flame radicals.

Chapt. IX   IONIZATION IN SEEDED FLAMES

We are then dealing with a case of partial equilibrium brought about by the balancing of a sufficiently rapid ionization and reverse neutralization process (see Sects II.3c-3 and 4a-2). The general causes for deviations from thermal ionization will be discussed in Sect. 3d. Here we stress only that thermal ionization can also exist if the dominant ionization process is chemical instead of physical; it is sufficient that the concentrations of the species involved in the ionization and neutralization reactions are in chemical equilibrium (cf. Sect. VI.4a sub iv).

When all species are in a state of thermal ionization, the flame is said to be in a state of *Saha equilibrium*.

Table IX.1 (page 823) presents typical examples of postulated ionization processes involving additives, classified according to the distinctions introduced above. The table includes the main types of ionization and reverse neutralization processes that can be of importance or have been studied in flames. For an explanation we refer to the Notes attached to the table. Some processes actually consist of a chain of several elementary steps (called: *multi-step process*; see also Table II.1 on page 108). For example, Fontijn (1972) has suggested that the chemical associative ionization process (viii) may proceed in two steps according to process 44 in Table II.1. Further, association of two H radicals can lead indirectly to alkali ionization, with chlorine acting as a catalyst, via the following three steps:

$Na + Cl \rightarrow Na^+ + Cl^-$ (forward process iv)

$Cl^- + H \rightarrow e^- + HCl$ (reverse process ix)

$HCl + H \rightarrow H_2 + Cl$ (reaction VIII.33)

$+$ ─────────────────────

$Na + 2H \rightarrow Na^+ + e^- + H_2$.   (xvi)

Each of these bimolecular reaction steps is balanced if enough chlorine is present to make this ionization process dominant over other processes. A steady-state balance between Na and Na^+ will then exist, which depends on the H concentration. When H is present in excess (see Sect. IV.5), the Na^+ concentration will rise above equilibrium, too; there is then no thermal ionization for Na. Under this condition one expects therefore that $[Na^+]$ and $[e^-]$ will initially rise when increasing amounts of chlorine are added. This rise, however, is followed by a fall in $[e^-]$ because of the increasing effect of Cl^- formation on the free-electron concentration through the balanced process (ix) with larger chlorine additions. The increasing formation of NaCl molecules through reaction VIII.32 also depresses the free-electron concentration. These expectations have been tested experimentally by Hayhurst and Sugden (1967).

As explained in Sect. II.4h multi-step processes should be well distinguished from (single-step) processes that occur via the formation of a temporary

SURVEY OF IONIZATION PROCESSES   Sect. IX.2a

collision complex. The lifetime of the latter complex is typically of the order of 10^{-13} s, which is much shorter than that of the intermediate species involved in a real multi-step process. When the reactants are 'heavy' particles (not electrons), the course of a single-step ionization process may be conveniently described by means of potential-energy curves (see Sect. II.4), as in the case of collisional excitation processes (see Sects VI.2a and 2b). For example, the forward electron-transfer process (iv) can be conceived as the result of a transition from a covalent (Na + Cl) state to an ionic (Na$^+$ + Cl$^-$) state through a (pseudo-)curve crossing (cf. Fig. II.6 on page 101; see Baede 1975, Olson 1977, and Los and Kleyn 1979). When Na and Cl are initially in their ground states, the temporary collision complex can dissociate into a pair of free Na$^+$ and Cl$^-$ ions if the available internal plus translational energy exceeds $E_{ion} - E_{aff}$, where E_{aff} is the electron affinity. The collision complex may also follow another 'path' that leads to the formation of an excited neutral atom (inelastic collision), or it may return to the 'entrance channel' (elastic collision).

Similar ionic-curve crossings are involved when an alkali atom becomes ionized upon collision with a di-atomic molecule such as Cl_2, HCl, CO, O_2 and N_2; these ionization processes have been studied by molecular-beam techniques (see, e.g., Lacmann and Herschbach 1970, Moutinho, Baede and Los 1970, Baede and Los 1971, and the reviews by Baede 1975, and Los and Kleyn 1979). With certain di-atomics, processes of type (i) may occur in two steps, e.g.: $Na + O_2 \to Na^+ + O_2^-$ through ionic curve-crossing, followed by collisional electron detachment: $O_2^- + X \to O_2 + e^- + X$, where X is an arbitrary flame molecule (see Hayhurst and Telford 1972a; see also Sect. 6b-4). From an energetic point of view ($E_{aff} \simeq 0.5$ eV for O_2; see Moutinho, Baede and Los 1970, and Baede 1975), the second step would be the more rapid one. Since N_2 has a negative electron affinity $E_{aff} \simeq -1.9$ eV and N_2^- does not exist for a much longer time than 10^{-15} s,[†] there is no two-step process of this kind for N_2.

Collisions of Na atoms in the ground state with molecules need not lead directly to Na ionization. If the Na atom is first transferred to a state of higher energy before it undergoes an ionizing collision with N_2, the overall process is in fact a multi-step, cumulative ionization process (see also Sect. 6a-1). The lifetime of the intermediate species, in casu an excited Na atom, is long enough (typically 10^{-8} s; see Sect. VI.4c-1) to distinguish it from the temporary collision complex considered above.

[†] See footnote ǂ on page 585.

Chapt. IX   IONIZATION IN SEEDED FLAMES

Previously, collisional ionization of alkalis was believed to occur in two steps involving Na^+H_2O as an intermediate species (see Mavrodineanu and Boiteux 1965, and Page 1973); a slow, rate-determining step: $Na + H_2O \rightarrow Na^+H_2O + e^-$ is followed by a fast step: $Na^+H_2O + N_2 \rightarrow Na^+ + H_2O + N_2$. The activation energy of the first step was assumed to be markedly lower than the atomic ionization energy; but this assumption had to be abandoned on the basis of later experiments (see Sugden 1965a). Although genuine Na^+H_2O ions do occur in the flame (and are not just an artefact produced in the sampling beam of the ion mass-spectrometer; see Hayhurst and Sugden 1966), the importance of this two-step process in the alkali ionization is nowadays denied (see Jensen and Padley 1967, Lawton and Weinberg 1969, and Gaydon and Wolfhard 1979; see also Sect. 6).

Ionization processes of type (vi) and (vii) involving chemical association may occur with curve-crossing (see Biondi *et al.* 1965) as well as without curve-crossing, namely through a Franck-Condon transition in which an electron is ejected, carrying with it the surplus energy (see Fig. IX.1 and Fig. II.3 on page 36; see also Fontijn 1972, 1974). The process can be either endo-ergic or exo-ergic depending on the relative positions of the $(M + X)$ and $(M^+ + X)$ potential-energy curves. The position of the $(M + X)$ curve depends, again, on the initial state in which the metal atom M is found.

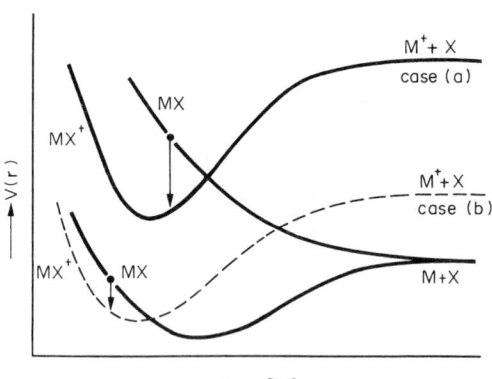

Fig. IX.1 Schematic potential-energy curves showing the potential energy, $V(r)$, of the collision complex as a function of distance, r, between metal atom M (or ion M^+) and collision partner X. Formation of MX^+ proceeds through a Franck-Condon transition (presented by a downward vertical arrow) in which an electron is ejected. In case (a) the minimum of the $(M^+ + X)$ curve (solid line) lies above the initial asymptote of the $(M + X)$ curve, and the process is endo-ergic. In case (b) (dashed curve) the process is exo-ergic. Only if sufficient energy is available can the MX^+ molecule redissociate into M^+ and X. (Redrawn after Fontijn 1974.)

SURVEY OF IONIZATION PROCESSES

Table IX.1
Classification and Examples of Flame Ionization Processes involving Additives[†]

No.	Examples*	Notes	Kind of process			Electron transition			Nature of process			
									chemical			phys
			ionization	neutralization	ionic re-arrangement	detachment	attachment	transfer	association	dissociation	exchange	collisional
i	$Na + N_2 \rightleftharpoons Na^+ + e^- + N_2$	1	x / x			x / x						x
ii	$Na + e^- \rightleftharpoons Na^+ + 2e^-$	2	x / x			x / x						x
iii	$Cl + e^- + N_2 \rightleftharpoons Cl^- + N_2$	7	x / x				x / x					x
iv	$Na + Cl \rightleftharpoons Na^+ + Cl^-$	3	x / x					x				x
v	$Na + SrOH^+ \rightleftharpoons Na^+ + SrOH$	4		x				x				x
vi	$Ca + OH \rightleftharpoons CaOH^+ + e^-$	5,14	x / x			x / x			x / x			
vii	$CaO + H \rightleftharpoons CaOH^+ + e^-$	5,14	x / x			x / x			x / x			
viii	$Na + CO + O \rightleftharpoons Na^+ + e^- + CO_2$	6	x / x			x / x			x / x			
ix	$HCl + e^- \rightleftharpoons Cl^- + H$	3,7	x / x				x / x		x / x			
x	$H_2O + HCl \leftarrow H_3O^+ + Cl^-$	8,13		x			x			x		
xi	$CaOH^+ + H \rightleftharpoons Ca^+ + H_2O$	9			x			x			x	
xii	$Na + H_3O^+ \rightarrow Na^+ + H_2O + H$	8,10			x			x		x		
xiii	$Pb + H_3O^+ \rightarrow PbOH^+ + H_2$	8,11			x			x			x	
xiv	$CrO + H_3O^+ \rightarrow CrOH^+ + H_2O$	8,11			x			x			x	
xv	$MnO + H_3O^+ \rightarrow Mn^+H_2O + OH$	8,12			x			x			x	

823

NOTES TO TABLE IX.1

† Examples are listed of processes that have been postulated for the formation and removal of atomic and molecular ions of metallic elements, and of halogen ions in flames. The processes are classified according to the entries at the top of the table, which are explained in the text.

* An arrow indicates in which direction the process commonly takes place when it is encountered or studied in flames. The appearance of arrows in both directions in some examples does not mean that the process is always balanced; the process corresponding to the upper arrow is classified by the crosses found in the sections of the boxes above the slanting dividing lines; the crosses below these lines relate to the same process in the opposite direction. The partners of each process may be in the ground state or in an excited state. Na, Ca and Cl are taken as being representative of an alkali, alkaline-earth, and halogen atom, respectively.

1. This is the dominant ionization and neutralization process for the alkali atoms (see Sects 5 and 6) and for Ga, In and Tl (see Kelly and Padley 1969) well away from the combustion zone and in absence of halogens (for a survey see Sugden 1971). It is also a dominant neutralization process for atoms with high ionization energy E_{ion}, such as Pb, Mn and Cr, that have been ionized above equilibrium in the combustion zone (see Jensen and Padley 1967, and Anthony et al. 1974).

2. Although free electrons have a high efficiency in ionizing atoms, this process will be important only if the electron concentration is exceptionally high or if the electrons are 'hot' (see Sect. IV.7). When hot electrons are present, the forward process (ii) will be dominant. In the literature opinions vary about the importance of this process for metal ionization in flames (see, e.g., Sugden 1965a, Calcote and Jensen 1966, and Hayhurst and Sugden 1966).

3. This process may become important when the halogen concentration in the flame is of the order of 1 mole percent or higher (see Hayhurst and Sugden 1967, and Burdett and Hayhurst 1977, 1979a; see Olson 1977 for theoretical calculations); it is part of a reaction chain that leads effectively to ionization of one alkali atom by recombination of a pair of H radicals (see reaction xvi in the text). Rate constants of the physical ionization process $K + Cl \rightarrow K^+ + Cl^-$ and the competitive dissociative chemi-ionization process $KCl + N_2 \rightarrow K^+ + Cl^- + N_2$ have been measured as a function of temperature in hydrogen flames by Burdett and Hayhurst (1977) (for a survey of similar measurements on alkali-halogen systems see Burdett and Hayhurst 1979a).

4. Since, in contrast to the alkaline-earth elements, ionization of Na proceeds relatively slowly, the electron-transfer process in the forward direction may be dominant and boost the rate of Na ionization.

5. This is one of the dominant formation processes of molecular alkaline-earth ions (see Jensen 1968, and Hayhurst and Kittelson 1974a); contrary to earlier assumptions (see Schofield and Sugden 1965, Jensen 1968, and Kelly and Padley 1971a) this process is not always rapid enough to be balanced everywhere in the flame (see Hayhurst and Kittelson 1972b). Similar associative ionization processes have been postulated for U, Mo and La compounds by Kelly and Padley (1969a).

SURVEY OF IONIZATION PROCESSES   Sect. IX.2a

6. This process has been postulated to explain the excess ionization rate of alkali atoms
    just above the combustion zone of CO flames, where excess O radicals persist (see
    Sect. 5); it may in fact be a two-step process (see text). A similar process involving
    the recombination of O+O, H+OH and H+H has been invoked to explain the excess chemi-
    ionization of K in the reaction zone of H_2 flames (see Semenov and Sokolik 1970).

7. This process, which is rapidly balanced, brings about a reduction in the free-electron
    concentration when a large amount of halogen is added as electron acceptor (see Hayhurst
    and Sugden 1966, and Bulewicz and Padley 1969); this reduction shifts the ionization
    equilibria of the metal atoms. The rate of reaction (ix) dominates that of the three-
    body electron-attachment process (iii) in H_2 flames at 1 atm by several orders of
    magnitude (see Hayhurst and Sugden 1966).

8. This reaction occurs in and just above the combustion zone of hydrocarbon flames and
    of H_2 flames with an admixture of hydrocarbon fuel (see also Sect. IV.7); it can
    easily lead to excess ionization of metal atoms or compounds with high ionization
    energies. The actual H_3O^+ concentration can be markedly affected by the occurrence of
    this process.

9. Through this process Ca^+ and Sr^+ are formed as secondary ions from the parent molecu-
    lar ion which is produced by reactions (vi) and (vii) (see Hayhurst and Kittelson 1974);
    the rapidly established partial equilibrium between the products and reactants provides
    a means for measuring the atomic hydrogen concentration from observations of the concen-
    tration ratio of the ionic species involved. Similar balanced processes may also occur
    with Sn (see Jensen 1969), Cr, Pb, Mn and Fe (see Hayhurst and Telford 1970).

10. This reaction also occurs for Tl but its importance for metals such as Pb, Cr and Mn
    having $E_{ion} > 6.3$ eV, making the reaction endo-ergic, is still under discussion (see
    Sugden 1963, Calcote and Jensen 1966, Hayhurst and Sugden 1966, Lawton and Weinberg
    1969, Sugden 1971, Page 1973, Hayhurst 1974, and Anthony et al. 1974).

11. Reactions (xiii) and/or (xiv) may be important for metals such as Pb, Cr, Mn and Cd
    with high E_{ion} (see Hayhurst and Telford 1970).

12. See Hayhurst and Telford (1970).

13. The rate constant of this neutralization process is about a tenth of that for the
    $H_3O^+ + e^-$ neutralization process considered in Sect. IV.7 (see Burdett, Hayhurst and
    Morley 1974).

14. The associative ionization reactions: $A^* + A^* \rightarrow A_2^+ + e^-$ and $A^* + A^0 \rightarrow A_2^+ + e^-$, where A^0
    and A^* denote an alkali atom in the ground-state and excited state, respectively,
    belong to this class. They have been studied for Na, K, Rb and Cs atoms in an atomic
    beam or vapour cell with the use of a dye laser or an alkali-discharge lamp as excita-
    tion source by Gounand et al. (1976), Klucharev et al. (1976, 1980), Korchevoi,
    Lukashenko and Khil'ko (1976), Bearman and Leventhal (1978), and de Jong and van der
    Valk (1979). Estimates made by van Dijk and Alkemade (1980a), based on the absolute
    cross section found in an atomic beam by de Jong and van der Valk (1979), have shown
    that the thermoneutral process: $Na(3P) + Na(3P) \rightarrow Na_2^+ + e^-$ is unimportant in flames when
    compared to collisional ionization process (i) from the same excited state. This con-
    clusion holds even when the 3P state is saturated and the flame temperature is as low
    as 1800 K if we assume that the Na density is below 10^{14} cm^{-3}.

2b FACTORS CONTROLLING THE PROBABILITY OF AN IONIZATION PROCESS

The (relative) importance of a given ionization process under given flame conditions depends on several factors. We shall discuss these factors here in a qualitative manner, using the examples shown in Table IX.1.

The rate of a collisional or chemical process, i.e. the number of processes taking place per unit time and per unit volume, is generally given by Eqs II.74 and 75 for a two- and three-body process, respectively. This rate depends on the concentrations of the partners involved and on the rate constant k which is usually a function of temperature. For a survey and quantitative discussion of ionization and neutralization rate constants found in flame experiments we refer to, e.g., Hayhurst (1974) and to Sects 5 and 6 where further references are given. Through the detailed-balance relation the rate constant of any process is directly related to that of the reverse process if the values of these constants do not deviate from those in thermodynamic equilibrium (see Sect. II.4a-2). The latter condition is not fulfilled when, for example, 'hot' electrons participate in ionization process (ii) in Table IX.1.

The importance of the *concentration factor* is obvious in processes where halogens or H_3O^+ ions are involved (see Notes 3 and 8 to Table IX.1). The presence of radical concentrations in excess of chemical equilibrium affects the rates of chemical ionization processes (vi)-(ix) and (xi), and shifts the steady-state balance of, e.g., reactions (ix) and (xi). It should be noted, however, that the ratio of the rates of ionization reactions (vi) and (vii) does not depend on the $[OH]/[H]$ ratio because of the partial equilibrium: $Ca + OH \rightleftharpoons CaO + H$. This partial equilibrium is based on the combination of the partial equilibria: $Ca + H_2O \rightleftharpoons CaO + H_2$ (see reaction VIII.17) and: $H + H_2O \rightleftharpoons OH + H_2$ (see Eq. IV.5). Consequently it is not possible to distinguish between the two ionization reactions (vi) and (vii) by varying $[OH]/[H]$. The relative importance of reaction (xiii) with respect to reaction (xiv) can however be varied by changing the degree of dissociation of the metal oxide concerned.

The boosting of Na ionization by process (v) upon addition of Sr, mentioned in Note 4, affects $[Na^+]$ only if thermal ionization does not exist for Na and/or Sr. In addition, the free electrons produced by the ionization of Sr enhance the rate of the Na^+ neutralization process (i). This second effect results in a depression of the $[Na^+]/[Na]$ ratio, also when the Na ionization is thermal (see Sect. 3b-2); this is in contrast to the first mentioned boosting effect.

Because the electron concentration is usually low compared to the other flame constituents, process (ii) is unimportant (see also Note 2). If it were important, the rate of ionization would be higher than first order with respect to the metal atoms. This holds because the concentration of the electrons produced by metal ionization will increase with metal concentration. Experimentally the ionization rate constant has been found to be independent of Na solution concentration in H_2 and CO flames at 1 atm pressure (see Hollander, Kalff and Alkemade 1963, and Sugden 1971). An extensive, general bibliography on ionization by electron impact is to be found in Kieffer (1976), and Beaty and Gallagher (1976).

The *number of reactants* participating in a process is also indicative of its probability. The three-body process (iii) is, for this reason, less significant for the production of Cl^- than the two-body process (ix) (see Note 7). A third body is required in the former process to stabilize the exo-ergic attachment of an electron to a Cl atom (see Sect. II.4h). A three-body reaction may actually consist of a chain of two- and/or three-body reactions (see the above discussion on processes viii and xvi). When a three-body multi-step process consists of rapidly balanced bimolecular processes, its overall rate will also be large.

With endo-ergic processes the rate constant contains an exponential *activation-energy factor* (see Eqs II.122 and 123); an activation energy > 1 eV will markedly impede the probability of the process and make it sensitively dependent on flame temperature. For alkali atoms in the ground state the activation energy of ionization process (i) is close to the ionization energy (see Sect.5). For excited alkali atoms the activation energy is expected to be lowered by the amount of the excitation energy. Since the electron affinity of halogen atoms, ranging from 3.3 eV (for I) to 3.8 eV (for Cl), is lower than the alkali ionization energy, forward process (iv) is endo-ergic.

When ionization is accompanied by an associative chemical reaction, the negative enthalpy of the neutral-neutral reaction will help by reducing the activation energy or even by making the process exo-ergic (see also Fig. IX.1). Ionization process (vi) is, for example, endo-ergic by only about 1 eV for Ca and Sr, and thermoneutral for Ba (see Jensen 1968, and Hayhurst and Kittelson 1974; for a more general discussion of chemi-ionization reaction enthalpies see Fontijn 1972, 1974). The positive change of about 1 eV in enthalpy for the ionic re-arrangement process (xii) is not believed to seriously affect the probability of this process for Pb atoms (see Lawton and Weinberg 1969; see also Note 10 to Table IX.1). Estimates of the activation energy of chemi-ionization processes may be based on a useful generalization made by Sugden (1965); where a reaction between an ion and a neutral, or between two oppositely charged species (including electrons) is exoergic and involves simple chemical re-arrangement, the appropriate rate constant will usually have negligible activation energy. The activation energy of the reverse endo-ergic process will then be virtually equal to the positive change in reaction enthalpy (see Sect. II.4b-2).

Considerations of the *steric factor* (see Sect. II.4b-2) have been used in

Chapt. IX   IONIZATION IN SEEDED FLAMES

the discussion on the importance of chemi-ionization process (vii) relative to process (vi) (see Hayhurst and Kittelson 1972, 1974a).

It is worth noting that cross sections for the exo-ergic ion-ion neutralization processes (iv) and (x) are of the order of 10 and 10^3 $Å^2$, respectively (see Hayhurst 1974). The large values of these cross sections are connected with the Coulomb attraction between the reactants. The relatively smaller value for the former process is due to a low probability of the ionic-curve crossing involved.

In the low-energy range ionization processes between alkalis and electronegative di-atomics like Cl_2 and O_2, involving ionic-curve crossings, may, somewhat surprisingly, have cross sections that favourably compete with those for ionization by electron impact (see Kistemaker 1973). This contradicts the frequently expressed generalization that processes induced by electrons are always more efficient than processes between 'heavy' particles.

Summarizing, we can state, following Sugden (1965), that bimolecular ionization reactions involving simple chemical re-arrangement with no activation energy have in general rate constants in the range 10^{-8} to 10^{-6} $cm^3 s^{-1}$ for pairs of oppositely charged ions, in the range 10^{-10} to 10^{-8} $cm^3 s^{-1}$ for ion-neutral pairs, and in the range 10^{-12} to 10^{-10} $cm^3 s^{-1}$ for neutral-neutral pairs. In the case of the alkaline-earth metals, ionization reactions (vi), (vii) and (xi) involving chemical rearrangement are faster than the collision processes of type (i) with inert flame molecules.

In a given flame and for a given metal, ionization processes of several types, in one or more steps, and with several collision partners often concur. Only a judicious selection of widely varying but known experimental conditions may enable us to unravel the contributions of each of these processes. As we have seen in Sect. II.4a-2, no information whatsoever about type and rate of a process can be obtained unless some deviation from full equilibrium exists in the system investigated. This will be shown in more detail in the following sections.

3. SOME USEFUL THEORETICAL RELATIONSHIPS

In this section we present some quantitative theoretical relationships that describe the dependence of the electric conductivity on the density of free charge carriers (Sect. 3a), and the dependence of this density on the total density of metallic element in the case of thermal ionization (Sect. 3b) as well as on the ionization rate constant in the case of ionization relaxation (Sect. 3c). In Sect. 3d we discuss the conditions for the existence of thermal ionization and possible

SOME USEFUL THEORETICAL RELATIONSHIPS   Sect. IX.3a

causes of deviations therefrom. These relationships will be used in the subsequent Sects 4-6 where methods of determining electron densities from conductivity measurements and experiments on ionization rate constants are described.

3a ELECTRIC CONDUCTIVITY[†]

The *complex conductivity* σ_c is defined as the ratio of complex current density and electric field strength at a given circular (or angular) frequency ω. The argument and the modulus of σ_c depend on the interaction of the charged particles with the flame gas molecules by collisions, which is characterized by the collision frequency ν (see Sect. 1). If several species of charged particles are present in the flame, the resulting conductivity is simply the sum of the conductivities corresponding to the separate species.

The real part σ' constitutes the ohmic conductivity; the imaginary part σ'' is connected with the relative permittivity ε_r according to $\sigma'' = \omega(\varepsilon_r - 1)/4\pi$ (in e.s. CGS units) or $\sigma'' = \omega\varepsilon_0(\varepsilon_r - 1)$ (in SI units).

A general method to compute σ_c has first been given by Margenau (1946) and is based on the Boltzmann transport equation. Under certain conditions — for instance if only elastic collisions between charged particles and neutral gas molecules occur — this theory leads to the following expression (see Rosen 1956) [*]

$$\sigma_c = -\frac{4\pi}{3}\frac{e^2\omega}{m}\int_0^\infty \frac{(\nu/\omega)-j}{\nu^2+\omega^2} v^3 \frac{\partial f_0}{\partial v} dv . \quad\quad (IX.1)$$

Here, f_0 represents the isotropic part of the velocity-distribution function of the charged particles (with charge e and mass m), the density of which is given by the normalization condition

$$n = 4\pi \int_0^\infty f_0 v^2 dv . \quad\quad (IX.2)$$

In particular, the dependence of the collision frequency ν on the particle velocity v must be known in order to determine σ_c. In the special case when ν is independent of v, Eq. IX.1 simplifies to

$$\sigma_c = (e^2 n/m\omega)[1+(\nu/\omega)^2]^{-1} [(\nu/\omega)-j] , \quad\quad (IX.3)$$

provided f_0 decreases more rapidly than v^{-3} as v approaches infinity. Now the real and imaginary parts of σ_c can be written in the form

---

[†] See, e.g., Lawton and Weinberg (1969) for a more extensive treatment.
[*] This and the following expressions IX.3-6 are formally the same in the Giorgi system as in the e.s. CGS system.

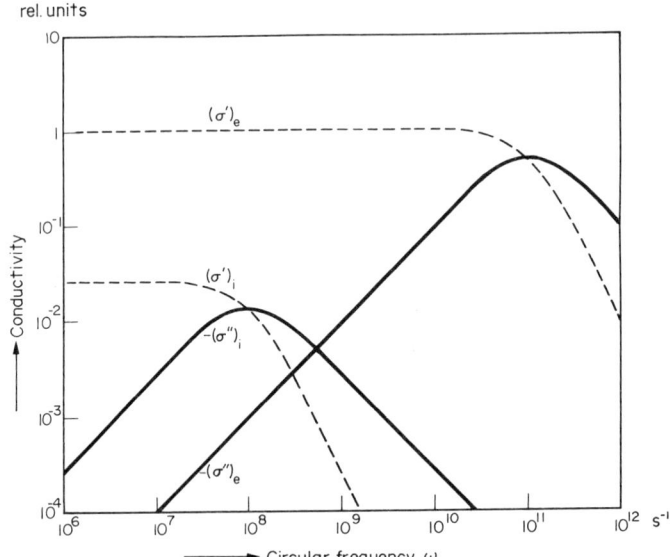

Fig. IX.2 Real part (σ') and imaginary part (σ'') of the conductivity of a flame plotted against the circular field frequency ω in the case of free electrons and free univalent ions (having an atomic weight of 20), respectively, assuming equal concentrations of electrons and ions, and a velocity-independent collision frequency $\nu_e = 10^{11}$ s^{-1} and $\nu_i = 10^8$ s^{-1}, respectively. For a CO-air flame consisting of 40 vol.% N_2, 35 vol.% CO and 22 vol.% CO_2 at a temperature of 1930 K and at 1 atm pressure Borgers (1978) calculated: $\sigma'_e = 1.7 \times 10^{-15} n_e$ Ω^{-1} cm^{-1} and $\sigma''_e = -1.1 \times 10^{-26} \omega n_e$ Ω^{-1} cm^{-1} (up to $\omega \simeq 10^8$ s^{-1}) from experimental data of the electron scattering cross section as a function of electron energy.

SOME USEFUL THEORETICAL RELATIONSHIPS   Sect. IX.3a

$$\sigma' = (e^2 n/m)[\nu/(\omega^2 + \nu^2)], \quad (IX.4)$$

$$\sigma'' = -(e^2 n/m)[\omega/(\omega^2 + \nu^2)]. \quad (IX.5)$$

In flames at 1 atm pressure one expects collision frequencies for free electrons (ν_e) that are of the order of 10^{11} s^{-1}; those for atomic ions (ν_i) are uncertain but might well be less by three orders of magnitude. Anyway, when a large fraction of the electrons are attached to form negative ions in the flame, the contribution to σ' made by the ions may become dominant.

With respect to σ'', the situation is quite different. At field frequencies $\omega \leq 10^8$ s^{-1} the contribution made by the electrons is small compared to that made by the ions, even when their concentrations are of comparable order of magnitude. Fig. IX.2 shows the high-frequency behaviour of σ' and σ'' for free electrons and free univalent ions (with relative atomic mass of 20) assuming a collision frequency of 10^{11} s^{-1} and 10^8 s^{-1}, respectively (see Borgers 1965).

The assumption of a velocity-independent collision frequency is approximately valid for electrons in collision with CO_2 but it is not a good assumption for electrons in collision with N_2 and CO (see Becker et al. 1971). For ions this assumption is in general doubtful (see Borgers 1978).

In the presence of a homogeneous magnetic field B parallel to the z axis, the real part of the conductivity measured in the x direction with the electric field vector parallel to the x axis[†] is given by

$$\sigma'_{xx} = \left(\frac{e^2 n_e}{2m_e}\right)\left[\frac{\nu_e}{\nu_e^2 + (\omega - \omega_c)^2} + \frac{\nu_e}{\nu_e^2 + (\omega + \omega_c)^2}\right]. \quad (IX.6)$$

Here ν_e is assumed to be constant and $\omega_c \equiv eB/m_e c = 1.8 \times 10^7 B$ (B in gauss; $e, c,$ and m_e in e.s. CGS units) is the (circular) *cyclotron-resonance frequency*. This equation applies to the case where the influence of the free electrons on σ' is dominant. For $B = 5 \times 10^4$ gauss we have: $\omega_c = 9 \times 10^{11}$ s^{-1} (or $\lambda_c \simeq 0.2$ cm). We note that for $\omega = \omega_c$ a maximum occurs in σ' (*cyclotron resonance*).

All the above expressions contain the factor $(e^2 n/m)$. In the case when the effect of the free electrons is dominant, we can express this factor in the (circular) *plasma frequency* ω_p given by (see Thorne 1974)[*]

$$\omega_p = \sqrt{4\pi e^2 n_e/m_e} = 5.64 \times 10^4 \sqrt{n_e}, \quad (IX.7)$$

where the numerical constant applies in the CGS system.

[†] In this case the conductivity is represented by a tensor.

[*] When working in the Giorgi system instead of the e.s. CGS system one has to replace $4\pi e^2$ by e^2/ε_0.

The *attenuation* of an electromagnetic wave in a homogeneous medium is related to σ'. When the attenuation is determined by the motion of the free electrons only, we have for the absorptivity k at not too high electron densities ($\omega_p/\omega < 0.3$) in good approximation

$$k \simeq 4\pi\sigma'/c \, , \qquad (IX.8)$$

where c is the velocity of light in vacuo. Using Eqs IX.4, 7 and 8 we find in the absence of a magnetic field

$$k \simeq (\omega_p^2/c) \left[\nu_e/(\omega^2 + \nu_e^2)\right] . \qquad (IX.9)$$

The attenuation, β, is given by

$$\beta \simeq 0.46 \, n_e \nu_e/(\omega^2 + \nu_e^2) \quad \text{(in dB/cm)} . \qquad (IX.10)$$

In the presence of a magnetic field we derive in a similar way from Eq. IX.6 (if $\omega^2 \gg \omega_p^2$)

$$\beta \simeq 0.23 \, n_e \nu_e \left[\frac{1}{\nu_e^2 + (\omega-\omega_c)^2} + \frac{1}{\nu_e^2 + (\omega+\omega_c)^2}\right] . \qquad (IX.11)$$

3b THERMAL IONIZATION

3b-1 *CASE OF A SINGLE METALLIC ELEMENT FORMING NO COMPOUNDS*

In the simplest case we have one metallic element that is present as a free atom (M) or ion (M+) only. Assuming that no other ions are present in the flame, i.e., $[e^-] = [M^+]$, we derive from the Saha equation (Eq. II.59)

$$\beta_i^2/(1-\beta_i) = K_i/([M]+[M^+]) \, , \qquad (IX.12)$$

where the *degree of ionization* β_i is defined by

$$\beta_i \equiv [M^+]/([M]+[M^+]) \qquad (IX.13)$$

and where K_i is the ionization constant at flame temperature. Introducing the total density of element: $[M]_t \equiv [M] + [M^+]$ we find from the above equations

$$\beta_i = \sqrt{\frac{K_i}{[M]_t} + \left(\frac{K_i}{2[M]_t}\right)^2} - \frac{K_i}{2[M]_t} . \qquad (IX.14)$$

The relative decrease in atomic density due to ionization is then found from

$$\frac{[M]}{[M]_t} = 1 - \beta_i . \qquad (IX.15)$$

From Eq. IX.14 we conclude that β_i increases with decreasing total element concentration and with increasing ionization constant; the latter is the greater, the higher the temperature or the lower the ionization energy are (see Sect. II.3b-4).

SOME USEFUL THEORETICAL RELATIONSHIPS   Sect. IX.3b

Equation IX.14 may be simplified for high and low element concentrations as follows

(1) $\beta_i \simeq \sqrt{K_i/[M]_t}$ or $[M^+] \simeq \sqrt{K_i[M]_t}$ and $[M] \simeq [M]_t(1-\sqrt{K_i/[M]_t})$ for $[M]_t \gg K_i$

(2) $\beta_i \simeq 1-[M]_t/K_i$ and $[M] \simeq [M]_t^2/K_i$ for $[M]_t \ll K_i$.

Consequently, ionization is, virtually, complete when $[M]_t \ll K_i$, whereas it is negligible when $[M]_t \gg K_i$. For $[M]_t = K_i$ we find $\beta_i = 0.61$. We also conclude from these asymptotic expressions that ionization results in a nonlinear relationship between atomic density $[M]$ and solution concentration (which is proportional to $[M]_t$). The analytical curve in atomic spectroscopy is concave as a result of ionization (see the textbooks on analytical spectroscopy cited in Chapt. I).

Some typical values of K_i at 2000 K for Li, Na, K, Rb and Cs (see Appendix A.4 for ionization energies) are in this order: 3.4×10^6, 1.5×10^7, 1.7×10^9, 4.4×10^9 and 2.3×10^{10} cm^{-3}, and of K_i at 2500 K in the same order: 4.3×10^9, 1.4×10^{10}, 5.3×10^{11}, 1.1×10^{12} and 4.3×10^{12} cm^{-3} (see Rubeška 1969). Some of these elements, however, form monohydroxides in flames (see Chapt. VIII) and reference should be made to the equations to be presented in Sect. 3b-3 in order to calculate β_i from K_i.

When upon addition of an excess of chlorine a fixed fraction, α, of the electrons produced by metal ionization is bound as Cl$^-$ ions because of the partial equilibrium of reaction (ix) in Table IX.1, we have to replace $[e^-]$ by $(1-\alpha)[M^+]$ in the Saha equation. In the above equations we must then substitute $K_i/(1-\alpha)$ for K_i. When α is close to 1, i.e. when practically all electrons are removed by Cl$^-$ formation, the metal will be fully ionized. We may expect α to be independent of $[M]_t$ as long as $[HCl] \gg [M]_t$ and $[H]$ is not affected by the metal.

When from some other origin free electrons are produced at a concentration $[e^-]_0$ far in excess over $[M]_t$, we may consider $[e^-] \simeq [e^-]_0$ as a constant in the Saha equation; the degree of ionization of M then becomes

$$\beta_i/(1-\beta_i) = K_i/[e^-]_0 , \qquad (IX.16a)$$

$$\beta_i = K_i/(K_i + [e^-]_0) . \qquad (IX.16b)$$

Then β_i and $[M]/[M]_t$ are independent of $[M]_t$ and the analytical curve is a straight line through the origin. When $[e^-]_0 \to \infty$, the metal ionization is fully suppressed.

3b-2   *IONIZATION INTERFERENCE BETWEEN TWO METALLIC ELEMENTS*

It has been known for a long time in analytical flame emission spectroscopy that the atomic emission of an ionizable element, e.g. K, is enhanced when another ionizable element, e.g. Na, is added to the solution. This has been explained as a *mutual ionization interference* caused by the suppression of the ionization of the former element by the electrons released by the second one, and vice versa (see Huldt 1948, Smit and Vendrik 1948, Vendrik 1949, Smit, Alkemade and Verschure 1951, de Montgareuil 1954, and Alkemade 1952, 1954). This effect has also been exploited in atomic spectroscopy to improve the sensitivity and the detection limit of elements that show a high degree of ionization in hot flames and for which no suitable ionic lines are available. Addition of, e.g., K or Cs (having low ionization energies) as electron donor to the nebulized solution is commonly used (see textbooks on analytical flame spectroscopy cited in Chapt. I; see also Fig. IX.15 in Sect. 6).

We can treat mutual ionization interference quantitatively by combining the Saha equations for the two elements, M and M', and by putting: $[e^-] = [M^+] + [M'^+]$. Denoting the corresponding ionization constants by K_i and K_i', and the total element densities by $[M]_t$ and $[M']_t$, we find the following implicit relationship for the degree of ionization, β_i, of M (see Alkemade 1954)

$$\frac{[M]_t + [M']_t}{[M]_t} = (1-\beta_i)\left[\frac{K_i}{[M]_t}\left\{\frac{K_i}{K_i'}(1-\beta_i)+\beta_i\right\}\bigg/\beta_i^2 - \left(\frac{K_i}{K_i'}-1\right)\right]. \quad (IX.17)$$

From this equation we derive an asymptotic relationship that holds if $\beta_i \ll 1$, $K_i' \gg K_i$ and $[M']_t \gg [M]_t$

$$\beta_i \simeq \tfrac{1}{2}(K_i/[M']_t)\left\{\sqrt{4([M']_t/K_i')+1}+1\right\}. \quad (IX.17a)$$

The underlying condition: $\beta_i \ll 1$ implies that: $K_i \ll [M']_t$ when $[M']_t \leqslant K_i'$, or that: $K_i \ll \sqrt{[M']_t K_i'}$ when $[M']_t \gg K_i'$. This asymptotic equation can be used to estimate how high we must make the total density, $[M']_t$, of the electron donor in order to suppress the ionization of M below, say, 1%. This equation can also be used to correct for any residual ionization of M when $[M']_t$ cannot be made high enough to suppress the ionization of M fully.

Such a correction is especially useful when one wants to determine the degree of ionization of M by measuring the relative increase in atomic metal signal upon addition of an electron donor in excess concentrations. If the ionization suppression is complete, the relative increase is simply equal to $[M]_t/[M] = (1-\beta_i)^{-1}$ (see Eq. IX.15).

3b-3 *CASE OF A METALLIC ELEMENT FORMING NEUTRAL AS WELL AS IONIZED ATOMS AND COMPOUNDS*

We now consider the case of an element, such as Sr, that forms neutral atoms and molecules, such as SrOH, SrO, $Sr(OH)_2$, as well as atomic and molecular ions, such as $SrOH^+$ (see Sect. 2a). This case includes the special case when an element, such as Cs, forms only one molecular compound, such as CsOH, but no molecular ion.

SOME USEFUL THEORETICAL RELATIONSHIPS   Sect. IX.3b

Suppose that element M is present as neutral atom (M), neutral molecules (MX, MY, ...), atomic ion (M$^+$) and molecular ion (MX$^+$). Let the last two species be the only ions present in the flame so that: $[M^+]+[MX^+] = [e^-]$. The total element concentration is given by: $[M]_t = [M]+[MX]+[MY] +\ldots [M^+]+[MX^+]$. The association factor $\phi_{MX} \equiv [MX]/[M]$ and total association factor $\phi_t \equiv ([MX]+[MY]+\ldots)/[M]$ are supposed to be independent of element concentration and ionization. This is a reasonable assumption when a (partial) equilibrium exists between M and MX, etc., and the total density of X, etc., in the flame largely exceeds $[M]_t$ (see Chapt. VIII). We again assume that ionization of M and MX is thermal, so that the Saha equation can be applied

$$[M^+][e^-]/[M] = K_i \, , \qquad (IX.18a)$$

$$[MX^+][e^-]/[MX] = K_i' \, . \qquad (IX.18b)$$

From these equations and the definition of ϕ_{MX} we simply arrive at

$$[MX^+] = \phi_{MX}(K_i'/K_i)[M^+] \equiv \alpha[M^+] \, , \qquad (IX.19a)$$

$$[e^-] = [MX^+] + [M^+] = (1+\alpha)[M^+] \, . \qquad (IX.19b)$$

Substituting this expression for $[e^-]$ in Eq. IX.18a and using: $[M]_t = (1+\phi_t)[M] + (1+\alpha)[M^+]$, we can express $[M^+]$ in $[M]_t$ as follows

$$\frac{[M^+]}{[M]_t} = \sqrt{\frac{K_i^0}{[M]_t(1+\alpha)} + \left(\frac{K_i^0}{2[M]_t}\right)^2} - \frac{K_i^0}{2[M]_t} \qquad (IX.20)$$

with

$$K_i^0 \equiv K_i/(1+\phi_t) \, . \qquad (IX.20a)$$

Equation IX.20 is similar to Eq. IX.14 with K_i being replaced by K_i^0, apart from the appearance of a factor $(1+\alpha)$ due to molecular ion formation. From the definition of K_i^0 in Eq. IX.20a, we see that the formation of neutral molecules MX, MY, ..., described by ϕ_t, has the same effect on the atomic ion production as a lowering of the atomic ionization constant K_i. Molecule formation thus results in a suppression of atomic ionization. Furthermore, we see from Eq. IX.20 that the formation of molecular ions, described by the $(1+\alpha)$ factor, suppresses the atomic ionization, too. This is because $[e^-]$ in Eq. IX.18a is enhanced by the formation of molecular ions.

The formation of neutral molecules represses the effect of mutual ionization interference on M (see Sect. 3b-2) because of the total mass balance: $[M] = [M]_t/(1+\phi_t) - [M^+]/(1+\phi_t)$ if we suppose: $[MX^+] \ll [M]_t$. A suppression of $[M^+]$ due to mutual ionization interference will benefit $[M]$ for a fraction $1/(1+\phi_t)$, which is but small when $\phi_t \gg 1$; the major part will benefit MX, etc. Molecule

835

formation has therefore a stabilizing effect on the density of neutral atoms.

The effect of ionization on $[M]/[M]_t$ follows directly from Eq. IX.20 and the total mass balance: $[M](1+\phi_t) = [M]_t - (1+\alpha)[M^+]$, and is described by

$$\frac{[M]}{[M]_t} = \left\{1 - (1+\alpha)\frac{[M^+]}{[M]_t}\right\} \bigg/ (1+\phi_t) . \qquad (IX.21)$$

The relative effect of ionization on the density of any of the metal compounds $[MX]$, $[MY]$ ... is the same as on $[M]$, because $[MX] = \phi_{MX}[M]$, etc., and ϕ_{MX} is independent of ionization. Consequently, the intensities of <u>atomic lines and molecular bands are equally affected by ionization</u>, if we disregard the possible effect of self-absorption on the line intensity.

The density of the molecular ions and of the electrons is found directly by a combination of Eqs IX.19a and 19b, respectively, with Eq. IX.20. One sees from Eq. IX.19a that the molecular ions dominate over the atomic ions when $\alpha \gg 1$, i.e. $\phi_{MX} \gg K_i/K_i'$. This case applies for Ca, Sr, and Ba where ϕ_{MOH} is usually considerably larger than unity (see Chapt. VIII), whereas the ionization energies of the monohydroxides are comparable to those of the corresponding atoms (see references in Table IX.8 in Sect. 5, and Appendix A.4). For these elements, however, deviations from thermal ionization may occur in flames with excess radicals (see Sect. 3d).

From Eq. IX.20 approximate expressions can be derived which hold in the case of very small or large values of $[M]_t$. For $[M]_t \ll \frac{1}{4} K_i (1+\alpha)/(1+\phi_t)$ we find by series expansion

$$\frac{[M^+]}{[M]_t} \simeq \frac{1}{1+\alpha}\left\{1 - \frac{[M]_t(1+\phi_t)}{K_i(1+\alpha)}\right\} . \qquad (IX.22a)$$

In the limit $[M]_t \to 0$, the element is fully ionized, that is:

$$[M^+] + [MX^+] \equiv (1+\alpha)[M^+] \to [M]_t .$$

For

$$[M]_t \gg \tfrac{1}{4} K_i (1+\alpha)/(1+\phi_t)$$

the first term under the square-root expression in Eq. IX.20 becomes dominant and we have approximately

$$\frac{[M^+]}{[M]_t} \simeq \sqrt{\frac{K_i}{[M]_t(1+\alpha)(1+\phi_t)}} . \qquad (IX.22b)$$

Combination of this equation with Eq. IX.19b yields finally

$$[e^-] \simeq \sqrt{K_i[M]_t(1+\alpha)/(1+\phi_t)} . \qquad (IX.22c)$$

The corresponding approximate expressions for $[M]/[M]_t$ are obtained by combining Eq. IX.21 with Eqs IX.22a and 22b, respectively.

SOME USEFUL THEORETICAL RELATIONSHIPS   Sect. IX.3c

3c IONIZATION RELAXATION

Ionization relaxation occurs when the dominant ionization and neutralization processes are too slow for a steady-state balance to be achieved. Ionization relaxation is found not only in the combustion zone, where the physical and chemical gas conditions change very rapidly (within 10 μs at 1 atm), but may also occur in the interzonal region. When the relaxation time is comparable to the travel time of the metal species in this region or when the temperature changes markedly with distance from the combustion zone, the local concentration of the ions may not correspond to thermal ionization (see Sect. 3d). The existence of such deviations from thermal ionization above the combustion zone, however, enables us to determine the rate constants of the dominant ionization or neutralization processes.

The general kinetic equation for the production and removal of ions of a given species in a stationary flame reads, when no fields are applied,

$$\frac{\partial n_i}{\partial t} = G - R - \text{div}(n_i \mathbf{v}) + \text{div}(D_a \text{ grad } n_i) \equiv 0 . \quad (IX.23)$$

Here $\partial n_i/\partial t$ represents the partial derivative of the ionic density with respect to time at a fixed place, that is, in a reference frame at rest; in a stationary flame this derivative must equal zero. G and R represent the local rates per unit volume and per unit time of formation and destruction of the considered ionic species by ionization, neutralization or ionic re-arrangement processes. In nonequilibrium conditions the reverse of the dominant ionization path need not be the dominant neutralization path; for example, ionization through process (xii) may be dominant, whereas neutralization may occur mainly through the inverse process (i) (see Table IX.1). The rates G and R can generally be expressed as products of densities of reactants and rate constants. The last two terms in the above expression for $\partial n_i/\partial t$ describe the contributions of convection and ambipolar diffusion to the change in local ionic density, respectively. The vector $\mathbf{v}$ is the local velocity of the flame gas considered in a reference frame at rest; D_a is the ambipolar diffusion constant which describes the joint transport of electrons and positive ions by diffusion. Although electrons move much more rapidly than the heavier ions, D_a is mainly determined by the properties of the slower partner because the electrons are tied to the ions by Coulombic forces (see Cambel, Duclos and Anderson 1963; see also Sect. X.4b). In the hard-sphere model D_a depends on temperature and total pressure, p_t, according to $\propto T^{\frac{3}{2}}/p_t$; for small ions a dependence $\propto T^2/p_t$ is more likely (see Ay, Ong and Sichel 1975).

When different ionic species are present together, Eq. IX.23 holds for each species separately; however, the expression for D_a of each species depends on the presence of the other species. In the following we shall, for simplicity's sake, ignore this complication.

Chapt. IX   IONIZATION IN SEEDED FLAMES

The density of ionic (and neutral) metal species can be solved as a function of position from the above differential equation at given boundary conditions if the composition, temperature and velocity of the flame gas are known functions of position. These flame properties can usually be considered as being independent of the state of metal ionization. When the main metal compounds formed are in (partial) equilibrium with the flame constituents (see Chapt. VIII), the degree of their dissociation can also be considered as being independent of the state of ionization. In solving the above differential equation, we should furthermore take into account the charge-balance equation ($n_i = n_e$) and the mass-balance equation that relates the total density of metallic element in the flame to the amount of element supplied per second.

It is clear that in the general three-dimensional case, where all relevant quantities and parameters are arbitrary functions of position, solutions of the above equation are hard to calculate. An attempt to solve this equation numerically for the concentration of H_3O^+ ions in premixed hydrocarbon flames has been made by Ay, Ong and Sichel (1975) on the basis of a semi-empirical model; in this model convection and ambipolar diffusion were taken into account besides ion formation and recombination.

Our interest is restricted to metal ionization above the combustion zone in thick, cylindrical flames where $|v|$ ($\equiv$ rise-velocity v_r) and D_a are constants and v is directed along the flame axis (= z axis) and where n_i is only a function of z. Then Eq. IX.23 reduces to

$$G - R = v_r dn_i/dz - D_a d^2 n_i/dz^2 \ . \qquad (IX.24)$$

As a further simplification we shall disregard the ambipolar diffusion term which is negligible outside the combustion zone for $p_t > 30$ Torr (see Lawton and Weinberg 1969). Choosing a reference frame moving with the flame gas at speed v_r we can substitute $z = v_r t$ where t is the travel time (= t_{tv} in Eq. IV.1) and obtain

$$G - R = dn_i/dt \ . \qquad (IX.25)$$

This equation will be worked out for three typically different border-line cases which are commonly met with in the experimental determination of rate constants (see Sect. 5).

Case A. We assume that the metallic species is present only as free atom (M) or free atomic ion (M⁺), that no other ions are present ($[M^+] = [e^-]$) and that the total element density $[M]_t$ ($\equiv [M] + [M^+]$) is constant. Ionization and neutralization are assumed to proceed through process (i) and its reverse (see Table IX.1).[†]

[†] A derivation in the case of a two-step ionization process is presented in Appendix A.9.

SOME USEFUL THEORETICAL RELATIONSHIPS   Sect. IX.3c

We then have for the ionization- and recombination rates (see Sect. II.4a-1)

$$G = k_i[M], \qquad (IX.26a)$$

$$R = k_r[M^+][e^-] . \qquad (IX.26b)$$

In a *single-component system* the apparent monomolecular rate constant k_i can be expressed by: $k_i = k_i'[Z]$ where $[Z]$ is the density of the collision partner. This monomolecular rate constant k_i describes the probability per second that a neutral atom M, in any quantum state, is split into an ion M^+, in any quantum state, and a free electron as a result of a single collision with one particle Z in any quantum state. An analogous definition holds for the apparent bimolecular rate constant k_r, which can be written as: $k_r = k_r'[Z]$.

In general, k_i' (or k_r') depends on the distribution of the relative velocity as well as on the distribution of the internal energy of the reactants involved in the ionization (or recombination) process. For k_i' we have an expression analogous to that for k_2 in Eq. II.79 if we replace $f(E_i)$ by the joint distribution function for the population of the energy levels of Z as well as of M. In thermodynamic equilibrium these distributions are invariably determined by the Maxwell-Boltzmann laws. The bimolecular rate constant k_i' (or termolecular rate constant k_r') of a given process is then a function of temperature T only and is called *thermal*. The thermal rate constants are denoted by $k_i'(T)$ and $k_r'(T)$.

In a system that is in thermodynamic equilibrium, however, we cannot measure rate constants (see Sect. II.4a-2). In order to measure ionization or recombination rate constants we need a system where some deviation from Saha equilibrium exists. But the very existence of a deviation from Saha equilibrium entails — in principle at least — also a deviation from the Maxwell-Boltzmann laws. In particular, the population distribution of the atomic levels will be affected by any unbalance between the ionization- and recombination rates (see Sect. II.4a-2). Since ionization from a high excited level will in general be more rapid than from a lower level, the value of k_i' will sensitively depend on the level-population distribution and thus indirectly on the extent of the deviation from Saha equilibrium. Under these circumstances k_i' depends on the actual concentrations of the neutral and ionized species and cannot be considered as a true constant of proportionality in Eq. IX.26a.

Similarly, the equality of k_r' to its thermal value $k_r'(T)$ may be questioned when there is no Saha equilibrium. Since recombination is an exo-ergic process with a negligible activation energy (see also Sect. IX.5), the excited-level population of the ionic species plays only a minor role. Any deviation of the ionic level population from Boltzmann's law, due to a deviation from Saha equilibrium, is thus of little consequence for k_r'. Since k_r' describes the probability of a single-step recombination process, resulting in a neutral atom in any level, it does not depend on the atomic level population either. We can thus assume a priori that k'_r is, in good approximation, a true constant of proportionality, independent of the actual level of ionization.

It will appear from the outcome of the flame experiments that, within the experimental error limits, the experimental k'_i is a true rate constant which is independent of the level of ionization. In fact, the measured k'_i values appear to be close to their thermal values $k'_i(T)$, which are defined for a system in thermodynamic equilibrium (for a further discussion see Sect. 6b-1). In the following we shall therefore consider k'_r as well as k'_i to be independent of the concentrations of M, M$^+$, e$^-$ and Z.

In a *multi-component system*, the overall, apparent monomolecular rate constant k_i (in s^{-1}) may be expressed as a sum over the contributions from different collision partners Z_j (cf. Eq. II.80 with $\overline{k_2^u} \triangleq k_i$ and $k_{2,j} \triangleq k'_{i,j}$)[†]

$$k_i = \sum_j k'_{i,j}[Z_j], \qquad (IX.27a)$$

where $k'_{i,j}$ are the specific, bimolecular rate constants. For k_r we have similarly

$$k_r = \sum_j k'_{r,j}[Z_j]. \qquad (IX.27b)$$

In the following we shall assume that the densities of the main flame constituents are independent of height z; a variation of k_i and k_r with height may, however, be induced by a height-dependent flame temperature.

Equation IX.25 can now be re-written as

$$d[M^+]/dt = k_i[M] - k_r[M^+]^2. \qquad (IX.28)$$

Since $[M^+] = [M]_t - [M]$ we also have

$$-d[M]/dt = k_i[M] - k_r([M]_t - [M])^2. \qquad (IX.29)$$

When the specific rate constants, $k'_{i,j}$ and $k'_{r,j}$, are thermal, i.e. when their values are the same as in thermodynamic equilibrium, we can apply the detailed-balance relation (see Eq. II.93): $k'_{i,j}(T)/k'_{r,j}(T) = K_i(T)$; here $K_i(T)$ is the ionization constant at flame temperature T. The T-dependent overall rate constants $k_i(T)$ and $k_r(T)$ then obey the detailed-balance relation, too

$$k_i(T)/k_r(T) = K_i(T). \qquad (IX.30)$$

This equation holds irrespective of the actual (equilibrium or nonequilibrium)

[†] See, however, Sects 6a-1 and 6b-1 for a critical discussion of this summation law. The validity of Eqs IX.27a and 27b is, however, not essential in the derivation of the relaxation expressions.

SOME USEFUL THEORETICAL RELATIONSHIPS   Sect. IX.3c

concentrations of the collision partners Z_j. The values of k_i and k_r separately still depend on the flame composition and should be called 'thermal' only if this composition conforms to the mass-action-law. However, a deviation from chemical equilibrium as such does not upset the state of thermal ionization as long as collisional ionization dominates over chemi-ionization and the <u>specific</u> rate constants $k'_{i,j}$ and $k'_{r,j}$ are thermal.[†]

Using Eq. IX.30 we obtain, after dividing both sides of Eq. IX.29 by [M],

$$-\frac{d \ln[M]}{dt} = k_i \left\{ 1 - \frac{([M]_t - [M])^2}{[M] \, K_i(T)} \right\} . \tag{IX.31}$$

For constant T the following asymptotic expressions can be derived from this equation. If the metal vapour is far removed from the thermal ionization state ($[M^+] \ll [M^+]_e$ where suffix e denotes the equilibrium value), we can ignore the second term between braces and obtain

$$\ln[M] \simeq -k_i t + \ln[M]_0 . \tag{IX.32}$$

Close to the thermal ionization state we can transform Eq. IX.28 by writing: $[M] \equiv [M]_e + \Delta$ and $[M^+] = [M^+]_e - \Delta$, where Δ is a small deviation from the equilibrium density, and obtain

$$-d\Delta/dt = k_i([M]_e + \Delta) - k_r([M^+]_e - \Delta)^2 \simeq (k_i + 2k_r[M^+]_e) \Delta . \tag{IX.33}$$

We have used here the equilibrium relation: $k_i[M]_e = k_r[M^+]_e^2$, which follows from Eq. IX.30 and the definition of K_i in Eq. II.59, while we have neglected in the last expression the quadratic term $k_r \Delta^2$. The solution is

$$\ln\left\{ ([M] - [M]_e)/[M]_t \right\} \simeq -(k_i + 2k_r[M^+]_e) \, t + C , \tag{IX.34a}$$

where C is an integration constant. Using again Eq. IX.30 and the definition of K_i we get

$$\ln\left\{ ([M] - [M]_e)/[M]_t \right\} \simeq -k_i(1 + 2[M]_e/[M^+]_e) t + C . \tag{IX.34b}$$

Consequently, in the limit of small deviations the approach towards thermal ionization is exponential in time and the time constant or *relaxation time* τ is given by

---

[†] On the other hand, the usage of thermal <u>rate constants</u> do not necessarily imply that the <u>state</u> of ionization (as defined in Sect. 2a) is thermal as well (cf. also Sect. II.4a). When the thermal rate constants are small, a deviation from thermal ionization may result from relaxation effects (see below). Note also that the adjective 'thermal' does not indicate a particular kind of process and should therefore not be identified with 'collisional' (cf. Sect. VI.4a sub iv).

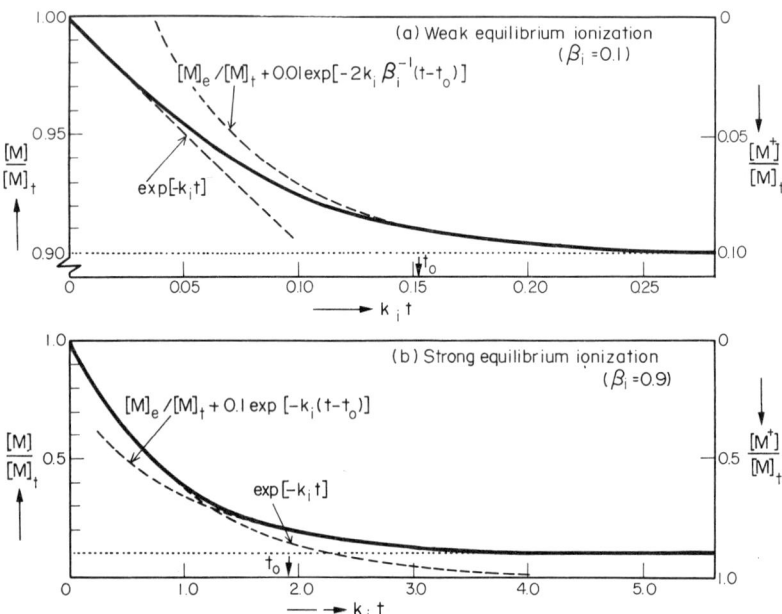

Fig. IX.3 Figures a and b show the dependence of atomic metal concentration [M] and ionic metal concentration [M$^+$] upon time t as a result of ionization relaxation in the case of weak $(\beta_i = 0.1)$ and strong $(\beta_i = 0.9)$ equilibrium ionization, respectively. It is assumed that at $t = 0$ the metal is present only as neutral atoms $([M] = [M]_t)$. The time axis is scaled by the monomolecular ionization rate constant k_i, which is assumed to be independent of t (isothermal flame). The full curves are numerically computed from Eq. IX.31. The dashed curves represent the asymptotic expressions obtained for $t \to 0$ and $t \to \infty$, respectively (see inserts). At $t = t_0$ [M]/[M]$_t$ has dropped to one percent above its equilibrium value $[M]_e/[M]_t$ ($\equiv 0.90$) in case a, and to twice its equilibrium value $[M]_e/[M]_t$ ($\equiv 0.10$) in case b.

(Computer calculations were performed by Mr G. Kuiper at the Physics Laboratory, University, Utrecht.)

SOME USEFUL THEORETICAL RELATIONSHIPS   Sect. IX.3c

$$\tau^{-1} = k_i(1 + 2[M]_e/[M^+]_e) \, . \tag{IX.35}$$

A 'small' deviation is here defined by the inequality: $\Delta \ll [M^+]_e$. In the *strong* ionization case ($[M^+]_e \gg [M]_e$), the relaxation time is about equal to k_i^{-1} and Δ need not be small relative to $[M]_e$. In the *weak* ionization case, the relaxation time is reduced by a factor of about $2[M]_e/[M^+]_e$; this factor is about equal to $2[M]_t/[M^+]_e = 2\beta_i^{-1}$, as $[M]_e \simeq [M]_t$ for weak ionization. According to Eq. IX.34a we can in this case write for the relaxation time also $2k_r[M^+]_e$. It should be realized that τ characterizes the rate of ionization relaxation only in the 'tail' of the relaxation curve describing $[M]$ or $[M^+]$ as a function of time. When the equilibrium degree of ionization is, e.g., 0.90, it takes a time $t_0 = 1.9\,k_i^{-1}$ for $[M]$ to drop from the initial value $[M]_0 \,(= [M]_t)$ to twice its equilibrium value $[M]_e \,(= 0.10\,[M]_t)$. After roughly this time the approximate Eq. IX.34a or 34b becomes applicable because we have then: $\Delta \leqslant 0.1\,[M]_t \simeq 0.1\,[M^+]_e$. When the equilibrium degree of ionization is 0.10, it takes a time $t_0 = 0.15\,k_i^{-1}$ for $[M]$ to drop from $[M]_t$ to one percent above its equilibrium value $[M]_e \,(= 0.90\,[M]_t)$. After roughly this time we can again apply Eq. IX.34a or 34b because then: $\Delta \leqslant 0.01\,[M]_t = 0.1\,[M^+]_e$. The course of $[M]/[M]_t$ or $[M^+]/[M]_t \,(= 1 - [M]/[M]_t)$ as a function of time in these two particular cases is shown in Figs IX.3a and b for given k_i.

Case B. We now suppose that as well as free atoms and atomic ions M forms a molecular compound MX and that the association factor $\phi \equiv [MX]/[M]$ is only a function of z or travel time t and is independent of the metal ionization. We also suppose that the flame temperature, the ionization and the recombination rate constants are independent of t. We furthermore assume that the state of ionization is far removed from Saha equilibrium so that we can neglect in Eq. IX.25 either the neutralization rate (Case B.1) or the ionization rate (Case B.2). The other assumptions remain the same as in the previous case.

Case B.1. Writing: $[M]_t = (1+\phi)[M] + [M^+]$, $G = k_i[M]$ and $R \equiv 0$ we get from Eq. IX.25

$$d[M^+]/dt \simeq k_i[M] = k_i([M]_t - [M^+])/(1+\phi) \, ,$$

or

$$-\frac{d\ln([M]_t - [M^+])}{dt} \simeq \frac{k_i}{1+\phi} \, . \tag{IX.36}$$

In order to integrate this equation we must know the function $\phi(t)$, that is, $\phi(z)$. When MX is an alkali hydroxide, ϕ varies with height because of the relaxation of excess radicals towards chemical equilibrium with increasing height (see Sect. IV.5c). In the range where $[H] \gg [H]_e$ we have: $[H]^{-1} = [H]_0^{-1} + At$ with $A > 0$ (see Eq. IV.14 and Fig. IV.8) and: $\phi(t) = \phi_0(1 + Bt)$ with $B > 0$ because Eq. VIII.20 applies here.

843

A similar linear dependence of ϕ on t may also apply effectively in a restricted height interval when [H] is not much larger than $[H]_e$. Using this expression for $\phi(t)$ we can integrate Eq. IX.36 and find (see Kelly and Padley 1972)

$$\ln\left\{[M]_t/([M]_t - [M^+])\right\} \simeq \left\{k_i/C(1+\phi_0)\right\}\ln(1+Ct) \qquad (IX.37a)$$

where $C \equiv \phi_0 B/(1+\phi_0)$ and $[M^+] = 0$ at $t = 0$. We can also derive from the last equation [M] as a function of time by substituting there $(1+\phi)[M]$ for $([M]_t - [M^+])$ and using the equality: $(1+\phi) = (1+\phi_0)(1+Ct)$; we then get

$$\ln\left\{[M]_t/[M](1+\phi_0)\right\} = \left\{1 + k_i/C(1+\phi_0)\right\}\ln(1+Ct). \qquad (IX.37b)$$

Case B.2. Writing: $R = k_r[M^+]^2$ while neglecting G we get from Eq. IX.25

$$d[M^+]/dt = -k_r[M^+]^2.$$

Integration yields directly

$$[M^+]^{-1} = k_r t + [M^+]_0^{-1}. \qquad (IX.38)$$

Case C. This case differs from case B in that the element now forms also a molecular ion MX^+ while it may form several neutral compounds $MX, MY, \ldots$. We shall take into account ionization and neutralization together and again assume that $[M]_t$ and the flame temperature are independent of position. This case has been applied in the study of alkaline-earth ionization in fuel-rich H_2 flames (see Sect. 5a) and we shall therefore specify from the outset the molecular ion as MOH^+ and the neutral compounds as MOH, $M(OH)_2$ and MO (see Chapt. VIII). The concentrations of ions and molecular compounds may differ from their equilibrium values because of relaxation in the attainment of ionization as well as chemical equilibria (see Sect. IV.5). These concentrations are therefore functions of distance z from the combustion zone or travel time t.

The primary ionization step is here the forward chemi-ionization reaction (vi) and/or (vii), both leading to MOH^+ (see Table IX.1). Atomic ions are formed from MOH^+ through process (xi), which is instantaneously balanced (see Note 9 of Table IX.1). Neutralization is assumed to occur through the dissociative $(MOH^+ + e^-)$ recombination process (vi) and/or (vii). For the net rate of production of atomic plus molecular ions we therefore have

$$dn_i/dt = k_6[M][OH] + k_7[MO][H] - k_r[MOH^+][e^-]. \qquad (IX.39)$$

In this equation n_i ($\equiv [MOH^+] + [M^+]) = [e^-]$ and k_6, k_7 are the second-order chemi-ionization rate constants of processes (vi) and (vii), respectively; k_r is the second-order ion-recombination rate constant of the <u>combined</u> inverse processes (vi) and (vii).

SOME USEFUL THEORETICAL RELATIONSHIPS   Sect. IX.3c

The two terms describing the chemi-ionization rate G can be combined by writing

$$G = k_6[M][OH]\left\{1 + (k_7/k_6)[MO][H]/[M][OH]\right\} =$$

$$k_6[M][OH]\left\{1 + (k_7/k_6)K_{17}K_3\right\} \equiv k_6[M][OH](1+\xi), \qquad (IX.40)$$

where K_{17} is the equilibrium constant of reaction VIII.17 and K_3 is the equilibrium constant of reaction (iii) in Sect. IV.5a ($K_3 \triangleq K_{iii}$ as defined by Eq. IV.5). In a given flame of uniform temperature ξ is a constant, independent of the actual radical concentrations and the state of ionization. The factors $[M]$ and $[OH]$ in the above equation depend on the radical concentrations and may thus vary with t. Since the ion concentrations are usually much smaller than $[M]_t$, the ratio $[M]/[M]_t$ can be expressed in $[H_2O]$, $[H_2]$ and $[H]$ and in the equilibrium constants of reactions VIII.12, VIII.17 and VIII.30 (see Hayhurst and Kittelson 1974a). The concentrations of the major constituents, $[H_2O]$ and $[H_2]$, are virtually equal to their equilibrium values and can be calculated as described in Sect. IV.4. $[H]$ may be present in excess over equilibrium; however, it can be determined as a function of z by one of the experimental methods outlined in Sect. IV.5b-2. Once $[H]$ is known, $[OH]$ follows from: $[OH]/[OH]_e = [H]/[H]_e$ (see Eq. IV.5) because $[H]_e$ and $[OH]_e$ can be calculated. Consequently we can rewrite Eq. IX.40 as

$$G = k_6(1+\xi)[M]_t \gamma^3 F(\gamma), \qquad (IX.41)$$

where $\gamma \equiv [H]/[H]_e$ and $F(\gamma)$ is a known function of γ, involving the equilibrium concentrations of H, H_2 and H_2O (see Hayhurst and Kittelson 1974a). The dependence of G on γ, and thus on z or t, is mainly governed by the cubic γ factor.

In the flames investigated $[M^+]$ appears to be much smaller than $[MOH^+]$. In the term describing the ion-recombination rate R we can therefore replace $[MOH^+]$ by n_i and get, using $[e^-]=n_i$,

$$R \simeq k_r n_i^2. \qquad (IX.42)$$

Inserting in Eq. IX.39 the expressions for G and R from Eqs IX.41 and 42 we finally get

$$dn_i/dt = k_6(1+\xi)[M]_t \gamma^3 F(\gamma) - k_r n_i^2, \qquad (IX.43)$$

which is the basic equation for determining the product $k_6(1+\xi)$ and k_r from experiments (see Sect. 5a). The separate values of k_6 and ξ (or k_7) cannot, however, be determined.

For low z- or t values, i.e. near the combustion zone, n_i will be relatively small and γ high; the first term in the right-hand side of Eq. IX.43 then dominates and dn_i/dz is positive. The reverse may hold for large z in cool

Chapt. IX IONIZATION IN SEEDED FLAMES

flames where γ^3 drops markedly with increasing z. A maximum in n_i is then found; the location of this maximum depends on $[M]_t$.

3d CONDITIONS FOR THERMAL IONIZATION AND CAUSES OF NONTHERMAL IONIZATION

Thermal ionization. According to the definition given in Sect. 2a an atomic or molecular species X is in a state of thermal ionization at a given position if

$$[X^+][e^-]/[X] = K_i(T), \quad (IX.44)$$

where $K_i(T)$ is the ionization constant at local flame temperature T. When the rates of the <u>dominant</u> ionization and neutralization process can be written as

$$G = k_i[X] \quad (IX.45a)$$

$$R = k_r[X^+][e^-], \quad (IX.45b)$$

thermal ionization thus implies that

$$k_i/k_r = K_i(T). \quad (IX.46)$$

It is not necessary for k_i and k_r themselves to be equal to their equilibrium values, $k_i(T)$ and $k_r(T)$. For example, when k_i and k_r are proportional to the density of one and the same collision partner (flame molecule or electron; see processes i and ii in Table IX.1), any deviation of this density from its equilibrium value will affect k_i and k_r alike and thus not their ratio or the state of ionization of X.

Ionization relaxation. Even when $k_i = k_i(T)$ and $k_r = k_r(T)$ and thus Eq. IX.46 is satisfied, a deviation from thermal ionization can exist as a result of ionization relaxation. When in a flame of uniform temperature and composition $[X^+] \neq [X^+]_e$ at $t = 0$, a deviation due to relaxation will exist for travel times $t \leqslant k_i^{-1}$ (cf. Eq. IX.32). In nonuniform flames deviations will persist also for longer travel times when the temperature gradient along the flame axis is so strong that the metal ionization lags continuously behind the local thermal ionization state determined by the local ionization constant. In other words, the curve depicting $[X^+]$ as a function of height z is then shifted with respect to the theoretical *Saha curve* describing the thermal ionization state as a function of z (see Case A in Sect. 5a).

When the temperature changes with height, K_i and k_i in Eq. IX.31 are to be considered as functions of z or t. If in a situation where this equation is applicable thermal ionization were to exist at any z, that is, if $[M^+] \equiv [M^+]_e$ and $[M] \equiv [M]_e$ at any t, the expression between braces on the right-hand side of this equation must be zero. However, when K_i and thus $[M]_e$ vary with t, the

SOME USEFUL THEORETICAL RELATIONSHIPS   Sect. IX.3d

left-hand side of this equation cannot be zero if [M] is to follow $[M]_e$ everywhere. Consequently, thermal ionization in a strict sense cannot co-exist with a height-dependent temperature. However, the inevitable deviation from thermal ionization in the presence of a given T-gradient (i.e. the deviation from unity of the second term between braces in Eq. IX.31) may be negligibly small, but still not zero, when k_i is sufficiently large; it is the product of this deviation and k_i that determines $d\ln[M]/dt$. In order for [M] to follow closely $[M]_e$ as a function of z in the presence of a T-gradient, k_i^{-1} need only be small compared to the time interval in which $[M]_e$ changes by, say, a factor e.

When in an actual situation where Eq. IX.31 is applicable the function $[M](z)$ happens to attain an extremum at some point z_0, a state of thermal ionization exists exactly at that point irrespective of the magnitude of k_i. This conclusion holds because at $z = z_0$ the left-hand side of Eq. IX.31 and therefore also the expression between braces in the right-hand side are strictly equal to zero. In other words, the actual $[M](z)$ curve and the theoretical Saha curve intersect at the very point where the former curve attains its extremum (see also Sect. 5a sub Case A).

Physical and chemical disequilibrium. Another cause of deviations from thermal ionization is the invalidity of Eq. IX.46 resulting from a physical or chemical disequilibrium which affects the magnitude of k_i and/or k_r for the dominant ionization or neutralization process. For example, when 'hot' electrons are present (see Sect. IV.7), we expect $k_i > k_i(T)$ when process (ii) in Table IX.1 is dominant. Another example of physical disequilibrium is found when one of the Na excitation levels is saturated by one- or two-photon absorption from an intense laser beam (see Sect. II.5a-2). The ionization rate constant will be strongly enhanced above its equilibrium value because the activation energy for collisional ionization from the excited state is reduced by the amount of the excitation energy and a significant fraction of the atoms is found in the saturated state (see also Sect. 2a).

When H_3O^+ ions are present in excess (see Sect. IV.7) and process (xii) is the dominant ionization path, the monomolecular rate constant k_i will exceed its thermal value by a factor $[H_3O^+]/[H_3O^+]_e$. Suprathermal metal ionization may then occur.

Chemical disequilibrium of the flame radicals H, OH and O (see Sect. IV.5) also causes nonthermal ionization when these radicals are involved in the dominant ionization or neutralization process. For example, excess O radicals existing near the combustion zone in CO flames make the product [CO][O] exceed its equilibrium value, whereas $[CO_2]$ is hardly affected. The value of k_i for the chemical ionization process (viii), which is proportional to $[CO][O]$, then exceeds $k_i(T)$, whereas k_r for the reverse process is hardly affected. This, again, may lead to excess ionization of the alkali atoms as will be discussed in Sect. 5a sub Case A.

A more complicated situation exists with regard to the ionization of Ca, Sr and Ba, which is dominated by processes (vi) and/or (vii) as a primary step, and by process (xi) as a secondary step for the formation of atomic ions. Assuming that these binary processes are balanced (which need not always be the case; see Sect. 3c sub Case C), the steady-state densities of the ions are governed by the relations

$$[MOH^+][e^-]/[M][OH] = K_{vi} , \qquad (IX.47)$$

$$[M^+][H_2O]/[MOH^+][H] = K_{xi} , \qquad (IX.48)$$

where K_{vi} and K_{xi} are the equilibrium constants of processes (vi) and (xi). Ionization process (vii), which in flames is indistinguishable from process (vi) (see Sect. 2b), leads to the same result and need not be considered separately. Furthermore, a partial equilibrium exists for: $M + H_2O \rightleftharpoons MOH + H$, at least in fuel-rich H_2 flames, which leads to (see Sect. VIII.4b and Eq. VIII.12)

$$[MOH][H]/[M][H_2O] = K_1 . \qquad (IX.49)$$

Combination of Eqs IX.47 and 49 immediately results in

$$[MOH^+][e^-]/[MOH] = (K_{vi}/K_1)([OH][H]/[H_2O]) . \qquad (IX.50)$$

Since $[H_2O] \simeq [H_2O]_e$ and $[H_2] \simeq [H_2]_e$ we find from Eq. IV.5 that:

$$[OH]/[OH]_e \simeq [H]/[H]_e \equiv \gamma$$

so that Eq. IX.50 can be rewritten as

$$[MOH^+][e^-]/[MOH] = K_i'(T) \gamma^2 . \qquad (IX.51a)$$

Here $K_i'(T)$ is the ionization constant of the process: $MOH \rightleftharpoons MOH^+ + e^-$, given by

$$K_i'(T) = K_{vi} K_{H_2O}/K_1 , \qquad (IX.51b)$$

where K_{H_2O} is the dissociation constant of the reaction: $H_2O \rightleftharpoons H + OH$. The deviation of MOH^+ from thermal ionization thus increases proportionally to γ^2; this factor may largely exceed unity in cool flames.

Equation IX.51b may be used to derive $K_i'(T)$ from a measured value of K_{vi} and known values of K_1 and K_{H_2O}; from $K_i'(T)$ the ionization energy of MOH can, in turn, be derived through the expression for $K_i'(T)$ from Statistical Mechanics (see Eq. II.60a). The ionization energy can also be derived directly from its relation with the reaction enthalpies corresponding with K_{vi}, K_1 and K_{H_2O}. The enthalpy of the chemi-ionization reaction (vi) can be determined from measurements

SOME USEFUL THEORETICAL RELATIONSHIPS   Sect. IX.3d

of K_{vi} in a similar way as the dissociation energy is determined from the measured dissociation constant (see Sect. VIII.5a; see also Jensen 1968, Kelly and Padley 1971a, and Hayhurst and Kittelson 1974a).

Combination of Eqs IX.48, 51a and 51b yields

$$[M^+][e^-]/[M] = K_{vi} K_{xi} K_{H_2O} \gamma^2 ,\qquad(\text{IX.52})$$

where $(K_{vi} K_{xi} K_{H_2O})$ equals the ionization constant of the process: $M \rightleftharpoons M^+ + e^-$. The deviation of M^+ from thermal ionization thus increases $\propto \gamma^2$, too.

Nonthermal ionization due to radical disequilibrium also occurs with the negative halogen ions. The density of, e.g., Cl^- ions is mainly governed by the balanced process (ix). We thus have

$$[Cl^-][H]/[HCl][e^-] = K_{ix} .\qquad(\text{IX.53})$$

According to Eq. VIII.33 we also have, because of the balanced reaction: $H_2 + Cl \rightleftharpoons HCl + H$,

$$[HCl][H]/[H_2][Cl] = K_7 .\qquad(\text{IX.54})$$

Combination of the last two equations yields

$$[Cl][e^-]/[Cl^-] = [H]^2/[H_2] K_7 K_{ix} .\qquad(\text{IX.55})$$

Since $[H_2] \simeq [H_2]_e$ it follows that the deviation from thermal ionization varies here again $\propto \gamma^2$.

Conclusion. It is not really possible to draw definite conclusions about the occurrence of thermal or nonthermal ionization in seeded flames. Much depends on the properties of the flame used (temperature and temperature-gradient, total pressure, kind of fuel gas, extent of radical disequilibrium, natural flame ionization), and on the kind of element (ionization and activation energy, total metal density, kind of ionic species and of compounds formed). Other important factors may be the presence of concomitants in the solution sprayed, especially of those that form electron acceptors or solid particles, the height of observation and the rise-velocity, the possibility of thermionic emission from solid particles and the presence of applied d.c. fields. In the primary combustion zone the state of metal ionization is in general nonthermal because of lack of physical and chemical equilibrium and because of the very short residence time (see Sect. 3c). As to the interzonal region one can only say that, on the whole, thermal ionization is more readily attained

(1)   in the hotter flames, where the ionization rate constants are larger and radical disequilibria less pronounced;

Chapt. IX   IONIZATION IN SEEDED FLAMES

(2)   for ions that are produced by a bimolecular chemical reaction in which
        (part of) the chemical energy released is used for ionization (see, e.g.,
        processes vi and vii); termolecular chemi-ionization (e.g., process viii)
        and collisional ionization (e.g., process i) occur generally at a markedly
        slower rate;

(3)   for elements with a low ionization energy, as they usually also have a low
        activation energy for ionization (see Sect. 5; compare the ionization ener-
        gies listed in Appendix A.4);

(4)   when ionization is catalyzed by the occurrence of an ionic re-arrangement
        process involving H_3O^+ ions (e.g., process xii) or $SrOH^+$ ions (process
        xi); the latter ion might also catalyze the Na ionization (see Schofield
        and Sugden 1965); the presence of H_3O^+ ions may, however, also lead to
        suprathermal ionization of species with high ionization energies, which
        would otherwise exhibit little or no ionization (see processes xii-xv);

(5)   further downstream from the combustion zone where deviations due to ioniza-
        tion relaxation or radical disequilibrium are less significant.

                              4. EXPERIMENTAL TECHNIQUES

        In this section we discuss various experimental techniques that have been
used for ionization measurements in flames. In Sect. 4a we present a survey of
experimental techniques, mentioning only their principal features; in Sect. 4b we
discuss the techniques separately and in more detail, in particular the underlying
formulas and experimental conditions. In our discussions we pay special attention
to the techniques that have been applied in the three special cases of ionization
rate constant measurements that will be discussed in Sect. 5.

        It should be emphasized that we discuss only techniques for stationary-state
measurements.

4a   SURVEY OF EXPERIMENTAL TECHNIQUES

        Table IX.2 lists the various experimental techniques used and summarizes
the specific information that can be derived herefrom (see also van Tiggelen 1968,
Lawton and Weinberg 1969, Jensen and Travers 1973, Miller 1976, and Gaydon and
Wolfhard 1979).

        In Table IX.2 techniques based on the measurement of electron noise tempera-
tures and d.c. current resistance, and on d.c. field effects (known as the
electric wind of Chattock) are not listed; these methods are not relevant for
the discussions in this chapter. The interested reader is referred for details
to Wilson (1944), and Lawton and Weinberg (1969).

EXPERIMENTAL TECHNIQUES   Sect. IX.4a

One should bear in mind that the various experimental methods yield quite different information (see Table IX.2): some methods tell us about the nature and concentration of the neutral particles, whereas others yield only free-electron concentrations. Other methods reveal the concentration and the kind of ions.

With *optical spectrometry* it is possible to detect and measure ionic concentrations either by direct observation of their optical spectra or by measuring the reduction in intensity of neutral atomic spectra caused by ionization.

The *ion mass spectrometer* samples and detects directly, i.e. without ionization chamber, positive and negative ions in flames at concentrations down to 10^6 ions cm^{-3} at flame pressures between 10^{-3} and 1 atm. The aim of sampling (through a small hole) is to obtain gas from a well-defined point in the flame and to quench any chemical reactions immediately, so that the composition of the sample is that of the gas in the flame at the sampling point. The ions are selected according to their charge/mass ratio usually by means of quadrupole filters; from the measured ion beam intensity the ionic concentrations are found.

Current-tension characteristics of *electrostatic probes* placed in flames have been widely used in attempts to determine positive-ion and free-electron concentrations as well as electron temperatures in flame plasmas. The technique of electrostatic probes has the advantage of good spatial resolution over a wide range of conditions while the probes are relatively easy to construct, but an interpretation of results obtained with them is fraught with difficulties. One can distinguish between *single* and *double probes*. This distinction is in the first instance purely geometric: the single probe is a highly asymmetric arrangement of a large reference electrode, well removed spatially from a much smaller electrode, the proper probe; the components of the double probe are normally of equal size and shape, and are situated in the same region of the flame plasma.

The single-probe characteristic contains information about $[M^+]$, $[e^-]$, electron temperature T_e, the floating potential and the plasma potential (see definitions in Sect. 4b).

Use of the floating double-probe technique removes some of the experimental difficulties associated with the single probe (see Sect. 4b), but at some cost: the double-probe technique yields information about $[M^+]$ and T_e only.

The disadvantage of using electrodes or probes in contact with the hot flame gases can be circumvented by the application of *electric high-frequency techniques*. The electric h.f. techniques can be distinguished according to whether radio-frequencies ($\approx$ 1 - 100 MHz) or microwave (MW) frequencies are applied.

In the *radio-frequency (r.f.) resonance technique* resonance of a r.f. circuit incorporating a flame plasma is used to determine free-electron concentrations. The method essentially involves measuring the change in Q-factor (or half-

IONIZATION IN SEEDED FLAMES

Table IX.2
Survey of Experimental Techniques Commonly Used in Flame Ionization Measurements

	Technique	Quantity measured directly	Quantity derived [+][*]	Absolute/ relative [†]	Freedom from flame disturbance ϕ	Spatial resolution ϕ	Minimum [‡] $[e^-]$, $[M^+]$ (cm^{-3})	Maximum [‡] $[e^-]$, $[M^+]$ (cm^{-3})	References
a	Optical spectrometry	Atomic (ionic) line intensity	B_i; $[M]$ or $[M^+]$, $[M]_t$	Relative; absolute	+	++	10^8	10^{11}	(1), (2)
b	Ion mass spectrometry	Mass/charge; beam intensity	Kind and concentration of ionic species	Relative	-	++	10^7	...	(3)-(8)
c	Electrostatic probe:								
c.1	Single probe	Positive or negative probe current (as function of probe tension)	$[M^+]/\sqrt{m_i}$, T_e; $[e^-]$; m_i/m_e	Relative (absolute from theory)	0/-	++	10^7	10^{13}	(9)-(11)
c.2	Double probe		$[M^+]/\sqrt{m_i}$; T_e		-	++	10^6	10^{13}	(12)-(15)
d	R.f. resonance	Damping and shift of resonance frequency	$[e^-]/\nu_e$; $([M^+]/m_i; \nu_i)$	Relative (absolute)	+	0	10^7	...	(16)-(18)
e	Cyclotron resonance	Peak and width of resonance curve	$[e^-]$; ν_e	Relative; absolute	+	0	...	...	(19)-(21)
f	Microwave:								
f.1	MW cavity resonance	Damping	$[e^-]$	Relative (absolute if ν_e is known)	0	+/0	10^8	10^{11}	(22), (23)
f.2	MW beam attenuation	Attenuation factor	$[e^-]$; ν_e		+	0	10^9	10^{12}	(24)-(26)

EXPERIMENTAL TECHNIQUES   Sect. IX.4a

NOTES TO TABLE IX.2

† For meaning of symbols see Sects. 2 and 3 or Appendix A.3.

* Quantity that follows from directly measured quantity without making specific assumptions about the system studied.

‡ Relative measurements calibrated by using a known external reference are not called absolute.

φ As to freedom from flame disturbance: + means no disturbance, 0 means some disturbance, − means much disturbance. As to spatial resolution: ++ means usually better than 1 mm; + means better than 1 cm; 0 means of the order of 1 cm; − means of the order of 10 cm or more.

‡ The lowest/highest electronic or ionic density that can be quantitatively detected is roughly indicated.

REFERENCES

1. Hollander, Kalff and Alkemade (1963)
2. Hollander (1964)
3. Hayhurst (1974)
4. Jensen and Miller (1970)
5. Hayhurst, Mitchell and Telford (1971)
6. Hayhurst and Telford (1977)
7. Hayhurst, Kittelson and Telford (1977)
8. Hayhurst and Kittelson (1977); Burdett and Hayhurst (1979)
9. Calcote (1962, 1963)
10. Soundy and Williams (1965)
11. Kelly and Padley (1969)
12. Bradley and Ibrahim (1975)
13. Travers and Williams (1965)
14. Bradley and Matthews (1967, 1967a)
15. Silla and Dougherty (1972)
16. Borgers (1965)
17. Knewstubb and Sugden (1958)
18. Borgers, Jongerius, Hollander and Alkemade (000)
19. Hofmann, Kohn and Schneider (1961)
20. Bulewicz (1962)
21. Bradley and Tse (1969)
22. Jensen and Padley (1966)
23. Sugden and Thrush (1951)
24. Jensen and Travers (1973)
25. Balwanz (1965)
26. Jensen (1965).

intensity width) or the shift of the resonance frequency resulting from the presence of charge carriers in the flame which forms a part of the di-electric of a capacitor or is embraced by an inductance coil in a resonant circuit.

The (*electron*) *cyclotron resonance technique* is closely related to the MW attenuation technique (see below). The flame is placed in a homogeneous magnetic field B. Microwave radiation propagates perpendicular to the B field with the E-vector directed also perpendicular to the B field. At fixed MW frequency one varies B and thus the cyclotron resonance frequency: $\omega_c = eB/m_e c$ (see Sect. 3a). One thus obtains resonance curves the integral of which is a direct measure for $[e^-]$, independent of ν_e. The width of the resonance curve is determined by $2\nu_e/\omega$ and yields the value of ν_e (see Eq. IX.6).

The *microwave techniques* have the advantage of greater simplicity than r.f. techniques. In the *microwave resonance technique* the damping of a cavity through the presence of electrons in a flame, which is (partially) contained in the cavity, is measured.

The *microwave attenuation technique* is irrelevant for the type of flame measurements to be discussed in Sect. 5 (see Jensen and Travers 1973, and Miller 1976). This technique is relatively simple and easy to calibrate if ν_e is known (see Eq. IX.10); relative measurements are feasible if ν_e is unknown. It has the disadvantages of poor spatial resolution and a poor detection limit (see Table IX.2). The value of ν_e can be found by measuring σ' as a function of ω.

4b   DETAILED DISCUSSION

Techniques that have been actually applied to the measurement of ionization rate constants to be considered in Sect. 5 will here be more extensively discussed than other techniques.

(a)   *Optical spectrometry*. Measurements of atomic and ionic spectral line intensities in emission or absorption yield relative or absolute densities of the species characterized by the wavelength of the line (see Sect. V.2). We present some examples:

(i)   If the absolute value of [M] is measured, $[M^+]$ (= $[e^-]$) follows indirectly, provided that only M^+ ions and M atoms are present under the flame conditions considered, and that the total density $[M]_t$ is known. The $[M]_t$ value can be determined by the experimental methods discussed in Sect. V.2.

(ii)   The β_i factor, being equal to $[M^+]/[M]_t$, can be derived (under the same experimental conditions as under i) from the measurement of relative [M] values with and without ionization suppression by addition of an electron donor element (see Sect. 3b-2).

EXPERIMENTAL TECHNIQUES   Sect. IX.4b

(iii)   If molecular formation also occurs, one can determine $[M^+]/[M]_t$ by measuring relative $[M]$ values with and without ionization suppression, if the total association factor ϕ_t is known. One therefore applies Eqs IX.20 and 20a, with $\alpha \equiv 0$ (see Sect. 3b-3).

The *advantages* of the optical spectrometric techniques (see also Table IX.2) are the following: no flame disturbance, good spatial resolution,[†] absolute calibration of densities $[M]$, $[M^+]$ and $[M]_t$ and hence of electron densities; no assumptions have to be made about collisional frequencies of electrons or ions. Furthermore the technique involves simple apparatus; accurate measurements of relative densities are feasible (for example, as a function of height of observation or of the concentration ratio of two elements). This technique works well down to low densities.

Among the *disadvantages* we mention that the determination of $[M^+]$ through the measurement of $[M]$ and $[M]_t$ is only feasible when β_i is not too small. Direct measurement of $[M^+]$ through ionic line intensities is only possible in a few cases.

In general: optical spectrometry is often a welcome technique as an additional means and/or check on the other, electric methods (see, e.g., Jensen and Padley 1966, Kelly and Padley 1972, and Borgers 1978).

(b)   *Ion mass spectrometry.* This technique makes use of a quadrupole mass filter which offers the following *advantages* for direct sampling from flames: the ion filtration is not strongly dependent upon incoming ion energy; mass selection and resolution can be readily varied; operating pressures in the beam-formation chamber may be kept relatively high because of short ion transmission path length, and there are no magnetic fields to perturb the flame studied. For experimental details the reader is referred to Jensen and Miller (1970), Hayhurst, Mitchell and Telford (1971), Hayhurst, Kittelson and Telford (1977), Hayhurst and Kittelson (1977), Hayhurst and Telford (1977), and Burdett and Hayhurst (1979). The method is applicable to mass numbers between about 10 and 400.

The design of the flame sampling system is a major problem; across this system the pressure must drop as rapidly as possible from that of the flame to a level (between about 10^{-7} and 10^{-6} atm) at which chemical reactions are effectively quenched. Usually a vertically positioned, cylindrical flame is examined through a coaxial sampling cone with a small hole at its tip; typical diameters of this hole are between 0.03 and 0.06 mm. This diameter is a compromise between significant

---

[†] When a pencil beam of flame radiation is detected and the flame is homogeneous over its cross section (as to n, T, etc.) good height resolution means good spatial resolution.

quenching of chemical reactions and prevention of orifice clogging. The cone used is mounted on a water-cooled plate; its tip protrudes through the boundary layer between this plate and the flame. The tip must become hot enough to prevent condensation of involatile flame products. The cone angle chosen represents a compromise between minimum flame disturbance and maximum pumping efficiency, the latter minimizing the number of molecule-wall collisions during sampling (see also Hayhurst 1974). Precautions are taken to ensure that there is no transmission of species that have undergone collisions in the boundary layer close to the wall of the cone. Overall, approximately 1% of the ions drawn through the orifice are registered by the detector. By varying the position of the burner with respect to the mass spectrometer orifice the flame locus under observation can be chosen. Hayhurst (1974) states that, with the set-up described, fully representative samples of flame gas can be obtained without chemical reactions altering the composition, provided that the relaxation times of these reactions are larger than 1 µs. However, the formation of hydrated ions during the sampling process is a serious complication. In the case of $H_3O^+ \cdot H_2O$ formation it has been reported by Hayhurst and Kittelson (1977) that the experimental ratio $[H_3O^+ \cdot H_2O]/[H_3O^+]$ found in a flame at 2000 K is about 500 times the calculated ratio.

No authors have so far attempted to determine absolute ion concentrations in flames by mass-spectrometric sampling. Instead, ion concentrations are usually obtained via comparison of observed ion detector currents with currents provided by alkali metal ions at known flame concentrations. The latter are either measured by other techniques or calculated for conditions where thermal ionization is known to be closely approached (see Sect. 3b). (Cs is mostly used on the assumption that there is complete ionization at low concentration.) In these determinations of ion concentrations it is frequently necessary to compromise between the conflicting requirements of high mass resolution and good sensitivity: in practice a mass resolution[†] of 155 could be achieved with a detection limit of below 10^7 ions cm^{-3} (see Jensen and Miller 1970).

Problems and *disadvantages* relating to these mass-spectrometric techniques are connected with the calibration procedure (see above) and the extent of flame disturbance (see Hayhurst, Kittelson and Telford 1977, and Hayhurst and Kittelson 1978). Furthermore one may wonder whether the sample that is taken at a certain position in the flame is representative for the situation at that position; in other words how do the presence of the orifice and the sampling process affect the local

---

[†] This mass resolution is defined as follows: when an ion of mass 155 was observed the ion current fell to 10% of its peak value at indicated mass numbers of 154.5 and 155.5.

flame conditions? Finally the question of hydration during the sampling process is an obvious complication (see above).

(c) *Electrostatic probe technique.* In Fig. IX.4 a typical arrangement is shown for *single-probe* measurements. The probe, usually a thin cylindrical wire or a small sphere, can be biased either positively or negatively with respect to the reference electrode also placed in the flame. In this way a current-tension (i_p-V_p) characteristic can be obtained (see Fig. IX.5). Over the region AB the probe has a sufficiently strong negative bias, with respect to the ionized gas, to repel electrons. Under these circumstances the probe current is almost solely due to positive ions. As V_p is made less negative, the current falls owing to a decrease in the flux of positive ions reaching the probe. In the region BC the electron current becomes appreciable and, at point C, is equal to the positive ion current.

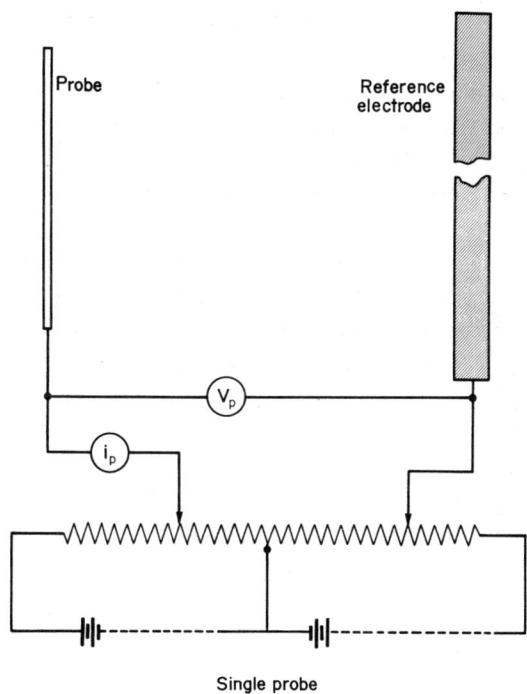

Fig. IX.4 Typical arrangement for single electric probe measurements. (From Lawton and Weinberg 1969.)

Chapt. IX   IONIZATION IN SEEDED FLAMES

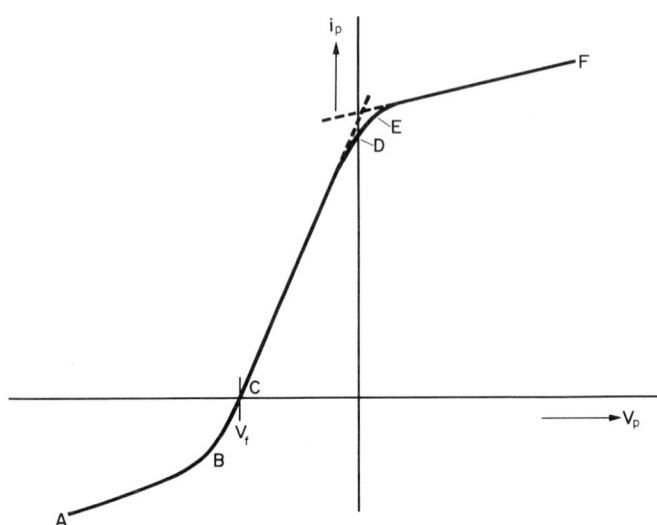

Fig. IX.5 Current-tension characteristics of a single probe in a flame. In practice the dashed lines intersect at $V_p = 0$. (From Lawton and Weinberg 1969.)

The potential of the probe at point C is called the *floating potential* V_f. In the region CD more electrons reach the probe than positive ions, causing the current to change direction. At the point D the probe is at *plasma potential*. At the point E the positive ion current is reduced to an insignificant level by the repulsive field, and the increase in net current in the region EF is due to an increase in the number of electrons reaching the probe under the action of the attractive field.

The interpretation of these characteristics to obtain reliable information about ion and electron concentrations and electron temperatures has posed severe theoretical problems (see Langmuir 1961). On the one hand, one has to cope with the mutual influencing of probe and flame plasma; on the other hand, one has to make sure that the conditions underlying the theoretical formulas used in the interpretation are realized in the experimental set-up.

For true single-probe behaviour to be exhibited it is necessary that the whole externally applied tension is effective in defining the potential-difference between probe and plasma and that little or none of this tension serves to increase the positive space-charge surrounding the reference electrode. This requirement implies (see Jensen and Travers 1973) that the current-collecting area for the reference electrode should be of the order of 10^5 greater than that of the probe: in other words, extremely small probes are necessary. Yanagi (1968, 1968a) making

EXPERIMENTAL TECHNIQUES Sect. IX.4b

use of a third, ring-shaped electrode has overcome the difficulty of maintaining a constant reference potential. Even if the reference potential remains well-defined, identification of the plasma potential from the current-tension characteristic may be difficult or uncertain if the electron drain on the plasma is high.

Calcote and co-workers (see, e.g., Calcote and King 1955, and Calcote 1962, 1963) applied a cylindrical single-probe technique to determine $[M^+]$, $[e^-]$ and T_e over a wide range of flame compositions, temperatures and pressures.

Soundy and Williams (1965), working with a negatively biased spherical probe in a shielded atmospheric-pressure H_2-O_2-N_2 flame, circumvented some problems of interpretation by in situ calibration of the probe current at a fixed probe bias against known levels of flame ionization. Use of this approach has since been extended by Kelly and Padley (1969, 1969a, 1970, 1971, 1972) in their rotating single-probe measurements in hydrogen flames at 1 atm (see Fig. IX.6).

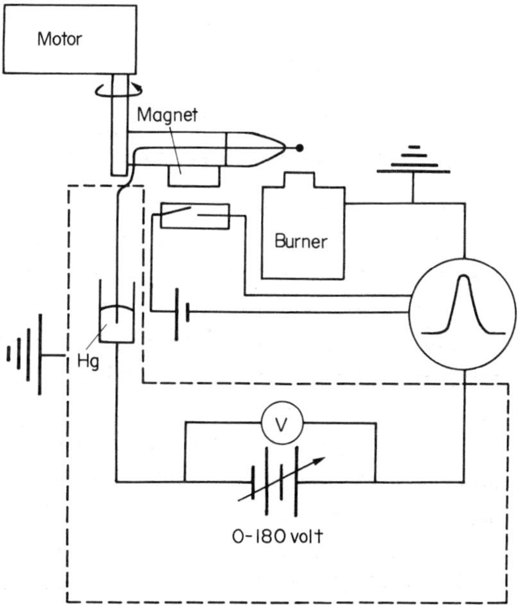

Fig. IX.6 Schematic diagram of probe assembly. (From Kelly and Padley 1969.)

Kelly and Padley used a probe consisting of a platinum sphere with radius 0.267 mm on the end of a piece of platinum wire with a much smaller radius. This probe rotated once per second in a horizontal circle. The residence time in the vertically burning flame was 0.02 s or less. At a fixed probe tension V_p of 100 V at a given height in a given flame the relation between the probe current i_p and the metal solution concentration c was determined experimentally for Na, K, Rb and Cs in a wide molarity range (10^{-5} to 1). It was found that for the metals

Chapt. IX   IONIZATION IN SEEDED FLAMES

studied i_p was proportional to $c^{1/1.3}$; MOH formation was accounted for here. At $V_p = 100\,V$ the probe equation used by Kelley and Padley was

$$[M^+] = A'' \, i_p^{1.3} \, V_p^{-0.65} \, T^{-2.9} \,. \tag{IX.56}$$

This relationship proved to be valid for the whole $[M^+]$ range considered (see Table IX.2 on p. 852) i.e. from 2×10^7 up to 5×10^{12} ions cm^{-3}. The values of A' ($\equiv A'' \, T^{-2.9}$) where found by comparing the probe current i_p (at fixed V_p) and calculated values of $[M^+]$ under (near-)equilibrium conditions; the latter values were derived from experimental data about $[M]_t$, the ionization constant and ϕ ($\equiv [MOH]/[M]$) (see Sect. 3b-3).

It should be noted that the A' value was determined by making use of data from MW cavity measurements (see Jensen and Padley 1966) at known travel time (see Eq. IV.1). This A' value proved to vary only slightly (within less than a factor two) for the different alkali metals, but its value strongly depended on the probe size; the experimental relation between $[M^+]$ and i_p however did not depend on probe size or shape. The temperature-dependence of A' was established by Kelly and Padley (1971).

As to the *double-probe* technique: there is a great advantage to be gained from using two probes of comparable area when measuring electron temperatures. Excessive drain of charge from the plasma is avoided here; excessive drainage would result in a shift of plasma potential with probe tension and a consequent error in the measured T_e value (see Travers and Williams 1965). In the simplest case of identical probes the i_p-V_p characteristic is symmetric with respect to the point where the current is zero (the average potential of the probes being equal to the plasma potential).

In most cases the double-probe technique is used for determining T_e rather than $[M^+]$. The measurement of $[M^+]$ is difficult, but that of T_e is readily carried out (see Swift and Schwar 1970). We mention the double-probe measurements in low-pressure hydrocarbon-air and -oxygen flames by Travers and Williams (1965), in CO- and H_2 flames with added methane by Bradley and Matthews (1967, 1967a) and in CH_4-air flames by Bradley and Ibrahim (1975). In the interpretation of their results Bradley and co-workers used the boundary-layer theory of Su and Lam (1963). Bradley and Ibrahim (1975) have also reviewed the derivation of T_e in flame gases and compared the results with available theories.

In *summary*, electrostatic probes provide a reasonably reliable means of measuring positive ion concentrations in flames, and can also be used to measure T_e values. The validity of their use for determining free-electron concentrations is still a matter of dispute (see Jensen and Travers 1973). Electrostatic probe

EXPERIMENTAL TECHNIQUES

Sect. IX.4b

techniques offer good spatial resolution and sensitivity; in the case of rotating single probes the flame disturbance is (probably) negligible and calibration can be performed easily.

(d) *R.f. resonance technique*. With this technique one can determine the conductivities σ' and σ'' of a flame plasma (see Smith and Sugden 1952, Knewstubb and Sugden 1958, Borgers 1965, 1978, and Borgers *et al.* (000); see Sect. 3a). The flame is placed between two (usually plane) condenser plates which are mounted as closely as possible to the flame border to achieve a compromise between maximum sensitivity and minimum flame disturbance. In this way the flame forms part of the di-electric of a condenser. With a parallel r.f. coil one obtains a resonance circuit that can be tuned at a frequency of, say, 20 MHz. The resonance curve of this circuit is determined by means of a sweep generator the frequency of which is varied periodically at a low frequency in the relevant frequency interval. After rectifying the r.f. signal one obtains a periodically varying d.c. component that, again, is fed into an oscilloscope, the time base of which is synchronized with this periodic signal, or is stored digitally. The oscilloscope screen displays the resonance curve (see Fig. IX.7).

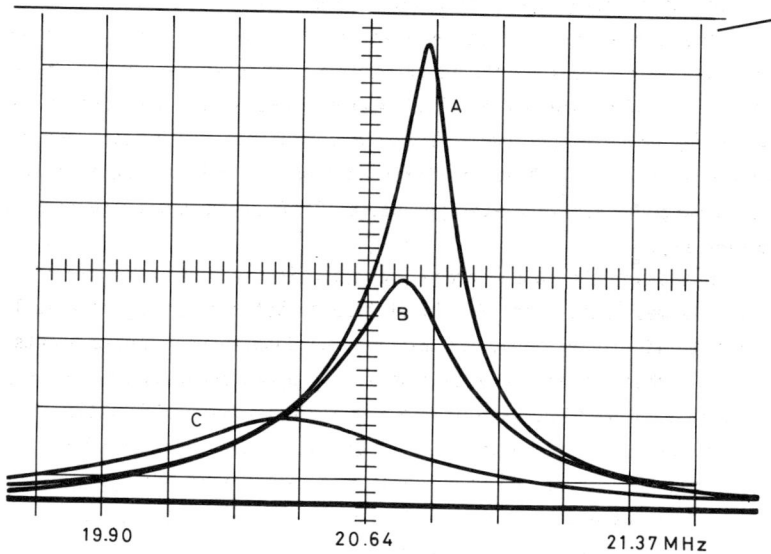

Fig. IX.7 Resonance curves displayed on an oscilloscope screen. A: metal-free $CO-O_2-N_2$ flame; B: same flame with 160 ppm Rb sprayed; C: ditto with 20,000 ppm Rb. (According to measurements by Dr A.J. Borgers at the Physical Laboratory, University, Utrecht.)

The introduction of ionizable metal vapour into the flame will bring about changes in σ' as well as σ''. Consequently the damping of the resonance circuit increases and with it the half-intensity width of its resonance curve while the resonance frequency is shifted (see Fig. IX.7). In principle one may derive the electron concentration from the magnitude of either effect, but a cumbersome calibration procedure is required. The question of calibrating the r.f. resonance method has not been answered completely. Borgers et al. (000) calibrated the changes in half-intensity width of the resonance curves, Δb, by means of additional optical spectrometric measurements of $[M]$ and/or $[M]_t$.

Advantages of the r.f. resonance technique are the good detection limit and the possibility of getting information about ion concentrations and electron mobilities. The *disadvantages* of the method are:

(i) poor spatial resolution;

(ii) a very complicated electronic set-up is needed, because a high degree of stability and absence of systematic errors (for example, due to nonlinearity of the rectifying process) are required;

(iii) an elaborate calibration procedure is needed to obtain absolute electron (and/or ion) densities; as an alternative, elaborate model calculations have been made to avoid the calibration procedure;[†]

(iv) the first 3-4 cm of the flame above the combustion zone cannot easily be investigated because of possible disturbances due to the presence of the metallic burner top (with ceramic burner tops however this problem can be overcome); the diameter of the condenser plates, the presence of a guard-ring condenser and the diameter of the flame are also factors that limit the usable height interval for r.f. resonance measurements;

(v) contributions of ions to the conductivity may have to be taken into account; these contributions are frequency-dependent (see Borgers 1978).

(e) *Cyclotron-resonance technique.* Here one measures the attenuation of the microwave radiation passing through the flame part studied. From Eq. IX.6 it follows that for $\omega = \omega_c$ a maximum in this attenuation occurs; the value of ω_c lies in the MW range, as ω_c is of the order of $10^{11}-10^{12}$ s^{-1} for B values of the order of 6-60 kgauss, which are common B values in this method (see Sect. 3a). Cyclotron

---

[†] When using a guard-ring condenser for improving the homogeneity of the electric field across the flame, the model calculations are somewhat simpler (see Borgers 1978 and Borgers et al. 000). The experimental calibration procedure is also simpler when only relative measurements are performed.

resonance is detectable only when ν_e is not too large; therefore this method has been applied in low-pressure flames, typically in the range $(1-5) \times 10^{-2}$ atm. The upper bound of ν_e depends on the achievable value of ω_c, that is, of B.

For further details the reader is referred to the literature: Hofmann, Kohn and Schneider (1961), Bulewicz (1962), Bulewicz and Padley (1962, 1963, 1963a and 1969), and Bradley and Tse (1969).

Advantages of this method are that it can provide information on $[e^-]$ and ν_e in low-pressure flames without disturbing the flame and that it requires no calibration. *Disadvantages* are that the spatial resolution is relatively poor and that the apparatus is expensive and difficult to construct.

(f1) *Microwave-cavity resonance technique.* With this technique a cylindrical flame is placed in a cylindrical or annular cavity; holes are drilled in the bottom and the lid (see, e.g., Padley and Sugden 1962, and Jensen and Padley 1966). Metal grids over these openings may prevent losses from causing too much damping. In order to prevent overheating of the cavity cooling is required or a rotating chopper may be used, which admits the flame gases only periodically; the presence of this chopper, however, causes flame disturbances.

The quantity σ' is found from the quality-factors Q_0 and Q_1 measured in the absence and presence of flame gas in the cavity, respectively, according to

$$\sigma' \simeq \frac{\omega g}{Q_0} \left(\frac{Q_0}{Q_1} - 1 \right), \qquad (IX.57)$$

where g is a geometrical factor that can be calculated for the particular experimental set-up used.

This technique has the *advantages* of greater simplicity than the r.f. resonance technique, the calibration is accurate, and the technique provides a favourable detection limit of 10^8 electrons per cm^3. The *disadvantages* are (see, e.g., Kelly and Padley 1969):

(i) its spatial resolution is limited by the air gap between the end plates and, to a lesser extent, by the presence of the holes in the cavity through which the flame must pass;

(ii) flame disturbance may be caused by the rotating chopper disk (see above);

(iii) the first 1 cm of flame gas above the combustion zone is not accessible for measurements.

(f2) *Microwave attenuation technique.* This technique has in practice been applied only to relatively large (rocket exhaust) flames, where electron concentrations typically range from 10^9 to 10^{12} electrons per cm^3 (see Balwanz 1965, Jensen 1965, Jensen and Travers 1973, and Miller 1976).

Chapt. IX   IONIZATION IN SEEDED FLAMES

5. EXPERIMENTAL RESULTS

In this section we present experimental data from some flame studies, viz. ionization experiments with metallic additives above the combustion zone. Details and results with regard to ionization- and ion-recombination rate constants and (specific) ionization cross sections will be discussed in Sect. 5a. Only a short survey about other flame experiments will be presented in Sect. 5b. A survey of ionization rate constants and equilibrium constants as a function of temperature, that are of interest in work with flames, has been presented by Jensen and Jones (1978).

5a. MEASUREMENTS OF IONIZATION AND ION-RECOMBINATION
     RATE CONSTANTS WITH METALLIC ADDITIVES

We focus primarily on three typical cases of ionization studies in flames, labelled A to C. These cases are distinguished according to the basic underlying assumptions specified in Sect. 3c. The main characteristics of these studies are summarized in Table IX.3, together with the references.

It is not our aim to give a complete account of results reported; we stress some typical results only. The formulas quoted in this section have been derived in Sect. 3c.

NOTES TO TABLE IX.3

† Several typical methods of determining ionization rate constants in laminar, premixed and shielded flames at 1 atm are compared as regards their underlying assumptions, measuring technique, experimental conditions, range of applications, etc. The experiments are classified in Cases A to C as described in Sect. 3c.

* See Table IX.1 on page 823.

‡ In Case A, k_r is expressed in k_i by a detailed-balance relation; in Case C, k_i and k_r are determined separately.

‡ Specific as to collision partner.

φ These processes are indistinguishable (see Sect. 3c).

Table IX.3
Synopsis of Assumptions and Experimental Conditions in the
Determination of Ionization Rate Constants in Flames[†]

	Case A	Case B.1	Case B.2	Case C
References	Hollander, Kalff and Alkemade (1963); Hollander (1964); Borgers (1965, 1978); Borgers et al. (000)	Kelly and Padley (1969, 1972); Jensen and Padley (1966); Ashton and Hayhurst (1973); Hayhurst and Telford (1972)	Anthony et al. (1974)	Hayhurst and Kittelson (1974, 1974a); Hayhurst (1974); Jensen (1968); Kelly and Padley (1971)
Flame	$CO-O_2-N_2$; $CO-O_2-Ar$	fuel-rich $H_2-O_2-N_2$, $-Ar, -CO, -CO_2, -H_2O$	fuel-rich $H_2-O_2-N_2$, $-Ar, -CO, -CO_2, -H_2O$, $+ 1\% C_2H_2$	fuel-rich $H_2-O_2-N_2$
Temperature range (K)	1900 - 2500	2000 - 2700	2000 - 2500	1820 - 2570
Vertical T gradient	allowed for	neglected, but small	neglected, but small	neglected, but small
Vertical $[M]_t$ gradient	negligible	negligible	negligible	negligible
Investigated metals	Na, K, Cs, Li	Tl, Li, Na, K, Cs, Ga, In	Li, Na, Pb, Mn, Cr, Ga, In, Tl	Ca, Sr
Metal compound allowed for	only LiOH	MOH	MOH	MOH, M(OH)$_2$, MO
Ionic species	M^+	M^+	M^+	M^+, MOH$^+$
Charge balance assumed	$[M^+] = [e^-]$	$[M^+] = [e^-]$	$[M^+] = [e^-]$	$[M^+] + [MOH^+] = [e^-]$
Chemi-ionization	corrected for, or negligible	assumed negligible	assumed negligible	assumed negligible
Measuring technique	atomic spectroscopy; r.f. resonance	rotating probe + atomic spectroscopy	rotating probe + atomic spectroscopy	ion mass spectrometry (+ atomic spectroscopy)
Assumed ionization (→) and neutralization (←) process *	(i) ⇄; (viii) ⇄	(i) →	(i) ←	(vi) + (vii) ⇄[ϕ]
Basic theoretical equation	IX.31	IX.37a,b	IX.38	IX.43
Rate constant(s) measured	ionization + neutralization[‡]	ionization	neutralization	ionization; neutralization
Specific[†] or overall rate constant	overall	specific	overall; (specific)	overall

865

Chapt. IX   IONIZATION IN SEEDED FLAMES

5a-1   *CASE A*

In the work of Hollander, Kalff and Alkemade (1963) experimental evidence has been found about the ionization relaxation of alkali metals in $CO-O_2$ flames diluted with N_2 or Ar, at 1 atm with temperatures ranging from 1900 to 2500 K. The profiles of atomic density [M] as a function of height z in the flame (or rise-time t) were measured by atomic emission spectrometry in the absence of self-absorption. A typical experimental height profile is presented in Fig. IX.8 together with the profile calculated with the help of the Saha equation for the known and constant total metal density $[M]_t$ and known local temperatures $T(z)$. From the comparison between the experimental and the Saha profiles, monomolecular, overall rate constants k_i were derived for Na, K and Cs as a function of T through Eq. IX.31. From the experimental $[M](z)$ curves the derivative $-d\ln[M]/dt$ was determined as a function of t by using $z = v_r t$ with measured rise-velocities v_r varying typically between 600 and 1300 cm s^{-1}. At the particular height where Saha equilibrium happens to exist, i.e. where $d[M]/dt = 0$ (see Fig. IX.8 and Sect. 3d) the absolute values of [M] (and $[M]_t$) are derived either by determining the ratio $[M]_t/[M]$ through ionization suppression, or by deriving $[M]_t$ from self-absorption measurements.

The $k_i(T)$ values obtained downstream from the temperature maximum for six N_2-diluted CO flames of different gas composition are plotted for Na, K and Cs in Fig. IX.9 as a function of the temperature at the point where $k_i(T)$ was determined. These $k_i(T)$ values were not corrected for the contribution of chemi-ionization (see below), but this correction proved to be at most 8%. The semi-logarithmic plot of $k_i(T)$ multiplied by a factor $T^{\frac{1}{2}}$ (see next paragraph) as a function of T^{-1} appears to be a straight line, the slope of which corresponds fairly well to E_{ion}/k (Arrhenius plot).

According to Eq. IX.27a one has:

$$k_i = \sum_j k'_{i,j}[Z_j]$$

where $k'_{i,j}$ are the specific, bimolecular rate constants. Using Eq. II.133 we obtain, if we write there $\mu_{M,j}$ for μ_j and $\sigma_{i,j}$ for the thermal phenomenological cross section $\langle \sigma_{n,j}\rangle(T)$ divided by the activation-energy factor $\exp[-E_{ion}/kT]$,

$$k_i(T) = \sum_j \left(\frac{8kT}{\pi \mu_{M,j}}\right)^{\frac{1}{2}} \sigma_{i,j}[Z_j]\exp[-E_{ion}/kT] \ . \qquad (IX.58)$$

The CO flames studied contain molecules of various kinds, but N_2 is present as a dominant constituent. When disregarding in first approximation the presence of different species, one may conclude from Eq. IX.58 that, provided σ_{i,N_2} is only slightly T-dependent, $k_i(T)\sqrt{T}$ as plotted in Fig. IX.9 should vary proportional to $\exp[-E_{ion}/kT]$. The implications of this approximation have been discussed by Hollander (1964) (see also Sect. 6a).

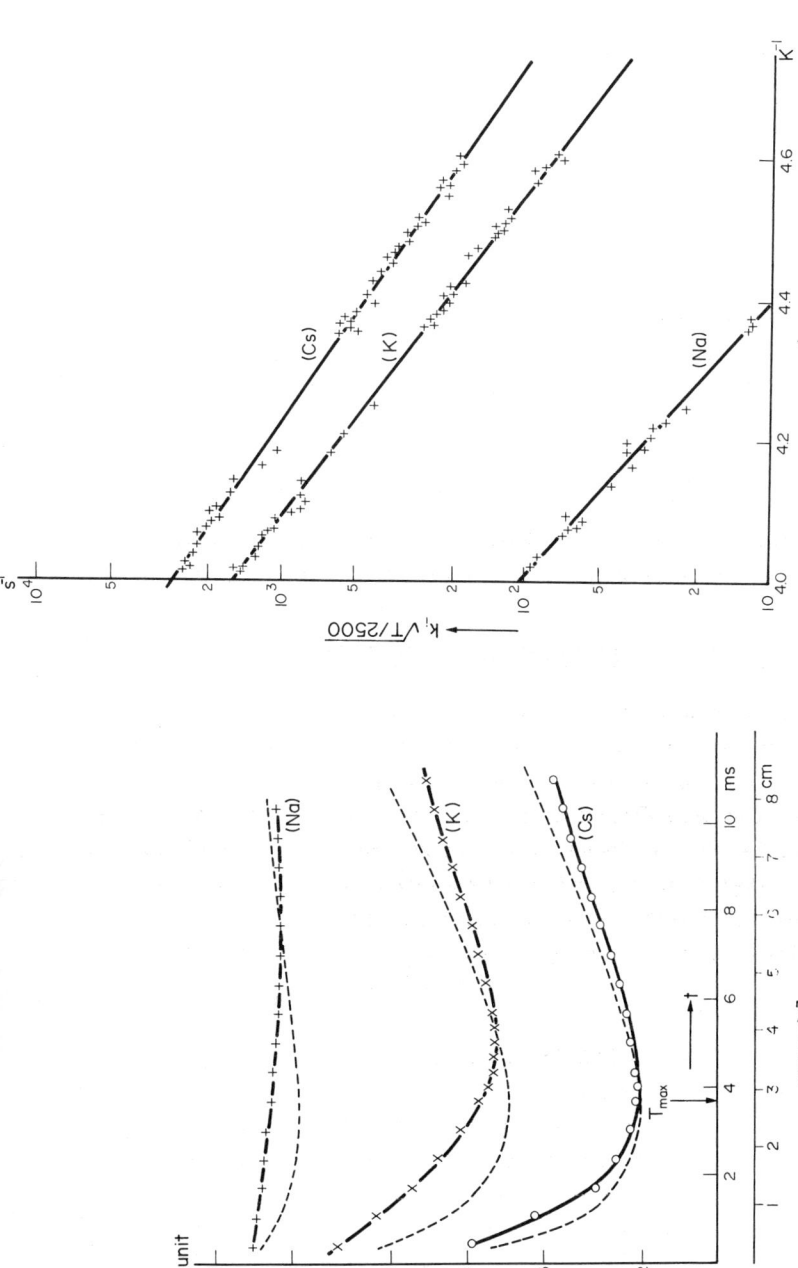

Fig. IX.8 The atomic Na, K and Cs concentration as a function of height z (or rise-time t) above the combustion zone in a $CO-O_2-N_2$ flame. Measured points are indicated by +, x and o, respectively. The dotted curves are the corresponding 'Saha-curves'. The place of maximum temperature is also indicated. (From Hollander, Kalff and Alkemade 1963.)

Fig. IX.9 The values of $k_i(T)$ multiplied by $[T/2500]^{\frac{1}{2}}$ as a function of reciprocal temperature for Na, K and Cs are shown for six N_2-diluted CO flames of different gas compositions. The solid (semi-logarithmic) straight lines represent the theoretical functions $\exp[-E_i/kT]$ with E_i = ionization energy of Na, K and Cs, respectively. (From Hollander, Kalff and Alkemade 1963.)

Chapt. IX   IONIZATION IN SEEDED FLAMES

For K and Cs, k_i values were also derived in one of the six CO flames ($T \approx 2300$ K) with N_2 replaced by Ar as diluent gas. The above T-dependence was again established in this flame. A decrease in $k_i(T)$ value as compared to the $CO-O_2-N_2$ flame at the same T was found for K and Cs; the extent of this decrease could be quantitatively understood by assuming that $\sigma_{i,Ar} \approx 0$ and that all molecular species have about the same specific cross section. We return to this subject in Sect. 6a-4.

Some typical $k_i(T)$ values are presented in Table IX.4. The $k_r(T)$ values derived through the detailed-balance relation from measured $k_i(T)$ values are presented in Table IX.5 (on page 874) and compared with the results from other authors.

In the CO flames used, the chemi-ionization process: $M + CO + O \rightarrow M^+ + e^- + CO_2$ (process viii in Table IX.1) can be expected to contribute to the measured overall ionization rate constant k_i; the chemical association energy released amounts to 5.5 eV and exceeds the ionization energy of Na, K and Cs. The contribution of this process was assessed by extending the ionization measurements into the region close to the combustion zone where O radicals are present in concentrations in excess of equilibrium (see Sect. IV.5). Supposing that the contribution of chemi-ionization to k_i is proportional to the product [CO][O] and ignoring the T-dependence of the proportionality constant, we have for the excess ionization rate constant k_i^{exc} [$\equiv k_i - k_i(T)$, where $k_i(T)$ is the thermal rate constant]

$$k_i^{exc} = A \{[CO][O] - [CO]_e[O]_e\}. \qquad (IX.59)$$

Since $[CO_2]$ is hardly affected by the disequilibrium of O radicals, the overall neutralization rate constant remains equal to its thermal value $k_r(T)$. The latter rate constant can therefore be expressed in $k_i(T)$ through the detailed-balance relation. Knowing T and deriving $k_i(T)$ from measurements higher up in the flame where CO and O are essentially in chemical equilibrium, one finds

$$k_i^{exc} = k_i(T) \frac{([M]_t - [M])^2}{[M] K_i(T)} - \frac{d \ln[M]}{dt} - k_i(T) . \qquad (IX.60)$$

A plot of k_i^{exc} versus $[CO][O] - [CO]_e[O]_e$ (a quantity that can be determined by the method outlined in Sect. IV.5b-4) shows a more or less straight line the initial slope of which yields the constant A in Eq. IX.59. Knowing A and the equilibrium product $[CO]_e[O]_e$ higher up in the flame where $k_i(T)$ is measured, we can correct $k_i(T)$ for the thermal rate constant, $k_i^{ch}(T) \equiv A[CO]_e[O]_e$, of the *chemi-ionization* process and find the thermal *collisional* rate constant $k_i^{coll}(T)$. Values of $k_i(T)$ and $k_i^{ch}(T)$ are presented in Table IX.4.

It is important to note that the monomolecular collisional rate constants appeared to be independent of the alkali concentrations when the latter was varied

EXPERIMENTAL RESULTS   Sect. IX.5a

by a factor of at least 10. This outcome proves that the ionization process does
not involve free electrons as collision partners (process ii in Table IX.1), as the
concentration of the latter varies with alkali concentration, too.

Table IX.4

Measured Values of $k_i(T)$ and $k_i^{ch}(T)$ in s^{-1} at the Point
of Maximum Temperature in $CO-O_2-N_2-$ and $CO-O_2-Ar$ Flames [†]

Flame	Temperature (K)	K ionization		Cs ionization	
		$k_i(T)$	$k_i^{ch}(T)$	$k_i(T)$	$k_i^{ch}(T)$
$CO-O_2-N_2$	2458	1190	60 - 100	2140	140 - 220
$CO-O_2-N_2$	2491	1520	100 - 140	2640	230 - 330
$CO-O_2-N_2$	2286	260	8 - 12	540	19 - 27
$CO-O_2-N_2$	2300	290	11 - 17	590	26 - 38
$CO-O_2-N_2$	2218	140	4 - 6	310	10 - 14
$CO-O_2-N_2$	2226	145	5 - 7	320	12 - 16
$CO-O_2-Ar$	2331	160	15 - 23	300	35 - 53

Recent measurements carried out by Borgers (1978), and Borgers et al.(000)
yielded $k_i(T)$ values for Li, Na, K and Cs in fuel-rich $CO-O_2-N_2$ flames at a uniform temperature of 1930 ± 10 K. These results are collected in Table IX.9 in Sect.
6a-2. These authors determined [e⁻] values as a function of z as well as of $[M]_t$
by making use of the r.f. resonance method (see Sect. 4). Additional curve-of-
growth (i.e. self-absorption) measurements were made for determining absolute $[M]_t$
values and for calibrating the r.f. resonance set-up. Furthermore, spectrometric
measurements on the behaviour of [M] were carried out to supplement the electric
measurements of [e⁻].

5a-2   *CASE B*

In these experiments two different experimental conditions are created (see
Table IX.3): in the first group (Case B.1) the flame conditions are such that
recombination can be neglected, i.e. $[M^+] \ll [M^+]_e$, whereas in the second group (Case
B.2) recombination processes are dominant.

In the experiments in *Case B.1* the alkali metals in fuel-rich H_2-O_2 flames
diluted with N_2, Ar, H_2O, CO and CO_2 are studied. A complication here is that the
metal hydroxide formation cannot be excluded; i.e. in the analysis of the data one
has either to take ϕ ($\equiv [MOH]/[M]$) as being independent of height in the interval

---

[†] From Hollander, Kalff and Alkemade (1963), and Hollander (1964).

considered, or ϕ has to be determined as a function of height through Eq. VIII.20 by measuring [H] by one of the methods outlined in Sect. IV.5b-2.

The basic rate equation that applies under the experimental conditions considered is Eq. IX.36 from Sect. 3c, which reads:

$$-\frac{d \ln([M]_t - [M^+])}{dt} \simeq \frac{k_i}{1+\phi} .$$

Integration of this expression is feasible if $\phi(t)$ is known. In Sect. 3c (under Case B.1) it is argued that a linear dependence of ϕ on t may apply either under condition that $[H] \gg [H]_e$ or in a restricted height interval when $[H] \simeq [H]_e$. Using a linear dependence of ϕ on t one obtains Eq. IX.37a, which reads:

$$\ln\{[M]_t/([M]_t - [M^+])\} \simeq \{k_i/C(1+\phi_0)\} \ln(1+Ct)$$

with $[M^+] = 0$ at $t = 0$. The final expression giving [M] as a function of t is Eq. IX.37b, which reads:

$$\ln\{[M]_t/[M](1+\phi_0)\} \simeq \{1 + k_i/C(1+\phi_0)\} \ln(1+Ct).$$

The measuring techniques involve rotating probe-, MW cavity- and optical spectrometric measurements (see Sect. 4). With the rotating probe technique $[M^+]$ is measured as a function of t; absolute $[M^+]$ values are obtained by calibrating the probe data against known levels of flame ionization, i.e. one introduces such small amounts of Cs into the flame that the Cs vapour is completely ionized (see, e.g., Kelly and Padley 1969, 1972). Conversion of solution concentrations into [M] or $[M]_t$ values in the flame is achieved by self-absorption measurements. With the MW cavity technique one obtains $[e^-]$ as a function of t, and absolute calibration of $[M]_t$ takes place in the same way (see, e.g., Jensen and Padley 1966). By optical spectrometry one determines [M] as a function of t and again $[M]_t$ values are found in absolute measure by self-absorption measurements (see, e.g., Kelly and Padley 1972).

From the probe measurements one obtains first-order plots (as to t) for the production of positive ions, that is, $\ln\{[M]_t/([M]_t - [M^+])\}$ as a function of t. The same plot is obtained from the MW cavity measurements, since $[e^-] = [M^+]$. The $k_i(T)$ values are found from these plots when corrections are made for the variation of ϕ as a function of t and ϕ_0 is known in absolute measure. The optical spectrometric technique yields [M] as a function of t, and under certain conditions $k_i(T)$ values can be obtained directly from this plot without knowledge of <u>absolute</u> [M] values (see Kelly and Padley 1972). In the latter case one applies Eq. IX.37b, in the other cases Eq. IX.37a. In Figs. IX.10 and 11 we present two examples of such plots.

EXPERIMENTAL RESULTS

Sect. IX.5a

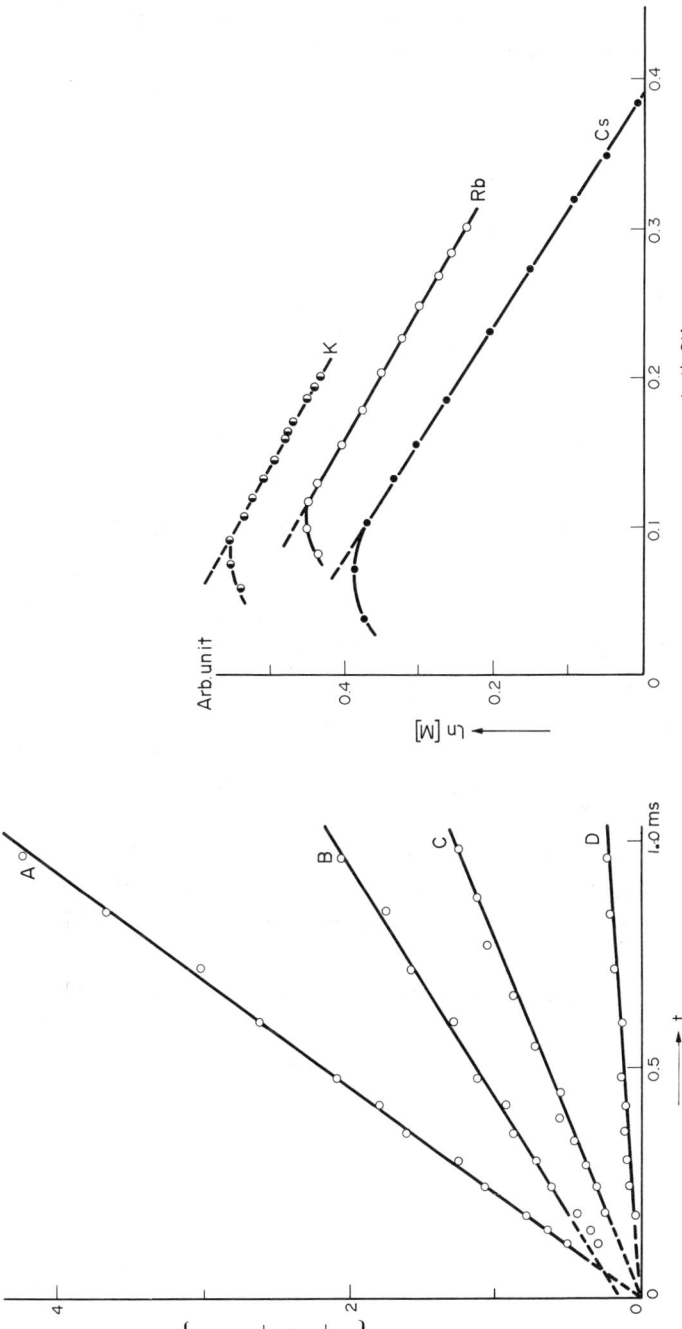

Fig. IX.10 Typical plots for the production of positive ions by Na (A and B) and Tl (C and D) in H_2-O_2-Ar flames from probe measurements. Flame A: $[H_2]/[O_2]/[Ar] = 4/1/3$ ($T = 2330$ K); B: = 2.9/1/5 (2260 K); C: = 2.6/1/3 (2720 K); and D: = 3.2/1/3 (2540 K). Notice that the experimental points for Na in flame B, on extrapolation to $t = 0$, indicate a finite level of ionization within the combustion zone. (From Kelly and Padley 1972.)

Fig. IX.11 The optical technique for determination of $k_1(T)$ for K, Rb and Cs. The (relative) atomic metal concentration [M] at any point (corresponding to a rise-time t) is given by the emission or absorption intensity of the first resonance doublets, here in a H_2-O_2-CO_2 flame (with $[H_2]/[O_2]/[CO_2] = 3.5/1/2$, $T = 2100$ K). Correction for MOH formation was made through Eq. IX.37b. The vertical displacements of the K, Rb and Cs plots have no significance, nor do in the present context the initial increases of [M] with increasing t very near (i.e. within 0.2 ms) the combustion zone. (From Kelly and Padley 1972.)

Chapt. IX   IONIZATION IN SEEDED FLAMES

In order to illustrate the derivation of $k_i(T)$ values from probe measurements as well as optical spectrometry, we discuss briefly some typical cases (see, e.g., Kelly and Padley 1972):

— Sodium and thallium in fuel-rich H_2-O_2-Ar flames (see caption of Fig. IX.10) form negligible amounts of MOH. In the derivation of $k_i(T)$ values from probe measurements we may use a simplified version of Eq. IX.37a, since $\phi_0 = 0$; the equation used is then

$$\ln\{[M]_t/([M]_t - [M^+])\} = k_i t \,, \tag{IX.61}$$

and $k_i(T)$ was found from the slope of the plots in Fig. IX.10.

— For K, Rb and Cs, $k_i(T)$ was derived from probe measurements as well as optical spectrometry. For the probe measurements Eq. IX.37a was applied, whereas Eq. IX.37b was used for the optical measurements. In Fig. IX.11 this optical technique is illustrated. Note that only the relative dependence of $[M]$ on t is needed here.

In the evaluation of k_i values with both measuring techniques it is necessary to have accurate values for $K_1(T)$ and $[H]$ (see Eq. VIII.20), since under the flame conditions considered the corrections for MOH formation are considerable. The values of $K_1(T)$ were taken from Kelly and Padley (1971), whereas $[H]$ and hence k_H (see Eq. IV.14) were derived according to the methods discussed in Sect. IV.5.

In order to stress the importance of the correction for MOH formation, we mention that in the case of K ionization the k_i value was found to be of the same order of magnitude as $K_1[H_2O]k_H$, or in terms of Eq. IX.37b: $k_i \approx C(1+\phi_0)$, i.e. ionization and MOH formation have about the same effect on the t-dependence of $\ln([M]_t/[M])$.

It was also established here (as in Case A) that the k_i values were independent of the metal concentration when the latter was varied by a factor of 50 to 100. Chemi-ionization contributions to k_i were not considered, so that the rate constants derived are total thermal rate constants, $k_i(T)$, rather than purely collisional rate constants $k_i^{col}(T)$.

Furthermore, Kelly and Padley (1972) derived for each metal specific ionization cross sections, $\sigma_{i,j}$, pertinent to collision partner Z_j by studying isothermal flames at different quantitative and qualitative gas compositions. Comparing Eqs IX.27a and IX.58 we have

$$k'_{i,j} = (8kT/\pi\mu_{M,j})^{\frac{1}{2}} \sigma_{i,j} \exp[-E_{ion}/kT] \,. \tag{IX.62}$$

The specific ionization cross sections are normalized with respect to the cross section for Ar by defining $\alpha_{M,j}$ according to

EXPERIMENTAL RESULTS   Sect. IX.5a

$$\sigma_{i,j}/\mu_{M,j}^{\frac{1}{2}} \equiv \alpha_{M,j}\sigma_{i,Ar}/\mu_{M,Ar}^{\frac{1}{2}}. \qquad (IX.63)$$

The authors found $\alpha_{M,j}$ values with a trial-and-error method from a set of linear equations. The specific $\sigma_{i,j}$ values from Kelly and Padley (1972) are discussed in Sect. 6a-4; here we quote the corresponding $\alpha_{M,j}$ values calculated by these authors: $\alpha_{M,H_2} = 1.7 \pm 0.3$; $\alpha_{M,N_2} = 1.7 \pm 0.3$; $\alpha_{M,CO} = 3.3 \pm 0.6$; $\alpha_{M,CO_2} = 3.3 \pm 0.6$; and $\alpha_{M,H_2O} = 5.0 \pm 1.0$. Within the error limits there was no dependence of $\alpha_{M,j}$ on the metal studied (lithium possibly excepted).

If the overall k_i values are corrected for varying flame gas composition via the known $\alpha_{M,j}$ values, all experimental points fit (on the average) the same straight line in the Arrhenius plot. Some of the experimental $k_i(T)$ values derived from the above measurements by Kelly and Padley are collected, after standardization as to gas composition, in Table IX.9 on page 887; in Table IX.5 the k_r values are presented that are calculated from k_i through the detailed-balance relation. It should be noted that also in older work on alkali metal ionization in H_2-O_2-diluent gas flames (see, e.g., Jensen and Padley 1966, and Hayhurst and Telford 1972) essentially the same formulas have been used, i.e., Eqs IX.37a and 37b.

NOTES TO TABLE IX.5

† Anthony et al. (1974) reported specific k_r values for Li; they deduced from their measurements k_r values of 2.5×, 3.4×, 4.1×, 3.0×, 3.9×, and 4.3×10⁻⁹ cm³ molecule⁻¹ s⁻¹ for the collision partners H_2, Ar, N_2, CO, CO_2, and H_2O, respectively.

†† Ashton and Hayhurst (1973) reported that overall k_r values for Li, Na, K, Rb and Cs were in the range $4.3 \times < k_r < 7.5 \times 10^{-9}$ cm⁻³ molecule⁻¹ s⁻¹ in H_2-O_2-N_2 flames at $T = 2250$ K.

‡ A temperature-dependence $k_r \propto T^{-2}$ follows from $k_i(T)$ and the detailed-balance relation; this T-dependence was experimentally verified by Jensen and Padley (1967) and by Anthony et al. (1974) (see also Ashton and Hayhurst 1973) for the alkali ionization.

* k_r is (*directly*) measured or (*indirectly*) derived from measured k_i values through a detailed-balance relation, or obtained in a *combined* measurement of k_i and k_r linked by detailed balance.

φ Specific k_r values can be derived from the specific k_i values reported by Kelly and Padley (1972).

§ k_r values derived at (1930 ± 10) K were converted here to 2400 K, for a better comparison, by using $k_r \propto T^{-2}$ (see footnote ‡).

Table IX.5

Overall Bimolecular Ion-Electron Recombination Rate Constants k_r
(in 10^{-9} cm^3 molecule^{-1} s^{-1}) in Flames at 1 atm with N$_2$ as Diluent[#]

Flame	CO/O$_2$/N$_2$	CO/O$_2$/N$_2$	H$_2$/O$_2$/N$_2$ +1% C$_2$H$_2$	H$_2$/O$_2$/N$_2$	H$_2$/O$_2$/N$_2$	H$_2$/O$_2$/N$_2$	H$_2$/O$_2$/N$_2$ +1% C$_2$H$_2$	H$_2$/O$_2$/N$_2$ +1% C$_2$H$_2$
Reference	1	2	3	4	5	6φ	6	7
Measuring technique	optical spectrometry	r.f. resonance + optical spectrometry	rotating probe (+ optical spectrometry)	optical spectrometry	MW cavity (+ optical spectrometry)	rotating probe (+ optical spectrometry)	rotating probe	rotating probe
Temperature (K)[†]	2400	2400 §	2370	2250	2250	2200	2200	2250 / 2370
Direct, indirect, or combined [*]	combined	combined	direct	direct	indirect	indirect	direct	direct
Li	-	47	9.0	-	-	-	-	3.0± (2370 K)
Na	8.7	25	-	6.4	8.5	7.3	-	8.9 (2370 K)
K	3.2	9.1	-	3.5	6.5	5.5	-	-
Rb	-	-	-	-	5.5	4.0	-	-
Cs	0.70	2.1	-	4.5	8.5	4.8	-	-
Tl	-	-	-	-	-	4.7	6.4	16
Ga	-	-	-	-	-	8.6	7.9	16.5
In	-	-	-	-	-	1.8	6.1	13
Pb	-	-	9.0	6.9	6.5	-	-	28
Mn	-	-	24	-	7.5	-	-	31
Cr	-	-	18	-	26	-	-	36

REFERENCES

1. Hollander, Kalff and Alkemade (1963); Hollander (1964)
2. Borgers (1978); Borgers et al. (1981)
3. Soundy and Williams (1965)
4. Hayhurst and Sugden (1965)
5. Jensen and Padley (1967)
6. Kelly and Padley (1969, 1972)
7. Anthony et al. (1974)

EXPERIMENTAL RESULTS

Sect. IX.5a

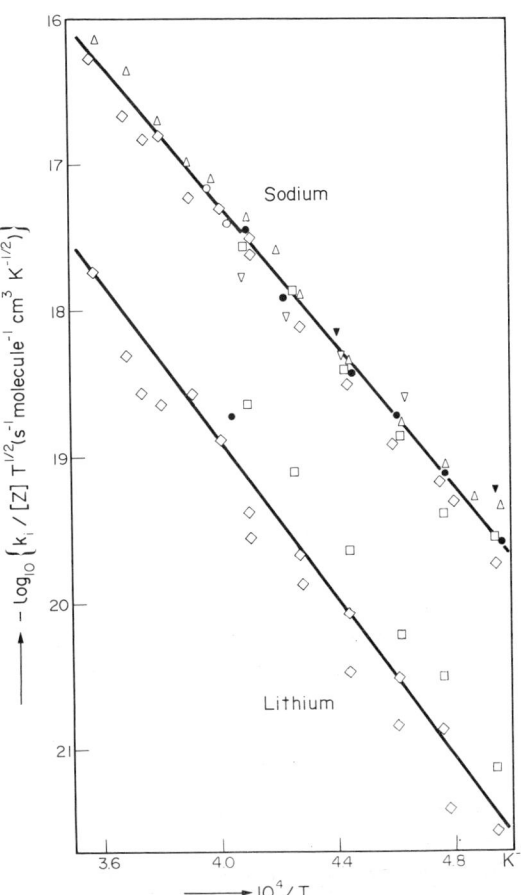

Fig. IX.12 Plots of $\log_{10}(k_i/[Z]T^{\frac{1}{2}})$ versus $1/T$ for Li and Na ($[Z]$ = total molecular concentration at 1 atm pressure). The points for Na have been displaced vertically upwardly by one unit for clarity. Results are from: o Ashton and Hayhurst (1973), ▽ Padley and Sugden (1962), ▼ Hayhurst and Sugden (1965), ▲ Schofield and Sugden (1965), ● Jensen and Padley (1966), △ Kelly and Padley (1969), □ Hayhurst and Telford (1972), ◊ Kelly and Padley (1972), and ■ Jensen (1965).
(From Ashton and Hayhurst 1973.)

In Fig. IX.12 we present some Arrhenius plots measured by several authors for Na and Li ionization (cf. Fig. IX.9). No correction for the different flame gas compositions was carried out; only data for H_2-O_2 flames with various diluents are shown.

In the experiments in *Case B.2* (Anthony et al. 1974) only ion recombination was considered. In H_2 flames containing a trace of hydrocarbon the degree of ionization of many metals may be boosted above equilibrium through process (xii) in Table IX.1. The H_3O^+ ions disappear very rapidly above the combustion zone, whereas $[M^+]$ remains appreciably above $[M^+]_e$ in the height interval studied. It was proved that for metals with ionization energy lower than 7.5 eV this ionization boost is relatively very large. The influence of process (xii) in Table IX.1 is relatively unimportant downstream from the combustion zone, and only ion recombination

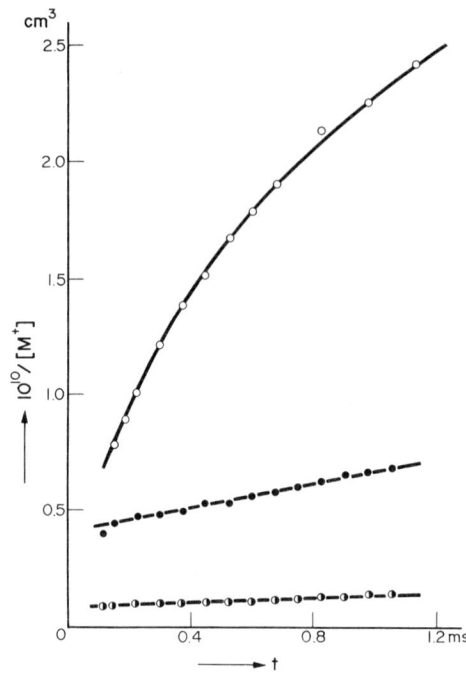

Fig. IX.13 Metal ion-electron recombination plots in a flame with $[H_2]/[O_2]/[N_2] = 4/1/3$ ($T = 2170$ K) with 1% C_2H_2 added, and for various sprayed metal nitrate solutions; ○, Cu; ●, Pb; and ◐, Li. (From Anthony et al. 1974.)

process (i) has to be considered.

The recombination rate constant k_r was found through Eq. IX.38 by determining the ionic concentration as a function of height by means of probe measurements. In the case of Li ionization, also specific $k_{r,j}$ values were considered. In Fig. IX.13 some typical plots are presented. In Table IX.5 these k_r values are presented together with various other k_r values, some of which were calculated from the corresponding k_i values according to Eq. IX.30.

5a-3   *CASE C*

This case applies when the alkaline-earth ionization is studied in (mainly) fuel-rich H_2-O_2-N_2 flames. The ionic species are M^+ and MOH^+, and the neutral compounds are specified as MOH, M(OH)$_2$ and MO (see Chapt. VIII). Ionization and neutralization are taken into account together, and $[M]_t$ and T are assumed to be independent of position. The concentrations of ions and molecular compounds may differ from their equilibrium values (see Sect. 3c under Case C) and are therefore functions of distance z (or travel time t).

We discuss here mainly the ion mass spectrometric measurements by Hayhurst and Kittelson (1974, 1974a) on Ca and Sr. (It should be noted that our presentation is somewhat different.) The main questions we consider are:

EXPERIMENTAL RESULTS   Sect. IX.5a

(i)   Which (dominant) primary ionization mechanism(s) is (are) responsible for
      the alkaline-earth ionization?

(ii) Which rate constant values apply in the primary ionization and recombina-
      tion mechanisms and which T-dependence applies for the rate constants?

One has postulated:

(a)   [M$^+$] and [MOH$^+$] are rapidly balanced through process (xi) (see Table IX.1).

(b)   The most likely neutralization mechanism is the reverse process (vi) and/or
      (vii); this neutralization mechanism is a second-order process, whereas
      all other possible mechanisms involve third-order processes.

The ion mass spectrometer measured ion currents for M$^+$ and MOH$^+$ separately.
The ion densities were measured as a function of axial distance z in each flame
(see Fig. IX.14). The measurements were carried out in a set of five fuel-rich

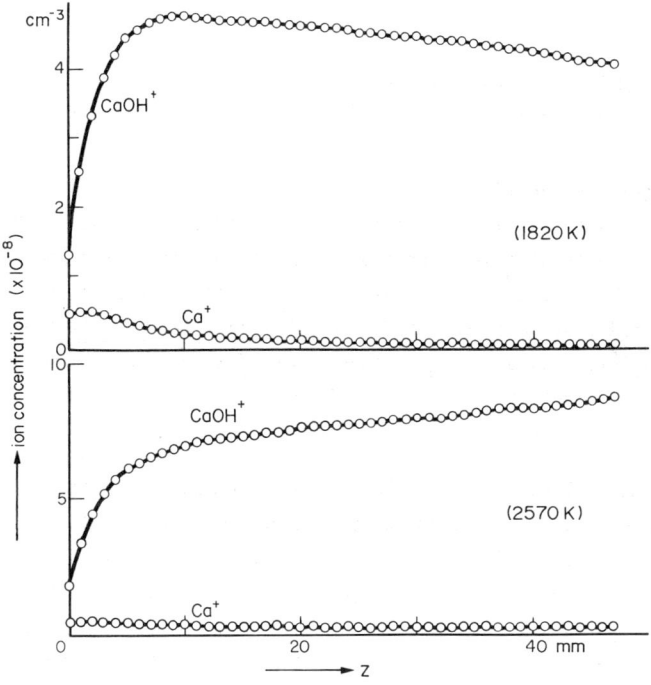

Fig. IX.14 Variation of measured ion concentrations with axial
distance z from the combustion zone in two H_2-O_2-N_2 flames of
different temperatures. In each case a 0.0008 M solution of
$CaCl_2$ was sprayed.   (From Hayhurst and Kittelson 1974a.)

Chapt. IX   IONIZATION IN SEEDED FLAMES

H_2-O_2-N_2 flames with different quantitative gas compositions (p_t = 1 atm); the temperature of the flames ranged from 1820 to 2870 K. In each flame the temperature was found to be uniform over the height interval considered. Solutions of MCl_2 at a molarity of about 10^{-3} (corresponding to $[M]_t \approx 10^{10}$ cm^{-3}) were sprayed. The ion currents were calibrated by comparison with known (by measurement or calculation) ionic densities in the flame. The necessary absolute calibration of $[M]$ and/or $[M]_t$ was realized by self-absorption measurements; $[M]_t$ proved to be independent of z. A complication arises as a result of the occurrence of hydrated-ion concentrations which are included in those of the parent ions. (Since ion hydration is a consequence of the sampling process rather than a feature of the flame itself, according to Hayhurst and Kittelson 1974, we will not discuss this subject in more detail.) The rise-velocity v_r was measured and proved to be constant along the flame axis; $[H]$ was measured as a function of z; the quantitative flame gas composition in each flame was computed as described in Sect. IV.4. Then all quantities are known for the evaluation of data from the ionic density profiles.

Answer to question (i). Various primary ionization mechanisms are possible, viz.

- physical processes of type (i) (see Table IX.1) with M or MOH;
- chemi-ionization processes: either chemical association processes (vi) + (vii) or chemical exchange processes such as: $MOH + H \rightarrow M^+ + H_2O + e^-$.

Processes (vi) and (vii) are energetically much more likely than the other ones if the activation energy involved is about equal to the endo-ergicity; however, the product of reactant concentrations may be $10^2 \times$ smaller in processes (vi) and (vii). For given $[M]_t$ the ionization rate G depends on the H concentration according to: $G \propto \gamma^n F(\gamma)$ where $\gamma \equiv [H]/[H]_e$ and $n = 1, 2, 3$ (only for processes vi and vii is $n = 3$; see also Eq. IX.41).

To test the value of exponent n operating under the prevailing flame conditions, Hayhurst and Kittelson (1974a) plotted $\log\{(dn_i/dz)/[M]_t\}$ versus $\log \gamma$ (in arbitrary units) under conditions of negligible ion recombination, while assuming that $F(\gamma)$ is only slightly dependent on z. The resulting slope proved to be described by: $n \simeq 3$. Less ambiguously (since ion recombination and dependence of $F(\gamma)$ on z were explicitly included), n was derived from the linear relationship between $\log\{n_i^2/F(\gamma)[M]_t\}$ and $\log \gamma$; the result was: $n = 3.03$. Here n_i and γ are taken at the maximum in the $n_i(z)$ curve which was measured for various $[M]_t$ values in the coolest flame ($T = 1820$ K) (cf. Fig. IX.14; see also Sect. 3d). (For each solution concentration the maximum is at a different point in the flame; a different value of γ applies therefore to each maximum.) In such a maximum $dn_i/dt = 0$ or $R = G$, i.e., $k_r n_i^2 = \text{constant} \times [M]_t \gamma^n F(\gamma)$ (cf. Eq. IX.43 for case $n = 3$).

EXPERIMENTAL RESULTS   Sect. IX.5a

The conclusion is that the indistinguishable processes (vi) and/or (vii) are the predominant primary ionization mechanisms.

Answer to question (ii). When the processes (vi) and/or (vii) are responsible for the ionization mechanisms studied, the theoretical expression for the ion production is given by Eq. IX.43. In this equation the unknown parameters are contained in $k_6(1+\xi)$, where $\xi \equiv (k_7/k_6)(K_{17}K_3)$ (see Eq. IX.40) and k_6 and k_7 are the second-order rate constants of primary ionization process (vi) and (vii), respectively.

The method for solving for k_r is as follows: either a numerical fitting procedure (by computer) is applied from Eq. IX.43 to differentiated experimental curves, dn_i/dz, or such a procedure is applied from the integrated Eq. IX.43 to the experimental curve $n_i(z)$. The results obtained by Hayhurst and Kittelson (1974a) are presented in Table IX.6.

Table IX.6
Overall Bimolecular Dissociative Neutralization
Rate Constant, k_r, for the Combined Processes:

$$MOH^+ + e^- \rightarrow \begin{array}{l} M + OH \\ MO + H \end{array}$$

Metal	Ca [†]	Sr [†]	Sr [*]
k_r in cm³ molecule⁻¹ s⁻¹ (at T = 2000 K)	$1.2 \times 10^{-7.0 \pm 0.3}$	$7.0 \times 10^{-8.0 \pm 0.2}$	7.0×10^{-8}
If $k_r(T)/k_r$ (2000 K) can be approximated by $(T/2000)^m$, m values are:	1.9 ± 1.5	-2.7 ± 1.1	-

From their experimental results Hayhurst and Kittelson (1974a) derived estimates of the upper limits for k_6 and k_7 by assuming that each reaction (process vi or vii) is in turn the sole path for ion production, i.e., $\xi = 0$ and $\xi \gg 1$, respectively (see Table IX.7).

[†] From Hayhurst and Kittelson (1974a).   [*] From Kelly and Padley (1971).

Table IX.7

Upper Limits for the Second-order Rate Constants[†] for Primary Ionization k_6 and k_7 (at $T \simeq 2000$ K), Expressed[*] in Terms of Pre-exponential Factor A and Activation Energy E_{Arrh}, and Endo-ergicity ΔH_0

Metal	Ca		Sr	
Rate constant	k_6	k_7	k_6	k_7
$-\log A$ [‡]	9.9 ± 1.3	10.6 ± 1.3	9.9 ± 0.8	10.7 ± 0.8
E_{Arrh} (kJ mol^{-1})	130 ± 40	58 ± 40	102 ± 35	63 ± 35
ΔH_0 (kJ mol^{-1})	124 ± 15	62 ± 15	95 ± 15	56 ± 15

[†] From Hayhurst and Kittelson (1974a).
[*] Compare Eq. II.124 on p.83.
[‡] The slight T-dependence of A is neglected as compared to the experimental scatter. A is expressed in cm^3 s^{-1}.

Alkali metal ionization and recombination processes also were studied in hydrocarbon flames by means of single-probe measurements by Calcote and King (1955, 1955a), and King (1957, 1958, 1962). The interested reader is referred to Shuler and Fenn (1963) for results and more detailed information.

5b SURVEY OF FURTHER IONIZATION EXPERIMENTS
    WITH ADDITIVES IN FLAMES

We summarize some further ionization experiments with additives in flames in Table IX.8. The interested reader is referred for further details to the literature cited there.

NOTES TO TABLE IX.8

[†] A = alkali atom; B = halogen atom; Z = flame species.

[*] Bates (1980) has recently investigated by computer-simulation 'experiments' the dependence of the neutralization rate constant on temperature and gas pressure as expected in flames.

[‡] Fehsenfeld *et al.* (1968) have pointed out flaws in the interpretation of some of the mobility measurements in flames. Gilardini (1972) reports nonflame results as well and has suggested recommended values.

Note: The influence of electric fields on flames and the ionization from solids in flames are not considered. The reader is referred to Lawton and Weinberg (1969).

Note: Hydrate formation (in the ion beam of the mass spectrometer) is discussed briefly in Sects 4 and 5a-3; hydration energies for CaOH$^+$, SrOH$^+$, Ca$^+$ and Sr$^+$ are reported in Hayhurst and Kittelson (1974).

Table IX.8
Survey of Further Ionization Experiments with Additives in Flames

Quantity studied	Species investigated	Experimental method	References
Rate constant of charge transfer from H_3O^+	Alkali; Tl, Cr, Pb, Mn, Cu, Fe, Cd	ion mass spectrometer	Hayhurst and Telford (1970)
Rate constant of neutralization by charge transfer and of reverse process*	$A^+ + B^- \rightleftharpoons A + B$	MW cavity resonance; ion mass spectrometer	Hayhurst and Sugden (1967); Hayhurst (1974); Burdett and Hayhurst (1977, 1979a)
Ionization energy ($MOH \rightarrow MOH^+ + e^-$)	CaOH, SrOH, BaOH	rotating probe (+ optical spectrometry)	Kelly and Padley (1971)
	Halogens	ion mass spectrometer	Hayhurst & Kittelson (1974a)
Electron affinity		cyclotron resonance	Bulewicz and Padley (1969)
	PO_2^-	ion mass spectrometer + MW cavity resonance	Miller (1978a)
Rate constant of dissociative chemi-ionization and of reverse process	$HB + e^- \rightleftharpoons H + B^-$	ion mass spectrometer	Burdett and Hayhurst (1974, 1977a)
	$AB + Z \rightleftharpoons A^+ + B^- + Z$	ion mass spectrometer	Burdett and Hayhurst (1977, 1979a)
Rate constant of neutralization by exchange reaction	$H_3O^+ + Cl^- \rightarrow H_2O + HCl$	ion mass spectrometer	Hayhurst (1974)
Mobility‡	e^-	MW cavity resonance	Belcher and Sugden (1950); Schneider and Hofmann (1959)
		cyclotron resonance	Schneider and Hofmann (1958); Bulewicz (1962); Schmidt and Kummer (1970); Quadbeck and Kummer (1976)

Chapt. IX   IONIZATION IN SEEDED FLAMES

6. DISCUSSION AND CONCLUSIONS

6a   CONCLUSIONS FROM AND INTERCOMPARISON OF IONIZATION AND RECOM-
      BINATION RATE CONSTANTS OF ALKALI ATOMS MEASURED IN FLAMES

We compare here the experimental rate constants for ionization and recombination of alkali atoms measured by various authors above the combustion zone in premixed flames and reported in Sect. 5a. It is noted that the experimental rate constants proved to be thermal (as defined in Sect. II.4a-2) whether there was Saha equilibrium or not (see Hollander, Alkemade and Kalff 1963, and Kelly and Padley 1972; see also Sect. 5a). This holds provided we correct the measured rate constants for possible contributions from excess chemi-ionization (see for this correction Sect. 5a-1 and Table IX.4). First we shall draw some conclusions about the ionization mechanism from the general trends in the observations.

6a-1   *SOME GENERAL CONCLUSIONS; THE ORIGINAL LADDER MODEL OF HOLLANDER ET AL.*

In addition to a chemi-ionization process of type (vii) (see Table IX.1 and Sect. 5a-1), two types of collisional processes (i) and (ii) have been considered to be operative in the ionization of alkali atoms in flames. We assume the absence of halogen species, which could add another ionization process of type (iii) or (iv) (see Burdett and Hayhurst 1977).

Process (ii) involving free electrons as collision partners in the ionization process can be ruled out because the unimolecular ionization rate constant $k_i(T)$ appears to be independent of $[e^-]$. This observation has been made independently by various authors (see Sect. 5a).

We have still to consider the detailed mechanism of collisional ionization process (i) induced by a neutral species (molecule or atom) of the flame gas. A clue to the mechanism of collisional ionization by a flame particle is found if we suppose, as a first trial, that the phenomenological cross section for collisional ionization is of the same order as that for gas-kinetic collisions mentioned in Sect. II.4b-1. Inserting a normal gas-kinetic value of 3-5 Å² for σ_0 in Eq. IX.58 and assuming that N_2 is the predominant collision partner in a flame of 1 atm pressure and a temperature of 2500 K, we calculate a unimolecular collisional rate constant k_i^{coll} that is about 3 orders of magnitude smaller than the experimentally found ionization rate constants (see Figs IX.9 and 12).

Hollander, Kalff and Alkemade (1963) were the first to note this discrepancy and to explain it qualitatively by taking into account the contribution of collisional ionization from excited alkali states (M^*) in addition to that from the ground-state (M^0). They in fact postulated that the largest contribution to the ionization rate comes from collisions between thermally excited alkali atoms and flame molecules Z (cf. process i in Table IX.1 were A may be in any excited state). The suggested ionization mechanism for a flame containing only M and Z

DISCUSSION AND CONCLUSIONS   Sect. IX.6a

particles reads thus (see also Page 1973):

$$M^0 + Z^\dagger \rightarrow M^+ + e^- + Z \quad (1)$$

$$M^0 + Z^\dagger \rightleftharpoons M^* + Z \quad (2)$$

$$M^* + Z^\dagger \rightarrow M^+ + e^- + Z \quad (3)$$

where M^* refers to any of a large number of <u>thermally populated</u> excitation levels of M. The symbol $Z^\dagger$ represents a flame particle whose internal energy and kinetic energy of relative motion together make up for the energy deficit of the endo-ergic excitation or ionization step. Each of the processes (1) - (3) is considered as occurring in a single, elementary step; the processes may occur in parallel or in series with each other. Because $[M]_t \ll [Z]$, we can here disregard collisions between two M atoms. Processes (2) and (3) constitute together a *ladder mechanism*, also called *cumulative ionization* (see Sects II.4b-2 and IX.2a). The reverse processes (1) and (3) apply for the recombination of M^+ and e^-.

The above authors now assume that the expression IX.58 for $k_i^{coll}(T)$ can be applied for any of the ionization steps (1) or (3) for each atomic energy level <u>separately</u>. They further assume for simplicity's sake that for each energy level the cross section occurring in this equation is the same (σ_0) and of gas-kinetic order. (In fact, the cross section may be expected to increase with increasing internal energy of M^* because of the larger electron orbit; this expectation would only strengthen the relevance of the ladder model.) We have only to replace E_{ion} by ($E_{ion} - E_{exc}$) in the activation-energy factor when we apply Eq. IX.58 to each excited level with excitation energy E_{exc}. The thermal phenomenological ionization rate constant, as determined in flames from relaxation experiments, which relates to the total atomic density $[M]$ ($\approx [M^0]$), is then found by summing the rate constants given by Eq. IX.58 for each energy level of M, after weighting them over the Boltzmann distribution. The sum of the internal and translational energies available in the reactants must (at least) equal the ionization energy E_{ion} (supposedly no extra activation energy is required). Because the distribution of these energies obeys the Maxwell-Boltzmann laws, Hollander, Kalff and Alkemade (1963) and Hollander (1964, 1968) arrived at the expression

$$k_i^{coll}(T) = \sigma_0 [Z]_t (8kT/\pi\mu)^{\frac{1}{2}} \exp[-E_{ion}/kT] \sum_{j=0}^{\infty} g_j / Q_M . \quad (IX.64)$$

Here g_j and Q_M are the statistical weight of the j-th energy level of M and the partition function of M, respectively (see Sect. II.3b-2). This expression holds when all M levels are thermally populated; this thermal population was considered to be a reasonable assumption, at least when (near-)Saha equilibrium exists (see however also Sect. 6b-1). Since Q_M is finite (cf. the discussion about the apparent

Chapt. IX   IONIZATION IN SEEDED FLAMES

divergence of Q_M in Sect. II.3b-2), the factor $\sum g_j/Q_M$ would be infinite if one sums the g_j's over all (i.e. over an infinite number of) energy levels of the isolated atom. Obviously a cut-off of the highest excitation levels near the ionization continuum must be introduced to keep $\sum g_j$ finite; this cut-off procedure is similar to that introduced for avoiding the divergence of the partition function. Anyway, even if we terminate the summation at an energy level situated at kT below the ionization limit of an isolated atom, we calculate a phenomenological rate constant that is several orders of magnitude greater than for gas-kinetic collisions.

The simplified equation IX.64 for $k_i^{coll}(T)$ implies in fact that collisional ionization from each energy level of M (excited or not) contributes to an equal extent to the total ionization rate. This might appear paradoxical, as the higher excited levels are much less populated than the ground-level. But the effect of this unfavourable population ratio is just compensated for by the larger number of Z molecules that have sufficient energy to ionize the excited atom. This compensation is possible because the energy required, that is $E_{ion}-E_{exc}$, decreases with increasing excitation energy E_{exc}, and because the thermal (internal and translational) energy distribution of Z includes the same exponential dependence on energy as the internal-energy distribution of M.

Direct measurement of the cross sections for single-step collisional ionization from a particular energy level may sustain the general assumptions made in the above model. Such measurements can be made in a crossed-molecular beam apparatus where the alkali beam is irradiated by a strong laser beam tuned at one of the optical transitions. In this way Na atoms in the 3P state in collision with 'vibrationally hot' N_2 molecules effusing from an oven appeared to have an ionization cross section of about 100 $Å^2$, the activation-energy factor (corresponding to the N_2-oven temperature) being singled out (see de Jong *et al.* 000).

In a crossed-beam experiment Schmidt, Haug and Rappenecker (1974) and Haug, Rappenecker and Schmidt (1974) found for ground-state K and Na atoms in collision with vibrationally excited N_2 molecules a lower and an upper limit for the ionization cross section of 30 and 150 $Å^2$, respectively. These limits depend on the assumed (nonthermal) vibrational-energy distribution. Vibrationally excited N_2 was formed in this case by a microwave discharge in a nitrogen flow.

The thermal phenomenological recombination rate constant in the above ladder model follows directly from $k_i^{coll}(T)$ by application of the detailed-balance relation, Eq. IX.30.

The general idea of a ladder mechanism has also been adopted by Padley and co-workers and several other British authors, experimentalists as well as theoreticians (see Jensen and Padley 1966,1967, Kelly and Padley 1969,1972, Hayhurst and Telford 1972a, Fowler and Preist 1972, Preist 1972, and Ashton and Hayhurst 1973).

DISCUSSION AND CONCLUSIONS   Sect. IX.6a

Their approach, however, differs from that originally suggested by Hollander *et al.* in that they take into consideration only collisional ionization from a multitude of energy levels <u>close</u> to (say, within kT) the ionization limit (see Kelly and Padley 1972). In their picture practically no activation energy is required to raise the highly excited atom to the continuum state. Various theoretical attempts have been made to obtain a realistic estimate of the number of discrete energy levels that actually contribute to the ionization process. Here again cut-off problems arise similar to those discussed in connection with the divergence of the partition function. For a further discussion we refer to Sect. 6b-1.

In the theoretical analysis by Fowler and Preist (see Fowler and Preist 1972, and Preist 1972) the Chew-Low approximation was used, which appeared to account qualitatively for the experimental ionization rate constants; these authors avoided the divergence of the partition function of hydrogen-like atoms by assuming Debije screening (see Sect. II.3b-4).

Preist (1972) argues that in plasmas with $[e^-] \geqslant 10^{11}$ cm^{-3} the 'spectrum' of electronic states near the ionization continuum will be altered due to the presence of free electrons. In his picture the number of bound states then becomes finite by the broadening of the levels due to the electrons; as a result adjacent levels will overlap and effectively the number of discrete states will decrease and become finite.

Since, however, in most ionization experiments in flames $[e^-]$ is 'fortunately' $\leqslant 10^{10}$ cm^{-3}, Preist's theoretical model is not applicable here. Moreover, since the experimental $k_i(T)$ values appear to be independent of $[e^-]$, his estimates of the number of atomic levels do not seem to apply to flame ionization processes.

There is a particular problem involved in applying the ladder model when the metal atoms are contained in a <u>pure</u> noble gas. Noble gas atoms are known to be very inefficient in (de-)exciting alkali atoms, at least as far as the lowest excited levels are concerned and under flame conditions (see Sect. VI.2a-1). The actual population of the excited levels will then be appreciably below the Boltzmann value (see Sect. VI.4c-2); the summation in Eq. IX.64 is then not valid and the actual k_i^{coll} value is much smaller than the value estimated from this equation. Besides, the cross section for ionization by noble gas atoms is theoretically expected to be negligibly small (see Sect. 6b-2).

The application of the above ladder model, be it in Hollander's interpretation or in that of Kelly and Padley, to *multi-component systems* is more complicated. In this model the single-step ionization processes (1) and (3) and the reverse processes should then be separately summed over the different collision partners while taking into account their concentration ratios. A situation may occur where a

Chapt. IX   IONIZATION IN SEEDED FLAMES

molecular species is present in a minor concentration, while an atomic species, such as Ar, is present as a major constituent. If the concentration of the former is still large enough to bring about an equilibrium population of the excited levels (molecules are in general efficient quenchers, see Sect. VI.2b), the other species [which is quite inefficient in the (de-)excitation step] could still contribute to the ionization rate, if its cross section for ionization of excited metal atoms were not negligibly small.

If we can assume that in a multi-component system the relevant atomic excitation levels are, virtually, populated according to the Boltzmann law (and this assumption could be tested by intensity measurements in the thermal emission spectrum!), the kinetics of the (de-)excitation process (2) is no longer important. The phenomenological ionization rate constant derived from ionization relaxation experiments is then determined only by the rate constants of the single-step ionizing collisions (1) and (3) for each energy level separately. When various partners Z_j are present, their single-step, binary ionizing collisions are independent and the overall rate is obtained by simply summing up their contributions, weighted according to their abundance ratios. This may justify the application of the linear relationship expressed by Eq. IX.27a, in combination with Eq. IX.62, to a multi-component system where Boltzmann equilibrium, but not necessarily Saha equilibrium, is (approximately) realized. We refer to Sect. 6b-1 for a further discussion.

6a-2   *INTERCOMPARISON OF STANDARDIZED EXPERIMENTAL* $k_i(T)$ *VALUES*

Standardized $k_i(T)$ values are collected in Table IX.9 (for details of standardization see Note * to the table). The contribution of chemi-ionization to the measured $k_i(T)$ values was assumed to be negligible in Refs (3) and (4), whereas a small correction for chemi-ionization was applied in Refs (1) and (2). The listed rate constants thus refer to purely collisional ionization.

The data in the first row of each column apply when one takes into account only the accidental errors. The data in the second row of the first two columns indicate the range within which the actual value of k_i may be situated, if one also accounts for possible systematic errors. Borgers, Jongerius, Ventevogel, Hollander and Alkemade (000) discussed possible errors that might explain the systematic discrepancy of about a factor of 3 between their recent k_i values for Na, K and Cs, and the corresponding values from Hollander, Kalff and Alkemade (1963). They were able to explain this discrepancy by a possible variation with height of the electric sensitivity with the r.f. resonance method in Ref. (2) or a slight unnoticed decrement of $[M]_t$ with height with the optical spectrometric method in Ref. (1).

When comparing the $k_i(T)$ values in Table IX.9 one may draw the following conclusions:

DISCUSSION AND CONCLUSIONS   Sect. IX.6a

Table IX.9

Comparison of Standardized*, Unimolecular Thermal Ionization Rate
Constants $k_i(T)$ at 1930 K and 1 atm in H_2- and CO Flames Diluted by N_2

Metal	Ionization rate constant k_i (s^{-1}) †			
	Hollander et al. (1)	Borgers et al. (2)	Kelly and Padley (3)	Ashton and Hayhurst (4)
Cs	$(0.14 \pm 0.02) \times 10^2$ $0.1 \times 10^2 - 0.4 \times 10^2$	$(0.47 \pm 0.04) \times 10^2$ $0.2 \times 10^2 - 0.7 \times 10^2$	$(1.1 \pm 0.2) \times 10^2$	$(1.2 \pm 0.3) \times 10^2$
K	4.6 ± 0.6 $4 - 14$	13.4 ± 0.4 $7 - 20$	9.0 ± 1.4	7.7 ± 2.1
Na	0.11 ± 0.02 $0.1 - 0.3$	0.298 ± 0.015 $0.2 - 0.4$	0.075 ± 0.012	0.062 ± 0.017
Li	—	0.125 ± 0.017 $0.10 - 0.16$	0.011 ± 0.004	0.014 ± 0.007
Method of measuring	Optical spectrometry	R.f. resonance and optical spectrometry	Rotating probe and optical spectrometry	Optical spectrometry
Flame type	$CO/O_2/N_2$		$H_2/O_2/N_2$	

† The errors quoted are only of accidental character; however, for each alkali element the second row of the first two columns presents the range within which the real value of k_i may be situated, if one accounts also for possible systematic errors [as estimated by Borgers (2) for the experimental results of the Utrecht group].

* All the results quoted are normalized to the temperature of 1930 K. In the conversion of the results from Reference (3) account is also taken of the flame gas composition (on the basis of the specific ionization efficiencies $\alpha_{M,j}$ — as defined in Sect. 5a by Eq. IX.63 — given by Kelly and Padley).

REFERENCES

1. Hollander, Kalff and Alkemade (1963); Hollander (1964).
2. Borgers (1978); Borgers, Jongerius, Hollander and Alkemade (1981). Borgers, Jongerius, Ventevogel, Hollander and Alkemade (1981).
3. Kelly and Padley (1972).
4. Ashton and Hayhurst (1973).

Chapt. IX   IONIZATION IN SEEDED FLAMES

- The k_i values from the various investigators, standardized to the same temperature, can be divided into two groups according to the kind of flame used: CO-flames and H_2-flames, respectively. In the first group the results from Hollander et al. and Borgers et al. agree only as to the <u>ratios</u> between the k_i values for the various alkalis. The results from Kelly and Padley agree with those from Ashton and Hayhurst, but these results together disagree with those derived from the CO-flames.

- When taking into account the possible systematic errors mentioned above one may conclude that the CO-results are (at least) mutually consistent within wide error limits.

- The ratio between available risetime-interval and ionization relaxation time (k_i^{-1}) was in the H_2-flames 2 or 3 orders worse than in the CO flame at comparable temperatures (see Refs 2 and 3). Nevertheless Kelly and Padley claimed an experimental error of no more than 50% in their k_i values.

- The large discrepancy between the values for Li of the two groups might be due to a residual systematic error in the LiOH correction with the H_2-flames, where the correction is much larger than in the CO-flames.

- The standardization of the results in the H_2-flames was carried out by accounting also for differences in the flame gas composition on the basis of the specific ionization efficiencies reported by Kelly and Padley (1972). It proved not to be possible to reconcile the values from the two groups by adopting a different but still realistic value for the ionization efficiency of CO_2.

We must conclude that, whereas the values for K from the two groups are consistent with each other, the values for Cs and Na show an unexplained discrepancy of about a factor of 2 or more, whereas those for Li disagree by almost an order of magnitude.

6a-3   *COMPARISON OF E_{ion} AND E_A; T-DEPENDENCE OF k_i AND k_r*

In Table IX.10 the available activation-energy values (E_A) of the collisional ionization from the literature are shown. We generally conclude that these E_A values are, within the experimental scatter, quite close to the corresponding ionization-energy values (E_{ion}). The E_A values were derived from the slope of the linear Arrhenius plots (cf. Figs IX.9 and IX.12). The equality $E_A \simeq E_{ion}$ proves that no extra activation energy is required above the ionization energy in the collisional ionization process and that there is no strong dependence on the relative velocity. This outcome confirms the generally accepted T-dependence of k_i (see the discussion following Eq. IX.58). Furthermore, the fact that unique E_A values

Table IX.10

Collisional-ionization Activation Energies E_A (kJ mol^{-1}), Derived from Arrhenius Plots of $\ln(k_i T^{\frac{1}{2}})$, Compared to Ionization Energy E_{ion} (kJ mol^{-1})

Metal \ Reference	1	2*	3	4*†	4†‡	5	E_{ion}
Li	–	–	–	(504 ± 20)	494 ± 20	490 ± 30	520.3
Na	490 ± 14	483 ± 24	486 ± 37	(475 ± 5)	467 ± 12	465 ± 17	495.6
K	419 ± 12	414 ± 16	464 ± 34	(413 ± 5)	394 ± 24	423 ± 9	419.0
Rb	–	406 ± 16	–	(387 ± 6)	363 ± 9	391 ± 14	402.8
Cs	377 ± 11	384 ± 24	373 ± 21	(379 ± 8)	370 ± 11	384 ± 13	375.5
Tl	–	–	–	(596 ± 8)	565 ± 8	–	589.0

* In this reference the factor $T^{\frac{1}{2}}$ was deleted in the Arrhenius plot, that is, $\ln k_i$ instead of $\ln(k_i T^{\frac{1}{2}})$ was plotted; see, however, also Note ‡.

† The authors have corrected their experimental k_i values for variations in flame gas composition by making use of known $\alpha_{M,j}$ values (see also Sect. 5a-2 and Eq. IX.63).

‡ The numbers in this column have been recalculated from Ref. (4) by including the $T^{\frac{1}{2}}$ factor in the Arrhenius plot.

REFERENCES

1. Hollander, Kalff and Alkemade (1963); Hollander (1964).
2. Jensen and Padley (1966).
3. Hayhurst and Telford (1972).
4. Kelly and Padley (1972).
5. Ashton and Hayhurst (1973).

equal to E_{ion} are found in different flames confirms the assumption that a free electron and a positive ion are formed in the primary ionization process. This conclusion may not hold when appreciable amounts of O_2 molecules would be present in the flame gases (see Sect. 6b-4).

The T-dependence of k_r can be found from that of k_i through the detailed-balance relation (Eq. IX.30) as: $k_r \propto T^{-2}$. This T-dependence has been directly verified by experiment (see Note ǂ to Table IX.5 on page 873).

From the data in Table IX.5 one can conclude that the k_r values measured directly are consistent with the k_r values calculated through the detailed-balance relation from observed $k_i(T)$ values.

6a-4 *THE EXPERIMENTAL IONIZATION CROSS SECTIONS*

From the relations between $k_i(T)$ and $k'_{i,j}$ (Eq. IX.27a) and between $k'_{i,j}$ and $\sigma_{i,j}$ (Eq. IX.62) one can either derive specific cross sections for each collision partner j apart (see Sect. 5; cf. also Kelly and Padley 1972) or an 'approximate' specific cross section for the dominant partner. The latter cross section refers to the value calculated on the simplifying assumption that the flame consists of only one collision partner (obviously the dominant species); here we consider only N_2 and Ar. The $\sigma_{i,j}$ values found in this way are phenomenological, i.e. they are averaged over the relative-velocity and internal-state distributions (see Sect. II.4b-3).

Table IX.11

Intercomparison of Ionization Cross Sections* (in 10^4 Å^2) Derived from Alkali Ionization Experiments in N_2-diluted Flames

Metal \ Reference	1	2	3	4	5
Li	-	-	0.85	0.63	7.1
Na	4.3	2.7	1.8	1.4	11.8
K	1.8	2.3	2.1	1.6	5.3
Rb	-	2.0	-	1.7	-
Cs	0.5	2.3	2.4	2.2	1.7

*'Approximate' specific cross sections for N_2 have been calculated through Eq. IX.62 from overall $k_i(T)$ values on the assumption that the flame consists of N_2 molecules only, except for those from Ref. 4 which refer to true specific σ_{i,N_2} values (see also Sect. 5a-2).

REFERENCES
1. Hollander, Kalff and Alkemade (1963).
2. Kelly and Padley (1969).
3. Hayhurst and Telford (1972).
4. Kelly and Padley (1972).
5. Borgers, Jongerius, Ventevogel, Hollander and Alkemade (000).

DISCUSSION AND CONCLUSIONS
Sect. IX.6a

In Table IX.11 a number of representative values are presented for N_2. An intercomparison of the results reveals an interesting discrepancy between Refs (1) and (5) on the one hand and Refs (2), (3) and (4) on the other: the σ_{i,N_2} values for Na-K-Cs from Refs (1) and (5) decrease in this sequence, whereas those from Refs (2), (3) and (4) increase in this sequence or remain more or less constant. The value of σ_{i,N_2} for Li is always found to be lower than that for Na.

The phenomenological cross sections listed in Table IX.11 are roughly 3 orders of magnitude larger than gas-kinetic cross sections and 2 orders larger than the single-step ionization cross sections found for Na and K in molecular beams with vibrationally excited N_2 (see Sect. 6a-1).

Specific $\sigma_{i,Ar}$ values for the alkali metals have been derived by Kelly and Padley (1972) from rate-constant measurements in isothermal flames with different gas compositions including Ar; these authors applied relations IX.27a and IX.62. From their analysis $\sigma_{i,Ar}$ values were found that were of the order of 8×10^3 to 12×10^3 Å^2 in the sequence from Li to Cs. These $\sigma_{i,Ar}$ values are between those found for H_2 and N_2, respectively.

Hollander, Kalff and Alkemade (1963) have determined the ratio of overall ionization rate constants in two isothermal CO-O_2 flames diluted by Ar and N_2, respectively. This ratio appears to be about twice as small as the ratio calculated from the specific ionization cross sections measured by Kelly and Padley (1972) and the known compositions of these flames. No explanation could be given for this discrepancy; we only note that the efficiency of noble gas atoms to ionize alkali atoms under flame conditions has been questioned by Bates (1976) on theoretical grounds.

The ionization cross sections listed in Table IX.11 are of the order of the squared mean distance between two neighbouring particles in a flame at 1 atm pressure (see footnote ǂ on page 535). This outcome, however, does not invalidate the binary collision model used in the interpretation of the flame ionization measurements, as these phenomenological cross sections do not relate to an elementary collision process (see Sect. 6a-1). They are formally obtained by dividing the phenomenological bimolecular rate constant by the average relative velocity and the activation-energy factor (see Eq. IX.62). They should thus be well distinguished from the cross sections (σ_0) for the ionization of an atom in a given state by a single, binary collision (see Eq. IX.64). The latter cross sections, which are indeed much smaller, have a direct physical meaning, whereas the former are rather an artefact with the same dimension but without a direct interpretation (see also Sect. 6a-1).

Chapt. IX   IONIZATION IN SEEDED FLAMES

6b   COMPARISON OF EXPERIMENT WITH THEORY

6b-1 *A THEORETICAL DISCUSSION OF THE LADDER MODEL AND THE
SUMMATION RULE IN MULTI-COMPONENT FLAMES*

The specific phenomenological cross sections, $\sigma_{i,j}$, which relate to a certain collision partner Z_j, are derived from the specific bimolecular rate constants $k'_{i,j}$ through Eq. IX.62. These rate constants are, in turn, found by measuring the overall, apparent monomolecular rate constant k_i (in s^{-1}) in a series of isothermal flames with different but known compositions. In multi-component flames the connection between k_i and the values of $k'_{i,j}$ for different collision partners with concentrations $[Z_j]$ is given by the *summation rule* in Eq. IX.27a. This rule implies a linear dependence of k_i on the $[Z_j]$'s.

The assumptions underlying the linear dependence of the overall ionization (and recombination) rate constant on the concentrations of the collision partners and, more generally, the original ladder-climbing model of the ionization process described in Sect. 6a-1 need a closer theoretical examination. The validity of the assumptions and implications of this model has been questioned by Bates (1976), who worked out a different version of the ladder model, which we shall call the 'bottleneck model' (see also Bates and Kingston 1961, Bates, Kingston and McWhirter 1962, 1962a, and Bates and Kingston 1964).

The crucial question in any ladder model is how the atomic level population is affected by the absence of Saha equilibrium, that is, by the unbalance between the ionization and recombination rates. As argued in Sect. 3c in connection with the validity of Eq. IX.26a, the bimolecular ionization rate constant k'_i is expected to depend on the population of the atomic levels. It is recalled that $k'_i[Z]$ describes the probability per second of a neutral atom with a given internal-state distribution to be split into an ion and a free electron as a result of a single collision with partner Z. In Sect. 3c we have argued that the termolecular recombination rate constant k'_r may be expected to be insensitive to the ionic level population and independent of the atomic level population. Then k'_r will be equal to its thermal value $k'_r(T)$, irrespective of the actual state of ionization. However, the equality of k'_i to its thermal value in the absence of Saha equilibrium is not obvious, as we shall now illustrate.

Suppose, for example, that the metal vapour relaxes to Saha equilibrium after it has initially been fully ionized. Suppose furthermore that a single-step recombination of an ion and an electron produces preferentially an atom in a high excited level rather than in the ground-level or a low excited level. Let the chance of a highly excited atom being again ionized be greater than its chance of being transferred to a low level upon collision. For simplicity's sake, we disregard radiative processes. Under these conditions the presence of ions and electrons

DISCUSSION AND CONCLUSIONS
Sect. IX.6b

in excess concentrations will lead to an atomic level population that is far above the Boltzmann population for the higher levels, but below it for the lower ones. The ionization rate constant will then be strongly suprathermal, that is, considerably larger than its thermal value, or $k_i \gg k_i(T)$. As a result a near-balance between the recombination rate $k_r(T)[\text{M}^+][\text{e}^-]$ and the ionization rate $k_i[\text{M}]$ is rapidly attained for a total atomic density [M] that is far below its Saha-equilibrium value $[\text{M}]_e$. This balance is not complete because the system is not in Saha equilibrium. There is thus a positive <u>net</u> recombination rate: $-d[\text{M}^+]/dt > 0$. But because the attainment of Saha equilibrium is a comparatively slow process, we have that this net recombination rate is small compared to $k_r(T)[\text{M}^+][\text{e}^-]$.

In his *bottleneck model* Bates (1976) has derived a theoretical expression for the collision-dominated net recombination rate in terms of the specific bimolecular rate constants for collisional de-excitation in a multi-component flame. The net recombination is rate-limited by the comparatively slow population redistribution of the atomic levels, which ultimately tends to a Boltzmann distribution. He conceives this population redistribution as a diffusion-like 'random walk' of the valence electron along the 'ladder' of atomic levels. When the ionic and electronic densities are initially far above their Saha equilibrium values, the valence electron cascades statistically down this ladder to the ground-level after it was initially produced in a highly excited state by ion-electron recombination.

One may expect that collisional transitions between the closely spaced high levels near the ionization limit are rapid, so that a partial Boltzmann equilibrium is established within this group of levels. This will not hold true for lower lying levels, which are therefore <u>under</u>populated in comparison with the levels near the ionization limit. These underpopulated lower levels constitute a 'bottleneck' for the cascading valence electron. The rate of passage of the valence electron through this bottleneck limits in effect the net recombination rate. The former rate depends on the 'width' of the bottleneck (which corresponds to the magnitude of the cross sections for inelastic collisions) as well as on the 'length' of the bottleneck (which corresponds to the number of the underpopulated levels involved). It is clear that the width and length of the bottleneck are dependent on the properties of the collision partner. Bates (1976) has considered in particular the role played here by the vibrational and rotational degrees of freedom of certain molecular partners in the (de-)excitation of the alkali atom. For a more general and detailed treatment of the role of a bottleneck in collisional-radiative recombination processes we refer to Stevefelt, Boulmer and Delpech (1975).

As long as such a bottleneck exists, the bimolecular rate constant k_i' will depend on the extent of the deviation from Saha equilibrium; it is then not a true constant of proportionality. However, while the concentrations of the atomic and

Chapt. IX   IONIZATION IN SEEDED FLAMES

ionic species approach their Saha-equilibrium values, the population distribution of all atomic levels will become closer to the Boltzmann distribution so that ultimately $k'_i \to k'_i(T)$.

Another consequence of this bottleneck is that in the presence of various collision partners, Z_j, the overall monomolecular ionization rate constant k_i can, in general, not be expressed linearly in the concentrations $[Z_j]$ (cf. Eq. IX.27a). This is because every specific, bimolecular rate constant $k'_{i,j}$ depends on the population distribution over the atomic levels; the latter again depends on the combined effect of the inelastic collisions with various collision partners. For this reason Bates (1976) has criticized the use of the summation rule in the interpretation of the flame experiments by Hollander, Kalff and Alkemade, and by Kelly and Padley.

It is noted that the recombination rate coefficient α as defined by Bates (1976) and Stevefelt, Boulmer and Delpech (1975) refers to the net recombination rate. This coefficient should thus be clearly distinguished from our recombination constant k_r which relates to a single-step recombination of an ion in any state and an electron to a neutral atom in any state.

The above bottleneck model of the recombination process has been exposed to some extent in order to better appraise — by contrast — the *improved version of Hollander's original model* (described in Sect. 6a-1) that we shall treat now. This version is believed to justify, at least in rough approximation, the interpretation of the ionization relaxation experiments given in Sects 3 and 5.

As a starting point we recapitulate the main features of the reported *experimental flame results:*

(i)   the ionization rate constants show an exponential dependence on T^{-1} with an activation energy equal to the ionization energy within an error interval of the order of kT (see Sect. 5a);

(ii)   the recombination rate constants derived from measured ionization rate constants through application of the detailed-balance relation are consistent with those directly determined (see Sect. 6a-3);

(iii)   there is no systematic difference between the rate constants measured in situations of a large or small, a positive or negative deviation from Saha equilibrium (see Sect. 5a);

(iv)   the specific ionization cross sections derived by application of summation rule Eq. IX.27a or 27b seem to reproduce reasonably well (but not exactly) most overall rate constants found in flames of widely different but known gas compositions;

DISCUSSION AND CONCLUSIONS   Sect. IX.6b

(v)   there is no evidence of systematic differences between the rate constants
        derived from atomic-density and ionic- or electronic-density measurements,
        respectively;

(vi)   the ionization relaxation times found are roughly of the order of 1 ms or
        longer;

(vii)   the phenomenological ionization cross sections found in flames for Na and
        K are roughly two orders larger than the single-step ionization cross
        sections found in molecular beams, which are in turn one order larger than
        gas-kinetic cross sections (see Sect. 6a-1).

In order to understand qualitatively these flame observations in terms of a modified
ladder mechanism we have to reconsider the proper meaning of the ionization and recombination rate constants, k_i and k_r, as measured in flames. This will be done
by making the following *assumptions*, that are believed to hold in crude approximation under the special conditions existing in atmospheric-pressure flames with a nonnegligible content of molecular species:

(1)   The (de-)excitation as well as ionization and recombination processes are
        collision-dominated.

(2)   There is a rapid <u>partial</u> equilibration of the relative population within
        the group (I) of excited levels that lie within a distance of, say, kT from
        the ionization continuum.

(3)   Between the excited atoms in group I and the ions and free electrons a quasi-
        Saha equilibrium is very rapidly established. For, the rate at which an
        atom in this group is ionized may be expected to be (at least) of the order
        of the gas-kinetic collision rate, which is about $3 \times 10^9 \, s^{-1}$. The number
        density of atoms in group I is, consequently, proportional to the product
        of the densities of the ions and free electrons.

(4)   The number density of atoms in group I is always small compared to the den-
        sity of ions.

(5)   There is also a rapid <u>partial</u> equilibration of the relative population in
        a group (III) of lower-lying levels including the ground-level.

(6)   The intermediate, nonequilibrated atomic levels are contained in a compara-
        tively narrow energy-band (II) located at a distance of the order of kT
        below the ionization continuum. The density of atoms in this *demarcation
        band* is small compared to the density of atoms in group I and group III.
        The possible variation of the location and effective width of this band
        with varying flame gas composition is small compared to kT. A similar

assumption is made for the dependence of the demarcation band on the actual deviation from overall Saha equilibrium.

(7) The 'flow' of atoms upwards or downwards over the demarcation band is (mainly) determined by single-step transitions from any level in group III to any level in group I or to the ionization continuum, and vice versa.

The *consequences* of these assumptions are the following:

(a) Whenever an atom is promoted from a level in group III to a level in group I, this will be followed quickly by an increase of the mean number of ions by about one unit in order to restore the quasi-Saha equilibrium (see assumptions 3 and 4). The time scale at which this is achieved is at most of the order of the gas-kinetic intercollision time and therefore much shorter than the ionization relaxation time measured in flames (see vi).

(b) Because of assumptions 4 and 6 we have that $dn_{III}/dt \simeq -dn_i/dt$, where n_{III} is the number density of atoms in group III and n_i the number density of ions (or free electrons). We note that dn_{III}/dt can be measured in flames by monitoring the population of the ground-level or some low excited level as a function of risetime t. Similarly we can measure dn_i/dt by monitoring the density of ions or free electrons.

(c) Because of consequence (a) the 'ionization rate constant', $(k_i)_{exp}$, as determined from flame experiments, relates actually to the rate at which atoms from group III cross the demarcation band II either by a transition to a level of group I or by an ionizing collision. The reverse holds true for the experimental recombination rate constant, $(k_r)_{exp}$. We note that the rate at which atoms from group I are transferred to a lower level belonging to group III is proportional to the density of atoms in group I. But this density is, in turn, proportional to the product of the ionic and electronic densities, according to assumption 3. So this rate can be formally expressed in the ionic and electronic densities in the same way as the true recombination rate. This justifies the formal introduction of the experimental recombination rate constant $(k_r)_{exp}$, which includes transitions from group I levels to group III levels.

(d) Because of assumptions 2, 5, 6 and 7 $(k_i)_{exp}$ and $(k_r)_{exp}$ are, virtually, independent of the deviation from overall Saha equilibrium. They are determined by the (partially equilibrated) population distributions of group III and I, respectively, and by the Maxwell-Boltzmann distribution of the collision partner. Since each of these relative distributions is governed by the flame temperature only, the values of these rate constants are thermal

DISCUSSION AND CONCLUSIONS							Sect. IX.6b

(that is, the same as in thermal equilibrium). As a result, the ratio $(k_i)_{exp}/(k_r)_{exp}$ can be simply proven to obey strictly the detailed-balance relation Eq. II.95, if also use is made of assumption 3.

(e) Because of assumptions 2, 5, 6 and 7 we can express $(k_i)_{exp}$ <u>linearly</u> in the concentrations of the flame constituents, as was presumed in Eq. IX.27a. For, in the crossing of the demarcation band II by a single-step collision different collision partners act <u>in parallel</u>. Note also that the existence of partial equilibrium within group III and the extent of this group were assumed to be, virtually, independent of the flame gas composition and governed by the flame temperature only. An analogous conclusion can be drawn for $(k_r)_{exp}$.

(f) The ladder model for ionization introduced in Sect. 6a-1 has to be modified according to the above re-interpretation of $(k_i)_{exp}$. In steps (2) and (3) (on page 883) M^* represents now an excited atom belonging to group III, while steps (1) and (3) include also transitions to levels of group I. Equation IX.64 remains then valid, in first approximation, if we replace

$$\sum_{j=0}^{\infty} \text{ by } \sum_{j=0}^{j_{II}},$$

where j_{II} corresponds to the bottom of the demarcation band, and Q_M by the partition function, Q_{III}, of group III levels. (Note that the divergence problem does here not arise.) In the derivation of the modified Eq. IX.64 we have also used the approximation that $\exp[-E_{II}/kT] \simeq \exp[-E_{ion}/kT]$, as the excitation energy, E_{II}, of the demarcation band, which lies about kT below the bottom of the ionization continuum, is virtually equal to E_{ion}. In the modification of Eq. IX.64 we have also used the (crude) approximation that the effective cross section for a single-step transition from a level in group III to any level in group I is a constant and of the same order of magnitude as σ_0, which relates to a direct ionizing collision. Taking for σ_0 a value of 100 Å^2 as suggested by molecular beam experiments (see Sect. 6a-1), we now estimate from the modified Eq. IX.64 and the measured values of $(k_i)_{exp}$ that

$$\sum_{j=0}^{j_{II}} g_j \approx 10^2$$

(see also observation vii).

On the basis of this model we can qualitatively explain the features (i) - (vii) of the flame experiments. With regard to observation (i) we recall that the experimental value of the activation energy for ionization is inaccurate by about kT.

Chapt. IX   IONIZATION IN SEEDED FLAMES

The essential point is that the observed ionization rate constant $(k_i)_{exp}$ refers now to the rate at which atoms cross the demarcation band from below, and not to the rate at which atoms cross the bottom of the ionization continuum. The latter rate is very high and its value is, in fact, irrelevant for the interpretation of the flame experiments [see (a)].

It should be pointed out that the demarcation band introduced here has not the character of a 'bottleneck' as in Bates (1976) theory. The latter theory is based on the net 'flow' of atoms through the bottleneck, which is brought about by a diffusion-like 'motion' of the valence electron on a ladder of excited levels. This contrasts with the former model where the atoms cross the (thin) demarcation band by a single jump from any level below this band to any level above it or to the continuum state, and vice versa (see assumption 7). The question which of the two models is the more realistic one, can ultimately be solved only if we know the detailed cross sections for the (de-)excitation and ionization processes as a function of the initial and final quantum states and in dependence on the kind of collision partner as well as on temperature.

6b-2   *THE IONIZATION EFFICIENCY OF Ar*

As discussed in Sect. 6a-4, Kelly and Padley (1972) derived specific $\sigma_{i,Ar}$ values for the alkali metals; these $\sigma_{i,Ar}$ values were of the same order of magnitude as for H_2 and N_2 perturbers. This outcome is in conflict, however, with theoretical arguments put forward by Bates (1976) for the inefficiency of noble-gas atoms to ionize alkali atoms under flame conditions. Ground-state Ar atoms possess only translational energy, which makes them inefficient in raising the (light) valence electron of the alkali atom to a higher state or the continuum state.

It should be noted, however, that if we assume a priori that $\sigma_{i,j}/\sqrt{\mu}$ is not markedly different among the molecular species present in $CO-O_2-N_2$ and $CO-O_2-Ar$ flames, the ratio of overall ionization rate constants measured in these flames points to a value $\sigma_{i,Ar} \approx 0$ (see Hollander, Kalff and Alkemade 1963). Anyway, this ratio does not agree with the ratio calculated from the specific ionization cross sections measured by Kelly and Padley (1972) and the known composition of these flames (cf. Sect. 6a-4). Further experiments are needed to answer definitively the question whether the contribution of Ar atoms to the alkali ionization rate in Ar-diluted flames is or is not negligibly small. Crossed atomic beam or shock tube experiments might be helpful to answer this question (see the pertinent discussion in de Jong *et al.* 000).

6b-3   *EXPECTED SEQUENCE OF $\sigma_{i,j}$ VALUES FOR DIFFERENT ALKALI ATOMS*

As was discussed in Sect. 6a-4, there are indications in the work of Kelly and Padley (1972) that the $\sigma_{i,j}$ values for various flame gas particles increase

DISCUSSION AND CONCLUSIONS   Sect. IX.6b

(slightly) in the sequence: Li-Na-K-Rb-Cs. In CO flames Hollander, Kalff and Alkemade (1963) found a reverse and more pronounced sequence: Cs-K-Na; this sequence has also been found in a similar flame by Borgers, Jongerius, Ventevogel, Hollander and Alkemade (1981). However, the possible occurrence of systematic errors might arouse doubts on the latter sequence (see Sect. 6a-2).

Bates (1976) expects on theoretical grounds that the ionization cross section should be rather insensitive to the species of alkali atom and that Cs, being less hydrogen-like than Na, should tend to have a somewhat larger cross section. From his point of view Bates concludes that the results of Kelly and Padley are theoretically the more acceptable ones.

We are still in need of a decisive flame experiment and more detailed theoretical calculations to solve this question.

6b-4   *THE ACTIVATION ENERGY AND THE IONIC-CURVE CROSSING MODEL OF IONIZATION*

Qualitative considerations. The activation energy E_A measured in flames for the alkalis and thallium appears to be about equal to or slightly less than the ionization energy, E_{ion}, for an isolated atom (see Table IX.10). In a real gas or plasma the <u>effective</u> ionization energy will be somewhat less than E_{ion} because the highest atomic levels merge into a continuum and Debije screening may play its role (see Sects II.3b-2 and II.3b-4). The experimental error limits and the systematic differences between the values obtained by different authors do not warrant drawing a definite conclusion about the (in-)equality of E_A and E_{ion}. Besides, the theoretical evaluation of the depression in ionization energy under flame conditions is still uncertain.

It should be noted that <u>thermal</u> ionization rate constants, $k_i(T)$, exist only in a system that is in thermodynamic equilibrium, including the internal-energy distribution of the metal atoms (cf. Sects II.4a-1 and IX.3c under Case A). But we can measure ionization and recombination rate constants only if there is at least some deviation from Saha equilibrium (see Sect. II.4a-2). But any deviation from Saha equilibrium may entail a deviation from Boltzmann's distribution law for the population of the atomic energy levels that are nearest to the ionization limit, in comparison with the ground-state population. These levels are readily populated and depopulated by recombination of ions and electrons, and by ionizing collisions, respectively. This deviation from Boltzmann's law has consequences for the interpretation of the experimental ionization rate constant $(k_i)_{exp}$ as well as of the experimental activation energy $(E_A)_{exp}$. In the improved ladder model of Hollander *et al.* (described in Sect. 6b-1) $(k_i)_{exp}$ relates in fact to single-step collisional transitions from atomic levels lying below a certain demarcation band to the ionization continuum as well as to atomic levels lying above this band and closely to the continuum. The reverse holds for the experimental recombination rate constant $(k_r)_{exp}$. In this model the experimental activation energy $(E_A)_{exp}$ corresponds to the height of the demarcation band, which was supposed to be thin. Since the distance of this band to the ionization continuum is expected to be of the order of kT, the value of $(E_A)_{exp}$ will not differ from E_{ion} by much more than kT. Under flame conditions and for alkali atoms kT/E_{ion} is less than 10%, which corresponds roughly to the relative inaccuracy of the experimental values of the activation energy. It is therefore not possible to draw a conclusion about the correctness of this model from the existing data for $(E_A)_{exp}$.

We recall that no systematic differences were observed in the values of $(k_i)_{exp}$ and $(E_A)_{exp}$ as determined in situations where a small or large, a positive or negative deviation from Saha equilibrium existed. Special assumptions had to be made in the above improved ladder model in order to explain this (see Sect. 6b-1). Consequently, the $(k_i)_{exp}$ values are close to what they should be in thermal equilibrium including Saha equilibrium. The experimental activation energy $(E_A)_{exp}$ therefore relates to the variation of the <u>thermal</u> experimental rate constant with flame temperature.

In the above model a quasi-Saha equilibrium was assumed to be rapidly established between the density of atoms in highly excited states near the ionization continuum on the one hand and the ionic and electronic densities on the other. The time scale at which this quasi-equilibrium is established is probably of the order of the gas-kinetic intercollision time. It is therefore not feasible to determine the rate constant for single-step transitions from these high-lying levels to the continuum, and vice versa, by flame relaxation experiments. This is precisely why a re-interpretation of the experimental ionization (and recombination) rate constants was made in the improved ladder model of Hollander. We note, however, that the equations for ionization relaxation presented in Sect. 3c remain formally the same. It is only the physical meaning of the rate constants occurring in these equations that had to be changed.

Theoretically it is by no means obvious that the activation energy is about equal to the (effective) ionization energy (see Fig. II.4 on page 84, where E_A for the endo-ergic process <u>exceeds</u> $\Delta G_0 \hat{=} E_{ion}$). For a qualitative theoretical understanding we need a microscopic model for the ionization steps (1) and (3) mentioned on page 883. Possible models have been briefly discussed in Sect. 2a, where it is concluded on the basis of molecular beam experiments that *ionic-curve crossing* is probably involved in the alkali ionization by di-atomic molecules (see also Fig. II.6 on page 101). If the first step in a collisional ionization process between two ground-state species A and B (A = metal atom; B = molecule with electron affinity E_{aff}) leads to an ion pair: $A^+ + B^-$, the activation energy for this step

DISCUSSION AND CONCLUSIONS   Sect. IX.6b

will be: $E_A = E_{ion} - E_{aff}$ (see same figure). This activation energy has to be supplied from the translational energy of relative motion; molecular beam measurements as a function of translational energy have shown that E_A may be lower or higher than E_{ion} according to whether E_{aff} is positive (as for O_2) or negative (as for N_2 and CO) (see references to the literature at the beginning of Sect. 2). In the case of a negative E_{aff}, the B⁻ ion is supposed to auto-ionize immediately, so that the production of A⁺ and e⁻ occurs in fact in a single-step process (as defined in Sect. II.4h). However, when E_{aff} is positive, the overall ionization process includes two steps: A + B → A⁺ + B⁻ followed by B⁻ + X → B + e⁻ + X (cf. forward process iv and reverse process iii in Table IX.1). Since in the case of O_2 we have: $E_{aff} \simeq +0.5$ eV (see footnote † on page 585), the activation energy for the formation of alkali ions by collision with O_2 would be expected to lie significantly below E_{ion}. On the other hand, the activation energy with N_2 and CO as collision partners would be expected to lie about 2 eV above E_{ion}, as in these cases $E_{aff} \simeq -2$ eV.

It seems, upon closer inspection of the experimental data in Sect. 5a, that the compositions of all flames investigated make the contribution of O_2 to the overall ionization rate too small to significantly affect the measured E_A values[†]. However, N_2 is found to play a major role in the ionization rate measurements in N_2-diluted flames. The E_A values found in these flames are certainly not (about) 2 eV above E_{ion}, as the molecular beam experiments would suggest. An explanation might be that under flame conditions both the metal atoms and N_2 molecules have a thermal internal-energy distribution. Vibrationally excited molecules could provide an 'ionization channel' in which the electron is detached from N_2^- while this ion is still under the influence of the near alkali ion; the energy of N_2^- in its lowest state may then be appreciably lower than the corresponding energy, $|E_{aff}|$, for a free N_2^- ion. On the other hand, there is then also the possibility that the alkali ion would recapture this electron, thus lowering the ionization probability.

In flames, process (1) mentioned on page 883, which involves only ground-state metal atoms, makes only a small contribution to the phenomenological ionization rate constant measured. According to the simplified ladder mechanism explained in Sect. 6a-1, process (3) involving metal atoms in a large number of excited levels dominates in the production of metal ions. Process (3) may be again postulated to proceed through ionic-curve crossing, just like process (1) for a ground-state atom[*].

[†] In a recent study of the opto-galvanic effect with Na atoms in the laser-saturated D level, van Dijk (1978) and van Dijk and Alkemade (1980a,b) noticed a considerable increase in ion production when their H_2-O_2-Ar flame was made oxygen-rich. They ascribed this effect to the lowering of the activation energy due to the presence of free O_2 molecules.

[*] When the initial internal energy of a molecular collision partner equals about the ionization energy, ionic-curve crossing need not be involved.

The aforementioned molecular beam measurements relate, however, to ground-state metal atoms (as well as ground-state molecules). Thus the results obtained cannot be directly compared with those of the flame experiments. In particular, the activation energy found in experiments where the ionization energy has to be supplied wholly from the translational degrees of freedom might well be different from the activation energy found in experiments where vibrational and atomic excitation energy essentially participate.

The situation is again different for Ar as collision partner, as this atom has only translational energy in flames. Besides, an ionic-curve crossing model does not seem to apply here, as can be concluded from the (de-)excitation experiments described in Sect. VI.2a-1.

It should be realized that the activation energy for any of the microscopic processes (3) involving a metal atom in a particular excitation level is lowered by the excitation energy E_{exc}. When a Boltzmann distribution exists for the population of the excitation levels, the contribution of each microscopic process (3) to the thermal phenomenological rate constant is weighted by a Boltzmann factor. The excitation energy occurring in the exponent of this factor adds again to the activation energy of the microscopic process. Therefore the contribution of each microscopic process (3) to the above rate constant contains the same exponential factor as that of process (1) for the ground-state. Consequently, the activation energy involved in the expression for this rate constant (see Eq. IX.64) is equal to the activation energy as counted from the ground-state.

As postulated before, the microscopic processes (1) and (3) considered in the statistical ladder-climbing model proceed through ionic-curve crossing when di-atomic molecules are involved. It is thus not meaningful to consider the ionic-curve crossing model as an alternative to the ladder-climbing model for explaining the anomalously large ionization cross sections found in flames [†].

Further atomic and molecular beam studies with velocity selection and internal-state selection of the metal atom as well as of the molecular collision partner are wanted.

Quantitative considerations. When a two-step ionization process: $M + Z \rightarrow M^+ + Z^-$ (with monomolecular rate constant k_1) followed by: $Z^- + X \rightarrow Z + e^- + X$ is operative along with a single-step process: $M + Z \rightarrow M^+ + e^- + Z$ (with monomolecular rate constant k_3), the kinetic equation for the removal of M atoms by ionizing collisions with Z reads according to Eq. C.9 in Appendix A.9

[†] Hayhurst and Telford (1972a) seem to consider these models as alternatives, however.

DISCUSSION AND CONCLUSIONS   Sect. IX.6c

$$-\frac{d\ln[M]}{dt} = (k_1 + k_3) \left\{ 1 - \frac{([M]_t - [M])^2}{C[M]K_3} \right\} . \qquad (IX.65)$$

Here $C \equiv 1 + [Z]/K_2$, and K_2 and K_3 are the temperature-dependent equilibrium constants defined by (cf. Eqs II.59 and 62)

$$K_2(T) = ([Z][e^-]/[Z^-])_{eq} , \qquad (IX.66a)$$

$$K_3(T) = ([M^+][e^-]/[M])_{eq} . \qquad (IX.66b)$$

Z is, e.g., an O_2 molecule with positive E_{aff} and X is an arbitrary flame molecule. Equation IX.65, which replaces Eq. IX.31, holds if the electron detachment process for Z^- is, virtually, equilibrated; equilibration may be expected when E_{aff} is small. In fact we then have: $[Z]/K_2 = [Z^-]/[e^-]$. When the latter ratio tends to zero, we have: $C \to 1$, and Eq. IX.65 becomes identical to the previous equation IX.31 with $k_1 + k_3 \triangleq k_i$. Assuming that k_1 and k_3 have activation energies equal to $(E_{ion} - E_{aff})$ and E_{ion}, respectively, the apparent activation energy of the combined rate constant $k_1 + k_3$, as determined experimentally through Eq. IX.65 (cf. Sect. 3c under Case A), lies between these extremes; its actual value depends on whether k_1 or k_3 dominates in the restricted T-range investigated. If for O_2 we have $k_1 \gg k_3$, the activation energy found for collisional ionization by O_2 will be about 0.5 eV lower than E_{ion}. This lowering might be detectable in cool, O_2-rich flames[†] (a low temperature favours k_1 with respect to k_3, just because of the difference in their activation energies).

6c CONSEQUENCES OF METAL IONIZATION IN FLAMES

6c-1 *CONSEQUENCES FOR OTHER METAL VAPOUR STUDIES*

Metal ionization in flames not only extends the range of interesting phenomena that can be fruitfully and rather simply investigated in this medium. It may also interfere with the measurement and interpretation of other phenomena. These interferences have been mentioned in the pertinent chapters. Disregard or misunderstanding of such interferences can lead to systematic errors in metal vapour studies. In some cases metal ionization can also be utilized to obtain information about the state of the flame gas. We shall here summarize these interferences and useful side effects of metal ionization. In this survey one should bear in mind that free electrons may be produced not only by metal ionization. They may also appear as a result of natural ionization, especially in the primary combustion zone and in hydrocarbon flames (see Sect. IV.7).

---

† See also footnote † on page 901.

Chapt. IX　　　　　　　　　　　　　　　　　　　　　IONIZATION IN SEEDED FLAMES

(i)　　Ionization of metal atoms may lead to the appearance in the flame spectrum of ion lines such as the 4215.5 Å Sr^+ line, the 3611.0 Å Y^+ line and the 2144.4 Å Cd^+ line (for a table of wavelengths of lines and bands observed in flames see Alkemade and Herrmann 1979). It may also lead to the appearance of an ion-recombination continuum (see Sect. II.2e). No bands of ionized metal compounds seem to have been observed in flames.

(ii)　　In principle, the flame temperature can be determined through Eq. II.60b by measuring the ionization constant <u>if</u> Saha equilibrium exists. In practice, this method has found only very limited application and is not recommended either.

(iii)　　H-radical concentrations can be determined by measuring the ratio of, e.g., Ca^+ to $CaOH^+$ concentrations (see Note 9 to Table IX.1).

(iv)　　Addition of metal species can affect the concentration of the important natural flame ion H_3O^+ (see Sect. IV.7 and Note 8 to Table IX.1).

(v)　　When in equilibrium the major fraction of a metallic element is present as a free ion, variations in the temperature will have a marked effect on the <u>atomic</u> metal content through the strong dependence of the ionization constant on temperature (see Eq. II.59 with $[e^-] = [A^+] \simeq [A]_t$, and Eq. II.60b). Since the degree of ionization as well as the Boltzmann factor increase with temperature, the combined effect of a variation in temperature on the <u>atomic emission</u> will be reduced.

(vi)　　In choosing a fully atomized standard element as a reference in atomic concentration measurements one should take into consideration the possible occurrence of ionization (see Sect. V.6).

(vii)　　When absolute atomic concentrations or a-parameters are to be derived from measurements of the curve-of-growth (in emission or absorption), the concave curvature in the low-concentration range due to ionization should be accounted for or eliminated (see Sect. V.2d).

(viii)　　When the temperature and thus the ionization constant are known, the total absolute concentration of a partially ionized element may be determined by measuring its degree of ionization in equilibrium. The reliability of this method is, however, affected by complications of various kinds (see Sect. V.4a).

(ix)　　Free electrons produced by metal ionization are not likely to play a significant role in the collisional (de-)excitation of metal species (see Sect. VI.2a-3). Anyway, deviations of the electron concentration from Saha equilibrium would not affect the excited-level populations even if the electrons were the dominant collision partners (see Sect. VI.4c-2 under i). This holds, provided the electrons are not 'hot' (see Sect. IV.7). A deviation from Saha equilibrium may, however, cause a deviation from Boltzmann equilibrium for the high excitation levels which are

DISCUSSION AND CONCLUSIONS   Sect. IX.6c

populated predominantly by recombination of an ion and a free electron (see process 12 in Table II.1 and the discussions in Sects VI.2a-3 and IX.6b-1). Excess ionization would then lead to suprathermal excitation.

(x) When an atomic excitation level is saturated by laser irradiance, the resulting considerable enhancement in the collisional ionization rate may affect the concentration of neutral atoms in the flame (see process 45 in Table II.1 on page 108 and Sect. IX.2a). This opto-galvanic effect should be accounted for when saturated-laser fluorescence is utilized for local probing of the metal concentration.

(xi) Stark broadening of spectral lines is not thought to be detectable in flames, even at comparatively high ion or electron concentrations (see Sect. VII.2a).

(xii) The (partial) dissociation and ionization equilibria are interwoven (see Sect. IX.3c under Cases B and C). Consequently, the atomic concentration of a metallic element forming ions as well as compounds may vary in a complicated way with height of observation when relaxation in the radical equilibria (see Sect. IV.5) occurs. This may hamper the determination of the dominant formation reaction of metal compounds (see Sect. VIII.4). Conversely, relaxation in the radical equilibria, affecting the metal compound formation, complicates the measurements of ionization relaxation rates (see Sect. IX.5a).

(xiii) Special interferences are found when halogen vapour is introduced into a flame with alkali vapour (see Sect. IX.2a). The halogen binds the alkali as halide through reaction VIII.32, depresses the free electron concentration by the formation of negative halogen ions through processes (iii) or (ix) in Table IX.1, and catalyzes the ionization of the alkali by radical recombination according to overall process (xvi) in Sect. IX.2a.

(xiv) The diffusion of the metal atoms is closely linked with the ambipolar diffusion of the metal ions when partial ionization occurs. This hampers the measurement of the atomic diffusion constant (see Sect. X.4b).

6c-2   *CONSEQUENCES IN ANALYTICAL FLAME SPECTROSCOPY*

A loss in spectrochemical sensitivity and thus a deterioration of the detection limit must obviously result from partial ionization when the atomic line or molecular band intensities are measured. The fractional loss is the same for the atoms as for the molecules since a constant ratio between their concentrations is usually maintained through the law of mass-action. In hot flames the ionization of, e.g., the alkaline-earth elements may be so strong that their ionic emission lines can be used for analysis. In general, the optimum choice of flame temperature and analysis line depends on the extent of ionization.

The lower the absolute metal concentration, the more serious the fractional loss of neutral species is expected to be. However, the possible presence of natural flame electrons or of electrons produced by the ionization of metal atoms due to impurities or concomitants in the sample may buffer this loss owing to the repressing effect of these free electrons on the degree of ionization (see Sect. 3b-2). Without such a buffering effect, micro-analysis of sodium in the part-per-billion range would be inconceivable when the ionization of sodium corresponds to Saha equilibrium. Ionization relaxation, as discussed in Sect. 3c, may also limit the extent of this atomic loss low in the flame.

Partial ionization also leads to the well-known concave curvature of the analytical curve for the atomic lines as well as for the molecular bands. This curvature results from the fact that the fractional loss of atoms or molecules generally increases with decreasing metal solution concentration (see Sect. 3b-1). On the other hand, ionic line intensities show a square-root dependence on the concentration in the range of high concentrations where $\beta_i \ll 1$; when $\beta_i \simeq 1$ for low solution concentrations, the ionic analytical curves should be nearly linear. The strong curvature at the atomic lines in the low concentration range would make hazardous the extrapolation of the analytical curve to smaller concentration values and calculations of detection limits. Fortunately, the aforementioned buffering action of natural flame ions or concomitant metal ions often straightens the analytical curve in the range near the detection limit (see Sect. 3b-1).

Mutual interferences caused by a reduction in the degree of ionization occur when two ionizable elements are present together in the sample (see Sect. 3b-2). At present this kind of interference is the best understood one and is easily recognized in practice. All atomic lines and molecular bands of the analyte are enhanced to the same relative extent when a second ionizable element is added, if we disregard self-absorption of the resonance emission lines. A characteristic feature of this kind of interference is the saturation observed when the concentration of the interfering element is increased to higher and higher values. This saturation is attained when β_i is virtually suppressed to zero for the analyte. This element is then present only as neutral species and cannot be further disturbed by a shift in the ionization equilibrium. The relative extent of ionization interference, for fixed concentration ratio of analyte and concomitant, generally increases with decreasing concentration of the analyte.

Figure IX.15 shows the effect of increasing concentrations of K on the absorbance of some atomic and ionic lines of alkaline-earth metals. The strong ionization effects observed are explained by the high temperature of the $C_2H_2-N_2O$ flame employed. As expected, a saturation sets in for the enhancement of the atomic lines just at that concentration of K where the ionic line absorbance practically

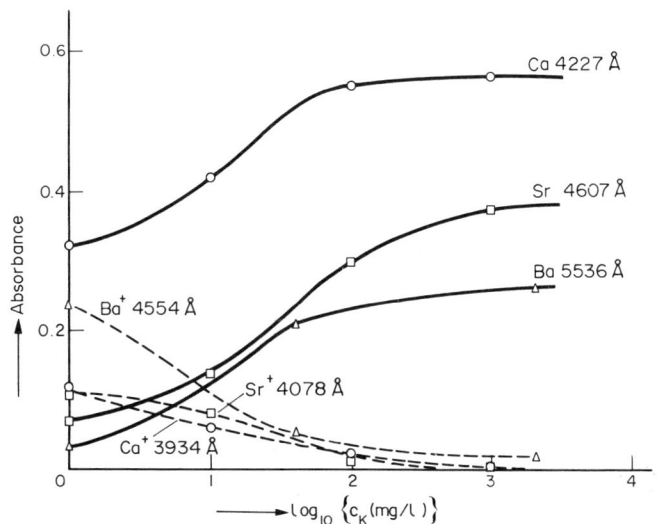

Fig. IX.15 Effect of increasing amounts of K on the absorbance of atomic and ionic spectral lines of Ca (5 mg/l), Sr (5.5 mg/l), and Ba (30 mg/l) in an C_2H_2-N_2O flame. (From Amos and Willis 1966.)

drops to zero. Similar observations have been made on the atomic and ionic emission lines of Eu in the same flame when KCl was added.

The stimulating influence of chlorine on the degree of metal ionization, owing to the removal of free electrons by Cl^- formation, explains the aggravation of the above disturbing ionization effects when halogen compounds are present in the solution in large concentrations (see Sect. 3b-1). Since the concentration of Cl^- ions depends on the concentration of free H radicals (see process ix in Table IX.1), any deviation from chemical equilibrium of the flame radicals thus affects the extent to which chlorine interferes with the metal ionization. Chlorine may also act as a catalyst in a series of reactions that effectively lead to the ionization of a Na atom in hydrogen flames. A discussion of this effect is given in Sect. 2a.

The optimum height of observation of an analysis line may depend on the variation of the degree of ionization with height due to a changing temperature and/or to ionization relaxation. Since this variation depends on the total amount of analyte introduced, the optimum height also depends on the analyte concentration in the solution.

For experimental methods of testing the occurrence of partial ionization of metal vapours in analytically useful flames see Alkemade (1970a).

CHAPTER X

Diffusion

1. INTRODUCTION

Diffusion processes tend to smooth local inhomogeneities in the metal vapour concentration in a flame. This smoothing effect may be a disadvantage in analytical flame spectroscopy because it may decrease the local concentration maximum and therefore the detectability of the metal. It may be an advantage, e.g., in measurements of ionization phenomena, where otherwise inhomogeneous clouds of atoms around vapourizing salt particles would make any realistic calculation impossible. For these and similar reasons and also because the subject is of interest in itself, it is important to know the diffusion coefficient of metal atoms and compounds in flames of various compositions and temperatures.

A great deal is known, theoretically and experimentally, about transport phenomena in gases, of which diffusion (= particle transport) forms a part. The Chapman-Enskog theory (see Chapman and Cowling 1939, and Hirschfelder, Curtiss and Bird 1954) yields a detailed description of these phenomena, from which transport coefficients can be calculated provided the interaction potential between the colliding particles is known. Generally, the type of potential function is chosen as a result of experience; the parameters involved are found by fitting the experimental data, especially the temperature-dependence of the coefficients, to the theoretical curves. This curve-fitting procedure is inevitable in bulk phenomena such as diffusion, since the results of single collisions are buried in weighted averages over the collision parameter and the relative velocity of the colliding particles.

In order to measure atomic and molecular diffusion coefficients the classical technique is to use a long tube divided in two parts by a stopcock. The two parts are filled with different gases. After opening the stopcock one measures the concentration of the gases at different times (of the order of minutes) and places

Chapt. X   DIFFUSION

Table X.1
Measured and Calculated Diffusion Coefficients (in $cm^2\ s^{-1}$) of Metal
Atoms and Compounds in Flames (at Elevated Temperatures) and in Noble
Gases (around Room Temperature), all at a Pressure of 1 atmosphere

Flame/gas	Temperature (K)	Species	D_{meas}	D_{calc}	Reference
$H_2-O_2-N_2$	2520	LiOH	11.8		1
$H_2-O_2-N_2$	1920	LiOH	6.8		1
$C_2H_2-O_2-N_2$	2440	LiOH	8.0		2
$C_2H_2-N_2O$	2860	Li	13.5	14.6	3
$H_2-O_2-N_2$	2520	^{23}Na	11.4	11.8	4
$H_2-O_2-N_2$	1920	^{23}Na	6.9	7.3	4
$C_2H_2-O_2-N_2$	2440	^{23}Na	9.9	9.1	2
$C_2H_2-O_2-N_2$	2040	^{23}Na	7.2	6.7	2
Ne	300	^{20}Na	0.35		5
Ne	428	^{23}Na	0.50		6
$H_2-O_2-N_2$	2520	K	8.1		1
$H_2-O_2-N_2$	1920	K	4.5		1
$C_2H_2-O_2-N_2$	2440	K	5.5		2
$H_2-O_2-N_2$	2520	Rb	6.7		1
$H_2-O_2-N_2$	1920	Rb	4.2		1
$C_2H_2-O_2-N_2$	2440	Rb	5.3		2
$C_2H_2-O_2-N_2$	2440	Rb		5.5	1
Ne	320	Rb	0.31		7
Ar	320	Rb	0.24		7
$H_2-O_2-N_2$	2520	Cs	5.6		1
$H_2-O_2-N_2$	1920	Cs	3.7		1
$C_2H_2-O_2-N_2$	2440	Cs/Cs$^+$	6.9		2
$C_2H_2-O_2-N_2$	2440	Cs	4.5		8
$C_2H_2-O_2-N_2$	2440	Cs$^+$	9.2 (ambipolar)		8
$H_2-O_2-N_2$	2520	Ca(OH?)	5.2		9
$H_2-O_2-N_2$	1920	Ca(OH?)	3.1		9
$C_2H_2-O_2-N_2$	2440	Ca(OH?)	3.5		2
$H_2-O_2-N_2$	2520	Sr(OH?)	4.75		9
$H_2-O_2-N_2$	1920	Sr(OH?)	2.7		9
$C_2H_2-O_2-N_2$	2440	Sr(OH?)	5.0		2
$H_2-O_2-N_2$	2520	Ba(OH?)	4.9		9
$H_2-O_2-N_2$	1920	Ba(OH?)	2.6		9
$C_2H_2-O_2-N_2$	2440	Ba(OH?)	2.5		2
$H_2-O_2-N_2$	2520	Cu	7.5		9
$H_2-O_2-N_2$	1920	Cu	4.5		9
$H_2-O_2-N_2$	2520	Tl	4.6		10
$H_2-O_2-N_2$	1900	Tl	2.8		10
$C_2H_2-N_2O$	2860	Pb	8.1	6.7	3

REFERENCES
1. Ashton and Hayhurst (1972);
2. Snelleman (1965);
3. L'vov *et al.* (1976);
4. Ashton and Hayhurst (1970a);
5. Coolen and Hagedoorn (1975), Coolen (1976);
6. Anderson and Ramsey (1963);
7. Franzen (1959);
8. Snelleman, unpublished work;
9. Ashton and Hayhurst (1976);
10. Hayhurst and Springett (1978).

in the tube. From these data the *diffusion coefficient* D, which is the mass transport through unit area per second, divided by the density-gradient perpendicular to that area, can be calculated.

The measurement and calculation of diffusion coefficients of metal vapours in gases at temperatures between 2000 and 3000 K is complicated by the following factors. Firstly, the pressure of metal vapours below 1000 K is mostly too low for the normal type of measurements described above. Only a few measurements with special techniques have been performed at temperatures of 300 and 400 K (see Table X.1). Consequently, interaction potentials of metal atoms are not known from low-temperature transport experiments. Therefore, measured diffusion coefficients of metal vapours at flame temperatures can only be compared with values calculated with the use of other molecular data such as the critical temperature of the metals or data from beam experiments involving metal atoms. The values obtained are often rather inaccurate. Secondly, the measurements of diffusion coefficients in flames are made in a stationary system of flowing gas in which the residence time of the vapour is only of the order of milliseconds. A different technique (called the point-source or line-source technique) involving the direct measurement of the spatial spread of the metal atoms diffusing in the flowing gas has to be used. Here the accuracy of the results is of the order of 5 - 10%, which is worse than that obtained in static systems. Within these limitations theoretical estimates and experimental results will be compared below.

The effect of thermal diffusion which describes diffusion transport under the influence of a temperature-gradient only, is of second order and, certainly above the combustion zone, does not play a role in flames (see Fristrom and Westenberg 1965).

2. THEORETICAL ESTIMATE OF DIFFUSION COEFFICIENTS

The expression for the binary diffusion coefficient D, in $cm^2 s^{-1}$, of a trace species 1 in a bulk gas of species 2 is in first approximation (correct to a few percent) of the Chapman-Enskog theory (see Hirschfelder, Curtiss and Bird 1954)

$$D = \tfrac{3}{16} kT/n\mu\Omega_{12}^{(1,1)}. \quad (X.1)$$

Here n is the total number of particles per cm^3, μ is the reduced mass and $\Omega_{12}^{(1,1)}$ is an integral expression for the product of the mean relative velocity of the particles and the collision cross section for diffusion. The Ω-integral contains the interaction potential pertaining to collisions between the two types of particles. When the particles are assumed to be rigid spheres, we have simply for the Ω-integral

Chapt. X   DIFFUSION

$$\Omega_{r.s.} = (\pi kT/2\mu)^{\frac{1}{2}} R_{12}^2, \qquad (X.2)$$

where R_{12} is the sum of the radii of the particles. When more realistic interaction potentials are used, one often employs 'reduced' Ω-integrals Ω^* (with dimension one) defined by

$$\Omega \equiv \Omega^* \Omega_{r.s.} . \qquad (X.3)$$

R_{12} is then taken as the particle separation at the (finite) zeropoint of the interaction potential. Generally, the modified Buckingham (exp;6) potential and the Lennard-Jones (12;6) potential are used (see Fig. X.1). The (exp;6) potential is given by

$$V_B = \frac{\varepsilon}{1-6/s} \cdot \left\{ \frac{6}{s} \exp\left[s\left(1-\frac{r}{r_m}\right)\right] - \left(\frac{r_m}{r}\right)^6 \right\}, \qquad (X.4)$$

where ε is the depth of the well in the potential, $r_m = 2^{\frac{1}{6}} R$ is the distance of the centres of the particles at which $V_B = -\varepsilon$ occurs, and s is a steepness-parameter

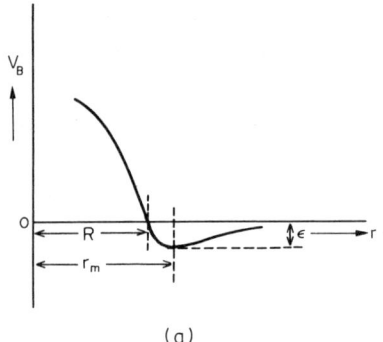

(a)

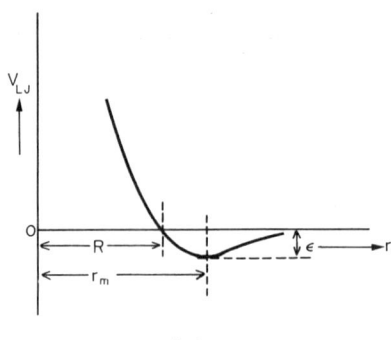

(b)

Fig. X.1 The shape of (a) the Buckingham (exp;6) potential, and (b) the Lennard-Jones (12;6) potential. The minimum $V = -\varepsilon$ of the potential is at $r = r_m$; the finite zeropoint $V = 0$ is at $r = R$.

THEORETICAL ESTIMATE OF DIFFUSION COEFFICIENTS   Sect. X.2

mostly taken to be 12. The Lennard-Jones potential is given by

$$V_{LJ} = \varepsilon\left\{\left(\frac{r_m}{r}\right)^{12} - 2\left(\frac{r_m}{r}\right)^6\right\}. \tag{X.5}$$

Values of Ω^* for these potentials are tabulated as a function of the reduced temperature $T^* \equiv kT/\varepsilon$ (see Hirschfelder, Curtiss and Bird 1954). The results for D are in good agreement with the measurements on binary systems consisting of noble gases and simple gases such as N_2 and H_2O, although the theory assumes spherical symmetry of the particles.

The remaining problem in the description of the diffusion of metal vapours in flames is to find values for R_{12} and ε_{12} relating to the collisions between the metal atoms (1) and the flame species (2). By using the combination rules

$$R_{12} = \frac{1}{2}(R_{11} + R_{22}) \tag{X.6}$$

and

$$\varepsilon_{12} = (\varepsilon_{11}\varepsilon_{22})^{\frac{1}{2}}, \tag{X.7}$$

the problem reduces to finding values for R_{11} and ε_{11} for like species, since values for R_{22} and ε_{22} for, e.g., N_2 and H_2O are available from viscosity measurements at lower temperatures. The quantities R_{11} and ε_{11} for metal atoms can be estimated, e.g., from the properties of the metal at the critical point (see Hirschfelder, Curtiss and Bird 1954) and from the dissociation energy of the metal molecule. Values for R_{11} and ε_{11} obtained in this way are indicated in Table X.2 by 'calc'. The corresponding values of D, calculated with the aid of Eqs. X.1,4, 5,6 and 7, are given in Table X.1 as D_{calc}.

Table X.2

Values of Particle Separation R_{11} and Depth of Potential Well ε_{11} as Estimated from Molecular Data (calc), Curve-fitting (bf) or Beam Experiments

Potential	Species	R_{11} (Å)	ε_{11}/k (K)	Method	Ref.
L.-J. (12;6)	LiOH-LiOH	2.2	740	bf	1
L.-J. (12;6)	Na-Na	2.4	500	calc	2
L.-J. (12;6)	Na-Na	2.1	750	bf	2
L.-J. (12;6)	Na-Na	2.4	3000	calc	3
Buck. (exp;6)	Na-Na	3.0	1500	calc	3
L.-J. (12;6)	K-K	2.8	840	bf	1
L.-J. (12;6)	Rb-Rb	3.8	58	bf	1
L.-J. (12;6)	Na-Na	3	2000	beam	4

REFERENCES

1. Ashton and Hayhurst (1972);   3. Snelleman (1965);
2. Ashton and Hayhurst (1970a); 4. Croucher and Clark (1969).

Chapt. X   DIFFUSION

Conversely, one may deduce values for the potential parameters R_{12} and ε_{12} from the measured temperature-dependence of D. Using a specified potential function one selects those values of R_{12} and ε_{12} that minimize the difference between the relation of D and the temperature T, as calculated with various values of R_{12} and ε_{12}, and the corresponding measured relation. The 'best-fit' values of R_{11} and ε_{11} derived from these selected values of R_{12} and ε_{12} with the aid of Eqs X.6 and 7 are indicated in Table X.2 by 'bf'. A discussion of the measurements of D as a function of T is given in Sect. X.4a.

3. MODEL CALCULATIONS OF ATOMIC DISTRIBUTIONS

3a   DIFFUSION FROM A POINT SOURCE IN A LAMINAR FLAME

In the point-source model (see, e.g., Wilson 1912, Ginsel 1933, Vendrik 1949, Walker and Westenberg 1958, Snelleman 1965, and Ashton and Hayhurst 1970) the metal atoms are assumed to be injected at a constant rate Q (s^{-1}) from a point source into a uniform laminar gas flow of velocity v, infinite cross section and constant temperature. When the direction of the gas flow is taken as the z axis and the point source is taken at $r \equiv (x^2+y^2+z^2)^{\frac{1}{2}} = 0$, the number density, $n_a(r,z)$, of metal atoms as a function of z and r follows from (cf. Eq. IX.23)

$$D \text{ div grad } n_a - v\frac{\partial n_a}{\partial z} = 0 \; , \qquad (X.8)$$

where D is the diffusion coefficient of the metal atoms in the gas. This equation holds when no other processes (such as ionization and compound formation) occur that may affect n_a. When it is assumed that $n_a(r,z) = 0$ at infinity, the solution of Eq. X.8 is

$$n_a(r,z) = (Q/4\pi Dr) \exp[v(z-r)/2D] \; . \qquad (X.9)$$

In emission or absorption measurements the quantity measured relates to the integral of the concentration along the line of viewing (y axis). Accordingly Eq. X.9 must be integrated over the y direction to yield the 'beam-integrated concentration' $N_a(x,z)$. On condition that $x/z < 1/5$ the integral is approximated within 2% by

$$N_a(x,z) \simeq Q/(4\pi Dvz)^{\frac{1}{2}} \exp[-vx^2/4Dz] \; . \qquad (X.10)$$

Since Eq. X.10 is also a solution of

$$D\frac{\partial^2 N_a}{\partial x^2} - v\frac{\partial N_a}{\partial z} = 0 \; , \qquad (X.11)$$

the approximation means, physically, that at some distance ($z > 5x$) above the source the diffusion transport in the z direction is negligible as compared to the convective transport in the z direction; the latter is represented by the second term in the left-hand sides of Eqs X.8 and 11.

Equation X.10 can be used in various ways to determine D. On the z axis, where $x=0$, Eq. X.10 reduces to

$$N_a(0,z) = Q/(4\pi Dvz)^{\frac{1}{2}} \ . \tag{X.12}$$

When no self-absorption occurs, the measured radiation intensity, $I(x,z)$, of a spectral line of the atom is proportional to N_a; thus we may write: $I(x,z) = \gamma N_a(x,z)$ where γ is a conversion factor. The source strength can be found by integration of $I(x,z)$ over x, yielding an integrated intensity I_0 at any z

$$I_0 = \gamma \int_{-\infty}^{\infty} N_a(x,z)\,dx = \gamma Q/v \ . \tag{X.13}$$

From Eqs X.12 and 13 we then obtain

$$I(0,z) = I_0 (v/4\pi Dz)^{\frac{1}{2}} \ . \tag{X.14}$$

The measured value of I/I_0 yields (in the absence of self-absorption) a value for D, provided the rise-velocity of the gas is known (see Sect. III.11). This method (I) was used by Ashton and Hayhurst (1970).

According to Eq. X.10 the horizontal intensity profile for fixed $z = z_0$ is given by

$$I(x,z_0) \propto \exp[-vx^2/4Dz_0] \ . \tag{X.15}$$

A plot of $\ln I(x,z_0)$ versus x^2 yields straight lines the slope of which is determined by D,v and z_0; this is the method (II) used by Vendrik (1949) in the determination of D for Li in a flame for known v and z_0. From the intersection of lines for different values of z_0 the location of the point source, that is, the point where $z=0$, can be determined.

Curves for <u>fixed</u> values of $N_a(x,z)$, that is, iso-intensity curves, can also be used to determine D (see Fig. X.2). Differentiation of such a curve, which is described by Eq. X.10 for a fixed value of $N_a(x,z)$,

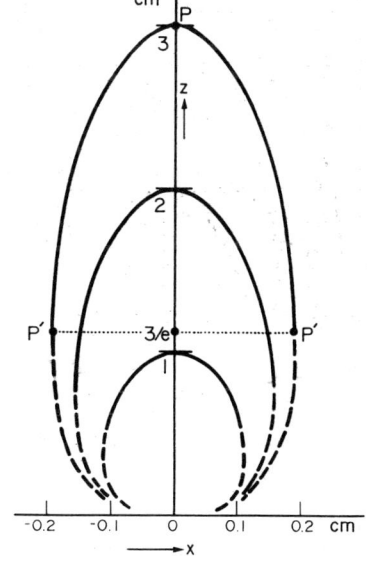

Fig. X.2 Iso-intensity lines observed when metal vapour is diffusing from a point source in the origin into a laminar gas flow moving in the positive z direction. The z coordinates of P and P' are related by: $z_{P'} = z_P/e$.

Chapt. X · DIFFUSION

with respect to z yields the relation between the coordinates x_m and z of the extreme points P' occurring in this figure

$$vx_m^2 = 2Dz \qquad (X.16)$$

Inserting Eq. X.16 into Eq. X.10 we find with $I = \gamma N_a(x_m, z)$ and $I_0 = \gamma Q/v$ (see Eq. X.13)

$$I(x_m, z) = I_0 (v/4\pi Dze)^{\frac{1}{2}} . \qquad (X.17)$$

Comparing Eqs X.14 and 17 we find that the intensity in point $z = z_0$ on the z axis (point P in Fig. X.2) equals the intensity in the horizontal extrema (outside the z axis) at $z = z_0/e$ (points P' in the same figure).

This relation permits an easy determination of D: one measures $I(0,z)$ for an arbitrary value of z and then looks at the height z/e for that value, x_m, of x for which $I(x_m, z/e) = I(0, z)$. Inserting the values of z and x_m, together with the rise-velocity, into Eq. X.16 one finds D. The advantage of this method is that the relation between $I(x, z)$ and $N_a(x, z)$ need only be single-valued, i.e. self-absorption does not interfere with the measurements. This method (III) was used by Snelleman (1965).

3b   DIFFUSION FROM A LINE SOURCE IN A LAMINAR FLAME

While investigating atom distributions in flames on slot burners (see Sect. III.2b-4) L'vov et al. (1976) used a diffusion model in which the source is extended in one horizontal direction (the y axis). The derivation of the equations for this linear source is analogous to that discussed in Sect. X.3a. For the number density integrated over the y direction at $x = 0$, which is proportional to the integral absorption $A_t^{(y)}$ in the low-density limit, we find in essentially the same approximation as before (viz. $vz/2D > 10$)

$$A_t^{(y)} \propto \int_0^l n_a(0, y, z) dy \equiv N_a(0, z) = 0.28 \, Q/(Dvz)^{\frac{1}{2}} , \qquad (X.18)$$

where l is the length of the flame. When the integral absorption is measured in the x direction (as is done in a measurement in which the light beam is perpendicular to the slot), $A_t^{(x)}$ is independent of z and y (within the boundaries of the flame), as we have

$$A_t^{(x)} \propto \int_{-\infty}^{\infty} n_a(x, y, z) dx \equiv N_a'(y, z) = Q/lv . \qquad (X.19)$$

Combining Eqs X.18 and 19 we obtain

$$A_t^{(y)}/A_t^{(x)} = 0.28 \, l(v/Dz)^{\frac{1}{2}} , \qquad (X.20)$$

from which D can be found (Method IV).

MODEL CALCULATIONS OF ATOMIC DISTRIBUTIONS   Sect. X.3c

3c   THE VALIDITY OF THE MODELS IN ACTUAL FLAME EXPERIMENTS

The assumptions that are used have been tested by Snelleman (1965) and Ashton and Hayhurst (1970). They are discussed here briefly.

1.   Laminarity of the cylindrical flame was assumed because the Reynolds Number[†] was calculated to be 500 for a nitrogen flow at 2500 K with a velocity equal to the rise-velocity of 10^3 cm s^{-1} in a circular tube with a radius equal to the flame radius of 1 cm. Experimentally this assumption was confirmed by photographing the luminous trail of soot particles that were formed in the central hole of the Méker burner used. The central exit port was fed separately from the rest of the flame with a very fuel-rich mixture. The soot particles, too heavy to diffuse noticeably, followed a straight flow line (see Fig. X.3).

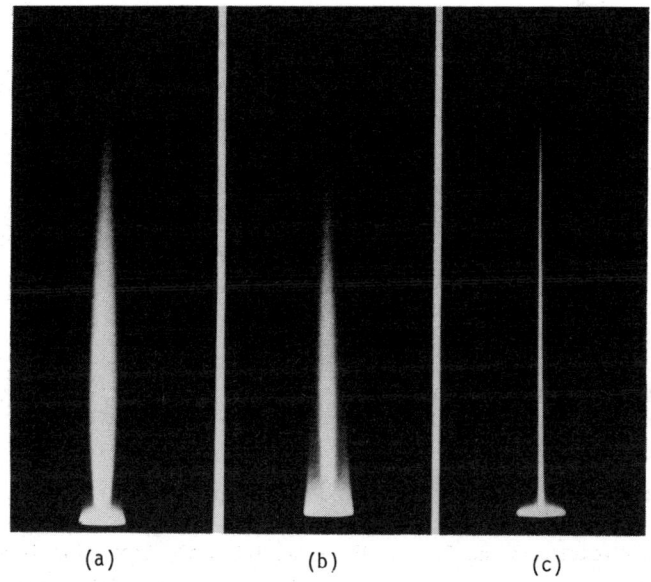

Fig. X.3 Pictures of a flame on a circular Méker burner. The central hole serves as a source for Na atoms (a) which diffuse rather quickly, BaO molecules (b) which diffuse considerably slower, and soot particles (c) which cannot be seen to diffuse at all and thereby demonstrate the laminarity of the gas flow in the flame.   (From Snelleman 1965.)

[†] The *Reynolds Number* Re is defined by: $Re \equiv vd\rho/\eta$ (v = velocity, d = diameter, ρ = density, η = viscosity) and has dimension one.

Chapt. X   DIFFUSION

2.   Computer calculations with reasonable values of z, D and v indicated that the cross section of the flames used was large enough to be considered infinite. In addition, calculations showed that, although the actual source has a diameter approximately equal to that of the exit hole ($\approx$ 1 mm), above about 5 mm the concentration pattern is equally well described by a virtual point source. The location of this virtual point source was found to be approximately at the burner top, i.e. a few mm below the tip of the combustion cone.

3.   The variation of flame temperature with height was measured and the emission measurements were corrected for this variation. Such measurements yielded the same results as measurements in absorption, in which the variation of temperature hardly plays a role. Horizontally, the flame was homogeneous in temperature over the width used. The direct influence of a temperature-variation on D is too weak ($D \propto T^2$, see Sect. X.4) to be noticeable in the cases considered.

4.   The constancy of I_0 corrected for T-variations (see Eq. X.13) as a function of height z above $z = 5$ mm was checked experimentally.

5.   The constancy of the rise-velocity v over the relevant height interval was checked experimentally; it was also ascertained that Eqs X.14 and 15 did hold above $z \simeq 5$ mm.

6.   In the case of a line source some corrections had to be introduced to allow for the horizontal expansion of the flame gases after leaving the slot burner (see Sect. X.4b).

4.   MEASUREMENTS OF DIFFUSION COEFFICIENTS IN FLAMES

4a   MEASUREMENTS IN SIMPLE SITUATIONS

The results of earlier flame measurements spread widely; e.g., for Na D ranged from 3.2 to 35 cm^2 s^{-1} at about 2100 K, probably because of incomplete atomization and/or inhomogeneous flames. In other cases ionization (with Cs) or compound formation (with Li) led to erroneous results. However, the later, extensive investigations by Snelleman (1965) and Ashton and Hayhurst (1970, 1970a, 1971, 1972, 1976) are in good agreement with each other, e.g., regarding the value for Na in N_2-diluted flames at 1 atm pressure and about 2400 K.

The method used in these later measurements involved a circular Méker burner with a separate inlet for the central hole (of about 1 mm diameter). This hole was fed with the same gas mixture as was fed through the surrounding holes, but with the addition of a sprayed solution. The visible spread of the metal vapour from this 'point' source differed for various elements, but was roughly 1 cm at 3 - 5 cm height for rise-velocities of about 600 cm s^{-1}, indicating that diffusion is not a priori negligible (see Fig. X.3). The flame was imaged on the slit of a spectrograph. A limited slit height of about 1 mm was used. With the aid of sledges the position of the flame, and thus the part of the flame imaged on the slit, was varied. The results of these and other measurements are given in Table X.1.

The temperature-dependence of the diffusion coefficient of Na was measured by Snelleman (1965) and Ashton and Hayhurst (1970, 1970a). In the first investigation the temperature of an $C_2H_2-O_2-N_2$ flame was varied from 2040 to 2440 K by varying the relative amount of nitrogen. The result was: $\beta = 1.75$ in the relation $D \propto T^\beta$. In the second the temperature of a $H_2-O_2-N_2$ flame was varied from 1920 to 2520 K by varying the relative amount of nitrogen and (to a smaller extent) the amount of hydrogen. The temperature-dependence was found to be: $\beta = 2.2$. The difference in these values of β may be partly caused by the varying amount of H_2 in the flames used in the second investigation, according to Ashton and Hayhurst (1970a).

When β has been determined, curve-fitting with the use of a chosen potential function yielded estimates of R_{12} and ε_{12}. Table X.2 gives some values for R_{11} and ε_{11} as deduced from these experiments with the use of the combination rules X.6 and 7. Also included in the table is an estimate of R_{11} and ε_{11} derived from the Na-N_2 beam experiments by Croucher and Clark (1969). Since these authors only determined the van der Waals constants $4\varepsilon_{12}R_{12}^6$, no unique values of ε and R can be deduced from their results.

Ashton and Hayhurst (1972) also measured the temperature-dependence of D for the other alkali atom. They found a value of $\beta = 2.0 \pm 0.2$. Curve-fitting yielded values for R_{11} and ε_{11} in the cases of Li+LiOH, Na and K, but the method failed for Rb and Cs, indicating very shallow wells in the potential. However, the resulting value of D is not very sensitive to the depth of the potential well and the authors obtained agreement with experimental results for Rb using the values of the parameters given in Table X.2. One may therefore conclude that the calculation of diffusion coefficients with the use of estimates of R and ε yields reasonably reliable results, but that, on the other hand, the deduction of the parameters of the potential function from diffusion experiments can hardly yield reliable figures.

4b   MEASUREMENTS IN MORE COMPLEX SITUATIONS

Some measurements and calculations have been made in situations in which not all assumptions of the point-source model or the line-source model were valid.

Some elements show considerable compound formation (see Chapt. VIII). When the ratio, ϕ^{-1}, of the atom and compound concentrations is constant in the flame volume considered, one can set up two diffusion equations (see Ashton and Hayhurst

1970). These equations are coupled since the atom/compound equilibrium is a dynamic one, i.e. transitions in both directions of, e.g., the reaction: $M+H_2O \rightleftharpoons MOH+H$ occur frequently. Therefore measurements on the intensity distribution of an atomic spectral line only yield an *effective diffusion coefficient*,

$$D_{\text{eff}} = \{D(M) + \phi D(MOH)\}/(1 + \phi),$$

intermediate between the separate values for the atom and its compound. The above assumption of a constant concentration ratio holds in a uniform flame volume, since the rate constants of the (binary) reactions generally are large enough for the partial equilibrium to be maintained in any place (see Chapt. VIII). In the case of Li the large fraction of LiOH would indicate that the value found pertains to the hydroxide. However, the value of D can still contain a considerable error, since the fraction of Li atoms is dependent on the H concentration which may vary with height (see Sect. IV.5b). For rather heavy atoms such as Rb and Ba the difference in D for the atom and the compound is expected to be small because of the small relative difference in mass. For lighter atoms such as Na and K the difference is larger, but, in view of the small fraction of compound, the value of D measured pertains to the atoms.

Ionization can, of course, be suppressed by the addition of a large amount of an easily ionizable element (see Sect. IX.3b-2 and Table X.1). When the ionization is not suppressed, one should apply more complete kinetic equations based on Eq. IX.23. In these equations the rate constant for ionization and the varying atom/ion ratio are taken into account. Furthermore, the diffusion of ions is ambipolar.

Ambipolar diffusion of ions occurs in a macroscopically neutral gas, i.e. a gas in which the ion density n_i and the electron density n_e are equal. When an atomic vapour is introduced into a flame and this vapour is (partly) ionized, the electrons formed will move much faster than the ions. As a consequence the negatively and positively charged particles tend to separate and an internal electric field is set up, which decreases the diffusion of the fast particles (the electrons) and increases the diffusion of the slow particles (the ions). One can show (see Lawton and Weinberg 1969) that to a good approximation: $D_a \simeq 2D_i$, where D_i is the diffusion coefficient of the ions in the absence of an external electric field and D_a is the ambipolar diffusion coefficient[†].

[†] Measurements (unpublished) of the ambipolar diffusion coefficient D_a of Cs and of the diffusion coefficient D of neutral Cs atoms were made in the Fysisch Laboratorium of the University, Utrecht, with the use of a diffusion equation containing both effects and the ionization rate constant. The results were: $D = 4.5 \text{ cm}^2 \text{ s}^{-1}$ (± 10%) and $D_a/D = 2.1$ (± 15%), confirming the often assumed equality: $D_i (= \frac{1}{2}D_a) = D$.

THE IMPORTANCE OF DIFFUSION IN METAL VAPOUR EXPERIMENTS   Sect. X.5

When a slot burner is used, the unburnt gas mixture as well as the metal vapour originate from a linear source. After combustion the flame gases expand considerably in a horizontal direction, invalidating the assumption of one-dimensional flow in the z direction. The divergence of the gas flow should then be taken into account. L'vov *et al.* (1976) applied corrections for this effect and derived (effective) diffusion coefficients for Li and Pb in an C_2H_2-N_2O flame from their absorption measurements as described in Sect. X.3b (see Table X.1). The results found seem rather high compared to the other results mentioned.

5. THE IMPORTANCE OF DIFFUSION IN METAL VAPOUR EXPERIMENTS

The measurements of diffusion coefficients give a reasonable basis for estimating the effect of diffusion in specific situations. In the case of thick (diameter $\geqslant$ 1 cm), cylindrical flames on Méker burners into which the metal salt is introduced over the whole cross section at the bottom of the flame, the favourable effect of diffusion is that the metal vapour-streams, originating from the single holes, merge to form quite a homogeneous distribution. Still, the losses of metal vapour to the cold ambient air seem to remain small. When such a flame is surrounded by a mantle flame with the same properties, into which no metal salt is introduced, the effect of diffusion of metal vapour into this mantle flame is larger than that of diffusion from the unshielded flame into the cold ambient air (see Snelleman 1965).

In the case of flames on slot burners diffusion greatly influences the distribution of atoms in the flame (see Sect. III.12).

APPENDIX A.1

Conversion Factors of Incoherent Units

Energy Units

	joule	erg	eV	cal
joule	1	10^7	6.24146×10^{18}	2.3901×10^{-1}
erg	10^{-7}	1	6.24146×10^{11}	2.3901×10^{-8}
eV	1.60219×10^{-19}	1.60219×10^{-12}	1	3.8294×10^{-20}
cal	4.1840	4.1840×10^7	2.6114×10^{19}	1

Energy Equivalents

Quantity	Value	Unit
eV/h	2.41797	10^{14} Hz
eV/hc	8.0655	10^3 cm^{-1}
eV/k	1.1604	10^4 K
eV/molecule	$\begin{cases}23.061\\96.485\end{cases}$	$\begin{cases}10^0 \text{ kcal/mol}\\10^0 \text{ kJ/mol}\end{cases}$
kcal/mol	4.3363	10^{-2} eV/molecule
kJ/mol	1.0364	10^{-2} eV/molecule

Pressure Units

	Pa	atm	Torr
Pa	1	9.8692×10^{-6}	7.50059×10^{-3}
atm	$\equiv 1.01325 \times 10^5$	1	760
Torr	1.33322×10^2	1.316×10^{-3}	1

Pressure Equivalents

Quantity	Value	Unit
1 atm at 1000 K	7.33887	10^{18} atoms per cm^3
1 Torr at 1000 K	9.65640	10^{15} atoms per cm^3

APPENDIX A.2
Some Fundamental Constants[1]

Quantity	Symbol	Value	Units SI	Units e.s.CGS	Uncertainty[2] (ppm)
Speed of light in vacuum	c	2.99792458	10^8 m.s^{-1}	10^{10} cm.s^{-1}	0.004
Elementary charge	e	$\begin{cases}1.6021892\\4.803242\end{cases}$	10^{-19} C —	— 10^{-10} e.s.u.	3 3
Planck constant	h	6.626176	10^{-34} J.s	10^{-27} erg.s	5
Avogadro constant	N_A	6.022045	10^{23} mol^{-1}	10^{23} mol^{-1}	5
Atomic mass unit ($\equiv \frac{1}{12}$ mass of nuclide ^{12}C)	u	1.6605655	10^{-27} kg	10^{-24} g	5
Electron rest mass	m_e	9.109534	10^{-31} kg	10^{-28} g	5
Proton rest mass	m_p	1.6726485	10^{-27} kg	10^{-24} g	5
Neutron rest mass	m_n	1.6749543	10^{-27} kg	10^{-24} g	5
Bohr radius	a_o	5.2917706	10^{-11} m	10^{-9} cm	0.8
Molar volume of ideal gas at s.t.p.[3]	V_m	2.241383	10^{-2} m^3.mol^{-1}	10^4 cm^3.mol^{-1}	30
Molar gas constant	R	8.31441	10^0 J.mol^{-1}.K^{-1}	10^7 erg.mol^{-1}.K^{-1}	30
Boltzmann constant	k	1.380662	10^{-23} J.K^{-1}	10^{-16} erg.K^{-1}	30
First radiation constant ($\equiv 2hc^2$)[4]	c_1	1.191062	10^{-16} W.m^{-2}	10^{-5} erg.cm^2.s^{-1}	5
Second radiation constant ($\equiv hc/k$)	c_2	1.438786	10^{-2} m.K	10^0 cm.K	30
Third radiation constant ($\equiv 8\pi hc$)[5]	c_3	4.992563	10^{-24} J.m	10^{-15} erg.cm	5
Permittivity of vacuum	ε_0	$\begin{cases}8.85418782\\1\end{cases}$	10^{-12} F.m^{-1} —	— 10^0	0.008 —

NOTES

[1] All fundamental constants are based on the ^{12}C system and are taken from: E.R. Cohen and B.N. Taylor, J. Phys. Chem. Ref. Data **2**, 663 (1973).

[2] The uncertainties denote one standard deviation.

[3] At $T_0 = 273.15$ K and $p_0 = 101325$ Pa (= 1 atm).

[4] Note that also another definition: $c_1 = 2\pi hc^2$ occurs in the literature.

[5] Introduced in this book.

APPENDIX A.3

List of Symbols and Terms Used[*]

Symbol(s)	Term / Meaning	Section[†]
A, A_n	pre-exponential (or frequency) factor in expression for rate constant (of process n)	II.4b-2
A, A_{qp}, $A_{2\to1}$	(Einstein) transition probability for spontaneous emission (from state q to state p, or from state 2 to state 1)	II.5a-1
A, A^{abs} (1)	(decadic) (internal) absorbance ($\equiv -\log_{10}\tau$)	App. A.5
$A(\lambda_0)$, $A^{abs}(\lambda_0)$ (2)	peak absorbance ($\equiv A^{abs}$ at line centre)	
$A(\lambda_m-\lambda)$	normalized instrumental profile	III.4b
A_G	total absorption ($\equiv 2\pi A_t$) ("Gesamtabsorption"; see Mitchell and Zemansky 1961) (3)	II.5c-2
A_t, $A_t^{(\lambda)}$ (2)	integral absorption (of spectral line) $[\equiv \int\alpha(\lambda)d\lambda]$	II.5c-2
A_t'	net integral absorption under (near-) saturation conditions	II.5c-2
A_t^*	$\equiv A_t$ at intersection of asymptotes of logarithmic COG	II.5c-2
a	a-parameter, damping parameter or line-broadening parameter $[\equiv(\delta\nu_L/\delta\nu_D)\sqrt{\ln 2}]$ (4)	II.5c-2
B (1) (5)	radiance	App. A.5

[*] This table lists the most important or the most frequently used quantities. The symbols and terms chosen conform as far as possible to international recommendations (see in particular IUPAC 1970, 1972, 1976 and 1976a). Alternative symbols are sometimes given for use in a particular context or for avoidance of confusion.
Numbers between parentheses refer to the Notes below.

[†] Number of section or appendix is given where further information can be obtained.

APPENDIX A.3   LIST OF SYMBOLS AND TERMS USED

Symbol(s)		Term / Meaning	Section
B_λ, $B_\lambda(\lambda)$	(2)	spectral radiance (as a function of λ)	App. A.5
$B_{\lambda 0}$	(2)	peak spectral radiance [$\equiv B_\lambda(\lambda_0)$]	
B_λ^b, $B_\lambda^b(\lambda,T)$	(2)	black body (= full radiator) spectral radiance (at λ and T)	II.3b-5
B, B_X		rotational constant (of linear molecule X)	II.2b-2
B_{pq}, $B_{1\to 2}$		(Einstein) transition probability for absorption from state p (or 1) to q (or 2)	II.5a-1
B_{qp}, $B_{2\to 1}$		(Einstein) transition probability for induced emission ($q\to p$ or $2\to 1$)	II.5a-1
b		impact parameter (6)	II.4e
COG		curve-of-growth	II.5c-2
C		capacity	
C_p		heat capacity at constant p	
c		solution concentration (of metallic element)	
c		velocity of light (in vacuo or air)	App. A.2
c_p		specific heat capacity at constant p ($\equiv C_p/m$)	
c_p'		molar heat capacity at constant p ($\equiv C_p$ per mole)	II.3b-3
c_1, c_2, c_3		radiation constants (occurring in Planck's radiation law): $c_1 \equiv 2hc^2$; $c_2 \equiv hc/k$; $c_3 \equiv 8\pi hc$	II.3b-5, App.A.2
D		diffusion coefficient	X.1
D		duplication factor	II.5c-3
D_0, $D_0(X)$		dissociation energy of 1 molecule or 1 mole of species X ($= \Delta E_0^0 = \Delta H_0^0$)	II.3b-3
$D_0^\star$		$\equiv D_0/hc$	
dx		fluctuation in variable x ($dx \equiv x - \bar{x}$); infinitesimal quantity	

APPENDIX A.3   LIST OF SYMBOLS AND TERMS USED

Symbol(s)	Term / Meaning	Section
E	energy (general term); irradiance	App. A.5
E_A, $(E_A)_n$	activation energy (of process n)	II.4b-2
E_{aff}	electron affinity	II.3b-4
E_{Arrh}	Arrhenius' energy of activation	II.4b-2
E_{exc}	excitation energy of particle	
E_{ion}	ionization energy of particle	
E_j, $(E_j)_X$	internal energy of j-th energy level of particle (of species X)	
E_k	kinetic energy of particle	
E_p	potential energy or particle	
E_λ (2)	spectral irradiance	App. A.5
e	elementary charge	App. A.2
$\vec{F}$, (F)	(quantum number of) total angular momentum including nuclear spin	II.2a-1
F_b	flow rate of burnt gas at T_f and 1 atm (volume of gas per unit time)	
F_l	solution aspiration rate (volume of solution per unit time)	
F_u, F_X, F_{air}	flow rate of unburnt gas at T_{room} and 1 atm (of species X or air) (volume of gas per unit time)	
f	frequency (electric)	
f_{mod}	frequency of a.c. modulation	
f, f_{ij}, $f_{1 \to 2}$	oscillator strength for absorption (i or 1 = lower state; j or 2 = upper state); f_{ji} for emission is defined by: $g_i f_{ij} = g_j f_{ji}$	II.5a-2
$f(x)$	distribution function of variable x	II.3b
G	free enthalpy or Gibbs free energy ($= U + pV - TS$)	
$G(v)$	vibrational energy term	II.2b-2

APPENDIX A.3 LIST OF SYMBOLS AND TERMS USED

Symbol(s)		Term / Meaning	Section
g		relative velocity of 2 particles	
g_0		statistical weight (or weight factor) of ground level	II.3b
g_j, g^*		statistical weight of j-th energy level or of an excited level	II.3b
$(g_j)_X$		statistical weight of j-th level of species X	II.3b
H		heat content or enthalpy ($= U + pV$)	
H_{fo}		heat of formation from elements in standard state (1 atm, 25 °C)	
h		Planck constant ($= 2\pi \hbar$)	App. A.2
$\hbar$		$\equiv h/2\pi$	
h		(entrance-)slit height of monochromator	
h_{obs}		height of observation above combustion zone	IV.1
I	(5)	intensity (in a loose sense)	App. A.5
I	(5)	radiant intensity	App. A.5
I_λ	(2)(5)	spectral intensity (in a loose sense)	App. A.5
I_λ	(2)(5)	spectral radiant intensity	App. A.5
I, I_X		principal moment of inertia of di-atomic molecule (X)	II.2b-2
I_A, I_B, I_C		principal moments of inertia of poly-atomic molecule	II.2b-2
$\vec{I}$, (I)		(quantum number of) nuclear spin	II.2a-1
i		electric current	
i_b		photocurrent due to background	
i_d		dark current (of photodetector)	
i_p		photocurrent	
$\vec{J}$, (J)		(quantum number of) total angular momentum of atom or molecule (without nuclear spin)	II.2a-1/ 2b-1

APPENDIX A.3 LIST OF SYMBOLS AND TERMS USED

Symbol(s)	Term / Meaning	Section
j	imaginary unit	
K	conversion factor ($\equiv$ number density of element in flame divided by its concentration in the solution)	III.2h
K_d, $K_d(T)$	dissociation constant (at T) expressed in number density	II.3b
K_d^*, $K_d^*(T)$	dissociation constant (at T) expressed in partial pressure	II.3b-3
K_i, $K_i(T)$	ionization constant (at T) expressed in unit of number density	II.3b-4
K_n, $K_n(T)$	equilibrium constant of process n (reaction or ionization) (at T)	II.3b-3
k	Boltzmann constant ($= R/N_A$)	App. A.2
k_a	association rate constant	II.4a-2
k_d	dissociation rate constant	II.4a-2
k_i	ionization rate constant	II.4a-2
k_n	rate constant for process or reaction n	II.4a-1
k_{-n}	rate constant for reverse process or reaction n	II.4a-1
k_r	recombination rate constant	II.4a-2
$k(\lambda)$ (2)	(napierian) absorptivity $[\equiv \alpha(\lambda)/l$ for $l \to 0]$	II.5a-2
$k'(\lambda)$ (2)	absorptivity under (near-)saturation conditions	II.5a-2
L	optical conductance	III.4b
$\vec{L}$, (L)	(quantum number of) resultant orbital angular momentum for atom	II.2a-1
l	length; (in particular, depth of flame in direction of optical axis); absorption path length	

APPENDIX A.3　　　　　　　　　　　　　LIST OF SYMBOLS AND TERMS USED

Symbol(s)	Term / Meaning	Section
M_X, $M(X)$	molar mass of X ($M = 12$ g for ^{12}C)	
$(M_r)_X$, $M_r(X)$	relative atomic or molecular mass of X ($M_r = 12$ for ^{12}C)	
m	mass (in general)	
m, m_X	mass of atom or molecule (X)	
m_e	electron mass	App. A.2
N	number of particles	
N_A	Avogadro constant	App. A.2
$\vec{N}$, (N)	(quantum number of) total angular momentum of di-atomic molecule without electronic and nuclear spin	II.2b-1
n	refractive index	
n	(number) density of particles ($\equiv$ number of particles per unit flame volume); also called: concentration	
n_a, n_X, $[X]$	density of free atoms of <u>element</u> X	
n_e, $[e^-]$	density of free electrons	
n_i, n_{X^+}, $[X^+]$	density of free atomic ions of <u>element</u> X	
n_j, $n_j(X)$, $[X^*]$	density of <u>species</u> X in j-th level or in an excited level	
n_0, $n_0(X)$, $[X^0]$	density of <u>species</u> X in ground level	
n_m	density of molecules	
n_t, $n_t(X)$, $[X]_t$	total density of <u>element</u> X in different forms (atom + ion + molecule) or total density of <u>free atom</u> X in different levels	
n_X, $[X]$	density of <u>species</u> X (e.g., X = Na, Na*, Na$^+$ or NaOH)	
P	steric factor	II.4b-2
P, $P(x)$	population factor of excited level $\{\equiv [M^*]/[M^*]_e\}$ (at x)	II.4a-3
p	pressure	

APPENDIX A.3   LIST OF SYMBOLS AND TERMS USED

Symbol(s)	Term / Meaning	Section
p_t	total pressure of gas mixture	
p_X, $p(X)$	partial pressure of species X	II.3b-3
Q	electric charge	
Q	radiant energy	App. A.5
Q_X, $Q(X)$	internal partition function of species X (with zero energy corresponding to ground level)	II.3b-2
$q(\lambda)$	quantum efficiency of photocathode at wavelength λ	III.5b
R	ohmic resistance	
R	(molar) gas constant	App. A.2
$\vec{R}$, (R)	(quantum number of) angular momentum of di-atomic molecule due to nuclear rotation	II.2b-1
RMS	root-mean-square value ($\equiv \sqrt{\overline{x^2}}$)	
r	distance; radius	
S	entropy	
S	area	
$\vec{S}$, (S)	(quantum number of) resultant electron spin	II.2a-1
S_f	cross section of flame	
SNR	signal-to-noise ratio	III.14f-1
S_λ, $S_\lambda(\lambda)$ (2)	normalized spectral distribution function [$\int_0^\infty S_\lambda(\lambda)d\lambda \equiv 1$]	II.5a-1
s	(entrance-)slit width of monochromator	
s, s_X	symmetry number of molecule (X)	II.3b-2
$s(\lambda)$ (2)	responsivity of photodetector or photocathode at λ ($\equiv$ photocurrent divided by incident radiant flux)	

APPENDIX A.3 LIST OF SYMBOLS AND TERMS USED

Symbol(s)	Term / Meaning	Section
T	absolute temperature (in K) (12)	
T_f	flame temperature (in K)	III.10
T_r	radiance temperature (in K)	
t	Celsius temperature (in °C) (12)	
t	time	
t_{tv}, t	travel time ($\equiv$ time needed for flame gas element to travel from base of flame to the centre of observation volume); also called: rise time	IV.1
U	internal energy of thermodynamic system	
V	volume	
V	electric potential; tension (= potential difference)	
$V(r)$	interaction potential, or potential (function) (= potential energy of interaction as a function of distance between two particles) (8)	II.2b-1/4d
v	vibrational quantum number	II.2b-2
v, v_X	velocity of particle (X) (7)	
v_b	burning velocity (of flame front)	IV.1
v_n	rate (or "velocity") of process or reaction n	II.4a-1
v_{-n}	rate (or "velocity") of reverse process or reaction n	II.4a-1
v_r	(vertical) rise-velocity of flame gas	III.11
v_{rel}	relative velocity	
v_u	flow speed of unburnt gas mixture in exit port(s) of burner	III.2
$W_x(f)$	spectral noise power of fluctuating quantity $x(t)$ at frequency f	III.5c
x	spatial coordinate; general variable	
x_e	anharmonicity constant (of molecular vibration)	II.2b-2

932

APPENDIX A.3 LIST OF SYMBOLS AND TERMS USED

Symbol(s)		Term / Meaning	Section
Y, Y_q		(quantum) efficiency of fluorescence	II.5c-4
Y_p		power efficiency of fluorescence	II.5c-4
Y_t		total quantum efficiency of fluorescence [$\equiv Y_q$ when upper level is (in)directly populated by absorption of several lines]	
y		spatial coordinate; general variable	
z		spatial coordinate; general variable	
z		degree of saturation (of optical transition)	VI.2c-5
$\alpha, \alpha(\lambda)$	(2)	absorption factor or absorptance ($\equiv \Phi_A/\Phi_0$) (as a function of λ for monochromatic radiation beam)	App. A.5
β_a		fraction atomized ($\equiv n_a/n_t$)	V.1
β_d		degree of dissociation $\{\equiv [M]/([M] + [MX])\}$	II.3b-3
β_i		degree of ionization $\{\equiv [M^+]/([M] + [M^+])\}$	II.3b-4
β_s		fraction of desolvated aerosol	III.9a
β_v		fraction of volatilized aerosol	III.9c
ΔE		energy difference or discrepancy	
Δf		noise bandwidth (of measuring instrument)	III.5c
ΔG_T		change in free enthalpy at T	II.3b-3
$\Delta G(v+\tfrac{1}{2})$		$\equiv G(v+1) - G(v)$	VIII.2a
ΔH_T		heat of reaction (or dissociation) at constant p and T	II.3b-3
ΔU_T		energy of reaction (or dissociation) at constant V and T	II.3b-3
$\Delta \lambda$	(2)	wavelength difference	

APPENDIX A.3   LIST OF SYMBOLS AND TERMS USED

Symbol(s)		Term / Meaning	Section
$\Delta\lambda_{eff}$	(2)	effective width of (any) spectral profile	II.5b-1
$\Delta\lambda_M$	(2)	monochromator bandwidth ($\equiv$ effective width of instrumental profile)	III.4b
$\delta\lambda$	(2)	half-intensity width ($\equiv$ full width at half peak height of spectral profile)	II.5b-1
$\delta\lambda_{abs}$	(2)	half-intensity width of absorption line	
$\delta\lambda_C$	(2)(9)	collisional half-intensity width	II.5b-2
$\delta\lambda_D$	(2)	Doppler half-intensity width	II.5b-2
$\delta\lambda_L$	(2)	half-intensity width of Lorentzian line profile owing to the combined effects of collision and natural broadening	II.5b-3
$\delta\lambda_N$	(2)	natural half-intensity width	II.5b-2
$\delta\lambda_S$	(2)	collisional line shift	VII.2b
$\varepsilon(\lambda)$	(2)	emission factor at λ [$\equiv B_\lambda/B_\lambda^b(T)$ with T = temperature of thermal radiator]	II.5c-3
ε_a		efficiency of atomization	III.2h
ε_n		efficiency of nebulization	III.8a
ε, (ε_0)		permittivity (of vacuum)	(App. A.2)
ε_r		relative permittivity of medium	
ζ		number of moles of burnt gases per mole unburnt mixture	III.11
η		viscosity	
$\kappa(\lambda)$	(2)	atomic absorptivity or atomic absorption cross section [$\equiv k(\lambda)/n$]	II.5a-2 (App. A.5)
$\bar{\Lambda}$, (Λ)		(quantum number of) resultant orbital angular momentum of electrons along internuclear axis of di-atomic molecule	II.2b-1
λ		wavelength	

APPENDIX A.3   LIST OF SYMBOLS AND TERMS USED

Symbol(s)		Term / Meaning	Section
λ_0		wavelength of atomic line centre	
$\lambda_{X,Y}$		mean free path of X with respect to collisions with Y	II.4b-1
μ		reduced mass of two particles	II.3a
ν		optical frequency	
ν, (ν_e, ν_i)		collision frequency (of electron or ion)	II.4b-1
ν_c		frequency (in continuous spectrum)	
ν_d		frequency (in discrete spectrum)	
ν_0		frequency of atomic line centre	
ν_v		molecular vibrational frequency	
ρ		mass density	
ρ	(1)	volume density of radiant energy	App. A.5
ρ_D		Debije shielding distance	II.3b-4
ρ_λ, $\rho_\lambda(\lambda)$	(2)	spectral volume density of radiant energy (as a function of λ)	App. A.5
$\rho_{\lambda 0}$	(2)	$\equiv \rho_\lambda(\lambda_0)$	
ρ_λ^b, $\rho_\lambda^b(\lambda,T)$	(2)	spectral volume density of radiant energy of black body (= full radiator) (at λ and T)	II.3b-5
$\vec{\Sigma}$, (Σ)		(quantum number of) component of $\vec{S}$ along internuclear axis of di-atomic molecule	II.2b-1
σ		wave number ($\equiv 1/\lambda$)	
σ, (σ_0)		cross section [$\hat{=} \pi(r_1+r_2)^2$] (in gas-kinetic collision)	II.4b-1
σ_{eff}, σ_n		effective cross section (of process n)	II.4b-3
$\langle\sigma_{eff}\rangle$, $\langle\sigma_n\rangle$		phenomenological effective cross section ($\equiv k_n/\bar{g}$) (of process n)	II.4b-3
σ_m, $(\langle\sigma_m\rangle)$		(phenomenological) mixing cross section	VI.2a-2/ 2b-3

APPENDIX A.3   LIST OF SYMBOLS AND TERMS USED

Symbol(s)			Term / Meaning	Section
σ_q, $(\langle \sigma_q \rangle)$			(phenomenological) quenching cross section	VI.2a-1/ 2b-2
σ_x			standard deviation of variable x	III.5c
τ, $\tau(\lambda)$		(2)	transmission factor or transmittance (at λ) ($\equiv \Phi_t/\Phi_0$)	App. A.5
τ		(10)	relaxation time; lifetime	II.4a-3
τ_c			time constant or RC-time	
τ_i			integration time	
τ_m			measuring time	III.14f
τ_r		(11)	radiative or optical lifetime	II.5a-2
τ_{ic}			intercollision time or collisional lifetime	II.4b-1
Φ		(5)	radiant flux	App. A.5
Φ_0			incident radiant flux	
Φ_t			transmitted radiant flux	
Φ_λ, $\Phi_\lambda(\lambda)$		(2)	spectral radiant flux (at λ)	App. A.5
$\Phi_{\lambda 0}$			$\equiv \Phi_\lambda(\lambda_0)$	
ϕ			association factor $\{\equiv [MX]/[M]\}$	VIII.1
Ω		(5)	solid angle	
$\vec{\Omega}$, (Ω)			(quantum number of) total electronic angular momentum along internuclear axis of di-atomic molecule ($\equiv \vec{\Lambda} + \vec{\Sigma}$)	II.2b-1
ω			circular or angular frequency ($\equiv 2\pi f$ or $2\pi \nu$)	
ω_e			$\equiv \nu_v/c$	II.2b-2

APPENDIX A.3   LIST OF SYMBOLS AND TERMS USED

NOTES

(1) Write $A(\lambda)$, etc., when referring to monochromatic radiation at λ.

(2) λ may be replaced by ν (with appropriate change of unit).

(3) Not used in this book.

(4) Mitchell and Zemansky (1961) define in Eq. 39: $a \equiv (\delta\nu_N/\delta\nu_D)\sqrt{\ln 2}$, and in Eq. 96: $a' \equiv [(\delta\nu_N + \delta\nu_C)/\delta\nu_D]\sqrt{\ln 2}$. Van der Held (1932) defines $a \equiv 2 \times [a$-parameter according to Mitchell and Zemansky].

(5) Subscript A, E, F or S refers to absorbed, emitted, fluorescent or scattered radiation etc., respectively.

(6) In the COG theory (Sect. II.5c-2) b stands for $\pi\delta\nu_D/\sqrt{\ln 2}$.

(7) Occasionally symbol v is also used for <u>relative</u> velocity.

(8) In Chapt. VII, $V(r)$ stands for the <u>difference</u> of the interaction potentials of the upper and lower state of an optical transition.

(9) In Chapt. VII, subscripts C' and C'' are used to distinguish between diabatic and adiabatic collision broadening, respectively.

(10) Occasionally also used for: sampling time (Sect. III.14) and: correlation time (Sect. II.5b-2).

(11) Occasionally also used for: response time (Sect. III.6b).

(12) The name "degree" (abbrev.: deg) is occasionally used for: temperature <u>interval</u>.

<u>Mathematical Symbols</u> are as usual (but here $\simeq$ means "about equal to").

<u>Explanation of General Affixes</u> [†]

A	= absorption
b	= (as superscript) black body; (as subscript) background
bg	= background
c	= spectral continuum; crossing point of potential energy curves
E	= emission
e	= equilibrium
e, el	= electron(ic)
eff	= effective
exc	= excited state; excitation
F	= fluorescence
f	= flame
g	= gas
i, ion	= ionization; ionized
l	= liquid; spectral line
ls	= light source (used as background)

[†] Some of the listed affixes may also have another, more special meaning (see the above "List of Symbols and Terms Used").

APPENDIX A.3 — LIST OF SYMBOLS AND TERMS USED

m	= maximum; molecular
o	= ground state; (as subscript to thermodynamic quantities) state at $0\,K$; (as superscript to thermodynamic quantities) state at standard pressure; (as subscript to radiation quantities) incident; (as subscript to kinetic quantities) gas-kinetic collision
r	= recombination (of charged species); rotational
ref	= reference
rel	= relative
S	= scattering
s	= solid state; saturation; signal; radiation (standard) source
st	= stationary
T	= (as subscript to thermodynamic quantities) state at temperature T
t	= total; (as subscript to radiation quantities) transmitted
thr	= threshold
tr	= translational
v	= vibrational
Δ	= difference
*	= excited state; complex conjugate
+	= (as superscript) positively charged
−	= (as superscript) negatively charged
−	= (as bar) average value
[X]	= number density of species X

APPENDIX A.4
Ionization Energies and Spectral Line Characteristics of Some Metal Atoms

APP. A.4 IONIZATION ENERGIES AND SPECTRAL LINE CHARACTERISTICS OF SOME METAL ATOMS

Element	Ionization energy* (eV)	Wavelength⌀ (in air) (Å)	Transition levels[†] symbols	Transition levels[†] energies (eV)		gf-values[≠]	References
Ag	7.58	3280.7	$5^2S_{1/2} - 5^2P^0_{3/2}$	0	-3.78	0.9	1, 2
Al	5.99	3944.0	$3^2P^0_{1/2} - 4^2S_{1/2}$	0	-3.14	0.25	3, 4, 5
		3092.7/8	$3^2P^0_{3/2} - 3^2D$	0.014	-4.02	0.9	4, 6
Au	9.22	2428.0	$6^2S_{1/2} - 6^2P^0_{3/2}$	0	-5.10	0.4 -0.8	7, 8, 9
Ba	5.21	5535.5	$6^1S_0 - 6^1P^0_1$	0	-2.24	1.6	10, 11, 12
Ba⁺	10.00	4554.0	$6^2S_{1/2} - 6^2P^0_{3/2}$	0	-2.72	1.5 -2.0	10, 12, 13
		4934.1	$6^2S_{1/2} - 6^2P^0_{1/2}$	0	-2.51	0.7	10, 12, 13
Be	9.32	2348.6	$2^1S_0 - 2^1P^0_1$	0	-5.28	1.0 -1.4	14, 15, 16
Bi	7.29	3067.7	$6^4S^0_{3/2} - 7^4P_{1/2}$	0	-4.04	0.3 -0.5	1, 17
Ca	6.11	4226.7	$4^1S_0 - 4^1P^0_1$	0	-2.93	1.5 -1.8	3, 18, 19
Ca⁺	11.87	3933.7	$4^2S_{1/2} - 4^2P^0_{3/2}$	0	-3.15	1.3	5, 12, 20
		3968.5	$4^2S_{1/2} - 4^2P^0_{1/2}$	0	-3.12	0.7	5, 20
Cd	8.99	2288.0	$5^1S_0 - 5^1P^0_1$	0	-5.42	1.0 -1.4	6, 17, 58, 61
Co	7.86	3453.5	$b^4F_{9/2} - y^4G^0_{11/2}$	0.43	-4.02	3.2 -6.0	6, 21, 22
Cr	6.77	3578.7	$a^7S_3 - y^7P^0_4$	0	-3.46	2.1 -2.7	2, 4, 23, 66
		4254.3	$a^7S_3 - z^7P^0_4$	0	-2.91	0.78	4, 65, 66
Cs	3.89	8521.1	$6^2S_{1/2} - 6^2P^0_{3/2}$	0	-1.45	1.45-1.6	24, 25, 26, 27
		8943.5	$6^2S_{1/2} - 6^2P^0_{1/2}$	0	-1.39	0.7 -0.8	24, 25, 26, 27
Cu	7.73	3247.5	$4^2S_{1/2} - 4^2P^0_{3/2}$	0	-3.82	0.86	29, 64
		3274.0	$4^2S_{1/2} - 4^2P^0_{1/2}$	0	-3.79	0.43	29, 63

APP. A.4 IONIZATION ENERGIES AND SPECTRAL LINE CHARACTERISTICS OF SOME METAL ATOMS

Element	Ionization energy* (eV)	Wavelength ∅ (in air) (Å)	Transition levels [+] symbols	Transition levels [+] energies (eV)		gf-values [≠]		References
Fe	7.87	2483.3	$a^5D_4 - x^5F_5^0$	0	-4.99	3.1	-3.9	6, 30
		3719.9	$a^5D_4 - z^5F_5^0$	0	-3.33	0.25	-0.42	31, 32, 33
Ga	6.00	2874.2	$4^2P_{1/2}^0 - 4^2D_{3/2}$	0	-4.31	0.3	-0.5	1, 2, 34
		4033.0	$4^2P_{1/2}^0 - 5^2S_{1/2}$	0	-3.07	0.15	-0.26	1, 2, 35
Hg	10.44	2536.5	$6^1S_0 - 6^3P_1^0$	0	-4.89	0.34		6
In	5.79	3039.4	$5^2P_{1/2}^0 - 5^2D_{3/2}$	0	-4.08	0.5	-0.7	1, 2, 4
		4101.8	$5^2P_{1/2}^0 - 6^2S_{1/2}$	0	-3.02	0.2	-0.3	1, 36
K	4.34	7664.9	$4^2S_{1/2} - 4^2P_{3/2}^0$	0	-1.617	1.4		3, 37
		7699.0	$4^2S_{1/2} - 4^2P_{1/2}^0$	0	-1.610	0.68		3
		4044.1	$4^2S_{1/2} - 5^2P_{3/2}^0$	0	-3.06	0.012	-0.026	3, 38, 39
Li	5.39	6707.8	$2^2S_{1/2} - 2^2P^0$	0	-1.85	1.4	-1.6	6, 40, 41
		3232.6	$2^2S_{1/2} - 3^2P^0$	0	-3.83	0.02		42
Mg	7.65	2852.1	$3^1S_0 - 3^1P_1^0$	0	-4.35	1.5	-1.8	3, 20, 43
Mn	7.43	2794.8	$a^6S_{5/2} - y^6P_{7/2}^0$	0	-4.43	3.5		4, 44, 66
		4030.8	$a^6S_{5/2} - z^6P_{7/2}^0$	0	-3.07	0.32	-0.38	45, 46, 66
Na	5.14	5889.95	$3^2S_{1/2} - 3^2P_{3/2}^0$	0	-2.104	1.3		3, 47
		5895.92	$3^2S_{1/2} - 3^2P_{1/2}^0$	0	-2.102	0.66		3
		3302.3	$3^2S_{1/2} - 4^2P_{3/2}^0$	0	-3.75	0.028		3
		8194.8	$3^2P_{3/2}^0 - 3^2D_{5/2}$	2.10	-3.62	9		6
Ni	7.63	2320.0	$a^3F_4 - y^3G_5^0$	0	-5.34	0.9		48
Pb	7.42	2833.1	$6^3P_0 - 7^3P_1^0$	0	-4.37	0.2		49
		2170.0	$6^3P_0 - 6^3D_1^0$	0	-5.71	0.4		1, 49, 50

APP. A.4 IONIZATION ENERGIES AND SPECTRAL LINE CHARACTERISTICS OF SOME METAL ATOMS

Rb	4.18	7800.2	$5^2S_{1/2} - 5^2P^0_{3/2}$	0	1.35	26
		7947.6	$5^2S_{1/2} - 5^2P^0_{1/2}$	0	0.65	26, 51
Sc	6.54	3907.5	$a^2D_{3/2} - y^2F^0_{5/2}$	0	2.7	6
Sn	7.34	2246.0	$^3P_0 - {}^3D^0_1$	0	0.4	49
Sr	5.69	4607.3	$5^1S_0 - 5^1P^0_1$	0	1.5 − 2.1	18, 52, 53, 54, 55, 62
Sr⁺	11.03	4077.7	$5^2S_{1/2} - 5^2P^0_{3/2}$	0	1.5	12, 55, 56
		4215.5	$5^2S_{1/2} - 5^2P^0_{1/2}$	0	0.7	12
Ti	6.82	3642.7	$a^3F_3 - y^3G^0_4$	0.02 − 3.42	1.8	6
Tl	6.11	2767.9	$6^2P^0_{1/2} - 6^2D_{3/2}$	0	0.7	1, 2
		3775.7	$6^2P^0_{1/2} - 7^2S_{1/2}$	0	0.3	1, 2
		5350.5	$6^2P^0_{3/2} - 7^2S_{1/2}$	0.97 − 3.28	0.5	1, 57
V	6.74	3184.0	$a^4F_{3/2} - x^4G^0_{5/2}$	0.04 − 3.93	2.1	66
Zn	9.39	2138.6	$4^1S_0 - 4^1P^0_1$	0	1.2 − 1.5	58, 59, 60
Zr	6.84	3601.2	$a^3F_4 - x^3G^0_5$	0.15 − 3.60	2.0	6

NOTES

* Ionization energy data are derived from Moore (1970), where more precise values can be found.

∅ Wavelength data are taken from Corliss and Bozman (1962).

† Transition-level data are taken from Moore (1959).

⧣ The gf-values (g = statistical weight of the lower level; f = oscillator strength for the absorption transition; see Sect. II.5a-2) are taken from the references in the last column.

APP. A.4 IONIZATION ENERGIES AND SPECTRAL LINE CHARACTERISTICS OF SOME METAL ATOMS

REFERENCES (gf-VALUES)

1. Cunningham and Link (1967)
2. Lawrence, Link and King (1965)
3. Wiese, Smith and Miles (1969)
4. Penkin (1964)
5. Smith and Liszt (1971)
6. Corliss and Bozman (1962)
7. Penkin and Slavenas (1963)
8. Moise (1966)
9. Levin and Budick (1966)
10. Miles and Wiese (1970)
11. Erdevdi and Shimon (1976)
12. Gallagher (1967)
13. Arnesen et al. (1975)
14. Wiese, Smith and Glennon (1966)
15. Bergström et al. (1969)
16. Andersen, Jessen and Sörensen (1969)
17. L'vov (1965)
18. Letfus (1966)
19. Odintsov (1963)
20. Anderson et al. (1970)
21. Morozova and Startsev (1964)
22. Allen and Asaad (1957)
23. Bucka et al. (1966)
24. Fabry (1976)
25. Exton (1976)
26. Link (1966)
27. Koenig (1971)
28. Bielski (1975)
29. Corliss (1970)
30. Corliss and Tech (1968)
31. Brzozowski et al. (1976)
32. Wolnik, Berthel and Wares (1970)
33. Margoshes and Scribner (1962)
34. Penkin and Shabanova (1965)
35. Ottinger and Ziock (1961)
36. Chen and Smith (1959)
37. Zimmermann (1975)
38. Hinnov and Kohn (1957)
39. Heavens (1961)
40. Marr and Creek (1968)
41. Anderson and Zilitis (1963)
42. D'Ans-Lax (1970)
43. Zhuvikin, Penkin and Shabanova (1976)
44. Hefferlin and Gearhart (1964)
45. Luther and Walter (1966)
46. Huldt and Lagerqvist (1952)
47. Mashinskii (1970)
48. L'vov (1970)
49. Penkin and Slavenas (1963a)
50. Bell and King (1961)
51. Altman and Kazantsev (1970)
52. Lurio, de Zafra and Goshen (1964)
53. Dickie et al. (1973)
54. Bucka and Schüssler (1961)
55. Kelly, Koh and Mathur (1974)
56. Ostrovskii and Penkin (1961)
57. Penkin, Ruzov and Shabanova (1973)
58. Baumann and Smith (1970)
59. Landman and Novick (1964)
60. Andersen and Sörensen (1973)
61. Lurio and Novick (1964)
62. Hansen (1978)
63. Krellmann, Siefart and Weihreter (1975)
64. Hannaford and McDonald (1978)
65. Measures, Drewell and Kwong (1977)
66. Younger et al. (1978)

APPENDIX A.5

General Radiant Quantities[†]

Term	Symbol	Definition	Dimension
(radiant) energy [*]	Q	energy in the form of radiation	energy
radiant flux, radiant power [*]	Φ	$\Phi = Q/t$ (t = time)	power
radiant intensity [*]	I	$I = \Phi/\Omega$ (Ω = solid angle)	$\frac{\text{power}}{\text{solid angle}}$
intensity [*]	J	relative, loose expression referring to any radiant quantity without specification (see Sect. II.5c-1)	unspecified
radiance [*]	B	$B = \Phi/S_\perp \Omega$ ($S_\perp$ = area of radiating surface element projected on a plane perpendicular to radiation direction)	$\frac{\text{power}}{\text{area} \times \text{solid angle}}$
(radiant) energy density [*]	ρ	$\rho = Q/V$ (V = volume)	$\frac{\text{energy}}{\text{volume}}$
irradiance [*]	E	$E = \Phi/S$ (S = irradiated surface)	$\frac{\text{power}}{\text{area}}$
absorption factor, absorptance	$\alpha, \alpha(\lambda)$ [ø]	$\alpha = \Phi_A/\Phi_0$ (Φ_0 = incident flux; Φ_A = absorbed flux)	1
transmission factor, transmittance	$\tau, \tau(\lambda)$ [ø]	$\tau = \Phi_t/\Phi_0$ (Φ_t = transmitted flux)	1
(decadic) (internal) absorbance	$A, A^{abs}, A(\lambda)$ [‡]	$A^{abs} = -\log_{10} \tau$	1
integral absorption (of spectral line)	$A_t, A_t^{(\lambda)}$	$A_t^{(\lambda)} = \int_{\text{line}} \alpha(\lambda)\, d\lambda$	length
(napierian) absorptivity [⧧]	$k(\lambda)$	$k(\lambda) = \alpha(\lambda)/l$ for $l \to 0$ (l = absorption path length)	(length)$^{-1}$
atomic absorptivity	$\kappa(\lambda)$	$\kappa(\lambda) = k(\lambda)/n_p$ (n_p = density of atoms in absorbing state)	(length)^2
emission factor [§]	$\varepsilon(\lambda)$	$\varepsilon(\lambda) = B_\lambda/B_\lambda^b(T)$ (B_λ = spectral radiance of thermal radiator at temper. T; $B_\lambda^b(T)$ is corresponding value for full radiator or black body)	1

APPENDIX A.5   GENERAL RADIANT QUANTITIES

NOTES

† Terms and symbols conform in the main to the IUPAC Nomenclature for Spectrochemical Analysis, Part I and III (IUPAC 1971 and 1976a), the IUPAC Manual of Symbols and Terminology for Physicochemical Quantities and Units (IUPAC 1970), and the Intern. Lighting Vocabulary, Public. C.I.E. no.17 (E-1.1.)(1970). When frequencies are used instead of wavelengths, λ is to be replaced by ν. Note that the dimensions of $A_t^{(\lambda)}$ and $A_t^{(\nu)}$ are different.

‡ This quantity refers to absorption in the flame, with the exclusion of losses in the optical system (see also explanation of the concept of absorption in Sect. II.5a-2). The second symbol is preferred whenever confusion with the Einstein probability of spontaneous emission (same symbol A) is likely to occur.

‡ Often called 'absorption coefficient' in the literature. However, according to IUPAC (1970) the latter term designates the (internal) decadic absorbance per unit length.

* When this quantity is defined <u>per unit of wavelength</u> or <u>frequency</u>, it is preceded by the adjective '<u>spectral</u>'. The appropriate symbol is then obtained by adding λ or ν as a suffix, as in Q_λ or Q_ν. We have, for example, the relation: $dQ = Q_\lambda d\lambda$. The units of Q and Q_λ are different. Note that the units of, e.g., Q_λ and Q_ν are also different. These spectral quantities themselves are usually a function of λ or ν, which may be made explicit by writing: $Q_\lambda(\lambda)$, etc.

ø When applied to monochromatic radiation, this quantity may be defined as a function of wavelength and written, for example, as $\alpha(\lambda)$ (not: α_λ).

§ There is much confusion in the literature as to the nomenclature of 'emission factor'. For example, Penner (1959) and the Intern. Lighting Vocabulary denote this quantity by 'emissivity' (with the same symbol). For gaseous radiating sources, this quantity depends on the <u>size</u> of the source, as do the absorption and transmission factors. Terms with the ending '-ivity' (as in 'absorptivity') should denote properties that are independent of the size of the source (see IUPAC 1971). Note that Richter (1968) defines 'emission coefficient' with symbol ε_λ as the spectral power radiated per unit of volume and solid angle.

APPENDIX A.6
(ad Sect. III.4b)

The Concept of Optical Conductance

The spatial and angular extent of a radiation beam passing through an optical train is restricted by the dimensions and positions of the apertures present (e.g., radiating surface, window, lenses, stops, slits and receiver). With narrow apertures wavelength-dependent diffraction effects must be taken into account. The property of an optical system or single component to 'conduct' a radiation beam is expressed by the *optical conductance*, L. The optical conductance of a (monochromatic) *radiation beam* is operationally defined by (see also Eq. III.2)

$$\Phi_1 = BL\tau. \qquad (A.1)$$

Here Φ_1 is the radiant flux of the beam passing through the system, B is the uniform beam radiance at the entrance of the system, while the transmission factor τ takes into account the optical losses in the system.[†] The <u>maximum</u> value of Φ_1 that can be accommodated by an optical system, for given values of B and τ, determines through Eq. A.1 the optical conductance of the *system*. We shall here present some examples in which diffraction effects are neglected and the medium between the optical parts of the system has a refractive index of unity.[*]

A.6-1 RADIATING SURFACE S OF ARBITRARY SHAPE IN COMBINATION WITH STOP D

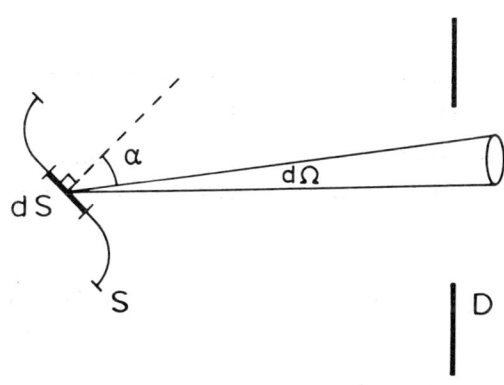

Fig. A.1 See text.

The light flux passing through stop D, according to the definition of B, is given by

$$\Phi_1 = \tau \int_S \int_\Omega B \cos\alpha \, dS \, d\Omega, \qquad (A.2)$$

where the integration extends over the whole surface area S and the whole solid angle Ω subtended by D (see also Fig. A.1). Comparing Eqs A.1 and A.2 and remembering that B is assumed to be uniform and isotropic, we get

$$L = \int_S \int_\Omega \cos\alpha \, dS \, d\Omega \qquad (A.3)$$

[†] See Appendix A.5 for a definition of these radiant quantities.

[*] A more general treatment is found in Appendix B of IUPAC (1972).

APPENDIX A.6   THE CONCEPT OF OPTICAL CONDUCTANCE

A.6-2   FLAT SURFACE S RADIATING INTO CONE C

We assume that the cone has an axis perpendicular to S and a semi-angle u (see Fig. A.2). Application of Eq. A.3 leads to

$$L = S\pi \sin^2 u = S\Omega \tfrac{1}{2} \sin^2 u / (1 - \cos u), \quad (A.4)$$

where Ω is the solid angle subtended by the cone. When S can freely radiate into the half-space ($\Omega = 2\pi$ or $u = \pi/2$) we have from Eq. A.4

$$L = \pi S. \quad (A.5)$$

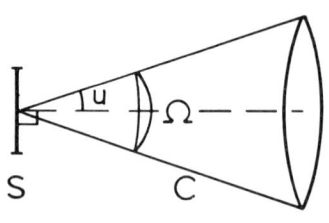

Fig. A.2   See text.

Note that in this extreme case L is <u>half</u> the product of S and solid angle $\Omega = 2\pi$. We also see from Eq. A.4 by series expansion that the equation

$$L \approx S\Omega \quad (A.6)$$

is a good approximation only if u is sufficiently small. For u smaller that 1/4 rad (or 15°) the error is less than 1.5%, whereas for $u = \pi/6$ (or 30°) the error is still as small as 7%. This approximate expression has in fact been used in Sect. III.4b where instead of a radiating surface we considered an illuminated slit or stop.

A.6-3   SPHERE RADIATING INTO FREE OUTSIDE SPACE

Since each surface element dS of the sphere with radius r_0 contributes to L by an amount $dL = \pi dS$ (compare Eq. A.5), the total L is given by

$$L = 4\pi^2 r_0^2. \quad (A.7)$$

We conclude by proving the invariance of L of the beam that passes through an optical train. When a beam passes through several apertures (including lenses) in succession, without optical losses, the radiance B is invariant. This is a generalization of a well-known theorem in Optics according to which object and image have the same radiance (see, e.g., Born and Wolf 1970, and Keitz 1971). We suppose here that the refractive index of the medium in which the radiance is considered is constant. It then follows immediately from Eq. A.1 that the optical conductance of the beam is invariant too, because Φ_1, as measured over the whole cross section of the passing beam, is conserved in the absence of optical losses.

APPENDIX A.7   DERIVATION OF EXPRESSIONS FOR SIGNAL-TO-NOISE RATIO (SNR)

APPENDIX A.7
(ad Sect. III.14f)

Derivation of Expressions for Signal-to-Noise Ratio (SNR) [†]

A.7-1   *GENERAL SNR EXPRESSION FOR PAIRED D.C. METER READINGS*

The same assumptions and notation will be used here as in Sect. III.14f-2 and in Table III.5.

Following Alkemade (1975) and Alkemade, Snelleman *et al.* (1978) we find from Eq. III.71 for the variance of the difference of two, paired background readings separated by sampling time τ

$$\sigma^2_{\Delta x} \equiv \overline{\{dx_b(t_0+\tau) - dx_b(t_0)\}^2} =$$
$$= \overline{dx_b(t_0+\tau)^2} + \overline{dx_b(t_0)^2} - \overline{2\,dx_b(t_0+\tau)\cdot dx_b(t_0)} =$$
$$= 2\{\sigma^2_b - \Psi_x(\tau)\} = 2\{\Psi_x(0) - \Psi_x(\tau)\}, \qquad (B.1)$$

where a bar denotes the average value. Here $\Psi_x(\tau)$ is the *autocorrelation function* of the fluctuations in the background deflection $dx_b(t)$, defined by

$$\Psi_x(\tau) \equiv \overline{dx_b(t_0+\tau)\cdot dx_b(t_0)} . \qquad (B.2)$$

Use has been made of the assumed statistical stationarity of the fluctuation process, i.e. the fluctuating background current, having constant statistical properties, is admitted to the meter long before readings are taken. Consequently it holds that $\overline{dx_b(t_0+\tau)^2} = \overline{dx_b(t_0)^2} = \sigma^2_b$, where σ^2_b is the time-independent variance of $dx_b(t)$. The final expression in Eq. B.1 follows from the identity: $\sigma^2_b = \Psi_x(0)$ [see definition of $\Psi_x(0)$ in Eq. B.2].

In order to calculate $\sigma^2_{\Delta x}$ we express the autocorrelation function in the spectral noise power, $W_{i_b}(f)$, of the background-current fluctuations and in the characteristics of the amplifier-readout system. We proceed by using the *Wiener-Khintchine theorem* which generally relates the autocorrelation function to the spectral noise power through a Fourier transformation. Defining $W_x(f)$ as the spectral noise power of $dx_b(t)$ we have according to this well-known theorem in noise theory

$$\Psi_x(\tau) = \int_0^\infty W_x(f)\cos(2\pi f\tau)\,df . \qquad (B.3)$$

Note that the right-hand side of this equation is only a function of τ, as the limits of the integration over f are zero and infinity. Since the measuring system is assumed to be linear, we find the spectral noise power, $W_x(f)$, of the meter

---

[†] A general introduction to the noise concepts and relations used in this Appendix may be found in, e.g., Blackman and Tukey (1959), MacDonald (1962), and van der Ziel (1970).

APPENDIX A.7   DERIVATION OF EXPRESSIONS FOR SIGNAL-TO-NOISE RATIO (SNR)

fluctuations by simply multiplying the spectral noise power, $W_{i_b}(f)$, of the background-<u>current</u> fluctuations by the *squared frequency response*, $G(f)^2$, of the amplifier-readout system. $G(f)$ represents the ratio of the amplitude of the meter oscillation to that of an a.c. input current as a function of frequency f. The square of this function enters here, because the spectral noise power relates to the <u>squared</u> fluctuations (see also Sect. III.5c). Thus we have

$$W_x(f) = W_{i_b}(f) G(f)^2 , \qquad (B.4)$$

so that Eq. B.3 can be re-written as

$$\Psi_x(\tau) = \int_0^\infty W_{i_b}(f) G(f)^2 \cos(2\pi f \tau) \, df . \qquad (B.5)$$

Using the latter expression we get finally from Eq. B.1

$$\sigma_{\Delta x}^2 = 2 \int_0^\infty W_{i_b}(f) G(f)^2 (1 - \cos 2\pi f \tau) \, df . \qquad (B.6)$$

It is seen that in general $\sigma_{\Delta x}^2$ is a function of sampling time τ. When $\tau \to 0$, then $\sigma_{\Delta x}$ becomes zero, too; this is, of course, to be expected as $\Delta x = 0$ for $\tau = 0$.

The signal deflection, $x_s(t_0 + \tau)$, due to a constant signal current, i_s, that is suddenly applied to the input at time t_0, can be generally written as

$$x_s(t_0 + \tau) = G i_s S(\tau) . \qquad (B.7)$$

Herein G is the d.c. sensitivity of the amplifier-readout system (see also Eq. III.72) and $S(\tau)$ the *normalized step response* of the system (cf. definition of step response in Sect. III.14f-2 and Eq. III.67). We have by definition: $S(\infty) \equiv 1$.

Using the above expressions for the noise (Eq. B.6) and the signal deflection (Eq. B.7) we find from Eq. III.70 for the SNR in the case of paired d.c. meter readings

$$(\text{SNR})_1^{\text{pair}} = \frac{G i_s S(\tau)}{[2 \int_0^\infty W_{i_b}(f) G(f)^2 (1 - \cos 2\pi f \tau) \, df]^{\frac{1}{2}}} . \qquad (B.8)$$

Since the d.c. sensitivity G equals $G(0)$, a simplification is obtained by introducing the *normalized frequency response*: $g(f) \equiv G(f)/G(0)$. Equation B.8 then reduces to

$$(\text{SNR})_1^{\text{pair}} = \frac{i_s S(\tau)}{[2 \int_0^\infty W_{i_b}(f) g(f)^2 (1 - \cos 2\pi f \tau) \, df]^{\frac{1}{2}}} . \qquad (B.9)$$

In general, both the numerator and denominator in the latter expression are a function of sampling time τ, and so the SNR will be too. Equation B.9 thus enables us to optimize τ and the instrumental characteristics, described by $S(\tau)$ and $g(f)$,

APPENDIX A.7   DERIVATION OF EXPRESSIONS FOR SIGNAL-TO-NOISE RATIO (SNR)

with respect to SNR for given input noise spectrum $W_{i_b}(f)$.

The normalized step response, $S(\tau)$, in the time domain and the normalized frequency response, $g(f)$, in the conjugated frequency domain are interrelated and cannot be chosen independently of each other. A system that responds quickly to a sudden variation of the input signal responds also to a broad band of a.c. frequencies. For a fixed, linear transmission system the impulse response (i.e. the time response to a unit impulse or Dirac delta function at the input) and the complex frequency response or 'transfer function' (see also below under A.7-5) are each other's Fourier transforms (see, e.g., Blackman and Tukey 1959). The absolute value of the complex frequency response is identical to the function $g(f)$ as defined above. The impulse response is again simply related to the step response, since the impulse can formally be considered as the derivative of a step function (see ibidem).

A.7-2 *SNR EXPRESSION FOR PAIRED READINGS ON A D.C. METER WITH EXPONENTIAL STEP RESPONSE IN THE CASE OF 'WHITE' NOISE*

We assume $W_{i_b}(f) \equiv W_0$ to be independent of frequency and the step response of the meter which is damped by a RC-filter to be given by Eq. III.67. We thus have for the normalized step response

$$S(\tau) = 1 - \exp[-\tau/\tau_c] \quad \text{with} \quad \tau_c = RC = \tau_r/2\pi. \qquad (B.10)$$

The normalized frequency response $g(f)$ corresponding to this step response is given by

$$g(f)^2 = \frac{1}{1+(2\pi\tau_c f)^2} = \frac{1}{1+f^2\tau_r^2}. \qquad (B.11)$$

Inserting the latter two expressions into Eq. B.9 we obtain for the SNR after some algebra

$$(SNR)_1^{pair} = i_s(1-\exp[-2\pi\tau/\tau_r])/\left\{2\int_0^\infty \frac{W_0(1-\cos 2\pi f\tau)}{1+f^2\tau_r^2}df\right\}^{\frac{1}{2}} =$$

$$= i_s\sqrt{\tau_r/\pi W_0}\sqrt{1-\exp[-2\pi\tau/\tau_r]}. \qquad (B.12)$$

For fixed response time τ_r, the latter expression increases monotonously with increasing τ and reaches asymptotically the maximum value

$$\text{Max }(SNR)_1^{pair} = i_s\sqrt{\tau_r/\pi W_0} \quad \text{for} \quad \tau \to \infty. \qquad (B.13)$$

The SNR already approaches this maximum within 0.2% when $\tau \approx \tau_r$.

Substituting the shot-noise expression given by Eq. III.63 for W_0 in Eq. B.13, one immediately obtains Eq. III.77, which was derived in Sect. III.14f-2 in an approximative way on condition that τ is large.

APPENDIX A.7   DERIVATION OF EXPRESSIONS FOR SIGNAL-TO-NOISE RATIO (SNR)

A.7-3 *SNR EXPRESSION FOR PAIRED INTEGRATOR READINGS IN THE CASE OF 'WHITE' NOISE*

When we let τ_r ($= 2\pi RC$) tend to infinity, e.g., by choosing R very large, the measuring system behaves as a capacitor, which integrates the input current. The difference, Δx, between two consecutive meter readings, τ seconds apart, then corresponds to the time-integral of the current over an integration period $\tau_i = \tau$.

The integrator output, y_b, for the fluctuating background current is given by Eq. III.80. The SNR obtained with (single) integrator readings can be derived from Eq. B.12 by substituting τ_i for τ and by taking the limit for $\tau_r \to \infty$. Using the approximation: $1 - \exp[-2\pi\tau_i/\tau_r] \simeq 2\pi\tau_i/\tau_r$ for $\tau_i/\tau_r \ll 1$ in Eq. B.12, we find for the SNR

$$(\text{SNR})_2 = i_s\sqrt{(\tau_r/\pi W_0)(2\pi\tau_i/\tau_r)} = i_s\sqrt{2\tau_i/W_0} \ . \tag{B.14}$$

In practice it may be again advantageous to apply <u>paired</u> integrator readings, i.e. to measure the <u>difference</u>, Δy, between the integrator outputs for two consecutive integration periods (cf. Eq. III.81 and the pertinent discussion). Because the background noise spectrum is 'white', the background-current fluctuations at two distinct times are statistically uncorrelated. The fluctuations in the <u>integrated</u> background current for two, non-overlapping time intervals are then also uncorrelated. Consequently, we obtain the SNR for paired integrator readings by square addition of the standard deviations of two single integrator readings. This means that we must divide the right-hand side of Eq. B.14 by $\sqrt{2}$ in order to find $(\text{SNR})_2^{\text{pair}}$

$$(\text{SNR})_2^{\text{pair}} = i_s\sqrt{\tau_i/W_0} \ . \tag{B.15}$$

In the special case of shot noise this equation reduces to Eq. III.88 after substitution of $2e\bar{i}_b$ for W_0. The result expressed by Eq. B.15 is, however, more generally valid for any 'white' noise.

A.7-4 *SNR EXPRESSION FOR PAIRED READINGS ON A D.C. METER WITH EXPONENTIAL STEP RESPONSE IN THE CASE OF FLICKER NOISE*

In this case the variance, $\sigma^2_{\Delta x}$, for paired readings is directly found from Eq. B.6 by substituting there $(A\bar{i}_b^2/f)$ for $W_{i_b}(f)$ and $G^2/(1+f^2\tau_r^2)$ for $G(f)^2 \equiv G^2 g(f)^2$ (cf. Eq. B.11). This leads immediately to

$$\sigma^2_{\Delta x} = 2G^2 A \bar{i}_b^2 \int_0^\infty \frac{(1-\cos 2\pi f\tau)}{f(1+f^2\tau_r^2)} df \ . \tag{B.16}$$

The SNR for paired readings is found by making the same substitutions in Eq. B.8 and using Eq. B.10 for $S(\tau)$. This immediately leads to Eq. III.90.

APPENDIX A.7 DERIVATION OF EXPRESSIONS FOR SIGNAL-TO-NOISE RATIO (SRN)

A.7-5 *SNR EXPRESSION FOR PAIRED INTEGRATOR READINGS IN THE CASE OF FLICKER NOISE*

In contrast to the case of 'white' noise, the fluctuations of the background current, i_b, in two consecutive integration periods are now statistically correlated. The simple method used in A.7-3 can thus not be applied here for the derivation of the SNR in the case of paired readings.

We shall proceed by treating the integrator output y_b, with fixed integration perion τ_i (see Eq. III.80), itself as a stochastic function of time (t_0). We can then derive the variance $\sigma^2_{\Delta y}$ from the spectral noise power, $W_y(f)$, of the fluctuating time function $y_b(t_0)$ in the same way as we derived $\sigma^2_{\Delta x}$ from $W_x(f)$ associated with the time function $x_b(t)$. To achieve this we have only to replace the index x by y in Eqs B.1 and B.3, and equalize $\tau = \tau_i$, as the paired integrator readings take place τ_i seconds apart. By combining the latter two equations we then get

$$\sigma^2_{\Delta y} = 2 \int_0^\infty W_y(f)\,(1 - \cos 2\pi f \tau_i)\,df. \qquad (B.17)$$

We can express $W_y(f)$ in the spectral noise power of the amplified background-current fluctuations (amplification factor G is assumed to be independent of frequency; i.e. no filters present). The latter spectral noise power is then simply equal to: $G^2 W_{i_b}(f)$, where the index i_b refers to the background current at the amplifier input. The relation between $W_y(f)$ and $G^2 W_{i_b}(f)$ is (see below)

$$W_y(f) = G^2 W_{i_b}(f)\,(1 - \cos 2\pi f \tau_i)/2\pi^2 f^2. \qquad (B.18)$$

Filling in the expression for $W_y(f)$ in Eq. B.17 and writing:

$$W_{i_b}(f) = A \bar{i}_b^2 / f,$$

we finally get in the case of flicker noise

$$\sigma^2_{\Delta y} = (G^2 A \bar{i}_b^2 / \pi^2) \int_0^\infty \frac{(1 - \cos 2\pi f \tau_i)^2}{f^3}\,df. \qquad (B.19)$$

Introducing $z \equiv 2\pi f \tau_i$ as an integration variable with dimension unity we can rewrite Eq. B.19

$$\sigma^2_{\Delta y} = 4 G^2 A \bar{i}_b^2 \tau_i^2 \int_0^\infty \frac{(1 - \cos z)^2}{z^3}\,dz. \qquad (B.20)$$

The integral can be proved to equal exactly: $\ln 2 = 0.694\ldots$ (see Gradshteyn and Ryzhik 1965) so that

$$\sigma^2_{\Delta y} = 2.78\, G^2 A \bar{i}_b^2 \tau_i^2. \qquad (B.21)$$

Using the latter expression and Eq. III.79 for the integrated signal y_s, we find for the SNR of paired integrator readings in the case of flicker noise

951

APPENDIX A.7   DERIVATION OF EXPRESSIONS FOR SIGNAL-TO-NOISE RATIO (SNR)

$$(SNR)_2^{pair} \equiv \frac{y_s}{\sigma_{\Delta y}} = 0.60\,(i_s/\bar{i}_b\sqrt{A}) \ . \tag{B.22}$$

We note that the SNR is independent of integration time τ_i (cf. discussion in Sect. III.14f-3). Comparing Eqs B.22 and III.93 it appears that the SNR for paired integrator readings virtually equals the maximum SNR for paired d.c. meter readings at optimal $\tau \simeq \frac{1}{8}\tau_r$.

The validity of Eq. B.18 can be proved by deriving the *complex transfer function*, i.e. the ratio of the a.c. response, $B(t)$, of the integrator circuit to a complex a.c. input signal $A(t) = A_m \exp[j\omega t]$, with $\omega \equiv 2\pi f$ and $j =$ imaginary unit (see Alkemade, Snelleman et al. 1978).[†] The integrator output, in complex notation, is then found to be

$$B(t) = \int_{t-\tau_i}^{t} A_m \exp[j\omega t']dt' = A_m \exp[j\omega t]\,(1-\exp[-j\omega\tau_i])/j\omega =$$

$$= A(t)(1 - \exp[-j\omega\tau_i])/j\omega \ . \tag{B.23}$$

The transfer function $T(f) \equiv B(t)/A(t)$ is thus

$$T(f) = (1 - \exp[-j\omega\tau_i])/j\omega \ . \tag{B.24}$$

The absolute value of $T(f)$ equals the ratio of the amplitude of the a.c. integrator output to that of the a.c. input current. We have from Eq. B.24

$$|T(f)| = \sqrt{2(1-\cos 2\pi f \tau_i)}/2\pi f \ . \tag{B.25}$$

The ratio of the spectral noise power of the integrator output to that of the input current is given by $|T(f)|^2 = (1 - \cos 2\pi f \tau_i)/2\pi^2 f^2$; this result has been used to derive Eq. B.18.

The same result has been obtained by Robben (1971) who used the general relation between the impulse response and the complex transfer function, mentioned at the end of Subsection A.7-1. In this way he also obtained an expression for $\sigma_{\Delta y}^2$, which agreed with our combined Eqs B.17 and B.18.

Analogous expressions have been derived by Alkemade, Snelleman, Boutilier and Winefordner (1980) for the signal-to-noise ratio with paired readings in the case of *multiplicative noise* (arising, e.g., in emission spectroscopy when the source temperature fluctuates, or when the amplification factor of the photomultiplier tube or of the electronic amplifier fluctuates).

[†] Several other derivations can be given for this equation, e.g., by expressing $y(t)$ in a convolution integral, by deriving an expression for the autocorrelation function of $y(t)$ and applying Fourier transformation, and by deriving directly the Fourier transform of $y(t)$ from its very definition.

APPENDIX A.8

Spectrograms of the Visible Bands of Ca, Sr and Ba in Flames

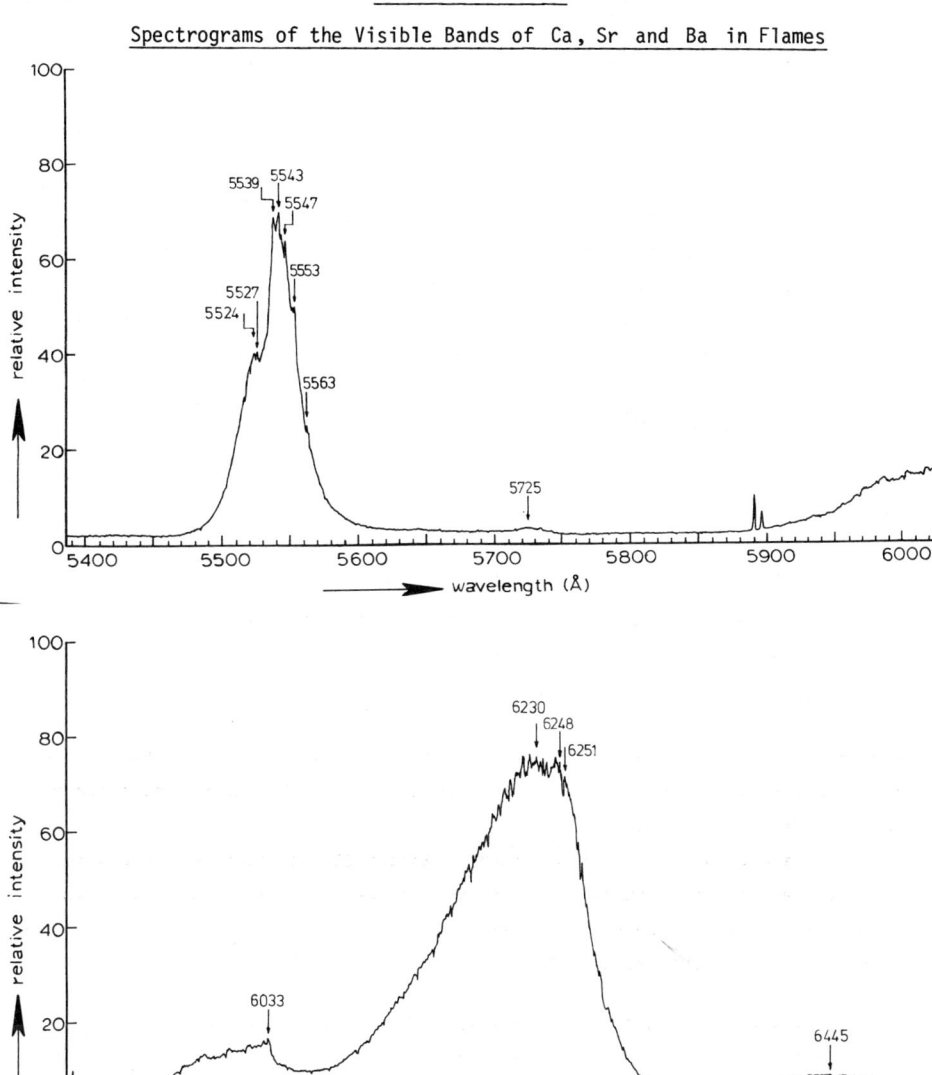

Spectrogram 1 Ca spectrum obtained by nebulizing an aqueous $CaCl_2$ solution into a C_2H_2-air flame of about 2450 K, showing the CaOH bands in the visible region. The relative intensity was not corrected for the spectral response of the spectrometer. The monochromator bandwidth was about 1.1 Å. The flame background intensity amounted to about 2 scale units. (From van der Hurk 1974.)

APPENDIX A.8   SPECTROGRAMS OF THE VISIBLE BANDS OF Ca, Sr AND Ba IN FLAMES

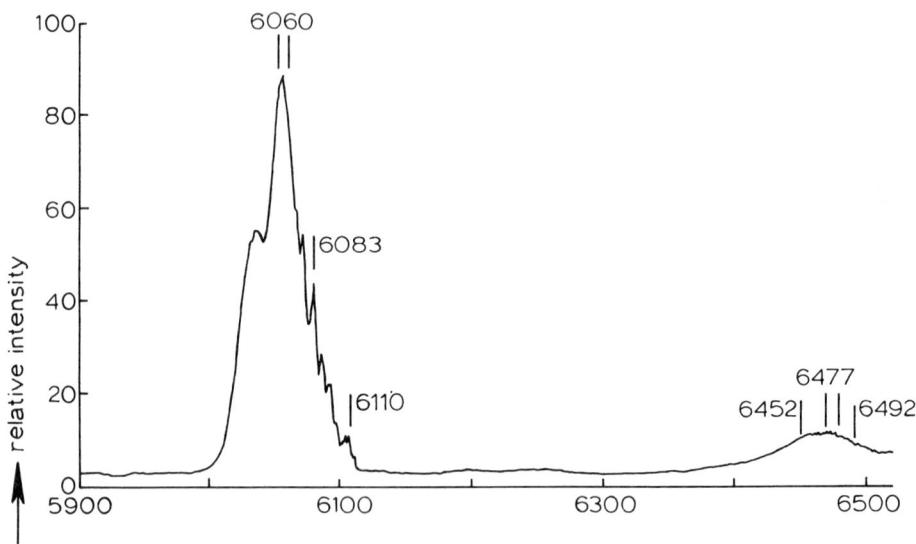

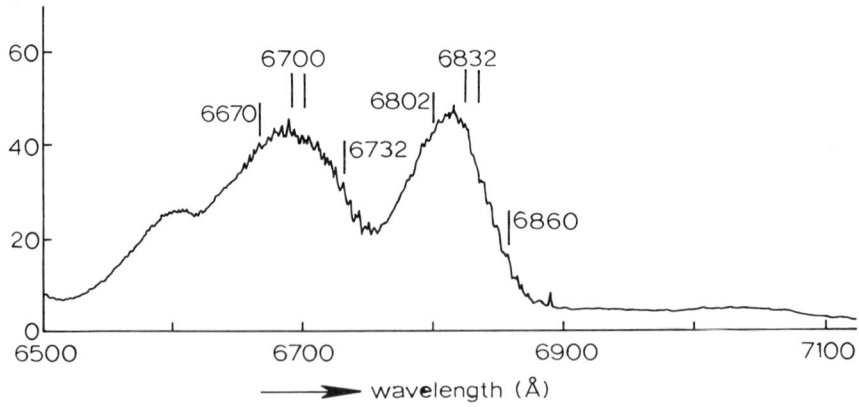

Spectrogram 2   Sr spectrum obtained by nebulizing an aqueous $SrCl_2$ solution into a C_2H_2-air flame of about 2450 K, showing the SrOH bands in the visible region. The relative intensity was not corrected for the spectral response of the spectrometer. The monochromator bandwidth was about 1.1 Å. The flame background intensity amounted to about 2 scale units. The well-resolved structure of the band peaks has been partially lost in the reproduction of the original recording.   (From van der Hurk 1974.)

APPENDIX A.8   SPECTROGRAMS OF THE VISIBLE BANDS OF Ca, Sr AND Ba IN FLAMES

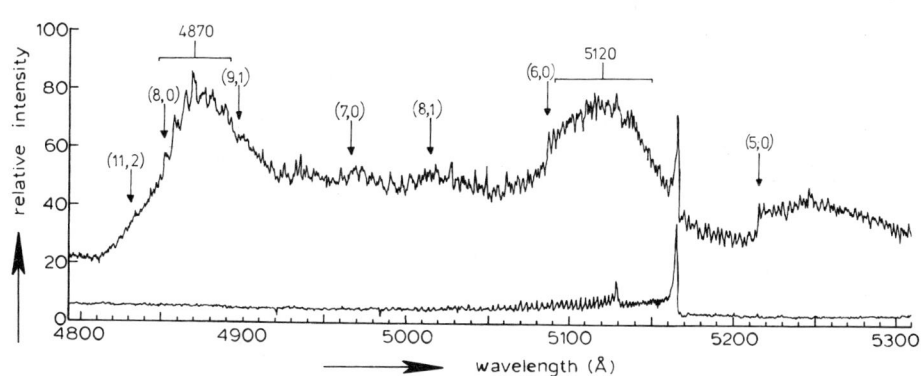

Spectrogram 3   Ba spectrum obtained by nebulizing an aqueous $BaCl_2$ solution into a fuel-rich C_2H_2-air flame of 2446 K. The relative intensity was not corrected for the spectral response of the spectrometer. The monochromator bandwidth was about 0.55 Å. The lower curve shows the flame background spectrum with the (0,0) and (1,1) bands of C_2 at 5165 and 5129 Å, respectively. The diffuse BaOH bands at 4870 and 5120 Å are indicated; the BaO band heads are indicated by their vibrational quantum numbers (v', v'').
(From van der Hurk 1974.)

APPENDIX A.8   SPECTROGRAMS OF THE VISIBLE BANDS OF Ca, Sr AND Ba IN FLAMES

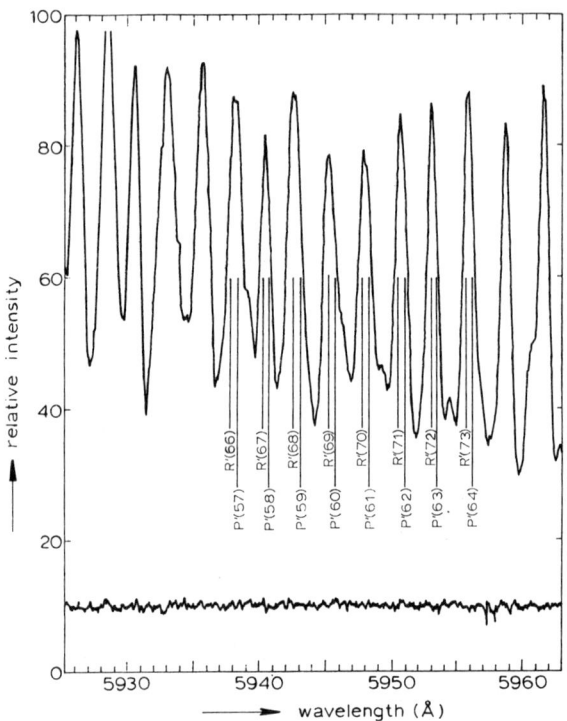

Spectrogram 4 Part of the (2,1) emission band of BaO under the same observation conditions as in spectrogram 3 in a slightly more oxygen-rich flame. The lower curve shows the flame background spectrum. The assignment of the rotational lines was based on data from Lagerqvist, Lind and Barrow (1950). (From van der Hurk 1974.)

APPENDIX A.9 DERIVATION OF RELAXATION EQUATION INCLUDING NEGATIVE-ION FORMATION

APPENDIX A.9
(ad Sect. IX.6b-4)

Derivation of Relaxation Equation including Negative-Ion Formation

We consider the following simultaneous ionization processes:

$$M + Z \underset{\rightarrow}{\overset{(\leftarrow)}{}} M^+ + Z^- \quad , \quad (i)$$

$$Z^- + X \underset{\rightarrow}{\overset{\leftarrow}{}} Z + e^- + X \, , \quad (ii)$$

$$M + Z \underset{\rightarrow}{\overset{(\leftarrow)}{}} M^+ + e^- + Z \, . \quad (iii)$$

M is a metal atom in any state of excitation or in the ground state (cf. Sect. IX.6a-1); Z is a specific flame molecule with positive electron affinity, so that Z^- is a stable species; X is an arbitrary flame particle. The processes (i) and (iii) are supposed to be slow and to be responsible for the ionization relaxation; process (ii) is supposed to proceed so quickly that it is partially balanced. The latter supposition is a reasonable one when the electron affinity of Z is small, say below 1 eV. The monomolecular rate constants of the forward processes (i) and (iii) are k_1 ($\equiv k'_1[Z]$) and k_3 ($\equiv k'_3[Z]$), respectively; the bimolecular rate constants of the reverse processes (i) and (iii) are k'_{-1} and k_{-3} ($\equiv k'_{-3}[Z]$). All these rate constants are supposed to be thermal. The equilibrium constants of the three processes are K_1, K_2 and K_3, respectively.

The rate equation for ionization relaxation is given by

$$-d[M]/dt = k'_1[M][Z] + k'_3[M][Z] - k'_{-1}[M^+][Z^-] - k'_{-3}[M^+][e^-][Z] \, . \quad (C.1)$$

Because of the partial equilibrium of process (ii) we have

$$[Z][e^-]/[Z^-] = K_2 \, , \quad (C.2)$$

and because the rate constants are thermal we can write (cf. Eqs II.93 and IX.30)

$$k'_1/k'_{-1} = K_1 \equiv ([M^+][Z^-]/[M][Z])_{eq} \, , \quad (C.3)$$

$$k'_3/k'_{-3} = K_3 \equiv ([M^+][e^-]/[M])_{eq} \, . \quad (C.4)$$

Using Eqs C.2 - 4 and the equality: $K_1 K_2 = K_3$, we find from Eq. C.1, after some algebra,

$$-d[M]/dt = (k'_1 + k'_3)[M][Z](1 - [M^+][e^-]/[M] K_3) \, . \quad (C.5)$$

From the charge-balance relation

$$[M^+] = [e^-] + [Z^-] \quad (C.6)$$

and from the relation between $[e^-]$ and $[Z^-]$ as given by Eq. C.2 we obtain

APPENDIX A.9   DERIVATION OF RELAXATION EQUATION INCLUDING NEGATIVE-ION FORMATION

$$[M^+] = [e^-](1+[Z]/K_2) \,. \tag{C.7}$$

Defining the total concentration, $[M]_t$, of element M by: $[M]_t \equiv [M^+] + [M]$, we find from Eqs C.5 and C.7

$$-d[M]/dt = (k_1' + k_3')[M][Z]\{1-([M]_t-[M])^2/(1+[Z]/K_2)[M]K_3\} \,. \tag{C.8}$$

After dividing both sides of this equation by [M] and after substituting k_1 and k_3 for $k_1'[Z]$ and $k_3'[Z]$, respectively, we finally get

$$-d\ln[M]/dt = (k_1+k_3)\{1-([M]_t-[M])^2/(1+[Z]/K_2)[M]K_3\} \,. \tag{C.9}$$

When at $t=0$ all metal is present as free atoms, we have from Eq. C.9

$$-\{d\ln[M]/dt\}_{t=0} = k_1+k_3 \,. \tag{C.10}$$

When for $t \to \infty$ Saha equilibrium has been established (in a flame with uniform temperature), we have

$$0 = 1-([M]_t-[M])^2/(1+[Z]/K_2)[M]K_3 \,, \tag{C.11}$$

or

$$[M^+]^2/[M] = (1+[Z]/K_2)K_3 \,. \tag{C.12}$$

Bibliography

EXPLANATIONS

In the text, references to the literature are identified by the authors' surnames and year of publication. When there are more than three authors, only the first author is mentioned, followed by "et al". In the Bibliography, the full names of all authors are given, except when there are more than six authors; the abbreviation "et al" is then used again. Publications without authors' names are identified by the (abbreviated) names of the organization concerned.
Volume numbers are given between square brackets.
000 as a date means that the year of publication is (as yet) unknown.
ä, ö, and ü are printed as ae, oe and ue, respectively.
Names beginning with "Mc" or "Mac" are written with a space, e.g. "Mc Donald" instead of "McDonald".

Acquista N. and Abramowitz S.	1969	J. Chem. Phys. [51] 2911
Acquista N., Abramowitz S. and Lide D.R.	1968	J. Chem. Phys. [49] 780
Alder J.F., Thompson K.C. and West T.S.	1970	Anal. Chim. Acta [50] 383
Alder J.F., Thompson K.C. and West T.S.	1972	Chemia Analityzna [17] 1091
Aldous K.M., Alder D., Dagnall R.M. and West T.S.	1970	Lab. Pract. [19] 589
Aldous K.M., Bailey B.W. and Rankin J.M.	1972	Anal. Chem. [44] 191
Aleksakhin I.S. and Zapesochnyi I.P.	1967	5th Int. Conf. on the Physics of Electronic and Atomic Collisions, Leningrad, USSR, 1967; "Abstracts of papers", Nauka, Leningrad, p.499
Aleksakhin I.S., Zapesochnyi I.P., Garga I. and Starodub V.P.	1975	Opt. Spectry. [38] 126
Aleksandrov E.B., Baranov A.V. and Kulyasov V.N.	1978	Opt. Spectry. [44] 624
Alexander C.A., Ogden J.S. and Levy A.	1963	J. Chem. Phys. [39] 3057
Alger D., Kirkbright G.F. and Troccoli O.E.	1973	Appl. Spectrosc. [27] 177
Alkemade C.Th.J.	1952	Physica [18] 933
Alkemade C.Th.J.	1954	"A contribution to the development and understanding of flame photometry", Ph.D. Thesis, Utrecht, Netherl.
Alkemade C.Th.J.	1959	8th Colloq. Spectrosc. Int., Luzern, Verlag Sauerlaender, Aarau, Switzerland, p.162
Alkemade C.Th.J.	1963	10th Colloq. Spectrosc. Int., College Park (Md) 1962; Spartan Books, Washington (DC), p.143
Alkemade C.Th.J.	1965	see: Proc. 10th Symp. (Int.) on Comb. (1965), p.33 Discussion on: Zeegers and Alkemade (1965a)
Alkemade C.Th.J.	1966	Anal. Chem. [38] 1252
Alkemade C.Th.J.	1968	Appl. Opt. [7] 1261
Alkemade C.Th.J.	1969	see: Dean and Rains (Eds.) (1969), Ch.4, p.101
Alkemade C.Th.J.	1970	Pure Appl. Chem. [23] 73
Alkemade C.Th.J.	1970a	see: Mavrodineanu (Ed.) (1970), Ch.1, p.1
Alkemade C.Th.J.	1973	Proc. Soc. Anal. Chem. [10] 130
Alkemade C.Th.J.	1975	5th Int. Conf. on Atomic Spectroscopy, Melbourne, Australia, 1975; "Abstracts of papers" no. C22, The Australian Academy of Science

BIBLIOGRAPHY

Alkemade C.Th.J. — 1977 — 20th Colloq. Spectrosc. Int. and 7th Int. Conf. on Atomic Spectroscopy, Praque, 1977; "Invited lectures", Praque, Sbornik VSCHT, p.93
Alkemade C.Th.J. — 1980 — Spectrochim. Acta B [35] 671
Alkemade C.Th.J. — 1981 — Appl. Spectrosc. [35] 1
Alkemade C.Th.J. and Herrmann R. — 1979 — "Fundamentals of analytical flame spectroscopy", Adam Hilger, Bristol, UK
Alkemade C.Th.J., Hollander Tj., Honings K.E.J., Koenders M.A. and Zijlstra R.J.J. — 1979 — Spectrochim. Acta B [34] 85
Alkemade C.Th.J., Hollander Tj. and Kalff P.J. — 1965 — Combust. Flame [9] 101
Alkemade C.Th.J., Hollander Tj., Snippe H. and Zijlstra R.J.J. — 1981 — Spectrochim. Acta B [36] 77
Alkemade C.Th.J. and Hooymayers H.P. — 1966 — Combust. Flame [10] 306
Alkemade C.Th.J., Hooymayers H.P., Lijnse P.L. and Vierbergen T.J.M.J. — 1972 — Spectrochim. Acta B [27] 149
Alkemade C.Th.J. and Jeuken M.E. — 1957 — Z. Anal. Chem. [158] 401
Alkemade C.Th.J. and Laven W.J. — 1957 — Appl. Sci. Res. B [6] 337
Alkemade C.Th.J. and Milatz J.M.W. — 1955 — Appl. Sci. Res. B [4] 289; J. Opt. Soc. Am. [45] 583
Alkemade C.Th.J., Snelleman W., Boutilier G.D., Pollard B.D., Winefordner J.D. and Chester T.L. et al — 1978 — Spectrochim. Acta B [33] 383
Alkemade C.Th.J., Snelleman W., Boutilier G.D. and Winefordner J.D. — 1980 — Spectrochim. Acta B [35] 261
Alkemade C.Th.J. and Voorhuis M.H. — 1958 — Z. Anal. Chem. [163] 91
Alkemade C.Th.J. and Wijchers T. — 1977 — Anal. Chem. [49] 2111
Alkemade C.Th.J. and Zeegers P.J.Th. — 1971 — see: Winefordner (1971a), Ch.1, p.3
Allen C.W. — 1963 — "Astrophysical quantities" 2nd ed., The Athlone Press, London
Allen C.W. and Asaad A.S. — 1957 — Mon. Not. Roy. Astron. Soc. [117] 36
Allen J.E., Anderson W.R., Crosley D.R. and Fansler T.D. — 1979 — see: Proc. 17th Symp. (Int.) on Comb. (1979), p.797
Allkins J.R. — 1975 — Anal. Chem. [47] 752A
Altman E.L. and Kazantsev S.A. — 1970 — Opt. Spectry. [28] 432

Amaee B. and Bottcher C. — 1978 — J. Phys. B [11] 1249
Amos M.D. and Thomas P.E. — 1965 — Anal. Chim. Acta [32] 139
Amos M.D. and Willis J.B. — 1966 — Spectrochim. Acta [22] 1325

Andersen T., Desesquelles J., Jensen K.A. and Soerensen G. — 1970 — J. Quant. Spectrosc. Radiat. Transfer [10] 143
Andersen T., Jessen K.A. and Soerensen G. — 1969 — Phys. Rev. [188] 76
Andersen T. and Soerensen G. — 1973 — J. Quant. Spectrosc. Radiat. Transfer [13] 369
Anderson E.M. and Zilitis V.A. — 1963 — Opt. Spectry. [16] 211
Anderson L.W. and Ramsey A.T. — 1963 — Phys. Rev. [132] 712
Anderson P.W. — 1949 — Phys. Rev. [76] 647
Anderson P.W. and Talman J.D. — 1955 — Proc. "Conf. on Broadening of Spectral Lines", University of Pittsburgh, Pittsburgh (Pa)
Anderson P.W. and Talman J.D. — 1956 — Bell Telephone Lab., Techn. Publ. no.3117, Murray Hill, N.J.
Anderson R.W., Aquilanti V. and Herschbach D.R. — 1969 — Chem. Phys. Lett. [4] 5
Anderson R.W., Goddard T.P., Parravano C. and Warner J. — 1976 — J. Chem. Phys. [64] 4037
Anderson R.W. and Herschbach D.R. — 1975 — J. Chem. Phys. [62] 2666
Andreev E.A. — 1972 — High Temp. [10] 637
Andreev E.A. — 1973 — Chem. Phys. Lett. [23] 516
Andreev E.A. and Voronin A.I. — 1969 — Chem. Phys. Lett. [3] 488
Andrews G.E. and Bradley D. — 1972 — Combust. Flame [18] 133
Anthony E.J., Bulewicz E.M., Kelly R. and Padley P.J. — 1974 — J. Chem. Soc., Faraday Trans.,I [70] 1067
Antonov E.E. and Korchevoi Yu. P. — 1976 — High Temp. [14] 1032

Appelblad O. and Lagerqvist A. — 1974 — Phys. Scripta [10] 307
Apt J. and Pritchard D.E. — 1976 — Phys. Rev. Lett. [37] 91
Apt J. and Pritchard D.E. — 1979 — J. Phys. B [12] 83

Arecchi F.T., Berne A. and Sona A. — 1966 — Phys. Rev. Lett. [17] 260
Armstrong B.H. — 1967 — J. Quant. Spectrosc. Radiat. Transfer [7] 61
Arnesen A., Bengtsson A., Hallin R., Kandela S., Noreland T. and Lidholt R. — 1975 — Phys. Lett. A [53] 459
Arnold J.O., Whiting E.E. and Lyle G.C. — 1969 — J. Quant. Spectrosc. Radiat. Transfer [9] 775
Arrington C.A., Brennen W., Glass G.P., Michael J.V. and Niki H. — 1965 — J. Chem. Phys. [43] 1489

Arrington C.A., Brennen W., Glass G.P., Michael J.V. and Niki H.	1965a	J. Chem. Phys. [43] 525
Ashton A.F. and Hayhurst A.N.	1970	Trans. Faraday Soc. [66] 824
Ashton A.F. and Hayhurst A.N.	1970a	Trans. Faraday Soc. [66] 833
Ashton A.F. and Hayhurst A.N.	1971	Trans. Faraday Soc. [67] 2348
Ashton A.F. and Hayhurst A.N.	1972	Trans. Faraday Soc. [68] 652
Ashton A.F. and Hayhurst A.N.	1973	Combust. Flame [21] 69
Ashton A.F. and Hayhurst A.N.	1976	J. Chem. Soc., Faraday Trans.,I [72] 208
Atakan A.K. and Jacobson H.C.	1972	J. Quant. Spectrosc. Radiat. Transfer [12] 289
Atakan A.K. and Jacobson H.C.	1973	Phys. Rev. A [7] 1452
Attard M.C.	1966	"Energies of electrons in plasmas", Ph.D. Thesis, Oxford, UK
Avni R. and Alkemade C.Th.J.	1960	Mikrochim. Acta [3] 460
Ay J., Ong R.S.B. and Sichel M.	1975	Combust. Science Technol. [11] 19
Babeliowsky T.P.	1962	"Thermochemical investigation above 1000 K with the mass-spectrometer" (in dutch), Ph.D. Thesis, Amsterdam
Babrov H.J. and Tourin R.H.	1968	Appl. Opt. [7] 2171
Bacherikov V.V., Vernyi A.E. and Kudryavtsev V.V.	1974	Izmeritel'naya Tekhnika no 12, 60 (transl. Measurement Techniques, p.1894)
Baede A.P.M.	1975	in: "Advances of chemical physics", vol.30, (Ed. K.P. Lawley), Wiley, New York, p.463
Baede A.P.M. and Los J.	1971	Physica [52] 422
Baestlein Ch., Baumgartner G. and Brosa B.	1969	Z. Physik [218] 319
Balducci G., De Maria G., Guido M. and Piacente V.	1971	J. Chem. Phys. [55] 2596
Balducci G., De Maria G., Guido M. and Piacente V.	1972	J. Chem. Phys. [56] 3422
Balfour W.J. and Lindgren B.	1978	Can. J. Phys. [56] 767
Balwanz W.W.	1965	see: Proc. 10th Symp. (Int.) on Comb. (1965), p.685
Baranger M.	1958	Phys. Rev. [111] 481
Baranger M.	1958a	Phys. Rev. [111] 494
Baranger M.	1958b	Phys. Rev. [112] 855
Baranger M.	1962	see: Bates (Ed.) (1962), p.517
Barker J.R. and Weston R.E.	1973	Chem. Phys. Lett. [19] 235
Barker J.R. and Weston R.E.	1976	J. Chem. Phys. [65] 1427
Barnes R.H., Moeller C.E., Kircher J.F. and Verber C.M.	1973	Appl. Opt. [12] 2531
Baronavski A.P. and Mc Donald J.R.	1977	J. Chem. Phys. [66] 3300
Baronavski A.P. and Mc Donald J.R.	1977a	Appl. Opt. [16] 1897
Barr G.	1934	J. Sci. Instr. [11] 321
Barrat J.P., Casalta D., Cojan J.L. and Hamel J.	1966	J. Phys. (Paris) [27] 608
Barrat-Rambosson M. and Kucal H.	1978	J. Phys. B [11] 2491
Barrow R.F. (Ed.)	1979	"Diatomic molecules; A critical bibliography of spectroscopic data", vol. 1, 2 and 3, Editions du Centre National de la Recherche Scientifique, Paris
Bashkin S. and Stoner J.O.	1975	"Atomic energy levels and Grotrian diagrams, vol.I: H-P"; North-Holland, Amsterdam; American-Elsevier, New York
Bastiaans G.J. and Hieftje G.M.	1974	Anal. Chem. [46] 901
Batalli-Cosmovici G. and Michel K.W.	1972	Chem. Phys. Lett. [16] 77
Bates D.R. (Ed.)	1962	"Atomic and molecular processes", Academic Press, New York-London
Bates D.R.	1962a	Discussions Faraday Soc. [33] 7
Bates D.R.	1976	Proc. Roy. Soc., Ser. A [348] 427; Erratum: idem Ser. A [354] 537 (1977)
Bates D.R.	1980	J. Phys. B [13] 205
Bates D.R. and Bederson B. (Eds.)	1975	"Advances in atomic and molecular physics", vol. 11, Academic Press, New York-London
Bates D.R. and Bederson B. (Eds.)	1976	"Advances in atomic and molecular physics", vol. 12, Academic Press, New York-London
Bates D.R. and Bederson B. (Eds.)	1977	"Advances in atomic and molecular physics", vol. 13, Academic Press, New York-London
Bates D.R. and Estermann I. (Eds.)	1965	"Advances in atomic and molecular physics", vol. 1, Academic Press, New York-London
Bates D.R. and Estermann I. (Eds.)	1966	"Advances in atomic and molecular physics", vol. 2, Academic Press, New York-London
Bates D.R. and Estermann I. (Eds.)	1967	"Advances in atomic and molecular physics", vol. 3, Academic Press, New York-London

BIBLIOGRAPHY

Bates D.R. and Estermann I. (Eds.)	1968	"Advances in atomic and molecular physics", vol. 4, Academic Press, New York-London
Bates D.R. and Estermann I. (Eds.)	1969	"Advances in atomic and molecular physics", vol. 5, Academic Press, New York-London
Bates D.R. and Estermann I. (Eds.)	1972	"Advances in atomic and molecular physics", vol. 8, Academic Press, New York-London
Bates D.R. and Estermann I. (Eds.)	1973	"Advances in atomic and molecular physics", vol. 9, Academic Press, New York-London
Bates D.R., Johnston H.C. and Stewart I.	1964	Proc. Phys. Soc. [84] 517
Bates D.R. and Kingston A.E.	1961	Nature [189] 652
Bates D.R. and Kingston A.E.	1964	Proc. Phys. Soc. [83] 43
Bates D.R., Kingston A.E. and Mc Whirter R.W.P.	1962	Proc. Roy. Soc., Ser. A [267] 297
Bates D.R., Kingston A.E. and Mc Whirter R.W.P.	1962a	Proc. Roy. Soc., Ser. A [270] 155
Batrakov R.I.	1970	Opt. Technol. [37] 392
Bauer E.	1967	5th Int. Conf. on the Physics of Electronic and Atomic Collisions, Leningrad, USSR, 1967; "Abstracts of papers", Nauka, Leningrad, p.236
Bauer E.	1969	see: Hochstim (Ed.) (1969a), Ch.10, p.381
Bauer E., Fisher E.R. and Gilmore F.R.	1969	Research Paper P-471, Institute for Defense Analyses, Science and Technology Division, Arlington (Va)
Bauer E., Fisher E.R. and Gilmore F.R.	1969a	J. Chem. Phys. [51] 4173
Baulch D.L., Drysdale D.D., Duxbury J. and Grant S.J.	1976	"Evaluated kinetic data for high temperature reactions", vol. 3, Butterworths, London
Baulch D.L., Drysdale D.D. and Horne D.G.	1970	High Temperature Reaction Rate Data no 5, july 1970
Baulch D.L., Drysdale D.D., Horne D.G. and Lloyd A.C.	1969	High Temperature Reaction Rate Data no 4, dec. 1969
Baulch D.L., Drysdale D.D., Horne D.G. and Lloyd A.C.	1972	"Evaluated kinetic data for high temperature reactions", vol. 1, Butterworths, London
Baulch D.L., Drysdale D.D., Horne D.G. and Lloyd A.C.	1973	"Evaluated kinetic data for high temperature reactions", vol. 2, Butterworths, London
Baulch D.L., Drysdale D.D. and Lloyd A.C.	1968	High Temperature Reaction Rate Data no 1, may 1968
Baulch D.L., Drysdale D.D. and Lloyd A.C.	1968a	High Temperature Reaction Rate Data no 2, nov. 1968
Baulch D.L., Drysdale D.D. and Lloyd A.C.	1969	High Temperature Reaction Rate Data no 3, april 1969
Baumann S.R. and Smith W.H.	1970	J. Opt. Soc. Am. [60] 345
Baylis W.E.	1969	J. Chem. Phys. [51] 2665
Baylis W.E.	1977	Can. J. Phys. [55] 1924
Baylis W.E.	1978	see: Hanle and Kleinpoppen (Eds.) (1978), Ch.6, p.207
Baylis W.E.	1979	see: Hanle and Kleinpoppen (Eds.) (1979), Ch.28, p.1227
Baylis W.E., Walentynowicz E., Phaneuf R.A. and Krause L.	1973	Phys. Rev. Lett. [31] 741
Beahn T.J., Condell W.G. and Mandelberg H.I.	1966	Phys. Rev. [141] 83
Bearman G.H. and Leventhal J.J.	1978	Phys. Rev. Lett. [41] 1227
Beaty E.C. and Gallagher J.W.	1976	"Bibliography of low energy electron and photon cross section data", JILA Information Center Report no 15, Boulder (Col)
Beckel Ch.L., Shafi M. and Engelke R.	1971	J. Mol. Spectry. [40] 519
Becker K.H. and Bayes K.D.	1968	J. Chem. Phys. [48] 653
Becker K.H. and Haaks D.	1972	Z. Naturforsch. A [28] 249
Becker K.H., Haaks D. and Tatavczyk T.	1974	Z. Naturforsch. A [29] 829
Becker R., Kummer J., Nicol K. and Schmidt W.	1971	Z. Physik [241] 380
Behmenburg W.	1964	J. Quant. Spectrosc. Radiat. Transfer [4] 177
Behmenburg W.	1968	Z. Astrophys. [69] 368
Behmenburg W.	1970	Habilitationsschrift, Saarbruecken, W.Germ.
Behmenburg W.	1979	see: Hanle and Kleinpoppen (Eds.) (1979), Ch.27, p.1187
Behmenburg W. and Kohn H.	1964	J. Quant. Spectrosc. Radiat. Transfer [4] 163
Behmenburg W., Kohn H. and Mailaender M.	1964	J. Quant. Spectrosc. Radiat. Transfer [4] 149
Belcher H. and Sugden T.M.	1950	Proc. Roy. Soc., Ser. A [202] 17
Bell G.D. and King R.B.	1961	Astron. J. [133] 718
Bell J.C. and Bradley D.	1970	Combust. Flame [14] 225
Bell R.J.	1972	"Introductory Fourier transform spectroscopy", Academic Press, New York
Belles F.E. and Lauver M.R.	1964	J. Chem. Phys. [40] 445
Bellisio J.A. and Davidovits P.	1970	J. Chem. Phys. [53] 3474
Bellisio J.A., Davidovits P. and Kindlmann P.J.	1968	J. Chem. Phys. [48] 2376
Belyaev Y.I., Ivantsov L.M., Karyakin A.V., Pham Hung Phi and Shemet V.V.	1968	J. Anal. Chem. USSR [23] 855
Ben Reuven A.	1966	Phys. Rev. [141] 34; [145] 7
Bendat J.S.	1958	"Principles and application of random noise theory", Wiley, New York
Bendat J.S. and Piersol A.G.	1966	"Measurement and analysis of random data", Wiley, New York

Bender P.L., Crosley D.R., Palmer D.R. and Zare R.N.	1967	5th Int. Conf. on the Physics of Electronic and Atomic Collisions, Leningrad, USSR, 1967; "Abstracts of papers", Nauka, Leningrad, p.510
Benetti P., Omenetto N. and Rossi G.	1971	Appl. Spectrosc. [25] 57
Bengtsson E. and Olsson E.	1931	Z. Physik [72] 163
Bennett B.G. and Dalby F.W.	1964	J. Chem. Phys. [40] 1414
Bennett H.E.	1966	Appl. Opt. [5] 1265
Benson F.A.	1958	Trans. Illum. Engl. Soc. (London) [23] 127
Benson S.W.	1960	"The foundations of chemical kinetics", McGraw-Hill, New York
Berdowski W., Shiner T. and Krause L.	1967	Appl. Opt. [6] 1683
Bergstrom I., Bromander J., Buchta R., Lundin L. and Martinson I.	1969	Phys. Lett. A [28] 721
Berman P.R.	1975	Appl. Phys. [6] 283
Berman P.R. and Lamb W.E.	1971	Phys. Rev. A [4] 319
Bernstein R.B. (Ed.)	1979	"Atom-molecule collision theory", Plenum, New York
Berry R.S.	1963	see: Shuler and Fenn (Eds.) (1963), Ch.1, p.3
Bevan M.J. and Husain D.	1975	J. Photochem. [4] 51
Bevan M.J. and Husain D.	1976	J. Phys. Chem. [80] 217
Bezemer J.	1976	"Optical pyrometry, employed in a new method to measure the melting points of Pt and Rh", Ph.D. Thesis, Utrecht, Netherl.
Bielski A.	1975	J. Quant. Spectrosc. Radiat. Transfer [15] 463
Bingel W.A.	1967	"Theorie der Molekuelspektren", Verlag Chemie, Weinheim, W.Germany
Biondi M.A., Connor T.R., Weller C.S. and Kasner W.H.	1965	see: Proc. 10th Symp. (Int.) on Comb. (1965), p.579
Bjerre A.	1967	5th Int. Conf. on the Physics of Electronic and Atomic Collisions, Leningrad, USSR, 1967; "Abstracts of papers", Nauka, Leningrad, p.238
Bjerre A.	1968	Theor. Exp. Chem. [4] 372
Bjerre A. and Nikitin E.E.	1967	Chem. Phys. Lett. [1] 179
Blackburn M.B., Mermet J.M. and Winefordner J.D.	1978	Spectrochim. Acta A [34] 847
Blackman R.B. and Tukey J.W.	1959	"The measurement of power spectra", Dover, New York
Blades A.T.	1971	Can. J. Chem. [49] 2476
Blades A.T.	1976	Can. J. Chem. [54] 2919
Bleekrode R.	1966	"Absorption and emission spectroscopy of CC, CH and OH in low-pressure oxyacetylene flames", Ph.D. Thesis, Amsterdam
Bleekrode R.	1968	Phys. Lett. A [27] 673
Bleekrode R.	1968a	Appl. Spectrosc. [22] 536
Bleekrode R.	1970	see: Mavrodineanu (Ed.) (1970), Ch.7, p.411
Bleekrode R. and Benthem W. Van	1969	J. Chem. Phys. [51] 2757
Bleekrode R. and Nieuwpoort W.C.	1965	J. Chem. Phys. [43] 3680
Bleekrode R. and Nieuwpoort W.C.	1965a	Appl. Opt., Supplement no.2 "Chemical lasers", p.179
Blickensderfer R.P., Breckenridge W.H. and Moore D.S.	1975	J. Chem. Phys. [63] 3681
Bloembergen N. and Levenson M.D.	1976	see: Shimoda (Ed.) (1976), Ch.8, p.315
Blue G.D., Green J.W., Ehlert Th.C. and Margrave J.L.	1963	Nature [199] 804
Boers A.L.	1963	"Apparatus for mass spectrometric analysis of flame gases" (in dutch with summary in english), Ph.D. Thesis, Utrecht Netherl.; see also Vriens, Boers and Smit (1965)
Boers A.L., Alkemade C.Th.J. and Smit J.A.	1956	Physica [22] 358
Boeschoten F., Milatz J.M.W. and Smit C.	1954	Physica [20] 139
Bojovic V. and Antic-Jovanovic	1972	Spectrochim. Acta B [27] 385
Bolshov M.A., Zybin A.V. and Koloshnikov V.G.	1979	Opt. Spectry. [46] 233
Bolshov M.A., Zybin A.V., Koloshnikov V.G. and Koshelev K.N.	1977	Spectrochim. Acta B [32] 279
Bonczyk P.A. and Shirley J.A.	1979	Combust. Flame [34] 253
Bonhoeffer K.F. and Haber T.	1928	Z. Phys. Chem. A [137] 263
Bonner T.W.	1932	Phys. Rev. [40] 105
Borgers A.J.	1965	see: Proc. 10th Symp. (Int.) on Comb. (1965), p.627
Borgers A.J.	1978	"Use of a radio-frequency resonance circuit in studies of alkali ionization in flames", Ph.D. Thesis, Utrecht, Neth.
Borgers A.J., Jongerius M.J. and Hollander Tj.	1980	Appl. Spectrosc. [34] 46
Borgers A.J., Jongerius M.J., Hollander Tj. and Alkemade C.Th.J.	1981	J. Chem. Soc., Faraday Trans.,I [11] 1075

BIBLIOGRAPHY

Borgers A.J., Jongerius M.J., Ventevogel W.J., Hollander Tj. and Alkemade C.Th.J.	1981	J. Chem. Soc., Faraday Trans.,I [11] 1083
Born M. and Wolf E.	1959	"Principles of optics", Pergamon, Oxford-New York; 4th ed. 1970; 5th ed. 1975
Born M. and Wolf E.	1970	see Born and Wolf (1959)
Born M. and Wolf E.	1975	see Born and Wolf (1959)
Boss C.B. and Hieftje G.M.	1978	Appl. Spectrosc. [32] 377
Botha J.P. and Spalding D.B.	1954	Proc. Roy. Soc., Ser. A [225] 71
Bottcher C.	1975	Chem. Phys. Lett. [35] 367
Bottcher C. and Sukumar C.V.	1977	J. Phys. B [10] 2853
Bouchareine P. and Connes P.	1963	J. Rad. Phys. [24] 134
Bouckaert R., D'Olieslager J. and De Jaegere S.	1972	Anal. Chim. Acta [58] 347
Boumans P.W.J.M.	1966	"Theory of spectrochemical excitation", Hilger and Watts, London
Boumans P.W.J.M.	1968	Spectrochim. Acta B [23] 559
Boumans P.W.J.M.	1972	in: "Analytical emission spectroscopy", (Ed. E.L. Grove Part 2, M. Dekker, New York, Ch.6, p.1
Boumans P.W.J.M. and Brouwer G.	1972	Spectrochim. Acta B [27] 247
Bousquet P.	1969	"Spectroscopie instrumentale", Dunod, Paris
Boutilier G.D., Blackburn M.B., Mermet J.M., Weeks S.J., Haraguchi H. and Winefordner J.D. et al	1978	Appl. Opt. [17] 2291
Boutilier G.D., Bradshaw J.D., Weeks S.J. and Winefordner J.D.	1977	Appl. Spectrosc. [31] 307
Boutilier G.D., Pollard B.D., Winefordner J.D., Chester T.L. and Omenetto N.	1978	Spectrochim. Acta B [33] 401
Boutilier G.D., Winefordner J.D. and Omenetto N.	1978	Appl. Opt. [17] 3482
Bower N.W. and Ingle J.D.	1979	Spectrochim. Acta B [34] 275
Bowman C.T. and Seery D.J.	1968	Combust. Flame [12] 611
Bowman J.A., Sullivan J.V. and Walsh A.	1966	Spectrochim. Acta [22] 205
Bowman M.R., Gibson A.J. and Sandford M.C.W.	1969	Nature [221] 456
Bradley D. and Ibrahim S.M.A.	1975	Combust. Flame [24] 169
Bradley D. and Matthews K.J.	1967	see: Proc. 11th Symp. (Int.) on Comb. (1967), p.359
Bradley D. and Matthews K.J.	1967a	Phys. Fluids [10] 1336
Bradley D. and Matthews K.J.	1968	J. Mech. Eng. Sci. [10] 299
Bradley D. and Sheppard C.G.W.	1970	Combust. Flame [15] 323
Bradley J.N. and Tse R.S.	1969	Nature [222] 474
Bradshaw J., Bower J., Weeks S., Fujiwara K., Omenetto N. and Haraguchi H. et al	1979	see: Hastie (Ed.) (1979), p.1079
Bradshaw J., Nikdel S., Reeves R., Bower J., Omenetto N. and Winefordner J.D.	1980	see: Crosley (Ed.) (1980), p.199
Brandenberger J.R.	1971	Rev. Sci. Instrum. [42] 1535
Bransden B.H.	1970	"Atomic collision theory", Benjamin, New York
Bratzel M.P., Dagnall R.M. and Winefordner J.D.	1970	Anal. Chim. Acta [52] 175
Brault J.W. and White O.R.	1971	Astron. Astrophys. [13] 169
Breckenridge W.H., Blickensderfer R.P. and Fitz Patrick J.	1976	J. Phys. Chem. [80] 1963
Breckenridge W.H., Blickensderfer R.P., Fitz Patrick J. and Oba D.	1979	J. Chem. Phys. [70] 4751
Breckenridge W.H., Donovan R.J. and Kim Malmin O.	1979	Chem. Phys. Lett. [62] 608
Breckenridge W.H. and Fitz Patrick J.	1976	J. Phys. Chem. [80] 1955
Breckenridge W.H. and Renlund A.M.	1978	J. Phys. Chem. [82] 1474
Breckenridge W.H. and Renlund A.M.	1978a	J. Phys. Chem. [82] 1484
Breckenridge W.H. and Renlund A.M.	1979	J. Phys. Chem. [83] 303
Breckenridge W.H. and Renlund A.M.	1979a	J. Phys. Chem. [83] 1145
Breene R.G.	1961	"The shift and shape of spectral lines", Pergamon, Oxford-New York
Brennen W. and Carrington T.	1967	J. Chem. Phys. [46] 7
Brewer L. and Brackett E.	1961	Chem. Rev. [61] 425
Brewer L. and Porter R.F.	1954	J. Chem. Phys. [22] 1867
Brewer L. and Rosenblatt G.M.	1969	in: "Advances in high-temperature chemistry, II Dissociation energies and free-energy functions of gaseous monoxides", (Ed. L. Eyring), Academic Press, New York, p.1
Brewer L. and Searcy A.W.	1956	Ann. Rev. Phys. Chem. [7] 259
Brewer L., Somayajulu G.R. and Brackett E.	1963	Chem. Rev. [63] 111

Brinkley S.R. and Lewis B.	1952	"The thermodynamics of combustion gases: General considerations", Bureau of Mines, Report of Investigations 4806, US Dept of the Interior, Washington (DC)
Britske M.E. and Sukach Yu.S.Saveleva A.N.	1973	Industr. Lab. [39] 578 (transl. from the russian)
Broida H.P., Schiff H.J. and Sugden T.M.	1961	Trans. Faraday Soc. [57] 259
Broida H.P. and Shuler K.E.	1957	J. Chem. Phys. [27] 933
Brokaw R.S.	1967	see: Proc. 11th Symp. (Int.) on Comb. (1967), p.1063
Brown R.J. and Parsons M.L.	1978	Spectrochim. Acta B [33] 777
Browner R.F., Patel B.M., Glenn T.H., Rietta M.E. and Winefordner J.D.	1972	Spectrosc. Lett. [5] 311
Browner R.F. and Winefordner J.D.	1973	Spectrochim. Acta B [28] 263
Bruce C.F. and Hannaford P.	1971	Spectrochim. Acta B [26] 207
Brus L.	1970	J. Chem. Phys. [52] 1716
Brzozowski J., Erman P., Lyyra M. and Hayden-Smith Wm.	1976	Phys. Scripta [14] 48
Bucka H., Budick B., Goshen R.J. and Marcus S.	1966	Phys. Rev. [144] 96
Bucka H. and Schuessler H.J.	1961	Ann. Physik [7] 225
Buell B.E.	1963	Anal. Chem. [35] 372
Bulewicz E.M.	1958	Ph.D. Thesis, Cambridge, UK
Bulewicz E.M.	1962	J. Chem. Phys. [36] 385
Bulewicz E.M.	1967	Combust. Flame [11] 297
Bulewicz E.M., James C.G. and Sugden T.M.	1956	Proc. Roy. Soc., Ser. A [235] 89
Bulewicz E.M., Mc Ewan M.J. and Phillips L.F.	1967	Combust. Flame [11] 63
Bulewicz E.M. and Padley P.J.	1961	Combust. Flame [5] 331
Bulewicz E.M. and Padley P.J.	1962	J. Chem. Phys. [36] 2231
Bulewicz E.M. and Padley P.J.	1963	see: Proc. 9th Symp. (Int.) on Comb. (1963), p.638
Bulewicz E.M. and Padley P.J.	1963a	see: Proc. 9th Symp. (Int.) on Comb. (1963), p.647
Bulewicz E.M. and Padley P.J.	1969	Trans. Faraday Soc. [65] 186
Bulewicz E.M. and Padley P.J.	1971	Trans. Faraday Soc. [67] 337
Bulewicz E.M. and Padley P.J.	1971a	see: Proc. 13th Symp. (Int.) on Comb. (1971), p.73
Bulewicz E.M. and Padley P.J.	1973	Spectrochim. Acta B [28] 125
Bulewicz E.M., Phillips L.F. and Sugden T.M.	1961	Trans. Faraday Soc. [57] 921
Bulewicz E.M. and Sugden T.M.	1956	Trans. Faraday Soc. [52] 1481
Bulewicz E.M. and Sugden T.M.	1956a	Trans. Faraday Soc. [52] 1475
Bulewicz E.M. and Sugden T.M.	1958	Trans. Faraday Soc. [54] 830
Bulewicz E.M. and Sugden T.M.	1958a	The Chemical Society, London, Special Publication no.9, 81
Bulewicz E.M. and Sugden T.M.	1958b	Trans. Faraday Soc. [54] 1855
Bulewicz E.M. and Sugden T.M.	1959	Trans. Faraday Soc. [55] 720
Bulewicz E.M. and Sugden T.M.	1964	Proc. Roy. Soc., Ser. A [277] 143
Bulos B.R. and Happer W.	1969	6th Int. Conf. on the Physics of Electronic and Atomic Collisions, Cambridge, Mass., 1969; "Abstracts of papers", M.I.T. Press, Cambridge (Mass), p.602
Burdett N.A. and Hayhurst A.N.	1974	see: Proc. 15th Symp. (Int.) on Comb. (1974), p.979
Burdett N.A. and Hayhurst A.N.	1977	Chem. Phys. Lett. [48] 95
Burdett N.A. and Hayhurst A.N.	1977a	Proc. Roy. Soc., Ser. A [355] 377
Burdett N.A. and Hayhurst A.N.	1979	Combust. Flame [34] 119
Burdett N.A. and Hayhurst A.N.	1979a	Philos. Transactions Royal Soc. of London A [290] 299
Burdett N.A., Hayhurst A.N. and Morley C.	1974	Chem. Phys. Lett. [25] 596
Burger J.C., Gillies W. and Yamasaki G.K.	1969	see: Mavrodineanu (Ed.) (1970), Ch.12, p.625
Burns R.P.	1966	J. Chem. Phys. [44] 3307
Burns R.P., De Maria G., Drowart J. and Inghram M.G.	1963	J. Chem. Phys. [38] 1035
Burriel-Marti F. and Ramirez-Munoz J.	1957	"Flame photometry, a manual of methods and applications", Elsevier, Amsterdam. Reprinted 1958, 1960, 1964
Burrow P.D.	1960	Ph.D. Thesis, University of California, Berkeley (Calif)
Burrow P.D. and Davidovits P.	1968	Phys. Rev. Lett. [21] 1789
Burrows K.M. and Horwood J.F.	1963	Spectrochim. Acta [19] 17
Busch G.E. and Wilson K.R.	1972	J. Chem. Phys. [56] 3638
Butaux J., Schuller F. and Lennuier R.	1972	J. Phys. (Paris) [33] 635
Butler L.R.P. and Brink J.A.	1971	see: Dean and Rains (Eds.) (1971), p.21
Bykhovskii V.K. and Nikitin E.E.	1964	Opt. Spectry. [16] 111
Cagnac B., Grynberg G. and Biraben F.	1973	J. Phys. (Paris) [34] 845
Cahill J.E.	1980	Intern. Laboratory [10] 64
Calcar R.A. Van	1980	"An experimental investigation into power broadening of the Na-D lines in a flame", Ph.D. Thesis, Utrecht, Neth.
Calcar R.A. Van, Ven M.J.M. Van De, Uitert B.K. Van, Biewenga K.J., Hollander Tj. and Alkemade C.Th.J.	1979	J. Quant. Spectrosc. Radiat. Transfer [21] 11
Calcote H.F.	1962	see: Proc. 8th Symp. (Int.) on Comb. (1962), p.184

BIBLIOGRAPHY

Calcote H.F.	1963	see: Proc. 9th Symp. (Int.) on Comb. (1963), p.622
Calcote H.F.	1965	see: Wilsted (Ed.) (1965), p.1
Calcote H.F. and Jensen D.E.	1966	in: "Advances in chemistry series", no. 58: "Ion-molecule reactions in the gasphase", Amer. Chem. Soc., p.291
Calcote H.F. and King I.R.	1955	see: Proc. 5th Symp. (Int.) on Comb. (1955), p.423
Calcote H.F. and King I.R.	1955a	J. Chem. Phys. [23] 2203
Calcote H.F., Kurzius S.C. and Miller W.J.	1965	see: Proc. 10th Symp. (Int.) on Comb. (1965), p.605
Callaway J. and Bauer E.	1965	Phys. Rev. A [140] 1072
Callear A.B.	1965	Appl. Opt., Supplement no. 2 "Chemical lasers", p. 145
Callear A.B.	1967	in: "Photochemistry and reaction kinetics", (Eds. P.G. Ashmore, F.S. Dainton and T.M. Sugden), The University Press, Cambridge, UK, p.133
Callear A.B. and Hedges R.E.M.	1970	Trans. Faraday Soc. [66] 615
Callear A.B. and Mc Gurk J.C.	1972	J. Chem. Soc., Faraday Trans.,I [68] 289
Callear A.B. and Wood P.M.	1971	Trans. Faraday Soc. [67] 2862
Cambel A.B., Duclos D.P. and Anderson T.P.	1963	"Real gases", Academic Press, New York-London
Cann M.W.P.	1969	Appl. Opt. [8] 1645
Carabetta R. and Kaskan W.E.	1967	see: Proc. 11th Symp. (Int.) on Comb. (1967), p.321
Carabetta R. and Kaskan W.E.	1968	J. Phys. Chem. [72] 2483
Carleton N. (Ed.)	1974	"Astrophysics, Part A: Optical and infrared", Academic Press, New York-London
Carrington C.G. and Gallagher A.	1974	Phys. Rev. A [10] 1464
Carrington T.	1959	J. Chem. Phys. [30] 1087
Carrington T.	1962	see: Proc. 8th Symp. (Int.) on Comb. (1962), p.257
Carrington T. and Polanyi J.C.	1972	in: "Chemical kinetics, phys. chem.", Ser.1, vol. 9 (Ed. J.C. Polanyi), MTP Int. Rev. of Science, Univ. Park Press, Baltimore (Md), p.163
Carter G.M., Pritchard D.E., Kaplan M. and Ducas T.W.	1975	Phys. Rev. Lett. [35] 1144
Cartwright J.S., Sebens C. and Slavin W.	1966	Atom. Absorption Newsl. [5] 22
Cath P.G.	1970	see: Mavrodineanu (1970), Ch.4, p.145
Celotta R.J., Bennett R.A., Hall J.L., Siegel M.W. and Levine J.	1972	Phys. Rev. A [6] 631
Chan C. and Daily J.W.	1978	see: Hastie (Ed.) (1979), p.847
Chan C. and Daily J.W.	1979	Meeting of Western States section of Combustion Institute, Provo (Utah) 1979; Paper no. 79-20
Chan C. and Daily J.W.	1979a	J. Quant. Spectrosc. Radiat. Transfer [21] 527
Chang A.M. and Pritchard D.E.	1979	J. Chem. Phys. [70] 4524
Chapman G.D. and Krause L.	1966	Can. J. Phys. [44] 753
Chapman S. and Cowling T.G.	1939	"Mathematical theory of non-uniform gases", The Universi Press, Cambridge, UK
Chappell W.R., Cooper J., Smith E.W. and Dillon T.	1971	J. Stat. Phys. [3] 401
Charton M. and Gaydon A.G.	1958	Proc. Roy. Soc., Ser. A [245] 84
Chen C.L. and Phelps A.V.	1973	Phys. Rev. A [7] 470
Chen S.T. and Gallagher A.	1978	Phys. Rev. A [17] 551
Chen S.Y. and Garrett R.O.	1966	Phys. Rev. [144] 59
Chen S.Y. and Garrett R.O.	1966a	Phys. Rev. [144] 66
Chen S.Y. and Garrett R.O.	1967	Phys. Rev. [155] 38
Chen S.Y., Gilbert D.E. and Tan D.K.L.	1969	Phys. Rev. [182] 51
Chen S.Y., Gilbert D.E. and Tan D.K.L.	1969a	Phys. Rev. [184] 51
Chen S.Y. and Henry P.K.	1973	J. Quant. Spectrosc. Radiat. Transfer [13] 41; 385
Chen S.Y. and Smith A.	1959	Physica [25] 1289
Chen S.Y. and Takeo M.	1957	Rev. Mod. Phys. [29] 20
Chen S.Y. and Wilson R.A.	1961	Physica [27] 497
Chenevier M. and Lombardi M.	1972	Chem. Phys. Lett. [16] 154
Chenevier M. and Lombardi M.	1974	6th EGAS Conf., Berlin-W, "Book of abstracts", paper no. 29
Cheron B.	1975	J. Phys. (Paris) [36] 17 and 29
Chester J.E., Dagnall R.M. and Taylor M.R.G.	1970	Anal. Chim. Acta [51] 95
Chester T.L., Haraguchi H., Knapp D.O., Messman J.D. and Winefordner J.D.	1976	Appl. Spectrosc. [30] 410
Child M.S.	1979	see: Bernstein (Ed.) (1979), Ch.13, p.427
Christian G.D. and Feldman F.J.	1970	"Atomic absorption spectroscopy: applications in agriculture, biology and medicine", Wiley-Interscience, New York
Clampitt N.C. and Hieftje G.M.	1972	Anal. Chem. [44] 1211
Clouston J.G., Gaydon A.G. and Glass I.I.	1958	Proc. Roy. Soc., Ser. A [248] 429
Clouston J.G., Gaydon A.G. and Hurle I.R.	1959	Proc. Roy. Soc., Ser. A [252] 143
Clyne M.A.A. and Thrush B.A.	1962	Proc. Roy. Soc., Ser. A [269] 404
Clyne M.A.A. and Thrush B.A.	1962a	Discussions Faraday Soc. [33] 139
Clyne M.A.A. and Thrush B.A.	1963	see: Proc. 9th Symp. (Int.) on Comb. (1963), p.177

Cohen E.R., Dermond J.W.M., Layton T.W. and Rollet J.S.	1955	Rev. Mod. Phys. [27] 363
Cohen-Tannoudji C.	1962	Ann. Phys. (Paris) [7] 423
Cohen-Tannoudji C.	1977	in:"Frontiers in laser spectroscopy", vol.1 (Eds.R.Balian, S. Haroche and S. Liberman), 27th Ecole d'ete de physique theoretique, Les Houches, France, 1975; North-Holland, Amsterdam, Part I, p.3
Coker D.T. and Ottaway J.M.	1971	Nature (London) Phys. Sci. [230] 156
Coker D.T., Ottaway J.M. and Pradhan N.K.	1971	Nature (London) Phys. Sci. [233] 69
Colin R., Drowart J. and Verhaegen G.	1965	Trans. Faraday Soc. [61] 1364
Colin R., Goldfinger P. and Jeunehomme M.	1964	Trans. Faraday Soc. [60] 306
Comite Int. Des Poids Et Mesures	1968	Metrologia [5] 35
Connes P.	1956	Rev. Opt. [35] 37
Cooke D.O., Dagnall R.M. and West T.S.	1971	Anal. Chim. Acta [54] 381
Cooke D.O., Dagnall R.M. and West T.S.	1971a	Anal. Chim. Acta [56] 17
Coolen F.C.M.	1976	"Fluorescence studies on 20Na and excited neon atoms in proton-induced plasmas", Ph.D. Thesis, Eindhoven, Neth.
Coolen F.C.M. and Hagedoorn H.L.	1975	Physica C [79] 402
Cooney R.P., Boutilier G.D. and Winefordner J.D.	1977	Anal. Chem. [49] 1048
Copley G., Kibble B.P. and Krause L.	1967	Phys. Rev. [163] 34
Copley G. and Krause L.	1969	Can. J. Phys. [47] 533
Copley G. and Krause L.	1969a	Can. J. Phys. [47] 1881
Coppens P., Smoes S. and Drowart J.	1967	Trans. Faraday Soc. [63] 2140
Corliss C.H.	1967	"Revision of the NBS tables of spectral-line intensities below 2450 A", NBS Monograph 32,Suppl.,Washington (DC); also in Spectrochim. Acta B [23] 117 (1967)
Corliss C.H.	1970	J. Res. Nat. Bur. Std. (US) A [74] 781
Corliss C.H. and Bozman W.R.	1962	"Experimental transition probabilities for spectral lines of 70 elements", NBS Monograph 53, Washington (DC)
Cosmovici C.B., D'Anna E., D'Innocenzo A., Leggieri G., Perrone A. and Dirscherl R.	1977	Chem. Phys. Lett. [47] 241
Cottereau M.J. and Stepowski D.	1980	see: Crosley (Ed.) (1980), p.131
Cotton D.H. and Jenkins D.R.	1968	Trans. Faraday Soc. [64] 2988
Cotton D.H. and Jenkins D.R.	1969	Trans. Faraday Soc. [65] 1537
Cotton D.H. and Jenkins D.R.	1970	Spectrochim. Acta B [25] 283
Cotton D.H. and Jenkins D.R.	1971	Trans. Faraday Soc. [67] 730
Cottrell T.L.	1958	"The strengths of chemical bonds", Butterworths, London
Coulson C.A. and Zalewski K.	1962	Proc. Roy. Soc., Ser. A [268] 437
Cova S. and Longoni A.	1979	see: Omenetto (Ed.) (1979), Ch.7, p.411
Cowan R.D. and Dieke G.H.	1948	Rev. Mod. Phys. [20] 418
Cozens J.R.	1965	J. Electron. [19] 61
Cozens J.R. and Engel A. Von	1964	Adv. in Electronics and Electron Physics [20] 99
Crosley D.R. (Ed.)	1980	"Laser probes for combustion research", ACS symposium series 134, American Chemical Society, Washington (DC)
Croucher D.J. and Clark J.L.	1969	J. Phys. B [2] 603
Cummings G.A. and Hutton E.	1966	Combust. Flame [10] 195
Cundall R.B.	1969	in:"Transfer and storage of energy by molecules", vol. 1, (Eds. G.M. Burnett and A.M. North), Wiley, London, p.1
Cunningham P.T. and Link J.K.	1967	J. Opt. Soc. Am. [57] 1000
Curry S.M., Happer W., Tam A.C. and Yabuzaki T.	1978	Phys. Rev. Lett. [40] 67
Cuvellier J., Berlande J., Benoit C., Perrin M.Y., Mestdagh J.M. and Mesmay J. De	1979	J. Phys. B [12] L461
Cuvellier J., Fournier P.R., Gounand F. and Berlande J.	1973	Compt. Rend. B [276] 855
Cuvellier J., Fournier P.R., Gounand F., Pascale J. and Berlande J.	1975	Phys. Rev. A [11] 846
Czajkowski M. and Krause L.	1965	Can. J. Phys. [43] 1259
Czajkowski M. and Krause L.	1974	Can. J. Phys. [52] 2228
Czajkowski M. and Krause L.	1976	Can. J. Phys. [54] 603
Czajkowski M., Krause L. and Skardis G.M.	1973	Can. J. Phys. [51] 1582
Czajkowski M., Mc Gillis D.A. and Krause L.	1966	Can. J. Phys. [44] 741
Czajkowski M., Skardis G.M. and Krause L.	1973a	Can. J. Phys. [51] 334
Dagnall R.M., Smith D.J., Thompson K.C. and West T.S.	1969	Analyst (London) [94] 871
Dagnall R.M., Thompson K.C. and West T.S.	1967	Analyst (London) [92] 506
Dagnall R.M., Thompson K.C. and West T.S.	1967a	Atom. Absorption Newsl. [6] 117

BIBLIOGRAPHY

Dagnall R.M., Thompson K.C. and West T.S.	1968	Analyst (London) [93] 518
Dagnall R.M., Thompson K.C. and West T.S.	1969	Analyst (London) [94] 643
Dagnall R.M. and West T.S.	1968	Appl. Opt. [7] 1287
Daily J.W.	1976	Appl. Opt. [15] 955
Daily J.W.	1978	Appl. Opt. [17] 225
Daily J.W.	1979	Appl. Opt. [18] 360
Daily J.W. and Chan C.	1978	Combust. Flame [33] 47
Dalton M.L.	1965	Appl. Opt. [4] 603
Damburg R.J. and Fabrikant I.I.	1975	9th Int. Conf. on the Physics of Electronic and Atomic Collisions, Seattle (Wash) 1975; "Abstracts of papers", p.1117
D'Ans-Lax	1970	"D'Ans-Lax Taschenbuch fuer Chemiker und Physiker", Band III "Eigenschaften von Atomen und Molekuelen", (Eds. K. Schaefer und C. Synowietz), 3rd ed. Springer, Berlin
Darwent B. De B.	1970	"Bond dissociation energies in simple molecules", NSRDS-NBS 31, Washington (DC)
Dashevskaya E.I.	1979	Opt. Spectry. [46] 236
Dashevskaya E.I. and Nikitin E.E.	1967	Opt. Spectry. [22] 473
Dashevskaya E.I. and Nikitin E.E.	1969	6th Int. Conf. on the Physics of Electronic and Atomic Collisions, Cambridge (Mass), 1969 "Abstracts of papers"; MIT Press, Cambridge (Mass), p.652
Dashevskaya E.I., Nikitin E.E. and Reznikov A.I.	1970	J. Chem. Phys. [53] 1175
Dashevskaya E.I., Nikitin E.E. and Reznikov A.I.	1970a	Opt. Spectry. [29] 540
Dashevskaya E.I., Nikitin E.E., Voronin A.I. and Zembekov A.A.	1970	Can. J. Phys. [48] 981
Dashevskaya E.I., Voronin A.I. and Nikitin E.E.	1969	Can. J. Phys. [47] 1237
David D.J.	1964	Spectrochim. Acta [20] 1185
Davies D. Kenneth	1967	J. Appl. Phys. [38] 4713
Davies M.G.	1968	"Investigation of the excitation mechanism of the OH radical in hydrogen flames", Ph.D. Thesis, University of Tennessee, Knoxville (Tenn)
Davison W.D.	1968	J. Phys. B [1] 139
De Galan		see: Galan, L. de
De Jong		see: Jong, A.A. de
De Olivares		see: Olivares, D.R. de
De Vos		see: Vos, J.C. de
Dean A.M. and Kistiakowsky G.B.	1970	J. Chem. Phys. [53] 830
Dean J.A.	1960	"Flame photometry", McGraw-Hill, New York
Dean J.A. and Carnes W.J.	1962	Anal. Chem. [34] 192
Dean J.A. and Rains T.C. (Eds.)	1969	"Flame emission and atomic absorption spectrometry, vol.1, Theory", M. Dekker, New York
Dean J.A. and Rains T.C. (Eds.)	1971	"Flame emission and atomic absorption spectrometry, vol.2, Components and techniques", M. Dekker, New York
Dean J.A. and Rains T.C. (Eds.)	1975	"Flame emission and atomic absorption spectrometry, vol. 3 Elements and matrices", M. Dekker, New York
Dean J.A. and Stubblefield C.B.	1961	Anal. Chem. [33] 382
Deech J.S. and Baylis W.E.	1971	Can. J. Phys. [49] 90
Deech J.S., Pitre J. and Krause L.	1971	Can. J. Phys. [49] 1976
Deleage J.P., Kunth D., Testor G., Rostas F. and Roueff E.	1973	J. Phys. B [6] 1892
Demayo A., Hunter L.W. and Kruus P.	1968	Can. J. Chem. [46] 3151
Demtroeder W.	1962	Z. Physik [166] 42
Denton M.B. and Malmstadt H.V.	1971	Appl. Phys. Lett. [18] 485
Denton M.B. and Malmstadt H.V.	1972	Anal. Chem. [44] 241
Despain A.M. and Bell J.W.	1970	in: Aspen Int. Conf. on Fourier Spectroscopy, Aspen (Col) p.397
Deventer J.M.M. Van	1980	"Differential cross sections for the scattering of Na[S(1/2)] and Na[P(3/2)] by Xe[S(0)] (Determination of ground-state and excited state potentials)", Ph.D. Thesis, Utrecht, Netherl.
Dicke R.H.	1953	Phys. Rev. [89] 472
Dickens P.G., Linnett J.W. and Sovers O.	1962	Discussions Faraday Soc. [33] 52
Dickie L.O., Kelly F.M., Koh T.K., Mathur M.S. and Suk F.C.	1973	Can. J. Phys. [51] 1088
Diederichsen J. and Wolfhard H.G.	1956	Trans. Far. Soc. [52] 1102; Proc. Roy. Soc., Ser. A [236] 89
Dieke G. and Crosswhite H.	1962	J. Quant. Spectrosc. Radiat. Transfer [2] 97
Dijk C.A. Van	1977	Opt. Communic. [22] 343
Dijk C.A. Van	1978	"Two-photon excitation of higher sodium levels and population transfer in a flame", Ph.D. Thesis, Utrecht, Netherl.

Dijk C.A. Van and Alkemade C.Th.J.	1980	J. Quant. Spectrosc. Radiat. Transfer [23] 445
Dijk C.A. Van and Alkemade C.Th.J.	1980a	Combust. Flame [38] 37
Dijk C.A. Van and Alkemade C.Th.J.	1980b	see: Crosley (Ed.) (1980), p.183
Dijk C.A. Van, Alkemade C.Th.J. and Zeegers P.J.Th.	1978	Appl. Spectrosc. [32] 189
Dijk C.A. Van, Zeegers P.J.Th. and Alkemade C.Th.J.	1979	J. Quant. Spectrosc. Radiat. Transfer [21] 115
Dijk C.A. Van, Zeegers P.J.Th., Nienhuis G. and Alkemade C.Th.J.	1978	J. Quant. Spectrosc. Radiat. Transfer [20] 55
Dinning J.I.	1960	Anal. Chem. [32] 1475
Dixon R.N.	1963	Proc. Roy. Soc., Ser. A [275] 431
Dixon-Lewis G. and Greenberg J.B.	1975	J. Instit. Fuel [48] 132
Dixon-Lewis G. and Williams D.J.	1977	in: "Comprehensive chemical kinetics", vol.17 "Gas-phase combustion" (Eds. C.H. Bamford and C.F.H. Tipper), Elsevier, Amsterdam
Dixon-Lewis G., Wilson W.E. and Westenberg A.A.	1966	J. Chem. Phys. [44] 2877
Dmitrieva L.M., Dobrolizh B.V., Klucharev A.N. and Krivtsova N.V.	1977	Opt. Spectry. [42] 598
Dobrowolny M. and Engelman F.	1965	Nuovo Cimento [37] 965
Dodd J.N., Enemark E. and Gallagher A.	1969	J. Chem. Phys. [50] 4838
Doemeny L.J., Van Itallie F.J. and Martin R.M.	1969	Chem. Phys. Lett. [4] 302
Doherty G. and Jonathan N.	1964	Discussions Faraday Soc. [37] 73
Dolidze L.D. and Lebedev V.I.	1973	3rd Int. Congr. on Atomic Absorption and Atomic Fluorescence Spectrometry, Paris, 1971; "Collection of papers", Hilger, London, p.63
Donovan R.J. and Husain D.	1970	Chem. Rev. [70] 489
Douda B.E. and Bair E.J.	1974	J. Quant. Spectrosc. Radiat. Transfer [14] 1091
Dougherty G.J., Dunn M.R., Mc Ewan M.J. and Phillips L.F.	1971	Chem. Phys. Lett. [11] 124
Dougherty G.J., Mc Ewan M.J. and Phillips L.F.	1973	Combust. Flame [21] 253
Drake M.C. and Rosenblatt G.M.	1976	Chem. Phys. Lett. [44] 313
Drawin H.W.	1968	in:"Plasma diagnostics", (Ed. W. Lochte-Holtgreven), North-Holland, Amsterdam, Ch.14, p.842
Drowart J., Colin R. and Exsteen G.	1965	Trans. Faraday Soc. [61] 1376
Drowart J., Coppens P. and Smoes S.	1969	J. Chem. Phys. [50] 1046
Drowart J., De Maria G., Burns R.P. and Inghram M.G.	1960	J. Chem. Phys. [32] 1366
Drowart J., Exsteen G. and Verhaegen G.	1964	Trans. Faraday Soc. [60] 1920
Drowart J. and Goldfinger P.	1962	Ann. Rev. Phys. Chem. [13] 459
Drummond D.L. and Gallagher A.	1974	J. Chem. Phys. [60] 3426
Drummond G. and Barrow R.F.	1951	Trans. Faraday Soc. [47] 1275
Dueren R., Kwan K.Ch. and Pauly H.	1974	4th Int. Conf. on Atomic Physics, Heidelberg 1974, "Abstracts of papers", p.509
Dunken H., Pforr G., Mikkeleit W. and Geller K.	1964	Spectrochim. Acta B [20] 1531
Durie R.A., Johnson G.M. and Smith M.Y.	1971	Combust. Flame [17] 197
Earl B.L.	1973	"Photochemical studies of alkalihalide vapours", Ph.D. Thesis, Berkeley (Calif)
Earl B.L. and Herm R.R.	1973	Chem. Phys. Lett. [22] 95
Earl B.L. and Herm R.R.	1974	J. Chem. Phys. [60] 4568
Earl B.L., Herm R.R., Lin S.M. and Mims C.A.	1972	J. Chem. Phys. [56] 867
Eberhardt E.H.	1964	IEEE Trans. Nucl. Sci. [11] 48
Eckbreth A.C. and Hall R.J.	1978	see: Hastie (Ed.) (1979), p.943
Eckhard S. and Pueschel A.	1960	Z. Anal. Chem. [172] 334
Edelstein S.A.	1979	J. Chem. Phys. [70] 591
Edmonds F.N.	1968	J. Quant. Spectrosc. Radiat. Transfer [8] 1447
Edmondson H. and Heap M.P.	1970	Combust. Flame [15] 179

BIBLIOGRAPHY

Edwards H.E., Smith R.W. and Brinkley S.R.	1953	"The thermodynamics of combustion gases: Temperatures and composition of the combustion products of oxyacetylene flames", Bureau of Mines, Report of Investigations 4958, US Dept of the Interior, Washington (DC)
Edwards M.G.	1969	J. Phys. B [2] 719
Ehlers V.J. and Gallagher A.	1973	Phys. Rev. A [7] 1573
Ehrhardt H. and Willmann K.	1967	Z. Physik [204] 462
Elbel M.	1979	see: Hanle and Kleinpoppen (Eds.) (1979), Ch.29, p.1299
Elder P., Jerrick T. and Birkeland J.W.	1965	Appl. Opt. [4] 589
Elenbaas W.	1972	"Light sources", Philips Technical Library, McMillan, London
Elser R.C. and Winefordner J.D.	1971	Appl. Spectrosc. [25] 345
Elser R.C. and Winefordner J.D.	1972	Anal. Chem. [44] 698
Elwell W.T. and Gidley J.A.F.	1961	"Atomic absorption spectrophotometry", Pergamon, Oxford-London; 2nd ed. 1966
E.M.I. Photomultiplier Tubes	1970	Brochure P001/fP70, EMI Electronics Ltd, Electron Tube Div., Hayes (Middlesex) UK
E.M.I. Photomultiplier Tubes; Supplement	1972	Brochure P001 S/a72, EMI Electronics Ltd., Electron Tube Div., Hayes (Middlesex) UK
Emonds J.G.M.	1981	"A high-resolution Fourier spectrometer for the visible", Ph.D. Thesis, Utrecht, Netherl.
Enemark E.A. and Gallagher A.	1972	Phys. Rev. A [6] 192
Engel A. Von	1964	"Ionized gases", 2nd ed., Clarendon Press, Oxford, UK
Engel A. Von	1967	Brit. J. Appl. Phys. [18] 1661
Engel A. Von and Cozens J.R.	1964	Nature [202] 480
Engel A. Von and Cozens J.R.	1965	in: Wilsted (Ed.) (1965), p.123
Engleman R.Jr.	1969	J. Quant. Spectrosc. Radiat. Transfer [9] 391
Erdevdi N.M. and Shimon L.L.	1976	Opt. Spectry. [40] 443
Eremin A.V., Kulikovski A.A. and Naboko I.M.	1977	Chem. Phys. Lett. [45] 351
Ermisch W.	1966	Ann. Physik (7) [18] 271 and 379
Ernst R.R.	1965	Rev. Sci. Instrum. [36] 1689
Evans B.L. and Thompson K.T.	1969	J. Phys. E, ser. 2 [2] 327
Ewing J.J., Trainor D.W. and Yatsiv S.	1974	J. Chem. Phys. [61] 4433
Exton R.J.	1976	J. Quant. Spectrosc. Radiat. Transfer [16] 309
Exton R.J. and Snow W.L.	1978	J. Quant. Spectrosc. Radiat. Transfer [20] 1
Fabry M.	1976	J. Quant. Spectrosc. Radiat. Transfer [16] 127
Faddeyeva V.N. and Terentev N.M.	1961	"Mathematical tables", Pergamon, Oxford, UK
Falk H., Becker-Ross H. and Schiller H.	1971	Ann. Physik [26] 166
Fano U.	1963	Phys. Rev. [131] 259
Fano U. and Macek J.H.	1973	Rev. Mod. Phys. [45] 553
Farber M. and Srivastava R.D.	1973	Combust. Flame [20] 33
Farber M. and Srivastava R.D.	1973a	Combust. Flame [20] 43
Farber M. and Srivastava R.D.	1975	High Temp. Sci. [7] 74
Farber M. and Srivastava R.D.	1976	Combust. Flame [27] 99
Farr J.M. and Hindmarsh W.R.	1971	J. Phys. B [4] 568
Fassel V.A.	1966	Symp. on Flamespectrometric Methods of Analysis, Amer. Chem. Soc., Phoenix (Ariz) (unpublished)
Fassel V.A., Curry R.H. and Kniseley R.N.	1962	Spectrochim. Acta [18] 1127
Fassel V.A. and Golightly D.W.	1967	Anal. Chem. [39] 466
Fassel V.A., Myers R.B. and Kniseley R.N.	1963	Spectrochim. Acta [19] 1187
Fehsenfeld F.C., Evenson K.M. and Broida H.P.	1965	Rev. Sci. Instrum. [36] 294
Fehsenfeld F.C., Megill L.R. and Dropplmann L.K.	1968	J. Chem. Phys. [43] 3618
Fells I. and Harker J.H.	1967	J. Inst. Fuel [40] 477
Fenimore C.P.	1964	"Chemistry in premixed flames", Pergamon, Oxford, UK
Fenimore C.P. and Jones G.W.	1958	J. Phys. Chem. [62] 693
Fernandez M.A. and Bastiaans G.J.	1979	Appl. Spectrosc. [33] 145

Feugier A.	1970	Rev. Generale Thermique [9] 1045
Finn G.D. and Jefferies J.T.	1968	J. Quant. Spectrosc. Radiat. Transfer [8] 1675
Finn G.D. and Jefferies J.T.	1968a	J. Quant. Spectrosc. Radiat. Transfer [8] 1705
Fiorino J.A., Kniseley R.N. and Fassel V.A.	1968	Spectrochim. Acta B [23] 413
Fischer J. and Kropp R.	1960	Glastechn. Berichte [33] 380
Fishburne E.S., Bilwakesh R.R. and Edse R.	1967	Aerospace Res. Labs. Rep. ARL 67-0113, may 1967
Fisher E.H.	1977	Laser Focus [13] (no.11) 82
Fisher E.R. and Smith G.K.	1970	Chem. Phys. Lett. [6] 438
Fisher E.R. and Smith G.K.	1971	Appl. Opt. [10] 1803
Fisher E.R. and Smith G.K.	1972	Chem. Phys. Lett. [13] 448
Fite W.L.	1968	Conf. on Heavy Particle Collisions,Belfast, N.Irel., 1968; "Abstracts of papers"
Fitzgerald J.J. and Winefordner J.D.	1975	Revs. Anal. Chem. [2] 299
Fluorescence News	1969	march-april, [4] no.2, p.6
Foley H.M.	1946	Phys. Rev. [69] 616
Foner S.N. and Hudson R.L.	1953	J. Chem. Phys. [21] 1374
Fontijn A.	1966	J. Chem. Phys. [44] 1702
Fontijn A.	1972	in: "Progress in reaction kinetics", vol.6 (Ed. G. Porter) Pergamon, Oxford, UK, p.75
Fontijn A.	1974	Pure Appl. Chem. [39] 287
Fontijn A.	1977	Chem. Phys. Lett. [47] 142
Fontijn A. and Felder W.	1979	in:"Reactive intermediates in the gas phase. Generation and monitoring", (Ed. D.W. Setser), Academic Press, New York, Ch.2, p.59
Fontijn A., Felder W. and Houghton J.J.	1975	see: Proc. 15th Symp. (Int.) on Comb. (1974), p.775
Fontijn A., Felder W. and Houghton J.J.	1977	see: Proc. 16th Symp. (Int.) on Comb. (1977), p.871
Fontijn A., Kurzius S.C. and Houghton J.J.	1973	see: Proc. 14th Symp. (Int.) on Comb. (1973), p.167
Fontijn A., Kurzius S.C., Houghton J.J. and Emerson J.A.	1972	Rev. Sci. Instrum. [43] 726
Fontijn A., Sabadell A.J. and Ronco R.J.	1970	Anal. Chem. [42] 575
Foo P.D., Wiesenfeld J.R. and Husain D.	1975	Chem. Phys. Lett. [32] 443
Foo P.D., Wiesenfeld J.R., Yuen M.J. and Husain D.	1976	J. Phys. Chem. [80] 91
Forrester A.T.	1961	J. Opt. Soc. Am. [51] 253
Fowler G.N. and Preist T.W.	1972	J. Chem. Phys. [56] 1601
Fowler R.H. and Guggenheim E.A.	1965	"Statistical thermodynamics" (2nd ed. 1949, reprint 1965), University Press, Cambridge, UK
Fowler W.K. and Winefordner J.D.	1977	Anal. Chem. [49] 944
Fowles G.R.	1968	"Introduction to modern optics", Holt, Rinehart and Winston, New York
Fox R.L. and Jacobson H.C.	1969	Phys. Rev. [188] 232
Frank P. and Krauss L.	000	Z. Naturforsch., in preparation
Frank P. and Krauss L.	1974	Z. Naturforsch. A [29] 742
Frank P. and Krauss L.	1974a	Forschungsbericht 74-65, Institut fuer Reaktionskinetik der DFVLR, Stuttgard, W.Germ.
Frank P. and Krauss L.	1976	Z. Naturforsch. A [31] 1193
Franklin M., Baber C. and Koirtyohann S.R.	1976	Spectrochim. Acta B [31] 589
Franklin M.L., Horlick G. and Malmstadt H.V.	1969	Anal. Chem. [41] 2
Franzen W.	1959	Phys. Rev. [115] 850
Fraser L.M. and Winefordner J.D.	1971	Anal. Chem. [43] 1693
Fraser L.M. and Winefordner J.D.	1972	Anal. Chem. [44] 1444
French N.P.D. and Lawley K.P.	1977	Chem. Phys. [22] 105
Frie W.	1963	Ann. Physik, 7 Folge [10] 332
Frish S.E.	1970	"Spectroscopy of gas discharge plasma", Nauka, Leningrad, USSR
Frish S.E. and Bochkova O.P.	1963	Sov. Phys.-JETP [16] 237
Fristrom R.M.	1958	Combust. Flame [2] 103
Fristrom R.M.	1963	Science [140] 297
Fristrom R.M.	1963a	see: Proc. 9th Symp. (Int.) on Comb. (1963), p.560
Fristrom R.M. and Raezer S.	1956	Report APL/JHU-CF.2553, Appl. Phys. Lab., The Johns Hopkins University, Silver Spring (Md)
Fristrom R.M. and Westenberg A.A.	1965	"Flame structure", McGraw-Hill, New York
Friswell N.J. and Jenkins D.R.	1972	Combust. Flame [19] 197
Fujiwara K., Omenetto N., Bradshaw J.B., Bower J.N., Nikdel S. and Winefordner J.D.	1979	Spectrochim. Acta B [34] 317
Fukushima S.	1959	Mikrochim. Acta p.596

BIBLIOGRAPHY

Furuta N., Yoshimura E., Haraguchi H. and Fuwa K.	1978	Spectrochim. Acta B [33] 715
Fushiki Y. and Tsuchiya S.	1973	Chem. Phys. Lett. [22] 47
Futrelle R.P.	1972	Phys. Rev. A [5] 2162
Fuwa K.	1971	see: Winefordner (Ed.) (1971a), Ch.3, p.189
Gaillard-Cusin F. and James H.	1977	Combust. Flame [30] 211
Galan L. De	1969	Spectrochim. Acta B [24] 629
Galan L. De, Mc Gee W.W. and Winefordner J.D.	1967	Anal. Chim. Acta [37] 436
Galan L. De and Samaey G.F.	1970	Spectrochim. Acta B [25] 245
Galan L. De, Smith R. and Winefordner J.D.	1968	Spectrochim. Acta B [23] 521
Galan L. De and Wagenaar H.C.	1971	in: "Methodes physiques d'analyse", edition speciale du GAMS, Paris, 3ieme Congres Int. de Spectrometrie d'Absorption et de Fluorescence Atomique, Paris, p.10
Galan L. De and Winefordner J.D.	1966	Anal. Chem. [38] 1412
Galan L. De and Winefordner J.D.	1967	J. Quant. Spectrosc. Radiat. Transfer [7] 251
Galan L. De and Winefordner J.D.	1967a	J. Quant. Spectrosc. Radiat. Transfer [7] 703
Galan L. De and Winefordner J.D.	1968	Spectrochim. Acta B [23] 277
Galatry L.	1961	Phys. Rev. [122] 1218
Gallagher A.	1967	Phys. Rev. [157] 24
Gallagher A.	1968	Phys. Rev. [172] 88
Gallagher A.	1975	Phys. Rev. A [12] 133
Gallagher C.J. and Lapp M.	1967	13th Colloq. Spectrosc. Int., Ottawa, 1967; Hilger, London
Gallagher T.F. and Cooke W.E.	1979	Phys. Rev. A [19] 2161
Gallagher T.F., Cooke W.E. and Edelstein S.A.	1978	Phys. Rev. A [17] 125
Gallagher T.F., Cooke W.E. and Edelstein S.A.	1978a	Phys. Rev. A [17] 904
Gallagher T.F., Edelstein S.A. and Hill R.M.	1977	Phys. Rev. A [15] 1945
Gallagher T.F., Olson R.E., Cooke W.E., Edelstein S.A. and Hill R.M.	1977	Phys. Rev. A [16] 441
Garbuny M.	1965	"Optical physics", Academic Press, New York
Garga I.I., Aleksakhin I.S., Starodub V.P. and Zapesochnyi I.P.	1974	Opt. Spectry. [37] 482
Gatterer A., Junkes J., Salpeter E.W. and Rosen B.	1957	"Molecular spectra of metallic oxides", Citta del Vaticano
Gatzke J.	1963	Z. Phys. Chem. [223] 321
Gay J.C. and Schneider W.B.	1976	Z. Physik A [278] 211
Gaydon A.G.	1944	Proc. Roy. Soc., [183] 111
Gaydon A.G.	1946	Trans. Faraday Soc. [42] 292
Gaydon A.G.	1948	"Spectroscopy and combustion theory", 2nd ed., Chapman and Hall, London
Gaydon A.G.	1949	Nature [164] 22
Gaydon A.G.	1953	"Dissociation energies and spectra of diatomic molecules" 2nd ed., Chapman and Hall, London
Gaydon A.G.	1968	"Dissociation energies and spectra of diatomic molecules" 3rd ed., Chapman and Hall, London
Gaydon A.G.	1974	"The spectroscopy of flames", 2nd ed., Chapman and Hall, London
Gaydon A.G. and Hurle I.R.	1963	"The shocktube in high-temperature chemical physics", Chapman and Hall, London
Gaydon A.G. and Wolfhard H.G.	1970	"Flames; their structure, radiation and temperature", 3rd ed., Chapman and Hall, London; 4th ed. 1979
Gaydon A.G. and Wolfhard H.G.	1979	see: Gaydon and Wolfhard (1970)
Gedeon A., Edelstein S.A. and Davidovits P.	1971	J. Chem. Soc. [55] 5171
Gelbhaar B. and Hanle W.	1976	Z. Naturforsch. A [31] 87
Gelder Z. Van	1970	Spectrochim. Acta B [25] 669
Geltman S.	1969	"Topics in atomic collision theory", Academic Press, New York
Gerber C.R., Ishler N.H. and Borker E.	1951	Anal. Chem. [23] 684
Gersing E., Pauly H. and Schaedlich E.	1973	Faraday Discussions Chem. Soc. [55] 211
Gersten J.I. and Foley H.M.	1968	J. Opt. Soc. Am. [58] 7
Gibson J.H., Grossman W. and Cooke W.D.	1963	Anal. Chem. [35] 266
Gilardini A.	1972	"Low energy electron collisions in gases", Wiley, New York
Gilbert P.T.	1955	U.S. Patent Office no. 2714833
Gilbert P.T.	1963	10th Colloq. Spectrosc. Int., College Park, Maryland, 1962, Spartan Books, Washington (DC), p.171
Gilbert P.T.	1966	Symp. on Flamespectrometric Methods of Analysis, Amer. Chem. Soc., Phoenix (Ariz) (unpublished)
Gilbert P.T.	1970	see: Mavrodineanu (Ed.) (1970), Ch.5, p.181

...lbert P.T.	1971	"Flame spectra of the elements", in:"Handbook of chemistry and physics", (Ed. R.C. Weast), 51 ed., Chemical Rubber Co., Cleveland (Ohio), p.E-205
...lmore F.R.	1965	J. Quant. Spectrosc. Radiat. Transfer [5] 369
...lmore F.R., Bauer E. and Mc Gowan J.W.	1967	"A review of atomic and molecular excitation mechanisms in non-equilibrium gases up to 20,000 K", Memorandum RM-5202-ARPA, The Rand Corpor., S. Monica (Calif); J. Quant. Spectrosc. Radiat. Transfer [9] 157 (1969)
...lmore F.R., Bauer E. and Mc Gowan J.W.	1969	see: Gilmore, Bauer and McGowan (1967)
...nsel L.A.	1933	"Massentransport in Lichtbogen und Flammen und optische Bestimmung der Alkali-atomradien", Ph.D. Thesis, Utrecht
...asstone S., Laidler K.J. and Eyring H.	1941	"The theory of rate processes", McGraw-Hill, New York
...eason W.S. and Pertel R.	1971	Rev. Sci. Instrum. [42] 1638
...leditsch S.D. and Michael J.V.	1975	J. Phys. Chem. [79] 409
...olde M.F. and Thrush B.A.	1975	see: Bates and Bederson (Eds.) (1975), p.361
...oldman A.	1968	J. Quant. Spectrosc. Radiat. Transfer [8] 829
...ole J.L. and Zare R.N.	1972	J. Chem. Phys. [57] 5331
...olightly D.W., Kniseley R.N. and Fassel V.A.	1970	Spectrochim. Acta B [25] 451
...onzalez M.A., Karl G. and Watson P.J.S.	1972	J. Chem. Phys. [57] 4054
...ooderman W.J.	1944	J. Soc. Chem. Ind. [63] 351
...oodfellow G.I.	1967	Appl. Spectrosc. [21] 39
...ordeev S.V. and Shevtsov M.K.	1977	Opt. Spectry. [42] 577
...orokhov L.N., Gusarov A.V. and Panchevkov I.G.	1970	Zh. Fiz. Khim. [44] 269
...oulard R., Scala S.M. and Thomas R.N.	1968	"Radiative energy transfer", Pergamon, London
...ounand F., Cuvellier J., Fournier P.R. and Berlande J.	1976	J. Phys. (Paris) [37] L-169
...ounand F., Fournier P.R. and Berlande J.	1977	Phys. Rev. A [15] 2212
...ouy M.	1879	Compt. Rend. [88] 418
...radshteyn I.S. and Ryzhik I.M.	1965	"Tables of integrals, series and products", (transl.from the russian by A. Jeffrey), Academic Press, New York
...ranier J. and Granier R.	1973	J. Quant. Spectrosc. Radiat. Transfer [13] 473
...ranier J., Granier R. and Schuller F.	1975	J. Quant. Spectrosc. Radiat. Transfer [15] 619
...ranier R.	1969	Ph.D. Thesis, Paris
...ranier R., Granier J. and Schuller F.	1976	J. Quant. Spectrosc. Radiat. Transfer [16] 143
...ranzow A., Hoffman M.Z. and Lichtin N.N.	1969	J. Chem. Phys. [73] 4289
...ray A.L.	1974	Proc. Analyt. Div. Chem. Soc., july issue, p.182
...ray A.L.	1975	Proc. Analyt. Div. Chem. Soc., march issue, p.94
...ray A.L.	1975a	Analyst (London) [100] 289
...ray A.L.	1975b	Anal. Chem. [47] 600
...reen J.A. and Sugden T.M.	1963	see: Proc. 9th Symp. (Int.) on Comb. (1963), p.607
...reen R.B., Keller R.A., Luther G.G., Schenck P.K. and Travis J.C.	1976a	Appl. Phys. Lett. [29] 727
...reen R.B., Keller R.A., Schenck P.K., Travis J.C. and Luther G.G.	1976	J. Am. Chem. Soc. [98] 8517
...reen R.B., Travis J.C. and Keller R.A.	1976	Anal. Chem. [48] 1954
...reenstein H. and Bates Jr. C.W.	1975	J. Opt. Soc. Am. [65] 33
...reig J.R.	1965	Brit. J. Appl. Phys. [16] 957
...riem H.R.	1964	"Plasma spectroscopy", McGraw-Hill, New York
...rimley R.T., Burns R.P. and Inghram M.G.	1961	J. Chem. Phys. [35] 551
...rove E.L., Scott C.W. and Jones F.	1965	Talanta [12] 327
...rove R.E., Wu F.Y. and Ezekiel S.	1977	Phys. Rev. A [15] 227
...uenther R. and Janish G.	1971	Chemie-Ing.-Techn. [43] 975
...uido M., Gigli G. and Balducci G.	1972	J. Chem. Phys. [57] 373
...unter J.W.D., Grant G.R. and Shaw S.A.	1970	Appl. Opt. [9] 251
...urvich L.V., Khachkuruzov G.A., Medvedev V.A. and Veits I.V.	1962	"Thermodynamic properties of individual materials" (in russian), vol. 1 and 2, Nauk, Moscow
...urvich L.V., Novikov M.M. and Ryabova V.G.	1965	Opt. Spectry. [18] 68
...urvich L.V. and Ryabova V.G.	1964	High Temp. [2] 190
...urvich L.V. and Ryabova V.G.	1964a	High Temp. [2] 486
...urvich L.V. and Ryabova V.G.	1965	Opt. Spectry. [18] 76
...urvich L.V., Ryabova V.G., Khitrov A.N. and Starovoitov E.M.	1971	High Temp. [9] 261
...urvich L.V. and Veits I.V.	1958	Izv. Akad. Nauk. SSSR, Ser. Fiz. [22] 673
...ussak L.A., Rjabikov O.B. and Semenov E.S.	1973	Kinet. Katal. [14] 843
...utman D. and Schott G.L.	1967	J. Chem. Phys. [46] 4576
...utsche B. and Herrmann R.	1974	Z. Anal. Chem. [269] 260

BIBLIOGRAPHY

Haar D. Ter	1954	"Elements of statistical mechanics", Rinehart, New York
Habitz P.	1980	Chem. Phys. [54] 131
Haensch T.W.	1972	Appl. Opt. [11] 895
Haensch T.W., Varsanyi F. and Schawlow A.I.	1971	Appl. Phys. Lett. [18] 108
Hafner H. and Kleinpoppen H.	1967	Z. Physik [198] 315
Hall A.R. and Pierson G.S.	1969	see: Proc. 12th Symp. (Int.) on Comb. (1969), p.1025
Halls D.J. and Townshend A.	1966	Anal. Chim. Acta [36] 278
Halstead C.J. and Jenkins D.R.	1967	Combust. Flame [11] 362
Halstead C.J. and Jenkins D.R.	1968	Chem. Phys. Lett. [2] 281
Halstead C.J. and Jenkins D.R.	1969	Trans. Faraday Soc. [65] 3013
Ham D.O.	1974	J. Chem. Phys. [60] 1802
Ham N.S. and Hannaford P.	1979	J. Phys. B [12] L199
Hambly A.N. and Rann C.S.	1969	see: Dean and Rains (Eds.) (1969), Ch.8, p.241
Hampson P.J. and Gilles P.W.	1971	J. Chem. Phys. [55] 3712
Hand C.W. and Kistiakowski G.B.	1962	J. Chem. Phys. [37] 1239
Hanle W. and Kleinpoppen H. (Eds.)	1978	"Progress in atomic spectroscopy",Part A, Plenum, New Yo
Hanle W. and Kleinpoppen H. (Eds.)	1979	"Progress in atomic spectroscopy",Part B, Plenum, New Yo
Hannaford P.	1979	21th Colloq. Spectrosc. Int. and 8th Int. Conf. on Atomic Spectroscopy, Cambridge, UK; "Keynote Lectures", Heyden, London, Ch.19, p.250
Hannaford P. and Lowe R.M.	1976	J. Phys. B [9] 2595
Hannaford P. and Lowe R.M.	1977	Phys. Rev. Lett. [38] 650
Hannaford P. and Mc Donald D.C.	1978	J. Phys. B [11] 1177
Hansen J.E.	1978	J. Phys. B [11] L579
Hanson H.G.	1955	J. Chem. Phys. [23] 1391
Hanson H.G.	1967	J. Chem. Phys. [47] 4773
Haraguchi H., Fowler W.K., Johnson D.J. and Winefordner J.D.	1976	Spectrochim. Acta A [32] 1539
Haraguchi H., Smith B., Weeks S., Johnson D.J. and Winefordner J.D.	1977a	Appl. Spectrosc. [31] 156
Haraguchi H. and Winefordner J.D.	1977	Appl. Spectrosc. [31] 195
Harber R.A. and Sonnek G.E.	1966	Appl. Opt. [6] 1039
Harker J.H. and Allen D.A.	1969	J. Inst. Fuel [42] 183
Harrison G.R.	1948	"Wavelength tables with intensities in arc, spark or discharge tube", Wiley, New York
Hartmann H. (Ed.)	1968	"Chemische Elementarprozesse", Springer, Berlin
Hassler J.C. and Polanyi J.C.	1967	Discussions Faraday Soc. [44] 182
Hastie J.W. (Ed.)	1979	"Characterization of high temperature vapors and gases", Proc. of the "10th Materials Research Symposium" at NBS, Gaithersburg (Md) 1978; NBS Special Publication no. 561, Washington (DC)
Haug R, Rappenecker G. and Schmidt C.	1974	Chem. Phys. [5] 255
Hayhurst A.N.	1974	IEEE Trans. Plasma Sci. PS-2, no.3, 115
Hayhurst A.N. and Kittelson D.B.	1972	Nature [236] 136
Hayhurst A.N. and Kittelson D.B.	1972a	J. Chem. Soc., Chem. Communic. 422
Hayhurst A.N. and Kittelson D.B.	1972b	Combust. Flame [19] 306
Hayhurst A.N. and Kittelson D.B.	1974	Proc. Roy. Soc., Ser. A [338] 155
Hayhurst A.N. and Kittelson D.B.	1974a	Proc. Roy. Soc., Ser. A [338] 175
Hayhurst A.N. and Kittelson D.B.	1977	Combust. Flame [28] 137
Hayhurst A.N. and Kittelson D.B.	1978	Combust. Flame [31] 37
Hayhurst A.N., Kittelson D.B. and Telford N.R.	1977	Combust. Flame [28] 123
Hayhurst A.N., Mitchell F.R.G. and Telford N.R.	1971	Int. J. Mass. Spectrom. Ion. Phys. [7] 177
Hayhurst A.N. and Springett M.J.	1978	J. Chem. Soc., Faraday Trans.,I [74] 715
Hayhurst A.N. and Sugden T.M.	1965	20th IUPAC Congress, Symp. on Low Temperature Plasmas, Moscow
Hayhurst A.N. and Sugden T.M.	1966	Proc. Roy. Soc., Ser. A [293] 36
Hayhurst A.N. and Sugden T.M.	1967	Trans. Faraday Soc. [63] 1375
Hayhurst A.N. and Telford N.R.	1970	Trans. Faraday Soc. [66] 2784
Hayhurst A.N. and Telford N.R.	1972	Nature [235] 114
Hayhurst A.N. and Telford N.R.	1972a	J. Chem. Soc., Faraday Trans.,I [68] 237
Hayhurst A.N. and Telford N.R.	1974	J. Chem. Soc., Faraday Trans.,I [70] 1999
Hayhurst A.N. and Telford N.R.	1975	J. Chem. Soc., Faraday Trans.,I [71] 1352
Hayhurst A.N. and Telford N.R.	1977	Combust. Flame [28] 67
Hearn A.G.	1963	Proc. Phys. Soc. [84] 11
Heavens O,S.	1961	J. Opt. Soc. Am. [51] 1058
Hedges R.E.M., Drummond D.L. and Gallagher A.	1972	Phys. Rev. A [6] 1519
Heek H.F. Van	1970	Spectrochim. Acta B [25] 107
Heerdt J.A. Ter	1979	"The high pressure xenon lamp as a source of radiation" (in dutch, with a summary in english), Ph.D. Thesis, Utrecht, Netherl.
Hefferlin R. and Gearhart J.	1964	J. Quant. Spectrosc. Radiat. Transfer [4] 9
Heierman J.H.	1937	"Absolute intensity measurements in alkali-spectra", (in dutch), Ph.D. Thesis, Utrecht, Netherl.

Heinrichs J.	1968	Phys. Rev. [176] 141
Heitler W.	1954	"The quantum theory of radiation", 3rd ed., University Press, Oxford, UK; repr. 1957, 1960
Held A. and Stephens R.	1975	Can. J. Spectrosc. [20] 10
Held E.F.M. Van Der	1932	"Measurement of the transition probability 2P-1S for Na by absolute intensity measurements in flames" (in dutch), Ph.D. Thesis, Utrecht, Netherl.
Hell A., Ulrich W.F., Shifrin N. and Ramirez-Munoz J.	1968	Appl. Opt. [7] 1317
Henry B.R. and Kasha M.	1968	Ann. Rev. Phys. Chem. [19] 161
Herbert F.	1974	J. Quant. Spectrosc. Radiat. Transfer [14] 943
Hercher M.	1967	Appl. Opt. [6] 947
Herman L. and Coulaud G.	1970	J. Quant. Spectrosc. Radiat. Transfer [10] 571
Herrmann R.	1971	see: Dean and Rains (Eds.) (1971), Ch.2, p.57
Herrmann R.	1974	Z. Klin. Chem. Klin. Biochem. [12] 393
Herrmann R. and Alkemade C.Th.J.	1960	"Flammenphotometrie", 2nd ed., Springer, Berlin
Herrmann R. and Alkemade C.Th.J.	1963	"Chemical analysis by flame photometry" (translated by P.T. Gilbert), Interscience, New York
Hertel I.V.	1980	in:"The role of the excited state in chemical physics II", (Ed. J.Wm. McGowan), Wiley, New York
Hertel I.V., Hofmann H. and Rost K.A.	1976	Phys. Rev. Lett. [36] 861
Hertel I.V., Hofmann H. and Rost K.A.	1977	Chem. Phys. Lett. [47] 163
Hertel I.V., Hofmann H. and Rost K.A.	1977a	Phys. Rev. Lett. [38] 343
Hertel I.V. and Stoll W.	1976	J. Appl. Phys. [47] 214
Hertel I.V. and Stoll W.	1977	see: Bates and Bederson (Eds.) (1977), p.113
Herzberg G.	1950	"Molecular spectra and molecular structure. I. Spectra of diatomic molecules", 2nd ed., Van Nostrand, London; repr. 1964
Herzberg G.	1954	"Molecular spectra and molecular structure. II. Infrared and Raman spectra of polyatomic molecules", 6th ed., Van Nostrand, London
Herzberg G.	1966	"Molecular spectra and molecular structure. III. Electronic spectra and electronic structure of polyatomic molecules", Van Nostrand, London
Herzberg G.	1971	"The spectra and structure of simple free radicals", Cornell Univ. Press, Ithaca
Hessel M.M., Broida H.P. and Drullinger R.E.	1975	J. Appl. Phys. [46] 2317
Heusinkveld W.A.	1966	Metrologia [2] 61
Heydtmann H., Polanyi J.C. and Taguchi R.I.	1971	Appl. Opt. [10] 1755
Hieftje G.M.	1971	Appl. Spectrosc. [25] 653
Hieftje G.M.	1972	Unpublished work
Hieftje G.M.	1972a	Anal. Chem. [44] (no.6) 81A
Hieftje G.M.	1972b	Anal. Chem. [44] (no.7) 69A
Hieftje G.M. and Bystroff R.I.	1975	Spectrochim. Acta B [30] 187
Hieftje G.M., Bystroff R.I. and Lim R.	1973	Anal. Chem. [45] 253
Hieftje G.M. and Malmstadt H.V.	1968	Anal. Chem. [40] 1860
Hieftje G.M. and Malmstadt H.V.	1969	Anal. Chem. [41] 1735
Hildenbrand D.L.	1968	J. Chem. Phys. [48] 2457
Hildenbrand D.L.	1968a	J. Chem. Phys. [48] 3657
Hildenbrand D.L.	1970	J. Chem. Phys. [52] 5712
Hildenbrand D.L.	1972	J. Chem. Phys. [57] 4556
Hildenbrand D.L.	1973	Chem. Phys. Lett. [20] 127
Hildenbrand D.L.	1975	Chem. Phys. Lett. [34] 352
Hildenbrand D.L. and Murad E.	1970	J. Chem. Phys. [53] 3403
Hill R.M.	1979	J. Quant. Spectrosc. Radiat. Transfer [21] 19
Hindmarsh W.R.	1959	Mon. Notic. Roy. Astron. Soc. [119] 11
Hindmarsh W.R.	1963	Phys. Lett. [7] 115
Hindmarsh W.R. and Farr J.M.	1969	J. Phys. B [2] 1388
Hindmarsh W.R. and Farr J.M.	1972	in: "Progress in quantum electronics", vol 2, Pt. 3, (Eds. J.H. Sanders and S. Stenholm), Pergamon, Oxford
Hindmarsh W.R., Petford A.D. and Smith G.	1967	Proc. Roy. Soc., Ser. A [296] 297
Hinkley E.D. (Ed.)	1976	"Laser monitoring of the atmosphere", Springer, Berlin
Hinnov E.	1957	J. Opt. Soc. Am. [47] 151
Hinnov E. and Kohn H.	1957	J. Opt. Soc. Am. [47] 156
Hirschfelder J.O., Curtiss F.C. and Bird R.B.	1954	"Molecular theory of gases and liquids", Wiley, New York
Hobbs R.S., Kirkbright G.F. and West T.S.	1971	Talanta [18] 859
Hochstim A.R. (Ed.)	1969	"Bibliography of chemical kinetics and collision processes", IFI/Plenum, New York-Washington
Hochstim A.R. (Ed.)	1969a	"Kinetic processes in gases and plasmas", Academic Press, New York
Hofacker G.L.	1968	Ber. Bunsengesellschaft [72] 969
Hofmann F.W. and Kohn H.	1961	J. Opt. Soc. Am. [51] 512

BIBLIOGRAPHY

Hofmann F.W., Kohn H. and Schneider J.	1959	Z. Naturforsch. A [14] 11
Hofmann F.W., Kohn H. and Schneider J.	1961	J. Opt. Soc. Am. [51] 508
Hofmann H.	1977	"Energietransferspektroskopie und Polarisationsabhaengigkeiten beim Quenchprozess laserangeregter Natriumatome durch einfache Molekuele", Ph.D. Thesis, Kaiserslautern, W.Germ.
Hollander Tj.	1964	"Self-absorption, ionization and dissociation of metal vapour in flames", Ph.D. Thesis, Utrecht, Netherl.
Hollander Tj.	1968	AIAA Journal, p.385
Hollander Tj. and Broida H.P.	1967	J. Quant. Spectrosc. Radiat. Transfer [7] 965
Hollander Tj. and Broida H.P.	1969	Combust. Flame [13] 63
Hollander Tj., Jansen B.J. and Alkemade C.Th.J.	1977	J. Quant. Spectrosc. Radiat. Transfer [17] 657
Hollander Tj., Jansen B.J., Plaat J.J. and Alkemade C.Th.J.	1970	J. Quant. Spectrosc. Radiat. Transfer [10] 1301
Hollander Tj., Kalff P.J. and Alkemade C.Th.J.	1963	J. Chem. Phys. [39] 2558
Hollander Tj., Kalff P.J. and Alkemade C.Th.J.	1964	J. Quant. Spectrosc. Radiat. Transfer [4] 577
Hollander Tj., Lijnse P.L., Franken L.P.L. and Zeegers P.J.Th.	1972	J. Quant. Spectrosc. Radiat. Transfer [12] 1067
Hollander Tj., Lijnse P.L., Jansen B.J. and Franken L.P.L.	1973	J. Quant. Spectrosc. Radiat. Transfer [13] 669
Holstein T.	1947	Phys. Rev. [72] 1212
Holstein T.	1950	Phys. Rev. [79] 744
Holstein T.	1951	Phys. Rev. [83] 1159
Homer J.B. and Kistiakowski G.B.	1966	J. Chem. Phys. [45] 1359
Hooymayers H.P.	1966	"Quenching of excited alkali atoms and hydroxyl radicals and related effects in flames", Ph.D. Thesis, Utrecht, Netherl.
Hooymayers H.P.	1968	Spectrochim. Acta B [23] 567
Hooymayers H.P. and Alkemade C.Th.J.	1966	J. Quant. Spectrosc. Radiat. Transfer [6] 847
Hooymayers H.P. and Alkemade C.Th.J.	1966a	J. Quant. Spectrosc. Radiat. Transfer [6] 501
Hooymayers H.P. and Alkemade C.Th.J.	1967	J. Quant. Spectrosc. Radiat. Transfer [7] 495
Hooymayers H.P. and Alkemade C.Th.J.	1969	Chem. Phys. Lett. [4] 277
Hooymayers H.P. and Lijnse P.L.	1969	J. Quant. Spectrosc. Radiat. Transfer [9] 995
Hooymayers H.P. and Nienhuis G.	1968	J. Quant. Spectrosc. Radiat. Transfer [8] 955
Horiguchi H. and Tsuchiya S.	1977	Bull. Chem. Soc. Japan [50] 1661
Horlick G.	1972	Appl. Spectrosc. [26] 395
Horlick G.	1976	Appl. Spectrosc. [30] 113
Horlick G. and Codding E.G.	1973	Anal. Chem. [45] 1490
Horlick G. and Yuen W.K.	1978	Appl. Spectrosc. [32] 38
Hornbeck G.A. and Hopfield H.S.	1949	J. Chem. Phys. [17] 982
Hosch J.W. and Piepmeier E.H.	1978	Appl. Spectrosc. [32] 444
Hougen J.T.	1970	"The calculation of rotational energy levels and rotational line intensities in diatomic molecules", NBS Monograph 115, US Govt. Printing Office, Washington (DC)
Hrycyshyn E.S. and Krause L.	1969	Can. J. Phys. [47] 223
Hrycyshyn E.S. and Krause L.	1969a	Can. J. Phys. [47] 215
Hrycyshyn E.S. and Krause L.	1970	Can. J. Phys. [48] 2761
Hsieh J.C. and Baird J.C.	1972	Phys. Rev. A [6] 141
Hsu D.S.Y. and Lin M.C.	1976	Chem. Phys. Lett. [42] 78
Huber K.P. and Herzberg G.	1979	"Molecular spectra and molecular structure. IV. Constants of diatomic molecules". van Nostrand, London
Hudson B.C. and Curnutte B.	1966	Phys. Rev. [148] 60
Huebner H.J.	1933	Ann. Physik [17] 781
Huldt L.	1948	"Eine spektroskopische Untersuchung des elektrischen Lichtbogens und der Azetylen-Luftflamme", Ph.D. Thesis, Uppsala, Sweden
Huldt L. and Knall E.	1956	Arkiv Fysik [11] 229
Huldt L. and Lagerqvist A.	1950	Arkiv Fysik [2] 233
Huldt L. and Lagerqvist A.	1951	Arkiv Fysik [3] 525
Huldt L. and Lagerqvist A.	1952	Arkiv Fysik [5] 91
Huldt L. and Lagerqvist A.	1954	Z. Naturforsch. A [9] 358
Hulpke E., Paul E. and Paul W.	1964	Z. Physik [177] 257
Human H.C.G.	1970	Ph.D. Thesis, Pretoria, South Africa
Human H.C.G. and Zeegers P.J.Th.	1975	Spectrochim. Acta B [30] 203
Human H.C.G., Zeegers P.J.Th. and Elst J.A. Van	1974	Spectrochim. Acta B [29] 111
Hummer D.G.	1965	Mem. R. Astr. Soc. [70] 1
Humphrey L.M., Gallagher T.F., Cooke W.E. and Edelstein S.A.	1978	Phys. Rev. A [18] 1383

Hunt B.L. and Sibulkin M.	1967	J. Quant. Spectrosc. Radiat. Transfer [7] 951
Hunziker H.E.	1971	IBM J. Res. Develop. [15] 10
Hurk J. Van Der	1974	"Origin and excitation energy of visible alkaline earth bands in flames", Ph.D. Thesis, Utrecht, Netherl.
Hurk J. Van Der, Hollander Tj. and Alkemade C.Th.J.	1973	J. Quant. Spectrosc. Radiat. Transfer [13] 273
Hurk J. Van Der, Hollander Tj. and Alkemade C.Th.J.	1974	J. Quant. Spectrosc. Radiat. Transfer [14] 1167
Hurk J. Van Der, Hollander Tj. and Alkemade C.Th.J.	1975	J. Quant. Spectrosc. Radiat. Transfer [15] 113
Hurle I.R.	1964	J. Chem. Phys. [41] 3911
Husain D.	1977	Ber. Bunsengesellschaft [81] 168
Husain D. and Littler J.G.F.	1974	Int. J. Chem. Kinet. [6] 61
Husain D. and Norris P.E.	1978	Chem. Phys. Lett. [53] 474
Inaba H.	1976	see: Hinkley (Ed.) (1976), Ch.5, p.153
Inaba H., Shimizu Y. and Tsuji Y.	1975	Jap. J. Appl. Phys. [14], Suppl. 14-1, p.23
Inghram M.G., Chupka W.A. and Berkowitz J.	1957	Mem. Soc. Roy. Sci. Liege [18] 513
Inghram M.G., Chupka W.A. and Porter R.F.	1955	J. Chem. Phys. [23] 2159
Ingle J.D. and Crouch S.R.	1971	Anal. Chem. [43] 1331
Ingle J.D. and Crouch S.R.	1972	Anal. Chem. [44] 777
Ingle J.D. and Crouch S.R.	1972a	Anal. Chem. [44] 785
Ingle J.D. and Crouch S.R.	1972b	Anal. Chem. [44] 1709
Ioli N., Strumia F. and Moretti A.	1971	J. Opt. Soc. Am. [61] 1251
Isaak G.R.	1961	Nature [4762] 373
Iupac	1970	"Manual of symbols and terminology for physicochemical quantities and units". Butterworths, London; Pure Appl. Chem. [21] 1. Revised and extended in: Information bulletin no.24:"Recommended names and symbols for light and related electromagnetic radiation", 1972,IUPAC
Iupac	1972	"Nomenclature, symbols, units and their usage in spectrochemical analysis. I."General atomic emission spectroscopy". Butterworths, London; Pure Appl. Chem. [30] 653; Spectrochim. Acta B [33] 219 (1978)
Iupac	1976	"Nomenclature, symbols, units and their usage in spectrochemical analysis.II. Data interpretation". Butterworths, London; Pure Appl. Chem. [45] 99; Spectrochim. Acta B [33] 241 (1978)
Iupac	1976a	"Nomenclature, symbols, units and their usage in spectrochemical analysis. III. Analytical flame spectroscopy and associated non-flame procedures". Butterworths, London; Pure Appl. Chem. [45] 105; Spectrochim. Acta B [33] 247 (1978)
Jacobson H.C.	1971	Phys. Rev. A [4] 1368
Jacquinot P.	1954	J. Opt. Soc. Am. [44] 761
Jacquinot P.	1976	see: Shimoda (Ed.) (1976), Ch.3, p.52
Jaffe H.	1962	"Atlas of analysis lines", Hilger, London
Jaffe J.H., Rank D.H. and Wiggins T.A.	1955	J. Opt. Soc. Am. [45] 636
James C.G. and Sugden T.M.	1953	Nature [171] 428
James C.G. and Sugden T.M.	1955	Proc. Roy. Soc., Ser. A [227] 312
James C.G. and Sugden T.M.	1955a	Nature [175] 333
James C.G. and Sugden T.M.	1955b	Nature [175] 252
James C.G. and Sugden T.M.	1958	Proc. Roy. Soc., Ser. A [248] 238
James J.F. and Sternberg R.S.	1969	"The design of optical spectrometers", Chapman and Hall, London
Janaf	1960	JANAF Thermochemical Tables, The Dow Chemical Co., Midland (Mich)
Janaf	1966	JANAF Thermochemical Tables, first addendum, The Dow Chemical Co., Midland (Mich)
Janaf	1971	JANAF Thermochemical Tables, 2nd ed. (Eds. D.R. Stull and H. Prophet), NSRDS, NBS no. 37, Washington (DC)
Janaf	1974	JANAF Thermochemical Tables, 1974 Supplement; Chase M.W., Cornutt J.L., Hu A.T., Prophet H., Syverud A.N. and Walker L.C., J. Phys. Chem. Ref. Data [3] 311 (1974)
Janaf	1975	JANAF Thermochemical Tables, 1975 Supplement; Chase M.W., Cornutt J.L., Hu A.T., Syverud A.N. and Walker L.C., J. Phys. Chem. Ref. Data [4] 1 (1975)
Janev R.K.	1976	see: Bates and Bederson (Eds.) (1976), p.1

BIBLIOGRAPHY

Jansen B.J.	1976	"Atomic spectral line profiles in flames; an experimental study", Ph.D. Thesis, Utrecht, Netherl.
Jansen B.J. and Hollander Tj.	1973	Proc. of Europhys. Study Conf. on "Spectral line broadening and related topics", Meudon, France
Jansen B.J. and Hollander Tj.	1977	Spectrochim. Acta B [32] 165
Jansen B.J., Hollander Tj. and Alkemade C.Th.J.	1977	J. Quant. Spectrosc. Radiat. Transfer [17] 187
Jansen B.J., Hollander Tj. and Alkemade C.Th.J.	1977a	J. Quant. Spectrosc. Radiat. Transfer [17] 663
Jansen B.J., Hollander Tj. and Alkemade C.Th.J.	1977b	J. Quant. Spectrosc. Radiat. Transfer [17] 695
Jansen B.J., Hollander Tj. and Franken L.P.L.	1974	Spectrochim. Acta B [29] 39
Jansen B.J., Hollander Tj. and Helvoort H. Van	1977	J. Quant. Spectrosc. Radiat. Transfer [17] 193
Jansson P.A. and Korb C.L.	1968	J. Quant. Spectrosc. Radiat. Transfer [8] 1399
Jefferies J.T.	1968	"Spectral line formation", Blaisdell, Waltham, Mass.
Jenkins D.R.	1966	Proc. Roy. Soc., Ser. A [293] 493
Jenkins D.R.	1967	Spectrochim. Acta B [23] 167
Jenkins D.R.	1968	Conf. on Heavy Particle Collisions, Belfast, 1968, "Abstracts of papers"
Jenkins D.R.	1968a	Trans. Faraday Soc. [64] 36
Jenkins D.R.	1968b	Proc. Roy. Soc., Ser. A [306] 413
Jenkins D.R.	1968c	Proc. Roy. Soc., Ser. A [303] 453
Jenkins D.R.	1968d	Proc. Roy. Soc., Ser. A [303] 467
Jenkins D.R.	1969	Proc. Roy. Soc., Ser. A [313] 551
Jenkins D.R.	1969a	Int. Conf. on Atomic Spectroscopy, Sheffield, UK, 1969; "Abstracts of papers", Hilger, London
Jenkins D.R. and Sugden T.M.	1969	see: Dean and Rains (Eds.) (1969), Ch.5, p.151
Jennings D.A., Braun W. and Broida H.P.	1973	J. Chem. Phys. [59] 4305
Jensen D.E.	1965	Ph.D. Thesis, Cambridge, UK
Jensen D.E.	1968	Combust. Flame [12] 261
Jensen D.E.	1969	J. Chem. Phys. [51] 4674
Jensen D.E.	1970	J. Phys. Chem. [74] 209
Jensen D.E. and Jones G.A.	1972	J. Chem. Soc., Faraday Trans.,I [68] 259
Jensen D.E. and Jones G.A.	1973	J. Chem. Soc., Faraday Trans.,I [69] 1448
Jensen D.E. and Jones G.A.	1978	Combust. Flame [32] 1
Jensen D.E., Jones G.A. and Mace A.C.H.	1979	J. Chem. Soc., Faraday Trans.,I [75] 2377
Jensen D.E. and Miller W.J.	1969	"Thermodynamic studies in metal containing flames", Report TP-223, Aerochem Research Labs, Princeton, N.J.
Jensen D.E. and Miller W.J.	1970	J. Chem. Phys. [53] 3287
Jensen D.E. and Padley P.J.	1966	Trans. Faraday Soc. [62] 2132
Jensen D.E. and Padley P.J.	1966a	Trans. Faraday Soc. [62] 2140
Jensen D.E. and Padley P.J.	1967	see: Proc. 11th Symp. (Int.) on Comb. (1967), p.351
Jensen D.E. and Travers B.E.L.	1973	Proc. IUPAC Int. Symp. on Plasma Chemistry, Kiel, W.Germ; Plasma Science [2] 34 (1974)
Johnson D.J., Plankey F.W. and Winefordner J.D.	1974	Anal. Chem. [46] 1898
Johnson G.M. and Smith M.Y.	1972	Spectrochim. Acta B [27] 269
Johnson R.W. and Schrenk W.G.	1964	Appl. Spectrosc. [18] 144
Johnson S.E.	1972	J. Chem. Phys. [56] 149
Johnston H.S.	1968	Ber. Bunsengesellschaft [72] 959
Jonah C.D., Zare R.N. and Ottinger Ch.	1972	J. Chem. Phys. [56] 263
Jones W.G. and Walsh A.	1960	Spectrochim. Acta [16] 249
Jong A.A. De, Kircz J.G., Alkemade C.Th.J. and Valk F. Van Der		to appear in: Chem. Phys (1981)
Jong A.A. De and Valk F. Van Der	1979	J. Phys. B [12] L561
Jongerius M.J.	1981	"A study of radiation redistribution and broadening of the Na-D lines due to Ar and N2 perturbers in flames and vapor cells", Ph.D. Thesis, Utrecht, Netherl.
Jongerius M.J., Bergen A.R.D. Van, Hollander Tj. and Alkemade C.Th.J.		to appear in: J. Quant. Spectrosc. Radiat. Transfer (1981)
Jongerius M.J., Bergen A.R.D. Van, Hollander Tj. and Alkemade C.Th.J.	1981	J. Quant. Spectrosc. Radiat. Transfer [25] 1
Jongerius M.J., Bij J.J. Van Der, Hollander Tj. and Alkemade C.Th.J.	1978	J. Quant. Spectrosc. Radiat. Transfer [20] 609
Jongerius M.J., Hollander Tj. and Alkemade C.Th.J.	1978	J. Quant. Spectrosc. Radiat. Transfer [20] 599
Jordan J.A. and Franken P.A.	1966	Phys. Rev. [142] 20
Kaiser H.	1970	Anal. Chem. [42] (no.2) 24A; idem [42] (no.4) 26A
Kaiser H. and Menzies A.C.	1968	"The limit of detection of a complete analytical procedure", Adam Hilger, London

Kalff P.J.	1971	"Alkaline earth compounds in flames and N2-alkali energy transfer in molecular beams", Ph.D. Thesis, Utrecht, Neth.
Kalff P.J. and Alkemade C.Th.J.	1972	Combust. Flame [19] 257
Kalff P.J. and Alkemade C.Th.J.	1973	J. Chem. Phys. [59] 2572; erratum in idem [69] 1698 (1974)
Kalff P.J., Hollander Tj. and Alkemade C.Th.J.	1965	J. Chem. Phys. [43] 2299
Kallend A.S.	1967	Combust. Flame [11] 81
Kallend A.S.	1967a	Trans. Faraday Soc. [63] 2442
Kallend A.S.	1972	Combust. Flame [19] 227
Kaneko M., Mori Y. and Tanaka I.	1968	J. Chem. Phys. [48] 4468
Karl G., Kruus P. and Polanyi J.C.	1967	J. Chem. Phys. [46] 224
Karl G., Kruus P., Polanyi J.C. and Smith I.W.M.	1967	J. Chem. Phys. [46] 244
Karl G. and Polanyi J.C.	1963	J. Chem. Phys. [38] 271
Karstensen F. and Schramm J.	1967	J. Opt. Soc. Am. [57] 654
Kashireninov O.E., Kuznetsov V.A. and Manelis G.B.	1977	AIAA Journal [15] 1035
Kaskan W.E.	1958	Combust. Flame [2] 229
Kaskan W.E.	1958a	J. Chem. Phys. [28] 729; idem [29] 1420
Kaskan W.E.	1959	J. Chem. Phys. [31] 944
Kaskan W.E.	1959a	Combust. Flame [3] 39
Kaskan W.E.	1965	see: Proc. 10th Symp. (Int.) on Comb. (1965), p.41
Kaufman F. and Parkes D.A.	1970	Trans. Faraday Soc. [66] 1579
Kaufman M., Wharton L. and Klemperer W.	1965	J. Chem. Phys. [43] 943
Keene J.P.	1963	Rev. Sci. Instrum. [34] 11
Keitz H.A.E.	1971	"Light calculations and measurements", McMillan, London
Keller R.A. and Travis J.C.	1979	see: Omenetto (Ed.) (1979), Ch.8, p.493
Kelly F.M., Koh T.K. and Mathur M.S.	1974	Can. J. Phys. [52] 1438
Kelly R. and Padley P.J.	1967	Nature [216] 258
Kelly R. and Padley P.J.	1969	Trans. Faraday Soc. [65] 355
Kelly R. and Padley P.J.	1969a	Trans. Faraday Soc. [65] 367
Kelly R. and Padley P.J.	1970	Trans. Faraday Soc. [66] 1127
Kelly R. and Padley P.J.	1971	Trans. Faraday Soc. [67] 740
Kelly R. and Padley P.J.	1971a	Trans. Faraday Soc. [67] 1384
Kelly R. and Padley P.J.	1972	Proc. Roy. Soc., Ser. A [327] 345
Kempter V.	1975	Adv. Chem. Phys. XXX 417
Kempter V., Koch W., Kuebler B., Mecklenbrauck W. and Schmidt C.	1974	Chem. Phys. Lett. [24] 117
Kempter V., Koch W. and Schmidt C.	1974	J. Phys. B [7] 1306
Kempter V., Kuebler B., Le Breton P., Lorek J. and Mecklenbrauck W.	1973	Chem. Phys. Lett. [21] 164
Kempter V., Kuebler B. and Mecklenbrauck W.	1974	J. Phys. B [7] 149
Kempter V., Mecklenbrauck W., Menzinger M. and Schlier Ch.	1971	Chem. Phys. Lett. [11] 353
Kempter V., Mecklenbrauck W., Menzinger M., Schuller G., Herschbach D.R. and Schlier Ch.	1970	Chem. Phys. Lett. [6] 97
Kenneth Davies D.	1967	J. Appl. Phys. [38] 4713
Keyes R.J. and Kingston R.H.	1972	Phys. Today [25] (no.3) 48
Khitrov A.N., Ryabova V.G. and Gurvich L.V.	1973	High Temp. [11] 1005
Kibble B.P., Copley G. and Krause L.	1967	Phys. Rev. [159] 11
Kibble B.P., Copley G. and Krause L.	1967a	Phys. Rev. [153] 9
Kibble B.P., Copley G. and Krause L.	1967b	5th Int. Conf. on the Physics of Electronic and Atomic Collisions, Leningrad, USSR, 1967; "Abstracts of papers", Nauka, Leningrad, p.508
Kieffer L.J.	1976	"Bibliography of low energy electron and photon cross section data", NBS Special Publication 426, Boulder (Col)
Kielkopf J.F.	1971	Spectrochim. Acta B [26] 371
Kielkopf J.F.	1976	J. Phys. B [9] 547
Kielkopf J.F., Davis J.F. and Gwinn J.A.	1970	J. Chem. Phys. [53] 2605
Kielkopf J.F. and Gwinn J.A.	1968	J. Chem. Phys. [48] 5570
Kimble H.J. and Mandel L.	1976	Phys. Rev. A [13] 2123
Kimble H.J. and Mandel L.	1977	Phys. Rev. A [15] 689
King I.R.	1957	J. Chem. Phys. [27] 817
King I.R.	1958	J. Chem. Phys. [29] 681
King I.R.	1962	J. Chem. Phys. [36] 553
King I.R.	1963	see: Shuler and Fenn (Eds.) (1963), p.197
King I.R. and Scheurig J.T.	1966	Rev. Sci. Instrum. [37] 1219
Kirkbright G.F., Peters M.K. and West T.S.	1967	Talanta [14] 789
Kirkbright G.F. and Sargent M.	1968	Analyst (London) [93] 552

BIBLIOGRAPHY

Kirkbright G.F. and Sargent M.	1970	Spectrochim. Acta B [25] 577
Kirkbright G.F. and Sargent M.	1974	"Atomic absorption and fluorescence spectroscopy", Academic Press, London
Kirkbright G.F., Semb A. and West T.S.	1968	Spectrosc. Lett. [1] 7
Kirkbright G.F. and Troccoli O.E.	1973	Spectrochim. Acta B [28] 33
Kirkbright G.F., Troccoli O.E. and Vetter S.	1973	Spectrochim. Acta B [28] 1
Kirkbright G.F. and West T.S.	1968	Appl. Opt. [7] 1305
Kirsten W.J. and Bertilsson G.O.B.	1966	Anal. Chem. [38] 648
Kistemaker J.	1973	in: "Invited papers of 11th Int. Conf. on Phenomena in Ionized Gases" (Eds. L. Pekarek and L. Laska), Prague, p.431
Klein L. and Margenau H.	1959	J. Chem. Phys. [30] 1556
Kleine R.	1973	Chemie-Ing.-Techn. [45] 300
Kley D. and Welge K.H.	1968	J. Chem. Phys. [49] 2870
Klucharev A.N., Lazarenko A.V. and Vujnovic V.	1980	J. Phys. B [13] 1143
Klucharev A.N. and Sepman V. Ya.	1976	Opt. Spectry. [40] 626
Knapp D.O., Omenetto N., Hart L.P., Plankey F.W. and Winefordner J.D.	1974	Anal. Chim. Acta [69] 455
Knewstubb P.F. and Sugden T.M.	1958	Nature [181] 474
Knewstubb P.F. and Sugden T.M.	1958a	Trans. Faraday Soc. [54] 372
Knewstubb P.F. and Sugden T.M.	1959	see: Proc. 7th Symp. (Int.) on Comb. (1959), p.247
Knewstubb P.F. and Sugden T.M.	1960	Proc. Roy. Soc., Ser. A [255] 520
Kniseley R.N., Butler C.C. and Fassel V.A.	1975	see: Dean and Rains (Eds.) (1975), Ch.5, p.116
Kniseley R.N., D'Silva A.P. and Fassel V.A.	1963	Anal. Chem. [35] 910
Kock M. and Richter J.	1969	Ann. Physik, 7 Folge [24] 30
Koenig E.	1971	Physica [62] 393
Koennen G.P., Haring A. and Vries A.E. De	1975	Chem. Phys. Lett. [30] 11
Kohn H.	1914	Ann. Physik [44] 749
Koirtyohann S.R. and Pickett E.E.	1968	Anal. Chem. [40] 2068
Koirtyohann S.R. and Pickett E.E.	1971	Spectrochim. Acta B [26] 349
Koizumi H. and Yasuda K.	1975	Anal. Chem. [47] 1679
Koizumi H. and Yasuda K.	1976	Spectrochim. Acta B [31] 237
Kondratiev V.N.	1959	see: Proc. 7th Symp. (Int.) on Comb. (1959), p.41
Kondratiev V.N. (Ed.)	1974	"Chemical bond energy, Ionization potentials", Nauka, Mosco
Kondratiev V.N. and Nikitin E.E.	1967	Russian Chem. Revs. [36] 872
Konjevic R. and Konjevic N.	1973	Spectrosc. Lett. [6] 177
Korchevoi Yu.P., Lukashenko V.I. and Khil'Ko I.N.	1976	Sov. Phys. Tech. Phys. [21] 1356
Korchevoi Yu.P. and Przonski A.M.	1967	Sov. Phys.-JETP [24] 1089
Kosasa K., Maruyama Y. and Urano Y.	1977	Jap. J. Appl. Phys. [16] 187
Kostkowski H.J. and Bass A.M.	1956	J. Opt. Soc. Am. [46] 1060
Kovacs I.	1969	"Rotational structure in the spectra of diatomic molecules", Hilger, London
Kovaleva T.A., Melamid A.E., Pertsev A.N. and Pisarevskii A.N.	1966	Instruments and experimental techniques, no.5, p.1025 (transl. from the russian)
Kraulinya E.K.	1964	Opt. Spectry. [17] 250
Kraulinya E.K.	1968	"The sensitized fluorescence of metal vapour mixtures". (collection of russian papers). Part 1. Publishing House of State Univ. of Latvin, Riga, USSR
Kraulinya E.K.	1969	"The sensitized fluorescence of metal vapour mixtures", (collection of russian papers). Part 2. Publishing House of State Univ. of Latvin, Riga, USSR
Kraulinya E.K., Kartasheva L.I. and Bryukhovetski A.P.	1975	9th Int. Conf. on the Physics of Electronic and Atomic Collisions, Seattle (Wash) 1975; "Abstracts of papers", University of Washington Press, Seattle (Wash), p.251
Kraulinya E.K. and Lezdin A.E.	1966	Opt. Spectry. [20] 304
Kraulinya E.K. and Lezdin A.E.	1977	Opt. Spectry. [42] 451
Kraulinya E.K., Liepa S.Ya. and Skudra A.Ya.	1976	Opt. Spectry. [40] 440
Krause H.F. and Datz S.	1976	Chem. Phys. Lett. [41] 339
Krause H.F., Datz S. and Johnson S.G.	1973	J. Chem. Phys. [58] 367
Krause H.F., Fricke J. and Fite W.L.	1972	J. Chem. Phys. [56] 4593
Krause L.	1966	Appl. Opt. [5] 1375
Krause L.	1972	in: "The physics of electronic and atomic collisions" (7th ICPEAC, Amsterdam 1971) (Eds. T.R. Govers and F.J. de Heer), North-Holland, Amsterdam, p.65

Krause L.	1975	in: "The excited state in chemical physics", Part 1 (Ed. J.Wm. McGowan), Interscience-Wiley, New York, Ch.4, p.268
Krause L., Siara I.N. and Dubois R.U.	1975	9th Int. Conf. on the Physics of Electronic and Atomic Collisions, Seattle (Wash) 1975; "Abstracts of papers", University of Washington Press, Seattle (Wash), p.521
Krauss M.	1968	J. Res. Nat. Bur. Std. (US) A [72] 553
Krellmann H., Siefart E. and Weihreter E.	1975	J. Phys. B [8] 2608
Kropp R.	1960	"Stoereinfluesse von Partnern in der Flammenspektrometrie sowie Methoden zu ihrer Eliminierung", Ph.D. Thesis, Frankfurt/M, W.Germ.
Kuang-Pang Li	1976	Anal. Chem. [48] 2050
Kuhl J., Marowsky G., Kunstmann P. and Schmidt W.	1972	Z. Naturforsch. A [27] 601
Kuhl J., Marowsky G. and Torge R.	1972	Anal. Chem. [44] 375
Kuhl J., Neumann S. and Kriese M.	1973	Z. Naturforsch. A [28] 273
Kuhn H.G.	1937	Proc. Roy. Soc., Ser. A [158] 212
Kuhn H.G.	1969	"Atomic spectra", 2nd ed., Longmans, London
Kuhn H.G. and London F.	1934	Phil. Mag. [18] 983
Kumar A. and Pandya T.P.	1970	Indian J. Pure Appl. Phys. [8] 42
Kumar L. and Callaway J.	1968	Phys. Lett. A [28] 385
Kunz C., Harteck P. and Dondes S.	1967	J. Chem. Phys. [46] 4157
Kunze P.	1931	Ann. Phys. (Leipzig) [8] 500
Kusch H.J., Roendigs G. and Wendt K.	1977	J. Quant. Spectrosc. Radiat. Transfer [17] 53
Kuznetsova L.A., Kuzmenko N.E., Kuzyakov Yu. Ya and Plastinin Yu.A.	1974	Sov. Phys.-Usp. [17] 405
Lacmann K. and Herschbach D.R.	1970	Chem. Phys. Lett. [6] 106
Lagerqvist A. and Huldt L.	1953	Z. Naturforsch. A [8] 493
Lagerqvist A. and Huldt L.	1954	Arkiv Fysik [8] 427
Lagerqvist A. and Huldt L.	1954a	Z. Naturforsch. A [9] 991
Lagerqvist A. and Huldt L.	1955	Arkiv Fysik [9] 227
Lagerqvist A., Lind E. and Barrow R.F.	1950	Proc. Phys. Soc. A [63] 1132
Laidler K.J.	1942	J. Chem. Phys. [10] 34
Laidler K.J.	1942a	J. Chem. Phys. [10] 43
Laidler K.J.	1949	J. Chem. Phys. [17] 221
Laidler K.J.	1955	"The chemical kinetics of excited states", Clarendon Oxford, UK
Laidler K.J.	1962	Discussions Faraday Soc. [33] 91
Laidler K.J.	1963	"Reaction kinetics, vol.1: Homogeneous gas reactions", Pergamon, Oxford-London
Laidler K.J. and Shuler K.E.	1951	Chem. Rev. [48] 153
Land P.L.	1971	Rev. Sci. Instrum. [42] 420
Landau L.D. and Lifshitz E.M.	1965	"Quantum-mechanics, non-relativistic theory", vol.3 of "Course of theoretical physics" (transl. from the russian), 2nd ed., Pergamon, Oxford, UK
Landolt-Boernstein	1950	"Physikalisch-chemische Tabellen", 6th ed., vol.1, p.1, Springer, Berlin
Lang W.	1966a	Z. Anal. Chem. [223] 241
Lang W.	1966b	Z. Anal. Chem. [219] 321
Langmuir J.	1961	"Collected works", vol.4, Pergamon, New York
Lapp M.	1974	see: Lapp and Penney (Eds.) (1974), p.107
Lapp M. and Penney C.M. (Eds.)	1974	"Laser Raman gas diagnostics", Plenum, New York
Lapple C.E.	1960	Chem. Eng. [1] 20th may
Lapworth K.C.	1976	J. Quant. Spectrosc. Radiat. Transfer [16] 357
Larkins P.L. and Willis J.B.	1974	Spectrochim. Acta B [29] 319
Larson G.F., Fassel V.A., Winge R.K. and Kniseley R.N.	1976	Appl. Spectrosc. [30] 384
Laudman A. and Novick R.	1964	Phys. Rev. A [134] 56
Lawrence G.M., Link J.K. and King R.	1965	Astrophys. J. [141] 293
Lawton J. and Weinberg F.	1969	"Electrical aspects of combustion", Clarendon, Oxford, UK
Learner R.C.M.	1962	Proc. Roy. Soc., Ser. A [269] 311
Lebedev V.I. and Dolidze L.D.	1969	Int. Conf.on Atomic Absorption Spectroscopy, Sheffield, UK, 1969; Abstracts of papers no.F-5, Hilger, London
Lecler D. and Laniepce B.	1976	J. Phys. (Paris) [37] 55
Lee C.M.	1974	Phys. Rev. A [10] 584
Lee P.H., Broida H.P., Braun W. and Herron J.T.	1973	J. Photochem. [2] 165
Leep D. and Gallagher A.	1975	9th Int. Conf. on the Physics of Electronic and Atomic Collisions, Seattle (Wash) 1975; "Abstracts of papers", University of Washington Press, Seattle (Wash), p.1111

BIBLIOGRAPHY

Leger A., Delmas B., Klein J. and Cheveigne S. De | 1976 | Revue Physique Appliquee [11] 309
Letfus V. | 1966 | Opt. Spectry. [21] 371
Levin L.A. and Budick B. | 1966 | Bull. Am. Phys. Soc., Ser. 2 [11] 455
Levine R.D. and Bernstein R.B. | 1972 | Chem. Phys. Lett. [15] 1
Levine R.D. and Bernstein R.B. | 1974 | "Molecular reaction dynamics", Clarendon, Oxford, UK
Levitt B.P. | 1965 | J. Chem. Phys. [42] 1038
Lewis B. and Elbe G. Von | 1961 | "Combustion, flames and explosions of gases", 2nd ed., Academic Press, New York-London
Lewis B., Pease R.N. and Taylor H.S. (Eds.) | 1956 | "Combustion processes", Princeton University Press, Princeton (NJ)
Lewis E.L., Mc Namara L.F. and Michels H.H. | 1971 | Phys. Rev. A [3] 1939

Lide D.R. and Kuczkowski | 1967 | J. Chem. Phys. [46] 4768
Light J.C., Ross J. and Shuler K.E. | 1969 | see: Hochstim (Ed.) (1969a)
Lijnse P.L. | 1972 | "Review of literature on quenching, excitation and doublet-mixing collision cross sections for the first resonance doublets of the alkalis", Report i398 of Physical Laboratory of Univ., Utrecht, Netherl.
Lijnse P.L. | 1973 | "Electronic-excitation transfer collisions in flames", Ph.D. Thesis, Utrecht, Netherl.
Lijnse P.L. | 1973a | Chem. Phys. Lett. [18] 73
Lijnse P.L. | 1973b | "Discussion of the present state-of-knowledge of electronic-excitation transfer collisions and a survey of literature", Report of Physical Laboratory of Univ., Utrecht, Netherl.
Lijnse P.L. | 1974 | J. Quant. Spectrosc. Radiat. Transfer [14] 1143
Lijnse P.L. | 1974a | J. Quant. Spectrosc. Radiat. Transfer [14] 1195
Lijnse P.L. and Elsenaar R.J. | 1972 | J. Quant. Spectrosc. Radiat. Transfer [12] 1115
Lijnse P.L. and Hornman J.C. | 1974 | J. Quant. Spectrosc. Radiat. Transfer [14] 1079
Lijnse P.L. and Maas C.J. Van Der | 1973 | J. Quant. Spectrosc. Radiat. Transfer [13] 741
Lijnse P.L., Zeegers P.J.Th. and Alkemade C.Th.J. | 1973 | J. Quant. Spectrosc. Radiat. Transfer [13] 1033
Lijnse P.L., Zeegers P.J.Th. and Alkemade C.Th.J. | 1973a | J. Quant. Spectrosc. Radiat. Transfer [13] 1301
Lin S.M. and Weston R.E. | 1976 | J. Chem. Phys. [65] 1443
Lindholm E. | 1942 | Ph.D. Thesis, Uppsala, Sweden
Lindholm E. | 1945 | Arkiv Math. Astron. Fysik A [32] 17
Link J.K. | 1966 | J. Opt. Soc. Am. [56] 1195
Lion K.S. | 1959 | "Instrumentation in scientific research", McGraw-Hill, New York
Lipson S.G. and Lipson H. | 1969 | "Optical physics", Cambridge Univ. Press, Cambridge, UK
Liu K. and Parson J.M. | 1976 | J. Chem. Phys. [65] 815
Liu M.B. and Wahlbeck P.G. | 1975 | J. Chem. Phys. [63] 1694

Lochte-Holtgreven W. | 1968 | "Plasma diagnostics", North-Holland, Amsterdam
Lochte-Holtgreven W. and Richter J. | 1968 | see: Lochte-Holtgreven (Ed.) (1968), Ch.5, p.250
Loesch H.J. and Brieger M. | 1977 | 10th Int. Conf. on the Physics of Electronic and Atomic Collisions, Paris, 1977; "Abstracts of papers", Commissariat a l'Energie Atomique, Paris, p.220
Longhurst R.S. | 1967 | "Geometrical and physical optics", Longmans, London
Los J. and Kleyn A.W. | 1979 | in: "Alkali halide vapors" (Eds. P. Davidovits and D.L. McFadden), Academic Press, New York, Ch.8, p.275
Losen J. and Behmenburg W. | 1973 | Z. Naturforsch. A [28] 1620
Loth C., Astier R. and Meyer Y.H. | 1972 | J. Phys. E, ser. 2 [5] 169
Lovett R.J. and Parsons M.L. | 1977 | Appl. Spectrosc. [31] 424
Lowe R.M. | 1971 | Spectrochim. Acta B [26] 201
Lowrance J.L. and Zucchino P. | 1974 | see: Carleton (Ed.) (1974), Ch.6.3, p.277

Lucht R.P. and Laurendeau N.M. | 1979 | Combust. Flame [34] 215
Lueck, K.C. and Thielen W. | 1978 | J. Quant. Spectrosc. Radiat. Transfer [20] 71
Lueck K.C. and Mueller F.J. | 1977 | J. Quant. Spectrosc. Radiat. Transfer [17] 403
Lue-Yung Chow Chiu | 1969 | 6th Int. Conf. on the Physics of Electronic and Atomic Collisions, Cambridge (Mass) 1969; "Abstracts of papers", M.I.T. Press, Cambridge (Mass), p.648
Lukaszewicz M. | 1974 | Ph.D. Thesis, Torun, Poland (in polish)
Lukaszewicz M. | 1975 | Bull.Acad.Pol.Sci.,Ser.Sci.Math.,Astr.,Phys. [23] 501
Luria M., Eckstrom D.J. and Benson S.W. | 1976 | J. Chem. Phys. [64] 3103
Lurio A. and Novick R. | 1964 | Phys. Rev. A [134] 608
Lurio A., Zafra R.L.De and Goshen R.J. | 1964 | Phys. Rev. A [134] 1198
Luther J. and Walter H. | 1966 | Phys. Lett. [23] 551

L'Vov B.V. | 1961 | Spectrochim. Acta [17] 761
L'Vov B.V. | 1965 | Opt. Spectry. [19] 282

L'Vov B.V.	1966	"Atomic absorption spectral analysis" (in russian), Nauka, Moscow
L'Vov B.V.	1970	"Atomic absorption spectrochemical analysis", (transl. from the russian by J.H. Dixon), Hilger, London
L'Vov B.V.	1970a	Opt. Spectry. [28] 8
L'Vov B.V.	1972	Meth. Phys. Anal. (GAMS) [8] 3
L'Vov B.V.	1972a	J. Quant. Spectrosc. Radiat. Transfer [12] 651
L'Vov B.V., Katskov D.A., Kruglikova L.P. and Polzik L.K.	1976	Spectrochim. Acta B [31] 49
Maassen J.D.M.	1976	Internal report "Atom in the radiation field" (in dutch), Inst. Theor. Phys., Univ. of Utrecht, Netherl.
Mac Donald D.K.C.	1962	"Noise and fluctuations: an introduction", Wiley, New York
Mack J.E., Mc Nutt D.P., Roesler F.L. and Chabbal R.	1963	Appl. Opt. [2] 873
Maeda M., Ishitsuka F., Matsumoto M. and Miyazoe Y.	1976	Opt. Communic. [17] 302
Magee J.L. and Ri T.	1941	J. Chem. Phys. [9] 638
Mahan B.H.	1975	J. Chem. Educ. [52] 299
Mahan G.D.	1972	Phys. Lett. A [39] 145; Phys. Rev. A [6] 1273
Mahan G.D. and Lapp M.	1969	Phys. Rev. [179] 19
Malmstadt H.V., Enke C.G. and Crouch S.R.	1974	"Electronic measurements for scientists", Benjamin, Menlo Park (Calif)
Malmstadt H.V., Enke C.G. and Toren E.C.	1962	"Electronics for scientists", Benjamin, Menlo Park (Calif)
Malmstadt H.V., Franklin M.L. and Horlick G.	1972	Anal. Chem. [44] (no.8) 63A
Mandelberg H.I.	1968	Conf. on Heavy Particle Collisions, Belfast, 1968, "Abstracts of papers"
Mandelstam S.L.	1939	Comptes Rendus (Doklady) de l'Academie des Sciences de l'URSS [22] 403
Mandelstam S.L.	1978	Spectrochim. Acta B [33] 577
Mandl A. and Hao-Lin Chen	1976	Phys. Rev. A [14] 264
Mansell R.E.	1970	Spectrochim. Acta B [25] 219
Mansfield J.M., Bratzel M.P., Norgordon H.O., Knapp D.O., Zacha K.E. and Winefordner J.D.	1968	Spectrochim. Acta B [23] 389
Mansfield J.M. and Winefordner J.D.	1968	Anal. Chim. Acta [40] 357
Margenau H.	1935	Phys. Rev. [48] 755
Margenau H.	1946	Phys. Rev. [69] 508
Margenau H. and Kestner N.R.	1971	"Theory of intermolecular forces", 2nd ed., Pergamon, Oxford, UK
Margenau H. and Landwehr G.W.	1958	Internal Report, Yale University, New Haven, 1958
Margenau H. and Lewis M.	1959	Rev. Mod. Phys. [31] 569
Margoshes M. and Scribner B.F.	1962	10th Colloq. Spectrosc. Int., College Park (Maryland), 1962; "Abstracts of papers", no.83
Maria G.De, Burns R.P., Drowart J. and Inghram M.G.	1960	J. Chem. Phys. [32] 1373
Marinkovic M. and Vickers T.J.	1970	Anal. Chem. [42] 1613
Marr G.V. and Creek D.M.	1968	Proc. Roy. Soc., Ser. A [304] 245
Martin M.D. and Thomas E.L.	1967	Phys. Lett. A [25] 637
Martin P.J., Clemens E., Zehnle L. and Kempter V.	1979	Z. Phys. A [289] 373
Mashinskii A.L.	1970	Opt. Spectry. [28] 1
Massey H.S.W., Burhop E.H.S. and Gilbody H.B.	1969	"Electronic and ionic impact phenomena", vol.1: "Collision of electrons with atoms", 2nd ed., Clarendon, Oxford, UK
Massey H.S.W., Burhop E.H.S. and Gilbody H.B.	1969a	idem. vol.2: "Collision of electrons with molecules and photo-ionization"
Massey H.S.W., Burhop E.H.S. and Gilbody H.B.	1971	idem. vol.3: "Slow collisions of heavy particles"
Massey H.S.W. and Gilbody H.B.	1974	idem. vol.4: "Recombination and fast collisions of heavy particles", Univ. Press, Oxford, UK
Matossi F., Mayer R. and Rauscher E.	1949	Phys. Rev. [76] 760
Mavrodineanu R.	1961	Spectrochim. Acta [17] 1016
Mavrodineanu R. (Ed.)	1970	"Analytical flame spectroscopy (selected topics)", Philips Technical Library, McMillan, London
Mavrodineanu R. and Boiteux H.	1965	"Flame spectroscopy", Wiley, New York
Mavrodineanu R. and Hughes R.C.	1968	Appl. Opt. [7] 1281
Mc Carroll B.	1970	Rev. Sci. Instrum. [41] 279
Mc Cartan D.G. and Farr J.M.	1976	J. Phys. B [9] 985
Mc Cartan D.G., Farr J.M. and Hindmarsh W.R.	1974	J. Phys. B [7] 208
Mc Cartan D.G. and Hindmarsh W.R.	1969	J. Phys. B [2] 1396
Mc Cartan D.G. and Lwin N.	1977	J. Phys. B [10] 17L
Mc Carthy W.J.	1971	in: Winefordner (Ed.) (1971a), Appendix, p.493
Mc Carthy W.J., Parsons M.L. and Winefordner J.D.	1967	Spectrochim. Acta B [23] 25

BIBLIOGRAPHY

Mc Cune J.E. 1963 Phys. Today [16] (no.4) 44
Mc Donald J.K. and Innes K.K. 1969 J. Mol. Spectry. [32] 501
Mc Donald J.R., Baronavski A.P., Pasternack L., Lemont S. and Harvey A.B. 1979 see: Hastie (Ed.) (1979), p.817
Mc Ewan M.J. and Phillips L.F. 1965 Combust. Flame [9] 420
Mc Ewan M.J. and Phillips L.F. 1966 Trans. Faraday Soc. [62] 1717
Mc Ewan M.J. and Phillips L.F. 1967 Combust. Flame [11] 63
Mc Gee W.W. and Winefordner J.D. 1967 J. Quant. Spectrosc. Radiat. Transfer [7] 261
Mc Gillis D.A. and Krause L. 1967 5th Int. Conf. on the Physics of Electronic and Atomic Collisions, Leningrad, USSR, 1967; "Abstracts of papers", Nauka, Leningrad, USSR, p.505
Mc Gillis D.A. and Krause L. 1967a Phys. Rev. [153] 44
Mc Gillis D.A. and Krause L. 1968 Can. J. Phys. [46] 25
Mc Gillis D.A. and Krause L. 1968a Can. J. Phys. [46] 1051
Mc Gillis D.A. and Krause L. 1969 Can. J. Phys. [47] 473

Measures R.M., Drewell N. and Kwong H.S. 1977 Phys. Rev. A [16] 1093
Medvedev V.A. 1961 Russian J. of Phys. Chem. [35] 729
Meggers W.F., Corliss C.H. and Scribner B.F. 1975 "Tables of spectral-line intensities", NBS Monograph 145, Parts 1 and 2; US Govt.Printing Office, Washington (DC)
Meitlis V.P. and Fishman V.M. 1972 High Temp. [10] 633
Menis O. and Rains T.C. 1970 see: Mavrodineanu (Ed.) (1970), Ch.2, p.47
Mentall J.E., Krause H.F. and Fite W.L. 1967 5th Int. Conf. on the Physics of Electronic and Atomic Collisions, Leningrad, USSR, 1967; "Abstracts of papers", Nauka, Leningrad, USSR, p.515
Mentall J.E., Krause H.F. and Fite W.L. 1967a Discussions Faraday Soc. [44] 157

Michael J.V. and Suess G.N. 1974 J. Phys. Chem. [78] 482
Michel R.G., Coleman J. and Winefordner J.D. 1978 Spectrochim. Acta B [33] 195
Michels A., Kluiver H. De and Seldam C.A. Ten 1959 Physica [25] 1231
Milatz J.M.W., Endt P.M., Alkemade C.Th.J. and Olink J.Th. 1948 Physica [14] 260
Miles B.M. and Wiese W.L. 1970 "Bibliography on atomic transition probabilities", US Govt. Printing Office, Washington (DC)
Miles B.M. and Wiese W.L. 1970a Atomic Data [1] 1
Miller W.J. 1968 "Ionization in combustion processes, oxidation and combustion reviews", Elsevier, Amsterdam
Miller W.J. 1972 J. Chem. Phys. [57] 2354
Miller W.J. 1976 AIAA 14th Aerospace Sciences Meeting Washington (DC) 1976, paper no. 76-135 published in "Progress in astronautics and aeronautics" (Ed. M. Summerfield), American Institute of Aeronautics and Astronautics, p.25
Miller W.J. 1978 see: Hastie (Ed.) (1979), p.443
Miller W.J. 1978a J. Chem. Phys. [69] 3709
Milne E.L. 1967 General Motors Corp. Defense Res. Lab. Report TR67-41
Milne E.L. 1970 J. Chem. Phys. [52] 5360
Milne T.A. and Greene F.T. 1965 see: Proc. 10th Symp. (Int.) on Comb. (1965), p.153
Milne T.A. and Greene F.T. 1966 J. Chem. Phys. [44] 2444
Mitchell A.C.G. and Zemansky M.W. 1961 "Resonance radiation and excited atoms", reprint, Univ. Press, Cambridge, UK
Mitchell D.G. 1970 Technicon Intern. Congress, November 1970, New York
Mizushima M. 1967 J. Quant. Spectrosc. Radiat. Transfer [7] 505

Moise N.L. 1966 Astrophys. J. [144] 774
Mollow B.R. 1969 Phys. Rev. [188] 1969
Mollow B.R. 1972 Phys. Rev. A [5] 1522
Mollow B.R. 1973 Phys. Rev. A [8] 1949
Montgareuil P.G. De 1954 "Contribution a l'etude des interactions chimiques dans les flammes", Ph.D. Thesis, Paris
Moore C.E. 1949 "Atomic energy levels, as derived from the analyses of optical spectra"; vol.1: H-V (1949); vol.2: Cr-Nb (1952); vol.3: Mo-La, Hf-Ac (1958), NBS Circular 467, US Govt. Printing Office, Washington (DC)
Moore C.E. 1950 "An ultraviolet multiplet table", Sect. 1-5, NBS Circular 488, US Govt. Printing Office, Washington (DC)
Moore C.E. 1959 "A multiplet table of astrophysical interest", NBS Technical Note 36, US Dept. Commerce, Washington (DC)
Moore C.E. 1970 "Ionization potentials and ionization limits derived from the analyses of optical spectra", NSRDS, NBS 34, Washington (DC)
Moore C.E. 1972 "Atomic energy levels", vol.1, 2 and 3, NSRDS, NBS 35, Washington (DC)

Moore W.J.	1962	"Physical chemistry", Prentice-Hall, Englewood Cliffs, N.J.
Moores D.L. and Norcross D.W.	1972	J. Phys. B [5] 1490
Mori Y.	1962	Bull. Chem. Soc. Japan [25] 1584
Morley C.	1980	18th Symp. (Int.) on Combustion, Waterloo, Can, 1980; "Abstracts of papers", The Combustion Institute, Pittsburgh (Pa), p.1
Morozov E.N. and Sosinskii M.L.	1973	Opt. Spectry. [35] 463
Morozova N.G. and Startsev G.P.	1964	Opt. Spectry. [17] 174
Morrissey B.W.	1975	J. Chem. Educ. [52] 296
Morten P.D., Freeman C.G., Claridge F.C. and Phillips L.F.	1974	J. Photochem. [3] 285
Morton G.A.	1968	J. Appl. Opt. [7] 1
Mossholder N.V., Fassel V.A. and Kniseley R.N.	1973	Anal. Chem. [45] 1614
Mossotti V.G. and Abercrombie F.N.	1971	16th Colloq. Spectrosc. Int., Heidelberg, 1971: Preprints vol.2, Hilger, London, p.441
Mott N.F. and Massey H.S.W.	1965	"The theory of atomic collisions", 3rd ed., Clarendon, Oxford, UK
Moutinho A.M.C., Baede A.P.M. and Los J.	1970	Physica [51] 432
Mukherjee N.R., Fueno T., Eyring H. and Ree T.	1962	see: Proc. 8th Symp. (Int.) on Comb. (1962), p.1
Muller C.H., Schofield K. and Steinberg M.	1978	see: Hastie (Ed.) (1979), p.855
Muller C.H., Schofield K. and Steinberg M.	1978a	Chem. Phys. Lett. [57] 364
Muller C.H., Schofield K. and Steinberg M.	1980	J. Chem. Phys. [72] 6620
Muller C.H., Schofield K., Steinberg M. and Broida H.P.	1979	see: Proc. 17th Symp. (Int.) on Comb. (1979), p.867
Murad E. and Hildenbrand D.L.	1975	J. Chem. Phys. [63] 1133
Murthy N.S. and Bagare S.P.	1978	J. Phys. B [11] 623
Murthy N.S., Bagare S.P. and Murthy B.N.	1978	J. Quant. Spectrosc. Radiat. Transfer [19] 455
Murthy N.S. and Prahllad U.D.	1978	J. Phys. B [11] 825
Myers B.F. and Bartle E.R.	1968	J. Chem. Phys. [48] 3935
Nadler M. and Kaskan W.E.	1970	J. Quant. Spectrosc. Radiat. Transfer [10] 25
Nakamura H.	1968	J. Phys. Soc. Japan [24] 1353
Naudeix D.	1971	Rev. Generale Thermique [10] 643
Naumann F. and Michel K.W.	1972	Z. Physik [255] 348
Neuhaus H.	1959	Arkiv Fysik [14] 551
Neumann K.K. and Knoche K.F.	1963	Chemie-Ing.-Techn. [35] 631
Newman R.N. and Page F.M.	1970	Combust. Flame [15] 317
Newman R.N. and Page F.M.	1971	Combust. Flame [17] 149
Nicholls R.W. and Stewart A.L.	1962	see: Bates (Ed.) (1962), Ch.2, p.47
Nienhuis G.	1973	Physica [66] 245
Nienhuis G.	1974	Physica [74] 157
Nienhuis G.	1977	Physica C [85] 151
Nienhuis G.	1978	Physica C [93] 393
Nienhuis G.	1978a	Physica C [95] 266
Nienhuis G.	1980	J. Phys. B [13] 2217
Nienhuis G. and Alkemade C.Th.J.	1976	Physica C [81] 181
Nienhuis G. and Schuller F.	1977a	Physica C [92] 397
Nienhuis G. and Schuller F.	1977b	Physica C [92] 409
Nienhuis G. and Schuller F.	1978	Physica C [94] 394
Niewitecka B. and Krause L.	1975	Can. J. Phys. [53] 1499
Niewitecka B., Skalinski T. and Krause L.	1974	4th Int. Conf. on Atomic Physics, Heidelberg, 1974, "Abstracts of papers", p.567
Nikitin E.E.	1965	J. Quant. Spectrosc. Radiat. Transfer [5] 435
Nikitin E.E.	1965a	J. Chem. Phys. [43] 744
Nikitin E.E.	1966	Combust. Flame [10] 381
Nikitin E.E.	1967a	High Temp. [5] 199
Nikitin E.E.	1967b	Opt. Spectry. [22] 379
Nikitin E.E.	1968	Chem. Phys. Lett. [2] 402
Nikitin E.E.	1968a	see: Hartmann (Ed.) (1968), p.43
Nikitin E.E.	1968b	Ber. Bunseges. Phys. Chem. [72] 949
Nikitin E.E.	1974	"Theory of elementary atomic and molecular processes in gases" (transl. from the russian by M.J.Kearsley), Clarendon, Oxford, UK
Nikitin E.E.	1975	in: "The excited state in chemical physics, I" (Ed. J.Wm. McGowan), Interscience-Wiley, New York, Ch.5, p.317
Nikitin E.E. and Bykhovskii V.K.	1964	Opt. Spectry. [17] 444

BIBLIOGRAPHY

Nikitin E.E. and Reznikov A.I.	1971	Chem. Phys. Lett. [8] 161
Nitis G.J., Svoboda V. and Winefordner J.D.	1972	Spectrochim. Acta B [27] 345
Norrish R.G.W. and Smith W.	1941	Proc. Roy. Soc., Ser. A [176] 295
Novikov M.M. and Gurvich L.V.	1967	Opt. Spectry. [22] 395
Odintsov A.I.	1963	Opt. Spectry. [14] 172
O'Haver T.C. and Winefordner J.D.	1968	Appl. Opt. [7] 1647
Okuyama M. and Zung J.T.	1967	J. Chem. Phys. [46] 1580
Olivares D.R. De	1976	"Studies into tunable-laser-excited atomic fluorescence spectrometry", Ph.D. Thesis, Bloomington (Ind)
Olivares D.R. De and Hieftje G.M.	1978	Spectrochim. Acta B [33] 79
Olivero J.J. and Longbothum R.L.	1977	J. Quant. Spectrosc. Radiat. Transfer [17] 233
Olson G.G.	1972	Optical Spectra, jan., , p.38
Olson R.E.	1977	Combust. Flame [30] 243
Olson R.E.	1977a	Phys. Rev. A [15] 631
Olson R.E., Smith F.T. and Bauer E.	1971	Appl. Opt. [10] 1848
Omenetto N.	1975	5th Int. Conf. on Atomic Spectroscopy, Melbourne, 1975, "Book of abstracts", paper no.6
Omenetto N.(Ed.)	1979	"Analytical laser spectroscopy", Wiley, New York
Omenetto N., Benetti P., Hart L.P., Winefordner J.D. and Alkemade C.Th.J.	1973	Spectrochim. Acta B [28] 289
Omenetto N., Benetti P. and Rossi G.	1972	Spectrochim. Acta B [27] 453
Omenetto N., Bower J., Bradshaw J., Dijk C.A. Van and Winefordner J.D.	1980	J. Quant. Spectrosc. Radiat. Transfer [24] 147
Omenetto N., Browner R. and Winefordner J.D.	1972	Anal. Chem. [44] 1683
Omenetto N., Epstein M.S., Bradshaw J.D., Bayer S., Horvath J.J. and Winefordner J.D.	1979	J. Quant. Spectrosc. Radiat. Transfer [22] 287
Omenetto N., Fraser L.M. and Winefordner J.D.	1973	Appl. Spectrosc. Revs. [7] 147
Omenetto N., Hart L.P., Benetti P. and Winefordner J.D.	1973	Spectrochim. Acta B [28] 301
Omenetto N., Hart L.P. and Winefordner J.D.	1972	Appl. Spectrosc. [26] 612
Omenetto N., Hatch N.N., Fraser L.M. and Winefordner J.D.	1973	Spectrochim. Acta B [28] 65
Omenetto N. and Rossi G.	1969	Spectrochim. Acta B [24] 95
Omenetto N. and Rossi G.	1970	Spectrochim. Acta B [25] 297
Omenetto N. and Winefordner J.D.	1972	Appl. Spectrosc. [26] 555
Omenetto N. and Winefordner J.D.	1979	see: Omenetto (Ed.) (1979), Ch.4, p.167
Omenetto N. and Winefordner J.D.	1979a	Prog. Analyt. Atom. Spectrosc. [2] 1
Omenetto N., Winefordner J.D. and Alkemade C.Th.J.	1975	Spectrochim. Acta B [30] 335
Omont A., Smith E.W. and Cooper J.	1972	Astrophys. J. [175] 185
Optical Transition Probabilities	1962	"Optical transition probabilities": A collection of russian articles from 1924 to 1960, transl. into english, Israel Program for Scientific Translations, (Ed. I. Meroz), Jerusalem
Optical Transition Probabilities	1963	"Optical transition probabilities": A collection of russian articles from 1932 to 1962, transl. into english, Israel Program for Scientific Translations, (Ed. I. Meroz), Jerusalem
Ostroumenko P.P. and Rossikhin V.S.	1965	Opt. Spectry. [19] 365
Ostrovskii Yu.I. and Penkin N.P.	1961	Opt. Spectry. [11] 307
Ottinger C., Scheps R., York G.W. and Gallagher A.	1975	Phys. Rev. A [11] 1815
Ottinger C. and Zare R.N.	1970	Chem. Phys. Lett. [5] 243

Ottinger C. and Ziock K.	1961	Z. Naturforsch. A [16] 720
Ower E. and Pankhurst R.C.	1966	"The measurement of air flow", 4th ed., Pergamon, Oxford, UK
Pace P.W. and Atkinson J.B.	1974	Can. J. Phys. [52] 1635
Pace P.W. and Atkinson J.B.	1974a	Can. J. Phys. [52] 1641
Padley P.J.	1959	Ph.D. Thesis, Cambridge, UK
Padley P.J.	1960	Trans. Faraday Soc. [56] 449
Padley P.J. and Sugden T.M.	1958	Proc. Roy. Soc., Ser. A [248] 248
Padley P.J. and Sugden T.M.	1959	Trans. Faraday Soc. [55] 2054
Padley P.J. and Sugden T.M.	1959a	see: Proc. 7th Symp. (Int.) on Comb. (1959), p.235
Padley P.J. and Sugden T.M.	1962	see: Proc. 8th Symp. (Int.) on Comb. (1962), p.164
Page F.M.	1973	in: "Physical chemistry of fast reactions.vol.1:Gas phase reactions of small molecules", (Ed. B.P. Levitt), Plenum, London, Ch.3, p.161
Page F.M. and Goode G.C.	1969	"Negative ions and the magnetron", Wiley-Interscience, London
Page F.M. and Woolley D.E.	1974	Combust. Flame [23] 121
Panchevkov I.G., Gusarov A.B. and Goroxov L.N.	1973	Zh. Fiz. Khim. [47] 101
Pao Y.-H. and Griffiths J.E.	1967	J. Chem. Soc. [46] 1671
Pao Y.-H., Zitter R.N. and Griffiths J.E.	1966	J. Opt. Soc. Am. [56] 1133
Parsons M.L., Mc Carthy W.J. and Winefordner J.D.	1966	Appl. Spectrosc. [20] 223
Parsons M.L. and Mc Elfresh P.M.	1971	"Flame spectroscopy. Atlas of spectral lines", Plenum, New York
Parsons M.L. and Mc Elfresh P.M.	1972	Appl. Spectrosc. [26] 472
Parsons M.L. and Winefordner J.D.	1966	Anal. Chem. [38] 1593
Pascale J. and Olson R.E.	1976	J. Chem. Phys. [64] 3538
Pascale J. and Stone P.M.	1976	J. Chem. Phys. [65] 5122
Pascale J. and Vandeplanque J.	1974	J. Chem. Phys. [60] 2278
Pasternack L., Baronavski A.P. and Mc Donald J.R.	1978	J. Chem. Phys. [69] 4830
Pearce R.W.B. and Gaydon A.G.	1976	"The identification of molecular spectra", Chapman and Hall, London
Pearce S.J., Galan L. De and Winefordner J.D.	1968	Spectrochim. Acta B [23] 793
Penkin N.P.	1964	J. Quant. Spectrosc. Radiat. Transfer [4] 41
Penkin N.P., Ruzov V.P. and Shabanova L.N.	1973	Opt. Spectry. [34] 588
Penkin N.P. and Shabanova L.N.	1965	Opt. Spectry. [18] 504
Penkin N.P. and Shabanova L.N.	1968	Opt. Spectry. [25] 446
Penkin N.P. and Slavenas I.Yu.Yu	1963	Opt. Spectry. [15] 3
Penkin N.P. and Slavenas I.Yu.Yu.	1963a	Opt. Spectry. [15] 83
Penner S.S.	1959	"Quantitative molecular spectroscopy and gas emissivities", Addison-Wesley, Reading, UK
Penner S.S. and Kavanagh R.	1953	J. Opt. Soc. Am. [43] 385
Penner S.S., Sulzmann K.G.P., Heffington W.M. and Parks G.E.	1978	J. Quant. Spectrosc. Radiat. Transfer [19] 173
Pepperl R.	1970	Z. Naturforsch. A [25] 927
Pforr G. and Klostermann K.	1965	Z. Chem. [5] 354
Phaneuf R.A.	1973	"Inelastic collisions of excited atoms and molecules at thermal energies: sensitized fluorescence of Cs and Rb atoms, and of Hg2 molecules", Ph.D. Thesis, Windsor, Can.
Phaneuf R.A., Skonieczny J. and Krause L.	1973	Phys. Rev. A [8] 2980
Phillips L.F.	1971	Rev. Sci. Instrum. [42] 1078
Phillips L.F. and Sugden T.M.	1961	Trans. Faraday Soc. [57] 914
Phillips L.F. and Sugden T.M.	1961a	Can. J. Chem. [38] 1804
Phillips W.D., Glaser C.L. and Kleppner D.	1977	Phys. Rev. Lett. [38] 1018
Phillips W.D., Serri J.A., Ely D.J. and Pritchard D.E.	1978	Phys. Rev. Lett. [41] 937
Picket R.C. and Anderson R.	1969	J. Quant. Spectrosc. Radiat. Transfer [9] 697
Piepmeier E.H.	1972	Spectrochim. Acta B [27] 431
Piepmeier E.H.	1972a	Spectrochim. Acta B [27] 445
Piepmeier E.H. and Malmstadt H.V.	1969	Anal. Chem. [41] 700
Pimbert M.	1972	J. Phys. (Paris) [33] 331

BIBLIOGRAPHY

Pimbert M., Rocchiccioli J.L. and Cuvellier J.	1970	Compt. Rend. B [270] 684
Pimbert M., Rocchiccioli J.L., Cuvellier J. and Pascale J.	1970	Compt. Rend. B [271] 415
Pimentel G.C.	1968	Appl. Opt. [7] 2155
Pinta M. (Ed.)	1971	"Spectrometrie d'absorption atomique, I, Problemes generaux", Masson, ORSTOM, Paris
Pinta M. (Ed.)	1971a	"Spectrometrie d'absorption atomique, II. Application a l'analyse chimique", Masson, ORSTOM, Paris
Pippard A.B.	1966	"The elements of classical thermodynamics", Univ. Press, Cambridge, UK
Pitre J. and Krause L.	1967	Can. J. Phys. [45] 2671
Pitre J. and Krause L.	1968	Can. J. Phys. [46] 125
Pitre J., Rae A.G.A. and Krause L.	1966	Can. J. Phys. [44] 731
Pitz R.W., Cattolica R., Robben F. and Talbot L.	1976	Combust. Flame [27] 313
Plankey F.W., Glenn T.H., Hart L.P. and Winefordner J.D.	1974	Anal. Chem. [46] 1000
Plass G.N.	1965	J. Opt. Soc. Am. [55] 104
Polanyi J.C.	1963	J. Quant. Spectrosc. Radiat. Transfer [3] 471
Polanyi J.C.	1965	Appl. Opt., Supplement no.2 "Chemical lasers", p.109
Polanyi J.C.	1972	Accounts Chem. Res. [5] 161
Poluektov N.S.	1961	"Techniques in flame photometric analysis" (transl. from the russian), Consultants Bureau, New York
Porter R.F.	1970	Combust. Flame [14] 275
Porter R.F., Chupka W.A. and Inghram M.G.	1955	J. Chem. Phys. [23] 1347
Posener D.W.	1959	Australian J. Phys. [17] 184
Preist T.W.	1972	J. Chem. Soc., Faraday Trans.,I [68] 661
Preobrazhenskii N.G.	1963	Opt. Spectry. [14] 183
Pringsheim P.	1949	"Fluorescence and phosphorescence", Interscience, New Yor
Pritchard D.E. and Apt J.	1975	9th Int. Conf. on the Physics of Electronic and Atomic Collisions, Seattle (Wash) 1975; "Abstracts of papers", University of Washington Press, Seattle (Wash), p.247
Pritchard D.E. and Carter G.M.	1975	9th Int. Conf. on the Physics of Electronic and Atomic Collisions, Seattle (Wash) 1975; "Abstracts of papers", University of Washington Press, Seattle (Wash), p.447
Pritchard H. and Harrison A.G.	1968	J. Chem. Phys. [48] 2827
Proc. 7th Symp. (Int.) on Comb.	1959	Proceedings of "7th Symposium (Int.) on Combustion", London and Oxford, UK, 1958, Butterworths, London
Proc. 8th Symp. (Int.) On Comb.	1962	Proceedings of "8th Symposium (Int.) on Combustion", Pasadena (Calif) 1960, Williams and Wilkins, Baltimore (Md)
Proc. 9th Symp. (Int.) on Comb.	1963	Proceedings of "9th Symposium (Int.) on Combustion", Ithaca (NY) 1962, Academic Press, New York
Proc. 10th Symp. (Int.) on Comb.	1965	Proceedings of "10th Symposium (Int.) on Combustion", Cambridge, UK, 1964, The Combustion Institute, Pittsburgh (Pa)
Proc. 11th Symp. (Int.) on Comb.	1967	Proceedings of "11th Symposium (Int.) on Combustion", Berkeley (Calif) 1966, The Combustion Institute, Pittsburgh (Pa)
Proc. 12th Symp. (Int.) on Comb.	1969	Proceedings of "12th Symposium (Int.) on Combustion", Poitiers, France, 1968, The Combustion Institute, Pittsburgh (Pa)
Proc. 13th Symp. (Int.) on Comb.	1971	Proceedings of "13th Symposium (Int.) on Combustion", Salt Lake City (Utah) 1970, The Combustion Institute, Pittsburgh (Pa)
Proc. 14th Symp. (Int.) on Comb.	1973	Proceedings of "14th Symposium (Int.) on Combustion", University Park (Penn) 1972, The Combustion Institute, Pittsburgh (Pa)
Proc. 15th Symp. (Int.) on Comb.	1974	Proceedings of "15th Symposium (Int.) on Combustion", Tokio 1974, The Combustion Institute, Pittsburgh (Pa)
Proc. 16th Symp. (Int.) on Comb.	1977	Proceedings of "16th Symposium (Int.) on Combustion", Cambridge (Mass) 1976, The Combustion Institute, Pittsburgh (Pa)
Proc. 17th Symp. (Int.) on Comb.	1979	Proceedings of "17th Symposium (Int.) on Combustion", Leeds, UK, 1978, The Combustion Institute, Pittsburgh (Pa)
Prugger H.	1969	Spectrochim. Acta B [24] 197
Prugger H., Grosskopf R. and Torge R.	1971	Spectrochim. Acta B [26] 191
Prunele E.De and Pascale J.	1979	J. Phys. B [12] 2511
Pueschel R.	1962	"Ueber die erreichbaren Na-Konzentrationen in turbulenten H2-O2 Flammen", Ph.D. Thesis, Giessen, W. Germ.
Pueschel R., Simon L. and Herrmann R.	1964	Optik [21] 441

METAL VAPOURS IN FLAMES

Pungor E.	1967	"Flame-photometry theory", Van Nostrand, London
Pungor E. and Hegedues A.J.	1960	Mikrochim. Acta p.87
Pyun C.W.	1968	J. Chem. Phys. [48] 1306
Quadbeck K. and Kummer J.	1976	Z. Physik A [277] 323
Quinn T.J. and Barber C.A.	1967	Metrologia [3] 19
Rae A.G. and Krause L.	1965	Can. J. Phys. [43] 1574
Rains T.C.	1969	see: Dean and Rains (Eds.) (1969), Ch.12, p.349
Ramirez-Munoz J.	1971	see: Winefordner (Ed.) (1971a), Ch.2, p.127
Rann C.S.	1967	J. Sci. Instr. [44] 227
Rann C.S.	1968	Spectrochim. Acta B [23] 827
Rann C.S.	1969	Spectrochim. Acta B [24] 685
Rann C.S. and Hambly A.N.	1965	Anal. Chem. [37] 879
Rasmuson J.O., Fassel V.A. and Kniseley R.N.	1973	Spectrochim. Acta B [28] 365; Erratum in Spectrochim. Acta B [31] 229 (1976)
Rasmuson J.O., Fassel V.A. and Kniseley R.N.	1976	see: Rasmuson, Fassel and Kniseley (1973)
Rauh E.G. and Ackermann R.J.	1975	Can. Metal. Quart. [14] 205
Rautian S.G. and Sobelman I.I.	1967	Sov. Phys.-Usp. [9] 701
Rawson R.A.G.	1966	Analyst (London) [91] 630
R.C.A. Photomultiplier Manual	000	RCA Photomultiplier Manual, PT-61, Radio Corpor. of America, Harrison (NJ)
Redfield D.	1961	Rev. Sci. Instrum. [32] 557
Reid R.H.G. and Dalgarno A.	1969	Phys. Rev. Lett. [22] 1029
Reid R.H.G. and Dalgarno A.	1970	Chem. Phys. Lett. [6] 85
Reid R.W. and Sugden T.M.	1962	Discussions Faraday Soc. [33] 213
Reid R.W. and Wheeler R.C.	1961	J. Phys. Chem. [65] 527
Reif I., Fassel V.A. and Kniseley R.N.	1973	Spectrochim. Acta B [28] 105
Reif I., Fassel V.A. and Kniseley R.N.	1974	Spectrochim. Acta B [29] 79
Reif I., Fassel V.A. and Kniseley R.N.	1975	Spectrochim. Acta B [30] 163
Reif I., Fassel V.A. and Kniseley R.N.	1976	Spectrochim. Acta B [31] 377
Reif I., Fassel V.A., Kniseley R.N. and Kalnicky D.J.	1978	Spectrochim. Acta B [33] 807
Reif I., Kniseley R.N. and Fassel V.A.	1970	Appl. Opt. [9] 2398
Rensburg H.C. and Zeeman B.P.	1968	Anal. Chim. Acta [43] 173
Rice O.K.	1961	J. Phys. Chem. [65] 1972
Rice P.A. and Ragone D.V.	1965	J. Chem. Phys. [42] 701
Richards W.G., Verhaegen G. and Moser C.M.	1966	J. Chem. Phys. [45] 3226
Richter J.	1968	see: Lochte-Holtgreven (Ed.) (1968), Ch.1, p.1
Riedmueller W., Brederlow G. and Salvat M.	1968	Z. Naturforsch. A [23] 731
Robaux O. and Roizen-Dossier B.	1970	Optica Acta [17] 733
Robb W.D.	1975	9th Int. Conf. on the Physics of Electronic and Atomic Collisions, Seattle (Wash) 1975; "Abstracts of papers", University of Washington Press, Seattle (Wash), p.1113
Robben F.	1971	Appl. Opt. [10] 776
Rocchiccioli J.L.	1972	Compt. Rend., Ser.B, [274] 787
Rodda S.	1953	"Photoelectric multipliers", McDonald, London
Roddier F.	1965	Ann. d'Astrophysique [28] 463
Roddier F.	1966	Ann. d'Astrophysique [29] 639
Rodgers C.D. and Williams A.P.	1974	J. Quant. Spectrosc. Radiat. Transfer [14] 319
Rodrigo A.B. and Measures R.M.	1973	IEEE J. Quantum Electron. [9] 972
Roessler F.	1968	Optica Acta [15] 257
Rolfe J. and Moore S.E.	1970	Appl. Opt. [9] 63
Rosen B.	1964	in: "Encyclopedia of physics, Spectroscopy I", Springer, Berlin, vol.27, p.221 (in french)
Rosen B.	1970	"Donnees spectroscopiques relatives aux molecules diatomiques", Pergamon, Oxford, UK
Rosen P.	1956	Phys. Rev. [103] 390
Rosenfeld J.L.J. and Sugden T.M.	1964	Combust. Flame [8] 44
Ross J., Light J.C. and Shuler K.E.	1969	see: Hochstim (Ed.) (1969a), Ch.8, p.281
Rossi G., Benetti P. and Omenetto N.	1971	Euratom Report Eur. 4598e
Rossi G. and Omenetto N.	1967	Appl. Spectrosc. [21] 329
Rost K.A.	1977	"Energieaustausch zwischen laserangeregten Atomen und einfachen Molekuelen", Ph.D Thesis, Kaiserslautern, W.Germ.
Rostas F. and Lemaire J.L.	1971	J. Phys. B [4] 555

BIBLIOGRAPHY

Roueff E.   1970   Astron. Astrophys. [7] 4
Roueff E.   1972   Phys. Lett. A [38] 8
Roueff E.   1972a   J. Phys. B [5] L79
Roueff E.   1972b   Ph.D. Thesis, Paris
Roueff E. and Launay J.M.   1977   J. Phys. B [10] L173
Rouse C.A. (Ed.)   1967   "Progress in high-temperature physics and chemistry" vol.1, Pergamon, Oxford, UK

Rubeska I.   1969   see: Dean and Rains (Eds.) (1969), Ch.11, p.317
Rubeska I.   1971   in: "Methodes physiques d'analyse", ed. spec. du GAMS: "3ieme Congres Int. de Spectrometrie d'Absorption et de Fluorescence Atomique", Paris, p.61
Rubeska I.   1975   Can. J. Spectrosc. [20] 156
Rubeska I. and Moldan B.   1969   "Atomic absorption spectrophotometry", Butterworths, London
Rukosueva A.V.   1964   Opt. Spectry. [16] 521
Russell B.J., Shelton J.P. and Walsh A.   1957   Spectrochim. Acta [8] 317
Rutgers G.A.W.   1971   Appl. Opt. [10] 2595

Ryabova V.G. and Gurvich L.V.   1964   High Temp. [2] 749
Ryabova V.G. and Gurvich L.V.   1965   High Temp. [3] 284
Ryabova V.G. and Gurvich L.V.   1965a   High Temp. [3] 604
Ryabova V.G., Gurvich L.V. and Khitrov A.N.   1971   High Temp. [9] 686
Ryabova V.G., Khitrov A.N. and Gurvich L.V.   1972   High Temp. [10] 669

Sadowski C.M., Schiff H.I. and Chow G.K.   1972   J. Photochem. [1] 23
Sakai H. and Stauffer F.R.   1964   J. Opt. Soc. Am. [54] 759
Sando K.M.   1974   Phys. Rev. A [9] 1103
Saturday K.A. and Hieftje G.M.   1977   Anal. Chem. [49] 2013
Sauerbrey G.   1972   Appl. Opt. [11] 2576
Saxon R.P., Olson R.E. and Liu B.   1977   J. Chem. Phys. [67] 2692

Schadee A.   1967   J. Quant. Spectrosc. Radiat. Transfer [7] 169
Schadee A.   1971   Astron. Astrophys. [14] 401
Schadee A.   1978   J. Quant. Spectrosc. Radiat. Transfer [19] 451
Schaefer F.P. (Ed.)   1973   "Dye lasers", Springer, Berlin
Schick H.L. (Ed.)   1966   "Thermodynamics of certain refractory compounds, vol. I, Discussion of theoretical studies", Sections 1-4, Academic Press, New York-London

Schieder R. and Walther H.   1974   Z. Physik [270] 55
Schieder R., Walther H. and Woeste L.   1972   Opt. Communic. [5] 337
Schmidt C., Haug R. and Rappenecker G.   1974   J. Atmospheric Terrestrial Phys. [36] 1809
Schmidt W.   1963   11th Colloq. Spectrosc. Int., Beograd
Schmidt W.   1970   Laser [4] 47
Schmidt W. and Kummer J.   1970   Z. Physik [234] 82
Schneider J. and Hofmann F.W.   1958   Phys. Rev. Lett. [1] 517
Schneider J. and Hofmann F.W.   1959   Phys. Rev. [116] 244
Schofield K.   1967   Chem. Rev. [67] 707
Schofield K.   1967a   Planet. Space Sci. [15] 643
Schofield K. and Broida H.P.   1968   in: "Methods of experimental physics, vol.7, Atomic and electron physics, atomic interactions", part B (Eds. B. Bederson and W.L. Fite), Academic Press, New York, Ch.8, p.189
Schofield K. and Sugden T.M.   1965   see: Proc. 10th Symp. (Int.) on Comb. (1965), p.589
Schofield K. and Sugden T.M.   1971   Trans. Faraday Soc. [67] 1054
Schuhknecht W.   1961   "Die Flammenspektralanalyse", Enke, Stuttgart, W.Germany
Schuhknecht W. and Schinkel H.   1958   Z. Anal. Chem. [162] 266
Schuller F.   1962   "Contribution a la theorie des perturbations spectrales dans les gaz comprimes", Ph.D. Thesis, Paris
Schuller F.   1971   J. Quant. Spectrosc. Radiat. Transfer [11] 725
Schuller F. and Behmenburg W.   1974   Phys. Reports C [12] 273
Schuller F. and Oksengorn B.   1969   J. Quant. Spectrosc. Radiat. Transfer [9] 185
Schuller F. and Oksengorn B.   1969a   J. Phys. (Paris) [30] 531
Schulz G. and Gutjahr H.J.   1968   Z. Angew. Phys. [25] 269
Schulz G. and Stopp W.   1967   Z. Physik [207] 470
Schulz G. and Stopp W.   1969   Z. Physik [210] 223
Schulz G.J.   1973   Rev. Mod. Phys. [45] 378 and 423
Schurer K.   1968   Appl. Opt. [7] 461
Schurer K. and Stoelhorst J.   1968   J. Sci. Instr. [44] 952
Schwar M.J.R. and Weinberg F.J.   1969   Nature [221] 357
Scott R.H. and Butler L.R.P.   1967   J. Sci. Instr. [44] 1050

Seiwert R.	1956	Ann. Physik [18] 54
Seiwert R.	1968	in: "Springer tracts in modern physics" (Ed. G. Hoehler), vol.47, p.143
Selzer P.M. and Yen W.M.	1976	Rev. Sci. Instrum. [47] 749
Semenov E.S. and Sokolik A.S.	1970	Combustion, Explosion and Shock Waves (engl. transl.) [6] 33
Shackleford W.L. and Penner S.S.	1966	J. Chem. Phys. [45] 1816
Sharp B.L. and Goldwasser A.	1976	Spectrochim. Acta B [31] 431
Sheinson R.S. and Williams F.W.	1973	Combust. Flame [21] 221
Shifrin N., Hell A. and Ramirez-Munoz J.	1967	18th Pittsburgh Conf. on Anal. Chem. and Appl. Spectrosc.
Shimoda K. (Ed.)	1976	"High-resolution laser spectroscopy", Springer, Berlin
Shimoda K.	1976a	see: Shimoda (Ed.) (1976), Ch.2, p.11
Shore B.W. and Menzel D.H.	1968	"Principles of atomic spectra", Wiley, New York
Shuler K.E.	1951	J. Chem. Phys. [19] 888
Shuler K.E.	1953	J. Phys. Chem. [57] 396
Shuler K.E., Carrington T. and Light J.C.	1965	Appl. Opt., Supplement 2 "Chemical lasers", p.81
Shuler K.E. and Fenn J.B. (Eds.)	1963	"Ionization in high-temperature gases", (selection of papers at the Amer. Rocket Soc. Conf. on "Ions in flames and rocket exhaust", Palm Springs (Calif) 1962; Academic Press, New York
Shull M. and Winefordner J.D.	1971	Anal. Chem. [43] 799
Siara I.N.	1972	"Inelastic collisions of excited atoms; 6 doublet P mixing and quenching in mixtures of rubidium with noble gases and molecules", Ph.D. Thesis, Windsor, Can
Siara I.N., Dubois R.U. and Krause L.	1974	4th Int. Conf. on Atomic Physics, Heidelberg, W.Germ. "Abstracts of papers", p.563
Siara I.N., Hrycyshyn E.S. and Krause L.	1972	Can. J. Phys. [50] 1826
Siara I.N. and Krause L.	1973	Can. J. Phys. [51] 257
Siara I.N., Kwong H.S. and Krause L.	1974	Can. J. Phys. [52] 945
Silla H. and Dougherty T.J.	1972	Combust. Flame [18] 65
Silver J.A., Blais N.C. and Kwei G.H.	1977	J. Chem. Phys. [67] 839
Silver J.A., Blais N.C. and Kwei G.H.	1979	J. Chem. Phys. [71] 3412
Silvester M.D. and Mc Carthy W.J.	1970	Spectrochim. Acta B [25] 229
Simon L.	1960	"Emissions-, Absorptions- und Temperaturmessungen an der H2-O2-Flamme des Beckman-Brenners", Ph.D. Thesis, Giessen, W.Germ.
Simon L.	1962	Optik [19] 621
Slater J.C.	1968	"Quantum theory of matter", 2nd ed., McGraw-Hill, New York
Slater J.C. and Kirkwood J.G.	1931	Phys. Rev. [37] 682
Slavin W.	1964	Atom. Absorption Newslett. no.24, p.15
Slavin W.	1966	Symp. on Flamespectrometric Methods of Analysis, Amer. Chem. Soc., Phoenix (Ariz) (unpublished)
Slavin W.	1967	At. Abs. Newslett. [6] 9
Slevin P.J., Muscat V.I. and Vickers T.J.	1972	Appl. Spectrosc. [26] 296
Smiley V.N.	1973	Proc. of "Int. Clean Air Conf.", Duesseldorf, W.Germ.
Smit C.	1961	"Development and investigation of a photon counter and measurement of optical excitation functions of helium", (in dutch with a summary in english), Ph.D. Thesis, Utrecht, Netherl.
Smit C. and Alkemade C.Th.J.	1963	Appl. Sci. Res. B [10] 309
Smit C., Alkemade C.Th.J. and Muntjewerff W.F.	1963	Physica [29] 41
Smit J.A.	1950	"The production and measurement of constant high temperatures up to 7000 K" (in dutch with a summary in english), Ph.D. Thesis, Utrecht, Netherl.
Smit J.A.	1966	Symp. no.26 "Abundance determination in stellar spectra", Int. Astron. Union, p.86
Smit J.A., Alkemade C.Th.J. and Verschure J.C.M.	1951	Biochim. Biophys. Acta [6] 508
Smit J.A. and Vendrik A.J.H.	1948	Physica [14] 505
Smith B., Winefordner J.D. and Omenetto N.	1977	J. Appl. Phys. [48] 2676
Smith E.W., Cooper J., Chappell W.R. and Dillon T.	1971	J. Quant. Spectrosc. Radiat. Transfer [11] 1547
Smith E.W., Cooper J., Chappell W.R. and Dillon T.	1971a	J. Quant. Spectrosc. Radiat. Transfer [11] 1567
Smith G.	1967	Proc. Roy. Soc., Ser. A [296] 288
Smith G.	1972	J. Phys. B [5] 2310
Smith G.	1975	J. Phys. B [8] 2273
Smith H. and Sugden T.M.	1952	Proc. Roy. Soc., Ser. A [211] 31
Smith H. and Sugden T.M.	1953	Proc. Roy. Soc., Ser. A [219] 204
Smith M.Y.	1972	Combust. Flame [18] 293

BIBLIOGRAPHY

Smith R.	1971	see: Winefordner (Ed.) (1971a), Ch.4, p.235
Smith R.W., Manton J. and Brinkley S.R.		"The thermodynamics of combustion gases: Temperatures of acetylene-air flames", Bureau of Mines, Report of Investigations 5035, US Dept of the Interior, Washington (DC)
Smith W.H. and Liszt H.S.	1971	J. Opt. Soc. Am. [61] 938
Smoes S., Mandy F., Auwera-Mahieu A.Van Der and Drowart J.	1972	Bull. Soc. Chim. Belge [81] 45
Smyly D.S., Townsend W.P., Zeegers P.J.Th. and Winefordner J.D.	1971	Spectrochim. Acta B [26] 531
Snelleman W.	1965	"A flame as a standard of temperature", Ph.D. Thesis, Utrecht, Netherl.
Snelleman W.	1967	Combust. Flame [11] 453
Snelleman W.	1968	Spectrochim. Acta B [23] 403
Snelleman W.	1968a	Metrologia [4] 117
Snelleman W.	1968b	Spectrochim. Acta B [23] 403
Snelleman W.	1969	see: Dean and Rains (Eds.) (1969), Ch.7, p.213
Snelleman W., Rains T.C., Yee K.W. and Cook H.D.	1970	Anal. Chem. [42] 394
Snelleman W. and Smit J.A.	1968	Metrologia [4] 123
Snider J.L.	1967	J. Opt. Soc. Am. [57] 1394
Snider N.S.	1965	J. Chem. Phys. [42] 548
Sobelman I.I.	1957	Fortschr. Phys. [5] 175
Sobelman I.I.	1972	"An introduction to the theory of atomic spectra" (transl. from the russian by T.F.J. Le Vierge), Pergamon, Oxford
Sobelman I.I.	1979	"Atomic spectra and radiative transitions", Springer, Berlin
Sobolev N.N., Mesheritscher E.M. and Rodin G.M.	1953	Abh. Sov. Physik III, Berlin, p.127
Sobolev N.N. et al	1962	Proc. of "5th Int. Conf. on Ionization Phenomena in Gases", Munich, 1961; (Ed. H. Maecker), North-Holland, Amsterdam, p.2122
Sosinskii M.L. and Morozov E.N.	1965	Opt. Spectry. [19] 352
Soundy R.G. and Williams H.	1965	see: Wilsted (Ed.)(1965), p.161
Speller E., Staudenmayer B. and Kempter V.	1979	Z. Physik A [291] 311
Spicer W.E. and Wooten F.	1963	"Photoemission and photomultipliers", Proc. IEEE, 1119
Stacey V. and Zare R.N.	1970	Phys. Rev. A [1] 1125
Stafford F.E. and Berkowitz J.	1964	J. Chem. Phys. [40] 2963
Stair R., Schneider W.E. and Jackson J.K.	1963	Appl. Opt. [2] 1151
Stamper J.H.	1965	J. Chem. Phys. [43] 759
Starr W.L.	1965	J. Chem. Phys. [43] 73
Starr W.L. and Shaw T.M.	1966	J. Chem. Phys. [44] 4181
Steiger R.P. and Cater E.D.	1975	High Temp. Sci. [7] 288
Stekelenburg L.H.M. Van	1943	"Measurement of transition probabilities in the spectrum of Ti" (in dutch with a summary in english), Ph.D. Thesis, Utrecht, Netherl.
Stephens R. and Ryan D.E.	1975	Talanta [22] 655, 659
Stepowski D. and Cottereau M.J.	1979	Appl. Opt. [18] 354
Sternberg J.C., Gallaway W.S. and Lones D.T.L.	1961	Gas Chromatogr. Intern. Symposium [3] 211
Stevefelt J., Boulmer J. and Delpech J.-F.	1975	Phys. Rev. A [12] 1246
Stevens B.	1967	"Collisional activation in gases", Pergamon, Oxford, UK
Stone J.M.	1963	"Radiation and optics", McGraw-Hill, New York
Strachan-Woods J.	1968	Appl. Spectrosc. [22] 799
Straubel H.	1953	Naturwiss. [40] 337
Straubel H.	1954	Z. Angew. Phys. [6] 264
Straubel H.	1955	Exp. Techn. Phys. [3] 89
Striganow A.R. and Sventitskii N.S.	1968	"Tables of spectral lines of neutral and ionized atoms", (transl. from the russian), Plenum, New York
Strumia F.	1967	Boll. Soc. Ital. Fis. [55] 80
Stumpf B., Becker K. and Schulz G.	1978	J. Phys. B [11] L639
Stupar J.	1969	Atom. Absorption Newsl. [8] 38
Stupavski M. and Krause L.	1968	Can. J. Phys. [46] 2127
Stupavsky M. and Krause L.	1969	Can. J. Phys. [47] 1249
Su C.H. and Lam S.H.	1963	Phys. Fluids [6] 1479
Suddendorf R.F. and Denton M.B.	1974	Appl. Spectrosc. [28] 8
Sugden T.M.	1956	Trans. Faraday Soc. [52] 1465
Sugden T.M.	1962	Ann. Rev. Phys. Chem. [13] 369
Sugden T.M.	1963	see: Shuler and Fenn (Eds.) (1963), p.145

Sugden T.M.	1965	see: Proc. 10th Symp. (Int.) on Comb. (1965), p.33 Discussion on: Zeegers and Alkemade (1965a)
Sugden T.M.	1965a	see: Wilsted (Ed.) (1965), p.43
Sugden T.M.	1965b	see: Proc. 10th Symp. (Int.) on Comb. (1965), p.539
Sugden T.M.	1971	in: "Reports of 10th Int. Conf. on Phenomena in Ionized Gases", Oxford, UK, p.437
Sugden T.M.	1972	Proc. 16th Colloq. Spectrosc. Int., Heidelberg, W. Germ., 1971, Hilger, London, p.211
Sugden T.M. and Knewstubb P.F.	1956	Research [9] (no.8) A1
Sugden T.M. and Schofield K.	1966	Trans. Faraday Soc. [62] 566
Sugden T.M. and Thrush B.A.	1951	Nature [168] 703
Sugden T.M. and Wheeler R.C.	1955	Discussions Faraday Soc. [19] 76
Sullivan J.V. and Walsh A.	1965	Spectrochim. Acta [21] 727
Sullivan J.V. and Walsh A.	1968	Appl. Opt. [7] 1271
Sulzmann K.G.P., Myers B.F. and Bartle E.R.	1965	J. Chem. Phys. [42] 3969
Sulzmann K.G.P., Myers B.F. and Bartle E.R.	1965a	J. Chem. Phys. [43] 1220
Swift J.D. and Schwar M.J.R.	1970	"Electrical probes for plasma diagnostics", Iliffe Books, London
Szudy J.	1970	Acta Phys. Pol. A [38] 779
Takeo M.	1970	Phys. Rev. A [1] 1143
Talmi Y., Crosmun R. and Larson N.M.	1976	Anal. Chem. [48] 326
Taran E.N., Nesterko N.A. and Tsikora I.L.	1973	Comb. Inst. European Symp. 1973; Academic Press, London, p.279
Taran E.N. and Tverdokhlebov V.I.	1966	High Temp. [4] 160
Tatum J.B.	1967	Astrophys. J. (Suppl. 124) [14] 21
Taylor H.S.	1979	Chem. Phys. Lett. [64] 17
Taylor R.L. and Bitterman S.	1969	Rev. Mod. Phys. [41] 26
Tebra W. and Visser C.A.	1972	Nucl. Instrum. Methods [98] 165
Ter Haar		see: Haar, D. ter
Ter Heerdt		see: Heerdt, J.A. ter
Thomas D.L.	1968	Combust. Flame [12] 541
Thomas P.E. and Pickering W.F.	1971	Talanta [18] 123
Thompson K.C. and Wildly P.C.	1970	Analyst (London) [95] 776
Thorn R.J.	1966	Ann. Rev. Phys. Chem. [17] 83
Thorne A.P.	1974	"Spectrophysics", Chapman and Hall, London
Thrash R.J., Weyssenhoff H.Von and Shirk J.S.	1971	J. Chem. Phys. [55] 4659
Thrush B.A.	1968	Ann. Rev. Phys. Chem. [19] 371
Thrush B.A.	1973	J. Chem. Phys. [58] 5191
Tiggelen A. Van (Ed.)	1968	"Oxydations et combustions" vol.1 and 2, Editions Techniques, Paris
Torrens I.M.	1972	"Interatomic potentials", Academic Press, Oxford, UK
Tourin R.H.	1966	"Spectroscopic gas temperature measurements", Elsevier, Amsterdam
Townes C.H. and Schawlow A.L.	1955	"Microwave spectroscopy", McGraw-Hill, London
Townsend W.P., Smyly D.S., Zeegers P.J.Th., Svoboda V. and Winefordner J.D.	1971	Spectrochim. Acta B [26] 595
Trainor D.W.	1976	J. Chem. Phys. [64] 4131
Trainor D.W.	1977	J. Chem. Phys. [67] 1206
Travers B.E.L. and Williams H.	1965	see: Proc. 10th Symp. (Int.) on Comb. (1965), p.657
Traving G.	1960	"On the theory of pressure broadening of spectral lines", Verlag G. Braun, Karlsruhe, W.Germ.
Traving G.	1968	see: Lochte-Holtgreven (Ed.) (1968), Ch.2, p.66
Treffers R.R.	1977	Appl. Opt. [16] 3103
Trigt C. Van	1968	J. Opt. Soc. Am. [58] 669
Trigt C. Van	1969	Phys. Rev. [181] 97
Trigt C. Van	1970	Phys. Rev. A [1] 1298
Trigt C. Van	1971	Phys. Rev. A [4] 1303
Trigt C. Van, Hollander Tj. and Alkemade C.Th.J.	1965	J. Quant. Spectrosc. Radiat. Transfer [5] 813

BIBLIOGRAPHY

Tsuchiya S.	1964	Bull. Chem. Soc. Japan [37] 828
Tsuchiya S.	1964a	Bull. Chem. Soc. Japan [37] 6
Tsuchiya S. and Kuratani K.	1964	Combust. Flame [8] 299
Tsuchiya S. and Suzuki I.	1969	J. Chem. Phys. [51] 5725
Tsuchiya S. and Suzuki I.	1971	Bull. Chem. Soc. Japan [44] 901
Tudorache S.S.	1977	Phys. Lett. A [61] 162
Tully J.C.	1973	J. Chem. Phys. [59] 5211
Turk G.C., Travis J.C., De Voe J.R. and O'Haver T.C.	1978	Anal. Chem. [50] 817
Turk G.C., Travis J.C., De Voe J.R. and O'Haver T.C.	1979	Anal. Chem. [51] 1890
Tyte D.C.	1967	Proc. Phys. Soc. [92] 1134
Uchida Y. and Hattori S.	1975	Oyo Butsuri [44] 52
Ueno Y. and Sato I.	1971	Bull. Chem. Soc. Japan [44] 637
Umemoto H., Tsunashima S. and Sato S.	1979	Chem. Phys. [43] 93
Unsoeld A.	1955	"Physik der Sternatmosphaeren", 2nd ed., Springer, Berlin
Unsoeld A.	1968	"Physik der Sternatmosphaeren", revised 2nd ed., Springer, Berlin
Uny G., Guea Lottin J.N., Tardif J.P. and Spitz J.	1971	Spectrochim. Acta B [26] 151
Uny G. and Spitz J.	1970	Spectrochim. Acta B [25] 391
Uy O.M. and Drowart J.	1969	Trans. Faraday Soc. [65] 3221
Uy O.M. and Drowart J.	1970	High Temp. Sci. [2] 293
Vaidya W.H.	1964	Proc. Roy. Soc., Ser. A [279] 572
Vallee B.L. and Baker M.R.	1955	Anal. Chem. [27] 320
Van Bergen		see: Bergen, A.R.D. van
Van Calcar		see: Calcar, R.A. van
Van Der Held		see: Held, E.F.M. van der
Van Der Hurk		see: Hurk, J. van der
Van Der Ziel		see: Ziel, A. van der
Van Deventer		see: Deventer, J.M.M. van
Van Dijk		see: Dijk, C.A. van
Van Gelder		see: Gelder, Z. van
Van Heek		see: Heek, H.F. van
Van Itallie F.J., Doemeny L.J. and Martin R.M.	1972	J. Chem. Phys. [56] 3689
Van Itallie F.J. and Martin R.M.	1972	Chem. Phys. Lett. [17] 447
Van Tiggelen		see: Tiggelen, A. van
Van Trigt		see: Trigt, C. van
Van Veen		see: Veen, N.J.A. van
Vanpee M., Kineyko W.R. and Caruso R.	1970	Combust. Flame [14] 381
Vasileva I.A., Deputatova L.V. and Nefedov A.P.	1975	Opt. Spectry. [39] 8
Vedeneyev V.I., Gurvich L.V., Kondratiev V.N., Medvedev V.A. and Frankevich Y.L.	1966	"Bond energies, ionization potentials and electron affinities", Arnold, London (transl. from the russian ed., Nauk, Moscow, 1962)
Veen N.J.A. Van, Vries M.S. De and Vries A.E. De	1979	Chem. Phys. Lett. [64] 213
Veen N.J.A. Van, Vries M.S. De and Vries A.E. De	1979a	Chem. Phys. Lett. [60] 184
Veillon C.	1971	see: Dean and Rains (Eds.) (1971), Ch.6, p.149
Veillon C., Mansfield J.M., Parsons M.L. and Winefordner J.D.	1966	Anal. Chem. [38] 204
Veillon C. and Margoshes M.	1968	Spectrochim. Acta B [23] 553
Veits I.V. and Gurvich L.V.	1956	Opt. Spectry. [1] 22
Vendrik A.J.H.	1949	"Optical investigation of alkali vapour in an acetylene flame" (in dutch with a summary in english), Ph.D. Thesis, Utrecht, Netherl.
Venghiatis A.A.	1968	Appl. Opt. [7] 1313

Vickers T.J., Cottrell C.R. and Breaky D.W.	1970	Spectrochim. Acta B [25] 437
Vickers T.J., Slevin P.J., Muscat V.I. and Farias L.T.	1972	Anal. Chem. [44] 930
Vickers T.J. and Vaught R.M.	1969	Anal. Chem. [41] 1476
Vidale G.L.	1960	US Dept. Comm., Office Tech. Serv., P.B. Rept. [148] 206, Washington (DC)
Vikis A.C., Torrie G. and Le Roy D.J.	1972	Can. J. Chem. [50] 176
Visser K., Hamm F.M. and Zeeman P.B.	1976	Appl. Spectrosc. [30] 72
Visser K., Hamm F.M. and Zeeman P.B.	1976a	Appl. Spectrosc. [30] 620
Von Engel A.		see: Engel A. von
Vos J.C. De	1954	Physica [20] 690
Vriens L.	1977	J. Appl. Phys. [48] 653
Vriens L., Boers A.L. and Smit J.A.	1965	Appl. Sci. Res. B [12] 65
Vriens L. and Smeets A.H.M.	1980	Phys. Rev. A [22] 940
Wade M.K., Czajkowski M. and Krause L.	1978	Acta Phys. Polonica A [54] 849
Wade M.K., Czajkowski M. and Krause L.	1978a	Can. J. Phys. [56] 891
Wagenaar H.C.	1976	"The influence of spectral line profiles upon analytical curves in atomic absorption spectrometry", Ph.D. Thesis, Delft, Netherl.
Wagenaar H.C. and De Galan L.	1973	Spectrochim. Acta B [28] 157
Wagenaar H.C. and De Galan L.	1975	Spectrochim. Acta B [30] 361
Wagenaar H.C., Novotny I. and De Galan L.	1974	Spectrochim. Acta B [29] 301
Wagenaar H.C., Pickford C.J. and De Galan L.	1974	Spectrochim. Acta B [29] 211
Wahlbeck Ph.G. and Gilles P.W.	1967	J. Chem. Phys. [46] 2465
Walentynowicz E., Phaneuf R.A., Baylis W.E. and Krause L.	1974	Can. J. Phys. [52] 584
Walentynowicz E., Phaneuf R.A. and Krause L.	1974	Can. J. Phys. [52] 589
Walker R.E. and Westenberg A.A.	1958	J. Chem. Phys. [29] 1139
Walker S. and Straw H.	1962	"Spectroscopy II", Chapman and-Hall, London; repr.1967
Walker S. and Straw H.	1967	see: Walker and Straw (1962)
Wallenstein R. and Haensch T.W.	1974	Appl. Opt. [13] 1625
Wallenstein R. and Haensch T.W.	1975	Opt. Communic. [14] 353
Walsh A.	1955	Spectrochim. Acta [7] 108
Walsh A.	1965	12th Colloq. Spectrosc. Int., Exeter, UK, Hilger, London, p.43
Walsh A.	1966	J. New Zealand Inst. Chem. [30] 7
Walsh A.	1974	Anal Chem. [46] 698A
Walsh A.	1975	Proc. Roy. Australian Chem. Instit. [42] 297
Walsh A.D.	1953	J. Chem. Soc. 2266
Walters J.P. and Malmstadt H.V.	1965	Anal. Chem. [37] 1484
Walther H.	1974	Phys. Scripta [9] 297
Walther H. and Hall J.L.	1970	Appl. Phys. Lett. [12] 239
Ward J., Cooper J. and Smith E.W.	1974	J. Quant. Spectrosc. Radiat. Transfer [14] 555
Ward J., Cooper J. and Smith E.W.	1974a	Physica [77] 372
Warnatz J.	1978	Ber. Bunsenges. Phys. Chem. [82] 643
Weeks S.J., Haraguchi H. and Winefordner J.D.	1978	J. Quant. Spectrosc. Radiat. Transfer [19] 633
Weide J.O. and Parsons M.L.	1972	Anal. Lett. [5] 363
Weinberg F.J.	1957	Proc. Roy. Soc., Ser. A [241] 132
Weinberg F.J.	1963	"Optics of flames", Butterworths, London
Wentink T. and Spindler R.J.	1972	J. Quant. Spectrosc. Radiat. Transfer [12] 129
West A.C., Fassel V.A. and Kniseley R.N.	1973	Anal. Chem. [45] 1586
West T.S.	1967	"Trace characterization, chemical and physical" (Eds. W.E. Meinke and B.F. Scribner), Report of a Symp. at Gaithersburg (Md) 1962, NBS Monograph no. 100, Washington (DC), p.215
West T.S. and Cresser M.S.	1973	Appl. Spectrosc. Revs. [7] 79
Westenberg A.A. and Fristrom R.M.	1965	see: Proc. 10th Symp. (Int.) on Comb. (1965), p.473
Wexler S. and Parks E.K.	1979	Ann. Rev. Phys. Chem. [30] 179
White W.B., Johnson S.M. and Dantzig G.B.	1958	J. Chem. Phys. [28] 751
Whiting E.E.	1968	J. Quant. Spectrosc. Radiat. Transfer [8] 1379
Whiting E.E. and Nicholls R.W.	1974	Astrophys. J. Suppl. [27], no. 235, p.1
Wiese W.L., Smith M.W. and Glennon B.M.	1966	"Atomic transition probabilities. vol.1: Hydrogen through Neon", NSRDS-NBS 4, Washington (DC)

BIBLIOGRAPHY

Wiese W.L., Smith M.W. and Miles B.M. 1969 "Atomic transition probabilities. vol.2: Sodium through Calcium", NSRDS-NBS 22, Washington (DC)
Wiesenfeld J.R. 1973 Chem. Phys. Lett. [21] 517
Wijchers T. 1981 "Fluorescence spectra and collisional energy transfer of laser excited YO-molecules in flames" (in dutch with a summary in english), Ph.D Thesis, Utrecht, Netherl.
Wijchers T., Dijkerman H.A., Zeegers P.J.Th. and Alkemade C.Th.J. 1980 Spectrochim. Acta B [35] 271
Wildly P.C. and Thompson K.C. 1970 Analyst (London) [95] 562
Wilkinson P.G. 1963 Astrophys. J. [138] 778
Williams F.A. 1965 "Combustion theory", Addison-Wesley, Reading (Mass)
Willis J.B. 1967 Spectrochim. Acta A [23] 811
Willis J.B. 1968 Appl. Opt. [7] 1295
Willis J.B. 1970 see: Mavrodineanu (Ed.) (1970), Ch.10, p.525
Willis J.B. 1971 Spectrochim. Acta B [26] 177
Willis J.B. and Amos M.D. 1966 Spectrochim. Acta [22] 1325
Wilson A.D. and Levine R.D. 1974 Mol. Phys. [27] 1197
Wilson H.A. 1912 "The electrical properties of flames and incandescent solids", University Press, London
Wilson H.A. 1912a Phil. Mag. [24] 118
Wilson H.A. 1931 Rev. Mod. Phys. [3] 156
Wilson H.A. 1944 "Modern physics", London
Wilson W.E. and Westenberg A.A. 1967 see: Proc. 11th Symp. (Int.) on Comb. (1967), p.1143
Wilsted H.D. (Ed.) 1965 "Fundamental studies of ions and plasmas", Pisa, Italy, 1965, AGARD Conf. Proc. Ser. no. 8, vol.1 and 2
Winefordner J.D. 1971 Accounts Chem. Res. [4] 259
Winefordner J.D. (Ed.) 1971a "Spectrochemical methods of analysis", Wiley-Interscience New York
Winefordner J.D. 1975 Chemtech. (february no.), p.123
Winefordner J.D., Avni R., Chester T.L., Fitzgerald J.J., Hart L.P. and Johnson D.J. 1976 Spectrochim. Acta B [31] 1
Winefordner J.D. and Elser R.C. 1971 Anal. Chem. [43] 24A
Winefordner J.D., Fitzgerald J.J. and Omenetto N. 1975 Appl. Spectrosc. [29] 369
Winefordner J.D., Mansfield C.T. and Vickers T.J. 1963 Anal. Chem. [35] 1607
Winefordner J.D. and Mansfield J.M. 1967 Appl. Spectrosc. Revs. [1] 1
Winefordner J.D., Parsons M.L., Mansfield J.M. and Mc Carthy W.J. 1967 Spectrochim. Acta B [23] 37
Winefordner J.D., Parsons M.L., Mansfield J.M. and Mc Carthy W.J. 1967a Anal. Chem. [39] 436
Winefordner J.D., Schulman S.G. and O'Haver T.C. 1972 "Luminescence spectrometry in analytical chemistry", Wiley-Interscience, New York
Winefordner J.D. and Smith R. 1970 see: Mavrodineanu (Ed.) (1970), Ch.11, p.599
Winefordner J.D. and Staab R.A. 1964 Anal. Chem. [36] 165
Winefordner J.D. and Vickers T.J. 1964 Anal. Chem. [36] 1947
Winefordner J.D. and Vickers T.J. 1964a Anal. Chem. [36] 161
Winefordner J.D. and Vickers T.J. 1972 Anal. Chem. Annual Rev. [44] 150R
Winefordner J.D., Vickers T.J. and Remington L. 1965 Anal. Chem. [37] 1216
Winge R.K., Fassel V.A. and Kniseley R.N. 1971 Appl. Spectrosc. [25] 636
Wittenberg G.K., Haun D.V. and Parsons M.L. 1979 Appl. Spectrosc. [33] 626
Wolnik S.J., Berthel R.O. and Wares G.W. 1970 Astron. J. [162] 1037
Woodward C. 1970 Anal. Chim. Acta [51] 548
Yaakobi B. 1972 J. Quant. Spectrosc. Radiat. Transfer [12] 1077
Yakovlev S.A. and Shishatskaya L.P. 1969 Sov. J. Opt. Tech. [36] 55
Yamada H.Y. 1968 J. Quant. Spectrosc. Radiat. Transfer [8] 1463
Yamamoto S., Takei T., Nishimura N. and Hasegawa S. 1976 Chemistry Lett. (Japan) 1413
Yamamoto S., Takei T., Nishimura N. and Hasegawa S. 1980 Bull. Chem. Soc. Japan [53] 860
Yanagi T. 1968 Jap. J. Appl. Phys. [7] 605
Yanagi T. 1968a Jap. J. Appl. Phys. [7] 656
Yang K., Paden J.D. and Hassell C.L. 1967 J. Chem. Phys. [47] 3824
Yokozeki A. and Menzinger M. 1976 Chem. Phys. [14] 427
York G., Scheps R. and Gallagher A. 1975 J. Chem. Phys. [63] 1052
Yoshimura E., Furuta N., Haraguchi H. and Fuwa K. 1977 Appl. Spectrosc. [31] 560
Young A.T. 1969 Appl. Opt. [8] 2431
Young A.T. 1971 Appl. Opt. [10] 1681
Young A.T. 1974 see: Carleton (Ed.) (1974), Ch.1, p.1

Young A.T.	1974a	see: Carleton (Ed.) (1974), Ch.2, p.95
Young C.	1965	Rep. ORA-05863, Univ. of Michigan, Ann Arbor (Mich)
Young C.	1965a	J. Quant. Spectrosc. Radiat. Transfer [5] 549
Youngbluth O.	1970	Appl. Opt. [9] 321
Younger S.M., Fuhr J.R., Martin G.A. and Wiese W.L.	1978	J. Phys. Chem. Ref. Data [7] 495
Yumlu V.S.	1967	Combust. Flame [11] 389
Yumlu V.S.	1968	Combust. Flame [12] 14
Zacha K.E., Bratzel M.P., Winefordner J.D. and Mansfield J.M.	1968	Anal. Chem. [40] 1733
Zaidel A.N. and Korennoi E.P.	1961	Opt. Spectry. [10] 299
Zaidel A.N., Prokofev V.K. and Raiski S.M.	1961	"Tables of spectrum lines" 2nd ed., Pergamon, London; "Spektraltabellen" 2nd ed., VEB Technik, Berlin
Zaidi H.R.	1972	Can. J. Phys. [50] 2792
Zapesochnyi I.P.	1967	High Temp. [5] 6
Zapesochnyi I.P. and Shimon L.L.	1965	Opt. Spectry. [19] 268
Zapesochnyi I.P. and Shimon L.L.	1966	Opt. Spectry. [21] 155
Zare R.N.	1964	J. Chem. Phys. [40] 1934
Zare R.N.	1974	Ber. Bunsenges. Phys. Chem. [78] 153
Zare R.N. and Herschbach D.R.	1963	Proc. IEEE [51] 173
Zeegers P.J.Th.	1966	"Recombination of radicals and related effects in flames", Ph.D. Thesis, Utrecht, Netherl.
Zeegers P.J.Th. and Alkemade C.Th.J.	1965	Combust. Flame [9] 247
Zeegers P.J.Th. and Alkemade C.Th.J.	1965a	see: Proc. 10th Symp. (Int.) on Comb. (1965), p.33
Zeegers P.J.Th. and Alkemade C.Th.J.	1965b	12th Colloq. Spectrosc. Int., Exeter, UK, Hilger, London, p.290
Zeegers P.J.Th. and Alkemade C.Th.J.	1970	Combust. Flame [15] 193
Zeegers P.J.Th., Smith R. and Winefordner J.D.	1968	Anal. Chem. [40] 26A
Zeegers P.J.Th., Townsend W.P. and Winefordner J.D.	1969	Spectrochim. Acta B [24] 243
Zeegers P.J.Th. and Winefordner J.D.	1971	Spectrochim. Acta B [26] 161
Zehnle L., Clemens E., Martin P.J., Schaeuble W. and Kempter V.	1978	J. Phys. B [11] 2133
Zembekov A.A. and Nikitin E.E.	1971	Chem. Phys. Lett. [9] 213
Zhitkevich V.F., Lyuty A.I., Nesterko N.A., Rossikhin V.S. and Tsikora I.L.	1963	Opt. Spectry. [14] 17
Zhuvikin G.V., Penkin N.P. and Shabanova L.N.	1976	Opt. Spectry. [41] 425
Ziel A. Van Der	1954	"Noise", Prentice Hall, New York
Ziel A. Van Der	1959	"Fluctuation phenomena in semiconductors", Butterworths, London
Ziel A. Van Der	1970	"Noise; sources, characterization, measurement", Prentice-Hall, Englewood Cliffs (NJ)
Ziel A. Van Der	1976	"Noise in measurements", Wiley-Interscience, New York
Zimmermann D.	1975	Z. Physik A [275] 5
Zizak G., Cignoli F. and Benecchi S.	1979	Appl. Spectrosc. [33] 179
Zmbov K.F.	1969	Chem. Phys. Lett. [4] 191
Zoller P.	1978	J. Phys. B [11] 805
Zung J.T.	1967	J. Chem. Phys. [46] 2064
Zwillenberg M.L.	1975	"A search for chemical laser action in low-pressure metal vapor flames", NASA-CR-146083 (US), Report 1214-T, Washington (DC)
Zworykin V.K.	1958	"Photoelektrische Zellen", Springer, Berlin

Subject Index

RECOMMENDATIONS FOR USE

1. As a rule, composite terms have been arranged alphabetically according to the first noun; however, when in a composite term the noun is nondescript or when this term is a standard expression, the expression is listed under the preceding adjective (e.g., Angular momentum; Refractive index; Primary combustion zone; Burning velocity). When a subdivision is required for a composite term, nouns and adjectives are listed alphabetically without distinction.

2. Whenever a composite standard expression beginning with a certain noun has already been listed separately, it is not repeated in the subdivision of that noun. For example, the term 'Absorption factor', which is listed separately, is not included in the subdivision of 'Absorption'.

3. If within a subdivision a reference is made to another term in that subdivision, the reference begins with a small letter; otherwise it begins with a capital.

4. When a frequently recurring term, such as 'Flame', 'Metal', or 'Intensity' is used in a loose sense only, no page reference is given.

5. The numbers that are in bold print indicate the principal references to the subject concerned.

6. The following abbreviations are used:

   f (after page number) = 'and following page(s)'

   $etc.$ = 'and similar terms'.

7. Chemical elements or compounds are listed under their chemical symbol; ions are identified by the sign of their charge. However, if a (composite) standard expression begins with the name of a chemical element, the expression is usually listed under the full name of the element, e.g. 'Sodium-D doublet' and 'Xenon (arc) lamp'.

   When a word is put in brackets, this indicates that the word is sometimes omitted in the text (see last example).

   The names of groups of elements or compounds, such as 'Alkalis' or 'Hydrocarbons', are given in full, as are the names of a few unfamiliar or complex compounds.

8. Symbols for physical quantities are generally not given in the Index. They can be found in Appendix A.3; references to this Appendix are to be found under the full names of these quantities. Exceptions are, e.g., 'a-Parameter' and 'f-Value', which are listed under A and F, respectively. The greek letter Ω is listed under O.

9. To find information on a particular subject in a particular case one should look under the references to the subject and to the case for page numbers that are common to both. For example, values of the dissociation energy of CaO are to be found on the pages listed under 'Dissociation energies (of molecule)' as well as under 'CaO'. If the reader wants information about, e.g., termolecular associative chemi-excitation reactions, he should look up in the subdivision of 'Reaction' the terms 'associative', 'chemi-excitation' and 'termolecular' and then consult the pages common to all. Finally, if the reader is interested, e.g., in the effect of flame temperature on the degree of ionization, he should consult the pages listed under 'Temperature (of flame or gas)' as well as 'Ionization, degree of'.

SUBJECT INDEX

AAS, atomic absorption spectroscopy;
  see Spectroscopy
Abel integral equation 327,483
Abel inversion 327f,334,341
Absolute-fluorescence method 563f,567f
Absolute-intensity method 481f,489f,492
Absorbance (decadic) 160f,335,388,906f,925,943
—, internal 925,943
—, net 173
—, peak 165,491f,509f,512f,925
Absorptance; see Absorption factor
Absorption 31,108,110,125f,144,159f,160,260f,334, 489,517f,525,532,619,628,702,709,737,812; see also Intensity, Self-absorption, Self-reversal
— by atom 110f; see also Absorption (general)
— cell; see Vapour cell
— coefficient 125,944
—, continuous; see Continuum
— curve-of-growth; see Curve-of-growth
— cross section; see Cross section, optical
—, integral; see Integral absorption
—, laser-amplified 369
— line profile; see Line profile
—, measurement of 173,257,300,368f,382
— by molecule 38,134f,159,186,194,438f,443f, 489,569,572f,596
—, net 173,203,657; see also Saturation
— path length 943
—, peak 128f,164f,167,192,196,368f,487,526, 722,748f
— probability; see Transition
— rate 621
—, ratio of 509f
—, saturated 174,202f; see also Saturation
— spectrum; see Spectrum
—, strength of 159f,317; see also Absorbance, etc.
—, total 167,925
—, two-photon 108,203,543,819
Absorption factor 56,160f,173,175,181,192,369, 375,382,387f,444,486f,504,654f,672,714,943
—, net 173,657
—, peak 165,196,368f,487,720,722
—, ratio of 509f
Absorption spectroscopy; see Spectroscopy
Absorptivity 110,123f,176,187,388,619,627,631, 655,661,679,929,943
—, atomic 129,161,382,934,943
—, integrated 134
— — over band 137f
— — over line 128f,137f
—, napierian 929,943
—, net 130f,133,140,173,203,626,631
—, peak 134,138
— for radio waves 832
Acetone 229,417,457
A.c. method for measurement of photocurrent 270
A.c. Stark effect 117,144
Activated complex; see Complex
Activation; see Excitation
Activation energy 82f,85,90,92,94,99,106,545,578, 581,584f,596f,647,666,759,763,785,791,822, 849f,880,883,885,888f,897,899f,927; see also Arrhenius energy of activation
—, apparent 903
— factor 82,86f,827,866,883,891
Adiabatic condition 57,59
Adiabaticity parameter 97
Aerosol 224,306f,519; see also Coagulation, Dispersion, Fraction, Nebulization
—, desolvation of; see Desolvation

Aerosol (continued)
—, dry 3,224,229,238
—, — introduction of 236f,309f
—, electrostatic precipitation of 520, 809
— loss 517f
—, scattering by 307f
—, solid-, generator 237
—, trapping of 303f
—, volatilization of; see Volatilization
—, wet 2,224,306f
AES, atomic emission spectroscopy; see Spectroscopy
AFS, atomic fluorescence spectroscopy; see Spectroscopy
Afterglow 572
— system 4
Ag 291,493,525,527,748f,771,773,939
Ag-comparison method; see Silver comparison method
AgH 524,768,779,794
AgO 766
Air-acetylene flame, etc.; see C_2H_2-air flame, etc.
Al 309,312f,506,725,748f,759,771f,939
— particle 330
Alcohol 229,672
AlH 768
Alignment 250f
Alkali(s) 20f,32f,41,101f,125,134,191,229,499, 508,514,519,524,531,534,537,543f,548f,552f, 555f,563,565f,573,577,580f,586,592f,598, 600f,633,648f,651f,702,704f,738,741,743, 747,771,775,780,793,813,823f,827,859f,869, 873f,880,882f,884f,888,890,898f,919; see also individual metals
— bromides 780
— chlorides 780
— dimers 762,825
— fluorides 780
— hydroxides 651,770,793f,805,843,869f
— iodides 780
— lines 688,939f
— —, resonance 621,993f
— monohalides 758,795f,905
— monoxides 792
— superoxides (MO_2) 775,793
Alkali-discharge lamp 825
Alkaline-earth(s) 21,31,33,38,194,389,499f,501, 557,639,702,743,746,759,775,780,824,828, 844,848f,905f; see also individual metals
— dihalides 783,796
— dihydroxides 779,794,808f
— halides 763,796
— hydroxychlorides 783
— hydroxyhalides 796
— monohalides 764,769,779f,801,806
— monohydroxides 38,651,763,768,775f,794,801, 808f,836
— oxides 236,787,791f,799f
AlO 757,759,766,772,787,801,806
Al_2O_3 238
— particle 312
AlOH 773
$Al(OH)_2$ 773
Amplification (of radiation) 131
Amplification factor 951
Amplifier 336,471,475; see also Lock-in
—, differential 255
—, electrometer 279
—, narrow-band 279
—, tuned 280

1000

Analysis (chemical) 8f,339f,479,508,819; *see also* Analytical curve, Detection limit, Flame spectroscopy
—, isotope 493
—, qualitative 8
—, quantitative 8
Analysis element; *see* Analyte
Analyte 7,491,**674**f,906f
— signal 325f,343f,348f,403f
Analytical curve 7,9f,158,160,165,171,186,200f, 392,502,508,**673**,747,753,832,906; *see also* Curve-of-growth
Angle of acceptance 341
Angular aperture 244,250,324,**395**f
Angular momentum 16f,60,102,106; *see also* Quantum number
— of nuclear rotation (in molecule) 931
—, orbital (of electron) 16,93,548
— —, resultant 16,25,96,929,934
— of relative (nuclear) motion 93,99,102
—, spin (of electron) 16,598
— —, exchange 531
— —, resultant 16,25,96,931
—, spin (of nucleus) 18,21,**33**,46f,928
—, total 552,928
— — atomic 18,927
— — electronic 17f,**25**,936
— — molecular 25f,930
Anharmonicity constant 27,932
Anisotropy (of radiation) 530f,535
Anti-Stokes fluorescence; *see* Fluorescence
α-Parameter 156f,162f,169f,188,201f,207,494,**496**f, 500,**504**f,508,689,709f,711,**718**f,**726**f,734, 742f,**747**f,749,904f,**925**
Appearance potential 758
Ar 221,293,406,456,463,534f,**537**,543f,549,551, **558**,572,604f,**693**f,**727**f,729,732,**735**f,**740**f,873, **890**f,**898**,902,910
—, quenching by 456
— as flame diluent 213,406,418,812; *see also* individual flames
Arc (electric) 3f,6,38,79,331,380,475,515; *see also* Carbon arc, Mercury (arc) lamp, Xenon (arc) lamp
Arrhenius' energy of activation 83f,763,927; *see also* Activation energy
Arrhenius equation 83f
Arrhenius plot 866,**873**f,**888**f
As 638,640
Aspiration (of solution), rate of 226f,234,302, 517,**927**
Association (of atoms *or* molecules) 49,69,108f; *see also* Rate constant, Reaction, Recombination
— factor 751f,764,**771**f,**776**f,**780**f,**787**f,796,799, 803,**805**f,835,**843**,870,**936**
— —, total 751,**808**f,835
— —, determination of 800f
Asymmetry; *see* Line profile
Atom(s) 15f,520,529,**536**f,**548**f,560f,599f,633,**635**f, 751f,**842**f,**882**f,**919**f; *see also* individual or groups of elements, Atomization, Energy, Radical
—, absorption by; *see* Absorption
—, aligned 17,569
—, diffusion coefficients of 910
—, emission by; *see* Emission
—, excitation energies of 120,571,**939**f
—, fluorescence by; *see* Fluorescence
—, free 520f,752f,773,**776**; *see also* Fraction atomized

METAL VAPOURS IN FLAMES

Atom(s) (*continued*)
—, internal energies of; *see* excitation energies
—, ionization energies of **939**f
—, oriented; *see* polarized
—, polarized 17,123,127,131,**531**
—, spectrum of 30f,32f,34,**939**f
—, structure of 15f; *see also* Energy level, Quantum state
Atomic absorption spectroscopy; *see* Absorption, Spectroscopy
Atomic beam; *see* Beam
Atomic emission spectroscopy; *see* Emission, Spectroscopy
Atomic fluorescence spectroscopy; *see* Fluorescence, Spectroscopy
Atomic number 15,554
Atomization 520f
—, anomalous 675
— efficiency; *see* Efficiency of atomization
Atomizer 9; *see also* Flame
—, electrothermal 4
—, nonflame 392,633
Attenuation (of electromagnetic wave) 832,862
Attenuation (of light) 338
Au 291,362,493,525,**526**,771,939
—, melting point of 362
AuH **768**,779,794
Autocorrelation function 143f,148f,151,154,472, 475,681,685,**693**,696,699,**947**,952
Autocorrelation technique 475
Avalanche multiplication 262
Avogadro constant 924

B 545,611
B⁻ 881
Ba 38,291,314,392,455,557,**617**,679,**735**f,**745**f, 759,779,**805**f,**809**f,827,836,**907**,910,920,**939**,**955**f
Ba⁺ 369,939
BaBr 769
Background; *see also* Flame background, Light source, Scattering, Signal-to-background ratio
— emission 355
— noise **346**f,**366**f,**381**f,**403**f,**947**f
Back mirror (for flame duplication) 191,**506**,508; *see also* Flame duplication
Back tracing 251
BaCl 535,**769**,800,**805**f
BaCl₂ **769**,806
BaF **769**
BaF₂ 769
BaHCl 806
BaI 769
Balance, detailed; *see* Detailed balance
Band (of molecule) 34,37,**134**f,137,180,443f,469, 484,489,535,571,640,753,778,**808**,906,**953**f; *see also* Flame, Spectrum
—, absorption 137,444; *see also* Absorption
— branch 36f
—, degradation of 37
—, emission 466,503,775,**953**f; *see also* Emission
—, fluorescence 444; *see also* Fluorescence
— head 37,443
—, infrared 466
— origin 36
— oscillator strength 137
— progression 36
—, resonance 37
— Schumann-Runge 466
— sequence 36,137
— strength 136f

1001

SUBJECT INDEX

Band (*continued*)
— system 36,181
— transition probability; *see* Transition
—, Vaidya 466,638
—, vibrational 24,467
— —, convergence limit of 755f
Bandwidth (noise); *see* Noise
Bandwidth (spectral) 241,248f; *see also* Effective width, Half-intensity-width, Resolution
— of monochromator 166,173,298,325, 343f, 376, 384, 404, 408, 486f, 490,495,504,512,**934**,953f
BaO 38,138,369,484,759,**766,772**,778,**791**,802,**809f**, **917,955**f
BaO^+ 762
BaOCl 806
BaOH 768,**776**,778,802,**809f**,**881,955**
$Ba(OH)_2$ **768**,**809f**
BaOHCl 806
BaOHOCl 806
Barrier; *see* Centrifugal b., Energy b., Potential b.
Be 639,939
Bead of salt 519f
Beam
—, atomic (*or* molecular) 6, 10, 38, 81, 298,**533f**, **538f, 546f, 571, 575f**, 603f,**758f**,821,891,900f, 911,**913**,919
—, crossed 4, 67, 102, **533f**, 549,562,568f,570, 575f, 594f, 618, 815, 825, **884**, 898
—, electronic 603, **617**
—, light 377f; *see also* Optical conductance, Radiation, Spectrometer
—, pencil 377f
—, probing 255, 371f,
—, reference 255, 371f,
—, splitting of 378f
Beating (of a.c. signals) 281
Beer plot 389, 711, 714f
Beer's law (*or* Beer-Lambert's) 160f, 201, 382, 389, 492f, 655, 714f
Bennett hole 132
BeO 639, **757**, **766**, 772, 792
Bi 537, 570, 611, 639, 649, **708**, 749, 939
Binary collision concept; *see* Collision
BiO 649,**766**, 772
Birge-Sponer extrapolation **756f**, 767
Black body 56, 60, 163, 181, 183,190,**299**,316,**362**, 622,926,935,943; *see also* Radiator, full
Bleaching (effect) 130, 631
Blow-off 216f, 421
BO **766**, 772
BO_2 34
B.O. approximation; *see* Born-Oppenheimer approximation
Bohr radius 924
Bohr's postulate 30
Boltzmann
— constant 273, **924**, 929
— equilibrium; *see* Equilibrium
— factor 46, 162, 578, 902, 904
— law 44f,51,68,77,119,162,175,315,598,625, 652f,657,668,883,886
— transport equation 829
Born-Oppenheimer approximation 24,**27**,46,91,95,136
Bottle, polythylene 230
Br_2 101
Broadening; *see* Line broadening
Broglie wavelength 582
Buffer; *see* Ionization
Buffering effect (on metal halide)
— of H halide 781f
— of metal compound 782f
— of metal ion 783

Bulk method(s) 534
Bulk system(s) 533
Bunsen burner; *see* Burner
Bunsen flame; *see* Flame
Burner 1, 211f, 413
—, Boling 220
—, Bunsen 213
—, capillary 213, 219
—, circular 220, 917f
—, direct-injection 211f,235,302,306,308,313f, 340,395,405,418,424,517
—, electrical preheating of 218
—, grid 213, **218f**
— head 213, 215, 217f, 221, 419
— —, clogging of 232
— —, cooling of 217, 219, 221, 456
— —, heat loss of 456
— —, temperature of 218
— housing 213, 215, 221, 517
—, low-pressure 221f
—, Méker 213f,307,**334f**,391,418,423,474,476, 917f, 921
— mounting 223f
—, multi-slot 213
—, porous-plate 213, **220**, 423
—, (exit) ports 213f, 216f, 219, 222, 419, 477
—, premix 211f
—, slot 213, **220f**, 332, **334f**,**419,425**,516,**916**, 918,921
— tip (*or* top) 213, 458, 862, 918
Burning velocity 215,219f,414f,420,932; *see also* Flame

C (radical) 635,639,**708**,775,792; *see also* Carbon
C^- 461
C^+ 461
C_2 420, 435f, 460f, 635, 640, 775, 792, **955**
C_2^- 461f
Ca 38, 295, 313f, 455, 544,557,617,**670f**,725 735f, 740, 745, 748f,**808**f,823,827,836,865,**876**f,**880**, **907**, **910**, 939, 953; *see also* Calcium
Ca^+ 443f, 725, 823, **877f**, **880**, 939
CaBr 769
CaCl **769**, 806
$CaCl_2$ 314, **769**
Cadmium-sulphide photoconductor 262
CaF **769**
CaF_2 **769**
CaI 769
Calcium level diagram 19
Calibration
— of intensity; *see* Radiation standard
— of metal concentration 303f,519f; *see also* Concentration (in flame)
— of wavelengths 248
Candoluminescence 39
CaO 92, **766**, **772f**, 778, **791**, 809, 823
CaOH 34, 194, 503, **768**, **776**, 778, **808f**, **881**, 953
$CaOH^+$ 442f, 823, **877f**, **880**
$Ca(OH)_2$ **768**, **808f**
Carbide particle 310
Carbon; *see also* C, C_2
— anode 366
— arc 289, 298
— particle **428f**, 461, 466, 917; *see also* Particle
Catalysis; *see* Recombination
Cavity; *see* Hohlraum, Microwave
CCl_4 239, 805
Cd 110,292,**299**,304,530,544,547,577,600f,**610f**, 616, 618, 630, 636, **640f**, 649, 825, 881, **939**
Cd^+ 904

Ce   635
Cell; *see* Vapour cell
Central region (*or* zone); *see* Interzonal region
Centrifugal barrier   28, **93f**, 99, 591
Centrifugal stretching   **28**
Centrifugal term   688
$C_2(F_5)_3$ 239
CH   420, 435, 436, **460f**, 466, 635, **636f**, 640, 753, 792
CH_3 462
CH_3^+ 461
CH_4 468, 565, 586, 595, 598, 604f
— -air flame   115, 725, 729, 860
— flame   466, 732
C_2H 462, **535f**, 792
C_2H_2 231, 457, 462, 535; *see also* Flame, Hydrocarbon flame
— -air flame   38, 115, 184f, 194, 197, 217f, 221f, 235, 261, 311, 320, 326, 335, **415**, 418, 420, 426f, 431, 447f, 453, 456f, 464f, 476, 500, **506f**, 514, 518, 523f, 564, 635, 651, 669f, 679, 717, 725, 730, **735f**, 744f, 748f, 779, 802f, **953f**
— flame   216, **433f**, 466, 535, 732, 792
— as fuel gas   213, 215, 220, 414
— -N_2O flame   220, 310, 315, 318, 320, 323, **415**, 420, 431f, **525f**, 564, 725, 748, 770, **773f**, **906f**, 910, 921
— -O_2 flame   30, 213, 312, 314, 318, **415**, 418, 422, 426, 432, 436, 465, 489, 519, 635, 671
— -O_2-Ar flame   744f
— -O_2-CO_2 flame   725, **735**, 744f
— -O_2-He flame   671, 744f
— -O_2-N_2 flame   221, 462, 671, 724f, 728, 743f, 749, **910**, 919
C_2H_4 568
$C_3H_3^+$ **461f**
C_3H_8; *see also* Flame, Hydrocarbon flame
— -air flame   217, **415**, 474, 514, 535, 744, 781
— flame   173, 216, 466, 535
— as fuel gas   215
— -N_2O flame   415
— -O_2 flame   415
Charge
— balance   838, **865f**, 957
— exchange reaction   461
— neutrality   813
— transfer, in (de-)excitation   584f, 588f, 592
— —, in ionization; *see* Process, electron transfer
— — state   100f, 584f, 588f
$CHCl_3$ 239
Chemi-excitation   468f, 529f, 634f, 779; *see also* Precursor, Rate, Rate constant, Reaction
Chemi-ionization   221, **461f**, **865f**; *see also* Rate, Rate constant, Reaction
Chemiluminescence   468f, 530, 532, 634f; *see also* Reaction
—, hard   **635f**, 674
—, nonthermal   443
—, soft   **641f**, 654, 674
— spectrum   759
— —, short-wavelength limit of   **759**
— suprathermal   10, 317, 340, 441, **502**, 523, 527, 580, 634f, 641f, 651, **663f**, 752, 792, 802, 812
Chemiquenching   153, **193**, 530, 569, 618, **634f**, 647; *see also* Rate, Rate constant, Reaction
CH_3I 446
Chimney   216, 221, 419
CHO; *see* HCO
CHO^+; *see* HCO^+
$C_2H_2O^+$ 463
$C_2H_3O^+$ 461
C_3HO^+ 461

METAL VAPOUR IN FLAMES

Chopper; *see* Light chopper
City-gas flame   466
Cl   41, 340, 438, **442f**, 467, 527, 797, **805**, **820**, 823, 833
Cl^- 783, 820, 823, 827, 833, **881**, 907
Cl_2 239, 467, 586, 759, **780f**, 821, 828
CN   435, 466, 753, 775, **792**
$C_2N_2-O_2$ flame   **415**, 417, 506
CO   41, 414, 434f, 439, 447, **506f**, **538f**, 547, 565f, **571f**, 578, **583f**, 591, 595f, 604f, 618, **635f**, **639f**, **642**, 659, **708**, 732, **745**, 786, 791, 811, 821f, 831, **868**, **873**, **901**; *see also* Flame
— -air flame   217, 220, 235, 330, **415**, 431, 461, 744f
— —, dry   236, **506f**, 523, 642, 781, 830
— flame   423, 426, 433f, **445f**, 460f, 468f, 523, 775, 791, 795, 799, 825, 860
— as fuel gas   213, 215, 220, 231, 423
— method   439, 445
— -N_2O flame   38, **415**, 431, 447, 706, 724, **730f**, 733, **735**, 778, 809, **811**
— -O_2 flame   **415**
— -O_2-Ar flame   **865f**, **869**, **891**
— -O_2-N_2 flame   38, 223, 432, 724, 749, 777, **861**, **865f**, **869**, **887f**
CO_2 319, **423f**, 432f, 439, 445f, 456, 464f, **468f**, 493, 534, **538**, 565, 568f, 576, 592, 595, **604f**, 636, **642**, 649, **727f**, 732, 740, 744f, 773, **786**, 791, 811, 823, 831, **873**, **888**
— as flame diluent   213
—, quenching by   456
C_2O 635, **636**
Co   291, 639, **939**
Coagulation (of drops)   215, **306**
Coalescence; *see* Coagulation
Coal gas   220
COG; *see* Curve-of-growth
Coherence   110, 113, **116f**, 176, 653, 655, 657
—, degree of   653
— length   143
— time   **143f**
Collection efficiency   **264f**
Collimator   242f, **251f**
— lens   **398f**
Collision(s)   93, 108, 145, **530f**, **560f**, **593f**, 599
—, adiabatic   147, 151, **555**, 595, 682, 684, **698f**, **700f**, 724; *see also* Cross section, Line broadening, Process, Rate, Rate constant, Reaction
— angle   533
—, bimolecular (*or* binary *or* two-body)   81, 105, **682**, **697**, **891**
—, binary; *see* bimolecular
— — concept   **682**, **690f**, 697
—, classification of   108
— complex   40, 91, **93f**, 96f, 104, 544f, 587f, 649, **821f**; *see also* Complex
— configuration   583
—, depolarizing; *see* Depolarization
—, diabatic   **147**, 553, **595**, 682, 684, **698f**, **700f**
— -dominated equilibrium   61
— — plasma   **626**
— — process   74
— duration   81, 544, 685, 687, 693
—, elastic   63, 91, 103, 108, 147, 422, 559, 699, 709, 821
— of the first kind   108, 112
— frequency   59, **81f**, **829f**, **935**; *see also* Electron, Ion
—, gas-kinetic   59, **80f**, 153, 900
—, head-on   93
—, inelastic   **63f**, 91, 102f, 148, 422, **535f**, 684, 821
—, inter-, time   81, 104, 148, 685, 693, 900, 936

SUBJECT INDEX

Collisions (*continued*)
- —, momentaneous 685, 687, 697
- —, near-adiabatic 552f
- —, nonadiabatic; *see* diabatic
- — partner 64, 68, 71, 534, 537f, 543f, 556f, 580f, 600, 604f, 680, 869, 873, 885, 890f, 897, 904; *see also* Perturber, Third body
- —, scattering; *see* Scattering
- — of the second kind 108, 112, 114, 126, 562
- —, superelastic 108
- —, termolecular (*or* ternary *or* three-body) 81, 105, 635
- —, ternary; *see* termolecular
- —, three-body; *see* termolecular
- —, trajectory 92, 96, 685, 693, 697
- —, two-body; *see* bimolecular

Combination rules (of potential parameters) 913, 919
Combinatory method 721
Combustion 414, 417, 435, 815f; *see also* Flame, Heat of combustion
- — cone; *see* Cone
- — engine 7
- —, quenched 414
- — reaction 60, 421, 434f
- —, secondary 60, 503
- — zones 1, 213, 222; *see also* Primary combustion zone, Secondary combustion zone

Comparison method 314, 520f, 527, 777, 800, 812
Complex 90, 92, 584; *see also* Collision complex, Quasi-molecule, Transition complex
- —, activated (*or* activation) 84, 89f
- — — theory 89f
- —, intermediate 545
- —, ionic 100, 544f
- —, long-lived 91, 555, 584
- — molecule 95, 583f
- —, short-lived 91, 562
- —, temporary 91

Composition of flame gases; *see* Flame
Concentration (in flame) 363, 504, 532, 826f, 930; *see also* Conversion factor, Flame, Number density
- — of metal species 2, 303f, 368, 387, 523, 660, 671, 734, 774, 842, 849, 852, 904f
- — —, absolute 303f, 333
- — —, calculation of 515
- — —, calibration of 303f, 519f
- — —, determination of 166f, 479f, 669
- — — —, absolute 171, 188f, 234, 481f, 513f, 869f, 878
- — — —, relative 508f, 511f, 521
- — — (spatial) distribution of 328, 333f, 482f, 485, 488, 495, 509, 518f, 653f, 914f, 921
- — —, integrated 914
- — —, specific 479
- — —, total 479
- — of natural ions 460f
- — profile 333; *see also* (spatial) distribution
- — of radicals 437f, 643f, 648; *see also* Radical
- — —, conversion of relative into absolute 448f
- — —, determination of 437f
- — —, excess 451f

Concentration (in solution) 346, 479f, 491, 497, 499f, 503, 505, 515f, 520, 527, 878, 906f, 926; *see also* Analysis, Analytical curve, Conversion factor, Detection limit, Flame spectroscopy, Interference
- —, characteristic 166

Concentration broadening; *see* Line broadening
Concomitant (in the solution) 674f, 849, 906
Conductance; *see* Optical conductance

Conductivity
- —, electric 465, 829f
- — — complex 829f, 861f
- —, ohmic 829

Cone 420f; *see also* Primary combustion zone
- — combustion 213, 215, 222
- —, primary 219
- —, inner *or* blue 419
- —, internal 419

Configuration; *see* Electron
- — mixing 18

Contamination
- — of concentration (in solution) 504
- — of salt solutions 229
- — of samples 227

Continuity of mass flow 516
Continuum (spectral) 39f, 105, 119, 190, 343f, 358, 363, 365, 367, 384f, 466f, 620; *see also* Flame background, Intensity
- —, absorption 40
- — —, long-wavelength limit of 40, 755f
- —, apparent 39
- —, association 108, 651
- — —, chemical 39f
- — —, short-wavelength limit of 40, 651f
- —, emission 39, 352, 444
- — —, thermal 540
- —, ion (-electron) recombination 39f, 108, 904
- —, photodissociation 40f
- — —, long-wavelength limit of 40f
- —, photo-ionization 40
- —, quasi- 39, 118, 466f
- —, radiation 118, 542
- —, recombination 467

Continuum of energy levels; *see* Energy level
Continuum (light) source 160, 166, 190, 196, 200f, 207, 245, 286f, 298f, 320f, 324f, 376, 403f, 410, 443, 486, 489, 491, 510f, 525, 628f, 712, 715, 719, 721, 749; *see also* individual lamps, Light source

Convection 837f
Convergence limit 755f
Conversion factor 497
- — of incoherent units 923
- — for number density in flame 235, 305, 929

Convolution (integral) 155f, 241, 248f, 689, 700f
- — of line profiles 681, 698

Cooling
- — of burner head 217, 219, 221, 456
- — of flame 458, 517
- — of (cathode of) photomultiplier 266, 275

Correlation (statistical) 355, 361, 387, 472, 474, 681, 951; *see also* Time correlation
- — technique 283
- — time 143, 149, 152, 937

Correlation diagram 649
Correlation rule 92
Corrosion 239
Counting (technique) 336
- —, photon 263
- —, pulse 266f, 270, 275

Cr 291, 309, 455, 545, 558, 638, 708, 725, 749, 824f, 865, 869f, 881, 939
- — halide 780
- — particle 315

CrO 194, 314f, 455, 766, 823
- — particle 314

CrOH$^+$ 823
Crossing (of potential curves *or* surfaces) 40, 93f, 100, 103, 544f, 553, 583f, 597, 602, 821f; *see also* No-crossing rule, Transition, Transmission coefficient

Crossing (*continued*)
—, avoided 94f
—, ionic-curve 99f, **539**, **584f**, 597, 648, 821, 828, **899f**
—, multiple 96, 99f
— point 93f, 101, 544f, 586f
—, probability for 96f
—, pseudo- 93f, 101
—, quasi- 95
— region 97
Cross section (effective) (for collision *or* process) 80f, **85f**, 99, 103, 535, 540, 569, 882, **884f**, 935
—, atomic absorption; *see* optical, for absorption
—, average 86
— for de-excitation; *see* quenching
— for diffusion 911
— for electronic-excitation transfer 537, **599f**, 602, **616**
— for excitation **537f**, 572, 576
— —, differential 547, 569
— — by electron impact 556f, 618
— for gas-kinetic collisions 80f, 85, 882, **935**
— for inelastic collisions 893f
— —, differential 534
— for ionization 828, **866f**, **872f**, **882f**, **890f**, 895, **898f**, 902
— —, single-step **884f**, 891
— for line-broadening collisions 149f, 724, **742f**
— —, adiabatic **689**
—, microscopic **85f**
— for mixing 935
— —, differential 534
— —, intermultiplet 603f
— —, intramultiplet **548f**, **593f**, **612f**
— —, mutual 555
— —, self- 602
—, optical, for absorption 129, 132, 149, **934**
— —, for line-broadening collisions; *see* line-broadening collisions
—, over-all 87, 542f
—, phenomenological 86, 563, 568, 577, 589, **603f**, **866f**, 882, **890f**, 895
— for quenching 99, 193, 406, 535, **537f**, **542f**, 548, **551f**, 561, 563, 565f, 568, 572, 576f, **581f**, 594f, 599, **603f**, **647f**, 935
— —, differential 562
— — by electron impact 558f
— —, self- 548, 555
— for scattering (of particles) 81
— —, differential 81, 102, 534, **568f**
— — of electrons 830
—, specific 533, **542f**, **566**, 572, 868, **872f**, **890f**, **898f**
— —, normalized 872f
—, thermal 87, 89, 99
—, total 86, 565, 618
Cs 33, 120, 238, 299, 449f, **506f**, 514, 537, 542, **545**, 554, **557f**, 565f, 579, **593f**, 597f, 601, **608f**, **613f**, 649, **694**, 737f, 742, 745, **749f**, 777, 833, 865f, **869f**, 887f, 899, 910, 918f, **920**, **939**
Cs⁺ 910
CsBr 769
CsCl 769
CsF 769
CsI 769
CsO₂ 775
CsOH 768, 776
Cu 125, 291f, 387, 438, 442f, 493f, 518, **525f**, 570, 609, 649, 670, 705, **748f**, 771, 779, 790, **881**, 910, **939**
— -halides 526
CuBr 769
CuCl 618, **769**, 781
CuCl₂ 769
CuF 769
CuH 34, 237, 369, 438, **443f**, 526, 650, **768**, 779, **789f**, **794f**, 802
CuI 618, **769**
CuO 526, **766**, 772, 779
CuOH 237, 526, 650, **768**, **775**, 779
Current-tension characteristic (of probe) 857f
Curve crossing; *see* Crossing
Curve-of-growth 157, **169f**, 192, 314, **495f**, 505, 508, 521, 628, 668, 671f, 689, **704f**, 733, **749f**, 869
— in absorption 502f, **718f**
—, asymptotes of 168f, 502
— —, intersection of 168f, **497f**, 705, **720f**, 749
—, distortion of 498, **663**, 709, 904
—, doublet 498
— in emission 184, 187f, 191, 502, 504, 523f, 526, **718f**
—, experimental 171f, 188, 191, **496**, 525, **734**
— in fluorescence 198, **718**, **721f**
—, inflection (point) of 171, 201
—, linear branch of 168, **170**, 189
—, square-root-branch of 168, 170, 172, 187, 189; *see also* Square-root-law, Square-root-region
c.w., continuous-wave; *see* Laser
Cyclotron-resonance 831, **862f**
— frequency **831**, 854
— technique **852**, 854, **862f**, 881

D₂ 565, 568f, **577f**, 590, 595f, 604f
Damping parameter **925**; *see also* α-Parameter
Dark current 266f. 270f, 359, 381, 387
—, reduction of 275f
De-activation; *see* De-excitation
Dead-time 274
Debije (*or* Debye)
— (shielding *or* screening) distance 55, **813**, 935
— length 55
— radius 55
— shielding (*or* screening) 45, **55**, 885, 899
Decay; *see* Energy level, population of, Relaxation
Deconvolution 250, 258, 389, 715
De-excitation 97, 193, **529f**, **536f**; *see also* Cross section, Process, Quenching, Rate, Rate constant, Reaction
Degeneracy 18f, **26f**, 122, 135, **702**, 804; *see also* Λ-Doubling
—, electronic 135
— factor **19**, 44, 48f
—, *l*- **20**
—, M_J- 179
—, magnetic 18
—, rotational 135
—, spatial 18, 20, **26**
Degree of dissociation; *see* Dissociation
Degree of ionization; *see* Ionization
De-ionization **500f**, **513f**; *see also* Ionization buffer
Delayed-coincidence method (*or* technique) 78, 540, 555
Delta-function, Dirac 249
Demarcation band (in collisional ionization) **895f**, **899f**
Demodulation 362
Density; *see* Mass density, Number density
Depletion (of excited-level population) 74
Depolarization (of radiation) 80, 195, **530f**, 580; *see also* M_J-Mixing
Depopulation (of excited state); *see* De-excitation
Derivative spectroscopy; *see* Spectroscopy

SUBJECT INDEX

Desolvation 2, **224**, 239, **306**f, 333, 457, **516**f; see also Fraction desolvated
Detailed balance **57**f, **68**f, 74, 88, 120f, 699, 708
—, optical **373**f
—, principle of 119, 539
— relation (for conjugated rate constants) **68**f, 73, 84, **88**f, **548**f, **558**, **576**, 599, 621, 654, 660, 664, 675, 826, 864, 868, **873**, **884**, 890, 897
—, total 74
Detectability 471; see also Detection limit
Detection; see also Measuring technique, Meter
—, a.c. 261
—, phase-sensitive 322
—, synchronous 257, 261, 274
Detection limit 9, 174, 202, 262, 271, **325**f, 346, **387**f, 392, **405**f, 416, 467, 479, 491, 629, 641, **673**f, 771, 774, 819, 834, 905f
Detector; see also Photocell, Photodetector
—, coherent 281
—, heterodyne 281
—, photo-electric 3, **261**f, **336**f
—, solid-state 343, 346
—, surface-junction 343
Detonation method 762
Deviation from equilibrium; see Disequilibrium
Dicke narrowing 153, 681, 699f, 736
Difference meter 371f
Difference potential; see Interaction potential
Difference technique 374f
Diffuser 400
Diffusion 85, **335**, 418, 459, 483, 518, 524, 905, **909**f
—, ambipolar 465, **837**f, 905, 920
— coefficient (or constant) 312, 333, 753, **837**f, 905, **909**f, **911**f, **914**f, 926
—, ambipolar 910, 920
—, binary 911f
—, effective 920f
—, measurement of 918
— of collision complex 587f
—, cross section for 911
— equations 919f
—, inward, of air; see Infusion
—, line-source model for **916**f, 919
—, point-source model for **914**f, 919
—, radical 517
— of radiation; see Radiation
—, thermal 911
Diffusion flame; see Flame
Dihalides **769**, 779, 783, 805; see also individual dihalides
Dihydroxides **768**, **775**f; see also individual dihydroxides
Dilution run **487**, 496
Dimers (of metallic elements); see individual dimers
Dipole-dipole interaction 579
—, resonant 555
Dipole matrix element **122**f, **135**f
Dipole moment (of molecules in liquid) 227
Direct-injection burner; see Burner
Direct-injection nebulizer; see Nebulizer
Direct-line fluorescence; see Fluorescence
Discharge lamp; see Gas-discharge lamp
Disequilibrium **58**f, 60, 65, 70, 75, 85, 89, 120f, 421, 435, 458, **532**f, 549, **655**f, **659**f, 708, **787**f; see also Relaxation
—, chemical **69**f, 72, 153, **434**, 634, **641**f, 660, **665**, 674, **783**, 775, 784, 841, **843**f, **847**f
—, dissociation 69, 71, 75, 654, 660, 791, **795**f
—, excitation 490, **839**f, **899**f, 904; see also Energy level population, Population factor

Disequilibrium (continued)
—, ionization 71, 559, 654, **659**f, **820**, 825, 837, **839**f, **846**f, **892**f, **895**f, **899**f, 904; see also Chemi-ionization
—, physical **847**f; see also radiative
—, radiative 61, 77, 177, 188, 456, 502, **620**f, **660**f, **671**f; see also Chemiluminescence
— — by depletion effect 75
— of radical (concentrations) 644, **776**f, 788, **791**f, 796, **799**f, **807**f, **847**f, 907
—, thermodynamic 819
Dispersion (in nebulization) 315, 365, 520; see also Aerosol
Dispersion (spectral)
—, angular **244**f
—, anomalous **122**, **125**, **129**f, 514
— function 144, 149, 152
Dissociation (of molecule) 2, 24, **49**f, 58, 418, **647**f, 763; see also (Dis-)Equilibrium, Photo-dissociation, Rate, Rate constant, Reaction
— abnormal 775
— constant 47, **49**, **52**f, 68, 751, 760, **773**f, 776, **797**f, 929
—, degree of **50**f, 54, 70, 432, 526, 751, 762, 773, 775, 784, 826, 933; see also Association factor
— energy (-ies) 28, 37, 40, **49**f, 53, 105, **434**, 440, 446, 525, **635**f, 648, **651**f, 752, **754**f, **764**f, **784**f, **789**, **796**f, **802**f, 913, 926
—, methods of determination of 754, **764**f, **796**f
— — —, appearance threshold **755**, **758**f, 767
— — —, calorimetric 760
— — —, chemical-kinetic **763**f
— — —, flame-spectrometric equilibrium **764**f, **796**f, **802**f
— — —, molecular **754**f
— — —, spectroscopic **755**f, **764**f
— — —, thermochemical (or thermal equilibrium) **755**, 760f, **764**f
—, heat of 51, 761, 933
Distribution (function) 58, 67
—, angular
— — of fluorescence 116
— — of scattered light 114, 116, 533
— — of scattered particles; see Cross section for scattering, differential
— — of concentration; see Concentration (in flame)
— — of drop-size 307, 312
— — of energy **58**f, **65**f, 85
—, Gauss 142, **152**
— — of intensity (spatial) 915f
— — of internal energy **44**f, **66**f, 71, 86, 102, 839, 884, 890, **894**f
— — laws (in equilibrium); see Equilibrium
—, Lorentz 142, 149, 155
—, Maxwell; see Maxwell law
— — of rotational energy 574, 584
— — spectral 120f, **138**f, **141**f, 176, 620, 653, 657, 660; see also Line profile, Spectrum
— —, normalized 139, 931
— — of temperature; see Temperature (of flame)
— — of translational energy; see (relative) velocity
— — of (relative) velocity 67, 79, 102, 117, **152**f, 316, 535, 549, **575**f, 603, 659, **708**f, 829, **839**, 890
— — of vibrational energy **571**f, **575**f, 584, **586**f, **590**f, 884
—, Voigt **156**f, 162, 169

D_2O 604

1006

Doppler
- broadening; see Line broadening
- effect 116, 131, **152**, 602, **696f**
- —, recoil **697**
- profile; see Line profile
- relation 133, **152f**
- shift 329
- temperature; see Temperature of flame
- width; see Half-intensity-width

Double-beam technique; see Spectrometer
Double-photon excitation, etc.; see Two-photon excitation, etc.
Double probe technique 464f, **851f**, **860**
—, floating 851
Doublet 20, 33, 37, 191, 204, 504, 508, 552, 555, **562f**, 564f, **594f**, 598, 633, 649, 728, 730; see also Multiplet
—, compound 33
- mixing; see Mixing
- splitting **552f**, 556, 649
Drain-off 226
Drift 256, 275, 279, 370, 376, 381
Drop(let) 495; see also Aerosol, Coagulation, Desolvation, Nebulization
- diameter; see size
—, fluctuation in supply of; see Noise, shot
- generator **238**, 311, 332, **520**
- —, single 327
- size 227, 239, **306f**; see also Distribution
—, water 307
Dry powder, introduction of **237**
Dual-beam-sampler **374**
Dual-beam (or wavelength) technique; see Spectrometer
Duplication
- curve **188f**, **505f**, **704f**, **718f**, 734, 749
- —, doublet **506f**
- factor **188f**, **505f**, **926**
- of flame thickness **506**; see also Flame duplication
- method 504f
- of solution concentration **505f**
Dy 635
Dye laser; see Laser
Dynode 267, 271, **275f**

Eddies (in flame) 418, 477
EDL; see Electrodeless-discharge lamp
Effective width (of spectral profile) **139**, 167, **248f**, 352, 365, 404, 491, 722, **934**; see also Bandwidth, Half-intensity-width
—, collisional **151**
—, Doppler **153**
- of instrumental profile **241f**
—, natural **142**
- of spectral selector 363; see also Resolution
- of Voigt profile **157**
Efficiency of atomization **234**, **934**
Efficiency of fluorescence **76f**, 130, 147, 173, 176, **192f**, **264f**, 324, 390, **392f**, 405, 411, 532, 558, **562f**, 580, **594**, 621, **624f**, 628, 633, 643, **646f**, 653, **660f**, **724**, 746
—, effective **627**
—, measurement of **393f**
—, power **193**, **933**
—, quantum **192f**, **262**, 326, **933**
- —, total **933**
Efficiency of nebulization **226f**, **234**, **236f**, **302f**, **494**, **503f**, **507**, **515f**, **521**, **523f**, **527**, **805**, **812**, **934**
Einstein (radiation) coefficient; see Transition probability
Einstein transition probability; see Transition probability

METAL VAPOURS IN FLAMES

Electric wind of Chattock 850
Electrodeless-discharge lamp **285**, **293f**, **299**, 301, 376, 389, 408, 630
—, microwave (MW) excited **293f**
—, radio-frequency excited **294**
Electromagnetic moment 762
Electron(s) 3, **15f**, 54, **532**, **556f**, 558, **813f**, **817f**, **823f**, **828f**, **850f**, **864f**, **882f**, **920**; see also Emission, Flame, Ionization, Process, Reaction
- acceptor **849**
- affinity 24, 54, 100, 463, **585f**, **821**, 827, 881, **900f**, 927, **957**
- attachment **54**, **818**, **825**
—, collision frequency of 81, 559, **814**, **831**, **852f**, **863**
- concentration, measurement of 514, **850f**
—, configuration of **16**, 19f
- core **16**
- detachment **54**, **818**
- donor **522**, **834**
—, hot **464f**, **559**, **659**, **824**, **847**, **904**
—, ion recombination **108f**, **817f**, **823**; see also Ion-recombination
- mobility **880f**
—, natural flame; see Flame
- noise temperature **850**
—, secondary **264f**
- spin 25; see also Angular momentum
- temperature **851**
- —, measurement of **851f**, **858f**
—, thermionic emission of 519
- transfer; see Process
—, valence **16**, **20**, **546**, **893**
Electronic image sensor **262**
Electron microdiffraction **310**
Electron spin resonance **420**
Electrostatic-probe technique **420**, **818**, **851f**, **857f**, **865**, **869f**; see also Double-probe technique, Single-probe technique
E.l.f. noise, Excess low-frequency noise; see Noise
Emission 31, **64**, 125, **489f**, **509**, **512**, **532**, **619**; see also Background, Chemiluminescence, Radiation, Transition
- by atoms **501**, **904**; see also Emission [general]
- coefficient **944**
—, continuous; see Continuum
- curve-of-growth; see Curve-of-growth
- of electrons from photocathode **264f**, **336**, **470**; see also Photodetector, Photomultiplier
- excess **445**; see also suprathermal
- factor **163**, **175**, **181f**, **190**, 652, **661**, **934**, **943**
- —, effective **620**
- field **266**
- by flame species; see Flame
—, induced **78**, **110**, **130f**, **132**, **144f**, **182**, **202**, **619f**, **624**, **657**
- — rate **619f**
—, infrathermal **812**
- intensity; see Intensity
- line profile; see Line profile
—, metal **471**
- method **334**
- by molecules **38**, **134f**, **179f**, **445f**, **484**, **501**, **640**, **775f**, **779**; see also Band
—, nonthermal **708f**; see also infra-(or supra-) thermal
- probability; see Transition
—, secondary, of electrons **263f**; **271f**; see also Photomultiplier
- —, of photons; see Fluorescence
—, spontaneous **63**, **110**, **145**, **619f**, **655**

1007

SUBJECT INDEX

Emission (*continued*)
- — rate 580, 620
- —, suprathermal 634, 642, 644f, 659, 663f, 777
- —, stimulated; *see* induced
- —, tertiary, of photons 195; *see also* Radiation diffusion
- —, thermal 163, **175f**, **179f**, **182f**, 317, 403, 410, 445f, 542, **668**
- —, thermionic, of electrons 266, 271, 275, 849; *see also* Dark current

Emission-absorption method 318f
Emission spectroscopy; *see* Spectroscopy
Emissivity **944**
Energy 927; *see also* Activation energy, Dissociation energy, Distribution, Excitation energy, Free energy, Gibbs free energy, Ionization energy, Reaction energy, Threshold energy
- — barrier 82, 84, 89, 759
- — density; *see* Radiant energy density
- — of dissociation 933
- —, electronic **18f**, **27**, 539
- —, internal
  - — (of atom) **18f**, 534, 536, **560f**, 927; *see also* Atom
  - — (of molecule) **27f**, 534, 536, **539**, **560f**, **570f**, 598, **659**; *see also* Molecule
  - — (of system) 50, 52, **932**
- —, kinetic; *see* Energy, translational
- —, potential; *see* Potential energy
- —, radiant; *see* Radiant energy
- — relaxation 58f; *see also* Relaxation
- —, rotational **27f**, 43, 46, 59, **571f**
- —, translational 20, **42f**, 59, 534, **536f**, 548, **556f**, 560, 572, 575f, **590f**, **595f**, 599, 650, 659, 708, 759, 927
- — units 923
- —, vibrational **27f**, 59, 100, 106f, 571, 573, 575f, 583, **586f**, 649
- — —, persistence of **573f**
- — —, quantum of **27**, 82, 180
- — —, zeropoint **27**, 47

Energy level(s) 19, 44, 46, 480; *see also* Multiplet, Quantum number, Quantum state, Term
- — of atom **18f**, 21, 117, **884**, **939f**; *see also* Atom
- —, broadening of; *see* Line broadening
- —, continuum of 20, 321, **885**
- —, degenerate **18f**, 44, 48, **702**; *see also* Degeneracy
- — diagram **19f**, 20, 29; *see also* individual atoms
- —, electronic (of molecule) 25, **26f**, 37, 560
- —, excitation (*or* excited) 20, 77, 566, 883
- —, ground **19f**, 895, 902
- —, high(er)-lying **536f**, **539f**, **545f**, **549f**, 552f, 554, 559, 561, **565f**, 579, 590, 594, 601, 635f, 659, **892f**, **895f**, 900, 904
- —, manifold of; *see* Quantum state
- —, metastable 33, 551, 555
- — of molecule **23f**, **26f**; *see also* Molecule
- —, non-degenerate 44, **702**
- —, population of 67, **539f**; *see also* Boltzmann law, Inversion, Population factor
- — — electronic 48
- — — excited 123, 130, 143, 147, 532, **622f**, **628f**, 635, 642f, **654f**, **658f**, 839, **885f**, **892f**, **895f**, 904
- — — —, decay (rate) of 540, 543, 570
- — — —, redistribution of 815, 819, 893
- — —, equilibrium; *see* population, thermal
- — — ground- 540
- — — infrathermal 441, 547, **885**, **893f**
- — — metastable 540
- — —, nonthermal 368, **635f**, **652f**, **893f**

Energy level(s) (*continued*)
- — —, partial equilibrium 542, 566, **593f**
- — — rotational 29, **47f**
- — —, stationary 74f, **540f**, 547, 580, **658**, 664
- — —, thermal **44f**, 580, 642, **652f**, 660, **883f**, **886**
- — — vibrational **23f**, 29, **48f**, 571, 659
- —, pumping of 540f, 543, 547
- —, resonance, first 565, 586, **591f**, 594
- —, second 565, 568
- —, rotational **26f**, 29, 37, 59, **560f**
- —, negative 26
- —, positive 26
- —, saturation of; *see* Saturation
- —, sign of 26, 35, 37
- —, symbol for **19f**, **22f**, **25**
- —, vibrational 23, **27f**, 29, **36f**, 59, **560**, 571, **577f**, **707**, **756f**

Enthalpy 50, 90, **928**
- — of activation 90
- — of formation; *see* Heat of formation
- —, free; *see* Free enthalpy
- — of reaction; *see* Reaction enthalpy, Reaction, heat of

Entrance slit **398f**
Entropy 50f, 90, **931**
Equilibration 58f, **61f**, 316, 532; *see also* Equilibrium, Relaxation
- —, chemical 456
- — of ionization 903
- —, partial **895f**
- — of reaction **786f**
- — time 316, 559
- — of velocity 316

Equilibrium 13, **41f**, **43f**, **50f**, **58**, **74**, 416, **422f**, 457, 481, **620f**, 634, 652, **657f**; *see also* Equilibration
- —, attainment of **58f**
- —, Boltzmann 120, 183
- —, partial 542, 547
- —, chemical 5, **49f**, **61**, 68, **425f**, 435, 447, 456, 469, 501, **641f**, 654, 774, 789, 820
- —, partial **61**, **72**
- —, constant 53, 69, 90, 105, **436**, **440f**, 453, 775f, **788f**, **799f**, **802f**, **805f**, 809, **903**, 929
- — for electron detachment 54
- — method **426f**
- —, deviation from; *see* Disequilibrium
- —, dissociation **49f**, 68, 760, 774, **791f**, **795f**, 905
- —, distribution (law) **42f**, 68, 70
- — for internal energy **44f**; *see also* Boltzmann law
- — for radiation **55f**; *see also* Planck radiation law
- — for translational energy 43; *see also* Maxwell law
- — for velocity **43f**, 153, 158; *see also* Maxwell law
- —, excitation; *see* equilibrium, Boltzmann
- —, ionization **53f**, **513f**, 825, **905f**; *see also* Saha
- —, partial **72f**
- —, Maxwell 421
- —, partial **57f**, **60f**, **436**, **442f**, 445, 453, 543, 641, 668, 775, 777, **787f**, **802**, 805, 820, **825f**, 833, 897, 920, 957
- —, physical 5, **61**, 654, 788
- — in pure flame **43f**, **58f**
- —, radiation 55f
- —, Saha 63, 501, 813, **820**, 839, 866, 883, 886, 900, 904, 958; *see also* ionization
- —, quasi- **895f**, 900
- —, thermal **61**, 74, 175, 315, 424, 481, 532, 656, 900
- —, local **61**, 316

1008

Equilibrium (*continued*)
—, thermodynamic 41f, 57, 60, 68, 70, 74, 119, 153, 175, 315, 625f, 653, 752, 784
— —, incomplete 61
— —, local 57f, 60f
—, water-gas 788
Equipartition (law) 42f, 52, 58, 61, 152, **788**
Equivalent width 167
Er 635
Eu 635, 907
Evaporation; *see* Desolvation, Volatilization
Excess low-frequency noise; *see* Noise
Exchange; *see* Angular momentum, spin, Reaction, Transfer
— interaction 555
Excitation 108f, **529**f, **534**f, 573, 599, 603; *see also* Emission, Fluorescence, Process, Reaction, Rate, Rate constant, Chemi-excitation, Photo-excitation
—, collisional **468**f
— efficiency 102, **667**
—, electronic **536**f, 659
— — transfer; *see* Transfer
— energy 19f, 24, 68, **189**f, 368, **524**, 559, 583, 636, 641, 648, 656, 658, **665**f, 674, 801, **806**f, 809, 829, 883f, 902, 927; *see also* Atom, Molecule
— — defect 543, 548, 555, 560, **577**f, 618
— —, effective lowering of 46
— function **557**f
— interference **674**f
— level; *see* Energy level
—, nonthermal 199, 441
— potential 20; *see also* energy
—, pulse 78, 123, 147, **540**, 543, 570, 580, 633
—, rotational 584
—, suprathermal **635**f, **641**f, **650**f, 659, **663**f, **673**f, 905
— temperature 62, 77, **317**f, 455, 458, 481, 657, **669**f
—, thermal 340, 387, 392, 441, 447, **652**f, 672
— threshold energy for **538**, 556
— by two-photon absorption 108,110f,203,543,819
—, vibrational 539, **571**f, **584**f, 596, 659
Exit slit 249
Expansion chamber; *see* Spray chamber
Expansion factor 516
Explosion method 762
Extinction 126

F_2 759
FAAS, flame atomic absorption spectroscopy; *see* Flame spectroscopy
FAFS, flame atomic fluorescence spectroscopy; *see* Flame spectroscopy
Family-effect **520**f; *see also* Flames, family of
FAS, flame absorption spectroscopy; *see* Flame spectroscopy
Fast-Fourier-Transform 475
Fast gas-flow reactor 787, 793
Fatigue (of photomultiplier) 276
Fe 248, 291, 318, 489, **526**, 557, **636**, 641, **670**f, 825, **881**, **940**
FeO 766, 772, 787
FES, flame emission spectroscopy; *see* Flame spectroscopy
FFS, flame fluorescence spectroscopy; *see* Flame spectroscopy
Field
— broadening; *see* Line broadening
—, electric 532, 850, 862
—, magnetic 260, 276, **531**, 714, 762, **831**f, 854, 862

Figure-of-merit (of flame noise) **472**f
Figure-of-merit (for monochromator) 241f, 245f
Fill gas 290
Filter
—, absorption, glass 246
—, didymium 331
—, frequency 269f
—, frequency-selective 475f
—, grey 338
—, interference 244, **246**, 343, 377
—, negative 261
—, nonlinear 173
—, RC- **269**, **949**
—, spectral 262
Fine-structure 21, 26, 33, 45, 165, 179, 186, 484, 489, 492f, 498f, 510, 513; *see also* Multiplet
Flame(s) 1f, 8, 38, 419, 465, 515, **533**f, 674; *see also* individual flames, Interzonal region, Primary and Secondary combustion zones
—, absorption by pure; *see* background
—, advantages of 4, 7, 36, 814
—, analysis 444
—, analytically useful **746**f
—, atmospheric-pressure 1, 5, 195, 212, **413**f, 419, 426, 437, 460, **535**, 891, 895, 910
—, atomic-hydrogen 417
—, augmented 6, 417
—, auxiliary 192, 260, 376, 444, 484, 494, 506
— background (spectrum), in absorption 382, 343f, **465**f
— —, in emission 5, 30, **39**, 60, **403**f, 407, 410, 416, 420, 423, 444f, 465f, 494, 674, **955**f
— —, in fluorescence **465**f
— — noise 467, **470**f, **476**, 491, 674
—, blow-off of **216**f, 421
—, bulbous 419
—, Bunsen **419**f, 422
—, burning velocities of 219f, **415**, 418, 422
—, carbon-rich 535
—, central 459
—, chemical 1, 417, 532
—, chemi-ionization in pure **460**f
—, chemiluminescence in pure 465f
—, classification of **417**f
—, composition of 5, 115, 414, 416, 522, 647, **728**, 789f, **800**f, 812, **888**f; *see also* equilibrium composition
—, conical 419
—, cool 451, 490, 652, 674, 753, 848
—, cooling of 458, 517
—, cylindrical 419
—, diffusion 753, 787
—, disadvantages of 5f
—, duplication **189**f, 191, 503, **505**f, 628; *see also* Back mirror, Duplication
— electrons, natural 416, 460f, 501, 906; *see also* Electron
—, emission by pure; *see* background
—, equilibrium composition of **424**f, **428**f
— —, calculation of 50, **424**f
— —, deviation from 434
—, family of **450**f, **522**f, 524, 643, **789**f, 794; *see also* Family effect
—, flashback of 214, 421
—, flat 419
—, flicker 470
—, fluorescence by pure; *see* background
— front 414, 419, 421
—, fuel-lean **417**f, 428, 435, 437, 441, 447, 642f, 773f, 777, 793, 901 903
—, fuel-rich 315,**417**f,428,**432**f,437,441,447f,451f, 503, 635, 640,642,674,753, **773**f, 783, **792**f, 809

SUBJECT INDEX

Flame(s) (*continued*)
—, heat losses in; *see* Heat loss
—, high-pressure 212, **419**
—, hot 451
—, ignition of 223, 414
—, incandescent 386, 641, 774
—, ionization, natural **460f**, 849; *see also* electron, ion
— ions, natural 416, **460f**, 514, 813; *see also* Ion
—, isothermal 67, 77, 88
—, laminar 212, **221f**, **332**, **418f**, 517, **864f**, **914f**, **917**
—, lifted **419**
—, lifting of 216f
—, low-pressure 61, 131, 195, 212, 311, **419**, 422, 437, 464, 489, 580, 641, 659f, 708, 863
—, luminous; *see* incandescent
—, mantle 212, **217f**, 332, 459, 485, 502, 517, 921
—, mixing ratio of; *see* Mixing ratio
—, noise 272, 424, 850; *see also* Noise
—, nonshielded; *see* unshielded
—, oxygen-rich (*or* oxidizing); *see* fuel-lean
—, planar **419**
—, premixed 311, **413f**, **418f**, 424, 475, **517f**, **527**, **864f**
—, (total) pressure of 6, 414, 416, **419**, 425f, 665f, 849
—, propagated 413f
—, radicals; *see* Radical
—, reactions in pure; *see* Reaction
—, rectangular **419**
—, reducing **418**, 432, 635, 774
—, relaxation in pure; *see* Relaxation
—, rise-velocities of 2, 223, 417, 866; *see also* Rise-velocity
—, seeding of **236f**
—, shape of **419**
— sheath 212, 214, 217, 406
—, sheathed 2, **218**, 423, 475
—, shielded 212, 217, **221f**, 456, **527**, 812, **864f**
—, sooting 60, 417f, 461, **917**
—, spectrum; *see* background
—, split **419**, 423
—, stationary 413f
—, stoichiometric 5, **415**, **417f**, 437, 642
—, structure of **419f**
—, temperatures of 1, 41f, 223, **418**, 433, 436, 449, 451f, **455f**, 720, 735, 744f, 748; *see also* Temperature (of flame)
— thickness; *see* Thickness (of flame)
—, turbulent 212, 307, 517f
—, unpremixed **418**, 423, 516f, 518, 675
—, unsheathed 475
—, unshielded 187, 448, 459, **517**, 921
Flame absorption spectroscopy; *see* Flame spectroscopy
Flame atomic absorption spectroscopy; *see* Flame spectroscopy
Flame detector 816
Flame emission spectroscopy; *see* Flame spectroscopy
Flame fluorescence spectroscopy; *see* Flame spectroscopy
Flame photometer; *see* Flame spectrometer
Flame photometry; *see* Flame spectroscopy
Flame spectrometer **209f**, 473
—, analytical 262
Flame-spectrometric equilibrium method; *see* Dissociation energy
Flame spectroscopy (-metry) 31, 37f, 45, 57, 125, **209**, 262, 272, 599; *see also* Spectroscopy
—, analytical 7f, 39, 212, 228, 234, 255, 274, 309, 336, **346**, 373, 413, 416f, 467, 471, 479, 502, 529, 600, 652, 723, **746f**, **753**, 774, 780, 816, 833, **905f**, 909

Flame spectroscopy (-metry) (*continued*)
— —, atomic absorption 8f, 160, 165, 171, 174, 221, 261f, 368, 382f, 491, 513, 747, 753, 773, 809
— —, atomic fluorescence 8f, **200f**, 207, 215, **262f**, 300, 308, **390f**, 395, 399, 405, 533, 580, 747, 753
— —, emission 8f, 262, **339f**, **392**, 423, 673f, 883
— —, special bibliography of 11f
— —, chemiluminescence 673f
— —, molecular 34
Flashback 222, 231
Flicker noise; *see* Noise
Floating potential 851, **858**
Flow (of gas)
— adjustment 232
— calibration **233f**
— rate of burnt gas **927**
— rate of unburnt gas 235, **927**
— regulation 230f
— speed of unburnt gas **932**
— system, low-pressure 571
—, unburnt gas 419f, 422
Flowing gas system 579
Flowmeter (for gas) **230f**; *see also* Gas meter, Rotameter
—, capillary-type 231
—, calibration of **233f**
—, —, by method of series connection 233
—, —, by substitution method 233f
—, liquid 231
—, standard 233f
Fluctuation; *see* Noise
Fluorescence **76f**, **112f**, 126, 174, **192f**, 319, 327, **390f**, 483, **530f**, **535**, 648; *see also* Curve-of-growth, Distribution, Excitation, Quenching, Spectroscopy
— by atoms 111, 633, 723; *see also* Fluorescence [general]
—, anti-Stokes **111f**, 319
—, cross- 112, 548
—, direct-line **111f**
—, efficiency of **76f**, **192f**; *see also* Efficiency of fluorescence
— excitation function **716f**
— by flame species; *see* Flame background
— induced by two-photon absorption; *see* Two-photon absorption
— intensity; *see* Intensity
—, isotropic **400**, **564**
—, laser-excited (*or* laser-induced) 436, 531, 535, 727f, **788**
— line profile; *see* Line profile
— measurement of 390f
— method **76f**, **319f**
—, —, absolute **76f**
—, —, relative **77**
— by molecules 194, 207, 444, 579; *see also* Flame background
—, nonresonance **109**, **111f**, 126, 203, **319**, 598, **633**
—, resonance **76f**, **109**, **112f**, **193**, **319f**, 390, 530, 621
—, saturated **202**, **334**, **905**; *see also* Saturation
— scattering 653, 655, 657
—, sensitized 79, 112, 548, 600, **603f**, 709
—, —, method of 600
— spectrum 679
—, stepwise line **111f**, **534**, **548f**, **555**, 593, **603f**
—, —, thermally assisted **112**
—, Stokes **111f**, 319
—, transient 194
—, yield factor of **76**; *see also* efficiency
Fluorescence spectroscopy; *see* Spectroscopy
Flux; *see* Radiant flux
Focal length 247

1010

Folding 155
Force constant (of interaction potential) 680f, 685f, 704, 707, 729, 739f, 743
Foreign gas broadening; see Line broadening
Fourier
— analysis 475, 693
— spectrometer 259
— technique 475f
— transform(ation) 144, 149, 152, 250, 259, 475, 685, 696, 947, 949, 952
Fraction
— atomized 234, 480, 513, 520f, 527, 635, 771, 933
— desolvated 234, 306, 518, 933
— —, local 306
— dissociated 502, 933
— ionized 499f, 933
— volatilized 234, 309, 503, 933
Franck-Condon
— factor 36, 136f, 181, 589; see also Sum rule
— principle 35f
— transition 822
Franck's rule 103, 555
Free enthalpy 50, 84, 427, 773, 927
— method 427f
Free path; see Mean free path
Free spectral range 247
Frequency
—, angular; see circular
—, circular 936
—, collision; see Collision
— criterion 729
—, -doubling; see Laser
—, electric 927
—, -factor 83f, 785, 925
— filter 269f, 475f
—, fundamental vibrational; see Wavenumber
—, modulation- ; see Modulation frequency
— -modulation ; see Modulation, wavelength-
—, optical 935, 944
— response; see Response
— of spectral line 30; see also Wavelength
f.s.; see Fine-structure
FSR ; see Free spectral range
Fuel (gas) 213, 215, 414, 417, 849; see also individual flames and gases
Fuel/oxidant ratio; see Mixing ratio
Full width at half maximum (FWHM) 139; see also Half-intensity-width
Fully atomized (standard) element; see Standard element
Fundamental constants 924
Furnace 4, 6, 9, 238, 762; see also Oven
f-Value; see Oscillator strength
FWHM ; see Full width at half maximum

Ga 314, 725, 748f, 824, 865, 869f, 940
GaBr 769
GaCl 769
GaF 769
GaH 768
GaI 769
Gain factor (of photomultiplier) 264f, 275
Galvanometer, a.c. 282f, 474
GaO 766, 778
GaOH 768, 778, 794
Gas(es); see also individual gases and flames, Fill gas, Flow, Flowmeter, Fuel, Inert gas, Noble gas, Oxidant gas
— chromatography 7, 816
—, discharge 818
—, handling of combustion 231
—, storage of 231

Gas(es) (continued)
— supply 231
— tank 231
Gas constant (molar) 924
Gas-discharge lamp 164f, 187, 191; see also individual lamps
—, alkali 825
—, electric 254
—, low-pressure 662
Gas meter, integrating 233
Ge 640
gf-Value 125, 939f
Gibbs free energy 50f, 760, 927
Gouy curve 505
Gouy factor 188f
Ground level; see Energy level
Ground state; see Quantum state
Guard ring condenser 862

H (radical) 314, 426, 434f, 437f, 440f, 446, 449f, 451, 454, 461, 467, 544f, 554f, 556, 641, 651, 664f, 680, 776f, 781f, 785f, 789f, 792, 795, 802, 803f, 809f, 811, 820, 823, 843f, 870f, 878f, 907, 920
— -halide 467
H_2 29, 414, 416, 423, 426, 434f, 453, 463, 560, 564f, 573, 576, 577f, 586, 590f, 595f, 604f, 636, 642, 649, 743f, 786, 820, 873, 891, 919; see also Flame
— -air flame 38, 195, 217, 219f, 258, 314, 320, 327, 333, 415, 431, 467, 525, 530, 641, 645f, 651, 670f, 775; see also Flame
— -CO-air flame 223
— diffusion flame (detector) 7, 418
— flame 194, 216, 309, 405, 416f, 420, 426, 433f, 446f, 452, 460f, 514, 523, 535, 666, 732f, 776f, 791, 793f, 824f, 844, 860
— as fuel gas 213f, 220, 231, 258, 414, 435
— -N_2O flame 415, 431, 466, 525
— -O_2-Ar flame 110f, 115, 145, 205, 219, 456, 466, 502, 525, 531, 558, 564, 594, 621, 632, 645f, 660f, 663, 667, 706, 717, 723f, 728, 730f, 733, 735, 740, 745, 750, 779, 865, 869f, 901
— -O_2-CO flame 865, 869f
— -O_2-CO_2 flame 865, 869f
— -O_2 flame 5, 213, 235, 308, 313f, 323, 415f, 467, 495, 591, 670f, 730f
— -O_2-H_2O flame 865, 869f
— -O_2-inert gas flame 217f
— -O_2-N_2 flame 219f, 235, 432, 453, 467, 523f, 526, 530, 558, 564, 642, 645, 706, 717, 723f, 728, 730f, 744f, 750, 781, 783, 788f, 792, 793f, 803, 805, 809, 859, 865, 869f, 876f, 887f, 910, 919
H_2^- 585
Hadamard spectrometer 260
Half-intensity-width (of spectral line) 139f, 152, 162f, 184f, 250, 296f, 299f, 363, 493f, 680f, 685f, 691f, 697f, 718f, 725, 734f, 737f, 746f, 934; see also Bandwidth, Effective width, Line broadening, Line profile
— of adiabatic (collision) broadening 724f, 741, 746
— of collision broadening 147, 149, 485, 488, 510, 512, 724f, 746, 749
— of diabatic (or quenching) broadening 701, 724f, 746
— of Doppler profile 150, 153, 162, 485, 708, 719, 746
— of Lorentz (or dispersion) profile 150f, 154, 701, 718f, 746
— of natural broadening 142, 701, 746
— of spectral line source 493f, 747
— of Voigt profile 150, 157f, 746

SUBJECT INDEX

Half-width 139; see also Half-intensity-width
Halides 764f, 779f, 795, 799; see also Buffering effect
Halogen(s) 2, 461, 779f, 795,799f, 823f, 826f, 881, 905; see also individual halogens
— atom 780
— compound, liquid 239
— hydride; see H-halides
—, introduction of 239
— ion 823f, 849, 905
— lamp 288
Hamiltonian 696
Hanle effect 195, 532, 551, 594
Harpoon model 101
HB 881
HBr 781
HCL; see Hollow-cathode lamp
HCl (liquid) 239
HCl (molecule) 780, 781, 805f, 820f, 873
HCN 792
HCO 420, 435, 466, 637f
HCO$^+$ 461f, 635
HD 565, 577f, 596f, 604f
He 293, 535, 537, 543f, 550f, 569f, 604f, 688, 727, 735, 737f, 740, 744f
Heat
— bath 41, 59
— of combustion 59
— content 928
— of dissociation 51, 761, 933
— of formation 428, 638f, 760, 928
—, latent 52
— loss 423, 456
— — to the burner 458
— of reaction 51, 427, 880, 933
—, specific 432, 760f, 803
— of sublimation 760
Heat capacity (at constant pressure) 52, 447, 926
—, molar 926
—, specific 428, 926
Heat-pipe-oven reactor 4
Heaviside's step function 350
Height of observation 223f, 253, 306, 315, 331, 414, 449, 518, 527, 643f, 672, 774, 779, 787f, 810, 812, 846, 849, 866f, 905, 907, 918, 928
Heisenberg uncertainty relation 97, 117, 141
HF 571, 780, 781
h.f.s.; see Hyper-fine-structure
Hg 292, 538, 544, 547, 549, 554, 557, 564, 566, 573, 576f, 582f, 594f, 600f, 610, 615f, 618, 649, 688, 693, 702, 706, 709, 737, 940
Hg$_2$ 602
HgH 569
HI 446, 781
High-frequency technique 851f
High-temperature fast-flow reactor 4
H$_2$O (liquid); see Water
H$_2$O (molecule) 237, 319, 426, 432f, 453, 456, 462f, 564f, 569f, 586, 595, 604f, 636, 642, 647f, 665, 681, 732, 740, 743f, 745, 770, 773, 777, 785f, 789f, 794, 803, 822f, 873, 913
—, quenching by 456
H$_3$O$^+$ 461f, 638f, 823, 826, 838, 847, 850, 875, 881, 904
Ho 635
Hohlraum 56, 119; see also Black body
Hole-burning, spectral 132f
Hollow-cathode discharge 260
Hollow-cathode lamp 10, 290f, 299f, 327, 376, 475, 493f, 630, 711
—, demountable 290
—, d.c. boosted 290, 301
—, van Gelder 291, 301

Hollow-cathode lamp (continued)
—, line broadening in 292
—, microwave excited (or boosted) 291f, 301
Holtsmark broadening 146, 680, 682
Hook effect 523f
Hönl-London factor 136f, 180f; see also Sum rule
Hund's (coupling) case a 25, 28f, 35, 37
Hund's (coupling) case b 25f, 29, 35, 37, 135
HPO 753
H$_2$WO$_4$ 770
Hydride 38, 764, 775f, 779, 786, 789, 794f
Hydrocarbon(s) 706
Hydrocarbon flame 423, 425, 435f, 445, 447, 460f, 464f,468f,523,598,635,640,674,777,791f,813, 825, 838, 860, 880; see also individual flames
Hydronium; see H$_3$O$^+$
Hydroxide(s) 432, 764f, 785, 793f, 848f, 865, 869f, 920; see also individual hydroxides
— ion 639, 844, 848f, 865
Hydroxychloride 783
Hydroxyl; see OH
Hydroxyl ion; see OH$^-$
Hyper-fine-structure 21, 33f, 44, 46, 165, 484, 489, 492f,498f,510,513,526, 704f, 726,743, 747,749f

I 446, 640
I$_2$ 586, 604f
Ignition 223
— temperature, spontaneous 414
Image dissector tube 259, 276
Image sensor, electronic 262
Impact
— approximation; see Line broadening
— parameter 93, 99, 102, 685f, 688, 693, 926
—, sudden 685, 687
Imprisonment of radiation; see Radiation
Impulse response; see Response
In 320,327,600,725, 748f, 794,824, 865, 869f, 940
InBr 753, 769
Incandescence 39, 314; see also Flame
InCl 340, 753, 769
Indium-antimonide photovoltaic cell 262
Induced emission; see Emission, Transition
Inert gas(es) 212f, 215, 217, 221, 414, 416f, 433f; see also individual flames and gases
InF 769
Informing power 259, 261
Infusion of secondary air (into flame) 423, 518, 799; see also Mixing (of flame gas)
InH 768
InI 753, 769
Inner cone 419; see also Primary combustion zone
Inner quantum number 18, 32
Inner zone; see Primary combustion zone
InO 766, 778
InOH 768, 778, 794
Integral absorption (method) 166f, 173,181, 182, 188, 196, 368f, 382, 384f, 486f, 489f, 492, 495f, 500, 504, 524f, 656f, 704f, 720, 722, 748, 916, 925, 943
—, net 173, 175, 206, 925
Integration time 347, 354, 357, 366, 936, 950f
Integrator 278, 354, 357, 385, 950f
—, box-car 278, 300, 379
Intensity 110f, 158f, 943
—, background 257
— of chemiluminescence 640f, 650
— of continuum 438f, 444f, 467f, 651
—, definition of 159
— distribution
— —, spatial 915f

1012

Intensity (*continued*)
- —, spectral; *see* Distribution, Line profile, Spectrum
- — of emission 138, **317**, **645f**, 777
- — of emission band 180f, **438f**, 443, **445f**, 753, 779,792,801,**806f**,809,836; *see also* Radiance
- — of emission line **175f**, **182f**, **188f**, **438f**, 441, 476, **481f**, 500, **509f**, 512, **522f**, 526, 625, 629, **643f**, **660f**, 672, 780f, **800f**, **805f**, 809, 836, 854,
- — of fluorescence 76, **192f**, **195f**, **198f**, **202f**,317, **319f**, 531, **540f**, **629f**
- —, (spectrally) integrated 249, **718f**
- —, measurement of
- — — absolute **362f**
- — — absorption 173, 257, 300, **368f**, 382
- — — emission **339f**, **362f**
- — — fluorescence 261
- — — relative **364f**
- —, noise; *see* Noise
- — profile; *see* distribution, spatial
- —, radiant 159, 363, **943**
- — —, spectral **468f**, **943f**
- — of radiation **654f**
- — —, infrathermal 625
- — —, nonthermal **655f**
- — —, thermal **175f**, 180f, **188f**, **654f**
- — of scattered light 114, 116
- —, spectral **943f**
- —, total 164

Interaction
- —, adiabatic 148
- — broadening 147
- —, long-range 103
- —, short-range 102
- —, van der Waals; *see* Interaction potential

Interaction potential 91f,102, 146, 534, 553, 587, 683, **685**, 691, **696f**, 701f, **732f**, 739, **909f**, **932**; *see also* Force constant, Potential barrier, Potential energy (function)
- —, (modified) Buckingham **912f**
- —, difference **685**, **689f**, **707**, 734, 739
- —, effective **702f**
- —, hard-sphere **687f**
- —, Lennard-Jones (L.-J.) **687f**, **691f**, **693f**, **707**, 729, 732, 734, **738f**, 741, **912f**
- —, Morse 742
- —, nonspherical 598
- —, semi-empirical 741
- —, spherically-symmetric **686f**, **739f**, 742
- —, van der Waals 496, 587, 591, 593, 598, **686f**, **689f**, 692f, 727, 734, 739

Intercollision time 81, 104, 148,685,**693f**,900,936
Intercombination line 32
Interconal zone; *see* Interzonal region
Interference 9, 395, 413; *see also* Matrix effect
- —, background **381f**
- —, excitation **674f**
- —, (mutual) ionization 5,10,819,**832f**,**903f**,**906f**
- —, lateral-diffusion 336
- —, quantum-mechanical 100, 114
- — by scattering 5, 203, 392
- —, spectral 5, 39, 240, 383, 392, 753, 819
- —, (solute-)volatilization 309, **313**

Interference filter 241, **397f**
Interferogram 259
Interferometer 241, 259
- —, Fabry-Pérot 174,184f,244,328,710f,**725f**,735f
- — —, confocal 246, 259
- — —, flat 246
- —, Michelson, field-widened 246, 259
- —, Pepsios 248
- —, scanning **710f**

Intermultiplet mixing; *see* Mixing (of energy levels)
Internal absorbance, *etc.*; *see* Absorbance, *etc.*
Internal cone 419; *see also* Primary combustion zone
Interzonal region **422f**, 463, **465f**, 792, 837, **849f**
Intramultiplet mixing; *see* Mixing (of energy levels)
Inverse-square law **337f**
Inversion (of energy level populations) 131, 547
IO 439, **445f**
- — -method 439, **445f**

Ion(s) 3, **520**, **638**, **813f**, **823f**,**830f**,**842f**,**882f**,**920**; *see also* individual *or* groups of ions, Fraction ionized, Ionization
- —, atomic; *see* Ion [general]
- —, collision frequency of **831**, **852f**
- —, hydrated 856, 878
- —, hydration energy of 880
- —, ionization by charge transfer from **823f**
- —, measurement of concentration of **850f**
- —, molecular 514, 752, **823f**, **834f**, 844
- —, natural flame; *see* Flame ion
- —, negative 54, **461f**, 501, 514, **817**,**831**,**851**,**957f**
- —, positive 55, **461f**, **851**, **870f**
- —, primary 461
- — (spatial) profiles 465
- —, secondary 462, **817**, 825

Ionic-curve crossing; *see* Crossing
Ionization 2, **53f**, 58, **104f**, 499, **504**, 507, **513f**, **521f**, 758, 777, **813f**, **828f**, **850f**, **864f**, **882f**, **903f**, 918, **957f**; *see also* Chemi-ionization, Cross section, Electron, (Dis-)Equilibrium, Ion, Photo-ionization, Process, Rate, Rate constant, Reaction, Relaxation
- —, boosting of **826**, **875**
- —, bottleneck model for **892f**
- — buffer(ing) **500f**, **834f**, **906**
- —, collisional **464f**
- — constant 54, 69, 105, **832f**, **848f**, **904**, 929
- — continuum 885, **895f**, **899**
- —, degree of 54, 70, **513f**, 752, **832f**, **842f**, **852f**, **904**, **906**, **933**
- — effect of electric field on 465, **816**, **850**
- — efficiency (specific) **872f**, **887f**, **898**
- — energy 24, 45, **54f**, 100, 460, **525f**, 585, **825**, **848f**, **867**, **881**, **883f**, **888f**, **899**, **927**, **939f**; *see also* Atom
- — —, effective 55, **899**
- — —, depression (*or* lowering) of 55, **899**
- —, excess; *see* suprathermal
- — of flame species; *see* individual ions, Flame ionization
- — of halogen **823f**; *see also* individual ions, Halogen ion
- — interference; *see* Interference
- —, ladder (climbing) model for
- — —, modified **894f**
- — —, original **882f**, **892f**, **901f**
- —, laser-enhanced (*or* laser-induced); *see* Optogalvanic effect
- — limit **19f**, 24, 45, 54, **884f**, **893**
- — —, effective 45
- —, measurement of; *see* Electron, Ion
- — of metal atoms *or* compounds; *see* individual *or* groups of ions
- —, natural; *see* Flame ionization
- —, nonthermal **820**, **846f**
- —, partial 524, **846f**
- —, suppression 238, 514, **833f**, **854f**, **866**, **920**; *see also* buffering
- —, suprathermal 460, **847f**, **905**

SUBJECT INDEX

Ionization (*continued*)
—, thermal 62f, **819**f, **826**, **832**f, **846**f
Ion mass spectrometer(-try) 4, 10, 461, 463, 817, **822**, **851**f, **855**f, **865**, **876**f, **881**
—, flame 10
Ion-recombination 53, 105, 108f, 559, **638**f, 659, **817**f, **823**, **875**f; see *also* Continuum, Neutralization, Process, Rate, Rate constant, Reaction
Irradiance **927**, **943**; see *also* Spectral irradiance
Isopropanol 64
Isotope(s) 15; see *also* individual isotopes, Analysis
— effect 21, **37** 560
— molecule **22**, 577
— shift (*or* splitting) 34, 493, 498
— substitution 38

j-Coupling scheme 18, **32**f
j-j Coupling 96
J-Mixing; see Mixing (of energy levels)

K 299, 455, 499f, **506**f, 514, **538**f, **542**f, 547, 549, 552f, 555, 565, 567, 572, **577**f, **586**, 591, **594**f, 601, **606**f, **612**f, **616**f, 642, 645f, **650**f, 667, 725, **730**, **732**, **738**, **743**f, **747**f, 777, **793**, **796**, **825**, **833**, **865**f, **869**f, **884**, **887**f, **906**f, **910**, **913**, **919**f, **940**
— -halide 780
KBr **769**
KCl 314, **441**f, **769**
K$_2$CrO$_4$ 770
KF **769**
KHVO$_3$ 751
KI 751, **769**
Kirchhoff's law 57, 120, **181**f, 319, 624, 654, 657, 679, **695**f, 699, **718**, **736**
K$_2$MoO$_4$ 770
Knudsen cell **761**
KO$_2$ **775**
KOH 652, **768**, **776**
Kr **537**, 612, **737**f
K$_2$VO$_3$ 751, 770
K$_2$WO$_4$ 770

La 635, 771, 824
Ladder-climbing model (for dissociation) 90
Ladder-climbing model (for ionization); see Ionization
Ladder-mechanism 82
Lamp; see individual lamps, Light source
Landau-Zener formula **97**f, **586**, **589**
Langmuir equation (for evaporation) **311**f
Langmuir-flame 417
LaO 489, **766**, **772**
Laser 107, **110**f, 120, **131**f, **202**f, 254, **296**f, 327, 334, 534, 536, 539, 549, 566, **570**f, 576, 624, **630**f, **818**, **901**
—, alkali **533**
—, Ar-ion 299
—, broad-band 203, **205**f, **630**, **632**
—, cavity 174
—, chemical 6
—, CO 572
—, continuous-wave (c.w.) **296**
—, dye 3, 10, 174, **296**, 369, **376**f, **390**, **392**, **466**, **534**f, **543**, **602**, **632**, **819**, **825**
—, c.w. 205, **300**f, **531**, **632**, **713**, **716**f
—, flash-lamp pumped 174, **572**, **632**, **717**
— —, Xe **297**f
—, N$_2$-pumped **297**f, **564**
— —, frequency-doubled 444, **535**, **543**, **819**
—, pulsed 145, 174, 205, **297**f, **301**, **535**, **543**, **564**, **572**, **632**, **716**f

Laser (*continued*)
— -enhanced chemistry **788**
— -enhanced ionization; see Optogalvanic effect
—, He-Ne 600
— modes 174
—, monochromatic; see narrow-band
—, narrow-band 206, 630
—, pulsed 144, 203, **296**, **383**, **390**, 547, **630**, **633**, **659**, **788**
— threshold 174
Latent heat 52
Λ-doubling 22, **26**, 37, 135, 179
Lennard-Jones (interaction) potential; see Interaction potential
Level; see Energy level
Li 33f, 237, 335, **438**, **440**f, 493, **506**f, **508**, **585**, **604**, **617**, **649**, **651**, **666**, **681**, **737**, **744**, **749**, **777**, **782**, **788**, **790**, **809**, **833**, **865**f, **869**f, **873**, **887**f, **910**, **918**f, **940**
— halide 780
LiBr **769**
LiCl **769**
Lidar 7
LiF **769**
Lifetime (of excited state) 724, **821**, **936**
—, actual 74, 123, **543**, **580**
—, apparent 555
—, collisional 81, **147**, **543**, **555**f, **936**
— of (collision) complex 91, **821**
—, optical; see radiative
—, (effective) radiative 79, 104, **122**f, 125, 152, **177**f, 193, **531**, **580**, **618**, **627**f, **936**
—, reduction of, by J-mixing 724
— —, by quenching 177
Lifetime method 78, **526**, **537**, **551**, **564**, **566**f, **603**f
Light; see *also* Radiation
Light chopper 258, 276, 279, 289, 323, **330**, 337, 359, **370**, **372**, **378**
—, air-turbine 254
—, double a.c. 339
—, electro-optical 254
—, rotating disc 253
—, square-pulse **371**
—, square-wave **371**
—, tuning fork 253
Light-guide **276**
— modulation; see Modulation
— scattering; see Scattering
— shift 117
— shutter 259, **330**f
— —, high-speed 254
Light source 8, **243**f, 251, **253**f, 260, 301, 343, **376**f, 392, 406, **628**f; see *also* individual lamps, Continuum source, Spectral line source
—, background 164, **323**f, 368, **370**, 373, 380, **385**f, **443**f, **473**, **491**
—, imaging of **251**f
—, low-pressure 176
—, monochromatic; see Narrow-line source
—, primary 196
LiH **649**
LiI **769**
Limit of detection; see Detection limit
Limits of inflammability **414**
Line (spectral) **343**f, 352, 358, 367, 481, 709, 753; see *also* Absorption, Emission, Fluorescence, Intensity, Spectrum, Transition
— asymmetry 492, 498
—, atomic 30f, **110**f, 138, **159**f, 178, **260**f, **384**f, **403**f, **652**f, **677**f, **752**, **906**f, **939**f
—, forbidden **738**

Line (spectral) (*continued*)
—, frequency of 30
—, intercombination 32
—, ionic 500f, 904 f
—, missing 37
—, molecular 34f, 134f, 677
— narrowing, collisional; *see* Dicke narrowing
—, nonresonance 32, 186, 489, 669
—, resonance 8, 32, 142, 162f, 171, 178, 186f, 193, 260f, 326, 484, 495f, 500f, 543f, 629, 632f, 655, 660f, 670, 679, 708f, 718
— —, first 32, 133, 142, 147, 747
—, reversed 187; *see also* Line reversal
—, rotational 135f, 138, 179f, 318, 443f, 491f, 526f, 956
— splitting 144; *see also* (Hyper-)Fine-structure
— strength 122f, 135f
—, wavelength(s) of; *see* Wavelength
—, wave number of; *see* Wave number
Linearity
— of measuring system 336f, 504, 507
— of nebulizer (*or* nebulization) 227f
— of photodetector 504
— of photomultiplier 266, 268
— of spectrometer 364
Linear momentum 60, 80, 81, 102, 107, 583, 708
Line broadening 138f, 221, 368, 533, 677f, 723f, 736f; *see also* Half-intensity-width, Line profile
—, collision 132, 145f, 154, 172, 491f, 653f, 679f, 682f, 710, 724f, 736, 739f, 749
— —, adiabatic 146f, 154, 580, 680, 682, 684, 687, 688f, 692f, 695f, 698f, 700f, 727, 742f, 937
— — —, effective 746
— — —, specific 743
— —, cross section for 149f, 689, 724f, 742f
— —, diabatic 146f, 154, 556, 680, 684, 700f, 937
— —, total 746
— — —, effective 746
— — —, specific 746
—, concentration 162f, 183f, 714, 719
—, Doppler 116, 132, 152f, 172, 491, 653f, 679, 695f, 700f, 723, 736
— —, extra 317, 502, 668, 673, 708f
—, field 146
—, fluorescence 115
—, foreign-gas 146, 680, 682, 686, 688f, 705f
—, Holtsmark 146, 680, 682; *see also* resonance
—, homogeneous 131f, 140, 206, 624, 631
—, inhomogeneous 131f, 624
—, interaction 147
—, intrinsic 39
—, J-mixing 680
—, Lorentz 146
—, natural 129, 132, 141f, 147, 152, 154, 679, 700f
—, nonadiabatic; *see* collision, diabatic
— parameter 925
—, power 144
—, pressure 146
—, quenching 148, 580, 680, 695f, 700f, 723f, 736
—, radiation 117, 144f
—, resonance 146, 498, 680f, 686, 736
—, saturation 140f
—, self- 146, 172, 680; *see also* resonance
—, self-absorption 163, 183f, 192, 498, 508, 711
—, Stark (effect) 146, 885, 905
— —, linear 680, 682
— —, quadratic 680f, 682, 686
—, theory for 138f, 679f, 722f, 736
— —, classical limit in 696f
— —, completed-collision assumption in 686f, 697

Line broadening (*continued*)
— —, density-criterion in 687, 690, 727, 729
— —, frequency-criterion in 687
— —, general-pressures 682f, 692f, 701, 729, 737, 739, 741
— —, generalized, pressure 682
— —, impact approximation in 115, 147, 149, 680, 682f, 685f, 692f, 697f, 701, 703f, 727f, 736f, 739
— —, Markov assumption in 686f
— —, nearest-neighbour approximation in 682, 690f
— —, quantummechanical 683f, 695f, 703
— —, quasi-static (QS) 682f, 690f, 693f, 703f, 707, 725, 738f
— — — regime in 698
— — —, semi- 694f
— —, semi-classical (SC) 683f, 693f, 703f, 736
— —, small-time limit in 696f
— —, statistic 682
— —, unified 682f
Line centre; *see* Line core
Line core 688f, 692, 694, 713, 716f, 725f, 733f, 736f, 739
Line profile 128, 138f, 249, 316, 365, 368, 652, 661f, 679f, 687, 691f, 723f, 739f, 749f; *see also* Half-intensity-width, Line core, Line wing, Satellite
— in absorption 114, 120f, 127, 140, 159f, 183f, 370, 376, 388f, 444, 534, 679, 694f, 712f, 718f, 735f, 748f
—, adiabatic 689, 736, 738
—, asymmetry 148, 172, 694
— — ratio 694, 726, 737, 739
—, central dip in 661f; *see also* Self-reversal
—, collisional 681, 739
—, combined 154f
—, convoluted 155
—, dispersion 142, 685, 687, 697f, 700, 725, 727f, 736f
—, Doppler 150f, 155, 316, 549, 681, 689, 708, 736
— —, integrated 153
— in emission 116, 120f, 140, 183f, 187, 192, 679, 695f, 710f, 718f, 735f, 748
— in fluorescence 114, 198, 717
— function 679
—, Gaussian 153, 700
—, instrumental (normalized) 241f, 248f, 258, 344, 925
—, integrated 153, 704f, 748
—, Lorentz(ian) 117, 129, 142, 150f, 154f, 168, 687
—, natural 116, 142f, 689
—, normalized 128, 679
—, quenching 689
—, total 681, 689, 699f, 713
—, true 138, 258
—, Voigt 150f, 155f, 168, 172, 492, 689, 699, 711, 718, 726f, 733f, 737f, 749
Line-reversal (method) 190f, 257, 285, 320f, 387, 481, 499, 537, 574f, 641, 669, 673
Line shift (collisional) 148f, 152, 164f, 172, 184, 382, 492f, 498, 510, 513, 677f, 685f, 691f, 697f, 711, 714, 717f, 722, 725f, 736f, 739f, 747, 934
—, adiabatic 740
Line source; *see* Spectral line source
Line width; *see* Effective (line) width, Half-intensity-width
Line wing(s) 148, 168, 172, 687, 690, 692f, 706, 712, 715f, 732f, 737f, 742; *see also* Satellite
—, far 682, 692, 729f, 738
—, near 692, 725f, 733f, 736
—, overlap of 728, 730
LiO 766, 772, 777, 792

SUBJECT INDEX

LiO_2 440, **775**
LiOH 237, 438, **440f**, **768**, 770, **776f**, 782, **788f**, 794, **802f**, 807, **865f**, 888, 910, 918f
— -method 438, **440f**, 448, 451, **807**, **809**, 913
Liquid bubble method 239
Local probing (or sensing) 391, 483, 905
Lock-in 374, 380; see also Rectifier
— amplifier **281f**, 284
— detector **378f**
— meter 258, 475
London's formula 741, 743
Long-wave(length) limit; see Continuum
Long-wavelength threshold 264
Lorentz broadening; see Line broadening
Loss(es)
— of aerosol **517f**
—, heat 423, **456**, 458
—, radiation **456**, 458
— of salt 519
Low-pressure flow system 571
L-S coupling (scheme) 16, 32
LTE, local thermodynamic equilibrium; see Equilibrium
Lu 635
Luminescence; see Chemiluminescence, Fluorescence, Photoluminescence
—, sensitized 112

Magneto-hydrodynamic power generator 6
Mantle flame; see Flame
Mass 930
— density **917**, **935**
— of electron, etc. 924
—, molar 930
— number 15
—, reduced 42, **935**
—, relative (of atom or molecule) 930
Mass-action-law **49f**,60,68,89,**315**,**751**,**771**,**781f**,**841**
Mass balance **425f**, 838
— equations **781f**
Massey parameter **97f**, 103, 544, 552, 596
Mass spectrometer(-try) 9, 420, **437**, **758f**, **761f**, 770, 778, 782, 796; see also Ion mass spectrometer(-try)
Matrix effect 395
Maxwell
— equilibrium; see Equilibrium
— (distribution) law 43, 59, 175, 315, 549, **839**, **883**
Mean free path **81f**, 153, **935**
Measuring technique (or system) **277f**; see also Detection, Meter
—, a.c. **279f**
—, d.c. **278f**
—, digital 274
Measuring time 347, 353, 357, 361, **936**
Melting point (of Au) 362
Mercury (arc) lamp **298f**, 301
—, high-pressure 285, **288f**
Metal atom; see individual or groups of elements, Atom
Metal compound; see individual or groups of molecules, Molecule
Metalloid vapour 2
Metal-vapour (studies) **2f**,**6**,**9**,**41**,**413**,**529**,**903f**,**921**
Metal-vapour discharge lamps 285, **630**
—, low-pressure **292f**
Meter; see also Detection, Measuring technique
—, a.c. **359f**, 378
—, d.c. **348f**, **947f**
—, synchronous 474

Methylidene; see CH
— oxide, see HCO
— — ion; see HCO^+
— trimer ion; see $C_3H_3^+$
Mg 304, 455, 526, 570, **600**, 609, **617**, **940**
MgBr 769
MgCl 769
$MgCl_2$ 310, 314
— $\cdot H_2O$ particle 310
MgF 769
MgH 26, **768**
$Mg(NO_3)_2$ 310
MgO **757**, **766**, 772
— particle 310
MgOH **768**, **777**
Micro-process; see Process
Microscopic reversibility 40, **64**, **88f**, 92, 101, 103, **537f**, 561, **572f**, 576, 581, 618, 634
—, principle of 536
Microwave (MW)
— attenuation technique **852**, 854, **863**
— discharge 445, **572**, 884
— excited (or boosted) discharge lamps **291f**, 301
— (-cavity) resonance technique **852**, 854, **860** **863**, **881**
Mirror
—, back 191, **506**, 508
—, butterfly 379
—, Cassegrain 408
—, spherical 408
Mist; see Aerosol
Mixing (of eigen- or quantum states) 95, 530
Mixing (of energy levels) 108, **534**, **618**; see also Cross section, Process, Rate constant
—, doublet 204, **530**, **548f**, **555**, **593**, **612f**, 663, **717**; see also intramultiplet
—, fine-structure; see doublet, intramultiplet
—, intermultiplet 530, **560f**, **603f**, **609**, 611; see also De-excitation, Quenching (of fluorescence)
—, intramultiplet **530f**, **548f**, **569**, **593f**, **612f**, **615**, **702f**, **723f**; see also doublet, triplet
—, J-; see doublet, intramultiplet
—, M_J- 108, **146**, **530f**, 581, 699, **702f**; see also Depolarization
—, mutual **555f**
—, self- **555f**
—, triplet **552**
Mixing (of flame gas with ambient air) **456**, **459**; see also Infusion
Mixing (of fuel and oxidant gases) 418, **774**
Mixing (of magnetic sublevels); see M_J-Mixing
Mixing (or fuel/oxidant) ratio 214, **414f**, **417f**, 428, **432f**, 451, 466, **522f**, 525
—, optimum 415
—, stoichiometric 415, **428f**
Mixture strength 417
M_J-Mixing; see Mixing (of energy levels)
Mn 304, 455, 725, **748f**, **824f**, **865**, **869f**, 881, 940
— -halide 780
MnBr 769
MnCl 769
MnF 769
$Mn^+ \cdot H_2O$ 823
MnI 769
MnO 766, **772f**, 823
MnOH 650, **768**, 779
Mo 310, 635, 638, 640, 774, **824**
Mode (of laser) 174
Mode (of vibration) 30, 48

Modulation (of light) 159, 282, 284, 337, **370**, **378**f, 402,**411**,472f,**494**,571; *see also* Light chopper
- frequency 78, 254f, **257**f, 272, 301, 323, 347, **358**f, **366**, 379, 395, 405, 471, 927
-, intensity- 78, 253, 274f, 289, 294, 301, 323, 345, **358**f, **379**f, **384**f
- -, square-wave 300
-, spectral selective 260f
- technique 345
- -, selective 260f
- waveform 359, 366
-, wavelength- (*or* frequency-) 255f,**258**,345,392
-, Zeeman- 261
Molecular beam; *see* Beam
Molecular constant 30, 761, 776, 797, **804**
Molecule(s) 514, 522, 529, 535, 537, 539, 546, 557, **560**f, **570**f, **580**f, **639**f, **650**f, **667**f, 686, 732, 742, **751**f, **783**f, **834**f, **843**f, 886, **918**f; *see also* individual *or* groups of molecules, Dimer(s) Energy, Quasi-molecule, Radical(s)
-, absorption by; *see* Absorption
-, asymmetric-top 30
-, complex 95; *see also* Complex
- configuration **805**, 807f
-, di-atomic 21f, 30, 34f, 43, 47, **90**f, 96, **135**f, 179, 535, 538, 567, 577, **582**f, **756**f, 763, **797**f, 821, 828, 900, 902
- -, heteronuclear 22, 46, 48f, 135, 179, **797**f
- -, homonuclear 21, 48f
-, diffusion coefficients of 910
-, dissociation (energies) of; *see* Dissociation
-, emission by; *see* Emission
-, fluorescence by; *see* Fluorescence
-, hot **465**
-, ion of; *see* Ion
-, ionization of; *see* Ionization
-, isotopic 22, 577; *see also* individual isotopes
-, polar 571
-, poly-atomic 21, 29f, 43, **47**f, **90**f, 535, 538, 758, 763, 785, **798**f; *see also* tri-atomic
- -, (co-)linear 22, 30, 46, **48**, 798f, **804**f
- -, nonlinear 22, 29, 47f, **798**f
-, saturated organic 565, 763
-, spectrum of; *see* Spectrum
-, stable 24
-, structure of 21f, 596, 754, **803**f, 807; *see also* Energy level, Quantum state
-, tri-atomic 34, 592
- -, bent **804**f
- -, linear 49, 92, **804**f
-, unstable 24
-, van der Waals **707**
Moment of inertia (of molecule) **28**
-, principal 29f, 799, **804**, 928
Monochromator 166, **242**f, 251, 321, 352, 382, 399, 407f, **494**f; *see also* Bandwidth (spectral)
-, double 253
-, grating 241, 244, **246**, 250, 253, 712
-, prism 241, 244, **246**, 326
-, resonance 260f
-, scanning 711, 715, 731
- -, rapid- 258
- slit; *see* Slit
Monohydrides 768
Monoxides; *see* Oxides
MoO 767
Morse (interaction) potential; *see* Interaction potential
Mössbauer spectroscopy 9
Multi-channel sampling 379
Multiple-pass system **377**f

Multiplet 19f, 45f, 48, **133**f, 179, 191, 530f, **603**f, 633; *see also* Doublet, Triplet
Multiplexing method **259**f
Multiplicity 17, **20**, 22, **26**, 136
Multiplier (phototube); *see* Photomultiplier

N 463
N_2 101, 151, 426, 453, 456, 463, 535, **538**f, 547, 551, **558**f, **564**f, **572**f, **577**f, **584**f, **590**f, **595**f, **604**f, 618, 621, 636, **727**f, 735, 738, **740**f, **743**f, **821**f, 831, **873**f, 884, **890**f. **901**, 913, 919
- as flame diluent 213, 215, 414, 418, 621, 732; *see also* individual flames
- flow system **572**
-, line shift by 493
-, quenching by 456
$^{15}N_2$ 577f
N_2^- 585, **587**f, 821
Na 33, 46, 110, 111, 151, 174, 204f, 299, 308, 313, 321, 326, 335, 369, 392, 438, **440**f, 493, 499,**502**, **506**f, 518, **523**f, 527, 534, **537**f, **543**f, 551, 555, **565**f, **568**f, **574**f, **577**f, **584**f, 590, 598, **600**f, **604**f, **612**f, **616**f, 649, 670, **709**, 717, **732**f, **735**, **740**f, **744**, **747**f, 773, **780**f, **788**, 793, 796, **820**, **822**f, 826, 833, 850, **865**, **869**f, 884, **887**f, 901, 910, 913, **918**f, 940; *see also* Sodium
- -comparison method; *see* Sodium comparison method
- -halide 780
Na^+ **523**, 545, 820, 823
^{20}Na 910
Na_2 523, 525, 751, 770
Na_2^+ 825
NaBr 769, 781
NaCl 438, **441**f, **769**f, **780**f, 820
- method 441f, 770
- particle 310f
NaCN 525
NaF 769, 780, 781
NaH_2 92
NaI 79, 709, **758**, **769**, 781
NaO 525, 767, **772**, 774
NaO_2 440, 444, 523, **775**
NaOH **440**f, **442**f, **523**f, **768**, 776, **788**
- method 438, **444**f
Narrow-line source 164f, 172. 196, 201, **246**, 249, **387**f, 491, 511, **629**f, 714
Na_2SO_4 770
Nb 635
Nd 635
Ne 293, 551, 607, **737**f, **910**
Near-resonance 106
Nebulization 302f; *see also* Aerosol, Dispersion, Drop, Efficiency of nebulization
- effect (on curve-of-growth) 503
-, linearity of **227**
- of organic solvents 229
Nebulizer 211f, **224**f, **228**, 302; *see also* Spray chamber, Spray nozzle
-, angular-type **225**
-, chamber-type **224**f, 302, 306f, 457, 519, 523
-, cleaning of 229
-, concentration range of 229
-, concentric-type **225**
-, controlled-flow 224
-, conversion factor for 234
-, corrosion effects of 229
-, direct-injection **229**; *see also* Burner, direct-injection
-, electrostatic **226**f
-, gravity-fed 224
-, linearity of 228f

SUBJECT INDEX

Nebulizer (*continued*)
— for low-pressure flames 229f
—, modulated 224
—, pneumatic 2, **223f**, 515
—, reflux-type 224
—, reproducibility of 228
—, split-ring **225**
—, suction 224
—, twin **238**
—, ultrasonic **226f**
Negative-filter (technique) 260f, 360
Neutralization 109, 817, **823f**; *see also* Ion recombination, Process, Rate, Rate constant, Reaction
— by charge transfer 823 881
— by electron detachment 823
NH 435, 466, 753, **792**
Ni 291, 314, 489, 639, **940**
— -halide 780
NiO 767
Nitrous oxide-acetylene flame, *etc.*; *see* C_2H_2-N_2O flame, *etc.*
NO 41, 426, 439, 445, 455, 460f, 470, 568, 570f, 576, 579, 584, 595, 753
— -method 445
NO$^+$ 461f
NO$_2$ 439, 445, 470, **538**, 759
N$_2$O 568, 576, 592
— as oxidant 213, 215, 220, 231, 418; *see also* individual flames
Noble gas 102, 214f, 292, 400, **537f**, **543f**, 547, 549, 551, **556f**, 563, **580f**, 600, 686, 702, 704, 706, 723, 737, 741, **885**, 891, **898**, 910, 913; *see also* individual gases
No-crossing rule 96, 100
Noise 268f; *see also* Background, Detection limit, Flame background, Precision, Signal-to-noise ratio
—, additive 275
—, analyzer 475f
—, bandwidth (effective) 269f, 347, 352, 360, 475, 933
—, dark-current 260, 270
—, excess low-frequency (e.l.f.) 272, 275 344f, 406, **471f**, *see also* flicker
— in flame emission signal 470f
—, flicker 254, **272**, 279f, **344f**, 348, **355f**, **361**, **367**, 375, 381, **385f**, 389, 395, 403, 405, 410, **473**, **950f**
—, fractional 473f
—, generation-recombination 346
—, impulse 283
—, $1/f$ 269; *see also* flicker
—, Johnson **272f**, 275, 465; *see also* Nyquist formula
—, low-frequency 286; *see also* excess low-frequency
—, multiplicative 275, **470**, **952**
— of photodetector 268, 273
—, power **345f**; *see also* Spectral noise power
—, quantizing 346
—, secondary-emission **271f**, 475
—, shot 268f, **272f**, 325, 344f, 348f, 351, 360, 367, **385f**, 389, 395, 403, 405, 410, 470f, 476; *see also* Poisson statistics
— — in aerosol supply 470
— spectrum **268f**, 290, **471f**, 475, 949
—, peaks in 476
—, white 269f
—, stationary, process 351
— temperature 465
— theory **268**, **947f**

Noise (*continued*)
—, white 269f, 274f, 362, 471, **949f**
—, zero-offset 280
Nonequilibrium; *see* Disequilibrium
Nonlinearity (of photodetectors) 383; *see also* Linearity
Nonresonance fluorescence; *see* Fluorescence
Nonresonance line; *see* Line
Nozzle; *see* Sprayer nozzle
Null-balance (method) 321, 336, **373f**, **378f**
—, optical 383
Number density (in flame) 44, 50, 480, **689f**, **930**; *see also* Concentration (in flame), Conversion factor
— and pressure 480
—, total 302, **930**
Nyquist formula 273
Nyquist noise; *see* Noise, Johnson

O (radical) 426, **432f**, 439, **445f**, **451f**, **461f**, **468f**, 470, 636f, 640f, 649, **771f**, **776**, **786**, **791f**, 797, 823, **825**, **868**
O$^-$ **445f**, **451f**, **461f**, 776
O$_2$ 414, 416, 426, **434f**, **445f**, **463f**, **538**, 560, **564f**, 570, 576, **578**, **585f**, 589, 592, **595f**, 604, 642, 648f, **649**, **759**, 761, **773f**, **787**, **792f**, 821, 828, 890, 901, 903
— as oxidant 213f, 231, 414, 418; *see also* individual flames
O$_2^-$ **461**, 821
O$_3$ 759
Observation height; *see* Height of observation
OH 26, 237, 314f, **333f**, **420f**, 426, **434f**, **437f**, **443f**, **451f**, 454, 460f, 467, 535, 636, 641, 645, **647f**, 650, 664f, 672, 725, **745**, **785f**, **789f**, 823
— absorption method **438f**, 443f
OH$^-$ **461**, 463
Ω-Integral **911f**
Optical conductance **242f**, 246, 260, 324, 343, 352, 365, 378, 386, 395, 404, 407, **929**, **945f**
Optical lifetime; *see* Lifetime
Optical pumping 531
Optical thickness **127f**, 140, 158, 162, 176, 183, 194, 388, 487f, 622
Optical wedge **371f**
Optogalvanic effect 10, 109, 819, 901, 905
Oscillator strength 10, **123f**, **134f**, 161, 320, 404, 443, **505**, **525f**, 527, 620, 812; *see also* gf-Value, Sum rule, Transition
— for absorption **124**, 927
—, determination of 125, 166f, 171, 480
—, duplication of **508**
— for emission 927
— of molecular band **137**
—, weighted **124**
Outer cone (*or* zone); *see* Secondary combustion zone
Oven 38, 540, 542; *see also* Furnace
Overlap (of interferometer orders) 247
Overlap (of lines)
—, apparent 134
—, degree of **172**
—, intrinsic 134, 172, 186
— of line wings 728, 730
— parameter **498f**, 704f
Overpopulation (of energy level); *see* Energy level, Excitation, suprathermal, Population factor
Oversaturation 311
Oxidant (gas) 213, 414, 417, **433f**; *see also* individual gases

1018

Oxides (of metal)   7, 335, 369, 635, **639f**, 708, 753, 758f, 761, **764f**, 771, 788, **791f**, 799; *see also* individual oxides, Particle
—, refractory   418
Oxy-acetylene flame, *etc*.; *see* C_2H_2-O_2 flame, *etc*.
Oxygen-carbon monoxide flame; *see* CO-O_2 flame

P   7, 309, 313
Paramagnetic-resonance method   **437**
Parity   18, 26, 31
Particle(s) (liquid *or* solid)   237f, 309f, 312, 329, 405, 774, 817, 849; *see also* individual chemical species, Desolvation, Distribution, Volatilization
—, aerosol   532; *see also* Aerosol
—, nonvolatilized   455
—, oxide   310, 314f
—, salt   518
—, soot; *see* Carbon
Particle-track method   **329**
Partition function   52, 162, 760f, 812
— of atom   44f, 509, 771, 797f, **883f**, 897
— —, divergency of   45, 55, **884f**
—, electronic   48
—, internal   44, 49f, 931
— of molecule   46, 48f, 509
— —, electronic   **764f**, 771, 797f
— —, rotational   46f, 798
— —, vibrational   47, 798
—, translational   50, 52
Partner; *see* Collision partner, Perturber, Reactant, Reaction partner, Third body
Pauli's exclusion principle   16
Pb   6, 335, 544, 547, 552, 598, 610, 615, 633, 643, 649f, 669, 670, 823f, 865, **869f**, 881, 910, 921, 940
PbO   767, 772
PbOH$^+$ 823
PdH   779
Peak-absorption method   491f, **495**, 809
PenRay lamp   293
Permeability   123
Permittivity   122, 829
— of vacuum   55
Persistence factor   106f, 649f
Perturber (in collision broadening)   682, 686, **689f**, 691f, 697f, 702, 704f, 723, **727**, 732, 737, 740f, 744
Phase-retardation plate   **377**
Phase-shift (of optical frequency)   682, 685, 687, 690f, **693**, 698
Phase-shift method   78, 537, **564**
Photocathode   264f, 272, **275f**, 380, 400
Photocell, solid-state   **262**
Photoconductive cell (*or* photoconductor)   **262**
Photocurrent   260, **270f**, 470f, 476
—, primary   473
Photodetector   255, **260f**, 471; *see also* individual photocells or photodetectors
— noise   268, 273
Photodiode
— array   334
—, silicon   262
—, solid-state   **262**
—, vacuum   264
Photodissociation   108, **529**, 534, 536, 552, 594, 603f, 618, 709, 754
— continuum   40f
— limit   40f, 79
— method   **758**
— technique   79, 564, 568
Photo-electron   264, 266
Photo-emission   264f

Photo-emissive (vacuum) tube; *see* Phototube
Photo-excitation   108, 319f, **529f**, 536, 549, 551, 571, **619f**, **626f**, 758, 788; *see also* Process, Rate, Rate constant
Photofragmentation spectroscopy   **758**
Photographic plate   261f; *see also* Informing power
Photography   330f; *see also* Schlieren photography
Photo-ionization   108
— continuum   40
Photoluminescence   109, 112, **530**; *see also* Fluorescence
Photolysis   108, **529**
—, flash   551
—, pulsed   537
Photomultiplier (tube)   263f, **273f**, 475
—, solar-blind   263, 407
Photon   30, 64
— counting technique   **263**; *see also* Counting, pulse
Phototransistor   **262**
Phototube   263f
—, multiplier; *see* Photomultiplier
—, single-stage   263f, 273
Pile-up effect   268, 336
— of electron pulses   274
Planck constant   30, 924
Planck radiation law   55f, 119f, 175, 182, 315, 319, 625, 653, **657**
Plasma   3f, 45, 55, 532, **626**, 813f, **885**
— flame   417
— frequency   **831**
— potential   **851**, **858f**
PO   466
PO_2^- 881
Pockels cell   254
Poisson statistics   269, 271, 354; *see also* Noise, shot
Polarizability   586, 589, 593, 598, 741
Polarization (of atom)   17, 123, 127, 131, 206, 531
— interaction   553
Polarization (of light *or* radiation)   32, 80, 115, 127, 131, 134, 194f, 276, 400, **530f**, 535, 569
—, circular   **531**, **714**
—, degree of   195, **530f**
—, partial   253
Polarization (in spectral apparatus)   253, 364
Polarizer   377, **714**
—, linear   338
Pollution research   445
Polymerization   461
Population; *see* Energy level
Population-depletion method   77f
Population factor   74, 77, **623f**, **626f**, **656f**, 660, 663f, 930
Population inversion   131, 547
Post-flame gas; *see* Interzonal region
Potential; *see also* Interaction potential
— barrier   49, 84
— difference; *see* Tension
—, electric   932
Potential (energy) curve   22f, 35f, 40, 91f, 101f, 544f, 547f, **553f**, **583f**, 754f, 821f; *see also* Interaction potential
—, adiabatic   94f, 591
—, attractive   587f
—, covalent   545, **587f**
—, crossing; *see* Crossing
—, diabatic   94f, 591
—, effective   28, **587f**
—, ionic   101, 545, **586f**
—, repulsive   93f, **587f**

SUBJECT INDEX

Potential energy (of collision complex) 91, 94, 102, 583f, 927; see also Interaction potential
Potential (energy) function 23, 909f, 919; see also Interaction potential
Potential (energy) hyper-surface 92
Potential (energy) surface 91f, 581f
—, adiabatic 95
—, covalent 100
—, diabatic 95
—, ionic 100
—, polar 100
Potential well 100, 707, 912f, 919
—, depth of 912f, 919
Power broadening; see Line broadening
Power efficiency of fluorescence; see Efficiency of fluorescence
Pr 635
Precision 471
Precursor (in chemi-excitation) 635f, 638
Predissociation 29, 37, 40, 754
Pre-exponential factor 83, 880, 925
Preheating zone 420, 461
Premix burner; see Burner
Pressure 50, 58, 540f
—, fractional 51
— gauge 230
— and number density 480
—, partial 50, 425f, 480, 931
—, total 496, 653, 762, 837, 931; see also Flame
— units 923
Pressure broadening; see Line broadening
Primary combustion zone 106, 221, 298, 311, 419f, 450f, 452, 460f, 464f, 502, 635, 637f, 641f, 659, 669f, 753, 774f, 792, 824f, 837, 849; see also Flame
—, ions in 871f
—, reactions in 434f
—, spectrum of 435, 465f
—, thickness of 419
Primary light source; see Light source
Primary radiation; see Radiation
Primary reaction zone; see Primary combustion zone
Principal moment of inertia 29f, 799, 804, 928
Principle of detailed balance; see Detailed balance
Principle of microscopic reversibility; see Microscopic reversibility
Probability of radiative transition; see Transition probability
Probe (technique); see Double-probe technique, Electrostatic-probe technique, Single-probe technique
Process(es) 63f, 74, 82, 104f, 108f, 576, 823; see also Collision, Reaction, Scattering
—, absorption; see Absorption
—, activation energy of 827; see also Activation energy
—, adiabatic 595, 597
—, associative; see Reaction
—, balanced 825
— —, partially 957
—, bimolecular (or binary or two-body) 65, 89, 105, 535
—, catalyzed 820
—, chemical; see Reaction
—, chemi-excitation; see Reaction
—, chemi-ionization; see Reaction
—, chemiluminescence; see Reaction
—, chemiquenching; see Reaction
—, classification of 103f, 108f

Process(es) (continued)
—, collisional 64, 66, 68f, 73, 85f, 88f, 94, 96, 100f, 103, 105f, 108f, 188, 193, 529f, 533f, 536f, 544f, 548f, 560f, 593f, 626, 647, 660, 818f, 823f, 828, 841, 850, 882f, 886f, 893f, 900, 904; see also Collision
—, collision-dominated 74, 895
—, cross section for; see Cross section
—, cumulative 109, 818, 821, 883f
—, de-excitation 64, 88, 96, 101, 106, 108, 112, 193, 529f, 532, 534f, 560f, 619f, 626, 634f, 886, 893f, 904
—, diabatic 93, 595, 598
—, dissociative; see Reaction
—, electron attachment 818, 823f, 827
— — detachment 818, 821, 823f, 903
— by electron impact 79, 83, 87, 556f, 659, 823f, 827f
—, electron transfer 103, 617, 818, 821f
—, elementary; see single-step
—, emission; see Emission
—, endo-ergic 70, 827
—, endothermic; see endo-ergic
—, energy of; see Energy
—, equilibrated; see balanced
—, exchange; see Reaction
—, excitation 64, 66, 68, 70, 79, 83, 85, 87, 101f, 108f, 188, 529f, 532f, 536f, 556f, 560f, 572, 581f, 617, 626, 654, 659f, 886, 893f, 904
— —, radiative; see photo-excitation
— — transfer; see Transfer
—, exo-ergic 70, 84, 827
—, exothermic; see exo-ergic
—, fluorescence; see Fluorescence
—, ionic re-arrangement 817, 823f, 837, 850, 957f
—, ionization 53, 63, 69f, 86, 96, 100f, 104f, 108f, 532, 536, 817f, 820f, 826f, 837f, 841, 850. 865f, 869, 882f, 886f, 890, 896f, 900f
—, ion-recombination 108, 536, 817f, 823f, 883f, 893f, 905
—, ladder 818; see also Ionization
—, line broadening; see Line broadening
—, micro-(scopic) 68, 71, 85, 88, 902
—, mixing 108, 530f, 534, 548f, 562, 593f, 597f, 702f, 723f; see also Mixing (of energy levels)
—, monomolecular; see unimolecular
—, multiphoton 110, 818
—, multi-step 66, 104, 109, 819f, 827
—, neutralization 108f, 817, 823f, 837f, 865f
—, nonadiabatic, see diabatic
—, one-step; see single-step
—, order of 66
—, over-all 66, 68, 104, 827, 900f
—, partner of; see reactant, Collision partner, Third body
—, Penning 818
—, photochemical; see Reaction
—, photodissociation; see Reaction
—, photo-excitation 64, 108f, 529f, 543; see also Absorption, Fluorescence
—, photo-ionization 108
—, photoluminescence; see Fluorescence
—, photolysis; see Reaction
—, physical 63, 104, 108f, 529f, 534f, 818, 823f; see also collisional, radiative
—, primary 890
—, product of 64, 85
—, quasi-resonance 602
—, quenching 77, 93f, 99f, 108, 193f, 530, 533, 536f, 544f, 548, 556f, 562, 570, 581f, 598, 647, 740

1020

Process(es) (*continued*)
—, radiation 104, 598; *see also* Absorption, Emission, Fluorescence, Radiation
—, radiative 69, 73f, 104, 108, 112, 193, 529f, 534, 619f, 626, 893
—, rate of; *see* Rate
—, rate constant of; *see* Rate constant
—, reactant of 64, 827; *see also* Reactant
—, recombination (of atoms *or* molecules); *see* Reaction
—, resonance 102f, 577
—, single-step 64f, 71, 104f, 108f, 821, 883f, 891, 894, 896f, 902
—, step-up 818
—, stepwise 109, 112, 529, 543
—, termolecular (*or* ternary *or* three-body) 65, 105
—, thermal 841
—, three-body; *see* termolecular
—, transfer 102
—, two-body; *see* bimolecular
—, two-photon 32, 108, 110, 114, 203, 390, 543, 819
—, two-step 104, 109, 821f, 901f
—, unbalanced 71, 74
—, unimolecular 65, 105
Profile; *see* Concentration profile, Distribution, Ion profile, Line profile
Propagation velocity (of flame front); *see* Burning velocity
Proportionality; *see* Linearity
Proton affinity 462
Pseudo-crossing; *see* Crossing
PtH 779
Pulse-height discriminator 267, 271
Pulse-height distribution 267
Pumping, optical 531
Pyrometry 762

Q-factor 851, 863
QS theory; *see* Line broadening theory, quasistatic
Quadrupole (mass) filter 851, 855
Quadrupole-quadrupole coupling 598
Quality factor; *see* Q-factor
Quantum efficiency of fluorescence; *see* Efficiency of fluorescence
Quantum efficiency of photocathode 264f, 931
Quantum number(s) 15f, 21f, 37, 44, 533, 552; *see also* Quantum state
— of angular momentum of relative (nuclear) motion 93, 99, 931
— of atom 15f, 31
—, azimuthal 15, 566
—, electron 15, 31
—, inner 18, 32
— of molecule 22f, 25f
—, orbital angular-momentum 15f, 31, 45, 741, 929, 934
— —, magnetic 15f
—, principal 15, 20, 31, 33, 543f, 565, 590, 741
—, rotational 25, 135f
—, spin 15f, 18, 21f, 32, 35, 45, 931, 935
— —, magnetic 15f
—, total atomic angular momentum 18, 31, 927
— —, magnetic 18, 31
—, total electronic angular momentum 17f, 31f, 35, 44, 928, 936
— —, magnetic 17f, 31f
—, total molecular angular momentum 25f, 35f, 928, 930
— —, magnetic 25f
—, vibrational 23f, 30, 35, 135f, 569, 571f, 756f, 932

Quantum state(s) 15, 19, 569f; *see also* Atom, Energy level, Lifetime, Molecule, Multiplet, Quantum number, Term
— of atom 15f
—, auto-ionizing 20, 818, 901
—, degenerate 18, 27, 683; *see also* Degeneracy
—, doublet 20
—, electronic 24, 619f
—, even 23, 35
—, excited (*or* excitation) 20, 739
—, ground 20, 24, 531, 739
—, high(er); *see* Energy level, high(er)-lying
—, magnetic sub-; *see* Zeeman sublevel
—, manifold of 566f, 618
—, metastable 122, 569f
—, mixing of 95, 670
— of molecule 21f, 26f, 589
—, negative 22
—, odd 23, 35
—, population of; *see* Energy level
—, positive 22
—, preparation of 533
— of quasi-molecule, adiabatic 553
—, resonance 20
—, Rydberg 543, 546f
—, selection of 533
—, sign of 26, 35
—, singlet 21, 26, 29, 544, 570, 639f, 791
—, symbols for 19f, 22f, 25
—, triplet 21, 544, 570, 639f, 791
—, unstable negative-ion 20
Quarter-wave plate 377, 714
Quartz halide lamp 288f
Quasi-continuum; *see* Continuum (spectral)
Quasi-crossing; *see* Crossing
Quasi-molecule 89, 91, 95, 98, 707
—, poly-atomic 100
—, tri-atomic 100
Quasi-resonance 102f, 602
Quasi-static theory; *see* Line broadening theory
Quenching (of chemical reaction) 851, 855f
Quenching (of fluorescence) 76f, 84, 98, 100f, 115, 147, 193, 221, 334, 447, 530f, 534, 536f, 551, 556f, 561f, 573, 593, 603, 618, 647f, 701; *see also* Chemiquenching, Cross section, De-excitation, Efficiency of fluorescence, Process, Rate, Rate constant, Reaction
— broadening; *see* Line broadening
—, self- 542
Quenching diameter (of burner ports) 214, 414

Rabi effect 117, 144
Radiance 56f, 178f, 186f, 190f, 245, 299f, 317, 325, 342f, 352, 363, 365, 385, 403f, 408f, 411, 485, 630, 652, 655f, 660f, 665, 669, 925, 943, 945; *see also* Spectral radiance
—, net 325
—, ratio of 509f, 512
— temperature; *see* Temperature (of flame)
Radiant density; *see* Radiant energy density
Radiant energy 943; *see also* Spectral radiant energy
Radiant energy density 121, 622f, 943; *see also* Spectral radiant energy density
—, integrated 620f, 624f, 630
Radiant flux 195f, 242, 296, 300, 343f, 363, 393, 403f, 943, 945; *see also* Spectral radiant flux
Radiant intensity 159, 363, 943; *see also* Intensity, Spectral radiant intensity
Radiant power; *see* Radiant flux

SUBJECT INDEX

Radiation 105, **529f**, 532, **943f**; *see also* Absorption, Chemiluminescence, (De-)Polarization, (Dis-)Equilibrium, Emission, Fluorescence, Intensity, Process, Spectrum
- — beam **945f**; *see also* Beam
- —, black-body 119, 175
- — coefficient; *see* Transition probability
- — constant 56, **924**
- —, first 56, **924**
- —, second 56, **924**
- —, third **924**
- —, continuous; *see* Continuum
- — damping 143
- — depletion 660
- — diffusion **176f**, 194
- —, equilibrium; *see* thermal
- — imprisonment 79, **176**, **194**, 658; *see also* trapping
- —, infrathermal 188, **659f**, **671f**
- — leak 316
- — loss 456, 458
- —, microwave 854, 862
- —, monochromatic 620
- —, nonequilibrium; *see* nonthermal
- —, nonthermal 177, 468, 481, 532, **652f**, **657f**, **668f**, **708f**
- —, primary 76, 112, 193, 308, 393, **411**
- —, quasicontinuous; *see* Continuum
- —, scattered; *see* Scattering (of light)
- —, secondary 76, 112
- — source; *see* Light source
- —, suprathermal **663f**, **673f**
- — standard **364f**, 486, 490
- —, primary **362f**
- —, secondary **362f**
- — temperature (effective) 145
- —, thermal, of atoms *or* molecules 61, **175f**, 316, 468, 532, 580, **652f**, **670f**
- —, composite **652f**
- — of solid particles 39
- — transfer; *see* transport
- — transport 60, 177, 194, 624, 628
- — trapping **79f**, **176f**, **548f**, 555, **626**
Radiative lifetime; *see* Lifetime
Radiator; *see also* Black body, Light source
- —, full 56, **181f**, 926, 935, 943
- —, thermal 57, **181f**, 190, 285, 943
Radical(s) 22, 41, 75, 194, 434f, 534, **636f**, **641**, **644**, **667f**, **674**, **751f**, **774**, **784f**, **847f**; *see also* individual radicals, Relaxation
- —, concentration of **437f**, **643f**, 648
- —, conversion of relative into absolute **448f**
- —, determination of **437f**
- —, excess 314, **451f**, **643f**, 666, 670, **674f**, **777f**, 782, 787, **794f**, 799, 802, 819f, **825f**
- — disequilibrium; *see* Disequilibrium
- — reactions; *see* Reaction
- —, recombination of; *see* Recombination (of atoms, *etc*.)
Radio-active element 305
Radio-frequency (r.f.) resonance
- — circuit **861f**
- — curve 854, **861f**
- —, width of 854, 862
- — frequency, shift of 854, 862
- — technique **851f**, **861f**, 865, 869, **886f**
Radius of first Bohr orbit 122, **924**
Rapid-scan method 258f
Rare-earths 369, 481, 774
Rate (of process *or* reaction) **64f**, 68, **70f**, **89f**, 106, 444, 787, 825, 932; *see also* Detailed balance

Rate (of process *or* reaction) (*continued*)
- —, chemi-excitation 649, **664f**
- —, chemi-ionization 826, 828, 844
- — chemiquenching 647
- — collisional 130, 416, 621, 625, 647, **666f**, 905
- — of collisions 80, **535f**, **635f**
- —, de-excitation 130, **619f**, 621, 625
- —, chemical; *see* chemiquenching
- — — by electron impact **558f**
- —, dissociation **786f**
- —, excitation 106, 621, 625, 649, **666f**
- —, chemical; *see* chemi-excitation
- — — by electron impact **558f**
- —, radiative; *see* photo-excitation
- —, ionization 86, 826, **837f**, 850, 884, **892f**, 905
- —, chemical; *see* chemi-ionization
- —, ion-recombination 4, **892f**; *see also* neutralization
- —, net **893f**
- —, neutralization 825, **837f**, 844; *see also* ion-recombination
- —, photo-excitation **619f**, 630, **664f**
- —, quenching 79, 532, 558, 580, 647, 658, 664
- —, chemical; *see* chemiquenching
- — of radiative process **619f**
- —, recombination (of atoms *or* radicals) 455
- —, thermal 79
- — of two-step process 649
- — of unbalanced process **892f**
Rate of aspiration; *see* Aspiration rate
Rate coefficient; *see also* Rate constant
- — for ion-electron recombination 894
Rate constant (of process *or* reaction) **64f**, **68f**, 80f, 88, 90, 102, 105, 532, 618, 640, 643, 929; *see also* Detailed-balance
- —, apparent unimolecular, *etc.* 67; *see also* unimolecular, *etc.*
- —, association 69, 929; *see also* recombination
- —, bimolecular 67, 69, 77, 89, 618, **839f**, 844, **866f**, **874**, **879f**, **891f**, 957
- —, bulk 66
- —, chemi-excitation **643f**, 649
- —, chemi-ionization 826, 828, 844, **868f**
- —, chemiquenching **647f**
- —, collisional 66, 76f, **147**, **539f**, **542**, **547**, **868f**, **882**
- — of collisions 66, 147
- —, de-excitation 67, 76f, 130, 533, 547; *see also* mixing, quenching
- —, chemical; *see* chemiquenching
- — — by electron impact **558f**
- —, dissociation 69, 929
- —, dissociative 879
- —, excess 868
- —, excitation 89, 533, **575f**, 658
- —, chemical; *see* chemi-excitation
- — — by electron impact **556f**
- —, radiative; *see* photo-excitation
- —, experimental **896f**
- —, ionization 66, 69, 580, 815, **826f**, **837f**, **846f**, **864f**, **868f**, 880, **882f**, **885f**, **890f**, **895f**, **899f**, **902f**, 920, 929, **957f**
- —, chemical; *see* chemi-ionization
- —, ion-recombination 463, **839f**, 844, **846f**, **873f**, **882f**, 890, 892, **894f**, 929; *see also* neutralization
- —, microscopic 66f, 71
- —, mixing 562
- —, intermultiplet 609, 611
- —, intramultiplet 548, 612, 615
- —, monomolecular; *see* unimolecular
- —, neutralization 825, **837f**, 844, 868, **879f**; *see also* ion-recombination

Rate constant (*continued*)
— , over-all 67,71,87,840,866f,873f,879,890,892f
— , phenomenological 66f,71,86,815,883f,891,901f
— , photo-excitation 619f; *see also* Transition probability
— , quenching 76f, 79, 89, 532, 537, 547, 562f, 594, 609, 611, 625, 647f; *see also* de-excitation
— — , chemical; *see* chemiquenching
— , radiative 110f, 144f, 562; *see also* Transition probability
— , reaction 435, 446, 452f, 784
— , recombination (of atoms *or* radicals) 69, 793, 884f; *see also* association
— , specific 67, 453, 840, 866f, 873f, 892f
— , standardized 885f
— , suprathermal 893f
— , termolecular 446, 649, 793, 839f, 892
— , thermal (specific) 67f, 70f, 83, 85, 88, 105, 539f, 839f, 868f, 883f, 892, 896f, 899f, 957
— , unimolecular 65f, 69, 77, 89, 539f, 562, 647f, 763, 839f, 866f, 882, 885f, 892f, 902f, 957
Rate of effusion 762
Rate of liquid up-take; *see* Aspiration rate
Ratio-meter 255, 371f, 374
Ratio technique 375
Rb 238, 335, 543, 546, 550f, 564f, 573, 579, 586, 594f, 601f, 607, 613, 616f, 737, 745, 777, 780, 833, 861, 889f, 910, 913, 919f, 941
RbBr 769
RbCl 769
RbF 769
RbI 769
RbOH 768, 776
RC-filter 269, 949
RC-time 263, 347, 936; *see also* Time constant
Re 635
Re-absorption 109, 176; *see also* Self-absorption
Reactant (of process *or* reaction) 64, 71, 85, 90, 674, 788, 827
Reaction(s) (chemical) 4, 60f, 64, 66, 69, 72, 84, 86, 89, 93, 103f, 108f, 419, 461, 529, 532, 534, 758f,783f,823; *see also* Process
— , activation energy of 827, 878; *see also* Activation energy
— , association 49, 106, 108f, 468, 664f, 818
— , associative 64, 68, 108f, 635, 820, 822f, 827, 844f, 878f
— — , three-body 468
— , attachment 463
— , balanced 61f, 72, 106, 419, 436, 440, 446, 452, 778,786f,789f,793f,797f,807f,811,820,825,849
— — , partially 61
— , bimolecular (*or* binary *or* two-body) 53,65,84, 90, 105f, 108f, 422, 436, 635f,648f,667f, 708f, 785f, 789f, 793f, 798f, 802f, 818, 827, 850, 879
— , chain(-branching) 434f
— , charge transfer 462
— , chemi-excitation 64, 68, 108f, 530, 634f, 638f, 650f; 708f; *see also* photolysis
— , chemi-ionization 63, 109, 638f, 688, 818, 820, 822f,826f,844f,847f,850,868f,876f,878,880,886
— , chemiluminescence 106, 109, 153, 445f, 468, 530, 534, 634f, 654, 663f, 667f, 708f
— , chemiquenching 108, 530, 569f, 581, 634f, 647f, 708f
— , classification of 103f, 108f, 823
— , collisional 82, 108
— , combustion 60, 421
— , de-excitation 64, 108, 634f; *see also* chemiquenching
— , diabatic 90

Reaction(s) (chemical) (*continued*)
— , dissociation 49, 69f, 90, 108, 758, 761f, 768f, 775f, 785f, 797f, 802, 818
— , dissociative 64, 108f, 462f, 569, 647f, 823f, 844f, 879
— by electron impact 758, 762, 768
— , elementary; *see* single-step
— , endo-ergic 70, 825
— , endothermic; *see* endo-ergic
— , energy of 53, 71, 84, 636, 638f, 708, 798, 809, 933; *see also* Dissociation energy
— , enthalpy of 84, 90, 636f, 648f, 803, 827, 848; *see also* heat
— , entropy change in 90
— , equilibrated; *see* balanced
— , exchange 53,108f,440,569,647f,651,667f,708f, 785f,789f,793f,798f,802f,811,818,823f, 878
— , excitation; *see* chemi-excitation
— , exo-ergic 70, 84, 414, 444
— , exothermic; *see* exo-ergic
— , formation 775, 777f, 783f, 791f
— , free enthalpy of 51, 84, 791
— , heat of 51, 427, 791; *see also* enthalpy
— , heterogeneous 3, 532
— , homogeneous 3
— , ionic re-arrangement 823f, 827, 850
— , ionization; *see* chemi-ionization
— , ion-recombination 109,463,817,823,844f,875f
— , molecularity of 66
— , multi-step 104, 109
— , neutralization 109, 462, 817, 823f, 828, 844f, 876f, 879
— , nonadiabatic; *see* diabatic
— , order of 66
— , over-all 66, 109, 827
— partner 65; *see also* reactant, Third body
— , path 84, 92, 752, 775, 787f
— , photochemical 108
— , photodissociation 108, 529, 758
— , photolysis 108, 529
— , polymerization 461
— , primary 877f
— , product 64, 105, 674
— , quenching; *see* chemiquenching
— , radiative 108
— , rate of; *see* Rate
— , rate constant of; *see* Rate constant
— , reactant of 64,71,90,827; *see also* Reactant
— , recombination 39, 53, 436, 444, 447, 449, 453, 461, 639, 641, 651, 785f, 789f, 792f; *see also* association, ion-recombination
— , single-step 104, 108f
— , suprathermal 708, 710
— , surface 314
— , termolecular (*or* ternary *or* three-body) 65, 105f, 108f,436,452,635f,641,664f, 667f, 827
— , three-body; *see* termolecular
— , two-body; *see* bimolecular
— , two-step 106, 109, 446, 635f, 649f,785f, 789f, 794f, 850
— , unbalanced 71f, 307, 419, 793, 802, 824
— , velocity of; *see* Rate
Reaction-free zone; *see* Interzonal region
Recombination (of atoms *or* radicals) 39, 75, 664f, 774, 820, 825, 905; *see also* Association, Continuum, Ion-recombination, Rate, Rate constant, Reaction
— , catalyzed 442, 445, 453f
— energy 40, 452, 642, 666; *see also* Dissociation energy
Recombination (of drops); *see* Coagulation

SUBJECT INDEX

Recording methods **258**f
Recoil effect 30, **697**, 699
Rectifier 336; *see also* Lock-in
—, frequency-sensitive **281**
—, phase-sensitive 337
—, synchronous **281**f, 284
Re-emission; *see* Fluorescence, Radiation
Reference-element technique **255**, 441
Reference solution 508
Reflux methods **303**f; *see also* Nebulizer, reflux-type
Refractive index 57, 118, **122**f, 129, 244, 422, 930, **945**f
— of air 30, **248**
Relaxation 2, 5, **57**f, 62, **72**f, 414, 422f, 551, **658**f; *see also* Equilibration
—, chemical 752, **786**f, **795**f; *see also* radical
— of excited-level population 73; *see also* Energy level
— for ionization 75, 209, **461**f, 514, **837**f, **846**f, 883, 886, **892**f, **900**, **905**f, **957**f
— method **76**f, 80
— of radical (concentrations) 75, **447**f, **451**f, **777**f, **843**, **905**
— for rotational energy 571
—, spin (-orbit) **530**f
— time **58**f, **74**f, 78, **131**f, 316, 414, 422, 464, 551, 574, 641, 658, 837, **841**f, **936**
—   — for energy level population 551
—   — for gas temperature **574**f
—, — ionization **888**, **895**f
—, — rotational 571
—, — vibrational **59**, 71, **574**f, 659
—, — translational (*or* velocity) 59
— for velocity of solid particles **331**
— for vibrational energy 71, 78, 106, **571**f
Residence time (in flame) 2, 75, 659, 911; *see also* Rise time
Resolution
—, frequency 476
—, height 855; *see also* spatial
—, mass **855**f
—, spatial **252**, 378, 437, 444, **852**f, 862f
—, spectral 34, 134, 138, 241f, **244**f, 248, 250, 260, **347**, 392, 710, **712**f
—   —, effective 261
—, — time 2, 78, 266f, 322
Resolving power, spectral **247**, 260
Resonance; *see also* Cyclotron resonance, Electron spin resonance, Quantum state, resonance, Radio-frequency resonance
Resonance (in inelastic collision); *see also* Near-resonance, Quasi-resonance
—, close **102**
— curve **601**
— defect **103**, 584, 597, 602
— effect **571**f, **577**f, 581, 584, 592, **596**f, **601**f
—, electronic-vibrational **577**, 579
— interaction **602**
— process **102**f, 577
—, adiabatic 104
—, diabatic 104
— separation **95**
— splitting **95**f
Resonance broadening; *see* Line broadening
Resonance detector **261**
Resonance fluorescence; *see* Fluorescence
Resonance line; *see* Line
Response
—, frequency- 948f
—   —, normalized **948**f
—, impulse **949**, 952
—, spectral **263**f, 310

Response (*continued*)
—, step **350**, **948**f
—, —, normalized **948**f
Response time **262**f, **278**, 325, **347**, **349**f, 353, 356, 359f, **366**, 403, 405, 411, **937**, **949**
Responsivity of photocathode (*or* photodetector) **262**f, 276, 325, 344, **931**
Reversal; *see also* Line, reversed, Line-reversal, Self-reversal
— dip 187, 198, 662
— point, **321**f, **324**f, 387
— temperature; *see* Temperature (of flame, *etc.*)
Reversibility **817**; *see also* Microscopic reversibility
Reynolds Number **917**
Rise time 2, 311, **414**, 424, 449, 454, 793, 838, **842**f, **870**f, **876**f, **932**; *see also* Residence time
— of electron pulse 267
Rise-velocity 223, 307, **329**f, **413**f, 516, 838, 849, **866**f, **915**f, **932**; *see also* Flame
Rocket
— exhaust 6, 816, 863
— fuel 6
Root-mean-square value **931**
Rotameter **231**f
Rotation (of molecule) 25, **28**f, 597, 798; *see also* Quantum number
Rotational constant **28**f, **36**f, 43, 181, 797. 926
Rotational energy, *etc.*; *see* Energy, *etc.*
Russell-Saunders coupling scheme **16**

S 7
S_2 753
Saddle point 92
Saha
— curve **846**f, **866**f
— equation; *see* law
— equilibrium; *see* Equilibrium
— law **53**f, 315, 460, **499**f, **513**f, 819, 832f
—, — departure from; *see* Disequilibrium
Sample 7, 9; *see also* Solution
—, reference 7
—, intermittent introduction of **226**
Sample-and-hold circuit 300
Sampling time **347**f, 353, **356**f, **359**f, 362, 366, 403, 411, **937**, **947**f
Satellite **705**f, 711, **733**, **738**, 742; *see also* Line profile
Saturation
— in ionization interference **906**f
— of photomultiplier **276**
— pressure 311
Saturation (of radiative transition) 114, **117**, 130f, **140**f, 161, **173**f, **202**f, 383, 483, 580, 626, **630**f, 815, 825, 847, **905**; *see also* Absorption, Bleaching, Fluorescence, Optogalvanic effect
— broadening; *see* Line broadening
— curve **204**f
—, degree of 173, **206**f, 377, 533, 556, **630**f, 933
— of doublet **632**f
— effect on alkali compound formation 515
— method 78
— of multiplet 599, 633
— parameter 78, 173, **203**f, **631**f
—, partial **203**f
— plateau **204**; *see also* curve
— by two-photon absorption 110, 390
Sb 638, 649, 708
Sc 635, 941
Scale expansion 346, **371**, **374**f, 379
Scattering (of light) 113f, 116f, 126, 202
— by aerosol **307**f, **394**f

Scattering (of light) (*continued*)
—, angular distribution of 114, 116, 533
—, elastic 102, 108, **112f**
—, inelastic 88, **113**
—, intensity of 114, 116
—, Mie **113**
—, nonresonant 114
—, polarization of 114, 117
—, Raman **113**, 115, 327
—, Rayleigh **113**, 115f, 126, 531, 717
—, resonant 113f
—, spectrum of 116f
— by unevaporated particles 314, 329, 383, **394f**, 403f, 410, 564
Scattering (of particles) 81, 88, 539, 739
— angle 534, 568f
—, angular distribution of; *see* Cross section for scattering, differential
—, elastic 534
— of electrons 589f, 592, 830
— —, elastic 546f, 554
— —, superelastic 573
—, inelastic 534
Scattering amplitude 100
Schlieren photography 422
Schumann-Runge bands 466
SC theory; *see* Line broadening theory, semi-classical
ScO 489, **767**, **772**
Secondary combustion zone 423f, 459, 465f, 489
Secondary-electron conduction image tube 262
Secondary emission; *see* Emission, Fluorescence
Secondary emission coefficient **264**
Secondary radiation; *see* Radiation
Secondary reaction zone; *see* Secondary combustion zone
Second-law method **761**, 798f, **803f**, 812
Selection
—, internal-state 815
— of observed area 250f
—, velocity; *see* Velocity
Selection rule(s) 60, 531
—, optical 31f, 35, 110, 122, 136, 541, 601
— —, general 31, 35
— —, special **32**, **35**
— — for two-photon process 32
— — for potential-curve crossing 96, 585
Selective-modulation technique **260f**
Selenium barrier-layer cell 262
Self-absorption 109, 138, 163, **176f**, **181f**, 188, **191f**, 194, **198f**, 202f, 295, 314, 318f, 326, 328, 393, 400, 472, **484f**, 489, 507, 509f, 512, 532, **626f**, 653, 661, 665, 668, **671f**, 673, 866, 870, 878, 916; *see also* Absorption, Curve-of-growth, Line broadening, Radiation diffusion, Self-reversal
Self-broadening; *see* Line broadening
Self-mixing 555f
Self-quenching 542
Self-reversal **187f**, 198, 261, 292f, 295, 494, **502f**, **507**, 653, 672; *see also* Self-absorption
— dip 187, 198, 662
Sensing, local; *see* Local probing
Sensitivity, analytical 747
Shell **16**
—, closed **16**, 18
—, sub- **16**, 31
Shift; *see* Isotope shift, Line shift
Shock-tube 6, 10, 78, **310f**, 435, 468, 533, 537, 544, 551, 558, **574f**, 659, **762**, 815, 898
Shock-wave; *see* Shock-tube
Short-wave(length) cut-off (*or* limit) 40, 470, 651f
Shot effect; *see* Noise, shot

Shot noise; *see* Noise, shot
Si 570, **611**, 649
Sign (of quantum state) **26**, **35**
Signal **948f**; *see also* Analyte signal
Signal-to-background ratio 226
Signal-to-noise ratio (SNR) 416, **258f**, 263, 268, 270f, 275, 278, 283, 300, **324**, 340, **346f**, **366f**, 374, **383f**, 391, **400f**, 485, **947f**
SiH **768**
Silica tube 212, 423
Silicon-diode array camera **262**
Silicon photodiode 262
Silicon vidicon **262**
Silver-comparison method **525f**
Single-beam technique 369, 380, 384
Single probe technique **851f**, **857f**, 880
—, rotating 463, **859f**, **865**, **869f**, **881**, **887**
Singlet 21, 26, 29, 33, **544**, 570, **639f**, **791**; *see also* Term
SiO **767**, **772**
Slit(s) (of monochromator)
—, entrance **242f**, 251f
—, exit **242f**, **255f**
— function **241**
— height **250**
— —, angular **243f**
—, matched 249
—, vibrating 256
— width 166, **352**, 386
— —, angular **243f**
Slope method (of temperature measurement) 318, 483, **671**, 673
Slot burner; *see* Burner
Sm 635, 759
SmCl 759
SmF 759
Smithells separator 419
Sn 46, 455, 552, **615**, 638, 640, 671f, **708**, 771, 774, 825, **941**
SnH 753, 779
SnO 455, **767**, **772**, 775, 779, **792**
SnOH 779
SO 41
SO_2 455, 565, 604f
Sodium-comparison method **439**, **523f**, 803, 811
Sodium-*D* doublet (*or* line) 21, 31, 33, 46, 114f, 129, 141, 145, 153, 168, 174, 205, 261, 476, 486, 493, 499, **524f**, **530f**, **537f**, 549, **551f**, **555f**, 563, **567f**, 572, 590f, 594, 597f, **626**, **632**, 648, **660f**, 663, 681, **705f**, 716f, **723f**, **727f**, 732f, 738, **740f**, 743, **749f**, 819
Sodium detector 534
Sodium flare 7
Sodium level diagram **19**
Solute 302
Solution 310f; *see also* Sample
—, metal salt, prepatation of 230
—, reference 399
—, stock 230
—, surface tension of 229
—, test 230
—, viscosity of 229
Solvent 302; *see also* Water
—, alcoholic 10
—, condensing system for 236
—, evaporation of 230; *see also* Desolvation
—, organic 228, 417, 641
— vapour 307
Soot 532; *see also* Carbon, Flame, sooting
Source of light; *see* Light source
Spark 38, 79, **298**
Spark-in-flame 417

SUBJECT INDEX

Specific heat 432, 760f, 803
Specific heat capacity 428, **926**
Specific heat of vaporization 307
Spectral band; see Band
Spectral bandwidth; see Bandwidth (spectral)
Spectral continuum; see Continuum (spectral)
Spectral continuum source; see Continuum source
Spectral intensity 943f; see also Spectral radiant intensity
Spectral irradiance 320, **632**, **927**, **943f**
Spectral light source; see Light source
Spectral line; see Line
Spectral line broadening; see Line broadening
Spectral line profile; see Line profile
Spectral line source 191f, 207, **285f**, **296f**, 376, 380f, 403f, 410, 443, 491f; see also individual lamps, Light source, Narrow-line source
Spectral line width; see Effective width, Half-intensity-width
Spectral noise power 268f, 272, 280, 348, 473, **475f**, 932, **947f**; see also Noise spectrum
Spectral overlap; see Overlap
Spectral profile; see Line profile
Spectral quantum response (of photocathode) 265
Spectral radiance 56f, 60, **179**, 182, **190f**, 324, **342f**, **352f**, 363, 365, 411, 661, **926**, **943f**
—, integrated 630, **655f**; see also Radiance
—, peak **926**
Spectral radiant density; see Spectral radiant energy density
Spectral radiant energy **943f**
Spectral radiant energy density 55, **118**, 121, 145, **622f**, **627f**, **943f**
Spectral radiant flux 57, **943f**
Spectral radiant intensity **468f**, **943f**
Spectral radiant power; see Spectral radiant flux
Spectral resolution; see Resolution
Spectral selector 5, **241f**; see also individual selectors
—, nondispersive 407
Spectral volume density of radiant energy; see Spectral radiant energy density
Spectrochemical analysis; see Analysis, Flame spectroscopy
Spectrogram 38, **953f**
Spectrograph(y) **259**, 340, 919
—, short-time 259
Spectrometer(-try) 8, 209, **240f**, **261f**, **277f**, 364, 379, **394f**, **854f**; see also Flame spectrometer, Mass spectrometry, Spectroscopy
—, bandwidth of; see Bandwidth (spectral)
—, calibration of; see Calibration
—, direct-reading 262
—, double-beam **254f**, 283, 360, 369, **371f**, 378, **381**, 386, 389, **714f**
— —, spatial 369
— —, spectral; see dual-beam
—, dual-beam (or -wavelength) **255f**, 360, 369, **373f**, 382, 386, 395
—, Fourier **259f**
—, grating 260
—, Hadamard **260**
—, instrumental profile of; see Line profile
—, multichannel 254, 262
—, multislit 259
—, resolution of; see Resolution, spectral
—, spectral response of **263f**, 310
—, stray light in; see Stray light
Spectrophotometer; see Spectrometer
Spectroscopic term; see Term
Spectroscopy; see also Flame spectroscopy, Mass spectrometry, Spectrometer

Spectroscopy (*continued*)
—, (atomic) absorption 3, 140, 164, 166, **368f**, 551, 555, 570
—, derivative **256f**, **322**, 360, 387
—, (atomic) emission 3, 261, **339f**
—, (atomic) fluorescence 3, 283, 296, **390f**, 474, 598, 626
—, Fourier **259f**
—, Hadamard **260**
—, infrared 260
—, microwave 141, 146
—, molecular 31, 34, 37, 261, 390
—, Mössbauer 9
—, multiplex 240
—, saturated fluorescence **515**
—, time-resolved 259
Spectrum 3, **321f**, 365, 418, **778f**; see also Flame spectroscopy, (Hyper-)fine-structure, Line, Mass spectrometry, Spectroscopy, Spectrometer
— of atom 30f, **32f**, 34, 774, **939f**
—, background 191, 257, 262
—, band 34, **36f**, **38f**, 194, 327, **465f**, **777f**, 802, **953f**; see also Band
—, chemiluminescence 759
—, continuous; see Continuum (spectral)
—, distortion of **248f**
—, flame background; see Flame background
—, fluorescence 579
—, line 242, 248
— of molecule 34f, 38, **953f**; see also band
—, noise; see Noise
—, quasi-continuous; see Continuum (spectral)
— of scattered radiation **116f**
—, transient **259**
—, true 249
Speed of light; see Velocity of light
Spin; see also Angular momentum, Quantum number
— -orbit coupling **649**
— -orbit precession 552
Spontaneous emission; see Emission
Spontaneous-ignition temperature **414**
Spray; see also Aerosol
—, density of 229
—, preheated 230
Spray chamber 2, **225**, 236, 306, 517; see also Nebulizer
—, (pre-)heating of 228, 236
Sprayer nozzle **224f**, 228
—, inlet of 226
Sputtering, cathode 260, 290, 537
Square-root-law (of curve-of-growth) 168, 172, 184, 314, 484, 488, 492, 498, **502f**, 505
Square-root-region (of curve-of-growth) 496, 498, 512
Sr 38, 185, 295, 369, 455, **500**, **563**, 570, 609, 633, 670, 679, **694**, 706, 710, **724f**, **729f**, 733, **735f**, **740f**, **743f**, **826f**, 836, **865**, **876f**, 880, 907, **910**, **941**, **954**
— -polymer 506
Sr⁺ **442f**, **500**, **880**, **904**, **941**
SrBr **769**
SrCl **769**, 783, 806
SrCl$_2$ **769**
SrF 753, **769**
SrF$_2$ **769**
SrI **769**
SrO 759, **767**, 772, **778**, **791**
Sr$_2$O$_2$ 770
SrOH 194, **768**, **776f**, **806f**, **823**, **881**, **954**
SrOH⁺ **442f**, **823**, **880**
Sr(OH)$_2$ **768**

SrOHCl 783
sr = steradian
Standard deviation 268f, 272f, 936
—, relative 268
Standard element, fully atomized 234, 480, 520f, 775, 801, 904
Standard lamp 365, 486; *see also* Radiation standard
Standard of wavelength 248
Stark broadening; *see* Line broadening
Stark effect 20, 146
Stark sublevel 20
State; *see also* Quantum state
—, intermediate 104
—, ionic 100
—, standard 51, 90
—, stationary 74f
— sum; *see* Partition function
Stationarity 74f
Statistical weight 44f, 48, 54, 68, 122, 133, 632, 883f, 928
Step response; *see* Response
Stepwise line fluorescence; *see* Fluorescence
Steric factor 82, 85f, 90, 106, 785, 827f, 930
Stern-Volmer formula (*or* relation) 76f, 193f, 563
Stimulated emission; *see* Emission, induced
Stirring device, ultrasonic 237
Stoichiometric (mixing) ratio 415, 428f
Stokes fluorescence; *see* Fluorescence
Stokes law 331
Stray light 253, 276, 364, 383; *see also* Scattering (of light)
Stripping effect 517
Stroboscope 330
Strong-collision model (for dissociation) 90
Sublimation method 237f
Subshell 16, 31
Summation rule (for collisional ionization) 840, 892f
Sum rule
— for Franck-Condon factors 137
— for Hönl-London factors 136
— for oscillator strengths 134
Suprathermal chemiluminescence, *etc.*; *see* Chemiluminescence, *etc.*
Surface tension 229
Symmetry (of collision complex) 93
Symmetry number 47, 49, 797f, 931

Tb 635
Te 639
Temperature (of flame *or* gas) 1, 5f, 13, 41f, 63, 178, 186, 190, 217, 220, 368, 405, 414f, 422, 432, 434, 443, 455f, 459, 464, 468, 486, 489, 492, 496, 502f, 509f, 512f, 514, 522, 524, 532, 549f, 553, 556, 567f, 575, 581, 590f, 595f, 598, 603f, 647f, 666f, 671f, 674, 690, 760f, 771f, 776, 789f, 798f, 806, 809, 812, 824, 837, 840, 849, 865f, 869, 873f, 877, 879, 888f, 904, 907, 910f, 918f, 932; *see also* Cooling, Flame temperatures
—, adiabatic 59, 415, 424f, 427f
—, average 459
—, calculated 428f, 455f
—, calculation of 455f
—, dissociation 62
—, (spatial) distribution of 318, 320, 326f, 333, 459, 482f, 652
—, Doppler 316
— —, effective 294
—, effective radiation 145
—, electron 63, 659, 851
—, (electronic) excitation 62, 77, 317f, 455, 458, 481, 657, 669f

Temperature (of flame *or* gas) (*continued*)
—, final 59, 62, 447
— fluctuations 387
— gradient 187, 453, 658, 846f, 849, 865, 911
—, inhomogeneities in 459
—, initial 455
—, initial rise of 447, 455f
—, ionization 62
—, local 60f, 115, 326f, 423, 455
— maximum 432, 448, 456
—, measurement of 257, 315f, 387, 673; *see also* specific methods
—, noise 465
—, radiance 145, 190f, 287f, 321, 323, 325, 365f, 386, 630, 932
—, (line-)reversal 318, 320f, 324f, 327, 457f, 641, 659
—, rotational 62, 180f, 318, 422, 572, 576, 669
—, translational 62f, 316f, 421, 572, 574, 653, 659
—, true 316f, 324, 457, 653f
—, two-line method for 317f, 459, 483
—, vibrational 62, 181, 318, 422, 455, 572, 574, 669
Temperature (general) 13, 41f, 49f, 58, 60, 557, 932
—, Celsius 932
—, definition of 315f
—, effective radiation 145
— interval 937
— lever 247
—, practical, scale 458
—, radiance; *see* Temperature (of flame)
—, thermodynamic 458
— —, scale 458
Temperature (special)
— of burner (head) 218
—, critical 911, 913
— furnace 575
—, initial, of unburnt gas mixture 432
— of lamp 365
—, reduced 913
— of resistor 273
Tension 932
Term, spectroscopic 20, 38, 133f, 530; *see also* Multiplet
— diagram 19, 21, 484
—, molecular 545
—, singlet 21, 633
— symbol 19f
—, triplet 21, 633
— value 21
—, vibrational 27f, 927
Thermal equilibrium, *etc.*; *see* Equilibrium, *etc.*
Thermochemical cycle 760
Thermocouple 326f, 422
Thermodynamic equilibrium, *etc.*; *see* Equilibrium, *etc.*
Thickness (of flame) 160f, 169f, 184f, 332f, 387, 481, 505f, 517f, 628, 661
Third body (effect) 69, 72, 105f, 436, 452f, 463, 467, 641, 827
Third-law method 761, 764, 798f, 803f, 809, 812
Third partner; *see* Third body
Three-body collision, *etc.*; *see* Collision, *etc.*, termolecular
Threshold 82
— energy 759
— wavelength 275
Throughput; *see* Optical conductance
Ti 635, 771, 941
TiO 767
Time; *see also* Correlation time, Integration time, Intercollision time, Measuring time, Relaxation time, Response time, Rise time, Sampling time

SUBJECT INDEX

Time (*continued*)
- constant 347f, **350**, **356f**, 936; see also RC-time
- -correlation 351, 539; see also Autocorrelation function
- resolution; see Resolution
- reversal 88

Time-of-flight technique 539, 569, 758

Tl 33, **320**, 327, 489, **526f**, **537**, **551f**, **555f**, 570, 579, **595f**, **609f**, **614f**, **642f**, **649f**, 709, 824f, **865**, **869f**, **889f**, **910**, **941**

TlBr **769**
TlCl **769**
TlF **769**
TlH **768**, 779, **794f**
TlI **769**
TlO **767**
TlOH **768**, 778
Tm 635

Tongue; see Primary combustion zone
Trajectory; see Collision
Transfer; see also Process, Reaction, Transport
- of charge *or* electron 818
- of electronic excitation energy 530, 542, 566, 579, 599f, **660**, 709
- -, electronic-translational energy 544
- -, electronic-vibrational energy 57, **571f**, **576f**, 582, **596f**
- of radiation; see Radiation transport

Transfer function 949, 952
- -, complex 952

Transient (effect *or* signal) 274, 278, **630**, 633

Transition 102; see also Collision, Cross section, Process
- absorption; see Absorption
- cascade 539, 551, 601
- collisional 111, 535, 540
- complex **89**; see also Complex
- -, diabatic 96f, 649
- elements 481; see also individual elements
- -, electric-dipole 31, 35, 121f, 702
- -, emission; see Emission
- -, Franck-Condon 822
- -, intermultiplet; see Mixing (of energy levels)
- -, intramultiplet; see Mixing (of energy levels)
- -, magnetic-dipole 122
- -, moment 122, 136
- -, optical; see radiative
- -, probability 117f, 136, 484
- - for absorption 117f, 134f, 926
- - -, spectral 121
- - of band 136
- - for induced emission 117f, 926
- - for spontaneous emission 65f, 76, 117f, 134f, 143, 177, 188, 193, 480, 569, 619, 746, **925**
- - -, effective 181, **627f**
- - -, total 137
- - -, weighted 124
- -, quenching; see Quenching
- -, radiationless 76, **93f**, 101, 112, 193, 468, 530; see also collisional
- -, allowed 96
- -, forbidden 96
- -, radiative 19, **30f**, 110f, 118f, 468, 470, 540, **939f**; see also Spectrum
- -, allowed 122; see also Selection rules
- -, electronic 30, **34f**, 135f
- -, forbidden 122; see also Selection rules
- -, rotational **34f**, 135f, 596
- -, saturated 114, 117f, 130f; see also Saturation
- -, vibrational **34f**, 135f, **575f**, **584f**

Transition (*continued*)
- -, two-photon; see Two-photon absorption

Translational energy, *etc*.; see Energy, *etc*.
Transmission coefficient 90, **96f**, 100, 545
Transmission factor 99, **160f**, 173, 243, 253, 325, 365, **370**, **943**, 945
Transmission function 241, 246
Transmission of photomultiplier window **265**
Transmittance; see Transmission factor

Transport
- of heat 60, 416
- of mass 60, 416, 911; see also Diffusion
- -, convective 914
- of radiation; see Radiation transport
- of salt 472

Trapping
- of aerosol 303f, **519f**
- of atoms (in substate) **531**
- of radiation; see Radiation

Travel time; see Rise time
Triplet 21, 33, 544, 570, **639f**, **791**; see also Mixing, Multiplet, Term
Tube, photo-emissive; see Phototube

Tungsten
- arc 323
- brush lamp **287**
- lamp 475
- -, calibrated **490**
- strip lamp **28f**, **299**, **301**, 321, **338**, **362**, **365**

Tunnelling (quantum-mechanical) 99
Turbulence 414, 418, 456
Turnover frequency **473f**, 477
Twin nebulizer **238**
Two-body collision, *etc*.; see Collision, *etc*., bimolecular
Two-hole method (for linearity check) **337**
Two-line method (for temperature measurement) 317f, 459, 483
Two-photon absorption 108f, 203, 543, 819
Two-photon excitation; see Excitation
Two-photon process; see Process
Two-slit method **331f**

U 455, 771, 824
Uncertainty principle; see Heisenberg uncertainty relation
UO **767**
Upper-atmosphere 7
U-shaped manometer 231

V 635, 640, **941**
Vaidya bands 466, 638
Valence electron; see Electron
Valve
- -, needle 230
- -, reducing 230

Van der Waals constant 919
Van der Waals (interaction) potential; see Interaction potential

Vapour cell 4f, 9, 80, 174, 531, **533f**, 537, 542, **548f**, 555, 561, **563f**, 570, 572, **577f**, 593, 595, 598, **602f**, 648, 692, 702, 706, 709, 716f, 723, 727, **737f**, **741f**, **762**, 825

Vapour cloud 260f
- -, pulsating 260
Vapour-discharge lamp; see individual lamps
Vaporization; see Desolvation, Volatilization
Variance 270
VCl_4 751
Velocity; see also Burning velocity, Rise velocity
- -, absolute 43, **837f**
- -, average 43

Velocity (*continued*)
—, distribution of; *see* Distribution, Maxwell law
— of light 924, 926
— of reaction *or* process 65, 106, 932; *see also* Rate
—, relative 42, 44, 99, 522f, 533, 549, **567f**, 581, 591f, 602f, 685, **689**, 691f, 928, 932, 937
—, —, average 44
— selection 67, **533f**, 568, 602, 815, 902
— of unburnt gas 419f, 456
Vibration (of molecule) 24, 28, 90, 100, 798; *see also* Quantum number
—, mode of 798
—, normal 30, 798
Vibrational energy, *etc.*; *see* Energy, *etc.*
Viscosity 229, **917**, 934
VO 767, 772, **792**
Voigt concept (*or* model) **700**, 733f
Voigt profile (*or* function); *see* Line profile
Volatility 773
Volatilization 2f, 225, 239, **309f**, **313f**, 333, 335f, 516, 527, 774
—, fraction of; *see* Fraction volatilized
—, heat of 313
—, incomplete 503, 505
—, rate of 311f
Vortices (in flame) 418, 477

Water (liquid) 237, **434f**, **457f**, 504, **776**; *see also* Aerosol, Desolvation, Drop, Nebulization
—, crystal 230
—, distilled 227, 229f
Wavelength 30, 34, 904, 934, **939f**, **944**; *see also* Spectrum
—, (de)Broglie 582
— calibration **248**
— modulation; *see* Modulation
— scanning 392
— —, repetitive **322**
Wave number 20, 30, **935**
—, fundamental vibrational **27**, **30**, 38, 797, 804

METAL VAPOURS IN FLAMES

Weight factor; *see* Statistical weight
Weisskopf radius 686f, **727f**
WF_6 751
Wien's displacement law 57
Wien's radiation law 57, 120, 182, **656f**
Wiener-Khintchine theorem 475, **947**
Wigner's (spin-conservation) rule 96, 544, 639, **785**, 791
WO_2 770
WO_3 770
Work function 266
—, photo-electric **264**

Xe 537, 544f, 549, 606f, 681, 694, 706, 737f, 742
Xenon flash lamp 288, 296
Xenon (arc) lamp 145, 197, **287f**, 301, 323, 387, 443, 630
—, high-pressure 258, **285**, **298f**, 366

Y 635
Y^+ 904
Yb 635, 759
YbF 759
YbO 759
Yield factor of fluorescence; *see* Fluorescence
YO 194, 489, 535

Zeeman
— effect 32, 260, 714f
— modulation; *see* Modulation
— scanning (technique) 184, 295, **374**, 376, **388f**, 713f, 725, 735f
— splitting 373
— sublevel (*or* substate) 20, **530f**, **702**
Zeropoint vibrational energy **27**, 47
Zero-suppression 346, **371**, 379
Zn 110, 292, 299f, 579, 600, **611**, **616**, 635, 725, **745**, **941**
ZnO **767**
Zr 941
ZrO **767**

1029

Other Titles in the Series in Natural Philosophy

Vol. 1. Davydov — Quantum Mechanis
Vol. 2. Fokker — Time and Space, Weight and Inertia
Vol. 3. Kaplan — Interstellar Gas Dynamics
Vol. 4. Abrikosov, Gor'kov and Dzyaloshinskii — Quantum Field Theoretical Methods in Statistical Physics
Vol. 5. Akun' — Weak Interaction of Elementary Particles
Vol. 6. Shklovskii — Physics of the Solar Corona
Vol. 7. Akhiezer et al. — Collective Oscillations in a Plasma
Vol. 8. Kirzhnits — Field Theoretical Methods in Many-body Systems
Vol. 9. Klimontovich — The Statistical Theory of Nonequilibrium Processes in a Plasma
Vol. 10. Kurth — Introduction to Stellar Statistics
Vol. 11. Chalmers — Atmospheric Electricity (2nd Edition)
Vol. 12. Renner — Current Algebras and their Applications
Vol. 13. Fain and Khanin — Quantum Electronics, Volume 1 — Basic Theory
Vol. 14. Fain and Khanin — Quantum Electronics, Volume 2 — Maser Amplifiers and Oscillators
Vol. 15. March — Liquid Metals
Vol. 16. Hori — Spectral Properties of Disordered Chains and Lattices
Vol. 17. Saint James, Thomas and Sarma — Type II Superconductivity
Vol. 18. Margenau and Kestner — Theory of Intermolecular Forces (2nd Edition)
Vol. 19. Jancel — Foundations of Classical and Quantum Statistical Mechanics
Vol. 20. Takahashi — An Introduction to Field Quantization
Vol. 21. Yvon — Correlations and Entropy in Classical Statistical Mechanics
Vol. 22. Penrose — Foundations of Statistical Mechanics
Vol. 23. Visconti — Quantum Field Theory, Volume 1
Vol. 24. Furth — Fundamental Principles of Theoretical Physics
Vol. 25. Zheleznyakov — Radioemission of the Sun and Planets
Vol. 26. Grindlay — An Introduction to the Phenomenological Theory of Ferroelectricity
Vol. 27. Unger — Introduction to Quantum Electronics
Vol. 28. Koga — Introduction to Kinetic Theory: Stochastic Processes in Gaseous Systems
Vol. 29. Galsiewicz — Superconductivity and Quantum Fluids
Vol. 30. Constantinescu and Magyari — Problems in Quantum Mechanics
Vol. 31. Kotkin and Serbo — Collection of Problems in Classical Mechanics
Vol. 32. Panchev — Random Functions and Turbulence
Vol. 33. Taipe — Theory of Experiments in Paramagnetic Resonance
Vol. 34. Ter Haar — Elements of Hamiltonian Mechanics (2nd Edition)

OTHER TITLES IN THE SERIES IN NATURAL PHILOSOPHY

- Vol. 35. Clarke and Grainger — Polarized Light and Optical Measurement
- Vol. 36. Haug — Theoretical Solid State Physics, Volume 1
- Vol. 37. Jordan and Beer — The Expanding Earth
- Vol. 38. Todorov — Analytical Properties of Feynman Diagrams in Quantum Field Theory
- Vol. 39. Sitenko — Lectures in Scattering Theory
- Vol. 40. Sobel'man — Introduction to the Theory of Atomic Spectra
- Vol. 41. Armstrong and Nicholls — Emission, Absorption and Transfer of Radiation in Heated Atmospheres
- Vol. 42. Brush — Kinetic Theory, Volume 3
- Vol. 43. Bogolybov — A Method for Studying Model Hamiltonians
- Vol. 44. Tsytovich — An Introduction to the Theory of Plasma Turbulence
- Vol. 45. Pathria — Statistical Mechanics
- Vol. 46. Haug — Theoretical Solid State Physics, Volume 2
- Vol. 47. Nieto — The Titius-Bode Law of Planetary Distances: Its History and Theory
- Vol. 48. Wagner — Introduction to the Theory of Magnetism
- Vol. 49. Irvine — Nuclear Structure Theory
- Vol. 50. Strohmeier — Variable Stars
- Vol. 51. Batten — Binary and Multiple Systems of Stars
- Vol. 52. Rousseau and Mathieu — Problems in Optics
- Vol. 53. Bowler — Nuclear Physics
- Vol. 54. Pomraning — The Equations of Radiation Hydrodynamics
- Vol. 55. Belinfante — A Survey of Hidden Variables Theories
- Vol. 56. Scheibe — The Logical Analysis of Wuantum Mechanics
- Vol. 57. Robinson — Macroscopic Electromagnetism
- Vol. 58. Gombas and Kisdi — Wave Mechanics and its Applications
- Vol. 59. Kaplan and Tsytovich — Plasma Astrophysics
- Vol. 60. Kovacs and Zsoldos — Dislocations and Plastic Deformation
- Vol. 61. Auvray and Fourrier — Problems in Electronics
- Vol. 62. Mathieu — Optics
- Vol. 63. Atwater — Introduction to General Relativity
- Vol. 64. Muller — Quantum Mechanics: A Physical World Picture
- Vol. 65. Bilenky — Introduction to Feynman Diagrams
- Vol. 66. Vodar and Romand — Some Aspects of Vacuum Ultraviolet Radiation Physics
- Vol. 67. Willett — Gas Lasers: Population Inversion Mechanisms
- Vol. 68. Akhiezer *et al.* — Plasma Electrodynamics, Volume 1. — Linear Theory
- Vol. 69. Glasby — The Nebular Variables
- Vol. 70. Bialynicki-Birula — Uantum Electrodynamics
- Vol. 71. Karpman — Non-linear Waves in Dispersive Media
- Vol. 72. Cracknell — Magnetism in Crystalline Materials
- Vol. 73. Pathria — The Theory of Relativity
- Vol. 74. Sitenko and Tartakovskii — Lectures on the Theory of the Nucleus
- Vol. 75. Belinfante — Measurement and Time Reversal in Objective Quantum Theory
- Vol. 76. Sobolev — Light Scattering in Planetary Atmospheres
- Vol. 77. Novakovic — The Pseudo-spin Method in Magnetism and Ferroelectricity
- Vol. 78. Novozhilov — Introduction to Elementary Particle Theory
- Vol. 79. Busch and Schade — Lectures on Solid State Physics
- Vol. 80. Akhiezer *et al.* — Plasma Electrodynamics, Volume 2
- Vol. 81. Soloviev — Theory of Complex Nuclei
- Vol. 82. Taylor — Mechanics: Classical and Quantum
- Vol. 83. Srinivasan and Parthasathy — Some Statistical Applications in X-Ray Crystallography
- Vol. 84. Rogers — A Short Course in Cloud Physics
- Vol. 85. Ainsworth — Mechanisms of Speech Recognition
- Vol. 86. Bowler — Gravitation and Relativity
- Vol. 87. Klinger — Problems of Linear Electron (Polaron) Transport Theory in Semiconductors
- Vol. 88. Weiland and Wilhelmson — Coherent Non-Linear Interaction of Waves in Plasmas

Vol. 89. Pacholczyk — Radio Galaxies
Vol. 90. Elgaroy — Solar Noise Storms
Vol. 91. Heine — Group Theory in Quantum Mechanics
Vol. 92. Ter Haar — Lectures on Selected Topics in Statistical Mechanics
Vol. 93. Bass and Fuks — Wave Scattering from Statistically Rough Surfaces
Vol. 94. Cherrington — Gaseous Electronics and Gas Lasers
Vol. 95. Sahade and Wood — Interacting Binary Stars
Vol. 96. Rogers — A Short Course in Cloud Physics 2nd Edition
Vol. 97. Reddish — Stellar Formation
Vol. 98. Patashinskii and Pokrovskii — Fluctuation Theory of Phase Transitions
Vol. 99. Ginzburg — Theoretical Physics and Astrophysics
Vol. 100. Constantinescu — Distributions and their Applications in Physics
Vol. 101. Gurzadyan — Flare Stars
Vol. 102. Lominadze — Cyclotron Waves in Plasma
Vol. 103. Alkemade — Metal Vapours in Flames
Vol. 104. Akhiezer and Peletminsky — Methods of Statistical Physics
Vol. 105. Klimontovich — Kinetic Theory of Non-Ideal Gases and Non-Ideal Plasmas
Vol. 106. Kadomtsev — Collective Phenomena in Plasmas
Vol. 107. Sitenko — Fluctuations and Nonlinear Interactions of Waves in Plasmas
Vol. 108. Sinai — Phase Transitions: Rigorous Results
Vol. 109. Davydov — Biology and Quantum Mechanics